Lecture Notes in Computer Science 5021

Commenced Publication in 1973
Founding and Former Series Editors:
Gerhard Goos, Juris Hartmanis, and Jan van Leeuwen

Editorial Board

David Hutchison
  Lancaster University, UK
Takeo Kanade
  Carnegie Mellon University, Pittsburgh, PA, USA
Josef Kittler
  University of Surrey, Guildford, UK
Jon M. Kleinberg
  Cornell University, Ithaca, NY, USA
Alfred Kobsa
  University of California, Irvine, CA, USA
Friedemann Mattern
  ETH Zurich, Switzerland
John C. Mitchell
  Stanford University, CA, USA
Moni Naor
  Weizmann Institute of Science, Rehovot, Israel
Oscar Nierstrasz
  University of Bern, Switzerland
C. Pandu Rangan
  Indian Institute of Technology, Madras, India
Bernhard Steffen
  University of Dortmund, Germany
Madhu Sudan
  Massachusetts Institute of Technology, MA, USA
Demetri Terzopoulos
  University of California, Los Angeles, CA, USA
Doug Tygar
  University of California, Berkeley, CA, USA
Gerhard Weikum
  Max-Planck Institute of Computer Science, Saarbruecken, Germany

Sean Bechhofer Manfred Hauswirth
Jörg Hoffmann Manolis Koubarakis (Eds.)

The Semantic Web: Research and Applications

5th European Semantic Web Conference, ESWC 2008
Tenerife, Canary Islands, Spain, June 1-5, 2008
Proceedings

Volume Editors

Sean Bechhofer
University of Manchester, School of Computer Science
Manchester M13 9PL, United Kingdom
E-mail: sean.bechhofer@manchester.ac.uk

Manfred Hauswirth
National University of Ireland, Galway
Digital Enterprise Research Institute (DERI), IDA Business Park
Lower Dangan, Galway, Ireland
E-mail: manfred.hauswirth@deri.org

Jörg Hoffmann
University of Innsbruck, STI
6020 Innsbruck, Austria
E-mail: joerg.hoffmann@sti2.at

Manolis Koubarakis
National and Kapodistrian University of Athens
Department of Informatics and Telecommunications
15784 Athens, Greece
E-mail: koubarak@di.uoa.gr

Library of Congress Control Number: 2008927186

CR Subject Classification (1998): H.4, H.3, C.2, H.5, I.2, K.4, D.2

LNCS Sublibrary: SL 3 – Information Systems and Application// incl. Internet/Web and HCI

ISSN   0302-9743
ISBN-10   3-540-68233-3 Springer Berlin Heidelberg New York
ISBN-13   978-3-540-68233-2 Springer Berlin Heidelberg New York

This work is subject to copyright. All rights are reserved, whether the whole or part of the material is concerned, specifically the rights of translation, reprinting, re-use of illustrations, recitation, broadcasting, reproduction on microfilms or in any other way, and storage in data banks. Duplication of this publication or parts thereof is permitted only under the provisions of the German Copyright Law of September 9, 1965, in its current version, and permission for use must always be obtained from Springer. Violations are liable to prosecution under the German Copyright Law.

Springer is a part of Springer Science+Business Media

springer.com

© Springer-Verlag Berlin Heidelberg 2008
Printed in Germany

Typesetting: Camera-ready by author, data conversion by Scientific Publishing Services, Chennai, India
Printed on acid-free paper   SPIN: 12271900   06/3180   5 4 3 2 1 0

Preface

This volume contains the papers from the technical programme of the 5th European Semantic Web Conference, ESWC 2008, that took place during June 1–5, 2008 in Tenerife, Islas Canarias, Spain.

ESWC 2008 was the latest in a series of annual, international events focusing on the dissemination and discussion of the latest research and applications of Semantic Web technologies. The call for papers saw over 270 submissions, a comparable figure to the previous year, indicating that the conference series has reached a certain level of maturity. The review process was organized using a two-tiered system. First, each submission was reviewed by at least three members of the Programme Committee. Submissions were also assigned to a Senior Programme Committee member, who led discussions between reviewers and provided a metareview and provisional decision. A physical Programme Committee meeting was then held, where the final decisions were made. Competition was as strong as ever, and the Programme Committee selected 51 papers to be presented at the conference.

In addition to the technical research paper track, a system demo track was included, with its own review process. Twenty-five demo papers were selected for publication. System demo authors were given the opportunity to present their work in dedicated sessions during the conference, while an evening reception was also devoted to the presentation of posters and demonstrations of systems.

As in past years, ESWC subscribed to the call to "eat our own dog food," with the publication of a rich set of semantic metadata describing the conference.

Three invited talks were given by distinguished scientists: Nigel Shadbolt (Garlik Ltd. and University of Southampton, UK) spoke about the use of Semantic Technology for the consumer; the challenge of Semantic Search was the subject of a talk from Ricardo Baeza-Yates (Yahoo! Research); and Claudio Gutierrez (Universidad de Chile) presented work on the state of the art in RDF Database theory.

The conference also included a programme of tutorials and associated workshops, held on the days preceeding the main conference track. A PhD symposium took place immediately before the conference, giving doctoral students within the Semantic Web community an opportunity to showcase their work and obtain valuable feedback from senior members of the community. This year, we also introduced a programme of panel discussions, bringing together experts to debate key issues in the Semantic Web research area.

As ever, the success of the conference was due to the hard work of a number of people. The members of the Organizing Committee, who were responsible for the selection of tutorials, workshops, panels and the PhD symposium, the Local Organizers and conference administration team, the sponsorship co-ordinator, webmaster and metadata co-ordinator are all listed on the following pages.

We would also particularly like to thank all those authors who submitted papers to ESWC 2008, all the members of the Programme Committee and the additional reveiwers who helped with the review process and contributed to ensuring the scientific quality of ESWC 2008. A quality review process is the cornerstone of a scientific conference, and we appreciate the time and effort spent by our Programme Committee.

ESWC2008 was supported by STI2 (Semantic Technology Institutes International), Universidad Politécnica Madrid and Ministerio de Educación y Ciencia. Sponsorship came from DERI Galway, from the EU Projects Okkam, NeOn, TAO, ACTIVE and Luisa, and from companies Ontos, Franz Inc., Yahoo! Research, Vulcan, Ximetrix and iSOCO. Support for recording presentations came from Jožef Stefan Institute in Slovenia.

We thank Springer for professional support and guidance during the preparation of these proceedings. Finally, we would like to thank the developers of the EasyChair conference management system (http://www.easychair.org/), which was used to manage the submission and review of papers and the production of this volume.

March 2008

Sean Bechhofer
Manfred Hauswirth
Jörg Hoffmann
Manolis Koubarakis

Conference Organization

General Chair:
  Manfred Hauswirth (DERI Galway, Ireland)
Programme Chairs:
  Sean Bechhofer (University of Manchester, UK)
  Manolis Koubarakis (National and Kapodistrian University of Athens, Greece)
Local Organizer:
  Oscar Corcho (UPM, Spain)
  Asunción Gómez-Pérez (UPM, Spain)
Tutorial Chair:
  Grigoris Antoniou (University of Crete and FORTH-ICS, Greece)
Workshop Chair:
  Heiner Stuckenschmidt (University of Mannheim, Germany)
Demo Chair:
  Jörg Hoffmann (STI Innsbruck, Austria)
Poster Chair:
  York Sure (SAP Research, Germany)
Panel Chair:
  Raphaël Troncy (CWI, The Netherlands)
Sponsor Chair:
  Raúl García Castro (UPM, Spain)
PhD Symposium Chair:
  Philippe Cudré-Mauroux (MIT, USA)
Semantic Web Technologies Co-ordinator:
  Chris Bizer (University of Berlin, Germany)
Webmaster and Conference Administrator:
  Katy Esteban-González (UPM, Spain)

Senior Programme Committee

Karl Aberer
Fabio Ciravegna
Dave de Roure
John Domingue
Carole Goble
Lynda Hardman
Ian Horrocks
Natasha Noy
Guus Schreiber
Michael Wooldridge

Richard Benjamins
Isabel Cruz
Stefan Decker
Jérôme Euzenat
Asunción Gómez-Pérez
Martin Hepp
Wolfgang Nejdl
Dimitris Plexousakis
Steffen Staab

Programme Committee

Harith Alani†
Yuan An
Jürgen Angele
Anupriya Ankolekar
Marcelo Arenas
Alessandro Artale†
Danny Ayers
Wolf-Tilo Balke
Nick Bassiliades
Abraham Bernstein†
Walter Binder†
Alex Borgida
John Breslin†
Paul Buitelaar†
Andrea Calì†
Mario Cannataro
Paul-Alexandru Chirita
Paulo Pinheiro da Silva
Olga De Troyer
Thierry Declerck
Ying Ding†
Wlodzimierz Drabent
Thomas Eiter
Boi Faltings
Enrico Franconi
Fabien Gandon†
Fausto Giunchiglia
John Goodwin†
Bernardo Cuenca Grau
Gianluigi Greco†
Peter Haase
Axel Hahn
Siegfried Handschuh
Pascal Hitzler†
Jane Hunter
Eero Hyvönen
Epaminondas Kapetanios
Vipul Kashyap†
Anastasios Kementsietsidis†
Hak Lae Kim†
Mieczyslaw M. Kokar
Ruben Lara

Jose Julio Alferes
Ion Androutsopoulos
Galia Angelova
Grigoris Antoniou
Lora Aroyo†
Yannis Avrithis
Christopher J. O. Baker
Cristina Baroglio
Zohra Bellahsene
Leopoldo Bertossi†
Kalina Bontcheva†
Paolo Bouquet
Francois Bry†
Christoph Bussler†
Diego Calvanese†
Silvana Castano
Oscar Corcho
Jos de Bruijn
Mike Dean
Rose Dieng-Kuntz
Jin Song Dong
Martin Dzbor
Michael Erdmann
Giorgos Flouris†
Gerhard Friedrich
Nick Gibbins
Christine Golbreich
Guido Governatori†
Mark Greaves†
Volker Haarslev†
Mohand-Said Hacid
Harry Halpin
Nicola Henze†
Andreas Hotho†
Carlos Hurtado
Arantza Illarramendi
Vangelis Karkaletsis†
Yevgeny Kazakov
Georgia Koutrika
Matthias Klusch†
Yiannis Kompatsiaris
Thibaud Latour

Georg Lausen
Yaoyong Li
Alexander Löser
Bertram Ludaescher
Zlatina Marinova
David Martin
Wolfgang May[†]
Peter Mika
Simon Miles
Dunja Mladenic[†]
Boris Motik
Saikat Mukherjee
Daniel Oberle
Daniel Olmedilla
Sascha Ossowski
Jeff Pan
Bijan Parsia[†] Terry Payne[†]
Marco Pistore
Line Pouchard
Wolfgang Prinz
Alan Rector
Ulrich Reimer
Riccardo Rosati
Lloyd Rutledge
Marta Sabou
Kai-Uwe Sattler
Sebastian Schaffert
Michael Schumacher[†]
David Shotton
Wolf Siberski
Elena Simperl[†]
Michael Sintek[†]
Spiros Skiadopoulos
Kavitha Srinivas
Heiko Stoermer
Ljiljana Stojanovic
Umberto Straccia
Rudi Studer
York Sure
Hideaki Takeda
Bernhard Thalheim
Christos Tryfonopoulos[†]
Victoria Uren[†]
Jacco van Ossenbruggen
Ubbo Visser

Nicola Leone
Francesca Alessandra Lisi
Joanne Luciano[†]
Jan Maluszynski
M. Scott Marshall
Mikhail Matskin
Diana Maynard
Alistair Miles
Riichiro Mizoguchi
Ralf Moeller
Enrico Motta[†]
Lyndon J B Nixon
Leo Obrst
Borys Omelayenko
Georgios Paliouras
Massimo Paolucci
H. Sofia Pinto
Axel Polleres[†]
Valentina Presutti
Jinghai Rao
Gerald Reif
Mark Roantree
Marie-Christine Rousset
Alan Ruttenberg
Vassilis Samoladas
Ulrike Sattler
Stefan Schlobach
Luciano Serafini
Pavel Shvaiko
Alkis Simitsis
Munindar P. Singh
Sergej Sizov
Derek Sleeman
Giorgos Stamou
George Stoilos
Michael Stollberg[†]
Michael Strube
Gerd Stumme
Vojtěch Svátek
Herman ter Horst
Robert Tolksdorf
Yannis Tzitzikas
Frank van Harmelen
Costas Vassilakis
Max Völkel

Raphael Volz[†]
Holger Wache
Peter Wood
Yiyu Yao

George Vouros
Graham Wilcock
Milena Yankova
Djamel A. Zighed

Additional Reviewers

Milan Agatonovic
Wee Tiong Ang
Alexander Artikis
Dominik Benz
Veli Bicer
Jürgen Bock
Amancio Bouza
Saartje Brockmans
Charis Charalambous
Juri Luca De Coi
Jerome Darmont
Renaud Delbru
Yu Ding
Nicola Fanizzi
Alberto Fernandez
Blaz Fortuna
Alexander Garcia-Castro
Chiara Ghidini
Alfio Gliozzo
Stephan Grimm
Andreas Harth
Ramon Hermoso
Thomas Hornung
Giovambattista Ianni
Qiu Ji
Rajaraman Kanagasabai
Pythagoras Karabiperis
Raman Kazhamiakin
Christoph Kiefer
Sheila Kinsella
Stasinos Konstantopoulos
Thomas Krennwallner
Joey Sik Chun Lam
Michel Leclere
Holger Lewen
Annapaola Marconi

Alia Amin
Darko Anicic
Matteo Baldoni
Piergiorgio Bertoli
Sebastian Blohm
Uldis Bojārs
Janez Brank
Sven Casteleyn
Philipp Cimiano
Alfredo Cuzzucrea
Brian Davis
Tina Dell'Armi
Pavlin Dobrev
Yuzhang Feng
Alfio Ferrara
Lorenzo Gallucci
Michael Gesmann
Georgios Giannakopoulos
Miranda Grahl
Gunnar Grimnes
Tom Heath
Michiel Hildebrand
Matthew Horridge
José Iria
Helene Jaudoin
Alissa Kaplunova
Atila Kaya
Maria Keet
Malte Kiesel
Pavel Klinov
Jacek Kopecký
Markus Krötzsch
Steffen Lamparter
Domenico Lembo
Davide Lorusso
Zlatina Marinova

[†] Also member of the Demo Programme Committee.

Rainer Marrone
Mihhail Matskin
Eduardo Mena
Malgorzata Mochol
Stefano Montanelli
Kreshnik Musaraj
Preslav Nakov
Andrew Newman
Vit Novácek
Eyal Oren
Mary Parmelee
Georgios Petasis
Valentina Presutti
Vladislav Ryzhikov
Katharina Reinecke
Quentin Reul
Marco Rospocher
Marta Sabou
Ahmad El Sayed
Francois Scharffe
Roman Schindlauer
Thomas Schneider
Joachim Selke
Arash Shaban-Nejad
Elena Simperl
Giorgos Stoilos
Ondřej Šváb-Zamazal
Stuart Taylor
Yannis Theoharis
Jonathan Tivel
Farouk Toumani
Panayiotis Tritakis
Vassileios Tsetsos
Nwe Ni Tun
Miroslav Vacura
Dimitris Vogiatzis
Sebastian Wandelt
Jiewen Wu
Elias Zavitsanos
Zhangbing Zhou
Mathieu d'Aquin

Alessandro Martello
Michael Meier
Iris Miliaraki
Fergal Monaghan
Thomas Moser
Knud Möller
Yannick Naudet
Deborah Nichols
Martin Francis O'Connor
Ruben Ortiz
Jorge Perez
Sergios Petridis
Guilin Qi
Paraskevi Raftopoulou
Vassiliki Rentoumi
Francesco Ricca
Sebastian Rudolph
Leonardo Salayandia
Simon Scerri
Simon Schenk
Michael Schmidt
Alexander Schutz
Sinan Sen
Kostyantyn Shchekotykhin
Vassilis Spiliopoulos
Fabian Suchanek
Martin Szomszor
VinhTuan Thai
Barbara Thönssen
Ioan Toma
Thanh Tran
Tuvshintur Tserendorj
Giovanni Tummarello
Christopher J. Tuot
Julien Velcin
Ingo Weber
Cathrin Weiss
Markus Zanker
Ilias Zavitsanos
Claudia d'Amato

XII   Organization

Sponsors

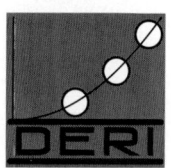  

A Paul G. Allen Company

 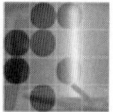

Table of Contents

Invited Talks

From Capturing Semantics to Semantic Search: A Virtuous Cycle 1
  Ricardo Baeza-Yates

Foundations of RDF Databases 3
  Claudio Gutierrez

Garlik: Semantic Technology for the Consumer 4
  Nigel Shadbolt

Agents

Semantic Web Technology for Agent Communication Protocols 5
  *Idoia Berges, Jesús Bermúdez, Alfredo Goñi, and
  Arantza Illarramendi*

xOperator – Interconnecting the Semantic Web and Instant Messaging
Networks 19
  Sebastian Dietzold, Jörg Unbehauen, and Sören Auer

Application Ontologies

An Ontology for Software Models and Its Practical Implications for
Semantic Web Reasoning 34
  Matthias Bräuer and Henrik Lochmann

A Core Ontology for Business Process Analysis 49
  Carlos Pedrinaci, John Domingue, and Ana Karla Alves de Medeiros

Applications

Assisting Pictogram Selection with Semantic Interpretation 65
  Heeryon Cho, Toru Ishida, Toshiyuki Takasaki, and Satoshi Oyama

KonneX^{SALT}: First Steps Towards a Semantic Claim Federation
Infrastructure 80
  Tudor Groza, Siegfried Handschuh, Knud Möller, and Stefan Decker

Building a National Semantic Web Ontology and Ontology Service
Infrastructure – The FinnONTO Approach 95
  Eero Hyvönen, Kim Viljanen, Jouni Tuominen, and Katri Seppälä

Creating and Using Geospatial Ontology Time Series in a Semantic
Cultural Heritage Portal 110
  Tomi Kauppinen, Jari Väätäinen, and Eero Hyvönen

Semantic Email as a Communication Medium for the Social Semantic
Desktop 124
  Simon Scerri, Siegfried Handschuh, and Stefan Decker

IVEA: An Information Visualization Tool for Personalized Exploratory
Document Collection Analysis 139
  VinhTuan Thai, Siegfried Handschuh, and Stefan Decker

Building a Semantic Web Image Repository for Biological Research
Images 154
  Jun Zhao, Graham Klyne, and David Shotton

Formal Languages

Mapping Validation by Probabilistic Reasoning 170
  Silvana Castano, Alfio Ferrara, Davide Lorusso,
  Tobias Henrik Näth, and Ralf Möller

Safe and Economic Re-Use of Ontologies: A Logic-Based Methodology
and Tool Support 185
  Ernesto Jiménez-Ruiz, Bernardo Cuenca Grau, Ulrike Sattler,
  Thomas Schneider, and Rafael Berlanga

dRDF: Entailment for Domain-Restricted RDF 200
  Reinhard Pichler, Axel Polleres, Fang Wei, and Stefan Woltran

Finite Model Reasoning in $DL\text{-}Lite$ 215
  Riccardo Rosati

Module Extraction and Incremental Classification: A Pragmatic
Approach for $\mathcal{EL}^+$ Ontologies 230
  Boontawee Suntisrivaraporn

Forgetting Concepts in DL-Lite 245
  Zhe Wang, Kewen Wang, Rodney Topor, and Jeff Z. Pan

Foundational Issues

An Entity Name System (ENS) for the Semantic Web 258
  Paolo Bouquet, Heiko Stoermer, and Barbara Bazzanella

A Functional Semantic Web Architecture 273
   Aurona Gerber, Alta van der Merwe, and Andries Barnard

Learning

Query Answering and Ontology Population: An Inductive Approach 288
   Claudia d'Amato, Nicola Fanizzi, and Floriana Esposito

Instance Based Clustering of Semantic Web Resources 303
   Gunnar AAstrand Grimnes, Peter Edwards, and Alun Preece

Conceptual Clustering and Its Application to Concept Drift and
Novelty Detection 318
   Nicola Fanizzi, Claudia d'Amato, and Floriana Esposito

Ontologies and Natural Language

Enriching an Ontology with Multilingual Information 333
   Mauricio Espinoza, Asunción Gómez Pérez, and Eduardo Mena

Rabbit: Developing a Control Natural Language for Authoring
Ontologies 348
   Glen Hart, Martina Johnson, and Catherine Dolbear

A Natural Language Query Interface to Structured Information 361
   Valentin Tablan, Danica Damljanovic, and Kalina Bontcheva

Distinguishing between Instances and Classes in the Wikipedia
Taxonomy 376
   Cäcilia Zirn, Vivi Nastase, and Michael Strube

Ontology Alignment

Two Variations on Ontology Alignment Evaluation: Methodological
Issues 388
   Laura Hollink, Mark van Assem, Shenghui Wang,
   Antoine Isaac, and Guus Schreiber

Putting Ontology Alignment in Context: Usage Scenarios, Deployment
and Evaluation in a Library Case 402
   Antoine Isaac, Henk Matthezing, Lourens van der Meij,
   Stefan Schlobach, Shenghui Wang, and Claus Zinn

CSR: Discovering Subsumption Relations for the Alignment of
Ontologies 418
   Vassilis Spiliopoulos, Alexandros G. Valarakos, and George A. Vouros

Query Processing

XSPARQL: Traveling between the XML and RDF Worlds – and
Avoiding the XSLT Pilgrimage 432
  *Waseem Akhtar, Jacek Kopecký, Thomas Krennwallner, and
  Axel Polleres*

Streaming SPARQL - Extending SPARQL to Process Data Streams 448
  Andre Bolles, Marco Grawunder, and Jonas Jacobi

The Creation and Evaluation of iSPARQL Strategies for
Matchmaking... 463
  Christoph Kiefer and Abraham Bernstein

Adding Data Mining Support to SPARQL Via Statistical Relational
Learning Methods.. 478
  Christoph Kiefer, Abraham Bernstein, and André Locher

A Semantic Web Middleware for Virtual Data Integration on the
Web.. 493
  Andreas Langegger, Wolfram Wöß, and Martin Blöchl

Graph Summaries for Subgraph Frequency Estimation 508
  *Angela Maduko, Kemafor Anyanwu, Amit Sheth, and
  Paul Schliekelman*

Querying Distributed RDF Data Sources with SPARQL 524
  Bastian Quilitz and Ulf Leser

Improving Interoperability Using Query Interpretation in Semantic
Vector Spaces 539
  *Anthony Ventresque, Sylvie Cazalens, Philippe Lamarre, and
  Patrick Valduriez*

Search

Hybrid Search: Effectively Combining Keywords and Semantic
Searches 554
  *Ravish Bhagdev, Sam Chapman, Fabio Ciravegna,
  Vitaveska Lanfranchi, and Daniela Petrelli*

Combining Fact and Document Retrieval with Spreading Activation for
Semantic Desktop Search 569
  Kinga Schumacher, Michael Sintek, and Leo Sauermann

Q2Semantic: A Lightweight Keyword Interface to Semantic Search 584
  Haofen Wang, Kang Zhang, Qiaoling Liu, Thanh Tran, and Yong Yu

Semantic Web Services

Conceptual Situation Spaces for Semantic Situation-Driven Processes ... 599
Stefan Dietze, Alessio Gugliotta, and John Domingue

Combining SAWSDL, OWL-DL and UDDI for Semantically Enhanced
Web Service Discovery 614
Dimitrios Kourtesis and Iraklis Paraskakis

Web Service Composition with User Preferences 629
Naiwen Lin, Ugur Kuter, and Evren Sirin

Enhancing Workflow with a Semantic Description of Scientific Intent ... 644
Edoardo Pignotti, Peter Edwards, Alun Preece, Nick Gotts, and Gary Polhill

WSMO Choreography: From Abstract State Machines to Concurrent
Transaction Logic 659
Dumitru Roman, Michael Kifer, and Dieter Fensel

WSMO-Lite Annotations for Web Services 674
Tomas Vitvar, Jacek Kopecký, Jana Viskova, and Dieter Fensel

Storage and Retrieval of Semantic Web Data

Semantic Sitemaps: Efficient and Flexible Access to Datasets on the
Semantic Web 690
Richard Cyganiak, Holger Stenzhorn, Renaud Delbru, Stefan Decker, and Giovanni Tummarello

On Storage Policies for Semantic Web Repositories That Support
Versioning 705
Yannis Tzitzikas, Yannis Theoharis, and Dimitris Andreou

User Interface and Personalization

Semantic Reasoning: A Path to New Possibilities of Personalization 720
Yolanda Blanco-Fernández, José J. Pazos-Arias, Alberto Gil-Solla, Manuel Ramos-Cabrer, and Martín López-Nores

An User Interface Adaptation Architecture for Rich Internet
Applications... 736
Kay-Uwe Schmidt, Jörg Dörflinger, Tirdad Rahmani, Mehdi Sahbi, Ljiljana Stojanovic, and Susan Marie Thomas

OntoGame: Weaving the Semantic Web by Online Games 751
Katharina Siorpaes and Martin Hepp

Demo Papers

SWING: An Integrated Environment for Geospatial Semantic Web
Services.. 767
  Mihai Andrei, Arne Berre, Luis Costa, Philippe Duchesne,
  Daniel Fitzner, Miha Grcar, Jörg Hoffmann, Eva Klien,
  Joel Langlois, Andreas Limyr, Patrick Maue, Sven Schade,
  Nathalie Steinmetz, Francois Tertre, Laurentiu Vasiliu,
  Raluca Zaharia, and Nicolas Zastavni

Semantic Annotation and Composition of Business Processes with
Maestro 772
  Matthias Born, Jörg Hoffmann, Tomasz Kaczmarek,
  Marek Kowalkiewicz, Ivan Markovic, James Scicluna,
  Ingo Weber, and Xuan Zhou

Learning Highly Structured Semantic Repositories from Relational
Databases: The RDBToOnto Tool 777
  Farid Cerbah

Cicero: Tracking Design Rationale in Collaborative Ontology
Engineering 782
  Klaas Dellschaft, Hendrik Engelbrecht, José Monte Barreto,
  Sascha Rutenbeck, and Steffen Staab

xOperator – An Extensible Semantic Agent for Instant Messaging
Networks 787
  Sebastian Dietzold, Jörg Unbehauen, and Sören Auer

LabelTranslator - A Tool to Automatically Localize an Ontology 792
  Mauricio Espinoza, Asunción Gómez Pérez, and Eduardo Mena

RKBExplorer.com: A Knowledge Driven Infrastructure for Linked Data
Providers 797
  Hugh Glaser, Ian C. Millard, and Afraz Jaffri

Semantic Browsing with PowerMagpie 802
  Laurian Gridinoc, Marta Sabou, Mathieu d'Aquin,
  Martin Dzbor, and Enrico Motta

Tagster - Tagging-Based Distributed Content Sharing................ 807
  Olaf Görlitz, Sergej Sizov, and Steffen Staab

The Web Service Modeling Toolkit 812
  Mick Kerrigan and Adrian Mocan

Mymory: Enhancing a Semantic Wiki with Context Annotations 817
  Malte Kiesel, Sven Schwarz, Ludger van Elst, and Georg Buscher

Pronto: A Non-monotonic Probabilistic Description Logic Reasoner 822
  Pavel Klinov

User Profiling for Semantic Browsing in Medical Digital Libraries 827
  Patty Kostkova, Gayo Diallo, and Gawesh Jawaheer

SWiM – A Semantic Wiki for Mathematical Knowledge Management ... 832
  Christoph Lange

Integrating Open Sources and Relational Data with SPARQL 838
  Orri Erling and Ivan Mikhailov

Previewing Semantic Web Pipes 843
  *Christian Morbidoni, Danh Le Phuoc, Axel Polleres,
  Matthias Samwald, and Giovanni Tummarello*

Demo: Visual Programming for the Semantic Desktop with Konduit 849
  *Knud Möller, Siegfried Handschuh, Sebastian Trüg, Laura Josan,
  and Stefan Decker*

SCARLET: SemantiC RelAtion DiscoveRy by Harvesting OnLinE
OnTologies 854
  Marta Sabou, Mathieu d'Aquin, and Enrico Motta

ODEWiki: A Semantic Wiki That Interoperates with the ODESeW
Semantic Portal 859
  *Adrián Siles, Angel López-Cima, Oscar Corcho, and
  Asunción Gómez-Pérez*

Simplifying Access to Large-Scale Health Care and Life Sciences
Datasets 864
  *Holger Stenzhorn, Kavitha Srinivas, Matthias Samwald, and
  Alan Ruttenberg*

GRISINO - An Integrated Infrastructure for Semantic Web Services,
Grid Computing and Intelligent Objects 869
  Ioan Toma, Tobias Bürger, Omair Shafiq, and Daniel Döegl

SemSearch: Refining Semantic Search 874
  Victoria Uren, Yuangui Lei, and Enrico Motta

The Combination of Techniques for Automatic Semantic Image
Annotation Generation in the IMAGINATION Application 879
  Andreas Walter and Gabor Nagypal

WSMX: A Solution for B2B Mediation and Discovery Scenarios 884
  Maciej Zaremba and Tomas Vitvar

Conceptual Spaces in ViCoS 890
  Claus Zinn

Author Index 895

From Capturing Semantics to Semantic Search: A Virtuous Cycle

Ricardo Baeza-Yates

Yahoo! Research, Barcelona, Spain

Abstract. Semantic search seems to be an elusive and fuzzy target to IR, SW and NLP researchers. One reason is that this challenge lies in between all those fields, which implies a broad scope of issues and technologies that must be mastered. In this extended abstract we survey the work of Yahoo! Research at Barcelona to approach this problem. Our research is intended to produce a virtuous feedback circuit by using machine learning for capturing semantics, and, ultimately, for better search.

Summary

Nowadays, we do not need to argue that being able to search the Web is crucial. Although people can browse or generate content, they cannot synthesize Web data without machine processing. But synthesizing data is what search engines do, based not only in Web content, but also in the link structure and how people use search engines. However, we are still far from a real semantic search. In [2] we argue that semantic search has not occurred for three main reasons. First, this integration is an extremely hard scientific problem. Second, the Web imposes hard scalability and performance restrictions. Third, there is a cultural divide between the Semantic Web and IR disciplines. We are simultaneously working in these three issues.

Our initial efforts are based in shallow semantics to improve search. For that we first need to develop fast parsers without losing quality in the semantic tagging process. Attardi and Ciaramita [1] have shown that this is possible. Moreover, Zaragoza *et al* [7] have shared a semantically tagged version of the Wikipedia. The next step is to rank sentences based on semantic annotations, and preliminary results in this problem are presented in [8]. Further results are soon expected, where one additional semantic source to exploit in the future is time [3].

The Semantic Web dream would effectively make search easier and hence, semantic search trivial. However, the Semantic Web is more a social rather than a technological problem. Hence, we need to help the process of adding semantics by using automatic techniques. A first source of semantics can be found in the Web 2.0. One example is the Flickr folksonomy. Sigurbjornsson and Van Zwol [5] have shown how to use collective knowledge (or the wisdom of crowds) to extend image tags, and also they prove that almost 80% of the tags can be semantically classified by using Wordnet and Wikipedia [6]. This effectively improves image search. A second source of implicit semantics are queries and the actions after them. In fact, in [4] we present a first step to infer semantic relations by defining equivalent, more specific, and related queries, which can represent an implicit folksonomy. To evaluate the quality of the results we used the Open Directory Project, showing that equivalence or specificity have precision of over 70% and 60%, respectively. For the cases that were not found in the ODP, a manually verified sample showed that the real precision was close to 100%. What happened was that the ODP was not specific enough to contain those relations, and every day the problem gets worse as we have more data. This shows the real power of the wisdom of the crowds, as queries involve almost all Internet users.

By being able to generate semantic resources automatically, even with noise, and coupling that with open content resources, we create a virtuous feedback circuit. In fact, explicit and implicit folksonomies can be used to do supervised machine learning without the need of manual intervention (or at least drastically reduce it) to improve semantic tagging. After, we can feedback the results on itself, and repeat the process. Using the right conditions, every iteration should improve the output, obtaining a virtuous cycle. As a side effect, in each iteration, we can also improve Web search, our main goal.

References

1. Attardi, G., Ciaramita, M.: Tree Revision Learning for Dependency Parsing. In: Proceedings of the HLT-NAACL 2007 Conference, Rochester, USA (2007)
2. Baeza-Yates, R., Mika, P., Zaragoza, H.: Search, Web 2.0, and the Semantic Web. In: Benjamins, R. (ed.) Trends and Controversies: Near-Term Prospects for Semantic Technologies, IEEE Intelligent Systems, January-February 2008, vol. 23(1), pp. 80–82 (2008)
3. Alonso, O., Gertz, M., Baeza-Yates, R.: On the Value of Temporal Information in Information Retrieval. ACM SIGIR Forum 41(2), 35–41 (2007)
4. Baeza-Yates, R., Tiberi, A.: Extracting Semantic Relations from Query Logs. In: ACM KDD 2007, San Jose, California, USA, August 2007, pp. 76–85 (2007)
5. Sigurbjornsson, B., Van Zwol, R.: Flickr Tag Recommendation based on Collective Knowledge. In: WWW 2008, Beijing, China (April 2008)
6. Overell, S., Sigurbjornsson, B., Van Zwol, R.: Classifying Tags using Open Content Resources (submitted for publication, 2008)
7. Zaragoza, H., Atserias, J., Ciaramita, M., Attardi, G.: Semantically Annotated Snapshot of the English Wikipedia v.0 (SW0) (2007), `research.yahoo.com/`
8. Zaragoza, H., Rode, H., Mika, P., Atserias, J., Ciaramita, M., Attardi, G.: Ranking Very Many Typed Entities on Wikipedia. In: CIKM 2007: Proceedings of the sixteenth ACM international conference on Information and Knowledge Management, Lisbon, Portugal (2007)

Foundations of RDF Databases*

Claudio Gutierrez

Department of Computer Science
Universidad de Chile
http://www.dcc.uchile.cl/cgutierr/

The motivation behind the development of RDF was, to borrow the words Tim Berners-Lee used for the Semantic Web, "to have a common and minimal language to enable to map large quantities of existing data onto it so that the data can be analyzed in ways never dreamed of by its creators." To bring to reality this vision, the processing of RDF data at big scale must be viable. This challenge amounts essentially to develop the theory and practice of RDF databases.

In this talk, we will present the current state of the theory of RDF databases, from the perspective of the work of our (Santiago, Chile) group. We will start discussing the RDF data model from a database perspective [3]. Then we will present the basis of querying RDF and RDFS data, and discuss the pattern matching paradigm for querying RDF [2]. (We will also briefly discuss the challenges of viewing RDF from a graph database perspective [1].) Finally we will discuss SPARQL, the recent W3C recommendation query language for RDF, its semantics, expressiveness and complexity [4].

References

1. Angles, R., Gutierrez, C.: Querying RDF Data from a Graph Database Perspective. In: Gómez-Pérez, A., Euzenat, J. (eds.) ESWC 2005. LNCS, vol. 3532, Springer, Heidelberg (2005)
2. Gutierrez, C., Hurtado, C., Mendelzon, A.O.: Foundations of Semantic Web Databases. In: ACM Symposium on Principles of Database Systems, PODS (2004)
3. Munoz, S., Perez, J., Gutierrez, C.: Minimal Deductive Systems for RDF. In: Franconi, E., Kifer, M., May, W. (eds.) ESWC 2007. LNCS, vol. 4519, Springer, Heidelberg (2007)
4. Perez, J., Arenas, M., Gutierrez, C.: The Semantics and Complexity of SPARQL. In: Cruz, I., Decker, S., Allemang, D., Preist, C., Schwabe, D., Mika, P., Uschold, M., Aroyo, L.M. (eds.) ISWC 2006. LNCS, vol. 4273, Springer, Heidelberg (2006)

---

* This research is supported by FONDECYT 1070348.

Garlik: Semantic Technology for the Consumer

Nigel Shadbolt

CTO Garlik Ltd
Richmond
UK
School of Electronics and Computer Science
University of Southampton
Highfield
Southampton, SO17 1BJ
United Kingdom
http://users.ecs.soton.ac.uk/nrs/

In under a decade the internet has changed our lives. Now we can shop, bank, date, research, learn and communicate online and every time we do we leave behind a trail of personal information. Organisations have a wealth of structured information about individuals on large numbers of databases. What does the intersection of this information mean for the individual? How much of your personal data is out there and more importantly, just who has access to it? As stories of identity theft and online fraud fill the media internet users are becoming increasingly nervous about their online data security. Also what opportunities arise for individuals to exploit this information for their own benefit?

Garlik was formed to give individuals and their families real power over the use of their personal information in the digital world. Garlik's technology base has exploited and extended results from research on the Semantic Web. It has built the world's largest, SPARQL compliant, native format, RDF triple store. The store is implemented on a low-cost network cluster with over 100 servers supporting a 24x7 operation. Garlik has built semantically informed search and harvesting technology and has used industrial strength language engineering technologies across many millions of people-centric Web pages. Methods have been developed for extracting information from structured and semi structured databases. All of this information is organised against a people-centric ontology with facilities to integrate these various fragments.

Garlik has received two substantial rounds of venture capital funding (as of March 2008), has established an active user base of tens of thousands of individuals, and is adding paying customers at an increasing rate. This talk reviews the consumer need, describes the technology and engineering, and discusses the lessons we can draw about the challenges of deploying Semantic Technologies.

Semantic Web Technology for Agent Communication Protocols

Idoia Berges[*], Jesús Bermúdez, Alfredo Goñi, and Arantza Illarramendi[**]

University of the Basque Country
{iberges003,jesus.bermudez,alfredo,a.illarramendi}@ehu.es
http://siul02.si.ehu.es

Abstract. One relevant aspect in the development of the Semantic Web framework is the achievement of a real inter-agents communication capability at the semantic level. The agents should be able to communicate and understand each other using standard communication protocols freely, that is, without needing a laborious a priori preparation, before the communication takes place.

For that setting we present in this paper a proposal that promotes to describe standard communication protocols using Semantic Web technology (specifically, OWL-DL and SWRL). Those protocols are constituted by communication acts. In our proposal those communication acts are described as terms that belong to a communication acts ontology, that we have developed, called COMMONT. The intended semantics associated to the communication acts in the ontology is expressed through social commitments that are formalized as fluents in the Event Calculus.

In summary, OWL-DL reasoners and rule engines help in our proposal for reasoning about protocols. We define some comparison relationships (dealing with notions of equivalence and specialization) between protocols used by agents from different systems.

Keywords: Protocol, Communication acts, agents.

1 Introduction

In the scenario that promotes the emergent Web, administrators of existing Information Systems, that belong to nodes distributed along the Internet network, are encouraged to provide the functionalities of those systems through agents that represent them or through Web Services. The underlying idea is to get a real interoperation among those Information Systems in order to enlarge the benefits that users can get from the Web by increasing the machine processable tasks.

---

[*] The work of Idoia Berges is supported by a grant of the Basque Government.
[**] All authors are members of the Interoperable DataBases Group. This work is also supported by the University of the Basque Country, Diputación Foral de Gipuzkoa (cosupported by the European Social Fund) and the Spanish Ministry of Education and Science TIN2007-68091-C02-01.

Although agent technology and Web Services technology have been developed in a separate way, there exists a recent work of several members from both communities trying to consolidate their approaches into a common specification describing how to seamlessly interconnect FIPA compliant agent systems [1] with W3C compliant Web Services. The purpose of specifying an infrastructure for integrating these two technologies is to provide a common means of allowing each to discover and invoke instances of the other [2]. Considering the previous approach, in the rest of this paper we will only concentrate on inter-agent communication aspects.

Communication among agents is in general based on the interchange of communication acts. However, different Information Systems have incorporated different classes of communication acts as their Agent Communication Language (ACL) to the point that they do not understand each other. Moreover, protocols play a relevant role in agents communication. A protocol specifies the rules of interaction between agents by restricting the range of allowed follow-up communication acts for each agent at any stage during a communicative interaction. It is widely recognized the interest of using standard communication protocols.

We advocate so that the administrators of the Information Systems proceed in the following way. When they wish to implement the agents that will represent their systems, they first select, from a repository of standard protocols (there can exist one or more repositories), those protocols that fulfill the goals of their agents. Sometimes a single protocol will be sufficient and other times it will be necessary to design a protocol as a composition of some other protocols. Next, they can customize the selected protocols before they incorporate them to the agents. In that setting, when agents of different Information Systems want to interoperate it will be relevant to reason about the protocols embedded in the agents in order to discover relationships such as equivalence or restriction between them. Moreover, once those relationships are discovered both agents can use the same protocol by replacing dynamically in one agent the protocol supported by the other. Finally, in our opinion it will be desirable to use a formal language to represent the protocols.

In this paper we present a proposal that promotes to describe standard communication protocols using Semantic Web Technology (OWL-DL and SWRL). In addition, communication acts that take part of the protocols are described as terms that belong to a communication acts ontology, that we have developed, called COMMONT (see more details about the ontology in [3]). The use of that ontology favours on the one hand, the explicit representation of the meaning of the communication acts and on the other hand, the customization of existing standard protocols by allowing the use of particular communication acts that can be defined as specializations of existing standard communication acts.

Terms of the COMMONT ontology are described using OWL-DL and we have adopted the so called *social approach* [4,5] for expressing the intended semantics of the communication acts included in the protocols. According to the social approach, when agents interact they become involved in social commitments or obligations to each other. Those commitments are public, and therefore they are suitable for an objective and verifiable semantics of agent interaction. Social commitments can be considered as *fluents* in the Event Calculus, which is a logic-based formalism for representing actions and their effects. Fluents are propositions that hold during time intervals. A formula in the Event Calculus is associated to a communication act for describing its social effects. The set of fluents that hold at a moment describes a state of the interaction. DL axioms and Event Calculus formulae apply to different facets of communication acts. DL axioms describe static features and are principally used for communication act interpretation purposes. Event Calculus formulae describe dynamic features, namely the social effects of communication acts, and are principally used for communication act operational contexts such as supervising conversations.

In summary the main contributions of the proposal presented in this paper are:

- It favours a flexible interoperation among agents of different systems by using standard communication protocols described through tools promoted by the W3C.
- It facilitates the customization of those standard communication protocols allowing to use communication acts in the protocols that belong to specific ACL of Information Systems. The particular communication acts are described in an ontology.
- It provides a basis to reason about relationships between two protocols in such a way that the following relations can be discovered: equivalence or restriction (and also considering a notion of specialization). Moreover, notice that our approach allows to get protocols classification in terms of the intended semantics of communication acts that appear in the protocols.
- It allows modeling the communication among agents without regarding only to the lower level operational details of how communication acts are interchanged but taking also into account the meaning of those acts.

The rest of the paper is organized as follows: Section 2 provides background on the communication ontology, that contains terms corresponding to communication acts that appear in the protocols, and on the semantics associated to those acts. Section 3 explains how protocols are described using Semantic Web Technology and presents the definitions of the relationships considered between protocols. Section 4 discusses different related works, and conclusions appear in the last section.

2 Two Basic Supports for the Proposal: The CommOnt Ontology and the Representation of the Semantics of Communication Acts

Among the different models proposed for representing protocols one which stands out is that of State Transition Systems (STS).

Definition 1. *A State Transition System is a tuple (S, s_0, L, T, F), where S is a finite set of states, $s_0 \in S$ is an initial state, L is a finite set of labels, $T \subseteq S \times L \times S$ is a set of transitions and $F \subseteq S$ is a set of final states.*

In our proposal we use STS where transitions are labeled with communication act classes described in a communication acts ontology called COMMONT. That is to say, the set of labels L is a set of class names taken from that ontology. Moreover, as mentioned before, the intended semantics associated to the communication acts in the ontology is expressed through predicates in the Event Calculus that initiate or terminate fluents. In our case, each state is associated to the set of fluents that holds at that moment.

In the following two subsections we present the main features of the COMMONT ontology and of the intended semantics associated to communication acts, respectively.

2.1 Main Features of the CommOnt Ontology

The goal of the COMMONT ontology is to favour the interoperation among agents belonging to different Information Systems. The leading categories of that ontology are: first, *communication acts* that are used for interaction by *actors* and that have different purposes and deal with different kinds of contents; and second, *contents* that are the sentences included in the communication acts.

The main design criteria adopted for the communication acts category of the COMMONT ontology is to follow the *speech acts* theory [6], a linguistic theory that is recognized as the principal source of inspiration for designing the most familiar standard agent communication languages. Following that theory every communication act is the sender's expression of an attitude toward some possibly complex proposition. A sender performs a communication act which is expressed by a coded message and is directed to a receiver. Therefore, a communication act has two main components. First, the attitude of the sender which is called the *illocutionary force* (F), that expresses social interactions such as informing, requesting or promising, among others. And second, the *propositional content* (p) which is the subject of what the attitude is about. In COMMONT this $F(p)$ framework is followed, and different kinds of illocutionary forces and contents leading to different classes of communication acts are supported. More specifically, specializations of illocutionary forces that facilitate the absorption of aspects of the content into the illocutionary force are considered.

COMMONT is divided into three interrelated layers: *upper*, *standards* and *applications*, that group communication acts at different levels of abstraction. Classes of the COMMONT ontology are described using the Web Ontology Language OWL-DL. Therefore, communication acts among agents that commit to COMMONT have an abstract representation as individuals of a shared universal class of communication acts.

In the upper layer, according to Searle's speech acts theory, five upper classes of communication acts corresponding to *Assertives*, *Directives*, *Commissives*, *Expressives* and *Declaratives* are specified. But also the top class `CommunicationAct`[1] is defined, which represents the universal class of communication acts. Every particular communication act is an individual of this class.

---

[1] This `type` style refers to terms specified in the ontology.

In COMMONT, components of a class are represented by properties. The most immediate properties of `CommunicationAct` are the content and the actors who send and receive the communication act. There are some other properties related to the context of a communication act such as the conversation in which it is inserted or a link to the domain ontology that includes the terms used in the content.

A standards layer extends the upper layer of the ontology with specific terms that represent classes of communication acts of general purpose agent communication languages, like those from KQML or FIPA-ACL. Although the semantic framework of those agent communication languages may differ from the semantic framework adopted in COMMONT, in our opinion enough basic concepts and principles are shared to such an extent that a commitment to ontological relationships can be undertaken in the context of the interoperation of Information Systems.

With respect to FIPA-ACL, we can observe that it proposes four primitive communicative acts [1]: *Confirm*, *Disconfirm*, *Inform* and *Request*. The terms `FIPA-Confirm`, `FIPA-Disconfirm`, `FIPA-Inform` and `FIPA-Request` are used to respectively represent them as classes in COMMONT. Furthermore, the rest of the FIPA communicative acts are derived from those mentioned four primitives. Analogously, communication acts from KQML can be analyzed and the corresponding terms in COMMONT specified. It is of vital relevance for the interoperability aim to be able of specifying ontological relationships among classes of different standards.

Finally, it is often the case that every single Information System uses a limited collection of communication acts that constitute its particular agent communication language. The applications layer reflects the terms describing communication acts used in such particular Information Systems. The applications layer of the COMMONT ontology provides a framework for the description of the nuances of such communication acts. Some of those communication acts can be defined as particularizations of existing classes in the standards layer and maybe some others as particularizations of upper layer classes. Interoperation between agents of two systems using different kinds of communication acts will proceed through these upper and standard layer classes.

Following we show some axioms in the COMMONT ontology. For the presentation we prefer a logic notation instead of the more verbose OWL/XML syntax.

$$\text{CommunicationAct} \sqsubseteq =1 \text{ hasSender.Actor} \sqcap \forall \text{hasReceiver.Actor} \sqcap$$
$$\forall \text{hasContent.Content}$$
$$\text{Request} \sqsubseteq \text{Directive} \sqcap \exists \text{hasContent.Command}$$
$$\text{Accept} \sqsubseteq \text{Declarative}$$
$$\text{Responsive} \sqsubseteq \text{Assertive} \sqcap \exists \text{inReplyTo.Request}$$

2.2 Semantics Associated to Communication Acts

Formal semantics based on mental concepts such as *beliefs*, *desires* and *intentions* have been developed for specifying the semantics of communication acts.

However, they have been criticized on their approach [4] as well as on their analytical difficulties [7]. We have adopted the so called social approach [5,8,9] to express the intended semantics of communication acts described in the COMMONT ontology. According to the social approach, when agents interact they become involved in social commitments or obligations to each other.

Definition 2. *A base-level commitment C(x, y, p) is a ternary relation representing a commitment made by x (the debtor) to y (the creditor) to bring about a certain proposition p.*

Sometimes an agent accepts a commitment only if a certain condition holds or, interestingly, only when a certain commitment is made by another agent. This is called a conditional commitment.

Definition 3. *A conditional commitment CC(x, y, p, q) is a quaternary relation representing that if the condition p is brought out, x will be committed to y to bring about the proposition q.*

Moreover, the formalism we use for reasoning about commitments is based on the Event Calculus. The basic ontology of the Event Calculus comprises *actions*, *fluents* and *time points*. It also includes predicates for saying what happens when (*Happens*), for describing the initial situation (*Initially*), for describing the effects of actions (*Initiates* and *Terminates*), and for saying what fluents hold at what times (*HoldsAt*). See [10] for more explanations.

Commitments (base-level and conditional) can be considered fluents, and semantics of communication acts can be expressed with predicates. For example:

- *Initiates(Request(s,r,P)*, CC*(r, s, accept(r,s,P), P), t)*
  A Request from *s* to *r* produces the effect of generating a conditional commitment expressing that if the receiver *r* accepts the demand, it will be commited to the proposition in the content of the communication act.
- *Initiates(Accept(s,r,P), accept(s,r,P), t)*
  The sending of an Accept produces the effect of generating the accept fluent.

Furthermore, some rules are needed to capture the dynamics of commitments. Commitments are a sort of fluents typically put in force by communication acts and that become inoperative after the appearance of other fluents. In the following rules *e(x)* represents an event caused by *x*. The first rule declares that when a debtor of a commitment that is in force causes an event that initiates the proposition committed, the commitment ceases to hold.

RULE 1: *HoldsAt(* C*(x, y, p), t)* ∧ *Happens(e(x), t)* ∧ *Initiates(e(x), p, t)* → *Terminates(e(x),*C*(x, y, p), t).*

The second rule declares that a conditional commitment that is in force disappears and generates a base-level commitment when the announced condition is brought out by the creditor.

RULE 2: *HoldsAt(* CC*(x, y, c, p), t)* ∧ *Happens(e(y), t)* ∧ *Initiates(e(y), c, t)* → *Initiates(e(y),*C*(x, y, p), t)* ∧ *Terminates(e(y),*CC*(x, y, c, p), t).*

Following we state some predicates that describe the semantics asociated to some of the communication acts of the upper level of the COMMONT ontology. This semantics is determined by the fluents that are initiated or terminated as a result of the sending of a message between agents.

- *Initiates(Assertive(s,r,P), P, t)*
- *Initiates(Commissive(s,r,C,Γ),* CC*(s,r,C,P), t)*
- *Initiates(Responsive(s,r,P, RA), P, t)*
  *Terminates(Responsive(s,r,P, RA),*C*(s,r,RA), t)*

Effects of these predicates can be encoded with SWRL rules. For instance, the predicate *Initiates(Request(s, r, P),* CC*(r, s, accept(r, s, P), P), t)* can be encoded as follows:
 Request(x) ∧ *hasSender(x,s)* ∧ *hasReceiver(x,r)* ∧ *hasContent(x,p)* ∧ *hasCommit(x,c)* ∧ *isConditionedTo(c,a)* ∧ *atTime(x, t)* → *initiates(x,c)* ∧ *hasDebtor(c,r)* ∧ *hasCreditor(c,s)* ∧ *hascondition(c,p)* ∧ *Acceptance(a)* ∧ *hasSignatory(a,r)* ∧ *hasAddressee(a,s)* ∧ *hasObject(a,p)* ∧ *atTime(c, t)*

3 Protocol Description

As mentioned in the introduction, our proposal promotes to describe standard protocols using Semantic Web technology. We use STS as models of protocols. More specifically, we restrict to deterministic STS (i.e. if $(s, l, s') \in T$ and $(s, l, s'') \in T$ then $s' = s''$). In order to represent protocols using OWL-DL, we have defined five different classes: `Protocol`, `State`, `Transition`, `Fluent` and `Commitment`, which respectively represent protocols, states, transitions in protocols, fluents and commitments associated to states.

We model those class descriptions with the following guidelines: A state has fluents that hold in that point and transitions that go out of it. A transition is labelled by the communication act that is sent and is associated to the state that is reached with that transition. A fluent has a time stamp that signals the moment it was initiated. An actual conversation following a protocol is an individual of the class `Protocol`. Following are some of the ontology axioms:

$$
\begin{aligned}
\texttt{Protocol} &\equiv \exists \texttt{hasInitialState.State} \sqcap \\
&\quad \forall \texttt{hasInitialState.State} \\
\texttt{State} &\equiv \forall \texttt{hasTransition.Transition} \sqcap \\
&\quad \exists \texttt{hasFluent.Fluent} \sqcap \\
&\quad \forall \texttt{hasFluent.Fluent} \\
\texttt{Transition} &\equiv =1\ \texttt{hasCommAct.CommunicationAct} \sqcap \\
&\quad =1.\texttt{hasNextState.State} \\
\texttt{FinalState} &\sqsubseteq \texttt{State} \sqcap \\
&\quad \forall \texttt{hasFluent.(Fluent} \sqcap \neg \texttt{Commitment)} \\
\texttt{Fluent} &\sqsubseteq =1\ \texttt{atTime} \\
\texttt{Commitment} &\sqsubseteq \texttt{Fluent} \sqcap =1\ \texttt{hasDebtor.Actor} \sqcap \\
&\quad =1\ \texttt{hasCreditor.Actor} \sqcap
\end{aligned}
$$

ConditionalCommitment ⊑ Fluent ⊓ =1 hasDebtor.Actor ⊓
=1 hasCreditor.Actor ⊓
=1 hasCondition.Fluent ⊓
=1 isConditionedTo.Fluent
=1 hasCondition.Fluent

The OWL-DL description of protocols reflects their static features and can be used to discover structural relationships between protocols. For instance, in Fig. 1 we show a simple protocol where agent A asks for time to agent B. The protocol description appears in the following:

Protocol AskTime

Fig. 1. Protocol AskTime

```
Asktime ≡ Protocol ⊓ ∃hasInitialState.S0
     S0 ≡ State ⊓ ∃hasTransition.T01 ⊓ ∃hasFluent.F0
     S1 ≡ State ⊓ ∃hasTransition.T12 ⊓ ∃hasFluent.F1
     S2 ≡ State ⊓ ∃hasTransition.T23 ⊓ ∃hasFluent.F2
     S3 ≡ FinalState ⊓ ∃hasFluent.F3
    T01 ≡ Transition ⊓ ∃hasCommAct.TimeRequest ⊓ ∃hasNextState.S1
    T12 ≡ Transition ⊓ ∃hasCommAct.TimeAccept ⊓ ∃hasNextState.S2
    T23 ≡ Transition ⊓ ∃hasCommAct.TimeInform ⊓ ∃hasNextState.S3
TimeRequest ≡ Request ⊓ =1 hasContent.TimeReq
 TimeAccept ≡ Accept ⊓ =1 hasContent.TimeReq
 TimeInform ≡ Responsive ⊓ =1 hasContent.TimeInfo ⊓ =1 inReplyTo.TimeRequest
```

However, dealing only with structural relationships is too rigid if a flexible interoperation among agents that use different standard protocols is promoted. For that reason, we propose to consider what we call *protocol traces*.

Definition 4. *A protocol trace is a sequence of time stamped fluents sorted in increasing order of time stamp.*

Notice that protocol traces are defined in terms of the semantics of communication acts, not in terms of the communication acts themselves; in contrast with many other related works (see section 4) that consider messages as atomic acts without considering their content, neither their semantics.

During a simulation of a protocol run we apply the SWRL rules that encode the semantics of the communication acts (see section 2.2) appearing in the run. Then, we can consider the sorted set of time stamped fluents that hold at a final state of the protocol. That set represents the effects of the protocol run. Following we show an example of the application of the rules to a run of protocol AskTime in Fig. 1.

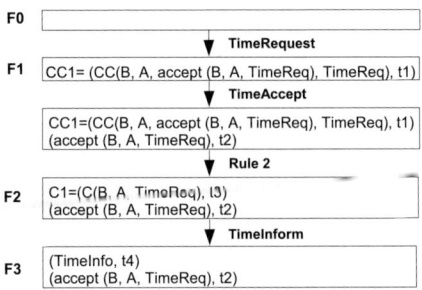

Fig. 2. Protocol fluents

In Fig. 2 we show which are the fluents associated to the states of the protocol and how they vary as a consequence of the communication acts that are sent and the rules described in section 2.2. We depart from a situation where the set of fluents is empty (F0). When the TimeRequest message is sent, due to the predicate *Initiates(Request(s, r, P), CC(r, s, accept(r,s,P), P), t)* the conditional commitment CC1 is initiated, which states that if agent B accepts to give information about the time, then it will be committed to do so; t_1 is the time stamp associated. By convention we sort time stamps by their subindexes, that is: $t_i < t_j$ if $i < j$. Then agent B agrees to respond by sending the TimeAccept message, and due to the predicate *Initiates(Accept(s,r,P), accept(s,r,P), t)*, the fluent *accept(B, A, TimeReq)* is initiated at time t_2. At this point, Rule 2 (see section 2.2) can be applied, so CC1 is terminated and the base commitment C1 is initiated at time t_3. Finally, agent B sends the TimeInform message, and because of the predicates *Initiates(Responsive(s,r,P, RA), P, t)* and *Terminates(Responsive(s,r,P, RA), C(s,r,RA), t)*, C1 is terminated and a new fluent, *TimeInfo*, is initiated at time t_4. So, at this point we can say that the fluents that hold at the final state of the protocol are $(accept(B, A, TimeReq), t_2)$ and $(TimeInfo, t_4)$.

Then, we say that the protocol trace $[(accept(B, A, TimeReq), t_2), (TimeInfo, t_4)]$ is *generated* by the protocol. We denote $\mathcal{T}(A)$ to the set of all protocol traces generated by a protocol A.

Now, we proceed with the definitions of relationships between protocols we are considering. Our relationships are not structure-based but effect-based. Intuitively, two protocols are equivalent if the same effects take place in the same relative order. Runs of a protocol are made up of communication acts, and fluents are the effects they leave.

Definition 5. *Protocol A is equivalent to protocol B if* $\mathcal{T}(A) = \mathcal{T}(B)$.

Sometimes, a protocol is defined by restrictions on the allowable communication acts at some states of a more general protocol. In those situations the application of those restrictions is reflected in the corresponding effects.

Definition 6. *Protocol A is a restriction of protocol B if* $\mathcal{T}(A) \subset \mathcal{T}(B)$.

Protocols for specific Information Systems may use specialized communication acts. Specialization can also be applied also to domain actions that can be represented by specialized fluents.

Definition 7. *A protocol trace t is a specialization of a protocol trace s, written $t \ll s$, if $\forall i.\ t(i) \sqsubseteq s(i)$ in an ontology of fluents.*

Definition 8. *Protocol A is a specialized-equivalent of protocol B if $\forall t \in \mathcal{T}(A)$. $\exists s \in \mathcal{T}(B).\ t \ll s$ and $\forall s \in \mathcal{T}(B).\ \exists t \in \mathcal{T}(A).\ t \ll s$.*

Definition 9. *Protocol A is a specialized-restriction of protocol B if $\forall t \in \mathcal{T}(A)$. $\exists s \in \mathcal{T}(B).\ t \ll s$.*

Notice that all those relationships can be easily discovered by straightforward algorithms supported by OWL-DL reasoners. Those reasoners deal with the ontology descriptions and rule engines that consider our semantic rules for generating protocol traces.

Moreover, sometimes we may be interested in comparing protocol traces independently of time stamps. That is, we may be interested in knowing if a protocol produces the same fluents as another, in whatever order. For example, in Fig. 3 we show two protocols that can be used to ask for information related to the vital signs temperature and pulse. In fact, for that purpose it is irrelevant the order in which the two requests are done.

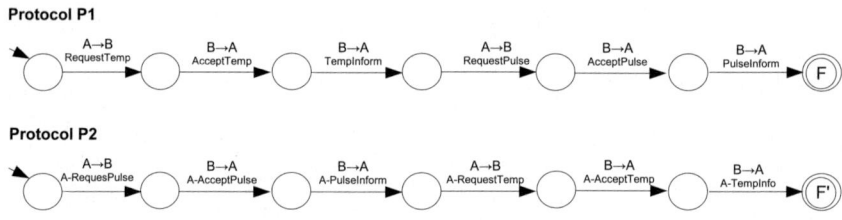

Fig. 3. Specialization of protocols

In protocol P1 we can find a general protocol, in which agent A makes a request about the temperature using the communication act `RequestTemp` for that purpose. Then, agent B accepts and replies with a `TempInform` message, which is used to give information about the temperature. Once agent A receives this information, it asks agent B information about the pulse using a `RequestPulse`. Finally agent B accepts and replies with a `PulseInform` message and the final state F is reached. On the other hand, in protocol P2 we can find the specific protocol used by the agents of a specific system, called AINGERU[2], to exchange information about vital signs. Protocol P2 may be a specialization of an standard protocol. First, agent A asks for the pulse, using the communication act `A-RequestPulse`. Then, agent B accepts and responds to the request using the `A-PulseInform` message. Next, agent A sends a `A-RequestTemp` message to ask about the temperature. Finally, agent B accepts and replies using the `A-TempInform` message and

---

[2] The A- prefix intends to label the AINGERU terminology.

reaches state F'. Following we show the OWL specification for the communication acts used in this example.

$$
\begin{aligned}
\text{RequestTemp} &\equiv \text{Request} \sqcap =1 \text{ hasContent.TempReq} \\
\text{AcceptTemp} &\equiv \text{Accept} \sqcap =1 \text{ hasContent.TempReq} \\
\text{TempInform} &\equiv \text{Responsive} \sqcap =1 \text{ hasContent.TempInfo} \sqcap =1 \text{ inReplyTo.RequestTemp} \\
\text{RequestPulse} &\equiv \text{Request} \sqcap =1 \text{ hasContent.PulseReq} \\
\text{AcceptPulse} &\equiv \text{Accept} \sqcap =1 \text{ hasContent.PulseReq} \\
\text{PulseInform} &\equiv \text{Responsive} \sqcap =1 \text{ hasContent.PulseInfo} \sqcap =1 \text{ inReplyTo.RequestPulse} \\
\text{A-RequestTemp} &\equiv \text{RequestTemp} \sqcap =1 \text{ theSystem.Aingeru} \sqcap =1 \text{ hasContent.A-TempReq} \\
\text{A-AcceptTemp} &\equiv \text{AcceptTemp} \sqcap =1 \text{ theSystem.Aingeru} \sqcap =1 \text{ hasContent.A-TempReq} \\
\text{A-TempInform} &\equiv \text{TempInform} \sqcap =1 \text{ theSystem.Aingeru} \sqcap =1 \text{ hasContent.A-TempInfo} \sqcap \\
& =1 \text{ inReplyTo.A-RequestTemp} \\
\text{A-RequestPulse} &\equiv \text{RequestPulse} \sqcap =1 \text{ theSystem.Aingeru} \sqcap =1 \text{ hasContent.A-PulseReq} \\
\text{A-AcceptPulse} &\equiv \text{AcceptPulse} \sqcap =1 \text{ theSystem.Aingeru} \sqcap =1 \text{ hasContent.A-PulseReq} \\
\text{A-PulseInform} &\equiv \text{PulseInform} \sqcap =1 \text{ theSystem.Aingeru} \sqcap =1 \text{ hasContent.A-PulseInfo} \sqcap \\
& =1 \text{ inReplyTo.A-RequestPulse}
\end{aligned}
$$

Notice that every communication act in protocol P2 is a subclass of its counterpart in protocol P1 (i.e. `A-RequestPulse` $\sqsubseteq$ `RequestPulse`, etc.) and correspondingly `A-PulseInfo` $\sqsubseteq$ `PulseInfo`, etc., is also satisfied.

Through a reasoning procedure analogous to that explained with the example of the AskTime protocol, we get the following sets of protocol traces:

$\mathcal{T}(P1) = \{[(accept(B, A, TempReq), t_2), (TempInfo, t_4), (accept(B, A, PulseReq), t_6), (PulseInfo, t_8)] \}$

$\mathcal{T}(P2) = \{[(accept(B, A, A\text{-}PulseReq), t_2), (A\text{-}PulseInfo, t_4), (accept(B, A, A\text{-}TempReq), t_6), (A\text{-}TempInfo, t_8)] \}$

Even if the structure of the protocols is not exactly the same, we can relate both protocols by a shallow notion of specialization from the following point of view. If we get abstracted from time stamps, we can see protocol traces as multi-sets. Let us denote *abstract-time(t)* to the multi-set formed by the fluents appearing in the protocol trace t, without any time stamp associated. Now, we define

$\mathcal{S}(A) = \{abstract\text{-}time(t) | t \in \mathcal{T}(A)\}$

Then, we are in condition to define analogous relationships to the previous five, but in a shallow mood.

Definition 10. 1. *Protocol A is shallow-equivalent to protocol B if $\mathcal{S}(A) = \mathcal{S}(B)$.*
2. *Protocol A is a shallow-restriction of protocol B if $\mathcal{S}(A) \subset \mathcal{S}(B)$.*
3. *A protocol trace t is a shallow-specialization of a protocol trace s, written $t \ll_s s$, if there is a map ϕ from abstract-time(t) to abstract-time(s) such that $\forall f \in$ abstract-time(t).$f \sqsubseteq \phi(f)$ in an ontology of fluents.*
4. *Protocol A is a shallow-specialized-equivalent of protocol B if $\forall t \in \mathcal{S}(A). \exists s \in \mathcal{S}(B). t \ll_s s$ and $\forall s \in \mathcal{S}(B). \exists t \in \mathcal{S}(A). t \ll_s s$.*
5. *Protocol A is a shallow-specialized-restriction of protocol B if $\forall t \in \mathcal{S}(A). \exists s \in \mathcal{S}(B). t \ll_s s$.*

Finally, using our proposal, we can conclude that protocols P1 and P2 are *shallow-specialized-equivalent*, although they use different communications acts and have different structure.

4 Related Works

Among the different related works that we can find in the specialized literature, the closer work is [11], where protocols are represented as transition systems and subsumption and equivalence of protocols are defined with respect to three state similarity funtions. We share some goals with that work, but the protocol description formalism used by them is not considered in the paper and there is no references to how protocol relationships are computed. In contrast, we describe protocols with a description logic language and protocol relationships can be computed by straightforward algorithms. It is worth mentioning that protocol relationships considered in that paper deserve study in our framework.

The works of [12] and [13] are quite similar one to each other. Both capture the semantics of communication acts through agents' commitments and represent communication protocols using a set of rules that operate on these commitments. Moreover those rule sets can be compiled as finite state machines. Nevertheless, they do not consider the study of relationships between protocols. In addition, in [14], protocols are also represented with a set of rules with terms obtained from an ontology, but their main goal is protocol development and, in order to reason about protocol composition, they formalize protocols into the π-calculus. Then, equivalence through bisimulation is the only process relationship considered. In [15], they also consider commitment protocols; however, their main focus is on combining them with considerations of rationality on the enactment of protocols. Our proposal could be complemented with their approach.

An alternative way to describe finite state machines with a description logic language is to take advantage of the relationship of that logic with Deterministic Propositional Dynamic Logic, see [16] for an example in the context of Web Services composition. The approach of that paper is very different in purpose from ours. Their states and transitions descriptions are not prepared to be confronted in a comparison. In constrast, our state and transition descriptions are carefully modelled as class descriptions such that semantics relationships between protocols can be captured.

Also in the context of Web Services, state transition systems are used in [17] for representing dynamic behaviour of services and they define some notions of compatibility and substitutability of services that can be easily translated to the context of compatibility of protocols. Relationships between their compatibility relations and our defined relationships deserve study.

In [18] protocols are defined as a set of permissions and obligations of agents participating in the communication. They use an OWL ontology for defining the terms of the specification language, but their basic reasoning is made with an ad hoc reasoning engine. We share their main goal of defining protocols in a general framework that allows reutilization. Nevertheless, they do not consider relationships between protocols.

The problem of determining if an agent's policy is conformant to a protocol is a very important one, but we are not treating that topic in this paper. Nevertheless, the topic is close to ours and it is worth mentioning the following papers that consider different notions of conformance: In [19], deterministic finite state machines are the abstract models for protocols, which are described by simple logic-based programs. Three levels of conformance are defined: weak, exhaustive and robust. They consider communication acts as atomic actions, in contrast to our semantic view. In [20] a nondeterministic finite state automata is used to support a notion of conformance that guarantees interoperabiliy among agents conformant to a protocol. Their conformance notion considers the branching structure of policies and protocols and applies a simulation-based test. Communication acts are considered atomic actions, without considering their semantics. In [21], communication acts semantics is described in terms of commitments but it is not used for the conformance notion. A third different notion of conformance is defined and, moreover, it is proved orthogonal to their proposed notions of coverage and interoperability.

Finally, [22] and [23] use finite state machines and Petri nets, respectively, but without taking into account the meaning of the communication acts interchanged, neither considering relationships between protocols.

5 Conclusions

Increasing machine-processable tasks in the Web is a challenge considered at present. In this line we have presented in this paper a proposal that favours the communication among agents that represent to different Information Systems accessible through the Web. The main contributions of the proposal are:

- The management of the semantics aspects when dealing with agent communication protocols.
- The provision of the possibility of customizing standard communication protocols and management of them.
- The use of standard Semantic Web tools to describe protocols.
- The support for discovering different kinds of relationships between protocols.

References

1. FIPA: FIPA communicative act library specification (July 2005), http://www.fipa.org/specs/fipa00037/SC00037J.html
2. Greenwood, D., Lyell, M., Mallya, A., Suguri, H.: The IEEE FIPA approach to integrating software agents and web services. In: International Conference on Autonomous Agents and Multiagent Systems AAMAS, Hawaii, USA, pp. 14–18 (2007)
3. Bermúdez, J., Goñi, A., Illarramendi, A., Bagüés, M.I.: Interoperation among agent-based information systems through a communication acts ontology. Inf. Syst. 32(8), 1121–1144 (2007)
4. Singh, M.P.: Agent Communication Languages: Rethinking the Principles. IEEE Computer 31(12), 40–47 (1998)
5. Singh, M.P.: A social semantics for agent communication languages. In: Issues in Agent Communication, pp. 31–45. Springer, Heidelberg (2000)
6. Austin, J.L. (ed.): How to do things with words. Oxford University Press, Oxford (1962)
7. Wooldridge, M.: Semantic Issues in the Verification of Agent Comunication Languages. Journal of Autonomous Agents and Multi-Agent Systems 3(1), 9–31 (2000)

8. Venkatraman, M., Singh, M.P.: Verifying compliance with commitment protocols. Autonomous Agents and Multi-Agent Systems 2(3), 217–236 (1999)
9. Fornara, N., Colombetti, M.: Operational specification of a commitment-based agent communication language. In: AAMAS 2002: Proceedings of the first international joint conference on Autonomous agents and multiagent systems, pp. 536–542. ACM Press, New York (2002)
10. Shanahan, M.: The event calculus explained. In: Veloso, M.M., Wooldridge, M.J. (eds.) Artificial Intelligence Today. LNCS (LNAI), vol. 1600, pp. 409–430. Springer, Heidelberg (1999)
11. Mallya, A.U., Singh, M.P.: An algebra for commitment protocols. Autonomous Agents and Multi-Agent Systems 14(2), 143–163 (2007)
12. Yolum, P., Singh, M.P.: Flexible protocol specification and execution: Applying event calculus planning using commitments. In: Proceedings of the 1st International Joint Conference on Autonomous Agents and MultiAgent Systems (AAMAS), July 2002, pp. 527–534. ACM Press, New York (2002)
13. Fornara, N., Colombetti, M.: Defining interaction protocols using a commitment-based agent communication language. In: AAMAS 2003: Proceedings of the second international joint conference on Autonomous agents and multiagent systems, pp. 520–527. ACM Press, New York (2003)
14. Desai, N., Mallya, A.U., Chopra, A.K., Singh, M.P.: Interaction protocols as design abstractions for business processes. IEEE Trans. Softw. Eng. 31(12), 1015–1027 (2005)
15. Yolum, P., Singh, M.: Enacting protocols by commitment concession. In: International Conference on Autonomous Agents and Multiagent Systems AAMAS, Hawaii, USA, pp. 116–123 (2007)
16. Berardi, D., Calvanese, D., Giacomo, G.D., Lenzerini, M., Mecella, M.: Automatic service composition based on behavioral descriptions. Int. J. Cooperative Inf. Syst. 14(4), 333–376 (2005)
17. Bordeaux, L., Salaün, G., Berardi, D., Mecella, M.: When are two web services compatible? In: Shan, M.-C., Dayal, U., Hsu, M. (eds.) TES 2004. LNCS, vol. 3324, pp. 15–28. Springer, Heidelberg (2005)
18. Kagal, L., Finin, T.: Modeling conversation policies using permissions and obligations. Autonomous Agents and Multi-Agent Systems 14(2), 187–206 (2007)
19. Endriss, U., Maudet, N., Sadri, F., Toni, F.: Logic-based agent communication protocols. In: Workshop on Agent Communication Languages, pp. 91–107 (2003)
20. Baldoni, M., Baroglio, C., Martelli, A., Patti, V.: A priori conformance verification for guaranteeing interoperability in open environments. In: Georgakopoulos, D., Ritter, N., Benatallah, B., Zirpins, C., Feuerlicht, G., Schoenherr, M., Motahari-Nezhad, H.R. (eds.) ICSOC 2006. LNCS, vol. 4652, pp. 339–351. Springer, Heidelberg (2007)
21. Chopra, A.K., Singh, M.P.: Producing compliant interactions: Conformance, coverage, and interoperability. In: Baldoni, M., Endriss, U. (eds.) DALT 2006. LNCS (LNAI), vol. 4327, pp. 1–15. Springer, Heidelberg (2006)
22. d'Inverno, M., Kinny, D., Luck, M.: Interaction protocols in agentis. In: Proceedings of the Third International Conference on Multi-Agent Systems (ICMAS 1998), pp. 261–268 (1998)
23. Mazouzi, H., Seghrouchni, A.E.F., Haddad, S.: Open protocol design for complex interactions in multi-agent systems. In: AAMAS 2002: Proceedings of the first international joint conference on Autonomous agents and multiagent systems, pp. 517–526. ACM Press, New York (2002)

xOperator – Interconnecting the Semantic Web and Instant Messaging Networks

Sebastian Dietzold[1], Jörg Unbehauen[2], and Sören Auer[1,3]

[1] Universität Leipzig, Department of Computer Science
Johannisgasse 26, D-04103 Leipzig, Germany
dietzold@informatik.uni-leipzig.de
[2] Leuphana - University of Lüneburg, Faculty III Environmental Sciences and Engineering, Volgershall 1, D-21339 Lüneburg
joerg@unbehauen.net
[3] University of Pennsylvania, Department of Computer and Information Science
Philadelphia, PA 19104, USA
auer@seas.upenn.edu

Abstract. Instant Messaging (IM) is in addition to Web and Email the most popular service on the Internet. With xOperator we present a strategy and implementation which deeply integrates Instant Messaging networks with the Semantic Web. The xOperator concept is based on the idea of creating an overlay network of collaborative information agents on top of social IM networks. It can be queried using a controlled and easily extensible language based on AIML templates. Such a deep integration of semantic technologies and Instant Messaging bears a number of advantages and benefits for users when compared to the separated use of Semantic Web technologies and IM, the most important ones being context awareness as well as provenance and trust. We showcase how the xOperator approach naturally facilitates contacts and calendar management as well as access to large scale heterogeneous information sources.

1 Introduction

With estimated more than 500 million users[1] Instant Messaging (IM) is in addition to Web and Email the most popular service on the Internet. IM is used to maintain a list of close contacts (such as friends or co-workers), to synchronously communicate with those, exchange files or meet in groups for discussions. Examples of IM networks are ICQ, Skype, AIM or the Jabber protocol and network[2]. The latter is an open standard and the basis for many other IM networks such as Google Talk, Meebo and Gizmo.

While there were some proposals and first attempts to bring semantic technologies together with IM (e.g. [9,5,12]) in this paper we present a strategy and implementation called xOperator, which deeply integrates both realms in order

---

[1] According to a sum up available at:
http://en.wikipedia.org/wiki/Instant_messaging#User_base
[2] http://www.jabber.org/

to maximise benefits for prospective users. The xOperator concept is based on the idea of additionally equipping an users' IM identity with a number of information sources this user owns or trusts (e.g. his FOAF profile, iCal calendar etc.). Thus the social IM network is overlaid with a network of trusted knowledge sources. An IM user can query his local knowledge sources using a controlled (but easily extensible) language based on Artificial Intelligence Markup Language (AIML) templates[13]. In order to pass the generated machine interpretable queries to other xOperator agents of friends in the social IM network xOperator makes use of the standard message exchange mechanisms provided by the IM network. After evaluation of the query by the neighbouring xOperator agents results are transferred back, filtered, aggregated and presented to the querying user.

Such a deep integration of semantic technologies and IM bears a number of advantages and benefits for users when compared to the separated use of Semantic Web technologies and IM. From our point of view the two most crucial ones are:

- **Context awareness.** Users are not required to world wide uniquely identify entities, when it is clear what/who is meant from the context of their social network neighbourhood. When asked for the current whereabout of Sebastian, for example, xOperator can easily identify which person in my social network has the name Sebastian and can answer my query without the need for further clarification.
- **Provenance and trust.** IM networks represent carefully balanced networks of trust. People only admit friends and colleagues to their contact list, who they trust seeing their online presence, not being bothered by SPAM and sharing contact details with. Overlaying such a social network with a network for semantic knowledge sharing and querying naturally solves many issues of provenance and trust.

The paper is structured as follows: after presenting envisioned usage scenarios and requirements in Section 2 we exhibit the technical xOperator architecture in Section 3. We report about a first xOperator evaluation according to different use cases in Section 4, present related work in Section 5. We draw conclusions and suggest directions for future work in Section 6.

2 Agent Communication Scenarios and Requirements

This section describes the three envisioned agent communication scenarios for xOperator. We will introduce some real-world application scenarios also later in Section 4. Figure 1 shows a schematic depiction of the communication scenarios. The figure is divided vertically into four layers.

The first two layers represent the World Wide Web. Mutually interlinked RDF documents (such as FOAF documents) reference each other using relations such as `rdf:seeAlso`.[3] These RDF documents could have been generated manually,

---

[3] The prefix `rdf`, `rdfs`, `foaf` and `ical` used in this paper represent the well known namespaces (e.g. `http://xmlns.com/foaf/0.1/` for `foaf`).

exported from databases or could be generated from other information sources. These can be, for example, mailing list archives which are represented as SIOC ontologies or personal calendars provided by public calendaring servers such as Google calendar. In order to make such information available to RDF aware tools (such as xOperator) a variety of transformation and mapping techniques can be applied. For the conversion of iCal calendar information for example we used Masahide Kanzaki's ical2rdf service[4].

The lower two layers in Figure 1 represent the Jabber Network. Here users are interacting synchronously with each other, as well as users with artificial agents (such as xOperator) and agents with each. A user can pose queries in natural language to an agent and the agent transforms the query into one or multiple SPARQL queries. Thus generated SPARQL queries can be forwarded either to a SPARQL endpoint or neighbouring agents via the IM networks transport protocol (XMPP in the case of Jabber). SPARQL endpoints evaluate the query using a local knowledge base, dynamically load RDF documents from the Web or convert Web accessible information sources into RDF. The results of SPARQL endpoints or other agents are collected, aggregated, filtered and presented to the user depending on the query as list, table or natural language response.

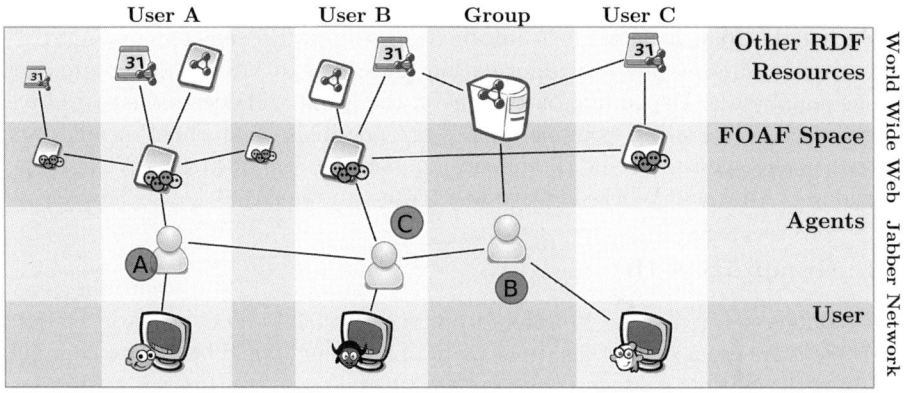

Fig. 1. Agent communication scenarios: (a) personal agent, (b) group agent, (c) agent network

The different communication scenarios are described in the following subsections:

2.1 Personal Agent (A)

This scenario is the most important one and also builds the foundation for the other two communication scenarios. A user of an Instant Messaging network installs his own personal agent and configures information sources he owns or

---
[4] http://www.kanzaki.com/courier/ical2rdf

trusts. For easy deployment the software representing the agent could be distributed together with (or as a plugin of) the IM network client (e.g. Pidgin). Information sources can be for example a FOAF profile of the user containing personal information about herself and about relationships to other people she knows and where to find further information about these. This information is represented in FOAF using the properties `foaf:knows` and `rdfs:seeAlso`. The following listing shows an excerpt from a FOAF profile.

```
:me a foaf:Person ;
  foaf:knows [
    a foaf:Person ;
    rdfs:seeAlso <http://eye48.com/foaf.rdf> ;
    foaf:name "Michael Haschke" ;
    foaf:nick "Haschek" ] .
```

Additionally this FOAF profile can link to other RDF documents which contain more information about the user and his activities. The RDF version of his calendar, for example, could be linked as follows:

```
:me rdfs:seeAlso <http://.../ical2rdf?u=http...> .
<http://.../ical2rdf?u=http...> a ical:Vcalendar ;
  rdfs:label "Haschek's Calendar" .
```

Such links span a network of information sources as depicted in Figure 1. Each user maintains his own information and links to information sources of his acquaintances. Depending on the query, the agent will access the respective resources. The following example queries are possible, when FOAF profiles are known to the agent: Tell me the phone / homepage / ... of Frank! What is the birthday of Michael? Where is Dave now? Who knows Alex?

2.2 Group Agent (B)

This communication scenario differs from the Personal Agent scenario in that multiple users get access to the same agent. The agent should be able to communicate with multiple persons at the same time and to answer queries in parallel. As is also depicted in Figure 1 the agent furthermore does not only access remote documents but can also use a triple store for answering queries. When used within a corporate setting this triple store can for example contain a directory with information about employees or customers. The triple store can be also used to cache information obtained from other sources and thus facilitates faster query answering. For agents themselves, however, the distinction between RDF sources on the Web and information contained in a local triple store is not relevant.

2.3 Agent Network (C)

This scenario extends the two previous ones by allowing communication and interaction between agents. The rationale is to exploit the trust and provenance characteristics of the Instant Messaging network: Questions about or related to acquaintances in my network of trust can best be answered by their respective agents. Hence, agents should be able to talk to other agents on the IM network. First of all, it is crucial that agents on the IM network recognize each other. A personal agent can use the IM account of its respective owner and can access the contact list (also called roster) and thus a part of its owner's social network. The agent should be able to recognise other personal agents of acquaintances in this contact list (auto discovery) and it should be possible for agents to communicate without interfering with the communication of their owners. After other agents are identified it should be possible to forward SPARQL queries (originating from a user question) to these agents, collect their answers and present them to the user.

3 Technical Architecture

First of all, the xOperator agent is a mediator between the Jabber Instant Messaging network [5] on one side and the World Wide Web on the other side. xOperator is client in both networks. He communicates anonymously (or using configured authentication credentials) on the WWW by talking HTTP with Web servers. On the Jabber network xOperator utilizes the Extensible Messaging and Presence Protocol (XMPP, [11]) using the Jabber account information provided by its owner. Jabber clients only communicate with the XMPP server associated with the user account. Jabber user accounts are have the same syntax as email

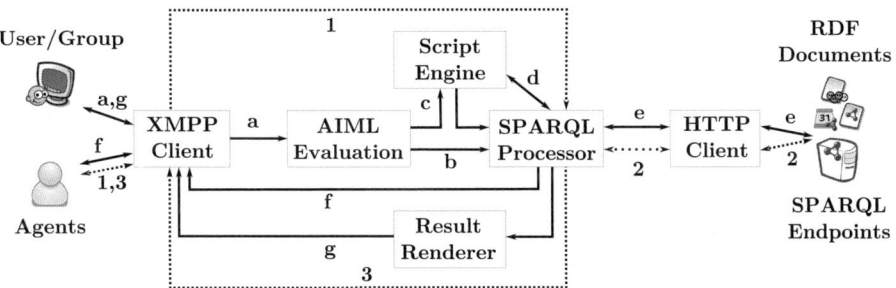

Fig. 2. Technical architecture of xOperator

addresses (e.g. soerenauer@jabber.ccc.de). The respective Jabber server cares about routing messages to the server associated with the target account or temporarily stores the message in case the target account is not online or its server is not reachable. Since 2004 XMPP is a standard of the Internet Engineering Task Force and is widely used by various services (e.g. Google Talk). Figure 2 depicts the general technical architecture of xOperator.

---

[5] Our implementation currently supports the Jabber network, but can be easily extended to other IM networks such as Skype, ICQ or MSN Messenger.

The agent works essentially in two operational modi:

1. (Uninterrupted line) Answer natural language questions posed by a user using SPARQL queries and respond to the user in natural language according to a predefined template. Questions posed by a user (a) are either directly mapped to a SPARQL query template (b) or SPARQL queries are generated by a query script (c), which might obtain additional information by means of sub queries (d). The resulting SPARQL query will be evaluated on resources of the user (e), as well as passed on the Jabber network to neighbouring agents for evaluation (f). All returned results are collected and prepared by a result renderer for presentation to the user (g). Algorithm 1 demonstrates the workings of xOperator.

Algorithm 1. Evaluation of XMPP user input

Input: User input I from XMPP
Output: Sendable Agent response
Data: set $S = A \cup D \cup E$ of agents, documents and endpoints
Data: set C of AIML categories
Data: set $R = \emptyset$ of results

1 **if** I *is an admin or extension command* **then return** `executeCommand` (I)
2 **else if** I *has no match in* C **then return** defaultmsg
3 **else if** I *has standard match in* C **then return** `aimlResult` (I, C)
4 **else**
5   **if** I *has SPARQL template match in* C **then**
6   Query = `fillPatterns` (`aimlResult` (I, C))
7   **else if** I *has query script match in* C **then**
8   Query = `runScript` (`aimlResult` (I, C))
9 **if** *Query* **then**
10   **foreach** $s \in S$ **do**
11          $R = R \cup$ `executeQuery`$(Query, s)$
12   **return** `renderResults` (R);
13 **else**
14   **return** error

---

2. (Dotted line) Receive SPARQL queries from neighbouring agents (1) on the IM network, evaluate these queries (2) on the basis of locally known RDF documents and SPARQL endpoints and send answers as XML SPARQL Result Set [3] back via XMPP (3).

In both cases the agent evaluates SPARQL queries by querying a remote SPARQL endpoint via HTTP GET Request according to the SPARQL HTTP Bindings [4] or by retrieving an RDF document as well via HTTP and evaluating the query by means of a local SPARQL query processor.

In the following we describe first the natural language component on the basis AIML templates and address thereafter the communication in the Jabber network.

3.1 Evaluation of AIML Templates

The Artificial Intelligence Markup Language (AIML, [13]) is an XML dialect for creating natural language software agents. In [6] the authors describe AIML to enable pattern-based, stimulus-response knowledge content to be served, received and processed on the Web and offline in the manner that is presently possible with HTML and XML. AIML was designed for ease of implementation, ease of use by newcomers, and for interoperability with XML and XML derivatives such as XHTML. Software reads the AIML objects and provides application-level functionality based on their structure. The AIML interpreter is part of a larger application generically known as a bot, which carries the larger functional set of interaction based on AIML. A software module called a responder handles the human-to-bot or bot-to-bot interface work between an AIML interpreter and its object(s). In xOperator AIML is used for handling the user input received through the XMPP network and to translate it into either a query or a call to a script for more sophisticated evaluations.

The most important unit of knowledge in AIML is the category. A category consists of at least two elements, a pattern and a template element. The pattern is evaluated against the user input. If there is a match, the template is used to produce the response of the agent. It is possible to use the star (*) as a placeholder for any word in a pattern. We have extended this basic structure in two ways:

Simple Query Templates: In order to enable users to create AIML categories on the fly we have created an extension of AIML. It allows to map natural language patterns to SPARQL query templates and to fill variables within those templates with parameters obtained from *-placeholders in the natural language patterns.

```
<category>
  <pattern>TELL ME THE PHONE OF *</pattern>
  <template>
    <external name="query"
      param="SELECT DISTINCT ?phone WHERE {...}" />
  </template>
</category>
```

Within the SPARQL template variables in the form of %%n%% refer to *-placeholder (n refers to the nth *-placeholder in the category pattern). The question for the phone number of a person, for example, can be represented with the following AIML template:

```
TELL ME THE PHONE OF *
```

A possible (very simple) SPARQL template using the FOAF vocabulary could be stored within the AIML category as follows:

```
SELECT DISTINCT ?phone WHERE
    { ?s foaf:name "%%1%%". ?s foaf:phone ?phone. }
```

On activation of a natural language pattern by the AIML interpreter the corresponding SPARQL templates variables are bound to the values of the placeholders and the resulting query is send independently to all known SPARQL endpoints and neighbouring agents. These answer independently and deliver result sets, which can complement each other, contain the same or contradictory results. The agent renders results as they arrive to the user, but filters duplicates and marks contradictory information. The agent furthermore annotates results with regard to their provenance.

This adoption of AIML is easy to use and directly extensible via the Instant Messaging client (cf. Sec. 3.2). However, more complex queries, which for example join information from multiple sources are not possible. In order to enable such queries we developed another AIML extension, which allows the execution of query scripts.

Query Scripts: Query scripts basically are small pieces of software, which run in a special environment where they have access to all relevant subsystems. They are given access to the list of known data sources and neighbouring agents. xOperator, for example, allows the execution of query scripts in the Groovy scripting language for Java.The execution of a query script results in the generation of a SPARQL query, which is evaluated against local information sources and passed to other agents as described in the previous section. We motivate and illustrate the workings of query scripts using an application scenario based on the FOAF space (cf. Figure 1). The FOAF space has the following characteristics:

- The `foaf:knows` relation points to other people known by this person.
- Other FOAF and RDF documents are linked through `rdfs:seeAlso`, allowing bots and agents to crawl through the FOAF space and to gather additional RDF documents like calendars or blog feeds.

To enable the agent to retrieve and evaluate additional information from sources which are referenced from the user's FOAF profile, a query script can contain subqueries, whose results are used within another query. Query scripts also enable the usage of special placeholders such as `now` or `tomorrow`, which can be populated for the querying of iCal calendars with the concrete values.

In order to extend the agent for other application domains or usage scenarios, xOperator allows to dynamically assign new query scripts to AIML categories. A query script is assigned to an AIML template by means of an `external` tag (as are also simple SPARQL templates). An example script implementing a subquery to retrieve relevant resources about a `foaf:person` is presented in Section 4.

3.2 Administration and Extension Commands

Users can easily change the configuration of their agents by using a set of administration and extension commands. These commands have a fix syntax and are executed without the AIML engine:

- `list ds, add ds {name} {uri}, del ds {name}`: With these commands, users can manage their trusted data sources. Each source is locally identified by a name which is associated to an URI.

- `list templates`, `add template {pattern} {query}`, `del template {pattern}`: With these template commands, users can manage simple query templates which are associateded by its AIML pattern.
- `query {sparql query}`: This command is used to send on-the-fly SPARQL queries to the xOperator. The query will be evaluated on every datastore and routed to every agent in the neighbourhood. The query results will be rendered by a default renderer.
- `list ns`, `add ns {prefix} {uri}`, `del ns {prefix}`: To easlily create on-the-fly SPARQL queries, users can manage namespaces. The namespaces will be added to the namespace section in the on-the-fly query.
- `help`: This is an entry point for the help system.

3.3 XMPP Communication and Behaviour

While the HTTP client of the agent uses standard HTTP for querying SPARQL endpoints and the retrieval of RDF documents, we extended XMPP for the mutual communication between the agents. This extension complies with standard extension routines of XMPP will be ignored by other agents. With regard to the IM network the following functionality is required:

- The owner of the agent should be able to communicate easily with the agent. He should be able to manage the agent using the contact list (roster) and the agent should be easily recognizable.
- The agent has to have access to the roster of its owner in order to identify neighbouring agents.
- It should be possible for other agents to obtain information about the ownership of an agent. His requests will not be handled by other agents for security reasons if he can not be clearly assigned to an owner.
- The agent should be only visible for his owner and neighbouring agents (i.e. agents of contacts of his owner) and only accept queries from these.

As a consequence from those requirements it is reasonable that the agent acts using the account of its owner (main account) for the communication with other agents, as well as an additional account (proxy account) for the communication with its owner[6]. Due to the usage of the main account other agents can trust the agents answers and easily track the provenance of query results. Figure 3 depicts the concept of using two different accounts for the communication with the owner and other agents. For unique identification of senders and recipients so called resource names (in the figure `Home`, `Work` and `Agent`) are used and simply appended to the account name.

We demonstrate the agent communication with two XMPP messages:

Agent Autodiscovery: Goal of the autodiscovery is the identification of agents among each other. For that purpose each agent sends a special message of type

---

[6] Technically, it is sufficient for the agent to use the owner's account which, however, could create confusing situations for the user when communicating with 'herself'.

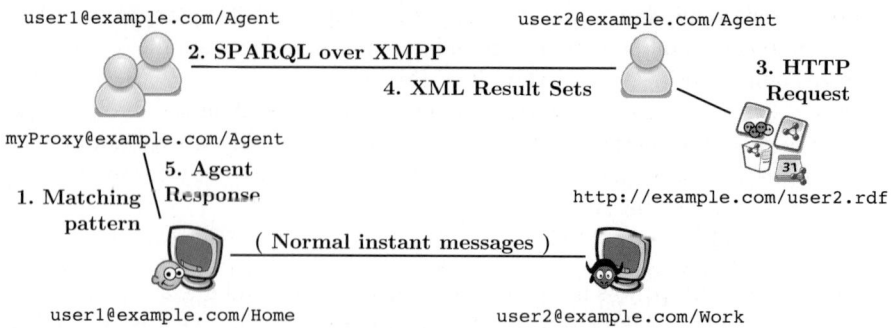

Fig. 3. XMPP Communication example

info/query (`iq`) to all known and currently active peers. Info/query messages are intended for internal communication and queries among IM clients without being displayed to the human users. An autodiscovery message between the two agents from Figure 3, for example, would look as follows:

```
<iq from="user1@example.com/Agent" type='get'
    to="user2@example.com/Agent" id='...'>
 <query xmlns='http://jabber.org/protocol/disco#info'/>
</iq>
```

A positive response to this feature discovery message from an xOperator agent would contain a feature with resource ID `http://www.w3.org/2005/09/xmpp-sparql-binding`. This experimental identifier/namespace was created by Dan Brickley for SPARQL / XMPP experiments (cf. Section 5). The response message to the previous request would look as follows:

```
<iq from='user2@example.com/Agent' type='result'
    to='user1@example.com/Agent' id='...' />
  <query xmlns='http://jabber.org/protocol/disco#info'>
    <identity
      category='client' name='xOperator' type='bot'/>
    <feature
      var='http://www.w3.org/2005/09/xmpp-sparql-binding'/>
    <!-- ... more here -->
  </query>
</iq>
```

Similar XMPP messages are used for sending SPARQL queries and retrieving results. The latter are embedded into a respective XMPP message according to the SPARQL Query Results XML Format[7].

Routing and Recall: Queries are propagated to all neighbouring xOperator agents. As currently there is no way of anticipating which agent could answer

---

[7] http://www.w3.org/TR/rdf-sparql-XMLres/

a question, asking all directly connected agents offers the best compromise between load and recall. Flooding the network beyond adjacent nodes would cause excessive load. Especially in the domain of personal information, persons or their respective agents directly related to the querying person or agent should be most likely to answer the query.

4 Evaluation

The xOperator concept was implemented in Java and is available as open-source software from: http://aksw.org/Projects/xOperator. The agent is able to log into existing accounts and can receive querying and configuration commands.

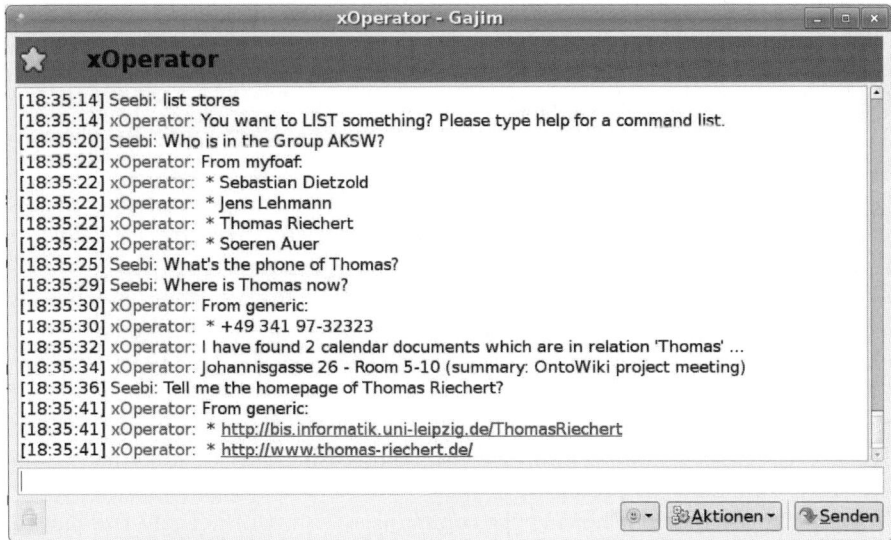

Fig. 4. Communication with the xOperator agent by means of an ordinary Jabber client

We evaluated our approach in a number of scenarios, which included various heterogeneous information sources and a different number of agents. As information sources we used FOAF profiles (20 documents, describing 50 people), the SPARQL endpoint of our semantic Wiki OntoWiki [2] (containing information about publications and projects), information stored in the LDAP directory service of our department, iCal calendars of group members from Google calendar (which are accessed using iCal2RDF) and publicly available SPARQL endpoints such as DBpedia [1]. Hence the resulting information space contains information about people, groups, organizations, relationships, events, locations and all information contained in the multidomain ontology DBpedia. We created a number of AIML categories, interacting with this information space. Some example patterns and corresponding timings for obtaining answers from the agent network in

Table 1. Average response time in seconds (client to client) of some AIML patterns used in three scenarios: (1) 20 documents linked from one FOAF profile, 1 personal agent with no neighbourhood (2) 20 documents linked from different FOAF profiles and spread over a neighbourhood of 5 agents (3) one SPARQL endpoint as an interface to a Semantic Wiki or DBpedia store with one group agent

	Template	Scenario 1	Scenario 2	Scenario 3
1	What is / Tell me (the) * of *	2.3	3.9	1.5
2	Who is member of *	3.5	4.3	1.6
3	Tell me more about *	3.2	5.6	1.1
4	Where is * now:	5.1	6.7	4.2
5	Free dates * between * and *	5.1	6.8	4.7
6	Which airports are near *	–	–	3.4

the three different network scenarios (personal agent, agent network and group agent) are summarized in Table 1.

The first three templates represent queries which are answered using simple SPARQL templates. Template 4 makes use of a reserved word (`now`), which is replaced for querying with an actual value. Template 5 is implemented by means of a query script which retrieves all available time slots from the calendars of a group of people and calculates the intersection thus offering suitable times to arrange meetings or events, where the attendance of all group members is required. Template 6 uses the DBpedia SPARQL endpoint in a group agent setting to answer questions about the geographic location of places (such as airports). These query templates are meant to give some insights in the wealth of opportunities for employing xOperator. Further, AIML templates can be created easily, even directly from within the IM client (using the administration and extension commands as presented in Section 3.2).

A typical user session showing the communication with the agent is depicted in Figure 4. The response timings indicate that the major factor are latency times for retrieving RDF documents or querying SPARQL endpoints. The impact of the number of agents in the agent network as well as the overhead required by the xOperator algorithm is rather small. The timings are furthermore upper bounds, since answers are presented to the user as they arrive. This results in intuitive perception that xOperator is a very responsive and efficient way for query answering.

Experiences during the evaluation have led to the following rules for creating patterns and queries in xOperator.

(1) Query as fuzzy as possible: Instant Messaging is a very quick means of communication. Users usually do not capitalize words and use many abbreviations. This should be considered, when designing suitable AIML patterns. If information about the person 'Sören Auer' should be retrieved, this can be achieved using the following graph pattern: `?subject foaf:name "Auer"`. However, information can be represented in multiple ways and often we have to deal with minor

misrepresentations (such as trailing whitespace or wrong capitalizations), which would result for the above query to fail. Hence, less strict query clauses should be used instead. For the mentioned example the following relaxed SPARQL clause, which matches also substrings and is case insensitive, could be used:

```
?subject foaf:name ?name.
FILTER regex(?name,'.*Auer.*','i')
```

(2) Use patterns instead of qualified identifiers for properties: Similar, as for the identification of objects, properties should be matched flexible. When searching for the `homepage` of 'Sören Auer' we can add an additional property matching clause to the SPARQL query instead of directly using, for example, the property identifier `foaf:homepage`):

```
?subject ?slabel ?spattern.
?subject ?property ?value.
?property ?plabel ?ppattern.
FILTER regex(?spattern,'.*Auer.*','i')
FILTER regex(?ppattern,'.*homepage.*','i')
```

This also enables multilingual querying if the vocabulary contains the respective multilingual descriptions. Creating fuzzy queries, of course, significantly increases the complexity of queries and will result in slower query answering by the respective SPARQL endpoint. However, since we deal with a distributed network of endpoints, where each one only stores relatively small documents this effect is often negligible.

(3) Use sub queries for additional documents: In order to avoid situations where multiple agents retrieve the same documents (which is very probable in a small worlds scenario with a high degree of interconnectedness) it is reasonable to create query scripts, which only distribute certain tasks to the agent network (such as the retrieval of prospective information sources or document locations), but perform the actual querying just once locally.

5 Related Work

Proposals and first prototypes which are closely related to xOperator and inspired its development are Dan Brickley's JQbus[8] and Chris Schmidt's SPARQL over XMPP[9]. However, both works are limited to the pure transportation of SPARQL queries over XMPP.

Quite different but the xOperator approach nicely complementing are works regarding the semantic annotation of IM messages. In [9] for example the authors present a semantic archive for XMPP instant messaging which facilitates search in IM message archives. [5] suggests ways to make IM more semantics aware by facilitating the classification of IM messages, the exploitation of semantically represented context information and adding of semantic meta-data to messages.

---

[8] http://svn.foaf-project.org/foaftown/jqbus/intro.html
[9] http://crschmidt.net/semweb/sparqlxmpp/

Comprehensive collaboration frameworks which include semantic annotations of messages and people, topics are, for example, CoAKTinG [12] and Haystack [7]. The latter is a general purpose information management tool for end users and includes an instant messaging component, which allows to semantically annotate messages according to a unified abstraction for messaging on the Semantic Web[10].

In [8] natural language interfaces (NLIs) are used for querying semantic data. The NLI used in xOpertor employs only a few natural language processing techniques, like stop word removal for better template matching. Generic templates would be possible to define, but as [8] shows user interaction is necessary for clarifying ambiguities. For keeping IM conversation as simple as possible, domain specific templates using AIML were chosen. Finally, in [6] the author enhanced AIML bots by generating AIML categories from RDF models. Different to xOperator, these categories are static and represent only a fixed set of statements.

6 Conclusions and Future Work

With the xOperator concept and its implementation, we have showed how a deeply and synergistic coupling of Semantic Web technology and Instant Messaging networks can be achieved. The approach naturally combines the well-balanced trust and provenance characteristics of IM networks with semantic representations and query answering of the Semantic Web. The xOperator approach goes significantly beyond existing work which mainly focused either on the semantic annotation of IM messages or on using IM networks solely as transport layers for SPARQL queries. xOperator on the other hand overlays the IM network with a network of personal (and group) agents, which have access to knowledge bases and Web resources of their respective owners. The neighbourhood of a user in the network can be easily queried by asking questions in a subset of natural language. By that xOperator resembles knowledge sharing and exchange in offline communities, such as a group of co-workers or friends. We have showcased how the xOperator approach naturally facilitates contacts and calendar management as well as access to large scale heterogeneous information sources. In addition to that, its extensible design allows a straightforward and effortless adoption to many other application scenarios such as, for example, sharing of experiment results in Biomedicine or sharing of account information in Customer Relationship Management.

In addition to adopting xOperator to new domain application we view the xOperator architecture as a solid basis for further technological integration of IM networks and the Semantic Web. This could include adding light-weight reasoning capabilities to xOperator or the automatic creation of AIML categories by applying NLP techniques. A more fine grained access control will be implemented in a future version. Instead of simply trusting all contacts on the roster, individual and group based policies can be created. An issue for further research is the implementation of a more sophisticated routing protocol, that allows query traversal beyond directly connected nodes without flooding the whole network.

References

1. Auer, S., Bizer, C., Kobilarov, G., Lehmann, J., Cyganiak, R., Ives, Z.G.: DBpedia: A Nucleus for a Web of Open Data. In: Proc. of ISWC/ASWC, pp. 722–735 (2007)
2. Auer, S., Dietzold, S., Riechert, T.: OntoWiki - A Tool for Social, Semantic Collaboration. In: Cruz, I., Decker, S., Allemang, D., Preist, C., Schwabe, D., Mika, P., Uschold, M., Aroyo, L.M. (eds.) ISWC 2006. LNCS, vol. 4273, pp. 736–749. Springer, Heidelberg (2006)
3. Beckett, D., Broekstra, J.: SPARQL Query Results XML Format. In: W3C Candidate Recommendation, World Wide Web Consortium (W3C) (April 2006)
4. Clark, K.G.: SPARQL Protocol for RDF. In: W3C Recommendation, World Wide Web Consortium (W3C) (2007)
5. Franz, T., Staab, S.: SAM: Semantics Aware Instant Messaging for the Networked Semantic Desktop. In: Semantic Desktop Workshop at the ISWC (2005)
6. Freese, E.: Enhancing AIML Bots using Semantic Web Technologies. In: Proc. of Extreme Markup Languages (2007)
7. Karger, D.R., Bakshi, K., Huynh, D., Quan, D., Sinha, V.: Haystack: A General-Purpose Information Management Tool for End Users Based on Semistructured Data. In: Proc. of CIDR, pp. 13–26 (2005)
8. Kaufmann, E., Bernstein, A.: How useful are natural language interfaces to the semantic web for casual end-users? In: Aberer, K., Choi, K.-S., Noy, N., Allemang, D., Lee, K.-I., Nixon, L., Golbeck, J., Mika, P., Maynard, D., Mizoguchi, R., Schreiber, G., Cudré-Mauroux, P. (eds.) ISWC 2007. LNCS, vol. 4825, pp. 281–294. Springer, Heidelberg (2007)
9. Osterfeld, F., Kiesel, M., Schwarz, S.: Nabu - A Semantic Archive for XMPP Instant Messaging. In: Semantic Desktop Workshop at the ISWC (2005)
10. Quan, D., Bakshi, K., Karger, D.R.: A Unified Abstraction for Messaging on the Semantic Web. In: WWW (Posters) (2003)
11. Saint-Andre, P.: Extensible Messaging and Presence Protocol (XMPP): Core. RFC 3920, The Internet Engineering Task Force (IETF) (October 2004)
12. Shum, S.B., De Roure, D., Eisenstadt, M., Shadbolt, N., Tate, A.: CoAKTinG: Collaborative advanced knowledge technologies in the grid. In: Proc. of 2nd Workshop on Adv. Collab. Env. at the HPDC-11 (2002)
13. Wallace, R.: Artificial Intelligence Markup Language (AIML). Working draft, A.L.I.C.E. AI Foundation (February 18, 2005)

An Ontology for Software Models and Its Practical Implications for Semantic Web Reasoning*

Matthias Bräuer and Henrik Lochmann

SAP Research CEC Dresden
Chemnitzer Str. 48, 01187 Dresden, Germany
{matthias.braeuer,henrik.lochmann}@sap.com

Abstract. Ontology-Driven Software Development (ODSD) advocates using ontologies for capturing knowledge about a software system at development time. So far, ODSD approaches have mainly focused on the unambiguous representation of domain models during the system analysis phase. However, the design and implementation phases can equally benefit from the logical foundations and reasoning facilities provided by the Ontology technological space. This applies in particular to Model-Driven Software Development (MDSD) which employs models as first class entities throughout the entire software development process. We are currently developing a tool suite called HybridMDSD that leverages Semantic Web technologies to integrate different domain-specific modeling languages based on their ontological foundations. To this end, we have defined a new upper ontology for software models that complements existing work in conceptual and business modeling. This paper describes the structure and axiomatization of our ontology and its underlying conceptualization. Further, we report on the experiences gained with validating the integrity and consistency of software models using a Semantic Web reasoning architecture. We illustrate practical solutions to the implementation challenges arising from the open-world assumption in OWL and lack of nonmonotonic queries in SWRL.

1 Introduction

In recent years, researchers and practitioners alike have started to explore several new directions in software engineering to battle the increasing complexity and rapid rate of change in modern systems development. Among these new paradigms is *Semantic Web Enabled Software Engineering (SWESE)*, which tries to apply Semantic Web technologies (such as ontologies and reasoners) in mainstream software engineering. This way, SWESE hopes to provide stronger logical foundations and precise semantics for software models and other development artifacts.

---
* This work is supported by the feasiPLe project (partly financed by the German Ministry of Education and Research (BMBF)) and by Prof. Assman from Chair of Software Technology at Dresden University of Technology.

The application of ontologies and Semantic Web technologies in software engineering can be classified along two dimensions [13]: the kind of knowledge modeled by the ontology and whether the approach tackles runtime or development time scenarios. If ontologies are employed during development time to capture knowledge of the software system itself (rather than the development infrastructure or process), we speak of *Ontology-Driven Software Development (ODSD)*.

So far, most approaches to ODSD have focused on using ontologies as unambiguous representations of domain (or conceptual) models during the initial phases of the software engineering process, i.e., requirements engineering and systems analysis. However, the design and implementation phases can equally benefit from the logical foundations and reasoning facilities provided by the Semantic Web. This applies in particular to *Model-Driven Software Development (MDSD)*, a development paradigm that employs models as first class entities throughout the entire development process [27].

MDSD advocates modeling different views on a system (e.g., data entities, processes, or user interfaces) using multiple domain-specific modeling languages (DSLs) [14]. This raises the need for sophisticated consistency checking between the individual models, decoupling of code generators, and automatic generation of model transformations [2]. We are currently developing a toolsuite called *HybridMDSD* that aims at leveraging Semantic Web technologies to address these challenges.

In this paper, we introduce HybridMDSD with its ontological foundation, an upper ontology for software models. Additionally, we concentrate on the challenges that arise from using Semantic Web technologies to validate the integrity and consistency of multiple software models. Our main contribution is an analysis of practical solutions to the implementation challenges posed by the open-world assumption in OWL and lack of nonmonotonic queries in SWRL. By doing so, we highlight the need for nonmonotonic extensions to the Semantic Web languages in the context of Ontology-Driven Software Development.

After briefly reviewing selected aspects related to ODSD in Sect. 2, we present our approach and its benefits in Sect. 3. Section 4 describes the main concepts of our newly developed upper ontology. Based on its axiomatization, Sect. 5 elaborates in detail on the challenges and solutions for Semantic Web reasoning over closed-world software models. In Sect. 6, we briefly highlight related work to place our method in context. Section 7 concludes on the paper.

2 Ontologies and Models

This section reviews the prevalent view on ontologies and software models in ODSD and the relation to the Semantic Web. This provides the foundation for highlighting the differences of our approach and discussing the resulting challenges in the following sections.

So far, ontology-driven software development has focused on using ontologies for domain representation in conceptual and business modeling [12]. This appears to stem from the view that ontologies and software models have differing, even opposing intentions. For instance, in a recent proposal to unify *Model-Driven Architecture (MDA)* and Semantic Web technologies [1], the authors point out that both ontologies and models are means to describe a domain of interest by capturing all relevant concepts and their relationships. However, a key difference between the two approaches is that a model is a *prescriptive* representation of a particular domain under *closed-world assumption (CWA)*. Essentially, this means that everything that is not explicitly contained in the model does not exist. Thus, models are ideally suited as exact specifications for software systems. Ontologies, by contrast, are *descriptive* and possibly incomplete representations of the "real world". They follow the *open-world assumption (OWA)* which means that a statement is not necessarily false if it cannot be proved true.

The closed-world assumption in software models is closely related to *nonmonotonic reasoning*. Nonmonotonic logics allow to make decisions based on incomplete knowledge, causing previously-drawn conclusions to be retracted when new information becomes available. An important property of nonmonotonic logics is strong negation or *negation as failure*, which allows to infer $\neg P$ if P cannot be proved. This is vital to validate integrity constraints in closed-world data models, but can cause problems in the open world of the Semantic Web. Since incomplete or changing knowledge is common in the web, nonmonotonicity could often lead to wrong conclusions. Also, implementing efficient nonmonotonic reasoners is difficult, because predicate nonmonotonic logic is undecidable. As a result, the Semantic Web is currently built on monotonic formalisms: the Web Ontology Language (OWL) corresponds to description logics, while the Semantic Web Rule Language (SWRL) is based on Horn clause rules.

3 HybridMDSD

Model-Driven Software Development facilitates the generation of executable software assets from technical abstractions of concrete domain knowledge. However, applying multiple domain-specific models to describe different views on the same system is still challenging, because it remains difficult to properly describe the semantic references and interdependencies between elements in different models and to maintain consistency. The HybridMDSD project tries to alleviate these challenges. In the following, we outline our approach and highlight its benefits for MDSD.

3.1 Approach

The core idea of our approach is to capture pure system-related knowledge of modeling languages and actively use this knowledge during language instantiation. To implement this, we establish a binding between the constructs of a modeling language and the concepts and relationships of an axiomatized upper ontology [20,4]. During system modeling, this binding is used to construct an ontology knowledge base that contains information about elements from different models and their interdependencies. We call this knowledge base a *semantic connector*, because it allows to create semantically sound references between different domain-specific models in a holistic view of the entire system.

3.2 Benefits

Mapping constructs of different modeling languages to a single reference ontology allows the generation of model transformations. Additionally, the semantic connector forms the basis for comprehensive ABox reasoning over the set of all semantically connected model elements. Thus, it permits integrity and consistency validation across the boundaries of individual models (cf. Sect. 5). This enables several modelers to work collaboratively on different models of a large system while maintaining semantic consistency between the individual views. In addition, domain-specific inference rules facilitate automatic adaptation of model instances in case of modifications to one or more connected models. Such modifications commonly occur during software evolution and especially in Software Product Line Engineering (SPLE), where a variable product feature may affect several system models [20].

4 An Ontology for Software Models

This section introduces the *Unified Software Modeling Ontology (USMO)*, a new upper ontology that is the basis for the semantic connector in the HybridMDSD project. USMO is the result of a careful analysis of existing foundational ontologies in the area of conceptual modeling, such as the *Bunge-Wand-Weber (BWW)* ontology [8], the *General Foundational Ontology (GFO)* [11] and the *Unified Foundational Ontology (UFO)* [12]. Specifically, we have compared the ontological foundations of conceptual models with those of software models. Our study revealed major differences in the corresponding ontological interpretation, which eventually prompted us to derive our own upper ontology. We have successfully employed this new ontology in a case study that involved the semantic integration of several domain-specific models [5]. A detailed description of all USMO concepts and relationships is beyond the scope of this paper, so we limit the discussion to those elements relevant in the following sections. A comprehensive coverage of the entire ontology — including its philosophical background and conceptualization of physical objects changing over time — can be found in [5].

4.1 Concepts and Relationships

In its current version, USMO totals 27 classes and 56 properties. In line with UFO, we divide the world of things into Entitys and Sets. In the context of software modeling, Entitys simply correspond to model elements. At the most general level, the Entity class therefore represents language constructs in modeling languages. Figure 1 depicts the top level concepts of USMO.

In our ontology, an Entity is simultaneously classified as (1) either an Actuality or a Temporality *and* (2) either a Universal or a Particular. The Actuality and

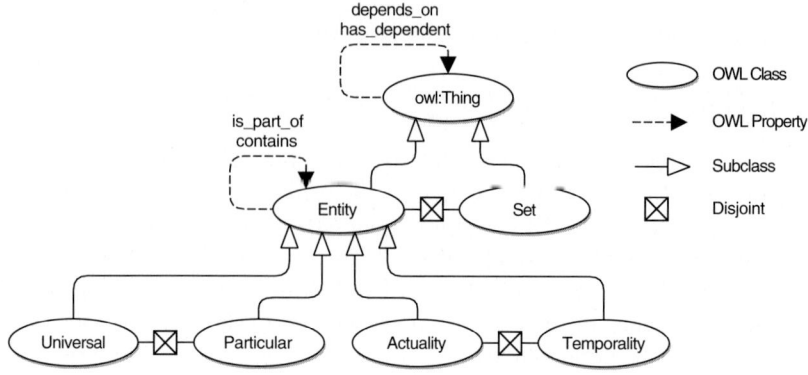

Fig. 1. Elementary concepts and relationships

Temporality concepts represent the philosophical notions of **Endurant** and **Perdurant**, respectively [9]. While an **Actuality** is wholly present at any time instant, **Temporalitys** are said to "happen in time". In the context of software models, this allows to ontologically interpret both structural and temporal elements.

Universals represent intensional entities that can be instantiated by **Particulars**. This distinction facilitates the ontological interpretation of both *type models* (e.g., class diagrams) and *token models* (e.g., object diagrams and behavioral models) [19]. Figure 2 illustrates the relationships between a **Universal** and its *extension*, which is the **Set** of instantiating **Particulars**. Note that USMO only supports two levels of ontological instantiation, so a **Universal** cannot instantiate other **Universals**.

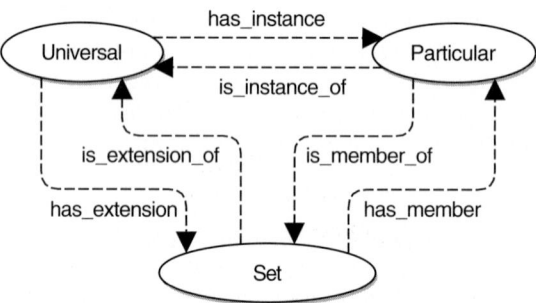

Fig. 2. The model of ontological instantiation

At the top level, we define two major relationships. The most important one is *existential dependency*, since the corresponding properties depends_on and has_dependent subsume all other USMO properties. The dependency relation therefore spans a directed, acyclic graph over all model elements captured in the semantic connector knowledge base. With the help of a reasoner, this design allows complex impact analysis across several domain-specific models. The second important relationship is *containment*, which adds the notion of transitive, non-shareable ownership.

To conclude the overview of our ontology, Fig. 3 shows the USMO conceptualization of universal structural elements, which are typically found in type models. A Schema is an Actuality Universal that classifies independent entities. By contrast, a Property cannot exist on its own and always depends on at least one Schema. There are two types of Propertys: a RelationalProperty relates at least two Schemas in a formal or physical relationship, while an IntrinsicProperty belongs to exactly one other Schema or Property. This allows to clearly specify the semantics of object-oriented language constructs like classes, associations, association classes and attributes.

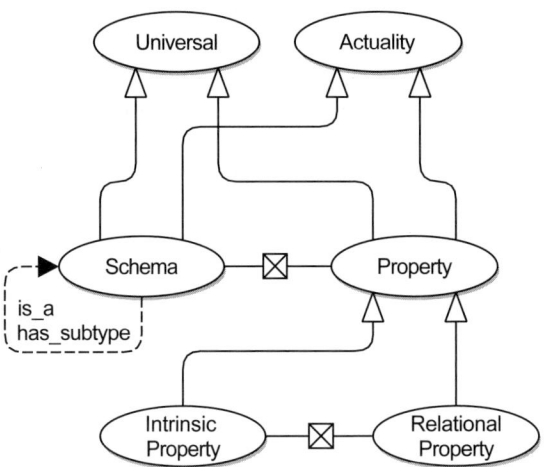

Fig. 3. An overview of universal structural concepts

4.2 Axiomatization

To facilitate inferencing and consistency checks, the semantics of each USMO concept are captured in a rich axiomatization. Since neither description logics nor Horn logic is a subset of the other, we are using a combination of both DL concept constructors and rules for this purpose. Currently, we employ only those constructors available in the OWL standard and emulate features from the recent OWL 1.1 proposal [23] with rules. This applies to axioms such as qualified number restrictions (e.g., $C \sqsubseteq \leqslant np.D$), role inclusion axioms ($R \circ S \sqsubseteq R$) and irreflexive, antisymmetric, or disjoint roles.

The entire concept taxonomy of USMO is defined using so-called *covering axioms* [15, p. 75]. This means that for each class C that is extended by a number of subclasses $D_1, ..., D_n$, we add an axiom

$$C \equiv D_1 \sqcup ... \sqcup D_n \qquad (1)$$

In addition, we declare all subclasses as disjoint, so an individual instantiating C must necessarily instantiate exactly one of $D_1, ..., D_n$.

Regarding the axiomatization with rules, we employ both *deductive* and *integrity* rules. Deductive rules (also known as derivation rules) assist the modeler by completing the knowledge base with knowledge that logically follows from asserted facts. They are always monotonic. A good example is the rule describing the semantics of the *instantiation* relationship between a **Particular** p and a **Universal** u:

$$\mathsf{is\_instance\_of}(p, u) \wedge \mathsf{has\_extension}(u, e) \rightarrow \mathsf{is\_member\_of}(p, e) \qquad (2)$$

Integrity rules, by contrast, describe conditions that must hold for the knowledge base to be in a valid state. These rules therefore ensure the *wellformedness* of individual models as well as the *consistency* of all domain-specific viewpoints that constitute a system description.

5 Semantic Web Reasoning over Software Models

As outlined in Sect. 3, we aim at leveraging the power of logical inference and Semantic Web reasoning for integrity and consistency checking of software models. This section describes the practical realization of this goal and discusses solutions to the implementation challenges arising from the open-world assumption in OWL and lack of nonmonotonic queries in SWRL.

5.1 Reasoning Architecture

In Sect. 4.2, we described how the semantics of our new upper ontology are specified using DL axioms and rules. To validate the integrity and consistency of software models using both DL and rule reasoning, we employ a three-layered reasoning architecture (Fig. 4). At the bottom layer, the Jena Semantic Web framework [16] parses OWL ontologies and knowledge bases serialized in RDF. On the middle layer, the Pellet DL reasoner [26] provides basic DL reasoning facilities like subsumption, satisfiability checking, and instance classification. At the top, the rule reasoner of the Jena framework evaluates both deductive rules and nonmonotonic consistency constraints. Each layer provides a view of the RDF graph to the layer above, so the Jena rule engine sees all statements entailed by the Pellet reasoner in addition to those asserted in the knowledge base.

The separation of DL and rule reasoning into two distinct layers is motivated by the following practical considerations: First, ontologies with complex TBox assertions like USMO (many disjunctions, inverse roles, and existential quantifications) require the power of tableau-based implementations available in dedicated DL reasoners such as Pellet. We experienced serious performance degradation when activating the OWL-DL integrity rules in the general-purpose rule engine of the Jena framework. Moreover, Jena rules are not *DL-safe* [21].

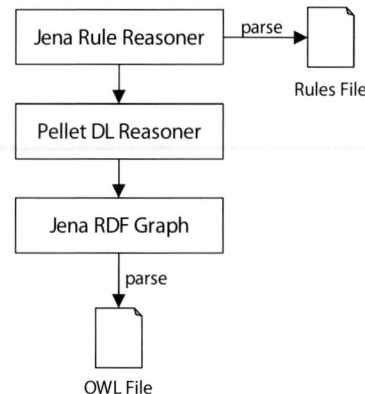

Fig. 4. The Reasoning Architecture in HybridMDSD

When used for DL reasoning, the reasoner creates *blank nodes* for existential quantification and minimum cardinality axioms. As a result, variables in USMO integrity constraints are bound to individuals that are not explicitly asserted in the ABox. This often results in wrongly reported integrity violations.

By contrast, the Pellet reasoner is DL-safe and only returns known individuals from the knowledge base. Unfortunately, Pellet alone does not suffice either. Since it solely processes rules encoded in SWRL, there is no support for non-monotonic querying over OWL ontologies. In particular, the lack of negation as failure renders it impossible to formulate and validate many USMO integrity constraints. The Jena rule engine does not have these limitations and supports complex nonmonotonic queries via extensible built-ins.

5.2 Simulating a Closed World

One of the major challenges of DL reasoning over software models is the open-world assumption of the Semantic Web. As outlined in Sect. 2, models of software systems usually represent a closed world. Thus, missing model elements might cause integrity constraint violations or inconsistencies between different viewpoints in multi-domain development scenarios. Under the OWA, these violations remain undetected since unknown information is not interpreted as false. Based on the axiomatization of our upper ontology for software models, this section provides a classification of the various problems and discusses possible solutions.

Essentially, we identify three types of USMO axioms negatively affected by the open-world assumption: (1) the covering axioms for "abstract" concepts, (2) existential property restrictions or cardinality axioms and (3) universal quantification axioms.

As an example for covering axioms, consider that every USMO Property is either an IntrinsicProperty or RelationalProperty:

$$\text{Property} \equiv \text{IntrinsicProperty} \sqcup \text{RelationalProperty} \qquad (3)$$

Now, an individual p solely asserted to be an instance of Property will *not* violate this constraint, because it trivially satisfies the disjunction [18]; it is just unknown which type of Property p exactly belongs to.

An example for existential quantification is the axiom describing a Property as a dependent entity that cannot exist by itself:

$$\text{Property} \sqsubseteq \exists \text{depends\_on.Schema} \qquad (4)$$

Due to the open-world semantics, a Semantic Web reasoner will *not* actually ensure that each Property depends on at least one Schema known in the knowledge base. Similar considerations apply to cardinality restrictions.

Finally, a typical axiom illustrating universal quantification is that Universals cannot depend on Particulars:

$$\text{Universal} \sqsubseteq \forall \text{depends\_on.Universal} \qquad (5)$$

Here, a reasoner will not validate the actual type of an individual i asserted as the object in a dependency relationship. Instead, i will be inferred to be of type Universal.

Theoretically, all above-listed axioms can be rewritten with the *epistemic operator* **K** [7] to gain the desired semantics. The **K** operator corresponds to $\Box$ (*Necessarily*) in modal logic and allows to close the world locally for a concept or role [18]:

$$\text{Property} \equiv \textbf{K}\text{IntrinsicProperty} \sqcup \textbf{K}\text{RelationalProperty} \qquad (6)$$
$$\text{Property} \sqsubseteq \exists \textbf{K}\text{depends\_on.Schema} \qquad (7)$$
$$\text{Universal} \sqsubseteq \forall \text{depends\_on.}\textbf{K}\text{Universal} \qquad (8)$$

Unfortunately, there is currently no support for the **K** operator in OWL, even though a corresponding extension has been suggested many times [18,10,22]. The only implementation known to us is part of the Pellet DL reasoner [6], but it is of prototypical nature and limited to the description logic $\mathcal{ALC}$ [18].

A method to simulate local closed worlds without explicit support for the **K** operator is documented in [22] and [24, pp. 85]. Its main idea is to use set-theoretic operations to determine the set of individuals that *possibly* violate an integrity constraint, in addition to those that are certainly invalid. The key observation is that for a given class C, we can partition all individuals into three distinct sets: (1) those that are known to be a member of C, (2) those that are known to be a member of $\neg C$, and (3) those that may or may not be a member of C. For instance, let C be the class of all individuals that are asserted to be instances of Property, but do not depend on a Schema, thereby causing an integrity violation. In an open world, simply enumerating C does not yield the expected result. The invalid Propertys can instead be found by subtracting everything in group 2 from the set of all Propertys. Obviously, a sufficient condition for individuals in group 2 is that they depend on at least one entity:

$$\text{DependsOnMinOne} \equiv \geqslant 1 \text{ depends\_on} \tag{9}$$

$$\text{InvalidProperty} = \text{Property} - \text{DependsOnMinOne} \tag{10}$$

Unfortunately, there is no DL concept constructor for expressing the difference of classes. Moreover, as observed in [22], this method of querying invalid entities is asymmetric to the usual instance retrieval for other classes. Hence, it does not suit the needs of an end user (i.e., the modeler) who wishes to validate the consistency of the knowledge base in a uniform way.

Since neither the **K** operator nor the set-theoretic approach are feasible in practice, our prototype explicitly "closes the world" before each validation run. In the first case (the covering axioms), this means to explicitly assert for each Property p that it is neither an IntrinsicProperty nor a RelationalProperty if the concrete type is unknown. This results in

$$p \in \text{Property} \sqcap \neg\text{IntrinsicProperty} \sqcap \neg\text{RelationalProperty} \tag{11}$$

and a reasoner will readily report the apparent inconsistency. An alternative approach is to declare the covering subclasses equivalent to an enumeration of all known members. The following listing exemplifies this method in pseudo Java code:

```
for each class C {
    get i_1,...,i_n with {i_1,...,i_n} ⊑ C
    assert C ≡ {i_1,...,i_n}
}
```

In the above example, this results in the following assertions:

$$\text{IntrinsicProperty} \equiv \{\} \tag{12}$$

$$\text{RelationalProperty} \equiv \{\} \tag{13}$$

A reasoner can now detect the inconsistency between an individual p declared to be an instance of Property and the unsatisfiability of the Property concept.

To validate the second type of problematic axioms, namely existential property restrictions or cardinality constraints, it is necessary to close the corresponding roles. This is achieved by asserting for every individual i that a particular role on i will have no more role fillers [3, p. 25]. The following listing illustrates a possible implementation:

```
for each individual i {
    for each class C with i ∈ C {
        if ( C is one of {∃p.X, ⩾ np, = np} ) {
            get n with i ∈ = np
            assert i ∈ ⩽ np
        }
    }
}
```

For each individual that belongs to a restriction class, we add a membership assertion to an anonymous class that limits the cardinality of the affected property to the number of current fillers. If this number violates an existential property restriction or cardinality constraint, a reasoner will detect the inconsistency between the different class assertions. Again, the counting of existing property assertions in line 4 is a nonmonotonic operation.

A major disadvantage of this approach is that any subsequent property assertions will render the knowledge base inconsistent. Therefore, if one of the represented models changes, we have to rebuild the entire ABox, which is expensive for large models and makes continuous validation of the knowledge base in the background difficult.

Compared to the problems discussed above, universal quantification axioms are relatively easy to validate. Since our upper ontology has been designed such that subclasses on the same level of the inheritance hierarchy are always pairwise disjoint, a reasoner will detect an inconsistency if fillers for universally quantified properties do not have the expected type.

5.3 Realizing Nonmonotonic Rules

Integrity rules are typically read like "if the body is true, then the head must also be true". Yet, to actually validate these rules and show the results to the user, a logical reformulation is necessary. To see why, consider the following rule that combines the irreflexivity and antisymmetry axioms of the dependency relationship:

$$\mathsf{depends\_on}(x,y) \rightarrow \neg\mathsf{depends\_on}(y,x) \tag{14}$$

Evidently, this rule is nonmonotonic as its head contains a negated atom. In this form, the rule is a statement about the conditions that hold in a consistent closed-world knowledge base **KB**. If we abbreviate the rule with R, we can thus state a new rule S as follows:

$$\neg R \rightarrow Inconsistent(\mathbf{KB}) \tag{15}$$

Actually, we are only interested in those (known) individuals that cause the knowledge base to be inconsistent, so we can reformulate S by inserting the old rule for R and using an auxiliary predicate that classifies an individual as invalid. We can then simply query the knowledge base for all instances of Invalid and present them to the user:

$$\neg(\mathsf{depends\_on}(x,y) \rightarrow \neg\mathsf{depends\_on}(y,x)) \rightarrow \mathsf{Invalid}(x) \wedge \mathsf{Invalid}(y) \tag{16}$$

Transforming this rule using simple logical equivalences yields:

$$\mathsf{depends\_on}(x,y) \wedge \mathsf{depends\_on}(y,x) \rightarrow \mathsf{Invalid}(x) \wedge \mathsf{Invalid}(y) \tag{17}$$

Obviously, we can rephrase every integrity rule R of the form $B_R \rightarrow H_R$ into a rule S of the form $B_S \rightarrow H_S$, where B_S is $B_R \wedge \neg H_R$. The intuitive reading of S is then "If there are individuals that cause the body of R to be true and the head of R to be false, then these individuals are invalid."

Not all nonmonotonic rules can be as easily transformed as in the example above. Some USMO integrity rules contain negated predicates in their body. We speak of *simple negation* in this case. As an example, consider the following structural integrity rule, which ensures that the containment relationships between entities span a tree rather than a graph:

$$\mathsf{contains}(e_1, e_3) \wedge \mathsf{contains}(e_2, e_3) \wedge e_1 \dot{\neq} e_2 \rightarrow \mathsf{contains}(e_1, e_2) \vee \mathsf{contains}(e_2, e_1) \quad (18)$$

Again, note the nonmonotonicity of this rule caused by the disjunction in the head. Applying the logical transformation outlined above yields:

$$\mathsf{contains}(e_1, e_3) \wedge \mathsf{contains}(e_2, e_3) \wedge e_1 \dot{\neq} e_2 \wedge \neg\mathsf{contains}(e_1, e_2) \wedge \neg\mathsf{contains}(e_2, e_1) \rightarrow$$
$$\mathsf{Invalid}(e_1) \wedge \mathsf{Invalid}(e_2) \wedge \mathsf{Invalid}(e_3) \quad (19)$$

The disjunction in the head is gone, but we have gained two negated predicates in the body of the new rule. Under an open-world assumption, we do not know for sure whether e_1 contains e_2 or vice versa. Without negation as failure semantics, this rule will never fire.

A more complex problem than simple negation is what we call *negated existential quantification*. Here, the rule body contains an atom that queries the knowledge base for the existence of an individual that matches a number of given criteria. This is required in several USMO integrity rules. As an example, consider the following rule which ensures that the extension of a Universal is not empty:

$$\mathsf{Universal}(u) \wedge \mathsf{has\_extension}(u, e) \rightarrow \exists p \, (\mathsf{Particular}(p) \wedge \mathsf{is\_member\_of}(p, e)) \quad (20)$$

Reformulating this rule according to our transformation guideline yields:

$$\mathsf{Universal}(u) \wedge \mathsf{has\_extension}(u, e) \wedge \neg\exists p \, (\mathsf{Particular}(p) \wedge \mathsf{is\_member\_of}(p, e)) \rightarrow$$
$$\mathsf{Invalid}(u) \quad (21)$$

Axioms like this one apparently require a nonmonotonic querying mechanism built into the rule language. Some USMO integrity rules even involve existential quantification over not just one, but several variables.

In our implementation, we realize nonmonotonic rules through dedicated built-ins for the Jena rule reasoner. Dealing with simple negation is easy since Jena already offers a nonmonotonic built-in `noValue` that allows to query the knowledge base for concept and role assertions. For instance, Axiom 19 can be expressed as shown in in the following listing:

```
1  [(?e1 usmo:contains ?e3),(?e2 usmo:contains ?e3),notEqual(?e1,?e2),
2    noValue(?e1 usmo:contains ?e2), noValue(?e2 usmo:contains ?e1) ->
3    (?e1 rdf:type sc:Invalid),(?e2 rdf:type sc:Invalid),
4    (?e3 rdf:type sc:Invalid) ]
```

By default, the Jena rule language is not expressive enough to formulate integrity constraints involving negated existential quantification. Fortunately, the set of available built-ins can easily be extended. We have written a custom built-in `notExists` that realizes the required semantics. The following listing exemplifies the usage of this built-in for Axiom 20:

```
1  [(?u rdf:type usmo:Universal), (?u usmo:has_extension ?e),
2     notExists(?p rdf:type usmo:Property, ?p usmo:is_member_of ?e) ->
3  (?u rdf:type sc:Invalid) ]
```

6 Related Work

To the best of our knowledge, the HybridMDSD project is the first attempt to leverage Semantic Web reasoning for consistency checking of multiple domain-specific software models based on semantics defined in a shared ontology.

Closely related are the works on UML model consistency validation with description logics by Simmonds et al. [25]. They, too, represent model elements as individuals in the ABox of a DL reasoning system. However, in contrast to our work, they do not employ an axiomatized upper ontology as a semantical foundation and concentrate solely on the UML metamodel, which results in very complex queries and hampers the reusability of their solution.

In the area of multi-domain modeling, Hesselund et al. have presented the *SmartEMF* system [14]. SmartEMF supports validating consistency constraints between several loosely coupled domain-specific languages by mapping all elements of a DSL's metamodel and instantiating models to a Prolog fact base. This yields the advantage of logical queries with closed-world semantics, but it does not offer the powerful features of a DL reasoner such as subsumption and instance classification.

Finally, both the *SemIDE* proposal by Bauer and Roser [2] and the *ModelCVS* project by Kappel et al. [17] aim to integrate different modeling languages based on common domain ontologies. Yet, both approaches focus exclusively on an integration on the metamodel level, which essentially means a restriction to TBox reasoning.

7 Conclusion

In this paper, we have analyzed the challenges in Ontology-Driven Software Development that arise from using Semantic Web technologies for representing and reasoning about models in the design and implementation phases. We have particularly focused on the conceptual mismatch between the open-world assumption of the Semantic Web and the closed world of software models in an MDSD process. To exemplify the resulting problems, we have introduced a new upper ontology that allows to semantically integrate different domain-specific software modeling languages. Based on the axiomatization of this ontology, we have illustrated different types of integrity constraints that require closed-world semantics and nonmonotonic reasoning. Our study of critical axioms comprised both DL concept constructors and Horn clause rules. Finally, we described a reasoning architecture that practically realizes the presented techniques.

Acknowledgment. The authors would like to thank Manuel Zabelt and the anonymous reviewers for their valuable comments on previous versions of this paper. Additionally, we would like to thank explicitly Prof. Uwe Assmann from Chair of Software Technology at Dresden University of Technology, who actively supports the HybridMDSD project.

References

1. Amann, U., Zschaler, S., Wagner, G.: Ontologies, Meta-models, and the Model-Driven Paradigm. In: Ontologies, Meta-models, and the Model-Driven Paradigm, pp. 249–273 (2006)
2. Bauer, B., Roser, S.: Semantic-enabled Software Engineering and Development. In: INFORMATIK 2006 - Informatik für Menschen, Lecture Notes in Informatics (LNI), vol. P-94, pp. 293–296 (2006)
3. Brachman, R.J., McGuiness, D.L., Patel-Schneider, P.F., Resnick, L.A.: Living with CLASSIC: When and How to Use a KL-ONE-Like Language. In: Sowa, J.F. (ed.) Principles of Semantic Networks: Explorations in the representation of knowledge, pp. 401–456. Morgan Kaufmann, San Mateo (1991)
4. Bräuer, M., Lochmann, H.: Towards Semantic Integration of Multiple Domain-Specific Languages Using Ontological Foundations. In: Proceedings of 4th International Workshop on (Software) Language Engineering (ATEM 2007) co-located with MoDELS 2007, Nashville, Tennessee (October 2007)
5. Bruer, M.: Design of a Semantic Connector Model for Composition of Metamodels in the Context of Software Variability. Diplomarbeit, Technische Universität Dresden, Department of Computer Science (November 2007)
6. Clark, L., Parsia: Pellet: The Open Source OWL DL Reasoner (2007), http://pellet.owldl.com
7. Donini, F.M., Lenzerini, M., Nardi, D., Nutt, W., Schaerf, A.: An epistemic operator for description logics. Artificial Intelligence 100(1-2), 225–274 (1998)
8. Evermann, J., Wand, Y.: Ontology based object-oriented domain modelling: fundamental concepts. Requirements Engineering 10(2), 146–160 (2005)
9. Gangemi, A., Guarino, N., Masolo, C., Oltramari, A., Schneider, L.: Sweetening Ontologies with DOLCE. In: Gómez-Pérez, A., Benjamins, V.R. (eds.) EKAW 2002. LNCS (LNAI), vol. 2473, Springer, Heidelberg (2002)
10. Grimm, S., Motik, B.: Closed World Reasoning in the Semantic Web through Epistemic Operators. In: OWL: Experiences and Directions (OWLED 2005), Calway, Ireland (November 2005)
11. Guizzardi, G., Herre, H., Wagner, G.: Towards Ontological Foundations for UML Conceptual Models. In: Meersman, R., Tari, Z., et al. (eds.) CoopIS 2002, DOA 2002, and ODBASE 2002. LNCS, vol. 2519, pp. 1100–1117. Springer, Heidelberg (2002)
12. Guizzardi, G., Wagner, G.: Applications of a Unified Foundational Ontology, ch XIII. In: Some Applications of a Unified Foundational Ontology in Business Modeling, pp. 345–367. IDEA Publisher (2005)
13. Happel, H.-J., Seedorf, S.: Applications of Ontologies in Software Engineering. In: 2nd International Workshop on Semantic Web Enabled Software Engineering (SWESE 2006), held at the 5th International Semantic Web Conference (ISWC 2006), Athens, GA, USA (November 2006)
14. Hessellund, A., Czarnecki, K., Wasowski, A.: Guided Development with Multiple Domain-Specific Languages. In: Engels, G., Opdyke, B., Schmidt, D.C., Weil, F. (eds.) MODELS 2007. LNCS, vol. 4735, Springer, Heidelberg (2007)

15. Horridge, M., Knublauch, H., Rector, A., Stevens, R., Wroe, C.: A Practical Guide To Building OWL Ontologies Using The Protégé-OWL Plugin and CO-ODE Tools Edition 1.0. The University Of Manchester (August 2004)
16. Jena – A Semantic Web Framework for Java. Version 2.5 (January 2007), http://jena.sourceforge.net
17. Kappel, G., Kramler, G., Kapsammer, E., Reiter, T., Retschitzegger, W., Schwinger, W.: ModelCVS - A Semantic Infrastructure for Model-based Tool Integration. Technical Report, Business Informatics Group, Wien Technical University (July 2005)
18. Katz, Y., Parsia, B.: Towards a Nonmonotonic Extension to OWL. In: OWL: Experiences and Directions (OWLED 2005), Calway, Ireland (November 2005)
19. Kühne, T.: Matters of (Meta-) Modeling. Software and Systems Modeling 5(4), 369–385 (2006)
20. Lochmann, H.: Towards Connecting Application Parts for Reduced Effort in Feature Implementations. In: Proceedings of 2nd IFIP Central and East European Conference on Software Engineering Techniques CEE-SET 2007 (WIP Track), Posen, Poland (October 2007)
21. Motik, B., Sattler, U., Studer, R.: Query Answering for OWL-DL with Rules. In: McIlraith, S.A., Plexousakis, D., van Harmelen, F. (eds.) ISWC 2004. LNCS, vol. 3298, Springer, Heidelberg (2004)
22. Ng, G.: Open vs Closed world, Rules vs Queries: Use cases from Industry. In: OWL: Experiences and Directions (OWLED 2005), Calway, Ireland (November 2005)
23. Patel-Schneider, P.F., Horrocks, I.: OWL 1.1 Web Ontology Language Overview. World Wide Web Consortium (W3C), W3C Member Submission (December 2006), http://www.w3.org/Submission/owl11-overview/
24. Racer Systems GmbH & Co. KG. RacerPro User's Guide Version 1.9 (December 2005)
25. Simmonds, J., Bastarrica, M.C.: A tool for automatic UML model consistency checking. In: Proceedings of the 20th IEEE/ACM international Conference on Automated Software Engineering (ASE 2005), Long Beach, CA, USA, Demonstration Session: Formal tool demo presentations, pp. 431–432. ACM Press, New York (2005)
26. Sirin, E., Parsia, B., Grau, B.C., Kalyanpur, A., Katz, Y.: Pellet: A Practical OWL-DL Reasoner. Journal of Web Semantics (submitted for publication, 2007)
27. Stahl, T., Völter, M.: Model-Driven Software Development: Technology, Engineering, Management, 1st edn. Wiley & Sons, Chichester (2006)

A Core Ontology for Business Process Analysis

Carlos Pedrinaci[1], John Domingue[1], and Ana Karla Alves de Medeiros[2]

[1] Knowledge Media Institute, The Open University, Milton Keynes, MK7 6AA, UK
{c.pedrinaci,j.b.domingue}@open.ac.uk
[2] Eindhoven University of Technology, P.O. Box 513, 5600MB, Eindhoven,
The Netherlands
a.k.medeiros@tue.nl

Abstract. Business Process Management (BPM) aims at supporting the whole life-cycle necessary to deploy and maintain business processes in organisations. An important step of the BPM life-cycle is the analysis of the processes deployed in companies. However, the degree of automation currently achieved cannot support the level of adaptation required by businesses. Initial steps have been performed towards including some sort of automated reasoning within Business Process Analysis (BPA) but this is typically limited to using taxonomies. We present a core ontology aimed at enhancing the state of the art in BPA. The ontology builds upon a Time Ontology and is structured around the process, resource, and object perspectives as typically adopted when analysing business processes. The ontology has been extended and validated by means of an Events Ontology and an Events Analysis Ontology aimed at capturing the audit trails generated by Process-Aware Information Systems and deriving additional knowledge.

1 Introduction

Many companies use information systems to support the execution of their business processes. Examples of such information systems are Enterprise Resource Planning, Customer Relationship Management, and Workflow Management Systems (WFMS). These systems usually generate events while executing business processes [1] and these events are recorded in logs. The competitive world we live in requires companies to adapt their processes in a faster pace. Therefore, continuous and insightful feedback on how business processes are executed becomes essential. Additionally, laws like the Sarbanes-Oxley Act force companies to show their compliance to standards. In short, there is a need for good analysis tools that can provide feedback about how business processes are actually being executed based on the observed (or registered) behaviour in event logs.

BPM results from the limitations exhibited by WFMS which mainly focus on the enactment of processes by generic engines and does not take into account the continuous adaptation and enhancement of existing processes. BPM acknowledges and aims to support the complete life-cycle of business processes which undoubtedly involves post-execution analysis and reengineering of process models. A key aspect for maintaining systems and the processes they support is the capability to analyse them. BPA is particularly concerned with the behavioural properties of enacted processes may it be at runtime, as in Business Process Monitoring, or post-execution as in Business Process Mining [1] or Reverse Business Engineering.

Due to its cyclic nature, BPM has however made more evident the existing difficulties for obtaining automated solutions from high-level business models, and for analysing the execution of processes from both a technical and a business perspective. The fundamental problem is that moving between the business-level and the IT-level is hardly automated [2]. Deriving an IT implementation from a business model is particularly challenging and requires an important and ephemeral human effort which is expensive and prone to errors. Conversely analysing automated processes from a business perspective, e.g., calculating the economical impact of a process or determining the performance of different departments in an organisation, is again an expensive and difficult procedure which typically requires a human in the loop. *Semantic Business Process Management* (SBPM), that is the combination of Semantic Web and Semantic Web Services technologies with BPM, has been proposed as a solution [2].

In this paper we present results obtained in the context of the European project SUPER (IST-026850) which aims at developing a SBPM framework, based on Semantic Web Services technology, that acquires, organises, shares and uses the knowledge embedded in business processes in order to make companies more adaptive. This semantic framework will support the four phases of the BPM life-cycle and the research presented in this paper provides the foundation for semantic BPA. *In particular we shall describe a core ontology for business process analysis which bridges the gap between low-level monitoring information and high-level business knowledge.* The remainder of the paper is organised as follows. Section 2 reviews existing research that makes use of semantic technologies and present a set of requirements and competency questions that semantic BPA technologies should address. Section 3 presents COBRA, a Core Ontology for Business pRocess Analysis, and Time Ontology which provides the basis for using temporal reasoning within BPA. Section 4 illustrates how COBRA can be applied to BPA. Section 5 presents our conclusions and describes future research to be carried out.

2 Semantics in Business Process Management

In the last years significant efforts have been devoted to integrating automated reasoning with the BPM domain, a field where the application of knowledge-based technologies appears to be the next evolutionary step [3]. These efforts can roughly be divided into top-down and bottom-up approaches. Top-down approaches make use of high-level conceptual models to structure and reason about Business Process Management activities. Among these approaches we find research on enterprise ontologies, models for resources consumption and provision, value flows, service bundling, etc. [2,4,5,6,7,8]. However, despite the variety of models and tools produced so far there is little uptake within the industry which is often due to the existing difficulty to provide and maintain good knowledge bases expressed in terms of these conceptual models.

On the other hand, bottom-up approaches integrate some sort of light-weight automated reasoning machinery with existing BPM solutions, see for instance [9, 10, 11]. These efforts are mainly dominated by researchers from the BPM area, where knowledge-based technologies have not been widely used so far. The focus has mainly been the annotation of data warehouses or the application of rule engines to control resources and ensure certain business policies are followed. Unfortunately, the information manipulated is mostly in syntactic formats which is hardly amenable to automated reasoning. In fact, most of the budget when applying so-called Business Intelligence solutions is typically devoted to the manual integration of data from BPM systems and this is often an ephemeral effort which has to be repeated over time. As a result the benefits gained by applying these techniques are largely limited. Still, as opposed to top-down approaches, the fact that these research efforts are grounded into deployed BPM systems increases their impact in the industry.

What can be distilled from the current state-of-the-art is that the existent epistemological gap between, on the one hand industry BPM solutions, and on the other hand knowledge-based research, hampers to an important extent the wider application of semantics in BPM. The research presented in this paper aims precisely at reducing this gap when it comes to analysing business process executions. In order to guide and validate our approach we present next a representative set of requirements and competency questions that we have identified based on existing practice within the BPM domain.

2.1 Requirements for Semantic Business Process Analysis

BPA is typically structured around three different views: (i) the *process view*; (ii) the *resource view*; and (iii) the *object view* [12]. The process view is concerned with the enactment of processes and is thus mainly focussed on the compliance of executed processes with respect to prescribed behaviours and Key Performance Indicators that can support business analysts in the examination and eventual optimisation of deployed processes [1]. Relevant information in this respect are (i) "the processes and activities currently running"; (ii) "which ones have been completed and whether they were successful or not"; (iii) "the execution time of the different business activities"; (iv) "which business activities have preceded which others", etc. The resource view is centred around the usage of resources within processes. In this perspective, the performance at different levels of granularity (individuals, organisational units, etc.), work distribution among the resources, and the optimisation of resources usage are the main aspects analysed. Typical questions in this perspective are for instance (i) "which resources were involved in which business activities"; (ii) "which actor was responsible for a certain process"; (iii) "which external providers appear to work more efficiently"; (iv) "what's the average number of orders processed by the sales department per month", etc. Finally, the object view focusses on business objects such as inquiries, orders or claims. This perspective is often adopted in order to better analyse the life-cycle of so-called Business Objects. In this perspective, business analysts often want answers to questions like (i) "what is the average cost per claim"; (ii) "which is the item we are currently selling the most (or the least)"; (iii) "what's the overall benefit we are obtaining per item"; (iv) "are critical orders processed in less than two hours", etc.

These three views are populated with statistical information such as the minimum, the average or the deviation of some parameter of interest, and correlations are typically established across them, e.g., "what is the average process execution time for processing each type of order?". Common to these scenarios where BPA techniques are applied is the underlying dependency with respect to time manipulation (e.g.,"are critical orders processed in less than two hours"), the need to navigate through different levels of abstraction (e.g., "what's the average number of orders processed by the sales department per month") and across the different perspectives, and the overall necessity to apply general purpose methods over domain specific data.

Therefore, to enhance the state-of-the-art of BPA we need a comprehensive conceptual model of the BPM domain that supports applying general purpose knowledge-based techniques over domain specific data, coupled with the capacity to navigate through different levels of abstraction across the process, resource, and object perspectives, and the ability to appropriately deal with temporal aspects. The next section is devoted to presenting a core ontology for supporting Business Process Analysis that aims to provide a generic and extensible conceptual model that can support the competency questions exposed above.

3 An Ontology for Business Process Analysis

In order to support the level of automation required by enterprises nowadays we need to enhance BPA with support for applying general purpose analysis techniques over specific domains in a way that allows analysts to use their particular terminology and existing knowledge about their domain. To this end we have defined the Core Ontology for Business pRocess Analysis. COBRA provides a core terminology for supporting BPA where analysts can map knowledge about some particular domain of interest in order to carry out their analyses. It is worth noting that COBRA does not aim to provide a fully-comprehensive conceptualisation for supporting each and every kind of analysis since the scope would simply be too big to be tackled appropriately in one ontology. Instead COBRA provides a pluggable framework based on the core conceptualisations required for supporting BPA and defines the appropriate hooks for further extensions in order to cope with the wide-range of aspects involved in analysing business processes. These extensions are currently been developed in SUPER as part of an ontological framework aimed at providing an extensive conceptualisation of the BPM domain ranging from process modelling to the definition of business strategies. Still, COBRA already provides a good basis for supporting the most typical analysis as described in the previous section.

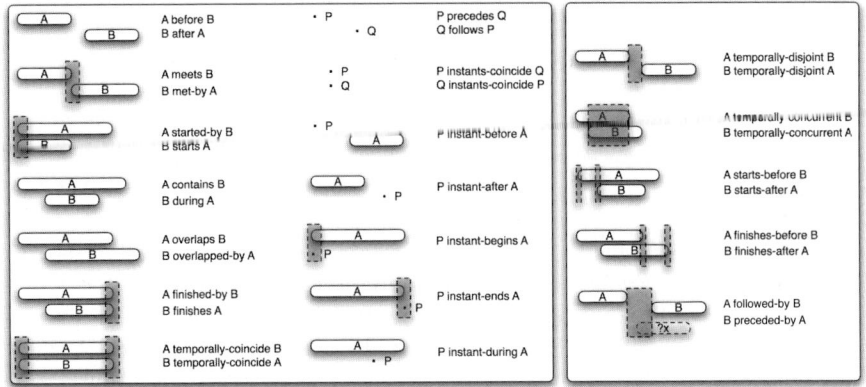

Fig. 1. Instants and Interval relations

COBRA has been developed using the Operational Conceptual Modelling Language (OCML) [13], which provides support for executing the definitions in the ontology as well as export mechanisms to other representations including OWL and WSML. COBRA builds upon two ontologies, namely Base Ontology and Time Ontology, and is currently enhanced with Events Ontology for capturing audit trails, and Events Analysis Ontology which provides a set of generic reusable rules and relations[1]. Base Ontology provides the definitions for basic modelling concepts such as tasks, relations, functions, roles, numbers, etc. The interested reader is referred to [13] for further information. The other ontologies will be described in the remainder of this section.

3.1 Time Ontology

COBRA builds upon Time Ontology that provides a temporal reference by means of which one can determine temporal relations between elements. The ontology defines three top-level concepts, namely *Time Instant*, *Time Interval*, and *Temporal Entity*. Time Instant is the main primitive element and it provides the means for identifying a point in time with precision up to the microsecond for we aim to support monitoring automated systems. Time Intervals are defined by means of the start and end instants and have therefore an associated duration which can be computed by means of a function that substracts the limiting instants. Temporal Entity, as opposed to the conceptualisation proposed in [14], represents entities that have a temporal occurrence, and are therefore different from Time Instant and Time Interval which are the base constructs that represent a particular point or period in time.

Using these core concepts we have implemented the interval relations defined by Allen [15], the additional instant-interval relations defined by Vilain [16], and useful functions for computing the duration of intervals or for obtaining the

---

[1] The ontologies can be found at http://kmi.open.ac.uk/people/carlos

current Time Instant. The left hand-side of Figure 1 illustrates these relations, whereby *A* and *B* represent Time Intervals, whereas *P* and *Q* represent Time Instants. The relations are self-explanatory, the interested reader is referred to [15] and [16] for further details. It is worth noting that we have renamed the equality relations for Time Intervals and Time Instants to *Temporally Coincide* and *Instants Coincide* respectively, for we believe it is counterintuitive to use the term "*equal*" for referring to different things that occur at the same time.

In addition to these relations we have also included for convenience a few typical disjunctions of Allen's algebra, e.g., *Before-Or-Meets*, and further relations which are relevant for BPA. The latter are depicted in the right-hand side of Figure 1. The new relations we have implemented are namely *Temporally Disjoint, Temporally Concurrent, Starts-Before, Starts-After, Finishes-Before, Finishes-After*. Two intervals are considered to be Temporally Disjoint if there is no interval shared between the two, which in Allen's interval algebra is equivalent to a disjunction between Before, After, Meets and Met-By. Temporally Concurrent is the inverse relation of Temporally Disjoint and it therefore holds when there exists some interval shared between the two concurrent intervals. Starts-Before, Starts-After, Finishes-Before and Finishes-After, which we believe are self-explanatory, are based on the numerical comparison between the start instant or end instant of the intervals.

Our Time Ontology considers two kinds of *Temporal Entities*, namely *Instantaneous Entity* and *Time Spanning Entity*. Instantaneous Entities are phenomena that occur at a specific point on time and whose duration can be neglected. By contrast, Time Spanning Entities are those that last over a period of time indicated by the *spansInterval* slot. The distinction between, on the one hand, Temporal Entities and, on the other hand, Time Instant and Time Interval allows us to apply the previous relations over a plethora of entities, i.e., every Temporal Entity. In addition to the previously mentioned relations we have included two which are specific to Time Spanning entities and are particularly useful for BPA, namely *Followed-By* and *Preceded-By*. A Time Spanning Entity *I* is *Followed-By* by another Time Spanning Entity *J* of kind *C*, if *J* is After *I* and there is no other Time Spanning Entity *X* of kind *C* which is After *I* and Starts-Before *J*. *Preceded-By* is the inverse relation of *Followed-By*.

One of the main characteristics of our Time Ontology is the use of *polymorphism*. Our ontology supports determining temporal relations about the primitive elements Time Instant and Time Interval, between these and Temporal Entities, and between any two Temporal Entities. To do so, the relations have been implemented on the basis of backward-chaining rules that perform the appropriate transformations between Temporal Entities and their primitive counterpart and then invoke the primitive relation. This polymorphism is a convenient feature of our conceptualisation in order to support BPA where temporal relations need to be evaluated between executions of activities, e.g., "was Activity A executed after Activity B?", executions of processes, e.g.,"has Process A been run concurrently with Process B", but also with respect to reference intervals or instants, e.g., "retrieve all the Processes executed in the last month".

3.2 Core Ontology for Business Process Analysis

We previously introduced that BPA is concerned with the analysis of the execution of business processes from several perspectives. In particular, we identified the *process view*, the *resource view*, and the *object view*. COBRA has therefore been structured around these very views in an attempt to enhance BPA with support for the automated reasoning, querying, and browsing of audit trails from different perspectives and at different levels of abstraction. The ontology is depicted in Figure 2 using an extended UML notation where arrows represent the *isA* relation, dashed arrows denote the *instanceOf* relation, and lines represent custom relations. Further notation extensions will be explained as the need arises during the description of the ontology.

The development of COBRA has been guided to an important extent by existing ontologies like the Enterprise Ontology [5], DOLCE [17], TOVE [4] and CIDOC [18]. COBRA distinguishes between Temporal Entities (see Section 3.1) and *Persistent Entities* which are disjoint. This terminology is borrowed from CIDOC [18] but is conceptually inline with DOLCE [17], whereby Endurant corresponds to Persistent Entity and Perdurant to Temporal Entity. In short, Temporal Entities are entities that have a temporal extent whereas Persistent Entities are essentially independent of time. COBRA uses this high-level categorisation as a foundational basis but it doesn't go however much further in the reuse of existing foundational ontologies for it aims at supporting analysis of processes and a complete grounding into this kind of ontologies would carry an important computational overhead. Instead, we provide a simple categorisation of Persistent Entities specifically tailored to our needs, though informed by DOLCE, whereby we contemplate Physical and Non-Physical Entities which are disjoint. Physical entities are those that have a mass.

Physical and Non-physical Entities are further refined into *Agentive* and *Non-Agentive*. The distinction between these classes which are obviously disjoint, is that Agentive Entities are those that can take an active part within some specific activity. Finally, we define *Agent* as the union of both Physical and Agentive Non-Physical Entities. We include for reference and self-containment a few concepts widely used within BPM. For instance, we include *Object, Person, Organisation, Software Agent,* and *Role*. The latter will be dealt with in more detail later on. COBRA, for its purpose is to provide core definitions for supporting business analysis, does not refine these classes any further. Instead they serve as placeholders for including additional conceptualisations such as Organisational Ontologies or domain-specific master data. By doing so we aim at reducing the ontological commitment, while we support the seamless integration of further specific conceptualisations. Finally, since sometimes one needs not specify a concrete instance but rather the type, e.g. "you require a computer", we have defined the meta-class Persistent Entity Type such that all the sub-classes of Persistent Entity are instances of Persistent Entity Type. This is depicted in Figure 2 by means of a double-headed arrow.

Core concepts in COBRA are *Business Activity* and *Business Activity Realisation*. A Business Activity is a Non-Agentive Non-Physical Entity (the isA

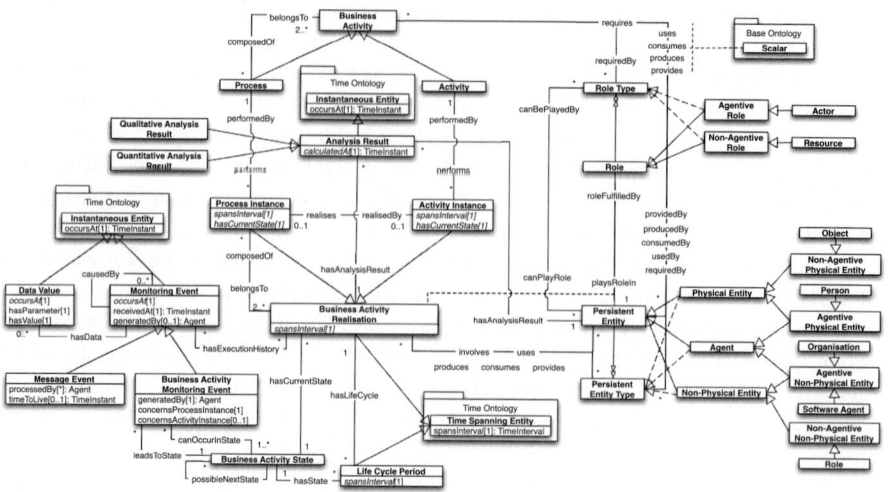

Fig. 2. Core Business Process Analysis Ontology

relation is not depicted in the figure for the sake of clarity) that represents the specification of any business activity at a high-level where aspects such as the control flow are abstracted away. We contemplate two kinds of Business Activities, namely *Process* and *Activity*, to reuse the terminology typically employed in the BPM domain [12]. Activity represents atomic Business Activities whereas Processes are *composedOf* at least two other Business Activities. Business Activity Realisations are Time Spanning Entities which represent the actual execution of Business Activities. Mirroring Business Activities, Process Instance and Activity Instance are the two kinds of Business Activity Realisations considered. Despite their name, which originates again from BPM literature, both are concepts which represent the actual executions of Processes and Activities respectively (see *performs* in Figure 2). In this way it is simple to move between fine-grained details concerning one single execution and aggregation details concerning all the executions of the same Business Activity. Additionally, we include the relation *realises* between Process Instance and Activity Instance in order to track the fact that what appears as an Activity for some Process might in fact be supported by a complex Process.

COBRA primarily characterises Business Activities from the perspective of the Persistent Entities involved since we aim to cover the Resource and Object views typically adopted in BPA. Our approach is based on the notion of *Role*[2]. Role, much like in the Descriptions and Situations ontology [19], is the function assumed or part played by a Persistent Entity in a particular Business Activity Realisation. This is defined by means of the ternary relation *playsRoleIn* which relates Roles, Persistent Entities and Business Activity Realisations. COBRA includes a simple categorisation of Roles into two disjoint types, *Agentive* and

---

[2] Role is duplicated in the figure for the sake of clarity.

Non-Agentive ones. Agentive Roles are those that can only be played by Agents whereas Non-Agentive Roles can, in principle, be played by any Persistent Entity. Further restrictions should be defined on a per Role basis. COBRA currently includes for self-containment an Agentive Role–*Actor*–and a Non-Agentive Role–*Resource*–which are of most use when analysing business processes. Again, Roles categorisation is to be extended for specific domains. Finally, we include the *Role Type* meta-class in order to support describing things like "we require an engineer". Persistent Entities are further characterised by a set of Role Types they can play within Business Activity Realisations. This allows to model for example that "Directors can play the Role Type Supervisor".

Given the notion of Role and how these relate to Persistent Entities, we can now fully describe Business Activity and Business Activity Realisation. Business Activities may *use, consume, produce*, and *provide* a set of Persistent Entity Types. The relationship *uses* may also be defined over specific Persistent Entities, e.g., "this Activity requires this specific machine", reason why we actually include two relations *usesPersistentEntity* and *usesPersistentEntityType*. Usage, like in the Enterprise Ontology, concerns Persistent Entities that can play a Resource Role, and which are not consumed during the execution of the business activity. In other words, the availability of the Resource will decrease during the execution of the Business Activity and will be restored to the original level at the end. For example, we use a screw-driver for screwing but as soon as we are done the screw-driver is available for others to use. Resource consumption is captured by means of the relationship *consumes*. This relationship is only applicable to those Persistent Entity Types which are not Agents. For situations where some things are required but not used or consumed, we provide the relation *requires*. Business Activities may *require* a set of Persistent Entities (e.g. "a particular document is required"), Persistent Entity Types (e.g. "one license is required to use the software"), and Role Types (e.g. "a coordinator is required") in order to be performed. The three scenarios are modelled as separate relations. The relationship *produces* captures the outcomes of a Business Activity and is applicable to Persistent Entity Types excepting Non-Agentive Non-Physical Entities for which we have devoted instead the relationship *provides*. These two relationships allow us to capture things like "this production process produces a robot" and "market analysis provides knowledge". These, excepting the relationship *provides*, are all ternary relationships that can be characterised by the quantity involved, see dashed line in Figure 2.

When it comes to Business Activity Realisations we capture their relation with Persistent Entities in a very similar manner. We do so by means of five relations–*involves, uses, consumes, produces*, and *provides*. Whereby *involves* is a super-relation of the others. We finally provide a ternary relation between Business Activity Realisation, Persistent Entity, and Role which allows us to capture the Role a Persistent Entity *plays in* a Business Activity Realisation (see *playsRoleIn* in Figure 2). Business Activity Realisations are the bridge between the high level conceptualisation of the BPM domain and the low-level monitoring information captured at runtime by the IT infrastructure. Thus,

Business Activity Realisations are further characterised by an *execution history*, a *life-cycle*, and the *current state* of the execution.

The execution history is a set of *Monitoring Events* relevant for monitoring the life-cycle of a Business Activity, see Figure 2. Monitoring Events are Instantaneous Entities generated by Agents. They are characterised by a reception timestamp which is to be filled by the logging infrastructure upon reception of an event. The main goal of this attribute is to support monitoring even in environments where clock synchronisation mechanisms are hardly applicable. Additionally, Monitoring Events can have a causality vector, i.e., the set of Monitoring Events that caused that particular event. This supports capturing the actual derivation of events by the monitoring infrastructure as necessary for Complex Event Processing. Finally, Monitoring Events might be characterised by additional associated data which is expressed as *Data Value* instances. These instances identify a particular parameter and the value associated to it.

Monitoring Events are further refined into *Message Events* and *Business Activity Monitoring Events*. The former accommodates Event-Based environments so that their execution can also be traced. The latter supports monitoring the life-cycle of Business Activity Realisations in Process-Aware Information Systems. Business Activity Monitoring Events therefore *concern* a specific Process Instance and, depending on the granularity of the event occurred, may also concern an Activity Instance. Similarly to the proposals in [20, 12, 19], Business Activity Monitoring Events are centred around the notion of state model. Every event identifies a particular transition within the state model, the transition being indicated by means of the *leadsToState* attribute. Conversely the canOccurInState attribute allows to ensure that the transitions are consistent with the prescribed state model or to detect anomalies within the execution history possibly due to missing events.

COBRA supports the definition of specific state models is a simple ontological form by means of the *Business Activity State* concept which has a set of *possibleNextStates*. Business Activity States are used to further characterise Business Activity Realisations with the *hasLifeCycle* and *hasCurrentState* slots. The former captures the overall life-cycle of Business Activity Realisations as a set of Life-Cycle Periods which are Time Spanning Entities whereby the executed business activity was in a particular state. The latter is a shortcut for avoiding heavy usage of temporal reasoning in order to obtain the current state. On the basis of these Life-Cycle Periods it is possible to revisit the complete life-cycle of a Business Activity Realisation in a suitable manner for interval-based temporal reasoning. Instead of prescribing a particular state model and the corresponding events COBRA remains agnostic from the domain-specific details. Still, we provide an extension, i.e., Events Analysis Ontology, with a set of generic event processing forward-chaining rules that can derive information based Business Activity Monitoring Events. These rules will be detailed in the next section.

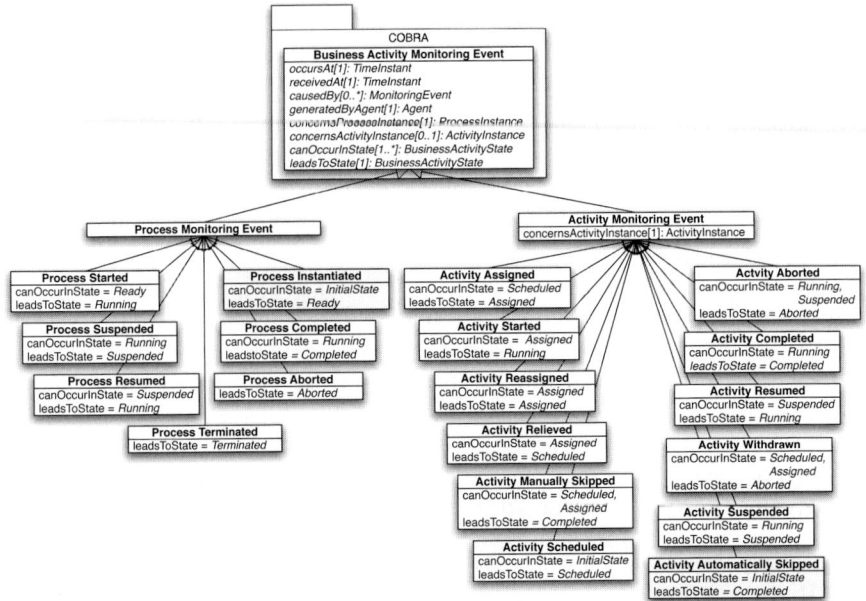

Fig. 3. Events Ontology

Finally, given that COBRA aims to support Business Process Analysis, both Persistent Entities and Business Activity Realisations are characterised by a set of *Analysis Results*. Thus one can capture results of previous analysis for all the relevant perspectives for BPA. Analysis Results are Instantaneous Entities of a *Quantitative* or *Qualitative* nature[3]. Being part of the core ontology for analysing business process, this allow us to reuse results across different types of BPA which paves the way for enhancing current analysis techniques [11]. For instance, metrics computed at runtime can be reused when performing RBE, mining results can be applied during monitoring, etc.

4 Events Processing

COBRA aims at providing a conceptual and extensible framework for supporting BPA. Thus, it purposely leaves many aspects, such as domain-specific data or infrastructure specific monitoring events unspecified. In order to apply COBRA to specific domains these particular details have to be modelled and integrated. As part of the overall BPA conceptual framework but also in order to validate and test our conceptualisation we have developed an ontology for capturing monitoring events from a plethora of BPM systems and a general purpose Events Analysis Ontology that allows to derive information in terms of COBRA from monitoring information. In the remainder of this section we shall describe first the Events Ontology and next the Events Analysis Ontology.

---

[3] Note that we have used slot renaming for occursAt.

4.1 Events Ontology

BPA takes the audit trails generated by the supporting IT infrastructure as a starting point, and attempts to derive information from the business perspective. Each of the supporting systems provides its own level of detail, in heterogeneous formats making it particularly difficult to integrate the audit trails generated as well as it complicates the creation of general purpose solutions. Common formats have been proposed as a solution to overcome this problem, e.g., MXML [20] or the Audit Trail Format by the Workflow Management Coalition (WFMC). Although these formats have proven their benefits, they are supported by technologies that are not suitable for automated reasoning. In order to overcome this, we have extended COBRA with a reference Events Ontology (EVO) that provides a set of definitions suitable to a large variety of systems and ready to be integrated within our core ontology for analysing business processes. EVO is however an optional module which can be replaced by other models if required.

EVO is based on the previously mentioned formats since they provide general purpose solutions that have shown to be suitable to capture logs generated by a plethora of systems. As prescribed by COBRA, EVO is centred around a state model that accounts for the status of processes and activities, see Figure 4. The figure shows the different states and possible transitions contemplated for both Process Instances and Activity Instances which we believe are self-explaining. Note that process abortion differs from process termination in that in the former any ongoing activity is allowed to finish [12]. The dark dot represents the initial state, arrows represent transitions, the smaller boxes depict states, and bigger boxes encapsulate (conceptual) families of states. The state model has been captured ontologically as shown in Figure 3, an enhanced with additional relations. For instance it is possible to determine whether an Activity Instance has been allocated–*isAllocated*–which is true for those that are either in state Running, Suspended, or Assigned. It is also possible to determine whether a Business Activity Realisation is active–*isActive*–which is equivalent to Running, or inactive–*isInactive*–which is true for the rest of the states.

The state model does not distinguish between Process Instances and Activity Instance. The reason for this is mainly to simplify some tasks, e.g. monitoring of active Business Activity Realisations. Still, this necessary distinction is preserved within the logs by means of the Business Activity Monitoring Events defined, see Figure 3. EVO includes two subclasses, namely *Process Monitoring Event* and *Activity Monitoring Event*. EVO currently captures seven Process Monitoring Events and twelve Activity Monitoring Events based on the state model in Figure 4. Process Monitoring Events capture the different transitions which are possible for Process Instances. A Process Instance can be *Instantiated, Started, Suspended, Resumed, Completed, Aborted* and *Terminated*. Activity Monitoring Events, in addition to the typical execution events, contemplate the distribution of work to Agents. Thus, there are events that capture the scheduling of activities, the *Assignment, ReAssignment*, or *Relief* of activities to specific agents. Additionally like MXML, EVO contemplates the possibility for skipping activities either manually or automatically, which lead to a correct completion. Finally,

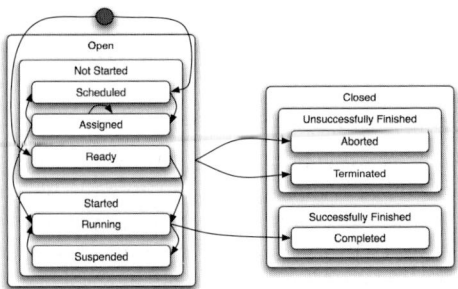

Fig. 4. Events Ontology State Model

EVO captures the abortion of activities by means of two events *Activity Aborted* and *Activity Withdrawn*. The distinction between the two lays in the fact that only started activities can be aborted.

4.2 Event Analysis Ontology

So far we have focussed on the conceptual models that capture the BPM domain spanning from the low-level details concerning audit trail information, to higher-level aspects such the roles played by agents in certain processes. In this section we focus on how, on the basis of this conceptual model and by capturing monitoring information ontologically, we derive knowledge about the enterprise that can then support business practitioners or even Knowledge-Based Systems in the analysis and eventual decision-making process.

OCML provides support for defining both backward and forward-chaining rules. In order to derive information upon reception of monitoring events we have defined a set of generic forward-chaining rules which are independent from the domain and the specific Monitoring Events defined. The goal is to provide reusable rules which can then be enhanced with domain specific ones to derive a richer knowledge-base. Additionally we have implemented a set of relations which are of most use when analysing processes. Some of these relations have been defined for COBRA in a generic manner, whereas others have been bundled with EVO for they are EVO-specific. The rules currently implemented support (i) deriving and checking the consistency of life-cycle information about Business Activity Realisations; (ii) updating the execution history of Business Activity Realisations; (iii) updating the relations between Process Instances and Activity Instances; (iv) tracking the Agents involved and; (v) updating the Roles played by Actors within Business Activities.

The first set of rules uses Business Activity Monitoring Events to update the current state of activity realisations, generate Life-Cycle Period instances, and contrast the transitions with the given state model. Basically, every event defines the end of a period and the beginning of a new one[4]. In this way, by simple

---

[4] The initial state is a special one which is predefined in COBRA.

updates over the life-cycle and with temporal reasoning we can support many of the monitoring competency questions previously exposed. To this end we provide a general purpose relation that holds when, given a Business Activity Realisation, a Time Instant, and a Business Activity State, the activity realisation was in the state given at that particular instant.

The second set of rules aim at correctly tracking the execution history for specific Business Activities so that they can later on be used within Business Process Mining algorithms. The third aspect is supported by a rule that tracks the coincidence of Process Instances and Activity instances within the same Business Activity Monitoring Event and derives the appropriate *composedOf* relation. Agents involvement is derived from the *generatedBy* slot in the events. Finally, whenever one of the Actors involved is the only one taking part in a Business Activity Realisation that can play a certain Role Type that was required, we can derive the role this Actor played. This last rule is bundled with EVO since it is necessary to know whether the business activity was completed before deriving this. The interested reader is referred to the ontologies for further details about the rules and relations currently implemented.

5 Conclusions and Future Work

BPM systems aim at supporting the whole life-cycle of business processes. However, BPM has made more evident the current lack of automation that would support a smooth transition between the business world and the IT world. Yet, moving back and forth between these two aspects is a bottleneck that reduces the capability of enterprise to adapt to ever changing business scenarios. As a consequence there is a growing need for integrating semantics within the BPM domain. A crucial branch of BPM where semantics have a clear and direct impact is Business Process Analysis, where in fact so-called Business Intelligence is appearing as a key enabler for increasing value and performance [3]. Important efforts but with limited success have been devoted to integrating semantics within BPA. The reason for this appears to be the fundamental gap between semantics technologies and the ones currently deployed within BPM solutions.

To reduce this gap we have defined COBRA, a core ontology for business process analysis. The research carried has been guided on a set of competency questions extracted from existing needs with the BPA domain. Our approach is based on a conceptualisation that links low-level monitoring details with high-level business aspects so as to bring this vital information to the business-level as required by business practitioners. This conceptual model is based on a Time Ontology and has been enhanced and validated by means of two extensions for logging monitoring information in a semantic manner, and eventually processing this information.

A key requirement underlying our work has been the need to produce a generic yet comprehensive conceptual infrastructure where additional extensions can be seamlessly plugged-in in order to better support BPA techniques. Future work will thus be devoted to extending our work along the vision previously presented in [11]. In particular, next steps will be devoted to the definition of a metrics computation engine that will support the computation of both generic and user defined metrics, and the implementation of a classification Problem-Solving Method for detecting process deviations. In parallel, we are working on an ontology-based user interface to the reasoning infrastructure as part of WSMO Studio (www.wsmostudio.org).

References

1. van der Aalst, W.M.P., Reijers, H.A., Weijters, A.J.M.M., van Dongen, B.F., de Medeiros, A.K.A., Song, M., Verbeek, H.M.W.: Business process mining: An industrial application. Information Systems 32(5), 713–732 (2007)
2. Hepp, M., Leymann, F., Domingue, J., Wahler, A., Fensel, D.: Semantic business process management: A vision towards using semantic web services for business process management. In: Lau, F.C.M., Lei, H., Meng, X., Wang, M. (eds.) ICEBE, pp. 535–540. IEEE Computer Society, Los Alamitos (2005)
3. Watson, H.J., Wixom, B.H.: The current state of business intelligence. Computer 40(9), 96–99 (2007)
4. Fox, M.S.: The tove project towards a common-sense model of the enterprise. In: Belli, F., Radermacher, F.J. (eds.) IEA/AIE 1992. LNCS, vol. 604, pp. 25–34. Springer, Heidelberg (1992)
5. Uschold, M., King, M., Moralee, S., Zorgios, Y.: The enterprise ontology. Knowledge Engineering Review 13(1), 31–89 (1998)
6. Geerts, G.L., McCarthy, W.E.: An accounting object infrastructure for knowledge-based enterprise models. IEEE Intelligent Systems 14(4), 89–94 (1999)
7. Gordijn, J., Akkermans, H.: Designing and evaluating e-business models. IEEE Intelligent Systems 16(4), 11–17 (2001)
8. Malone, T.W., Crowston, K., Herman, G.A.: Organizing Business Knowledge: The MIT Process Handbook. MIT Press, Cambridge (2003)
9. Castellanos, M., Casati, F., Dayal, U., Shan, M.C.: A comprehensive and automated approach to intelligent business processes execution analysis. Distributed and Parallel Databases 16(3), 239–273 (2004)
10. Grigori, D., Casati, F., Castellanos, M., Dayal, U., Sayal, M., Shan, M.C.: Business process intelligence. Computers in Industry 53(3), 321–343 (2004)
11. de Medeiros, A.K.A., Pedrinaci, C., van der Aalst, W.M.P., Domingue, J., Song, M., Rozinat, A., Norton, B., Cabral, L.: An Outlook on Semantic Business Process Mining and Monitoring. In: Proceedings of International IFIP Workshop On Semantic Web & Web Semantics (SWWS 2007) (2007)
12. zur Muehlen, M.: Workflow-based Process Controlling. In: Foundation, Design, and Implementation of Workflow-driven Process Information Systems. Advances in Information Systems and Management Science, vol. 6, Logos, Berlin (2004)
13. Motta, E.: Reusable Components for Knowledge Modelling. Case Studies in Parametric Design Problem Solving. Frontiers in Artificial Intelligence and Applications, vol. 53. IOS Press, Amsterdam (1999)
14. Hobbs, J.R., Pan, F.: Time ontology in owl (2006), available at
 http://www.w3.org/TR/owl-time/
15. Allen, J.F.: Maintaining knowledge about temporal intervals. Communications of the ACM 26(11), 832–843 (1983)
16. Vilain, M.B.: A system for reasoning about time. In: AAAI, pp. 197–201 (1982)

17. Masolo, C., Borgo, S., Gangemi, A., Guarino, N., Oltramari, A., Schneider, L.: WonderWeb Deliverable D17. The WonderWeb Library of Foundational Ontologies and the DOLCE ontology (2003), http://www.loa-cnr.it/Papers/DOLCE2.1-FOL.pdf
18. ICOM/CIDOC CRM Special Interest Group: CIDOC Conceptual Reference Model (2007), http://cidoc.ics.forth.gr/docs/cidoc_crm_version_4.2.2.pdf
19. Gangemi, A., Borgo, S., Catenacci, C., Lehmann, J.: Task taxonomies for knowledge content. Technical report, EU 6FP METOKIS Project D07 (2004)
20. van Dongen, B., de Medeiros, A., Verbeek, H., Weijters, A., van der Aalst, W.: The prom framework: A new era in process mining tool support. In: Ciardo, G., Darondeau, P. (eds.) ICATPN 2005. LNCS, vol. 3536, pp. 444–454. Springer, Heidelberg (2005)

Assisting Pictogram Selection with Semantic Interpretation

Heeryon Cho[1], Toru Ishida[1], Toshiyuki Takasaki[2], and Satoshi Oyama[1]

[1] Department of Social Informatics, Kyoto University, Kyoto 606-8501 Japan
cho@ai.soc.i.kyoto-u.ac.jp, {ishida,oyama}@i.kyoto-u.ac.jp
[2] Kyoto R&D Center, NPO Pangaea, Kyoto 600-8411 Japan
toshi@pangaean.org

Abstract. Participants at both end of the communication channel must share common pictogram interpretation to communicate. However, because pictogram interpretation can be ambiguous, pictogram communication can sometimes be difficult. To assist human task of selecting pictograms more likely to be interpreted as intended, we propose a *semantic relevance measure* which calculates how relevant a pictogram is to a given interpretation. The proposed measure uses pictogram interpretations and frequencies gathered from a web survey to define probability and similarity measurement of interpretation words. Moreover, the proposed measure is applied to categorized pictogram interpretations to enhance retrieval performance. Five pictogram categories are created using the five first level categories defined in the Concept Dictionary of EDR Electronic Dictionary. Retrieval performance among not-categorized interpretations, categorized and not-weighted interpretations, and categorized and weighted interpretations using semantic relevance measure were compared, and the categorized and weighted semantic relevance retrieval approach exhibited the highest F_1 measure and recall.

1 Introduction

Tags are prevalent form of metadata used in various applications today, describing, summarizing, or imparting additional meaning to the content to better assist content management by both humans and machines. Among various applications that incorporate tags, we focus on a pictogram email system which allows children to communicate to one another using pictogram messages[1,2]. Our goal is to support smooth pictogram communication between children, and to realize this, we focus on the pictogram selection stage where children select individual pictograms to create pictogram messages.

Pictogram is an icon which has a clear pictorial similarity with some object[3], and one who can recognize the object depicted in the pictogram can interpret the meaning associated with the object. Pictorial symbols, however, are not universally interpretable. A simple design like an arrow is often used to show direction, but there is no reason to believe that arrows suggest directionality to all people; they might also be taken as a symbol for war or bad luck[4].

Since the act of selecting a pictogram is done with a purpose of conveying certain meaning to the counterpart, the selected pictogram must carry intended meaning to both the sender and receiver of communication; that is, the selected pictogram must be relevant to the participants at both end of communication channel in order for the pictogram communication to be successful.

To assist pictogram selection, we propose a categorized usage of human-provided pictogram interpretations. Related research unifies the browsing by tags and visual features for intuitive exploration of image databases[5]. Our approach utilizes *categorized* pictogram interpretations together with the *semantic relevance measure* to retrieve and rank relevant pictograms for a given interpretation. We define *pictogram categories* by appropriating first level categories defined in the Concept Dictionary of EDR Electronic Dictionary[6]. We will show that categorized and weighted semantic relevance approach returns better result than not-categorized, not-weighted approaches.

In the following section, five pictogram categories are described, and characteristics in pictogram interpretation are clarified. Section 3 describes semantic relevance measure, and categorization and weighting of interpretation words. Section 4 presents precision, recall, and retrieval examples of four pictogram retrieval approaches. A prototype implementation is also presented. Finally, section 5 concludes this paper.

2 Ambiguity in Pictogram Interpretation

A twenty-five month pictogram web survey was conducted from October 1st, 2005 to November 7th, 2007 to collect free-answer English pictogram interpretation words or phrases from respondents living in the United States. Tallying the unique username–IP address pairs, a total of 1,602 respondents participated in the survey. Details of the earlier survey can be found in [7,8].

2.1 Polysemous Interpretation

From the pictogram web survey data, interpretations consisting of English words or phrases were tallied according to unique interpretation words. Phrasal expressions and misspellings were discarded. An example of tallied pictogram interpretation words is shown in Table 1. As shown, a pictogram can have multiple interpretations which include both similar and different-meaning interpretations. For example, words like *talking*, *talk*, *chatting*, *conversing*, *chat*, and *communicating* are all similar action-related words describing the act of speaking. Other action-related words are *date*, *flirt*, *sit*, *flirting*, *listening*, *love*, and *play*. On the other hand, when the focus shifts to the people depicted in the pictogram, the pictogram is interpreted as *friends* or *family*. Or it can be interpreted as some kind of place such as *church*, or as an emotional state such as *happy*. One way to organize mixed interpretations containing both similar and different meanings is to group them into categories. We use the Concept Dictionary in the EDR Electronic Dictionary[6] to group interpretation words into five first level categories

Table 1. An example of tallied pictogram interpretation words

PICTOGRAM	WORD	FREQ	RATIO	WORD	FREQ	RATIO
	talking	58	0.367	church	1	0.006
	talk	27	0.171	communication	1	0.006
	conversation	20	0.127	family	1	0.006
	friends	15	0.095	flirting	1	0.006
	chatting	13	0.082	friend	1	0.006
	conversing	5	0.032	friendly	1	0.006
	chat	2	0.013	happy	1	0.006
	communicating	2	0.013	listening	1	0.006
	date	2	0.013	love	1	0.006
	flirt	2	0.013	play	1	0.006
	sit	2	0.013			
	TOTAL FREQUENCY = 158,			**TOTAL RATIO = 0.999**		

defined in the dictionary. We borrow these five first level categories to define five *pictogram categories*.

The EDR Electronic Dictionary was developed for advanced processing of natural language by computers, and is composed of five types of dictionaries (Word, Bilingual, Concept, Co-occurrence, and Technical Terminology), as well as the EDR Corpus. The Concept Dictionary contains information on the approximately 410,000 concepts listed in the Word Dictionary and is divided according to information type into the Headconcept Dictionary, the Concept Classification Dictionary, and the Concept Description Dictionary. The Headconcept Dictionary describes information on the concepts themselves. The Concept Classification Dictionary describes the super-sub relations among the approximately 410,000 concepts. The "super-sub" relation refers to the inclusion relation between concepts, and the set of interlinked concepts can be regarded as a type of thesaurus. The Concept Description Dictionary describes the semantic (binary) relations, such as 'agent', 'implement', and 'place', between concepts that co-occur in a sentence[6]. We use the Headconcept Dictionary and the Concept Classification Dictionary to obtain super-concepts of the pictogram interpretation words. A record in the Headconcept Dictionary is composed of a record number, a concept identifier, an English headconcept, a Japanese headconcept, an English concept explication, a Japanese concept explication, and a management information[1]. Below shows two records containing the English headconcept "talk"[2]. Notice that there are two different concept identifiers (0dc0d6, 0dc0d7) for the same English headconcept "talk"[3]:

- CPH0144055 0dc0d6 talk JH "an informal speech" "JE" DATE="95/6/6"
- CPH0144056 0dc0d7 talk JH "a topic for discussion" "JE" DATE="95/6/6"

---

[1] A headconcept is a word whose meaning most closely expresses the meaning of the concept. A concept explication is an explanation of the concept's meaning.
[2] JH indicates Japanese headconcept and "JE" indicates Japanese explication.
[3] Overall, there are 13 concept identifiers matching the English headconcept "talk".

We obtain concept identifier(s) of a pictogram interpretation word by matching the interpretation word string to English headconcept string in the Headconcept Dictionary. Once the concept identifier(s) are obtained, we use the Concept Classification Dictionary to retrieve the first level categories of the concept identifier(s). A record in the Concept Classification Dictionary is composed of a record number, a concept identifier of the super-concept, a concept identifier of the sub-concept, and management information. Below shows two concept classification dictionary records containing the super-sub concept identifiers[4]:

- CPC0144500 444059 0dc0d6 DATE="95/6/7"
- CPC0419183 443e79 444059 DATE="95/6/7"

Note that there may be more than one super-concept (identifier) for a given concept (identifier) since the EDR Concept Dictionary allows multiple inheritance. By climbing up the super-concept of a given concept, we reach the root concept which is 3aa966 'concept'. Five categories defined at the first level, placed just below the root concept, will be used as five pictogram categories for categorizing pictogram interpretation words. The headings of the five categories are:

(a) Human or subject whose behavior (actions) resembles that of a human
(b) Matter or an affair
(c) Event/occurrence
(d) Location/locale/place
(e) Time

Superclasses in SUMO ontology[9] could be another candidate for defining pictogram categories, but we chose EDR because we needed to handle both English and Japanese pictogram interpretations. For brevity, we abbreviate the pictogram category headings as (a) AGENT, (b) MATTER, (c) EVENT, (d) LOCATION, and (e) TIME. Table 2 shows examples of nine pictograms' interpretation words categorized into the five pictogram categories. Each column contains interpretation words of each pictogram. Each cell on the same row contains interpretation words for each of the five categories. One interpretation word can be assigned to multiple categories, but in the case of Table 2, each word is assigned to a single *major category*. Major category is explained later in section 3.2. We now look at each category in detail.

(a) **AGENT.** Pictograms containing human figures can trigger interpretations explaining something about a person or people. Table 2 AGENT row contains words like *family, dancers, people, crowd, fortune teller,* and *magician* which all explain a specific kind of person or people.

(b) **MATTER.** Concrete objects or objective subjects are indicated. Table 2 MATTER row contains words like *good morning, good night, moon, good evening, dancing, chicken, picture, ballet, card, drama, crystal ball,* and *magic* which point to some physical object(s) or subject depicted in the pictogram.

---

[4] 444059 is the super-concept identifier of 0dc0d6, and 443e79 is the super-concept identifier of 444059.

Table 2. Polysemous interpretations (each column) and shared interpretations (boldface type)

	(1)	(2)	(3)	(4)	(5)	(6)	(7)	(8)	(9)
AGENT					**family**	**family** dancers		people **family** crowd	fortune teller magician
MATTER	good morning **good night**	moon good evening dancing **good night**	moon **good night**	chicken **picture**		**dancing** ballet		**card** drama **picture**	**card** crystal ball magic
EVENT	**talking** sleeping **happy** **play** friendly	**talking** **happy** **play**	sleeping dream peaceful	sleeping	**talking** date **play** friendly	dance jumping **play**	slide fun **play**	crying **happy** mixed	bowling guess **play**
LOCATION							playground	theater	
TIME	morning day bedtime	night evening	night bedtime	morning night evening					future

(c) **EVENT.** Actions or states are captured and described. Table 2 EVENT row contains words like *talking, sleeping, happy, play, friendly, dream, peaceful, date, dance, jumping, slide, fun, crying, mixed, bowling,* and *guess* which all convey present states or ongoing actions.

(d) **LOCATION.** Place, setting, or background of the pictogram are on focus rather than the object occupying the center or the foreground of the setting. Table 2 LOCATION row contains words like *playground* and *theater* which all indicate specific places or settings relevant to the central object(s).

(e) **TIME.** Time-related concepts are sometimes perceived. Table 2 TIME row contains words like *morning, day, bedtime, night, evening,* and *future* which all convey specific moments in time.

Categorizing the words into five pictogram categories elucidates two key aspects of polysemy in pictogram interpretation. Firstly, interpretations spread across different categories lead to different meanings. For example, interpretation words in Table 2 column (8) include AGENT category's *crowd* which describes a specific relationship between people, EVENT category's *crying* which describes an ongoing action, and LOCATION category's *theater* which describes a place for presenting a show, and they all mean very different things. This is due to the different focus of attention given by each individual.

Secondly, while interpretation words placed within the same category may contain similar words such as *sleeping* and *dream* (Table 2 column (3) row EVENT), or *dance* and *jumping* (Table 2 column (6) row EVENT), contrasting or opposite-meaning words sometimes coexist within the same category. For example, Table 2 column (4) row TIME contains both *morning* and *night*, which are contrasting time-related concepts, and column (1) row MATTER contains both *good morning* and *good night*, which are contrasting greeting words.

While the words in Table 2 column (2) row TIME are varied yet similar (*night* and *evening*), the words in column (4) row TIME are confusing because contrasting interpretations are given on the same viewpoint (*morning* and *night*). To summarize the above findings, it can be said that polysemy in pictogram interpretation is dependent on the interpreter's perspective; usually, interpretations differ across different categories or perspectives, but sometimes interpretations may vary even within the same category or perspective.

When a pictogram having polysemous interpretations is used in communication, there is a possibility that a sender and receiver might interpret the same pictogram differently. In the case of pictogram in Table 2 column (4), it could be interpreted quite differently as *morning* and *night* by the sender and receiver. One way to assist the sender to choose a pictogram with higher chance of conveying the intended message to the receiver is to display possible interpretations of a given pictogram. If various possible interpretations are presented, the sender can speculate receiver's interpretation before choosing and using the pictogram. For example, if the sender knows a priori that Table 2 pictogram (4) can be interpreted as both *morning* and *night*, s/he can guess ahead that it might be interpreted oppositely by the receiver, and avoid choosing the pictogram.

We will refer to this characteristic of one-to-many correspondence in pictogram-to-meaning and an associative measure of displaying possible pictogram interpretations as assisting selection of pictograms having polysemous interpretations.

2.2 Shared Interpretation

One pictogram may have various interpretations, but these interpretations are not necessarily different across different pictograms. Sometimes multiple pictograms share common interpretation(s) among themselves. Words indicated in boldface type in Table 2 are such interpretations shared by more than one pictogram: Table 2 (5), (6), and (8) share *family* (row AGENT); (1), (2), and (3) share *good night* (row MATTER); (1) and (5) share *friendly* (row EVENT); and (2), (3), and (4) share *night* (row TIME) and so forth.

The fact that multiple pictograms can share common interpretation implies that each one of these pictograms can be interpreted as such. The degree to which each is interpreted, however, may vary according to the pictogram. For example, Table 2 pictograms (1), (2), and (3) can all be interpreted as *good night* (row MATTER), but (1) can also be interpreted as *good morning* while (2) can also be interpreted as *good evening*. Furthermore, if we move down the table to row TIME, we see that (1) has *morning* as time-related interpretation while (2) and (3) have *night*. Suppose two people A and B each use pictogram (1) and (2) respectively to send a "good night" message to person C. Upon receiving the message, however, C may interpret the two messages as "good morning" for A and "good night" for B. Even though A and B both intend on conveying a "good night" message, it may not always be the case that C will interpret the two pictograms likewise. This is because the degree of interpretation may vary across similar-meaning pictograms; one reason may be due to other possible interpretations within the pictogram (as in *good morning* and *good evening*).

One way to assist the selection of pictograms among multiple similar-meaning pictograms is to rank those pictograms according to the degree of relevancy of a pictogram to a given interpretation. Presenting ranked pictograms to the user who selects the pictogram to be used in communication will allow the user to understand which pictogram is most likely to be interpreted as intended. In order to rank pictograms according to the interpretation relevancy, some kind of metric which measures the relevancy of a pictogram to an interpretation is needed. We will refer to this characteristic of one-to-many correspondence in meaning-to-pictogram and an associative measure of ranking pictograms according to interpretation relevancy as assisting selection of pictograms having shared interpretations.

3 Semantic Relevance Measure

We identified ambiguities in pictogram interpretation and possible issues involved in the usage of such pictograms in communication. Here, we propose a *semantic relevance measure* which outputs relevancy values of each pictogram when a pictogram interpretation is given. Our method presupposes a set of pictograms having a list of interpretation words and ratios for each pictogram.

3.1 Definition

We assume that pictograms each have a list of interpretation words and frequencies as the one shown in Table 1. Each unique interpretation word has a frequency. Each word frequency indicates the number of people who answered the pictogram to have that interpretation. The ratio of an interpretation word, which can be calculated by dividing the word frequency by the total word frequency of that pictogram, indicates how much support people give to that interpretation. For example, in the case of pictogram in Table 1, it can be said that more people support *talking* (58 out of 158) as the interpretation for the given pictogram than *sit* (2 out of 158). The higher the ratio of a specific interpretation word of a pictogram, the more that pictogram is accepted by people for that interpretation.

We define semantic relevance of a pictogram to be the measure of relevancy between a word query and interpretation words of a pictogram. Let $w_1, w_2, ..., w_n$ be interpretation words of pictogram e. Let the ratio of each interpretation word in a pictogram to be $P(w_1|e), P(w_2|e), ..., P(w_n|e)$. For example, the ratio of the interpretation word *talking* for the pictogram in Table 1 can be calculated as $P(talking|e) = 58/158$. Then the simplest equation that assesses the relevancy of a pictogram e in relation to a query w_i can be defined as follows:

$$P(w_i|e) \tag{1}$$

This equation, however, does not take into account the similarity of interpretation words. For instance, when "talking" is given as query, pictograms having similar interpretation word like "gossiping", but not "talking" fail to be measured as relevant when only the ratio is considered. To solve this, we need to define $similarity(w_i, w_j)$ between interpretation words in some way. Using the similarity, we can define the measure of *Semantic Relevance* or $SR(w_i, e)$ as follows:

$$SR(w_i, e) = \sum_j P(w_j|e) similarity(w_i, w_j) \tag{2}$$

There are several similarity measures. We draw upon the definition of similarity given by Lin[10] which states that similarity between A and B is measured by the ratio between the information needed to state the commonality of A and B and the information needed to fully describe what A and B are. Here, we calculate the similarity of w_i and w_j by figuring out how many pictograms contain certain interpretation words. When there is a pictogram set E_i having an interpretation word w_i, the similarity between interpretation word w_i and w_j can be defined as follows:

$$similarity(w_i, w_j) = |E_i \cap E_j|/|E_i \cup E_j| \tag{3}$$

$|E_i \cap E_j|$ is the number of pictograms having both w_i and w_j as interpretation words. $|E_i \cup E_j|$ is the number of pictograms having either w_i or w_j as interpretation words. Based on (2) and (3), the semantic relevance or the measure of relevancy to return pictogram e when w_i is input as query can be calculated as follows:

$$SR(w_i, e) = \sum_j P(w_j|e)|E_i \cap E_j|/|E_i \cup E_j| \qquad (4)$$

The resulting semantic relevance values will fall between one and zero, which means either a pictogram is completely relevant to the interpretation or completely irrelevant. Using the semantic relevance values, pictograms can be ranked from very relevant (value close to 1) to not so relevant (value close to 0). As the value nears zero, the pictogram becomes less relevant; hence, a cutoff point is needed to discard the less relevant pictograms. Setting an ideal cutoff point that satisfies all interpretations and pictograms is difficult, however, since all words contained in a pictogram, regardless of relation to each other, each influence the calculation. For example, let's say that we want to find a pictogram which can convey the meaning "friend" or "friends". Pictogram in Table 1 could be a candidate since it contains both words with a total ratio of 0.1. When the semantic relevance is calculated, however, the equation takes into account all the interpretation words including *talking* or *church* or *play*. Selecting a set of words relevant to the query would reduce the effect of less-relevant interpretation words affecting the calculation. Based on this prediction, we propose a semantic relevance calculation on categorized interpretations.

3.2 Word Categorization, Word Weighting, and Result Ranking

Word Categorization. Interpretation words are categorized into five pictogram categories described in section 2.1. Note that some headconcept(s) in the EDR Electronic Dictionary link to multiple concepts, and some concepts lead to multiple super-concepts (i.e. multiple inheritance). For example, in the case of the word (headconcept) *park*, three kinds of pictogram categories are obtained repeatedly: LOCATION category six times, MATTER category five times, and EVENT category four times. In such cases of multiple categories, we use all categories since we cannot accurately guess on the single correct category intended by each respondent who participated in the web survey.

Word Weighting. Although we cannot correctly decide on the single, intended category of a word, we can calculate the ratio of the pictogram category of each word. For example, in the case of *park*, the LOCATION category has the most number of repeated categories (six). Next is the MATTER category (five) followed by the EVENT category (four). In the case of the word *night*, the TIME category has most number of categories (seven) followed by EVENT (five) and MATTER (one). We can utilize such category constitution by calculating the ratio of the repeated categories and assigning the ratio as weights to the word in a given category. For example, the word *park* can be assigned to LOCATION, MATTER and EVENT category, and for each category, weights of 6/15, 5/15 and 4/15 can be assigned to the word. Same with *night*. The word *night* in the

TIME category will have the largest weight of 7/13 compared to EVENT (5/13) or MATTER (1/13). Consequently, the *major category* of *park* and *night* will be LOCATION and TIME respectively.

Result Ranking. Applying the semantic relevance calculation to categorized interpretations will return five semantic relevance values for each pictogram. We compare the highest value with the cutoff value to determine whether the pictogram is relevant or not. Once the relevant pictograms are selected, pictograms are then ranked according to the semantic relevance value of the query's major category. For example, if the query is "night", relevant pictograms are first selected using the highest semantic relevance value in each pictogram, and once candidate pictograms are selected, the pictograms are then ranked according to the semantic relevance value of the query's major category, which in this case is the TIME category. We use 0.5 cutoff value for the evaluation and prototype implementation described next.

4 Evaluation

Using the semantic relevance measure, retrieval tasks were performed to evaluate the semantic relevance measure and the categorized and weighted pictogram retrieval approach. Baseline for comparison was a simple string match of the query to interpretation words having a ratio greater than 0.5[5]. We also implemented a prototype web-based pictogram retrieval system (Fig. 1).

Comparison of Four Approaches. Four pictogram retrieval approaches were evaluated: (1) baseline approach which returns pictograms containing the query as interpretation word with ratio greater than 0.5; (2) semantic relevance approach which calculates semantic relevance value using not-categorized interpretations; (3) semantic relevance approach which calculates semantic relevance values using categorized interpretations; and (4) semantic relevance approach which calculates semantic relevance values using categorized and weighted interpretations. We wanted to see if (a) the fourth approach, the categorized and weighted approach, performed better than the rest; (b) the semantic relevance approach in general was better than the simple query match approach; (c) the categorized approach in general was better than the not-categorized approach.

Creation of Relevant Pictogram Set. A relevant pictogram set was created by five human judges who were all undergraduate students. There were 903 unique words for 120 pictograms, which meant that these words could be used as queries in the retrieval tasks. We performed retrieval tasks with these 903 words using the four approaches to filter out words that returned the same result among the four approaches, since those words would be ineffective in discerning the performance difference of the four approaches. A total of 399 words returned the same results for all four approaches. Another 216 words returned the same

---

[5] This is the same as selecting pictograms with $P(w_j|e) > 0.5$ where w_j equals query.

results for the three semantic relevance approaches. That left us with 288 words. Among the 288 words, words having more than 10 candidate pictograms, similar words (e.g. *hen, rooster*), singular/plural words (e.g. *girl, girls*), and varied tenses (e.g. *win, winning*) were filtered leaving 193 words to be judged for relevancy. For each of the 193 words, all pictograms containing the word were listed as candidate pictograms to be judged for relevancy.

A questionnaire containing each of the 193 words and candidate pictograms with ranked list of interpretation words[6] were given to five human judges, and they were to first judge whether each candidate pictogram can be interpreted as the given word, and then if judged relevant, write down the ranking among the relevant pictograms. The five judgments were analyzed by tallying the relevance judgments, and pictogram ranking was determined by calculating the averages and variances of the judgments[7]. After the five human judges' relevance judgments were analyzed, 30 words were additionally deleted since none of the candidate pictograms were judged as relevant. As a result, a ranked relevant pictogram set for 163 words was created and used in the evaluation[8].

Precision and Recall. Table 3 shows precision, recall, and F_1 measure of the four pictogram retrieval approaches. Each value is the mean performance value of 163 retrieval tasks performed[9]. A cutoff value of 0.5 was used for the three semantic relevance approaches. Based on the performance values listed in Table 3, we see that (a) the categorized and weighted semantic relevance approach performs better than the rest in terms of recall (0.70472) and F_1 measure (0.73757); (b) the semantic relevance approach in general performs much better than the simple query string match approach; and that (c) the categorized approach in general performs much better than the not-categorized approach.

It should be noted that the greatest gain in performance is achieved through the categorization of the interpretation words. By contrast, only a minimal gain is obtained through word-weighting as exhibited by the not-weighted vs. weighted performance values (e.g. 0.71492 vs. 0.73757 F_1 measure values). Through this

---

[6] The probability of each pictogram interpretation word was not displayed in the questionnaire, but was used to list the words with greater probability at the top.
[7] If three or more people judged relevant, the pictogram was judged relevant. Otherwise, the pictogram was discarded. Average ranking for each of the relevant pictogram was calculated. If average rankings were the same among multiple pictograms, variance was calculated and compared. The smaller the variance, the higher was the ranking.
[8] The composition ratio of the major category in 903, 288, 193 and 163 words were:
  - 903 words: [AGENT, 9%], [MATTER,27%], [EVENT,56%], [LOCATION, 4%], [TIME,3%]
  - 288 words: [AGENT,10%], [MATTER,26%], [EVENT,49%], [LOCATION, 8%], [TIME,6%]
  - 193 words: [AGENT, 9%], [MATTER,28%], [EVENT,47%], [LOCATION,10%], [TIME,6%]
  - 163 words: [AGENT, 9%], [MATTER,29%], [EVENT,47%], [LOCATION,10%], [TIME,6%]
[9] Mean precision value was calculated using the valid tasks that returned at least one result. The number of valid tasks for each approach was: QUERY MATCH = 9 tasks, *SR* WITHOUT CATEGORY = 49 tasks, *SR* WITH CATEGORY & NOT WEIGHTED = 139 tasks, and *SR* WITH CATEGORY & WEIGHTED = 153 tasks.

Table 3. Precision, recall, and F_1 measure of four pictogram retrieval approaches

APPROACH MEASURE (MEAN FOR 163 TASKS)	QUERY MATCH RATIO>0.5	SEMANTIC RELEVANCE		
		WITHOUT CATEGORY	WITH CATEGORY	
			NOT-WEIGHTED	WEIGHTED
PRECISION	1.00000	0.87755	0.79808	0.77364
RECALL	0.02853	0.18108	0.64746	0.70472
F_1 MEASURE	0.05547	0.30022	0.71492	0.73757

example, we confirmed that a simple classification or pre-categorization of words can contribute greatly to the improvement of retrieval performance.

Examples of Retrieved Results. Table 4 shows pictogram retrieval results of five queries, "doctor", "book", "cry", "playground", and "bedtime", on four different approaches: (1) ALL PICTOGRAMS WITH QUERY lists all pictograms containing the query as interpretation word. The pictograms are sorted from the largest interpretation ratio to the smallest; (2) HUMAN JUDGED AS RELEVENT lists relevant pictograms selected by five human judges upon seeing the candidate pictograms listed in (1). The pictograms are ranked with the most relevant pictogram starting from the left. The pictograms listed here are the relevant pictogram set of the given word; (3) QUERY MATCH RATIO > 0.5 lists all pictograms having the query as interpretation word with ratio greater than 0.5; (4) *SR* WITHOUT CATEGORY uses not-categorized interpretations to calculate the semantic relevance value; (5) *SR* WITH CATEGORY & NOT-WEIGHTED uses categorized interpretations to calculate five semantic relevance values for each pictogram; (6) *SR* WITH CATEGORY & WEIGHTED uses categorized and weighted interpretations to calculate five semantic relevance values for each pictogram. In the three semantic relevance approaches (4), (5), and (6), a cutoff value of 0.5 was used. Once the semantic relevance values were calculated, the pictograms were ranked according to the semantic relevance value of the major category. Images of the candidate pictograms that contain query as interpretation word are listed at the bottom five rows of Table 4.

For instance, for the query "book" (Table 4 third column), there are two relevant pictograms [059, 097] out of five, and we see that only one approach, *SR* WITH CATEGORY & WEIGHTED, succeeds in returning the first relevant pictogram [059]. Similar readings can be done on the remaining four queries.

Prototype Implementation. Figure 1 shows a screenshot of a web-based pictogram retrieval system which uses the interpretation words collected from the survey. A search result for query "night" using the categorized and weighted approach is displayed. Contrasting interpretations, such as *night* and *morning* at the bottom right, are elucidated once the interpretations are categorized.

Table 4. Comparison of four approachs' retrieved results for five queries (Note: *SR* is an abbreviation for *Semantic Relevance*)

QUERY APPROACH	DOCTOR	BOOK	CRY	PLAYGROUND	BEDTIME
ALL PICTOGRAMS WITH QUERY	043,044,090	059,105,097,101,102	007,006,050,073,072	039,113,118,105,038	003,001,009
HUMAN JUDGED AS RELEVANT	043,044	059,097	007,006	113,039,118,112	003,009
QUERY MATCH RATIO > 0.5					
SR WITHOUT CATEGORY	043		007	039	
SR WITH CATEGORY & NOT-WEIGHTED	043,044		007,006	039,113	
SR WITH CATEGORY & WEIGHTED	043,044	059	007,006	039,118,113	003
All pictograms containing DOCTOR	[043],	[059],	[09C]		
All pictograms containing BOOK	[059],	[105],	[097],	[102]	
All pictograms containing CRY	[007],	[006],	[050],	[101], [073],	[072]
All pictograms containing PLAYGROUND	[039],	[113],	[118],	[105],	[038]
All pictograms containing BEDTIME	[003],	[001],	[009]		

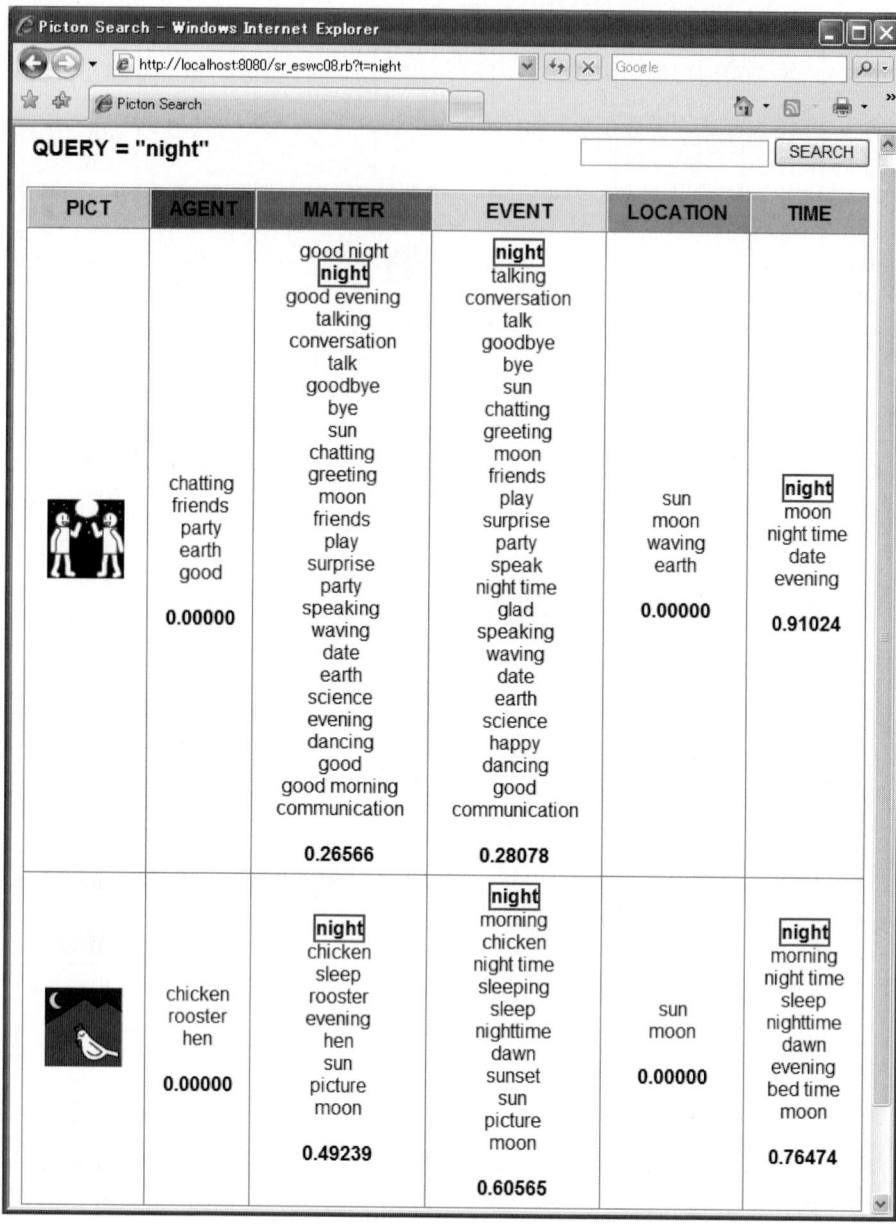

Fig. 1. A screenshot of web-based pictogram retrieval system prototype which uses the categorized and weighted semantic relevance approach with a 0.5 cutoff value. Results for the query "night" is displayed. Notice that contrasting interpretations, *night* and *morning* at the bottom right, become evident once the interpretations are categorized.

5 Conclusion

Pictograms used in a pictogram email system are created by novices at pictogram design, and they do not have single, clear semantics. To retrieve better intention-conveying pictograms using a word query, we proposed a semantic relevance measure which utilizes interpretation words and frequencies collected from a web survey. The proposed measure takes into account the probability and similarity in a set of pictogram interpretation words, and to enhance retrieval performance, pictogram interpretations were categorized into five pictogram categories using the Concept Dictionary in EDR Electronic Dictionary. The retrieval performance of (1) not-categorized, (2) categorized, and (3) categorized and weighted semantic relevance retrieval approaches were compared, and the categorized and weighted semantic relevance retrieval approach performed better than the rest.

Acknowledgements. This research was partially supported by the International Communications Foundation and the Global COE Program "Informatics Education and Research Center for Knowledge-Circulating Society."
All pictograms presented in this paper are copyrighted material, and their rights are reserved to NPO Pangaea.

References

1. Takasaki, T.: PictNet: Semantic infrastructure for pictogram communication. In: Sojka, P., Choi, K.S., Fellbaum, C., Vossen, P. (eds.) GWC 2006. Proc. 3rd Int'l WordNet Conf., pp. 279–284 (2006)
2. Takasaki, T.: Design and development of a pictogram communication system for children around the world. In: Ishida, T., R. Fussell, S., T. J. M. Vossen, P. (eds.) IWIC 2007. LNCS, vol. 4568, pp. 193–206. Springer, Heidelberg (2007)
3. Marcus, A.: Icons, symbols, and signs: Visible languages to facilitate communication. Interactions 10(3), 37–43 (2003)
4. Kolers, P.A.: Some formal characteristics of pictograms. American Scientist 57, 348–363 (1969)
5. Aurnhammer, M., Hanappe, P., Steels, L.: Augmenting navigation for collaborative tagging with emergent semantics. In: Cruz, I., Decker, S., Allemang, D., Preist, C., Schwabe, D., Mika, P., Uschold, M., Aroyo, L.M. (eds.) ISWC 2006. LNCS, vol. 4273, pp. 58–71. Springer, Heidelberg (2006)
6. National Institute of Information and Communications Technology (NICT): EDR Electronic Dictionary Version 2.0 Technical Guide
7. Cho, H., Ishida, T., Inaba, R., Takasaki, T., Mori, Y.: Pictogram retrieval based on collective semantics. In: Jacko, J.A. (ed.) HCI 2007. LNCS, vol. 4552, pp. 31–39. Springer, Heidelberg (2007)
8. Cho, H., Ishida, T., Yamashita, N., Inaba, R., Mori, Y., Koda, T.: Culturally-situated pictogram retrieval. In: Ishida, T., R. Fussell, S., T. J. M. Vossen, P. (eds.) IWIC 2007. LNCS, vol. 4568, pp. 221–235. Springer, Heidelberg (2007)
9. Niles, I., Pease, A.: Towards a standard upper ontology. In: FOIS 2001. Proc. 2nd Int'l Conf. on Formal Ontology in Information Systems (2001)
10. Lin, D.: An information-theoretic definition of similarity. In: ICML 1998. Proc. of the 15th Int'l Conf. on Machine Learning, pp. 296–304 (1998)

KonneX^{SALT}: First Steps Towards a Semantic Claim Federation Infrastructure

Tudor Groza, Siegfried Handschuh, Knud Möller, and Stefan Decker

DERI, National University of Ireland, Galway,
IDA Business Park, Lower Dangan, Galway, Ireland
{tudor.groza,siegfried.handschuh,knud.moeller,stefan.decker}@deri.org
http://www.deri.ie/

Abstract. Dissemination, an important phase of scientific research, can be seen as a communication process between scientists. They expose and support their findings, while discussing claims stated in related scientific publications. However, due to the increasing number of publications, finding a starting point for such a discussion represents a real challenge. At same time, browsing can also be difficult since the communication spans accross multiple publications on the open Web. In this paper we propose a semantic claim federation infrastructure, named KonneX^{SALT}, as a solution for both issues mentioned above: (i) finding claims in scientific publications, and (ii) providing support for browsing by starting with a claim and then following the links in an argumentation discourse network (ADN) (in our case, by making use of transclusion). In addition, we join the web of linked open data, by linking the metadata contained in KonneX^{SALT} with some of the known repositories of scientific publications.

1 Introduction

Dissemination, an important phase of scientific research, can be seen as a communication process between scientists. They expose and support their findings, while discussing claims stated in related scientific publications. This communication takes place over the course of several publications, where each paper itself contains a rhetorical discourse structure which lays out supportive evidence for the raised claims. Often this discourse structure is hidden in the semantics expressed by the publication's content and thus hard to discover by the reader.

Externalization, as defined by Nonaka and Takeuchi [1], represents *the process of articulating tacit knowledge into explicit concepts*. As such, it holds the key to knowledge creation. Consequently, the knowledge becomes crystallized, thus allowing it to be shared with and by others. Although made explicit, the externalized knowledge is dependent on the degree of formalization. In the case of the argumentation discourse based on claims, it can be a couple of keywords, or a weakly structured text, both possibly including direct references to the publications stating the actual claims.

In a previous paper [2], we have described SALT (Semantically Annotated LaTeX), an authoring framework for creating semantic documents and defined a

Web identification scheme (named *claim identification tree*) for claims in scientific publications. The goal of the framework was to define a clear formalization for externalizing the knowledge captured in the argumentation discourses. We defined a special LATEX markup syntax for annotating claims and arguments, and modeled the relationship between them by means of the SALT Rhetorical Ontology. The Web identification scheme allowed us to introduce a novel way for authoring and referencing publications, by giving authors the possibility of working at a fine-grained level, i.e., by citing claims within publications. The main goal was to support the creation of networks of claims, which span across multiple publications.

Having set the foundation, we have now reached the point where we can provide ways to use the externalized knowledge modelled as semantic metadata. In this paper, we introduce KonneX^{SALT}, a semantic claim federation infrastructure, designed with the goal of finding claims in scientific publications and providing support for browsing argumentative discussions, starting from a particular claim. KonneX^{SALT} was not conceived to be yet another search engine, but rather to represent a look-up service for externalized knowledge and realize efficiency through the minimization of the data to be indexed. By using latent semantic indexing, it also provides a means for discovering similarities among the managed claims. From a browsing point of view, KonneX^{SALT} defines Argumentative Discourse Networks (ADN) as realizations of associative trails [3]. As a consequence, it provides a method for federating claims with the help of semantic technologies. In addition, for improving the readability of the ADNs, it makes use of transclusion (*the inclusion of part of a document into another document by reference* [4]). Finally, KonneX^{SALT} contributes to the web of "linked open data" by linking the claims, and implicitly the publications hosting them, to publications referred to by social websites (e.g. Bibsonomy[1]) or managed by known repositories (e.g. DBLP[2]). Our ultimate goal is to transform KonneX^{SALT} in an open hub for linking scientific publications based on externalized knowledge.

The remainder of the paper is structured as follows: in Sect. 2 we present the relevant research performed in this field. Following, we introduce our use-cases (Sect. 3) and then, in Sect. 4, we provide background information to motivate the decisions taken for achieving our goals. In Sect. 5 we present the design and implementation of KonneX^{SALT}, and before concluding in Sect. 7, we discuss some ethical and technical challenges discovered during our research (Sect. 6).

2 Related Work

The relevant literature for our work can be split in two main categories: search engines focused on scientific publications and hypertext systems close to our browsing goals.

Table 1 shows a brief comparative overview of some of the known search engines for scientific publications and shows how KonneX^{SALT} compares to them

---
[1] http://www.bibsonomy.org/
[2] http://dblp.uni-trier.de/

Table 1. Scientific publication search engines overview

	Google Scholar	CiteSeer	Science-Direct	DBLP	KonneX^{SALT}
Focus	Full publications	References	Full publications	Shallow metadata	Shallow metadata, references, claims
Population	Crawling	Crawling	Manual	Manual	Author-driven
Full search	+	−	+	−	−
SW oriented	−	−	−	±	+
Linking	−	−	−	−	+
Openness	+	+	−	±	+

in a number of features. Traditional search engines focus on indexing the full content of the publications, inserted manually or via automatic crawling, but provide almost no means for creating/re-using semantic data (the only exception is DBLP, which offers an RDF dump). In contrast, our approach minimizes the quantity of data to be indexed and provides only semantically linked data.

On the browsing side, the roots of KonneX^{SALT} heavily reside on the work elaborated by visionaries like Vannevar Bush and Ted Nelson. In 1945, in his famous article "As we may think" [3], Bush described the Memex as a the first proto-hypertext system. The Memex was envisioned as an electronic device, which linked to a library, was able to display films or books from the library and when necessary, to follow cross-references from one piece to another. In addition to following links, Memex was also able to create links, based on a technology which combined electromechanical controls and microfilm cameras and readers. This is generally regarded as a main inspiration for and the first step towards today's hypertext.

In the Memex, Bush also introduced the concept of *associative trails*, which are realized in our approach by means of the argumentative discourse networks. Bush defines associative trails as *"a new linear sequence of microfilm frames across any arbitrary sequence of microfilm frames by creating a chained sequence of links, along with personal comments and 'side trails' "*. In a similar manner we (re)construct the argumentation discourse structure (the *new linear sequence of microfilm frames*), based on the claims stated in publications (*the arbitrary sequence of microfilm frames*), and following the positions and arguments expressed in papers referring to these claims (*chained sequence of links* ...).

In 1960, Ted Nelson founded Project Xanadu. It represented the first attempt to design and implement a hypertext system. Xanadu was describing the ideal approach of realizing a word processor which would allow versioning, visualization of difference between versions, and especially non-sequential writing and reading. Consequently, a reader was able to choose his own path through an electronic document, based on *transclusion*, or *zippered lists* [5], i.e., creating a document by embedding parts of other documents inside it. In KonneX^{SALT} we use transclusion in conjunction with the argumentative discourse networks to improve readability and the browsing process.

The Compendium methodology, developed as part of the ScholOnto project[3] was a major inspiration for our work. In [6], the authors describe the modeling foundation for capturing claims inside scientific publications, while [7] details the set of sensemaking tools which can be used for visualizing or searching claims in scholarly documents. As part of the work, transclusion is mentioned and handled in [8]. While our goals are very similar, there is a major difference in the modeling approach that we took and in our ultimate goal, i.e., to establish a new way of citing scientific publications not via usual references, but via claim identifiers.

In terms of navigation of argumentative discourse using hypertext systems, the main direction was given by Horst Rittel [9] when he introduced the IBIS (Issue-Based Information Systems) method. IBIS focuses on modeling *Issues* together with the *Positions* and *Arguments* in a decisional process. In 1987, Conklin et. al introduced a hypertext version of IBIS, called gIBIS, first by modeling team design deliberation [10] and then by exploring policy discussions [11]. Other approaches that make use of the IBIS method are for example the DILIGENT Argumentation Ontology [12] designed to capture the argumentative support for building discussions in DILIGENT processes, or Aquanet [13] designed to model the personal knowledge via claims and arguments.

3 Use-Cases

Due to the increasing number of scientific publications, finding relevant literature for a particular field represents a real challenge. At a more fine-grained level, if we look for specific claims, we are faced with an even more cumbersome task. One of the main reasons for this problem is the lack of a central access point to all publications. Information about publications is scattered over the Web. In theory, this would not really present an issue, if there was a uniform way of representing this information. In practice, however, each of the currently existing publishing sources (e.g. journals, conferences, publication repositories) handles the representation of the scientific publications in its own particular manner. Therefore, the same publications exist in different environments, maybe modelled from different aspects, but without any explicit links between the instances of the same publication.

For example, let us consider a publication X: (i) X might be listed in DBLP, together with the publication environment (e.g., the conference where it was presented) and some extra information about the authors of the publication (e.g. other papers they published), (ii) its citation record can be found using CiteSeer, (iii) independent reviews about the publication might be found on Revyu.com, (iv) while even more information about the authors might be found in OntoWorld.org. All these information sources provide different (partially overlapping) facets of the same publication, some in a traditional fashion, while others in the "Semantic Web" way, i.e., by describing the information in RDF (the last two). In the following, we detail the approach KonneX^{SALT} takes in providing solutions for both finding publications and linking them.

---

[3] http://kmi.open.ac.uk/projects/scholonto/

3.1 Search and Browse

Current solutions for finding scientific literature, such as Google Scholar or CiteSeer, follow a traditional direction, by crawling the open Web for publications, storing them locally and performing full-text indexing. Therefore the search operation includes in terms of cost all the previous three operations, while in terms of time consumption is expensive because the search space comprises the full text of all publications.

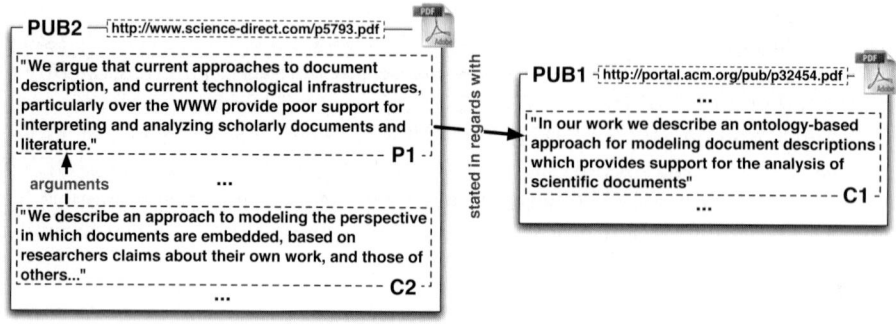

Fig. 1. Example of argumentative discussion in different publications

Contrary to this approach, KonneX^{SALT} tries to make the search operation as in-expensive and efficient as possible. In order to achieve this, we rely on the authors' incentive for externalizing their knowledge and on minimizing the search space only to this knowledge. As we show in Sect. 4, the benefit an author receives by annotating their publications with SALT is proportional to the effort they put in the annotation process. For the lowest possible level of annotation (i.e., claims and positions), the benefit is already clear: one can easily find the claims stated in one's publications and browse the argumentation line in the related work starting from such a claim. At the same time, the author themselves might find interesting connections between their claims and others, about which they might have not been aware. From an efficiency point of view, by indexing only the claims in publications, we restrain the search space to the minimum. This, combined with latent semantic indexing, provides us with a high level of accuracy, as well as the option of clustering similar claims.

As mentioned, in addition to the search functionality, the externalized knowledge gives us the possibility of enhancing the readability and browsing of the argumentative discourse spanning across multiple publications. Fig. 1 depicts an example of such a discourse: PUB1 claims C1, while PUB2 provides a position P1 in regards to the claim in PUB1 and arguments it with its own claim C2. Finding either of the claims in KonneX^{SALT} will result in the opportunity of browsing the entire discussion, as shown in Fig. 2.

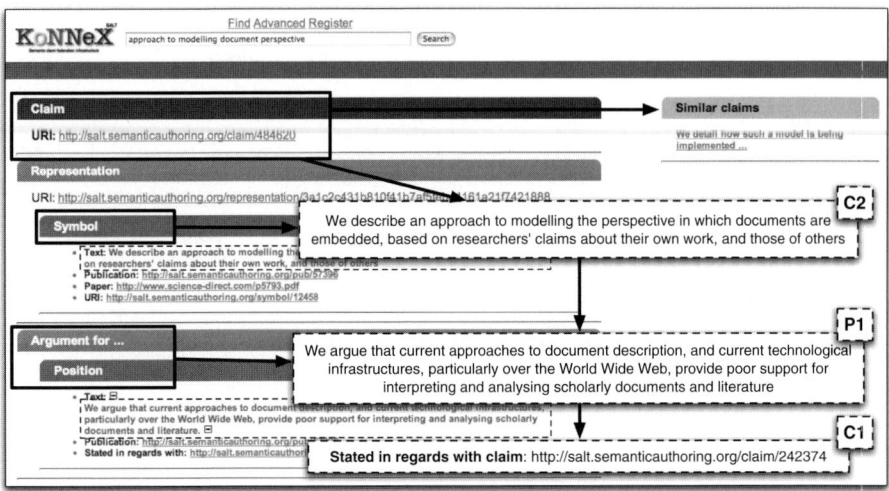

Fig. 2. Browsing a discussion in KonneX^{SALT}

3.2 Linking Open Data

As the Semantic Web grows, the focus in R&D moves more and more from producing semantic data towards the reuse of data. In this category we can also fit approaches for solving one of the collateral aspects of producing semantic data, i.e. replication. Different or overlapping aspects of the same data now exist on the Semantic Web. The Linking Open Data[4] initiative emerged from the necessity of creating bridges between those data islands. We join this effort by providing the means for linking publications in different environments. Fig. 3 depicts such an example, where the reader is advised to follow the links to find more information on this publication. The same linking information is also available as triples for reasoning purposes.

4 Background

As an intermediary step towards implementing the technical solutions to solve the use-cases we have just outlined, we will now provide some background information on the frameworks we use to achieve our goals and motivate our decisions. We start with a short description of SALT, then emphasize the evolution of the claim identification tree into the argumentative discourse network and finally introduce Latent Semantic Indexing (LSI).

4.1 SALT: Semantically Annotated LaTeX

SALT is a semantic authoring framework targeting the enrichment of scientific publications with semantic metadata. The result of a SALT process is a semantic

---

[4] http://esw.w3.org/topic/SweoIG/TaskForces/CommunityProjects/LinkingOpenData

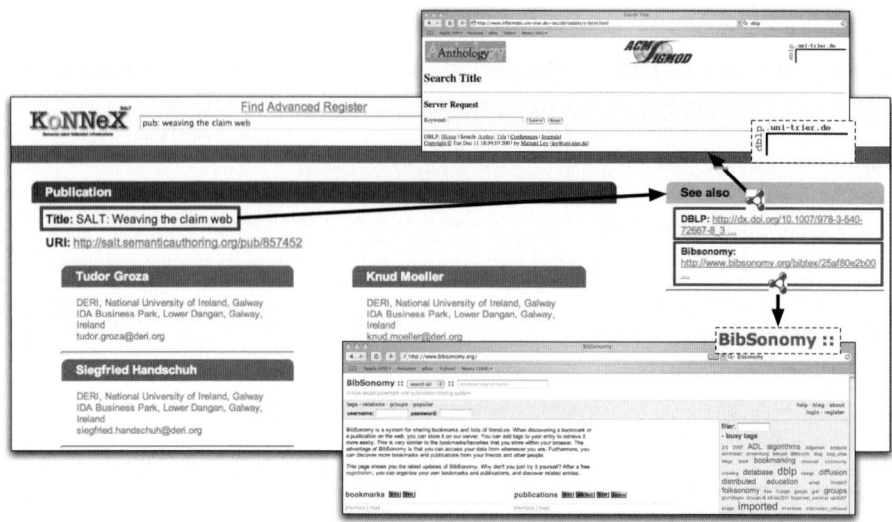

Fig. 3. Linked data in KonneX^{SALT}

document, built by using the PDF file format as a container for both the document content and the semantic metadata. SALT comprises two layers: (i) a syntactic layer, and (ii) a semantic layer. The syntactic layer represents the bridge between the semantic layer and the hosting environment, i.e., LATEX. It defines a series of new LATEX commands, while making use of some of the already existing ones to capture the logical structure of the document and the semantics in the publication. We chose LATEX because it is one of the most widely used writing environments in the scientific community. In a similar way, the SALT framework can be used also together with, for example, Microsoft Word.

The semantic layer is formed by a set of three ontologies: (i) The Document Ontology, which models the logical structure of the document, (ii) The Rhetorical Ontology, which models the rhetorical structure of the publication and (iii) The Annotation Ontology, which links the rhetorics captured in the document's content and the physical document itself. The Annotation Ontology also models the publication's shallow metadata. The Rhetorical Ontology has three sides: (i) one side is responsible for modeling the claims and their supporting argumens in scientific publications, as well as the rhetorical relations connecting them, (ii) a second side models the rhetorical block structure of the publication, while (iii) the last one, modeling the argumentative discourse over multiple publications, can be seen as a communication between different authors.

One of the main differences that distinguishes SALT from other semantic authoring solutions is its incremental approach. The SALT process was designed from the beginning to be flexible and entirely driven by the author. As shown in Fig. 4 the benefit that an author receives by using different parts of SALT is increasing proportionally with the amount of effort it is involved into the authoring process. While we do believe in a quasi-linear increase in benefit, note

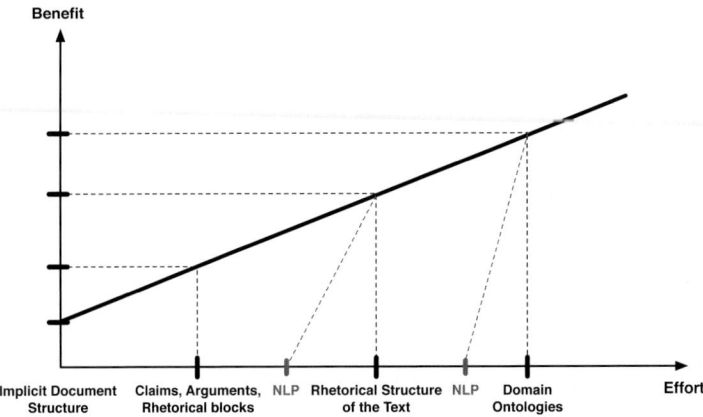

Fig. 4. SALT Process incremental approach

that this figure is only an approximation of the effort/benefit ratio we conjecture; the figure does not show an actual measurement.

For example, without investing any effort at all, the author still gains value from SALT, because it automatically extracts the shallow metadata of the publication together with its logical structure and references. Used within KonneX[SALT], using only this information would make it equivalent to CiteSeer or DBLP in terms of functionality. After this, adding simple annotations like the claims, arguments or rhetorical blocks the value brought to the user increases considerably. As a result, KonneX[SALT] provides specialized search and navigation of the argumentation discourse. Continuing in the same direction, the effort would increase, but in return we would improve the existing functionalities and maybe even discover unexpected features. At the same time, we would argue that by automating the process with the help on NLP techniques, the authors' effort would decrease substantially[5].

4.2 Argumentative Discourse Networks

In [2] we defined a semiotic approach for modeling claims and their textual representations, called *claim identification tree*. The tree has three levels: the root of the tree, representing a claim at an abstract level, a representation level linking multiple representations to the abstract claim and a symbolic level pointing to the actual textual claim in different publications on the Web.

Continuing in the same direction, we take the modeling of claims one step further and consider the associative trails (see Sect. 2) captured in the argumentative discourse between authors in different publications. Fig. 5 depicts the way in which the example introduced in the previous section is modeled as an argumentative discourse network.

---
[5] By introducing NLP techniques, the effort/benefit ratio would show a non-linear development.

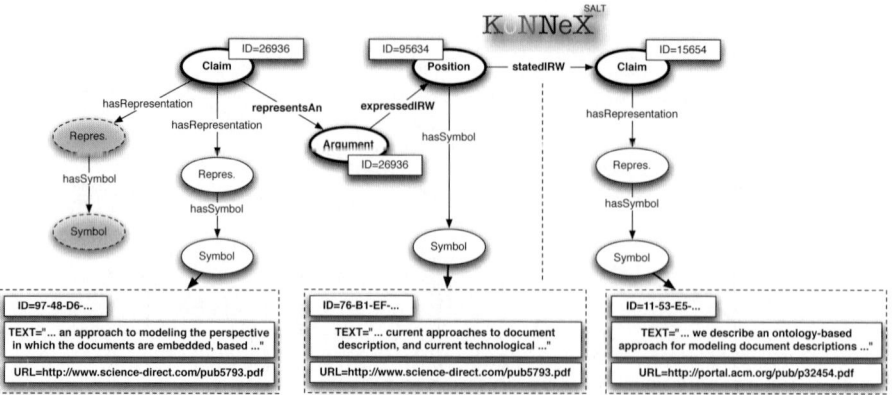

Fig. 5. Argumentative discourse network in KonneX^{SALT}

4.3 Latent Semantic Indexing

Latent Semantic Analysis (LSA) represents a technique used in natural language processing (NLP) which takes advantage of implicit higher-order structure in the association of terms with documents, in order to improve detection of relevant documents on the basis of terms found in queries [14]. Often, in its application in information retrieval, this technique is also called Latent Semantic Indexing (LSI). The basic LSI process represents documents as vectors in a multi-dimensional space and calculates relevancy rankings by comparing the cosine of the angles between each document vector and the original query vector.

Some of the proved limitations of LSA motivated us in using this technique for federating claims in KonneX^{SALT}: (i) long documents are poorly represented, due to the small scalar product and the large dimensionality – in our case each document is represented by a claim, i.e., a phrase or sentence with a maximum length of around 20 words; or, (ii) documents with similar context but different term vocabulary will not be associated – the different representations of the claim usually share a common (or similar) term vocabulary, due to the fact that they all describe the same abstract claim.

5 Semantic Claim Federation Infrastructure

This section gives an overview of the design of KonneX^{SALT}, including a brief description of the publication registration process, which is responbible for the linking data use-case described in Sect. 3.2 and of the data visualization issue in the context of a pure Semantic Web application.

Fig. 6 depicts the overall architecture of the semantic claim federation infrastructure. The core part is composed by the four managers (*Query*, *Registration*, *Index* and *Repository*), together with the associated indexes and the RDF repository. Our goal was to have a clean design and clearly distinguish between the

KonneX^{SALT}: First Steps Towards a Semantic Claim Federation Infrastructure

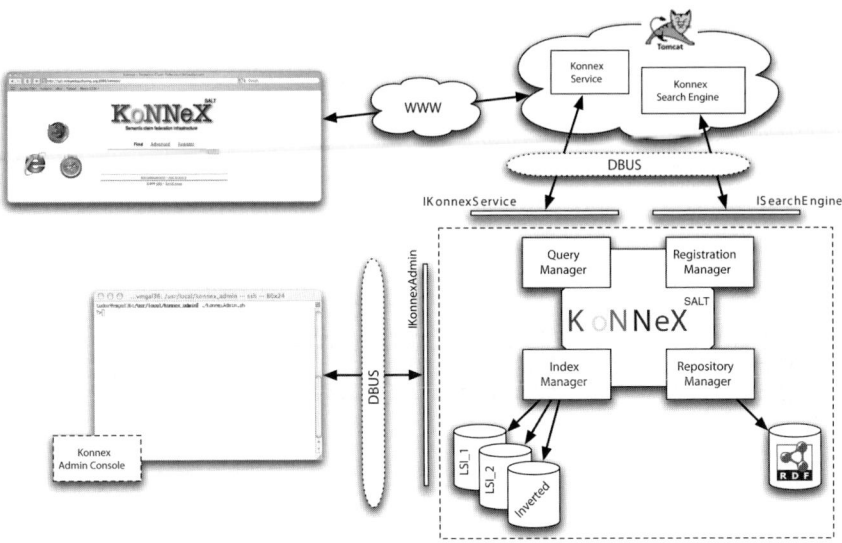

Fig. 6. Overall architecture of KonneX^{SALT}

functional modules placed outside the core. Thus, due to the existing overlaps between some of the exposed functionalities, we decided to use DBus [15] as a communication infrastructure. As a consequence, all the modules reside on the server, but each module deals with its own particular functionality. The core of KonneX^{SALT} acts a as daemon and exposes three different interfaces over DBus, thus limiting the access to a particular set of functionalities for a particular module.

The four managers comprising the core are: (i) the *Query Manager*, which acts as a query distribution and results merging hub for the Index Manager and the Repository Manager; (ii) the *Registration Manager*, which analyzes the publications to be registered, creates links to other publication repositories and populates the local indexes and RDF repository. For each publication, it extracts the SALT metadata and, based on the title-authors information, for linking purposes, it requests more data from DBLP and Bibsonomy. Due to possible errors that might appear in the shallow metadata of a publication, the results received from the two sources are filtered to a certain degree of similarity; (iii) the *Repository Manager*, which manages the RDF repository, which in our case is a wrapped Sesame[6] instance. Finally (iv) , the *Index Manager* orchestrates a set of three indexes: two latent semantic indexes (InfoMap [16] and Semantic Engine [17]) and one inverted index. The inverted index is used to provide the link from the textual representation to the root of the RDF graph modelling the claim. The final ranking of a query is computed as a weighted average over the rankings coming from the LSIs.

---

[6] http://www.openrdf.org/

Outside this core box there are three independent modules which provide separate functionalities to end-users: (i) the *Konnex Service*, which represents a RESTful Web service handling query requests, both full-text and URI-based. It can be used as an API for interacting with the semantic claim federation infrastructure, but also as a mechanism to retrieve URIs in usual web browsers. In the context of the URIs, we adhere to the Linking Open Data principle, and make them retrievable. (ii) The *Konnex Search Engine* acts as the main access point to the system for end-users. It is accompanied by a Web interface which allows searching for scientific claims and to register publications. (iii) The *Konnex Admin Console* is a utility for administrating the system. It provides capabilities for tuning the systems' parameters and querying the system's status.

As mentioned before, we argue that KonneX^{SALT} represents a pure Semantic Web application. The data flow between the core of the infrastructure and the other modules is strictly based on RDF graphs. Here, we also include the flow between the core and the end-user Web interface of the search engine. As a consequence, when retrieving a URI or performing text-based queries, the result will always be represented as an RDF graph. Rendering such an RDF graph for a human user in a Web browser is handled by using an XSL stylesheet. In this way, the burden of transformation is moved from the server to the client: the actual process of transformation is executed by the local web browser via its built-in XSLT processor, the RDF result having attached an XSL style sheet. This method helps us to avoid duplication of data and provides flexibility for both machines and humans, because it can be easily read by a human, and at the same time be analyzed by an application or fetched by a crawler.

6 Discussion

In this section we present two interesting issues which appeared during our research. First, we deal with the sensitive issue of copyright and transcopyright. Then we discuss how domain knowledge can improve search results and navigation in KonneX^{SALT}.

6.1 Copyright and Transcopyright

One of the main issues raised by implementing transclusion is the copyright of the included material. This was discussed for the first time by Ted Nelson and included in his list of 17 original rules of Xanadu [5]. The specific rules considering copyright were: (i) Permission to link to a document is explicitly granted by the act of publication; (ii) Every document can contain a royalty mechanism at any desired degree of granularity to ensure payment on any portion accessed, including virtual copies ("transclusions") of all or part of the document.

Nelson was the one to provide also a solution to the copyright issue, by defining *Transcopyright* [18], or the pre-permission for virtual republishing. He mainly proposed the introduction of a particular format to mark transcopyright, as an adaptation of the traditional copyright notice. For example, from (c) 1995 DERI to Trans(c) 1995 DERI.

In the case of KonneX^{SALT}, due to the author-driven approach, the copyright of the text used in transclusion has an implicit assumption: the author of the publication, by registering it with our system, grants us the right to cite and re-use the text for achieving our purposes. This assumption follows the common author-driven publishing trend, which traditional search engines also take advantage of, i.e., the authors themselves are the ones who expose their publications on the open Web and thus give everyone the right to read[7] the publications. Still, in order to maintain and exhibit the copyright of the material, we will adopt and implement a solution based on the approach proposed by Ted Nelson.

6.2 Improving Results Based on Domain Knowledge

The functionalities KonneX^{SALT} provides are driven by the way in which the authors use the SALT framework. As previously discussed in Sect. 4, the benefit an author gains by using SALT is proportional to the effort spent on creating the annotations. Consequently, the more features of SALT are used, the more functionalities KonneX^{SALT} can offer. An important aspect of the semantic authoring process is the use of domain knowledge. In a previous publication [19] we have described the support for annotating rhetorical elements in scientific publications with domain knowledge. In this section, we present an example where domain knowledge helps to improve the search results and the navigation of the argumentative discourse network.

Considering the bio-medical domain, we assume that the publications, were enriched with concepts present in a disease domain ontology. This could be performed either manually by the author, or semi-automatically by using a specialized tool. As a direct consequence, internally to KonneX^{SALT}, the rhetorical elements can be clustered not only based on linguistic information but also based on the attached domain concepts. Obviously, we would need an ontology import mechanism to make the system aware of the particular ontologies. In terms of enhancements, we foresee two new features:

- a special search syntax to take into account the concepts in the ontology. Such a feature could be used in two different ways: either to prune the search results based on the specified concepts (see the first example below), or to expand the search to a certain lens created around the specified concepts (the second example below).
  E.g. 1 :< *about* : *prot#P3456* & *about* : *prot#P6734* > (... *text* ...) — Search for *text* which has attached information only about the two proteins: P3456 and P6734.
  E.g. 2 :< *about* : *prot#P3456* | 1 > (... *text* ...) — Search for *text* which has attached information about protein P3456, but consider also the direct parent concept and direct subconcepts from the ontology hierarchy.
- attached concepts to the argumentative discourse network. The dual navigation (depicted in Fig. 7) could improve the understanding of the reader not

---

[7] In this particular case, when mentioning the read access, we also include citation, crawling and all other operations that re-use the text without modifying it.

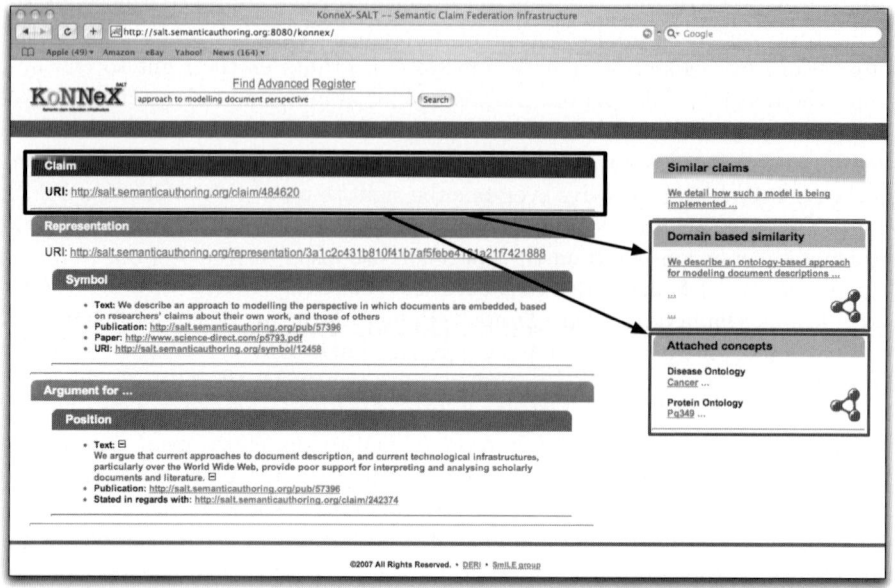

Fig. 7. Domain knowledge incorporated into the browsing process

only in the line of argumentation but also following the covered domain concepts. This feature would open up to the possibility of integrating a faceted browser as part of KonneX^{SALT}.

7 Conclusion

In this paper we presented KonneX^{SALT}, a semantic claim federation infrastructure geared towards finding claims in scientific publications and facilitating the navigation of argumentative discourse networks, starting from a root claim. Our main goal is to provide a look-up service for scientific claims and not yet another traditional search engine. Hence, we distinguish our approach from others by adopting an author-driven method of populating the publication repository. In addition, we give the reader the possibility of following the argumentation line by realizing associative trails in conjunction with on-demand transclusion. Finally, we implemented KonneX^{SALT} as a pure Semantic Web application (i.e. all the data is being strictly modeled as RDF) and follow the Linking Open Data principle by linking instances of our data with data elsewhere on the Semantic Web.

Future developments of KonneX^{SALT} include: (i) using domain knowledge as part of the argumentative discourse network, (ii) providing transclusion for rhetorical blocks, (iii) linking deeper into the Web of Open Data (e.g. authors' links), and (iv) introducing an advanced search method inside rhetorical blocks based on a special syntax. All the above mentioned features represent basic enhancements of the infrastructure. In addition, we are considering o provide a way for browsing the publications present in the repository, based on a faceted navigation methodology like the one offered by Exhibit [20].

Acknowledgments

The work presented in this paper was supported (in part) by the European project NEPOMUK No FP6-027705 and (in part) by the Lion project supported by Science Foundation Ireland under Grant No. SFI/02/CE1/I131. The authors would like to thank Diana Trif, Laura Dragan, Renaud Delbru, Alexander Schutz and Andreas Harth for their support and involvement in the the development of SALT.

References

1. Nonaka, I., Takeuchi, H.: The Knowledge Creating Company: How Japanese Companies Create the Dynamics of Innovation. Oxford University Press, Oxford (1995)
2. Groza, T., Möller, K., Handschuh, S., Trif, D., Decker, S.: SALT: Weaving the claim web. In: Aberer, K., Choi, K.-S., Noy, N., Allemang, D., Lee, K.-I., Nixon, L., Golbeck, J., Mika, P., Maynard, D., Mizoguchi, R., Schreiber, G., Cudré-Mauroux, P. (eds.) ISWC 2007. LNCS, vol. 4825, Springer, Heidelberg (2007)
3. Bush, V.: As We Think. The Atlantic Monthly 176(1), 101–108 (1945)
4. Nelson, T.H.: The Heart of Connection: Hypermedia Unified by Transclusion. Communications of the ACM 8, 31–33 (1996)
5. Nelson, T.H.: Literary Machines: The report on, and of, Project Xanadu concerning word processing, electronic publishing, hypertext, thinkertoys, tomorrow's intellectual.. including knowledge, education and freedom. Mindful Press, Sausalito, California, 1981 edition (1981) ISBN 089347052X
6. Mancini, C., Shum, S.B.: Modelling discourse in contested domains: a semiotic and cognitive framework. Int. J. Hum.-Comput. Stud. 64(11), 1154–1171 (2006)
7. Uren, V., Shum, S.B., Li, G., Bachler, M.: Sensemaking tools for understanding research literatures: Design, implementation and user evaluation. Int. Jnl. Human Computer Studies 64(5), 420–445 (2006)
8. Mancini, C., Scott, D., Shum, S.B.: Visualising Discourse Coherence in Non-Linear Documents. Traitement Automatique des Langues 47(1), 101–108 (2006)
9. Kunz, W., Rittel, H.: Issues as elements of information system. Working paper 131, Institute of Urban and Regional Development, University of California (1970)
10. Conklin, J., Begeman, M.L.: gIBIS: A Hypertext Tool for Team Design Deliberation. In: HYPERTEXT 1987: Proceeding of the ACM conference on Hypertext, pp. 247–251. ACM Press, New York (1987)
11. Conklin, J., Begeman, M.L.: gIBIS: A Hypertext Tool for Exploratory Policy Discussion. In: CSCW 1988: Proceedings of the 1988 ACM conference on Computer-supported cooperative work, pp. 140–152. ACM Press, New York (1988)
12. Tempich, C., Pinto, H.S., Sure, Y., Staab, S.: An Argumentation Ontology for Distributed, Loosely-controlled and evolvInG Engineering processes of oNTologies (DILIGENT). In: Gómez-Pérez, A., Euzenat, J. (eds.) ESWC 2005. LNCS, vol. 3532, pp. 241–256. Springer, Heidelberg (2005)

13. Marshall, C.C., Halasz, F.G., Rogers, R.A., William, C., Janssen, J.: Aquanet: A Hypertext Tool to Hold Your Knowledge in Place. In: HYPERTEXT 1991: Proceedings of the third annual ACM conference on Hypertext, pp. 261–275. ACM Press, New York (1991)
14. Deerwester, S., Dumais, S.T., Furnas, G.W., Landauer, T.K., Harshman, R.: Indexing by Latent Semantic Analysis. Journal of the Society for Information Science 41(6), 391–407 (1990)
15. Palmieri, J.: Get on D-BUS. Red Hat Magazine 3(1) (2005)
16. Infomap NLP Software: An Open-Source Package for Natural Language Processing (2007), http://infomap-nlp.sourceforge.net/
17. The Semantic Indexing Project: Creating tools to identify the latent knowledge found in text (2007),
http://www.hirank.com/semantic-indexing-project/index.html
18. Nelson, T.H.: Transcopyright: Dealing with the Dilemma of Digital Copyright. Educom Review 32(1), 32–35 (1997)
19. Groza, T., Handschuh, S., Möller, K., Decker, S.: SALT - Semantically Annotated for Scientific Publications. In: Franconi, E., Kifer, M., May, W. (eds.) ESWC 2007. LNCS, vol. 4519, Springer, Heidelberg (2007)
20. Huynh, D., Karger, D., Miller, R.: Exhibit: Lightweight Structured Data Publishing. In: Proceedings of the 16th International World Wide Web Conference, WWW 2007, Banff, Canada, May 8-12 (2007)

Building a National Semantic Web Ontology and Ontology Service Infrastructure –The FinnONTO Approach

Eero Hyvönen, Kim Viljanen, Jouni Tuominen, and Katri Seppälä

Semantic Computing Research Group (SeCo)
Helsinki University of Technology (TKK) and University of Helsinki
firstname.lastname@tkk.fi
http://www.seco.tkk.fi/

Abstract. This paper presents the vision and results of creating a national level cross-domain ontology and ontology service infrastructure in Finland. The novelty of the infrastructure is based on two ideas. First, a system of open source core ontologies is being developed by transforming thesauri into mutually aligned lightweight ontologies, including a large top ontology that is extended by various domain specific ontologies. Second, the ONKI Ontology Server framework for publishing ontologies as ready to use services has been designed and implemented. ONKI provides legacy and other applications with ready to use functionalities for using ontologies on the HTML level by Ajax and semantic widgets. The idea is to use ONKI for creating mash-up applications in a way analogous to using Google or Yahoo Maps, but in our case external applications are mashed-up with ontology support.

1 A National Ontology Infrastructure

The ambitious goal of the National Semantic Web Ontology project (FinnONTO 2003–2007)[1] [1] is to develop a semantic web infrastructure on a national level in Finland. The consortium behind the initiative—37 companies and public organizations—represents a wide spectrum of functions of the society, including libraries, health organizations, cultural institutions, government, media, and education. The project has produced a variety of scientific results, specifications, services, demonstrations, and applications:

1. *Metadata standards.* Nationally adapted standards for representing metadata in various application fields have been created, e.g. JHS 158[2] and [2].
2. *Core ontologies.* Several core ontologies[3] have been developed in order to initiate ontology development processes in Finland.

---
[1] http://www.seco.tkk.fi/projects/finnonto/
[2] http://www.jhs-suositukset.fi/suomi/jhs158, Public Recommendation for Geographic Metadata (in Finnish), Ministry of Internal Affairs.
[3] http://www.seco.tkk.fi/ontologies/

3. *Public ontology services.* An ontology library and web service framework ONKI[4] is being developed [3] to enable ontology usage in ontology development, content indexing, and information retrieval through public web and mash-up services.
4. *Tools for metadata creation.* A bottleneck limiting proliferation of the semantic web is production of metadata. For this purpose, semiautomatic content annotation tools have being developed [4,5].
5. *Tools for semantic portal building.* Tools for semantic search and recommendation based on amalgamating the multi-facet search paradigm with semantic web technologies have been developed [6,7,8].
6. *Pilot applications.* The framework is being evaluated by implementing a number of practical semantic portal applications in the domains of eCulture (MuseumFinland [9] and CultureSampo [10,11]), eHealth (HealthFinland [12,13], eGovernment [14], eLearning [15], and eCommerce [16].

The vision of FinnONTO from the outset has been the idea of developing a national "semantic content infrastructure" for the semantic web. The core of the infrastructure consists of 1) an ontology library system of mutually aligned central ontologies and 2) the ONKI ontology server framework [17,3] for cost-effective exploitation ontologies in legacy and other applications. The FinnONTO approach is based on international domain independent recommendations and best practices of W3C, such as RDF, SKOS, OWL etc., but builds, from the domain content perspective, upon existing national thesauri and vocabularies encompassing different application domains. It is argued that such a *cross-domain ontology system* is needed nationally on the web in same way as roads are needed for traffic and transportation, power plants and electrical networks are needed for energy supply, or GSM standards and networks are needed for mobile phones and wireless communication.

This paper presents the FinnONTO approach to developing a national cross-domain ontology infrastructure for the semantic web. In the following, the idea a creating a system of mutually aligned ontologies from thesauri is first elaborated. After this, the ONKI Ontology Service system is presented for application usage. The ontology framework is already operational on the web and is being used in creating the FinnONTO demonstration applications.

2 Creating a System of Mutually Aligned Ontologies

The traditional approach for harmonizing content indexing is to use keyword terms taken from shared vocabularies or thesauri [18,19]. In Finland, for example, lots of thesauri conforming to the ISO [2788] thesaurus standard [19] and its Finnish version SFS [5471] are in use, such as YSA[5], MASA [20], or Agriforest[6].

---

[4] http://www.seco.tkk.fi/services/onki/
[5] http://vesa.lib.helsinki.fi/
[6] http://www-db.helsinki.fi/triphome/agri/agrisanasto/Welcomeng.html

In order to reuse the effort already invested in developing thesauri, it was therefore decided to develop a method of minimal changes for transforming thesauri into ontologies. Thesauri-based class ontologies form the core of FinnONTO ontologies, but there are also other important ontologies being developed in the projects, especially the geographic ontology SUO [21] that consists currently of some 800,000 places in Finland (and over 4 million places abroad), the historical place ontology SAPO [22], and the actor ontology TOIMO based on some 120,000 actors and organizations of the Universal List of Artist Names vocabulary (ULAN)[7] with some Finnish extensions [23], and a Finnish version of the iconographic vocabulary ICONCLASS[8].

2.1 A Method for Transforming Thesauri into Ontologies

The idea of transforming thesauri into ontologies was also suggested in [24] and by the SKOS initiative[9]. However, in contrast to these initiatives, we stress that although a syntactic transformation into SKOS can be useful, it is not enough from a semantic viewpoint. The fundamental problem with a traditional thesaurus [18, 19], is that its semantic relations have been constructed mainly to help the indexer in finding indexing terms, and understanding the relations needs implicit human knowledge. Unless the meaning of the semantic relations of a thesaurus is made more explicit and accurate for the computer to interpret, the SKOS version is equally confusing to the computer as the original thesaurus, even if semantic web standards are used for representing it.

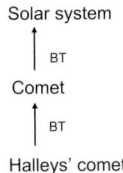

Fig. 1. An example of broader term hierarchy in the YSA thesaurus

For example, consider the example of Figure 1 taken from the YSA thesaurus, where BT indicates the "broader term" relation used in thesauri. We can easily understand its meaning but the machine is confused: Is Halley's Comet an individual or a class of them, such as Comet? Can there be many Halley's comets or only one? Is a comet a kind of solar system or a part of a solar system? Is it a part as a concept or are all individual comets a part of some solar system? Do comets have the properties of solar systems, e.g. own planets, based on the BT relation? Using the BT relations for term expansion, a search for "solar systems" would retrieve comets although comets are not solar systems.

---

[7] http://www.getty.edu/research/conducting_research/vocabularies/ulan/
[8] http://www.iconclass.nl/
[9] http://www.w3.org/2004/02/skos/

The idea of using ontologies is to define the meaning of indexing terms and concepts explicitly and accurately enough for the machine to use. This is essential in many application areas, such as semantic search, information retrieval, semantic linking of contents, automatic indexing, and in making contents semantically interoperable. With even a little extra work, e.g. by just systematically organizing concepts along subclass hierarchies and partonymies, substantial benefits can be obtained.

Therefore, our method for thesaurus-to-ontology transformation is not a syntactic one, but is done by refining and enriching the semantic structures of a thesaurus. As a general methodological guideline, criteria from DOLCE [25] were employed. However, the methodology is applied in a lightweight fashion in order to keep the job manageable.

The method used is based on making the following semantic refinements and extensions in the thesaurus structure:

1. *Missing links in the subclass-of hierarchy.* The BT relations do not (usually) structure the terms into a full-blown hierarchy but into a forest of separate smaller subhierarchies. In the case of YSA, for example, there were thousands of terms without any broader term. Many interesting relations between terms are missing in thesauri, especially concerning general terminology, where BT relations are not commonly specified in practice. As in DOLCE, the concepts on the top level were divided into three upper classes: 1) Abstract, Endurant, and Perdurant. A central structuring principle in constructing the hierarchies was to avoid multiple inheritance across major upper ontology categories.

2. *Ambiguity of the BT relation.* The semantics of the BT relation is ambiguous: it may mean either subclass-of relation, part-of relation (of different kinds, cf. [26]), or instance-of relation. This severely hinders the usage of the structure for reasoning [27]. For example, the BT relation cannot be used for property inheritance because this requires that the machine knows that BT means subclass-of and not e.g. part-of relation. In our method, existing BT relations were transformed into subclass-of and part-of relations, instance-of relations are not used. The difficult question of deciding what terms should be instances and what terms are classes is avoided by transforming even instance like terms, such as "Halleys' comet", into classes. The decision of whether to use classes or instances in annotations[10] is not made in the ontology but is left to the application developers for flexibility.

3. *Non-transitivity of the BT relation.* The transitivity of the BT relation chains is not guaranteed from the instance-class-relation point of view, when transforming BT relations into subclass-of relations. If x is an instance of a class A whose broader term is B, then it is not necessarily the case that x is an instance of B, although this a basic assumption in RDFS and OWL semantics [28] if BT is regarded as the subclass-of relation. For example, assume that x is a "make-up mirror", whose broader term is "mirror", and that its broader term is "furniture". When searching with the concept "furniture"

---

[10] Cf. http://www.w3.org/TR/swbp-classes-as-values/, W3C Working Group Note 5 April 2005.

one would expect that instances of furniture are retrieved, but in this case the result would include x and other make-up mirrors, if transitivity is assumed. This means e.g. that term expansion in querying cannot be used effectively based on the BT relation in a thesaurus, but can be used in our corresponding lightweight ontology.
4. *Ambiguity of concept meanings.* Lots of terms in our thesauri are ambiguous and cannot be related properly with each other in the hierarchy using the subclass-of relation. For example, in YSA there is the indexing term "child". This term has several meanings such as "a certain period of human life" or "a family relation". For example, George W. Bush is not a child anymore in terms of age but is still a child of his mother, Barbara Bush. The computer cannot understand this and is confused, unless the meanings of "child" are separated and represented as different concepts (with different URIs) in different parts of the ontology.

In FinnONTO, the central ontology developed is the General Finnish Ontology YSO[11] [29]. YSO is based on the general Finnish keyword thesaurus YSA[12] that contains some 23,000 terms divided into 61 domain groups, such as Physics, History, etc. YSA is maintained by the National Library of Finland. Since YSA is widely used in Finnish organizations, YSO is an important step in solving semantic interoperability problems in Finland. The ontology is trilingual. Swedish translations of the YSO ontology labels were acquired from the Allärs thesaurus[13], and a translation of the terms into English was produced by the City Library of Helsinki. This makes it possible in the future to align YSO with international English ontologies of the semantic web.

In ontologizing the YSA thesaurus lots of terms turned out to be ambiguous, i.e., they could not be placed in one place in the hierarchy. In such cases the term had to be split into several concepts in YSO. However, a lesson learned in our work was that the general ambiguous concept encompassing several meanings, say "child", can be useful for indexing purposes and should be available in YSO. For example, assume a painting depicting children playing in a park with their mothers watching. When selecting keywords (concepts) describing the subject, it would be tedious to the indexer to consider all the meaning variants of "childness" in YSO, while the single ambiguous indexing term "child" of YSA would encompass them properly in this case. We therefore included some useful ambiguous concepts, such as "child", in YSO as special *aggregate indexing concepts*. They lay outside of the subclass-hierarchies but can be defined, e.g., in terms of them by using Boolean class expressions as in OWL[14].

A principle used in transforming YSA was that each YSA term should have a counterpart in YSO. This makes it possible to use YSO for more accurate reasoning about content annotated using YSA. Since the original term meanings in YSA change when the term is connected into an ontology, the original YSA

---
[11] http://www.yso.fi/onto/yso/
[12] http://vesa.lib.helsinki.fi/
[13] http://vesa.lib.helsinki.fi/allars/index.html
[14] http://www.w3.org/TR/owl-ref/

terms had to be preserved in the YSO ontology as they are. YSO therefore consists of the following major parts: 1) a meaning preserving SKOS version of the original YSA, 2) an ontology of concepts corresponding to YSA terms, and 3) a mapping between the two structures.

The mapping makes it possible to explicitly tell the relation between YSO concepts and YSA terms. In our mapping schema, the relation between two concepts A and B is defined in terms of extensional overlap that can be expressed as two numerical values in the range [0,1]: 1) how much A overlaps B proportionally and 2) how much B overlaps A. This model is an adaptation of [22] where geographical overlap in area is considered. For example, if A is a subclass of B, then B overlaps A in meaning by 1, and A overlaps B by some value in the range [0,1]; equality means two overlap values 1, and partial overlaps can be expressed by selecting other values. In the first version of YSO equality of YSA and YSO concepts is used by default.

2.2 Aligning Ontologies

Thesauri are widely used for harmonizing content indexing. Different domain fields have thesauri of their own. The thesauri are typically developed by domain specific expert groups without much systematic collaboration with other fields. When using such thesauri in cross-domain environments, such as the web, semantic problems arise, e.g., due to ambiguity of literal word expressions. For example, in the finance domain the term "bank" has an obvious meaning as an institution, but when considering the nature or musical instrument domains, there are other meanings. In semantic web ontologies, the ambiguity problem is solved by dealing with unambiguous resources identified by URIs instead of literal words. However, support is needed for sharing the URIs across domains and users. If one needs to define the notion of "river bank", one should not mix this concept with "money bank". On the other hand, if one is defining the notion of "blood bank", it is possible to use the more general notion of "bank" and modify it, thus sharing this common notion with other kind of banks considered in other ontologies.

In focused domains and applications it may be possible to agree upon common ontological concepts, but in a larger cross-domain setting, this usually becomes more difficult. Different domains and applications may need different ontological representations even for the same real world objects, and different parties tend to have different philosophical opinions and needs on how to model the world. As a result, there is the danger that the global semantic web will not emerge but there will rather arise a set of isolated, mutually incompatible semantic web islands.

There are various complementary approaches for making semantic web ontologies interoperable. First, ontology mapping and alignment [30] can be used for mapping concepts with each other. Second, ontologies can share and be based on common foundational logical principles, like in DOLCE. This easily leads to complicated logical systems that may not scale up either epistemically or computationally to real word situations and practical usage. Third, horizontal top ontologies, such as the SUMO[15] can be created for aligning the concepts between vertical domain ontologies. Fourth, ontology engineering tools for creating ontologies in the first place as interoperable as possible can be created.

We adopted the idea that a shared top ontology is useful for enhancing semantic interoperability between various domain ontologies. In Finland the YSA thesaurus is widely used for content indexing in libraries, museums, and archives of various kinds both public and private. YSA can therefore be considered as a kind of semantic terminological "glue" between many other Finnish thesauri. Once the structure of the top ontology is defined, the same choices of hierarchical structures can be reused in many cases in the vertical ontologies that typically share lots of concepts with the top ontology. For example, when we created the cultural ontology MAO, based on the Finnish Cultural thesaurus MASA [20], about 2,000 out of MASA's 6,000 terms turned out to have a corresponding term in YSA. We now work e.g. on the Agriforest thesaurus[16], and on some other thesauri, where thousands of terms originate from YSA.

A simple method and a tool was created by which a Protégé[17] project is created containing an ontology O and YSO in their own namespaces. The classes of O are mapped with YSO classes having similar labels using equivalence, and for each class in O and YSO, subclasses from both ontologies are attached. Using this project, a human editor then checks the subclass-of chains along both ontologies by hand, and edits the structures. In contrast to the PROMPT Suite [31], our tool simply matches similar labels in subclass-of-hierarchies, and the focus is not to merge ontologies or create mappings, but rather to form an initial version of a double ontology project for maintaining an ontology O in accordance with another one (YSO).

Several ontologies have now been aligned with YSO pairwise. According to our experience, a lot of work is saved by reusing the work done with YSO in this way, and the resulting ontologies are aligned with each other at the same time with reasonable extra effort. Our goal is a system of mutually interlinked and aligned ontologies, as illustrated in Figure 2. In our vision, vertical domain ontologies add semantic depth to the top ontology. Interoperability is obtained by aligning the ontologies with each other.

3 Public Ontology Services

The Semantic Web is based on using shared ontologies for enabling semantically disambiguated data exchange between distributed systems on the web. This requires, from the ontology publisher's viewpoint, tools for publishing ontologies on the web to ensure the availability and acceptance of the ontologies. From the ontology user's viewpoint, support services are needed for utilizing ontologies easily and cost-effectively in the users' own systems that are typically legacy systems without ontology support. ONKI addresses both problems at the same

---

[15] http://suo.ieee.org/
[16] http://www-db.helsinki.fi/eviikki/Welcome_eng.html
[17] http://protege.stanford.edu/

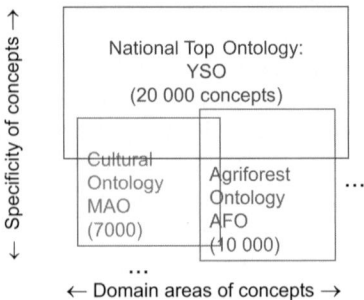

Fig. 2. YSO and related vertical ontologies intersect each other and share structures

time. It provides the publisher with a SKOS compatible server and an ontology browser whose functionalities, i.e., concept finding, semantic disambiguation, and concept fetching can be used in external applications as centralized ontology services [3]. A major contribution of ONKI is to provide these services as ready-to-use functionalities for creating "mash-up" applications very cost-efficiently. Two prototypes of the system—ONKI-SKOS [3] for all kinds of ontologies and ONKI-Geo [32] for geographical ontologies with a map mash-up interface—are operational on the web and are currently being used in several pilot applications.

3.1 ONKI Functionalities

One of the main lessons learned in our work on creating semantic portals [9, 14, 15, 12] is that metadata in data sources, such as museum databases, are often syntactically heterogeneous and contain typos, are semantically ambiguous, and are based on different vocabularies [33]. This results in lots of tedious syntactic correction, semantic disambiguation, and ontology mapping work when making the contents semantically interoperable, and when publishing them on the Semantic Web. A natural solution to this problem would be to enhance legacy cataloging and content management systems (CMS) with ontological annotation functions so that the quality of the original data could be improved and errors fixed in the content creation phase. However, implementing such ontological functions in existing legacy systems may require lots of work and thus be expensive, which creates a severe practical hindrance for the proliferation of the Semantic Web.

This relates to the more general challenge of the Semantic Web today: ontologies are typically published as files without much support for using them. The user is typically expected to open the files using some ontology editor, such as Protégé, have a closer look of the ontology, and then figure out whether it is of use to her. Once a suitable ontology is found, lots of programming effort is usually needed to utilize the ontology because it is likely to be used in a specific application and software environment. Even if the same ontological resources and structures could be shared by different content providers for interoperability,

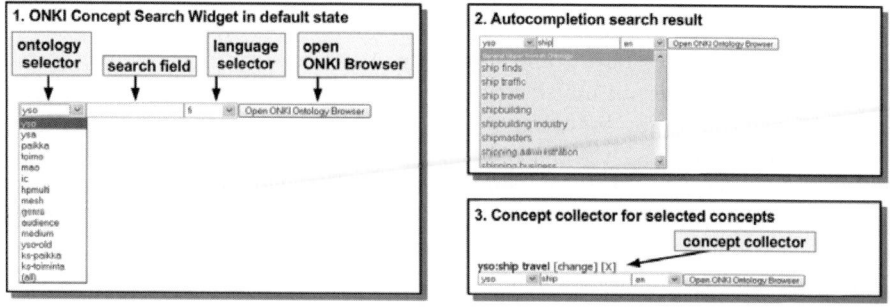

Fig. 3. ONKI Concept Search Widget

like in [9], it is not usually possible to share the *functionalities* of using the ontologies across applications. Instead, each application tends to re-implement generic functions for utilizing ontologies, such as semantic autocompletion and disambiguation [34], browsing and finding concepts, populating ontologies with new instances, etc. It is like re-creating map services from the scratch in different geographical web applications, rather than using available services such as Google Maps[18], Yahoo Maps[19], or Microsoft Live Search Maps[20].

The main functionalities of the ONKI service are 1) a *web widget for concept searching and fetching*, which is described in this section, and 2) *domain-specific ONKI Browsers*. The ONKI Browsers are user interfaces for searching and browsing ontologies, and they can be used independently or accessed via the web widget.

The general idea of the proposed mash-up approach is to provide the developer with a widget that can utilize ONKI services with minimal changes in the legacy system. In the case of an HTML-based legacy system, just a few lines of JavaScript code need to be added on the HTML page. In the case of other user interface technologies, the Web Service[21] interface can be used. The widget solves the problem of getting the right URIs into the application or a database; the actual usage of the acquired semantically correct data is the responsibility of the target application. Such a simple way for getting the URIs is crucial e.g. in various content creation systems for the semantic web, such as [33,12].

The ONKI web widget on an HTML form is illustrated in Figure 3. The widget enables the user, e.g. a content annotator, to find correct ontological concepts to describe the content to be annotated. The user is provided with searching and browsing capabilities to aid this task. When the correct concept is found, its URI and label can be fetched to the target application. In addition to annotating content, the web widget can be used for supporting other tasks where ontological concepts need to be searched and fetched, such as content searching.

---

[18] http://maps.google.com/
[19] http://maps.yahoo.com/
[20] http://maps.live.com
[21] http://www.w3.org/TR/ws-arch/

Part 1 of Figure 3 shows the default components of the widget. The ontology selector can be used to change the ontology used in search. At the moment, there are 14 different vocabularies and ontologies to choose from, including MeSH, Iconclass[22], the General Finnish Upper Ontology YSO[23] and HPMULTI[24]. The search field is used for inputting the query string. The language of concept labels used in matching the query string can be chosen by using the language selector. The choice of languages depends on the ontology selected. For example, for YSO, English and Swedish are supported in addition to Finnish, and the Finnish Geographic Ontology[25] can be used in Finnish, Swedish, and in three dialects of Sami languages spoken in Lapland. It is possible to use all languages simultaneously by selecting the option "all". The "Open ONKI Browser" button is used for opening the ONKI browser in a separate window.

The widget supports concept fetching to the target application. This can be done either by semantic autocompletion of the input field, or by pushing the "Open ONKI Browser" button for opening the ONKI Browser:

- *Using semantic autocompletion.* In part 2 of Figure 3, the user is typing a search string to the search field of the mash-up component. The system then dynamically performs a query after each input character (here "s-h-i-p-...") to the ONKI service, which returns the concepts whose labels match the string, given the language selection. The results of the query are shown in the web widget's result list below the input field. The desired concepts can be selected from the results. When selected, the concept's URI and label are fetched into the target application. In part 3 of Figure 3, the user has selected "ship travel" from the English version of the YSO ontology, and the concept's URI and label are stored onto the HTML page with the label shown, together with links `change` and `[X]`. In the case of a legacy application, which is not capable of handling URIs, only the labels of concepts can be fetched. By clicking the `change` link, it is possible to change the selected concept by using the ONKI browser (whose usage will be illustrated below). The annotation can be removed by clicking the link `[X]`.
- *Using ONKI Browser.* The alternative for using the autocompletion search, is to use the "Open ONKI Browser" button to search and browse concepts in a new ONKI Browser window. When the desired concept has been found by searching or browsing the ontology, the concept's URI and label are fetched into the application by pressing the "Fetch concept" button on the ONKI Browser page corresponding to the concept.

When the URI of a desired annotation concept is fetched, it is stored in a concept collector. The widget provides a default concept collector, but the concept collector can also be implemented in the target application. The default concept

---

[22] http://www.iconclass.nl/
[23] http://www.seco.tkk.fi/ontologies/yso/
[24] The European multilingual thesaurus on health promotion in 12 languages http://www.hpmulti.net/
[25] http://www.seco.tkk.fi/ontologies/suo/

collector shows the fetched concepts in the widget's user interface, and also stores them in hidden input fields. When the form is submitted, the hidden input fields can be processed by the target application. This way the URIs of the annotation concepts can be transferred, e.g., into the database of the application. The body of a HTTP POST request message used to submit the form's hidden fields to the target application server looks like the following example below where two URIs are selected as values of the "dc:subject" field.

```
dc:subject=http://www.yso.fi/onto/yso/p14629&
dc:subject=http://www.yso.fi/onto/yso/p8038
```

3.2 Implementation

The web widget is implemented as an easily integrable Ajax component. The widget uses HTML and JavaScript for the user interface components, and the Direct Web Remoting (DWR)[26] library for the asynchronous cross-domain Ajax communication between the web browser and the ontology server. The DWR library enables the straightforward use of the ontology server's Java methods in the JavaScript code in the web browser.

An input field in e.g. a cataloging system of a museum could be defined like this:

```
<input id="dc:subject"/>
```

The web widget can be integrated into this example system by adding the following lines of HTML/JavaScript code into the HTML page.

```
1) <script language="javascript" type="text/javascript"
   src="http://www.yso.fi/onki.js"></script>
```

```
2) <input id="dc:subject"
   onkeyup="onki['yso'].search()"/>
```

The code line 1) is used to load the needed ONKI library files and is typically added to the HEAD section of the HTML page. The code line 2) is added to the BODY section of the HTML page in the locations where the ONKI widget component is desired. The string "yso" in the code line 2) refers to the ontology server instance used in the search.

When a page containing the integration code is accessed, the ONKI JavaScript library files are loaded into the web browser. When loaded, the JavaScript library generates the user interface components of the web widget into the desired locations of the page. In this way, plain input text fields are transformed into ONKI web widgets.

The web widget can be customized e.g. by hiding the ontology or the language selection menus, the search field, or the "Open ONKI Browser" button. The

---

[26] http://getahead.org/dwr

desired customizations can be defined in the integration code[27]. The appearance of the web widget can be modified by CSS rules, e.g. by using the class attributes of the HTML elements of the widget, overriding the default ones.

We have defined an ONKI Java API that has to be implemented by the domain-specific ONKI servers for them to be used with the web widget. The API includes the following methods:

- *search(query, lang, maxHits, type, parent)* - for searching for ontological concepts.
- *getLabel(URI, lang)* - for fetching a label for a given URI in a given language.
- *getAvailableLanguages()* - for querying for supported languages of an ontology.

By implementing the shared API, different domain-specific ONKI servers can be used by one single web widget. The widget can even be used to query multiple ontology servers within the same query. In addition for the use of the web widget, the ONKI API has been published as a Web Service conforming to the SOAP[28] standard.

The ontologies in an ONKI service are also published as RDF files to support e.g. semantic web applications that perform complex ontological inferencing. Such applications need access to complete ontology files to be able to process the ontologies in their entirety. When a new version of an ontology is to be published in an ONKI service, the ontology file is tagged with a current date. This date forms part of the URI used for identifying the corresponding version of the ontology file, and this URI can be used as an URL for locating and downloading the RDF file. For every published ontology there is also a static URI which identifies the latest version of the ontology's source file.

4 Discussion and Related Work

FinnONTO aims to build an open source ontology infrastructure that consists of a set of mutually aligned ontologies from different application domains. The core of the system is a horizontal top ontology and related vertical domain ontologies extend its class hierarchy in different directions and application domains. This approach contributes to earlier work on creating ontology infrastructures that has focused either on developing domain independent standards on the international level (e.g. W3C activities), on developing single global ontologies, such as the Open Directory Project[29] or OpenCyc[30], or on developing ontologies within a particular application domain (e.g., SNOMED CT[31] in medicine and Getty Vocabularies[32] in cultural heritage). In contrast, FinnONTO tries to establish a

---

[27] http://www.yso.fi/onki/yso/app/annotation/integration-howto_en.html
[28] http://www.w3.org/TR/soap12-part1/
[29] http://dmoz.org/
[30] http://www.opencyc.org/
[31] http://www.ihtsdo.org/
[32] http://www.getty.edu/research/conducting_research/vocabularies/

comprehensive cross-domain ontology infrastructure, based on independent but mutually aligned ontologies. The experiment is conducted on a national level with a serious effort of actually starting to use ontologies in a larger scale.

The resulting ontology library is maintained as a centralized ONKI service[33] providing support for collaborative ontology publishing and ontology usage. A related approach is [35] presenting a distributed model for searching, reusing, and evolving ontologies. A novelty of the ONKI ontology services is that they can be used in external legacy and other applications as ready-to-use functionalities: we developed the idea of mash-up usage of ontologies in a way similar to Google Maps and other similar services. Another approach providing an integrable autocompletion widget for external systems is developed in [36].

We hope that the results of FinnONTO will encourage organizations to start developing machine (and human) "understandable" ontologies instead of thesauri targeted for human indexers, and that content providers will start using ontologies via ontology services such as ONKI. The FinnONTO framework is a practical approach to solving one of the most severe hindrances for the success of the Semantic Web: how to create good quality ontological metadata cost-efficiently. To test this, FinnONTO ontologies and ONKI ontology services will be maintained on a regular basis in a living lab environment for organizations to use in a national follow-up project of FinnONTO: Semantic Web 2.0 (FinnONTO 2.0, 2008–2010).

Our work is funded by the National Funding Agency for Technology and Innovation (Tekes) and a consortium of 38 companies and public organizations.

References

1. Hyvönen, E., Viljanen, K., Mäkelä, E., Kauppinen, T., Ruotsalo, T., Valkeapää, O., Seppälä, K., Suominen, O., Alm, O., Lindroos, R., Känsälä, T., Henriksson, R., Frosterus, M., Tuominen, J., Sinkkilä, R., Kurki, J.: Elements of a national semantic web infrastructure—case study Finland on the semantic web (invited paper). In: Proceedings of the First International Semantic Computing Conference (IEEE ICSC 2007), Irvine, California, pp. 216–223. IEEE Press, Los Alamitos (2007)
2. Suominen, O., Viljanen, K., Hyvönen, E., Holi, M., Lindgren, P.: TerveSuomi.fi:n metatietomäärittely (Metadata schema for TerveSuomi.fi), Ver. 1.0 (26.1.2007) (2007), http://www.seco.tkk.fi/publications/
3. Viljanen, K., Tuominen, J., Känsälä, T., Hyvönen, E.: Distributed semantic content creation and publication for cultural heritage legacy systems. In: Proceedings of the 2008 IEEE International Conference on Distibuted Human-Machine Systems, Athens, Greece, IEEE Press, Los Alamitos (2008)
4. Valkeapää, O., Alm, O., Hyvönen, E.: Efficient content creation on the semantic web using metadata schemas with domain ontology services (System description). In: Franconi, E., Kifer, M., May, W. (eds.) ESWC 2007. LNCS, vol. 4519, Springer, Heidelberg (2007)
5. Valkeapää, O., Hyvönen, E., Alm, O.: A framework for ontology-based adaptable content creation on the semantic web. J. of Universal Computer Science (2007)

---

[33] http://www.yso.fi/

6. Mäkelä, E., Hyvönen, E., Saarela, S., Viljanen, K.: OntoViews—a tool for creating semantic web portals. In: McIlraith, S.A., Plexousakis, D., van Harmelen, F. (eds.) ISWC 2004. LNCS, vol. 3298, Springer, Heidelberg (2004)
7. Viljanen, K., Känsälä, T., Hyvönen, E., Mäkelä, E.: Ontodella—A projection and linking service for semantic web applications. In: Bressan, S., Küng, J., Wagner, R. (eds.) DEXA 2006. LNCS, vol. 4080, Springer, Heidelberg (2006)
8. Mäkelä, E., Hyvönen, E., Saarela, S.: Ontogator—a semantic view-based search engine service for web applications. In: Cruz, I., Decker, S., Allemang, D., Preist, C., Schwabe, D., Mika, P., Uschold, M., Aroyo, L.M. (eds.) ISWC 2006. LNCS, vol. 4273, Springer, Heidelberg (2006)
9. Hyvönen, E., Mäkela, E., Salminen, M., Valo, A., Viljanen, K., Saarela, S., Junnila, M., Kettula, S.: MuseumFinland—Finnish museums on the semantic web. Journal of Web Semantics 3(2), 224–241 (2005)
10. Hyvönen, E., Ruotsalo, T., Häggström, T., Salminen, M., Junnila, M., Virkkilä, M., Haaramo, M., Kauppinen, T., Mäkelä, E., Viljanen, K.: CultureSampo—Finnish culture on the semantic web. The vision and first results. In: Semantic Web at Work—Proceedings of STeP 2006, Espoo, Finland (2006); To appear in: Klaus Robering (ed.) Information Technology for the Virtual Museum. LIT Verlag, 2008
11. Ruotsalo, T., Hyvönen, E.: An event-based approach for semantic metadata interoperability. In: Aberer, K., Choi, K.-S., Noy, N., Allemang, D., Lee, K.-I., Nixon, L., Golbeck, J., Mika, P., Maynard, D., Mizoguchi, R., Schreiber, G., Cudré-Mauroux, P. (eds.) ISWC 2007. LNCS, vol. 4825, Springer, Heidelberg (2007)
12. Hyvönen, E., Viljanen, K., Suominen, O.: HealthFinland—Finnish health information on the semantic web. In: Aberer, K., Choi, K.-S., Noy, N., Allemang, D., Lee, K.-I., Nixon, L., Golbeck, J., Mika, P., Maynard, D., Mizoguchi, R., Schreiber, G., Cudré-Mauroux, P. (eds.) ISWC 2007. LNCS, vol. 4825, Springer, Heidelberg (2007)
13. Suominen, O., Viljanen, K., Hyvönen, E.: User-centric faceted search for semantic portals. In: Franconi, E., Kifer, M., May, W. (eds.) ESWC 2007. LNCS, vol. 4519, Springer, Heidelberg (2007)
14. Sidoroff, T., Hyvönen, E.: Semantic e-goverment portals—a case study. In: Proceedings of the ISWC 2005 Workshop Semantic Web Case Studies and Best Practices for eBusiness (SWCASE 2005) (2005)
15. Känsälä, T., Hyvönen, E.: A semantic view-based portal utilizing Learning Object Metadata. In: 1st Asian Semantic Web Conference (ASWC 2006). Semantic Web Applications and Tools Workshop (August 2006)
16. Laukkanen, M., Viljanen, K., Apiola, M., Lindgren, P., Mäkelä, E., Saarela, S., Hyvönen, E.: Towards semantic web-based yellow page directory services. In: Proceeedings of the Third International Semantic Web Conference (ISWC 2004) (November 2004)
17. Komulainen, V., Valo, A., Hyvönen, E.: A tool for collaborative ontology development for the semantic web. In: Proceedings of International Conference on Dublin Core and Metadata Applications (DC 2005) (November 2005)
18. Foskett, D.J.: Thesaurus. In: Encyclopaedia of Library and Information Science, vol. 30, pp. 416–462. Marcel Dekker, New York (1980)
19. Aitchison, J., Gilchrist, A., Bawden, D.: Thesaurus construction and use: a practical manual. Europa Publications, London (2000)
20. Leskinen, R.L. (ed.): Museoalan asiasanasto. Museovirasto, Helsinki, Finland (1997)
21. Hyvönen, E., Lindroos, R., Kauppinen, T., Henriksson, R.: An ontology service for geographical content. In: Poster Proc. of the ISWC + ASWC 2007, Busan, Korea (2007)

22. Kauppinen, T., Hyvönen, E.: Modeling and reasoning about changes in ontology time series. In: Kishore, R., Ramesh, R., Sharman, R. (eds.) Ontologies in the Context of Information Systems, Springer, Heidelberg (2007)
23. Kurki, J., Hyvönen, E.: Relational semantic search: Searching social paths on the semantic web. In: Poster Proc. of the ISWC + ASWC 2007, Busan, Korea (2007)
24. van Assem, M., Menken, M.R., Schreiber, G., Wielemaker, J., Wielinga, B.: A method for converting thesauri to RDF/OWL. Springer, Heidelberg (2004)
25. Gangemi, A., Guarino, N., Masolo, C., Oltramari, A., Schneider, L.: Sweetening ontologies with DOLCE. In: Proc. of the 13th International Conference on Knowledge Engineering and Knowledge Management. Ontologies and the Semantic Web, Springer, Heidelberg (2002)
26. Fellbaum, C. (ed.): WordNet. An electronic lexical database. The MIT Press, Cambridge (2001)
27. Guarino, N., Welty, C.: Evaluating ontological decisions with ONTOCLEAN. Comm. of the ACM 45(2) (2001)
28. Antoniou, G., van Harmelen, F.: A semantic web primer. The MIT Press, Cambridge (2004)
29. Hyvönen, E., Valo, A., Komulainen, V., Seppälä, K., Kauppinen, T., Ruotsalo, T., Salminen, M., Ylisalmi, A.: Finnish national ontologies for the semantic web—towards a content and service infrastructure. In: Proceedings of International Conference on Dublin Core and Metadata Applications (DC 2005) (2005)
30. Hameed, A., Preese, A., Sleeman, D.: Ontology reconciliation. In: Staab, S., Studer, R. (eds.) Handbook on ontologies, pp. 231–250. Springer, Heidelberg (2004)
31. Noy, N.F., Musen, M.A.: The PROMPT Suite: Interactive tools for ontology merging and mapping. International Journal on Digital Libraries 59(6), 983–1024 (2003)
32. Hyvönen, E., Lindroos, R., Kauppinen, T., Henriksson, R.: An ontology service for geographical content. In: Poster Proceedings of the 6th International Semantic Web Conference (ISWC + ASWC 2007), Busan, Korea, Springer, Heidelberg (2007)
33. Hyvönen, E., Salminen, M., Kettula, S., Junnila, M.: A content creation process for the Semantic Web. In: Proceeding of OntoLex 2004: Ontologies and Lexical Resources in Distributed Environments, Lisbon, Portugal (2004)
34. Hyvönen, E., Mäkelä, E.: Semantic autocompletion. In: Mizoguchi, R., Shi, Z.-Z., Giunchiglia, F. (eds.) ASWC 2006. LNCS, vol. 4185, Springer, Heidelberg (2006)
35. Maedche, A., Motik, B., Stojanomic, L., Studer, R., Volz, R.: An infrastructure for searching, reusing and evolving distributed ontologies. In: Proceedings of the 12th International Conference on World Wide Web (WWW 2003), Budapest, ACM Press, New York (2003)
36. Hildebrand, M., van Ossenbruggen, J., Amin, A., Aroyo, L., Wielemaker, J., Hardman, L.: The design space of a configurable autocompletion component. Technical Report INS-E0708, CWI, Amsterdam (November 2007)

Creating and Using Geospatial Ontology Time Series in a Semantic Cultural Heritage Portal

Tomi Kauppinen[1], Jari Väätäinen[2], and Eero Hyvönen[1]

[1] Semantic Computing Research Group (SeCo)
Helsinki University of Technology
and University of Helsinki, Finland
firstname.lastname@tkk.fi
http://www.seco.tkk.fi/
[2] Geological Survey of Finland
firstname.lastname@gtk.fi
http://www.gtk.fi

Abstract. Content annotations in semantic cultural heritage portals commonly make spatiotemporal references to historical regions and places using names whose meanings are different in different times. For example, historical administrational regions such as countries, municipalities, and cities have been renamed, merged together, split into parts, and annexed or moved to and from other regions. Even if the names of the regions remain the same (e.g., "Germany"), the underlying regions and their relationships to other regions may change (e.g., the regional borders of "Germany" at different times). As a result, representing and finding the right ontological meanings for historical geographical names on the semantic web creates severe problems both when annotating contents and during information retrieval. This paper presents a model for representing the meaning of changing geospatial resources. Our aim is to enable precise annotation with temporal geospatial resources and to enable semantic search and browsing using related names from other historical time periods. A simple model and metadata schema is presented for representing and maintaining geospatial changes from which an explicit time series of temporal part-of ontologies can be created automatically. The model has been applied successfully to represent the complete change history of municipalities in Finland during 1865–2007. The resulting ontology time series is used in the semantic cultural heritage portal CULTURESAMPO to support faceted semantic search of contents and to visualize historical regions on overlaying maps originating from different historical eras.

1 Introduction

Geospatial ontologies define classes and individuals for representing e.g. geographic regions, their properties, and mutual relationships. By sharing ontological resources in different collections and application domains, interoperability in terms of geographical locations can be obtained, and intelligent end-user services such as semantic search, browsing, and visualization be facilitated [9,24,23]. For example, in the semantic portal

MUSEUMFINLAND[1] [10] a location partonomy[2] was used for annotating museum artifacts with metadata about the place of manufacture and the place of usage.

A lesson learned during this work was that geography changes rapidly, which makes it hard 1) to the content annotator to make correct references to spatiotemporal regions and 2) to the end-user to understand the changes in historical geography and, as a result, to formulate the queries. For example, many artifacts in MUSEUMFINLAND originate from regions that no longer exist and/or have not been a part of Finland but of Russia after the Second World War. Finding the right names for querying, understanding to which regions on the map the names refer to at different times, and understanding how old historical names relate to modern Finnish and Russian geography creates, at the same time, both a semantic challenge for the technology and an important part of useful content to learn when using the portal.

This paper addresses two essential needs from the end-user point of view when using historical geographic regions in a cultural heritage portal:

Ontology-based spatiotemporal search. It is necessary to be able to use both historical and modern regions as search concepts e.g. in a view-based, or multi-facet search [7,20,18,6]. The idea is that regions offer one view to a content and they can be used to select a subset of the content by specifying constraints. For example, selecting "Finland (1945-)" from a facet view would refer to a part of Europe relating to the modern post-war Finland.

Visualization of concepts. It is necessary for the end-user to be able to see where the historical regions are on the map in a proper temporal context. Moreover, there should be a way of visualizing the spatial relationship between the old regions and the modern ones on the maps in order to relate history with the world of today. Creation of several layers of information on maps is a common way to visualize maps [3] and related content at the same time. In our case, we decided to apply this idea to overlaying historical and modern maps to visualize spatiotemporal relationships of regions, and to display related cultural content on the maps. Such map visualizations also help in finding the right concepts for a search and for presenting the search results.

To successfully meet the above needs the following requirements can be set for the ontology creation and management:

1. Concepts representing the regions from different time intervals need to be identified by URIs, put into a valid spatial part-of hierarchy, and mapped with each other in the temporal dimension.
2. Essential geographical properties, such as coordinates of points or polygonal boundaries, time span, size, and names of the historical regions, need to be assigned to the URIs of the regions.

To meet these requirements, it is essential that a geospatial ontology used in a semantic cultural system can represent change in time [15]. In current historical geo-vocabularies

---

[1] http://www.museosuomi.fi
[2] This partonomy is a part-of hierarchy of individuals of the classes Continent, Country, County, City, Village, Farm, etc.

and ontologies, such as the Getty Thesaurus of Geographic Names (TGN)[3], historical regions may be found, but the aspect of change is usually missing. For example, in the TGN the historical city of "Rome" in Italy has an entry as an inhabited place, but its development from an Etruscan city of the 8th century BC to its declination in 330 AD is described only as a piece of literal text.

In this paper, we present a simple metadata schema and a model for representing geospatial changes and for maintaining them as an RDF repository. A method for constructing a time series of geospatial, temporal ontologies (an ontology time series) from the filled metadata schema is discussed, and a reasoning mechanism to infer properties (size), relationships, and mappings between spatiotemporal regions is then presented. To test and evaluate the approach, the system was used in a case study of creating a complete model of the changes of the Finnish municipalities in 1865–2007. Finally, we present how the resulting ontology time series has been applied to creating intelligent services and map-based visualizations in the semantic cultural heritage portal "CULTURESAMPO—Finnish Culture on the Semantic Web"[4] [12].

2 Modeling Geospatial Changes

2.1 Analysis of Change Types

We analyzed the kinds of regional changes of municipalities in Finland[5] between years 1865 and 2007. Table 1 lists the change types found and their quantities.

Table 1. Different types of regional changes between 1865 and 2007 in Finland

Change type	Quantity
Establishment (A region is established)	508
Merge (Several regions are merged into one)	144
Split (A region is split to several regions)	94
Namechange (A region changes its name)	33
Changepartof (Annexed (to a different country))	66
Changepartof (Annexed (from a different country))	1
Changepartof (Region moved to another city or municipality)	256
Total sum	1102

An example of a merge is depicted in Figure 1. In the year 1922 Nummi-Pusula was formed via the unification of two former municipalities, Nummi and Pusula. This means that the old notions of Nummi and Pusula became obsolete after 1922, and the new concept of Nummi-Pusula was introduced.

In Figure 2, there is an example of a split. Pirkkala was split into two municipalities, Pohjois-Pirkkala and Etelä-Pirkkala in 1922. In Figure 3 there are two examples of

---

[3] http://www.getty.edu/research/tools/vocabulary/tgn/
[4] http://www.kulttuurisampo.fi
[5] As collected by the Geological Survey of Finland.

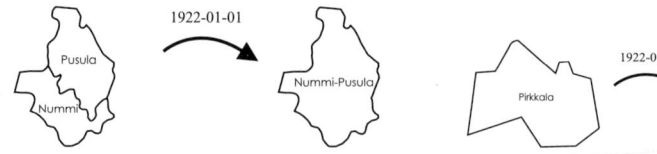
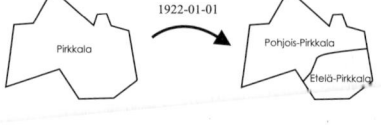

Fig. 1. An example of a merge

Fig. 2. An example of a split

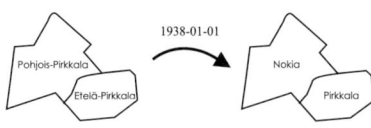
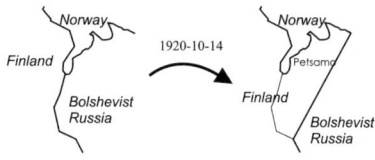

Fig. 3. Two examples of name changes

Fig. 4. An example of a change where a part of a region is moved

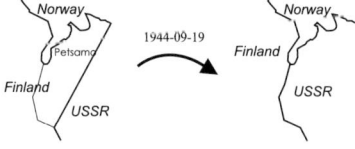
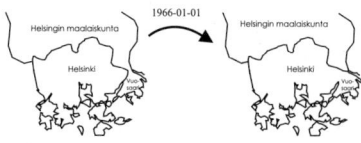

Fig. 5. Another example of a change in a partition hierarchy

Fig. 6. A third example of a change in a partition hierarchy

name changes. In year 1938 Pohjois-Pirkkala was renamed Nokia. At the same time, its neighbor Etelä-Pirkkala was renamed to Pirkkala. Finally, in Figures 4, 5 and 6 there are three different examples of changes in a partition hierarchy. Figure 4 depicts a change where Petsamo was annexed from Russia to Finland in 1920. Figure 5 depicts a change where Petsamo was annexed from Finland to USSR in year 1944. Finally, Figure 6 depicts a different change: Vuosaari was moved from Helsingin maalaiskunta to the city of Helsinki in year 1966.

These changes always change not only the sizes of the regions in question but also the partition hierarchy. This means that from year 1865[6] until 2007 there are 142 different kind of partition hierarchies of historical "Finlands".

Modeling all these different 142 temporal partition hierarchies of Finland, the resources and their mutual relationships, as separate ontologies by hand would be hard. Instead, we propose utilization of a simple schema for representing changes, and using an automated process for generating the different partition hierarchies.

2.2 A Schema for Representing and Maintaining Changes

The change types of Table 2 can be represented in terms of a few metadata fields (elements) listed in the Metadata Schema of Changes (see Table 2). An other metadata

---
[6] In year 1865 first municipalities were established in Finland.

Table 2. The Metadata Schema of Changes. Cardinalities are presented in the column C

Field	Definition	C	Value range
identifier	Identifier for a change (automatically generated)	1	Change ontology
date	The date of the change	1	W3CDTF (ISO8601)[7]
place	Place field	1	Location ontology
place type	The type of the place	1	Location ontology
change type	The type of the change	1	Change type ontology (either establishment, merge, split, namechange, or changepart, (see Fig. 1–6)
from	From where there are areas moving to in the change	1..*	Location ontology
to	To where there are areas moving to in the change	1..*	Location ontology
movedpart	Which part(s) are moving in the change (if they can be named) Note: Only used for changepartof	0..*	Location ontology
description	Description of a change	1	String

Table 3. The Metadata Schema of Current Places

Field	Definition	C	Value range
identifier	Identifier for a place	1	Location ontology
place name	Current place [8].	1	Location ontology
size	Size of the place in square kilometers	0..1	Double
partof	Which other administrational region this region is a part of	1	Location ontology
point	Representative point of the place	0..*	ISO 19107, WGS84
polygonal boundaries	Polygonal boundaries of the place	0..*	ISO 19107, WGS84
maps	Maps of the place	0..*	Map ontology

schema, the Metadata Schema of Current Places is meant for maintaining the contemporary places, like cities, municipalities and countries (see Table 3) and the Metadata Schema of Historical Places for properties such as boundaries of historical regions (see Table 4). Note that the last two schemas are very similar and could also be integrated.

Different fields of the Metadata Schema of Changes, such as *place*, *date*, *change type* and *from* and *to*-fields are filled up with the changes and resources they concern. For example, a change concerning the annexing of Petsamo from Finland to USSR on 1944-09-19 has an own instance conforming to the metadata schema, with the corresponding fields filled up (*from=Finland, to=USSR, movedpart=Petsamo, date=1944-09-19*, and so on). Notice that for each region modified by a change, a separate instance conforming to the metadata schema is created.

---
[8] For an implementation as an XML Schema Date, see http://www.w3.org/TR/xmlschema-2/

Table 4. The Metadata Schema of Historical Places

Field	Definition	C	Value range
identifier	Identifier for a place	1	Location ontology
place name	Place name	1..*	Location ontology
size	Size of the place in square kilometers	0..1	Double
partof	Which other administrational region this region is part of	0..1	Location ontology
point	Representative point of the place	0..*	ISO 19107, WGS84
polygonal boundaries	Polygonal boundaries of the place	0..*	ISO 19107, WGS84
measurement date	Date when the properties were valid (e.g. boundaries)	1	W3CDTF (ISO8601)
maps	Maps of the place	0..*	Map ontology

3 Creating an Ontology Time Series and Overlap Mapping

The previous section described three schemas used for creating an ontology time series. An ontology time series [15] defines a set of geospatial ontologies, including partonomy hierarchies for different time spans. This knowledge is represented in terms of RDF triples [1], where a resource (subject) is characterized by an identity (URI) and related property (predicate) values (object) in the form $< subject, predicate, object >$.

The following example motivates the creation of different temporal ontologies. Let us assume two RDF triples represented in a table, utilizing the name space *dcterms* of Dublin Core Metadata Terms [9] and another namespace *location*:

Subject	Predicate	Object
location:Monrepos	dcterms:isPartOf	location:Vyborg
location:Vyborg	dcterms:isPartOf	location:Russia

These triples could come from an RDF repository containing a traditional partonomy hierarchy that define the fact that the famous park Monrepos is a part of the city called Vyborg which in turn is a part of Russia. This is true for the ontology of the year 2007. However, the two RDF triples

Subject	Predicate	Object
location:Monrepos	dcterms:isPartOf	location:Viipuri
location:Viipuri	dcterms:isPartOf	location:Finland

define the historical fact that Monrepos is a part of Finland—this was true in 1921–1944. As we can see, these two sets of RDF triples would confuse a reasoner and the end-user, because *location : Monrepos* would be a part of two non-intersecting regions *location : Russia* and *location : Finland* (assuming that *dcterms:isPartOf* is transitive).

To overcome this problem, our ontology time series is populated with different *temporal parts* of places which are described by a metadata schema. Examples of temporal

---
[8] Present in year 2007 as of writing this paper.
[9] http://dublincore.org/documents/dcmi-terms/

parts of *location:Viipurin mlk* are *location:Viipurin mlk (1869-1905)*, *location:Viipurin mlk (1906-1920)*, *location:Viipurin mlk (1921-1944)* and *location:Vyborg(1944-)*. All these temporal parts have different polygonal boundaries, different sizes, and some of them are also in a different partonomy hierarchy. The ontology population process proceeds in the following way.

First, a place is created in the RDF repository (like *location:Viipurin mlk*). Based on the two sequential changes in the Metadata Schema of Changes for Viipurin mlk, that happened e.g. in 1906 and in 1921, a temporal part *location:Viipurin mlk (1906-1920)* is created and added to the union of [10] *location:Viipurin mlk*. Similarly, by examining the next two sequential changes concerning the place, additional temporal parts (like *location:Viipurin mlk (1921-1944)*) are created. If there are no more changes for that place, then the place has ceased to exist (like *location:USSR (1944-1991)*) or it is a contemporary one (like *location:Helsinki (1966-)*). Whether the place is a contemporary one is checked from the Metadata Schema of Current Places.

Second, the properties for the temporal parts of places are retrieved from the Metadata Schema of Current Places and from the Metadata Schema of Historical Places, depending whether the place in question is an existing one or has ceased to exist. This phase creates RDF triples representing, for example, the polygonal boundaries, the center point, the size, and partonomical relationships of the temporal part of the place. For example, two different partonomy hierarchies of our previous example of Monrepos is defined by four triplets

Subject	Predicate	Object
location:Monrepos(1921-1944)	dcterms:isPartOf	location:Viipuri(1921-1944)
location:Viipuri(1921-1944)	dcterms:isPartOf	location:Finland(1921-1944)
location:Monrepos(1991-)	dcterms:isPartOf	location:Vyborg(1991-)
location:Vyborg(1991-)	dcterms:isPartOf	location:Russia(1991-)

In addition, there are triples defining that different temporal parts of Monrepos belong to the same union of *location:Monrepos*, and triples defining different properties for temporal parts.

A temporal ontology [15] includes all temporal parts (of places) of some time span. For example, a temporal ontology of the year 1926 would include *location:Viipurin mlk (1921-1944)* because the year 1926 is within the range 1921–1944. Furthermore, the ontology contains all the partonomical relationships of those temporal parts that are valid during its time span.

Next, when all the places, their temporal parts and properties are created in the ontology time series, a model of changes is created based on the fields of the Metadata Schema of Changes. In each change there is something *before* the change (like *location:Viipurin mlk (1869-1905)*) and something *after* the change (like *location:Viipurin mlk (1906-1920)*). This is expressed with properties *before* and *after*. In practice, the following types of RDF triples are added to the repository:

---

[10] owl:unionOf is used, http://www.w3.org/TR/owl-guide/

Subject	Predicate	Object
change:change15	change:before	location:Viipurin mlk(1869-1905)
change:change15	change:after	location:Viipurin mlk(1906-1920)
change:change15	change:after	location:Nuijamaa(1906-1944)
change:change15	change:date	"1906-01-01"
change:change15	change:changetype	change:split

These triples are used as an input for a previously published method [14,15] to create *a global overlap table* between different temporal parts in the ontology time series. This table tells how much each place overlaps with the others. The repository is filled by following kind of triples based on the global overlap table calculation:

Subject	Predicate	Object
overlapping:overlap31	overlapping:overlaps	1.0
overlapping:overlap31	overlapping:overlappedBy	0.3131
overlapping:overlap31	overlapping:argument1	location:Viipurin mlk(1869-1905)
overlapping:overlap31	overlapping:argument2	location:Nuijamaa(1906-1944)

For example, since the size of Nuijamaa (1906-1944) is 407 square kilometers and the size of Viipurin mlk (1869-1905) is 1300 square kilometers, Nuijamaa overlaps Viipurin mlk by value $407/1300 = 0.3131$ and is overlappedBy by Viipurin mlk by value $407/407 = 1.0$ after the split (cf. the example above).

The Figure 7 illustrates the global overlap table by depicting overlaps with colors between a selected set of regions. The black color indicates a full 100% overlap between the temporal parts and the white color a 0% overlapping, accordingly. Different shades of grey indicate the level of overlapping: the darker the box, the greater is the overlap. From this illustration it is easy to see the mutual asymmetric overlaps between the temporal parts, and that the overlap-relation in this case is fairly complicated.

4 Creation of a Finnish Spatio-Temporal Ontology

The metadata schemas and methods described in the previous sections were implemented to create a *Finnish Spatio-temporal Ontology*, an ontology time series of Finnish municipalities over the time interval 1865–2007.

The metadata schemas were implemented as a spreadsheet tables[11] for easy editing. Figure 8 shows a screenshot of the Metadata Schema of Changes. Different schema fields, such as *place, date, change type, from*, and *to*, are filled up with resources and values. For example, the split of Viipurin mlk (1869-1905) into Nuijamaa (1906-1944) and Viipurin mlk (1906-1920) is seen on the row 1197, and the annexing of Viipuri from Finland to USSR on 1944-09-19 is on the row 1195. Most changes have a natural language explanation of the event.

The methods for creating an ontology time series from the metadata schemas were implemented using Java and Jena Semantic Web Framework[12] [2]. The resulting RDF

---
[11] We used the freely available OpenOffice Calc (http://www.openoffice.org/)
[12] http://jena.sourceforge.net/

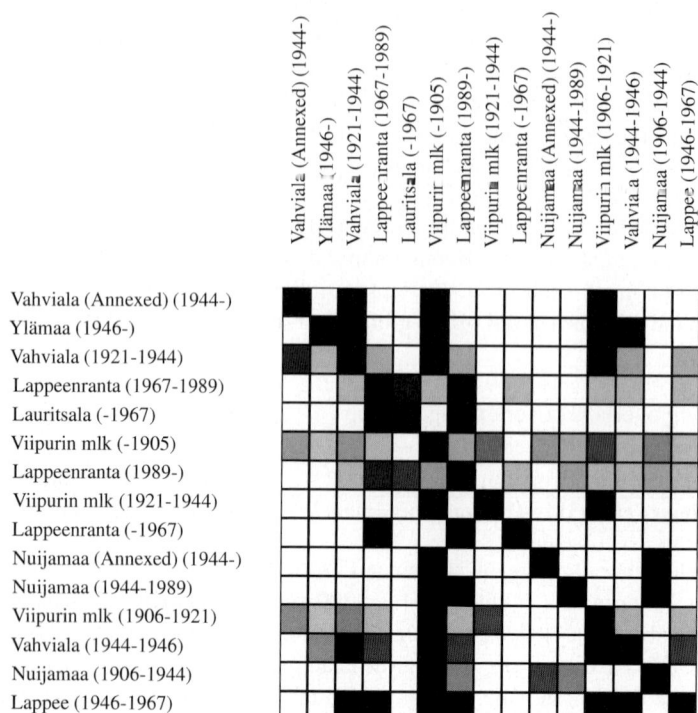

Fig. 7. Overlaps between temporal parts of places visualized using colored boxes. The black color indicates a full 100% overlap between the temporal regions and the white color a 0% overlap, accordingly. Different shades of grey indicate the level of overlap between regions: the darker the box, the greater is the overlap between the regions.

ontologytimeseries-2008-03-14.ods - OpenOffice.org Calc

File Edit View Insert Format Tools Data Window Help

	D	K	Y	Z	AA	AF	AI	AJ	
1	**Place**	**Date**	**Change type**	**From 1**	**From 2**	**Movedpart**	**To 1**	**To 2**	**Description**
1194	Viipuri	1403	Establishment				Viipuri		Viipuri sai ka
1195	Viipuri	1944-09-19	Changepart	Finland		Viipuri	USSR		Luovutetuilla
1196	Viipurin mlk	1869	Establishment				Viipurin mlk		
1197	Viipurin mlk	1906	Split	Viipurin mlk			Nuijamaa	Viipurin mlk	Nuijamaa its
1198	Viipurin mlk	1921	Split	Viipurin mlk			Vahviala	Viipurin mlk	
1199	Viipurin mlk	1944-09-19	Changepart	Finland		Viipurin mlk	USSR		Luovutetuilla
1200	Viitasaari	1868	Establishment				Viitasaari		Viitasaari pe
1201	Viitasaari	1934	Changepartof	Viitasaari		Viitasaari: Vuoksenkoski	Kannonkoski		Viitasaaresta
1202	Viljakkala	1874	Establishment				Viljakkala		Perustettu 18
1203	Viljakkala	2007	Merge	Viljakkala	Ylöjärvi		Ylöjärvi		Viljakkala yh
1204	Vilppula	1904	Changepartof	Keuruu		Suuri osa Keuruunkylää	Vilppula		
1205	Vilppula	1904	Changepartof	Ruovesi		"osia"	Vilppula		
1206	Vilppula	1922	Split	Vilppula			Mänttä	Vilppula	Mänttä erosi

Places / Areal sizes

Sheet 1 / 3 TAB_Taul1 100% STD * Sum=1906

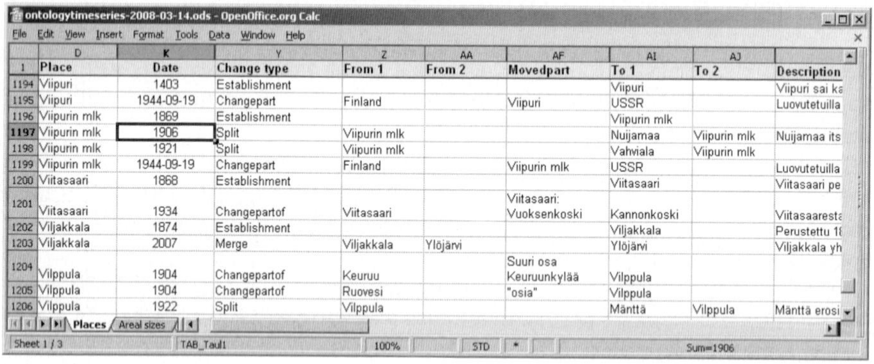

Fig. 8. A set of changes collected as a spreadsheet table

repository contains 1105 different changes and 976 different temporal parts of 616 different historical and modern places, meaning each place has on average 1.58 temporal parts. For example, the place resource *location:Viipurin mlk* got the temporal parts *location:Viipurin mlk (1869-1905)*, *location:Viipurin mlk (1906-1920)*, *Viipurin mlk (1921-1943)*, and *location:Viipurin mlk (1944-)*. The temporal parts and their partonomy hierarchies in the RDF repository constitute 142 different temporal ontologies between the years 1865 and 2007, each of which is a valid model of the country during its own time span.

5 Applications for Spatiotemporal Search and Visualization

Two case applications were created to utilize the resulting ontology time series in real application scenarios. The first one uses partition hierarchies of different time spans in faceted search facilitating ontology-based spatio-temporal search. Both historical and modern regions can be used as search categories. To illustrate this, in Figure 9 two categories corresponding to temporal parts of *location:Helsinki*, namely *location:Helsinki (1966-)* and *location:Helsinki (1640-1946)* are selected in a search facet, and related items from cultural collections are retrieved.

This functionality is included in the semantic CULTURESAMPO portal [12] that currently contains over 32 000 distinct cultural objects. The annotations of the objects were enriched automatically by comparing the time span and place of each annotation with those of the temporal parts of places. If they overlapped and place names matched, then the annotation was enriched accordingly. CULTURESAMPO also allows for searching with places on a map as illustrated in Figure 11. By clicking a place on a map, the items annotated with that place are retrieved and shown on the right side of the map. Furthermore, the user can formulate a search query as a polygon by pointing out *n* points on a map. All the places that have a point inside that polygon are retrieved and the content related to those places are listed on the right side of the page.

Our second application [13] utilizes the ontology time series in visualizing historical and modern regions on top of maps and satellite images. This answers to the need for visualizing spatiotemporal places: it is necessary for the end-user to be able see where the historical regions are on the map in a proper temporal context. Figure 12 illustrates the application. Historical regions, i.e. temporal parts of places, can be selected from a drop-down menu on the left. Here a temporal part *location:Viipuri(1920-1944)* of *location:Viipuri* is selected. As a result, the polygonal boundaries of Viipuri (1920–1944) are visualized on a contemporary Google Maps satellite image, map, or on a historical map. In addition, modern places from ONKI-Geo [11] that are inside the polygonal boundaries of the historical region are retrieved in a mash-up fashion, and can be used to browse the map. The content related to *location:Viipuri(1920-1944)* is listed in this view on the right. Furthermore, content from historical regions that overlap *location:Viipuri(1920-1944)* are listed as recommendations. The overlappings are looked up from the global overlap table.

Historical maps can be shown on top of the contemporary maps, as depicted in Figure 10. In the middle, a contemporary satellite Google Maps image of the city of Viipuri in the Karelia region is shown. In the middle, a smaller rectangular area

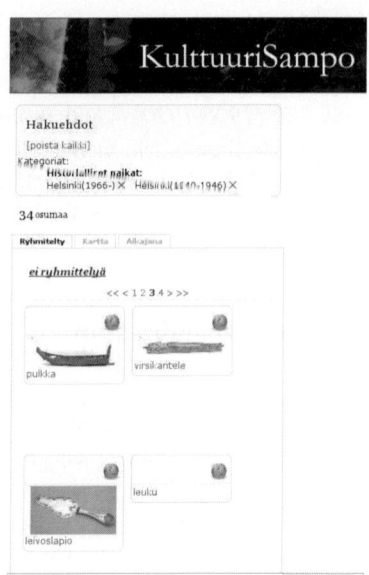

Fig. 9. Temporal parts of places used as a search constraint in CultureSampo

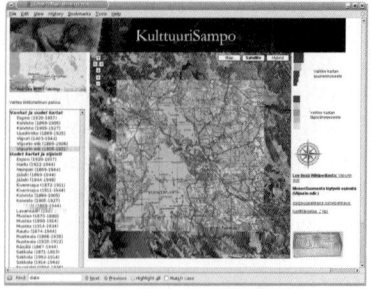

Fig. 10. Using multiple maps simultaneously. A historical Karelian map depicting the city of Viipuri is shown semi-transparently on top of a modern satellite image provided by the Google Maps service. Temporal parts of places on the left can be used to select different maps. The search for cultural artefacts can be constrained in this view by pointing our n points on a map.

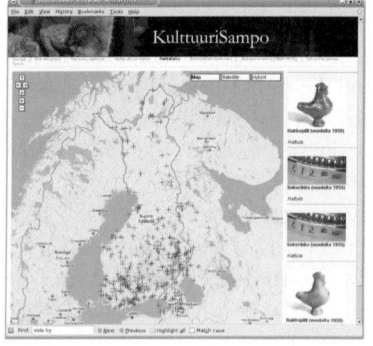

Fig. 11. A search with regions without temporal extensions

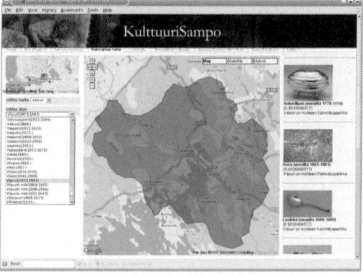

Fig. 12. Temporal parts are used to visualize polygonal boundaries of historical regions in CULTURESAMPO and for searching historical artifacts

is shown with a semi-transparent[13] old Karelian map that is positioned correctly and is of the same scale as the Google Maps image. This smaller view shows the old Viipuri, an old Finnish city that nowadays is a part of Russia. The place cannot be found in current maps as it was, which makes it difficult for modern users to locate the place geographically. In order to move around the user is able to use the zooming and

---

[13] We use transparency libraries provided by http://www.kokogiak.com/ which allow the alteration of the level of transparency.

navigation functions of Google Maps and the historical view is automatically scaled and positioned accordingly.

To provide the historical maps, we used a set of old Finnish maps from the early 20th century covering the area of the annexed Karelia region before the World War II. The maps were digitized and provided by the National Land Survey of Finland[14]. In addition, a geological map of the Espoo City region in 1909, provided by the Geological Survey of Finland[15], was used. This application is also included in the CULTURE-SAMPO portal [12].

6 Conclusions

6.1 Contributions

This paper presented an analysis of change types in historic regions, a model of changes based on the analysis and an ontology time series from the model, and a practical tool for maintaining the RDF repository of changes. We have also succesfully applied an existing method [15] to create a global overlap table from the repository of changes. We have evaluated the usability of the resulting ontological structure—the ontology time series—in two real life applications for information retrieval and for visualization in a semantic cultural heritage portal.

These applications can be used for teaching where historic regions have been and how they are related with each other in a partonomy hierarchy. The visualization is made using a rich set of historic maps, modern maps, satellite images, and polygonal boundaries. In addition, the applications can be used for retrieving historical cultural content related to the regions. The relationship is explicated for the user indicating whether the content has been found, used, manufactured, or located in a specific region.

Old maps and names on them could be of substantial benefit when using visualization in annotating or searching content in cultural heritage systems. The idea of using overlaid transparent maps is useful when comparing geo-information from different eras (e.g., how construction of cities has evolved) or from different thematic perspectives (e.g., viewing a geological map on top of a satellite image). We believe that map-based views of historic locations together with rich, precisely, and spatio-temporally annotated cultural content offer a good use case of semantic web technologies for solving real life interoperability and information retrieval problems.

6.2 Related Work

Traditions in ontology versioning [17] and ontology evolution [19] are interested in finding mappings between different ontology versions, doing ontology refinements and other changes in the conceptualization [16,22], and in reasoning with multi-version ontologies[8]. In ontology mapping research, there have been efforts to do mappings based on probabilistic frameworks [21]. Means for handling inconsistencies between ontology versions [5] have been developed. Methods for modeling temporal RDF have been proposed recently [4].

---

[14] http://www.maanmittauslaitos.fi/default.asp?site=3
[15] http://en.gtk.fi

In contrast to these works, our approach is merely about the evolution of ontology time series that is due to changes in the underlying domain. Hence it should not be confused with ontology versioning, database evolution, or ontology evolution even if changes are considered in all of these approaches as well. Each temporal member ontology in a time series is a valid, consistent model of the world within the time span it concerns, and may hence be used correctly in e.g. annotation.

6.3 Future Work

In the future, we would like to investigate whether the methods and tools presented in this paper could be generalized to other domains, where concepts overcome changes affecting their extensions, properties, or positions in ontological hierarchies and structures.

Acknowledgements

Our research is a part of the National Semantic Web Ontology Project in Finland[16] (FinnONTO, 2003–2007 and 2008–2010) funded by the Finnish Funding Agency for Technology and Innovation (Tekes) and a consortium of 38 companies and public organizations.

References

1. Brickley, D., Guha, R.V.: RDF Vocabulary Description Language 1.0: RDF Schema W3C Recommendation, February 10, 2004. Recommendation, World Wide Web Consortium (February 2004)
2. Carroll, J.J., Dickinson, I., Dollin, C., Reynolds, D., Seaborne, A., Wilkinson, K.: Jena: Implementing the semantic web recommendations. Technical Report HPL-2003-146, HP Labs (December 24, 2003)
3. de Berg, M., van Kreveld, M., Overmars, M., Schwarzkopf, O.: Computational geometry: algorithms and applications, 2nd edn. Springer, New York (2000)
4. Gutierrez, C., Hurtado, C., Vaisman, A.: Temporal RDF. In: Gómez-Pérez, A., Euzenat, J. (eds.) ESWC 2005. LNCS, vol. 3532, pp. 93–107. Springer, Heidelberg (2005)
5. Haase, P., van Harmelen, F., Huang, Z., Stuckenschmidt, H., Sure, Y.: A framework for handling inconsistency in changing ontologies. In: International Semantic Web Conference, Galway, Ireland, November 6-10, 2005, pp. 563–577. Springer, Heidelberg (2005)
6. Hearst, M., Elliott, A., English, J., Sinha, R., Swearingen, K., Lee, K.-P.: Finding the flow in web site search. CACM 45(9), 42–49 (2002)
7. Hildebrand, M., van Ossenbruggen, J., Hardman, L.: /facet: A browser for heterogeneous semantic web repositories. In: The Semantic Web - Proceedings of the 5th International Semantic Web Conference 2006, November 5-9, 2006, pp. 272–285 (2006)
8. Huang, Z., Stuckenschmidt, H.: Reasoning with multi-version ontologies: A temporal logic approach. In: International Semantic Web Conference, pp. 398–412 (2005)
9. Hyvönen, E.: Semantic portals for cultural heritage. In Manuscript draft for a book chapter (September 2007)

---

[16] http://www.seco.tkk.fi/projects/finnonto/

10. Hyvönen, E., Junnila, M., Kettula, S., Mäkelä, E., Saarela, S., Salminen, M., Syreeni, A., Valo, A., Viljanen, K.: Finnish Museums on the Semantic Web. User's perspective on museumfinland. In: Selected Papers from an International Conference Museums and the Web 2004 (MW 2004), Arlington, Virginia, USA (2004)
11. Hyvönen, E., Lindroos, R., Kauppinen, T., Henriksson, R.: An ontology service for geographical content. In: Poster+Demo Proceedings, ISWC+ASWC-2007, Busan, Korea (2007)
12. Hyvonen, E., Ruotsalo, T., Häggström, T., Salminen, M., Junnila, M., Virkkilä, M., Haaramo, M., Kauppinen, T., Mäkelä, E., Viljanen, K.: CultureSampo—Finnish culture on the Semantic Web: The vision and first results. In: Semantic Web at Work — Proceedings of the 12th Finnish Artificial Intelligence Conference STeP 2006, Helsinki, Finland, vol. 1 (2006)
13. Kauppinen, T., Deichstetter, C., Hyvönen, E.: Temp-o-map: Ontology-based search and visualization of spatio-temporal maps. In: Franconi, E., Kifer, M., May, W. (eds.) ESWC 2007. LNCS, vol. 4519, Springer, Heidelberg (2007)
14. Kauppinen, T., Hyvönen, E.: Modeling coverage between geospatial resources. In: Posters and Demos of the 2nd European Semantic Web Conference ESWC 2005. European Semantic Web Conference, Heraklion, Greece, May 29 - June 1 (to appear, 2005)
15. Kauppinen, T., Hyvönen, E.: Ontologies: A Handbook of Principles, Concepts and Applications in Information Systems. In: Modeling and Reasoning about Changes in Ontology Time Series, Springer, Heidelberg (2007)
16. Klein, M.: Change Management for Distributed Ontologies. PhD thesis, Vrije Universiteit Amsterdam (August 2004)
17. Klein, M., Fensel, D.: Ontology versioning on the Semantic Web. In: Proceedings of the International Semantic Web Working Symposium (SWWS), Stanford University, California, USA, July 30–August 1, 2001, pp. 75–91 (2001)
18. Mäkelä, E., Hyvönen, E., Sidoroff, T.: View-based user interfaces for information retrieval on the semantic web. In: Proceedings of the ISWC 2005 Workshop End User Semantic Web Interaction (November 2005)
19. Noy, N., Klein, M.: Ontology evolution: Not the same as schema evolution. Knowledge and Information Systems 5 (2003)
20. Oren, E., Delbru, R., Decker, S.: Extending faceted navigation for rdf data. In: International Semantic Web Conference, November 5-9, 2006, pp. 559–572 (2006)
21. Pan, R., Ding, Z., Yu, Y., Peng, Y.: A bayesian network approach to ontology mapping. In: International Semantic Web Conference 2005, Galway, Ireland, November 6-10, 2005, pp. 563–577. Springer, Heidelberg (2005)
22. Stojanovic, L.: Methods and Tools for Ontology Evolution. PhD thesis, University of Karlsruhe, Germany (2004)
23. Stuckenschmidt, H., Harmelen, F.V.: Information Sharing on the Semantic Web. Springer, Heidelberg (2004)
24. Visser, U.: Intelligent information integration for the Semantic Web. Springer, New York (2004)

Semantic Email as a Communication Medium for the Social Semantic Desktop

Simon Scerri, Siegfried Handschuh, and Stefan Decker

Digitial Enterprise Research Institute,
National University of Ireland Galway,
IDA Business Park, Galway, Ireland
{Simon.Scerri,Siegfried.Handschuh,Stefan Decker}@deri.org

Abstract. In this paper, we introduce a formal email workflow model based on traditional email, which enables the user to define and execute ad-hoc workflows in an intuitive way. This model paves the way for semantic annotation of implicit, well-defined workflows, thus making them explicit and exposing the missing information in a machine processable way. Grounding this work within the Social Semantic Desktop [1] via appropriate ontologies means that this information can be exploited for the benefit of the user. This will have a direct impact on their personal information management - given email is not just a major channel of data exchange between desktops, but it also serves as a virtual working environment where people collaborate. Thus the presented workflow model will have a concrete manifestation in the creation, organization and exchange of semantic desktop data.

Keywords: Semantic Email, Workflow Patterns, Speech Act Theory, Semantic Web, Social Semantic Desktop.

1 Introduction

Despite sophisticated collaboration environments being around for a long time already, email is still the main means for distributed collaboration. People still use it for maintaining to-do lists, tracking, documentation, organization, etc. The major reasons for this may be grounded in the ease of use, the negligible learning effort to be able to use it, the universal availability, and that literally everyone on the Internet uses it and can be contacted via email. Email is the social communication backbone of the Internet and is a great example of the fact that often not the number of functionalities decide on the success of a technology, but its flexibility and how people use it (a similar example from the telecom world, would be text messaging which already produces larger revenues than other mobile services).

However, the many ways in which people use email are not well supported as the uses of email are beyond its original intended design [2]. Functionalities such as deadlines and reminders, tasks and task tracking, prioritizing etc. are missing or only supported to a limited degree. Emails in a conversation and their content are related by mere identifiers or mail subjects and simple extensions ("Re:", "Fwd:", etc.). What would actually be needed would be support for simple action item management, i.e., task definition, time-lines, dependencies of tasks, checks if the action items have been completed, etc., in a seamless way. This could well be achieved by identifying and placing patterns of communication into a structured form.

From a technical perspective this can be seen as support for ad-hoc workflows [3], i.e., it helps users coordinate and track action items that involve multiple steps and users. Users can leverage ad-hoc workflows to better manage less-structured processes in a better way than via traditional rigid workflow systems. In this paper, we introduce a workflow model, based on traditional email, which enables the user to define and execute ad-hoc workflows in an intuitive way. By eliciting and semantically annotating implicit, well-defined workflows we collect missing information in a semantic, i.e., machine processable way. By grounding this work within the Semantic Desktop through ontologies, this information can be exploited for the benefit of the user via a tool that supports the user with personal information management.

In the Semantic Desktop paradigm [1], personal information such as address book data, calendar data, email data, folder structures and file metadata, etc. is lifted onto an RDF representation. Hence, information items on the desktop are treated as Semantic Web resources and ontologies and vocabularies allow the user to express such desktop data formally. The NEPOMUK Project developed an infrastructure of ontologies for this purpose[1], including NRL - a representational language [4] resembling RDF/S best practice. Other ontologies tackle different aspects of the semantic desktop. One such ontology in development is the Personal Information Model Ontology (PIMO) [5], which acts as a formal representation of the structures and concepts within a knowledge workers mental model. An instance of this model will include representations and inter-relationships between information elements on the user's desktop. The Information Element Ontologies provide a basis for these representations and are able to represent anything from files (NFO), contacts (NCO), calendar data (NCal) and so forth. The social aspect of the semantic desktop is dependent on data exchange and communication. Messages exchanged between desktops are rarely self-contained, and commonly refer to items in the user's PIMO like people, projects, events and tasks. The purpose of the NEPOMUK Message Ontology (NMO) is to introduce the message concept to the semantic desktop so that these links can be made explicit. These messages are exchanged between desktops within a workflow. Thus, the workflow model will have a concrete manifestation in the creation and organisation of semantic desktop data.

Our workflow model supports standard workflow patterns, i.e., templates for typical interactions, as defined by van der Aalst et al.[2] with a standard notation and well defined operational semantics. This semantics lays the foundation for automatic execution (this is beyond the scope of this paper). While our model incorporates a set of such templates, this template library can be extended by the user. Semantic annotations in conjunction with workflows not only support the user, but in the long-run also can be used in business environments for semantic business process mining, i.e., the identification of hidden, currently unsupported workflows and workflow consistency and compliance checking.

---

[1] http://www.semanticdesktop.org/ontologies/
[2] http://www.workflowpatterns.com/

The paper is organized as follows. In Section 2 we discuss related work. In Section 3 we present the main contribution of this work - a workflow model that supports and gives semantics to Email ad-hoc workflows. In Section 4 we provide a proof of concept for our work by applying our model to four common email workflows. We also introduce our prototype Semantic Email application. Finally in Section 5 we give an outline of our future work before concluding.

2 Related Work

The problems of Email and its impact on personal information management has been discussed a large number of times, most notably by Whittaker et.al.[2][6]. Giving the Email process a semantics via some formal standard would enable machines to understand it. As a result, machines could guide and support the user in the communicative process, automate the flow between trivial states of the communication, and in general participate actively in the communication alongside the user. This idea was extensively dealt-with in the research which first introduced the notion of Semantic Email [7]. Here a broad class of Semantic Email Processes was defined – to manage simple but tedious processes that are handled manually in non-semantic email. An email process was modelled as a set of updates to a dataset, via logical and decision-theoretic models. This research is suitable for a number of processes that are commonly executed in email collaboration. The processes need to be pre-meditated and have to follow fixed templates for each individual process (e.g. a template for meeting scheduling). However, email is frequently multi-purpose, and thus multiple processes should be able to execute within one email. A template-only based system would only support one process per email. Our rationale favours a system where the nature of the ad-hoc workflow can be elicited from the text of an email in a new thread. Workflows defined via a template selection would then complement processes detected on-the-fly. In [7] the possibility of establishing interactions between semantic email and other semantic web applications to support more sophisticated reasoning techniques was pointed out. Grounding Semantic Email within the Semantic Desktop enables the realization of this prospect.

In order to be able to elicit the nature of a process from the email content, an improved semantic email system needs to resort to some linguistic analysis that is, at least partially, automatic. A relevant standard theory on which this analysis can be based is Speech Act Theory [8]. The theory has been very influential in modeling electronic communicative patterns. The core idea behind is that every explicit utterance (or text) corresponds to one or more implicit actions. The theory was applied to the Email domain a number of times: in particular to ease task management for email-generated tasks [9][10] as well as for email classification [11][12][13]. In previous related work, efforts in this area where aligned with the concept of semantic email. This resulted in sMail - a conceptual framework for Semantic Email Communication based on Speech Act Theory [14]. This conceptual framework is also influenced by research that investigated the sequentiality of speech acts in conversational structure [15] as well as their fulfillment[16]. In this paper we will model the workflow of email collaboration based on this framework, using formal workflow patterns that serve as a conceptual basis for process technology.

The main contributions of this framework are the definitions and conceptualization of the *Email Speech Act Model* and the *Email Speech Act Process Model*. An Email Speech Act is defined as a triple (v,o,s) - where v denotes a *verb*, o an *object* and s a *subject*. The Email Speech Act Model contains instances for these speech act parameters. Verbs (*Request, Commit, Propose, Suggest, Deliver, Abort and Decline*) define the action of the speech act. The Object refers to the object of the action. Objects are classified in *Data* - representing something which can be represented within email (*Information, Resource, Feedback*); and *Activities* - representing external actions occurring outside of email (*Task, Event*). The subject is only applicable to speech acts having an activity as their object, and it represents who is involved in that activity – i.e., the *Initiator* (e.g. "Can I attend?"), the *Participant* (e.g. "You will write the document") or *Both* (e.g. "Let's meet tomorrow").

The Email Speech Act Process Model considers each speech act as a separate process. In essence, it outlines the expected reaction from both initiator and participant of a speech act, on sending it and on receiving it respectively. It assigns the *Initiator Expected Reaction* [IEA] and the *Participant Expected Reaction* [PER] to each speech act combination, and is applied to the Speech Act Model as (v,o,s) [IEA] → [PER]. The IEA refers to the status or action of the speaker, or in this case, the initiator on sending a speech act (*Expect, None*). The PER refers to the reaction expected from the hearer, or in this case the participant upon receiving and acknowledging a speech act (*Reply, Perform, None*).

3 A Behavioural Model for the Email Process

The sMail Conceptual Framework [14] on which we base this work refers to an Email Speech Act Workflow (ESAW) which models sequences of speech acts within email messages in email threads. Thus this model is equivalent to the *Email Speech Act Process Flow Model*. The ESAW was however never formally defined. In this work we explicitly model it in a standardised workflow language. Its workflow patterns can

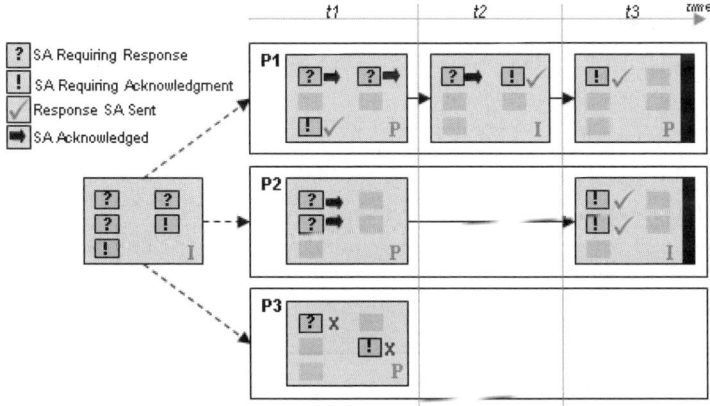

Fig. 1. Email Breakdown into seven ESAWs within three 1-1 transactions

be given semantics through their translation to YAWL[3] and subsequently Petri Nets[4]. Given this semantics, Semantic Email is not merely fancy email that is able to transport semantic content via Semantic Web technologies. The Email process itself is given semantics, by breaking it down into a number of speech act processes, each of which can execute according to a formal workflow.

In order to model our ESAW, every email is conceptually broken down into a number of 1-1 transactions between the *Initiator* and the *Participant* agents. The initiator refers to the agent that starts the email thread, whereas the participant is any other agent implicated in the email. Both agents can play the role of a sender or a recipient of an email in a thread in their own time. If the initiator addresses an email to n participants, then this amounts to n transactions. Each transaction can have zero or more speech acts which can be modeled using n distinguishable ESAWs. This concept is demonstrated in Fig. 1 – a timeline demonstrating how an email is broken down into separate, independent 1-1 transactions (cases where these transactions are not independent will be discussed later). Although the email received by the three participants (P1, P2, P3) is identical, the speech acts in the email are addressed to different participants. The first transaction between the Initiator and P1 consists of three speech acts (i.e. 3 out of the 5 speech acts are addressed to P1). Each speech act can execute along an independent ESAW. At time interval $t1$, P1 terminates one of these workflows, but sends control back to the initiator for the other two workflows. The activities in Fig. 1 are marked P or I depending on which agent has control. At time $t2$, the initiator terminates the second of these two, and returns control to P1 for the first. At time $t3$ P1 terminates the last ESAW and as a direct result this 1-1 transaction terminates. The same happens with the transaction between the initiator and P2, although at a different time. However the third transaction does not terminate by time $t3$ since P3 has stalled both ESAWs by not reacting to them.

The ESAW we present is grounded on key research in the area of control flow workflow patterns. The Workflow Patterns Initiative [3] was established with the aim of delineating the fundamental requirements that arise during business process modelling on a recurring basis and describe them in an imperative way. The contribution of this work is a set of patterns describing the control-flow perspective of workflow systems. Our ESAW uses 9 patterns out of the 40 documented patterns. Table 1 enlists these patterns (excluding the 10th which is our customized version of the 1st) together with a short description[5] and the graphical notation, used based on UML 2.0 Activity Diagram notations[6].

The workflow is graphically represented in Fig. 2. A swimlane splits the figure vertically to distinguish between the initiator and the participant(s) in separate 1-1 transactions. Whereas the initiator is always the agent that starts the workflow, its termination can depend on the initiator, the participant, or both. This is reflected in the workflow figure given the termination symbol is outside of both agent spaces. We will now give a brief walkthrough description of the workflow. The workflow will be further explained through a number of Semantic Desktop use-cases in Section 4.

---

[3] http://www.yawl-system.com/
[4] http://www.informatik.uni-hamburg.de/TGI/PetriNets/
[5] The given descriptions are neither formal nor complete. Refer to [3] for the complete specifications.
[6] http://www.uml.org/

Table 1. Patterns used in the ESAW

Name	Description	Graphical Notation
Exclusive Choice (XOR-split)	The thread of control is immediately passed to exactly one outgoing branch based on the outcome of a logical expression associated with each.	
Simple Merge (XOR-join)	Convergence of two or more exclusive branches into a single branch. The active incoming branch passes control over to the outgoing branch.	
Multi-Choice (OR-split)	The thread of control is passed to one or more of the outgoing branches based on the outcome of logical expressions associated with each.	
Parallel Split (AND-split)	Branch diverges into two or more parallel branches each of which executes concurrently.	
Structured Synchronizing Merge	The thread of control is passed to the subsequent branch when each active incoming branch has been enabled.	
Multi-Merge	The thread of control is passed to the subsequent branch immediately when just one of the active incoming branches is enabled.	
Structured Loop: Post Test	Executes an activity or sub-process repeatedly. Post-test is evaluated at the end of the loop to determine whether it should continue.	
Recursion	The ability of an activity to invoke itself during its execution or an ancestor in terms of the overall decomposition structure with which it is associated.	
Persistent Trigger	An activity is triggered by a signal from the external environment. These triggers are persistent and are retained by the workflow until they can be acted on.	
Simplified XOR-Split	The majority of exclusive choices in the workflow have two common choices. These are abstracted with this symbol and expanded later in Fig. 2.	

An agent can initiate a speech act sequence by sending one of the following speech acts (shown below with their brief description and their intended effect).

1. (Suggest, Activity, *): Activity suggestion – No reply or action required
2. (Deliver, Data, Ø): Deliver unrequested data – No reply or action required
3. (Propose, Activity, *): Dependent (1-n) – n replies required
4. (Abort, Activity, *): Notification of an aborted activity – Action possible
5. (Commit, Activity, *): Notification of a commitment – Action possible
6. (Request, Activity, *): Independent (1-1) – Reply required, action possible
7. (Request, Data, Ø): Data request – A reply is required

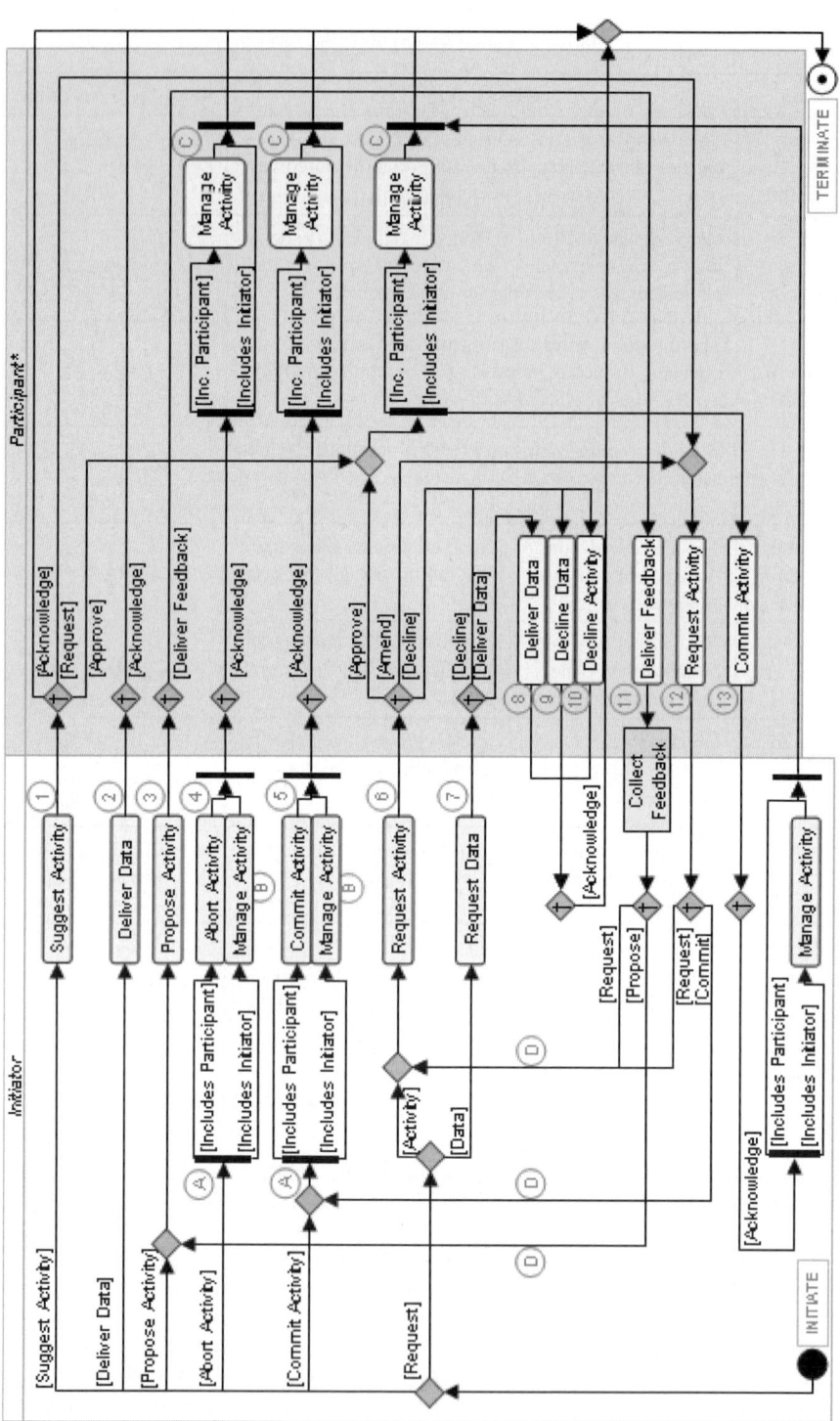

Fig. 2. Speech Act Process Flow Model

– where '*' denotes any subject (i.e. the event or task implicates the initiator, participant or both). These seven speech acts are marked 1-7 in Fig. 2. The ESAW shows that the choice of some of these initiative speech acts (marked A), namely (Abort, Activity, *) and (Commit, Activity, *) results in a multi-choice split. If the subject of the activity includes the initiator (e.g. "I will do something!") the respective path is executed. If it includes the participant (e.g. "You will do something!") the alternative path executes. If both are included (e.g. "We will do something!"), both paths execute simultaneously. If the activity implicated an action by the participant, the process continues simply by sending the speech act. If the activity implicated an activity by the initiator, the initiator is expected to manage the generated activity (marked B), e.g. represent it on the semantic desktop as an event or task.

On receiving a speech act (grey half of Fig. 2), the participant is presented with a choice of activities according to the type of speech act received. Whereas some of these choices lead to the termination of the ESAW, e.g. 'Acknowledge', others execute a more interesting path. Some lead to activity management for the participant (marked C), whereas a number of paths make it necessary for the participant to send one of six reply speech act to the initiator. These six speech acts are marked 8-13 in Fig. 2. A brief description of these reply speech acts and of their context (in reply to [i.r.t]) is given below:

8. (Deliver, Data, Ø): Deliver requested data – i.r.t. request for data
9. (Decline, Data, Ø): Decline requested data – i.r.t. request for data
10. (Decline, Activity, *): Decline participation – i.r.t. an activity request
11. (Deliver, Feedback, Ø): Deliver feedback – i.r.t. activity proposal
12. (Request, Activity, *): Same as before – amending an activity request
13. (Commit, Activity, *): Same as before – i.r.t. an activity request

On sending these reply speech acts, control is returned to the initiator (white half of Fig. 2). At this point, some paths lead to the termination of the workflow (e.g. via Acknowledge) whereas other paths restart the loop by returning control to the participant via another speech act process (marked D).

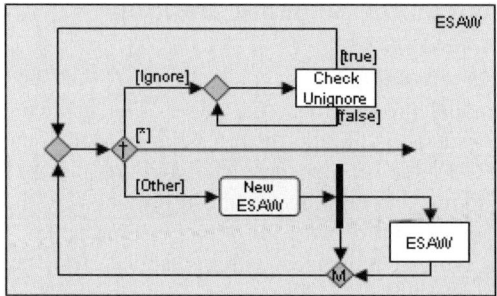

Fig. 3. *Other* and *Ignore* choices in simplified XOR-split

As we mentioned earlier, the workflow in Fig. 2 has been simplified. In reality, each exclusive choice that an agent must consider after receiving a speech act (simplified XOR-split in Table 1) has two extra choices – 'Ignore' and 'Other'. Fig. 3 shows the behaviour of these choices.

The 'Ignore' path leads to a structured loop pattern that uses a post test which continuously checks for a reactivation signal. If this is detected it leads back to the start of the subactivity. The 'Other' path is more interesting since it uses the recursion pattern. Given this path is enabled, a new ESAW is initiated as a sub-workflow of the current ESAW. In the meantime, control is returned to the start of the subactivity via a structured synchronizing merge. This means that via the 'Other' path, n sub-workflows can be initiated. The parent ESAW does not need any of these sub-workflows to terminate in order to regain control. However, it does need all n sub-workflows which branched-off to terminate in order to terminate itself.

We also mentioned that there are cases when a speech act addressed to n participants does not branch in n simultaneous independent workflows. This is the case with speech acts with the *Propose* verb. In this case, the workflow will stall until the initiator gets the responses of all n participants. In Fig. 2 this behaviour is represented with a 'Collect Feedback' activity. This sub-activity is expanded in Fig. 4. Here we see that when each of the n participants delivers the required feedback to the initiator, a signal is fired. On the initiator's side, each time a signal fires a post-check structured loop checks whether all participants have submitted their feedback. If this is the case the initiator can decide on which action to take, depending on the desired level of consensus in the cumulative feedback.

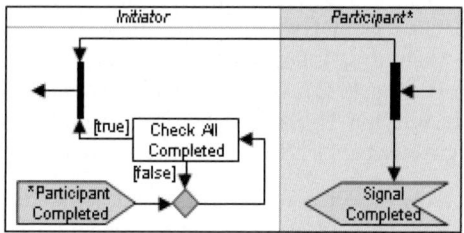

Fig. 4. *Propose* Speech Acts

4 Semantic Email on the Semantic Desktop

We have implemented a prototype of an application [17] that supports semantic email and the workflow model presented in this paper. The application, called *Semanta* (Semantic Email Assistant), is an add-in to a popular Mail User Agent[7]. The sMail models discussed in 2.3 have been encapsulated and are available within a Semantic Email Ontology[8]. Linking this ontology to those provided by the Semantic Desktop was a step towards email and desktop information unification. *Semanta* uses these ontologies to enhance email messages with semantics. Annotation of the email is performed both automatically via a speech act extraction web service, and manually via an appropriate wizard. The metadata is transported alongside the email content via a specific RDF MIME extension in the email headers. In this section we will provide

---
[7] Currently only Microsoft Outlook is supported but an add-in for Mozilla Thunderbird is in the pipeline.
[8] http://smile.deri.ie/ontologies/smail.rdf

a proof of concept for our ESAW and demonstrate how it can benefit the social aspect of the social semantic desktop. We will do this via four common use-cases that take place between the semantic desktops of four employees in a hypothetical company. The employees communicate via email, supported by *Semanta*. The employees have different levels of authority - Marilyn is a director, James a manager whereas Frank and Grace are regular employees.

4.1 Data Request

Marilyn requires a recent photograph of two employees - James and Frank. She selects an email Request Template where she asks the two contacts to deliver her the required data – a resource. The email has one speech act, a (Request, Resource, Ø), addressed to both participants. As shown in Fig. 5 at time interval $t1$, the speech act is broken down into two separate processes, which follow the ESAW independently:

i. P1 (Marilyn-Frank): When Frank (P) receives the email from Marilyn (I) at time $t1$, *Semanta* will recognize its purposes and will mark it as having pending action items until Frank follows a path in the ESAW which terminates the workflow. When Frank, clicks on the annotated speech act *Semanta* presents him the following choices: *Deliver, Decline, Ignore* and *Other*, as defined in the ESAW. He selects the Deliver option and, supported by the knowledge stored in his PIMO, he finds an appropriate photo in seconds. On selecting it, *Semanta* generates an automatic reply email. This is locally linked to the previous email in the thread, as well as to the instance of the photo in Frank's PIMO. Some auto-generated text in the email is automatically annotated with the speech act (Deliver, Resource, Ø). Frank does not need to add anything to the conversation so *Semanta* sends the reply. When Marilyn receives the reply at $t2$, *Semanta* shows her the following options: *Acknowledge, Ignore, Other*. She acknowledges the receipt and given the nature of the speech act, *Semanta* prompts her to save the file. When she does, an instance of the file is created in her PIMO and linked to the email message that introduced it to her desktop. This workflow instance terminates.

ii. P2 (Marilyn-James): At time interval $t1$, James does not have a more recent photo available. He selects the Decline option, and *Semanta* prompts him to provide the justification as input text to an automatic reply email. This is

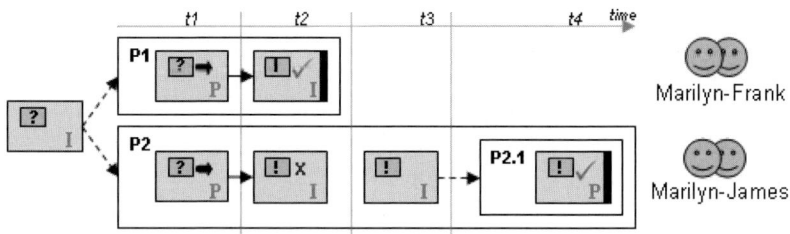

Fig. 5. Data Request and Task Delegation

automatically annotated as (Decline, Resource, Ø). When Marilyn receives the reply at *t2*, she is assisted with choosing from *Acknowledge*, *Ignore* or *Other*. She selects the latter, since she wants to tell James that he should get a new photo as soon as possible. *Semanta*'s wizard assists her with creating and annotating this reply (Figure 6). The results is a (Commit, Task, Participant) addressed to James. On sending it at *t3*, she sets off a new sub-workflow (P2.1). Before this new workflow terminates, the parent workflow, P2, cannot terminate.

As opposed to previous work [7], even after selecting a seemingly fixed data request template, the ensuing workflow is still to a great extent ad-hoc. For example, when James or Frank got the data request, they could have replied to Marilyn with a request for information asking why she needs their photo – and wait for the answer, before actually delivering the resource.

4.2 Task Delegation

At time interval *t4* (Sub-workflow P2.1 in Fig. 5) James receives Marilyn's latest email with the speech act committing him to a task. *Semanta* assists him with choosing from Acknowledge, Ignore and Other. He selects the first option, thus accepting the task. The subject of the speech act in Marilyn's email is the Participant, so according to the ESAW, given that James is the participant in the transaction, he has to manage the generated activity (activity management is not shown in the time graphs). *Semanta* prompts him with the activity manager (Figure 6). Unless he chooses to dismiss the generated task, the *Semanta* can support him with adding the task as an object on his semantic desktop (i.e task list). In his PIMO, the task would be linked to the email in which it was generated and to the contacts involved, i.e. Marilyn. When James is done managing the task, the workflow terminates.

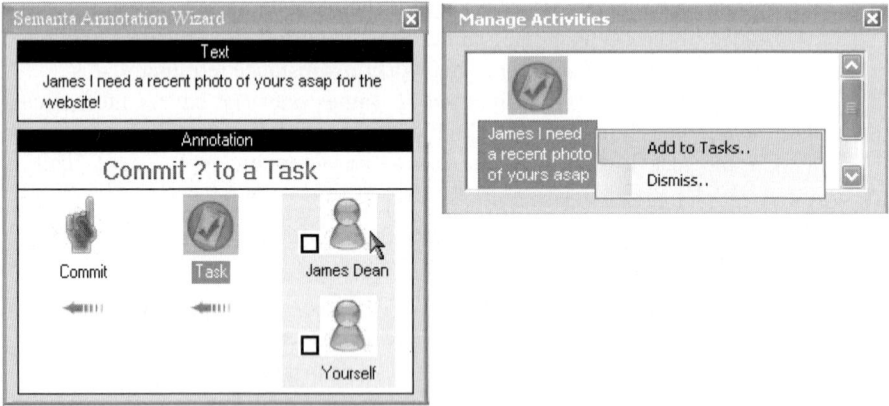

Fig. 6. *Semanta*'s Annotation Wizard and Activity Manager

4.3 Appointment Scheduling

James wants to organize a meeting with Grace and Frank. At time interval zero, he looks into his calendar to find an appropriate date and time. From his calendar application, James generates a meeting proposal for Tuesday at 11am, addressing it to Grace and Frank. *Semanta* generates an automatic email with one speech act. Since James is implicated in the event, the speech act equivalent to this proposal is (Propose, Event, Both). James sends the proposal, thus initiating two parallel instance of the ESAW as shown in Fig. 7. Until James gets the requested feedback, the email item will show in a list of items pending a response. If James feels that too much time has elapsed, *Semanta* can assist him with sending a reminder.

i. P1: When Grace receives the email, at time $t1$, *Semanta* recognizes the speech act within it as being a meeting proposal. She is shown her calendar and the commitments for the proposed day. *Semanta* then shows her the following options: *Deliver Feedback, Ignore, Other*. Since Grace is free at the proposed time, she expresses her availability and delivers the feedback.
ii. P2: Unlike Grace, Frank has another commitment. In his feedback at $t1$, he informs James that on Tuesday he is only available after 1pm.

Given the semantics of a Propose, the two workflow instances generated by James are not independent. In fact, the separate parallel instances of the ESAW merge back together once both participants have delivered their feedbacks as stated in Fig. 4 and seen in Fig. 7 at time interval $t2$. James is now shown two options: Propose – to propose again and Request – to call for the meeting. Since James requires everyone to attend the meeting and no consensus was reached, he proposes another time for the meeting – Tuesday at 2pm. When he sends the new proposal at time $t3$, the cycle is repeated, and the control flow splits again into two dependent activity instances:

i. P1: Grace is available at the new time and delivers her positive feedback at $t4$.
ii. P2: Frank is also available and also delivers the positive feedback at $t4$.

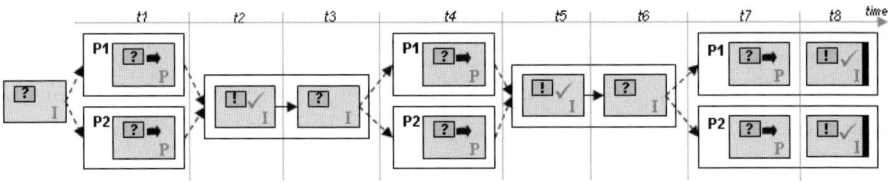

Fig. 7. Appointment Scheduling

At time $t5$, James gets both Grace's and Frank's feedback. Since consensus was reached this time around, he chooses the request option in *Semanta*'s user prompt. This translates into a (Request, Event, Both) addressed to both Frank and Grace. After sending at $t7$, the workflow splits again into two, this time independent, instances:

i. P1: At *t7*, Grace gets the request and she can choose from: *Approve, Amend, Decline, Ignore* or *Other*. She approves her participation and given the ESAW, the workflow follows two parallel paths. In the first *Semanta* assists Grace with managing the email-generated activity. In the second path, *Semanta* sends an automatic email to James with a (Commit, Event, Both) speech act. On receipt at *t8*, James acknowledges the commitment to the event. Since the subject of the speech act includes him, *Semanta* provides him with the possibility of adding the event to his desktop. If he does, a representation for the event is created in the semantic desktop's RDF repository, with a link to the source email amongst other metadata. This exposes the email-generated event to all semantic desktop applications. The workflow instance terminates.

ii. P2: At *t7*, Frank repeats the process performed by Grace, and is supported by *Semanta* with adding the event to his semantic desktop and sending a (Commit, Event, Both) back to James. On getting this at *t8*, James again acknowledges the commitment and is prompted to manage the generated event. This time he chooses to dismiss the event, since he already has a representation for this event on his desktop. The workflow instance terminates.

4.4 Event Announcement

James writes an email to inform Marilyn of the upcoming meeting so if she wants to, she can attend. *Semanta*'s Speech Act Extraction service recognizes a sentence as being either a suggestion or a request related to a new event involving Marilyn. James confirms that this is a suggestion, and *Semanta* creates a (Suggest, Event, Participant) addressed to Marilyn. On sending, an instance of the ESAW is created - P1 in Fig. 8, and control passes over to the Marilyn at time interval *t1*. *Semanta* offers to show her the known commitments for the said time and day. When considering the speech act, Marilyn is presented with the following options: *Acknowledge, Approve, Request, Other* and *Ignore*. Since the speech act is a suggestion, she does not have to take any action or even reply – in which case she would just acknowledge it. However she decides that she can attend the event, and chooses the Approve option. Given her selection, the workflow follows two parallel paths. Since the subject of the speech act she approved included herself, the first path sees *Semanta* assisting her with managing the generated event. The second sees *Semanta* generating a (Commit, Event, Participant) to be sent in an auto-reply to James. When James acknowledges this at *t2*, the path terminates because, being the initiator, he is not included in the subject of the speech act. Given both paths terminated, the workflow instance terminates at time interval *t2*.

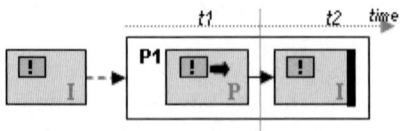

Fig. 8. Event Announcement

5 Conclusions and Outlook

In this paper, we have introduced the *Email Speech Act Process Flow Model* (ESAW) – a workflow model based on traditional email, which enables the user to define and execute ad-hoc workflows in an intuitive way. Given this model we can semantically annotate implicit, well-defined workflows, thus making them explicit and exposing the missing information in a machine processable way. Grounding this work within the Semantic Desktop through appropriate ontologies means that this information can be exploited by various semantic desktop applications for the benefit of the user. This will have a direct impact on their personal information management

Our main contribution is the formal description of the ad-hoc workflows of email communication. To the best of our knowledge, this was never formally described. We believe that defining this model using formal work flow patterns is equivalent to providing a formal semantics for email collaboration. After having defined the ESAW workflow model, we provided the corresponding implementation, namely the *Semanta* Email Client add-in. We believe that the human effort required to support *Semanta* with the semi-automatic annotation of email is minimal compared to the support provided by *Semanta* throughout the complete execution of the ensuing workflow. Alternatively our one-click semantic email process templates are still an improvement over previous work, since even after selecting a seemingly fixed template the workflow is still to a great extent ad-hoc.

In the near future we would like to perform two kinds of evaluation for our work. First we will perform a statistical evaluation by considering a corpus of real email messages. We will manually apply the ESAW model to conversations in this corpus, by breaking them down into a set of ESAW workflow patterns. This will determine the applicability of our workflow model to real conversations and point out any particular scenarios that the current model does not support. This evaluation will also serve as a study into the sequentiality of activities in email conversations, via probabilistic graphical models like Bayesian networks represented as transition diagrams. From a more formal point of view, we will analyse properties of the ESAW model mathematically, in an attempt to prove them as well as to determine the complexity of various patterns. Similar work has been done in follow-up work to [7].

Secondly we will investigate ways to evaluate the benefits of using *Semanta* to exchange semantic email, i.e. trial-based evaluation. We are currently improving *Semanta* and integrating further its functionalities within the Semantic Desktop. The ESAW we presented is completely extensible and new patterns can easily be introduced in the workflow. From an application point of view, *Semanta* can easily be extended to support custom workflow patterns required by a user or an organization. From another perspective, given its flexible nature support for the ESAW can be implemented on top of any existing protocol and email client. We will investigate the possibilities for integration of ESAW within widely-used email services like Gmail[9].

We believe that the here presented workflow model will have a concrete manifestation in the creation and organisation of semantic desktop data. Semantic annotations in conjunction with workflows not only support the user, but in the long-run also can be used in business environments for semantic business process mining,

---

9 http://mail.google.com/mail/

i.e., the identification of hidden, currently unsupported workflows and workflow consistency and compliance checking.

Acknowledgments. We would like to give our special thanks to Prof. Manfred Hauswirth and Armin Haller for sharing their knowledge on workflow patterns. The work presented in this paper was supported (in part) by the Lion project supported by Science Foundation Ireland under Grant No. SFI/02/CE1/I131 and (in part) by the European project NEPOMUK No FP6-027705.

References

1. Groza, T., Handschuh, S., Moeller, K., Grimnes, G., Sauermann, S., Minack, E., Mesnage, C., Jazayeri, M., Reif, G., Gudjonsdottir, R.: The NEPOMUK Project - On the way to the Social Semantic Desktop. In: 3rd International Conference on Semantic Technologies (ISEMANTICS 2007), Graz, Austria (2007)
2. Whittaker, S., Bellotti, V., Gwizdka, J., Email and PIM: Problems and Possibilities. In: CACM07 (2007).
3. Voorhoeve, M., Van der Aalst, W.: Ad-hoc workflow: problems and solutions. In: 8th International Workshop on Database and Expert Systems Applications (DEXA07) (1997)
4. Sintek, M., van Elst, L., Scerri, S., Handschuh, S.: Distributed Knowledge Representation on the Social Semantic Desktop: Named Graphs, Views and Roles in NRL. In 4th European Semantic Web Conference, Innsbruck, Austria (2007)
5. Sauermann, L., van Elst, L., Dengel, A.: PIMO - a Framework for Representing Personal Information Models. In: I-Semantics' 07, pp. 270–277. (2007)
6. Whittaker, S., Sidner, C.: Email overload: exploring personal information management of email. In: CHI '96 Conference, Vancouver, BC Canada, April, pp. 276–283. (1996)
7. McDowell, L., Etzioni, O., Halevey, A., Levy, H.: Semantic Email. In: 14th international conference on World Wide Web, New York, NY, USA (2004)
8. Searle, J.: Speech Acts. Cambridge University Press (1969).
9. Khoussainov, R., Kushmerick, N.: Email task management: An iterative relational learning approach. In: Conference on Email and Anti-Spam, CEASŠ2005 (2005).
10. Corston-Oliver, S., Ringger, E., Gamon, M., Campbell, R.: Task-focused summarization of email. In: Text Summarization Branches Out workshop, ACL 2004 (2004).
11. Khosravi, H., Wilks, Y.: Routing email automatically by purpose not topic. Natural Language Engineering, Vol. 5, pp. 237–250. (1999)
12. . Goldstein, J., Sabin, R.E.: Using Speech Acts to Categorize Email and Identify Email Genres. In: System Sciences, HICSS2006 (2006)
13. Carvalho, V, Cohen, W.: On the collective classification of email speech acts. In: SIGIR-2005, pp. 345–352. (2005)
14. Scerri, S., Mencke, M., Davis, B., Handschuh, S.: Evaluating the Ontology underlying sMail Ű the Conceptual Framework for Semantic Email Communication. In: 6th Language Resources and Evaluation Conference (LREC 2008), Marrakech, Morocco (2008)
15. Jose, P.: Sequentiality of speech acts in conversational structure. In: Journal of Psycholinguistic Research. Vol.17, Number 1, pp. 65Ű88. (1988)
16. Singh, M.: A Semantics for Speech Acts. In: Annals of Mathematics and Artificial Intelligence. Vol. 8, Numbers 1-2, pp. 47Ű71. (2006)
17. Scerri, S.: Semanta - Your Personal Email Semantic Assistant. In: Demo Session at the International Conference on Intelligent User Interfaces (IUI08). Island of Madeira, Canary Islands, Spain. (2008)

IVEA: An Information Visualization Tool for Personalized Exploratory Document Collection Analysis

VinhTuan Thai, Siegfried Handschuh, and Stefan Decker

Digital Enterprise Research Institute (DERI),
National University of Ireland, Galway
firstname.lastname@deri.org

Abstract. Knowledge work in many fields requires examining several aspects of a collection of documents to attain meaningful understanding that is not explicitly available. Despite recent advances in document corpus visualization research, there is still a lack of principled approaches which enable the users to personalize the exploratory analysis process. In this paper, we present IVEA (**I**nformation **V**isualization for **E**xploratory Document Collection **A**nalysis), an innovative visualization tool which employs the PIMO (Personal Information Model) ontology to provide the knowledge workers with an interactive interface allowing them to browse for information in a personalized manner. Not only does the tool allow the users to integrate their interests into the exploration and analysis of a document collection, it also enables them to incrementally enrich their PIMO ontologies with new entities matching their evolving interests in the process, and thus benefiting the users not only in their future experiences with IVEA but also with other PIMO-based applications. The usability of the tool was preliminarily evaluated and the results were sufficiently encouraging to make it worthwhile to conduct a larger-scale usability study.

Keywords: Personal Information Management, Semantic Desktop, PIMO ontology, Exploratory Document Collection Analysis, Information Visualization, Coordinated Multiple Views, Human-Computer Interaction.

1 Introduction

Apart from the need to retrieve information from documents that are relevant to certain topics of interest, oftentimes knowledge workers also need to explore and analyze a collection of documents as a whole to gain further understanding. This particular kind of information-seeking activity can be referred to as *exploratory data analysis* or *information analysis* and is commonly carried out in science, intelligence and defense, or business fields [1]. Unlike the information retrieval activity, the information analysis activity aims to provide the users with an overall picture of a text collection as a whole on various dimensions instead of presenting them with the most relevant documents satisfying some search criteria. The insights obtained from the exploration and analysis of a text collection can enable the knowledge workers to understand the distribution of topics, to find clusters of similar documents, or to identify trends or linkages between different entities [1]. Information visualization is an effective mechanism to support the information analysis task and has also been widely employed in many data mining and knowledge discovery tools to identify previously unknown useful information.

While there are many existing tools reported in the Information Visualization literature that support document corpus visualization (such as [1,2,3,4]), most are based only on the text of the documents in the corpus and hence present findings that are independent of the users' interests. Their approaches focus on the entity extraction process in order to identify the main entities (e.g. topics, people, locations) within a document collection and then visualize different linkages between the identified entities onto a 2D or 3D user interface. While the automatic extraction of entities is helpful, the visual exploration process cannot be aligned with the knowledge workers' interests, especially when a number of entities contained within their spheres of interests are of particular importance to their knowledge work. As a result, the knowledge workers cannot have a personal viewpoint over the entities and relationships that they wish to focus on to gain insights by using the existing tools. There is a clear need to link various important concepts and structures within parts of the knowledge workers' mental models with the document collection visual exploration and analysis activity. To achieve this, it is necessary that these concepts and structures can be externalized to some formal representation.

The above requirement motivates us to investigate existing work within the Semantic Desktop area where several integrated Personal Information Management environments have been introduced [5]. In the Semantic Desktop paradigm, desktop items are treated as Semantic Web resources and formal ontologies "allow the user to express personal mental models and form the semantic glue interconnecting information and systems" [5]. One such ontology has been designed and developed within the Gnowsis project and is based upon the *Personal Information Model* (PIMO) [6]. Since the PIMO ontology acts as a "formal representation of the structures and concepts" within a knowledge worker's mental model [6], it can be employed as a means to integrate her personal interests into an exploratory visualization tool. This would give the PIMO-based visualization tool clear advantages over the keyword-based visualization tools in that: (1) it can align the exploration process with the users' interests and hence can offer the users the control over which aspects of a text collection they wish to focus on, (2) it can utilize the hierarchical structure (class-subclasses, class-instances) within the PIMO ontology to provide the users with the flexibility to explore a text collection at different levels of detail.

In this context, we propose an innovative visualization tool called IVEA, which leverages upon the PIMO ontology and the Coordinated Multiple Views technique to support the personalized exploration and analysis of document collections. IVEA allows for an interactive and user-controlled exploration process in which the knowledge workers can gain meaningful, rapid understanding about a text collection via intuitive visual displays. Not only does the tool allow the users to integrate their interests into the visual exploration and analysis activity, it also enables them to incrementally enrich their PIMO ontologies with entities matching their evolving interests in the process. With the newly added entities, the PIMO ontology becomes a richer and better representation the users' interests and hence can lead to better and more personalized exploration and analysis experiences in the future. Furthermore, not only can IVEA be beneficial to its targeted task, it provides an easy and incremental way that requires minimal effort from the users to keep their PIMO ontologies in line with their continuously changing interests. This, indirectly, can also benefit other PIMO-based applications.

The remainder of this paper is organized as follows: Section 2 details different visual aspects of IVEA. Its system architecture is described in Section 3. In Section 4, we present a usability study and a discussion on the results. Related work is highlighted in Section 5. Section 6 concludes the paper and outlines future work.

2 Proposed Solution

2.1 Design Desiderata

In order to design a suitable visualization solution, various factors need to be taken into consideration, including: (1) *the nature of the task* (personalized exploratory analysis, integration of newly discovered entities to evolving spheres of interests), (2) *the type of data* (contents of documents in a collection rather than those of a single metadata file describing it), (3) *the target users* (knowledge workers rather than computer users in general), and (4) *their typically available work environment* (2D display interface rather than advanced 3D environment). Furthermore, the design of advanced visual interfaces can also benefit from the well-known visual information-seeking mantra: *"Overview first, zoom and filter, then details-on-demand"* proposed by Shneiderman in his seminal paper [7]. Taking into account all those aspects, desirable capabilities for such a visualization solution would be:

- Provide an overall view showing the degrees of relevance of all documents in a collection with respect to the personal interests represented by the PIMO ontology.
- Enable the users to interactively explore and analyze a text collection at different levels of detail by filtering the set of documents based on the PIMO concepts and instances and their hierarchical relationships.
- Provide a detailed view on demand which focuses on the distribution of entities of interest within each document.
- Suggest the users with entities potentially matching their evolving interests and enable them to enrich the PIMO ontology with the selected entities.

2.2 Visual Interface

To cater for the above-mentioned needs, we employ the Coordinated Multiple Views (CMV) technique [8] to design IVEA, as shown in Fig. 1. The use of the CMV technique enables the users to understand a text collection better via different visual displays highlighting different aspects of it, as well as via the interactions with and coordinations between those displays. Interested readers are encouraged to view the demo screencast available online[1].

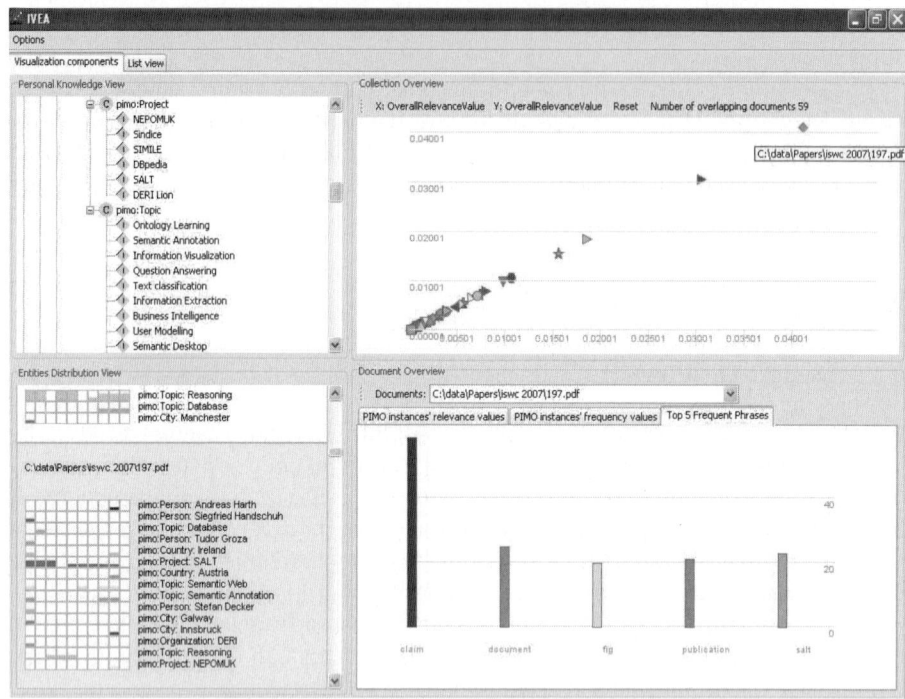

Fig. 1. IVEA interface

IVEA's visual interface consists of four views as follows:

- **Personal Knowledge View**: The tree structure, as shown on the upper-left corner of Fig. 1, is used to display the concepts, instances and their hierarchical relationships within the PIMO ontology. Although this tree structure does not display many other relationships that exist between PIMO concepts and instances, it is sufficient to serve as an anchor keeping the users informed of their entities of interest. This view acts as the basis for some of the interactions in IVEA.

---

[1] http://smile.deri.ie/projects/ivea/demo/demo1.html

- **Collection Overview:** The scatter plot, as shown on the upper-right corner of Fig. 1, is used as the overall view to display documents matching the users' interests. On the scatter plot, each document is represented by a dot and its file name is shown in the dot's tooltip text. The coordinate of a dot on the scatter plot is determined by the relevance values of the dot's corresponding document with respect to the classes or instances set on the x and y axes. The initial display of the scatter plot, as shown in Fig. 1, uses the overall relevance values of documents in the collection with respect to the set of all PIMO instances on both axes. Based on this initial display, the users can, for example, see that in the collection being explored, 59 documents overlap with their interests and that the document "*C:\data\Papers\iswc 2007\197.pdf*", represented by the rightmost dot on the scatter plot, is most relevant, based on its coordinate. More details about that particular document can be obtained immediately from the coordinated document overview and entities distribution view as described shortly. Moreover, the dimension of either of the two axes can be changed to reflect how relevant the documents are with respect to the concepts or instances placed on it. The dots' colors and shapes are used to differentiate the associated documents. The dot's size can be customized to accommodate for text collections of different sizes.

- **Document Overview:** Bar charts are used to display detailed views on various characteristics of each document in the collection. Three different bar charts are used on the lower-right corner of Fig. 1. The first bar chart shows the PIMO instances appearing in a document together with the relevance values of that document with respect to them. The second one displays the matching PIMO instances based on their frequencies. The third bar chart shows the top frequent phrases or terms in a document. For instance, in Fig. 1, the top frequent phrases bar chart of the document "*197.pdf*" mentioned above shows that the word "*claim*" appears most frequently.

- **Entities Distribution View:** This view, as shown in the lower-left corner of Fig. 1, is based on the TileBars, originally reported in [9], and some of its variants in [10]. It is used to display the matching PIMO instances within each fragment of a document. The rows are associated with PIMO instances whose labels are placed next to them on the right. The number of column in this view is the number of fragments to divide a document up to, whose value can be set by the users. Each document is split into sentences and they are put into fragments such that each fragment contains an approximately equal number of sentences. The height of the colored area in a cell is determined by the frequency of the corresponding PIMO instance in that particular fragment. By using this view, the users can quickly be informed of the relative locations, in which certain entities of interest appear, together with their respective frequencies. For instance, Fig. 1 shows that the instance "*SALT*" of the concept "*pimo:Project*" appears more often in the first three fragments than in any other parts of the document "*197.pdf*".

2.3 Interactions, Manipulations and Coordinations

In line with the visual information-seeking mantra [7], IVEA provides the users with the freedom to interact and control the visual displays in the following manners:

- **Filtering:** It is essential that the users are able to filter out uninteresting documents to focus only on a restricted subset of a text collection. In IVEA, the users can directly manipulate the overall view displayed in the scatter plot by dragging concepts or instances from the personal knowledge view and dropping them onto the labels of the x and y axes to indicate the dimensions upon which the relevance values of documents are to be measured. IVEA instantly updates the scatter plot by executing dynamic queries for the relevance values of all documents with respect to the instance or the aggregated relevance values with respect to the concept placed on the axes. In Fig. 2, the x-axis highlights documents relevant to *"Semantic Desktop"* while the y-axis highlights documents relevant to *"pimo:Person"*. The relevance value of a document with respect to *"pimo:Person"* is dynamically measured as the aggregated relevance value of that document with respect to all instances of the concept *"pimo:Person"* in the PIMO ontology. Hence, the scatter plot can show, among others, documents referring to both the topic *"Semantic Desktop"* and one or more persons who are of specific interest to the users (documents plotted above both axes). The example also illustrates that IVEA can take advantage of the hierarchical relationships between entities within the PIMO ontology to allow for the rapid exploration of a text collection at different levels of detail.
- **Details-on-demand:** Once the overall view has been restricted to a specific set of documents, the users can investigate the details of each document within that set. Clicking on a dot in the scatter plot will open up the corresponding document in its associated application. Furthermore, to interactively link the collection overview with the detailed views, coordinations between the scatter plot and the bar charts as well as the Entities Distribution View are provided in such a way that hovering the mouse over a dot in the scatter plot will display the corresponding bar charts and highlight the TileBars of the document represented by that dot. This interaction is also illustrated in Fig. 2.
- **PIMO ontology enrichment:** An innovative feature of IVEA is its capability to enable the users, with minimal efforts, to enrich their PIMO ontologies while exploring a text collection. While investigating a particular document to see how it overlaps with their spheres of interests, the users are presented with the most frequent phrases[2] in that document. The top frequent phrases and their corresponding frequencies are displayed in a histogram as shown in Fig. 3. If any of the presented phrases is of specific interest to the users, they are just two-click away from adding it to the PIMO ontology simply by dragging its respective column in the histogram and dropping on a concept

---

[2] Candidate phrases are nouns or noun chunks consisting of at most 3 words.

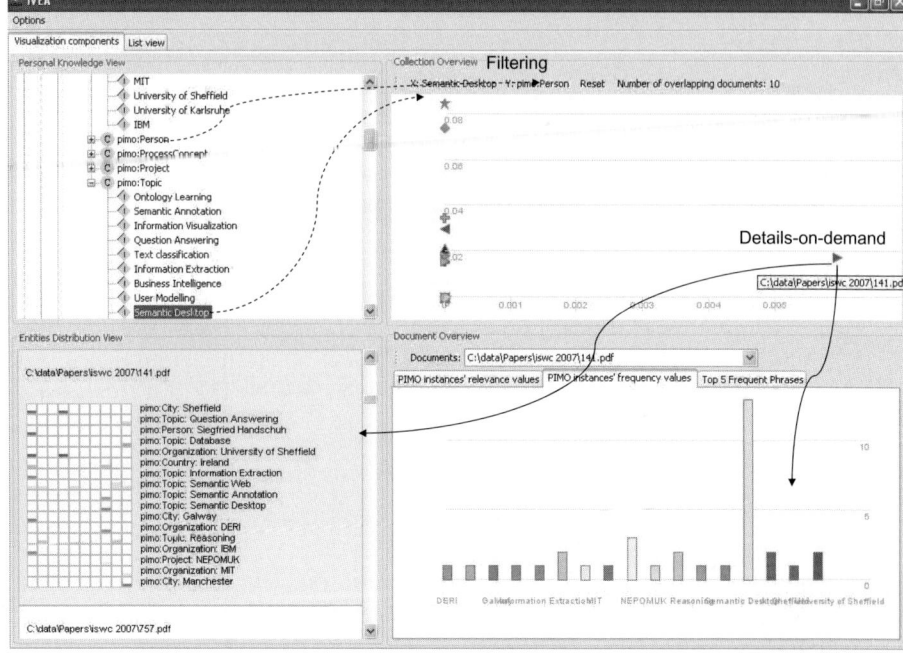

Fig. 2. Interactive Filtering and Details-on-demand

in the PIMO ontology. The users have the option to add the selected phrase as a subclass or instance of the targeted concept. Fig. 3 and 4 illustrate how the users can enrich the PIMO ontology by adding *"conference"* as a subclass of *"pimo:Event"*. We believe that this is of benefit to the users as it allows them to update their PIMO ontologies with new entities on-the-fly. Consequently, they can better explore a text collection when their spheres of interests are better represented. Besides, an extended PIMO ontology is useful not only for IVEA but also for other PIMO-based applications.

3 System Architecture

The system architecture is shown in Fig. 5. Documents in a collection are first analyzed by the text processing component, which is based on various natural language processing resources (Tokenizer, Sentence Splitter, POS Tagger, JAPE Transducer) provided by the GATE [11] API. Each document is split into sentences to identify the fragments' boundaries for use in the Entities Distribution View. Furthermore, to suggest the users with entities potentially matching their interest, most frequent noun chunks in each document together with their frequencies are extracted.

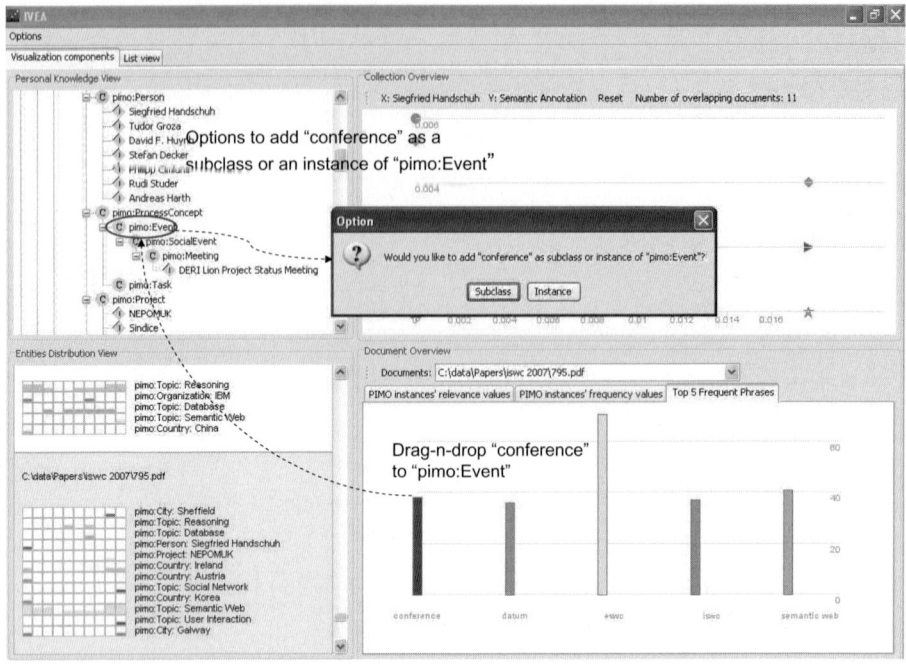

Fig. 3. Adding new concepts or instances to the PIMO ontology

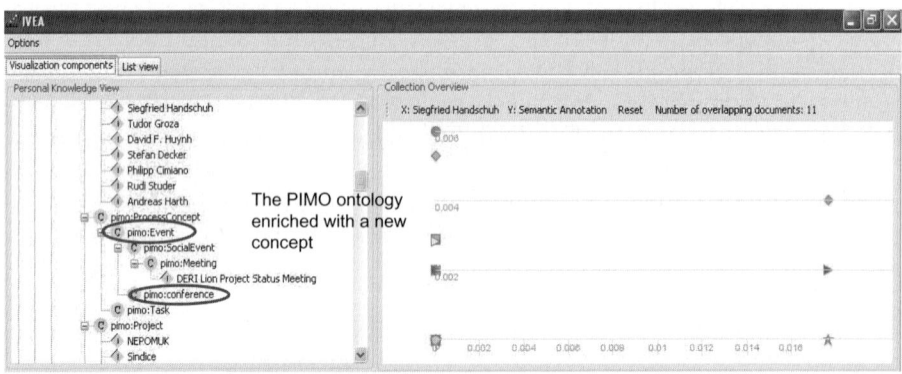

Fig. 4. The PIMO ontology added with a new concept

The analyzed text is then stored into a Lucene[3] index. A Boolean query (with the default OR operator) consisting of all PIMO instances is used to retrieve documents that are of interest to the users. The term weight in a document of an instance is used as the relative relevance value of that document with respect

---

[3] http://lucene.apache.org/

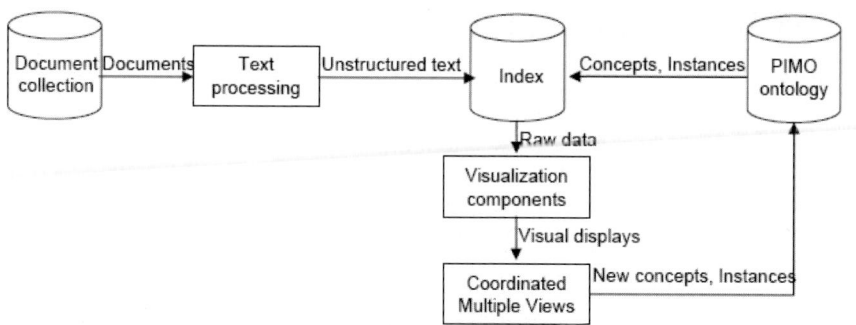

Fig. 5. System architecture

to that instance. In Lucene, a variant of the well-known TF.IDF term weight function is used, which takes into account the frequency of a term locally (in a document) and globally (in the whole collection), as well as the length of a document itself [12]. The relative relevance value of a document with respect to a class is the aggregated relevance value of that document with respect to all of its direct instances and recursively, all of its subclasses.

The above relevance values and the frequencies of PIMO instances in fragments of each document are used as raw data for the visualization components. The implementation of the visual displays is based on the prefuse library [13] and Java Swing components.

4 Usability study

Given a text collection on the desktop, most users are used to keyword-based search interface to look for documents satisfying some criteria using a desktop search engine. To explore a text collection this way, the users may have to perform many searches, each with a query containing multiple Boolean clauses. As such, we implemented a baseline interface as shown in Fig. 6 and performed a small-scale usability study to (1) get initial indications about the users' performances by using IVEA versus by using the baseline interface, and (2) identify IVEA's potential usability problems.

4.1 Description

Six researchers participated in the study. They all had prior knowledge about ontology and Boolean query operators. They were assumed to have the same sphere of interest, whose concepts and structures were encoded into a predefined PIMO ontology. This PIMO ontology acted as the basis for exploring and analyzing a test collection consisting of 62 research papers in the Semantic Web domain. Although the collection's size was not large, it was suitable for this initial study since it could

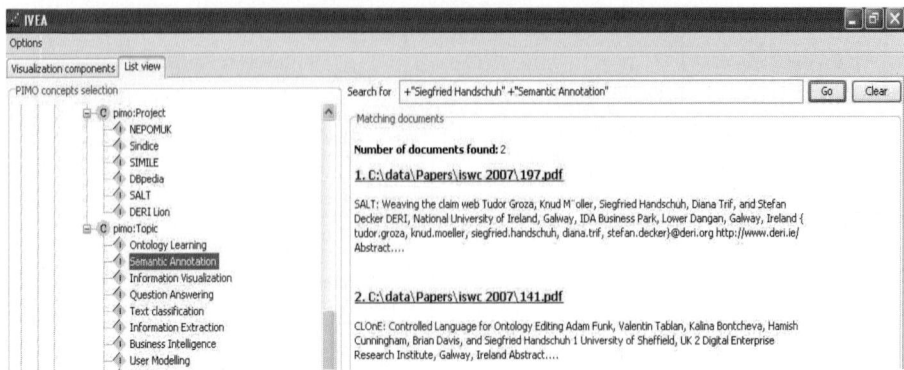

Fig. 6. Baseline interface

be representative of some real-world situation, e.g. when a researcher wishes to understand different characteristics of a document collection consisting of scientific papers published at a conference in a particular year.

The subjects were asked to perform the same set of 4 simple tasks and 4 complex tasks by using first the baseline interface and then the visual interface, followed by filling out the follow-up questionnaire. The following measures were recorded based on the subjects' responses:

- **Effectiveness**: calculated as the accuracy and completeness with which the users accomplished a task's goal [14].
- **Temporal efficiency**: calculated as the ratio between the effectiveness achieved in a task and the time taken to complete that task [14].
- **Users' satisfaction**: determined by the ratings responded to the follow-up questionnaire consisting of 12 Likert-scale questions (the ratings are on a range from -2 (strongly disagree/very bad) to 2 (strongly agree / very good)) and 1 open-ended question seeking suggestions on improving IVEA's design.

Further details about the tasks, the questionnaire and their respective results can be found online[4]. In the next section, we elaborate our findings and discuss what can be deduced from them.

4.2 Findings

We are aware that the study described above is not a summative usability evaluation, which would (1) involve a larger number of participants who are representative of the target users population (knowledge workers) and hence may have limited knowledge about ontology and Boolean query operators, (2) require a larger test collection, (3) use a more extensive set of tasks, and (4) require that each interface be used first by half of the subjects. Nevertheless, at this stage

---

[4] http://smile.deri.ie/projects/ivea/usability/preliminary.html

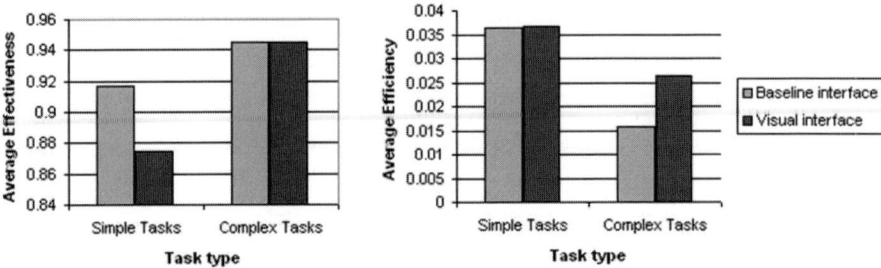

Fig. 7. Average Effectiveness and Temporal Efficiency by Task Type and Interface Used

of IVEA's design and development, the outcomes of the conducted study provide us with initial suggestive indications of its performances as well as potential usability problems.

With respect to the participants' performances, the average effectiveness and temporal efficiency are shown in Fig. 7.

In terms of effectiveness, it is interesting to note that, on average, the participants achieved a better score on the set of simple tasks by using the baseline interface than by using IVEA. We believe that this could be due to the effect of experience, whereby all the participants were very familiar with using keyword-based search interface to find out simple facts but had no prior experience with IVEA before. Meanwhile, they achieved the same score on the set of complex tasks by using the two interfaces, and the score was higher than those achieved on the set of simple tasks. This could be attributed to the order in which the task types were introduced (from simple to complex) whereby the subjects learned from the mistakes they made when using both interfaces for the first time while performing the simple tasks, and they became more experienced with both interfaces' features as they moved along to carry out the complex tasks.

When the task completion time was taken into consideration, Fig. 7 shows that the temporal efficiencies of the participants are comparable on the set of simple tasks by using the two interfaces. Meanwhile, on the set of complex tasks, the participant were considerably more efficient by using IVEA than by using the baseline interface. Moreover, the overall temporal efficiency of the participants on the set of complex tasks is lower than that of the participants on the set of simple tasks on both of the interfaces. The results suggest that our work is toward the right direction, however we are aware that a further usability study on a large scale is necessary.

With respect to the participants' satisfaction, their responses on the questionnaire are shown on Fig. 8. Among the 6 participants, a particularly well-liked aspect of IVEA was that its design and layout were appealing, whereas the ratings were a bit low for its self-descriptiveness and being complex for the task. We believe that with proper documentation, these setbacks can be mitigated. Some of the participants suggested to display 10 frequent phrases instead of the default

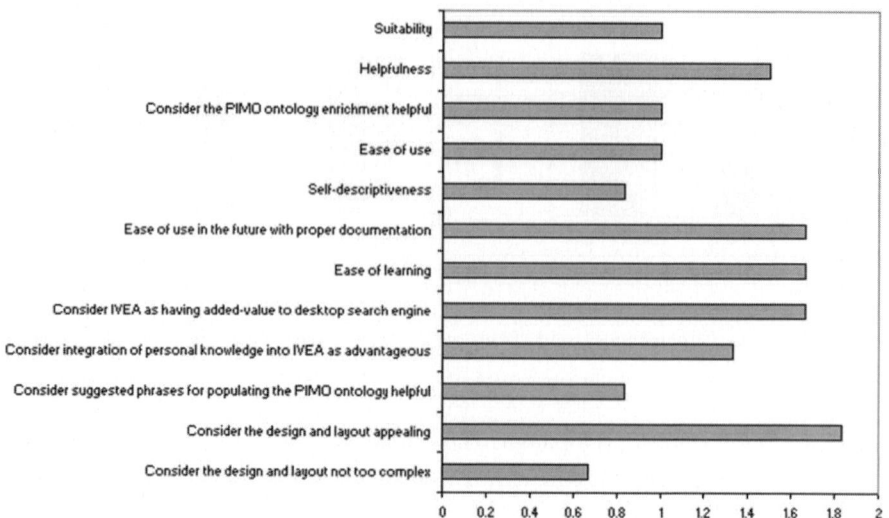

Fig. 8. Questionnaire responses

setting of 5. In fact, this number could be customized, but the lack of detailed manual rendered the participants uninformed of this feature. The participants also provided us with valuable design feedback such as:

- Highlight the most relevant document on the scatter plot.
- Group the visual items in the document view and the Entities Distribution View by class.
- Display the text surrounding the PIMO instances in the tooltip text of the TileBars' cells.
- Enable the Collection Overview to display at more than 2 dimensions.

To sum up, the study provided us with useful suggestive indications and valuable feedback to improve the design of IVEA in the future. In the next section, we review related work.

5 Related Work

Many research efforts in the area of Semantic Desktop [15,16,17,18] have introduced semantically-enabled Personal Information Management environments on the desktop. Although many useful functionalities are provided in these environments, we are not aware of any visualization tool that enables the users to explore and analyze a text collection according to their interests.

In the Information Visualization literature, a number of exploratory visualization tools are described in [19]. These tools, however, visualize different aspects of a domain based on the contents of *a single file*, which contains manually cleaned metadata describing a set of documents within that domain [19]. They do not enable exploring a text collection based on the documents' contents, and the considerable efforts required to produce the metadata file make it impractical for wide-spread use. In addition, several document corpus visualization tools have been developed (e.g. [1,2,3,4,20]). However, they are based only on the textual contents of the documents and provide no mechanism to integrate and utilize an ontology representing the users' knowledge or interests. Hence, they lack some of the capabilities that ontology-based visualization tools offer, e.g. they do not allow the users to personalize the dimensions in which insights are to be gained. Nor do they enable the addition of newly discovered entities into the users' knowledge. It is also worth noting Aduna's AutoFocus, an ontology-based visualization tool which displays search results for documents on the desktop as clusters of populated concepts [21]. While allowing for the exploration of a document collection based on the extracted metadata and search keywords, in AutoFocus no deep insights or comparisons can be attained when the degrees of relevance of documents with respect to each entity within the users' spheres of interests are not available. Many other ontology-based visualization tools are designed specifically for certain domains and hence have different objectives than ours.

Also related to our work are faceted search and browsing tools, e.g. Flamenco [22] or Exhibit [23], which allow the users to navigate along different conceptual dimensions of a given document collection or a structured dataset. Although these tools are very effective in supporting browsing and exploratory tasks, they were designed either to work on document collection whose rich metadata categories are available or only to be used for browsing structured data.

6 Conclusions and Future Work

In this paper, we introduced IVEA, an innovative information visualization tool which supports the task of personalized exploratory document collection analysis. IVEA leverages upon the PIMO ontology as a formal representation of the users' interests and combined various visual forms in an innovative way into an interactive visualization tool. The integration of the PIMO ontology into IVEA allows the users to rapidly explore text collections at different levels of detail. To cater for the knowledge workers' evolving spheres of interests, IVEA enables them to enrich their PIMO ontologies with new concepts and instances on-the-fly. A small-scale usability study was carried out and the results were sufficiently encouraging to make it worthwhile to conduct a larger-scale usability study.

In future work, we plan to improve IVEA's functionalities based on the useful feedback gathered. In addition, we intend to employ anaphora resolution techniques so that co-references of PIMO instances (e.g. "European Central Bank", "ECB", "the bank", "it") can be correctly identified. Furthermore, the users should be able to tag documents with PIMO concepts and instances. IVEA will also need to be capable of handling more types of textual documents than just PDF files. Once the final version of IVEA is developed, a full-scale summative evaluation is to be carried out to gauge its usability.

As a final remark, it is our hope that the benefits that IVEA can bring about are not confined within its targeted task. If its final version has a high usability, it will demonstrate, to a certain extent, that the PIMO ontology is very useful and hence will motivate more users to start populating their own PIMO ontologies and use them to describe desktop items. This activity is certainly of particular importance to the realization of Semantic Desktop environments.

Acknowledgments. We thank Anthony Jameson for his constructive comments on a draft of this paper, and the anonymous reviewers for the insightful remarks. We are also grateful to the researchers who participated in the usability study. This work is supported by the Science Foundation Ireland(SFI) under the DERI-Lion project (SFI/02/CE1/1131) and partially by the European Commission 6th Framework Programme in context of the EU IST NEPOMUK IP - The Social Semantic Desktop Project, FP6-027705.

References

1. Shaw, C.D., Kukla, J.M., Soboroff, I., Ebert, D.S., Nicholas, C.K., Zwa, A., Miller, E.L., Roberts, D.A.: Interactive Volumetric Information Visualization for Document Corpus Management. Int. J. on Digital Libraries 2, 144–156 (1999)
2. Zhu, W., Chen, C.: Storylines: Visual exploration and analysis in latent semantic spaces. Computers & Graphics 31, 338–349 (2007)
3. Fortuna, B., Grobelnik, M., Mladenic, D.: Visualization of Text Document Corpus. Informatica 29, 497–504 (2005)
4. Grobelnik, M., Mladenic, D.: Visualization of News Articles. Informatica 28, 32 (2004)
5. Sauermann, L., Bernardi, A., Dengel, A.: Overview and Outlook on the Semantic Desktop. In: Decker, S., Park, J., Quan, D., Sauermann, L. (eds.) Proc. of Semantic Desktop Workshop at the ISWC, Galway, Ireland, November 6, 2005, vol. 175 (2005)
6. Sauermann, L., van Elst, L., Dengel, A.: PIMO - a Framework for Representing Personal Information Models. In: Pellegrini, T., Schaffert, S. (eds.) Proc. of I-Semantics 2007, JUCS, pp. 270–277 (2007)
7. Shneiderman, B.: The Eyes Have It: A Task by Data Type Taxonomy for Information Visualizations. In: IEEE Visual Languages, pp. 336–343 (1996)
8. Roberts, J.C.: State of the Art: Coordinated & Multiple Views in Exploratory Visualization. In: CMV 2007: Proceedings of the 5th Int'l Conf. on Coordinated and Multiple Views in Exploratory Visualization, pp. 61–71. IEEE Computer Society, Washington, DC, USA (2007)
9. Hearst, M.A.: TileBars: visualization of term distribution information in full text information access. In: CHI 1995: Proc. of the SIGCHI Conf. on Human factors in computing systems, pp. 59–66. ACM Press/Addison-Wesley Publishing Co., New York (1995)
10. Reiterer, H., Mussler, G., Mann, T., Handschuh, S.: INSYDER - An Information Assistant for Business Intelligence. In: SIGIR 2000: Proc. of the 23rd Annual Int'l ACM SIGIR 2000 Conf. on Research and Development in Information Retrieval, pp. 112–119. ACM press, New York (2000)

11. Cunningham, H., Maynard, D., Bontcheva, K., Tablan, V.: GATE: A framework and graphical development environment for robust NLP tools and applications. In: Proceedings of the 40th Anniversary Meeting of the Association for Computational Linguistics (2002)
12. Hatcher, E., Gospodnetic, O.: Lucene in Action. Manning Publications Co., Greenwich (2004)
13. Heer, J., Card, S.K., Landay, J.A.: prefuse: a toolkit for interactive information visualization. In: CHI 2005: Proc. of the SIGCHI Conf. on Human factors in computing systems, pp. 421–430. ACM Press, New York (2005)
14. ISO: ISO9241-11:1998: Ergonomic requirements for office work with visual display terminals (VDTs) – Part 11: Guidance on usability (1998)
15. Sauermann, L., Grimnes, G.A., Kiesel, M., Fluit, C., Maus, H., Heim, D., Nadeem, D., Horak, B., Dengel, A.: Semantic Desktop 2.0: The Gnowsis Experience. In: Proc. of the ISWC Conf., pp. 887–900 (2006)
16. Karger, D.R., Bakshi, K., Huynh, D., Quan, D., Sinha, V.: Haystack: A General-Purpose Information Management Tool for End Users Based on Semistructured Data. In: CIDR, pp. 13–26 (2005)
17. Cheyer, A., Park, J., Giuli, R.: IRIS: Integrate. Relate. Infer. Share. In: Decker, S., Park, J., Quan, D., Sauermann, L. (eds.) Proc. of Semantic Desktop Workshop at the ISWC, Galway, Ireland, November 6, 2005, vol. 175 (2005)
18. Groza, T., Handschuh, S., Moeller, K., Grimnes, G., Sauermann, L., Minack, E., Mesnage, C., Jazayeri, M., Reif, G., Gudjonsdottir, R.: The NEPOMUK Project – On the way to the Social Semantic Desktop. In: Proc. of the 3rd International Conference on Semantic Technologies (I-SEMANTICS 2007), Graz, Austria (2007)
19. Plaisant, C., Fekete, J.D., Grinstein, G.: Promoting Insight-Based Evaluation of Visualizations: From Contest to Benchmark Repository. IEEE Transactions on Visualization and Computer Graphics 14, 120–134 (2008)
20. Olsen, K.A., Korfhage, R.R., Sochats, K.M., Spring, M.B., Williams, J.G.: Visualization of a document collection: the VIBE system. Information Processing and Management 29, 69–81 (1993)
21. Fluit, C., Sabou, M., van Harmelen, F.: Ontology-based Information Visualisation: Towards Semantic Web Applications. In: Geroimenko, V. (ed.) Visualising the Semantic Web, 2nd edn., Springer, Heidelberg (2005)
22. Yee, K.P., Swearingen, K., Li, K., Hearst, M.: Faceted metadata for image search and browsing. In: CHI 2003: Proc. of the SIGCHI Conf. on Human factors in computing systems, pp. 401–408. ACM, New York (2003)
23. Huynh, D., Karger, D., Miller, R.: Exhibit: Lightweight Structured Data Publishing. In: 16th Int'l WWW Conf., Banff, Alberta, Canada, ACM, New York (2007)

Building a Semantic Web Image Repository for Biological Research Images

Jun Zhao, Graham Klyne, and David Shotton

Department of Zoology
University of Oxford
South Parks Road, Oxford
OX1 3PS United Kingdom
{jun.zhao,graham.klyne,david.shotton}@zoo.ox.ac.uk

Abstract. Images play a vital role in scientific studies. An image repository would become a costly and meaningless data graveyard without descriptive metadata. We adapted EPrints, a conventional repository software system, to create a biological research image repository for a local research group, in order to publish images with structured metadata with a minimum of development effort. However, in its native installation, this repository cannot easily be linked with information from third parties, and the user interface has limited flexibility. We address these two limitations by providing Semantic Web access to the contents of this image repository, causing the image metadata to become programmatically accessible through a SPARQL endpoint and enabling the images and their metadata to be presented in more flexible faceted browsers, jSpace and Exhibit. We show the feasibility of publishing image metadata on the Semantic Web using existing tools, and examine the inadequacies of the Semantic Web browsers in providing effective user interfaces. We highlight the importance of a loosely coupled software framework that provides a lightweight solution and enables us to switch between alternative components.

1 Introduction

Images are semantic instruments for capturing aspects of the real world, and form a vital part of the scientific record for which words are no substitute. In the digital age, the value of images depends on how easily they can be located, searched for relevance, and retrieved. Images are usually not self-describing. Rich, well-structured descriptive image metadata thus carries high value information, permitting humans and computers to comprehend and retrieve images, and without them an image repository would become little more than a meaningless and costly data graveyard. A public image repository should thus publish its images along with such metadata.

In this paper, we present the *Drosophila* Testis Gene Expression Database, FlyTED [1], which presently contains images of expression patterns of more than

---
[1] http://www.fly-ted.org/

five hundred genes that are expressed in the testis of the fruitfly *Drosophila melanogaster*, both in normal wild type and in five meiotic arrest mutant strains, revealed by the technique of *in situ* hybridisation. These images were created as part of an ongoing research effort [1] by Helen White-Cooper and her team, who have also provided us with user requirements. Each image in FlyTED is described by the following *domain-specific* metadata: 1) the gene name, 2) the strain name, and 3) the gene expression pattern, in addition to the metadata about experimental details. This database aims to enable scientists to search for images of particular genes and compare their expression patterns between wild type flies and mutant strains.

To avoid building yet another purpose-built database, and to create a working database rapidly with the minimum development effort, we developed FlyTED by adapting an existing open source repository software system, EPrints[2]. The initial implementation of FlyTED was presented to our biological colleagues and received positive feedback. However, because EPrints is designed for text publications and is not Semantic Web-enabled, it is not effective in making its metadata programmatically accessbile and linkable to other *Drosophila* data resources, such as the Berkeley Drosophila Genome Project (BDGP) database of gene expression images in *Drosophila* embryos[3], or the global database of *Drosophila* genomic information, FlyBase[4]. Furthermore, although the built-in EPrints user interface can be customised to improve data presentation using technologies such as Cascading Style Sheets (CSS), it is not trivial to support some of the advanced image browsing functionalities requested by our researchers, such as filtering images first by gene name, then by mutant name, and then by expression pattern.

These limitations of metadata dissemination and user interface motivated us to enhance FlyTED with Semantic Web technologies, in order to make its images and metadata Semantic Web accessible. This has enabled the FlyTED image metadata to be queryable through the SPARQL protocol, and the images to be accessible through simple HTTP requests, enabling the use of faceted Semantic Web browsing interfaces.

The goals of this paper are to:

- Share our experience of enhancing an image repository developed using a conventional repository software package with Semantic Web technologies, which may be useful to others holding legacy data in conventional databases.
- Show the benefits obtained by making this image repository Semantic Web accessible.
- Discuss lessons we learnt through this experiment, summarised as the following:
    - Available Semantic Web tools can be employed to support Semantic Web access to data held in pre-existing Web applications.
    - The creation of effective user interfaces still remains challenging.

---

[2] http://www.eprints.org/
[3] http://www.fruitfly.org/
[4] http://www.flybase.org/

- A software framework that permits loose coupling between components shows its advantages, by enabling the reuse of existing toolkits and by facilitating switching between alternative software implementations.

2 Background

In this section, we provide background information about FlyTED and the EPrints 3.0 software system.

2.1 FlyTED

The spatial and temporal expression patterns of genes provide important knowledge for biologists in understanding the development and functioning of organelles, cells, organs and entire organisms. Biological researchers in the Department of Zoology at the University of Oxford are working towards this goal by determining expression data for approximately 1,500 genes involved in spermatogenesis in the testis of the fruitfly *Drosophila melanogaster*, representing ∼10% of the genes in the entire genome [1]. Comparative studies are being carried out in both wild type flies and five meiotic arrest mutant fly strains in which sperm development is defective. It is hoped that such studies can assist in the understanding of and the development of treatments for human male infertility.

In this work, for each gene, at least one *in situ* gene expression image is acquired from a wild type fly, and possibly one or more are taken from each strain. In addition, images are sometimes acquired of expression patterns of mutants of these genes in wild type flies, where such mutants are available. Thousands of images were accumulated by our biological researchers during the initial months of their research project. These images were kept in a file system organised by the date of creation and described using Excel spreadsheets. Each row in the spreadsheets describes one gene expression image by the following metadata terms (columns):

- The `GeneName` associated with this image.
- The `Strain` name of the fly from which the image was acquired.
- The gene `ExpressionPattern` being revealed in the image, as defined by the *Drosophila* anatomy ontology[5].
- Other domain-specific metadata, such as the number of the microscope slide on which the *in situ* hybridisation specimen was mounted for image acquisition.

Without organizing these images using structured metadata, it was extremely difficult for researchers to locate images from the file directories by their domain-specific metadata. A proper image repository was needed to assist researchers in uploading, storing, searching and publishing images with appropriate domain-specific metadata.

---
[5] http://obofoundry.org/cgi-bin/detail.cgi?fly_anatomy

2.2 EPrints

EPrints is an open source software package for building open access digital repositories that are compliant with the Open Archives Initiative Protocol for Metadata Harvesting (OAI-PMH) [2]. The latest release, EPrints 3.0, was chosen for setting up FlyTED for the following reasons:
- It is one of the well-established repository software systems for archiving digital items, including theses, reports and journal publications, with ~240 installations worldwide.
- It has built-in support for the OAI-PMH protocol, a simple HTTP-based protocol which allows repository metadata to be harvested by any OAI-PMH compliant parties.
- It has a built-in user interface, which makes it fairly quick to set up the repository and present it to users.
- It has previously been adapted by the Southampton SERPENT project[6] to publish images using domain-specific metadata.

2.3 Image Ingest

EPrints was designed as a digital text archive. It has good support for using Dublin Core (DC) metadata for describing and searching for digital items. However, to use EPrints to store and publish our biological research images along with domain-specific metadata, we needed to take the following steps:
- Preprocess researchers' Excel spreadsheets into a collection of free-text image metadata files, one for each image, using a Python script. Each file contains all the domain-specific metadata terms used by the researchers to describe their gene expression images. These metadata terms from different columns of the original spreadsheets are line separated in the metadata files.
- Modify the underlying EPrints relational database schema to accommodate some of the domain-specific metadata needed for both browsing and searching images, *i.e.*, the gene name, strain name and expression pattern.
- Ingest images and their metadata files into EPrints using a customised script written in Perl. This script achieves two goals: 1) extract from the image metadata files the metadata terms needed for searching and browsing images and store these metadata in the EPrints relational database when uploading images; 2) store the images as well as the individual metadata files for the images as binary objects within the EPrints database. Both the images and their image metadata files become Web-accessible.

The Perl and Python image ingest scripts are available at our SVN repository[7]. In addition to this customisation of the underlying EPrints database, we also needed to customise the user interface in order to permit searching and browsing for images using domain-specific metadata, and to enable the metadata within the search results to be displayed in the most useful manner. The details of this interface customisation can be found in [3].

---
[6] http://archive.serpentproject.com/
[7] https://milos2.zoo.ox.ac.uk/svn/ImageWeb/FlyTED/Trunk/

3 Problem Statement

This EPrints FlyTED repository clearly fulfils our initial goal of publishing images and their metadata on the Web. We can provide a working system for scientists and their community quickly and easily. However, it has two limitations:
- The OAI-PMH protocol fails to provide an effective programmatic access to the images themselves and to all their domain-specific metadata, in a form that can be linked with other data sources that are not OAI-PMH compliant.
- It is difficult to configure the user interface to permit researchers to browse, search and retrieve images by their domain-specific metadata for complex tasks.

3.1 Metadata Accessibility

The first limitation is caused by the nature of the OAI-PMH protocol, through which metadata in FlyTED can be programmatically accessed. OAI-PMH [2] is developed by the Open Archives Initiative for harvesting the metadata descriptions of resources in a digital archive. It is widely supported by repository software systems, including EPrints, DSpace[8], and Fedora[9].

It is compulsory for OAI-PMH compliant repositories to return XML format DC metadata for OAI-PMH requests. However, exposing domain-specific metadata in a semantic-rich data format is not natively supported by repository software systems, and would require a significant amount of work from repository administrators to implement. Furthermore, as a harvesting protocol, OAI-PMH does not allow *querying* of resources based on their (domain-specific) metadata values. To overcome these shortcomings of OAI-PMH, we chose to deploy a SPARQL endpoint over FlyTED.

SPARQL [4], a W3C-recommended RDF query language and protocol, has two distinct advantages over OAI-PMH for providing accessible metadata:
- SPARQL permits query selection by domain-specific metadata values. For example, a SPARQL query "?image hasPatternIn Mid_elongation-stage _spermatid" would query for all entities, locally designated by ?image, that have an ExpressionPattern in Mid_elongation-stage_spermatid, which is a type of Elongation-stage_spermatid. OAI-PMH cannot do this.
- SPARQL query results are in RDF, which can be linked to the wider Semantic Web of data, including the metadata resources of BDGP that have also been exposed via a SPARQL endpoint[10].

3.2 User Interface Issues

We customised the EPrints user interface using CSS to improve the presentation of images. However, this interface was still unable to fulfil all the requirements

---
[8] http://www.dspace.org/
[9] http://www.fedora.info/
[10] http://spade.lbl.gov:2021/sparql

of our users, and we found it difficult to perform further customisations without significant modification of the software source code.

Each image in FlyTED is described by at least three key metadata properties, the GeneName, one of several Strain names, and one or more ExpressionPattern keyword(s). Scientists need to use these domain specific metadata to examine images from different perspectives, for example, comparing expression patterns among all the images of the same gene in different strains. Furthermore, scientists need groups of images to be presented as thumbnails, so that they can obtain an overview of them and make comparisons between them.

The following list of requirements were articulated by our biologist colleagues for browsing and searching for images, ordered by their priorities:

- **R1:** Browse images by the gene name, strain name, or expression pattern keyword, preferably with images presented as thumbnails.
- **R2:** Search for all images of a gene that show the same expression pattern, *e.g.*, searching for all images of gene CG10396 with expression in Mid_elongation-stage_spermatid.
- **R3:** Search for all images of a gene that show the same *set* of expression patterns, *e.g.*, searching for all images of gene CG10396 with expression in both Mid_primary_spermatocyte and Mid_elongation-stage_spermatid.
- **R4:** Browse images of a single gene, first by the different strains in which this gene was imaged, and then by the expression patterns exhibited in those images.
- **R5:** Browse images from a particular strain by the (set of) expression pattern(s) shown in the images.
- **R6:** Find all the images of a particular strain NOT showing a certain set of expression patterns.

EPrints supports requirements R1-3, but not the others. It allows users to construct conjunctive conditions (such as both Mid_primary_spermatocyte and Mid_elongation-stage_spermatid) in the search interface, but not in the browsing interface.

EPrints provides *views* for users to browse repository records that are grouped by the value of their metadata properties. These views are dynamic web pages automatically generated by EPrints' built-in Perl scripts. They are capable of sorting and presenting repository resources by each different value of a property, but not by *a set of* values. For example, EPrints can provide a "Strain View" which groups images by each of the six different strain names, but it cannot provide a view grouping images by different sets of expression patterns, as required by R4-5. R6 is supported neither in the search interface nor in the browsing interface. To support requirements R4-6 using EPrints alone, we would have needed to put substantial effort into modifying EPrints' Perl source code, which would have consumed valuable human resources and led to a less sustainable software package.

Although R4-6 are less frequently required than R1-3, they are nevertheless essential to enable researchers to organise and integrate these images effectively, for example, by comparing the expression patterns of different mutants of the same

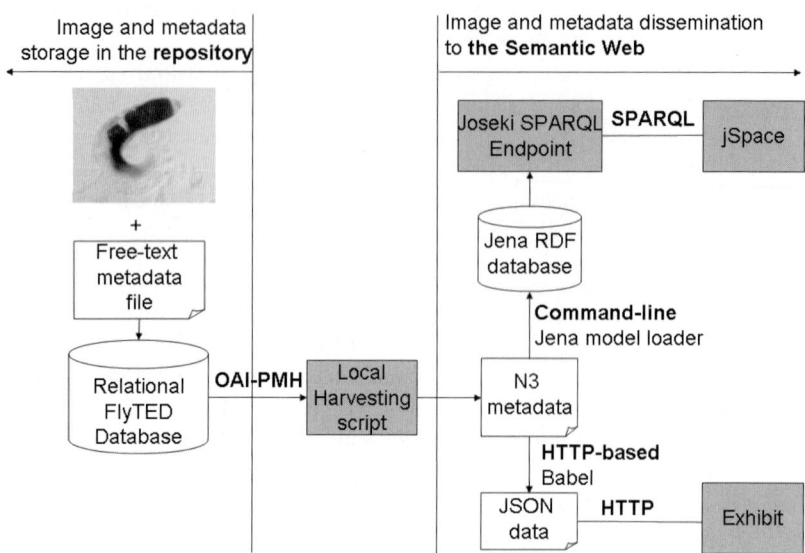

Fig. 1. The software framework of building a Semantic Web accessible image repository

gene or reviewing the group of genes expressed at different locations throughout spermatogenesis.

Requirements R4-5 closely correlate to the vision proposed by existing Semantic Web faceted browsing tools [5,6,7]. These tools provide user interfaces either for RDF data (such as Exhibit [6]) or for resources exposed through SPARQL endpoints (such as jSpace [7]). This means that by making our image metadata Semantic Web accessible, as either RDF metadata or a SPARQL endpoint, we would obtain an opportunity of exploring alternative and more flexible user interfaces for our images and associated metadata.

4 Publishing Semantic Web Accessible Metadata

Figure 1 shows the software framework that enables our images and their domain-specific metadata to become accessible to the Semantic Web. The gap between the EPrints FlyTED repository and the Semantic Web is bridged by a local harvesting script, which harvests domain-specific image metadata from FlyTED through OAI-PMH, and writes this metadata in Notation 3 (N3) format to a local disk. We then disseminate this RDF metadata to the Semantic Web, directly or through a SPARQL endpoint, enabling our image metadata to be accessible to any Semantic Web data resources and applications.

4.1 Metadata Harvesting

The local harvesting script extracted image metadata from the FlyTED repository and constructed semantic metadata in three steps:

- Harvesting of DC metadata of all the repository records using OAI-PMH.
- Analysis of the DC metadata of each record to extract the value of `dc:identifier` for two URLs: one pointing to the image and the other to its metadata file.
- Retrieval of the image metadata file from the FlyTED repository through the HTTP protocol using the metadata file URL, and construction of statements about this image in RDF.

As said in Sect. 2.3, both the images in FlyTED and their image metadata files are made Web-accessible. The meaning of each metadata term in the image metadata files had been well understood when we first constructed these files from our researchers' Excel spreadsheets during the image ingest. This understanding guided us in the creation of classes and properties that make RDF statements about each image using these metadata terms.

The harvesting script was written in Java, and built upon the open source OAIHarvester2 APIs from the Online Computer Library Center (OCLS)[11]. This harvesting script, and also the image ingest scripts for EPrints (written in Perl and Python), are available at our SVN repository[12].

4.2 Creating the SPARQL Endpoint

We built a SPARQL endpoint over FlyTED[13] using the Jena/Joseki[14] toolkit. This approach shows the following two advantages:

- The ability to use a lightweight HTTP-based protocol: the local harvesting script, described above, achieved harvesting metadata from the relational repository database through the simple HTTP-based OAI-PMH protocol.
- The ability to use a lightweight toolkit: the Jena model loader, a command-line tool from Jena, supports loading any RDF data into a Jena database without requiring any code writing, and its generic software interface means that it can be executed in any programming language. This makes our framework more sustainable.

This FlyTED SPARQL endpoint not only allows us to expose our image metadata through a programmatic interface that permits querying of metadata but also provides an interface for developers to execute SPARQL queries that are not included in the list of our users' requirements.

To summarize, this software framework for enabling our images and their metadata Semantic Web accessible (Figure 1) shows the following advantages:

- It avoids defining tightly constrained interfaces between components that can be inflexible and fragile when faced with evolving needs.
- It creates a lightweight software environment using simple RESTful (Representational State Transfer) [8] interfaces and HTTP-based protocols.

---

[11] http://www.oclc.org/research/software/oai/harvester2.htm
[12] https://milos2.zoo.ox.ac.uk/svn/ImageWeb/FlyTED/Trunk/
[13] http://www.fly-ted.org/sparql/
[14] http://www.joseki.org/

- It minimises the cost of development effort by adopting or adapting existing tools and services.
- It maximises the opportunity of replacing or updating any element of the technology used.

The images and their metadata, thus made available on the Semantic Web, can now be explored using Semantic Web faceted browsers, which present data in more flexible ways than does EPrints, without requiring any software installation.

5 Faceted Image Browsing

A faceted browser [5,6,7] presents categories of a knowledge domain to a user, which assists the user in filtering information by selecting or combining categories. This type of interaction enables users to query and manipulate information in an intuitive manner without having to construct logically sophisticated queries, which requires specialised knowledge about query languages and the underlying data model. This freedom is what is greatly valued by biological researchers: provision of a flexible interface for exploring their datasets without imposing additional cognitive demands. Two faceted browsers, Exhibit [6] and jSpace [7], were evaluated for accessing our image data.

5.1 The Exhibit Approach

Exhibit is a lightweight server-side data publishing framework that allows people with basic HTML knowledge to create Web pages with rich, dynamic visualisation of structured data, and with faceted browsing [6]. To 'exhibit' FlyTED, we needed to create the dataset that could be consumed by Exhibit, and the HTML page to present these data.

Our local harvesting script exports image metadata from the FlyTED repository into N3 format, which we then translate into JSON format[15] using Babel[16]. The initial HTML code was copied from Exhibit's tutorial and then customised based on the feedbacks from our users. Figure 2 shows the first 10 thumbnails of the 38 wild type gene expression images that reveal expressions in both Mid_primary_spermatocyte and Mid_elongation-stage_spermatid. Clicking on an image caption will take users to a pop-up bubble that displays metadata details about this image, including a live URL of this image record in the FlyTED repository.

5.2 The jSpace Approach

jSpace [7] is a faceted browser that has been inspired by mSpace [5] (which was not available for evaluation at the time of writing). It is a client-side Java Web Start application, and it supports browsing of RDF data presented by SPARQL

---
[15] http://www.json.org/
[16] http://simile.mit.edu/babel/

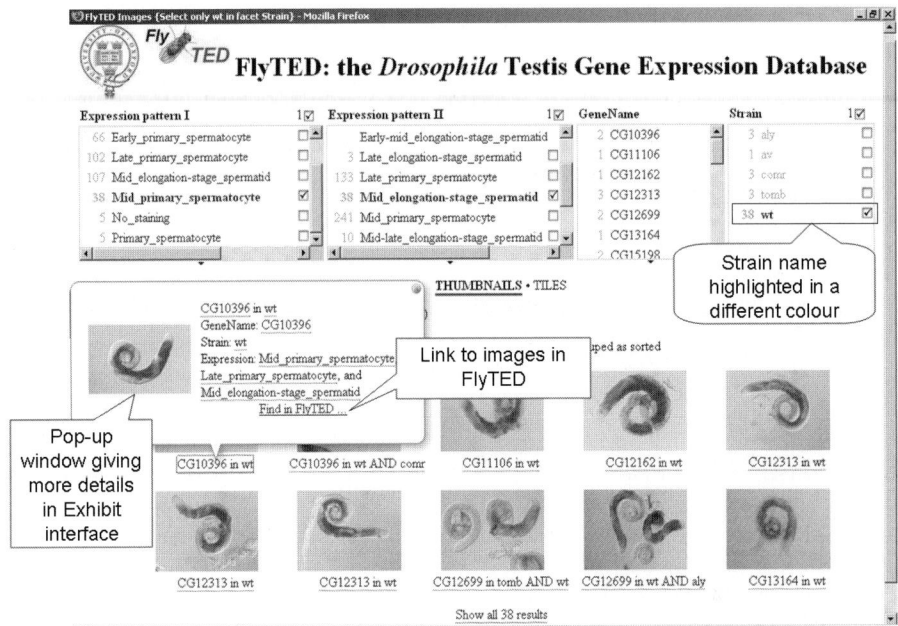

Fig. 2. Presenting FlyTED in Exhibit

endpoints, or kept in local files or Sesame [9] RDF databases. We used jSpace to access images and their domain metadata through the FlyTED Joseki SPARQL endpoint. To present FlyTED in jSpace we needed to:
- Publish our metadata as a SPARQL endpoint, which could then be accessed by jSpace using the SPARQL protocol.
- Create a jSpace model file, which defines the facets used for filtering and presenting images.
- Create a *Web view builder* to load images from the FlyTED repository Web site.

Out of the box, jSpace does not support browsing multimedia data content, but it can load and display Web pages which contain multimedia content using its Web view builder API. By default, this builder takes the string value of a currently selected resource, searches for that term on Google restricted to site:wikipedia.org, and navigates to the first hit. For our experiment, we created an alternative customised Web view builder, which takes the URL of a currently selected image and then navigates to the FlyTED repository site using that image's URL. Figure 3 shows jSpace being used to browse gene expression images recorded from wild type flies that reveal expressions in both Mid_primary_spermatocyte and Mid_elongation-stage_spermatid. Selecting an image URI in the image column brings, in this example, the expression image of gene CG10396 in wild type flies retrieved from FlyTED. Compared to Exhibit, the major limitation of jSpace is that links displayed in the retrieved

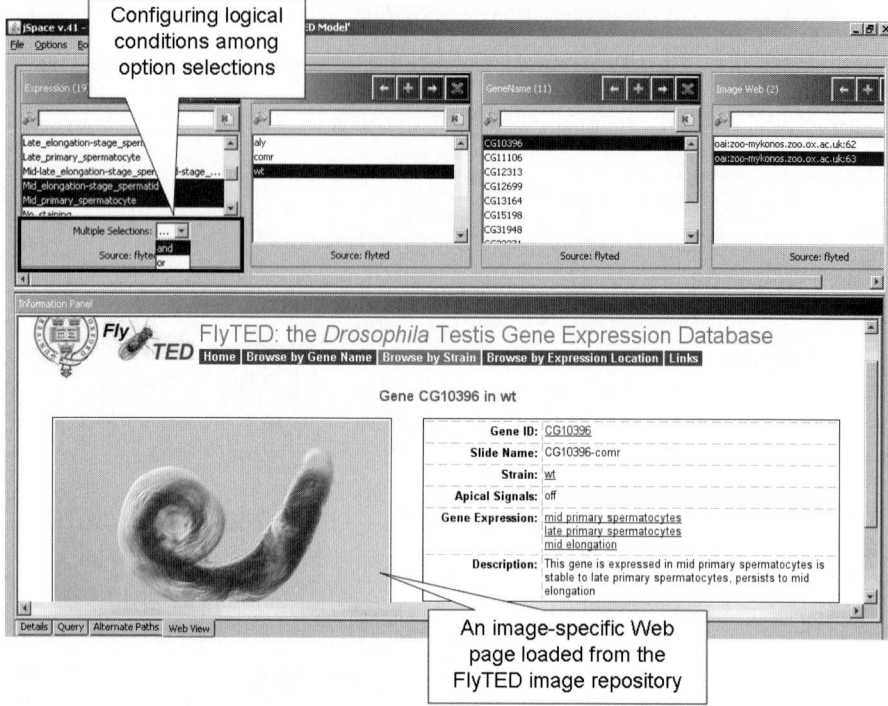

Fig. 3. Presenting FlyTED in jSpace

FlyTED web page are not active and that a view showing thumbnails of all retrieved images side by side is not available. But jSpace has better support for constructing logical conditions between option selections (see Figure 3).

5.3 Functionality Measurement

Table 1 compares the support for our users' image browsing requirements given by EPrints, Exhibit and jSpace. The result shows that the faceted browsers provide more flexible interfaces than the conventional repository interface, which can only support R1-3. However, a list of inadequacies remain to be addressed by the faceted browsers, which include:

– Support for users in the construction of complex conditions between multiple selections, including AND, OR (possible in jSpace) and NOT operators.
– Support for users' customisation of their sorting conditions, for example, presenting the *wild type* strain at the top of the strain categorisation, rather than providing only the *ad hoc* string index in alphabetical order.
– Support for the organisation of facet conditions in a tree structure, so that users can zoom in/out within one category.
– Support for image browsing by presenting a group of images as thumbnails (possible in Exhibit), or presenting an enlarged view of the image when the user hovers the mouse over the thumbnail image.

Table 1. Functionality comparison between EPrints, Exhibit and jSpace

Requirements	EPrints	Exhibit	jSpace
R1: Browse images by the gene name, strain name, or expression pattern keyword, preferably with images presented as thumbnails	Yes	Yes	Partial
R2: Search for all images of a gene that show the same expression pattern	Yes	Yes	Yes
R3: Search for all images of a gene that show the same *set* of expression patterns	Yes	Yes	Yes
R4: Browse images of a single gene, first by the different strains in which this gene was imaged, and then by the expression patterns exhibited in those images	No	Partial	Yes
R5: Browse images from a particular strain by the (set of) expression patterns shown in the images	No	Partial	Yes
R6: Find all the images of a particular strain NOT showing a certain set of expression patterns	No	No	No

5.4 Performance

Figure 4 shows the performance achieved for loading and browsing images and their metadata in datasets of varying sizes in Exhibit and jSpace respectively. The test was performed using a laptop of 1GB memory. Our goal is to compare how these two tools perform for browsing an image dataset of moderate size on an average personal desktop where most scientists would work on.

There are currently around 26,000 RDF triples for ~1500 images in our Jena RDF repository and in the JSON metadata file. The size of our dataset is likely to grow at least three times by the end of the biological image gathering.

The evaluation aimed to discover how these two tools performed for typical image browsing tasks shown in Figure 2 and 3 as the size of data grows. The results show that:1) Exhibit takes on average 10 times longer to load the dataset into a Web browser than jSpace (Figure 4(a)) sending the query and receiving the SPARQL response from the local FlyTED endpoint, and 2) Exhibit takes 2-4 times less time to browse images and their metadata than jSpace (Figure 4(b)).

Exhibit loads all the metadata into memory during initialisation. This makes its subsequent image browsing more responsive, but means that coping with large datasets becomes more difficult. The loading time in Exhibit grows linearly with the growth of data and its scalability problem is known to its developers. In jSpace, each selection of an option leads to a HTTP call between the client and the SPARQL endpoint. The scalability of jSpace thus relates to how many triples are returned from the client for each SPARQL query.

5.5 Development Cost

Both tools required no installation and were easy to start up. The existing examples on both tools' web sites provided sufficient information that could be copied and pasted to create either the model files required for jSpace or the

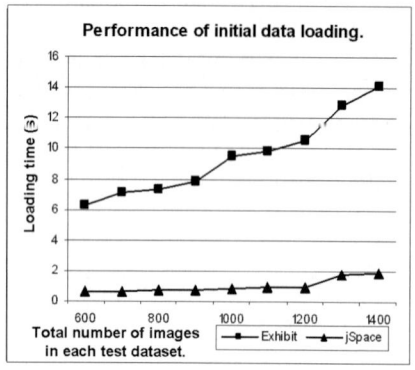 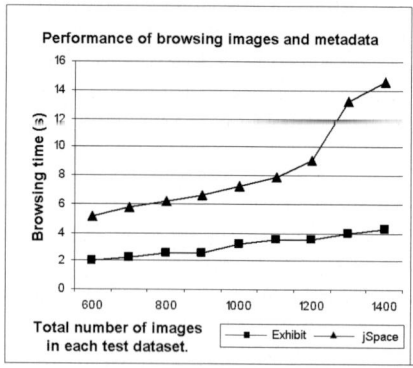

(a) Loading time (LT) of Exhibit and jSpace. LT_Exhibit=Load the JSON data into Web browser; LT_jSpace =Send the query and receive the SPARQL response.

(b) Browsing time (BT) of Exhibit and jSpace. BT_Exhibit=$\sum$ response time of the Web Browser for each selection of an option; BT_jSpace =$\sum$ SPARQL response time of each selection.

Fig. 4. The performance comparison between Exhibit and jSpace

HTML pages and JSON data needed for Exhibit. jSpace required moderate extra effort to build the customised Web view builder, but its facet widgets provide more flexibility for users. Total effort required: 3 working days for jSpace and 2 working days for Exhibit.

5.6 Summary

We compared the costs and benefits of two Semantic Web faceted browsers, Exhibit and jSpace, for accessing images in the FlyTED image repository. The experiments showed that these tools provided more flexible user interfaces than does the conventional EPrints repository software. However, it also revealed gaps in the functionality of these tools, which fell short both in satisfying user requirements and in tool performance.

6 Related Work

Semantic digital library systems [10] describe repository resources with rich semantics so that they can be integrated with other resources and consumed by machines. This is closely in line with our goal of building a Semantic Web accessible image repository. Some existing digital libraries systems (such as Jerome DL [11]) also provide a "social semantic web library", which builds a social network among library users to promote sharing of resources and knowledge. However, these systems, compared to existing digital repository systems, are still in their early stage. Their stability and scalability are yet to be tested, and their components are often tightly coupled. The ability of extending existing digital repository software to provide the extra functionalities of semantic digital libraries is yet to be verified.

Researchers in Southampton have experimented with using mSpace to browse a knowledge repository of heterogenous digital collections, including publication information from their university EPrints repositories and information about researchers from their web sites or funding bodies [12]. Their previous experience encouraged this work.

7 Conclusions

This paper reports our experience in building a Semantic Web accessible image repository by combining an existing repository software package with Semantic Web tools. This approach bridged the gap between conventional repositories and the Semantic Web. The latter provides facilities for disseminating repository resources so that they can be processed along with information from third parties, and for visualising these resources in more flexible user interfaces.

The contributions of this paper are threefold:

- It shows the feasibility of building a Semantic Web accessible image repository using existing tools and simple HTTP-based protocols. This saved us from having to build a repository software system from scratch to achieve the desired functionalities.
- It demonstrates that although existing Semantic Web faceted browsers do provide more flexible user interfaces, they have limitations in supporting a real-world scientific usage. Some of the missing functionalities are likely to be required in different application contexts, such as supporting logical combinations of conditions in one facet; while others are required by the challenges of presenting image data, such as loading multiple images and presenting them side by side.
- It illustrates the significant advantages of employing a lightweight software framework: saving development effort by reusing existing toolkits; and providing the flexibility of replacing components with alternative tools, which increases the sustainability of this framework and enables us to experiment with different approaches.

The provision of the FlyTED repository information in a Semantic Web-accessible form is our first step. We now plan to integrate this information with other related data resources in a data web [13,14], in order to provide a unified platform for scientists to access and integrate these distributed and heterogeneous resources. We also anticipate continuing our experiment with faceted browsers, and thereby to contribute requirements to the browser developers from real scientific studies that involve frequent interactions with image data.

Acknowledgements

The FlyTED Database was developed with funding from the UK's BBSRC (Grant BB/C503903/1, Gene Expression in the *Drosophila* Testis, to Drs Helen White-Cooper and David Shotton). This work was supported by funding from the JISC (Defining Image Access Project to Dr David Shotton; http://imageweb.zoo.ox.ac.uk/wiki/index.php/DefiningImageAccess), and from BBSRC (Grant BB/E018068/1, The FlyData Project: Decision Support and Semantic Organisation of Laboratory Data in *Drosophila* Gene Expression Experiments, to Drs David Shotton and Helen White-Cooper). Help from Michael Grove of jSpace and from the developers of Exhibit is gratefully acknowledged.

References

1. Benson, E., Gudmannsdottir, E., Klyne, G., Shotton, D., White-Cooper, H.: FlyTED - The *Drosophila* Testis Gene Expression Database. In: Proc. of the 20th European Drosophila Research Conference, Vienna, Austria (2007)
2. Lagoze, C., de Sompel, H.V., Nelson, M., Warner, S.: The Open Archives Initiative Protocol for Metadata Harvesting - Version 2 (June 2002), http://www.openarchives.org/OAI/openarchivesprotocol.html
3. Shotton, D., Zhao, J., Klyne, G.: JISC Defining Image Access Project Final Report - Images and Repositories: Present status and future possibilities. Section 12: Project software developments. Project report, University of Oxford (August 2007), http://imageweb.zoo.ox.ac.uk/pub/2007/DefiningImageAccess/FinalReport/
4. Prud'hommeaux, E., Seaborne, A.: SPARQL query language for RDF, W3C recommendation (January 2008), http://www.w3.org/TR/rdf-sparql-query/
5. schraefel, m.c., Wilson, M.L., Russell, A., Smith, D.A.: mSpace: Improving information access to multimedia domains with multimodal exploratory search. Communication of the ACM 49(4), 47–49 (2006)
6. Huynh, D.F., Karger, D.R., Miller, R.C.: Exhibit: Lightweight structured data publishing. In: Proc. of the 16th International Conference on World Wide Web, Banff, Canada, pp. 737–746 (2007)
7. jSpace: a tool for exploring complex information spaces, http://clarkparsia.com/jspace/
8. Fielding, R.T.: Architectural styles and the design of network-based software architectures. ch. 5: Representational state transfer (REST). Ph.D. Thesis, Department of Information and Computer Science, University of California, Irvine (2000)
9. Broekstra, J., Kampman, A., van Harmelen, F.: Sesame: a generic architecture for storing and querying RDF and RDF Schema. In: Proc. of the 1st International Semantic Web Conference, Sardinia, Italy, June 2002, pp. 54–68 (2002)
10. Kruk, S.R., Decker, S., Haslhofer, B., Knežević, P., Payette, S., Krafft, D.: Semantic digital libraries (May 2007), http://wiki.corrib.org/index.php/SemDL/Tutorial/WWW2007
11. Kruk, S.R., Decker, S., Zieborak, L.: JeromeDL - Adding Semantic Web technologies to digital libraries. In: Andersen, K.V., Debenham, J., Wagner, R. (eds.) DEXA 2005. LNCS, vol. 3588, pp. 716–725. Springer, Heidelberg (2005)

12. schraefel, m.c., Smith, D.A., Carr, L.A.: mSpace meets EPrints: a case study in creating dynamic digital collections. Technical report, University of Southampton (January 2006)
13. FlyWeb: Data web for linking laboratory image data with repository publications, http://imageweb.zoo.ox.ac.uk/wiki/index.php/FlyWeb_project
14. Shotton, D.: Data webs for image repositories. In: World Wide Science: Promises, Threats and Realities, Oxford University Press, Oxford (in press, 2008)

Mapping Validation by Probabilistic Reasoning *

Silvana Castano[1], Alfio Ferrara[1], Davide Lorusso[1],
Tobias Henrik Näth[2], and Ralf Möller[2]

[1] Università degli Studi di Milano,
DICo, 10235 Milano, Italy
{castano,ferrara,lorusso}@dico.unimi.it
[2] Hamburg University of Technology,
Institute of Software Systems,
21079 Hamburg, Germany
{r.f.moeller,tobias.naeth}@tu-harburg.de

Abstract. In the semantic web environment, where several independent ontologies are used in order to describe knowledge and data, ontologies have to be aligned by defining mappings among the elements of one ontology and the elements of another ontology. Very often, mappings are not derived from the semantics of the ontologies that are compared. Rather, mappings are computed by evaluating the similarity of the ontologies terminology and/or of their syntactic structure. In this paper, we propose a new mapping validation approach. The approach is based on the notion of probabilistic mappings and on the use of probabilistic reasoning techniques to enforce a semantic interpretation of similarity mappings as probabilistic and hypothetical relations among ontology elements.

1 Introduction

In the semantic web environment, as well as in any information system where two or more independent ontologies are used in order to describe knowledge and data, some kind of ontology interoperability is needed. It can be realized either by integrating the different ontologies into a new one or by aligning them with a mechanism for translating a query over an ontology A into a query over another ontology B. In both cases, two (or more) ontologies have to be aligned by defining correspondences among the elements (e.g., concepts, properties, instances) of one ontology and the elements of another ontology. Such correspondences are called *mappings*. Usually a mapping is featured by a relation holding between the two elements mapped and a measure in the range [0,1], which denotes the strength of the relation between the two elements. In literature, many ontology matching approaches and systems have been proposed for automatically discovering ontology mappings [1]. In most of the proposed approaches, mappings are not derived from the semantics of the ontologies that are compared, but, rather, from an evaluation of the similarity of the terminology used in the two ontologies

---

* This paper has been partially funded by the BOEMIE Project, FP6-027538, 6th EU Framework Programme.

or of their syntactic structure. The reason is that mapping discovery is not really a deductive problem, if the two ontologies are really independent. In fact, the information about the correspondence of an element of the first ontology with an element of the second one is not implicitly contained in any of the two. However, mappings derived from linguistic/syntactic approaches cannot be interpreted as semantic relations among the ontology elements as it should be in order to use them for query processing. On the other hand, some approaches for detecting mappings with the support of logical reasoning have been proposed [2]. In this case, a set of semantic relations is derived from mappings, but all the relations are interpreted as ontology axioms. However, a certain level of uncertainty is intrinsically due to the mechanism used for mapping discovery. In fact, ontology mapping tools derive such mappings from uncertain sources or techniques (e.g., string matching, graph matching) or they rely on external lexical sources (e.g., WordNet) that are not specifically tailored for the domain at hand. This uncertainty leads to mapping inaccuracy. A way to overcome inaccuracy is to manually validate each mapping not only per se (i.e., to state that the mapping is correct with respect to the meaning of the involved elements), but also with respect to all the other mappings in the set. Of course, this activity is time consuming and it would however be affected by a subjectivity component, as it is performed by human experts. In all these cases, a post-processing activity on ontology mappings is required in order to validate mappings with respect to the semantics of the ontologies involved and, at the same time, by maintaining the uncertain nature of mappings. In this paper, we propose a new mapping validation approach for interpreting similarity-based mappings as semantic relations, by coping also with inaccuracy situations. The idea is to see two independent ontologies as a unique distributed knowledge base and to assume a semantic interpretation of ontology mappings as probabilistic and hypothetical relations among ontology elements. We present and use a probabilistic reasoning tool in order to validate mappings and to possibly infer new relations among the ontologies.

The paper is organized as follows: in Section 2, we present our approach and we show an example of mapping acquisition. In Section 3, we present the P-$\mathcal{SHIQ}$(D) formalism and we show how mappings can be represented as probabilistic knowledge. In Section 4, we discuss probabilistic reasoning and its use for mapping validation. In Section 5, we discuss some related work concerning our approach. Finally, in Section 6, we give our concluding remarks.

2 Ontology Mappings

Our approach to mapping validation is articulated in four phases.

1. *Ontology mapping acquisition.* In this phase, we acquire mappings produced using an ontology matching system; the matching system can rely on syntactic, linguistic, structural, or even semantic matching techniques.
2. *Probabilistic interpretation of mappings.* In this phase, acquired mappings are interpreted as probabilistic constraints holding among the elements of the

two ontologies; in other words, the mapping is associated with a probability that the relation expressed by the mapping is true.
3. *Probabilistic reasoning over mappings.* In this phase, the merged ontology obtained by combining the two initial ontologies using the semantic relations derived from probabilistic mappings is validated by means of a probabilistic reasoning system.
4. *Mapping validation and revision.* In this phase, mappings are revised according to the reasoning results; mappings causing incoherences within the merged ontology are discarded while new mappings are inferred from the valid mappings.

In this paper, the main contribution regards the representation of mappings as probabilistic knowledge and the reasoning-driven mapping validation (phases 2, 3, and 4), which are addressed in more detail in Sections 3, and 4, respectively. In the remainder of this section, we first introduce the notion of probabilistic mapping and then we describe an example of acquisition of linguistic mappings generated by our ontology matching system HMatch 2.0 [3].

2.1 Probabilistic Mappings

Given two ontologies O_1 and O_2, a mapping set $\mathcal{M}_{O_1}^{O_2}$ is a collection of mappings $m_{i,j,r}$ of the form:

$$m_{i,j,r} = \langle r(i,j), v \rangle$$

where $i \in O_1$ and $j \in O_2$ are two elements of O_1 and O_2, respectively, r is a binary relation holding between i and j, and $0 \leq v \leq 1$ is a value denoting the strength of the relation r. Such kind of mappings can be automatically defined by a wide range of available techniques and by using one of the ontology matchmaking tools proposed in literature. Goal of the mappings is to define which is the kind of semantic relation which holds between an element of O_1 and an element of O_2. The relation r associated with a mapping depends on the matching technique used. In most cases, it denotes a generic correspondence between two elements (e.g., similarity), while, in other cases, it represents a terminological relationship between the names of i and j (e.g., synonymy, hypenymy, hyponymy). The value v usually denotes the strength of the belief in the proposition $r(i,j)$. In order to take into account the intrinsic inaccurate nature of a mapping, we define a probabilistic mapping as follows.

Definition 1. *Probabilistic Mapping. Given two ontology elements i and j, a probabilistic mapping $pm_{i,j,r}$ between i and j is a mapping $m_{i,j,r}$ whose value v is the probability that the mapping relation $r(i,j)$ is true, that is:*

$$pm_{i,j,r} = \langle r(i,j), v \rangle, such\ that\ Pr(r(i,j)) = v$$

In defining probabilistic mappings out of generic mappings, we introduce the hypothesis that the strength of the relation holding between two mapped elements can be expressed in terms of probabilities. In particular, we want to measure the probability of the assertion $r(i,j)$ to be true.

2.2 Mapping Acquisition

In mapping acquisition, we work under the assumption that the mechanism used for ontology matching is compatible with a probabilistic interpretation of the resulting mappings. In this paper, we show how this can be achieved using a linguistic matching technique based on the WordNet lexical system. In particular, we rely on the linguistic matching technique of HMatch 2.0, which determines the linguistic mapping between two ontology elements (i.e., concepts, properties, individuals) on the basis of their names and of the terminological relationships holding between in WordNet. The mapping degree $v \in [0,1]$ is calculated through a Linguistic Affinity function (LA), which returns a value proportional to the probability that a terminological relationship holds between two terms. Given two elements i and j of two ontologies O_1 and O_2, the interaction between HMatch 2.0 and the lexical system WordNet produces a set of linguistic mappings of the form:

$$m_{i,j,r} = \langle r(T_i, T_j), v \rangle$$

where:

- T_i and T_j are terms used as names for i and j, respectively.
- $r \in \{\text{SYN}, \text{BT}, \text{NT}\}$ is a terminological relationship defined in WordNet for T_i and T_j, where SYN denotes synonymy, BT denotes hypernymy, and NT denotes hyponymy.
- v is the mapping value in the range [0,1] associated with r.

The LA function works on the synsets of WordNet. In WordNet, each possible meaning of a term is represented by a set of synonyms, called *synset*. Given two terms T_i and T_j, we search in WordNet the set R of terminological relationships holding between at least one synset of T_i and at least one synset of T_j. When linguistic mappings are interpreted as probabilistic mappings, many terms can have more than one meaning and we do not know which specific meaning has been used in the ontologies to be compared. If we find a terminological relationship $r_i(s_l, s_k) \in R$ between two synsets s_l of T_i and s_k of T_j, the probability $Pr(r_i(s_l, s_k))$ is equal to $Pr(Pr(s_l) \wedge Pr(s_k))$, where $Pr(s_l)$ is the probability that the intended meaning of T_i in its ontology is expressed by the synset s_l, while $Pr(s_k)$ is the probability that the intended meaning of T_j in its ontology is expressed by the synset s_k, respectively. In general, given a term T, the probability $Pr(s_t)$ that s_t is the synset which denotes the intended meaning of T depends on the total number N_T of synsets for T in WordNet, such that $Pr(s_t) = \frac{1}{N_T}$. Thus, the probability $Pr(r_i(s_l, s_k))$ of the terminological relationship r_i to be true is defined as:

$$Pr(r_i(s_l, s_k)) = Pr(Pr(s_l) \wedge Pr(s_k)) = \frac{1}{N_{T_i}} \cdot \frac{1}{N_{T_j}}$$

Example. Let us consider the two simple ontologies of Figure 1. In order to define mappings between concepts described as classes, we consider their names.

Ontology 1	Ontology 2
hotel ⊑ building ⊓ ∃hosts.tourist	accommodation ≡ ∃hosting.traveler motel ⊑ accommodation hostel ⊑ accommodation camping ⊑ accommodation

Fig. 1. Example of two ontologies in the touristic domain

Consider the terms hotel and hostel. In WordNet, hotel is associated only with a synset $s(hotel)_1$, defining the hotel as "a building where travelers can pay for lodging and meals and other services"; hostel is associated with two synsets $s(hostel)_1$ and $s(hostel)_2$, stating that a hostel is a "hotel providing overnight lodging for travelers" and a "inexpensive supervised lodging", respectively. By means of the linguistic component of HMatch 2.0, we retrieve the relationship $BT(s(hotel)_1, s(hostel)_1)$. Our intuition is that such a relation is correct if the intended meaning of the term hotel is expressed by $s(hotel)_1$ and if the intended meaning of the term hostel is expressed by $s(hostel)_1$. In the first case, since hotel has only one meaning, the probability is equal to one. In the second case, since hostel has two possible meanings, the probability of $s(hostel)_1$ to be the correct synset is equal to $1/2 = 0.5$. Then, the comprehensive probability of the relation between hotel and hostel is calculated as follows:

$$Pr(BT(s(hotel)_1, s(hostel)_1)) = \frac{1}{1} \cdot \frac{1}{2} = 0.5$$

In order to define linguistic mappings between concepts expressed by a property restriction of the form $(\exists \mid \forall \mid \geq n \mid \leq n \mid = n)R.C$, the quantifier must be equal and a linguistic mapping between the two properties R and the two concepts C must hold. In particular, given a terminological relationship r_1 between the properties of two restrictions and a terminological relationship r_2 between the concepts of two restrictions, we establish a terminological relationship r_3 between the two restrictions by choosing r_3 as the most restrictive terminological relationship between r_1 and r_2. The probability associated with r_3 is calculated as $Pr(r_3) = Pr(r_1) \cdot Pr(r_2)$. For example, take into account the concepts $\exists hosts.tourist$ and $\exists hosting.traveler$. Given the mappings $SYN(hosts, hosting)$ with probability 1.0 and $NT(tourist, traveler)$ with probability 1.0, we obtain the mapping:

$$Pr(NT(\exists hosts.tourist, \exists hosting.traveler)) = 1.0$$

When the two ontologies of Figure 1 are aligned, we collect all the mappings available among the concepts of the two ontologies, that are shown in Figure 2 together with their probabilities.

3 Formalizing Probabilistic Mappings Using P-$\mathcal{SHIQ}$(D)

In this section, we present P-$\mathcal{SHIQ}$(D), the probabilistic description logics we use in order to formalize probabilistic mappings as semantic relations.

Mapping relation	Probability
BT(hotel, motel)	1.0
NT(∃hosts.tourist, ∃hosting.traveler)	1.0
NT(tourist, traveler)	1.0
BT(hotel, hostel)	0.5
BT(building, motel)	0.25
BT(building, hostel)	0.125

Fig. 2. Example of probabilistic mappings between the ontologies of Figure 1

3.1 A Probabilistic Extension of $\mathcal{SHIQ}(D)$

In order to extend a description logic for dealing with probabilistic knowledge an additional syntactical and semantical construct is needed. The additional construct is called a *conditional constraint*. This extension of description logics has been first formalized in [4] for $\mathcal{SHOQ}(D)$. Later the extension was adapted to $\mathcal{SHIF}(D)$ and $\mathcal{SHOIN}(D)$ in [5], due to the fact that probabilistic reasoning problems are reduced to solving linear programs and standard satisfiability tests regarding the underling DL. We use $\mathcal{SHIQ}(D)$ as underling DL, hence the name P-$\mathcal{SHIQ}(D)$. For Syntax and Semantic of $\mathcal{SHIQ}(D)$ the reader is referred to [6]. A conditional constraint consists of a statement of conditional probability for two concepts C, D as well as a lower bound l and an upper bound u on that probability. It is written as follows: $(D|C)[l, u]$ Where C can be called the *evidence* and D the *hypothesis*. To gain the ability to store such statements in a knowledge base it has to be extended to a probabilistic knowledge base $\mathcal{PKB}$. Additionally to the TBox $\mathcal{T}$ of a description logic knowledge base we introduce the PTBox $\mathcal{PT}$, which consists of $\mathcal{T}$ and a set of conditional constraints $\mathcal{D}_g$, and a PABox $\mathcal{PA}$ holding sets of conditional constraints $\mathcal{D}_o$ for each probabilistic individual o. We also define the set of probabilistic individuals I_p, which contains all individuals o for which some probabilistic knowledge is available and therefore a set $\mathcal{D}_o$. In [4] there is no ABox declared, knowledge about so called classical individuals is also stored inside the TBox using nominals. $\mathcal{D}_g$ therefore represents statistical knowledge about concepts and $\mathcal{D}_o$ represents degrees of belief about the individual o. To be able to define the semantics for a description logic with probabilistic extension the interpretation $\mathcal{I} = (\Delta^\mathcal{I}, \cdot)$ has to be extended by a probability function μ on the domain of interpretation $\Delta^\mathcal{I}$. The extended interpretation is called the *probabilistic interpretation* $\mathcal{Pr} = (\mathcal{I}, \mu)$. Each individual o in the domain $\Delta^\mathcal{I}$ is mapped by the probability function μ to a value in the interval [0,1] and the values of all $\mu(o)$ have to sum up to 1 for any probabilistic interpretation $\mathcal{Pr}$. With the probabilistic interpretation $\mathcal{Pr}$ at hand the probability of a concept C, represented by $\mathcal{Pr}(C)$, is defined as sum of all $\mu(o)$ where $o \in C^\mathcal{I}$. The probabilistic interpretation of a conditional probability $\mathcal{Pr}(D|C)$ is given as $\frac{\mathcal{Pr}(C \sqcap D)}{\mathcal{Pr}(C)}$ where $\mathcal{Pr}(C) > 0$. A conditional constraint $(D|C)[l, u]$ is *satisfied* by $\mathcal{Pr}$ or $\mathcal{Pr}$ models $(D|C)[l, u]$ if and only if $\mathcal{Pr}(D|C) \in [l, u]$. We will write this as $\mathcal{Pr} \models (D|C)[l, u]$. A probabilistic interpretation $\mathcal{Pr}$ is said to *satisfy* or *model* a terminology axiom T, written $\mathcal{Pr} \models T$, if and only if $\mathcal{I} \models T$.

A set $\mathcal{F}$ consisting of terminological axioms and conditional constraints, where F denotes the elements of $\mathcal{F}$, is satisfied or modeled by $\mathcal{P}r$ if and only if $\mathcal{P}r \models F$ for all $F \in \mathcal{F}$. The *verification* of a conditional constraint $(D|C)[l,u]$ is defined as $\mathcal{P}r(C) = 1$ and $\mathcal{P}r$ has to be a model of $(D|C)[l,u]$. We also may say $\mathcal{P}r$ *verifies* the conditional constraint $(D|C)[l,u]$. On the contrary the *falsification* of a conditional constraint $(D|C)[l,u]$ is given if and only if $\mathcal{P}r(C) = 1$ and $\mathcal{P}r$ does **not** satisfy $(D|C)[l,u]$. It is also said that $\mathcal{P}r$ falsifies $(D|C)[l,u]$. Further a conditional constraint F is said to be *tolerated* under a Terminology $\mathcal{T}$ and a set of conditional constraints $\mathcal{D}$ if and only if a probabilistic interpretation $\mathcal{P}r$ can be found that verifies F and $\mathcal{P}r \models \mathcal{T} \cup \mathcal{D}$. With all these definitions at hand we are now prepared to define the *z-partition* of a set of generic conditional constraints $\mathcal{D}_g$. The z-partition is build as ordered partition $(\mathcal{D}_0, \ldots, \mathcal{D}_k)$ of $\mathcal{D}_g$, where each part $\mathcal{D}_i$ with $i \in \{0, \ldots, k\}$ is the set of all conditional constraints $F \in \mathcal{D}_g \setminus (\mathcal{D}_0 \cup \cdots \cup \mathcal{D}_{i-1})$, that are tolerated under the generic terminology $\mathcal{T}_g$ and $\mathcal{D}_g \setminus (\mathcal{D}_0 \cup \cdots \cup \mathcal{D}_{i-1})$. If the z-partition can be build from a PABox $\mathcal{PT} = (\mathcal{T}, \mathcal{D}_g)$, it is said to be *generically consistent* or *g-consistent*. A probabilistic knowledge base $\mathcal{PKB} = (\mathcal{PT}, (\mathcal{P}_o)_{o \in I_p})$ is *consistent* if and only if $\mathcal{PT}$ is g-consistent and $\mathcal{P}r \models \mathcal{T} \cup \mathcal{D}_o$ for all $o \in I_p$.

We use the z-partition for the definition of the lexicographic order on the probabilistic interpretations $\mathcal{P}r$ as follows: A probabilistic interpretation $\mathcal{P}r$ is called *lexicographical preferred* to a probabilistic interpretation $\mathcal{P}r'$ if and only if some $i \in \{0, \ldots, k\}$ can be found, that $|\{F \in D_i \mid \mathcal{P}r \models F\}| > |\{F \in D_i \mid \mathcal{P}r' \models F\}|$ and $|\{F \in D_j \mid \mathcal{P}r \models F\}| = |\{F \in D_j \mid \mathcal{P}r' \models F\}|$ for all $i < j \leq k$.

We say a probabilistic interpretation $\mathcal{P}r$ of a set $\mathcal{F}$ of terminological axioms and conditional constraints is a *lexicographically minimal model* of $\mathcal{F}$ if and only if no probabilistic interpretation $\mathcal{P}r'$ is lexicographical preferred to $\mathcal{P}r$. By now the meaning of *lexicographic entailment* for conditional constraints from a set $\mathcal{F}$ of terminological axioms and conditional constraints under a PTBox $\mathcal{PT}$ is given as: A conditional constraint $(D|C)[l,u]$ is a *lexicographic consequence* of a set $\mathcal{F}$ of terminological axioms and conditional constraints under a PTBox $\mathcal{PT}$, written $\mathcal{F} \parallel\!\sim (D|C)[l,u]$ under $\mathcal{PT}$, if and only if $\mathcal{P}r(D) \in [l,u]$ for every lexicographically minimal model $\mathcal{P}r$ of $\mathcal{F} \cup \{(C|\top)[1,1]\}$. Tight lexicographic consequence of $\mathcal{F}$ under $\mathcal{PT}$ is defined as $\mathcal{F} \parallel\!\sim_{tight} (D|C)[l,u]$ if and only if l is the infimum and u is the supremum of $\mathcal{P}r(D)$. We define $l = 1$ and $u = 0$ if **no** such probabilistic interpretation $\mathcal{P}r$ exists. Finally we define lexicographic entailment using a probabilistic knowledge base $\mathcal{PKB}$ for generic and assertional conditional constraints F. If F is a generic conditional constraint, then it is said to be a lexicographic consequence of $\mathcal{PKB}$, written $\mathcal{PKB} \parallel\!\sim F$ if and only if $\emptyset \parallel\!\sim F$ under $\mathcal{PT}$ and a tight lexicographic consequence of $\mathcal{PKB}$, written $\mathcal{PKB} \parallel\!\sim_{tight} F$ if and only if $\emptyset \parallel\!\sim_{tight} F$ under $\mathcal{PT}$. If F is an assertional conditional constraint for $o \in I_P$, then it is said to be a lexicographic consequence of $\mathcal{PKB}$, written $\mathcal{PKB} \parallel\!\sim F$, if and only if $\mathcal{D}_o \parallel\!\sim F$ under $\mathcal{PT}$ and a tight lexicographic consequence of $\mathcal{PKB}$, written $\mathcal{PKB} \parallel\!\sim_{tight} F$ if and only if $\mathcal{D}_o \parallel\!\sim_{tight} F$ under $\mathcal{PT}$.

3.2 Interpreting Mappings as Probabilistic Knowledge

The basic idea behind the formalization of mappings as probabilistic knowledge is to transform probabilistic mappings into conditional constraints. In order to do that, we assume as hypothesis that a mapping relation r can be seen as a corresponding constraint over the ontology concepts. As an example, we say that, given a terminological relation $BT(A, B)$ with probability p, it can be interpreted as the probability that an instance of B is also an instance of A. In general, given the set R of relations produced by a matching system, we introduce a set R' of rules for the translation of each kind of relation in R into a corresponding conditional constraint. Taking into account the example of Section 2, we start from the set of relations $R = \{SYN, BT, NT\}$ and we define the following translation rules: i) $Pr(SYN(A, B)) = p \rightarrow (A|B)[p,p] \wedge (B|A)[p,p]$; ii) $Pr(BT(A, B)) = p \rightarrow (A|B)[p,p]$; iii) $Pr(NT(A, B)) = p \rightarrow (B|A)[p,p]$. The goal of the translation process is to assume a hypothetical representation of probabilistic mappings as semantic relations in P-$\mathcal{SHIQ}$(D). Given two $\mathcal{SHIQ}$(D)-compatible ontologies O_1 and O_2, we build a new P-$\mathcal{SHIQ}$(D) ontology $O_{1,2}$ by merging the axioms of O_1 and O_2 and by augmenting the resulting ontology with conditional constraints derived from mappings. Recalling the example of Section 2, the resulting ontology is shown in Figure 3.

TBox	PTBox
hotel$\sqsubseteq$building $\sqcap$ $\exists$hosts.tourist	$(hotel|motel)[1.0, 1.0]$
accommodation$\equiv$$\exists$hosting.traveler	$(\exists hosting.traveler|\exists hosts.tourist)[1.0, 1.0]$
motel$\sqsubseteq$accommodation	$(traveler|tourist)[1.0, 1.0]$
hostel$\sqsubseteq$accommodation	$(hotel|hostel)[0.5, 0.5]$
camping$\sqsubseteq$accommodation	$(building|motel)[0.25, 0.25]$
	$(building|hostel)[0.125, 0.125]$

Fig. 3. P-$\mathcal{SHIQ}$(D) representation of the ontologies of Figure 1

4 Probabilistic Reasoning Over Mappings

Now we will recap the techniques required to reason in P-$\mathcal{SHIQ}$(D). The motivation of this discussion is to present algorithmic issues on the one hand, and point to implementation issues on the other. Subsequently, we describe our approach to mapping validation by using the proposed ContraBovemRufum reasoning System (CBR).

4.1 Reasoning with P-$\mathcal{SHIQ}$(D)

As previously mentioned the proposed approach for solving probabilistic reasoning problems relies on use of standard reasoning services provided by a description logic reasoner in order to build linear programs which may be handed over to a solver in order to decide the solvability of the programs. Although this approach seems attractive usually severe performance problems can be expected

$$\sum_{r \in \mathcal{R}_\mathcal{T}(\mathcal{F}), r \models \neg D \sqcap C} -l y_r + \sum_{r \in \mathcal{R}_\mathcal{T}(\mathcal{F}), r \models D \sqcap C} (1-l) y_r \geq 0 \quad \text{(for all } (D|C)[l,u] \in \mathcal{F})\text{(1a)}$$

$$\sum_{r \in \mathcal{R}_\mathcal{T}(\mathcal{F}), r \models \neg D \sqcap C} u y_r + \sum_{r \in \mathcal{R}_\mathcal{T}(\mathcal{F}), r \models D \sqcap C} (u-1) y_r \geq 0 \quad \text{(for all } (D|C)[l,u] \subset \mathcal{T})\text{(1b)}$$

$$\sum_{r \in \mathcal{R}_\mathcal{T}(\mathcal{F})} y_r = 1 \tag{1c}$$

$$y_r \geq 0 \quad \text{(for all } r \in \mathcal{R}_\mathcal{T}(\mathcal{F}))\tag{1d}$$

Fig. 4. Constraints of the linear program

for expressive logics. The complexity of the used algorithms lies within NP or worse depending on the underlying DL as it has been analyzed in [5]. Despite these results we think it is still worth investigating the approach to develop optimization techniques to bring down average case complexity.

In order to deciding satisfiability the first objective is to build a set $\mathcal{R}_\mathcal{T}(\mathcal{F})$. It contains the elements r, which map conditional constraints $F_i = (D_i|C_i)[l_i, u_i]$, elements of a set of conditional constraints $\mathcal{F}$, onto one of the following terms $D_i \sqcap C_i$, $\neg D_i \sqcap C_i$ or $\neg C_i$ under the condition, that the intersection of our r is not equal to the bottom concept given the terminology $\mathcal{T}$, written $\mathcal{T} \not\models r(F_1) \sqcap \cdots \sqcap r(F_n) \sqsubseteq \bot$. In the following we will write $\sqcap r$ instead of $r(F_1) \sqcap \cdots \sqcap r(F_n)$ as an abbreviation. With the set $\mathcal{R}_\mathcal{T}(\mathcal{F})$ at hand we are able to set up linear programs to decide the satisfiability of the terminology $\mathcal{T}$ and a finite set of conditional constraints $\mathcal{F}$. The constraints of the linear program are displayed in Figure 4. We say that $\mathcal{T} \cup \mathcal{F}$ is satisfiable if and only if the linear program with the constraints 1a-d is solvable for variables y_r, where $r \in \mathcal{R}_\mathcal{T}(\mathcal{F})$. This means, that in the objective function all coefficients preceding the variables y_r are set to 1. We further need to introduce the meaning of $r \models C$ which is used as index of the summation in 1a and 1b. It is an abbreviation for $\emptyset \models \sqcap r \sqsubseteq C$. So the coefficient preceding the variables y_r is set in linear constraints 1a and 1b if either $r \models \neg D \sqcap C$ or $r \models D \sqcap C$ may be proven.

Why is the creation of linear programs reasonable? Consider the following: By definition a conditional constraint is satisfied if $u \geq Pr(D|C) \geq l \Leftrightarrow uPr(C) \geq Pr(D \sqcap C) \geq lPr(C)$. This may lead us to linear constraints 1a and 1b. Lets focus on the upper bound, whose derivation is displayed in Figure 5. The derivation for the lower bound 1a follows analogously. The linear constraints 1c and 1d reflect that all $\mu(o)$ have to sum up to 1 and all $\mu(o) \in [0,1]$ Lets have a look at the number of subsumbtion tests which need to be performed to set up the system of linear constraints. At a first glance, finding a $\models$ under each sum, one might say four tests per variable and conditional constraint. Looking closer we discover that only two are required because they are identical for lower and upper bound. But even this may be optimised further. Considering that the $\sqcap r$ represents a map of all conditional constraints on $D_i \sqcap C_i$, $\neg D_i \sqcap C_i$ or $\neg C_i$ and they are tested on subsumbtion against $D \sqcap C$ and $\neg D \sqcap C$ we observe

$$u \sum_{r \in \mathcal{R}_\mathcal{T}(\mathcal{F}), r \models C} y_r \geq \sum_{r \in \mathcal{R}_\mathcal{T}(\mathcal{F}), r \models D \sqcap C} y_r \Leftrightarrow \quad (2a)$$

$$u \sum_{r \in \mathcal{R}_\mathcal{T}(\mathcal{F}), r \models (\neg D \sqcap C) \sqcup (D \sqcap C)} y_r \geq \sum_{r \in \mathcal{R}_\mathcal{T}(\mathcal{F}), r \models D \sqcap C} y_r \Leftrightarrow \quad (2b)$$

$$u \sum_{r \in \mathcal{R}_\mathcal{T}(\mathcal{F}), r \models \neg D \sqcap C} y_r + u \sum_{r \in \mathcal{R}_\mathcal{T}(\mathcal{F}), r \models D \sqcap C} y_r \geq \sum_{r \in \mathcal{R}_\mathcal{T}(\mathcal{F}), r \models D \sqcap C} y_r \Leftrightarrow \quad (2c)$$

$$\sum_{r \in \mathcal{R}_\mathcal{T}(\mathcal{F}), r \models \neg D \sqcap C} u y_r + \sum_{r \in \mathcal{R}_\mathcal{T}(\mathcal{F}), r \models D \sqcap C} (u-1) y_r \geq 0 \quad (2d)$$

Fig. 5. Upper bound derivation

that only if the first subsumbtion test of $\sqcap r \sqsubseteq D \sqcap C$ failed the second one is necessary. Therefore significantly reducing the number of required tests per variable and conditional constraint. With the tool at hand to decide satisfiability, we may also decide if a conditional constraint may be tolerated by a set of conditional constraints $\mathcal{F}$. To verify a constraint we add a conditional constraint $(C|\top)[1, 1]$. With the extended set the linear program is generated and solved. If an unfeasible solution is computed the conditional constraint is conflicting. If an optimal solution is found, the conditional constraint is tolerated. Now the z-partition of a set of conditional constraints is computable. How to compute tightest probability bounds for given evidence C and conclusion D in respect to a set of conditional constraints $\mathcal{F}$ under a terminology $\mathcal{T}$? The task is named *tight logical entailment* and denoted $\mathcal{T} \cup \mathcal{F} \models_{tight} (D|C)[l, u]$. Given that $\mathcal{T} \cup \mathcal{F}$ is satisfiable, a linear program is set up for $\mathcal{F} \cup (C|\top)[1, 1] \cup (D|\top)[0, 1]$. The objective function is set to $\sum_{r \in R, r \models D} y_r$. So the coefficient in front of the variables y_r are set 1 if $r \models D$. The tight logical entailed lower bound l is computed by minimising, respectively the upper bound u by maximising the linear program. In order to compute tight probabilistic lexicographic entailment for given evidence C and conclusion D under a $\mathcal{PKB}$ the following steps have to be taken:

1. Compute the z-partition of $\mathcal{D}_g$ in order to be able to generate a lexicographic ordering
2. Compute lexicographic minimal sets $\mathcal{D}'$ of conditional constraints of $\mathcal{D}_g$ as elements of $\overline{\mathcal{D}}$.
3. Compute the tight logical entailment $\mathcal{T} \cup \mathcal{F} \cup \mathcal{D}' \models_{tight} (D|C)[l, u]$ for all $\mathcal{D}' \in \overline{\mathcal{D}}$.
4. Select the minimum of all computed lower bounds and the maximum of all upper bounds.

The 2. step needs some explanation since a new task "compute *lexicographic minimal sets*" is introduced. In order to define a lexicographic minimal set $\mathcal{D}'$, a preparatory definition is required. A set $\mathcal{D}' \subset \mathcal{D}_g$ *lexicographic preferable* to

$\mathcal{D}'' \subset \mathcal{D}_g$ if and only if some $i \in \{0,\ldots,k\}$ exists such that $|\mathcal{D}' \cap \mathcal{D}_i| > |\mathcal{D}'' \cap \mathcal{D}_i|$ and $|\mathcal{D}' \cap \mathcal{D}_i| > |\mathcal{D}'' \cap \mathcal{D}_i|$ for all $i < j \leq k$. With the lexicographic order introduced onto the sets $\mathcal{D}'$ the definition of lexicographic minimal is given as: $\mathcal{D}'$ is lexicographic minimal in $\mathcal{S} \subseteq \{S | S \subseteq \mathcal{D}_g\}$ if and only if $\mathcal{D}' \in \mathcal{S}$ and no $\mathcal{D}'' \in \mathcal{S}$ is lexicographic preferable to $\mathcal{D}'$.

We implemented the algorithms presented by [4,5] in the Java programming language into the ContraBovemRufum System. For the reasoning tasks RacerPro with its JRacer interface is used. As solvers a native Java solver by Opsresearch and the Mosek linear program solver have been integrated. For application programmers two different sets of interfaces are provided. The first set contains the ProbabilisticKBInterface, which provides all operations related to setting up and modifying PTBox and PABox, and the ProbabilisticEntailmentInterface, which offers the probabilistic inference operations to decide consistency for $\mathcal{PT}$ and $\mathcal{PKB}$ as well as probabilistic subsumption and probabilistic instance checking. The second set of interfaces handles the configuration of the services. Using the SolverConfiguration interface the selection of the solver may be changed at runtime. The ReasonerConfiguration interface makes the settings for the reasoner. With the EntailmentConfiguration interface the algorithm used to compute the entailment may be chosen at runtime. Currently tight logical entailment and the two available algorithms for tight lexicographic entailment are supported.

4.2 Mapping Validation and Revision

After mapping acquisition and representation as probabilistic constraints, we want to use tight lexicographic entailment in order to check if one (or more) constraints used to represent mappings causes incoherence within the PTBox of the ontology obtained by merging the two ontologies. A mapping can produce contradictions in two ways: i) the constraint itself is contradictory with respect to some of the axioms already present in the two original ontologies; ii) the constraint is correct, but the probability associated with it is not compatible with the probability constraints already present in the ontologies. Moreover, incoherence can be caused by two different situations: i) the mapping is not compatible with one or more axioms already present in the merged ontology; ii) the mapping is not compatible with other mappings introduced in the merged ontology. In the first case, incompatible mappings are discarded in order to preserve the original knowledge provided by the ontologies at hand. In the second case, we need a heuristic for choosing the incompatible mapping(s) to be discarded. In our approach, we propose to keep the mappings featured by higher probabilities to be true under the assumption that relations which are more probable are also more reliable. In order to implement this approach, we adopt the mapping validation procedure depicted in Figure 6.

The procedure takes as input a mapping set together with a merged ontology. As a first step, the mapping set M is ordered by descending ordering from the most probable to the less probable mapping. In such a way, we first insert the more probable mappings and, in case of incoherence, we discard always the latter (i.e., less probable) mapping(s). Then, for each mapping, we translate the

```
mapping_validation(M,O):
  input: a mapping set M, a merged ontology O
  output: a validated mapping set M'
  begin
    M := sort M w.r.t. the probability
         associated with each mapping m_i ∈ M;
    for m_i ∈ M:
      c_i := translate m_i into a conditional constraint;
      add c_i to O;
      check coherency of O with CBR;
      if O is coherent then:
        continue;
      else:
        remove c_i from O;
      end if
  end
```

Fig. 6. Mapping validation and revision procedure

mapping into a conditional constraint as shown in Section 3 and we insert such a constraint into the P-$\mathcal{SHIQ}$(D) ontology. We use the CBR reasoning system in order to check the coherence of the ontology augmented with the conditional constraint. In the coherence check fails, we discard the mapping and we delete the conditional constraint from the ontology. The final result is that the ontology is augmented only with mappings which do not cause incoherence problems into the PTBox.

Example. As an example, we consider mappings of Figure 2 and the merged ontology of Figure 3. The first four mappings do not cause any incoherence within the PTBox, then the corresponding conditional constraints are added to the ontology. The mapping $BT(building, motel)$ with probability 0.25 is not compatible with other mappings. In particular, it happens that an instance of motel is also an instance of hotel with probability 1.0, due to the first mapping. Moreover, the concept hotel is subsumed by the concept building. Then, an instance of motel must be an instance of building with probability $p \geq 1.0$. According to the mapping validation procedure, this mapping is discarded. We note that

Table 1. Mappings after validation and revision

Discarded constraints	Revised constraints
$(building\|motel)[0.25, 0.25]$	$(\exists hosts.tourist\|motel)[1.0, 1.0]$
$(building\|hostel)[0.125, 0.125]$	$(\exists hosts.tourist\|hostel)[0.5, 1.0]$
	$(building\|motel)[1.0, 1.0]$
	$(building\|hostel)[0.5, 1.0]$
	$(accommodation\|hotel)[1.0, 1.0]$
	$(accommodation\|\exists hosts.tourist)[1.0, 1.0]$
	$(\exists hosts.tourist\|accommodation)[1.0, 1.0]$

such a mapping is also useless, since the relation between motel and building is already implicitly expressed by the axioms of the first ontology and by the first mapping. Analogously, we discard also the mapping $BT(building, hostel)$ with probability 0.125. Finally, we can check which conditional constraints can be inferred by the CBR system by considering the validated mappings. The final result of our example is summarized in Table 1.

5 Related Work

Related work concerns both ontology mapping validation and probabilistic reasoning techniques. The problem of automatically discovering ontology/schema mappings has been studied both in the field of ontologies [1] and in the field of databases [7]. In the literature, some approaches have been proposed for validating mappings with respect to the semantics of the involved ontologies, but in most of the cases these approaches do not deal with uncertain mappings. An example of work in this direction is the *S-Match* system [2]. The key idea behind S-Match is to compute an initial set of relations between concepts of two taxonomies by means of external sources (i.e. WordNet); then, the problem of matching is translated into a validity check problem on the concept taxonomy. However, the validation approach of S-Match does not take into account the uncertain nature of mappings which are instead translated into crisp relations. Another important step towards mapping validation is represented by the theoretical study proposed in [8], which differs from previous works in that logical reasoning about mappings is performed. Mappings are translated into logical constraints and represented by means of Distributed Description Logics [9], an extension of classical DLs for distributed knowledge bases. An implementation of this approach has been discussed in [10]. However, also in this case, mapping uncertainty is not supported. Other relevant work has been done in using fuzzy reasoning for handling uncertainty in ontologies [11]. A fuzzy approach is different from our proposal in that a different understanding of the semantics of mappings is applied. However, especially for instance matching, the use of fuzzy reasoning can be integrated in our approach. In [12] the authors propose a mapping validation approach based on the idea of representing mappings as probabilistic rules. With respect to this work, we present a linguistic matching procedure which is compatible with the probabilistic interpretation of mappings instead of applying probabilities to every kind of results produced by matching tools. In recent years, the problem of modeling uncertainty has become one focus of interest also in description logic research. First probabilistic extensions started by modelling a degree of overlap between concepts via probabilities at the TBox level [13] as well as believes about certain individuals at the ABox level [14]. In P-CLASSIC [15] an approach was presented which integrates Bayesian networks as underlying reasoning formalism. Further work on integration of description logic and Bayesian networks has been presented in [16]. Efforts to integrate several approaches of uncertainty representation into one coherent framework have been made in [17]. The presented extension differs from the aforementioned ones, due to its ties to default logic, a non-monotonic reasoning method developed by

Reiter [18]. Here lexicographic entailment for defaults [19] was adapted to probabilistic description logics which provides non-monotonic inheritance properties along subclass relationships.

6 Concluding Remarks

In this paper, we have proposed a mapping validation approach with the following original contributions: i) *separation of concern*: the mapping validation process is independent from the mapping discovery process, in order to enable interoperability with existing matching systems; *uncertainty-compliance*: a notion of probabilistic mapping is introduced to capture the uncertain nature of the correspondences stated by mappings; *well-founded approach*: the probabilistic extension P-$\mathcal{SHIQ}$(D) of $\mathcal{SHIQ}(D)$ is used to represent probabilistic mappings and a sound reasoning system for P-$\mathcal{SHIQ}$(D) is presented and used for working with probabilistic mappings and formally validate them.

Our approach to mapping validation is adopted in the European BOEMIE project, where ontology mappings are used as a support for ontology evolution. In the context of BOEMIE, an evaluation of this validation approach is being performed on linguistic mappings between the athletics domain ontology of BOEMIE and a set of external knowledge sources. Goal of the evaluation is to compare the quality of mappings in terms of precision and accuracy before and after the validation process. Our ongoing and future work is mainly devoted to: i) use of validated mappings for ontology evolution; ii) investigation on optimization techniques for the probabilistic reasoning algorithms implemented in the CBR System, in order to improve the systems average case performance; iii) investigation on the problem of automatically detecting the minimal set of mappings causing incoherence during validation, in order to substitute the current heuristic-based approach to mapping revision with a reasoning-based approach.

References

1. Euzenat, J., Shvaiko, P.: Ontology Matching. Springer, Heidelberg (2007)
2. Giunchiglia, F., Shvaiko, P., Yatskevich, M.: S-Match: an Algorithm and an Implementation of Semantic Matching. In: Bussler, C.J., Davies, J., Fensel, D., Studer, R. (eds.) ESWS 2004. LNCS, vol. 3053, pp. 61–75. Springer, Heidelberg (2004)
3. Castano, S., Ferrara, A., Montanelli, S.: Matching ontologies in open networked systems: Techniques and applications. Journal on Data Semantics (JoDS) V (2006)
4. Giugno, R., Lukasiewicz, T.: P-$\mathcal{SHOQ}(\mathbf{D})$: A probabilistic extension of $\mathcal{SHOQ}(\mathbf{D})$ for probabilistic ontologies in the semantic web. In: Flesca, S., Greco, S., Leone, N., Ianni, G. (eds.) JELIA 2002. LNCS (LNAI), vol. 2424, pp. 86–97. Springer, Heidelberg (2002)
5. Lukasiewicz, T.: Expressive probabilistic description logics. Artif. Intell. 172, 852–883 (2008)
6. Haarslev, V., Möller, R.: Expressive abox reasoning with number restrictions, role hierarchies, and transitively closed roles. In: Proc. of the 7th International Conference on Principles of Knowledge Representation and Reasoning (KR 2000), Breckenridge, Colorado, USA, pp. 273–284 (2000)

7. Rahm, E., Bernstein, P.A.: A survey of approaches to automatic schema matching. VLDB Journal 10, 334–350 (2001)
8. Stuckenschmidt, H., Serafini, L., Wache, H.: Reasoning about ontology mappings. Technical report, ITC-IRST, Trento (2005)
9. Borgida, A., Serafini, L.: Distributed description logics: Assimilating information from peer sources. Journal on Data Semantics 1, 153–184 (2003)
10. Serafini, L., Tamilin, A.: Drago: Distributed reasoning architecture for the semantic web. In: Gómez-Pérez, A., Euzenat, J. (eds.) ESWC 2005. LNCS, vol. 3532, pp. 361–376. Springer, Heidelberg (2005)
11. Stoilos, G., Stamou, G., Pan, J., Tzouvaras, V., Horrocks, I.: Reasoning with very expressive fuzzy description logics. Journal of Artificial Intelligence Research 30, 133–179 (2007)
12. Cali, A., Lukasiewicz, T., Predoiu, L., Stuckenschmidt, H.: A framework for representing ontology mappings under probabilities and inconsistencies. In: Aberer, K., Choi, K.-S., Noy, N., Allemang, D., Lee, K.-I., Nixon, L., Golbeck, J., Mika, P., Maynard, D., Mizoguchi, R., Schreiber, G., Cudré-Mauroux, P. (eds.) ISWC 2007. LNCS, vol. 4825, Springer, Heidelberg (2007)
13. Heinsohn, J.: Probabilistic description logics. In: Proc. of the 10th Conf. on Uncertainty in Artificial Intelligence, Seattle, Washington, pp. 311–318 (1994)
14. Jaeger, M.: Probabilistic reasoning in terminological logics. In: Proceedings of the Fourth International Conference on Knowledge Representation and Reasoning (KR 1994), pp. 305–316 (1994)
15. Koller, D., Levy, A.Y., Pfeffer, A.: P-classic: A tractable probabilistic description logic. In: AAAI/IAAI, pp. 390–397 (1997)
16. Ding, Z., Peng, Y.: A Probabilistic Extension to Ontology Language OWL. In: HICSS 2004: Proc. of the 37th Annual Hawaii International Conference on System Sciences (HICSS 2004), Washington, DC, USA (2004)
17. Haarslev, V., Pai, H.I., Shiri, N.: A generic framework for description logics with uncertainty. In: ISWC-URSW, pp. 77–86 (2005)
18. Reiter, R.: A logic for default reasoning. Artif. Intell. 13, 81–132 (1980)
19. Lehmann, D.: Another perspective on default reasoning. Annals of Mathematics and Artificial Intelligence 15, 61–82 (1995)

Safe and Economic Re-Use of Ontologies: A Logic-Based Methodology and Tool Support

Ernesto Jiménez-Ruiz[1], Bernardo Cuenca Grau[2], Ulrike Sattler[3],
Thomas Schneider[3], and Rafael Berlanga[1]

[1] Universitat Jaume I, Spain
{berlanga,ejimenez}@uji.es
[2] University of Oxford, UK
berg@comlab.ox.ac.uk
[3] University of Manchester, UK
{sattler,schneider}@cs.man.ac.uk

Abstract. Driven by application requirements and using well-understood theoretical results, we describe a novel methodology and a tool for modular ontology design. We support the user in the *safe* use of imported symbols and in the *economic* import of the relevant part of the imported ontology. Both features are supported in a well-understood way: safety guarantees that the semantics of imported concepts is not changed, and economic import guarantees that no difference can be observed between importing the whole ontology and importing the relevant part.

1 Motivation

Ontology design and maintenance require an expertise in both the domain of application and the ontology language. Realistic ontologies typically model different aspects of an application domain at various levels of granularity; prominent examples are the National Cancer Institute Ontology (NCI)[1] [1], which describes diseases, drugs, proteins, etc., and GALEN[2], which represents knowledge mainly about the human anatomy, but also about other domains such as drugs.

Ontologies such as NCI and GALEN are used in bio-medical applications as *reference ontologies*, i.e., ontology developers reuse these ontologies and customise them for their specific needs. For example, ontology designers use concepts[3] from NCI or GALEN and refine them (e.g., add new sub-concepts), generalise them (e.g., add new super-concepts), or refer to them when expressing a property of some other concept (e.g., define the concept Polyarticular_JRA by referring to the concept Joint from GALEN).

One of such use cases is the development within the Health-e-Child project of an ontology, called JRAO, to describe a kind of arthritis called JRA (Juvenile

---

[1] Online browser: http://nciterms.nci.nih.gov/NCIBrowser/Dictionary.do, latest version: ftp://ftp1.nci.nih.gov/pub/cacore/EVS/NCI_Thesaurus
[2] http://www.co-ode.org/galen
[3] We use the Description Logic terms "concept" and "role" instead of the OWL terms "class" and "property".

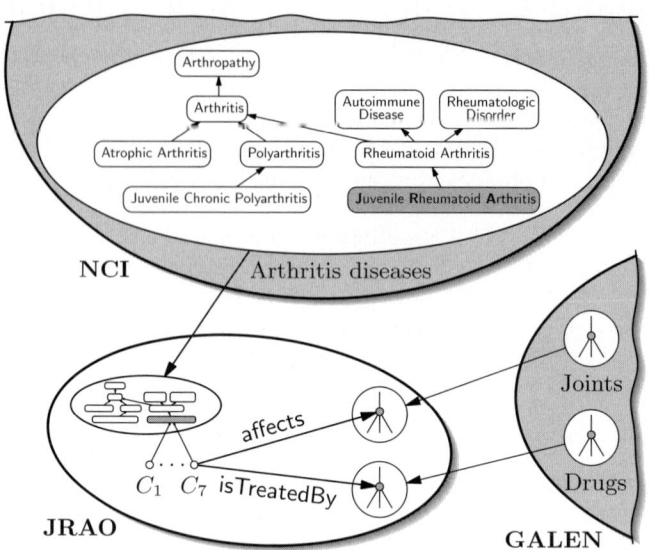

Fig. 1. Constructing the ontology JRAO reusing fragments of GALEN and NCI

Rheumatoid Arthritis).[4] Following the ILAR[5], JRAO describes the kinds of JRA. Those are distinguished by several factors such as the joints affected or the occurrence of fever, and each type of JRA requires a different treatment. GALEN and NCI contain information that is relevant to JRA, such as detailed descriptions of the human joints as well as diseases and their symptoms. Figure 1 gives a fragment of NCI that defines JRA. It also shows our reuse scenario, where $C_1, \ldots, C_7$ refer to the kinds of JRA to be defined in JRAO.

The JRAO developers want to reuse knowledge from NCI and GALEN for three reasons: (a) they want to save time through reusing existing ontologies rather than writing their own; (b) they value knowledge that is commonly accepted by the community and used in similar applications; (c) they are not experts in all areas covered by NCI and GALEN.

Currently, GALEN, NCI, and JRAO are written in OWL DL [2], and thus they come with a logic-based semantics, which allows for powerful reasoning services for classification and consistency checking. Thus, ontology reuse should take into account the semantics and, more precisely, should provide the following two guarantees. First, when reusing knowledge from NCI and GALEN, the developers of JRAO do not want to change the original meaning of the reused concepts. For example, due to (b) and (c) above, if it follows from the union of JRAO and NCI that JRA is a genetic disorder, then it also follows from NCI alone. Second, only small parts of large ontologies like NCI and GALEN are relevant

---

[4] See http://www.health-e-child.org. This project aims at creating a repository of ontologies that can be used by clinicians in various applications.
[5] Int. League of Associations for Rheumatology http://www.ilarportal.org/

to the sub-types of JRA. For efficiency and succinctness, the JRAO developers want to import only those axioms from NCI and GALEN that are relevant for JRAO. By importing only fragments of NCI and GALEN, one should not lose important information; for example, if it follows from the union of the JRAO and NCI that JRA is a genetic disorder, then this also follows from the union of JRAO and the chosen fragment of NCI.

Our scenario has two main points in common with other ontology design scenarios: the ontology developer wants to reuse knowledge without damaging it, and also to import only the relevant parts of an existing ontology. To support these scenarios whilst providing the two above guarantees, a logic-based approach to reuse is required. Current tools that support reuse, however, do not implement a logic-based solution and thus do not provide the above guarantees—and neither do existing guidelines and "best practices" for ontology design.

In this paper, we propose a methodology for ontology design in scenarios involving reuse which is based on a well-understood logic-based framework [3]. We describe a tool that implements this methodology and report on experiments.

2 Preliminaries on Modularity

Based on the application scenario in Section 1, we define the notions of a *conservative extension*, *safety*, and *module* [4, 3]. For simplicity of the presentation, we restrict ourselves to the description logic $\mathcal{SHIQ}$, which covers most of OWL DL [2]. Therefore the ontologies, entailments and signatures we consider are relative to $\mathcal{SHIQ}$. The results mentioned in this section, however, can be extended to $\mathcal{SHOIQ}$ and therefore OWL DL [3]. In this and the following section, we have omitted a few technical details, for instance proofs of the propositions. They can be found in a technical report available at http://www.cs.man.ac.uk/~schneidt/publ/safe-eco-reuse-report.pdf.

2.1 The Notions of Conservative Extension and Safety

As mentioned in Section 1, when reusing knowledge from NCI and GALEN, the developer of JRAO should not change the original meaning of the reused concepts. This requirement can be formalised using the notion of a *conservative extension* [4, 5]. In the following, we use Sig() to denote the signature of an ontology or an axiom.

Definition 1 (Conservative Extension). *Let $\mathcal{T}_1 \subseteq \mathcal{T}$ be ontologies, and $\mathbf{S}$ a signature. We say that $\mathcal{T}$ is an $\mathbf{S}$-conservative extension of $\mathcal{T}_1$ if, for every axiom α with $\mathsf{Sig}(\alpha) \subseteq \mathbf{S}$, we have $\mathcal{T} \models \alpha$ iff $\mathcal{T}_1 \models \alpha$; $\mathcal{T}$ is a conservative extension of $\mathcal{T}_1$ if $\mathcal{T}$ is an $\mathbf{S}$-conservative extension of $\mathcal{T}_1$ for $\mathbf{S} = \mathsf{Sig}(\mathcal{T}_1)$.*

Definition 1 applies to our example as follows: $\mathcal{T}_1 = $ NCI is the ontology to be reused, $\mathcal{T}$ is the union of JRAO and NCI, $\mathbf{S}$ represents the symbols reused from NCI, such as JRA and Rheumatologic_Disorder, and α stands for any axiom over the reused symbols only, e.g., JRA $\sqsubseteq$ Rheumatologic_Disorder.

Definition 1 assumes that the ontology to be reused (e.g. NCI) is static. In practice, however, ontologies such as NCI are under development and may evolve beyond the control of the JRAO developers. Thus, it is convenient to keep NCI separate from the JRAO and make its axioms available on demand via a reference such that the developers of the JRAO need not commit to a particular version of NCI. The notion of *safety* [3] can be seen as a stronger version of conservative extension that abstracts from the particular ontology to be reused and focuses only on the reused *symbols*.

Definition 2 (Safety for a Signature). *Let $\mathcal{T}$ be an ontology and $\mathbf{S}$ a signature. We say that $\mathcal{T}$ is safe for $\mathbf{S}$ if, for every ontology $\mathcal{T}'$ with $\mathsf{Sig}(\mathcal{T}) \cap \mathsf{Sig}(\mathcal{T}') \subseteq \mathbf{S}$, we have that $\mathcal{T} \cup \mathcal{T}'$ is a conservative extension of $\mathcal{T}'$.*

2.2 The Notion of Module

As mentioned in Section 1, by importing only a fragment of NCI and GALEN, one should not lose important information. This idea can be formalised using the notion of a *module* [3]. Intuitively, when checking an arbitrary entailment over the signature of the JRAO, importing a module of NCI should give exactly the same answers as if the whole NCI had been imported.

Definition 3 (Module for a Signature). *Let $\mathcal{T}_1' \subseteq \mathcal{T}'$ be ontologies and $\mathbf{S}$ a signature. We say that $\mathcal{T}_1'$ is a module for $\mathbf{S}$ in $\mathcal{T}'$ (or an $\mathbf{S}$-module in $\mathcal{T}'$) if, for every ontology $\mathcal{T}$ with $\mathsf{Sig}(\mathcal{T}) \cap \mathsf{Sig}(\mathcal{T}') \subseteq \mathbf{S}$, we have that $\mathcal{T} \cup \mathcal{T}'$ is a conservative extension of $\mathcal{T} \cup \mathcal{T}_1'$ for $\mathsf{Sig}(\mathcal{T})$.*

The notions of safety and module are related as follows:

Proposition 4 ([3], Safety vs. Modules). *If $\mathcal{T}' \setminus \mathcal{T}_1'$ is safe for $\mathbf{S} \cup \mathsf{Sig}(\mathcal{T}_1')$, then $\mathcal{T}_1'$ is an $\mathbf{S}$-module in $\mathcal{T}'$.*

2.3 Locality Conditions

The decision problems associated with conservative extensions, safety and modules—i.e., whether $\mathcal{T}$ is an $\mathbf{S}$-conservative extension of $\mathcal{T}_1$, whether $\mathcal{T}$ is safe for $\mathbf{S}$, or whether $\mathcal{T}_1'$ is an $\mathbf{S}$-module in $\mathcal{T}$—are undecidable for $\mathcal{SHOIQ}$ [6, 3]. Sufficient conditions for safety have been proposed: if an ontology satisfies such conditions, then we can guarantee that it is safe, but the converse does not necessarily hold [3]. By means of Proposition 4, such conditions could be used for extracting modules. A particularly useful condition is *locality* [3]: it is widely applicable in practice and it can be checked syntactically.

As mentioned in Section 1, when using a symbol from NCI or GALEN, the JRAO developers may refine it, extend it, or refer to it for expressing a property of another symbol. The simultaneous refinement and generalisation of a given "external" symbol, however, may compromise safety. For example, JRAO cannot simultaneously contain the following axioms:

$$\text{Polyarticular\_JRA} \sqsubseteq \text{JRA} \quad (\bot\text{-local}) \quad (1)$$
$$\text{Juvenile\_Chronic\_Polyarthritis} \sqsubseteq \text{Polyarticular\_JRA} \quad (\top\text{-local}) \quad (2)$$

where the underlined concepts are reused from NCI, see Figure 1. These axioms imply Juvenile_Chronic_Polyarthritis $\sqsubseteq$ JRA, and therefore an ontology containing axioms (1) and (2) is not safe w.r.t. $\mathbf{S} = \{$JRA, Juvenile_Chronic_Polyarthritis$\}$. Thus, when designing sufficient conditions for safety, we are faced with a fundamental choice depending on whether the ontology designer wants to reuse or generalise the reused concepts. Each choice leads to a different locality condition.

The following definition introduces these conditions and refers to Figure 2. In this figure, $A^\dagger$ and $R^\dagger$ stand for concept and role names not in $\mathbf{S}$. The letters C and R denote arbitrary concepts and roles.

Definition 5 (Syntactic $\bot$-Locality and $\top$-Locality). *Let $\mathbf{S}$ be a signature. An axiom α is $\bot$-local w.r.t. $\mathbf{S}$ ($\top$-local w.r.t $\mathbf{S}$) if $\alpha \in \mathsf{Ax}(S)$, as defined in Figure 2 (a) ((b)). An ontology $\mathcal{T}$ is $\bot$-local ($\top$-local) w.r.t. $\mathbf{S}$ if α is $\bot$-local ($\top$-local) w.r.t. $\mathbf{S}$ for all $\alpha \in \mathcal{T}$.*

(a) $\bot$-Locality   Let $C^\dagger \in \mathbf{Con}(\bar{\mathbf{S}})$, $C^\mathbf{s}_{(i)} \in \mathbf{Con}(\mathbf{S})$

$\mathbf{Con}(\bar{\mathbf{S}}) ::= A^\dagger \mid \neg C^\mathbf{s} \mid C \sqcap C^\dagger \mid C^\dagger \sqcap C \mid \exists R.C^\dagger \mid \geqslant n\, R.C^\dagger \mid (\exists R^\dagger .C) \mid (\geqslant n\, R^\dagger .C)$
$\mathbf{Con}(\mathbf{S}) ::= \neg C^\dagger \mid C^\mathbf{s}_1 \sqcap C^\mathbf{s}_2$
$\mathsf{Ax}(\mathbf{S}) ::= C^\dagger \sqsubseteq C \mid C \sqsubseteq C^\mathbf{s} \mid R^\dagger \sqsubseteq R \mid \mathsf{Trans}(R^\dagger)$

(b) $\top$-Locality   Let $C^\mathbf{s} \in \mathbf{Con}(\mathbf{S})$, $C^\dagger_{(i)} \in \mathbf{Con}(\bar{\mathbf{S}})$

$\mathbf{Con}(\mathbf{S}) ::= \neg C^\dagger \mid C \sqcap C^\mathbf{s} \mid C^\mathbf{s} \sqcap C \mid \exists R.C^\mathbf{s} \mid \geqslant n\, R.C^\mathbf{s}$
$\mathbf{Con}(\bar{\mathbf{S}}) ::= A^\dagger \mid \neg C^\mathbf{s} \mid C^\dagger_1 \sqcap C^\dagger_2 \mid \exists R^\dagger .C^\dagger \mid \geqslant n\, R^\dagger .C^\dagger$
$\mathsf{Ax}(\mathbf{S}) ::= C^\mathbf{s} \sqsubseteq C \mid C \sqsubseteq C^\dagger \mid R \sqsubseteq R^\dagger \mid \mathsf{Trans}(R^\dagger)$

Fig. 2. Syntactic locality conditions

Axiom (2) is $\top$-local w.r.t. $\mathbf{S} = \{$Juvenile_Chronic_Polyarthritis$\}$, and Axiom (1) is $\bot$-local w.r.t. $\mathbf{S} = \{$JRA$\}$. Note that the locality conditions allow us to refer to a reused concept for expressing a property of some other concept; for example, the axiom Polyarticular_JRA $\sqsubseteq\, \geqslant 5$ affects.Joint is $\bot$-local w.r.t. $\mathbf{S} = \{$Joint$\}$.

Both $\top$-locality and $\bot$-locality are sufficient for safety:

Proposition 6 ([3], Locality Implies Safety). *If an ontology $\mathcal{T}$ is $\bot$-local or $\top$-local w.r.t. $\mathbf{S}$, then $\mathcal{T}$ is safe for $\mathbf{S}$.*

Propositions 4 and 6 suggest the following definition of modules in terms of locality.

Definition 7. *Let $\mathcal{T}_1 \subseteq \mathcal{T}$ be ontologies, and $\mathbf{S}$ a signature. We say that $\mathcal{T}_1$ is a $\bot$-module ($\top$-module) for $\mathbf{S}$ in $\mathcal{T}$ if $\mathcal{T} \setminus \mathcal{T}_1$ is $\bot$-local ($\top$-local) w.r.t. $\mathbf{S} \cup \mathsf{Sig}(\mathcal{T}_1)$.*

We illustrate these notions by an example. Figure 3 (a) shows an example ontology (TBox). The set of external symbols is $\mathbf{S}_0 = \{A, t_1, t_2\}$. In order to extract

$\mathcal{T} = \{\ A \sqsubseteq A_2,\quad A_2 \sqsubseteq \forall s_2.E_2,\quad A_1 \sqsubseteq A,\quad \forall s_1.E_1 \sqsubseteq A_1,$
$\phantom{\mathcal{T} = \{\ }A_2 \sqsubseteq \exists r_2.C_2,\quad A_2 \sqsubseteq \forall t_2.F_2,\quad \exists r_1.C_1 \sqsubseteq A_1,\quad \forall t_1.F_1 \sqsubseteq A_1\ \}$
$\phantom{\mathcal{T} = \{\ }A_2 \sqsubseteq \forall r_2.D_2,\quad\quad \forall r_1.D_1 \sqsubseteq A_1$

(a) The TBox

Consideration	Consequence
(1) $A \in \mathbf{S}_0$, $A_2 \notin \mathbf{S}_0$ $\Rightarrow$ $A \sqsubseteq A_2 \notin \mathsf{Ax}(\mathbf{S}_0)$	$\mathbf{S}_1 = \mathbf{S}_0 \cup \{A_2\}$
(2) $A_2 \in \mathbf{S}_1$, $C_2 \notin \mathbf{S}_1$ $\Rightarrow$ $\exists r_2.C_2 \in \mathbf{Con}(\bar{\mathbf{S}}_1)$ $\Rightarrow$ $A_2 \sqsubseteq \exists r_2.C_2 \notin \mathsf{Ax}(\mathbf{S}_1)$	$\mathbf{S}_2 = \mathbf{S}_1 \cup \{r_2, C_2\}$
(3) $A_2 \in \mathbf{S}_2$, $r_2 \in \mathbf{S}_2$, $D_2 \notin \mathbf{S}_2$ $\Rightarrow$ $\exists r_2.\neg D_2 \notin \mathbf{Con}(\bar{\mathbf{S}}_2)$ $\Rightarrow$ $\neg\exists r_2.\neg D_2 \notin \mathbf{Con}(\bar{\mathbf{S}}_2)$ $\Rightarrow$ $A_2 \sqsubseteq \forall r_2.D_2 \notin \mathsf{Ax}(\mathbf{S}_2)$	$\mathbf{S}_3 = \mathbf{S}_2 \cup \{D_2\}$
(4) $A_2 \in \mathbf{S}_3$, $s_2 \notin \mathbf{S}_3$, $E_2 \notin \mathbf{S}_3$ $\Rightarrow$ $\exists s_2.\neg E_2 \in \mathbf{Con}(\bar{\mathbf{S}}_3)$ $\Rightarrow$ $\neg\exists s_2.\neg E_2 \in \mathbf{Con}(\bar{\mathbf{S}}_3)$ $\Rightarrow$ $A_2 \sqsubseteq \forall s_2.E_2 \in \mathsf{Ax}(\mathbf{S}_3)$	$\mathbf{S}_4 = \mathbf{S}_3$
(5) analogous to (3)	$\mathbf{S}_5 = \mathbf{S}_4 \cup \{F_2\}$
$\mathbf{S}_5 = \{A, A_2, C_2, D_2, F_2, r_2, t_1, t_2\}$ The $\bot$-module consists of $\mathbf{S}_5$ and all axioms $\alpha \in \mathcal{T}$ with $\mathsf{Sig}(\alpha) \subseteq \mathbf{S}_5$.	

(b) Extracting the $\bot$-module

Fig. 3. An example illustrating $\bot$- and $\top$-modules

the $\bot$-module, we extend $\mathbf{S}_0$ stepwise as in Figure 3 (b). The $\top$-module is obtained analogously and consists of $\mathbf{S}'_5 = \{A, A_1, C_1, D_1, F_1, r_1, t_1, t_2\}$ and all axioms $\alpha \in \mathcal{T}$ with $\mathsf{Sig}(\alpha) \subseteq \mathbf{S}'_5$.

It is clear that $\bot$-modules and $\top$-modules are modules as in Definition 3:

Proposition 8 ([3], Locality-based Modules are Modules). *Let $\mathcal{T}_1$ be either a $\bot$-module or a $\top$-module for $\mathbf{S}$ in $\mathcal{T}$ and let $\mathbf{S}' = \mathbf{S} \cup \mathsf{Sig}(\mathcal{T}_1)$. Then $\mathcal{T}_1$ is an $\mathbf{S}'$-module in $\mathcal{T}$.*

These modules enjoy an important property which determines their scope: suppose that $\mathcal{T}_1$ ($\mathcal{T}_2$) is a $\bot$-module ($\top$-module) for $\mathbf{S}$ in $\mathcal{T}$, then $\mathcal{T}_1$ ($\mathcal{T}_2$) will contain all super-concepts (sub-concepts) in $\mathcal{T}$ of all concepts in $\mathbf{S}$:

Proposition 9 ([3], Module Scope). *Let $\mathcal{T}$ be an ontology, X, Y be concept names in $\mathcal{T} \cup \{\top\} \cup \{\bot\}$, $\alpha := (X \sqsubseteq Y)$, $\beta := (Y \sqsubseteq X)$, and $\mathcal{T}_X \subseteq \mathcal{T}$ with $X \in \mathbf{S}$. Then the following statements hold.*
 (i) *If $\mathcal{T}_X$ is a $\bot$-module in $\mathcal{T}$ for $\mathbf{S}$, then $\mathcal{T}_X \models \alpha$ iff $\mathcal{T} \models \alpha$.*
 (ii) *If $\mathcal{T}_X$ is a $\top$-module in $\mathcal{T}$ for $\mathbf{S}$, then $\mathcal{T}_X \models \beta$ iff $\mathcal{T} \models \beta$.*

For example, if we were to reuse the concept JRA from NCI as shown in Figure 1 by extracting a $\bot$-module for a signature that contains JRA, such a module would contain all the super-concepts of JRA in NCI, namely Rheumatoid_Arthritis, Autoimmune_Disease, Rheumatologic_Disorder, Arthritis, and Arthropathy. Since such a fragment is a module, it will contain the axioms necessary for entailing those subsumption relations between the listed concepts that hold in NCI.

Finally, given $\mathcal{T}$ and $\mathbf{S}$, there is a unique minimal $\bot$-module and a unique minimal $\top$-module for $\mathbf{S}$ in $\mathcal{T}$ [3]. We denote these modules by $\mathsf{UpMod}(\mathcal{T}, \mathbf{S})$ and $\mathsf{LoMod}(\mathcal{T}, \mathbf{S})$. This is motivated by the alternative terms "upper/lower module" that refer to the property from Proposition 9. Following a similar approach as exemplified in Figure 3(b), the modules can be computed efficiently [3].

3 A Novel Methodology for Ontology Reuse

Based on our scenario in Section 1 and the theory of modularity summarised in Section 2, we propose a novel methodology for designing an ontology when knowledge is to be borrowed from several external ontologies. This methodology provides precise guidelines for ontology developers to follow, and ensures that a set of logical guarantees will hold at certain stages of the design process.

3.1 The Methodology

We propose the working cycle given in Figure 4. This cycle consists of an *offline phase*—which is performed independently from the current contents of the external ontologies—and an *online phase*—where knowledge from the external ontologies is extracted and transferred into the current ontology. Note that this separation is not strict: The first phase is called "offline" simply because it does not need to be performed online. However, the user may still choose to do so.

The Offline Phase starts with the ontology $\mathcal{T}$ being developed, e.g., JRAO. The ontology engineer specifies the set $\mathbf{S}$ of symbols to be reused from external ontologies, and associates to each symbol the external ontology from which it will be borrowed. In Figure 4 this signature selection is represented in the *Repeat* loop: each $\mathbf{S}_i \subseteq \mathbf{S}$ represents the external symbols to be borrowed from a particular ontology $\mathcal{T}'_i$; in our example, we have $\mathbf{S} = \mathbf{S}_1 \uplus \mathbf{S}_2$, where $\mathbf{S}_1$ is associated with NCI and contains JRA, and $\mathbf{S}_2$ is associated with GALEN and contains symbols related to joints and drugs. This part of the offline phase may involve an "online" component since the developer may browse through the external ontologies to choose the symbols she wants to import.

Next, the ontology developer decides, for each $\mathbf{S}_i$, whether she wants to refine or generalise the symbols from this set. For instance, in the reuse example shown in Figure 1, the concept JRA from NCI is refined by the sub-concepts abbreviated $C_1, \ldots, C_7$. In both cases, the user may also reference the external symbols via roles; in our example, certain types of JRA are defined by referencing concepts in GALEN (e.g., joints) via the roles affects and isTreatedBy. As argued in Section 1, refinement and generalisation, combined with reference, constitute the main possible intentions when reusing external knowledge. Therefore it is reasonable for the user, both from the modelling and tool design perspectives, to declare her intention. This step is represented by the *For* loop in Figure 4.

At this stage, we want to ensure that the designer of $\mathcal{T}$ does not change the original meaning of the reused concepts, independently of what their particular meaning is in the external ontologies. This requirement can be formalised using the notion of safety introduced in Section 2:

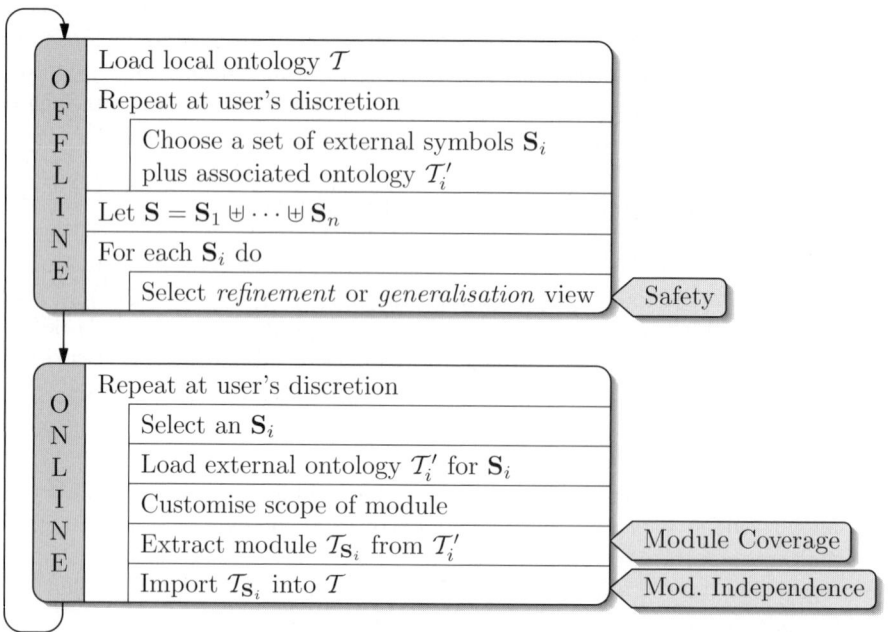

Fig. 4. The two phases of import with the required guarantees

Definition 10 (Safety Guarantee). *Given an ontology $\mathcal{T}$ and signatures $\mathbf{S}_1, \ldots, \mathbf{S}_n$, $\mathcal{T}$ guarantees safety if $\mathcal{T}$ is safe for $\mathbf{S}_i$ for all $1 \leq i \leq n$.*

In the next subsection, we will show how to guarantee safety.

In the Online Phase, the ontology engineer imports the relevant knowledge from each of the external ontologies. As argued in Section 1 and 2, we aim at extracting only those fragments from the external ontologies that are relevant to the reused symbols, and therefore the extracted fragments should be modules in the sense of Definition 3.

As shown in Figure 4, the import for each external ontology $\mathcal{T}_i'$ is performed in four steps. First, $\mathcal{T}_i'$ is loaded; by doing so, the ontology engineer commits to a particular version of it. Second, the scope of the module to be extracted from $\mathcal{T}_i'$ is customised; in practice, this means that the ontology engineer is given a view of $\mathcal{T}_i'$ and enabled to extend $\mathbf{S}_i$ by specifying requirements such as: "The module has to contain the concept 'joint', all its direct super-concepts and two levels of its sub-concepts". In the third step, the actual fragment of $\mathcal{T}_i'$ is extracted. At this stage, we should ensure that the extracted fragment is a module for the customised signature according to Definition 3. Therefore, the following guarantee should be provided for each external ontology and customised signature:

Definition 11 (Module Coverage Guarantee). *Let* $\mathbf{S}$ *be a signature and* $\mathcal{T}'_\mathbf{S}, \mathcal{T}'$ *be ontologies with* $\mathcal{T}'_\mathbf{S} \subset \mathcal{T}'$ *such that* $\mathbf{S} \subseteq \mathsf{Sig}(\mathcal{T}'_\mathbf{S})$. *Then,* $\mathcal{T}'_\mathbf{S}$ *guarantees coverage of* $\mathbf{S}$ *if* $\mathcal{T}'_\mathbf{S}$ *is a module for* $\mathbf{S}$ *in* $\mathcal{T}'$.

Finally, the actual module $\mathcal{T}_{\mathbf{S}_i}$ is imported. The effect of this import is that the ontology $\mathcal{T}$ being developed evolves to $\mathcal{T} \cup \mathcal{T}_{\mathbf{S}_i}$; as a consequence, the safety guarantee with respect to the remaining external ontologies might be compromised. Such an effect is obviously undesirable and hence the following guarantee should be provided after the import:

Definition 12 (Module Independence Guarantee). *Given an ontology* $\mathcal{T}$ *and signatures* $\mathbf{S}_1, \mathbf{S}_2$, *we say that* $\mathcal{T}$ *guarantees module independence if, for all* $\mathcal{T}_1$ *with* $\mathsf{Sig}(\mathcal{T}) \cap \mathsf{Sig}(\mathcal{T}_1) \subseteq \mathbf{S}_1$, *it holds that* $\mathcal{T} \cup \mathcal{T}_1$ *is safe for* $\mathbf{S}_2$.

3.2 Achieving the Logical Guarantees

In order to provide the necessary guarantees of our methodology, we will now make use of the locality conditions introduced in Section 2.3 and some general properties of conservative extensions, safety and modules.

The Safety Guarantee. In Section 2.3, we argue that the simultaneous refinement and generalisation of an external concept may compromise safety. To preserve safety, we propose to use two locality conditions: $\perp$-locality, suitable for refinement, and $\top$-locality, suitable for generalisation. These conditions can be checked syntactically using the grammars defined in Figure 2, and therefore they can be easily implemented in a tool. In order to achieve the safety guarantee at the end of the offline phase, we propose to follow the procedure sketched in Figure 5, where "locality of $\mathcal{T}$ for $\mathbf{S}_i$ according to the choice for $\mathbf{S}_i$" means $\perp$-locality if $\mathbf{S}_i$ is to be refined and $\top$-locality if $\mathbf{S}_i$ is to be generalised.

Input:
  $\mathcal{T}$: Ontology
  $\mathbf{S}_1, \ldots, \mathbf{S}_n$: disjoint signatures
  a choice among refinement and generalisation for each $\mathbf{S}_i$
Output:
  ontology $\mathcal{T}_1$ that guarantees safety

1: $\mathcal{T}_1 := \mathcal{T}$
2: **while** exists $\mathbf{S}_i$ such that $\mathcal{T}$ **not** local according to the selection for $\mathbf{S}_i$ **do**
3:    check locality of $\mathcal{T}_1$ w.r.t. $\mathbf{S}_i$ according to the choice for $\mathbf{S}_i$
4:    **if** non-local **then**
5:        $\mathcal{T}_1 :=$ repair $\mathcal{T}_1$ until it is local for $\mathbf{S}_i$ according to the choice for $\mathbf{S}_i$
6:    **end if**
7: **end while**
8: **return** $\mathcal{T}_1$

Fig. 5. A procedure for checking safety

It is immediate to see that the following holds upon completion of the procedure: for each $\mathbf{S}_i$, if the user selected the refinement (generalisation) view, then $\mathcal{T}$ is $\bot$-local ($\top$-local) w.r.t. $\mathbf{S}_i$. This is sufficient to guarantee safety, as given by the following proposition:

Proposition 13. *Let $\mathcal{T}$ be an ontology and $\mathbf{S} = \mathbf{S}_1 \uplus \ldots \uplus \mathbf{S}_n$ be the union of disjoint signatures. If, for each $\mathbf{S}_i$, either $\mathcal{T}$ is $\bot$-local or $\top$-local w.r.t. $\mathbf{S}_i$, then $\mathcal{T}$ guarantees safety.*

The Module Coverage Guarantee. The fragment extracted for each customised signature in the online phase must satisfy the module coverage guarantee. As seen in Section 2.3, $\bot$-locality and $\top$-locality can also be used for extracting modules in the sense of Definition 3. Given an external ontology $\mathcal{T}'$ and customised signature $\mathbf{S}_i$, the scope of the $\bot$-module and $\top$-module is determined by Proposition 9: as shown in Figure 3, the $\bot$-module will contain all the super-concepts in $\mathcal{T}'$ of the concepts in $\mathbf{S}_i$, whereas the $\top$-module will contain all the sub-concepts.

The construction in Figure 3 also shows that the extraction of $\bot$-modules or $\top$-modules may introduce symbols not in $\mathbf{S}_i$, and potentially unnecessary. To make the module as small as possible, we proceed as follows, given $\mathcal{T}'_i$ and $\mathbf{S}_i$: first, extract the minimal $\bot$-module M for $\mathbf{S}$ in $\mathcal{T}'_i$; then, extract the minimal $\top$-module for $\mathbf{S}$ in M. The fragment obtained at the end of this process satisfies the module coverage guarantee as given in the following proposition:

Proposition 14. *Let $\mathcal{T}'_2 = \mathsf{LoMod}(\mathsf{UpMod}(\mathcal{T}', \mathbf{S}), \mathbf{S})$. Then $\mathcal{T}'_2$ guarantees coverage of $\mathbf{S}$ in $\mathcal{T}'$.*

The Module Independence Guarantee. When a module is imported in the online phase (see Figure 4), module independence should be guaranteed—that is, importing a module for a signature $\mathbf{S}_i$ from an external ontology $\mathcal{T}'_i$ into $\mathcal{T}$ should not affect the safety of $\mathcal{T}$ w.r.t. to the other external ontologies. The following proposition establishes that this guarantee always holds provided that the imported module in $\mathcal{T}'_i$ does not share symbols with the remaining external ontologies $\mathcal{T}'_j$. In practice, this is always the default situation since different reference ontologies have different namespaces and therefore their signatures are disjoint. The following proposition is an immediate consequence of the syntactic definition of $\bot$-locality and $\top$-locality.

Proposition 15. *Let $\mathcal{T}$ be an ontology and $\mathbf{S}_1, \mathbf{S}_2$ disjoint signatures. If, for each $i = 1, 2$, $\mathcal{T}$ is $\bot$-local or $\top$-local w.r.t. $\mathbf{S}_i$, then $\mathcal{T}$ guarantees module independence.*

4 The Ontology Reuse Tool

We have developed a Protégé 4^6 plugin that supports the methodology presented in Section 3. The plugin and user manual can be downloaded from http://krono.act.uji.es/people/Ernesto/safety-ontology-reuse.

[6] Ontology Editor Protégé 4: http://www.co-ode.org/downloads/protege-x/

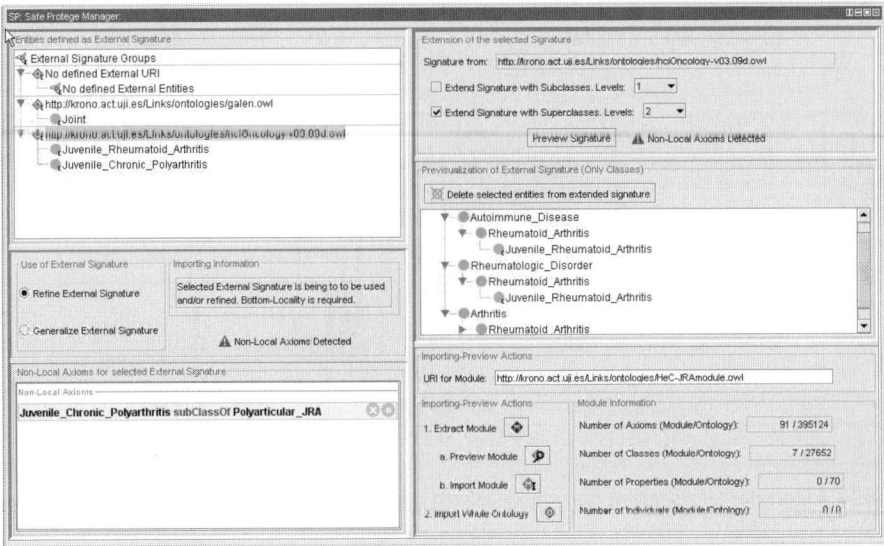

Fig. 6. Protégé Safe Manager Interface

The Offline Phase. The first step of the offline phase involves the selection of the external entities. Our plugin provides functionality for declaring entities as external as well as for defining the external ontology URI (or signature subgroup) for the selected entities; this information is stored in the ontology using OWL 1.1 annotations [7] as follows: we use an ontology annotation axiom per external ontology, an entity annotation axiom to declare an entity external, and an entity annotation axiom per external entity to indicate its external ontology. The set of external entities with the same external ontology URI can be viewed as one of the $\mathbf{S}_i$. Finally, the UI of the plugin also allows for the specification, for each external ontology, whether it will be refined or generalised. Once the external entities have been declared and divided into groups, the tool allows for safety checking of the ontology under development w.r.t. each group of external symbols separately. The safety check uses $\bot$-locality ($\top$-locality) for signature groups that adopt the refinement (generalisation) view. The non-local axioms to be repaired are appropriately displayed.

Figure 6 shows the *Safe Protégé Manager* component with the set of signature subgroups in the top left corner, and the non-local axioms in the bottom left corner. Note that, in this phase, our tool does allow the user to work completely offline, without the need of extracting and importing external knowledge, and even without knowing exactly from which ontology the reused entities will come from. Indeed, the specification of the URI of the external ontologies is optional at this stage, and, even if indicated, such URI may not refer to a real ontology, but it may simply act as a temporary name.

The Online Phase. In the online phase, the user chooses external ontologies and imports axioms from them. At this stage, the groups of external symbols to

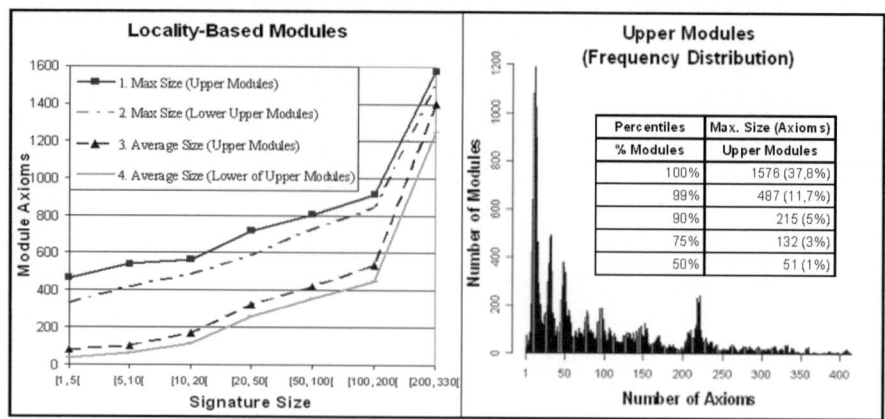

Fig. 7. Module information obtained from Galen. (left) Module average size against max size. (right) Frequency distributions for module sizes.

be imported should refer to the location of a "real" external ontology. Once an external signature group S_i has been selected for import, the selected signature can be customised by adding super-concepts and sub-concepts of the selected symbols. The tool provides the functionality for previewing the concept hierarchy of the corresponding external ontology for this purpose. Once the specific signature group under consideration has been customised, a module for it can be extracted. The module is computed using the procedure in Proposition 14; the user can compute the module, preview it in a separate frame, and either import it or cancel the process and come back to the signature customisation stage. The user is also given the option to import the whole external ontology instead of importing a module. Note that currently the import of a module is done "by value", in the sense that the module becomes independent from the original ontology: if the external ontology on the Web evolves, the previously extracted module will not change.

The right hand side of the *Safe Protégé Manager* component (see Figure 6) includes the set of necessary components to support the proposed steps for the online phase.

5 Evaluation

So far, we have demonstrated our tool to various ontology developers[7] who have expressed great interest, and we are currently working on a proper user study. In the following, we describe the experiments we have performed to prove that locality-based modules are reasonably sized compared to the whole ontology.

---

[7] Thanks to Elena Beißwanger, Sebastian Brandt, Alan Rector, and Holger Stenzhorn for valuable comments and feedback.

For each concept name A in NCI or Galen[8], we have proceeded as follows. (i) For each pair (u, ℓ) between $(0,0)$ and $(3,3)$, construct the signature $\mathbf{S}(A, u, \ell)$ by taking A, its super-concepts in the next u levels and its sub-concepts in the next ℓ levels. (ii) Extract $\mathcal{T}_1' = \mathsf{UpMod}(\mathcal{T}', \mathbf{S}_A)$ and $\mathcal{T}_2' = \mathsf{LoMod}(\mathcal{T}_1', \mathbf{S}_A)$, see Proposition 14. From now on, we will refer to the modules $\mathcal{T}_1'$ and $\mathcal{T}_2'$ mentioned above as "Upper Module (UM)" and "Lower of Upper Module (LUM)".

We have grouped the extracted modules according to the size of the input signature in order to evaluate its impact on the size of the modules. Figure 7 shows the obtained results for Galen (3161 entities and 4170 axioms), where the size of a module is the number of its axioms. The following conclusions can be drawn from the empirical results: first, the modules obtained are small on average since 99% of UMs have at most 487 axioms ($\sim 12\%$ of the size of Galen), and 99% of the LUMs contain at most and 386 axioms ($\sim 9\%$ of the size of Galen) for initial signatures $\mathbf{S}(A, u, \ell)$ containing between 1 and 330 entities; second, the growth in the size of the modules w.r.t. the size of the initial signature is smooth and linear up to initial signatures containing 100 entities. We have similar findings for NCI v3 (27,722 entities and 395,124 axioms, most of them annotations), NCI v7 (63,754 entities and 867,374 axioms, most of them annotations) and SNOMED (entities and 389,541 axioms and 389,544 entities).

In addition to the "synthetic" experiments, we have undertaken some real experiments in the context of the Health-e-Child project's user scenario, see Section 1. The experiments focus on JRA and Cardiomyopathies (CMP)—a group of diseases that are central to the project. Using our tool, members of the project have manually selected signatures that are relevant to JRA and Cardiomyopathies from both Galen and NCI, expanded these signatures as in the case of the synthetic tests, and extracted the corresponding modules. For example, in the case of JRA in Galen, the initial signature and expanded signature consisted of 40 and 131 entities. The following tables show the sizes of all signatures and modules extracted from (a) Galen and (b) NCI.

(a) Disease	JRA	JRA	CMP	(b) Disease	JRA	JRA	CMP
Signature size	40	131	77	Signature size	48	356	124
# axioms	490	1151	614	# axioms	300	1258	537
# concepts	296	663	334	# concepts	193	613	283
# roles	69	116	56	# roles	17	21	10

6 Related Work

Ontology Engineering Methodologies: Several ontology engineering methodologies can be found in the literature; prominent examples are Methontology [8], On-To-Knowledge (OTK) [9], and ONTOCLEAN [10]. These methodologies, however, do not address ontology development scenarios involving reuse. Our proposed methodology is complementary and can be used in combination with them.

---

[8] We have used a fragment of GALEN expressible in OWL:
http://krono.act.uji.es/Links/ontologies/galen.owl/view

Ontology Segmentation and Ontology Integration Techniques: In the last few years, a growing body of work has been developed addressing Ontology Modularisation, Ontology Mapping and Alignment, Ontology Merging, Ontology Integration and Ontology Segmentation, see [11, 12, 13] for surveys. This field is diverse and has originated from different communities.

In particular, numerous techniques for extracting fragments of ontologies. Most of them, such as [14, 15, 16], rely on syntactic heuristics for detecting relevant axioms. These techniques do not attempt to formally specify the intended outputs and do not provide any guarantees.

Ontology Reuse techniques: There are various proposals for "safely" combining modules; most of these proposals, such as $\mathcal{E}$-connections, Distributed Description Logics and Package-based Description Logics propose a specialised semantics for controlling the interaction between the importing and the imported modules to avoid side-effects, for an overview see [17]. In contrast, in our paper we assume that reuse is performed by simply building the logical union of the axioms in the modules under the standard semantics; instead, we provide the user with a collection of reasoning services, such as safety testing, to check for side-effects. Our paper is based on other work on modular reuse of ontologies [18, 19, 6, 5] which enables us to provide the necessary guarantees. We extend this work with a methodology and tool support.

7 Lessons Learned and Future Work

We have described a logic-based approach to the reuse of ontologies that is both *safe* (i.e., we guarantee that the meaning of the imported symbols is not changed) and *economic* (i.e., we import only the module relevant for a given set of symbol and we guarantee that we do not lose any entailments compared to the import of the whole ontology). We have described a methodology that makes use of this approach, have implemented tool support for it in Protégé, and report on experiments that indicate that our modules are indeed of acceptable size.

In the future, we will extend the tool support so that the user can "shop" for symbols to reuse: it will allow to browse an ontology for symbols to reuse and provide a simple mechanism to pick them and, on "check-out", will compute the relevant module. Moreover, we are working on more efficient ways of module extraction that make use of already existing computations. Next, we plan to carry out a user study to learn more about the usefulness of the interface and how to further improve it. Finally, our current tool support implements a "by value" mechanism: modules are extracted and added at the user's request. In addition, we would like to support import "by reference": a feature that checks whether the imported ontology has changed and thus a new import is necessary.

Acknowledgements

This work was partially supported by the PhD Fellowship Program of the *Generalitat Valenciana*, by the *Fundació Caixa Castelló-Bancaixa*, and by the UK EPSRC grant no. EP/E065155/1.

References

[1] Golbeck, J., Fragoso, G., Hartel, F.W., Hendler, J.A., Oberthaler, J., Parsia, B.: The National Cancer Institute's Thésaurus and Ontology. JWS 1(1), 75–80 (2003)
[2] Horrocks, I., Patel-Schneider, P.F., van Harmelen, F.: From $\mathcal{SHIQ}$ and RDF to OWL: The making of a web ontology language. JWS 1(1), 7–26 (2003)
[3] Cuenca Grau, B., Horrocks, I., Kazakov, Y., Sattler, U.: Modular reuse of ontologies: Theory and practice. J. of Artificial Intelligence Research 31, 273–318 (2008)
[4] Ghilardi, S., Lutz, C., Wolter, F.: Did I damage my ontology? A case for conservative extensions in description logics. In: Doherty, P., Mylopoulos, J., Welty, C. (eds.) KR 2006, pp. 187–197. AAAI Press, Menlo Park (2006)
[5] Lutz, C., Walther, D., Wolter, F.: Conservative extensions in expressive description logics. In: IJCAI 2007, pp. 453–459. AAAI Press, Menlo Park (2007)
[6] Cuenca Grau, B., Horrocks, I., Kazakov, Y., Sattler, U.: A logical framework for modularity of ontologies. In: IJCAI 2007, pp. 298–304. AAAI Press, Menlo Park (2007)
[7] Motik, B., Patel-Schneider, P.F., Horrocks, I.: OWL 1.1 Web Ontology Language Structural Specification and Functional-Style Syntax. W3C Member Submission (2007)
[8] Fernandez, M., Gomez-Perez, A., et al.: Methontology: From ontological art towards ontological engineering. In: AAAI, Stanford, USA (1997)
[9] Sure, Y., Staab, S., Studer, R.: On-to-knowledge methodology. In: Staab, S., Studer, R. (eds.) Handbook on Ontologies, Springer, Heidelberg (2003)
[10] Guarino, N., Welty, C.: Evaluating ontological decisions with ontoclean. Commun. ACM 45(2), 61–65 (2002)
[11] Kalfoglou, Y., Schorlemmer, M.: Ontology mapping: The state of the art. The Knowledge Engineering Review 18, 1–31 (2003)
[12] Noy, N.F.: Semantic integration: A survey of ontology-based approaches. SIGMOD Record 33(4), 65–70 (2004)
[13] Noy, N.F.: Tools for mapping and merging ontologies. In: Staab, S., Studer, R. (eds.) Handbook on Ontologies. International Handbooks on Information Systems, pp. 365–384. Springer, Heidelberg (2004)
[14] Noy, N., Musen, M.: The PROMPT suite: Interactive tools for ontology mapping and merging. Int. J.of Human-Computer Studies 6(59) (2003)
[15] Seidenberg, J., Rector, A.L.: Web ontology segmentation: analysis, classification and use. In: Proc. of WWW 2006, pp. 13–22. ACM, New York (2006)
[16] Jiménez-Ruiz, E., Berlanga, R., Nebot, V., Sanz, I.: Ontopath: A language for retrieving ontology fragments. In: Proc. of ODBASE. LNCS, pp. 897–914. Springer, Heidelberg (2007)
[17] Cuenca Grau, B., Kutz, O.: Modular ontology languages revisited. In: Proc.of the Workshop on Semantic Web for Collaborative Knowledge Acquisition (2007)
[18] Cuenca Grau, B., Horrocks, I., Kazakov, Y., Sattler, U.: Just the right amount: extracting modules from ontologies. In: Williamson, C.L., Zurko, M.E., Patel-Schneider, P.F., Shenoy, P.J. (eds.) WWW, pp. 717–726. ACM, New York (2007)
[19] Cuenca Grau, B., Horrocks, I., Kazakov, Y., Sattler, U.: Ontology reuse: Better safe than sorry. In: DL 2007. CEUR Workshop Proceedings, vol. 250 (2007)

dRDF: Entailment for Domain-Restricted RDF[*]

Reinhard Pichler[1], Axel Polleres[2], Fang Wei[1], and Stefan Woltran[1]

[1] Institut für Informationssysteme, Technische Universität Wien, Austria
{pichler,wei,woltran}@dbai.tuwien.ac.at
[2] Digital Enterprise Research Institute (DERI), National University of Ireland, Galway
axel.polleres@deri.org

Abstract. We introduce domain-restricted RDF (dRDF) which allows to associate an RDF graph with a fixed, finite domain that interpretations for it may range over. We show that dRDF is a real extension of RDF and discuss impacts on the complexity of entailment in dRDF. The entailment problem represents the key reasoning task for RDF and is well known to be NP-complete. Remarkably, we show that the restriction of domains in dRDF raises the complexity of entailment from NP- to Π_2^P-completeness. In order to lower complexity of entailment for both domain-restricted and unrestricted graphs, we take a closer look at the graph structure. For cases where the structure of RDF graphs is restricted via the concept of bounded treewidth, we prove that the entailment is tractable for unrestricted graphs and coNP-complete for domain-restricted graphs.

1 Introduction

The Resource Description Framework [18] provides means to publish and share metadata on the Web in a machine readable form. One of the features of RDF is to express incomplete metadata by so-called blank nodes, which allow to make statements about unknown resources, such as "I know *somebody* called 'Tim Berners-Lee' (but I don't know the URI identifying him)". In a sense, blank nodes can be viewed as existential variables in the data. In certain circumstances however, it is conceivable that more refined statements could be made about this "*somebody*". Normally, an RDF graph is interpreted over an infinite set of resources. However, one often has a concrete set of resources in mind when writing RDFs. For instance, we want to be able to say: "I don't know the URI identifying Tim, but I know that it is one of the URI's listed at: http://www.example.org/w3c-people", i.e. we want to assign blank nodes only to certain URI's from a restricted, finite set, but we just do not know which one.

In this paper, we introduce and investigate so-called *domain-restricted RDF (dRDF)* graphs which allow to define exactly such restrictions. Domain-restricted RDF graphs are graphs for which interpretations are bound to a fixed, finite domain.

Example 1. The RDF graphs in Fig. 1 model collaboration links between various people. In the figure and subsequent examples, we use $\_:b_1, \_:b_2, ..., \_:b_n$ to denote blank nodes, quoted strings for literals of L, and colon separated pairs of alphanumeric strings

---

[*] The work of Axel Polleres has been supported by the European FP6 project inContext (IST-034718) and by Science Foundation Ireland under the Lion project (SFI/02/CE1/I131).

G_1	G_2	G_3
(_:b_1,foaf:name,"Fang"), (_:b_2,foaf:name,"Stefan"), (_:b_3,foaf:name,"Reini"), (_:b_1,:worksWith,_:b_2), (_:b_2,:worksWith,_:b_3)	(_:b_1,foaf:name,"Stefan"), (_:b_2,foaf:name,"Reini"), (_:b_3,foaf:name,"Fang"), (_:b_1,:worksWith,_:b_2), (_:b_3,:worksWith,_:b_1), (_:b_1,:worksWith,_:b_3), (_:b_4,foaf:name,"Axel"), (_:b_1,:worksWith, _:b_4)	(_:b_2,foaf:name,"Stefan"), (_:b_1,foaf:name,"Axel"), (_:b_2,:worksWith,_:h_1)

Fig. 1. Fictitious collaboration graphs published by Fang, Stefan W. and Stefan D

where the prefix may be empty for QNames/URIs.[1] Graphs are sets of triples, as usual. The two fictitious graphs G_1 and G_2 describe metadata we assume to be published by two of the authors of this paper working at TU Vienna, Fang and Stefan. Fang's graph only talks about current employees of TU Vienna, Stefan's graph talks about current and past employees of TU Vienna, whereas G_3 denotes collaboration links of Stefan Decker, who talks in his graph only about current DERI employees. Even if we assume that lists of URIs to denote these domains [2] are published at some Web referenceable address, current RDF does not provide means to allow the respective publishers of the graphs $G_1 - G_3$ to express or reference the domain they are talking about. dRDF fills exactly this gap. □

The key reasoning task for RDF is deciding whether the information in one RDF graph is subsumed by what is said by another RDF graph – the RDF entailment problem. Entailment should intuitively be affected by restricting the domain of a graph. For instance, the graph G_3 is subsumed by G_2 modulo blank node renaming. Nevertheless, since these graphs talk about different domains, a reasoning engine aware of these domain restrictions should not conclude entailment here.

It is well known that blank nodes raise the complexity of the entailment problem to NP-completeness [14]. A major goal of this work is to search for realistic restrictions which might ensure tractability of the entailment problem. We thus study two kinds of restrictions: one is the restriction to a fixed, finite domain (i.e., dRDF) mentioned above. The other one is the restriction of the graph structure of the (RDF or dRDF) graphs. More precisely, we investigate the entailment problem for graphs having *bounded treewidth*, which can be thought of as a generalization of acyclicity. It has been successfully applied to graph-related problems in many areas [3,8] where otherwise intractable problems have been proved to become tractable if the underlying graph structure has bounded treewidth.

One may expect that both kinds of restrictions decrease the complexity of the entailment problem. Somewhat surprisingly, we will show that the restriction to finite domains

---

[1] We use QNames in the sense of RDF notations such as Turtle [2], where e.g. foaf:name, :axel, or :worksWith stand for full URIs, but we leave out the actual namespace prefixes here, as they do not matter for illustration.

[2] Complete lists of URIs denoting all employees of TU Vienna, DERI, etc. should be easy to obtain. Institutes typically already do publish this data, see e.g. http://www.deri.ie/about/team/ or http://www.dbai.tuwien.ac.at/staff/. It would be easy to write e.g. a GRDDL [9] transformation for those pages which creates lists of unique identifiers for their respective team members.

does not help at all. In contrast, it even increases the complexity of entailment up to the second level of the polynomial complexity hierarchy, viz. to Π_2^P-completeness. On the other hand, we will show that the restriction to RDF graphs of bounded treewidth indeed makes the entailment problem tractable. We will present a polynomial-time algorithm for this case. Actually, also for dRDF graphs, bounded treewidth decreases the complexity of entailment by one level in the polynomial hierarchy; we thus end up with coNP-completeness rather than Π_2^P-completeness.

Our complexity results are summarized as follows. Note that the case of infinite resources and no restriction on the treewidth is well known to be NP-complete [14].

	finite domain-restricted graphs	unrestricted graphs
bounded treewidth	**coNP-complete**	**in P**
unbounded treewidth	Π_2^P**-complete**	NP-complete

The remainder of this paper is organized as follows. In Section 2 we will first review the formal definitions of RDF's syntax and semantics and introduce *domain-restricted RDF (dRDF)* along the way. In this section we will also prove some important theoretical properties concerning general RDF entailment vs. dRDF entailment. The complexity of the entailment problem in case of domain-restricted RDF is dealt with in Section 3. The effect of bounded treewidth without or with domain-restriction is investigated in Section 4 and Section 5, respectively. We wrap up the paper with an outlook to related and future works and draw conclusions in Sections 6 and 7. For an extended version of this paper including more detailed proofs we refer to [22].

2 Preliminaries

In this paper, we exclusively deal with *simple* RDF entailment, i.e., without giving any special semantics to the RDF(S) vocabulary. For short, we shall therefore use the term "RDF entailment" throughout this paper in order to refer to "simple RDF entailment". For the definition of the syntax and semantics of RDF graphs, we find the notation given in [14] more convenient than the one used for defining the standard semantics in [10]. It should be noted that for *simple* interpretations which we consider here both approaches are equivalent, apart from the fact that plain literals are ignored in [14]. It can be easily verified that our complexity results also hold if we stick literally to the definitions in [10].

2.1 RDF Graphs and Domain-Restricted RDF Graphs

We consider an infinite set U (RDF URI references), an infinite set B (blank nodes, also referred to as variables), and an infinite set L (RDF literals). An *RDF triple* is a triple of the form $(v_1, v_2, v_3) \in (U \cup B) \times U \times (U \cup B \cup L)$. In such a triple, v_1 is called the *subject*, v_2 the *predicate*, and v_3 the *object*. The union of the sets U and L is often denoted by UL, and likewise, $U \cup B \cup L$ is often denoted by UBL.

An *RDF graph* (or simply a *graph*) is a set of RDF triples. A subgraph is a subset of a graph. The *vocabulary* of a graph G, denoted by UL_G, is the set of elements of UL

occurring in triples of G. A graph is ground if it has no blank nodes. RDF graphs are often represented as edge-labeled, directed graphs where a triple (a, b, c) is represented by $a \xrightarrow{b} c$.

A *map* is a function $\mu\colon UBL \to UBL$ preserving URIs and literals, i.e., $\mu(v) = v$ for all $v \in UL$. We define $\mu(G) := \{(\mu(s), \mu(p), \mu(o)) \mid (s, p, o) \in G\}$. A graph G' is an instance of G if there exists a map μ with $G' = \mu(G)$. With some slight ambiguity we say that there exists a map $\mu\colon G_1 \to G_2$ if there is a map $\mu\colon UBL \to UBL$, such that $\mu(G_1)$ is a subgraph of G_2. Let G_1 and G_2 be graphs. The *union* $G_1 \cup G_2$ is the set-theoretical union of their sets of triples.

Let $D \subseteq UL$ be a non-empty set of URI references and literals and G be an RDF graph. A *domain-restricted RDF graph (dRDF graph)* is a pair $\langle G, D \rangle$. Graphs such that $|D| = n$ is finite are also called *finitely restricted* (or simply *restricted* for short); graphs with $D = UL$ are also called *unrestricted* graphs. Sightly abusing notation, instead of $\langle G, UL \rangle$ we also write G to denote unrestricted graphs.

2.2 Semantics of (Domain-Restricted) RDF Graphs

A simple *interpretation* $I = (Res, Prop, Lit, \varepsilon, IS, IL)^3$ of an RDF graph G over vocabulary UL_G is defined by (1) a non-empty set of resources Res (also called the domain of I) and of properties $Prop$, (2) a distinguished subset $Lit \subseteq Res$, (3) an extension $\varepsilon(pr) \subseteq Res \times Res$ for every property $pr \in Prop$, and (4) mappings $IS\colon U_G \to Res \cup Prop$ and $IL\colon L \to Lit$.

We write $I(.)$ to denote the valuation under the interpretation I. We have $I(u) := IS(u)$ for a URI u and, $I(l) := IL(l)$ for a literal l. A triple (s, p, o) has the value "true" in I if $IS(p) \in Prop$ and $(I(s), I(o)) \in \varepsilon(IS(p))$; otherwise (s, p, o) has the value "false". For a ground graph G, we have $I(G) =$ "true" if every triple of G is true in I.

Blank nodes in non-ground graphs are interpreted as existentially quantified variables. Let $A\colon B \to Res$ be a *blank node assignment* (or an *assignment*, for short), and let I be an interpretation. Then we write $[I + A]$ to denote the interpretation I extended by the blank node assignment A. Clearly, $[I + A](b) = A(b)$ for blank nodes $b \in B$, while $[I + A](a) = I(a)$ for $a \in UL$. A non-ground graph G is true in I, if there exists an assignment $A'\colon B \to Res$, s.t. every triple of G is true in $[I + A']$. If a graph G is true in an interpretation I, then we say that *I is a model of G* or *I satisfies G*.

We say that an RDF graph G_1 *entails* the graph G_2, if every interpretation I which satisfies G_1 also satisfies G_2. If this is the case, we write $G_1 \models G_2$. This leads us to the *RDF entailment problem*: Given two RDF graphs G_1, G_2, does $G_1 \models G_2$ hold? This problem is well known to be NP-complete [14]. We may assume w.l.o.g. that $UL_{G_2} \subseteq UL_{G_1}$, since otherwise $G_1 \not\models G_2$ clearly holds (i.e., we can easily construct an interpretation I which satisfies G_1 but not G_2).

Interpretations for a dRDF graph $\langle G, D \rangle$ restrict general RDF interpretations in the following sense. Given an interpretation $I = (Res, Prop, Lit, \varepsilon, IS, IL)$ and a set $D \subseteq UL$ we call the interpretation $I = (Res \cap D, Prop, Lit \cap D, \varepsilon, IS', IL')$ with $IS' = IS_{Res \cap D}$ and $IL' = IL_{Res \cap D}$ the *D-restriction* of I, also written I_D. Note that

---

[3] As mentioned above, we are following the notation from [14]. Clearly, $Res, Prop, Lit, \varepsilon, IS,$ and IL correspond to $IR, IP, LV, IEXT, IS,$ and IL, respectively, in [10].

we do not restrict the domain of $Prop$ in I_D. Since the purpose of domain-restrictions is mainly to restrict the values which blank nodes may take, we do not need to restrict properties—blank nodes are not allowed in property position in RDF anyway.

We define *d-models* as before with the only difference that for any interpretation I its D-restriction is considered. I.e., given an interpretation I and a dRDF graph $\langle G, D \rangle$, if G is true in I_D, then we say that *I is a d-model of* $\langle G, D \rangle$ or *I d-satisfies* $\langle G, D \rangle$.

Finally, we say that a dRDF graph $\langle G_1, D_1 \rangle$ *d-entails* $\langle G_2, D_2 \rangle$ (by overloading $\models$ we write $\langle G_1, D_1 \rangle \models \langle G_2, D_2 \rangle$), if for any interpretation I s.t. I_{D_1} satisfies G_1, I_{D_2} also satisfies G_2. Obviously, if D_1 contains an element not existing in D_2, then this condition can never be fulfilled. Indeed, if $c \in D_1 \setminus D_2$, then we can easily construct a D_1-model of G_1 (where every URI in G_1 is mapped to c) which is not a D_2-model of G_2. Conversely, if D_2 contains elements not existing in D_1, then these elements play no role for d-entailment, i.e., we have $\langle G_1, D_1 \rangle \models \langle G_2, D_2 \rangle$ iff $\langle G_1, D_1 \rangle \models \langle G_2, D_1 \cap D_2 \rangle$. Therefore, in the sequel, we shall restrict ourselves w.l.o.g. to the case $D_1 = D_2$.

Example 2 (Example 1 cont'd). Getting back to the graphs in Fig. 1, it is easy to see that $G_2 \models G_3$ and that $G_1 \models G_2'$, where G_2' is the graph obtained from G_2 by removing the last three statements of G_2. As mentioned earlier, Fang's graph G_1 talks only about people working at TU Vienna, i.e., it is restricted to the fixed domain $D_1 = \{$"Fang", "Stefan", "Reini"$\} \cup D_{TUV}$ where D_{TUV} is a fixed, finite list of URIs which gives identifiers to all current TU Vienna employees and contains for instance the URIs :fangwei, :stefanwoltran, and :reinhardpichler. This list may be huge and instead of looking up all the real identifiers there, Fang still uses blank nodes as in the example for publishing her metadata. But in order to indicate the fact that her graph talks about a finite domain she publishes the dRDF graph $\langle G_1, D_1 \rangle$. Likewise, Stefan publishes his collaboration links as graph $\langle G_2, D_2 \rangle$. Stefan's graph is restricted to $D_2 = D_1 \cup D_{TUVold}$ where D_{TUVold} is a finite list of identifiers of former TU Vienna members that also contains the URI :axelpolleres, for example. Both $\langle G_1, D_1 \rangle \models \langle G_2', D_2 \rangle$ and $\langle G_2, D_2 \rangle \models G_3$ hold. However, G_3 is in fact none of the authors' but Stefan Decker's collaboration graph at DERI and restricted to the domain $D_3 = \{$"Stefan", "Axel"$\} \cup D_{DERI}$ where D_{DERI} is the (again finite) list of identifiers of DERI employees that contains among others the URIs :axelpolleres and :stefandecker, but none of the other previously mentioned URIs. Obviously, $\langle G_2, D_2 \rangle \not\models \langle G_3, D_3 \rangle$ despite the fact $\langle G_2, D_2 \rangle \models G_3$. □

2.3 Properties of (Domain-Restricted) Entailment

Before we have a closer look at the complexity of this restricted form of the entailment problem, let us discuss some fundamental properties of (domain-restricted) entailment.

Proposition 1. *Let G_1, G_2 be graphs and D a finite domain. Then $G_1 \models G_2$ implies $\langle G_1, D \rangle \models \langle G_2, D \rangle$ while the converse is, in general, not true.*

Proof. Clearly, entailment implies d-entailment, since every d-model is also a model. To see that the converse is, in general, not true, consider the following counter-example: Let $G_1 = \{(a, p, b), (a, p, c), (b, p, c)\}$ and $G_2 = \{(x, p, x)\}$ where $a, b, c, p \in U$ and

$x \in B$. Moreover, let $D = \{d_1, d_2\}$. Then $\langle G_1, D \rangle \models \langle G_2, D \rangle$ holds: Indeed, with $|D| = 2$, any d-model I of $\langle G_1, D \rangle$ assigns the same value d_i (for some $i \in \{1,2\}$) to two URIs out of $\{a, b, c\}$. Hence, G_2 is true in $[I + A]$ with $A(x) = d_i$. □

Proposition 2. *Let G_1, G_2 be graphs and D a finite domain with $|D| \geq |UT_{G_1 \cup G_2}|$. Then $G_1 \models G_2$ iff $\langle G_1, D \rangle \models \langle G_2, D \rangle$.*

Proof. The "only if" direction immediately follows from Proposition 1. The basic idea of the 'if'-direction is that, for any interpretation, only the "active domain" (i.e, the elements in Res which are actually used for interpreting the elements in $ULG_{1 \cup G_2}$) is relevant. For details, see [22]. □

Intuitively, Proposition 2 states that entailment and d-entailment coincide for a sufficiently large domain D.

We conclude this section by showing that w.l.o.g. several simplified assumptions may be made, both for the entailment problem and the d-entailment problem.

A *Skolemization* of a graph G is a ground instance of G which maps every blank node in G to some "fresh" URI reference. These fresh URI references are called the Skolem vocabulary. The Skolemization of G is denoted as $sk(G)$. The usefulness of Skolemizations is due to the following property:

Lemma 1. *Let G_1, G_2 be graphs and let $sk(G_1)$ be a Skolemization of G_1, s.t. the Skolem vocabulary is disjoint from both G_1 and G_2. Moreover, let D be a finite domain. Then the following equivalences hold: $G_1 \models G_2 \Leftrightarrow sk(G_1) \models G_2$ and $\langle G_1, D \rangle \models \langle G_2, D \rangle \Leftrightarrow \langle sk(G_1), D \rangle \models \langle G_2, D \rangle$.*

Proof. The correctness of this lemma in case of ordinary entailment is shown in [10]. The case of d-entailment can be shown by exactly the same arguments. □

In other words, for both ordinary entailment and d-entailment, we may assume w.l.o.g. that the graph G_1 is ground. After having restricted the syntax, we show that also the set of models to be inspected by an (ordinary or d-) entailment test can be significantly restricted. In [10], entailment testing is reduced to *Herbrand models*. However, in case of domain-restricted graphs, we can of course not be sure that the Herbrand universe is contained in the finite domain D. We thus have to generalize the idea of Herbrand models to *minimal* models.

Definition 1. *We call a model I of an RDF graph G (resp. a dRDF graph $\langle G, D \rangle$) a minimal model of G (resp. $\langle G, D \rangle$), if the extensions $\varepsilon(pr)$ in I are chosen minimal (for every $pr \in Prop$) s.t. G is true in I. In other words, for every property $pr \in Prop$, a minimal model is characterized by the following relation*

$$\varepsilon(pr) = \{(I(s), I(o)) \mid (s, p, o) \in G_1 \text{ and } IS(p) = pr\}.$$

Clearly, every Herbrand model is a minimal model while the converse is, in general, not true. The following lemma states that, for (d-) entailment testing, we may restrict ourselves to minimal models of G_1.

Lemma 2. *Let G_1, G_2 be graphs, s.t. $ULG_2 \subseteq ULG_1$ and G_1 is ground. Moreover, let D denote a finite domain. Then the following equivalences hold:*
(a) $G_1 \models G_2$ iff every minimal model I of G_1 satisfies G_2.
(b) $\langle G_1, D \rangle \models \langle G_2, D \rangle$ iff every minimal model I of G_1 with $Res \subseteq D$ satisfies G_2.

Proof. The restriction to minimal models of G_1 (resp. $\langle G_1, D \rangle$) is based on the following observation: Suppose that G_1 (or $\langle G_1, D \rangle$) is true in some interpretation I. It clearly remains true if we restrict ε to ε' with $\varepsilon'(pr) = \varepsilon(pr) \cap \{(I(s), I(o)) \mid (s, p, o) \in G_1$ and $IS(p) = pr\}$. In case (b), the restriction to interpretations I with $Res \subseteq D$ is obvious since, in a d-interpretation, Res is restricted to a subset of D anyway. □

3 Complexity of d-Entailment

We are now ready to investigate the complexity of d-entailment testing. It turns out that it is one level higher in the polynomial hierarchy than without domain restrictions.

The Π_2^P upper bound is easily established via the lemmas from Section 2.3.

Lemma 3. *The d-entailment problem is in Π_2^P.*

Proof. Recall that, by Lemma 1, we may assume w.l.o.g. that G_1 is ground. Then the complementary problem "$\langle G_1, D \rangle$ does not d-entail $\langle G_2, D \rangle$" can be decided by the following Σ_2^P-algorithm.

1. Guess an interpretation I over the vocabulary of $G_1 \cup G_2$, s.t. G_1 is true in I.
2. Check that for all assignments A for the blank nodes in G_2, the graph G_2 is false in $[I + A]$. Clearly, this check can be done by a coNP-oracle. □

The proof of the Π_2^P lower bound is much more involved. Due to space limitations, we can only give a rough sketch here. For details, see [22]. The proof goes by a reduction from a restricted form of the so-called *H-subsumption problem*. H-subsumption was introduced in the area of automated deduction as a powerful technique of redundancy elimination (cf. [12]). Given two clauses C, C', and a Herbrand universe H, $C \leq_{ss}^{H} C'$ holds, iff, for each substitution ϑ of the variables in C' to H, there exists a substitution μ of the variables in C to H, such that $C\mu \subseteq C'\vartheta$. In this paper we are only interested in the case that H is a *finite* domain of constants. In [21], it was shown that the H-subsumption problem is Π_2^P-complete even if C and C' consist of unnegated atoms only. However, we need a strongly restricted version of H-subsumption: In particular, we have to restrict the H-subsumption problem to the setting, where no constants are allowed to occur in the clauses and where all predicates are binary. We call such problems total, binary H-subsumption problems (TBH-subsumption, for short). Of course, it is a priori by no means clear that TBH-subsumption is still Π_2^P-hard. Hence, the Π_2^P-hardness proof essentially consists of two parts: the problem reduction from TBH-subsumption to d-entailment and the Π_2^P-hardness proof of TBH-subsumption.

Lemma 4. *The TBH-subsumption problem can be reduced in polynomial time to the d-entailment problem.*

Proof. Consider an instance $C \leq_{ss}^H C'$ of the TBH-problem over some finite universe H. In C, C', all predicates are binary and all arguments of the atoms in C and C' are first-order variables. W.l.o.g., the clauses C and C' have no variables in common. Moreover, all predicates in C also occur in C' (since otherwise $C \not\leq_{ss}^H C'$ trivially holds) and all predicates in C' also occur in C (since literals in C' with a predicate symbol not occurring in C' play no role at all in the H-subsumption test—they can never be matched by literals in C). We define the dRDF graphs $\langle G_1, D\rangle$ and $\langle G_2, D\rangle$ with $D = H$, $G_1 = \{(s,p,o) \mid p(s,o) \in C'\}$, and $G_2 = \{(s,p,o) \mid p(s,o) \in C\}$, s.t. the vocabulary of $G_1 \cup G_2$ is given as $U := \{s, p, o \mid p(s,o) \in C'\}$ and $L = \emptyset$. Moreover, we have $B = \{s, o \mid p(s,o) \in C\}$. In other words, G_1 is ground while G_2 contains only blank nodes. Clearly, this reduction is feasible in polynomial time. The correctness (i.e., $C \leq_{ss}^H C' \Leftrightarrow \langle G_1, D\rangle \models \langle G_2, D\rangle$) is shown in [22]. □

Lemma 5. *The TBH-subsumption problem is Π_2^P-hard.*

Proof. The proof is highly involved and very technical. It proceeds in three steps: First, Π_2^P-hardness is shown for clauses using only ternary predicates over a 2-element Herbrand universe $H = \{a, b\}$. This result is then extended to an arbitrary, finite H with $|H| \geq 2$. Finally, it is shown that Π_2^P-hardness still holds even if the clauses are built up from binary predicates only and $|H| \geq 4$; details are fully worked out in [22]. □

Putting the Lemmas 3–5 together, we immediately get the following result.

Theorem 1. *The d-entailment problem is Π_2^P-complete.*

In other words, the complexity of entailment increases from NP- to Π_2^P-completeness if we restrict the domain. This unexpected effect can be explained as follows. RDF entailment with unrestricted domain admits a syntactical characterization: $G_1 \models G_2$ iff there exists a map from G_2 to G_1. The proof of the "only if" direction of this equivalence crucially depends on an argument via Herbrand interpretations of G_1 (see the interpolation lemma and its proof in [10]). Of course, this argument is no longer valid for dRDF-graphs if the domain D is smaller than the Herbrand universe (note that the counter-example given in the proof of Proposition 1 is based on a similar idea).

4 Efficient Entailment Through Bounded Treewidth

In this section we first define the treewidth of an RDF graph, then show that the entailment problem of $G_1 \models G_2$ can be solved in polynomial time, if G_2 has bounded treewidth. Recall that an RDF triple has the form $(v_1, v_2, v_3) \in (U \cup B) \times U \times (U \cup B \cup L)$. Let us denote those triples (v_1, v_2, v_3) where v_1 and v_3 are two distinct variables as *blank triples*. Moreover, we speak of *semi-blank triples* if only one of v_1, v_3 is a variable or if v_1 and v_3 are identical variables.

It is interesting to observe that the intractability of the RDF entailment problem $G_1 \models G_2$ depends *only* on the blank triples in G_2. To see this, consider the ground and semi-blank triples in G_2: finding a map of any ground triple is merely an existence test of the triple in G_1. Thus all the ground triples can be tested independently from each other. Now let us assume that G_2 contains only semi-blank triples with k distinct variables. Assume further that $|G_1| = m$ and $|G_2| = n$. To test $G_1 \models G_2$, we first partition all the triples of G_2 into k disjoint sub-graphs $P_1, \ldots, P_k$, s.t. two triples belong to the same sub-graph if and only if they contain the same variable. For each i, let n_i denote the cardinality $|P_i|$ of P_i. Clearly, $n_1 + \cdots + n_k = n$. We can then check the entailment of the sub-graphs one by one. For each $P_{i(1 \leq i \leq k)}$, the variable in P_i can be mapped to m possible values. Because there is only one variable in P_i, for each map μ, we have to execute the existence test $\mu(P_i) \subseteq G_1$, which takes maximum mn_i steps. Thus in summary, the total cost of the entailment test is $\mathcal{O}(m^2 n)$.

However, if the graph G_2 contains blank triples, it is possible that the variables are intertwined s.t. no variable can be tested independently, thus the number of possible maps is exponential in the size of the variables occurring in blank triples. Treewidth is a well-known metric on graphs that measures how tree-like a graph is. Many intractable problems become tractable, if the treewidth of the underlying structure is bounded.

We shall now show that the entailment problem $G_1 \models G_2$ becomes tractable if the graph G_2 has bounded treewidth. Recall the syntactical characterization of entailment [10,14]: $G_1 \models G_2$ iff there exists a map from G_2 to G_1. Hence, the entailment problem for unrestricted RDF graphs comes down to a special case of conjunctive query containment where all predicates are binary. Hence, the notion of treewidth and the tractability of conjunctive query containment in case of bounded treewidth (see e.g. [5]) naturally carry over to RDF graphs and the entailment problem. However, we prefer to give a *native* definition of tree decomposition for RDF graphs here, so that the RDF intuition is preserved. Likewise, we explicitly present an entailment algorithm in terms of the RDF terminology rather than by just referring to conjunctive queries.

We start by giving the definitions of tree decomposition and treewidth for an RDF graph. By the above considerations, we assume that the RDF graph does not contain any ground triple. We denote all the variables occurring in G as B_G.

Definition 2. *A tree decomposition $\mathcal{T}$ of an RDF graph G is defined as $\langle T, (B_i)_{i \in T} \rangle$ where T is a tree and each B_i is a subset of B_G with the following properties:*

1. *Every $b \in B_G$ is contained in some B_i.*
2. *For every blank triple $(v_1, v_2, v_3) \in G$, there exists an $i \in T$ with $\{v_1, v_3\} \subseteq B_i$.*
3. *For every $b \in B_G$, the set $\{i \mid b \in B_i\}$ induces a subtree of T.*

The third condition is usually referred to as the *connectedness condition*. The sets B_i are called the *blocks* of $\mathcal{T}$. The *width* of the tree decomposition $\langle T, (B_i)_{i \in T} \rangle$ is defined as $\max\{|B_i| \mid i \in T\} - 1$. The *treewidth* of an RDF graph G (denoted as $tw(G)$) is the minimal width of all tree decompositions of G. For a given $w \geq 1$, it can be decided in linear time whether some graph has treewidth $\leq w$. Moreover, in case of a positive answer, a tree decomposition of width w can be computed in linear time [4].

Example 3. Consider the graph G_2 given in Fig. 2. The undirected graph and the tree decomposition are depicted in Fig. 3. The treewidth of G_2 is 2. □

Below, we describe an algorithm which, given the tree decomposition of G_2, tests $G_1 \models G_2$ *in polynomial time*. The intuition behind the algorithm is as follows: we first construct *partial* maps from the nodes on the tree decomposition into G_1 (denoted

G_1	G_2
(_:b_1,:worksWith,_:b_2), (_:b_2,:worksWith,_:b_3), (_:b_1,:worksWith,_:b_3), (_:b_2,:worksWith,_:b_5), (_:b_1,:worksWith,_:b_5), (_:b_3,:worksWith,_:b_6), (_:b_3,:worksIn,"TUV"), (_:b_5,:worksIn,"DERI")	(_:b_1,:worksWith,_:b_2), (_:b_2,:worksWith,_:b_3), (_:b_1,:worksWith,_:b_3), (_:b_2,:worksWith,_:b_4), (_:b_1,:worksWith,_:b_4), (_:b_2,:worksWith,_:b_5), (_:b_4,:worksWith,_:b_6), (_:b_3,:worksIn,"TUV"), (_:b_5,:worksIn,_:b_7)

Fig. 2. RDF graphs for Example 3

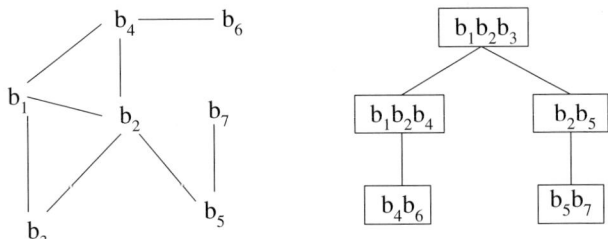

Fig. 3. Undirected graph of G_2 from Fig. 2 and the tree decomposition of G_2

as M_i in the algorithm below), then successively merge those partial maps which are consistent with each other. If at last the merging succeeds, $G_1 \models G_2$ holds, otherwise not. Note that the connectedness property of the tree decomposition allows us to merge such partial maps in a bottom up manner on the tree (by using the semi-join operation of relational algebra), in polynomial time. We thus carry over ideas proposed in [5] for testing conjunctive query containment to the RDF entailment problem.

Polynomial Time Algorithm. Let $\langle T, (B_i)_{i \in T} \rangle$ be the tree decomposition of the RDF graph G_2 with treewidth k. Given a node i in T, S_i is denoted as the union of all the blocks in the sub-tree rooted at i. The induced sub-graph $G[S_i]$ contains all the triples (v_1, v_2, v_3) in G_2, such that either v_1 or v_3 belongs to S_i. We maintain for each node i in T a relation M_i. In the algorithm below, $\ltimes$ is the natural semi-join operator.

The *Polycheck* algorithm for checking $G_1 \models G_2$ consists of the following steps:

1. For each node i in T, generate the sub-graph G'_i which contains all the triples (v_1, v_2, v_3) such that $\{v_1, v_3\} \subseteq B_i \cup UL_{G_2}$ and $\{v_1, v_3\} \cap B_i \neq \emptyset$.
2. Initialize the relation M_i as follows: for each map μ from G'_i to G_1, the tuple $\mu(B_i)$ is in M_i.
3. Process the tree nodes bottom-up as follows: Suppose i is a tree node in T all of whose children have been processed. For each child j of i, we set $M_i := M_i \ltimes M_j$.
4. Let r be the root of T. Then $G_1 \models G_2$ if and only if M_r is not empty.

Example 4. Let us continue with Example 3. With the given tree decomposition of G_2, we illustrate in Fig. 4 how the *Polycheck* algorithm works when testing $G_1 \models G_2$.

Step 1: We need to generate the sub-graphs $G'_1, \ldots, G'_5$ for the nodes 1–5 of the tree decomposition. For instance, G'_4 is the sub-graph consisting of only one triple (_:b_4,: $worksWith$, _:b_6).

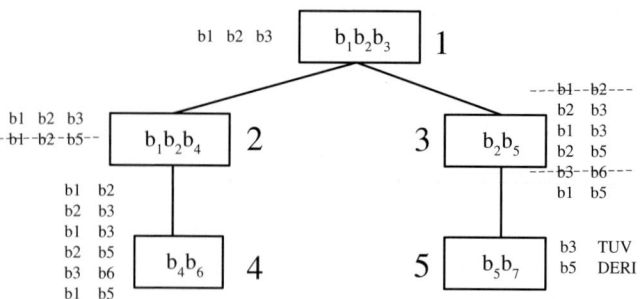

Fig. 4. Bottom up processing on the tree decomposition

Step 2: Next we generate the *partial* maps $M_{i\,(1 \leq i \leq 5)}$, which are given as the tables beside the tree nodes. Note that following the convention of relational databases, the variable names at each block give the relation schema for that block and every row of a table is called a tuple. For the time being, let us ignore the dotted lines drawn over the tuples. Now consider M_4. For every map μ with $\mu((\_:b_4,\,:\,worksWith,\,\_:b_6) \in G_1$, we insert the tuple $\mu(\_:b_4,\_:b_6)$ into M_4. It is easy to verify that there are six distinct maps from G'_4 to G_1, thus M_4 consists of six tuples.

Step 3: We execute semi-joins along the bottom-up traversal of the tree decomposition. The tables at the leaf nodes remain unchanged. Let us consider the semi-join operation $M_2 \ltimes M_4$. By definition, the result of the semi-join is the set of those tuples t in M_2 for which there is a tuple t' in M_4, s.t. t and t' coincide on their common attributes (in our case b_4). Such a t' is called a *partner* of t. Now let us consider the first tuple (b_1, b_2, b_3) in M_2. In this case, b_3 is the value for the common attribute b_4. A partner tuple (b_3, b_6) in M_4 is found, thus the tuple (b_1, b_2, b_3) remains in M_2. However, for the second tuple (b_1, b_2, b_5), there does not exist any partner tuple in M_4. Therefore the tuple (b_1, b_2, b_5) is deleted by the semi-join operation.

Step 4: Finally, the only tuple in M_1 remains after the semi-join operation with both M_2 and M_3, thus the entailment test succeeds. □

Theorem 2. *The algorithm* Polycheck *correctly decides whether* $G_1 \models G_2$.

Proof. We use induction on the number of nodes processed in the tree, with the following hypothesis: After node i is processed, tuple $t \in M_i$ if and only if there is a map μ from $G[S_i]$ to G_1 such that $\mu(B_i) = t$. Thus when the root r has been processed, M_r is non-empty if and only if there is a map μ from $G[S_r]$ to G_1. Because $G[S_r]$ is G_2, we can therefore conclude that $G_1 \models G_2$.

The induction hypothesis holds for the leaves, because of step 2, and the induced sub-graph $G[S_l]$ of any leaf node l is the the graph G'_l we defined in the step 1.

For the induction, assume that we have processed all the children $j_1, \ldots, j_r$ of node i. Suppose that $t \in M_i$ holds *before* the processing of node i. Let ϕ be the map of G'_i to G_1 s.t. $\phi(B_i) = t$. If $t \in M_i$ still holds *after* the processing of i (i.e., the semi-join operations with all the child nodes), then for each $j_{k\,(1 \leq k \leq r)}$, there is a tuple $t_k \in M_{j_k}$, that agrees with t on the variables in $B_i \cap B_{j_k}$. By the induction hypothesis, there is a map ϕ_k from $G[S_{j_k}]$ to G_1, such that $\phi_k(B_{j_k}) = t_k$.

It remains to show that the maps $\phi, \phi_1, \ldots, \phi_r$ are consistent. Assume that v occurs in the blocks B_{j_α} and B_{j_β} of two children of i. According to the connectedness condition, v occurs in B_i too. Since ϕ, ϕ_{j_α} and ϕ_{j_β} agree on the common variables, ϕ, ϕ_{j_α} and ϕ_{j_β} are consistent. Let $\mu := \phi \cup \phi_1 \cup \ldots \cup \phi_r$. Then μ is clearly a map from $G[S_i]$ to G_1 such that $\mu(B_i) = t$.

Conversely, assume there is a map μ from $G[S_i]$ to G_1 such that $\mu(B_i) = t$, we show that t is in M_i after the processing of i. Clearly t is in M_i before the processing of i, because G'_i as defined in step 1 is a sub-graph of $G[S_i]$. Let $\phi_1, \ldots, \phi_r$ be the projection μ onto the variables in $S_{j_1}, \ldots, S_{j_r}$, and let ϕ be the projection of μ onto the variables in B_i. By the induction hypothesis, there is a tuple $t_i \in M_{j_k}$, where $1 \leq k \leq r$, such that $\phi_k(B_{j_k}) = t_k$. After step 2, there is a tuple $t \in M_i$, such that $\phi(B_i) = t$. Since t agrees with the tuples $t_1, \ldots, t_r$ on all common attributes, t is in M_i after the processing of i. □

Theorem 3. *The entailment problem of $G_1 \models G_2$ can be decided in polynomial time if G_2 has bounded treewidth.*

Proof. Suppose that $|G_1| = n$, $|G_2| = m$ and $tw(G_2) = k - 1$. Step (1): For each tree node i, we need to scan all the triples in G_2 to generate the subgraph G'_i of G_2. Since the size of the tree decomposition of G_2 is not more than m, we have an m^2 upper bound. Step (2): For block B_i with size k, there are n^k possible tuples to be checked. For each tuple t, we generate the map μ from B_i to t. If $\mu(G_i) \subseteq G_1$, then t is added to M_i. Thus, the cost for the initialization of all the nodes is mn^k. Step (3): Each semi-join operation of two k-ary relations takes n^{2k} (using primitive nested loops), thus the total cost is mn^{2k}. In summary, we get the upper bound $\mathcal{O}(m^2 + mn^{2k})$ on the time complexity of the algorithm *Polycheck*. □

To summarize, the entailment problem $G_1 \models G_2$ is intractable, only if the blank triples in G_2 are cyclic. We note that, in practice, an RDF graph contains rarely blank nodes, and even less blank triples. Hence, most of the real RDF graphs are acyclic or have low treewidth such as 2, and the entailment can be tested efficiently with the above algorithm. For instance, all the graphs in Fig. 1 are acyclic and thus have $tw \leq 1$.

5 Bounded Treewidth and d-Entailment

In the previous section, we have seen for RDF graphs that bounded treewidth significantly decreases the complexity of entailment. We shall now prove a similar result for d-entailment, where bounded treewidth again has a positive impact on the complexity.

Lemma 6. *The d-entailment problem of $\langle G_1, D \rangle \models \langle G_2, D \rangle$ is in coNP if G_2 has bounded treewidth.*

Proof. Suppose that $tw(G_2)$ is bounded by some constant. Recall that, by Lemma 1, we may assume w.l.o.g. that G_1 is ground. Then the complementary problem of testing $G_1 \not\models G_2$ can be decided by the following NP-algorithm:

1. Guess an interpretation I over the vocabulary of $G_1 \cup G_2$, s.t. G_1 is true in I.
2. Check that there exists no assignments A for the blank nodes in G_2, s.t. the graph G_2 is true in $[I + A]$.

The check in step 2 comes down to an ordinary entailment test $G'_1 \not\models G'_2$ with $G'_1 := \{(I(s), I(p), I(o)) \mid (s, p, o) \in G_1\}$ and $G'_2 := \{(I(s), I(p), I(o)) \mid (s, p, o) \in G_2\}$, where we stipulate $I(z) = z$ for the variables z in G_2. We clearly, have $tw(G_2) = tw(G'_2)$. Hence, by Theorem 3, the check $G'_1 \not\models G'_2$ is feasible in polynomial time. □

Lemma 7. *The d-entailment problem of $\langle G_1, D \rangle \models \langle G_2, D \rangle$ is coNP-hard for bounded treewidth of G_2. It remains coNP-hard even if $tw(G_2) = 0$ (i.e., the graph induced by the blank nodes consists of isolated nodes only).*

Proof. We prove the coNP-hardness by reducing the well-known NP-complete problem of graph ℓ-colorability with $\ell \geq 3$ to the complementary problem $\langle G_1, D \rangle \not\models \langle G_2, D \rangle$.

Let $G = (V, E)$ be a graph with vertices V and edges E. We define two RDF graphs G_1 and G_2 as $G_1 := \{(u, e, v) \mid (u, v) \text{ is an edge in } E\}$ and $G_2 := \{(x, e, x)\}$ for some blank node x. Clearly, $tw(G_2) = 0$. Moreover, this reduction is feasible in polynomial time. It remains to show the correctness of this reduction, which can be seen as follows: By definition, G is ℓ-colorable iff there exists a mapping ϑ, assigning different colors to any two adjacent vertices $u, v \in V$. Obviously, such an assignment exists iff there exists an interpretation I sending all triples (u, e, v) in G_1 to values $(I(u), I(e), I(v))$ with $I(u) \neq I(v)$. This, in turn, is the case iff there exists no blank node assignment A, s.t. $(A(x), A(x)) \in \varepsilon(I(e))$. □

In summary, we thus have the following exact complexity classification.

Theorem 4. *The d-entailment problem of $\langle G_1, D \rangle \models \langle G_2, D \rangle$ is coNP-complete if G_2 has bounded treewidth.*

6 Related and Future Work

Our results touch upon many related issues on RDF reasoning and Semantic Web reasoning in general. First of all, we point out that the peculiarities of reasoning with open and restricted domains raised by dRDF are closely related to similar issues discussed in the context of reasoning with rules and ontologies [6].

Next, we have only discussed *simple* (d)RDF entailment in this paper. As for future works, it will be interesting to see, which implications restrictions on the domain will have when higher entailment regimes such as RDF entailment, RDFS entailment, D-entailment, or entailments in OWL variants are considered. Standard fragments of OWL are well-known to be syntactic variants of decidable Description Logics [1], i.e. OWL Light is reducible to SHIF(D) and OWL DL is reducible to SHOIN(D) [16]. We plan to investigate how (finite) domain-restrictions on the data affect the complexity of entailment in these languages, see also [1, Chapter 5]. Alternatively to finitely restricting the domain of interpretations for the whole graph it seems that restricting the blank nodes in an RDF graph to a finite, enumerated class (using OWL's `oneOf` constructor) could have similar effects, when we extend our considerations towards OWL. We are currently investigating respective generalizations of the definition of dRDF graphs. As for rule extensions on top of ontology languages, we remark that RDF(S) and some non-standard fragments of OWL entailment can be reduced to sets of

Datalog rules [7,13,17,27] and thus combined straightforwardly with datalog rules on top. Note however, that subsumption of arbitrary Datalog programs is undecidable [26]. Issues get even more involved, when standard OWL fragments or (non-monotonic) rule languages are added on top of these languages (see [11,15,25] and references therein) since in the unrestricted case, even the satisfiability problem for rule-extended ontologies problem becomes undecidable. However, here domain-restrictions may turn out to be actually a good thing, since those cases become decidable for finite domains. A complete investigation of complexity classes for such combinations is on our agenda. In the context of (non-monotonic) rules extensions, let us mention that restricting the domain of interpretations is also related to restricting the scope of negation to closed sets of rules in non-monotonic rule languages for the Web, see [23] for further details.

As for related results on finding tractable fragments of RDF, Muñoz et al. [19] define a syntactic subclass of RDFS with $\mathcal{O}(n \log n)$ bounds for entailment (without blank nodes though), which our results complement.

Deciding whether a SPARQL [24] query has an answer is an extension of simple RDF entailment which is PSPACE complete in general but also NP-complete in many cases [20]. We expect that our results propagate to tractable fragments of SPARQL over unrestricted RDF as well as over dRDF graphs, which to define is on our agenda.

Bounded treewidth is a well-established method for identifying tractable subclasses of otherwise intractable problems. It has been successfully applied to a great variety of graph-related problems like network reliability, job scheduling, compiler optimization, model checking, etc. (see e.g., [3,8]). To the best of our knowledge though, bounded treewidth has not yet been considered in the context of Semantic Web reasoning.

7 Conclusions

Entailment checking is the key reasoning task for RDF. In this work, we have investigated how the complexity of deciding entailment in RDF is affected by two restrictions. Firstly, we introduced dRDF, a variant of RDF which allows to associate an RDF graph with a fixed, finite domain that interpretations for it may range over. We have demonstrated that such restrictions are useful in environments where someone wants to make RDF statements over closed contexts such as enterprises or institutions. Secondly, we investigated restrictions of the graph structure of (d)RDF graphs. Particularly, we investigated the effect of restricting the structure of RDF graphs to bounded treewidth, which considerably lowered the complexity of entailment checking. As related works show, there are many promising directions for applying our results, such as finding further tractable algorithms for fragments of SPARQL, or applying respective restrictions beyond simple RDF entailment.

References

1. Baader, F., Calvanese, D., McGuinness, D.L., Nardi, D., Patel-Schneider, P.F. (eds.): The Description Logic Handbook. Cambridge (2003)
2. Beckett, D.: Turtle - Terse RDF Triple Language (November 2007), available at http://www.dajobe.org/2004/01/turtle/
3. Bodlaender, H.L.: A tourist guide through treewidth. Acta Cybern. 11(1-2), 1–22 (1993)

4. Bodlaender, H.L.: A linear-time algorithm for finding tree-decompositions of small treewidth. SIAM J. Comput. 25(6), 1305–1317 (1996)
5. Chekuri, C., Rajaraman, A.: Conjunctive query containment revisited. Theor. Comput. Sci. 239(2), 211–229 (2000)
6. de Bruijn, J., Eiter, T., Polleres, A., Tompits, H.: On representational issues about combinations of classical theories with nonmonotonic rules. In: Lang, J., Lin, F., Wang, J. (eds.) KSEM 2006. LNCS (LNAI), vol. 4092, Springer, Heidelberg (2006)
7. de Bruijn, J., Heymans, S.: RDF and logic: Reasoning and extension. In: Proc. WebS 2007, IEEE Computer Society Press, Los Alamitos (2007)
8. Downey, R.G., Fellows, M.R.: Parameterized Complexity. Springer, Heidelberg (1999)
9. Conolly, D. (ed.): Gleaning Resource Descriptions from Dialects of Languages (GRDDL). W3C recommendation (September 2007)
10. Hayes, P. (ed.): RDF Semantics. W3C Recommendation (February 2004)
11. Eiter, T., Ianni, G., Polleres, A., Schindlauer, R., Tompits, H.: Reasoning with rules and ontologies. In: Reasoning Web 2006, Springer, Heidelberg (2006)
12. Fermüller, C.G., Leitsch, A.: Hyperresolution and automated model building. J. Log. Comput. 6(2), 173–203 (1996)
13. Grosof, B., Horrocks, I., Volz, R., Decker, S.: Description logic programs: Combining logic programs with description logic. In: Proc. WWW 2003, ACM Press, New York (2003)
14. Gutiérrez, C., Hurtado, C.A., Mendelzon, A.O.: Foundations of semantic web databases. In: Proc. PODS 2004, ACM Press, New York (2004)
15. Heymans, S., de Bruijn, J., Predoiu, L., Feier, C., Van Nieuwenborgh, D.: Guarded hybrid knowledge bases. Theory and Practice of Logic Programming (to appear, 2008)
16. Horrocks, I., Patel-Schneider, P.F.: Reducing OWL entailment to description logic satisfiability. In: Fensel, D., Sycara, K.P., Mylopoulos, J. (eds.) ISWC 2003. LNCS, vol. 2870, Springer, Heidelberg (2003)
17. Ianni, G., Martello, A., Panetta, C., Terracina, G.: Faithful and effective querying of RDF ontologies using DLV^{DB}. In: Proc. ASP 2007 (2007)
18. Klyne, G., Carroll, J.J. (eds.): Resource Description Framework (RDF): Concepts and Abstract Syntax. W3C Recommendation (February 2004)
19. Muñoz, S., Pérez, J., Gutiérrez, C.: Minimal deductive systems for RDF. In: Franconi, E., Kifer, M., May, W. (eds.) ESWC 2007. LNCS, vol. 4519, pp. 53–67. Springer, Heidelberg (2007)
20. Pérez, J., Arenas, M., Gutierrez, C.: Semantics and complexity of SPARQL. In: Cruz, I., Decker, S., Allemang, D., Preist, C., Schwabe, D., Mika, P., Uschold, M., Aroyo, L.M. (eds.) ISWC 2006. LNCS, vol. 4273, Springer, Heidelberg (2006)
21. Pichler, R.: On the complexity of H-subsumption. In: Gottlob, G., Grandjean, E., Seyr, K. (eds.) CSL 1998. LNCS, vol. 1584, Springer, Heidelberg (1999)
22. Pichler, R., Polleres, A., Wei, F., Woltran, S.: Entailment for domain-restricted RDF (ext.version). Tech. Report DBAI-TR-2008-59, available at http://www.dbai.tuwien.ac.at/research/report/dbai-tr-2008-59.pdf
23. Polleres, A., Feier, C., Harth, A.: Rules with contextually scoped negation. In: Sure, Y., Domingue, J. (eds.) ESWC 2006. LNCS, vol. 4011, Springer, Heidelberg (2006)
24. Prud'hommeaux, E., Seaborne, A. (eds.): SPARQL Query Language for RDF. W3C Proposed Recommendation (November 2007)
25. Rosati, R.: $\mathcal{DL}+log$: Tight integration of description logics and disjunctive datalog. In: Proc. KR 2006, AAAI Press, Menlo Park (2006)
26. Shmueli, O.: Decidability and expressiveness aspects of logic queries. In: Proc. PODS 1987, ACM Press, New York (1987)
27. ter Horst, H.J.: Completeness, decidability and complexity of entailment for RDF Schema and a semantic extension involving the OWL vocabulary. JWS 3(2-3), 79–115 (2005)

Finite Model Reasoning in *DL-Lite*

Riccardo Rosati

Dipartimento di Informatica e Sistemistica
Sapienza Università di Roma
Via Ariosto 25, 00185 Roma, Italy
rosati@dis.uniroma1.it

Abstract. The semantics of OWL-DL and its subclasses are based on the classical semantics of first-order logic, in which the interpretation domain may be an infinite set. This constitutes a serious expressive limitation for such ontology languages, since, in many real application scenarios for the Semantic Web, the domain of interest is actually finite, although the exact cardinality of the domain is unknown. Hence, in these cases the formal semantics of the OWL-DL ontology does not coincide with its intended semantics. In this paper we start filling this gap, by considering the subclasses of OWL-DL which correspond to the logics of the *DL-Lite* family, and studying reasoning over finite models in such logics. In particular, we mainly consider two reasoning problems: deciding satisfiability of an ontology, and answering unions of conjunctive queries (UCQs) over an ontology. We first consider the description logic $DL\text{-}Lite_R$ and show that, for the two above mentioned problems, finite model reasoning coincides with classical reasoning, i.e., reasoning over arbitrary, unrestricted models. Then, we analyze the description logics $DL\text{-}Lite_F$ and $DL\text{-}Lite_A$. Differently from $DL\text{-}Lite_R$, in such logics finite model reasoning does not coincide with classical reasoning. To solve satisfiability and query answering over finite models in these logics, we define techniques which reduce polynomially both the above reasoning problems over finite models to the corresponding problem over arbitrary models. Thus, for all the *DL-Lite* languages considered, the good computational properties of satisfiability and query answering under the classical semantics also hold under the finite model semantics. Moreover, we have effectively and easily implemented the above techniques, extending the *DL-Lite* reasoner QuOnto with support for finite model reasoning.

1 Introduction

The semantics of OWL-DL [3] and its fragments [8] are based on the classical semantics of first-order logic, in which the interpretation domain may be either a finite or an infinite set. This constitutes a serious expressive limitation for these ontology languages, since in many real application scenarios for the Semantic Web, the domain of interest is actually finite, although the exact cardinality of the domain is unknown. Hence, in these cases the formal semantics of the OWL-DL ontology does not coincide with its intended semantics.

We illustrate the above problem through two simple examples (in the following examples, the ontologies are expressed using the $DL\text{-}Lite_F$ language, which corresponds to a fragment of OWL-DL: $DL\text{-}Lite_F$ is formally introduced in Section 2).

Example 1. Let $\mathcal{O}$ be the following ontology about employees:

$$Employee \sqsubseteq \exists isHelpedBy \tag{1}$$
$$\exists isHelpedBy^- \sqsubseteq Employee \tag{2}$$
$$\exists isHelpedBy^- \sqsubseteq HighSalary \tag{3}$$
$$(funct\ isHelpedBy^-) \tag{4}$$
$$Employee(Paul) \tag{5}$$

which formalizes the following knowledge about the concepts *Employee*, *HighSalary* and the role *isHelpedBy*:

- every employee has some colleague who is committed to help her/him to perform some special task (assertion (1));
- those who are committed to help employees are also employees (assertion (2)) and have a high salary (assertion (3));
- an employee can commit to help at most one of her/his colleagues (assertion (4));
- Paul is an employee (assertion (5)).

Now, on the one hand, it can be shown that the ontology $\mathcal{O}$ does not entail (under the standard OWL-DL semantics) *HighSalary(Paul)*: indeed, consider the following interpretation $\mathcal{I}$ over the countably infinite domain $\{d_0, \ldots, d_n, \ldots\}$:

- $Paul^\mathcal{I} = d_0$;
- $d_i \in Employee^\mathcal{I}$ for each $i \geq 0$;
- $d_i \in HighSalary^\mathcal{I}$ for each $i \geq 1$;
- $\langle d_i, d_{i+1} \rangle \in isHelpedBy^\mathcal{I}$ for each $i \geq 0$.

It is immediate to verify that $\mathcal{I}$ satisfies the ontology $\mathcal{O}$ and that *HighSalary(Paul)* is not satisfied in $\mathcal{I}$.

On the other hand, *every finite model for $\mathcal{O}$ satisfies HighSalary(Paul)*. In fact, if the domain is finite, then the chain of employees induced by the ontology on the binary relation *isHelpedBy* must be finite, so the only possible way to close such a chain is to "come back" to the initial employee, i.e., Paul, who is the only employee who does not help someone yet. Consequently, in every finite model for $\mathcal{O}$, Paul helps some colleague, hence he has a high salary.

Now, for the above ontology $\mathcal{O}$ it seems very natural to assume that the domain of interest is always finite (although not exactly known), i.e., it is unreasonable to assume as possible the existence of an infinite number of employees. Hence, in this case we would like to conclude from the above ontology that Paul has a high salary. However, all current OWL reasoners will not derive such a conclusion. ∎

Example 2. Let $\mathcal{O}$ be the following ontology about peer networks:

$$EUpeer \sqsubseteq \exists hasNAmirror \tag{6}$$
$$\exists hasNAmirror^- \sqsubseteq NApeer \tag{7}$$
$$NApeer \sqsubseteq \exists hasEUmirror \tag{8}$$

$$\exists hasEUmirror^- \sqsubseteq EUpeer \qquad (9)$$
$$\exists hasEUmirror^- \sqsubseteq AlwaysOnline \qquad (10)$$
$$(funct\ hasNAmirror^-) \qquad (11)$$
$$(funct\ hasEUmirror^-) \qquad (12)$$
$$EUpeer(\mathtt{dis.uniroma1.it}) \qquad (13)$$

which formalizes the following knowledge about the concepts *EUpeer*, *NApeer*, *AlwaysOnline* and the roles *hasNAmirror*, *hasEUmirror*:

- every European peer has a mirror who is a North American peer (assertions (6) and (7));
- every North American peer has a mirror who is a European peer (assertions (8) and (9));
- peers who are mirrors of North American peers are always on-line (assertion (10));
- a peer can be the mirror of at most one North American peer (assertion (11)) and of at most one European peer (assertion (12));
- dis.uniroma1.it is a European peer (assertion (13)).

In a way similar to the previous example, it can be shown that $\mathcal{O}$ does not entail (under the standard OWL-DL semantics) *AlwaysOnline*(dis.uniroma1.it), while every finite model for $\mathcal{O}$ satisfies *AlwaysOnline*(dis.uniroma1.it). Similarly to the previous example, also for the above ontology $\mathcal{O}$ it seems very natural to assume that the domain of interest is always finite. Hence, we would like to conclude that peer dis.uniroma1.it is always on-line, but all current OWL reasoners will not derive such a conclusion. ∎

The above examples highlight the fact that OWL-DL ontologies (under the classical first-order semantics) lack the ability to properly handle finite domains of interest. In this respect, it is also important to point out that database applications are always based on a finite domain assumption: such an assumption should be taken into account in ontology-based access to database information sources, which is going to be one of the prominent applications of the Semantic Web technology [12].

Finite model reasoning has actually been studied in past research in Description Logics (DLs) [2], which constitute the logical basis of OWL-DL. In particular, besides previous studies summarized in [2], recent research has mainly focused on $\mathcal{ALCQI}$, a large fragment of OWL-DL [4,10]: for such a logic, an EXPTIME-completeness result for finite model reasoning has been established [10]. However, such results on finite model reasoning in DLs only consider "classical" DL reasoning tasks (concept subsumption, knowledge base satisfiability) and do not take *query answering* into account. Moreover, none of the currently available (OWL)DL reasoners supports finite model reasoning.

Thus, with respect to finite model reasoning in OWL-DL ontologies, we are still missing: (i) a thorough computational analysis, in particular for tractable fragments of OWL-DL [8]; (ii) an analysis of query answering over finite models; (iii) the implementation of ontology-management systems supporting finite model reasoning.

In this paper we start filling this gap, by considering the subclasses of OWL-DL which correspond to the logics of the *DL-Lite* family of Description Logics [6], and study reasoning over finite models in such logics. In particular, we consider three reasoning problems: deciding entailment of intensional assertions (TBox entailment), deciding satisfiability of the ontology, and answering queries (specifically, unions of conjunctive queries) over an ontology.

Our contributions can be summarized as follows:

1. We first consider the description logic *DL-Lite$_R$* and show that, for all the above mentioned problems, finite model reasoning coincides with classical reasoning, i.e., reasoning over arbitrary, unrestricted models.
2. Then, we analyze the description logics *DL-Lite$_F$* and *DL-Lite$_A$*. Differently from *DL-Lite$_R$*, in such logics finite model reasoning does not coincide with classical reasoning. To solve TBox entailment, satisfiability and query answering over finite models in these logics, we define techniques which reduce polynomially all the above reasoning problems over finite models in such logics to the corresponding problem over arbitrary models. This allows us to show that, for all the *DL-Lite* languages considered, the good computational properties of TBox entailment, satisfiability and query answering under the classical semantics also hold under the finite semantics.
3. Finally, we have effectively and easily implemented the above techniques to provide the *DL-Lite* reasoner QuOnto with support for finite model reasoning. To the best of our knowledge, such an extension of QuOnto constitutes the first ontology management system providing an automated support to finite model reasoning.

2 The Description Logics *DL-Lite$_F$* and *DL-Lite$_R$*

2.1 Syntax

We start from three mutually disjoint alphabets: an alphabet of concept names, an alphabet of role names, and an alphabet of constant (or individual) names. We call *basic concept* an expression of the form $B ::= A \mid \exists P \mid \exists P^-$, where A is a concept name and P is a role name, and we call *basic role* an expression of the form $R ::= P \mid P^-$, where P is a role name.

A *DL-Lite$_R$* TBox assertion is an expression of one of the following forms:

- $B_1 \sqsubseteq B_2$ (concept inclusion assertion) where B_1, B_2 are basic concepts;
- $R_1 \sqsubseteq R_2$ (role inclusion assertion) where R_1, R_2 are basic roles;
- $B_1 \sqsubseteq \neg B_2$ (concept disjointness assertion) where B_1, B_2 are basic concepts;
- $R_1 \sqsubseteq \neg R_2$ (role disjointness assertion) where R_1, R_2 are basic roles;

A *DL-Lite$_F$* TBox assertion is an expression of one of the following forms:

- $B_1 \sqsubseteq B_2$ (concept inclusion assertion) where B_1, B_2 are basic concepts;
- $B_1 \sqsubseteq \neg B_2$ (concept disjointness assertion) where B_1, B_2 are basic concepts;
- $(funct\ R)$ (functionality assertion) where R is a basic role.

A *DL-Lite$_R$* TBox is a set of *DL-Lite$_R$* TBox assertions, while a *DL-Lite$_F$* TBox is a set of *DL-Lite$_F$* TBox assertions.

A *membership assertion* is a ground atom, i.e., an expression of the form $A(a)$, $P(a, b)$ where A is a concept name, P is a role name, and a, b are constant names.

An ABox is a set of membership assertions.

A *DL-Lite$_R$ knowledge base* (KB) is a pair $\mathcal{K} = \langle \mathcal{T}, \mathcal{A} \rangle$ where $\mathcal{T}$ is a *DL-Lite$_R$* TBox and $\mathcal{A}$ is an ABox. Analogously, a *DL-Lite$_F$* KB is a pair $\mathcal{K} = \langle \mathcal{T}, \mathcal{A} \rangle$ where $\mathcal{T}$ is a *DL-Lite$_F$* TBox and $\mathcal{A}$ is an ABox.

We call *DL-Lite$_F^-$* TBox a *DL-Lite$_F$* TBox without disjointness assertions. Analogously, We call *DL-Lite$_R^-$* TBox a *DL-Lite$_R$* TBox without disjointness assertions.

We now introduce queries. A *union of conjunctive query (UCQ)* is an expression of the form

$$\{x_1, \ldots, x_n \mid conj_1 \vee \ldots \vee conj_m\}$$

where $n \geq 0$, $m \geq 1$, and each $conj_i$ is an expression of the form $(\exists y_1, \ldots, y_j . a_1 \wedge \ldots \wedge a_h)$, where $j \geq 0$, $h \geq 1$, each a_i is an atom, i.e., an expression of the form $A(t)$ or $P(t, t')$ where A is a concept name, P is a role name and t, t' are either constants or variables from $\{x_1, \ldots, x_n, y_1, \ldots, y_j\}$. When $n = 0$ (i.e., the above expression is a first-order sentence) the UCQ is called a *Boolean* UCQ.

2.2 Semantics

The semantics of a DL is given in terms of interpretations, where an *interpretation* $\mathcal{I} = (\Delta^\mathcal{I}, \cdot^\mathcal{I})$ consists of a non-empty *interpretation domain* $\Delta^\mathcal{I}$ and an *interpretation function* $\cdot^\mathcal{I}$ that assigns to each concept C a subset $C^\mathcal{I}$ of $\Delta^\mathcal{I}$, and to each role R a binary relation $R^\mathcal{I}$ over $\Delta^\mathcal{I}$. In particular, we have:

$$A^\mathcal{I} \subseteq \Delta^\mathcal{I}$$
$$P^\mathcal{I} \subseteq \Delta^\mathcal{I} \times \Delta^\mathcal{I}$$
$$(P^-)^\mathcal{I} = \{(o_2, o_1) \mid (o_1, o_2) \in P^\mathcal{I}\}$$
$$(\exists R)^\mathcal{I} = \{o \mid \exists o'.(o, o') \in R^\mathcal{I}\}$$
$$(\neg B)^\mathcal{I} = \Delta^\mathcal{I} \setminus B^\mathcal{I}$$
$$(\neg R)^\mathcal{I} = \Delta^\mathcal{I} \times \Delta^\mathcal{I} \setminus R^\mathcal{I}$$

An interpretation $\mathcal{I} = (\Delta^\mathcal{I}, \cdot^\mathcal{I})$ is called *finite* if $\Delta^\mathcal{I}$ is a finite set.

An interpretation $\mathcal{I}$ is a *model* of $B \sqsubseteq C$, where C is either a basic concept or the negation of a basic concept, if $B^\mathcal{I} \subseteq C^\mathcal{I}$. Similarly, $\mathcal{I}$ is a *model* of $R \sqsubseteq E$, where R is a basic role and E is either a basic role or the negation of a basic role, if $R^\mathcal{I} \subseteq E^\mathcal{I}$.

To specify the semantics of membership assertions, we extend the interpretation function to constants, by assigning to each constant a a *distinct* object $a^\mathcal{I} \in \Delta^\mathcal{I}$. Note that this implies that we enforce the *unique name assumption* on constants [2]. An interpretation $\mathcal{I}$ is a model of a membership assertion $A(a)$, (resp., $P(a, b)$) if $a^\mathcal{I} \in A^\mathcal{I}$ (resp., $(a^\mathcal{I}, b^\mathcal{I}) \in P^\mathcal{I}$).

Given an (inclusion, or membership) assertion ϕ, and an interpretation $\mathcal{I}$, we denote by $\mathcal{I} \models \phi$ the fact that $\mathcal{I}$ is a model of ϕ. Given a (finite) set of assertions Φ, we denote by $\mathcal{I} \models \Phi$ the fact that $\mathcal{I}$ is a model of every assertion in Φ. A *model of a KB* $\mathcal{K} = \langle \mathcal{T}, \mathcal{A} \rangle$ is an interpretation $\mathcal{I}$ such that $\mathcal{I} \models \mathcal{T}$ and $\mathcal{I} \models \mathcal{A}$. A finite interpretation that is a model of a KB $\mathcal{K}$ is called *finite model* of $\mathcal{K}$.

The reasoning tasks we are interested in are UCQ entailment, KB satisfiability, and TBox entailment. More precisely:

- A KB $\mathcal{K}$ *entails* a Boolean UCQ q, denoted by $\mathcal{K} \models q$, if all models of $\mathcal{K}$ are also models of q (where an interpretation $\mathcal{I}$ is a model of q if $\mathcal{I}$ satisfies the first-order sentence q according to the standard notion of satisfiability in first-order logic). $\mathcal{K}$ finitely entails a UCQ q, denoted by $\mathcal{K} \models_{fin} q$, if all finite models of $\mathcal{K}$ are also models of q;
- a KB is *satisfiable* if it has at least one model, and is *finitely satisfiable* if it has at least one finite model.
- a TBox $\mathcal{T}$ *entails* a TBox assertion ϕ, denoted by $\mathcal{T} \models \phi$, if all models of $\mathcal{T}$ are also models of ϕ, while $\mathcal{T}$ *finitely entails* ϕ, denoted by $\mathcal{T} \models_{fin} \phi$, if all finite models of $\mathcal{T}$ are also models of ϕ.

Since in the following we will focus on UCQ entailment, from now on when we speak about UCQs we always mean *Boolean* UCQs. We recall that answering arbitrary (i.e., non-Boolean) UCQs can be easily reduced to UCQ entailment (for more details see e.g. [9]).

3 Finite Model Reasoning in *DL-Lite$_R$*

In this section we show that, in *DL-Lite$_R$*, finite model reasoning coincides with classical reasoning. We start by proving such property for *DL-Lite$_R^-$* KBs (i.e., *DL-Lite$_R$* KBs without disjointness assertions) and for UCQ entailment.

Lemma 1. *Let $\mathcal{T}$ be a DL-Lite$_R^-$-TBox, and let q be a UCQ. Then, for every ABox $\mathcal{A}$, $\langle \mathcal{T}, \mathcal{A} \rangle \models_{fin} q$ iff $\langle \mathcal{T}, \mathcal{A} \rangle \models q$.*

Proof. The proof is a consequence of Theorem 4 of [11], since a *DL-Lite$_R$* KB corresponds to a database with inclusion dependencies (interpreted under open-world assumption). More precisely, given a *DL-Lite$_R$* KB $\mathcal{K} = \langle \mathcal{T}, \mathcal{A} \rangle$, we build a database instance (i.e., a set of facts) $\mathcal{D}$ and a set of inclusion dependencies $\mathcal{C}$ as follows:

- for every concept name A (respectively, role name R) occurring in $\mathcal{K}$, the database schema contains a unary relation A (respectively, a binary relation R)
- for every concept inclusion in $\mathcal{T}$ of the form $A_1 \sqsubseteq A_2$, $\mathcal{C}$ contains the inclusion dependency $A_1[1] \subseteq A_2[1]$;
- for every concept inclusion in $\mathcal{T}$ of the form $A \sqsubseteq \exists R$ (respectively, $A \sqsubseteq \exists R^-$), $\mathcal{C}$ contains the inclusion dependency $A[1] \subseteq R[1]$ (respectively, $A[1] \subseteq R[2]$);
- for every concept inclusion in $\mathcal{T}$ of the form $\exists R \sqsubseteq A$ (respectively, $\exists R^- \sqsubseteq A$), $\mathcal{C}$ contains the inclusion dependency $R[1] \subseteq A[1]$ (respectively, $R[2] \subseteq A[1]$);
- for every concept inclusion in $\mathcal{T}$ of the form $\exists R_1 \sqsubseteq \exists R_2$ (respectively, $\exists R_1 \sqsubseteq \exists R_2^-$), $\mathcal{C}$ contains the inclusion dependency $R_1[1] \subseteq R_2[1]$ (respectively, $R_1[1] \subseteq R_2[2]$);
- for every concept inclusion in $\mathcal{T}$ of the form $\exists R_1^- \sqsubseteq \exists R_2$ (respectively, $\exists R_1^- \sqsubseteq \exists R_2^-$), $\mathcal{C}$ contains the inclusion dependency $R_1[2] \subseteq R_2[1]$ (respectively, $R_1[2] \subseteq R_2[2]$);
- for every role inclusion in $\mathcal{T}$ of the form $R_1 \sqsubseteq R_2$ (respectively, $R_1 \sqsubseteq R_2^-$), $\mathcal{C}$ contains the inclusion dependency $R_1[1,2] \subseteq R_2[1,2]$ (respectively, $R_1[1,2] \subseteq R_2[2,1]$);

– for every role inclusion in $\mathcal{T}$ of the form $R_1^- \sqsubseteq R_2$ (respectively, $R_1^- \sqsubseteq R_2^-$), $\mathcal{C}$ contains the inclusion dependency $R_1[2,1] \subseteq R_2[1,2]$ (respectively, $R_1[2,1] \subseteq R_2[2,1]$).

Finally, for every membership assertion of the form $A(a)$ (respectively, $R(a,b)$) in the ABox $\mathcal{A}$, the database instance $\mathcal{D}$ contains the fact $A(a)$ (respectively, $R(a,b)$).

We now recall the definition of [11] of semantics of the pair $(\mathcal{C}, \mathcal{D})$: we denote by $sem(\mathcal{C}, \mathcal{D})$ the set of database instances $\{\mathcal{B} \mid \mathcal{B} \supseteq \mathcal{D}$ and $\mathcal{B}$ satisfies $\mathcal{C}\}$, where each $\mathcal{B}$ is a (possibly infinite) set of facts. Moreover, we denote by $sem_f(\mathcal{C}, \mathcal{D})$ the subset of $sem(\mathcal{C}, \mathcal{D})$ where each $\mathcal{B}$ is a *finite* set of facts.

It is now immediate to verify that the set of models of $\mathcal{K}$ is in one-to-one correspondence with $sem(\mathcal{C}, \mathcal{D})$: more precisely, every database instance $\mathcal{B}$ in $sem(\mathcal{C}, \mathcal{D})$ corresponds to a model $\mathcal{I}(\mathcal{B})$ of $\mathcal{K}$ where $\mathcal{I}(\mathcal{B})$ is the interpretation defined as follows: for each concept name C and for each constant a, $a \in C^{\mathcal{I}(\mathcal{B})}$ iff $C(a) \in \mathcal{B}$, and for each role name R and pair of constant a, b, $\langle a, b \rangle \in R^{\mathcal{I}(\mathcal{B})}$ iff $R(a,b) \in \mathcal{B}$.

This in turn immediately implies that for every UCQ q, $\langle \mathcal{T}, \mathcal{A} \rangle \models q$ iff q is true in all database instances of $sem(\mathcal{C}, \mathcal{D})$, and $\langle \mathcal{T}, \mathcal{A} \rangle \models_{fin} q$ iff q is true in all database instances of $sem_f(\mathcal{C}, \mathcal{D})$. Since from Theorem 4 of [11] q is true in all database instances of $sem(\mathcal{C}, \mathcal{D})$ iff q is true in all database instances of $sem_f(\mathcal{C}, \mathcal{D})$, the thesis follows. □

We now extend the above result to arbitrary *DL-Lite$_R$* KBs (i.e., KBs whose TBox may also contain disjointness assertions).

Given a *DL-Lite$_R$* TBox $\mathcal{T}$, let $D(\mathcal{T})$ denote the set of disjointness assertions occurring in $\mathcal{T}$, and let $\mathcal{T}^- = \mathcal{T} - D(\mathcal{T})$, i.e., $\mathcal{T}^-$ denotes the TBox obtained from $\mathcal{T}$ by eliminating all disjointness assertions.

Given a disjointness assertion ϕ, we denote by $Q(\phi)$ the Boolean conjunctive query defined as follows:

- if $\phi = A_1 \sqsubseteq \neg A_2$, then $Q(\phi) = \exists x. A_1(x) \wedge A_2(x)$;
- if $\phi = A \sqsubseteq \neg \exists R$ or $\phi = \exists R \sqsubseteq \neg A$, then $Q(\phi) = \exists x, y. A(x) \wedge R(x,y)$;
- if $\phi = A \sqsubseteq \neg \exists R^-$ or $\phi = \exists R^- \sqsubseteq \neg A$, then $Q(\phi) = \exists x, y. A(x) \wedge R(y,x)$;
- if $\phi = \exists R_1 \sqsubseteq \neg \exists R_2$, then $Q(\phi) = \exists x, y, z. R_1(x,y) \wedge R_2(x,z)$;
- if $\phi = \exists R_1 \sqsubseteq \neg \exists R_2^-$ or $\phi = \exists R_2^- \sqsubseteq \neg \exists R_1$, then $Q(\phi) = \exists x, y, z. R_1(x,y) \wedge R_2(z,x)$;
- if $\phi = \exists R_1^- \sqsubseteq \neg \exists R_2^-$, then $Q(\phi) = \exists x, y, z. R_1(y,x) \wedge R_2(z,x)$.

Informally, $Q(\phi)$ is satisfied in an interpretation $\mathcal{I}$ iff $\mathcal{I}$ does not satisfy the disjointness assertion ϕ.

Furthermore, given a set of disjointness assertions Φ, we define $Q(\Phi)$ as the following UCQ:

$$Q(\Phi) = \bigvee_{\phi \in \Phi} Q(\phi)$$

Lemma 2. *Let $\mathcal{T}$ be a DL-Lite$_R$-TBox, and let q be a UCQ. Then, for every UCQ q, $\langle \mathcal{T}, \mathcal{A} \rangle \models q$ iff $\langle \mathcal{T}^-, \mathcal{A} \rangle \models q \vee Q(D(\mathcal{T}))$ and $\langle \mathcal{T}, \mathcal{A} \rangle \models_{fin} q$ iff $\langle \mathcal{T}^-, \mathcal{A} \rangle \models_{fin} q \vee Q(D(\mathcal{T}))$.*

Proof. It is immediate to verify that $D(\mathcal{T})$ corresponds to a first-order sentence ϕ which is equivalent to $\neg Q(D(\mathcal{T}))$, therefore from the deduction theorem it follows that $\langle \mathcal{T}, \mathcal{A} \rangle \models_{fin} q$ iff $\langle \mathcal{T}^-, \mathcal{A} \rangle \models_{fin} q \vee Q(D(\mathcal{T}))$. Moreover, since $\mathcal{T}^-$ is a $DL\text{-}Lite_R^-$ TBox, from Theorem 1 it follows that $\langle \mathcal{T}^-, \mathcal{A} \rangle \models_{fin} q \vee Q(D(\mathcal{T}))$ iff $\langle \mathcal{T}^-, \mathcal{A} \rangle \models q \vee Q(D(\mathcal{T}))$. □

As an immediate consequence of Lemma 2 and of Lemma 1, we obtain the following property.

Theorem 1. *Let $\mathcal{T}$ be a DL-Lite$_R$-TBox, and let q be a UCQ. Then, for every ABox $\mathcal{A}$, $\langle \mathcal{T}, \mathcal{A} \rangle \models_{fin} q$ iff $\langle \mathcal{T}, \mathcal{A} \rangle \models q$.*

Then, we turn our attention to KB satisfiability. It can easily be shown that the technique for reducing KB unsatisfiability in *DL-Lite$_R$* to UCQ entailment (see Lemma 16 of [6]) is also correct when we restrict to finite models. This fact and Theorem 1 imply the following property.

Theorem 2. *Let $\mathcal{K}$ be a DL-Lite$_R$ KB. Then, $\mathcal{K}$ is finitely satisfiable iff $\mathcal{K}$ is satisfiable.*

Moreover, it can also be shown that the technique for reducing TBox entailment in *DL-Lite$_R$* to KB unsatisfiability (see Theorem 22 and Theorem 23 of [6]) is also correct when we restrict to finite models. Consequently, the two previous theorems imply the following property.

Theorem 3. *Let $\mathcal{T}$ be a DL-Lite$_R$-TBox, and let ϕ be a DL-Lite$_R$ TBox assertion. Then, $\mathcal{T} \models_{fin} \phi$ iff $\mathcal{T} \models \phi$.*

4 Finite Model Reasoning in *DL-Lite$_F$*

In this section we study finite model reasoning in *DL-Lite$_F$*. First, we remark that, differently from the case of *DL-Lite$_R$*, in *DL-Lite$_F$* UCQ entailment over finite models differs from UCQ entailment over arbitrary models, as illustrated by both Example 1 and Example 2. So, since we cannot simply establish an equivalence between classical reasoning and finite model reasoning as in the case of *DL-Lite$_R$*, we must look for new reasoning methods to solve finite model reasoning in *DL-Lite$_F$*.

We start by considering *DL-Lite$_F^-$* KBs (i.e., *DL-Lite$_F$* KBs without disjointness assertions) and define inference rules for TBox assertions in *DL-Lite$_F^-$*.

4.1 Finite TBox Inference Rules in *DL-Lite$_F^-$*

In the following, R denotes a basic role expression (i.e., either P or P^- where P is a role name), while R^- denotes the inverse of R, i.e., if $R = P$ (R is a role name), then $R^- = P^-$, while if $R = P^-$, then $R^- = P$.

Definition 1. *Given a DL-Lite$_F^-$ TBox $\mathcal{T}$, $finClosure_F(\mathcal{T})$ denotes the closure of $\mathcal{T}$ with respect to the following inference rules:*

1. *(inclusion-rule)* if $B_1 \sqsubseteq B_2$ and $B_2 \sqsubseteq B_3$ then conclude $B_1 \sqsubseteq B_3$;
2. *(functionality-inclusion-cycle-rule)* if there is an inclusion-functionality cycle, i.e., a sequence of TBox assertions of the form

$$(\mathit{funct}\ R_1),\ \exists R_2 \sqsubseteq \exists R_1^-,\ (\mathit{funct}\ R_2),\ \exists R_3 \sqsubseteq \exists R_2^-,\ \ldots$$
$$\ldots,\ (\mathit{funct}\ R_k),\ \exists R_1 \sqsubseteq \exists R_k^-$$

(where each R_i is a basic role expression, i.e., either P or P^-), then conclude

$$(\mathit{funct}\ R_1^-),\ \exists R_1^- \sqsubseteq \exists R_2,\ (\mathit{funct}\ R_2^-),\ \exists R_2^- \sqsubseteq \exists R_3, \ldots$$
$$\ldots,\ (\mathit{funct}\ R_k^-),\ \exists R_k^- \sqsubseteq \exists R_1$$

It is immediate to verify that the above inference rules are not sound with respect to classical TBox entailment (i.e., entailment over unrestricted models). On the other hand, we now prove that the above inference rules are sound with respect to TBox entailment over finite models.

Lemma 3. *Let $\mathcal{T}$ be a DL-Lite$_F^-$ TBox and let ϕ be a DL-Lite$_F^-$ TBox assertion. If $\mathit{finClosure}_F(\mathcal{T}) \models \phi$ then $\mathcal{T} \models_{\mathit{fin}} \phi$.*

Proof. The proof is a consequence of the axiomatization for unary inclusion dependencies and functional dependencies presented in [7]. □

We call a TBox *f-closed* if it is closed with respect to the two inference rules above.

4.2 Query Answering over Finite Models in DL-Lite$_F^-$

We now turn our attention to query answering over finite models in DL-Lite$_F^-$ KBs. Notice that Lemma 3 does not imply that query answering over finite models can be solved by simply augmenting the TBox with the new assertions implied by Definition 1. However, we now show that this strategy is actually complete, i.e., we can answer (unions of) conjunctive queries over finite models of a DL-Lite$_F^-$ KB $\mathcal{K} = \langle \mathcal{T}, \mathcal{A} \rangle$ by first generating the augmented TBox $\mathcal{T}'$ obtained from $\mathcal{T}$ by adding the new assertions implied by Definition 1, and then computing the certain answers to queries over the KB $\langle \mathcal{T}', \mathcal{A} \rangle$ according to the unrestricted semantics.

Theorem 4. *Let $\mathcal{T}$ be a DL-Lite$_F^-$-TBox, and let q be a UCQ. Then, for every ABox $\mathcal{A}$, $\langle \mathcal{T}, \mathcal{A} \rangle \models_{\mathit{fin}} q$ iff $\langle \mathit{finClosure}_F(\mathcal{T}), \mathcal{A} \rangle \models q$.*

Proof (sketch). The proof is rather involved and is based on the notion of *chase* of a *DL-Lite* KB [6]. More precisely, we modify the notion of (generally infinite) chase of a *DL-Lite$_F$* KB, denoted by $\mathit{chase}(\mathcal{K})$ and presented in [6], thus defining the notion of *finite chase of degree k* of $\mathcal{K}$, denoted by $\mathit{finChase}_k^F(\mathcal{K})$.

Informally, $\mathit{chase}(\mathcal{K})$ is an ABox built starting from the initial ABox $\mathcal{A}$ and applying a chase expansion rule that adds new membership assertions to the ABox until all inclusions in $\mathcal{T}$ are satisfied. In this expansion process new constants are introduced in the ABox, and in the presence of cyclic concept inclusions this process may produce an infinite ABox. The important semantic properties of $\mathit{chase}(\mathcal{K})$ are the following: (i) $\mathit{chase}(\mathcal{K})$ is isomorphic to an interpretation $\mathcal{I}(\mathit{chase}(\mathcal{K}))$ which is a model of $\mathcal{K}$; (ii) UCQ entailment over $\mathcal{K}$ can be solved by simply evaluating the query over $\mathcal{I}(chase(\mathcal{K}))$.

We modify the above chase procedure and define a chase procedure, called $finChase_k^F(\mathcal{K})$ (parameterized with respect to a positive integer k), that always terminates, thus returning a finite ABox. Then, we prove that $finChase_k^F(\mathcal{K})$ only partially preserves the above semantic properties of the chase with respect to finite model semantics. More precisely, we prove that:

(A) $finChase_k^F(\mathcal{K})$ is isomorphic to an interpretation $\mathcal{I}(finChase_k^F(\mathcal{K}))$ which is a model of $\mathcal{K}$, *under the assumption that the TBox $\mathcal{T}$ of $\mathcal{K}$ is f-closed*;
(B) for every positive integer k, entailment of a UCQ *of length less or equal to k* (i.e., all of whose conjunctions have a number of atoms less or equal to k) can be decided by simply evaluating the query over $\mathcal{I}(finChase_k^F(\mathcal{K}))$.

With the notion of $finChase_k^F(\mathcal{K})$ in place, we can prove the theorem as follows. First, let $\mathcal{K} = \langle \mathcal{T}, \mathcal{A} \rangle$ and $\mathcal{K}' = \langle finClosure_F(\mathcal{T}), \mathcal{A} \rangle$. Soundness is trivial: if $\mathcal{K} \models_{fin} q$, then $\mathcal{K}' \models_{fin} q$, and since the set of finite models of $\mathcal{K}'$ is a subset of the set of models of $\mathcal{K}'$, it follows that $\mathcal{K}' \models q$. To prove completeness, assume that $\mathcal{K}' \not\models q$ and let k be the length of q. Now, by property (A) above, $\mathcal{I}(finChase_k^F(\mathcal{K}'))$ is a finite model of $\mathcal{K}'$, and by Lemma 3, $\mathcal{I}(finChase_k^F(\mathcal{K}'))$ is a model of $\mathcal{K}$. Finally, since $\mathcal{K}' \not\models q$, from the above property (B) it follows that $\mathcal{I}(finChase_k^F(\mathcal{K}')) \not\models q$, which in turn implies that $\mathcal{K} \not\models_{fin} q$. □

4.3 Query Answering over Finite Models in *DL-Lite$_F$*

It is now immediate to extend the above theorem to *DL-Lite$_F$*, since it is possible to encode disjointness assertions in the UCQ, as illustrated already in Section 3. Given a *DL-Lite$_F$* TBox $\mathcal{T}$, let $D(\mathcal{T})$ denote the set of disjointness assertions occurring in $\mathcal{T}$, and let $\mathcal{T}^- = \mathcal{T} - D(\mathcal{T})$, i.e., $\mathcal{T}^-$ denotes the *DL-Lite$_F^-$* TBox obtained from $\mathcal{T}$ by eliminating all disjointness assertions.

Theorem 5. *Let $\mathcal{T}$ be a DL-Lite$_F$-TBox, and let q be a UCQ. Then, for every ABox $\mathcal{A}$, $\langle \mathcal{T}, \mathcal{A} \rangle \models_{fin} q$ iff $\langle finClosure_F(\mathcal{T}^-), \mathcal{A} \rangle \models q \vee Q(D(\mathcal{T}))$.*

4.4 Finite KB Satisfiability and TBox Entailment

We now focus on KB satisfiability. Again, we start by showing that, differently from the case of *DL-Lite$_R$*, in *DL-Lite$_F$* finite KB satisfiability does not coincide with classical KB saisfiability. In particular, there are *DL-Lite$_F$* KBs that only admit infinite models.

Example 3. Let $\mathcal{T}$ be the following *DL-Lite$_F$* TBox:

$$A \sqsubseteq \exists R \qquad A \sqsubseteq \neg \exists R^- \qquad \exists R^- \sqsubseteq \exists R \qquad (funct\ R^-)$$

and let $\mathcal{A}$ be the ABox $\mathcal{A} = \{A(a)\}$. It is easy to see that the KB $\mathcal{K} = \langle \mathcal{T}, \mathcal{A} \rangle$ is not finitely satisfiable, while $\mathcal{K}$ is satisfiable (i.e., there are models for $\mathcal{K}$ but they are all infinite). ∎

To compute finite KB satisfiability, it is possible to show that the technique for reducing KB unsatisfiability in *DL-Lite$_F$* to UCQ entailment (see Lemma 16 of [6]) is also correct when we restrict to finite models. This fact and Theorem 5 imply that we can reduce finite KB satisfiability in *DL-Lite$_F$* to standard KB satisfiability. Formally, the following property holds:

Theorem 6. *Let $\mathcal{K}$ be a DL-Lite$_F$ KB. Then, $\mathcal{K}$ is finitely satisfiable iff $\langle \mathcal{T} \cup finClosure_F(\mathcal{T}^-), \mathcal{A} \rangle$ is satisfiable.*

It also turns out that finite TBox entailment in *DL-Lite$_F$* can be reduced to finite KB unsatisfiability. In fact, it is easy to show that Theorem 22 and Theorem 24 of [6] hold also when restricting to finite models, while Theorem 25 of [6] holds for finite models under the assumption that the TBox $\mathcal{T}$ is f-closed. Consequently, the following property holds.

Theorem 7. *Let $\mathcal{T}$ be a DL-Lite$_F$-TBox, and let ϕ be a DL-Lite$_F$ TBox assertion. Then, $\mathcal{T} \models_{fin} \phi$ iff $\mathcal{T} \cup finClosure_F(\mathcal{T}^-) \models \phi$.*

5 Complexity Results

We now study the computational complexity of finite model reasoning in *DL-Lite$_R$* and *DL-Lite$_F$*.

First, in the case of *DL-Lite$_R$*, the theorems shown in Section 3 immediately imply that, for the reasoning tasks considered in this paper, the complexity of finite model reasoning and of classical reasoning coincide. Hence, the complexity results for *DL-Lite$_R$* reported in [6] also holds when restricting to finite models.

We now analyze complexity in the case of *DL-Lite$_F$*. First, we show the following property.

Lemma 4. *Given a DL-Lite$_F^-$ TBox $\mathcal{T}$ and a DL-Lite$_F^-$ TBox assertion ϕ, $\mathcal{T} \models_{fin} \phi$ can be decided in polynomial time with respect to the size of $\mathcal{T} \cup \{\phi\}$.*

Moreover, by definition, $finClosure_F(\mathcal{T})$ is only composed of *DL-Lite$_F^-$* TBox assertions using concept and role names occurring in $\mathcal{T}$, and the number of possible *DL-Lite$_F^-$* TBox assertions using such names is quadratic with respect to the size of $\mathcal{T}$. Thus, from the above lemma, it follows that $finClosure_F(\mathcal{T})$ can be computed in polynomial time with respect to the size of $\mathcal{T}$. Furthermore, it is immediate to see that $Q(D(\mathcal{T}))$ can also be computed in polynomial time with respect to the size of $\mathcal{T}$.

Therefore, from the theorems shown in Section 4, it follows that the complexity results for *DL-Lite$_F$* reported in [6] also holds when restricting to finite models.

The above results are formally summarized as follows:

Theorem 8. *Deciding UCQ entailment over finite models in DL-Lite$_R$ and DL-Lite$_F$ is:*

- *in LOGSPACE with respect to the size of the ABox;*
- *in PTIME with respect to the size of the KB;*
- *NP-complete with respect to the size of the KB and the query.*

Theorem 9. *Let $\mathcal{K}$ be either a DL-Lite$_R$ or a DL-Lite$_F$ KB. Deciding whether $\mathcal{K}$ is finitely satisfiable is:*

- *in LOGSPACE with respect to the size of the ABox;*
- *in PTIME with respect to the size of the TBox;*
- *in PTIME with respect to the size of the KB.*

Theorem 10. *Finite entailment of an assertion ϕ with respect to a TBox $\mathcal{T}$ in DL-Lite$_R$ and DL-Lite$_F$ can be decided in PTIME with respect to the size of $\mathcal{T} \cup \{\phi\}$.*

6 Extension to *DL-Lite$_A$*

In this section we extend the previous results for finite model reasoning to the case of *DL-Lite$_A$*. Due to space limitations, in the present version of the paper we are not able to introduce *DL-Lite$_A$* in detail (see [5]) and just sketch the way in which we have extended the previous results to the case of *DL-Lite$_A$* KBs.

Informally, a *DL-Lite$_A$* TBox is a TBox which admits all the TBox assertions allowed in both *DL-Lite$_F$* and *DL-Lite$_R$* with the following limitation: a functional role cannot be specialized, i.e., if the assertion $(funct\ R)$ or $(funct\ R^-)$ is in the TBox, then there is no role inclusion assertion of the form $R' \sqsubseteq R$ or of the form $R' \sqsubseteq R^-$ in the TBox. Moreover, *DL-Lite$_A$* allows for defining concept *attributes*, i.e, binary relations whose ranges, called value-domains, are concepts interpreted over a domain that is disjoint from the interpretation domain of ordinary concept and roles. A *DL-Lite$_A$* TBox allows for value-domain inclusion/disjointness assertions and for attribute inclusion/disjointness assertions. We refer the reader to [5] for more details.

First, we denote by *DL-Lite$_A^-$* the version of *DL-Lite$_A$* that does not allow for (concept or role or value-domain or attribute) disjointness assertions.

Given a *DL-Lite$_A^-$* TBox $\mathcal{T}$, we denote by $finClosure_A(\mathcal{T})$ the closure of $\mathcal{T}$ with respect to the following inference rules:

1. *(binary-inclusion-rule)* if $R_1 \sqsubseteq R_2$ is either a role inclusion assertion or an attribute inclusion assertion, then conclude $\exists R_1 \sqsubseteq \exists R_2$ and $\exists R_1^- \sqsubseteq \exists R_2^-$;
2. *(transitivity-rule)* if $B_1 \sqsubseteq B_2$ and $B_2 \sqsubseteq B_3$ then conclude $B_1 \sqsubseteq B_3$;
3. *(functionality-inclusion-cycle-rule)* if there is an inclusion-functionality cycle, i.e., a sequence of TBox assertions of the form

$$(funct\ R_1),\ \exists R_2 \sqsubseteq \exists R_1^-,\ (funct\ R_2),\ \exists R_3 \sqsubseteq \exists R_2^-,\ \ldots$$
$$\ldots,\ (funct\ R_k),\ \exists R_1 \sqsubseteq \exists R_k^-$$

(where each R_i is a basic role expression, i.e., either P or P^-), then conclude

$$(funct\ R_1^-),\ \exists R_1^- \sqsubseteq \exists R_2,\ (funct\ R_2^-),\ \exists R_2^- \sqsubseteq \exists R_3,\ \ldots$$
$$\ldots,\ (funct\ R_k^-),\ \exists R_k^- \sqsubseteq \exists R_1$$

Observe that, with respect to $finClosure_F$, to compute $finClosure_A$ we basically just add inference rules for deriving unary (i.e., concept and value-domain) inclusions from binary (i.e., role and attribute) inclusions.

We are now able to prove that the reduction of query answering over finite models to query answering over unrestricted models proved in Section 4 for $DL\text{-}Lite_F$ can be extended to $DL\text{-}Lite_A$.

Theorem 11. *Let $\mathcal{T}$ be a $DL\text{-}Lite_A$-TBox, and let q be a UCQ. Then, for every ABox $\mathcal{A}$, $\langle \mathcal{T}, \mathcal{A} \rangle \models_{fin} q$ iff $\langle finClosure_A(\mathcal{T}^-), \mathcal{A} \rangle \models q \vee Q(D(\mathcal{T}))$.*

Proof. First, consider a $DL\text{-}Lite_A^-$ TBox $\mathcal{T}$. Let $U(\mathcal{T})$ denote the set of unary (i.e., concept and value-domain) inclusions and functionality assertions in $\mathcal{T}$.

We first modify the procedure for computing $finChase_k^F$, illustrated in the proof of Theorem 4, thus producing a new terminating chase procedure $finChase_k^A$. Then, we prove that: (A) $finChase_k^A(\mathcal{K})$ is isomorphic to an interpretation $\mathcal{I}(finChase_k^A(\mathcal{K}))$ that is a model of $\mathcal{K}$; (B) for every positive integer k, entailment of a UCQ of length less or equal to k can be decided by simply evaluating the query over $\mathcal{I}(finChase_k^A(\mathcal{K}))$. From the above two properties, the thesis follows with a proof analogous to the proof of Theorem 4.

Finally, let us consider the case when $\mathcal{T}$ is an arbitrary $DL\text{-}Lite_A$ TBox, i.e., when $\mathcal{T}$ also contains disjointness assertions. In this case, the thesis is proved in the same way as in the proof of Theorem 5. □

From the above theorem, and in the same way as in the case of $DL\text{-}Lite_F$, we are able to derive the following properties.

Theorem 12. *Let $\mathcal{K}$ be a $DL\text{-}Lite_A$ KB. Then, $\mathcal{K}$ is finitely satisfiable iff $\langle \mathcal{T} \cup finClosure_A(\mathcal{T}^-), \mathcal{A} \rangle$ is satisfiable.*

Theorem 13. *Let $\mathcal{T}$ be a $DL\text{-}Lite_A$-TBox, and let ϕ be a $DL\text{-}Lite_A$ TBox assertion. Then, $\mathcal{T} \models_{fin} \phi$ iff $\mathcal{T} \cup finClosure_A(\mathcal{T}^-) \models \phi$.*

Finally, from the above results it follows that the computational properties expressed by Theorem 8, Theorem 9, and Theorem 10 extend to the case of $DL\text{-}Lite_A$.

7 Implementation

In this section we show that the techniques presented in this paper allow for an efficient and effective implementation of finite model reasoning services for $DL\text{-}Lite$ ontologies.

We have implemented the above techniques in QuOnto [1]. QuOnto is a $DL\text{-}Lite_A$ reasoner whose main purpose is to deal with very large instances (i.e., ABoxes). To this aim, in QuOnto the ABox is stored and managed by a relational database system (RDBMS). The main reasoning services provided by QuOnto are the "extensional" ones, i.e., KB satisfiability and query answering (it allows for posing UCQs over the ontology).

Computation in QuOnto is divided into an off-line and an on-line phase. Off-line processing concerns purely intensional tasks, i.e., upload of the TBox, classification of concept and roles, etc. On-line processing concerns the tasks involving the ABox, i.e., KB satisfiability and query answering. In particular, query answering is divided into a query rewriting phase, in which the query is reformulated in terms of an SQL query (through a reasoning step which exploits the knowledge expressed by the TBox), and a query evaluation phase, in which the generated SQL query is evaluated over the ABox by the RDBMS, and the answers are presented to the user.

To provide support for finite model reasoning, we have extended QuOnto as follows:

- during off-line processing, the system computes $finClosure_A(\mathcal{T})$, and the assertions in $finClosure_A(\mathcal{T}) - \mathcal{T}$ are stored in an auxiliary TBox $\mathcal{T}'$;
- during on-line processing, at every service request (KB satisfiability or query answering) the user may choose between the classical semantics and the finite model semantics;
- if finite model semantics is selected, then the system executes its reasoning service method (KB satisfiability or query answering) using the TBox $\mathcal{T} \cup \mathcal{T}'$, otherwise it executes its method on the TBox $\mathcal{T}$.

Notice that the main additional computation requested is the computation of $finClosure_A(\mathcal{T})$, which is only executed during off-line processing. Hence, finite model reasoning does not imply any significant overhead during on-line processing. Moreover, even the off-line overhead caused by the computation of $finClosure_A(\mathcal{T})$ is usually not very significant, since such a computation is in the worst case quadratic in the number of functional roles of $\mathcal{T}$ (which are usually only a small fraction of the total number of roles in $\mathcal{T}$). This very nice computational behaviour has been confirmed by our experiments. We have thus included these functionalities for finite model reasoning in the next version of QuOnto, which is going to be publicly released in 2008.

8 Conclusions

Comparison with Related Work. We remark that $DL\text{-}Lite_R$ and $DL\text{-}Lite_F$ are fragments of expressive Description Logics for which finite model reasoning has been studied in the past [4,2,10]. In particular, decidability of finite KB satisfiability and of finite TBox entailment for both $DL\text{-}Lite_F$ and $DL\text{-}Lite_R$ is a consequence of such previous results. However, the complexity of such tasks in these two logics was not known, while the PTIME upper bound for finite TBox entailment in $DL\text{-}Lite_F^-$ is implied by the results in [7]. Furthermore, nothing was known about decidability and complexity of answering conjunctive queries and unions of conjunctive queries over finite models in such logics.

Future Work. From the theoretical viewpoint, our aim is to extend the present computational analysis towards other species of OWL-DL. In particular, we are interested in the tractable Description Logics which could become part of the standardization process of OWL 1.1 [8]. From a both theoretical and practical viewpoint, we would like to explore the idea of extending the ontology specification language with constructs that allow the user to associate single subparts of the ontology with either finite or unrestricted interpretation.

Acknowledgments. The author wishes to thank Maurizio Lenzerini for inspiring the present work. The author also warmly thanks the anonymous reviewers for their comments, in particular for pointing out the relationship between Lemma 3 and the results in [7]. This research has been partially supported by FET project TONES (Thinking ONtologiES), funded by the EU under contract number FP6-7603, by project HYPER, funded by IBM through a Shared University Research (SUR) Award grant, and by MIUR FIRB 2005 project "Tecnologie Orientate alla Conoscenza per Aggregazioni di Imprese in Internet" (TOCAI.IT).

References

1. QuOnto web site, http://www.dis.uniroma1.it/~quonto
2. Baader, F., Calvanese, D., McGuinness, D., Nardi, D., Patel-Schneider, P.F. (eds.): The Description Logic Handbook: Theory, Implementation and Applications. Cambridge University Press, Cambridge (2003)
3. Bechhofer, S., van Harmelen, F., Hendler, J., Horrocks, I., McGuinness, D.L., Patel-Schneider, P.F., Stein, L.A.: OWL Web Ontology Language reference. W3C Recommendation (February 2004), available at http://www.w3.org/TR/owl-ref/
4. Calvanese, D.: Finite model reasoning in description logics. In: Proc. of KR 1996, pp. 292–303 (1996)
5. Calvanese, D., De Giacomo, G., Lembo, D., Lenzerini, M., Poggi, A., Rosati, R.: MASTRO-I: Efficient integration of relational data through dl ontologies. In: Proceedings of the 2007 International Workshop on Description Logic (DL 2007). CEUR Electronic Workshop Proceedings (2007)
6. Calvanese, D., De Giacomo, G., Lembo, D., Lenzerini, M., Rosati, R.: Tractable reasoning and efficient query answering in Description Logics: The DL-Lite family. J. of Automated Reasoning 39, 385–429 (2007)
7. Cosmadakis, S.S., Kanellakis, P.C., Vardi, M.: Polynomial-time implication problems for unary inclusion dependencies. J. of the ACM 37(1), 15–46 (1990)
8. Cuenca Grau, B.: Tractable fragments of the OWL 1.1 Web Ontology Language, http://www.cs.man.ac.uk/~bcg/tractable.html
9. Glimm, B., Horrocks, I., Lutz, C., Sattler, U.: Conjunctive query answering for the description logic SHIQ. In: Proc. of the 20th Int. Joint Conf. on Artificial Intelligence (IJCAI 2007), pp. 399–404 (2007)
10. Lutz, C., Sattler, U., Tendera, L.: The complexity of finite model reasoning in description logics. Information and Computation 199, 132–171 (2005)
11. Rosati, R.: On the decidability and finite controllability of query processing in databases with incomplete information. In: Proc. of PODS 2006, pp. 356–365 (2006)
12. Shadbolt, N., Hall, W., Berners-Lee, T.: The semantic web revisited. IEEE Intelligent Systems, 96–101 (May-June 2006)

Module Extraction and Incremental Classification: A Pragmatic Approach for $\mathcal{EL}^+$ Ontologies*

Boontawee Suntisrivaraporn

Theoretical Computer Science, TU Dresden, Germany
meng@tcs.inf.tu-dresden.de

Abstract. The description logic $\mathcal{EL}^+$ has recently proved practically useful in the life science domain with presence of several large-scale biomedical ontologies such as SNOMED CT. To deal with ontologies of this scale, standard reasoning of classification is essential but not sufficient. The ability to extract relevant fragments from a large ontology and to incrementally classify it has become more crucial to support ontology design, maintenance and re-use. In this paper, we propose a pragmatic approach to module extraction and incremental classification for $\mathcal{EL}^+$ ontologies and report on empirical evaluations of our algorithms which have been implemented as an extension of the CEL reasoner.

1 Introduction

In the past few years, the $\mathcal{EL}$ family of description logics (DLs) has received an increasing interest and been intensively studied (see, e.g., [1,2,3,8]). The attractiveness of the $\mathcal{EL}$ family is twofold: on the one hand, it is computationally tractable, i.e., subsumption is decidable in polytime; on the other hand, it is sufficiently expressive to formulate many life science ontologies. Examples include the Gene Ontology, the thesaurus of the US National Cancer Institute (NCI), the Systematized Nomenclature of Medicine, Clinical Terms (SNOMED CT), and large part (more than 95%) of the Galen Medical Knowledge Base (GALEN). We lay emphasis on SNOMED CT which comprises about four hundred thousand axioms and is now a standardized clinical terminology adopted by health care sectors in several countries [13].

Being a standard ontology, SNOMED has been designed to comprehensively cover a whole range of concepts in the medical and clinical domains. For this reason, it is often the case that only a small part is actually needed in a specific application. The ability to automate extraction of meaningful sub-ontologies that cover all relevant information is becoming important to support re-use of typically comprehensive standardized ontologies. Several techniques for syntactic module extraction have been proposed [9,11,6], since semantic extraction is highly complex [6]. Though (deductive) conservative extension could be used as a sufficient condition for extracting a module, it is unfortunately too expensive

---

* Supported by DFG-Project under grant BA 1122/11-1 and EU-Project TONES.

(ExpTime-complete already in $\mathcal{EL}$ with GCIs [8]). In Section 3 of the present paper, we define a new kind of module, called *reachability-based modules*, which is motivated by a once-employed optimization technique in the CEL system and which can be extracted in linear time. Also, we propose an algorithm for extracting modules of this kind and show some interesting properties.

Despite being classifiable by modern DL reasoners, design and maintenance of large-scale ontologies like SNOMED CT requires additional reasoning support. This is due to the fact that an ontology under development evolves continuously, and the developer often has to undergo the long process of *full classification* after addition of a few new axioms. Though classification of SNOMED requires less than half an hour (see [2] or Table 1 in the present paper), the ontology developer is not likely willing to wait that long for a single change. In the worst case, she may end up not using automated reasoning support which could have helped identify potential modeling errors at an early stage. In Section 4, we propose a *goal-directed* variant of the $\mathcal{EL}^+$ classification algorithm developed in [3] which can be used for testing subsumption queries prior to full classification. Section 5 presents an extension of the algorithm in [3] to cater for two ontologies: the permanent ontology $\mathcal{O}_p$ which has been carefully modeled, and axioms of which are not supposed to be modified; and, the temporary ontology $\mathcal{O}_t$ that contains new axioms currently being authored. The extended algorithm reuses information from the previous classification of $\mathcal{O}_p$ and thus dispense with the need of the full classification of $\mathcal{O}_p \cup \mathcal{O}_t$. We call reasoning in this setting *restricted incremental classification*.

All algorithms proposed in this paper have been implemented in the CEL reasoner [2] and various experiments on realistic ontologies have been performed. The experiments and their promising results are discussed in Section 6.

For interested readers, proofs omitted from the present paper can be found in the associated technical report [12].

2 Preliminaries

The present paper focuses on the sub-Boolean DL $\mathcal{EL}^+$ [3], which is the underlying logical formalism of the CEL reasoner [2]. Similar to other DLs, an $\mathcal{EL}^+$ signature is the disjoint union $\mathbf{S} = \mathsf{CN} \cup \mathsf{RN}$ of the sets of concept names and role names. $\mathcal{EL}^+$ *concept descriptions (or complex concepts)* can be defined inductively as follows: each concept name $A \in \mathsf{CN}$ and the top concept $\top$ are $\mathcal{EL}^+$ concept descriptions; and, if C, D are $\mathcal{EL}^+$ concept descriptions and $r \in \mathsf{RN}$ is a role name, then concept conjunction $C \sqcap D$ and existential restriction $\exists r.C$ are $\mathcal{EL}^+$ concept descriptions. An $\mathcal{EL}^+$ *ontology* $\mathcal{O}$ is a finite set of *general concept inclusion (GCI)* axioms $C \sqsubseteq D$ and *role inclusion (RI)* axioms $r_1 \circ \cdots \circ r_n \sqsubseteq s$ with C, D $\mathcal{EL}^+$ concept descriptions and r_i, s role names. Concept equivalences and (primitive) concept definitions are expressible using GCIs, whereas RIs can be used to express various role axioms, such as reflexivity ($\epsilon \sqsubseteq r$), transitivity ($r \circ r \sqsubseteq r$), right-identity ($r \circ s \sqsubseteq r$), and role hierarchy ($r \sqsubseteq s$) axioms. Figure 1 illustrates an example in the medical domain. For convenience, we write $\mathsf{Sig}(\mathcal{O})$ (resp., $\mathsf{Sig}(\alpha)$, $\mathsf{Sig}(C)$)

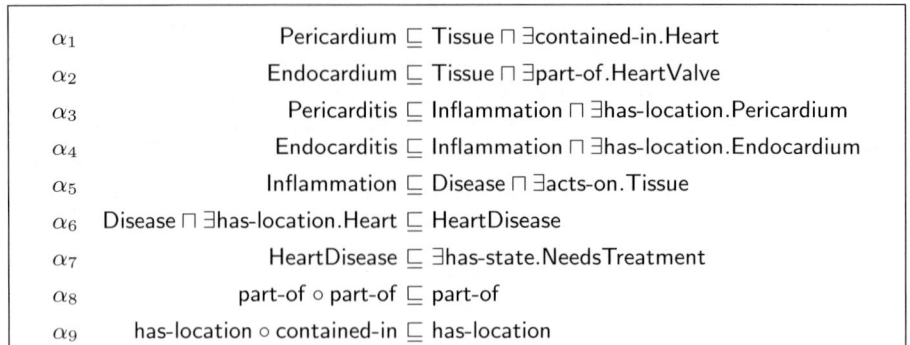

Fig. 1. An example $\mathcal{EL}^+$ ontology $\mathcal{O}_{ex}$

to denote the signature of the ontology $\mathcal{O}$ (resp., the axiom α, the concept C), i.e., concept and role names occurring in it. Also, let $\mathsf{CN}^\top(\mathcal{O})$ denote the set of $\top$ and concept names occurring in $\mathcal{O}$.

The main inference problem for concepts is *subsumption query*: given an ontology $\mathcal{O}$ and two concept descriptions C, D, check if C is subsumed by (i.e., more specific than) D w.r.t. $\mathcal{O}$, written $C \sqsubseteq_\mathcal{O} D$. From our example ontology, it is not difficult to draw that $\mathsf{Pericarditis} \sqsubseteq_{\mathcal{O}_{ex}} \exists\mathsf{has\text{-}state.NeedsTreatment}$. The identification of subsumption relationships between *all* pairs of concept names occurring in $\mathcal{O}$ is known as *ontology classification*.

The semantics of $\mathcal{EL}^+$ ontologies, as well as of subsumption, is defined by means of interpretations in the standard way, and we refer the reader to [12,1].

3 Modules Based on Connected Reachability

In this section, we introduce a new kind of module based on *connected reachability*, and propose an algorithm for extracting the modules of this kind. We also show that, in the DL $\mathcal{EL}^+$, our modules indeed correspond to modules based on syntactic locality first introduced in [6]. We start by giving the general definition of module:

Definition 1 (Modules for an axiom and a signature). *Let $\mathcal{O}$ be an $\mathcal{EL}^+$ ontology, and $\mathcal{O}'$ a (possibly empty) set of axioms from $\mathcal{O}$. We say that $\mathcal{O}'$ is a module in $\mathcal{O}$ for an axiom α (for short, α-module in $\mathcal{O}$) if: $\mathcal{O}' \models \alpha$ iff $\mathcal{O} \models \alpha$.*

We say that $\mathcal{O}'$ is a module for a signature $\mathbf{S}$ if, for every axiom α with $\mathsf{Sig}(\alpha) \subseteq \mathbf{S}$, we have that $\mathcal{O}'$ is an α-module in $\mathcal{O}$.

Intuitively, a module of an ontology $\mathcal{O}$ is a subset $\mathcal{O}' \subseteq \mathcal{O}$ that preserves an axiom of interest or the axioms over a signature of interest. Observe that this is a very generic definition, in the sense that the whole ontology is itself a module. In the following, we are interested in certain sufficient conditions that not only help extract a module according to Definition 1 but also guarantee relevancy of the extracted axioms. Note that if $\mathcal{O} \models \alpha$, a justification (minimal axiom set that has the consequence) is a minimal α-module in $\mathcal{O}$. A justification covers one axiom, not the axioms over a signature, thus it is normally expensive to obtain and involve standard inference reasoning, such as subsumption. For this reason, various syntactic approaches to extracting ontology fragments have been proposed in the literature [9,11,6]. In [6], Cuenca Grau et al. introduced a kind of module based on so-called syntactic locality for $\mathcal{SHOIQ}$. Though $\mathcal{EL}^+$ is not a sublanguage of $\mathcal{SHOIQ}$ due to RIs, the definition from [6] can be straightforwardly adjusted to suit $\mathcal{EL}^+$ as shown below:

Definition 2 (Locality-based modules). *Let $\mathcal{O}$ be an $\mathcal{EL}^+$ ontology, and $\mathbf{S}$ a signature. The following grammar recursively defines* $\mathbf{Con}^\perp(\mathbf{S})$:

$$\mathbf{Con}^\perp(\mathbf{S}) ::= A^\perp \mid (C^\perp \sqcap C) \mid (C \sqcap C^\perp) \mid (\exists r.C^\perp) \mid (\exists r^\perp.C)$$

with r is a role name, C a concept description, $A^\perp, r^\perp \notin \mathbf{S}$, and $C^\perp \in \mathbf{Con}^\perp(\mathbf{S})$.

An $\mathcal{EL}^+$ axiom α is syntactically local w.r.t. $\mathbf{S}$ if it is one of the following forms: (1) RI $R^\perp \sqsubseteq s$ where $R^\perp$ is either a role name $r^\perp \notin \mathbf{S}$ or a role composition $r_1 \circ \cdots \circ r_n$ with $r_i \notin \mathbf{S}$ for some $i \leq n$, or (2) GCI $C^\perp \sqsubseteq C$ where $C^\perp \in \mathbf{Con}^\perp(\mathbf{S})$. We write local($\mathbf{S}$) *to denote the collection of all $\mathcal{EL}^+$ axioms that are syntactically local w.r.t. $\mathbf{S}$.*

If $\mathcal{O}$ can be partitioned into $\mathcal{O}'$ and $\mathcal{O}''$ s.t. every axiom in $\mathcal{O}''$ is syntactically local w.r.t. $\mathbf{S} \cup \mathsf{Sig}(\mathcal{O}')$, then $\mathcal{O}'$ is a locality-based module *for $\mathbf{S}$ in $\mathcal{O}$.*

Now we consider the optimization techniques of "reachability" that are used to heuristically determine obvious subsumption and non-subsumption relationships. The reachability heuristic for non-subsumption can easily be exploited in module extraction for $\mathcal{EL}^+$ ontologies. To obtain a more satisfactory module size, however, we introduce a more appropriate (i.e., stronger) reachability notion and develop an algorithm for extracting modules based on this notion.

Definition 3 (Strong/weak reachability). *Let $\mathcal{O}$ be an $\mathcal{EL}^+$ ontology, and $A, B \in \mathsf{CN}^\top(\mathcal{O})$. The strong (weak) reachability graph $\mathcal{G}_s(\mathcal{O})$ $(\mathcal{G}_w(\mathcal{O}))$ for $\mathcal{O}$ is a tuple (V_s, E_s) $((V_w, E_w))$ with $V_s = \mathsf{CN}^\top(\mathcal{O})$ $(V_w = \mathsf{CN}^\top(\mathcal{O}))$ and E_s (E_w) the smallest set containing an edge (A, B) if $B = \top$ or $A \sqsubseteq D \in \mathcal{O}$ s.t. B is a conjunct in D (if $B = \top$ or $C \sqsubseteq D \in \mathcal{O}$ s.t. $A \in \mathsf{Sig}(C)$ and $B \in \mathsf{Sig}(D)$).*

We say that B is strongly reachable *(*weakly reachable*) from A in $\mathcal{O}$ if there is a path from A to B in $\mathcal{G}_s(\mathcal{O})$ $(\mathcal{G}_w(\mathcal{O}))$.*

Observe that B is strongly reachable from A in $\mathcal{O}$ implies $A \sqsubseteq_\mathcal{O} B$, while $A \sqsubseteq_\mathcal{O} B$ implies that B is weakly reachable from A in $\mathcal{O}$.

The weak reachability graph $\mathcal{G}_w(\mathcal{O})$ for $\mathcal{O}$ can be extended in a straightforward way to cover all the symbols in $\mathcal{O}$, i.e., also role names. Precisely, we define the extension as $\mathcal{G}'_w(\mathcal{O}) := (\mathsf{Sig}(\mathcal{O}) \cup \{\top\}, E'_w)$ with $(x, y) \in E'_w$ iff $y = \top$ or there is an axiom $\alpha_L \sqsubseteq \alpha_R \in \mathcal{O}$ s.t. $x \in \mathsf{Sig}(\alpha_L)$ and $y \in \mathsf{Sig}(\alpha_R)$. A module for $\mathbf{S} = \{A\}$ in an ontology $\mathcal{O}$ based on extended weak reachability can be extracted as follows: construct $\mathcal{G}'_w(\mathcal{O})$, extract all the paths from A in $\mathcal{G}_w(\mathcal{O})$, and finally, accumulate axioms responsible for the edges in those paths. However, this kind of module is relatively large, and many axioms are often irrelevant.

For example, any GCIs with Disease appearing on the left-hand side, such as Disease ⊓ ∃has-location.Brain ⊑ BrainDisease, would be extracted as part of the module for **S** = {Pericarditis}. This axiom is irrelevant since Pericarditis does not refer to Brain and thus BrainDisease. Such a module would end up comprising the definitions of all disease concepts. To rule out this kind of axioms, we make the notion of reachability graph stronger as follows: All symbols appearing on the left-hand side (e.g., Disease, has-location and Brain) are viewed as a connected node in the graph, which has an edge to each symbol (e.g., BrainDisease) on the right-hand side of the axiom. The connected node is reachable from x iff all symbols participating in it are reachable from x. In our example, since Brain is not reachable from Pericarditis, neither is BrainDisease. Therefore, the axiom is not extracted as part of the refined module.

Definition 4 (Connected reachability and modules). *Let $\mathcal{O}$ be an $\mathcal{EL}^+$ ontology, $\mathbf{S} \subseteq \mathsf{Sig}(\mathcal{O})$ a signature, and $x, y \in \mathsf{Sig}(\mathcal{O})$ concept or role names. We say that x is* connectedly reachable *from $\mathbf{S}$ w.r.t. $\mathcal{O}$ (for short, $\mathbf{S}$-reachable) iff $x \in \mathbf{S}$ or there is an axiom (either GCI or RI) $\alpha_L \sqsubseteq \alpha_R \in \mathcal{O}$ s.t. $x \in \mathsf{Sig}(\alpha_R)$ and, for all $y \in \mathsf{Sig}(\alpha_L)$, y is reachable from $\mathbf{S}$.*

We say that an axiom $\beta_L \sqsubseteq \beta_R$ is connected reachable *from $\mathbf{S}$ w.r.t. $\mathcal{O}$ (for short, $\mathbf{S}$-reachable) if, for all $x \in \mathsf{Sig}(\beta_L)$, x is $\mathbf{S}$-reachable. The reachability-based* module *for $\mathbf{S}$ in $\mathcal{O}$, denoted by $\mathcal{O}_\mathbf{S}^{\mathsf{reach}}$, is the set of all $\mathbf{S}$-reachable axioms.*

Intuitively, x is connectedly reachable from $\{y\}$ w.r.t. $\mathcal{O}$ means that y syntactically refers to x, either directly or indirectly via axioms in $\mathcal{O}$. If x, y are concept names, then the reachability suggests a potential subsumption relationship $y \sqsubseteq_\mathcal{O} x$. Note, in particular, that axioms of the forms $\top \sqsubseteq D$ and $\epsilon \sqsubseteq r$ in $\mathcal{O}$ are connectedly reachable from any signature because $\mathsf{Sig}(\top) = \mathsf{Sig}(\epsilon) = \emptyset$, and therefore occur in every reachability-based module. In our example, $\mathcal{O}_{\{\mathsf{Pericarditis}\}}^{\mathsf{reach}}$ contains axioms $\alpha_1, \alpha_3, \alpha_5 - \alpha_7$ and α_9. We now show some properties of connected reachability and reachability-based modules that are essential for establishing the subsequent lemma and theorem:

Proposition 1 (Properties of reachability and $\mathcal{O}_\mathbf{S}^{\mathsf{reach}}$). *Let $\mathcal{O}$ be an $\mathcal{EL}^+$ ontology, $\mathbf{S}, \mathbf{S}_1, \mathbf{S}_2 \subseteq \mathsf{Sig}(\mathcal{O})$ signatures, x, y, z symbols in $\mathsf{Sig}(\mathcal{O})$, and A, B concept names in $\mathsf{CN}(\mathcal{O})$. Then, the following properties hold:*

1. *If $\mathbf{S}_1 \subseteq \mathbf{S}_2$, then $\mathcal{O}_{\mathbf{S}_1}^{\mathsf{reach}} \subseteq \mathcal{O}_{\mathbf{S}_2}^{\mathsf{reach}}$.*
2. *If x is $\{y\}$-reachable and y is $\{z\}$-reachable, then x is $\{z\}$-reachable.*
3. *If x is connected reachable from $\{y\}$ w.r.t. $\mathcal{O}$, then $\mathcal{O}_{\{x\}}^{\mathsf{reach}} \subseteq \mathcal{O}_{\{y\}}^{\mathsf{reach}}$*
4. *$x \in \mathbf{S} \cup \mathsf{Sig}(\mathcal{O}_\mathbf{S}^{\mathsf{reach}})$ if, and only if, x is $\mathbf{S}$-reachable w.r.t. $\mathcal{O}$.*
5. *If B is not connected reachable from $\{A\}$ w.r.t. $\mathcal{O}$, then $A \not\sqsubseteq_\mathcal{O} B$.*

The converse of Point 5 is not true in general, for instance, Pericarditis involves Tissue, but the corresponding subsumption does not follow from the ontology. This suggests that we could use connected reachability as a heuristic for answering negative subsumption, in a similar but finer way as in weak reachability.

We outline our algorithm for extracting the reachability-based module given a signature **S** and an ontology $\mathcal{O}$ in Algorithm 1. Similar to the technique developed

Algorithm 1. extract-module

Input: $\mathcal{O}$: $\mathcal{EL}^+$ ontology; **S**: signature
Output: $\mathcal{O}_\mathbf{S}$: reachability-based module for **S** in $\mathcal{O}$

1: $\mathcal{O}_\mathbf{S} \leftarrow \emptyset$
2: queue $\leftarrow$ active-axioms(**S**)
3: **while not** empty(queue) **do**
4: $\quad (\alpha_L \sqsubseteq \alpha_R) \leftarrow$ fetch(queue)
5: $\quad$ **if** $\text{Sig}(\alpha_L) \subseteq \mathbf{S} \cup \text{Sig}(\mathcal{O}_\mathbf{S})$ **then**
6: $\quad\quad \mathcal{O}_\mathbf{S} \leftarrow \mathcal{O}_\mathbf{S} \cup \{\alpha_L \sqsubseteq \alpha_R\}$
7: $\quad\quad$ queue $\leftarrow$ queue $\cup$ (active-axioms(Sig(α_R)) $\setminus \mathcal{O}_\mathbf{S}$)
8: **return** $\mathcal{O}_\mathbf{S}$

in [3], we view the input ontology $\mathcal{O}$ as a mapping active-axioms : $\text{Sig}(\mathcal{O}) \to \mathcal{O}$ with active-axioms(x) comprising all and only axioms $\alpha_L \sqsubseteq \alpha_R \in \mathcal{O}$ such that x occurs in α_L. The main differences, compared to the $\widehat{\mathcal{O}}$ mapping in [3] (also used in Section 4), are that active-axioms does not assume the input ontology to be in normal form, and that it is defined for both concept and role names. The intuition is that every axiom $\alpha \in$ active-axioms(x) is "active" for x, in the sense that y could be connectedly reachable via α from x for some $y \in \text{Sig}(\mathcal{O})$. For convenience, we define active-axioms(**S**) := $\bigcup_{x \in \mathbf{S}}$ active-axioms(x) for a signature $\mathbf{S} \subseteq \text{Sig}(\mathcal{O})$.

It is easy to see that each axiom Algorithm 1 extracts to $\mathcal{O}_\mathbf{S}$ is **S**-reachable. The fact that all **S**-reachable axioms are extracted to $\mathcal{O}_\mathbf{S}$ can be proved by induction on connected reachability.

Proposition 2 (Algorithm 1 produces $\mathcal{O}_\mathbf{S}^{\text{reach}}$). Let $\mathcal{O}$ be an $\mathcal{EL}^+$ ontology and $\mathbf{S} \subseteq \text{Sig}(\mathcal{O})$ a signature. Then, Algorithm 1 returns the reachability-based module for **S** in $\mathcal{O}$.

In fact, connected reachability can be reduced to propositional Horn clause implication. The idea is to translate each $\mathcal{EL}^+$ axiom $\alpha_L \sqsubseteq \alpha_R$ into the Horn clause $l_1 \wedge \cdots \wedge l_m \to r_1 \wedge \cdots \wedge r_n$ where $l_i \in \text{Sig}(\alpha_L)$ and $r_i \in \text{Sig}(\alpha_R)$. Given a signature **S** and a symbol x, x is **S**-reachable iff x is implied by $\bigwedge_{y \in \mathbf{S}} y$ w.r.t. the Horn clauses. The Dowling-Gallier algorithm [4] can check this in linear time.

In the following, we show a tight relationship between our reachability-based modules and the (minimal) locality-based modules.

Theorem 1 ($\mathcal{O}_\mathbf{S}^{\text{reach}}$ is the minimal locality-based module). Let $\mathcal{O}$ be an $\mathcal{EL}^+$ ontology, and $\mathbf{S} \subseteq \text{Sig}(\mathcal{O})$ a signature. Then, $\mathcal{O}_\mathbf{S}^{\text{reach}}$ is the minimal locality-based module for **S** in $\mathcal{O}$.

So, Algorithm 1 can be used to extract a locality-based module in an $\mathcal{EL}^+$ ontology. The main difference, in contrast to the algorithm used in [6,5], is that our algorithm considers only "active" axioms for α_R when a new axiom $\alpha_L \sqsubseteq \alpha_R$ is extracted. Also, testing whether an $\mathcal{EL}^+$ axiom $\alpha = (\alpha_L \sqsubseteq \alpha_R)$ is non-local w.r.t. a signature $\mathbf{S} \cup \text{Sig}(\mathcal{O}_\mathbf{S})$ boils down to testing **S**-reachability of α, which is a simpler operation of testing set inclusion $\text{Sig}(\alpha_L) \subseteq^? \mathbf{S} \cup \text{Sig}(\mathcal{O}_\mathbf{S})$. This is due to the fact that any concept description and role composition α_L, with $x \in \mathsf{Sig}(\alpha_L)$ interpreted as the empty set, is itself interpreted as the empty set. This observation could be used to optimize module extraction for ontologies in expressive description logics.

It has been shown for $\mathcal{SHOIQ}$ that locality-based modules for $\mathbf{S} = \{A\}$ in $\mathcal{O}$ preserves the subsumption $A \sqsubseteq B$ for any $B \in \mathsf{CN}(\mathcal{O})$ [6]. This property could have been transferred to our setting as a corollary of Theorem 1 if $\mathcal{EL}^+$ were a sublanguage of $\mathcal{SHOIQ}$. Despite this not being the case, it is not hard to show that reachability-based modules in $\mathcal{EL}^+$ also enjoy the property:

Lemma 1 ($\mathcal{O}_A^{\mathsf{reach}}$ **preserves** $A \sqsubseteq_\mathcal{O} B$)**.** *Let $\mathcal{O}$ be an $\mathcal{EL}^+$ ontology, $A \in \mathsf{CN}(\mathcal{O})$, and $\mathcal{O}_{\{A\}}^{\mathsf{reach}}$ the reachability-based module for $\mathbf{S} = \{A\}$ in $\mathcal{O}$. Then, for any $\alpha = A \sqsubseteq B$ with $B \in \mathsf{CN}(\mathcal{O})$, $\mathcal{O} \models \alpha$ iff $\mathcal{O}_{\{A\}}^{\mathsf{reach}} \models \alpha$.*

4 Goal-Directed Subsumption Algorithm

In general, the techniques developed for module extraction have a number of potential applications, including optimization of standard reasoning, incremental classification, explanation, and ontology re-use. An obvious way to exploit module extraction to speed up standard reasoning, such as subsumption $\phi \sqsubseteq_\mathcal{O}^? \psi$, is to first extract the module $\mathcal{O}_{\{\phi\}}^{\mathsf{reach}}$ for $\mathbf{S} = \{\phi\}$ in $\mathcal{O}$, and then query the subsumption $\phi \sqsubseteq_{\mathcal{O}_{\{\phi\}}^{\mathsf{reach}}}^? \psi$, i.e., w.r.t. the module instead of the original ontology. Based on the assumption that modules are relatively much smaller than the ontology, this optimization should be highly effective. In this section, however, we argue that module extraction actually does not help speed up standard reasoning in $\mathcal{EL}^+$. This stems from the deterministic and goal-directed nature of the reasoning algorithm for deciding subsumption in $\mathcal{EL}^+$, which is in contrast to non-deterministic tableau-based algorithms for expressive logics, such as $\mathcal{SHOIQ}$.

In fact, with small modifications to the $\mathcal{EL}^+$ classification algorithm (first introduced in [1] for $\mathcal{EL}^{++}$ and later refined for implementation in [3]), we obtain a subsumption testing algorithm. The modified algorithm does not actually have to perform steps irrelevant to the subsumption in question – *the goal*. We call this variant the *goal-directed subsumption algorithm*.

Algorithm 2 outlines the modified core procedure **goal-directed-process** to replace **process** of Figure 3 in [3]. The procedure **process-new-edge**, as well as essential data structures, i.e., $\hat{\mathcal{O}}$, **queue**, R, S, remains intact. In particular, we view the (normalized) input ontology $\mathcal{O}$ as a mapping $\hat{\mathcal{O}}$ from concepts (appearing on the left-hand side of some GCI) to sets of queue entries. Here, $\mathbf{B}$ denotes the set of all concept names appearing in the conjunction $B_1 \sqcap \cdots \sqcap B_n$.

The main difference is the initialization of S, thus of **queue**. Since we are interested in the particular subsumption $\phi \sqsubseteq \psi$, we "activate" only ϕ by initializing $S(\phi)$ with $\{\phi, \top\}$ and **queue**(ϕ) with $\hat{\mathcal{O}}(\phi) \cup \hat{\mathcal{O}}(\top)$. We activate a concept name B *only* when it becomes the second component of a tuple added to some $R(r)$ and has not been activated previously (see lines 8-9 in **goal-directed-process** of Algorithm 2). Thereby, $S(B)$ and **queue**(B) are initialized accordingly. Queues

Algorithm 2. Goal-directed subsumption algorithm

Procedure subsumes($\phi \sqsubseteq \psi$)
Input: ($\phi \sqsubseteq \psi$): target subsumption
Output: 'positive' or 'negative' answer to the subsumption
1: activate(ϕ)
2: **while not** empty(queue(A)) for some $A \in \mathsf{CN}(\mathcal{O})$ **do**
3:     $X \leftarrow$ fetch(queue(A))
4:     **if** goal-directed-process($A, X, \phi \sqsubseteq \psi$) **then**
5:        **return** 'positive'
6: **return** 'negative'

Procedure goal-directed-process($A, X, \phi \sqsubseteq \psi$)
Input: A: concept name; X: queue entry; ($\phi \sqsubseteq \psi$): target subsumption
Output: 'positive' or 'unknown' answer to the subsumption
1: **if** $X = \mathbf{B} \to B$, $\mathbf{B} \subseteq S(A)$ and $B \notin S(A)$ **then**
2:     $S(A) := S(A) \cup \{B\}$
3:     queue(A) := queue(A) $\cup \widehat{\mathcal{O}}(B)$
4:     **for** all concept names A' and role names r with $(A', A) \in R(r)$ **do**
5:        queue(A') := queue(A') $\cup \widehat{\mathcal{O}}(\exists r.B)$
6:     **if** $A = \phi$ and $B = \psi$ **then**
7:        **return** 'positive'
8: **if** $X = \exists r.B$ and $(A, B) \notin R(r)$ **then**
9:     activate(B)
10:     process-new-edge(A, r, B)
11: **return** 'unknown'

are processed in the same fashion as before except that ϕ and ψ are now being monitored (Line 6), so that immediately after ψ is added to $S(\phi)$, the algorithm terminates with the positive answer (Line 7). Otherwise, goal-directed-process terminates normally, and the next queue entry will be fetched (Line 3 in subsumes? of Algorithm 2) and processed (Line 4). Unless 'positive' is returned, queues processing is continued until they are all empty. In this case, the algorithm returns 'negative.'

It is important to note that the goal-directed algorithm activates only concept names relevant to the target subsumption $\phi \sqsubseteq \psi$, i.e., those reachable via $R(\cdot)$ from ϕ. The subsumer sets of concept names that do not become activated are not populated. Moreover, axioms that are involved in rule applications during the computation of subsumes?($\phi \sqsubseteq \psi$) are those from the reachability-based module $\mathcal{O}^{\text{reach}}_{\{\phi\}}$ in $\mathcal{O}$. The following proposition states this correlation:

Proposition 3 (subsumes?($\phi \sqsubseteq \psi$) only requires axioms in $\mathcal{O}^{\text{reach}}_{\phi}$). *Let $\mathcal{O}$ be an ontology in $\mathcal{EL}^+$ normal form, and $\mathcal{O}^{\text{reach}}_{\{\phi\}}$ the reachability-based module for $\mathbf{S} = \{\phi\}$ in $\mathcal{O}$. Then, subsumes?($\phi \sqsubseteq \psi$) only requires axioms in $\mathcal{O}^{\text{reach}}_{\{\phi\}}$.*

Intuitively, the proposition suggests that our goal-directed subsumption algorithm inherently takes into account the notion of connected reachability, i.e., it applies rules only to relevant axioms in the reachability-based module. In fact, the preprocessing overhead of extracting the relevant module $\mathcal{O}_{\{\phi\}}^{\text{reach}}$ for the subsumption query $\phi \sqsubseteq_{\mathcal{O}}^{?} \psi$ makes the overall computation time for an individual subsumption query longer. This has been empirically confirmed in our experiments (see the last paragraph of Section 6).

Despite what has been said, module extraction is still useful for, e.g., ontology re-use, explanation, and full-fledged incremental reasoning [5].

5 Duo-Ontology Classification

Unlike tableau-based algorithms, the polynomial-time algorithm in [1,3] inherently classifies the input ontology by making all subsumptions between *concept names* explicit. This algorithm can be used to query subsumption between concept names occurring in the ontology, but complex subsumptions, such as

Inflammation $\sqcap$ $\exists$has-location.Heart $\sqsubseteq_{\mathcal{O}_{\text{ex}}}^{?}$ HeartDisease $\sqcap$ $\exists$has-state.NeedsTreatment

cannot be answered directly. First, the ontology $\mathcal{O}_{\text{ex}}$ from Figure 1 has to be augmented to $\mathcal{O}'_{\text{ex}} := \mathcal{O}_{\text{ex}} \cup \{A \sqsubseteq$ Inflammation $\sqcap$ $\exists$has-location.Heart, HeartDisease $\sqcap$ $\exists$has-state.NeedsTreatment $\sqsubseteq B\}$ with A, B new concept names, and then the subsumption test $A \sqsubseteq_{\mathcal{O}'_{\text{ex}}}^{?} B$ can be carried out to decide the original complex subsumption. Since A, B are new names not occurring in $\mathcal{O}_{\text{ex}}$, our complex subsumption holds iff $A \sqsubseteq_{\mathcal{O}'_{\text{ex}}} B$. This approach is effective but inefficient unless only one such complex subsumption is queried for each ontology. Constructing and normalizing the augmented ontology every time each subsumption is tested is not likely to be acceptable in practice, especially when the background ontology is large. For instance, normalization of SNOMED CT takes more than one minute.

In this section, we propose an extension to the refined algorithm (henceforth referred to as *the original algorithm*) developed in [3] to cater for a *duo-ontology* $\mathcal{O} = (\mathcal{O}_p \cup \mathcal{O}_t)$ with $\mathcal{O}_p$ a *permanent* $\mathcal{EL}^+$ ontology and $\mathcal{O}_t$ a set of *temporary* GCIs. Intuitively, $\mathcal{O}_p$ is the input ontology of which axioms have been read in and processed before, while $\mathcal{O}_t$ contains temporary GCIs that are asserted later. The main purpose is to reuse the information made available by the preprocess and classification of $\mathcal{O}_p$. Once $\mathcal{O}_p$ has been classified, the classification of $\mathcal{O}_p \cup \mathcal{O}_t$ should not start from scratch but rather use the existing classification information together with the new GCIs from $\mathcal{O}_t$ to do incremental classification.

In our extension, we use two sets of the core data structures $\widehat{\mathcal{O}}(\cdot), R(\cdot), S(\cdot)$, but retain a single set of queues queue$(\cdot)$. The mappings $\widehat{\mathcal{O}}_p, R_p, S_p$ are initialized and populated exactly as in the original algorithm, i.e., $\widehat{\mathcal{O}}_p$ encodes axioms in $\mathcal{O}_p$, and R_p, S_p store subsumption relationships inferred from $\mathcal{O}_p$. Similarly, the mapping $\widehat{\mathcal{O}}_t$ encodes axioms in $\mathcal{O}_t$, but R_t, S_t represent additional inferred subsumptions drawn from $\mathcal{O}_p \cup \mathcal{O}_t$ that are not already present in R_p, S_p, respectively. The extended algorithm is based on the tenet that description logics are monotonic, i.e., $\mathcal{O}_p \models \alpha$ implies $\mathcal{O}_p \cup \mathcal{O}_t \models \alpha$. There may be an additional consequence β such that $\mathcal{O}_p \not\models \beta$ but $\mathcal{O}_p \cup \mathcal{O}_t \models \beta$. Our algorithm stores such a consequence β in a separate set of data structures, namely R_p, S_p. Analogously to the original algorithm, queue

Algorithm 3. Processing queue entries in duo-ontology classification

Procedure process-duo(A, X)
Input: A: concept name; X: queue entry;
1: **if** $X = \mathbf{B} \to B$, $\mathbf{B} \subseteq S_p(A) \cup S_t(A)$ **and** $B \notin S_p(A) \cup S_t(A)$ **then**
2:     $S_t(A) := S_t(A) \cup \{B\}$
3:     queue$(A) :=$ queue$(A) \cup \widehat{\mathcal{O}}_p(B) \cup \widehat{\mathcal{O}}_t(B)$
4:     **for** all A' and r with $(A', A) \in R_p(r) \cup R_t(r)$ **do**
5:         queue$(A') :=$ queue$(A') \cup \widehat{\mathcal{O}}_p(\exists r.B) \cup \widehat{\mathcal{O}}_t(\exists r.B)$
6: **if** $X = \exists r.B$ and $(A, B) \notin R_p(r) \cup R_t(r)$ **then**
7:     process-new-edge(A, r, B)

Procedure process-new-edge-duo(A, r, B)
Input: A, B: concept names; r: role name;
1: **for** all role names s with $r \sqsubseteq^*_{\mathcal{O}_p} s$ **do**
2:     $R_t(s) := R_t(s) \cup \{(A, B)\}$
3:     queue$(A) :=$ queue$(A) \cup \bigcup_{\{B' | B' \in S_p(B) \cup S_t(B)\}} (\widehat{\mathcal{O}}_p(\exists s.B') \cup \widehat{\mathcal{O}}_t(\exists s.B'))$
4:     **for** all concept name A' and role names u, v with $u \circ s \sqsubseteq v \in \mathcal{O}_p$ and $(A', A) \in R_p(u) \cup R_t(u)$ and $(A', B) \notin R_p(v) \cup R_t(v)$ **do**
5:         process-new-edge-duo(A', v, B)
6:     **for** all concept name B' and role names u, v with $s \circ u \sqsubseteq v \in \mathcal{O}_p$ and $(B, B') \in R_p(u) \cup R_t(u)$ and $(A, B') \notin R_p(v) \cup R_t(v)$ **do**
7:         process-new-edge-duo(A, v, B')

entries are repeatedly fetched and processed until all queues are empty. Instead of the procedures process and process-new-edge, we use the extended versions for duo-ontology classification as outlined in Algorithm 3.

The behavior of Algorithm 3 is identical to that of the original one [3] if $\mathcal{O}_p$ has not been classified before. In particular, $\widehat{\mathcal{O}}_p(\cdot) \cup \widehat{\mathcal{O}}_t(\cdot)$ here is equivalent to $\widehat{\mathcal{O}}(\cdot)$ in [3] given that $\mathcal{O} = (\mathcal{O}_p \cup \mathcal{O}_t)$. Since no classification has taken place, $S_p(A) = R_p(r) = \emptyset$ for each concept name A and role name r. Initialization and processing of queues are done in the same manner with the only difference that inferred consequences are now put in R_t and S_t.

If $\mathcal{O}_p$ has been classified (thus, S_p, R_p have been populated), then a proper initialization has to be done w.r.t. the previously inferred consequences (i.e., S_p, R_p) and the new GCIs (i.e., $\widehat{\mathcal{O}}_t$). To this end, we initialize the data structures by setting:

- for each role name $r \in \mathsf{RN}(\mathcal{O})$, $R_t(r) := \emptyset$;
- for each *old* concept name $A \in \mathsf{CN}(\mathcal{O}_p)$, $S_t(A) := \emptyset$ and
  queue$(A) := \bigcup_{X \in S_p(A)} \widehat{\mathcal{O}}_t(X) \cup \bigcup_{\{(A,B) \in R_p(r), X \in S_p(B)\}} \widehat{\mathcal{O}}_t(\exists r.X)$;
- for each *new* concept name $A \in \mathsf{CN}(\mathcal{O}_t) \setminus \mathsf{CN}(\mathcal{O}_p)$, $S_t(A) := \{A, \top\}$
  queue$(A) := \widehat{\mathcal{O}}_t(A) \cup \widehat{\mathcal{O}}_t(\top)$.

After initialization, queue processing is carried out by Algorithm 3 until all queues are empty. Observe the structural analogy between these procedures and the original ones in [3]. Observe also the key difference: information is always retrieved from both sets of data structures, e.g., $S_p(A) \cup S_t(A)$ in Line 1, while modifications are only made to the temporary set of data structures, e.g., $S_t(A) := S_t(A) \cup \{B\}$ in Line 2. The correctness of Algorithm 3 can be shown following the correctness proofs structures of the original algorithm (see the submitted journal version of [3]) w.r.t. additional subsumption consequences obtained during incremental classification.

Lemma 2 (Correctness of Algorithm 3). *Let $\mathcal{O} = (\mathcal{O}_p \cup \mathcal{O}_t)$ be a duo-ontology, and S_p, R_p be the results after the original algorithm terminates on $\mathcal{O}_p$. Then, the extended algorithm (Algorithm 3), applied to $\mathcal{O}_t$, incrementally classifies $\mathcal{O}_t$ against $\mathcal{O}_p$ (i.e., classifies $\mathcal{O}$) in time polynomial in the size of $\mathcal{O}$. That is, $B \in S_p(A) \cup S_t(A)$ iff $A \sqsubseteq_\mathcal{O} B$ for all $A, B \in \mathsf{CN}(\mathcal{O})$.*

In our example, we may view $\mathcal{O}_{\mathsf{ex}}$ as the permanent ontology $\mathcal{O}_p$ and the two new GCIs as the temporary ontology $\mathcal{O}_t$. We can then run the extended algorithm on $\mathcal{O}_p \cup \mathcal{O}_t$ and reuse existing information in S_p and R_p, if any. After termination, our complex subsumption boils down to the set membership test $B \in^? S_p(A) \cup S_t(A) = S_t(A)$. To decide subsequent subsumption queries, only $\mathcal{O}_t, R_t, S_t$, and queue need to be initialized, leaving the background ontology $\mathcal{O}_p$ and possibly its classification information R_t, S_t intact.

Interestingly, this algorithm can be used effectively in certain scenarios of incremental classification. Consider $\mathcal{O}_p$ as a well-developed, permanent ontology, and $\mathcal{O}_t$ as a small set of temporary axioms currently being authored. Obviously, if the permanent ontology is large, it would be impractical to reclassify from scratch every time some new axioms are to be added. Algorithm 3 incrementally classifies $\mathcal{O}_t$ against $\mathcal{O}_p$ and its classification information. If the inferred consequences are satisfactory, the temporary axioms can be committed to the permanent ontology by merging the two sets of data structures. Otherwise, axioms in $\mathcal{O}_t$ and their inferred consequences could be easily retracted, since these are segregated from $\mathcal{O}_p$ and its consequences. To be precise, we simply dump the values of $\mathcal{O}_t(\cdot), R_t(\cdot)$ and $S_t(\cdot)$, when the temporary axioms are retracted.

6 Experiments and Empirical Results

This section describes the experiments and results of the three algorithms we proposed in this paper: module extraction, goal-directed subsumption query, and duo-ontology classification, which have been implemented and integrated as new features into the CEL reasoner [2] version 1.0b. All the experiments have been carried out on a standard PC: 2.40 GHz Pentium-4 processor and 1 GB of physical memory. In order to show interesting characteristics of reachability-based modules and scalability of subsumption and incremental classification in $\mathcal{EL}^+$, we have selected a number of large-scale medical ontologies. Our test suite comprises SNOMED CT, NCI, and the $\mathcal{EL}^+$ fragments[1] of FULLGALEN and

---

[1] FULLGALEN is precisely based on $\mathcal{SHIF}$ dispensed with negation, disjunction, and value restriction. The DL $\mathcal{EL}^+$ can indeed express most of its axioms, namely 95.75%, and we obtained this fragment for experimental purposes by dropping role inverse and functionality axioms.

Table 1. $\mathcal{EL}^+$ ontology test suite

Ontologies	♯Concepts/roles	♯Concept/role axioms	Class. time (sec)	Positive subs. (%)
$\mathcal{O}^{\text{NotGalen}}$	2 748 / 413	3 937 / 442	7.36	0.6013
$\mathcal{O}^{\text{FullGalen}}$	23 136 / 950	35 531 / 1 016	512.72	0.1648
$\mathcal{O}^{\text{Nci}}$	27 652 / 70	46 800 / 140	7.01	0.0441
$\mathcal{O}^{\text{Snomed}}$	379 691 / 62	379 691 / 13	1 671.23	0.0074

NotGalen, denoted respectively by $\mathcal{O}^{\text{Snomed}}$, $\mathcal{O}^{\text{Nci}}$, $\mathcal{O}^{\text{FullGalen}}$, and $\mathcal{O}^{\text{NotGalen}}$.[2] The FullGalen ontology shall not be confused with the original version of Galen, the latter of which is almost 10 times smaller and commonly used in DL benchmarking. The sizes of our test suite ontologies are shown in the second and third columns of Table 1. The last but one column shows the time CEL needs to classify each ontology, while the last presents in percentage the ratio of positive subsumption relationships between concept names. Observe that all ontologies have a very low ratio of positive subsumption (less than 1%); in particular, less than a ten-thousandth of all potential subsumptions *actually* hold in $\mathcal{O}^{\text{Snomed}}$.

Modularization: For each ontology $\mathcal{O}$ in the test suite and each concept name $A \in \mathsf{CN}(\mathcal{O})$, we extracted the reachability-based module $\mathcal{O}_A^{\text{reach}}$. Statistical data concerning the sizes of modules and times required to extract them are presented in Table 2. Observe that it took a tiny amount of time to extract a single module based on connected reachability, with the maximum time less than four seconds. However, extracting large the number of modules (i.e., one for each concept name) required considerably more time and even longer than classification. This was nevertheless the first implementation that was not highly optimized. Several optimization techniques could be employed in module extraction, especially recursive extraction as suggested by Point 3 of Proposition 1 and the counting techniques from [4]. To empirically support Theorem 1, we have compared our modularization algorithm to that from [5,6]. As expected, the results of both algorithms coincide w.r.t. $\mathcal{O}^{\text{NotGalen}}$ and $\mathcal{O}^{\text{Nci}}$, while we were unable to obtain locality-based modularization results w.r.t. the other two ontologies.[3]

Interestingly, module extraction reveals important structural dependencies that reflect complexity of the ontology. Though very large, concepts in $\mathcal{O}^{\text{Nci}}$ and $\mathcal{O}^{\text{Snomed}}$ are loosely connected w.r.t. reachability which makes it relatively easy to classify. In contrast, $\mathcal{O}^{\text{FullGalen}}$ contains more complex dependencies[4], thus is hard to classify.

Duo-ontology classification: As mentioned before, there are at least two applications of Algorithm 3, viz., complex subsumption query and (restricted)

---
[2] Obtainable at http://lat.inf.tu-dresden.de/~meng/toyont.html.
[3] Due to memory exhaustion with 0.8 GB of Java heap space.
[4] Based on the statistical data analysis, there are two clearly distinct groups of concepts in $\mathcal{O}^{\text{FullGalen}}$: the first with module sizes between 0 and 523 (med. 39; avg. 59.29) and the second between 14 791 and 15 545 (med. 14 792; avg. 14 829). Surprisingly, there is no module of size between those of these two groups.

Table 2. Module extraction (time in second; size in number of axioms)

Ontologies	Extraction time				Module size (%)		
	median	average	maximum	total	median	average	maximum
$\mathcal{O}^{\text{NotGalen}}$	< 0.01	~ 0.00	0.01	2.38	35 (1.27)	68.64 (2.50)	495 (18.00)
$\mathcal{O}^{\text{FullGalen}}$	0.01	0.04	0.85	960	178 (0.77)	7092 (30.65)	15 545 (67.18)
$\mathcal{O}^{\text{Nci}}$	< 0.01	~ 0.00	0.17	3.43	12 (0.026)	28.97 (0.062)	436 (0.929)
$\mathcal{O}^{\text{Snomed}}$	< 0.01	~ 0.01	3.83	3744	18 (0.005)	30.31 (0.008)	262 (0.069)

Table 3. Incremental classification (in second)

♯Temp. axioms	$\mathcal{O}^{\text{NotGalen}}$		$\mathcal{O}^{\text{FullGalen}}$		$\mathcal{O}^{\text{Nci}}$		$\mathcal{O}^{\text{Snomed}}$	
($|\mathcal{O}_t|$)	C. time	IC. time	C. time	IC. time	C. time	IC. time	C. time	IC. time
0.2%	6.53	1.75	486.19	56.94	5.10	2.00	1 666.43	55.86
0.4%	6.50	1.88	484.89	59.37	4.81	2.15	1 663.51	57.97
0.6%	6.48	2.45	482.13	62.34	4.78	2.37	1 661.49	68.58
0.8%	6.43	2.88	466.97	80.52	4.70	2.54	1 652.84	83.27
1.0%	6.38	4.46	450.61	109.81	4.59	3.19	1 640.11	93.89

incremental classification. For complex subsumption query, we have adopted the "activation" idea from Algorithm 2 to quickly answer the query. To perform meaningful experiments, it is inevitable to involve a domain expert to obtain sensible test data. Though we have done so w.r.t. $\mathcal{O}^{\text{Snomed}}$, the numbers of complex subsumption queries and additional axioms are very small compared to the ontology size.[5] For this reason, we have developed our test strategy as follows: for each ontology $\mathcal{O}$ and various numbers n, we have (i) partitioned $\mathcal{O}$ into $\mathcal{O}_p$ and $\mathcal{O}_t$ such that $\mathcal{O}_t$ contains $n\%$ of GCIs from $\mathcal{O}$; (ii) classified $\mathcal{O}_p$ normally; finally, (iii) incrementally classified $\mathcal{O}_t$ against $\mathcal{O}_p$. The average computation times for several runs of (ii) and (iii) are shown in the left and right columns of each ontology in Table 3, respectively. It requires only 4% (resp., 15%, 35%, and 38%) of the total classification time for $\mathcal{O}^{\text{Snomed}}$ (resp., for $\mathcal{O}^{\text{FullGalen}}$, $\mathcal{O}^{\text{Nci}}$, and $\mathcal{O}^{\text{NotGalen}}$) to incrementally classify up to 1% of all axioms, i.e., about four-thousand axioms in the case of $\mathcal{O}^{\text{Snomed}}$.

Subsumption: To evaluate our goal-directed algorithm, we have run subsumption tests between *random* pairs of concept names without any heuristics.[6] Average/maximum querying times (in second) are 0.09/1.51 for $\mathcal{O}^{\text{NotGalen}}$, 124.01/254.31 for $\mathcal{O}^{\text{FullGalen}}$, 0.0034/0.44 for $\mathcal{O}^{\text{Nci}}$, and 0.0183/3.32 for $\mathcal{O}^{\text{Snomed}}$.

---

[5] On average, a typical complex subsumption query against $\mathcal{O}^{\text{Snomed}}$ took 0.00153 *milli*seconds, while incremental classification of one axiom needed 48.74 seconds.

[6] Since there are about *144 billion pairs* of concept names in the case of $\mathcal{O}^{\text{Snomed}}$ and some subsumption queries against $\mathcal{O}^{\text{FullGalen}}$ took a few minutes, performing subsumption queries between *all* pairs would not be feasible. Therefore, one thousand random pairs of subsumption were tested against $\mathcal{O}^{\text{FullGalen}}$, and one million random pairs against each of the other ontologies.

Notice that subsumption requires a negligible amount of time and not much more than extracting a module in the case of $\mathcal{O}^{\text{NCI}}$ and $\mathcal{O}^{\text{SNOMED}}$. Observe also that subsumption querying times are roughly proportional to module sizes, which reflects the nature of the goal-directed algorithm as stated by Proposition 3.

7 Related Work

Recently, various techniques for extracting fragments of ontologies have been proposed in the literature. An example is the algorithm proposed in [11] which was developed specifically for Galen. The algorithm traverses in definitional order and into existential restrictions but does not take into account other dependencies, e.g., role hierarchy and GCIs. If applied to our example ontology $\mathcal{O}_{\text{ex}}$, the algorithm extracts only α_1, α_3 and α_5 as its segmentation output for Pericarditis. This is obviously not a module because we lose the subsumption Pericarditis $\sqsubseteq_{\mathcal{O}_{\text{ex}}}$ HeartDisease. Another example is the Prompt-Factor tool [9] which implements an algorithm that, given an ontology $\mathcal{O}$ and a signature $\mathbf{S}$, retrieves a subset $\mathcal{O}_1 \subseteq \mathcal{O}$ by retrieving to $\mathcal{O}_1$ axioms that contain symbols in $\mathbf{S}$ and extending $\mathbf{S}$ with $\text{Sig}(\mathcal{O}_1)$ until a fixpoint is reached. This is similar to our modules based on *weak* reachability, but it does not distinguish symbols occurring on lhs and rhs of axioms. In our example, the tool will return the whole ontology as output for $\mathbf{S} = \{\text{Pericarditis}\}$, even though several axioms are irrelevant. As we have shown, modules based on syntactic locality [6] are equivalent to our reachability-based modules relative to $\mathcal{EL}^+$ ontologies. Since reachability is much simpler to check, our algorithm has proved more efficient.

Incremental classification and reasoning have received much attention in the recent years. In [7,10], the so-called model-caching techniques have been investigated for application scenarios that only ABox is modified. A technique for incremental schema reasoning has recently been proposed in [5]: it utilizes modules to localize ramifications of changes and performs additional reasoning only on affected modules. The framework supports full-fledged incremental reasoning in the sense that arbitrary axioms can be retracted or modified, and as such it is worthwhile to investigate how its techniques can be integrated into our duo-ontology classification algorithm. All above-mentioned works focus on expressive languages. Here, however, we developed a very specific approach to (restricted) incremental classification in $\mathcal{EL}^+$. Since the technique exploits the facts that the original $\mathcal{EL}^+$ algorithm maintains completed subsumer sets, it is not obvious how this may benefit tableau-based algorithms for expressive DLs.

8 Conclusion

In this paper, we have introduced a new kind of module (based on connected reachability) and proposed an algorithm to extract them from $\mathcal{EL}^+$ ontologies. We have shown that these are equivalent to locality-based modules w.r.t. $\mathcal{EL}^+$ ontologies and empirically demonstrated that modules can be extracted in reasonable time and are reasonably small. Also, we have proposed a goal-directed variant of the algorithm in [3] for testing subsumption prior to classification and have extended this algorithm to cater for a duo-ontology which can be utilized to answer complex subsumption queries and to do (restricted) incremental classification. Our empirical results have evidently confirmed that the proposed algorithms are practically feasible in large-scale ontology applications.

Despite not being directly useful to speed up standard reasoning in $\mathcal{EL}^+$, modularization obviously benefits ontology re-use and explanation. As future work, we shall study the effectiveness of using modules to optimize axiom pinpointing, which is the cornerstone of explanation support.

Acknowledgement. The author would like to acknowledge Franz Baader and Carsten Lutz for their valuable suggestions and Christian H.-Wiener for his willingness in comparing the two modularization approaches.

References

1. Baader, F., Brandt, S., Lutz, C.: Pushing the $\mathcal{EL}$ envelope. In: Proc. of IJCAI 2005, Morgan Kaufmann, San Francisco (2005)
2. Baader, F., Lutz, C., Suntisrivaraporn, B.: CEL—a polynomial-time reasoner for life science ontologies. In: Proc. of IJCAR 2006, Springer, Heidelberg (2006)
3. Baader, F., Lutz, C., Suntisrivaraporn, B.: Efficient reasoning in $\mathcal{EL}^+$. In: Prof. of DL (2006), J. of Logic, Language and Information (to appear)
4. Dowling, W.F., Gallier, J.: Linear-time algorithms for testing the satisfiability of propositional horn formulae. J. of Logic Programming 1(3), 267–284 (1984)
5. Cuenca Grau, B., Halaschek-Wiener, C., Kazakov, Y.: History matters: Incremental ontology reasoning using modules. In: Aberer, K., Choi, K.-S., Noy, N., Allemang, D., Lee, K.-I., Nixon, L., Golbeck, J., Mika, P., Maynard, D., Mizoguchi, R., Schreiber, G., Cudré-Mauroux, P. (eds.) ISWC 2007. LNCS, vol. 4825, Springer, Heidelberg (2007)
6. Cuenca Grau, B., Horrocks, I., Kazakov, Y., Sattler, U.: Just the right amount: Extracting modules from ontologies. In: Proc. of WWW 2007, ACM Press, New York (2007)
7. Haarslev, V., Möller, R.: Incremental query answering for implementing document retrieval services. In: Proc. of DL 2003 (2003)
8. Lutz, C., Wolter, F.: Conservative extensions in the lightweight description logic $\mathcal{EL}$. In: Pfenning, F. (ed.) CADE 2007. LNCS (LNAI), vol. 4603, Springer, Heidelberg (2007)
9. Noy, N., Musen, M.: The PROMPT suite: Interactive tools for ontology mapping and merging. International Journal of Human-Computer Studies (2003)
10. Parsia, B., Halaschek-Wiener, C., Sirin, E.: Towards incremental reasoning through updates in OWL-DL. In: Proc. of Reasoning on the Web Workshop (2006)
11. Seidenberg, J., Rector, A.: Web ontology segmentation: Analysis, classification and use. In: Proc. of WWW 2006, ACM Press, New York (2006)
12. Suntisrivaraporn, B.: Module extraction and incremental classification: A pragmatic approach for $\mathcal{EL}^+$ ontologies. LTCS-Report. TU Dresden, Germany (2007), see http://lat.inf.tu-dresden.de/research/reports.html
13. The systematized nomenclature of medicine, clinical terms (SNOMED CT). The International Health Terminology Standards Development Organisation (2007), http://www.ihtsdo.org/our-standards/

Forgetting Concepts in DL-Lite

Zhe Wang[1], Kewen Wang[1], Rodney Topor[1], and Jeff Z. Pan[2]

[1] Griffith University, Australia
[2] The University of Aberdeen, UK

Abstract. To support the reuse and combination of ontologies in Semantic Web applications, it is often necessary to obtain smaller ontologies from existing larger ontologies. In particular, applications may require the omission of many terms, e.g., concept names and role names, from an ontology. However, the task of omitting terms from an ontology is challenging because the omission of some terms may affect the relationships between the remaining terms in complex ways. We present the first solution to this problem by adapting the technique of forgetting, previously used in other domains. Specifically, we present a semantic definition of forgetting for description logics in general, which generalizes the standard definition for classical logic. We then introduce algorithms that implement forgetting in both DL-Lite TBoxes and ABoxes, and in DL-Lite knowledge bases. We prove that the algorithms are correct with respect to the semantic definition of forgetting, and that they run in polynomial time.

1 Introduction

Ontologies are required for Semantic Web applications as they provide a shared representation of the terms and the relationships between terms in particular application domains. Such Semantic Web applications will normally require access to only some of the terms, i.e., concept and role names, in available ontologies. However the tasks of restricting attention to relevant terms in, or omitting irrelevant terms from, a given ontology is challenging because the omission of some terms generally affects the relationships between the remaining terms. Accordingly, the ontological engineering tasks of combining and transforming large ontologies have received extensive attention recently.

Current technologies, however, provide only limited support for such operations. The web ontology language OWL allows users to *import* one ontology into another using the ⟨owl : imports⟩ statement. Most ontology editors allow the reuse of another ontology by including it in the model that is being designed. For example, Protégé allows user to include other projects. However, these approaches to ontology reuse have at least two limitations: (1) Some ontologies are very large and ontology engineers might need to import only a part of the available ontologies. For instance, the medical ontology UMLS [1] contains around 300,000 terms and only some of these will be relevant to any particular application. (2) The ability to deal with inconsistency and contradiction caused by merging ontologies is very limited in these approaches. Hence, researchers

---

[1] http://www.nlm.nih.gov/research/umls

have proposed alternative approaches to address these problems in reusing and merging ontologies. In the literature, the problem of reusing portions of large ontologies is referred to as *partial use of ontologies*.

Consider the following scenario. Suppose we need an ontology about *Cancer* but only have a large ontology *Medicine*, which describes many diseases and treatments. It would not be efficient to adopt and use the whole ontology *Medicine*, since we only need a small part of its contents. A better strategy is to discard those terms that are not required (such as *Odontia*) and use the resulting restriction of the ontology.

Another scenario occurs when ontology engineers are constructing a complex ontology. The engineers may wish to conveniently delete terms which are unnecessary or poorly defined. This task is relatively easy in traditional database systems, but in an ontology, where concepts are more closely related to each other, simply removing a concept or role name may destroy the consistency of the ontology and may cause problems with subsequent reasoning. More sophisticated methods for deleting (omitting/hiding) information in ontologies are required.

In this paper we address the issue of partial ontology reuse by employing the technique of *forgetting*, which has been previously been thoroughly studied in classical logic [13,12] and logic programming [6,14]. Informally, forgetting is a particular form of reasoning that allows a piece of information (say, p) in a knowledge base to be discarded or hidden in such a way that future reasoning on information irrelevant to p will not be affected. Forgetting has proved to be a very useful tool in many tasks such as query answering, planning, decision-making, reasoning about actions, knowledge update and revision in classical logic [13,12] and logic programming [6,14]. However, to the best of our knowledge, forgetting has not previously been applied to description logic ontology reuse.

In particular, we study new techniques for forgetting from knowledge bases in the DL-Lite family of description logics. This family was recently proposed by Calvanese *et al.* [2,3,4]. Logics in this family are particularly attractive because they are expressive enough for many purposes and have polynomial time reasoning algorithms in the worst case, in contrast to more common description logics which have exponential time reasoning algorithms. Indeed, logics in the DL-Lite family have been proved to be the maximal logics allowing efficient conjunctive query answering using standard database technology.

The main contributions of the paper include the following.

1. We provide a semantic definition of forgetting from DL-Lite TBoxes, which also applies to other description logics.
2. We prove that the result of forgetting from TBoxes in languages of the DL-Lite family can always be expressed in the same languages, and present algorithms to compute TBox forgetting for such languages.
3. We introduce a definition of forgetting for DL-Lite ABoxes, which preserves conjunctive query answering, and use this to define forgetting for arbitrary DL-Lite knowledge bases.
4. Finally, we prove our algorithms are correct and that they run in polynomial time.

The rest of the paper is organized as follows. Some basics of DL-Lite are briefly recalled in Section 2. We present the semantic definition of forgetting in arbitrary description logic

(DL) TBoxes in Section 3 and show the DL forgetting has the same desirable properties that classical forgetting has. In Section 4, we introduce our algorithms for computing the result of forgetting in DL-Lite TBoxes, and show the algorithms are correct with respect to the semantic definition. We note that the forgetting algorithms are simple, run in polynomial time, and do not make other reasoning processes more complex. The forgetting technique and the results are then extended to DL-Lite ABoxes and knowledge bases in Section 5, and detailed examples are presented to show how forgetting algorithms work. Finally, Section 6 concludes the paper and discusses future work.

2 Preliminaries

Description logics (DLs) are a family of concept-based knowledge representation formalisms, equipped with well-defined model-theoretic semantics [1]. The DL-Lite family is a family of lightweight ontology languages that can express most features in UML class diagrams but still have low reasoning overheads [2]. Besides standard reasoning tasks like subsumption between concepts and satisfiability of knowledge bases, the issue of answering complex queries is especially considered. The DL-Lite family consists of a core language DL-Lite$_{core}$ and some extensions, among which two main extensions are DL-Lite$_{\mathcal{F},\sqcap}$ and DL-Lite$_{\mathcal{R},\sqcap}$.

The DL-Lite$_{core}$ language has the following syntax:

$$\begin{aligned} B &\longrightarrow A \mid \exists R \\ C &\longrightarrow B \mid \neg B \\ R &\longrightarrow P \mid P^- \end{aligned}$$

where A is an atomic concept and P is an atomic role (with P^- as its inverse). B is called a *basic concept*, R a *basic role* and C is called a *general concept*.

A DL-Lite$_{core}$ TBox is a set of inclusion axioms of the form

$$B \sqsubseteq C$$

A DL-Lite ABox is a set of membership assertions on atomic concepts and atomic roles:

$$A(a), P(a,b)$$

where a and b are constants.

A DL-Lite knowledge base (KB) is a tuple KB $\mathcal{K} = \langle \mathcal{T}, \mathcal{A} \rangle$, where $\mathcal{T}$ is a DL-Lite TBox and $\mathcal{A}$ is a DL-Lite ABox.

The semantics of a DL is given by interpretations. An interpretation $\mathcal{I}$ is a pair $(\Delta^{\mathcal{I}}, \cdot^{\mathcal{I}})$, where $\Delta^{\mathcal{I}}$ is a non-empty set called the *domain* and $\cdot^{\mathcal{I}}$ is an interpretation function which associates each atomic concept A with a subset $A^{\mathcal{I}}$ of $\Delta^{\mathcal{I}}$ and each atomic role P with a binary relation $P^{\mathcal{I}} \subseteq \Delta^{\mathcal{I}} \times \Delta^{\mathcal{I}}$. For DL-Lite$_{core}$, the interpretation function $\cdot^{\mathcal{I}}$ can be extended to complex descriptions:

$$\begin{aligned} (P^-)^{\mathcal{I}} &= \{(a,b) \mid (b,a) \in P^{\mathcal{I}}\} \\ (\exists R)^{\mathcal{I}} &= \{a \mid \exists b.(a,b) \in R^{\mathcal{I}}\} \\ (\neg B)^{\mathcal{I}} &= \Delta^{\mathcal{I}} \setminus B^{\mathcal{I}} \end{aligned}$$

An interpretation $\mathcal{I}$ is a model of $B \sqsubseteq C$ iff $B^\mathcal{I} \subseteq C^\mathcal{I}$, *that is*, all instances of concept B are also instances of concept C.

An interpretation $\mathcal{I}$ is a model of $A(a)$ (resp., $P(a,b)$) if $a^\mathcal{I} \in A^\mathcal{I}$ (resp., $(a^\mathcal{I}, b^\mathcal{I}) \in P^\mathcal{I}$).

An interpretation $\mathcal{I}$ is a model of a KB $\langle \mathcal{T}, \mathcal{A} \rangle$, if $\mathcal{I}$ is a model of all axioms of $\mathcal{T}$ and assertions of $\mathcal{A}$. A KB $\mathcal{K}$ is consistent if it has at least one model. Two KBs that have the same models are said to be equivalent. A KB $\mathcal{K}$ logically implies an axiom (or assertion) α, denoted $\mathcal{K} \models \alpha$, if all models of $\mathcal{K}$ are also models of α.

Although DL-Lite is a simple language, it is useful because it is sufficiently expressive and because conjunctive query evaluation over DL-Lite KBs is extremely efficient. A conjunctive query (CQ) $q(x)$ over a KB $\mathcal{K}$ is an expression of the form

$$\{x \mid \exists y.conj(x,y)\}$$

where x, y are lists of variables, $conj(x,y)$ is a conjunction of atoms, and atoms have the form $A(s)$ or $P(s,t)$, where A is an atomic concept, P is an atomic role, and s and t are either individual names or variables. Given an interpretation $\mathcal{I}$, $q^\mathcal{I}$ is the set of tuples of domain elements such that, when assigned to x, $\exists y.conj(x,y)$ is true in $\mathcal{I}$. Given a CQ $q(x)$ and a KB $\mathcal{K}$, the answer to q over $\mathcal{K}$ is the set $\text{ans}(q, \mathcal{K})$ of tuples a of constants in $\mathcal{K}$ such that $a^\mathcal{I} \in q^\mathcal{I}$ for every model $\mathcal{I}$ of $\mathcal{K}$.

DL-Lite$_{\mathcal{F}, \sqcap}$ extends DL-Lite$_{core}$ by allowing for conjunction of concepts in the left-hand side of inclusions and role functionality axioms in TBoxes. The extended syntax is:

$$B \longrightarrow A \mid \exists R$$
$$D \longrightarrow B \mid D_1 \sqcap D_2$$
$$C \longrightarrow B \mid \neg B$$
$$R \longrightarrow P \mid P^-$$

and TBoxes contain axioms of the form:

$$D \sqsubseteq C \quad \text{and} \quad (\mathit{funct}\ R).$$

Given an interpretation $\mathcal{I}$, we define $(D_1 \sqcap D_2)^\mathcal{I} = D_1^\mathcal{I} \cap D_2^\mathcal{I}$. Then $\mathcal{I}$ is a model of $(\mathit{funct}\ R)$ iff $(a, b_1) \in R^\mathcal{I}$ and $(a, b_2) \in R^\mathcal{I}$ implies $b_1 = b_2$.

DL-Lite$_{\mathcal{R}, \sqcap}$ extends DL-Lite$_{core}$ by allowing for conjunctions of concepts in the left-hand side of inclusions axioms, role complements, and role inclusion axioms in TBoxes. In this case the extended syntax is:

$$B \longrightarrow A \mid \exists R$$
$$D \longrightarrow B \mid D_1 \sqcap D_2$$
$$C \longrightarrow B \mid \neg B$$
$$R \longrightarrow P \mid P^-$$
$$S \longrightarrow R \mid \neg R$$

and TBoxes contain axioms of the form:

$$D \sqsubseteq C \quad \text{and} \quad R \sqsubseteq S.$$

Given an interpretation $\mathcal{I}$, we define $(\neg R)^\mathcal{I} = \Delta^\mathcal{I} \times \Delta^\mathcal{I} \setminus R^\mathcal{I}$. Then $\mathcal{I}$ is a model of $R \sqsubseteq S$ iff $R^\mathcal{I} \subseteq S^\mathcal{I}$.

We note that the data complexity of DL-Lite$_{core}$ is log-space whereas the data complexity of the two extensions (as defined here) is polynomial time.

Example 2.1. The following is a simple DL-Lite$_{core}$ knowledge base "Library", which describes the resources, users and lending policies of a library. This knowledge base has a TBox $\mathcal{T}$ consisting of the following axioms:

$\exists onLoanTo \sqsubseteq LibItem, \exists onLoanTo^- \sqsubseteq Member,$
$Member \sqsubseteq Person, Visitor \sqsubseteq Person, Visitor \sqsubseteq \neg Member,$
$LibItem \sqsubseteq \exists hasCatNum, \exists hasCatNum^- \sqsubseteq CatNum,$
$\exists hasCatNum \sqcap Missing \sqsubseteq \neg LibItem$

and an ABox $\mathcal{A}$ consisting of the following assertions:

$LibItem(SWPrimer), onLoanTo(DLHandBook, Jack).$

Here, $LibItem$ denotes library items, $onLoanTo$ denotes the loan relationship between library items and people, $CatNum$ denotes catalogue numbers, and $hasCatNum$ denotes the relationship between library items and their catalogue numbers. Note that not every person is a member or visitor.

If we regarded this example as a knowledge base of DL-Lite$_{\mathcal{F},\sqcap}$, we could add the TBox axioms $(funct\ onLoanTo)$ and $(funct\ hasCatNum)$.

3 Forgetting Concepts from DL-Lite TBoxes

In this section, we define the operation of forgetting about a concept A in a TBox $\mathcal{T}$. Informally, the TBox that results from forgetting about A in $\mathcal{T}$ should: (1) not contain any occurrence of A, (2) be weaker than $\mathcal{T}$, and (3) give the same answer to any query that is irrelevant to A. We will first give a semantic definition of forgetting in DL-Lite, investigate its properties and in the next section introduce algorithms for computing the result of forgetting in different languages of the family DL-Lite.

Let $\mathcal{L}$ be a DL-Lite language. The signature $\mathcal{S}_\mathcal{L}$ of $\mathcal{L}$ is the set of concept and role names in $\mathcal{L}$. We will omit the subscript if there is no confusion caused. Our semantic definition of forgetting in DL-Lite is an adaption of the corresponding definition for classical logic.

Let A be an atomic concept name in $\mathcal{L}$ and $\mathcal{I}_1$ and $\mathcal{I}_2$ interpretations of $\mathcal{L}$. We define $\mathcal{I}_1 \sim_A \mathcal{I}_2$ iff $\mathcal{I}_1$ and $\mathcal{I}_2$ agree on all atomic and concept role names except possibly A:

1. $\mathcal{I}_1$ and $\mathcal{I}_2$ have the same domain ($\Delta^{\mathcal{I}_1} = \Delta^{\mathcal{I}_2}$), and interpret every individual name the same ($a^{\mathcal{I}_1} = a^{\mathcal{I}_2}$ for every individual name a).
2. For every concept name A_1 distinct from A, $A_1^{\mathcal{I}_1} = A_1^{\mathcal{I}_2}$.
3. For every role name P, $P^{\mathcal{I}_1} = P^{\mathcal{I}_2}$.

Clearly, $\sim_A$ is an equivalence relation, and we say $\mathcal{I}_1$ is A-equivalent to $\mathcal{I}_2$.

Definition 3.1. *Let $\mathcal{T}$ be a TBox in $\mathcal{L}$ and A an atomic concept in $\mathcal{T}$. A TBox $\mathcal{T}'$ on the signature $\mathcal{S} \setminus \{A\}$ is a* result of forgetting *about A in $\mathcal{T}$ if any interpretation $\mathcal{I}'$ is a model of $\mathcal{T}'$ if and only if there is a model $\mathcal{I}$ of $\mathcal{T}$ such that $\mathcal{I} \sim_A \mathcal{I}'$.*

It follows from the above definition that the result of forgetting about an atomic concept A in a TBox $\mathcal{T}$ is unique in the sense that, if both $\mathcal{T}'$ and $\mathcal{T}''$ are results of forgetting about A in $\mathcal{T}$, then they are equivalent. So we will use forget($\mathcal{T}, A$) to denote the result of forgetting about A in $\mathcal{T}$ throughout the paper.

Obviously, forget($\mathcal{T}, A$) does not contain any occurrence of A and is weaker than $\mathcal{T}$. However, the definition of forgetting guarantees that forget($\mathcal{T}, A$) and $\mathcal{T}$ are equivalent under query answering on $\mathcal{S} \setminus \{A\}$.

Note that the above definition of forgetting can be applied to other description logics.

Example 3.1. Consider the TBox $\mathcal{T}$ in Example 2.1. Suppose the library now wishes to allow nonmembers to borrow library items but still wishes to prevent visitors from borrowing library items, *i.e*, suppose the library wishes to forget about atomic concept *Member* in $\mathcal{T}$. From the definition, forget($\mathcal{T}, Member$) now consists of the following axioms:

$\exists onLoanTo \sqsubseteq LibItem, \exists onLoanTo^- \sqsubseteq Person,$
$Visitor \sqsubseteq Person, \exists onLoanTo^- \sqsubseteq \neg Visitor,$
$LibItem \sqsubseteq \exists hasCatNum, \exists hasCatNum^- \sqsubseteq CatNum,$
$\exists hasCatNum \sqcap Missing \sqsubseteq \neg LibItem.$

We believe this definition correctly captures the informal operation of forgetting a concept from a TBox.

Definition 3.1 clearly captures our informal understanding of forgetting. However, we have not yet shown that the result of forgetting about a concept always exists or how to compute the result of forgetting. In the next section, we introduce algorithms for computing the result of forgetting in different DL-Lite languages. From the soundness and completeness of these algorithms, we can immediately conclude that the result of forgetting about concepts exists for every TBox in DL-Lite.

Theorem 3.1. *Let $\mathcal{T}$ be a TBox in a DL-Lite language $\mathcal{L}$ and A an atomic concept. Then the result of forgetting about A in $\mathcal{T}$ always exists and is in $\mathcal{L}$.*

Forgetting in DL-Lite has other important properties. In particular, it preserves reasoning relative to TBoxes.

Proposition 3.1. *Let $\mathcal{T}$ be a TBox in $\mathcal{L}$ and A an atomic concept. Let $\mathcal{T}' = $ forget($\mathcal{T}, A$). Then, for any ABox $\mathcal{A}$ on $\mathcal{S} \setminus \{A\}$, we have:*

- *The knowledge base $\langle \mathcal{T}, \mathcal{A} \rangle$ is consistent iff $\langle \mathcal{T}', \mathcal{A} \rangle$ is consistent.*
- *For any inclusion axiom α not containing A, $\mathcal{T} \models \alpha$ iff $\mathcal{T}' \models \alpha$.*
- *For any membership assertion β not containing A, $\langle \mathcal{T}, \mathcal{A} \rangle \models \beta$ iff $\langle \mathcal{T}', \mathcal{A} \rangle \models \beta$.*
- *For any conjunctive query q not containing A, ans($q, \langle \mathcal{T}, \mathcal{A} \rangle$) = ans($q, \langle \mathcal{T}', \mathcal{A} \rangle$).*

It is straightforward to generalize Definition 3.1 to the operation of simultaneously forgetting about a *set* of concept names.

We can forget about a set of concept names by forgetting one by one since, if A_1 and A_2 are concept names, it is easy to show that

$$\text{forget}(\text{forget}(\mathcal{T}, A_1), A_2) \equiv \text{forget}(\text{forget}(\mathcal{T}, A_2), A_1).$$

This property allows us to define the result of forgetting a *set* $S = \{A_1, \ldots, A_n\}$ of concept names by

$$\mathsf{forget}(\mathcal{T}, S) \equiv \mathsf{forget}(\ldots(\mathsf{forget}(\mathcal{T}, A_1), \ldots), A_n).$$

4 Computing the Result of Forgetting in DL-Lite

In Example 3.1, the result of forgetting about $Member$ in the TBox $\mathcal{T}$ was obtained from $\mathcal{T}$ by simple syntax transformations. In this section, we introduce algorithms for computing the result of forgetting in different languages of DL-Lite. We prove that our algorithms are sound and complete with respect to the semantic definition of forgetting in the previous section. Our algorithms show that the result of forgetting in a DL-Lite TBox can always be obtained using simple syntax-based transformations.

A language of DL-Lite$_{core}$ (resp. DL-Lite$_{\mathcal{F},\sqcap}$, DL-Lite$_{\mathcal{R},\sqcap}$) is denoted $\mathcal{L}_{core}$ (resp. $\mathcal{L}_{\mathcal{F},\sqcap}$, $\mathcal{L}_{\mathcal{R},\sqcap}$). We first introduce the forgetting algorithm for DL-Lite$_{core}$ as Algorithm 1, then extend it to algorithms for DL-Lite$_{\mathcal{F},\sqcap}$ and DL-Lite$_{\mathcal{R},\sqcap}$. The basic idea of Algorithm 1 is to first transform the given TBox into a standard form and then remove all occurrence of A.

Algorithm 1 (Computing the result of forgetting in DL-Lite$_{core}$)
Input: A TBox $\mathcal{T}$ in $\mathcal{L}_{core}$ and an atomic concept A.
Output: $\mathsf{forget}(\mathcal{T}, A)$
Method:
Step 1. Remove axiom $A \sqsubseteq A$ from $\mathcal{T}$ if it is present.
Step 2. If axiom $A \sqsubseteq \neg A$ is in $\mathcal{T}$, remove each axiom $A \sqsubseteq C$ or $B \sqsubseteq \neg A$ from $\mathcal{T}$, and replace each axiom $B \sqsubseteq A$ in $\mathcal{T}$ by $B \sqsubseteq \neg B$.
Step 3. Replace each axiom $B \sqsubseteq \neg A$ in $\mathcal{T}$ by $A \sqsubseteq \neg B$.
Step 4. For each axiom $B_i \sqsubseteq A$ ($1 \leq i \leq m$) in $\mathcal{T}$ and each axiom $A \sqsubseteq C_j$ ($1 \leq j \leq n$) in $\mathcal{T}$, where each B_i is a basic concept and each C_j is a general concept, if $B_i \sqsubseteq C_j$ is not in $\mathcal{T}$ already, add $B_i \sqsubseteq C_j$ to $\mathcal{T}$.
Step 5. Return the result of removing every axiom containing A in $\mathcal{T}$.

Fig. 1. Forgetting in a DL-Lite$_{core}$ TBox

Example 4.1. Consider the TBox $\mathcal{T}$ in Example 2.1 again. Algorithm 1 replaces the axioms

$$\exists onLoanTo^- \sqsubseteq Member, \ Member \sqsubseteq Person \ \text{and} \ Visitor \sqsubseteq \neg Member$$

by the axioms

$$\exists onLoanTo^- \sqsubseteq Person \ \text{and} \ \exists onLoanTo^- \sqsubseteq \neg Visitor,$$

which gives the same result as Example 3.1.

If the library wants to completely eliminate lending restrictions, then the result of forgetting about the concept $Visitor$ can be obtained by removing the following (lending restriction) axioms:

$$Visitor \sqsubseteq Person \ \text{and} \ \exists onLoanTo^- \sqsubseteq \neg Visitor.$$

We can now show that Algorithm 1 is sound and complete with respect to the semantic definition of forgetting in DL-Lite$_{core}$ TBoxes.

Theorem 4.1. *Let T be a TBox in $\mathcal{L}_{core}$ and A an atomic concept appearing in T. Then Algorithm 1 always returns* forget(T, A).

Given the forgetting algorithm for a DL-Lite$_{core}$ TBox, we can extend it to compute the results of forgetting in DL-Lite$_{\mathcal{F},\sqcap}$ and DL-Lite$_{\mathcal{R},\sqcap}$ TBoxes.

Recall that both DL-Lite$_{\mathcal{F},\sqcap}$ and DL-Lite$_{\mathcal{R},\sqcap}$ extend DL-Lite$_{core}$ by allowing conjunctions of basic concepts in the left-hand side of inclusion axioms. Therefore, to extend Algorithm 1 to these two extensions, we need a method of handling such conjunctions.

For inclusion axioms in which the concept A occurs negatively, *i.e.*, axioms of the form of $D \sqsubseteq \neg A$, where D is a conjunction, we cannot transform it into an equivalent axiom $A \sqsubseteq \neg D$, as we did previously, since this is not an axiom in the DL-Lite language. However, the following two lemmas give useful properties of conjunctions that can be applied in the extended algorithm.

Lemma 4.1. *Let B, B' be basic concepts and D a conjunction of basic concepts, then the two axioms $B \sqcap D \sqsubseteq \neg B'$ and $B' \sqcap D \sqsubseteq \neg B$ are equivalent.*

Lemma 4.2. *Suppose T is a TBox in $\mathcal{L}$ and $T = \{ D_i \sqsubseteq A \mid i = 1, \ldots, m \} \cup \{ A \sqcap D'_j \sqsubseteq C_j \mid j = 1, \ldots, n \} \cup T_A$, where $m, n \geq 0$, A is an atomic concept, each D_i is a nonempty conjunction, each D'_j is a (possibly empty) conjunction, each D_i, D'_j and C_j does not contain A, and T_A is a set of axioms that do not contain A. Then $T' = \{D_i \sqcap D'_j \sqsubseteq C_j \mid i = 1, \ldots, m, j = 1, \ldots, n\} \cup T_A = $ forget(T, A). (If m or n is 0, then T' is the empty set).*

Because we are currently concerned only with forgetting concepts, the extensions to functional role axioms and role inclusion axioms are not relevant to our algorithm. Hence, the same algorithm implements concept forgetting for both DL-Lite$_{\mathcal{F},\sqcap}$ and DL-Lite$_{\mathcal{R},\sqcap}$ TBoxes. This algorithm is now given as Algorithm 2. Throughout Algorithm 2, D denotes a possibly empty conjunction and D' denotes a nonempty conjunction.

From the definition of Algorithm 2 it is easy to see that the following result holds.

Proposition 4.1. *Let T be a TBox T in $\mathcal{L}_{\mathcal{F},\sqcap}$ or $\mathcal{L}_{\mathcal{R},\sqcap}$ and A an atomic concept name appearing in T. Then Algorithm 2 always terminates and takes polynomial time in the size of T.*

It is an immediate corollary that Algorithm 1 has the same properties.

From Lemma 4.1 and Lemma 4.2, it can easily be shown that Algorithm 2 is sound and complete with respect to the semantic definition of forgetting for DL-Lite. In other words, we have following theorem.

Theorem 4.2. *Let T be a TBox T in $\mathcal{L}_{\mathcal{F},\sqcap}$ or $\mathcal{L}_{\mathcal{R},\sqcap}$ and A an atomic concept name appearing in T. Then Algorithm 2 always returns* forget(T, A).

Example 4.2. Recall the TBox T in Example 2.1. If we want to forget about concept name $LibItem$ in T, then Algorithm 2 returns the following axioms, which comprise forget$(T, LibItem)$:

$\exists onLoanTo \sqsubseteq \exists hasCatNum, \exists onLoanTo^- \sqsubseteq Member,$

Algorithm 2 (Computing the result of forgetting in DL-Lite$_{core}$ extensions)
Input: A TBox $\mathcal{T}$ in $\mathcal{L}_{\mathcal{F},\sqcap}$ or $\mathcal{L}_{\mathcal{R},\sqcap}$ and an atomic concept name A.
Output: forget$(\mathcal{T}, A)$
Method:
Step 1. Remove each axiom $A \sqcap D \sqsubseteq A$ in $\mathcal{T}$.
Step 2. If axiom $A \sqsubseteq \neg A$ is in $\mathcal{T}$, remove each axiom $A \sqcap D \sqsubseteq C$ or $D' \sqsubseteq \neg A$, and replace each axiom $B \sqcap D' \sqsubseteq A$ by $D' \sqsubseteq \neg B$ and $B \sqsubseteq A$ by $B \sqsubseteq \neg B$.
Step 3. Replace each axiom $A \sqcap D \sqsubseteq \neg A$ in $\mathcal{T}$ by $D \sqsubseteq \neg A$.
Step 4. Replace each axiom $B \sqcap D \sqsubseteq \neg A$ in $\mathcal{T}$ by $A \sqcap D \sqsubseteq \neg B$.
Step 5. For each axiom $D'_i \sqsubseteq A$ ($1 \leq i \leq m$) in $\mathcal{T}$ and each axiom $A \sqcap D_j \sqsubseteq C_j$ ($1 \leq j \leq n$) in $\mathcal{T}$ if $D'_i \sqcap D_j \sqsubseteq C_j$ is not in $\mathcal{T}$, simplify $D'_i \sqcap D_j$ and add $D'_i \sqcap D_j \sqsubseteq C_j$ to $\mathcal{T}$.
Step 6. Return the result of removing every axiom containing A in $\mathcal{T}$.

Fig. 2. Forgetting in DL-Lite$_{core}$ extension TBoxes

$Member \sqsubseteq Person$, $Visitor \sqsubseteq Person$, $Visitor \sqsubseteq \neg Member$,
$\exists hasCatNum^- \sqsubseteq CatNum$,
$\exists hasCatNum \sqcap \exists onLoanTo \sqsubseteq \neg Missing$.

5 Forgetting Concepts from DL-Lite Knowledge Bases

Given the semantic definition of forgetting in TBoxes, we want to extend the notion of forgetting to ABoxes and DL-Lite Knowledge Bases (KB). However, DL-Lite languages are not closed under ABox forgetting, i.e., the classical forgetting result in an DL-Lite ABox is not expressible in the same language. A simple example would be, a DL-Lite KB with one axiom $A \sqsubseteq \neg B$ in TBox and one assertion $A(a)$ in ABox. To forget about concept name A in ABox, we have to remove the assertion $A(a)$. However, the original KB has a logical consequence $\neg B(a)$, that is, a cannot be interpreted as a member of concept B, which can not be expressed by standard DL-Lite ABoxes. Once concept A is forgotten, we also lose the information about a and B.

Adopting classical forgetting into DL-Lite ABoxes might lead outside of the original language. However, we argue that, for ABoxes, the classical forgetting is not necessary. Since ABoxes are used for maintain membership information and for query-answering, an ABox forgetting operation, as long as it preserves membership relation (rather than non-membership relation) and query-answering, should be good enough for ontology uses.

Given a Knowledge Base (KB) $\mathcal{K} = \langle \mathcal{T}, \mathcal{A} \rangle$ in DL-Lite language $\mathcal{L}$, the result of forgetting about a concept name A in $\mathcal{K}$ should be a KB $\mathcal{K}'$ in $\mathcal{L}$ such that (1) $\mathcal{K}'$ does not contain any occurrence of A and (2) $\mathcal{K}$ and $\mathcal{K}'$ are equivalent w.r.t. any conjunctive query on the signature $\mathcal{L} \setminus \{A\}$.

Definition 5.1. *Let $\mathcal{K} = \langle \mathcal{T}, \mathcal{A} \rangle$ be a knowledge base in a DL Lite language $\mathcal{L}$ and A an atomic concept in $\mathcal{A}$. An ABox $\mathcal{A}'$ on the signature $\mathcal{S} \setminus \{A\}$ is a result of forgetting about A in $\mathcal{A}$ with respect to $\mathcal{K}$ if, for any conjunctive query q on the signature of $\mathcal{S} \setminus \{A\}$, $\mathsf{ans}(q, \langle \mathcal{T}, \mathcal{A} \rangle) = \mathsf{ans}(q, \langle \mathcal{T}, \mathcal{A}' \rangle)$ holds.*

Example 5.1. Consider the following ABox from Example 2.1:

$$\mathcal{A} = \{LibItem(SWPrimer), onLoanTo(DLHandBook, Jack)\}$$

To forget about atomic concept *LibItem*, we can't just remove the membership assertion *LibItem(SWPrimer)*, because we would lose information about *SWPrimer*. According to the TBox axiom, we know that *SWPrimer* must have a catalogue number, so we can replace *LibItem(SWPrimer)* with *hasCatNum(SWPrimer, z)* where z denotes the catalogue number of *SWPrimer*. This is the result of forgetting about *LibItem* in $\mathcal{A}$.

From the above example we can see that a major issue in forgetting about an atomic concept A in an ABox $\mathcal{A}$ is how to preserve subsumption relations between different concepts. For example, if $A \sqsubseteq B$ is in TBox $\mathcal{T}$ and $A(a)$ is in $\mathcal{A}$, then $B(a)$ is derivable from $\mathcal{K} = \langle \mathcal{T}, \mathcal{A} \rangle$. However, if we simply delete $A(a)$ from $\mathcal{A}$, we will be unable to derive $B(a)$ from the resulting knowledge base. Hence, we have to propagate selected information before a concept name can be forgotten. For convenience, we slightly extend our definition of ABoxto allow variables in assertions. An assertion with variables represents a scheme (in particular, we use the symbol '_' to represent non-distinguished variables) in ABox [8].

To forget about $A(a)$ from DL-Lite$_{core}$ ABox $\mathcal{A}$, an intuitive way is to remove $A(a)$ and add all $B(a)$ to $\mathcal{A}$ such that B subsumes A. However, as we will show later, it is sufficient to add each $\Phi(a)$, where Φ is explicitly asserted to subsume A.

New membership assertions generated from $A(a)$ form a set $f(A, a) = \{ \Phi(a) \mid A \sqsubseteq \Phi \text{ in } \mathcal{T} \}$, where Φ is a concept description in DL-Lite, and

$$\Phi(a) = \begin{cases} B(a) & \text{if } \Phi = B \text{ is an atomic concept,} \\ P(\_, a) & \text{if } \Phi = \exists P^-, \\ P(a, \_) & \text{if } \Phi = \exists P. \end{cases}$$

The result of forgetting about a concept name A in $\mathcal{A}$, denoted forget$_\mathcal{K}(\mathcal{A}, A)$, is a new ABox obtained by replacing every assertion $A(a) \in \mathcal{A}$ with the corresponding set $f(A, a)$ of assertions. When the DL-Lite KB $\mathcal{K}$ is clear from the context, we can omit it from forget$_\mathcal{K}(\mathcal{A}, A)$ and write forget$(\mathcal{A}, A)$ instead.

In a DL-Lite$_{\mathcal{F},\sqcap}$ or DL-Lite$_{\mathcal{R},\sqcap}$ ABox, if we want to forget about an atomic concept A, then axioms of the form $A \sqcap D \sqsubseteq \Phi$ must also be considered, *i.e.*, if $A(a)$ needs to be removed and $\mathcal{K} \models D(a)$, then $\Phi(a)$ must be added into the new knowledge base.

Let $\mathcal{A}$ be an ABox in $\mathcal{L}_{\mathcal{F},\sqcap}$ or $\mathcal{L}_{\mathcal{R},\sqcap}$ and A an atomic concept name. Then, as in the case of DL-Lite$_{core}$, we can define $\Phi(a)$ and thus $f(A, a) = \{ \Phi(a) \mid A \sqcap D \sqsubseteq \Phi \text{ in } \mathcal{T} \text{ and } \mathcal{K} \models D(a) \}$, where D is a concept conjunction ($\mathcal{K} \models D(a)$ is true when D is empty).

An important observation is that ABox forgetting does not coincides with classical forgetting, *i.e.*, for some model $\mathcal{I}'$ of the result of forgetting $\mathcal{A}'$, there may not exist a model $\mathcal{I}$ of $\mathcal{A}$ such that $\mathcal{I} \sim_A \mathcal{I}'$.

For simplicity, we will only consider consistent knowledge bases in the following theorem.

Theorem 5.1. *Let A be an atomic concept name in a consistent knowledge base $\mathcal{K} = \langle \mathcal{T}, \mathcal{A} \rangle$ in DL-Lite language $\mathcal{L}$. For any conjunctive query q on the signature $\mathcal{S}_\mathcal{L} \setminus \{A\}$, we have*

$$\mathsf{ans}(q, \mathcal{K}) = \mathsf{ans}(q, \langle \mathcal{T}, \mathsf{forget}_\mathcal{K}(\mathcal{A}, A) \rangle).$$

Having defined forgetting for both ABoxes and TBoxes, we are now able to define forgetting in a DL-Lite knowledge base.

Definition 5.2. *Let $\mathcal{K} = \langle \mathcal{T}, \mathcal{A} \rangle$ be a DL-Lite knowledge base and A a concept in DL-Lite language $\mathcal{L}$. The result of forgetting about A in $\mathcal{K}$ is defined to be*

$$\mathsf{forget}(\mathcal{K}, A) = \langle \mathsf{forget}(\mathcal{T}, A), \mathsf{forget}_\mathcal{K}(\mathcal{A}, A) \rangle,$$

where $\mathsf{forget}_\mathcal{K}(\mathcal{A}, A) = \mathcal{A} \setminus \{A(a) \mid a \text{ is a constant in } \mathcal{L}\} \cup \{\Phi(a) \mid A \sqcap D \sqsubseteq \Phi \text{ in } \mathcal{T} \text{ and } \mathcal{K} \models D(a)\}$.

It is important to note that the forgetting operation in ABoxes is with respect to the original knowledge base (especially, the original TBox). However, forgetting in the TBox does not depend on the ABox. The computation process of forgetting about a concept in a DL-Lite knowledge base is shown in Algorithm 3. In this algorithm, D denotes a possibly empty conjunction.

Algorithm 3 (Computing the result of forgetting in an DL-Lite knowledge base)
Input: A knowledge base $\mathcal{K} = \langle \mathcal{T}, \mathcal{A} \rangle$ in $\mathcal{L}$ and an atomic concept A.
Output: $\mathsf{forget}(\mathcal{K}, A)$.
Method:
Step 1. For each assertion in $\mathcal{A}$ of the form $A(a)$,
- for every axiom $A \sqcap D \sqsubseteq B$ in $\mathcal{T}$, if $\mathcal{K} \models D(a)$ (true when D is empty), add $B(a)$ to $\mathcal{A}$;
- for every axiom $A \sqcap D \sqsubseteq \exists P$ in $\mathcal{T}$, if $\mathcal{K} \models D(a)$, add $P(a, \_)$ to $\mathcal{A}$;
- for every axiom $A \sqcap D \sqsubseteq \exists P^-$ in $\mathcal{T}$, if $\mathcal{K} \models D(a)$, add $P(\_, a)$ to $\mathcal{A}$;
Step 2. Remove all assertions containing A in $\mathcal{A}$, giving $\mathcal{A}'$. (As shown before, $\mathcal{A}' = \mathsf{forget}_\mathcal{K}(\mathcal{A}, A)$)
Step 3. Let $\mathcal{T}'$ be the result of forgetting about A in $\mathcal{T}$, ignoring $\mathcal{A}'$.
Step 4. Return the knowledge base $\mathcal{K}' = \langle \mathcal{T}', \mathcal{A}' \rangle$.

Fig. 3. Forgetting in a DL-Lite knowledge base

From Proposition 3.1 and Theorem 5.1, it is not hard to see following theorem holds.

Theorem 5.2. *Let $\mathcal{K} = \langle \mathcal{T}, \mathcal{A} \rangle$ be a consistent knowledge base in DL-Lite language $\mathcal{L}$ and A an atomic concept in $\mathcal{K}$. For any conjunctive query q on $\mathcal{S}_\mathcal{L} \setminus \{A\}$, we have*

$$\mathsf{ans}(q, \mathcal{K}) = \mathsf{ans}(q, \mathsf{forget}(\mathcal{K}, A))$$

6 Related Work

An alternative approach to partial reuse of ontologies is based on the notion of *conservative extensions* [7], a notion that has been well investigated in classical logic. An ontology $\mathcal{T}$ is a conservative extension of its subontology $\mathcal{T}' \subseteq \mathcal{T}$ w.r.t a signature $\mathcal{S}$ if and only if $\mathcal{T}$ and $\mathcal{T}'$ are equivalent under query answering on $\mathcal{S}$. The theory of conservative extensions can be applied in ontology refinement and ontology merging [7]. Since the goal of partial use of ontology is to obtain a smaller ontology from a larger ontology, a dual theory of conservative extensions, called *modularity of ontology* is also explored in [10,11]. Unfortunately, given an ontology $\mathcal{T}$, there does not exist a conservative extension in many cases. Moreover, it is exponential to decide if there is a conservative extension/module for a given ontology in common description logics such as $\mathcal{ALC}$. A more recent approach is introduced in [9]. Unlike modularity approach, this approach focuses on ABoxes update. A shortcoming of this approach is that DL-Lite languages are not closed under their ABox update operation. Eiter *et al* [5] have also attempted to define forgetting for OWL ontologies by transforming an OWL ontology into a logic program. However, this approach only works for some OWL ontologies.

7 Conclusion

We have presented the first complete account of forgetting from DL-Lite description logics. This process is a key operation in reusing and combining ontologies, which are based on such description logics. We have presented a semantic definition of forgetting from DL-Lite TBoxes and ABoxes and correct, efficient implementations of the semantics.

Natural next steps are to extend these approaches to other description logic families, to investigate forgetting roles as well as concepts, to find more efficient algorithms for forgetting a set of concepts (or roles) simultaneously, and to implement and apply the methods to practical problems.

Acknowledgments

The authors would like to thank three anonymous referees for their helpful comments. This work was partially supported by the Australia Research Council (ARC) Discovery Projects DP0666107.

References

1. Baader, F., Calvanese, D., McGuinness, D., Nardi, D., Patel-Schneider, P.: The Description Logic Handbook. Cambridge University Press, Cambridge (2002)
2. Calvanese, D., De Giacomo, G., Lembo, D., Lenzerini, M., Rosati, R.: DL-Lite: Tractable description logics for ontologies. In: Proceedings of the 20th National Conference on Artificial Intelligence (AAAI 2005), pp. 602–607 (2005)

3. Calvanese, D., De Giacomo, G., Lembo, D., Lenzerini, M., Rosati, R.: Data complexity of query answering in description logics. In: Proceedings of the 10th International Conference on Principles of Knowledge Representation and Reasoning (KR 2006), pp. 260–270 (2006)
4. Calvanese, D., De Giacomo, G., Lembo, D., Lenzerini, M., Rosati, R.: Tractable reasoning and efficient query answering in description logics: The DL-Lite family. J. Autom. Reasoning 39(3), 385–429 (2007)
5. Eiter, T., Ianni, G., Schindlauer, R., Tompits, H., Wang, K.: Forgetting in managing rules and ontologies. In: Proceedings of the IEEE/WIC/ACM International Conference on Web Intelligence (WI 2006), pp. 411–419 (2006)
6. Eiter, T., Wang, K.: Forgetting and conflict resolving in disjunctive logic programming. In: Proceedings of the 21st National Conference on Artificial Intelligence (AAAI 2006), pp. 238–243 (2006)
7. Ghilardi, S., Lutz, C., Wolter, F.: Did i damage my ontology? a case for conservative extensions in description logics. In: Proceedings, Tenth International Conference on Principles of Knowledge Representation and Reasoning, pp. 187–197 (2006)
8. De Giacomo, G., Lenzerini, M., Poggi, A., Rosati, R.: On the update of description logic ontologies at the instance level. In: Proceedings of the 21th National Conference on Artificial Intelligence (AAAI 2006) (2006)
9. De Giacomo, G., Lenzerini, M., Poggi, A., Rosati, R.: On the approximation of instance level update and erasure in description logics. In: Proceedings of the 22th National Conference on Artificial Intelligence (AAAI 2007), pp. 403–408 (2007)
10. Grau, B., Kazakov, Y., Horrocks, I., Sattler, U.: A logical framework for modular integration of ontologies. In: Proceedings of the 20th International Joint Conference on Artificial Intelligence (IJCAI 2007), pp. 298–303 (2007)
11. Cuenca Grau, B., Kazakov, Y., Horrocks, I., Sattler, U.: Just the right amount: Extracting modules from ontologies. In: Proceedings of the 16th International World Wide Web Conference (WWW 2007), pp. 717–726 (2007)
12. Lang, J., Liberatore, P., Marquis, P.: Propositional independence: Formula-variable independence and forgetting. J. Artif. Intell. Res (JAIR) 18, 391–443 (2003)
13. Lin, F., Reiter, R.: Forget it. In: Proceedings of the AAAI Fall Symposium on Relevance, New Orleans (LA), pp. 154–159 (1994)
14. Wang, K., Sattar, A., Su, K.: A theory of forgetting in logic programming. In: Proceedings of the 20th National Conference on Artificial Intelligence, pp. 682–687. AAAI Press, Menlo Park (2005)

An Entity Name System (ENS) for the Semantic Web*

Paolo Bouquet[1], Heiko Stoermer[1], and Barbara Bazzanella[2]

[1] Dipartimento di Ingegneria e Scienza dell'Informazione – University of Trento
Via Sommarive, 14 – 38050 Trento, Italy
[2] Dipartimento di Scienze della Cognizione e della Formazione – University of Trento
Via Matteo del Ben, 5 38068 Rovereto (TN), Italy
bouquet@disi.unitn.it, stoermer@disi.unitn.it,
b.bazzanella@email.unitn.it

Abstract. In this paper, we argue that implementing the grand vision of the Semantic Web would greatly benefit from a service which can enable the reuse of globally unique URIs across semantic datasets produced in a fully decentralized and open environment. Such a service, which we call *Entity Name System* (ENS), stores pre–existing URIs and makes them available for reuse mainly – but not only – in Semantic Web contents and applications. The ENS will make the integration of semantic datasets much easier and faster, and will foster the development of a whole family of applications which will exploit the data level integration through global URIs for implementing smart semantic-based solutions.

1 Introduction

In a note from 1998, Tim Berners-Lee describes the grand vision of the Semantic Web as follows:

> Knowledge representation is a field which currently seems to have the reputation of being initially interesting, but which did not seem to shake the world to the extent that some of its proponents hoped. It made sense but was of limited use on a small scale, but never made it to the large scale. This is exactly the state which the hypertext field was in before the Web [...]. The Semantic Web is what we will get if we perform the same globalization process to Knowledge Representation that the Web initially did to Hypertext [http://www.w3.org/DesignIssues/RDFnot.html].

We understand this parallel as follows. Like the WWW provided a global space for the seamless integration of small hypertexts (or local "webs of documents")

---

* This work is partially supported by the by the FP7 EU Large-scale Integrating Project **OKKAM – Enabling a Web of Entities** (contract no. ICT-215032). For more details, visit http://fp7.okkam.org. The authors are also very grateful to Claudia Niederee and Rodolfo Stecher for their support in distilling the ENS core concepts.

into a global, open, decentralized and scalable *web of documents*, so the Semantic Web should provide a global space for the seamless integration of semantic repositories (or "local semantic webs") into a global, open, decentralized and scalable *web of knowledge bases*.

Today, as a result of many independent research projects and commercial initiatives, relatively large and important knowledge repositories have been made available which actually are (or can be easily tranformed into) "local semantic webs", namely sets of statements connecting to each others any type of resource through properties which are defined in some schema or vocabulary. DBpedia, GeoNames, DBLP, MusicBrainz and the FOAF profiles are only a few examples of knowledge bases which have been made available in semantic web formats (RDF/OWL); but any social network, digital library metadata collection, commercial catalog and in principle any relational database could be easily (and mostly syntactically) transformed into a "local semantic web" by exposing its data on the Web in RDF/OWL. Apparently, the necessary building blcks of the Semantic Web are available. So why is the integration of these local "semantic webs" not progressing as expected?

The argument we propose in this paper is the following. The integration of local "webs of documents" into the WWW was largely made possible by a key enabling factor: the introduction of a global and unique addressing mechanism for referring to and locating/retrieving resources. This addressing space relies on the existence of a service like the Domain Name System (DNS[1]) which maps any Uniform Resource Locator (URL) into a physical location on the Internet. This is how we be sure that, for example, a document with a suitable URL will be always and unmistakably located and retrieved, and that a `href` link to that resource (through its URL) will always be resolved to the appropriate location (even when the physical location of the resource may have changed, e.g. it was moved to another machine with a different IP address). The integration of "local semantic webs" is based on a very powerful generalization of what can be addressed on the web: from information objects (like HTML pages, documents, servers, etc.) to any type of object, including concrete entities (like people, geographical locations, events, artifacts, etc.) and abstract objects (like concepts, relations, ontologies, etc.). This leads to a higher level concept of integration: if two (or more) independent "local semantic webs" make statements about the same entity e_1, then these statements should be connected to each other, this way combining the knowledge about e_1 provided separately in the two sources. For this seamless integration of "local semantic webs" to become possible, it is required that independent data sources address (i.e. refer to) the same resource through the same URI. But this is not the case today, and a new URI is minted for a resource every time it occurs in a RDF/OWL knowledge base.

As we will show, there are two general views on how to address this issue. The *ex post* view is based on the idea that the multiplicity of URIs for the same entity is not bad *per se*, and that an appropriate solution is the creation of identity statements between any URIs which have been created for the same entity; this

---

[1] See http://www.ietf.org/rfc/rfc1034.txt

approach is well exemplified by the Linked Data initiative[2]. The *ex ante* approach is based on the idea that the proliferation of URIs for the same entity should be limited from the outset, and that a suitable solution should support the widest possible use (and – most importantly – reuse) of globally unique URIs. Though the two solutions are in principle not mutually exclusive, in another paper [12] we presented some arguments for preferring the *ex ante* over the *ex post* view. In this paper we present a technical solution for supporting the *ex ante* view based on an *Entity Name System* (ENS) for the Semantic Web, an open and global service which can be used within existing applications to support the creators/editors of semantic web content to (re)use the same globally unique URI for referring to the same entity in a systematic way.

2 An Ordinary Day on the Semantic Web

Imagine an ordinary day on the Semantic Web:

- the University of Trento exports in RDF its bibliographic database;
- the 5th European Semantic Web Conference (ESWC2008) makes available the metadata about authors and participants as part of the Semantic Web Conference (SWC) initiative;
- participants at ESWC2008 upload their pictures on `http://www.flickr.com/` and tag them;
- some participants at ESWC2008 attend a talk on FOAF and decide to create their FOAF profiles at `http://www.ldodds.com/foaf/foaf-a-matic` and publish them on their web servers;
- ...

At the end of the day, a lot of related material has been created. In principle, the newly created RDF content should allow Semantic Web programs to answer questions like: "Find me which of my friends is attending ESWC2008", "Find me pictures of Fausto's friends who attended ESWC2008", "Find me the papers published by people of the University of Trento (or their friends) accepted at ESWC2008", and so on. But, unfortunately, this can't be done. And the reason is that every time an entity (Fausto, ESWC2008, University of Trento, ...) is mentioned in one of the data sets, it is referred to through a different URI. And this does not allow RDF graph merging based on the fact that the same resource is referred to by the same URI. This is an instance of the so–called problem of identity and reference on the Semantic Web.

This scenario is quite typical of how Semantic Web content is produced today. Nearly like hypertexts in the pre-WWW era, any tool for creating semantic content mints new URIs for every resource, and this seriously hinders the bootstrapping of this global knowledge space called Semantic Web as it was originally envisioned. More and more attention is payed to reusing existing vocabularies or

---

[2] See `http://linkeddata.org/`

ontologies, but statements about specific resources (instances, individuals) cannot be automatically integrated, as there is nothing practically supporting the (desirable) practice of using a single global URI for every resource, and reusing it whenever a new statement about it is made through some content creation application.

The ENS we will discuss in the next section is our proposed approach and solution for addressing this issue in a systematic and scalable way.

3 ENS: A Prototytpe Architecture

Our current prototype implementation of an ENS service is called OKKAM[3] and represents one node in a federated architecture, which is depicted as a cloud in the center of Figure 1. The aim of the OKKAM prototype is to provide a basic set of ENS functionality, i.e. searching for entities, adding new entities and creating new identifiers. The identifiers that OKKAM issues are *absolute URIs* in the sense of RFC3986 [1], which makes them viable global identifiers for use in all current (Semantic) Web data sources; they furthermore are valid UUIDs, i.e. identifiers that guarantee uniqueness across space and time[4], which prevents accidental generation of duplicates and thus also enables their use as primary keys, e.g. in relational data sources.

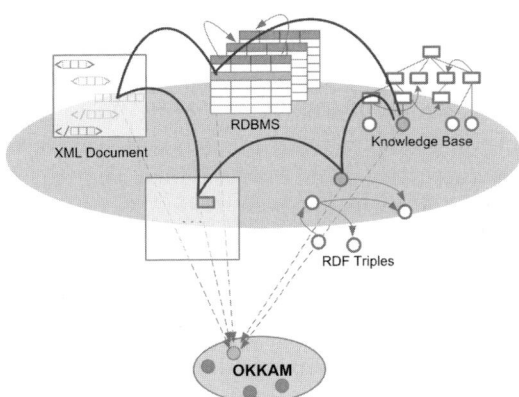

Fig. 1. The ENS providing entity identifiers across system boundaries

What is illustrated in Figure 1, and currently implemented as a single node, is planned to become a distributed system that is fully in line with the distributed nature of the (Semantic) Web.

---

[3] As a variation of the **Ockham's razor**, we propose the **Okkam's razor** as a driving principle: *"entity identifiers should not be multiplied beyond necessity"* ...
[4] See http://java.sun.com/j2se/1.5.0/docs/api/java/util/UUID.html for details.

3.1 Interaction with Okkam

A critical feature of an ENS is to provide a means for searching for the identifier of an entity. This step is strictly connected to the algorithm that supports the population of the system's repository with new entities. Indeed, when a query is submitted to the system, it has to decide if the query corresponds to an entity already stored (and return the information about it) or if a new entity has to be generated.

The standard use-case for the *okkamization*[5] of content goes as follows. A client application (such as FOAF-O-MATIC or OKKAM4P, presented in Section 5) accesses the OKKAM API, and presents (if available) a list of top candidates which match the description for the entity provided within the client application. If the entity is among these candidates, the client agent (human or software) uses the associated OKKAM identifier in the respective information object(s) *instead* of a local identifier. If the entity cannot be found, the client application can create a new entry for this entity in OKKAM and thus cause an identifier for the entity to be issued and used as described before.

3.2 Matching and Ranking in Okkam

The problem of searching for an entity in an ENS can be viewed as the problem of matching an entity description Δ from an external source against the set EP of all entity profiles stored in the OKKAM entity repository[6]. The setting of our problem is thus very similiar to what Pantel et al. describe about their *Guspin* system [2]: due to the high level of heterogeneity on the schema level (or in our case, the absence of such a level), we will pursue a purely data-driven approach for entity matching. For our first prototype, we have implemented an exemplary matching and ranking algorithm, whose objective is to provide a first solution of the matching problem described above, and can serve as a baseline and benchmark for future developments.

Matching and ranking in OKKAM is a two-step process: first, a set of candidate matches is retrieved from the storage backend, which, in the second step, is ranked with respect to the input query. With this approach we try to alleviate the problem that while storage backends such as relational databases perform extremely well in its main purpose, the production of ranked query results is not a "native" feature and thus hard to achieve. Furthermore, it allows us to apply methods for ranking that such storage backends simply do not provide.

Due to the differences between the matching problem in OKKAM and much of the related work, we decided to pursue an approach that is both schema-independent (entities in OKKAM are not described with a fixed schema) and type-independent (entities are untyped). The solution we came up with is to see

---

[5] We call *okkamization* the process of assigning an OKKAM identifier to an entity that is being annotated in any kind of content, such as an OWL/RDF ontology, an XML file, or a database, to make the entity globally identifiable.

[6] Note that Δ and EP are "compatible" for matching in the sense that every element $E \in EP$ contains a Δ by definition.

the EntityDescription Δ_e of an entity as a type of document which we can compare against the EntityDescription Δ_i that was provided in the input query. By computing a similarity between the two, and doing so for all candidate matches, we are able to provide a ranked query result.

The resulting algorithm, called `StringSimilarityRank`, is the following (with Δ_e being denoted by de and Δ_i by di):

```
d = concatenate(valuesOf(di)) forall candidates
   c = concatenate(valuesOf(de))
   s = computeSimilarity(d,c)
   rankedResult.store(s)
rankedResult.sort()
```

The function `valuesOf()` returns the value parts of the name/value pairs that form part of Δ, while `concatenate()` creates a single string from a set of strings; the combination of the two creates a "document" that can be matched against another, which is performed by the function `computeSimilarity()`.

To compute the similarity between two descriptions, we have selected the Monge-Elkan algorithm [3] as the result of extensive testing and evaluation of different algorithms [4]. The matching results that can be achieved with this approach are satisfactory as a baseline, as will be evident from Sect. 4.

This matching approach is completely general and "a-semantic", in that it neither uses background knowledge nor any kind of type-specific heuristics to perform the described matching. This is in strong contrast with other approaches for matching that are currently pursued e.g. in the Linked Data community, which heavily rely on different kinds knowledge to perform such a match, and as a consequence, require a special heuristic to be developed for different schemas or entity types. One example is the matching of FOAF profiles based on the inverse functional property of the email hash, which is a highly specialized combination of knowledge about the entity type, its schematic representation, and the available data. While we believe that in the mid-term a well-designed set of specialized algorithms embedded in an adaptive system is a very promising approach, for the current prototype we explicitly pursued the goal of implementing an algorithm that is completely independent of any such knowledge, and thus can be used to any type of entity, in any representation.

4 An Experiment in ABox Integration

To illustrate one possible application of the OKKAM infrastructure, we performed an ontology integration experiment with the Semantic Web data which cover information about papers, schedules, attendees, etc. of the two recent Semantic Web conferences, namley ISWC2006 and ISWC2007[7].

While this is not the "typical" application of OKKAM, as it is an *ex-post* alignment which we do not propagate as best practice, we set up this experiment

---

[7] These datasets are available at http://data.semanticweb.org; for future reference, we made available a copy of the datasets at http://okkam.dit.unitn.it/swonto/

to test and improve the performance of the current OKKAM prototype and of the implemented methods for entity matching and for ranking results.

The aim of the experiment is to perform fully automated object consolidation on entities of type *foaf:Person*, to evaluate several aspects of this process, and consequently, to establish a threshold for entity identity on which processes such as automatic alignment can rely. In the following, we evaluate three steps:

1. Establishing threshold t_{fp}, which can be considered a "good" value below which a best match found by OKKAM should be considered a false positive.
2. Establishing a golden standard g to evaluate the results of the merging process grounded on the threshold t_{fp}.
3. Performing an unsupervised ontology merge and analyzing the results.

4.1 Establishing an Identity Threshold

In OKKAM, deciding whether an entity e matches a query q relies on a similarity threshold t_{fp} below which e should be considered a false positive.

To fix this threshold, we ran the system on a set of example queries (all person-entities from the ISWC2006, ESWC2006 and ISWC2007 metadata sets). For each query, the system returns an OKKAM URI and a corresponding similarity value. Subsequently we checked manually the performance of the system, comparing the data available about the source URI with the Okkam URI to verify whether the match was correct or false.

Subsequently we evaluate how the performance of the system changes, varying the threshold on the range of similarity calsses ($t_1 = s_1, ..., t_j = s_j$) and for each class we compute the contingency table (see Table 1), including values for True Positive (TP_j), True Negative (TN_j), False Positive (FP_j) and False Negative (FN_j).

Table 1. Contingency table

S_j	Expert assigns YES	Expert assigns NO
System assigns YES	TP_j	FP_j
System assigns NO	FN_j	TN_j

Here, TP_j (True Positive with respect to the threshold t_j) is the number of entities correctly identified by the system when the threshold is t_j, TN_j is the number of entities that the system correctly did not identify when the threshold is t_j, FP_j is the number of the entities that have been incorrectly identified by the system when the threshold is t_j and FN_j is the number of entities that the system incorrectly did not identify.

The first evaluation that we performed was comparing the trend of TP with respect to FP. This analysis is motivated by the aim of our investigation to find the threshold that results in a minimum of FP but preserves a good level of TP. In general, if the number of FP is too high, the negative effects are two-fold: on

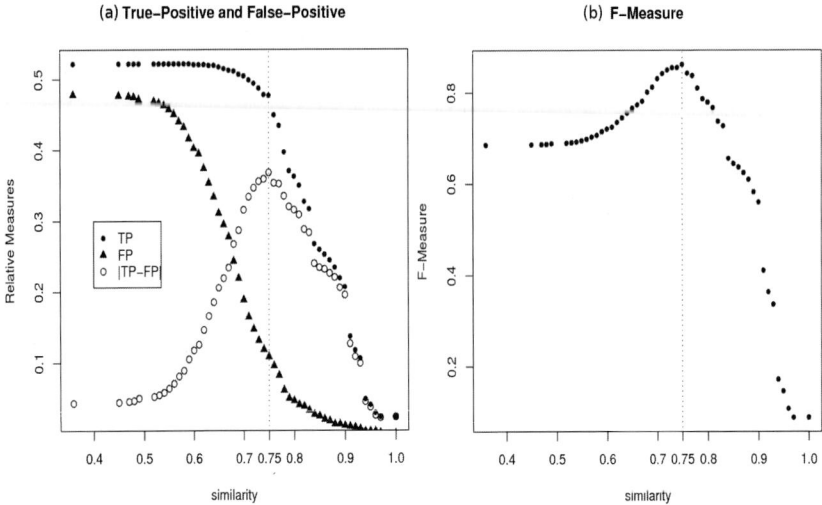

Fig. 2. Evaluation Measures

the one hand the results returned by the system would be polluted by irrelevant information, while on the other hand if the same threshold is used to perform the entity-merging, two different entities would be collapsed. The latter is a very undesirable circumstance because it leads to the loss of one of the two entities, and in the assignment of wrong information to the other (wrong merge/purge).

In order to determine an acceptable TP-FP trade-off we adopt a distance measure between TP and FP (the absolute value of the difference between TP and FP, $|TP - FP|$ or *Manhattan distance*) to establish the value of similarity in respect to this distance is maximized. In Figure 2(a) we plot TP and FP and the absolute value $|TP - FP|$. The graph shows that FP decrease more rapidly compared to TP when the similarity increses and the trend of difference $|TP - FP|$ shows a like normal distribution with a peak in correspondence to the maximum on a level of similarity equal to 0.75. On this level the system presents TP=0.47 and FP=0.10.

In order to confirm our result we evaluated the performance of the system measuring its effectiveness by means of Precision (P), Recall (R) and F-Measure (F)[8]. For each similarity class we calculate these evaluation measures to find which similarity value ensures the best performance of the system. We present the results relative to the F-Measure that give an overall description of the performance. In Figure 2(b) we show how the F-Measure varies as a function of similarity. We can see that the F-Measure increases up to a level of similarity equal to 0.75 and then dicreases rapidly. This evidence confirms the same result of the first analysis, indicating as the best threshold $t_{fp} = 0.75$. On this level we register a value of the F-Measure equal to 0.86, corresponding to P=0.81 and

---

[8] See http://en.wikipedia.org/wiki/F-measure for an overview of these performance measures from the field of Information Retrieval.

Table 2. Performance for $t = t_{fp} = 0.75$

$t_{fp} = 0.75$	TP	TN	FP	FN	P	R	FM
	0.47	0.36	0.10	0.04	0.81	0.91	0.86

R=0.91. Table 2 summarizes the performance of the system when the threshold is $t_{fp} = 0.75$.

4.2 Evaluating the Ontology Merge

In order to evaluate the performance of the system with respect to the results of the merging process, we have to define a benchmark that we consider as the golden standard in our analysis.

For this purpose we took into account two (ISWC2006 and ISWC2007) out of three Semantic Web ontologies considered in the first phase of our evaluation analysis.

As a first step, we compare manually the two data sets to detect which URLs in the two datasets refer to the same real world entities (persons). This comparison returns the number ($g = 48$) of entities that the system should be able to consolidate, and represents the golden standard of our analysis.

In the second step of the analysis we perform an automatic merge of the same data sets (ISWC2006 and ISWC2007), and compare this merge to the golden standard. In Table 3 we report the results of our analysis respect to three exemplary thresholds which we examined.

If we consider the first data column in Table 3 in which we have the results respect to a value of $t_{fp} = 0.75$, we notice that the correct mappings amount to 46, which – compared to the golden standard of $g = 48$ – shows that the system returns almost all the correct mappings. However the number of false positives is still quite high and it reduces precision to $P = 0.65$. In other words, the system recognises some wrong mappings, which requires us to search for another (more safe) threshold that guarantees a lower number of FP, but preserving a satisfying number of TP. Table 3 shows that $t_{fp} = 0.90$ increases precision

Table 3. Results of the merging process

	$t_{fp} = 0.75$	$t_{fp} = 0.90$	$t_{fp} = 0.91$	Golden standard
Total Positives	70	68	43	48
True Positive	46	45	25	48
True Negative	380	385	405	403
False Positive	24	20	20	0
False Negative	1	1	1	0
Precision	0.66	0.69	0.56	1
Recall	0.98	0.98	0.96	1
F-Measure	0.78	0.81	0.7	1

without sacrificing substantially TP, while $t_{fp} = 0.91$ leads to a degeneration of the results.

Summing up, our experiment showed that it is possible to move from two data sources that should set-theoretically present a certain overlap but syntactically do not[9], to a situation where good recall of matches can be reached through an alignment against OKKAM. The approach presented here requires no ad-hoc implementations or knowledge about the representation of the entities, as opposed to other approaches, such as [5] or the ones described in Section 6.

5 Two ENS-enabled Applications

To illustrate the viability and the usefulness of the approach, we have developed two exemplary applications that both have been strategically selected from the area of content creation, and serve – in contrast to the experiment described in Sect. 4 – as means to achieve an *a-priory* alignment of identifiers that we propagate in our approach. The reason for this selection was the fact that the success of the ENS approach depends entirely on a certain saturation of suitable content ("critical mass"), and in effect on the availability of tools for the creation of such content.

5.1 Okkam4P

The first tool is called OKKAM4P [6], a plugin for the widely-used ontology editor Protégé. This plugin enables the creator of an ontology to issue individuals with identifiers from OKKAM, instead of assigning local identifiers that bear the risk of non-uniqueness on a global scale. The choice for this tool was made based on two criteria, namely the target audience being rather 'expert' users of the Semantic Web, and, secondly, the very wide usage of the Protégé editor, which makes it a promising candidate for a rapid distribution of the tool.

Based on the data about an individual provided in the KB developed by the user. The plugin queries OKKAM to see whether an identifier already exists which can be assigned to the newly created individual, otherwise a new identifier is created and returned.

Access to the plugin is given through the context menu of the individual, as depicted in Figure 3. The plugin then guides the user through the search and selection process and finally replaces the local identifier for the entity with the one retrieved from the ENS. The result is an OWL ontology that is equipped with globally unique and re-usable identifiers and thus enables vastly simplified, automatic integrations with high precision. The plugin is available at the following URL: http://www.okkam.org/projects/okkam4p.

5.2 Foaf-O-Matic

The second application is called FOAF-O-MATIC [7], a WWW-based service for the creation of okkamized FOAF[10] profiles. Indeed, FOAF is in our opinion one

---

[9] In fact, the two data sources present an overlap of zero identifiers for person entities.
[10] http://www.foaf-project.org

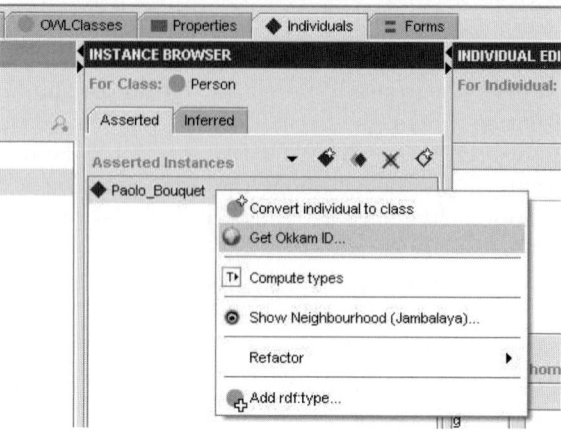

Fig. 3. Assigning a global identifier to an individual

of the few real success stories of the Semantic Web so far, as it is one of the few applications that really contributed to the creation of a non-toy amount of RDF data, with the special restriction that the agreement on URIs for persons is extremely low [5]. As content creation tools for FOAF are mostly rather prototypical, we decided to create a completely new application that both serves the user with state-of-the-art technology and at the same time creates okkamized FOAF profiles.

As we have discussed in [7], what is currently missing from FOAF is a reliable and pervasive way to identify "friends". The focus of the new application is to allow users to integrate OKKAM identifiers within their FOAF document in a user-friendly way. In this way, it will be possible to merge more precisely a wider number of FOAF graphs describing a person's social networks, enhancing the integration of information advancing toward the goal of the FOAF initiative.

A view of the application layout is given in Figure 4: it includes functions to re-use existing FOAF profiles (1), a form for describing oneself (2), the list of friends (3), and the form for adding friends (4) which initiates the ENS search process. The application is deployed and usable at the following URL: http://www.okkam.org/foaf-O-matic.

6 Related Work

The problem of *recognizing* that an entity named in some content (e.g. in an RDF graph) is the same as an entity stored in the ENS repository (the *entity matching* problem) is obviously related to well-known problems in several disciplines[11] (e.g. named entity recognition, coreference, object consolidation, entity

---

[11] See [4] for a more detailed discussion of related work.

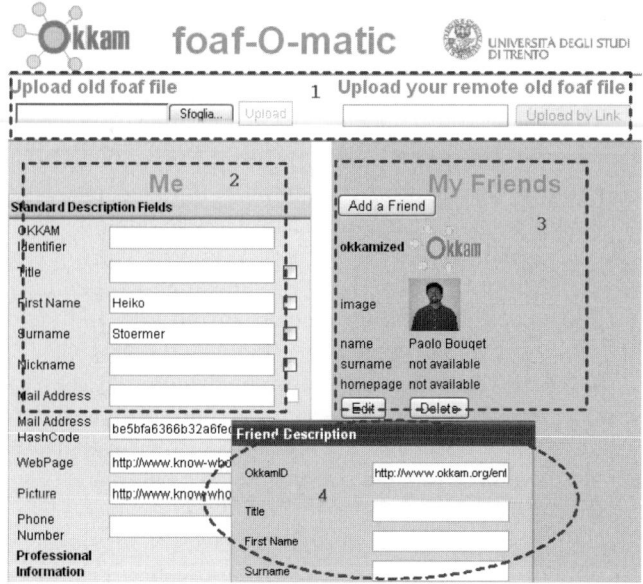

Fig. 4. FOAF-O-MATIC The main interface of FOAF-O-MATIC

resolution). In the areas of database and information integration, there is a substantial amount of related work that deals with the problem of detecting whether two records are the same or describe the same object. However, the matching problem in OKKAM is different for the following reasons:

1. the description of the entity that is searched for can be generated by client applications that are of very different nature. Some (like a text editor) may only provide a simple query string, while others (like ontology editors) may provide (semi-) structured descriptions. It is thus not foreseeable which key/value pairs a description contains;
2. the set of entity profiles stored in OKKAM is untyped, semi-structured and may as well contain arbitrary values. This aspect, combined with the previous one, makes the ideal OKKAM solution very different from most record–linkage approaches, as they rely on fixed (and/or identical) schemas, whereas OKKAM does not;
3. the objective is not deduplication (or Merge/Purge etc.) but rather the production of a ranked list of candidate matches within a time frame of a fraction of second. For this reason, unoptimized approaches that perform deduplication by iterating over entity profiles in a serial fashion must be avoided.

As part of the solution, we will investigate how we can automatically build a contextual profile for entities named in content specified in different formats (e.g. text, HTML or XML files, RDF/OWL databases, relational databases) and how such a profile can be used for matching the entity against the profile available in an ENS server.

For dealing with the proliferation of indentifiers in the Semantic Web, there are currently at least two major approaches which are very relevant.

Jaffri et al. [8], in their work resulting from the ReSIST project, recently came to a conclusion not very different from the one to we advoceated in [9,10], namely that the problem of the proliferation of identifiers (and the resulting coreference issues) should be addressed on an infrastructural level; consequently they propose what they call a *Consistent Reference Service*. While we share this general view, their point about URIs potentially changing "meaning" depending on the context in which they are used, is philosophically disputable: the fact that several entities might be *named* in the same way ("Spain" the football team, "Spain" the geographic location) must not lead to the conclusion that they can be considered *the same* under certain circumstances[12]. Furthermore, their implementation of "coreference bundles" for establishing identity between entitites is in fact very similar to a collection of `owl:sameAs` statements (see below).

Another notable approach is the effort of the *Linking Open Data Initiative*[13], which has the goal to "connect related data that wasn't previously linked". The main approach pursued by the initiative is to establish `owl:sameAs` statements between resources in RDF. While the Linked Data community has made a huge effort to interlink a large number of datasets, our view is that this approach is not optimal to realize the vision of the Semantic Web as a large, decentralized knowledge base. First of all, a very large number of `owl:sameAs` statements is necessary, and it grows with the number of different URIs which are available for the same entity; second, querying distributed datasets cannot be done by simple SPARQL queries, as it must be combined with some form of reasoning (unless all the implied identity statements are computed beforehand and stored with the data); third, it sounds quite unrealistic that users will spend time in creating identity statements about their local data. However, we would like to stress that the ENS approach and the Linked Data initiative are not at all mutually exclusive, as OKKAM identifiers can be easily linked to other non-okkamized datasets through `owl:sameAs` statements, and `owl:sameAs` statements can be used in OKKAM to generate aliases for an OKKAM identifier. See [12] for a more thoroughly discussion of the relationship between the two approaches.

7 Challenges and Conclusions

In the paper, we presented the idea and the results of a test on ontology integration with a prototype of the ENS. However, designing, implementing and making available the ENS on a global scale involves some very difficult scientific and technological challenges. Here we list some challenges, and discuss how we plan to address them in the FP7 project OKKAM.

---

[12] See e.g. Kripke [11].
[13] See http://linkeddata.org/

In the previous section we alrady discussed the *entity matching* problem. A second issue has to do with bootstrapping the service. This problem has two dimensions. First, we need to make sure that the ENS is pre-populated with a significant number of entities, so that there is a reasonable chance that people will find a URI to reuse in their applications; this will be done by implementing tools for importing entities (and their profiles) from existing sources. Second, and even more important, we need to make sure that the interaction with the service is integrated in the largest possible number of common applications for creating content. In Section 5, we described two simple examples of how we imagine this interaction should happen; however, it is our plan to extend the idea also to non–Semantic Web tools, like office applications or web-based authoring environments (including forums, blogs, multimedia tagging portals, and so on). This approach should make interaction with the ENS very easy, sometimes even transparent, and will slowly introduce the good practice of OKKAMizing any new content which is created on the Web.

A third big issue has to do with scalability of the proposed solution. Indeed, the number of entities which people might want to refer to on the Web is huge, and the number of requests that the ENS might be exposed to can be extremely high. For this reason, the architecture we envisage for the ENS is distributed and decentralized.

Last but not least, there are two non-technical related issues. The first has to do with acceptance: how will we convince people to adopt the ENS? Especially, we need to make sure that the benefits outnumber the concerns by proving the advantages of the service in a few very visible and popular domains. The second issue has indeed to do with the general problem of guaranteeing privacy and security of the ENS. As to this respect, it is important that we do not raise the impression that the ENS is about storing lots of information about entities. The profiles which we will store will be minimal, and will serve only to support reasonably robust matching techniques. Also, we need to make sure that people have some degree of control on what can be stored in a profile, what cannot, and on what can be stored for improving matching but should never be returned as the result of a query to the ENS.

We are aware that the challenges are quite ambitious, but in our opinion the ENS may become the enabling factor which will make possible for *knowledge-representation-on-the-web* to "shake the world" as it has never done before.

References

1. Berners-Lee, T., Fielding, R., Masinter, L.: RFC 3986: Uniform Resource Identifier (URI): Generic Syntax. IETF (Internet Engineering Task Force) (2005), http://www.gbiv.com/protocols/uri/rfc/rfc3986.html
2. Pantel, P., Philpot, A., Hovy, E.H.: Matching and Integration across Heterogeneous Data Sources. In: Proceedings of the 7th Annual International Conference on Digital Government Research, DG.O 2006, San Diego, California, USA, May 21-24, 2006, pp. 438–439 (2006)
3. Monge, A.E., Elkan, C.: An Efficient Domain-Independent Algorithm for Detecting Approximately Duplicate Database Records. In: DMKD (1997)

4. Stoermer, H.: OKKAM: Enabling Entity-centric Information Integration in the Semantic Web. PhD thesis, University of Trento (2008), http://eprints.biblio.unitn.it/archive/00001389/
5. Hogan, A., Harth, A., Decker, S.: Performing object consolidation on the semantic web data graph. In: i3: Identity, Identifiers, Identification. Proceedings of the WWW 2007 Workshop on Entity-Centric Approaches to Information and Knowledge Management on the Web, Banff, Canada, May 8 (2007)
6. Bouquet, P., Stoermer, H., Xin, L.: Okkam4P - A Protégé Plugin for Supporting the Re-use of Globally Unique Identifiers for Individuals in OWL/RDF Knowledge Bases. In: Proceedings of the Fourth Italian Semantic Web Workshop (SWAP 2007), Bari, Italy, December 18-20 (2007), http://CEUR-WS.org/Vol-314/41.pdf
7. Bortoli, S., Stoermer, H., Bouquet, P.: Foaf-O-Matic - Solving the Identity Problem in the FOAF Network. In: Proceedings of the Fourth Italian Semantic Web Workshop (SWAP 2007), Bari, Italy, December 18-20 (2007), http://CEUR-WS.org/Vol-314/43.pdf
8. Jaffri, A., Glaser, H., Millard, I.: Uri identity management for semantic web data integration and linkage. In: 3rd International Workshop On Scalable Semantic Web Knowledge Base Systems, Springer, Heidelberg (2007)
9. Bouquet, P., Stoermer, H., Giacomuzzi, D.: OKKAM: Enabling a Web of Entities. In: i3: Identity, Identifiers, Identification. Proceedings of the WWW 2007 Workshop on Entity-Centric Approaches to Information and Knowledge Management on the Web. CEUR Workshop Proceedings, Banff, Canada, May 8, 2007 (2007), online http://CEUR-WS.org/Vol-249/submission_150.pdf ISSN 1613-0073
10. Bouquet, P., Stoermer, H., Mancioppi, M., Giacomuzzi, D.: OkkaM: Towards a Solution to the "Identity Crisis" on the Semantic Web. In: Proceedings of SWAP 2006, the 3rd Italian Semantic Web Workshop. CEUR Workshop Proceedings, Pisa, Italy, December 18-20, 2006 (2006), online http://ceur-ws.org/Vol-201/33.pdf ISSN 1613-0073
11. Kripke, S.: Naming and Necessity. Basil Blackwell, Boston (1980)
12. Bouquet, P., Stoermer, H., Cordioli, D., Tummarello, G.: An Entity Name System for Linking Semantic Web Data. In: Proceedings of LDOW 2008 (2008), http://events.linkeddata.org/ldow2008/papers/23-bouquet-stoermer-entity-name-system.pdf

A Functional Semantic Web Architecture

Aurona Gerber, Alta van der Merwe, and Andries Barnard

Meraka Institute and University of South Africa (Unisa),
Pretoria, South Africa
aurona.gerber@meraka.org.za,
{vdmeraj,barnaa}@unisa.ac.za
http://www.meraka.org.za
http://www.unisa.ac.za

Abstract. A layered architecture for the Semantic Web that adheres to software engineering principles and the fundamental aspects of layered architectures will assist in the development of Semantic Web specifications and applications. The most well-known versions of the layered architecture that exist within literature have been proposed by Berners-Lee. It is possible to indicate inconsistencies and discrepancies in the different versions of the architecture, leading to confusion, as well as conflicting proposals and adoptions by the Semantic Web community. A more recent version of a Semantic Web layered architecture, namely the CFL architecture, was proposed in 2007 by Gerber, van der Merwe and Barnard [23], which adheres to software engineering principles and addresses several of the concerns evident from previous versions of the architecture. In this paper we evaluate this recent architecture, both by scrutinising the shortcomings of previous architectures and evaluating the approach used for the development of the latest architecture. Furthermore, the architecture is applied to usage scenarios to evaluate the usefulness thereof.

Keywords: Semantic Web Architecture, Software Engineering, System Architecture, Layered Architecture.

1 Introduction

The establishment of the architecture of any information system is one of the crucial activities during the design and implementation thereof. There is general consensus in literature that an architecture (at least) depicts the structure of a system within a specific context. This depicted structure should portray the components that a system comprises of, as well as the relationships between the identified components. One of the main purposes of a system architecture is the provision of an agreed-upon functional description of system structure, components and component interactions. It is thus plausible to state that an architecture for the Semantic Web is crucial to its eventual realisation and that it is therefore necessary to attach indisputable meaning to the specification of the architecture for the languages of the Semantic Web.

The most well-known versions of the layered architecture that exist within literature have been proposed by Berners-Lee, and the literature offers no description or specification of meaning for any of these. It is possible to indicate inconsistencies and discrepancies in the different versions of this reference architecture, leading to confusion, as well as conflicting proposals and adoptions by the Semantic Web community. In addition, none of the current formal initiatives by the W3C address the Semantic Web architecture specifically, which could be regarded as an omission.

A layered architecture for the Semantic Web that adheres to software engineering principles and the fundamental aspects of layered architectures will assist in the development of Semantic Web specifications and applications. Furthermore, several of the current research and implementation issues associated with the implementation of the Semantic Web could potentially be resolved.

A more recent version of a Semantic Web layered architecture, namely the CFL architecture, was proposed by Gerber, van der Merwe and Barnard [23]. They claim that their abstracted CFL architecture of the Semantic Web adheres to software engineering principles and addresses several of the concerns evident from previous versions of the architecture.

In this paper we evaluate this recent architecture, both by scrutinising the shortcomings of previous architectures and evaluating the approach used for the development of the latest architecture. Furthermore, the architecture is applied to usage scenarios to evaluate the usefulness thereof. A similar approach was used in the construction of one of the most significant layered architectures in popular use today, notably the ISO/OSI reference model for network protocols. We reach the conclusion that the approach indeed has merit in resolving current issues with regards to the architecture of the languages of the Semantic Web. However, the proposed version needs to be refined through consensus by all role players, including the W3C.

This paper starts with a background in section 2 describing the four versions of the Semantic Web layered architecture as proposed by Berners-Lee [6, 9, 10, 11, 12, 13]. This discussion includes an investigation into some inconsistencies prevalent from these architectures. Section 3 contains a discussion of the proposed CFL architecture for the Semantic Web. Section 4 evaluates the CFL architecture by applying it to usage scenarios and section 5 concludes this paper.

2 Background: The Semantic Web Architecture

Since the publication of the original Semantic Web vision of Berners-Lee, Hendler and Lassila [15], Berners-Lee proposed four versions of a Semantic Web architecture, for the purposes of this discussion called the reference architecture. All of the reference architecture versions were presented by Berners-Lee in *presentations*, they were never published in literature or included as part of a W3C Recommendation. The different versions are depicted as V1 to V4 in Figure 1.

In support of the founding vision of the Semantic Web [15], a first version (V1) of a layered architecture was introduced in 2000 by Berners-Lee [6]. As

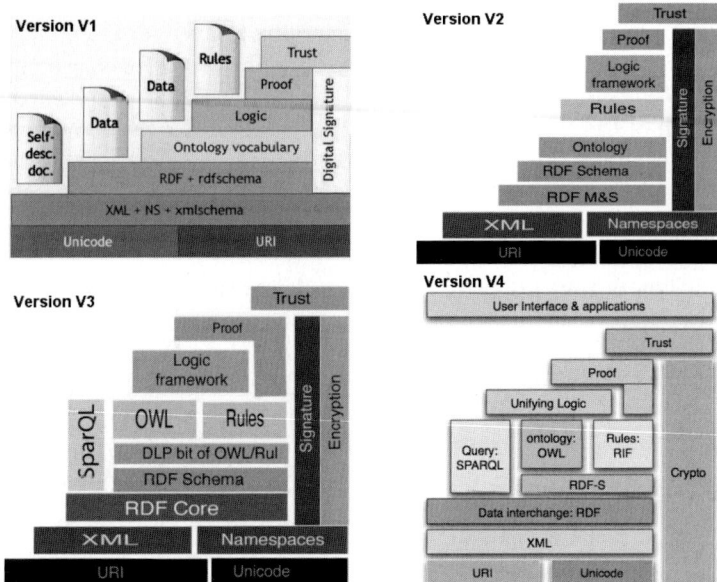

Fig. 1. The four versions of the Semantic Web reference architecture (V1-V4) proposed by Berners-Lee [6, 9, 10, 11, 12, 13]

participant in the ongoing activities of the W3C, Berners-Lee proposed a second version (V2) of the Semantic Web reference architecture in 2003 as part of a presentation at the SIIA Summit [30]. V2 was furthermore presented as part of two presentations in 2003 [9, 10, 11]. Berners-Lee proposed a third version (V3) at WWW2005 [12], and in his keynote address at AAAI2006 in July 2006, Berners-Lee introduced the latest (V4) version of the Semantic Web reference architecture [13].

In different publications, researchers argued that the the Semantic Web architectures cannot be described as *architectures* since Semantic Web *languages* or *W3C technologies* are depicted. They argue that this is not the nature of an architecture and that the term *architecture* is generally used to depict system functionality at different conceptual levels. The proposed versions of the Semantic Web architecture does not depict functionality and should be described as a *stack* or even *layered cake* [1, 24].

In order to respond to this argument, a definition for the term *system architecture* was compiled from an extensive literature analysis. There is general consensus in literature that an architecture *at least* depicts the structure of a system within a specific context [3, 4, 19, 20]. This structure should portray the components that a system comprise of, as well as the relationships between the identified components. System architects depicting the organisation of system components often make use of several identified architectural patterns, one of which is the *layered architecture* [2, 4, 17, 28]. In a layered architecture the principal elements or components are arranged in the form of a stack where each layer resides above a lower layer. Generally, a layer represents a grouping of elements that provides related services. A higher layer may use either only various services defined by the immediate lower layer (closed architecture) or services by all of the lower layers (open architecture). However, the lower layers are unaware of higher layers [2, 17, 18, 28], and are not allowed to access functionality provided by upper layers. This implies a strict ordering of access to the functionality provided by components in a layered architecture in one direction only [3].

In conclusion of the proposition that the Semantic Web layered architecture versions are regarded as *architectures*, we provide an example of a similar, widely disseminated architecture, namely the ISO/OSI (International Standards Organisation / Open Systems Interconnect) layered architecture. This architecture specifies the functionality required to define *protocols* necessary for *network interoperability* between applications [32]. The Semantic Web layered architecture has purpose similar to that of the the ISO/OSI layered architecture in that it aims to depict the *languages* necessary for *data interoperability* between applications. The ISO/OSI model is regarded as an *architecture*, and thus it is proposed that the Semantic Web model should be regarded in the same way as an *architecture* for the purpose of this discussion.

If the Semantic Web architectures are regarded as *architectures*, we have to define the context, components and relations. Within the *context* of the *languages required for meta-data specification*, the proposed versions of the reference architecture [6, 9, 12, 13] depict the organisation of the *language* components for the Semantic Web. The layering of the languages provides the *structure*. Thus, the general definition of an *system architecture* is adhered to. The fact that *functionality* is not always depicted on layers within the current versions should be regarded as an omission. This omission provides one of the motivations for the development of a *functional* architecture for the Semantic Web. In addition, an investigation into the structure of the reference architecture versions highlighted some discrepancies and irregularities.

☐ **Side-by-side layers**

In all four versions depicted in Figure 2 *URI* and *Unicode* are depicted on the bottom layer as two separate blocks or side-by-side layers (refer to (1) in Figure 2). It can be assumed that the intended implication is that the two technologies both reside on the bottom layer. However, this is contentious and inconsistent when referring to the presentation of other layers containing more than one technology such as V1 that depicts the *RDF* and *RDF Schema* technologies in one layer (*RDF + rdfschema*). In V2, *XML* and *Namespaces* are also depicted as side-by-side layers, similar to *URI* and *Unicode* on the bottom layer. However, to be consistent with the assumptions of layering, the layering of this version should imply that *XML* only uses *URI*, whereas *Namepaces* uses only *Unicode* as a preceding layer. Based on our knowledge of XML and Namespaces, this is not the case even though it is depicted as such. The same discussion holds for several examples in all versions.

☐ **Triangular structure of the layered architecture**
In all four versions depicted in Figure 2 the layers are staggered into a triangular structure with the lower layers wider than upper layers (refer to (2) in Figure 2). It is not clear whether this means that the upper layers use only *part* of what is provided by lower layers, or whether a lower layer specifies additional functionality that is not used for the purpose of the Semantic Web.

☐ **Mixing technologies and functionality descriptions in the naming of layers**
It is not clear what the layers in the four versions represent since certain layers are labelled using *technologies* whilst others are labelled using *functionality descriptions* (refer to (3) in Figure 2). In all versions the bottom layers represent technologies while higher layers are labelled with functionality descriptions.

☐ **Vertical layers**
All versions of the architecture in Figure 2 depict vertical layers such as *Digital Signatures* in V1 and *Crypto* in V4 (refer to (4) in Figure 2). The precise meaning of any of these vertical layers is not specified in literature. It is possible to speculate that these layers are included in all the other layers, or that these layers reside alongside the other layers, or even that these layers only depict technologies building on top of their lower layers and excluding the upper layers. Furthermore, it is possible to remove any *security related* vertical layer in any of the versions of the Semantic Web layered architecture without compromising the integrity of the system as regards to the *languages* of the Semantic Web.

In summary, the versions of the reference architecture Berners-Lee [6, 9, 12, 13] have not been formally documented and all depict certain inconsistencies and

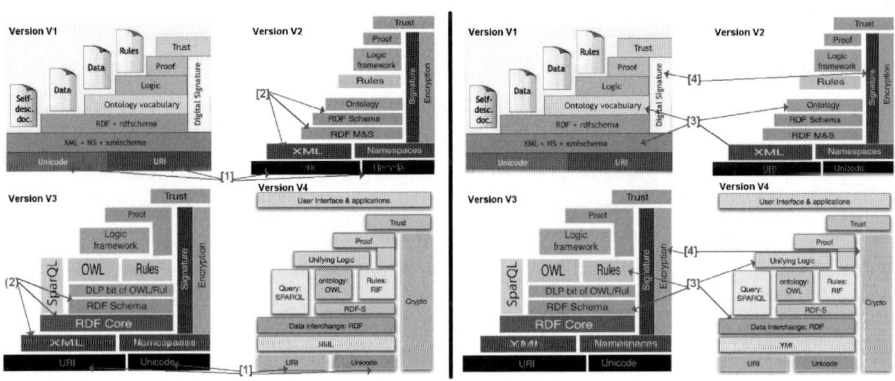

Fig. 2. Versions V1-V4: side-by-side layers (1), triangular structure (2), functionality vs technology (3) and vertical layers (4)

discrepancies. This often leads to confusion, as well as conflicting proposals and adoptions of technology and functionality by the Semantic Web community [25, 26, 29]. It is not unreasonable to propose that the architecture of any system is one of the primary aspects to consider during design and implementation thereof, and the proposed architecture of the Semantic Web is thus crucial to its eventual realisation. It is necessary to attach indisputable meaning to the specification of the architecture for the languages of the Semantic Web. In addition, because there is no precise functionality and interface definition, applications implementing Semantic Web technologies using these architectures will not be able to interoperate. The interoperation of Semantic Web applications is crucial for the eventual realisation of the founder vision.

In is interesting to note at this stage that, as part of the ongoing W3C design initiatives, Berners-Lee argues for the inclusion of certain software engineering design principles into W3C design efforts, and he identifies four principles namely simplicity, modularity, decentralisation and tolerance [7]. *Simplicity* strives to use only a few basic elements to achieve the required results. *Modularity* suggests a system design adhering to *loose coupling* and *tight cohesion* of system elements. *Decentralisation* avoids any common point that may be a single point of complete failure and *Tolerance* specifies liberal requirements and conservative implementations. In spite of these design principles, none of the current formal initiatives by the W3C address the Semantic Web architecture specifically [5, 7, 8, 14, 16, 31].

An approach to develop a layered architecture from software engineering principles was proposed by Gerber et al. [23], referred to as the Comprehensive, Functional, Layered (CFL) architecture for the Semantic Web. *Comprehensive* implies that the architecfture is based on software engineering principles, *functional* means that the architecture depicts required functionality and *layered* refers to the architecture structure [23].

3 The CFL Architecture for the Semantic Web

The proposed CFL architecture (Figure 3) for the Semantic Web varies from the previously suggested versions (V1-V4) of the architectures mainly because it adheres to the evaluation criteria for layered architectures that are based on established software engineering principles [21, 22, 23]. In addition, it is noticeable that the CFL architecture abstracts and depicts related functionalities rather than the W3C technologies used to instantiate these functionalities. A detailed discussion on the development of the CFL architecture can be found in Gerber et al. [23].

From Figure 3 a reader may conclude that the proposed CFL architecture depicts only a simplification of the original architecture. The *abstraction* of the functionality from the original versions of the Semantic Web layered architecture does indeed entail a simplification of the model, but this simplification introduces several advantages such as a mode understandable and universal model that facilitates debate. One of the most significant advantages is that such an

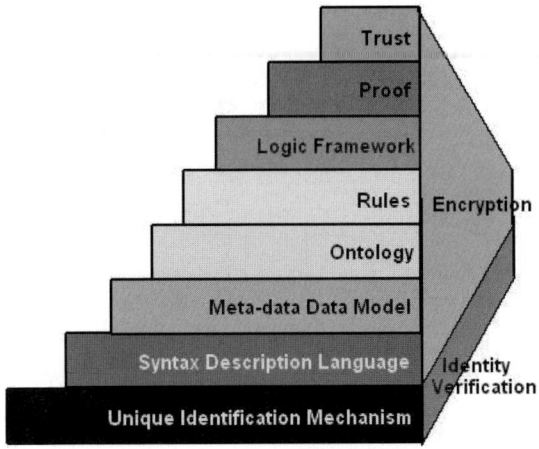

Fig. 3. The proposed CFL architecture for the Semantic Web comprises two orthogonal architecture stacks, the *language* stack and the *security* stack

abstraction enables the use of diverse and non-related technologies to implement the functionality of a specific layer. In other words, different technologies could be used to instantiate the functionality of a specific layer in the proposed Semantic Web layered architecture.

To determine whether this CFL architecture is indeed usefull for standardisation of Semantic Web technologies, as well as the resolution of the layering debate, the CFL architecture is evaluated using usage scenarios.

4 Evaluation of Usefulness

The first usage scenario was obtained from literature and the proposed CFL architecture is applied to solve some of the issues with regard to the layering of the Semantic Web in section 4.1. In addition, the current V4 Semantic Web reference architecture [13] is investigated as being an instantiation of the proposed CFL architecture (section 4.2). Lastly, a practical application example was developed using the proposed CFL architecture in section 4.3.

4.1 Scenario: The Two-tower Architecture of Horrocks, Parsia, Patel-Schneider and Hendler[25]

Originally, versions V2 and V3 of the reference architecture were discussed by Kifer, Bruijn, Boley and Fensel [27], who argued strongly in support of *multiple independent, but interoperable, stacks of languages*. They argued that a single stack architecture as depicted in V2 (Figure 1) is *unrealistic* and *unsustainable* and they regarded the side-by-side layering of *OWL* and *Rules* in V3 (Figure 1) as an implementation of the *interoperable stacks of languages* concept. They therefore supported the *multi-stack architecture* as depicted by V3. In particular, they discuss

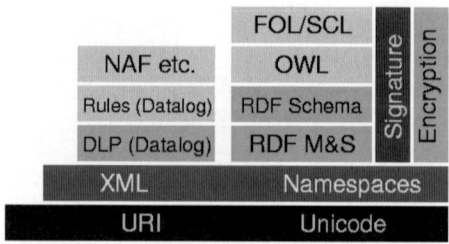

Fig. 4. The two-tower architecture of Horrocks, Parsia, Patel-Schneider and Hendler [25]

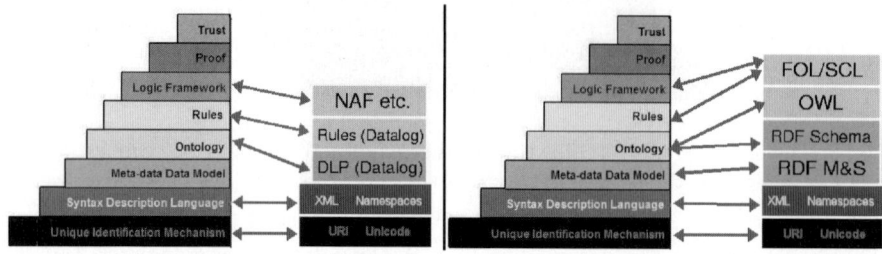

Fig. 5. The two-tower architecture as instantiations of the Semantic Web CFL architecture

the incorporation of the Datalog rule-based languages as *Rules*, which could not be layered on top of OWL, as a separate stack alongside the stack of OWL-based languages with their prescribed Description Logics.

Horrocks, Parsia, Patel-Schneider and Hendler [25] entered the debate about the positioning of *Rules* with their *Semantic Web Architecture: Stack or Two Towers?* article. They further argued that a realistic architecture for the Semantic Web had to be based on multiple independent, but interoperable, stacks of languages. In particular, they pointed out that DLP/Datalog and RDFS/OWL are not semantically compatible and cannot be layered as proposed in the V3 version of the present Semantic Web layered architecture (Figure 1). They consequently proposed a two-tower architecture (Figure 4) as the solution for the acceptance of both RDFS/OWL and DLP/Datalog as Semantic Web languages or technologies [25].

This two-tower architecture of Horrocks et al. [25] depicts two possible instantiations of the Semantic Web layered architecture. In order to demonstrate the value of the proposed CFL architecture, the instantiations in the two towers of Horrocks et al. [25] are related to the CFL architecture demonstrating the inclusion of different technologies. The relation of the two towers to the CFL architecture for the Semantic Web is depicted in Figure 5.

In *Tower 1* on the left-hand side in Figure 4, Horrocks et al. omitted an explicit data representation layer. Datalog was developed as a rule and query language for deductive databases and syntactically Datalog is a subset of Prolog, hence it does not constitute a data representation layer. To improve the composition of Tower 1, a *meta-data data model* could be added. Thus, the adapted Semantic Web architecture supports more than one technology to be used in this layer. In addition, the Ontology functionality is implemented by DLP (Description Logic Programs) where these DLP extend Datalog to include knowledge representation. The Rules functionality is implemented with Datalog Rules and NAF. This relationship is depicted in Figure 5.

In contrast to Tower 1, Tower 2 on the right-hand side in Figure 4 implements the meta-data data model functionality by means of RDF. The Ontology layer functionality is implemented using RDF Schema and OWL. In this scenario, the contention is that the FOL/SCL layer provides for Rules and to a limited extent, Logic Framework functionality. This instantiation is depicted in Figure 5.

Note that the purpose of this scenario is not to resolve the layering debate of the different technologies. In this paper, we evaluate the usefulness of the CFL architecture. The architecture defines *functionality* that has to be implemented by the different layers, and which allows for the acceptance of diverse technology *instantiations* that implement the requisite functionality. This is aptly demonstrated in this usage scenario where both the *towers* are accommodated by the proposed Semantic Web CFL architecture, even though they represent different semantic bases.

4.2 Scenario: The V4 Semantic Web Architecture

In this section, the latest version (V4) of the layered architecture as proposed by Berners-Lee [13] is investigated as being an instantiation of the proposed CFL architecture for the Semantic Web. Figure 6 depicts the mapping of the depicted W3C technologies to the functionality layers of the CFL architecture. The mapping irregularities, which is due to the inconsistencies in the V4 architecture as discussed in section 2.

The *Unique Identification Mechanism* layer maps to Unicode and URI as before. URI essentially uses Unicode and therefore the meaning of the side-by-side layering of Unicode and URI in V4 is unclear. However, it is plausible to state that URI making use of Unicode is an instantiation of the layer 1 *unique identification mechanism* functionality of the CFL architecture. XML instantiates the *Syntax Description Language* layer and the *meta-data data model* is instantiated with RDF.

The *Ontology* layer of the CFL architecture is instantiated with either RDF Schema (depicted as RDF-S in V4) or RDF Schema and OWL (RDF-S and OWL). If RDF-S as instantiation is used, RIF is the instantiation of the *Rules* layer. *RIF* is an acronym for *Rule Interchange Format* and it is a draft specification for a rule language using RDF Schema as its ontology basis.

Above RIF, no technologies are depicted in V4 and therefore *instantiations* cannot be discussed. *Unifying Logic* maps to *Logic Framework*, and *Proof* and *Trust* remains *Proof* and *Trust*. It is unclear what is meant by both *Unifying Logic* and *Proof* residing on top of *Rules:RIF*. It is possible to reason that both reside alongside each other above *Rules:RIF* or it might mean that the layered

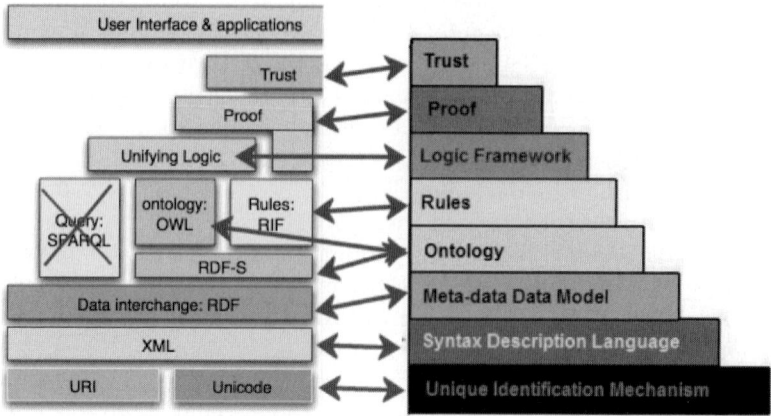

Fig. 6. The V4 version of the Semantic Web reference architecture [13] as instantiation of the adapted architecture

architecture is *open* in this instance so that *Proof* might access functionality of *RIF* directly without having to access the *Unifying Logic* layer. However, in general the technologies from URI and Unicode through to RIF can be regarded as an instantiation of the adapted, functional architecture.

In contrast, if the *Ontology* layer is instantiated with RDF Schema together with OWL (RDF-S and OWL), a deviation is depicted in the layers because of the omission of a *Rules* layer above *OWL*. It is however foreseen that a rule language will be required above an OWL ontology, hence the SWRL initiative of the W3C (RIF builds on RDF Schema only). It is plausible to speculate that this omission will have to be rectified in future. The CFL architecture thus allows for diverse technology specifications for the instantiation of a layer.

It can be argued that the latest version of the Semantic Web reference architecture also presents two towers of possible technology instantiations, and that both can be regarded as instantiations of the CFL architecture.

4.3 Application Usage Scenario

In this section the usefulness of the CFL architecture is illustrated by simulating two applications using two different meta-data data models.

In this usage scenario, the proposed system is a distributed management application for book publishers using Semantic Web technologies of the bottom three layers as depicted in Figure 7. The implementation or instantiation of the technology at the layer of communication should be similar. In this scenario, the meta-data data model could be either an entity relationship model, or an RDF diagram. Because the system implements up to and including Layer 3, the layer 3 models have to be implemented using the same technology.

The important aspect is that there is a *choice* of meta-data data models when the proposed CFL architecture for the Semantic Web is used. In contrast, the original reference architecture would have limited the choice to RDF.

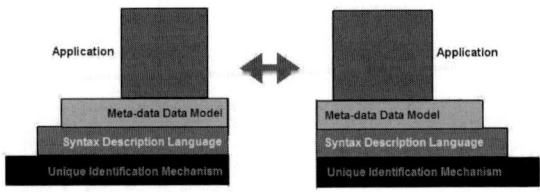

Fig. 7. Applications using the first three layers of the proposed Semantic Web CFL architecture

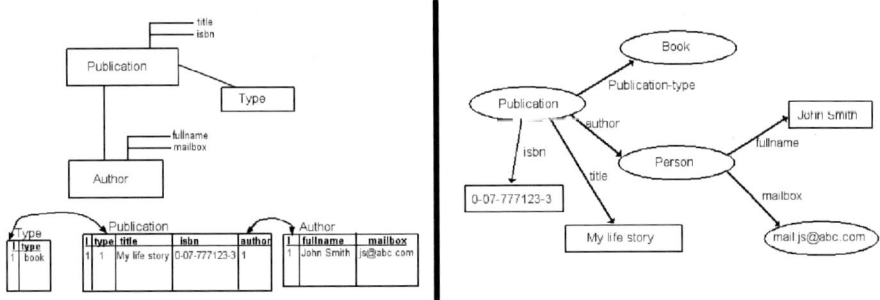

Fig. 8. An entity relationship diagram with a database implementation (left) and a RDF diagram (right) modelling the *publication* data of the scenario

In the scenario data model, a *publication*, which is of a specific *type* (a book in this case), has a *title* (*'My life story'*), an *isbn* (0-07-777123-3) and an *author*. The *author* is a *Person* with a *fullname* (John Smith) and a *mailbox* (mail:js@abc.com).

The Layer 3 implementation will be a data model using either an ER (entity relationship) diagram or an RDF diagram. A possible ER diagram is depicted on the left in Figure 8. Similarly, this data can be modelled in an RDF diagram. An RDF diagram modelling the same data is depicted on the right in Figure 8.

Figure 8 therefore depict two possible choices of meta-data data modelling technologies that can be used in this scenario as the proposed CFL architecture does not prescribe the technology. As long as the implementation adheres to the principles of layering and the specified functionality of the layer it implements, any meta-data data modelling technology can be used. Both models use a Layer 2 that is instantiated with XML using various serialisation techniques, and a Layer 1 that decodes to Unicode URIs. The Layer 1 implementation for both systems uses URI to uniquely identify the resources under discussion. In addition, the systems encode the serialised XML using Unicode to ensure that it is uniquely interpreted across the world.

In the usage scenario in this section an application is implemented using the proposed CFL architecture for the Semantic Web. Developers are able to choose between alternative technologies to implement the different layers of the

CFL architecture for the Semantic Web effectively. In this usage scenario, it is demonstrated how two possible meta-data data-modelling technologies can be used to implement Layer 3.

5 Conclusion

Within this paper the different versions of the Semantic Web reference architecture proposed by Berners-Lee Berners-Lee [6, 9, 12, 13] were discussed. The versions primarily depict technologies adopted by the W3C as standards or W3C Recommendations. This correlates with the mandate of the W3C that includes the development of *specifications* for the Web. However, when compared to definitions and criteria for architectures from software engineering principles, it is possible to identify several inconsistencies and discrepancies in these proposed versions. This often leads to confusion, as well as conflicting proposals and adoptions of technology and functionality by the Semantic Web community. Furthermore, Semantic Web applications based on technologies using these architectures will not be able to interoperate because there is no precise functionality and interface definition.

The design of the architecture of any system is a crucial activity during its development and it is necessary to attach precise and indisputable meaning to the system components, their functionlity and relationships with the associated interfaces. Similarly, it is necessary to attach indisputable meaning to the architecture for the languages of the Semantic Web.

When an architecture is designed, it is necessary to adhere to software engineering design principles. These principles imply that a system architecture should represent functional system components within a specific context with clearly defined purposes and interfaces. An example of such an architecture is the ISO/OSI layered architecture for network interoperability.

Similarly to the ISO/OSI model description and development approach, Gerber et al. [23] proposed a CFL architecture for the Semantic Wen that aims to present a structure that permits the meta-data languages for the Semantic Web to be viewed as logically composed of a succession of layers, each wrapping the lower layers and isolating them from the higher layers [32]. The CFL architecture depicts a simplification of the original architecture versions proposed by Bernes-Lee as a result of the *abstraction* of required functionality of language layers. Gerber et al. [23] argues that an abstracted layered architecture for the Semantic Web with well-defined functionalities will assist with the resolution of several of the current Semantic Web research debates such as the layering of language technologies. A functional architecture can accommodate diverse technologies for implementation of the required functionalities of each layer as long as the layer implementation adheres to interface specifications.

It is acknowledged that all issues with regard to the layering of the Semantic Web languages have not been resolved by this approach. However, the establishment of an approach and a first iteration CFL architecture for the Semantic Web is an important and necessary step towards realising the eventual notion of the Semantic Web. The CFL architecture for the Semantic Web should assist in the development of Semantic Web specifications and applications, or even W3C Recommendations, and several of the current research and implementation issues associated with the implementation of the Semantic Web could potentially be resolved. The CFL architecture might even assist the W3C to include all technology developments rather than to make a choice for the adoption of only one standard to the exclusion of other technologies. This position is similar to the position adopted by the network community with the acceptance of the ISO/OSI layered architecture as architectural framework for network interoperability standards.

In order to evaluate the usefulness of the proposed CFL architecture with its associated claims, the architecture was applied to various usage scenarios. In all of the scenarios, the CFL architecture proved to be of value.

It is plausible to speculate that the present versions of the Semantic Web reference architecture were developed to depict the progression pertaining to the development of W3C technologies. The definition of a general, accepted layered architecture with functional components (CFL architecture) could possibly assist the W3C to develop different technology specifications for the implementation of a specific layer functionality. This approach would *include*, rather than *exclude* technologies that would have as benefit a more rapid adoption of the Semantic Web vision. In addition, the definition of W3C specifications requires an appropriate level of abstraction to be able to penetrate sufficiently into the implementation domain, and it is the contention of the author that this can be provided by the CFL architecture for the Semantic Web as proposed in this paper.

References

[1] Antoniou, G., von Harmelen, F.: A Semantic Web Primer. MIT Press, Cambridge (2004)
[2] Bachman, C.: Personal Chronicle: Creating Better Information Systems, with Some Guiding Principles. IEEE Transactions on Knowledge and Data Engineering, 17–32 (1989)
[3] Bachmann, F., Bass, L., Carriere, J., Clements, P., Garlan, D., Ivers, J., Nord, R., Little, R.: Software Architecture Documentation in Practice: Documenting Architectural Layers. Technical Report CMU/SEI-2000-SR-004, Carnegie Mellon Software Engineering Institute (2005)
[4] Bass, L., Clements, P., Kazman, R.: Software Architecture in Practice. Addison Wesley Professional, Reading (2003) (Last accessed 12/8/2006)
http://www.aw-bc.com/catalog/academic/product/0,4096,0321154959,00.html
[5] Berners-Lee, T.: Axioms of Web Architecture: Metadata Architecture (1997) (Last accessed 2/10/2006), W3C Web site
http://www.w3.org/DesignIssues/Metadata.html
[6] Berners-Lee T.: Semantic Web - XML 2000. W3C Web site (2000) (Last accessed 11/8/2006),
http://www.w3.org/2000/Talks/1206-xml2k-tbl/slide10-0.html

[7] Berners-Lee, T.: Axioms of Web Architecture: Principles of design (2002) (Last accessed 5/10/2006), W3C Web site
http://www.w3.org/DesignIssues/Principles.html
[8] Berners-Lee, T.: Web Architecture from 50,000 feet (2002) (Last accessed 4/10/2006), W3C Website
http://www.w3.org/DesignIssues/Architecture.html
[9] Berners-Lee, T.: The Semantic Web and Challenges (2003) (Last accessed 10/10/2006), W3C Web site Slideshow
http://www.w3.org/2003/Talks/01-sweb-tbl/Overview.html
[10] Berners-Lee, T.: Standards, Semantics and Survival. SIIA Upgrade, pp. 6–10 (2003)
[11] Berners-Lee, T.: WWW Past and Future (2003) (Last accessed 10/10/2006), W3C Web site
http://www.w3.org/2003/Talks/0922-rsoc-tbl/slide30-0.html
[12] Berners-Lee, T.: WWW 2005 Keynote. W3C Web site (2005) (Last accessed 10/10/2006),
http://www.w3.org/2005/Talks/0511-keynote-tbl/
[13] Berners-Lee, T.: Artificial Intelligence and the Semantic Web: AAAI 2006 Keynote (2006) (Last accessed 12/10/2006), W3C Web site
http://www.w3.org/2006/Talks/0718-aaai-tbl/Overview.html
[14] Berners-Lee, T., Bray, T., Connolly, D., Cotton, P., Fielding, R., et al.: Architecture of the World Wide Web, Volume One (2004), W3C Web site
http://www.w3.org/TR/2004/REC-webarch-20041215/
[15] Berners-Lee, T., Hendler, J., Lassila, O.: The Semantic Web. The Scientific American 5(1), 36 (2001) (Last accessed 20/9/2006),
http://www.scientificamerican.com/2001/0501issue/0501berbers-lee.html
[16] Bos, B.: What is a good standard? An essay on W3C's design principles (2003) (Last accessed 2/10/2006), W3C Web site
http://www.w3.org/People/Bos/DesignGuide/introduction
[17] Bruegge, B., Dutoit, A.H.: Object-oriented Software Engineering using UML, Patterns, and Java, 2nd edn. Prentice-Hall, Englewood Cliffs (2004)
[18] Fowler, M.: Patterns of Enterprise Application Architecture. Addison-Wesley, Boston (2003)
[19] Garlan, D.: Software architecture: a roadmap. In: ICSE 2000: Proceedings of the Conference on The Future of Software Engineering, ACM Press, New York (2000)
[20] Garlan, D., Shaw, M.: An Introduction to Software Architecture (1994) (Last accessed 15/9/2006), http://www.inf.ed.ac.uk/teaching/courses/seoc1/2005_2006/resources/intro_softarch.pdf
[21] Gerber, A., Barnard, A., van der Merwe, A.: Design and Evaluation Criteria for Layered Architectures. In: Proceedings of the MSVVEIS Workshop hosted at the 8th International Conference on Enterprise Informatiopn Systems, Paphos, Cyprus, pp. 163–172 (2006) ISBN 972-8865-49-8
[22] Gerber, A., Barnard, A., van der Merwe, A.: A Semantic Web Status Model. In: Proceedings of the Ninth World Conference on Integrated Design & Process Technology, San Diego, California, IEEE, Los Alamitos (2006)
[23] Gerber, A., van der Merwe, A., Barnard, A.: Towards a Semantic Web Layered Architecture. In: Proceedings of IASTED International Conference on Software Engineering (SE 2007), Innsbruck, Austria, pp. 353–362 (2007) ISBN 978-0-88986-641-6

[24] Hendler, J.: Agents and the Semantic Web. IEEE Intelligent Systems 16, 30–37 (2001)
[25] Horrocks, I., Parsia, B., Patel-Schneider, P., Hendler, J.: Semantic Web Architecture: Stack or Two Towers? In: Fages, F., Soliman, S. (eds.) PPSWR 2005. LNCS, vol. 3703, Springer, Heidelberg (2005)
[26] Horrocks, I., Patel-Schneider, P.F.: Three theses of representation in the semantic web. In: WWW 2003: Proceedings of the 12th international conference on World Wide Web, pp. 39–47. ACM Press, New York (2003)
[27] Kifer, M., Bruijn, J., Boley, H., Fensel, D.: A Realistic Architecture for the Semantic Web. In: Adi, A., Stoutenburg, S., Tabet, S. (eds.) RuleML 2005. LNCS, vol. 3791, pp. 17–29. Springer, Heidelberg (2005)
[28] Nutt, G.J.: Centralized and Distributed Operating Systems. Prentice-Hall International Editions (1992)
[29] Patel-Schneider, P.F.: A Revised Architecture for Semantic Web Reasoning. In: Fages, F., Soliman, S. (eds.) PPSWR 2005. LNCS, vol. 3703, pp. 32–36. Springer, Heidelberg (2005)
[30] SIIA (Last accessed 21/9/2006), Website http://www.siia.net/
[31] W3C. W3C Technical Architecture Group (TAG) (2006) (Last accessed 4/10/2006), W3C Web site http://www.w3.org/2001/tag/
[32] Zimmermann, H.: OS1 Reference Model - The ISO Model of Architecture for Open Systems Interconnection. IEEE Transactions on Communications 71, 1334–1340 (1980) (Last accessed 26/9/2006), http://www.comsoc.org/livepubs/50_journals/pdf/RightsManagement_eid=136833.pdf

Query Answering and Ontology Population: An Inductive Approach

Claudia d'Amato, Nicola Fanizzi, and Floriana Esposito

Department of Computer Science, University of Bari
{claudia.damato,fanizzi,esposito}@di.uniba.it

Abstract. In order to overcome the limitations of deductive logic-based approaches to deriving operational knowledge from ontologies, especially when data come from distributed sources, inductive (instance-based) methods may be better suited, since they are usually efficient and noise-tolerant. In this paper we propose an inductive method for improving the instance retrieval and enriching the ontology population. By casting retrieval as a classification problem with the goal of assessing the individual class-memberships w.r.t. the query concepts, we propose an extension of the *k-Nearest Neighbor* algorithm for OWL ontologies based on an *entropic* distance measure. The procedure can classify the individuals w.r.t. the known concepts but it can also be used to retrieve individuals belonging to query concepts. Experimentally we show that the behavior of the classifier is comparable with the one of a standard reasoner. Moreover we show that new knowledge (not logically derivable) is induced. It can be suggested to the knowledge engineer for validation, during the ontology population task.

1 Introduction

Classification and query answering for retrieving resources in a knowledge base (KB) are important tasks. Generally these activities are performed by means of logical approaches that may fail when data comes from distributed sources, and is therefore exposed to inconsistency problems. This has given rise to alternative methods, such as *non-monotonic, paraconsistent* [8], *approximate* reasoning (see the discussion in [9]).

Inductive methods are known to be be quite efficient and more noise-tolerant, hence they seem suitable for contexts where knowledge is intended to be acquired from distributed sources. In this paper we propose an inductive *instance-based* method for *concept retrieval* [1] and *query answering* that may suggest new assertions which could not be logically derived, providing also a measure of their likelihood which may help dealing with the uncertainty caused by the inherent incompleteness of the KBs in the Semantic Web.

Namely, instance retrieval and query answering can be cast as classification problems, i.e. assessing the class-membership of the individuals in the KB w.r.t. some query concepts. Reasoning by analogy, similar individuals should likely belong to the extension of similar concepts. Moving from such an intuition, an instance-based framework for retrieving resources contained in ontological KBs has been devised, to inductively infer (likely) consistent class-membership assertions that may be not logically derivable. As such, the resulting assertions may enrich the KBs since the method can also provide a likelihood measure for its outcomes. Then the time-consuming ontology population task can be facilitated since the knowledge engineer only has to validate such new knowledge, as also argued in [2].

Logic-based approaches to (approximate) instance retrieval have been proposed in the literature [13, 11]. We intend to apply inductive forms of reasoning borrowed from *machine learning*. Specifically, we propose an extension of the well-known *Nearest Neighbor* search (henceforth, *NN*) [14] to the standard representations of the SW (RDF through OWL). Our analogical approach is based on a dissimilarity measures for resources in these search space. The procedure retrieves individuals belonging to query concepts, by analogy with other training instances, namely on the grounds of the classification of the nearest ones (w.r.t. the dissimilarity measure). This approach may be quite efficient because it requires checking class-membership for a limited set of training instances yielding a decision on the classification of new instances.

From a technical viewpoint, extending the NN setting to the target representations founded in Description Logics (DL) [1], required suitable metrics whose definition could not be straightforward. In particular, a theoretical problem is posed by the *Open World Assumption* (OWA) that is generally made on the semantics of SW ontologies, differently from the typical standards of databases where the *Closed World Assumption* (CWA) is made. Moreover, the NN algorithms are devised for simple classifications where classes are assumed to be pairwise disjoint, which is quite unlikely in the Semantic Web context where an individual can be instance of more than one concept. Furthermore, dissimilarity measures that can cope with the semantics of expressive representations are necessary.

Most of the existing measures focus on concept (dis)similarity and particularly on the (dis)similarity of atomic concepts within hierarchies or simple ontologies (see the discussion in [3]). Conversely, for our purposes, a notion of dissimilarity between *individuals* is required. Recently, dissimilarity measures for specific description logics concept descriptions have been proposed [3, 4]. Although they turned out to be quite effective for the inductive tasks of interest, they are still partly based on structural criteria (a notion of normal form) which determine their main weakness: they are hardly scalable to deal with standard ontology languages.

In order to overcome these limitations, an extension of a semantic pseudo-metrics [7] is exploited. This language-independent measure assesses the dissimilarity of two individuals by comparing them on the grounds of their behavior w.r.t. a committee of features (concepts), namely those defined in the KB or that can be generated to this purpose[1]. In the former measures, all the features have

---
[1] The choice of optimal committees may be performed in advance through randomized search algorithms [7].

the same importance in determining the dissimilarity. However, it may well be that some features have a larger discriminating power w.r.t. the others. In this case, they should be more relevant in determining the dissimilarity value. Moving from this observation, we propose an extension of the measures presented in [7], where each feature of the committee is weighted on the grounds of the *quantity of information* that it conveys. This weight is then determined as an *entropy* measure, also used in attribute selection when building decision trees. The rationale is that the more general a feature (or its negation) is (i.e. low entropy) the less likely it may be usable for distinguishing the two individuals and vice versa.

The measure has been integrated in the NN procedure [4] and the classification of resources (individuals) w.r.t. a query concept has been performed through a voting procedure weighted by the neighbors' similarity. The resulting system allowed for an experimentation of the method on performing instance retrieval and query answering with a number ontologies drawn from public repositories. Its predictions were compared to assertions that were logically derived by a deductive reasoner. The experiments showed that the classification results are comparable (although slightly less complete) and also that the classifier is able to induce new knowledge that is not logically derivable. The experimentation also compared the outcomes obtained by the former measure, extended in this paper. Such a comparison showed that the measure presented in this paper may improve the classification results.

The paper is organized as follows. The basics of the instance-based approach applied to the standard representations are recalled in Sect. 2. The next Sect. 3 presents the semantic dissimilarity measures adopted in the retrieval procedure. Sect. 4 reports the outcomes of the experiments performed with the implementation of the procedure. Possible developments are finally examined in Sect. 5.

2 Resource Retrieval as Nearest Neighbor Search

2.1 Representation and Inference

In the following sections, we assume that concept descriptions are defined in terms of a generic sub-language based on OWL-DL that may be mapped to *Description Logics* with the standard model-theoretic semantics (see the handbook [1] for a thorough reference).

A *knowledge base* $\mathcal{K} = \langle \mathcal{T}, \mathcal{A} \rangle$ contains a *TBox* $\mathcal{T}$ and an *ABox* $\mathcal{A}$. $\mathcal{T}$ is a set of axioms that define concepts. $\mathcal{A}$ contains factual assertions concerning the resources, also known as individuals. Moreover, the *unique names assumption* may be made on the ABox individuals, that are represented by their URIs. The set of the individuals occurring in $\mathcal{A}$ will be denoted with $\mathsf{Ind}(\mathcal{A})$.

As regards the inference services, like all other instance-based methods, our procedure may require performing *instance-checking* [1], which roughly amounts to determining whether an individual, say a, belongs to a concept extension, i.e. whether $C(a)$ holds for a certain concept C. Note that because of the OWA, a reasoner may be unable to give a positive or negative answer to a class-membership query. This service is provided proof-theoretically by a reasoner.

2.2 The Method

Query answering boils down to determining whether a resource belongs to a (query) concept extension. Here, an alternative inductive method is proposed for retrieving the resources that likely belong to a query concept. Such a method may also be able to provide an answer even when it may not be inferred by deduction, Moreover, it may also provide a measure of the likelihood of its answer.

In *similarity search* [14] the basic idea is to find the most similar object(s) to a query one (i.e. the one that is to be classified) with respect to a similarity (or dissimilarity) measure. We review the basics of the k-NN method applied to the Semantic Web context [4] context.

The objective is to induce an approximation for a discrete-valued target hypothesis function $h : IS \mapsto V$ from a space of instances IS to a set of values $V = \{v_1, \ldots, v_s\}$ standing for the classes (concepts) that have to be predicted. Note that normally $|IS| \ll |\mathsf{Ind}(\mathcal{A})|$ i.e. only a limited number of training instances is needed especially if they are prototypical for a region of the search space. Let x_q be the query instance whose class-membership is to be determined. Using a dissimilarity measure, the set of the k nearest (pre-classified) training instances w.r.t. x_q is selected: $NN(x_q) = \{x_i \mid i = 1, \ldots, k\}$.

In its simplest setting, the k-NN algorithm approximates h for classifying x_q on the grounds of the value that h is known to assume for the training instances in $NN(x_q)$, i.e. the k closest instances to x_q in terms of a dissimilarity measure. Precisely, the value is decided by means of a weighted majority voting procedure: it is simply the most *voted* value by the instances in $NN(x_q)$ weighted by the similarity of the neighbor individual.

The estimate of the hypothesis function for the query individual is:

$$\hat{h}(x_q) := \operatorname*{argmax}_{v \in V} \sum_{i=1}^{k} w_i \delta(v, h(x_i)) \quad (1)$$

where δ returns 1 in case of matching arguments and 0 otherwise, and, given a dissimilarity measure d, the weights are determined by $w_i = 1/d(x_i, x_q)$.

Note that the estimate function $\hat{h}$ is defined extensionally: the basic k-NN method does not return an intensional classification model (a function or a concept definition), it merely gives an answer for the instances to be classified.

It should be also observed that this setting assigns a value to the query instance which stands for one in a set of pairwise disjoint concepts (corresponding to the value set V). In a multi-relational setting this assumption cannot be made in general. An individual may be an instance of more than one concept.

The problem is also related to the CWA usually made in the knowledge discovery context. To deal with the OWA, the absence of information on whether a training instance x belongs to the extension of the query concept Q should not be interpreted negatively, as in the standard settings which adopt the CWA.

Rather, it should count as neutral (uncertain) information. Thus, assuming the alternate viewpoint, the multi-class problem is transformed into a ternary one. Hence another value set has to be adopted, namely $V = \{+1, -1, 0\}$, where the three values denote, respectively, membership, non-membership, and uncertainty, respectively.

The task can be cast as follows: given a query concept Q, determine the membership of an instance x_q through the NN procedure (see Eq. 1) where $V = \{-1, 0, +1\}$ and the hypothesis function values for the training instances are determined as follows:

$$h_Q(x) = \begin{cases} +1 & \mathcal{K} \models Q(x) \\ -1 & \mathcal{K} \models \neg Q(x) \\ 0 & \textit{otherwise} \end{cases}$$

i.e. the value of h_Q for the training instances is determined by the entailment[2] the corresponding assertion from the knowledge base.

Note that, being based on a majority vote of the individuals in the neighborhood, this procedure is less error-prone in case of noise in the data (e.g. incorrect assertions) w.r.t. a purely logic deductive procedure, therefore it may be able to give a correct classification even in case of (partially) inconsistent knowledge bases.

It should be noted that the inductive inference made by the procedure shown above is not guaranteed to be deductively valid. Indeed, inductive inference naturally yields a certain degree of uncertainty. In order to measure the likelihood of the decision made by the procedure (individual x_q belongs to the query concept denoted by value v maximizing the argmax argument in Eq. 1), given the nearest training individuals in $NN(x_q, k) = \{x_1, \ldots, x_k\}$, the quantity that determined the decision should be normalized by dividing it by the sum of such arguments over the (three) possible values:

$$l(class(x_q) = v | NN(x_q, k)) = \frac{\sum_{i=1}^{k} w_i \cdot \delta(v, h_Q(x_i))}{\sum_{v' \in V} \sum_{i=1}^{k} w_i \cdot \delta(v', h_Q(x_i))} \qquad (2)$$

Hence the likelihood of the assertion $Q(x_q)$ corresponds to the case when $v = +1$.

3 A Semantic Pseudo-Metric for Individuals

As mentioned in the first section, various attempts to define semantic similarity (or dissimilarity) measures for concept languages have been made, yet they have still a limited applicability to simple languages [3] or they are not completely semantic depending also on the structure of the descriptions [4]. Moreover, for our purposes, we need a function for measuring the similarity of individuals rather than concepts. It can be observed that individuals do not have a syntactic structure that can be compared. This has led to lifting them to the concept description level before comparing them (recurring to the notion of the *most*

---

[2] We use $\models$ to denote entailment, as computed through a reasoner.

specific concept of an individual w.r.t. the ABox [1], yet this makes the measure language-dependent. Besides, it would add a further approximations as the most specific concepts can be defined only for simple DLs.

For the NN procedure, we intend to exploit a new measure that totally depends on semantic aspects of the individuals in the knowledge base.

3.1 The Family of Measures

The new dissimilarity measures are based on the idea of comparing the semantics of the input individuals along a number of dimensions represented by a committee of concept descriptions. Indeed, on a semantic level, similar individuals should behave similarly with respect to the same concepts. Following the ideas borrowed from [12], totally semantic distance measures for individuals can be defined in the context of a knowledge base.

More formally, the rationale is to compare individuals on the grounds of their semantics w.r.t. a collection of concept descriptions, say $\mathsf{F} = \{F_1, F_2, \ldots, F_m\}$, which stands as a group of discriminating *features* expressed in the OWL-DL sub-language taken into account.

In its simple formulation, a family of distance functions for individuals inspired to Minkowski's norms L_p can be defined as follows [7]:

Definition 3.1 (family of measures). *Let $\mathcal{K} = \langle \mathcal{T}, \mathcal{A} \rangle$ be a knowledge base. Given a set of concept descriptions $\mathsf{F} = \{F_1, F_2, \ldots, F_m\}$, a family of dissimilarity functions $d_p^\mathsf{F} : \mathsf{Ind}(\mathcal{A}) \times \mathsf{Ind}(\mathcal{A}) \mapsto [0,1]$ is defined as follows:*

$$\forall a, b \in \mathsf{Ind}(\mathcal{A}) \quad d_p^\mathsf{F}(a,b) := \frac{1}{|\mathsf{F}|} \left[\sum_{i=1}^{|\mathsf{F}|} w_i \mid \delta_i(a,b) \mid^p \right]^{1/p}$$

where $p > 0$ and $\forall i \in \{1, \ldots, m\}$ the dissimilarity function δ_i is defined by:

$$\forall (a,b) \in (\mathsf{Ind}(\mathcal{A}))^2 \quad \delta_i(a,b) = \begin{cases} 0 & F_i(a) \in \mathcal{A} \land F_i(b) \in \mathcal{A} \text{ or} \\ & \neg F_i(a) \in \mathcal{A} \land \neg F_i(b) \in \mathcal{A} \\ 1 & F_i(a) \in \mathcal{A} \land \neg F_i(b) \in \mathcal{A} \text{ or} \\ & \neg F_i(a) \in \mathcal{A} \land F_i(b) \in \mathcal{A} \\ 1/2 & \text{otherwise} \end{cases}$$

or, model theoretically:

$$\forall (a,b) \in (\mathsf{Ind}(\mathcal{A}))^2 \quad \delta_i(a,b) = \begin{cases} 0 & \mathcal{K} \models F_i(a) \land \mathcal{K} \models F_i(b) \text{ or} \\ & \mathcal{K} \models \neg F_i(a) \land \mathcal{K} \models \neg F_i(b) \\ 1 & \mathcal{K} \models F_i(a) \land \mathcal{K} \models \neg F_i(b) \text{ or} \\ & \mathcal{K} \models \neg F_i(a) \land \mathcal{K} \models F_i(b) \\ 1/2 & \text{otherwise} \end{cases}$$

Note that the original measures [7] correspond to the case of uniform weights.

The alternative definition for the projections, requires the entailment of an assertion (instance-checking) rather than the simple ABox look-up; this can make the measure more accurate yet more complex to compute unless a KBMS is employed maintaining such information at least for the concepts in F.

In particular, we will consider the measures $d_1^\mathsf{F}(\cdot,\cdot)$ or $d_2^\mathsf{F}(\cdot,\cdot)$ in the experiments.

As regards the weights employed in the family of measures, they should reflect the impact of the single feature concept w.r.t. the overall dissimilarity. As mentioned, this can be determined by the quantity of information conveyed by a feature, which can be measured as its entropy. Namely, the extension of a feature F w.r.t. the whole domain of objects may be probabilistically quantified as $P_F = |F^\mathcal{I}|/|\Delta^\mathcal{I}|$ (w.r.t. the canonical interpretation $\mathcal{I}$). This can be roughly approximated with: $P_F = |\mathsf{retrieval}(F)|/|\mathsf{Ind}(\mathcal{A})|$. Hence, considering also the probability $P_{\neg F}$ related to its negation and that related to the unclassified individuals (w.r.t. F), denoted P_U, we may give an entropic measure for the feature:

$$H(F) = -\left(P_F \log(P_F) + P_{\neg F} \log(P_{\neg F}) + P_U \log(P_U)\right).$$

These measures may be normalized for providing a good set of weights for the distance measures.

3.2 Discussion

It is easy to prove [7] that these functions have the standard properties for pseudo metrics (i.e. semi-distances [14]):

Proposition 3.1 (pseudo-metric). *For a given a feature set* F *and* $p > 0$, d_p^F *is a pseudo-metric.*

Proof. It is to be proved that:

1. $d_p(a,b) \geq 0$
2. $d_p(a,b) = d_p(b,a)$
3. $d_p(a,c) \leq d_p(a,b) + d_p(b,c)$

1. and 2. are trivial. As for 3., noted that

$$(d_p(a,c))^p = \frac{1}{m^p} \sum_{i=1}^{m} w_i \mid \delta_i(a,c) \mid^p$$

$$\leq \frac{1}{m^p} \sum_{i=1}^{m} w_i \mid \delta_i(a,b) + \delta_i(b,c) \mid^p$$

$$\leq \frac{1}{m^p} \sum_{i=1}^{m} w_i \mid \delta_i(a,b) \mid^p + \frac{1}{m^p} \sum_{i=1}^{m} w_i \mid \delta_i(b,c) \mid^p$$

$$\leq (d_p(a,b))^p + (d_p(b,c))^p \leq (d_p(a,b) + d_p(b,c))^p$$

then the property follows for the monotonicity of the root function.

It cannot be proved that $d_p^\mathsf{F}(a,b) = 0$ iff $a = b$. This is the case of *indiscernible* individuals with respect to the given set of features F. To fulfill this property several methods have been proposed involving the consideration of equivalent classes of individuals or the adoption of a supplementary meta-feature F_0 determining the equality of the two individuals: $\delta_0(a,b) = 0$ if $a^\mathcal{I} = b^\mathcal{I}$ otherwise $\delta_0(a,b) = 1$.

Compared to other proposed dissimilarity measures [3, 4], the presented functions do not depend on the constructors of a specific language, rather they require only (retrieval or) instance-checking for computing the projections through class-membership queries to the knowledge base.

The complexity of measuring he dissimilarity of two individuals depends on the complexity of such inferences (see [1], Ch. 3). Note also that the projections that determine the measure can be computed (or derived from statistics maintained on the knowledge base) before the actual distance application, thus determining a speed-up in the computation of the measure. This is very important for algorithms that massively use this distance, such as all instance-based methods.

The measures strongly depend on F. Here, we make the assumption that the feature-set F represents a sufficient number of (possibly redundant) features that are able to discriminate really different individuals. The choice of the concepts to be included – *feature selection* – is beyond the scope of this work (see [7] for a randomized optimization procedure aimed at finding optimal committees). Experimentally, we could obtain good results by using the very set of both primitive and defined concepts found in the knowledge base.

Of course these approximate measures become more and more precise as the knowledge base is populated with an increasing number of individuals.

4 Experimentation

4.1 Experimental Setting

The NN procedure integrated with the pseudo-metric proposed in the previous section has been tested in a number of retrieval problems. To this purpose, we selected several ontologies from different domains represented in OWL, namely: SURFACE-WATER-MODEL (SWM), NEWTESTAMENTNAMES (NTN) from the Protégé library[3], the Semantic Web Service Discovery dataset[4] (SWSD), the University0.0 ontology generated by the Lehigh University Benchmark[5] (LUBM), the BioPax glycolysis ontology[6] (BioPax) and the FINANCIAL ontology[7]. Tab. 1 summarizes details concerning these ontologies.

For each ontology, 20 queries were randomly generated by composition (conjunction and/or disjunction) of (2 through 8) primitive and defined concepts in each knowledge base. Query concepts were constructed so that each offered both positive and negative instances among the ABox individuals. The performance

---

[3] http://protege.stanford.edu/plugins/owl/owl-library
[4] https://www.uni-koblenz.de/FB4/Institutes/IFI/AGStaab/Projects/xmedia/dl-tree.htm
[5] http://swat.cse.lehigh.edu/projects/lubm
[6] http://www.biopax.org/Downloads/Level1v1.4/biopax-example-ecocyc-glycolysis.owl
[7] http://www.cs.put.poznan.pl/alawrynowicz/financial.owl

Table 1. Facts concerning the ontologies employed in the experiments

Ontology	DL language	#concepts	#object prop.	#data prop.	#individuals
SWM	$\mathcal{ALCOF}(D)$	19	9	1	115
BioPAX	$\mathcal{ALCHF}(D)$	28	19	30	323
LUBM	$\mathcal{ALR^+HI}(D)$	43	7	25	555
NTN	$\mathcal{SHIF}(D)$	47	27	8	676
SWSD	$\mathcal{ALCH}$	258	25	0	732
Financial	$\mathcal{ALCIF}$	60	17	0	1000

of the inductive method was evaluated by comparing its responses to those returned by a standard reasoner[8] as a baseline.

Experimentally, it was observed that large training sets make the distance measures (and consequently the NN procedure) very accurate. In order to make the problems more difficult, we selected limited training sets (TS) that amount to only 4% of the individuals occurring in each ontology. Then the parameter k was set to $\log(|TS|)$ depending on the number of individuals in the training set. Again, we found experimentally that much smaller values could be chosen, resulting in the same classification.

The simpler distances (d_1^F) were employed from the *original* family (uniform weights) and *entropic* family (weighted on the feature entropy), using all the concepts in the knowledge base for determining the set F with no further optimization.

4.2 Results

Standard Measures. Initially the standard measures precision, recall, F_1-measure were employed to evaluate the system performance, especially when selecting the positive instances (individuals that should belong to the query concept). The outcomes are reported in Tab. 2. For each knowledge base, we report the average values obtained over the 20 random queries as well as their standard deviation and minimum-maximum ranges of values.

As an overall consideration we may observe that generally the outcomes obtained adopting the extended measure improve on those with the other one and appear also more stable (with some exceptions). Besides, it is possible to note that precision and recall values are generally quite good for all ontologies but SWSD, where especially recall is significantly lower. Namely, SWSD turned out to be more difficult (also in terms of precision) for two reasons: a very limited number of individuals per concept was available and the number of different concepts is larger w.r.t. the other knowledge bases. For the other ontologies values are much higher, as testified also by the F-measure values. The results in terms of precision are also more stable than those for recall as proved by the limited variance observed, whereas single queries happened to turn out quite difficult as regards the correctness of the answer.

The reason for precision being generally higher is probably due to the OWA. Indeed, in a many cases it was observed that the NN procedure deemed some

---

[8] We employed Pellet v. 1.5.1. See http://pellet.owldl.com

Table 2. Experimental results in terms of standard measures: averages ± standard deviations and [min,max] intervals

	Original Measure		
	precision	recall	F-measure
SWM	89.1 ± 27.3 [16.3;100.0]	84.4 ± 30.6 [11.1;100.0]	78.7 ± 30.6 [20.0;100.0]
BioPax	99.2 ± 1.9 [93.8;100.0]	97.3 ± 11.3 [50.0;100.0]	97,8 ± 7.4 [66.7;100.0]
LUBM	100.0 ± 0.0 [100.0;100.0]	71.7 ± 38.4 [9.1;100.0]	76.2 ± 34.4 [16.7;100.0]
NTN	98.8 ± 3.0 [86.9;100.0]	62.6 ± 42.8 [4.3;100.0]	66.9 ± 37.7 [8.2;100.0]
SWSD	74.7 ± 37.2 [8.0;100.0]	43.4 ± 35.5 [2.2;100.0]	54.9 ± 34.7 [4.3;100.0]
Financial	99.6 ± 1.3 [94.3;100.0]	94.8 ± 15.3 [50.0;100.0]	97.1 ± 10.2 [66.7;100.0]
	Entropic Measure		
	precision	recall	F-measure
SWM	99.0 ± 4.3 [80.6;100.0]	75.8 ± 36.7 [11.1;100.0]	79.5 ± 30.8 [20.0;100.0]
BioPax	99.9 ± 0.4 [98.2;100.0]	97.3 ± 11.3 [50.0;100.0]	98,2 ± 7.4 [66.7;100.0]
LUBM	100.0 ± 0.0 [100.0;100.0]	81.6 ± 32.8 [11.1;100.0]	85.0 ± 28.4 [20.0;100.0]
NTN	97.0 ± 5.8 [76.4;100.0]	40.1 ± 41.3 [4.3;100.0]	45.1 ± 35.4 [8.2;97.2]
SWSD	94.1 ± 18.0 [40.0;100.0]	38.4 ± 37.9 [2.4;100.0]	46.5 ± 35.0 [4.5;100.0]
Financial	99.8 ± 0.3 [98.7;100.0]	95.0 ± 15.4 [50.0;100.0]	96.6 ± 10.2 [66.7;100.0]

individuals as relevant for the query issued while the DL reasoner was not able to assess this relevance and this was computed as a mistake while it may likely turn out to be a correct inference when judged by a human agent.

Because of these problems in the evaluation with the standard indices, especially due to the cases on unknown answers from the reference system (the reasoner) we thought to make this case more explicit by measuring both the rate of inductively classified individuals and the nature of the mistakes.

Alternative Measures. Due to the OWA, cases were observed when, it could not be (deductively) ascertained whether a resource was relevant or not for a given query. Hence, we introduced the following indices for a further evaluation:

- *match rate*: number of individuals that got exactly the same classification ($v \in V$) by both the inductive and the deductive classifier with respect to the overall number of individuals (v vs. v);

- *omission error rate*: amount of individuals for which inductive method could not determine whether they were relevant to the query or not while they were actually relevant according to the reasoner (0 vs. ±1);
- *commission error rate*: number of individuals (analogically) found to be relevant to the query concept, while they (logically) belong to its negation or vice-versa (+1 vs. −1 or −1 vs. +1);
- *induction rate*: amount of individuals found to be relevant to the query concept or to its negation, while either case is not logically derivable from the knowledge base (±1 vs. 0);

Tab. 3 reports the outcomes in terms of these indices. Preliminarily, it is important to note that, in each experiment, the commission error was quite low or absent. This means that the inductive search procedure is quite accurate, namely it did not make critical mistakes attributing an individual to a concept that is disjoint with the right one. Also the omission error rate was generally quite low, yet more frequent than the previous type of error.

The usage of all concepts for the set F of d_1^{F} made the measure quite accurate, which is the reason why the procedure resulted quite conservative as regards inducing new assertions. In many cases, it matched rather faithfully the reasoner decisions. From the retrieval point of view, the cases of induction are interesting because they suggest new assertions which cannot be logically derived by using a deductive reasoner yet they might be used to complete a knowledge base [2], e.g. after being validated by an ontology engineer. For each candidate new assertion, Eq. 2 may be employed to assess the likelihood and hence decide on its inclusion (see next section).

If we compare these outcomes with those reported in other works on instance retrieval and inductive classification [4], where the highest average match rate observed was around 80%, we find a significant increase of the performance due to the accuracy of the new measure. Also the elapsed time (not reported here) was much less because of the different dissimilarity measure: once the values for the projection functions are pre-computed, the efficiency of the classification, which depends on the computation of the dissimilarity, was also improved.

As mentioned, we found also that a choice for smaller number of neighbors could have been made for the decision on the correct classification was often quite easy, even on account of fewer (the closest) neighbors. This yielded also that the likelihood of the inferences made (see Eq. 2) turned out quite high.

Likelihood and Top-k Answers. A further investigation concerned the likelihood of the inductively answers provided by the NN procedure. In Tab. 4, we report the average likelihoods computed (for all queries per ontology) during the previous experiments in case of induction of new consistent assertions (see Eq. 2), when the reasoner was not able to assess the membership. The first line reports the averages when answers were given based on the normalization of the likelihood over the 3 possible values. As expected, they are even higher when only the two cases +1 or −1 (membership, non-membership) are considered (see second line). As mentioned, since the distance measure accurately selected

Table 3. Results with alternative indices: averages ± standard deviations and [min,max] intervals

	ORIGINAL MEASURE			
	match	commission	omission	induction
SWM	93.3 ± 10.3 [68.7;100.0]	0.0 ± 0.0 [0.0;0.0]	2.5 ± 4.4 [0.0;16.5]	4.2 ± 10.5 [0.0;31.3]
BIOPAX	99.9 ± 0.2 [99.4;100.0]	0.2 ± 0.2 [0.0;0.06]	0.0 ± 0.0 [0.0;0.0]	0.0 ± 0.0 [0.0;0.0]
LUBM	99.2 ± 0.8 [98.0;100.0]	0.0 ± 0.0 [0.0;0.0]	0.8 ± 0.8 [0.0;0.2]	0.0 ± 0.0 [0.0;0.0]
NTN	98.6 ± 1.5 [93.9;100.0]	0.0 ± 0.1 [0.0;0.4]	0.8 ± 1.1 [0.0;3.7]	0.6 ± 1.4 [0.0;6.1]
SWSD	97.5 ± 3.7 [84.6;100.0]	0.0 ± 0.0 [0.0;0.0]	1.8 ± 2.6 [0.0;9.7]	0.8 ± 1.5 [0.0;5.7]
FINANCIAL	99.5 ± 0.8 [97.3;100.0]	0.3 ± 0.7 [0.0;2.4]	0.0 ± 0.0 [0.0;0.0]	0.2 ± 0.2 [0.0;0.6]
	ENTROPIC MEASURE			
	match	commission	omission	induction
SWM	97.5 ± 3.2 [89.6;100.0]	0.0 ± 0.0 [0.0;0.0]	2.2 ± 3.1 [0.0;10.4]	0.3 ± 1.2 [0.0;5.2]
BIOPAX	99.9 ± 0.2 [99.4;100.0]	0.1 ± 0.2 [0.0;0.06]	0.0 ± 0.0 [0.0;0.0]	0.0 ± 0.0 [0.0;0.0]
LUBM	99.5 ± 0.7 [98.2;100.0]	0.0 ± 0.0 [0.0;0.0]	0.5 ± 0.7 [0.0;1.8]	0.0 ± 0.0 [0.0;0.0]
NTN	97.5 ± 1.9 [91.3;99.3]	0.6 ± 0.7 [0.0;1.6]	1.3 ± 1.4 [0.0;4.9]	0.6 ± 1.7 [0.0;7.1]
SWSD	98.0 ± 3.0 [88.3;100.0]	0.0 ± 0.0 [0.0;0.0]	1.9 ± 2.9 [0.0;11.3]	0.1 ± 0.2 [0.0;0.5]
FINANCIAL	99.7 ± 0.2 [99.4;100.0]	0.0 ± 0.0 [0.0;0.1]	0.0 ± 0.0 [0.0;0.0]	0.2 ± 0.2 [0.0;0.6]

Table 4. Results (percentages) concerning the likelihood of the answers when the reasoner is not able to assess the class membership

	SWM	BIOPAX	LUBM	NTN	SWSD	FINANCIAL
3-valued case	76.26	99.99	99.99	98.36	76.27	92.55
2-valued case	100.0	99.99	99.99	98.36	76.27	92.55

very similar neighbors, seldom tight cases occurred during the majority votes of the NN, hence the observed likelihood of the answers turned out quite high on average.

We also took into account the top-10 (positive) answers provided by the inductive procedure for the various queries, ranked according to the likelihood of the decision. Most of the values amounted to 100%. In order to assess the

Table 5. Average differences of likelihood values (%) observed comparing the NN procedure to the reasoner on the top-10 (positive) answers

	SWM	BioPax	LUBM	NTN	SWSD	Financial
likelihood diff.	0.0	0.2	0.0	0.3	2.5	0

accuracy of such answers, we compared their related likelihood values to those of the deductive decisions made by the reasoner. Namely, we assigned a maximum likelihood of 100% to the decisions on membership (and 0% to the non-membership answer, if any) while 50% was assigned to cases when the reasoner was uncertain on the answer. The pairwise difference of likelihood are averaged over the top-10 answers of the various queries per each ontology. In Tab 5, we report such average difference. As expected such difference values are quite low, reflecting the fact that the top-ranked answers are also the most accurate ones.

5 Conclusions and Outlook

This paper explored the application of a distance-based procedure for semantic search applied knowledge bases represented in OWL. We extended a family of semantic dissimilarity measures based on feature committees [7] taking into account the amount of information conveyed by each feature based on an estimate of its entropy. The measure were integrated in an distance-based search procedure that can be exploited for the task of approximate instance retrieval which can be demonstrated to be effective even in the presence of incomplete (or noisy) information.

One of the advantages of the measures is that their computation can be very efficient in cases when statistics (on class-membership) are maintained by the KBMS [10]. As previously mentioned, the subsumption relationships among concepts in the committee is not explicitly exploited in the measure for making the relative distances more accurate. The extension to the case of concept distance may also be improved. Hence, scalability should be guaranteed as far as a good committee has been found and does not change also because of the locality properties observed for instances in several domains (e.g. social or biological networks).

The experiments made on various ontologies showed that the method is quite effective, and while its performance depends on the number (and distribution) of the available training instances, even working with quite limited training sets guarantees a good performance in terms of accuracy. Moreover, even if the measure accuracy embedded into the system depends on the chosen feature set, the high accuracy registered for almost all considered data sets shows that the method can be applied to any domain and its performances are not connected to a particular domain. Besides, the procedure appears also robust to noise since it seldom made commission errors in the experiments carried out so far.

Various developments for the measure can be foreseen as concerns its definition. Namely, since it is very dependent on the features included in the committee, two immediate lines of research arise: 1) reducing the number of concepts saving those concepts which are endowed of a real discriminating power; 2) learning optimal sets of discriminating features, by allowing also their composition employing the specific constructors made available by the representation language of choice [7]. Both these objectives can be accomplished by means of machine learning techniques especially when ontologies with a large set of individuals are available. Namely, part of the entire data can be drawn in order to learn optimal feature sets, in advance with respect to the successive usage.

As mentioned, the distance measures are applicable to other instance-based tasks which can be approached through machine learning techniques. The next step has been plugging the measure in flat or hierarchical clustering algorithms where clusters would be formed grouping instances on the grounds of their similarity assessed through the measure [6, 5].

References

[1] Baader, F., Calvanese, D., McGuinness, D., Nardi, D., Patel-Schneider, P. (eds.): The Description Logic Handbook. Cambridge University Press, Cambridge (2003)

[2] Baader, F., Ganter, B., Sertkaya, B., Sattler, U.: Completing description logic knowledge bases using formal concept analysis. In: Veloso, M. (ed.) Proceedings of the 20th International Joint Conference on Artificial Intelligence, Hyderabad, India, pp. 230–235 (2007)

[3] Borgida, A., Walsh, T.J., Hirsh, H.: Towards measuring similarity in description logics. In: Horrocks, I., Sattler, U., Wolter, F. (eds.) Working Notes of the International Description Logics Workshop, Edinburgh, UK. CEUR Workshop Proceedings, vol. 147 (2005)

[4] d'Amato, C., Fanizzi, N., Esposito, F.: Reasoning by analogy in description logics through instance-based learning. In: Tummarello, G., Bouquet, P., Signore, O. (eds.) Proceedings of Semantic Web Applications and Perspectives, 3rd Italian Semantic Web Workshop, SWAP 2006, Pisa, Italy. CEUR Workshop Proceedings, vol. 201 (2006)

[5] Fanizzi, N., d'Amato, C., Esposito, F.: Evolutionary conceptual clustering of semantically annotated resources. In: Proceedings of the IEEE International Conference on Semantic Computing, ICSC 2007, Irvine, CA, IEEE, Los Alamitos (2007)

[6] Fanizzi, N., d'Amato, C., Esposito, F.: A hierarchical clustering procedure for semantically annotated resources. In: Basili, R., Pazienza, M.T. (eds.) AI*IA 2007. LNCS (LNAI), vol. 4733, pp. 266–277. Springer, Heidelberg (2007)

[7] Fanizzi, N., d'Amato, C., Esposito, F.: Induction of optimal semi-distances for individuals based on feature sets. In: Calvanese, D., Franconi, E., Haarslev, V., Lembo, D., Motik, B., Turhan, A.-Y., Tessaris, S. (eds.) Working Notes of the 20th International Description Logics Workshop, DL 2007, Bressanone, Italy. CEUR Workshop Proceedings, vol. 250 (2007)

[8] Haase, P., van Harmelen, F., Huang, Z., Stuckenschmidt, H., Sure, Y.: A framework for handling inconsistency in changing ontologies. In: Gil, Y., Motta, E., Benjamins, V.R., Musen, M.A. (eds.) ISWC 2005. LNCS, vol. 3729, pp. 353–367. Springer, Heidelberg (2005)

[9] Hitzler, P., Vrandečić, D.: Resolution-based approximate reasoning for OWL DL. In: Gil, Y., Motta, E., Benjamins, V.R., Musen, M.A. (eds.) ISWC 2005. LNCS, vol. 3729, pp. 383–397. Springer, Heidelberg (2005)

[10] Horrocks, I.R., Li, L., Turi, D., Bechhofer, S.K.: The Instance Store: DL reasoning with large numbers of individuals. In: Haarslev, V., Möller, R. (eds.) Proceedings of the 2004 Description Logic Workshop, DL 2004. CEUR Workshop Proceedings, vol. 104, pp. 31–40 (2004)
[11] Möller, R., Haarslev, V., Wessel, M.: On the scalability of description logic instance retrieval. In: Parsia, B., Sattler, U., Toman, D. (eds.) Description Logics. CEUR Workshop Proceedings, vol. 189 (2006)
[12] Sebag, M.: Distance induction in first order logic. In: Džeroski, S., Lavrač, N. (eds.) ILP 1997. LNCS, vol. 1297, pp. 264–272. Springer, Heidelberg (1997)
[13] Wache, H., Groot, P., Stuckenschmidt, H.: Scalable instance retrieval for the semantic web by approximation. In: Dean, M., Guo, Y., Jun, W., Kaschek, R., Krishnaswamy, S., Pan, Z., Sheng, Q.Z. (eds.) WISE-WS 2005. LNCS, vol. 3807, pp. 245–254. Springer, Heidelberg (2005)
[14] Zezula, P., Amato, G., Dohnal, V., Batko, M.: Similarity Search – The Metric Space Approach. In: Advances in database Systems, Springer, Heidelberg (2007)

Instance Based Clustering of Semantic Web Resources

Gunnar AAstrand Grimnes[1], Peter Edwards[2], and Alun Preece[3]

[1] Knowledge Management Department, DFKI GmbH
Kaiserslautern, Germany
[2] Computing Science Department
University of Aberdeen, UK
[3] Computer Science Department
Cardiff University, UK
gunnar.grimnes@dfki.de, pedwards@csd.abdn.ac.uk,
a.d.preece@cs.cf.ac.uk

Abstract. The original Semantic Web vision was explicit in the need for intelligent autonomous agents that would represent users and help them navigate the Semantic Web. We argue that an essential feature for such agents is the capability to analyse data and learn. In this paper we outline the challenges and issues surrounding the application of clustering algorithms to Semantic Web data. We present several ways to extract instances from a large RDF graph and computing the distance between these. We evaluate our approaches on three different data-sets, one representing a typical relational database to RDF conversion, one based on data from a ontologically rich Semantic Web enabled application, and one consisting of a crawl of FOAF documents; applying both supervised and unsupervised evaluation metrics. Our evaluation did not support choosing a single combination of instance extraction method and similarity metric as superior in all cases, and as expected the behaviour depends greatly on the data being clustered. Instead, we attempt to identify characteristics of data that make particular methods more suitable.

1 Introduction

Currently on the Semantic Web there is an imbalance in the number of data-produces compared to the number of consumers of RDF data. The original Semantic Web vision [1] envisaged intelligent autonomous agents that could navigate the Semantic Web and perform information gathering tasks for their users, but such agents have not yet been realised. We argue that autonomous Semantic Web agents need to be capable of more sophisticated data-processing techniques than the deductive reasoning offered by existing Semantic Web inference engines, and to be able to solve personal tasks for a user on the heterogeneous and noisy Semantic Web an agent must be able to learn. We have in previous work been exploring how traditional machine learning techniques may be applied to Semantic Web data [2,3] and in this paper we focus with the particularities arising from attempting to cluster Semantic Web data. Clustering is the process of classifying similar object into groups, or more precisely, the partitioning of a data-set into subsets called clusters. The process is applicable for a large range of problems, and would make a valuable tool for an autonomous Semantic Web agent.

In this paper we identify two challenges that need to be tackled for applying traditional clustering mechanisms to Semantic Web resources: firstly, traditional clustering methods are based on *instances* not on a large interconnected graph as described by RDF; How do we extract a representation of *an instance* from such a RDF graph? Secondly, having extracted the instances, how do you computer the distance between two such instances?

We discuss three approaches to instance extraction in Section 2.1 and three distance measures in Section 2.2. We test our methods by using a simple Hierarchical Agglomerative Clustering (HAC) [4] algorithm. Although many more performant and modern clustering algorithms exist, we are not primarily interested in absolute performance, but rather the relative performance between our different approaches. HAC is stable, generates clustering solutions with exclusive cluster membership and generally well understood and is thus well suited for such an empirical comparison. The result of the HAC algorithm is a tree of clusters, with the similarity between clusters increasing with the depth of the tree. The tree can easily be cut to generate any number of separate clusters by iteratively cutting the weakest link of the tree until the desired number of clusters is reached.

We test our clustering techniques on three different RDF data-sets: the first consisting of computer science papers from the Citeseer citation database[1], where we generated RDF descriptions of each using the BibTeX description. The papers are categorised into 17 different subject areas of computing science, and exemplifies a typical relational database to RDF conversion: the data is shallow, there is no real ontology and most properties take literal values, and very few properties relating two resources. This dataset has 4220 instances. The second dataset is a Personal Information Model (PIMO) from a user of the Semantic Desktop system Gnowsis [5] – this data-set is based on a well-defined and rich ontology and exemplifies a data-set built on Semantic Web best practises. Both the ontology and the instances are partially generated by an end-user, but using a sophisticated tool assuring that the data is consistent and semantically sound. This dataset has 1809 instances in 48 classes. The third data-set is a crawl of FOAF documents (Friend of a Friend[2]) consisting of 3755 person instances, this data-set is identical to the one used in [2]. For the citeseer and PIMO datasets existing classifications of the instances are available, allowing a supervised evaluation. We present the results of several common metrics for evaluating clustering solution in Section 3, and Section 4 makes concluding remarks and outlines the future plans for this work.

2 Clustering RDF Instances

When applying traditional clustering techniques to RDF data there are two issues that require consideration:

1. Instance extraction – What exactly constitutes an individual in an RDF graph? The main virtue of RDF is that all data be interconnected, and all relations can be represented and made explicit. However, this means that a rich data-set might just appear

---

[1] Citeseer: http://citeseer.ist.psu.edu/, Our dataset can be downloaded from http://www.csd.abdn.ac.uk/~ggrimnes/swdataset.php
[2] The FOAF Project, http://www.foaf-project.org/

as one large graph, rather than separate instances. By instance extraction we mean the process of extracting the subgraph that is relevant to a single resource from the full RDF graph. The possibilities range from considering only the immediate attributes, to more sophisticated analysis of the whole graph, based on the graph topology, ontological information or information-theory metrics.
2. Distance Measure – How is the distance between two RDF instances computed? Assuming that the output from the instance extraction is a smaller RDF graph with the resource as root node, how may one compare two such graphs? They might partially overlap, or they may only have edges with common labels. Again the alternatives range from naive approaches to methods based on graph-theory or the ontological background information.

In addition to these two points, a Semantic Web capable of autonomous clustering requires a method for determining the appropriate number of clusters for a data-set. However, this problem is not unique to Semantic Web clustering nor is the process changed significantly by the processing of RDF, and we will not discuss it further in this paper.

2.1 Instance Extraction

The problem of instance extraction is best illustrated with an example. Consider the example RDF graph shown on the left in Figure 1. We would like to extract the parts of the graph relevant to the resource *ex:bob*. Bob is related to several other RDF resources in this graph; he knows some other people, he works for a company and he is the author of publications. Several of these other resources might constitute important parts of what defines Bob as a person. The challenge is to find a reasonable subset of the graph surround Bob, that still includes any important information without being too large. The problem of extracting instance subgraphs is not exclusive to clustering. For instance, the SPARQL specification[3] includes the *DESCRIBE* directive which should return *a single result RDF graph containing RDF data about resources*. However, the exact form of this description is not specified in the current SPARQL specification, and should instead be determined by the SPARQL query processor. Another area where these methods are applicable is programming user-interfaces for RDF enabled applications, where it is often required to extract an informative representation of a resource for presentation to an end-user. We will discuss three domain-independent approaches that have been implemented for this paper, and we will revisit the example graph for each one. In addition we will discuss some possible domain-dependent methods that improve on the domain-independent methods, but require additional semantic infra-structure support that is currently unavailable on the Semantic Web.

Immediate Properties. Our first and most direct approach for extracting instance representations from RDF graphs is considering only the immediate properties of resources. This method is illustrated on the right of Figure 1 where the extracted part of the graph around the *ex:bob* resource has been highlighted. This seems like an intuitive approach if viewing RDF from an object-oriented programming perspective, it is simple to implement and it runs very quickly. However, the approach has the obvious drawback that

---

[3] The SPARQL Specification: http://www.w3.org/TR/rdf-sparql-query/

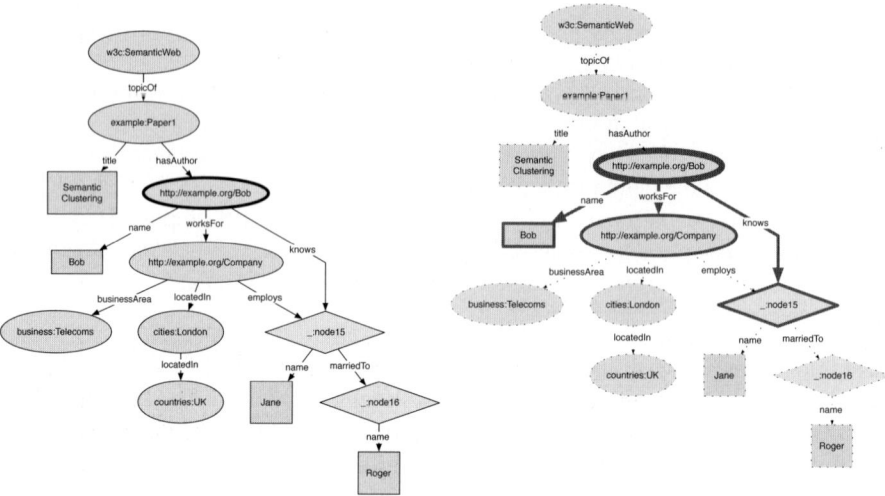

Fig. 1. An Example RDF Graph and Naive Instance Extraction

Fig. 2. CBD and DLT Instance Extraction Examples

much of the information that is relevant to the resource is lost, indeed large parts of the graph may not be included in any instance graph.

Concise Bounded Description. An improvement over simple immediate properties can be achieved by taking the types of nodes in the graph into account. Concise Bounded Description (CBD)[4] is one such improvement and is defined as a recursive algorithm

---

[4] Concise Bounded Description: http://www.w3.org/Submission/CBD/

for extracting a subgraph around an RDF resource. Informally it is defined as the resource itself, all it's properties, and all the properties of any blank nodes connected to it, formally the CBD of graph G is defined recursively as follows:

$$CBD_G(x) = \{<x,p,o> \mid <x,p,o> \in G\}$$
$$\cup \{<s,p,o> \mid <s,p,o> \in G$$
$$\wedge \exists s',p'(<s',p',s> \in CBD_G(X) \wedge bn(s))\}$$

where $bn(x)$ is true if x is a blank node.

The CBD extraction from the example graph is shown on the left in Figure 2, where the extracted graph is highlighted. Note how all info about _:node15 and _:node16 is included, but no further triples from *http://example.org/Company*. An alternative CBD definition also considers backward links, where triples having the resource in question as an object are also included. CBD is a much better option than just considering the immediate properties of a resource, however it also has several disadvantages: since the behaviour of CBD depends on the use of blank nodes in the data it is not strictly domain independent. For instance, some applications discourage the use of blank nodes because of the difficulty in implementing their semantics correctly, in this case CBD is no better than immediate properties. Conversely, in data where no natural identifiers exist there might be a large ratio of blank nodes to labelled resources, and the CBD might only be bound by the size of the full graph.

Depth Limited Crawling. An alternative to limiting the sub-graph by node-type is to limit the sub-graph by depth and simply traverse the graph a certain number of steps from the starting node. This has the advantage that it is stable over any input data making it easier to limit the size of the subgraph. For our initial experiments with clustering of FOAF data [2] we found that traversing two edges forward and one edge backwards from the root node provides a good tradeoff between size and the information contained in the subgraph, although this is clearly data-dependent and may need adjusting for special cases. The sub-graph extracted from the example graph is shown on the right in Figure 2. Note how this is the only method where also the publications that link to *Bob* as the author are included.

Since the results of the instance extraction algorithms vary greatly depending on the input data, we analyse the graphs extracted from our three data-sets. The mean and standard deviation of the number of triples in the graphs extracted is shown in Table 1. Note that the results for the Citeseer dataset are the same for all extraction methods, the reason being that the only properties linking two resources in this dataset is *rdf:type* relating publications to their class, and the classes have no further properties. All other properties in this dataset are literal properties. Note also that the naïve approach and CBD is identical for the PIMO data-set, as the PIMO language disallows the use of blank nodes, so CBD does not do any further traversal. The table also clearly illustrates the weakness of CBD when dealing with large crawled data, since FOAF encourages the use of blank nodes to represent people it is possible to traverse large parts of the FOAF crawl through blank nodes alone. In fact, the largest sub-graph extracted from the FOAF graph using CBD contained 20108 triples.

Table 1. Mean and standard deviation of size of extracted instance graphs

Dataset	Naive	CBD	DLT
Citeseer	18.22 / 5.31	18.22 / 5.31	18.22 / 5.31
PIMO	6.84 / 1.49	6.84 / 1.49	828.04 / 5.12
FOAF	13.95 / 19.14	1299.26 / 2983.97	145.06 / 153.58

2.2 Distance Measure

The choice of distance measure is the most important factor for a clustering setup, the definition of what constitutes close or remote instances affects everything about the behaviour of the clusterer. Commonly clustering has been performed on instance represented as numerical vectors and viewing the vectors as coordinates in an N-dimensional space allows the use of geometric distance measures, such as Euclidean or Manhattan distance.

However, transforming RDF instances represented as graphs into coordinates in a Euclidean space is not trivial, and thus standard distance metrics do not readily apply to RDF instances. In this section we present several methods for solving the problem of finding the distance between pairs of RDF instances, each instance consisting of a root node (the RDF resource) and a (sub)-graph considered relevant to this instance, known as R_n and G_n respectively. We consider one distance measure inspired by traditional feature vector based clustering, one based on the topology of the RDF graphs and one based on the ontological background information available.

Feature Vector Based Distance Measures. As a base-line approach to an RDF similarity measure we wanted to adapt traditional vector-based similarity to comparing RDF instances. The first step is to transform an RDF graph and root node pair into a feature vector representation. Since the method should be compatible with all instance extraction methods outlined above we needed to go beyond the obvious solution of mapping the direct RDF properties of the resource to features in the vector. Instead we map paths in the RDF graph to features, and let the value of the feature be a set of all the nodes reachable through that path. First we defined the set of nodes reachable from a root-node X in a graph G:

$$reachable(X, G) = \{n| <x, p, n> \in G\} \cup \{n'|n' \in R_G(n)\} \quad (1)$$

The feature vector for a single instance n is then defined as:

$$FV(n) = \{shortestPath(R_n, x, G_n)|x \in reachable(R_n, G_n)\} \quad (2)$$

where $shortestPath(R, X, G)$ is the shortest path from node R to node X in graph G, and a path is defined as a list of edges to traverse (i.e. RDF properties). The feature vector for a set of instances is simply the union of all the individual feature sets. Consider again the example graph in Figure 1, the features for the instance *ex:bob* would be as follows:

Table 2. Longest and average path-lengths

Dataset	Naive	CBD	DLT
Citeseer	1.00 / 0.98	2.00 / 1.02	2.0 / 1.20
PIMO	1.00 / 0.97	1.00 / 0.97	3.0 / 1.87
FOAF	1.00 / 1.00	26.00 / 19.40	3.0 / 1.90

[name, worksFor, knows, worksFor→businessArea, worksFor→locatedIn, knows→name, knows→marriedTo, worksFor→locatedIn→locatedIn, knows→marriedTo→name]

Note how the path worksFor→employs is not included as the same resource (_:node15) can be reacher by the shorter path of knows. As the feature vector can get very long for complicated data-set it is often desirable to only include the most frequent paths. In our experiments we used only the 1000 paths most frequently included when considering all instances, i.e. we used the same feature vector of length 1000 for all instances, but not all features have values for all instances. Table 2 shows the maximum and average length for the feature vectors extracted from our three data-sets when using each of the three instance extraction methods above; once again the weakness of CBD combined with FOAF is clear, but no other extraction method or dataset result in prohibitively long vectors. As mentioned above, the values of each feature is the set of RDF nodes reachable through that path[5]. It is necessary to use set-valued features because each RDF property might be repeated several times. One could imagine that cardinality constraints from an ontology could be used to restrict certain paths to single valued features, however, such constraints are unlikely to remain unbroken on a noisy heterogeneous Semantic Web, and we are primarily interested in methods that do not depend on data correctly conforming to an ontology. Again, consider the example graph from Figure 1, the values of the feature vector for the instance *ex:bob* would be:

[{''bob''}, {ex:TheCompany}, {_:node15}, {business:Telecoms}, {cities:London}, {''Jane''}, {_:node16}, {countries:UK}, {''Roger''}]

Note how in this basic examples every feature is a single valued set and no features are missing. In a more complex scenario many of the feature-values might be the empty set where the RDF properties in the path are not available. Computing the distance between two feature vectors is then done using a straight-forward similarity measure inspired by the Dice coefficient: The distance between instance X and Y over feature-vector FV is defined as:

$$simFV(X, Y, FV) = \frac{1}{|FV|} \sum_{f \in FV} \frac{2 * |X_f \cap Y_f|}{|X_f| + |Y_f|} \qquad (3)$$

---

[5] Note that this is not quite as query intensive as it may seem, by collapsing the common sub-paths in all paths to a tree, an efficient algorithm can be devised that makes only the minimal set of queries.

where X_f and Y_f is the set value of feature f for X and Y respectively. The $simFV$ takes a value between 0 and 1 where 1 means that the instances have exactly the same properties and 0 means no shared properties.

Graph Based Distance Measures. Montes-y-Gómez et. al developed a similarity measures for comparing two conceptual graphs [6]. Conceptual graphs are a data-structure commonly used for natural language processing. They consist of a network of concept nodes, representing entities, attributes, or events and relation nodes between them. The similarity metric incorporates a combination of two complementary sources of similarity: the *conceptual similarity* and the *relational similarity*, i.e. the overlap of nodes and the overlap of edges within the two graphs.

Conceptual graphs and RDF instances are structurally sufficiently similar that the this similarity metric is also appropriate for comparing RDF graphs. For each instance to be clustered the extracted sub-graph is used for the comparison and the root node is ignored. Space reasons prevent us from presenting the details of the metric here, please refer to Montes-y-Gómez et. al's paper for the details and to our previous work for details on applying this metric to RDF data [2].

Ontologically Based Distance Measures. There have been several efforts for developing an ontology based similarity metric for RDF data. Most of these originate from the fields of ontology mapping and are focussing on similarity between whole ontologies, or classes within. For instance, [7] compares two ontologies using an approach based on conceptual graphs, and [8] presents a metric designed especially for OWL-Lite. Maedche and Zacharias have developed a similarity metric explicitly for instance data, but with heavy emphasis on the ontological background information [9] . Their metric assumes the existence of a well-defined ontology and conforming instance data. However, real Semantic Web data is often noisy and inconsistent and some minor adjustments to the algorithm are required to allow for handling of more realistic data. The ontological similarity metric is a weighted combination of 3 dimensions of similarity:

$$simOnt(X,Y) = \frac{t \times TS(X,Y) + r \times RS(X,Y) + a \times AS(X,Y)}{t + r + a} \quad (4)$$

Where TS is the taxonomy similarity, RS the relational similarity, AS the attribute similarity, and t, r, a the weights associated with each. The taxonomy similarity considers only the class of the instances being compared and is determined by the amount of overlap in the super-classes of each instance's class. The attribute similarity focuses on the attributes of the instances being compared, i.e. the properties that take literal values (known as data-type properties in OWL), and is based on using some type-specific distance measure for comparing the literal values for matching predicates (i.e. Levenshtein edit distance for string literals and simple numeric difference for numbers, etc.) Finally, the relational similarity is based on the relational properties of a resource, i.e. the properties that link this resource to other resources (and not to literals). This is defined recursively, so for finding the similarity between two linked resources the ontological similarity measure is applied again. To make it possible to compute the similarity measure the recursion is only followed until a certain depth, after which the relational similarity is ignored and the combination of only attribute similarity and taxonomy similarity is returned. For our experiments we stopped recursing after 3 levels. Again we do not have the space to present the algorithm in full, and we will only briefly cover our modifications that allowed us to relax some assumptions made about the input data which are unlikely to hold true for real-life Semantic Web data, namely:

1. The ontologies must be strictly hierarchical, i.e. no multi-class inheritance.
2. Resources must be members of exactly one class.
3. Range and domain must be well defined for all properties.
4. Range and domain constraints must never violated.
5. There must be a clear distinction between datatyped properties and object properties, i.e. a property takes either literal objects or other resources.

Both point one and two are natural simplifying restrictions many applications chose to enforce internally, although both multi-class inheritance and multiple types are supported by both RDFS and OWL. Point three is not required by RDF, but would be a logical effect of using well-engineered ontologies. Point four is a valid assumption if one assumes a semi-closed data-base view of the Semantic Web (i.e. the database-semantic web view). Point five is another natural assumption to make since in OWL the distinction is made between datatype properties and object properties (Maedche and Zacharias call these attributes and relations). However, for RDFS based ontologies this is not the case.

To enable $simOnt$ to work with our data-sets we made the following modifications to the algorithm:

1. For the definition of the taxonomy similarity we let $SC(x)$ mean the set of superclasses of ALL types of x.
2. We do not rely on range/domain alone for finding applicable literal-/relational properties, we also add the properties that are used with literals/resource in the data available. This also solves the missing range/domain declaration problem.

Note that this distance measure does not make use of the extracted sub-graph, only the ontological information associated with the root-node.

3 Comparison

Clustering is a task that is notoriously hard to evaluate. Recent overviews over available methods can be found in [10]. Ultimately clustering is a data-analysis exercise, and the true value of a clustering solution comes from understanding of the data gained by a human data-analyst, but in this paper we are interested in automated methods for checking the quality of a clustering solution. For two of our data-sets the preexisting classifications may be used for evaluation, but since there is no way to train a clustering algorithms to look for particular classification of the data, a supervised evaluation with regard to this classification is not necessarily a good measure of the performance of the clusterer. I.e. the clustering algorithm might have chosen to partition the data along an axis different to partitioning in the given "correct" classification, meaning that although the supervised quality measures might be very poor, the resulting clustering might be an equally correct classification. In this paper we conduct a supervised evaluation for the PIMO and citeseer using the following supervised quality metrics:

- Entropy — A supervised measure that looks at how the different classes are distributed within each cluster. If each cluster only contains instances from a single class the entropy is zero. The entropy is trivially optimised by having a single cluster containing all instances.
- Purity — A supervised measure that quantifies the degree of which a cluster contains instances from only one class. As for entropy, this is trivially optimised by having a single cluster containing all instances.
- F-Measure — The F-measure in information retrieval is an harmonic mean of precision and recall, and can also be used a supervised quality metric for clustering [11]. For a perfect clustering solution with an F-Measure of one there must be a cluster for each class, containing exactly the members of the corresponding class. This is unlikely for an unguided clustering solution, and one can therefore not compare F-measure values for clustering with what is considered a good performance for classification algorithms.
- Hess Precision/Recall — Another view of precision and recall for clustering was introduced in a recent PhD thesis by Andreas Hess [12] . Instead of considering individual instances when defining precision and recall, Hess considers all pairs of objects that were clustered. The pairs are then classified whether they occur in the same class in the correct classification as in the clustering solution, and precision and recall is then defined as normal for each pair. Hess observes that the precision metric is biased towards small clusters and the recall towards large clusters.

For the FOAF and PIMO data-set we also carried out an unsupervised evaluation, using the basic Sum of Squared Errors and Zamir's Quality Metric (ZQM) [13], a function aiming to represent a trade-off between small well-defined clusters and larger mixed clusters.

3.1 Results

We evaluated all combinations of the instance extraction methods and similarity metric presented above on the PIMO and Citeseer data-set. For both data-sets we specified the number of clusters to generate based on the preexisting classification and calculated the entropy, purity, F-measure and Hess's pair-wise precision and recall based on this classification. The results are shown in Table 3 and the F-Measure and Hess-Recall are shown graphically in Figure 3 (for space reasons plots were only included for the most interesting metrics).

For the FOAF and PIMO data-set we used the jump-method as described by Sugar and James [14] to find a natural number of clusters for each data-set and user the unsupervised evaluation metrics to analyse the resulting clusters. The number of clusters found and evaluation metrics are shown in Table 4, the size of clusters found and the ZQM are also shown graphically in Figure 4.

Note first that recreating the preexisting classification is a very challenging problem, especially for the Citeseer dataset where the RDF representation of the papers is very simple and the information is not really sufficient for recreating the subject-area classification. Note also that it might also be possible to achieve overall better results using

---

[6] Please note the non-zero origins of F-Measure graphs and different scale for the two Hess Recall graphs.

Table 3. Results for Supervised Evaluation of PIMO and Citeseer Data-sets

Data-set	Extract.	Sim.	Entr.	Purity	F-Meas.	Hess-P	Hess-R
citeseer	N/A	ont	0.96	0.11	0.12	0.07	0.22
citeseer	naive	cg	0.94	0.12	0.13	0.07	0.36
citeseer	dlt	cg	0.94	0.12	0.13	0.07	0.38
citeseer	cbd	cg	0.94	0.13	0.13	0.07	0.53
citeseer	dlt	fv	0.97	0.09	0.12	0.07	0.78
citeseer	naive	fv	0.97	0.09	0.12	0.06	0.89
citeseer	cbd	fv	0.97	0.09	0.12	0.07	0.89
pimo	ont	ont	0.32	0.59	0.42	0.38	0.28
pimo	naive	cg	0.32	0.60	0.42	0.38	0.28
pimo	dlt	cg	0.31	0.60	0.44	0.40	0.29
pimo	cbd	cg	0.31	0.59	0.43	0.39	0.29
pimo	dlt	fv	0.30	0.62	0.44	0.40	0.31
pimo	naive	fv	0.33	0.59	0.42	0.39	0.33
pimo	cbd	fv	0.33	0.58	0.43	0.40	0.37

Table 4. Results for Unsupervised Evaluation of FOAF and PIMO Data-sets

Data-set	Extract.	Sim.	SSE	ZQM	# Clusters.
pimo	naive	fv	61.74	0.06	101.
pimo	naive	cg	120.97	0.08	104.
pimo	cbd	fv	59.67	0.06	117.
pimo	cbd	cg	126.38	0.15	70.
pimo	dlt	fv	35.77	0.04	128.
pimo	dlt	cg	3.91	0.04	96.
pimo	N/A	ont	32.11	0.5	16.
foaf	naive	fv	47.09	0.06	120.
foaf	naive	cg	65.7	0.09	86.
foaf	cbd	fv	77.95	0.18	56.
foaf	cbd	cg	60.62	21.43	2.
foaf	dlt	fv	35.98	0.05	113.
foaf	dlt	cg	135.69	0.13	95.
foaf	N/A	ont	29.14	0.52	14.

a more sophisticated clustering algorithm than HAC, but as mentioned previously, this was not the focus of this work.

Looking first at the unsupervised results, observer that for the PIMO data-set the number of clusters detected are far higher than the number of classes in the existing classification (The PIMO data-set has 48 classes). This holds for all methods apart from the ontological similarity measure. This can be explained by the idiosyncratic *Person* class in the PIMO data-set, containing 387 instances. We believe that further sub-divisions of the instances of this class are possible, making a larger number of clusters more natural for the PIMO dataset. Something that is not clear from the result table or graphs is that although many clusters are generated, many of them are singleton clusters, this holds even for the supervised clustering results where the *right* number of clusters is specified. The feature-vector distance measure was especially prone to creating a small number of large clusters, for the Citeseer dataset all experiments with the feature-vector distance metric the largest clusters contained over 85% of the instances and only 2 or 3 clusters had more than a single instance. For the PIMO dataset every method generated one large cluster with roughly half of the instances, caused again by the abnormal

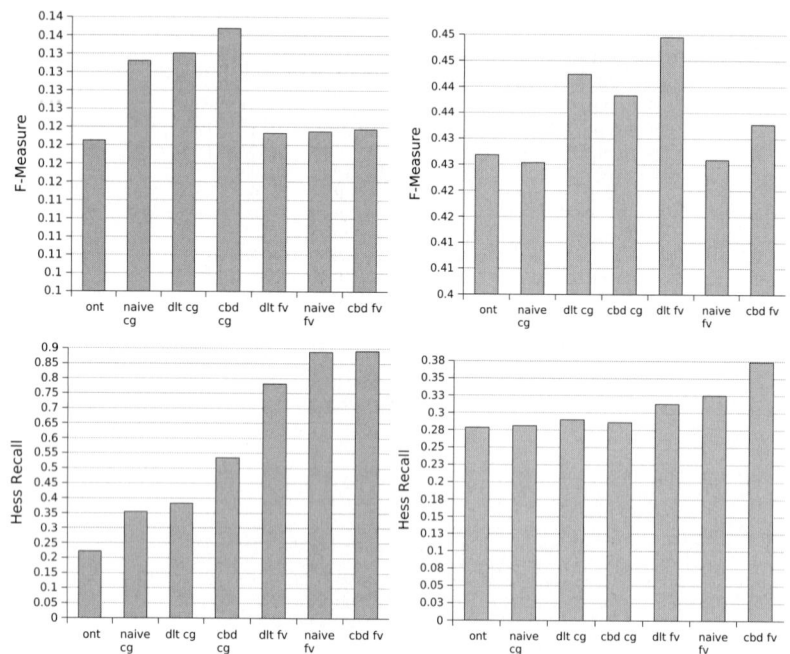

Fig. 3. F-Measure and Hess-Recall results for the Citeseer (left) and PIMO (right) data-sets[6]

Person class, the PIMO instances in this class are automatically generated from the user's address-book, and these all have identical structure.

For the FOAF data-set the number of clusters detected varies widely between the different methods. The two values that stand out are for the run using CBD and the conceptual graph similarity measure, which is extremely low, only two clusters were found. As mentioned before, the FOAF data-set exposes a problem with CBD due to the large number of blank-nodes and for some instances very large sub-graph are extracted. In combination with CG these huge instance graphs cause very unusual results. For the feature vector similarity metric the problem is not so severe, as the large instance graphs will only contribute a few features, and these are likely to be excluded from the 1500 most common features used in the end.

Considering the supervised results, we know that the PIMO dataset has many more classes than the Citeseer dataset (48 to 17), but fewer instances (1809 to 4220) and this is clearly reflected in the scores for entropy and purity, the PIMO clustering solutions has many classes that contain only a single instance which gives a high purity score, whereas for the Citeseer data-set each cluster is likely to have a mix of instances and the entropy is therefore high. Note also that for the Citeseer data-set the choice of instance extraction mechanism is largely irrelevant because the data is so shallow (there are no sub-graphs deeper than a single level).

Looking at the graphs in Figure 3 we can see that surprisingly the feature-vector method achieved high Hess-Recall scores for both data-sets, however, this is a side-effect of the large clusters created by this method and is consistent with Hess'

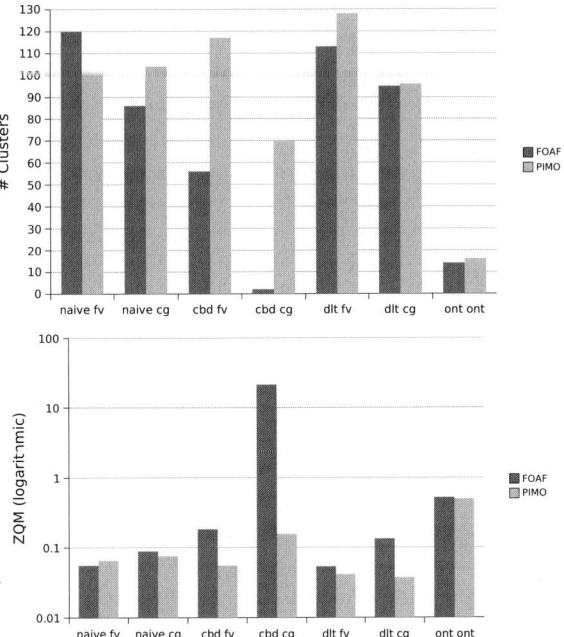

Fig. 4. Cluster Sizes and Zamir Quality Measure FOAF and PIMO data-sets

observation. Looking at the F-measure for the feature-vector it is consistently low for the Citeseer data-set, but performs rather well for the PIMO data-set, referring back to Table 2 we can see that the Citeseer data-set is just not rich enough to allow for paths of any length and any real distinction between the instances is lost.

The ontological similarity measure performs consistently well in these experiments, but not significantly better than the others. It does however seem to result in a much lower number of clusters being generated, which also results in a higher ZQM score due to the part of ZQM favouring a small number of well-defined clusters. The ontological similarity measure is theoretically well founded and makes the most use of the semantics out of the similarity measures presented here, and we had therefore expected it to perform better, but it may be that our data-set have insufficient ontological data to really make the most of this metric. However, it must be noted that the run-time performance of this metric is much worse than the two other metrics that were investigated; the recursive nature often makes it necessary to compute the similarity between a much larger number of nodes than the instances that are going to be clustered in the first place: For instance, clustering a small set of only 8 FOAF person nodes, described by only 469 triples, 178 nodes were visited. On larger graphs this number can grow very quickly. On the other hand, an advantage of the ontological similarity measure is that extraction of instance-graphs from the large RDF graph is not required, removing one layer of complexity from the clustering process.

In summary, the results do not support a clear conclusion of what is the *best* combination of extraction method and distance metric, and the best performing combination varies for each data-set. The DLT extraction method may be the best choice for a data-independent extraction method, as although it may occasionally excluded parts of the data-graph it has no obvious weakness like CBD has for FOAF. For similarity measure the ontological similarity does well for unsupervised clustering, generating a small number of well-defined clusters, but at the cost of a longer running-time. For supervised clustering it performs worse since not enough clusters are generated to mirror the preexisting classification.

4 Conclusion and Future Work

In this paper we have identified two important challenges for performing clustering of Semantic Web data, firstly extracting instance from a large RDF graph and secondly computing the distance between these instances. We evaluated our approaches using three datasets representing typical data one might find on the Semantic Web. Our evaluation did not yield one combination of instance extraction method and distance measure that outperforms the others, but highlights a strong dependency on the structure of the data for choosing the optimal methods.

An additional method of instance extraction we would like to investage is the creation of a hybrid method for instance extraction, combining CBD and depth-limited traversal, where one always traverses the graph a certain minimum number of steps regardless of the node-type, then follow only blank-nodes up to a certain maximum level. Another approach would be to consider the frequency of the predicates when crawling, for instance an algorithm could be conceived that would only include triples not included in any other immediate sub-graph.

With regard to the evaluation, it is clearly very difficult to do a domain-independent evaluation like we attempted here. Although many metrics are available, the supervised metrics may not measure the quality we are interested in, and the unsupervised metrics are often biased towards one particular shape of clusters, i.e. either many small clusters or few large ones. A more interesting evaluation method we would like to explore is to use classification algorithms to learn a classifier for each of the identified clusters, and the clustering quality becomes the accuracy of the classifiers learned.

Finally, we believe that we have shown that clustering of RDF resources is an interesting area, where many questions remain unanswered. Our long-term goal is to create a tool-box of learning techniques that an autonomous Semantic Web Agent may use to understand the world; and we believe clustering could form a central tool for such an agent. Further investigation is needed to determine how shallow data-analysis could help an agent chose the optimal instance extraction methods, distance metrics and clustering algorithms.

References

1. Berners-Lee, T., Hendler, J., Lassila, O.: The Semantic Web. Scientific American 284, 28–37 (2001)
2. Grimnes, G.A., Edwards, P., Preece, A.: Learning Meta-Descriptions of the FOAF Network. In: McIlraith, S.A., Plexousakis, D., van Harmelen, F. (eds.) ISWC 2004. LNCS, vol. 3298, pp. 152–165. Springer, Heidelberg (2004)

3. Edwards, P., Grimnes, G.A., Preece, A.: An Empirical Investigation of Learning from the Semantic Web. In: ECML/PKDD, Semantic Web Mining Workshop, pp. 71–89 (2002)
4. Johnson, S.C.: Hierarchical clustering schemes. Psychometrika 2, 241–254 (1967)
5. Sauermann, L., Grimnes, G.A., Kiesel, M., Fluit, C., Maus, H., Heim, D., Nadeem, D., Horak, B., Dengel, A.: Semantic desktop 2.0: The gnowsis experience. In: Cruz, I., Decker, S., Allemang, D., Preist, C., Schwabe, D., Mika, P., Uschold, M., Aroyo, L.M. (eds.) ISWC 2006. LNCS, vol. 4273, Springer, Heidelberg (2006)
6. Montes-y-Gómez, M., Gelbukh, A., López-López, A.: Comparison of Conceptual Graphs. In: Cairó, O., Cantú, F.J. (eds.) MICAI 2000. LNCS, vol. 1793, pp. 548–556. Springer, Heidelberg (2000)
7. Dieng, R., Hug, S.: Comparison of personal ontologies represented through conceptual graphs. In: Proceedings of ECAI 1998, pp. 341–345 (1998)
8. Euzenat, J., Valtchev, P.: An integrative proximity measure for ontology alignment. In: Proceedings of the 1st Intl. Workshop on Semantic Integration. CEUR, vol. 82 (2003)
9. Maedche, A., Zacharias, V.: Clustering ontology-based metadata in the semantic web. In: Elomaa, T., Mannila, H., Toivonen, H. (eds.) PKDD 2002. LNCS (LNAI), vol. 2431, pp. 348–360. Springer, Heidelberg (2002)
10. Strehl, A.: Relationship-based Clustering and Cluster Ensembles for High-dimensional Data Mining. PhD thesis, The University of Texas at Austin (2002)
11. Larsen, B., Aone, C.: Fast and effective text mining using linear-time document clustering. In: KDD 1999: Proceedings of the fifth ACM SIGKDD international conference on Knowledge discovery and data mining, pp. 16–22. ACM Press, New York (1999)
12. Heß, A.: Supervised and Unsupervised Ensemble Learning for the Semantic Web. PhD thesis, School of Computer Science and Informatics, University College Dublin, Dublin, Ireland (2006)
13. Zamir, O., Etzioni, O., Madani, O., Karp, R.M.: Fast and intuitive clustering of web documents. In: KDD, pp. 287–290 (1997)
14. Sugar, C.A., James, G.M.: Finding the Number of Clusters in a Data Set - An Information Theoretic Approach. Journal of the American Statistical Association 98, 750–763 (2003)

Conceptual Clustering and Its Application to Concept Drift and Novelty Detection

Nicola Fanizzi, Claudia d'Amato, and Floriana Esposito

Dipartimento di Informatica – Università degli Studi di Bari
Campus Universitario, Via Orabona 4, 70125 Bari, Italy
{fanizzi,claudia.damato,esposito}@di.uniba.it

Abstract. The paper presents a clustering method which can be applied to populated ontologies for discovering interesting groupings of resources therein. The method exploits a simple, yet effective and language-independent, semi-distance measure for individuals, that is based on their underlying semantics along with a number of dimensions corresponding to a set of concept descriptions (discriminating features committee). The clustering algorithm is a partitional method and it is based on the notion of medoids w.r.t. the adopted semi-distance measure. Eventually, it produces a hierarchical organization of groups of individuals. A final experiment demonstrates the validity of the approach using absolute quality indices. We propose two possible exploitations of these clusterings: concept formation and detecting concept drift or novelty.

1 Introduction

In the perspective of automatizing the most burdensome activities for the knowledge engineer, such as ontology construction, matching and evolution, they may be assisted by supervised or unsupervised methods crafted for the standard representations adopted in the Semantic Web (SW) context and founded in *Description Logics* (DLs).

In this work, we investigate unsupervised learning for populated ontologies. Specifically, we focus on the problem of *conceptual clustering* [27] of semantically annotated resources, that amounts to grouping them into clusters according to some criteria (e.g. similarity). The benefits of conceptual clustering in the context of knowledge bases maintenance are manifold. Clustering resources enables the definition of new emerging concepts (*concept formation*) on the grounds of those already defined (intensionally or extensionally) in the knowledge base; supervised methods can then exploit these clusters to induce new concept definitions or to refining existing ones (*ontology evolution*); intensionally defined groupings may speed-up the task of search and *discovery* [6]; a hierarchical clustering also suggests criteria for *ranking* the retrieved resources.

Essentially, most of the existing clustering methods are based on the application of similarity (or density) measures defined over a fixed set of attributes of the domain objects. Classes of objects are taken as collections that exhibit low interclass similarity (density) and high intraclass similarity (density). More rarely these methods are able to exploit (declarative) forms of *prior* or *background knowledge* to characterize the clusters with intensional definitions. This hinders the interpretation of the outcomes of these methods which is crucial in the SW perspective that should enforce semantic interoperability through knowledge sharing and reuse. Thus, specific conceptual clustering methods have to be taken into account, such as those focussing on the definition of groups of objects through conjunctive descriptions based on selected attributes [27]. More recent related works are based on similarity measures for clausal spaces [18], yet the expressiveness of these representations is incomparable w.r.t. DLs [3]. Also the underlying semantics is different since the *Open World Assumption* (OWA) is made on DL languages, whereas the *Closed World Assumption* (CWA) is the standard in machine learning and data mining.

Regarding dissimilarity measures in DL languages, as pointed out in a seminal paper [4], most of the existing measures focus on the similarity of atomic concepts within hierarchies or simple ontologies. Moreover, they have been conceived for assessing *concept* similarity, whereas for accomplishing inductive tasks, such as clustering, a notion of similarity between *individuals* is required. Recently, dissimilarity measures for specific DLs have been proposed [5]. Although they turned out to be quite effective for the inductive tasks, they are still partly based on structural criteria which makes them fail to fully grasp the underlying semantics and hardly scale to more complex ontology languages that are commonly adopted in the SW context.

Therefore, we have devised a family of dissimilarity measures for semantically annotated resources, which can overcome the aforementioned limitations. Namely, we adopt a new family of measures [8] that is suitable for a wide range of languages since it is merely based on the discernibility of the individuals with respect to a fixed set of features (henceforth a *committee*) represented by concept definitions. These measures are not absolute, yet they depend on the knowledge base they are applied to. Thus, the choice of the optimal feature sets may require a preliminary feature construction phase. To this purpose we have proposed randomized procedures based on *genetic programming* or *simulated annealing* [8, 9].

Regarding conceptual clustering, the expressiveness of the language adopted for describing objects and clusters is equally important. Former alternative methods devised for terminological representations, pursued logic-based approaches for specific languages [17, 10]. Besides of the language-dependency limitation, it has been pointed out that these methods may suffer from noise in the data. This motivates our investigation on similarity-based clustering methods which should be more noise-tolerant and language-independent.

Thus we propose a multi-relational extension of effective clustering techniques, which is tailored for the SW context. It is intended for grouping similar resources w.r.t. the novel measure. The notion of *means* characterizing partitional algorithms descending from (BISECTING) K-MEANS [15] originally developed for numeric or ordinal features, is replaced by the notion of *medoids* [16] as central individuals in a cluster. Hence we propose a BISECTING AROUND MEDOIDS algorithm, which exploits the aforementioned measures [8].

The clustering algorithm produces hierarchies of clusters. An evaluation of the method applied to real ontologies is presented based on internal validity indices such as the silhouette measure [16]. Then, we also suggest two possible ways for exploiting the outcomes of clustering: concept formation and detect concept drift or novelty detection. Namely, existing concept learning algorithms for DLs [14, 20] can be used to produce new concepts based on a group of examples (i.e. individuals in a cluster) and counterexamples (individuals in disjoint clusters, on the same hierarchy level). Besides, we provide also a method to detect interesting cases of concepts that are evolving or novel concepts which are emerging based on the elicited clusters.

The paper is organized as follows. Sect. 2 recalls the basics of the representation and the distance measure adopted. The clustering algorithm is presented and discussed in Sect. 3. After Sect. 4, concerning the related works, we present an experimental evaluation of the clustering procedure in Sect. 5. Conclusions are finally examined in Sect. 6.

2 Semantic Distance Measures

In the following, we assume that resources, concepts and their relationship may be defined in terms of a generic ontology representation that may be mapped to some DL language with the standard model-theoretic semantics (see the handbook [1] for a thorough reference).

In this context, a *knowledge base* $\mathcal{K} = \langle \mathcal{T}, \mathcal{A} \rangle$ contains a *TBox* $\mathcal{T}$ and an *ABox* $\mathcal{A}$. $\mathcal{T}$ is a set of concept definitions. $\mathcal{A}$ contains assertions (facts, data) concerning the world state. Moreover, normally the *unique names assumption* is made on the ABox individuals[1] therein. The set of the individuals occurring in $\mathcal{A}$ will be denoted with $\mathsf{Ind}(\mathcal{A})$. As regards the inference services, like all other instance-based methods, our procedure may require performing *instance-checking*, which amounts to determining whether an individual, say a, belongs to a concept extension, i.e. whether $C(a)$ holds for a certain concept C.

2.1 A Semantic Semi-distance for Individuals

For our purposes, a function for measuring the similarity of individuals rather than concepts is needed. It can be observed that individuals do not have a syntactic structure that can be compared. This has led to lifting them to the concept description level before comparing them (recurring to the approximation of the *most specific concept* of an individual w.r.t. the ABox) [5].

We have developed a new measure whose definition totally depends on semantic aspects of the individuals in the knowledge base [8]. On a semantic level, similar individuals should behave similarly with respect to the same concepts. The computation of the similarity of individuals is based on the idea of comparing their semantics along a number of dimensions represented by a committee of concept descriptions. Following the ideas borrowed from ILP [25], we propose the definition of totally semantic distance measures for individuals in the context of a knowledge base.

The rationale of the new measure is to compare individuals on the grounds of their behavior w.r.t. a given set of hypotheses, that is a collection of concept descriptions, say $\mathsf{F} = \{F_1, F_2, \ldots, F_m\}$, which stands as a group of discriminating *features* expressed in the language taken into account.

---

[1] Individuals can be assumed to be identified by their own URI.

In its simple formulation, a family of distance functions for individuals inspired to Minkowski's distances can be defined as follows:

Definition 2.1 (family of measures). *Let $\mathcal{K} = \langle \mathcal{T}, \mathcal{A} \rangle$ be a knowledge base. Given a set of concept descriptions $\mathsf{F} = \{F_1, F_2, \ldots, F_m\}$, a family of functions*

$$d_p^{\mathsf{F}} : \mathsf{Ind}(\mathcal{A}) \times \mathsf{Ind}(\mathcal{A}) \mapsto [0, 1]$$

is defined as follows:

$$\forall a, b \in \mathsf{Ind}(\mathcal{A}) \quad d_p^{\mathsf{F}}(a, b) := \frac{1}{m} \left(\sum_{i=1}^{m} \mid \pi_i(a) - \pi_i(b) \mid^p \right)^{1/p}$$

where $p > 0$ and $\forall i \in \{1, \ldots, m\}$ the projection function π_i is defined by:

$$\forall a \in \mathsf{Ind}(\mathcal{A}) \quad \pi_i(a) = \begin{cases} 1 & \mathcal{K} \models F_i(x) \\ 0 & \mathcal{K} \models \neg F_i(x) \\ 1/2 & \text{otherwise} \end{cases}$$

It is easy to prove that these functions have the standard properties for semi-distances:

Proposition 2.1 (semi-distance). *For a fixed feature set and $p > 0$, d_p is a semi-distance, i.e. given any three instances $a, b, c \in \mathsf{Ind}(\mathcal{A})$, it holds that:*

1. $d_p(a, b) \geq 0$ and $d_p(a, b) = 0$ if $a = b$
2. $d_p(a, b) = d_p(b, a)$
3. $d_p(a, c) \leq d_p(a, b) + d_p(b, c)$

It cannot be proved that $d_p(a, b) = 0$ iff $a = b$. This is the case of *indiscernible* individuals with respect to the given set of hypotheses F.

Compared to other proposed distance (or dissimilarity) measures [4], the presented function does not depend on the constructors of a specific language, rather it requires only retrieval or instance-checking service used for deciding whether an individual is asserted in the knowledge base to belong to a concept extension (or, alternatively, if this could be derived as a logical consequence).

In the perspective of integrating the measure in ontology mining algorithms which massively use it, such as all instance-based methods, it should be noted that the π_i functions ($\forall i = 1, \ldots, m$) can be computed in advance for the training instances, thus determining a speed-up in the overall computation.

2.2 Feature Set Optimization

The underlying idea for the measure is that similar individuals should exhibit the same behavior w.r.t. the concepts in F. Here, we make the assumption that the feature-set F represents a sufficient number of (possibly redundant) features that are able to discriminate really different individuals.

Experimentally, we could obtain good results by using the very set of both primitive and defined concepts found in the ontology (see Sect. 5). However, the choice of the concepts to be included – *feature selection* – may be crucial, for a good committee may discern the individuals better and a possibly smaller committee yields more efficiency when computing the distance.

We have devised a specific optimization algorithm founded in *genetic programming* and *simulated annealing* (whose presentation goes beyond the scope of this work) which are able to find optimal choices of discriminating concept committees [9]. Namely, since the function is very dependent on the concepts included in the committee of features F, two immediate heuristics can be derived: 1) control the number of concepts of the committee, including especially those that are endowed with a real discriminating power; 2) finding optimal sets of discriminating features, by allowing also their composition employing the specific constructors made available by the representation language of choice.

3 Hierarchical Clustering for Individuals in an Ontology

The conceptual clustering procedure that we propose can be ascribed to the category of the heuristic partitioning algorithms such as K-MEANS [15]. For the categorical nature of the assertions on individuals the notion of mean is replaced by the one of medoid, as in PAM (*Partition Around Medoids* [16]). Besides the procedure is crafted to work iteratively to produce a hierarchical clustering.

The algorithm implements a top-down bisecting method, starting with one universal cluster grouping all instances. Iteratively, it creates two new clusters by bisecting an existing one and this continues until the desired number of clusters is reached. This algorithm can be thought as levelwise producing a dendrogram: the number of levels coincides with the number of clusters.

Each cluster is represented by one of its individuals. As mentioned above, we consider the notion of medoid as representing a cluster center since our distance measure works on a categorical feature-space. The medoid of a group of individuals is the individual that has the lowest dissimilarity w.r.t. the others. Formally, given a cluster $C = \{a_1, a_2, \ldots, a_n\}$, the medoid is defined:

$$m = \text{medoid}(C) = \operatorname*{argmin}_{a \in C} \sum_{j=1}^{n} d(a, a_j)$$

The proposed method can be considered as a hierarchical extension of PAM. A bipartition is repeated level-wise producing a dendrogram. Fig. 1 reports a sketch of our algorithm. It essentially consists of two nested loops: the outer one computes a new level of the resulting dendrogram and it is repeated until the desired number of clusters is obtained (which corresponds to the final level; the inner loop consists of a run of the PAM algorithm at the current level.

Per each level, the next worst cluster is selected (SELECTWORSTCLUSTER() function) on the grounds of its quality, e.g. the one endowed with the least average inner similarity (or cohesiveness [27]). This cluster is candidate to being splitted. The partition is constructed around two medoids initially chosen (SELECTMOSTDISSIMILAR() function) as the most dissimilar elements in the cluster and then iteratively adjusted in

```
clusterVector HIERARCHICALBISECTINGAROUNDMEDOIDS(allIndividuals, k, maxIterations)
input   allIndividuals: set of individuals
        k: number of clusters;
        maxIterations: max number of inner iterations;
output clusterVector: array [1..k] of sets of clusters

begin
level ← 0;
clusterVector[1] ← allIndividuals;
repeat
        ++level;
        cluster2split ← SELECTWORSTCLUSTER(clusterVector[level]);
        iterCount ← 0;
        stableConfiguration ← false;
        (newMedoid1,newMedoid2) ← SELECTMOSTDISSIMILAR(cluster2split);
        repeat
            ++iterCount;
            (medoid1,medoid2) ← (newMedoid1,newMedoid2);
            (cluster1,cluster2) ← DISTRIBUTE(cluster2split,medoid1,medoid2);
            newMedoid1 ← MEDOID(cluster1);
            newMedoid2 ← MEDOID(cluster2);
            stableConfiguration ← (medoid1 = newMedoid1) ∧ (medoid2 = newMedoid2);
        until stableConfiguration ∨ (iterCount = maxIterations);
        clusterVector[level+1] ← REPLACE(cluster2split,cluster1,cluster2,clusterVector[level]);
until (level = k);
end
```

Fig. 1. The HIERARCHICAL BISECTING AROUND MEDOIDS Algorithm

the inner loop. In the end, the candidate cluster is replaced by the newly found parts at the next level of the dendrogram.

The inner loop basically resembles to a 2-MEANS algorithm, where medoids are considered instead of means that can hardly be defined in symbolic computations. Then, the standard two steps are performed iteratively:

distribution given the current medoids, distribute the other individuals to either partition on the grounds of their distance w.r.t. the respective medoid;

medoid re-computation given the bipartition obtained by DISTRIBUTE(), compute the new medoids for either cluster.

The medoid tend to change at each iteration until eventually they converge to a stable couple (or when a maximum number of iterations have been performed).

An adaptation of a PAM algorithm has several favorable properties. Since it performs clustering with respect to any specified metric, it allows a flexible definition of similarity. This flexibility is particularly important in biological applications where researchers may be interested, for example, in grouping correlated or possibly also anti-correlated elements. Many clustering algorithms do not allow for a flexible definition of similarity, but allow only Euclidean distance in current implementations.

In addition to allowing a flexible distance metric, a PAM algorithm has the advantage of identifying clusters by the medoids. Medoids are robust representations of the cluster centers that are less sensitive to outliers than other cluster profiles, such as the cluster means of K-MEANS. This robustness is particularly important in the common context that many elements do not belong exactly to any cluster, which may be the case of the membership in DL knowledge bases, which may be not ascertained given the OWA.

The representation of centers by means of medoids has two advantages. First, it presents no limitations on attributes types, and, second, the choice of medoids is dictated by the location of a predominant fraction of points inside a cluster and, therefore, it is less sensitive to the presence of outliers. In K-MEANS a cluster is represented by its centroid, which is a mean (usually weighted average) of points within a cluster. This works conveniently only with numerical attributes and can be more negatively affected by a single outlier.

3.1 Evolution: Automated Concept Drift and Novelty Detection

As mentioned in the introduction conceptual clustering enables a series of further activities related to dynamic settings: 1) concept drift [28]: i.e. the change of known concepts w.r.t. the evidence provided by new annotated individuals that may be made available over time; 2) novelty detection [26]: isolated clusters in the search space that require to be defined through new emerging concepts to be added to the knowledge base.

The algorithms presented above are suitable for an online unsupervised learning implementation. Indeed as soon as new annotated individuals are made available these may be assigned to the *closest* clusters (where closeness is measured as the distance to the cluster medoids or to the minimal distance to its instances). Then, new runs of the clustering algorithm may yield a modification of the original clustering both in the clusters composition and in their number.

Following [26], the clustering representing the starting concepts is built based on the clustering algorithm. For each cluster, the maximum distance between its instances and the medoid is computed. This establishes a decision boundary for each cluster. The union of the boundaries of all clusters represents a global decision boundary which defines the current model.

A new unseen example that falls inside this global boundary is consistent with the model and therefore considered *normal* and may be classified according to the current clustering; otherwise, a further analysis is needed. A single individual located externally w.r.t. the boundary should not be considered as novel *per se*, since it could somehow simply represent noise. Due to lack of evidence, these individuals are stored in a short-term memory, which is to be monitored for detecting the formation of new clusters that might indicate two conditions: novelty and concept drift. Namely, the clustering algorithm applied to individuals in the short-term memory generates candidate clusters. In order to validate a candidate cluster w.r.t. the current model (clustering), the algorithm in Fig. 2 can be applied.

The candidate cluster emerging from the short-memory candCluster is considered *abnormal*[2] (for the aims of drift or novelty detection) when the mean distance between

---

[2] This aims at choosing clusters whose density is not lower than that of the model.

```
(decision, newClustering) DRIFT_NOVELTY_DETECTION(currModel, candCluster)
input:   currModel: current clustering;
         candCluster: candidate cluster;
output:  (decision, newClustering);

begin
  $m_{CC}$ := medoid(candCluster);
  for each $C_j \in$ currModel do
     $m_j$ := medoid($C_j$);
  overallAvgDistance := $\frac{1}{|currModel|} \sum_{C_j \in currModel} \left[ \frac{1}{|C_j|} \sum_{a \in C_j} d(a, m_j) \right]$;
  candClusterAvgDistance := $\frac{1}{|candCluster|} \sum_{a \in CCluster} d(a, m_{CC})$;
  if overallAvgDistance $\geq$ candClusterAvgDistance then
      begin // abnormal candidate cluster detected
        $\overline{m}$ := medoid($\{m_j \mid C_j \subset$ currModel$\}$); // global medoid
        thrDistance := $\max_{m_j \in currModel} d(\overline{m}, m_j)$;
        if $d(\overline{m}, m_{CC}) \leq$ thrDistance then
            return (drift, replace(currModel, candCluster))
        else
            return (novelty, currModel $\cup$ candCluster)
      end
  else return (normal, integrate(currModel, candCluster))
end
```

Fig. 2. Concept drift and novelty detection algorithm

medoids and the respective instances, averaged over all clusters of the current model (overallAvgDistance) is greater than the average distance from the new individuals to the medoid of the candidate cluster (candClusterAvgDistance).

Then a threshold distance thrDistance for distinguishing between concept drift and novelty is computed as the maximum distance between the medoids of clusters in the original model m_j's and the global medoid[3] $\overline{m}$. When the distance between the overall medoid and the candidate cluster medoid m_{CC} does not exceed the threshold distance then a concept drift case is detected and the candidate cluster can replace an old cluster in the current clustering model. This may simply amount to reassigning the individuals in the drifted cluster to the new clusters or it may even involve a further run of the clustering algorithm to restructure the clustering model. Otherwise (novelty case) the clustering is simply extended with the addition of the new candidate cluster. Finally, when the candidate cluster is made up of normal instances these can be integrated by assigning them to the closest clusters.

The main differences from the original method [26], lie in the different representational setting (simple numeric tuples were considered) which allows for the use of off-the-shelf clustering methods such as k-MEANS [15] based on a notion of centroid

---
[3] Clusters which are closer to the boundaries of the model are more likely to appear due to a drift occurred in the normal concept. On the other hand, a candidate cluster appearing to be far from the normal concept may represent a novel concept.

which depend on the number of clusters required as a parameter. In our categorical setting, medoids substitute the role of means (or centroids) and, more importantly, our method is able to detect an optimal number of clusters autonomously, hence the influence of this parameter is reduced.

3.2 From Clusters to Concepts

Each node of the tree (a cluster) may be labeled with an intensional concept definition which characterizes the individuals in the given cluster while discriminating those in the twin cluster at the same level. Labeling the tree-nodes with concepts can be regarded as solving a number of supervised learning problems in the specific multi-relational representation targeted in our setting. As such it deserves specific solutions that are suitable for the DL languages employed.

A straightforward solution may be found, for DLs that allow for the computation of (an approximation of) the *most specific concept* (MSC) and *least common subsumer* (LCS) [1], such as $\mathcal{ALN}$, $\mathcal{ALE}$ or $\mathcal{ALC}$. This may involve the following steps: given a cluster of individuals node_j

– **for each** individual $a_i \in \text{node}_j$ **do**
   compute $M_i \leftarrow \text{MSC}(a_i)$ w.r.t. $\mathcal{A}$;
– **let** $\text{MSCs}_j \leftarrow \{M_i \mid a_i \in \text{node}_j\}$;
– **return** $\text{LCS}(\text{MSCs}_j)$

However, the use of this generalizing operator may be criticized for the sensitiveness to the presence of outliers and for the excessive specificity of the resulting concepts which may result in poorly predictive descriptions w.r.t. future unseen individuals. yet this also depends on the degree of approximation of the MSC's.

As an alternative, algorithms for learning concept descriptions expressed in DLs may be employed [19, 14]. Indeed, concept formation can be cast as a supervised learning problem: once the two clusters at a certain level have been found, where the members of a cluster are considered as positive examples and the members of the dual cluster as negative ones. Then any concept learning method which can deal with these representations may be utilized for this new task.

4 Related Work

The unsupervised learning procedure presented in this paper is mainly based on two factors: the semantic dissimilarity measure and the clustering method. To the best of our knowledge in the literature there are very few examples of similar clustering algorithms working on complex representations that are suitable for knowledge bases of semantically annotated resources. Thus, in this section, we briefly discuss sources of inspiration for our procedure and some related approaches.

4.1 Relational Similarity Measures

As previously mentioned, various attempts to define semantic similarity (or dissimilarity) measures for concept languages have been made, yet they have still a limited applicability to simple languages [4] or they are not completely semantic depending also on the structure of the descriptions [5]. Very few works deal with the comparison of individuals rather than concepts.

In the context of clausal logics, a metric was defined [23] for the Herbrand interpretations of logic clauses as induced from a distance defined on the space of ground atoms. This kind of measures may be employed to assess similarity in *deductive databases*. Although it represents a form of fully semantic measure, different assumptions are made with respect to those which are standard for knowledgeable bases in the SW perspective. Therefore the transposition to the context of interest is not straightforward.

Our measure is mainly based on Minkowski's measures [29] and on a method for distance induction developed by Sebag [25] in the context of *machine learning*, where *metric learning* is developing as an important subfield. In this work it is shown that the induced measure could be accurate when employed for classification tasks even though set of features to be used were not the optimal ones (or they were redundant). Indeed, differently from our unsupervised learning approach, the original method learns different versions of the same target concept, which are then employed in a voting procedure similar to the Nearest Neighbor approach for determining the classification of instances.

A source of inspiration was also *rough sets* theory [24] which aims at the formal definition of vague sets by means of their approximations determined by an indiscernibility relationship. Hopefully, these methods developed in this context will help solving the open points of our framework (see Sect. 6) and suggest new ways to treat uncertainty.

4.2 Clustering Procedures

Our algorithm adapts to the specific representations devised for the SW context a combination of the distance-based approaches (see [15]). Specifically, in the methods derived from K-MEANS and K-MEDOIDS each cluster is represented by one of its points.

PAM, CLARA [16], and CLARANS [22] represent early systems adopting this approach. They implement iterative optimization methods that essentially cyclically relocate points between perspective clusters and recompute potential medoids. Ester et al. [7] extended CLARANS to deal with very large spatial databases.

Further comparable clustering methods are those based on an *indiscernibility relationship* [13]. While in our method this idea is embedded in the semi-distance measure (and the choice of the committee of concepts), these algorithms are based on an iterative refinement of an equivalence relationship which eventually induces clusters as equivalence classes.

Alternatively evolutionary clustering approaches may be considered [9] which are also capable to determine a good estimate of the number of clusters [11, 12]. The UNC algorithm is a more recent related approach which was also extended to the hierarchical clustering case H-UNC [21].

As mentioned in the introduction, the classic approaches to conceptual clustering [27] in complex (multi-relational) spaces are based on structure and logics. Kietz & Morik proposed a method for efficient construction of knowledge bases for the BACK representation language [17]. This method exploits the assertions concerning the roles available in the knowledge base, in order to assess, in the corresponding relationship,

Table 1. Ontologies employed in the experiments

ontology	DL	#concepts	#obj. prop.	#data prop.	#individuals
FSM	$\mathcal{SOF}(D)$	20	10	7	37
S.-W.-M.	$\mathcal{ALCOF}(D)$	19	9	1	115
TRANSPORTATION	$\mathcal{ALC}$	44	7	0	250
FINANCIAL	$\mathcal{ALCIF}$	60	17	0	652
NTN	$\mathcal{SHIF}(D)$	47	27	8	676

those subgroups of the domain and ranges which may be inductively deemed as disjoint. In the successive phase, supervised learning methods are used on the discovered disjoint subgroups to construct new concepts that account for them. A similar approach is followed in [10], where the supervised phase is performed as an iterative refinement step, exploiting suitable refinement operators for a different DL, namely $\mathcal{ALC}$.

System OLINDDA [26] is one of the first methods exploiting clustering for detecting concept drift and novelty. Our method improves on it both in the representation of the instances and in being based on an original clustering method which is not parametrized on the number of clusters.

5 Experimental Evaluation of the Clustering Procedure

An experimental session was planned in order to prove the method feasible. It could not be a comparative experimentation since, to the best of our knowledge no other hierarchical clustering method has been proposed which is able to cope with DLs representations (on a semantic level) except [17, 10] which are language-dependent and produce non-hierarchical clusterings.

For the experiments, a number of different ontologies represented in OWL were selected, namely: FSM, SURFACE-WATER-MODEL, TRANSPORTATION and NEWTESTAMENTNAMES from the Protégé library[4], the FINANCIAL ontology[5] employed as a testbed for the PELLET reasoner. Table 1 summarizes important details concerning the ontologies employed in the experimentation.

A preliminary phase, may regard the selection of the features for the metric. Experimentally, we noted that the optimization affected the efficiency of the distance computation more than the metric sensitiveness. Thus we decided to employ the whole set of named concepts in the KB as features.

As pointed out in several surveys on clustering, it is better to use a different criterion for clustering (e.g. for choosing the candidate cluster to bisection) and for assessing the quality of a cluster. For the evaluation we employed standard validity measures for clustering: the mean square error (WSS, a measure of cohesion) and the *silhouette* measure [16]. Besides, we propose a the extension of Dunn's validity index for clusterings produced by the hierarchical algorithm [2]. Namely, we propose a modified version of

---

[4] http://protege.stanford.edu/plugins/owl/owl-library
[5] http://www.cs.put.poznan.pl/alawrynowicz/financial.owl

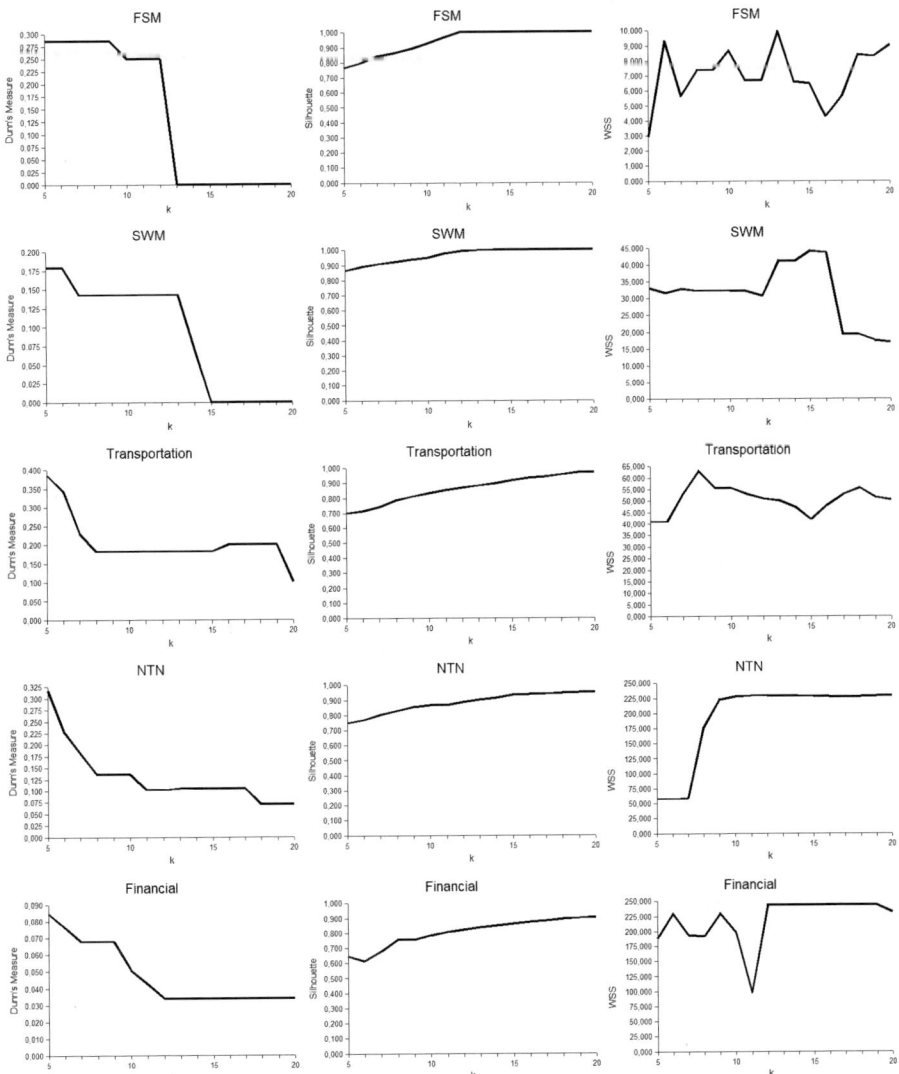

Fig. 3. Outcomes of the experiments: Dunn's, Silhouette, and WSS index graphs

Dunn's index to deal with medoids. Let $P = \{C_1, \ldots, C_k\}$ be a possible clustering of n individuals in k clusters. The index can be defined:

$$V_{GD}(P) = \min_{1 \leq i \leq k} \left\{ \min_{\substack{1 \leq j \leq k \\ i \neq j}} \left\{ \frac{\delta_p(C_i, C_j)}{\max_{1 \leq h \leq k} \{\Delta_p(C_h)\}} \right\} \right\}$$

where δ_p is the Hausdorff distance for clusters[6] derived from d_p and the cluster diameter measure Δ_p is defined:

$$\Delta_p(C_h) = \frac{2}{|C_h|} \left(\sum_{c \in C_h} d_p(c, m_h) \right)$$

which is more noise-tolerant w.r.t. other standard measures.

For each populated ontology, the experiments have been repeated for varying numbers k of clusters (5 through 20). In the computation of the distances between individuals (the most time-consuming operation) all concepts in the ontology have been used for the committee of features, thus guaranteeing meaningful measures with high redundancy. The PELLET reasoner[7] was employed to compute the projections. An overall experimentation of 16 repetitions on a dataset took from a few minutes to 1.5 hours on a 2.5GHz (512Mb RAM) Linux Machine.

The outcomes of the experiments are reported in Fig. 3. For each ontology, we report the graph for Dunn's, Silhouette and WSS indexes, respectively, at increasing values of k (number of clusters searched, which determines the stopping condition).

Particularly, the decay of Dunn's index may be exploited as a hint on possible cut points (the *knees*) in the hierarchical clusterings (i.e. optimal values of k).

It is also possible to note that the silhouette values, as absolute clustering quality measures, are quite stably close to the top of the scale (1). This gives a way to assess the effectiveness of our algorithms w.r.t. others, although applied to different representations.

Conversely, the cohesion coefficient WSS may vary a lot, indicating that for some level the clustering found by the algorithm, which proceeds by bisection of the worst cluster in the previous level, is not the natural one, and thus is likely to be discarded.

6 Conclusions and Outlook

This work has presented a clustering method for the standard (multi-)relational representations adopted in the SW research. Namely, it can be used to discover interesting groupings of semantically annotated resources in a wide range of concept languages.

The method exploits a novel dissimilarity measure, that is based on the resource semantics w.r.t. a number of dimensions corresponding to a committee of features represented by a group of concept descriptions (discriminating features). The algorithm, is an adaptation of the classic bisecting k-means to complex representations typical of the ontology in the SW.

Currently we are working on extensions of the metric based on weighting the discernibility power of the features based on information theory (entropy). Besides, we are also investigating evolutionary clustering methods both for performing the optimization of the feature committee and for clustering individuals automatically discovering an optimal number of clusters [9].

---

[6] The metric δ_p is defined, given any couple of clusters (C_i, C_j), $\delta(C_i, C_j) = \max\{d_p(C_i, C_j), d_p(C_j, C_i)\}$, where $d_p(C_i, C_j) = \max_{a \in C_i}\{\min_{b \in C_j}\{d_p(a, b)\}\}$.
[7] http://pellet.owldl.com

Finally, we plan to perform further experiments to evaluate the quality of the clustering and of the induced (new) concepts, although it may be questionable to assess this objectively. The output of our method is thought to be validated by a domain expert. then, a knowledge engineer may foresee to adopt these methods to validate knowledge bases under construction while the experts are collaborating on the task.

References

[1] Baader, F., Calvanese, D., McGuinness, D., Nardi, D., Patel-Schneider, P. (eds.): The Description Logic Handbook. Cambridge University Press, Cambridge (2003)
[2] Bezdek, J.C., Pal, N.R.: Some new indexes of cluster validity. IEEE Transactions on Systems, Man, and Cybernetics 28(3), 301–315 (1998)
[3] Borgida, A.: On the relative expressiveness of description logics and predicate logics. Artificial Intelligence 82(1-2)
[4] Borgida, A., Walsh, T.J., Hirsh, H.: Towards measuring similarity in description logics. In: Horrocks, I., Sattler, U., Wolter, F. (eds.) Working Notes of the International Description Logics Workshop, Edinburgh, UK. CEUR Workshop Proceedings, vol. 147 (2005)
[5] d'Amato, C., Fanizzi, N., Esposito, F.: Reasoning by analogy in description logics through instance-based learning. In: Tummarello, G., Bouquet, P., Signore, O. (eds.) Proceedings of Semantic Web Applications and Perspectives, 3rd Italian Semantic Web Workshop, SWAP 2006, Pisa, Italy. CEUR Workshop Proceedings, vol. 201 (2006)
[6] d'Amato, C., Staab, S., Fanizzi, N., Esposito, F.: Efficient discovery of services specified in description logics languages. In: Di Noia, T., et al. (eds.) Proceedings of Service Matchmaking and Resource Retrieval in the Semantic Web Workshop at ISWC 2007, vol. 243, CEUR (2007)
[7] Ester, M., Kriegel, H.-P., Sander, J., Xu, X.: A density-based algorithm for discovering clusters in large spatial databases. In: Proceedings of the 2nd Conference of ACM SIGKDD, pp. 226–231 (1996)
[8] Fanizzi, N., d'Amato, C., Esposito, F.: Induction of optimal semi-distances for individuals based on feature sets. In: Working Notes of the International Description Logics Workshop, DL 2007, Bressanone, Italy. CEUR Workshop Proceedings, vol. 250 (2007)
[9] Fanizzi, N., d'Amato, C., Esposito, F.: Randomized metric induction and evolutionary conceptual clustering for semantic knowledge bases. In: Silva, M., Laender, A., Baeza-Yates, R., McGuinness, D., Olsen, O., Olstad, B. (eds.) Proceedings of the ACM International Conference on Knowledge Management, CIKM 2007, Lisbon, Portugal, ACM Press, New York (2007)
[10] Fanizzi, N., Iannone, L., Palmisano, I., Semeraro, G.: Concept formation in expressive description logics. In: Boulicaut, J.-F., Esposito, F., Giannotti, F., Pedreschi, D. (eds.) ECML 2004. LNCS (LNAI), vol. 3201, pp. 99–113. Springer, Heidelberg (2004)
[11] Ghozeil, A., Fogel, D.B.: Discovering patterns in spatial data using evolutionary programming. In: Koza, J.R., Goldberg, D.E., Fogel, D.B., Riolo, R.L. (eds.) Genetic Programming 1996: Proceedings of the First Annual Conference, Stanford University, CA, USA, pp. 521–527. MIT Press, Cambridge (1996)
[12] Hall, L.O., Özyurt, I.B., Bezdek, J.C.: Clustering with a genetically optimized approach. IEEE Trans. Evolutionary Computation 3(2), 103–112 (1999)
[13] Hirano, S., Tsumoto, S.: An indiscernibility-based clustering method. In: Hu, X., Liu, Q., Skowron, A., Lin, T.Y., Yager, R., Zhang, B. (eds.) 2005 IEEE International Conference on Granular Computing, pp. 468–473. IEEE, Los Alamitos (2005)

[14] Iannone, L., Palmisano, I., Fanizzi, N.: An algorithm based on counterfactuals for concept learning in the semantic web. Applied Intelligence 26(2), 139–159 (2007)
[15] Jain, A.K., Murty, M.N., Flynn, P.J.: Data clustering: A review. ACM Computing Surveys 31(3), 264–323 (1999)
[16] Kaufman, L., Rousseeuw, P.J.: Finding Groups in Data: an Introduction to Cluster Analysis. John Wiley & Sons, Chichester (1990)
[17] Kietz, J.-U., Morik, K.: A polynomial approach to the constructive induction of structural knowledge. Machine Learning 14(2), 193–218 (1994)
[18] Kirsten, M., Wrobel, S.: Relational distance-based clustering. In: Page, D.L. (ed.) ILP 1998. LNCS, vol. 1446, pp. 261–270. Springer, Heidelberg (1998)
[19] Lehmann, J.: Concept learning in description logics. Master's thesis, Dresden University of Technology (2006)
[20] Lehmann, J., Hitzler, P.: A refinement operator based learning algorithm for the alc description logic. In: The 17th International Conference on Inductive Logic Programming (ILP). LNCS, Springer, Heidelberg (2007)
[21] Nasraoui, O., Krishnapuram, R.: One step evolutionary mining of context sensitive associations and web navigation patterns. In: Proceedings of the SIAM conference on Data Mining, Arlington, VA, pp. 531–547 (2002)
[22] Ng, R., Han, J.: Efficient and effective clustering method for spatial data mining. In: Proceedings of the 20th Conference on Very Large Databases, VLDB 1994, pp. 144–155 (1994)
[23] Nienhuys-Cheng, S.-H.: Distances and limits on herbrand interpretations. In: Page, D.L. (ed.) ILP 1998. LNCS, vol. 1446, pp. 250–260. Springer, Heidelberg (1998)
[24] Pawlak, Z.: Rough Sets: Theoretical Aspects of Reasoning About Data. Kluwer Academic Publishers, Dordrecht (1991)
[25] Sebag, M.: Distance induction in first order logic. In: Džeroski, S., Lavrač, N. (eds.) ILP 1997. LNCS, vol. 1297, pp. 264–272. Springer, Heidelberg (1997)
[26] Spinosa, E.J., Ponce de Leon Ferreira de Carvalho, A., Gama, J.: OLINDDA: A cluster-based approach for detecting novelty and concept drift in data streams. In: Proceedings of the 22nd Annual ACM Symposium of Applied Computing, SAC 2007, Seoul, South Korea, vol. 1, pp. 448–452. ACM, New York (2007)
[27] Stepp, R.E., Michalski, R.S.: Conceptual clustering of structured objects: A goal-oriented approach. Artificial Intelligence 28(1), 43–69 (1986)
[28] Widmer, G., Kubat, M.: Learning in the presence of concept drift and hidden contexts. Machine Learning 23(1), 69–101 (1996)
[29] Zezula, P., Amato, G., Dohnal, V., Batko, M.: Similarity Search – The Metric Space Approach. In: Advances in Database Systems, Springer, Heidelberg (2007)

Enriching an Ontology with Multilingual Information

Mauricio Espinoza[1], Asunción Gómez-Pérez[1], and Eduardo Mena[2]

[1] UPM, Laboratorio de Inteligencia Artificial, 28660 Boadilla del Monte, Spain
asun@fi.upm.es, mespinoza@delicias.dia.fi.upm.es
[2] IIS Department, Univ. of Zaragoza, María de Luna 1, 50018 Zaragoza, Spain
emena@unizar.es

Abstract. Organizations working in a multilingual environment demand multilingual ontologies. To solve this problem we propose LabelTranslator, a system that automatically localizes ontologies. Ontology localization consists of adapting an ontology to a concrete language and cultural community.

LabelTranslator takes as input an ontology whose labels are described in a source natural language and obtains the most probable translation into a target natural language of each ontology label. Our main contribution is the automatization of this process which reduces human efforts to localize an ontology manually. First, our system uses a translation service which obtains automatic translations of each ontology label (name of an ontology term) from/into English, German, or Spanish by consulting different linguistic resources such as lexical databases, bilingual dictionaries, and terminologies. Second, a ranking method is used to sort each ontology label according to similarity with its lexical and semantic context.

The experiments performed in order to evaluate the quality of translation show that our approach is a good approximation to automatically enrich an ontology with multilingual information.

Keywords: Ontology localization, Multilingual ontologies.

1 Introduction

The Semantic Web offers the most appropriate scenario for exploiting the potentialities of ontologies due to the large amount of information which is to be exposed and accessed. However, most of the ontologies on the Web are in English and although there are a few exceptions (like EuroWordNet [13]), it is necessary to guarantee that the same knowledge be recognizable in different natural languages.

Currently, more and more organizations working in multilingual environments demand ontologies supporting different natural languages. In the framework of the NeOn project [7], all case studies have expressed the need for multilingual ontologies. One case study is led by the Food and Agriculture Organization of the United Nations (FAO), an international organization with a huge amount of multilingual resources, some of them in more than ten languages. The second use case is concerned with the pharmaceutical industry in Spain, and requires ontologies in the different languages spoken in the country. Consequently, the inclusion of multilingual data in ontologies is not an option but a must.

In this paper we introduce *LabelTranslator*, a system that automatically localizes ontologies in English, Spanish and German. The Ontology Localization Activity (OLA) consists of adapting an ontology to a concrete language and culture community, as defined in [6]. We describe here the features and design aspects of the current prototype of our system and discuss some of the innovations we are planning for a future prototype. *LabelTranslator* takes as input an ontology whose labels are expressed in a source natural language, and obtains the most probable translation of each ontology label into a target natural language.

The rest of this paper is as follows. In Section 2 we describe the main components of our approach. In Section 3 we show how the system obtains the translation of each ontology label from different linguistic resources. The translation ranking method used to rank the translations of each ontology label according to its context can be found in Section 4. Some experimental results are presented in Section 5. Related work can be found in Section 6. Finally, conclusions and future work appear in Section 7.

2 Overview of the System

As a motivating example, let us consider the extract of the sample university ontology shown in Figure 1. Let us suppose that the user wants to translate the term *chair* from English into Spanish. According to the domain of the sample ontology, the correct translation of the selected term should be in the sense of the position professor, nor in the sense of a place where one person can sit down and nor an instrument of execution by electrocution, etc.

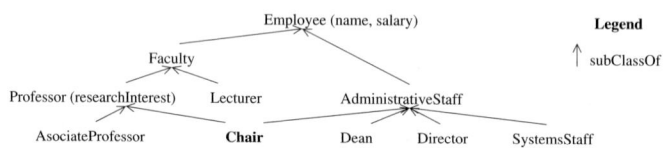

Fig. 1. Extract of the sample university ontology

In the following, we first show the main components of our system and then we describe how these components with interact each other.

2.1 Structural Overview

There are three important aspects to be considered to automatically localize monolingual ontologies to other natural languages and to allow users to access to multilingual information: 1) obtaining the possible translations for ontology labels, 2) disambiguating label senses, and 3) ranking translations. Certainly, all multilingual results have to be appropriately stored and presented to the user. Figure 2 shows the main components of our system and illustrates the process for enriching an ontology with linguistic information:

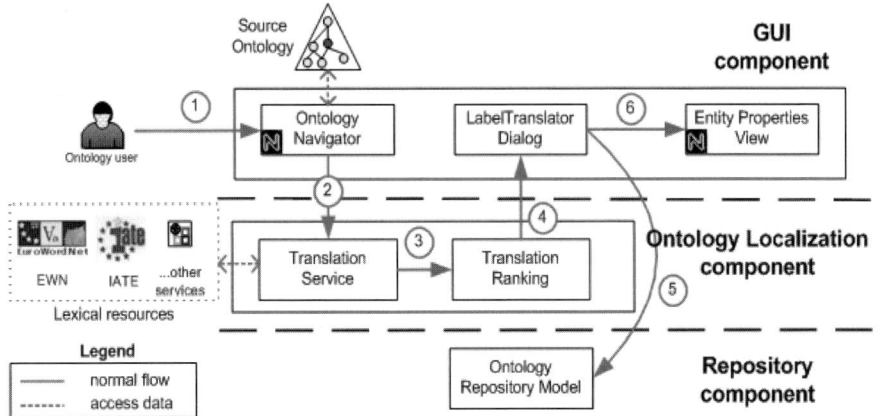

Fig. 2. Main components of *LabelTranslator* plug-in

- **GUI component:** This component controls the GUI in order to show the multilingual results appropriately. Once invoked, *LabelTranslator* uses some views[1] of the Neon ToolKit to load the ontology and store the multilingual results respectively. Due to space limitations, the user interfaces used by our system are not shown here, but can be found in [4].
- **Ontology Localization Component:** This component is responsible of obtaining the most probable translation for each ontology label. It relies on two advanced modules. The first, *translation service*, automatically obtains the different possible translations of an ontology label by accessing to different linguistic resources. This service also uses a compositional method in order to translate compound labels (multi-word labels). A more detailed description of the translation service can be found in Section 3.

  The second module, *translation ranking*, sorts the different translations according to the similarity with its lexical and semantic context. The method relies on a relatedness measure based on glosses to disambiguate the translations. This is done by comparing the senses associated to each possible translation and their context. More details about the ranking method can be found in Section 4.
- **Repository Component:** This component captures all the linguistic information associated with concepts. *LabelTranslator* supports the linguistic model [10] designed for the representation of multilingual information in ontologies. In the current version of our system, the link we establish between

---

[1] In the NeOn ToolKit a view is typically used to navigate a hierarchy of information, open an editor, or display properties for the active editor.

the terms in the ontology and their associated translations is characterized by simple references between concepts and labels (as offered by the standard owl:comment and rdfs:label properties). The representation of the multilingual information obtained by our tool is out of scope of this paper, however, a detailed description of this multilingual model can be found in [10].

The current version of LabelTranslator has been implemented as a NeOn plug-in, but it can easily become an independent module, for example a web service.

2.2 Functional Overview

This section briefly describe the interaction of the components shown in the Figure 2. A more detailed description of the execution process of our system can be found in [4]. The main activities can be summarized as follows:

1. the user chooses the label of the ontology term(s) to be translated.
2. the *translation service* access bilingual or multilingual linguistic resources to look for the possible translations of the selected label(s).
3. the *ranking method* compares the obtained translation possibilities against the label(s) in the original language and its ontological context.
4. the user confirms the translation proposed by the ranking method, or he/she chooses the translation that better fits in the ontological context.
5. the system updates the ontology model with the selected translation(s) and additional linguistic information.
6. finally, the linguistic information associated with ontology term(s) is shown to the user.

More details on the features of the *ontology localization component*, are given in the following sections.

3 Translation Service

In this section we provide the details that show how the system obtains the different translations of an ontology label (which can name different kinds of ontology terms: concepts, properties or relations) using different linguistic resources.

The *translation service* takes as input an ontology label l described in a source language and returns a set of possible translations $T = \{t_1, t_2, ..., t_n\}$ in a target language. The current prototype supports translations among English, Spanish, and German. In order to discover the translations of each ontology label, the system accesses different lexical resources: 1) remote lexical databases as EuroWordNet [13], 2) multilingual dictionaries as GoogleTranslate[2], Wiktionary[3], Babelfish[4], and FreeTranslation[5], and 3) other lexical resources as IATE[6]. A *cache* stores previously translations to avoid accessing the same data twice.

---

[2] http://www.google.com/translate_t
[3] http://en.wiktionary.org/wiki/
[4] http://babelfish.altavista.com/
[5] http://ets.freetranslation.com
[6] http://iate.europa.eu/iatediff/SearchByQueryLoad.do?method=load

The algorithm used by the translation service is summarized in the following: 1) If the selected ontology label is already available in the target language in our cache, then LabelTranslator just displays it, with all the relevant available information, 2) If the translation is not stored locally, then it accesses remote repositories to retrieve possible translations. A compositional method may be needed to translate compound labels (explained in the Section 3.2). If no results are obtained from the two previous steps, then the user can enter his/her own translation (together with the definition).

In our approach, the translation of an ontology label denoted by t, is a tuple $\langle trs, senses \rangle$, where trs is translated label in the specific target language, and $senses$ is a list of semantic senses extracted from different knowledge pools. In the following we briefly describe the task of automatically retrieving the possible semantic senses of a translated label.

3.1 Semantically Representating a Sense

In order to discover the senses of each translated label (t_i), we have considered the approach proposed in a previous work [12]. Our system takes as input a list of words (each t_i), discovers their semantics in run-time and obtains a list of senses extracted from different ontology pools; it deals with the possible semantic overlapping among senses. We summarize here the key characteristic of the sense discovering process:

1. To discover the semantic of the input words, the system relies on a pool of ontologies instead of just a single ontology.
2. The system builds a sense (meaning) with the information retrieved from matching terms in the ontology pool.
3. Each sense is represented as a tuple $s_k = <s, grph, descr>$, where s is the list of synonym names[7] of word k, $grph$ describes the sense s_k by means of the hierarchical graph of hypernyms and hyponyms of synonym terms found in one or more ontologies, and $descr$ is a description in natural language of such a sense.
4. As matching terms could be ontology classes, properties or individuals, three lists of possible senses are associated with each word k: S_k^{class}, S_k^{prop} and S_k^{indv}.
5. Each word sense is enhanced incrementally with the synonym senses (which also searches the ontology pool).
6. A sense alignment process integrates the word sense with those synonym senses representing the same semantics, and discards the synonym senses that do not enrich the word sense.

A more detailed description of this process can be found in [12]. In order to perform cross-language sense translations, the external resources are limited to those resources that have multilingual information like EuroWordNet; however

---

[7] The system extracts the synonym names of a term by consulting the synonym relationships defined in the ontology of such a term.

other resources can be used too. For example, a specific domain resource for the FAO (Food and Agricultural Organization) is Agrovoc[8], which could cover the vocabulary missed in EuroWordNet. The multilingual retrieval of a word sense (synset) in EuroWordNet is done by means of the InterlingualIndex (ILI), that serves as a link among the different wordnets. For example, when a synset, e.g. "chair" with the meaning "the position professor", is retrieved from the English wordnet, its synset ID is mapped through the ILI to the synsets IDs of the same concept in the different languages-dependent wordnets,(German, Spanish, etc.) that describe the same concept, but naturally contain the word description in its specific language. A similar retrieval process is used in the case of multilingual ontologies, but using the references between concepts and labels as offered by the standard owl:comment and rdfs:label properties.

Coming back to the example of section 3, in Figure 3 we show the translations of the ontology label "chair" from English into Spanish; our prototype finds eight translations, but due to space limitations we only show three. Notice that $t3$ has the desired semantics according to the similarity with the lexical and semantic ontology context (see figure 1 in section 2).

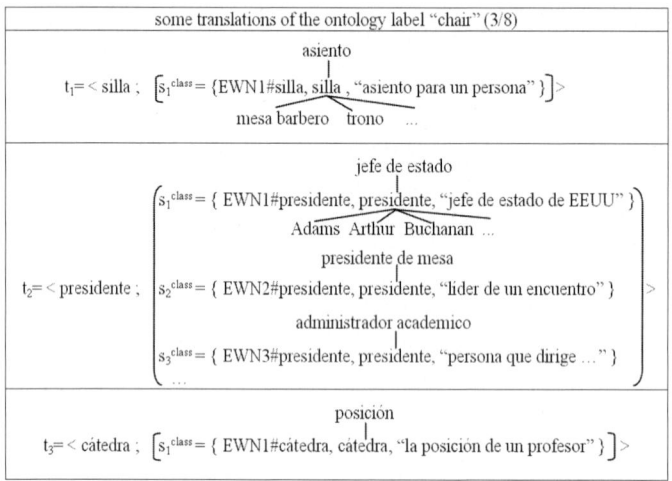

Fig. 3. Some translations of the ontology label "chair" into Spanish

3.2 Compositional Method to Translate Compound Labels

Compound labels which have an entry in linguistic ontologies such as EuroWordNet [13] (for example "jet lag", "travel agent" and "bed and breakfast") are treated in our system as single words. Others like "railroad transportation", which have no entry in the previous resources, are translated using a compositional method. This method split the label into tokens ("railroad" and

---

[8] http://www.fao.org/aims/ag_download.htm

"transportation" in the example); the individual components are translated and then combined into a compound label in the target language. Care is taken to combine the components respecting the word order of the target language. A set of lexical templates derived from different ontologies are used to control the order of translation. The main steps of the algorithm are:

1. The compound label is normalized, e.g., rewriting in lowercase, hyphens are removed, it is split into tokens, etc.
2. A set of possible translations is obtained for each token of the compound label using the translation service.
3. Since translations between languages do not keep the same word order, the algorithm creates candidate translations in the target language using lexical templates[9]. Each lexical template contains at least a pair of patterns, namely 'source' and 'target' patterns. A source pattern is a template to be compared with the *tagged compound label*[10], described in the source language, while the target pattern is used to generate the label in the target language. If no applicable template is found, the compound label is translated using the translation service directly.
4. All the candidate labels that fulfill the target pattern are returned as candidate translations of the compound label.

The senses of each candidate translation are discovered using the sense discovering process described in the section 3.1. If not results are obtained, the method tries to discover the senses of each token separately.

In the following we describe the process to learn the lexical templates used to control the order of translation of compound labels.

Learning Lexical Templates from Ontological Labels

We believe that lexical templates used to translate compound labels are a necessary component to produce high quality translations because 1) it guarantees grammatical output and, 2) it makes sure that the structural source language meaning is preserved. In our approach, we used a semi-automatic process to obtain the lexical templates. As we explained before, each lexical template is composed of source and target patterns. The ontology labels used to learn the source patterns were extracted from different domain ontologies expressed in English, German, or Spanish. Each label was tokenized and tagged using the language independent part-of-speech tagger proposed in [11]. On the other hand, the labels used to learn the target patterns were extracted either from the multilingual information associated with each ontological term or by means of a manual translation process. The same process used to annotate part of speech (POS) in the labels of the source patterns was used to annotate the labels of

---
[9] The notion of lexical template proposed in this paper refers to text correlations found between a pair of languages.
[10] We use TreeTagger [11] in order to annotate the compound labels with part-of-speech and lemma information.

Table 1. Some lexical templates to translate a compound label from English into Spanish

Templates (4/25)	Samples of source and target patterns	
	English	Spanish
$[J_1 \ N_2]en \rightarrow [N_2 \ J_1]es$	spatial region→	región espacial
	industrial product→	producto industrial
	natural hazard→	peligro natural
$[N_1 \ N_2]en \rightarrow [N_2 \langle pre \rangle N_1]es$	transport vehicle→	vehículo de transporte
	knowledge domain→	dominio del conocimiento
	research exploration→	exploración de la investigación
$[J_1 \ VB_2]en \rightarrow [VB_2 \langle pre \rangle J_1]es$	remote sensing→	detección remota;
		detección a distancia
$[J_1 N_2 N_3]en \rightarrow [N_2 \langle pre \rangle N_3 \ J_1]es$	associated knowledge domain→	dominio de conocimiento asociado

J: adjective; N: noun; VB: verb

the target patterns. The empirical results collected during the learning of lexical templates are briefly described below:

- *Existing ontologies share the same lexical patterns.* For instance, approximately 60% of the labels that describe an ontological concept makes use of an adjective followed by a noun (e.g. spatial region, industrial product, natural hazard, etc.). Other labels use as lexical pattern ($\approx$ 30%) a noun followed by another noun (e.g., transport vehicle, knowledge domain, etc.).
- *Ontology labels usually have less than four tokens.* Approximately 85% of labels fulfill this. Thus, for the current prototype we only focus on the definition of lexical templates for compound labels of two o three tokens.

A repository is used to store all the lexical templates obtained for each pair of languages. Due to space limitations, in Table 1 we show only a sample list of the lexical templates learned to translate compound labels from English into Spanish.

As an illustrating example of the compositional method, we show in Figure 4 the steps of the algorithm when collecting Spanish translations for the English compound label "AssociateProfessor", which was introduced in our motivating example (see Figure 1). Our prototype finds ten translations for the token "associate" and one for "professor" (normalized in the first step). In the next step, our tool searches a lexical template (in our repository) to create candidate translations. In the template found, $[J_1 \ N_2]en$ represents the source pattern in English whilst $[N_2 \ J_1]es$ represents the target pattern in Spanish. In both cases, numbers represent the position of each token of the compound label. Notice that, in the last step the candidate translations "profesor socio" (professor member) and "profesor compañero" (professor mat) are discarded because they do not fulfill the target pattern.

4 Translation Ranking Method

In this section we explain the *ranking method*, which sorts the list of translations according to similarity with the context of the label to be translated. The method

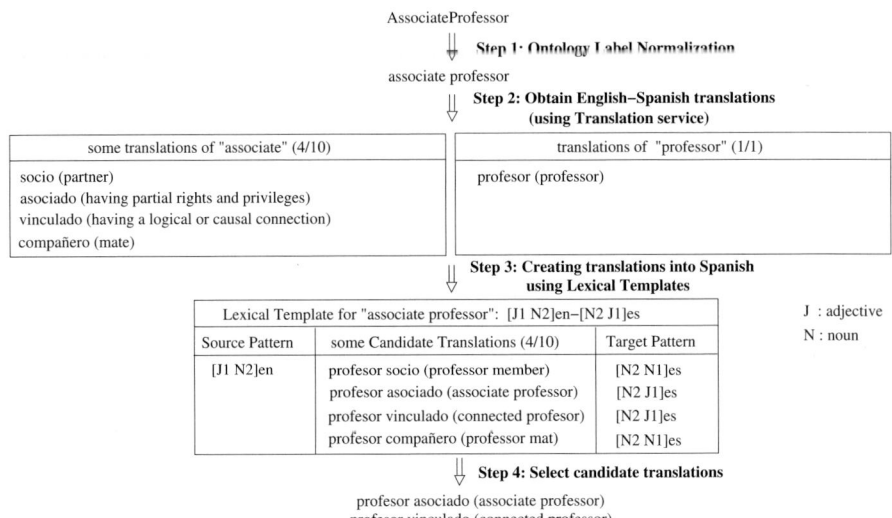

Fig. 4. Algorithm to translate the compound label "AssociateProfessor" into Spanish

takes as input the set of translations T obtained in the previous step. From this set of translations, the ranking method uses a disambiguation algorithm described in [9] to sort the translations. Once all the translations are ranked, the method allows two operation modes:

- Semi-automatic mode: It shows a list with all the possible translations sorted decreasingly. The method proposes the most relevant translation to be selected first although the user can change this default selection.
- Automatic mode: It automatically selects the translation with the highest score.

Next, we first describe how the system obtains the context of each ontology label, and then we describe the disambiguation algorithm used to sort the translations according to similarity with their context.

4.1 Determining the Context of an Ontology Term

We defined context as the information/knowledge that can be used additionally to perform some task. In our approach, the context of an ontology term is used to disambiguate the lexical meaning of a ontology term. To determine the context of an ontology term, the system retrieves the labels of the set of terms associated with the term under consideration. The list of context labels, denoted by C, comprises a set of names which can be direct label names and/or attributes label names, depending on the type of term that is being translated.

In order to mitigate risks associated with system performance, the *ranking method* limits the number of context labels used to disambiguate the translated label. Every context label $c \in C$ is compared with the ontology label l using a measure based on Normalized Google Distance [1] (NGD). NGD measures the semantic relatedness between any two terms, considering the relative frequency in which two terms appear in the Web within the same documents. Those labels with the higher values of similarity are chosen (maximum 3). To discover the senses of each context label (denoted by S_c), the system performs the same process used to discover the senses of each translated label (as explained in the previous section).

In Figure 5, on the left, the dashed area represents all the context labels found for the ontology label "chair". Our prototype finds five labels, but only selects three (see the dotted area) to disambiguate the term. In the table on the right, we show for each type of ontology term (concept, attribute, or relation) the context labels that could be extracted. For instance, for the concept "chair" the system retrieves its hypernyms, hyponyms, attributes, and sibling concepts.

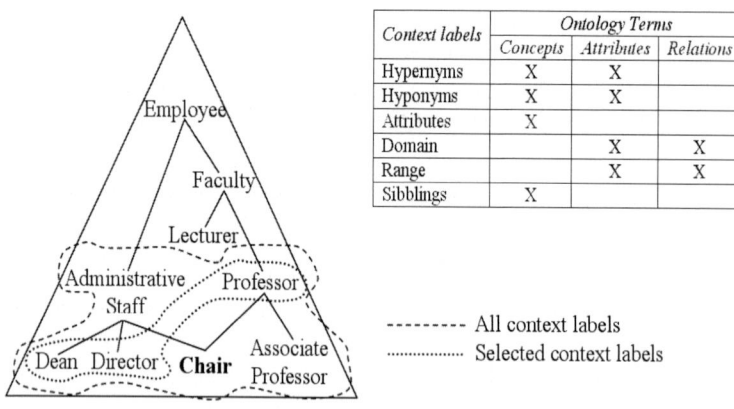

Fig. 5. Context of the ontology label "chair"

4.2 Disambiguating the Senses of the Translations

In some works [8,9] the glosses are considered as a very promising means of measuring relatedness, since they can be used: 1) to make comparisons between concepts semantically different, and 2) to discover relations of which no trace is present in the resource they come from. For the current version of the prototype, the ranking method relies on a measure based on glosses proposed in [9] to sort the translations according to their context. However, we recognize that glosses are by necessity short and may not provide sufficient information on their own to make judgments about relatedness. Therefore, we make use of the hierarchical graph of the sense to extend the gloss with the relatedness glosses of their ontological terms.

We carry out disambiguation in relation to the senses of each translated label and the senses of the context labels. In the following we describe the method: let us suppose that the ontology label l after executing the translation process has yielded n translations: $T = \{t_1, t_2, ..., t_n\}$. For each translation the system retrieves its corresponding senses, for example the first translated label (t_1) to be disambiguated has n senses $S_{t_1} = \{s_{t_1}^1, s_{t_1}^2, ..., s_{t_1}^n\}$. We use the notation TSC (translation sense collection) in order to group the senses of all translated labels.

$$TSC = \{S_{t_1} \cup S_{t_2} \cup ... \cup S_{t_n}\}$$

where S_{t_j}, $t_j \in T$, represents all senses corresponding to j^{th} translated label.

Now, suppose that the ontology label l has the context C which comprises several labels: c_1, c_2, c_3. Each of these context labels has a list of corresponding senses, for instance, c_i has m senses: $S_{c_i} = \{s_{c_i}^1, s_{c_i}^2, ..., s_{c_m}^n\}$. We use the notation CSC (context sense collection) in order to group the senses of each context label.

$$CSC = \{S_{c_1} \cup S_{c_2} \cup S_{c_3}\}$$

where S_{c_j}, $c_j \in C$, represents all senses corresponding to j^{th} context label.

The goal of the disambiguation algorithm is to select one of the senses from the set TSC as the most appropriate sense of the translation of label l. The algorithm performs word sense disambiguation by using a measure of semantic relatedness that is given by:

$$\max_{j=1}^{|TSC|}(SenseScore(TSC_j, CSC))$$

where TSC_j is the representation of one of the senses of each translated label. The chosen sense is the one with the greater value of SenseScore, defined as:

$$SenseScore(TSC_j, CSC) = \sum_{k=1}^{|CSC|}(Similarity(TSC_j, CSC_k))$$

where CSC_k is the representation of each sense of the different context labels.

In order to compute the similarity between the senses of each context and the translated label, the method applies an overlap scoring mechanism. Details about this process are available in [9] as it is not the main goal of this paper.

In our example, "cátedra" (cathedral) in the sense of "the position of professor" is ranked as first translation of the ontology label "chair". Once the right sense has been selected, the system updates the linguistic information of the corresponding ontological term.

5 Experimental Evaluation

The ontology corpus used for the evaluation was selected from the set of KnowledgeWeb [2] ontologies used to manage EU projects. The corpus statistics are given in Table 2. In order to illustrate the utility of the tool to a broad community of ontology users, we are working in the evaluation of prominent ontologies such as those used in the use cases the NeOn project. The results obtained are ongoing work.

Table 2. Ontologies corpus statistics

Ontology Domain	Number of Ontological Terms			% Compound labels	
	concepts	attributes	relations	≤3 tokens	>3 tokens
Documentation-Meeting	42	61	22	44%	25.6%
Person&Project	25	18	12	47.2%	10.9%
Organization	10	7	11	46.4%	7.1%
Office	20	12	8	12.5%	0%
University	30	10	12	17.3%	0%

5.1 Experimental Results

The goal of the experiments is to evaluate some aspects of the translation ranking method (described in Section 4) which tries to select the most appropriate translation for each ontology label. In particular, we evaluated three aspects of the algorithm 1) the quality of the output when the algorithm automatically suggest an translation, 2) the quality of all the set of translations, and 3) the quality of translation of the compound labels.

The manual evaluation that we decided to apply was done by Spanish speakers with a good level of English. In all the experiments a reference translation (gold standard) provided by the evaluators was used. The "gold standard" allows users to compare the quality of the translations provided by an expert with the translations provided by the algorithm. Next, we give a short overview of each experiment and show the obtained results.

Experiment 1: Accuracy. In order to evaluate the quality of the output of the ranking method in automatic operation mode we propose a measure of accuracy. The accuracy measures the capacity of the algorithm of translation to get in an automatic way a correct translation according to context. To measure the accuracy of the algorithm, we counted the number of times the first translation was correct.

Experiment 2: Precision and Recall. The previous evaluation does not allow checking the completeness of the translations since it does not observe the behavior of all the translated labels. Thus, we have measured *precision* as the number of correct translations of all the translations provided by the system and divided by the total number of translations provided by the system. To measure the *recall*, we divided the number of correct translations of all the translations provided by the system into the number of correct translations (provided by the gold standard).

Experiment 3: Adequacy and Fluency. In order to measure the quality of the translation of compound labels we propose a subjective 1-5 score for adequacy and fluency. The adequacy measures the capability of the translation algorithm to determinate the quantity in which the meaning of a correct translation is preserved. On the other hand, the fluency measures the capability of the algorithm to determine how good the corresponding language is. In this experiment, each evaluator assigned fluency and adequacy ratings for each translated

Table 3. Results obtained in the three experiments

Ontology Domain	Spanish					German				
	Accu.	Prec.	Rec.	Adeq.	Flu.	Accu.	Prec.	Rec.	Adeq.	Flu.
Documentation	0.51	0.47	0.39	68%	75%	-	-	-	-	-
Person&Project	0.73	0.35	0.81	89%	93%	-	-	-	-	-
Organization	0.81	0.41	0.78	87%	95%	0.73	0.33	0.64	73%	69%
Office	0.79	0.49	0.77	93%	95%	0.78	0.34	0.74	67%	76%
University	0.80	0.36	0.87	96%	93%	0.71	0.23	0.71	69%	73%

label. Each score ranges from one to five (with one being the poorest grade and five the highest). The adequacy and fluency scores of two evaluators for each sentence were averaged together, and an overall average adequacy and average fluency score was calculated for each evaluated ontology.

In Table 3 we show the results achieved by the prototype in each experiment. The values are organized by target language. All the percentages of adequacy and fluency shown in this table correspond to those translations punctuated with a value greater than 4. The experimental results show that our system is a good approximation to enhance the linguistic expressivity of existing ontologies. For example, in average our system suggest the correct translation 72% of the times. Also, the values of recall suggest that a high percentage of correct translations are part of the final translations shown to the user in the semi-automatic operation mode. Moreover, the obtained results in each metric help us to analyze which components need improvement. The main limitations discovered are:

- *Translation service is highly dependent on the types of resources used and their domain coverage.* The worst values of precision and recall were obtained by the documentation ontology, because the domain of this ontology is covered by the resources used for the translation only partially.
- *The lack of learning of new lexical patterns limits the scalability of our tool.* The percentages of adequacy and fluency obtained for English-German compound label translations are in general lower than the percentages of the English-Spanish ones. Our explication is that a major effort was put (in the current version) for the learning of templates between English-Spanish languages. However, this situation can be improved by allowing users to provide, in runtime, new lexical templates when these do not exist yet in any repository.

6 Related Work

Our work enhances the work presented in [3], where a system for supporting the multilingual extension of ontologies expressed in just one natural language was proposed. This tool is used to support "the supervised translation of ontology labels". Therefore, the tool offers a semi-automatic strategy. In our approach we have implemented an automatic method to reduce human intervention while enriching an ontology with linguistic information.

In [8] they propose a framework for adding linguistic expressivity to conceptual knowledge, as represented in ontologies. They use two lexical resources for the linguistic or multilingual enrichment; WordNet, for the linguistic enrichment of ontologies with English labels, and DICT dictionaries, for the linguistic and multilingual enrichment of ontologies. In this work, they do not describe the process to translate compound ontology labels, which are often not contained in linguistic ontologies as WordNet. In our work, we use a compositional method which first searches for translation candidates of a compound label and then builds the translations for the candidates using lexical templates.

In [5] a method to give support to multilingual ontology engineering is developed. In this work some software tools have been used for supporting the process of term extraction and translation. In particular the translation process requires sentence aligned parallel text, tokenized, tagged and lemmatized. In our opinion, obtaining a corpus aligned is not a simple task. Unlike this work, we rely on some multilingual translation services and extend them by using lexical templates.

7 Conclusions and Future Work

In this paper we have presented LabelTranslator, a system that automatically localize ontologies, providing translations across different natural languages. The main features of this tool are the following:

1. It uses a translation mining service to obtain from different linguistic resources the possible translations of each ontological label. A compound label is translated using a compositional method that searches candidate translations of each lexical component and then builds a translation using lexical templates.
2. It uses a disambiguation method that ranks the possible translations of each ontology label. A gloss overlap scoring mechanism is used to calculate the similarity between two senses.

The experimental results obtained encourage us to tackle further improvements and tests on our system.

Acknowledgements

This work is supported by the European Commission's Sixth Framework Program under the project name: Lyfecycle support for networked ontologies (NeOn) (FP6-027595), the National Project "GeoBuddies" (TSI2007-65677C02), and the spanish CICYT project TIN2007-68091-C02-02.

References

1. Cilibrasi, R.L., Vitányi, P.M.: The Google Similarity Distance. IEEE Transactions on Knowledge and Data Engineering 19(3), 370–383 (2007)
2. Corcho, O., López-Cima, A., Gómez-Pérez, A.: The odesew 2.0 semantic web application framework. In: Proceedings of the 15th International Conference on World Wide Web, WWW 2006, Edinburgh, Scotland, May 23-26, 2006, pp. 1049–1050. ACM Press, New York (2006)

3. Declerck, T., Gómez-Pérez, A., Vela, O., Gantner, Z., Manzano-Macho, D.: Multilingual lexical semantic resources for ontology translation. In: Proceedings of LREC 2006 (2006)
4. Espinoza, M., Gómez-Pérez, A., Mena, E.: Labeltranslator - a tool to automatically localize an ontology. In: Proc. of 5th European Semantic Web Conference (ESWC 2008), Tenerife (Spain) (June 2008)
5. Kerremans, K., Temmerman, R.: Towards multilingual, termontological support in ontology engineering. In: Proceedings Workshop on Terminology, Ontology and Knowledge représentation, Lyon, France, January 22-23 (2004)
6. Suarez-Figueroa, M.C. (coordinator): NeOn Development Process and Ontology Life Cycle. NeOn Project Deliverable 5.3.1 (2007)
7. Neon Project (2006), http://www.neon-toolkit.org/
8. Pazienza, M.T., Stellato, A.: Exploiting linguistic resources for building linguistically motivated ontologies in the semantic web. In: Second Workshop on Interfacing Ontologies and Lexical Resources for Semantic Web Technologies (OntoLex 2006), held jointly with LREC 2006, Genoa (Italy), May 24-26 (2006)
9. Pedersen, T., Banerjee, S., Patwardhan, S.: Maximizing Semantic Relatedness to Perform Word Sense Disambiguation. Research Report UMSI 2005/25, University of Minnesota Supercomputing Institute (March 2005)
10. Peters, W., Montiel-Ponsoda, E., de Cea, G.A.: Localizing ontologies in owl. In: OntoLex 2007 (2007)
11. TreeTagger (1997), http://www.ims.uni-stuttgart.de/projekte/corplex/
12. Trillo, R., Gracia, J., Espinoza, M., Mena, E.: Discovering the semantics of user keywords. Journal on Universal Computer Science. Special Issue: Ontologies and their Applications (2007) ISSN 0948-695X
13. Vossen, P.: Eurowordnet: a multilingual database for information retrieval. In: Proceedings of the workshop on Cross-language Information Retrieval, Zurich (1997)

Rabbit: Developing a Control Natural Language for Authoring Ontologies*

Glen Hart, Martina Johnson, and Catherine Dolbear

Ordnance Survey of Great Britain, Romsey Road, Maybush, Southampton SO16 4GU England
{glen.hart,catherine.dolbear}@ordnancesurvey.co.uk,
martina.johnson@gmail.com

Abstract: The mathematical nature of description logics has meant that domain experts find them hard to understand. This forms a significant impediment to the creation and adoption of ontologies. This paper describes Rabbit, a Controlled Natural Language that can be translated into OWL with the aim of achieving both comprehension by domain experts and computational preciseness. We see Rabbit as complementary to OWL, extending its reach to those who need to author and understand domain ontologies but for whom descriptions logics are difficult to comprehend even when expressed in more user-friendly forms such as the Manchester Syntax. The paper outlines the main grammatical aspects of Rabbit, which can be broadly classified into declarations, concept descriptions and definitions, and elements to support interoperability between ontologies. The paper also describes the human subject testing that has been performed to date and indicates the changes currently being made to the language following this testing. Further modifications have been based on practical experience of the application of Rabbit for the development of operational ontologies in the domain of topography.

"Owl," said Rabbit shortly, "you and I have brains. The others have fluff. If there is any thinking to be done in this Forest - and when I say thinking I mean thinking - you and I must do it." A. A. Milne

1 Introduction

Ordnance Survey, Great Britain's national mapping agency, is currently in the process of building a topographic ontology to express the content of its topographic database. Ordnance Survey's aim is to enable the semi-automation of data integration, product repurposing and quality control. We are devising a methodology that enables domain experts working with ontology engineers to construct ontologies that have both a conceptual, human readable aspect, and a computation aspect that is interpretable by machines [1]. This methodology is based on the notion that ontologies are best constructed through a close collaboration between domain expert and ontology engineer. A key part of the methodology is that it enables the first stages of ontology authoring to be conducted using a controlled natural language (CNL) based on English (Rabbit) that allows the domain expert to easily understand the ontology

---

* © Crown Copyright 2007 Reproduced by permission of Ordnance Survey.

whilst supporting all the OWL DL [2] language features. It thus provides a means for the domain expert and ontology engineer to communicate effectively, and also enables other domain experts to understand the content and thus verify it. The computational aspect of the ontology, expressed using OWL, is treated as a compiled assembler code representation of the conceptual ontology as written in Rabbit.

This paper introduces Rabbit, gives examples of the language constructs, illustrated using portions of the topographic ontology that we are building, and shows how they are mapped to OWL. The paper also describes the human subject testing that we are performing and how this is helping to modify the language.

Rabbit is intended for use with a software tool. At the time of writing, a plug-in for Protégé[1] is being implemented that will assist domain experts to author Rabbit and will automatically translate the Rabbit sentences to OWL-DL. [13].

2 Related Work

Ever since OWL was conceived there have been concerns that its form makes it inaccessible to all but those with a good understanding of mathematics [3]. It is therefore difficult for it to be used by domain experts to author or validate ontologies. This in turn creates a serious impediment to the adoption of OWL and semantic web technologies in general since there are far too few people with both domain knowledge and the knowledge to be able to use languages such as OWL in a competent and reliable manner. There have been a number of attempts to resolve this issue through the creation of grammars for OWL that attempt to make it more understandable. Such grammars include the Manchester Syntax [3] that attempts to replace the abstract symbology of description logic. For example the statement:

River ⊑ BodyOfWater ⊓ ∃flowsIn.Channel ⊓ ∃hasCurrent.Current

is represented in the Manchester Syntax as:
Class: River
  subClassOf:
    BodyOfWater *and* flowsIn **some** Channel *and* hasCurrent **some** Current

Whilst this is significantly more readable than the pure mathematical representation, the rather formal nature of the description and the odd language constructs (such as "hasCurrent some Current") will be off-putting to the average domain expert and more complex examples will cause them to struggle to understand what it means.

Other approaches are to use CNLs of English, examples being ACE [4] and Processable English (PENG) [5] both of which provide grammars based on constrained English to represent First Order Logic (FOL) and both have now Description Logic (DL) subsets [6] and [7], the PENG DL version being recently dubbed the "Sydney Syntax" or SOS [10]. Another CNL is CLoNE [12]. CLoNE enables OWL ontologies to be authored in a grammatically relaxed fashion, for example allowing multiple classes to be expressed in a single sentence. ACE, SOS and CLoNE do provide significantly more readable representations than the DL form of OWL. Many of the language constructs in these approaches are similar or indeed

---

[1] http://protege.stanford.edu

identical to language constructs in Rabbit. ACE and SOS are related in that SOS can trace its origins to ACE, but the Rabbit language structures were independently developed and thus the similarities can be seen as convergent evolution. The DL version of ACE differs from Rabbit in that the former has been developed as a way to express Description Logic (OWL DL) in English whereas Rabbit was developed as part of a methodology where comprehension for the domain expert took priority and the language was then "back-fitted" to OWL. SOS sits somewhere between the two: it has a lineage that can be traced back to ACE but shares many of the design aspirations of Rabbit. Both Rabbit and SOS attempt to be minimal languages sufficient to enable ontologies to be authored. An example of the difference between the Rabbit/SOS approach and ACE and CLoNE is that in ACE (for example) it is possible to write complex constructs such as: France is a country and Paris is a city. Both Rabbit and SOS require these two separate facts to be expressed as discrete statements. CLoNE differs from Rabbit in its relaxed approach to grammar. Rabbit is designed to work within Protégé, while CLoNE by contrast, relies on the author writing CLoNE sentences that are directly interpreted. As a result CLoNE includes sentence structures that are designed to modify the ontology, for example: "Forget that Journals have Articles" – a sentence that deletes this axiom "Journals have Articles".

The Rabbit/SOS approach not only prevents the construction of sentences containing unrelated facts but also means that individual sentences tend to be shorter and thus more understandable. Such short sentences are of course possible in ACE and CLoNE but their grammars do nothing to discourage the construction of large more complex sentences.

Rabbit differs from both ACE and SOS through the addition of language elements that are not implementations of description logic but which enable Rabbit to represent whole ontologies. As an example, language elements within Rabbit enable one Rabbit ontology to reference concepts defined in other Rabbit ontologies, a feature that does not exist within either ACE or SOS. As indicated above CLoNE supports different types of meta-statements.

The authors believe ACE, SOS and Rabbit in particular have much in common and each can learn from the approaches of the others. All three research groups are currently involved in an OWL-ED taskforce [11] looking at defining a CNL for OWL. The first task of this taskforce is a comparison of the three languages with the longer term aim of finding a consensus that can be used to develop a CNL for OWL that can be formally adopted.

3 Rabbit – Motivation and Design Principles

The methodology we have developed to author ontologies gives the domain expert the prominent role; but we acknowledge the importance of the knowledge engineer in the process and importance of each to complement and support the other. Our research has been focused on developing a language that overcomes some of the limitations described above: namely, it should be easily readable and writable by domain experts; easy for them to digest, and allow them to express what they need to in order to describe their domain. It should also be translatable into OWL. We have named this language Rabbit, after Rabbit in Winnie the Pooh, who was really cleverer than OWL. To this end, we have involved domain experts from the outset in the core language design decisions.

The fundamental principles underlying the design of Rabbit are:

1. To allow the domain expert, with the aid of a knowledge engineer and tool support, to express their knowledge as easily and simply as possible and in as much detail as necessary.
2. To have a well defined grammar and be sufficiently formal to enable those aspects that can be expressed as OWL to be systematically translatable.
3. To be comprehensible by domain experts with little or no knowledge of Rabbit.
4. To be independent of any specific domain.

We regard Rabbit as the authoritative source of the ontology. OWL is very important as it is an established standard with tool support. For example we use OWL reasoners to flag inconsistencies and these are then fed back to Rabbit for correction. Our original intention was for Rabbit to enable domain experts alone to author ontologies. However, practice showed that whilst domain experts could build ontologies, these ontologies often contained many modelling errors not related to the language but the modelling processes. None-the-less Rabbit still enables the domain expert to take the lead and to author ontologies with guidance in modelling techniques from a knowledge engineer. Rabbit also enables other domain experts to verify the ontology.

4 Language Elements

Rabbit contains three main types of language element. Those used to express axioms, those used to introduce (or declare) concepts and relationships, and those used to import or reference other Rabbit ontologies. In the following sections the form of Rabbit used is the revised form following initial human subject tests. Due to lack of space in this paper we do not present the complete Rabbit grammar, so what follows are illustrative examples.

4.1 Declarations

Rabbit allows an author to explicitly declare concepts, relationships and instances. Concepts are introduced using the form:

<Concept> is a concept [, plural <Plural>].
E.g.
River is a concept. (plural defaults to Rivers.)
Sheep is a concept, plural Sheep.

Homonyms are dealt with using brackets. For example a Pool can either be a pool of water on land or a pool of water within a river. These are quite different physical phenomena. The former is more common and so we would introduce it as:

Pool is a concept.
whereas the later would be introduced as:
Pool (River) is a concept.

Where it is unclear which was more common both would use the bracketed form.

In Owl these statements translate to: River → Thing with the annotation "River" as a label. (Our notation here uses → to indicate subclass.) Homonyms will all share the same annotation rdf:label so Pool and Pool (River) will both be annotated as "Pool" but the class names will be Pool and Pool_River respectively.

Relationships and instances are introduced in a similar way:

<relationship> is a relationship.
<instance> is a <concept>.
E.g.
next to is a relationship.
Derwent Water is a Lake.

4.2 Concept Descriptions, Definitions and Axioms

A concept is described (necessary conditions) or defined (necessary and sufficient conditions) by a collection of axioms relating to that concept. Each axiom is represented by a single Rabbit sentence. The simplest sentence form is where one concept it associated with another through a simple relationship.

For example:

Every <concept> is a kind of <concept>. (Subsumption.)
Every <concept> <relationship> <concept>.

E.g.
Every House is a kind of Building.
Every River contains Water.

Such statement s translate to OWL as follows:
House → Building.
River → contains some Water.

A number of modifiers exist to enable cardinality and lists to be supported.
For example:
Every River Stretch has part at most two Confluences.
Every River flows into exactly one of River, Lake or Sea.

Concept descriptions comprise a series of axioms relating to the same concept. A definition comprises a group of axioms that make up the necessary and sufficient conditions. The axioms are grouped by an introductory statement and then follow as a list:

A School is defined as:
  Every School is a kind of Place;
  Every School has purpose Education;
  Every School has part a Building that has purpose Education.
Which translates into OWL as:

School ≡ Place and (hasPurpose some Education) and (hasPart some (Building and hasPurpose some Education))

Rabbit at present contains a sentence form that cannot be translated into OWL, although its omission does not invalidate the OWL ontology it does make the OWL representation less complete. This is the ability of Rabbit to modify an axiom by adding "usually". For example:

A Postal Address usually contains a Road Name.

As OWL is unable to express the existential quantifier for elements on the left hand side of an expression, this sentence cannot be converted to OWL. The reason for including it is to enable the domain expert to record frequent but not mandatory relationships, the absence of which would make certain definitions seem strange. Indeed without the use of "Usually" the only things that could be said about a British Postal address are:

A Postal Address contains exactly 1 Post Town Name.
A Postal Address contains exactly 1 Postcode.

All the other aspects of a postal address that we normally "expect" are not mandatory. The inclusion of such a feature is contentious. And, to a certain degree it has been included to be deliberately contentious to create a debate about the need for certain non-OWL support statements that nonetheless increase a domain expert's ability to accurately define a domain.

4.3 Intersection and Union

Rabbit implements intersection in a number of different ways in order to promote the development of short sentences, rather than encourage the authoring of long sentence structures that could become hard to understand. Most common amongst these mechanisms is that all Rabbit sentences that refer to the same concept are converted to OWL as one long conjunction comprising all the Rabbit sentences. Within a sentence, "and" is supported, but in practice has rarely been used. In fact in the authors' experiences, only once over three ontologies that in total include over 800 core concepts. Rabbit also supports "that" which is interpreted exactly the same as "and" and which has been used far more often than "and".

For example:

Every Almshouse has part a Building that has purpose Housing of Elderly People.

Here we encourage the use of "that" rather than "and" as it both sounds better, and we also believe it will discourage long chains of conjunctions that would be better treated as separate Rabbit sentences.

Another mechanism used to implement intersection is the use of "of", "for" and "by" (again all semantically equivalent in their OWL translation). They are used in the sentences of the structure:

Every <concept1> <relationship> <concept2> [of | for | by] <concept3>.
e.g.
Every School has purpose Education of Children.
this translates into OWL as:
School → hasPurpose some (Education and appliesTo some Child)

Here "of", "for" and "by" are translated to "and appliesTo" in OWL with appliesTo being a predefined Rabbit relationship.

"Or" is supported in both inclusive and exclusion forms. By default Rabbit treats "or" (and ",") as exclusive. So:

Every River flows into a Sea or Lake.
Is interpreted in OWL as:
River ➙ flowsInto some ((River or Lake) and not (River and Lake)).

However, we found that the exclusive or was very rarely used in ontology modelling, and where it did appear, usually indicated some mistake in the content of our model. In the above example, we would more accurately designate "flows into" as a functional property, and hence not need to consider the exclusive or.

Inclusive Or is implemented through the use of the modifier "One or more of":
Every Mission has purpose one or more of Christian Worship or Charitable Activities.
which in OWL is:
Mission ➙ hasPurpose some (ChristianWorship or CharitableActivity).

4.4 Ontology Referencing

No ontology can be an island unto itself: the point of ontologies is interoperability, and therefore they need mechanisms to include concepts and relationships from other ontologies.

OWL achieves this through the owl:import statement, although since it operates at the level of the whole ontology, it is a fairly crude approach.

The equivalent Rabbit construct is:
Use ontologies:

<reference1> from <url 1>;

...

<reference n> from <url n>.
e.g.
Use ontologies:
Wildfowl from http://www.ordnancesurvey.co.uk/ontology/v1/Wildfowl.rbt
Transport from http://www.ordnancesurvey.co.uk/ontology/v1/Transport.rbt
Concepts and relationships taken from these ontologies are then identified using the notation:

<imported concept> [<reference>]
for example
Every Duck Pond is a kind of Pond.
Every Duck Pond contains Ducks [Wildfowl].
This indicates that the concept Duck is obtained from the Wildfowl ontology.
Since repeatedly referencing Ducks [Wildfowl] can be a bit cumbersome, Rabbit also enables a local name to be applied:

Refer to Duck [Wildfowl] as Duck.
This produces the following OWL:
Duck ➙ Thing
Duck ≡ <http://www.ordnancesurvey.co.uk/ontology/v1/Wildfowl #Duck>))
As this creates a new local concept, it is then possible to extend the definition of the imported concept if required by adding further axioms to the local concept.

4.5 Property Restrictions - An Area of Weakness

All CNLs appear to struggle when it comes to implementing property characteristics. such as symmetry or transitivity. Fundamentally this is because these constructs are not well aligned to the way that people normally think about things and hence there are no easily understood natural language analogues. Probably the worst example is transitivity. In Rabbit such characteristics are defined as follows:

> The relationship <relationship> is transitive.
> e.g.
> The relationship "is part of" is transitive.

However as discussed below, human subject testing has shown that this is very poorly understood by people, as are similar constructions. ACE and SOS have both taken a different approach. They attempt to define such relationships through example. SOS express the transitive relationship as follows:

> If X <relationship> Y and Y <relationship> Z then X <relationship> Z.
> If X is part of Y and Y is part of Z then X is part of Z.

It may well be that such expressions are more understandable and we are planning to include such sentences in the next phase of human subject testing. However, we strongly suspect that such structures will still present domain experts with significant problems with interpretation, and take much longer to input when faced with authoring a large ontology. We are prepared to admit that there is no good solution within the language and so will have to be managed through training, tool support and guidance by the knowledge engineer.

5 Human Subject Testing

As Rabbit was developed with domain experts who have no training in description logic in mind, it was designed to resemble natural language as closely as possible. In order to test whether the resulting constructs in Rabbit are understandable, human comprehension tests were conducted. We were interested in whether people with no prior knowledge about Rabbit and no training in computer science, would be able to understand and correctly interpret and author sentences in Rabbit.

We are conducting a series of multiple-choice questionnaires to flag up any constructs that are ambiguous or otherwise problematic with a view to modifying Rabbit. Here we describe the results of the first set of questionnaires which only investigates the comprehension of Rabbit. Later experiments will investigate the ease by which people may author ontologies.

5.1 Version of Rabbit Tested

The earlier version of Rabbit that was tested differs from the grammar described so far in this paper in two important ways. First axioms were defined in a way that referred to the subject as A <concept> (or An <concept>) rather than the current form of Every <concept>. For example:

A Building has part a Roof.
rather than:
Every Building has part a Roof.

It should also be noted that although we have at present adopted "every" the next round of human subject testing will also test using "all" as well as "every":

All Buildings have part a Roof.

Secondly, Rabbit supported two forms of sentence: productive and receptive sentence types. Productive sentences are those that are used by an author to express the ontology. Receptive sentences are those designed to enable a tool to express statements made by an author in a different manner to confirm to the author that what was written was what was meant. Although we have not abandoned the development of receptive sentences, work on them is currently on hold so that the completion of the productive sentences may be advanced. Hence they have not been mentioned above. However such sentences were included in the first phase of human subject testing.

5.2 Stimuli

Thirty-one sentences were constructed, 20 of which were receptive Rabbit sentences, 11 of which were productive sentences. The software used for the tests did not allow for the randomisation of sentences across participants, so two surveys were constructed with a different random order. Randomisation of sentences across participants would have allowed us to check that there was no order effect, i.e. that the order in which sentences were presented did not have an effect on how participants were answering. By constructing two surveys with a different sentence order, the results can be checked for order bias. Each sentence was presented twice to check whether participants were merely guessing.

In order to ensure that participants used as little background subject knowledge as possible to choose their answers, the sentences were constructed using fictional words. The relationships were based on a biological ontology for mayflies where for example the word *Mayfly* was substituted by the fictional "Acornfly" and the genus *Ephemeroptera* was substituted by "Halucinoptera". Participants were advised that they were not expected to know anything about acornflies, and that the questionnaire was designed for people who do not know anything about acornflies.

For each Rabbit sentence, 3 or 4 answer choices were created. Where possible, the answer choices were created to indicate why participants were getting the answer wrong. For example, for the Rabbit sentence 'An Halucinopetera is a kind of insect' if participants choose the answer 'The specific thing called Halucinopetera is an insect' then we can infer it is clear that the participant thinks that "Halucinopetera" is an instance, rather than a concept. The order of answer choices for each sentence was randomised across participants to ensure there was no order bias.

5.3 Participants

All participants were students at the universities of Newcastle, Sheffield, Edinburgh and UCL. Participants had no university-level training in computer science, mathematics or related areas to ensure that there was no possibility of participants having any knowledge of description logic and were required to speak English as a first language. Each participant could opt to receive £10 in vouchers for participation in the study by indicating their name and address at the end of the questionnaire. In total 223 students completed the questionnaire.

There were two groups of participants; the productive first group which was presented with productive sentences first, and the receptive first group which was presented with receptive sentences first. This was to ensure there was no order bias for productive and receptive sentences. There were 122 participants in the productive first group and 101 participants in the receptive first group.

5.4 Materials

The questionnaires were written using the software SurveyMonkey (www.surveymonkey.com). As mentioned above, four different questionnaires were created, two for the productive group, two for the receptive group. The questionnaires were accessible from a URL. All responses were recorded by the software.

5.5 Procedure

Participants were presented with a Rabbit sentence and the answer choices. Participants' task was to choose the sentence they thought was the closest interpretation of the Rabbit sentence. They were advised that only one answer was the correct one.

For each sentence, along with the answer choices, participants were also given the option to answer "Unsure" but were encouraged to use this option as little as possible. If they did choose this option, participants were asked to explain why they were unsure in the box provided.

An example is show in figure 1:

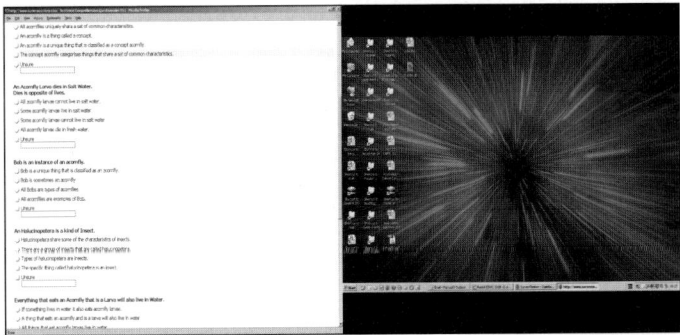

Fig 1. An example of a productive sentence test question

The correct answer is the first interpretation. The second one is wrong, because the Bob has to be an acornfly and cannot be anything else; the third one is wrong because there is only one Bob; and the fourth interpretation is wrong because the sentence is not saying anything about any other acornflies, it is specifically talking about Bob.

Participants were also told that there may be some unfamiliar uppercase letters in the sentences, but that they should not worry about this.

5.6 Results and Discussion

Thirteen of the sentences were answered correctly by 75 per cent of participants. These sentences were deemed sufficiently understandable by most participants, and have not been further analysed. It is not possible to discuss all 18 other sentences in detail here; instead, several typical examples will be discussed.

In this first set of questionnaires the Rabbit sentences were constructed such that singular "An Acornfly" was used to signify 'All Acornflies'. However, the results of the questionnaire show that participants did not necessarily interpret "An Acornfly" as being all acornflies . This became particularly evident in participants' comments as to why they were unsure about the answer.

> "it's just talking about one acornfly, so cannot generalise [to all]"
> "doesn't say if it's 1 or all"
> "is [the] question referring to the plural or the individual?"

It is clear that this is confusing several participants. In a second phase of the human subject testing that is currently being undertaken we are comparing three different options: 1) the plural, i.e. 'All Acornflies live in Water', 2) using 'every', i.e. 'Every Acornfly lives in Water' (which is the current preferred option and mimics SOS and ACE) and 3) the singular (the same as in the first phase of testing) 'An Acornfly lives in Water'. We can then determine whether to continue using the singular form and advise people in advance that it applies to the whole set, or whether the plural or using 'every' would be better understood.

Jargon from Description Logics and knowledge representation, such as the words 'instance' (Bob is an instance of an Acornfly), 'concept' (Acornfly is a concept), 'transitive' (Contained in is transitive), 'subject' (The is life stage of relationship can only have a Life Stage as a subject) and 'object' (The final moult into relationship can only have a Pictago as an object) were not sufficiently understood (average 66.4; 33; 26.8; 36.5; 52.4 resp.). These words need to be defined and explained to ontology authors in advance.

When using relative clauses (e.g. Everything that eats an Acornfly that is a Larva will also live in Water) it was not clear to participants (average 45.5 per cent) which noun the relative clause referred to. Adding commas or parentheses is possible in Rabbit and it is thought that this will make this relationship much clearer. This supposition will be tested in the second phase of human subject testing which is currently underway. It should be noted however that in our experience of ontology building, these types of sentence, which translate to General Concept Inclusion Axioms or Complex Role Inclusion Axioms in OWL, are only needed very occasionally,

For the second phase of the human subject testing, we have modified those sentences that were understood by less than 75 per cent of participants based on their responses or comments. We will re-test these sentences in a further set of questionnaires. Furthermore, in this next phase we will also be testing not just Rabbit but also Manchester Syntax with both being used to express the same knowledge. This will enable a direct comparison to be made between these two methods of expressing OWL.

6 Conclusions

This paper has provided a summary of the Rabbit CNL syntax for OWL. The language is developing through a combination of practical application in building real ontologies and through human subject testing. Certain language elements have been modified following a first round of human subject tests and a second round of testing is now being developed. The language is designed to overcome what is seen to be a major impendent to the general adoption of OWL: its inability to be easily understood by domain experts. Rabbit does this by providing easy-to-construct constrained English sentences that can be directly translated into OWL. Rabbit itself cannot alone provide sufficient support to a domain expert to author an ontology unassisted and so Rabbit is being developed in conjunction with a methodology that encourages close working between domain experts and knowledge engineers [1].

Acknowledgements: We would like to acknowledge the work of Hayley Mizen who first "took the plunge" and Fiona Hemsley-Flint for her help and thoughts – both domain experts bemused by OWL.

References

1. Mizen, M., Hart, G., Dolbear, C.: A Two-Faced Approach to Developing a Topographic Ontology. In: Proceedings of the GIS Research UK, 14th Annual Conference 2006, pp. 227–231 (2006)
2. W3C, OWL Web Ontology Language Guide, W3C Recommendation (February 10, 2004), http://www.w3.org/TR/owl-guide/
3. Horridge, M., Drummond, N., Goodwin, J., Rector, A., Stevens, R., Wang, H.: The Manchester OWL Syntax
4. Fuchs, N.E., Höfler, S., Kaljurand, K., Rinaldi, F., Schneider, G.: Attempto Controlled English: A Knowledge Representation Language Readable by Humans and Machines. In: Eisinger, N., Małuszyński, J. (eds.) Reasoning Web. LNCS, vol. 3564, Springer, Heidelberg (2005)
5. Schwitter, R.: English as a formal Specification Language. In: Hameurlain, A., Cicchetti, R., Traunmüller, R. (eds.) DEXA 2002. LNCS, vol. 2453, pp. 228–232. Springer, Heidelberg (2002)
6. Kaljurand, K., Fuchs, N.E.: Mapping Attempto Controlled English to OWL DL. In: 3rd European Semantic Web Conference. Demo and Poster Session. Budva, Montenegro, June 12 (2006)
7. Schwitter, R., Tilbrook, M.: Let's Talk in Description Logic via Controlled Natural Language. In: Proceedings of the Third International Workshop on Logic and Engineering of Natural Language Semantics (LENLS 2006) in Conjunction with the 20th Annual Conference of the Japanese Society for Artificial Intelligence, Tokyo, Japan, June 5-6, pp. 193–207 (2006)
8. Kalyanpur, A., Halaschek-Wiener, C., Kolovski, V., Hendler, J.: Effective NL paraphrasing of ontologies on the semantic web (Technical Report), http://www.mindswap.org/papers/nlpowl.pdf

9. W3C, Simple part-whole relations in OWL Ontologies, W3C Editor's Draft 11 (August 2005), http://www.w3.org/2001/sw/BestPractices/OEP/SimplePartWhole/simple-part-whole-relations-v1.5.html
10. Cregan, A., Schwitter, R., Meyer, T.: Sydney OWL Syntax - towards a Controlled Natural Language Syntax for OWL 1.1. In: OWLED 2007, OWL: Experiences and Directions, Third International Workshop, Innsbruck, Austria, June 6-7 (2007)
11. http://code.google.com/p/owl1-1/wiki/OwlCnl
12. Funk, A., Tablan, V., Bontcheva, K., Cunningham, H., Davis, B., Handschuh, S.: CLOnE: Controlled Language for Ontology Editing. In: Aberer, K., Choi, K.-S., Noy, N., Allemang, D., Lee, K.-I., Nixon, L., Golbeck, J., Mika, P., Maynard, D., Mizoguchi, R., Schreiber, G., Cudré-Mauroux, P. (eds.) ISWC 2007. LNCS, vol. 4825, Springer, Heidelberg (2007)
13. Denaux, R., Holt, I., Corda, I., Dimitrova, V., Dolbear, C., Cohn, A.G.: ROO: A Tool to Assist Domain Experts with Ontology Construction. In: ESWC 2008 (submitted, 2008)

A Natural Language Query Interface to Structured Information

Valentin Tablan, Danica Damljanovic, and Kalina Bontcheva

Department of Computer Science
University of Sheffield
Regent Court, 211 Portobello Street
S1 4DP, Sheffield, UK
{v.tablan,d.damljanovic,k.bontcheva}@dcs.shef.ac.uk

Abstract. Accessing structured data such as that encoded in ontologies and knowledge bases can be done using either syntactically complex formal query languages like SPARQL or complicated form interfaces that require expensive customisation to each particular application domain. This paper presents the QuestIO system – a natural language interface for accessing structured information, that is domain independent and easy to use without training. It aims to bring the simplicity of Google's search interface to conceptual retrieval by automatically converting short conceptual queries into formal ones, which can then be executed against any semantic repository.

QuestIO was developed specifically to be robust with regard to language ambiguities, incomplete or syntactically ill-formed queries, by harnessing the structure of ontologies, fuzzy string matching, and ontology-motivated similarity metrics.

Keywords: Searching, Querying, User Interfaces, Conceptual Search.

1 Introduction

Structured information in various guises is becoming ubiquitous on today's computers. This may include a user's contacts list, their calendar of events, other structured files such as spreadsheets or databases. In addition to this, unstructured textual content may refer or add information to entities from a user's structured information space. For example a project meeting mentioned in the calendar relates to textual files such as the agenda for the meeting, the various documents relevant to the particular project, the contact information of the people who will be attending the meeting, etc. All this information can be modelled as, or mapped onto, an ontology in order bring the benefits of all the technologies developed for the semantic web to the user's desktop.

We believe that text interfaces have a role to play because they are familiar to end users, benefit from very good support both on the desktop and in web interfaces, and are easily available on all types of devices. In previous work

[1,2] we have presented the CLOnE system (originally named CLIE) that provides a textual interface for editing a knowledge base (**KB**) through the use of an open-vocabulary, general purpose controlled language. That was designed as an interface for manual intervention in the process of generating ontological data from either structured information, through direct mapping, or from unstructured text, through semantic annotation. Continuing the same line of development, the work discussed in this paper focuses on providing access to the information stored in a KB by means of natural language queries.

Most knowledge stores provide facilities for querying through the use of some formal language such as SPARQL or SeRQL. However, these have a fairly complex syntax, require a good understanding of the data schema and are error prone due to the need to type long and complicated URIs. These languages are homologous to the use of SQL for interrogating traditional relational databases and should not be seen as an end user tool.

Different methods for user-friendly knowledge access have been developed previously. Some provide a graphical interface where users can browse an ontology, others offer a forms-based interface for performing semantic search based on an underlying ontology whilst hiding the complexity of formal languages. The most sophisticated ones provide a simple text box for a query which takes Natural Language (**NL**) queries as input.

The evaluation conducted in [3] contains a usability test with 48 end-users of Semantic Web technologies, including four types of query language interfaces. They concluded that NL interfaces were the most acceptable, being significantly preferred to menu-guided, and graphical query language interfaces. Despite being preferred by users, Natural Language Interface (**NLI**) system are not very frequent due to the high costs associated with their development and customisation to new domains, which involves both domain experts and language engineers.

We present QuestIO (**Quest**ion-based **I**nterface to **O**ntologies), a NLI system for accessing structured information from a knowledge base. The main impetus for this work is the desire to create a user-friendly way of accessing the information contained in knowledge stores, which should be easy to use, requiring little or no training.

The QuestIO application is open-domain (or customisable to new domains with very little cost), with the vocabulary not being pre-defined but rather automatically derived from the data existing in the knowledge base. The system works by converting NL queries into formal queries in SeRQL (though other query languages could be used). It was developed specifically to be robust with regard to language ambiguities, incomplete or syntactically ill-formed queries, by harnessing the structure of ontologies, fuzzy string matching, and ontology-motivated similarity metrics.

The following section presents some background information regarding user interfaces for knowledge access, putting this work into context. Next the design and implementation of the QuestIO system are presented, followed by an evaluation of its coverage, portability and scalability. Finally, we conclude and present some plans for future development.

2 Context

Tools for accessing data contained in ontologies and knowledge bases are not new, several have been implemented before using different design approaches which reach various levels of expressivity and user-friendliness.

A popular idea is adding search and browsing support to ontology editing environments. For instance Protégé [4] provides the Query Interface, where one can specify the query by selecting some options from a given list of concepts and relations. Alternatively, advanced users can type a query using a formal language such as SPARQL. Facilities of this type give the maximum level of control but are most appropriate for experienced users.

While typing queries in formal languages provides the greatest level of expressivity and control for the user, it is also the least user-friendly access interface. Query languages have complex syntax, require a good understanding of the representation schema, including knowledge of details like namespaces, class and property names. All these contribute to making formal query languages difficult to use and error prone. The obvious solution to these problems is to create some additional abstraction level that provides a user friendly way of generating formal queries, in a manner similar to the many applications that provide access to data stored in standard relational databases.

One step toward user-friendliness is creating a forms-bases graphical interface where the information request can be expressed by putting together a set of restrictions. Typically these restrictions are provided as ready-made building blocks that the user can add to create a complex query. Systems like this can either be customised for a particular application domain or general purpose. A good example of forms-based interface that offers both generic and domain specific options is provided by the KIM knowledge management platform [5].

Interfaces of this type are well suited for domain specific uses; when customised to a particular application, they provide the most efficient access path to the information. They can take advantage of good support for forms in graphical interfaces and benefit from a long tradition of forms-based interface design. Their downside, however, becomes apparent when moving to a general purpose search interface, in which case they can be either too restrictive or too complex with either too many input fields or too many alternative options.

Probably due to the extraordinary popularity of search engines such as Google, people have come to prefer search interfaces which offer a single text input field where they describe their information need and the system does the required work to find relevant results. While employing this kind of interface is straightforward for full text search systems, using it for conceptual search requires an extra step that converts the user's query into semantic restrictions like those expressed in formal search languages. A few examples of such query interfaces are discussed next.

SemSearch [6] is a concept-based system which aims to have a Google-like Query Interface. It requires a list of concepts (classes or instances) as an input query (e.g. the 'news:PhD Students' query asks for all instances of class News that are in relation with PhD Students). This approach allows the use of a simple text field as input but it requires good knowledge of the domain ontology and provides no way of specifying the desired relation between search terms, which can reduce precision when there are several ontology properties applicable.

Another notable example, and one of the most mature from this family of systems, is AquaLog [7]. It uses a controlled language for querying ontologies with the addition of a learning mechanism, so that its performance improves over time in response to the vocabulary used by the end users. The system works by converting the natural language query into a set of ontology-compatible triples that are then used to extract information from a knowledge store. It utilises shallow parsing and WordNet, and so requires syntactically correct input. It seems geared mainly towards queries containing up to two triples and expressed as questions (e.g., who, what).

Orakel [8] is a natural language interface (**NLI**) to knowledge bases. The key advantage is support for compositional semantic construction which helps it support questions involving quantification, conjunction and negation. These advanced features come at a cost, however, as the system requires a mandatory customisation whenever it is ported to a new application domain. Due to the expertise required for performing the customisation, this is a fairly expensive process.

ONLI (Ontology Natural Language Interaction) [9] is a natural language question answering system used as front-end to the RACER reasoner and to nRQL, RACER's query language. ONLI assumes that the user is familiar with the ontology domain and works by transforming the user's natural language queries into nRQL. No details are provided regarding the effort required for re-purposing the system.

Querix [10] is another ontology-based question answering system that translates generic natural language queries into SPARQL. In case of ambiguities, Querix relies on clarification dialogues with users.

To summarise, existing language-based query interfaces either support keyword-like search or require full-blown, correctly phrased questions. In this paper, we argue that it is technologically possible and advantageous to marry both kinds of approaches into one robust system, which supports both interaction styles.

3 The QuestIO System

Like many of the systems discussed in the previous section, QuestIO works by converting natural language input into a formal semantic query. Because we use Sesame[1] as a knowledge store, the system is configured to generate SeRQL as a query language. The same architecture and most of the implementation can be used to generate queries in other formal languages.

The driving principles behind the design for the QuestIO system were that:

– it should be easy to use, requiring as little user training as possible, ideally none;

---

[1] Sesame is an open-source RDF repository. More details are available on its homepage at http://www.openrdf.org/

- it should be open domain, with no need for customisation, or with customisation being done automatically or through other inexpensive means;
  it should be robust, able to deal with all kinds of input, including ungrammatical text, sentence fragments, short queries, etc.

In order to be robust, the system needs to attempt to make sense of whatever input it gets; it cannot rely on syntax, grammatically correct queries or enough context to perform disambiguation through linguistic analysis. Unlike other similar systems, our approach puts more weight on leveraging the information encoded in the ontology, and only uses very lightweight linguistic processing of the query text. This ensures that any textual input can be ingested successfully while the bulk of the question analysis work is based on the contents of the ontology, which is a larger resource and more likely to have been well engineered.

When a query is received, some of the contained words will match ontology concepts, while the textual segments that remain unmatched can be used to predict property names and for disambiguation. The sequence of concepts and property names is then converted into a formal query that is executed against the knowledge store. Throughout the process, a series of metrics are used to score the possible query interpretations, allowing the filtering of low scoring options, thus reducing ambiguity and limiting the search space.

3.1 Initialisation of the System

When the system is initialised, it processes the domain ontology. Of great importance to the functioning of this system is the ability to recognise textual references to resources from the ontology or the knowledge base. This is done by automatically creating a *gazetteer* when initialising the system with a given knowledge base. In traditional natural language processing applications, a *gazetteer* is a large list of known words or phrases that need to be recognised in text. These typically include various types of names, such as locations, organisations, or people, and a variety of domain dependent terms. In our case we build a gazetteer by automatically extracting lexicalisations from the knowledge base.

A lexicalisation is a textual form used to refer to a particular concept or entity. Our approach is based on the observation that most ontologies and knowledge bases contain a large amount of textual data that can be used to extract a domain vocabulary. The textual elements that we use include:

- The local part of URIs.
- The values of `rdfs:label` properties.
- The values of a custom list of properties, which can be specified as a customisation option.
- The values of datatype properties that can be converted to a string. This was introduced as as a catch-all case that is intended to capture custom-named properties use to encode entity names – for example properties like 'has alias'. This is also useful because it links characteristics of entities (encoded as property values) to the entity itself, which is sometimes necessary to identify ontology instances that have no names or labels, e.g. identifying an

instance of a 'Deadline' class based on its date value. Identifying all datatype property values in the query can lead to noise but that is mitigated by giving this kind of match a lower score than the previous, more direct, matches.

Because the strings used in an ontology sometimes use different orthography conventions, we attempt to normalise them by:

– recognising and splitting words that use underscores as a form of spacing;
– recognising and splitting CamelCase words;
– finding the morphological root (i.e. noninflected form) for all constituent words;
– for complex noun phrases, we derive lexicalisations based on their constituents, e.g. from "ANNIE JAPE Transducer" we derive "JAPE transducer" and "transducer" as additional ways of referring to the same concept.

For instance, if there is an ontology instance with a local name of "*ANNIEJapeTransducer*", and with assigned property `rdfs:label` with value "*Jape Transducer*", and with assigned property `rdfs:comment` with value "*A module for executing Jape grammars*", the gazetteer will contain following the strings:

– "*ANNIEJapeTransducer*" – the value of the local name;
– "*ANNIE Jape Transducer*" – local name after camel case splitting;
– "*Jape Transducer*" – the value of `rdfs:label`;
– "*A module for execute Jape grammar*" – the value of comment, after morphological normalisation. Note that morphological normalisation is also applied to the words in the input query, thus matching is possible regardless of inflection.

All the lexicalisations thus extracted are stored in a gazetteer that is able to identify mentions of classes, properties, instances, and property values associated with instances. More details regarding the linguistic processing involved can be found in [11].

3.2 Run-Time Operation

When a query is received, the system performs the following steps:

1. linguistic analysis;
2. ontological gazetteer lookup;
3. iterative transformation until a SeRQL query is obtained;
4. execute the query against the knowledge base and display the results.

The linguistic analysis stage performs morphological analysis of the text by running a tokeniser, part-of-speech tagger and a morphological analyser. All these types of linguistic analysis are very lightweight and robust as they do not depend on grammatical structure such as syntax. Because of this, they will complete successfully on any kind of input, be it fully formed sentences or simple fragments. We use a morphological analyser to apply the same kind of normalisation as that used for the lexicalisations from the ontology. This allows us to later match query terms with concepts in the ontology, regardless of the way they are inflected.

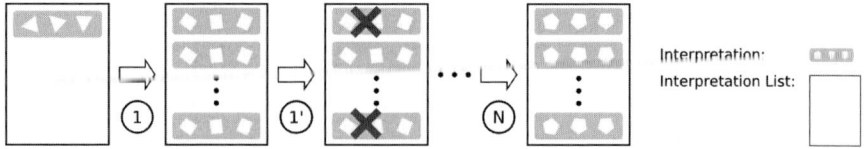

Fig. 1. Information Flow through the System

The second processing phase consists of running the ontological gazetteer built at initialisation time over the morphologically normalised query text. This creates annotations for all mentions of ontological resources that the gazetteer can identify – classes, properties, instances, and values of datatype properties.

Thirdly an iterative transformation process is started for converting the input free text into a formal query. It goes through several steps, each of which starts with a list of candidate interpretations and converts those into more complex ones. The information flow through the system is depicted in Figure 1. At the beginning of the process, the list of candidate interpretations is initialised with a simple interpretation containing the input text and the annotations created in the previous steps. In each iteration, the interpretations currently in the candidate list are transformed into more detailed ones. In cases of ambiguity, it is possible that the number of candidate interpretations grows during a transformation step (see the step numbered (1) in the diagram). All candidate interpretations are scored according to a set of metrics that will be detailed later and, in order to keep the number of alternatives under control, candidates that score too low can be eliminated as shown in the step (1') in the figure.

Expressed in pseudo-code, the top level algorithm is the following:

```
1  annotate input text with morphological data;
2  annotate input text with mentions of ontology resources;
3  create an interpretation containing the input text and annotations;
4  initialise the candidates list with the interpretation;
5  do
6    apply all possible transformations to the current candidates;
7    eliminate candidates that score too low;
8  until(no more transformation possible)
9  generate SeRQL using top scoring interpretation(s)
```

Following are some details about each of the steps, referred to using line numbers.

Step 1 includes breaking the input text into separate tokens, determining the part of speech for each token and annotating it with its morphological root.

Step 2 deals with identifying in the input text mentions of resources from the ontology. The matching is done using a normalised form that takes care of multi-word names and morphological variation.

Cycle 5 – 8 performs repeated transformations of the candidate interpretations until no more are possible.

Step 9 uses the highest-scoring interpretation to generate a SeRQL query.

In order to create a modular implementation, two abstractions are used: *Interpretation* and *Transformer*. Interpretations are used as a container for information – the contents of different interpretation types can be quite different. To begin with, a single Interpretation is created which holds the input text, together with all the annotations resulted from the morphological pre-processing. As the process progresses ever more complex interpretations are generated. Similarly, a *Transformer* represents an algorithm for converting a type of interpretation into another. The implementation of the top level work-flow is quite independent of the types on interpretations known or the kinds of transformers available; the cycle at lines 5-8 simply applies all known types of transformers to the current list of interpretations.

Creating a particular implementation of the system requires defining a set of interpretations and transformers (as Java classes that implement particular interfaces). In our current implementation, an interpretation is defined by the following elements:

- A list of interpretation elements which hold the actual data. The actual type of data held varies, becoming more and more detailed as the iterative process continues. We are currently using ten types of different interpretation elements of various complexity.
- A list of text tokens representing the original query text. Each interpretation element is aligned with a sequence of one or more input tokens. This is used to keep track of how each text segment was interpreted.
- A back-link to the interpretation that was used to derive the current one. This can be used to back-track through the interpretation steps and justify the reasoning that led to the final result.
- A score value – a dynamically calculated numeric value used to sort the interpretations list based on confidence.

The current implementation has seven types of transformers, which are used during the iteration steps for converting interpretations. They perform various operations to do with associating input text segments to ontology resources.

3.3 Identifying Implicit Relations

One of the main functions performed by the transformers is to identify relations which are not explicitly stated in the input query. After the ontological gazetteer is used to locate explicit references to ontology entities, the remaining text segments between those references are used to infer relations. Relations between query terms are essentially homologous with object properties in OWL, so our system attempts to match snippets from the input query to properties in the ontology. The process starts by identifying a list of possible candidates based on the definitions in the ontology: from two consecutive ontological references, we identify the ontology classes they belong to. References to classes are used directly; in the case of instance references we identify the class of the instance; for property values we first find the associated instance and then its class. If the reference is to a property name, we consider that to be an explicit relation mention and this process does not take place. Once we have the two classes, we build a list of properties that could conceivably apply given the range and domain restrictions in the ontology.

The list of applicable properties is then used to generate candidate interpretations that are scored using the following metrics:

String Similarity Score. This score is based on the assumption that the words used for denoting a relation in the query might be quite similar to some of the lexicalisations of the candidate properties. The actual numeric value is calculated using the *Levenshtein distance metrics* which represents the distance between two strings as the minimum number of operations needed to transform one string into the other, where an operation is an insertion, deletion, or substitution of a single character. Resulting scores are normalised to range from 0 to 1.

Specificity Score. This score is based on the sub-property relation in the ontology definition. The assumption behind using this is that more specific properties are preferable to more general ones. Specificity score is determined by the distance of a property from its farthermost super-property that has no defined super-properties. This makes more specific properties, i.e. that are placed at deeper levels in the property hierarchy, score higher. Its value is normalised by dividing it with the maximum distance calculated for the properties on the ontology level.

Distance Score. This metric was developed to compensate for the fact that in many ontologies the property hierarchy is rather flat. It is trying to infer an implicit specificity of a property based on the level of the classes that are used as its domain and range. Consider the following example: a top level class named "`Entity`" with a subclass "`GeographicalLocation`" and the properties "`partOf`" (with range and domain "`Entity`") and "`subregionOf`" (with range and domain "`GeographicalLocation`"). Although not expressed explicitly using sub-property relations, the property "`subregionOf`" can be seen as more specific than "`partOf`". Assuming that the system is trying to find the most appropriate relation between two instances of types "`City`" and "`Country`" (both subclasses of "`GeographicalLocation`") then, all other things being equal, the "`subregionOf`" property should be preferred.

The general case is illustrated in Figure 2, where classes are represented as circles, inheritance relations are shown as vertical arrows and ontological object properties are presented as horizontal arrows. Given two classes, one considered to be domain, the other range, the distance score for a candidate property represents the length of the path travelled from the domain class to the range class when choosing the property. These distances are represented by the thick dotted lines in the figure. Once calculated, the distances are normalised, through division by double the maximum depth of the ontology and inverted – as the shortest path represents the most specific property and should yield the highest score.

These similarity metrics are ontology-motivated and are largely comparable to those used in the AquaLog system. The final score associated with the candidate properties is a weighted sum of the three different atomic measures.

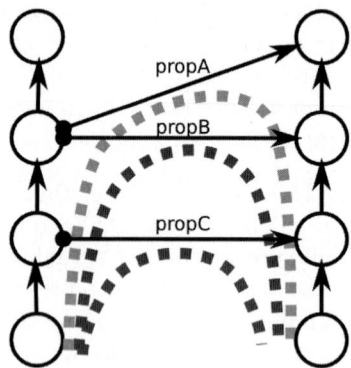

Fig. 2. Distance Score

3.4 Creating Queries

After the implicit relations have been identified, the final interpretation of the input query is presented as a list of explicit references to ontology resources interspersed with references to properties. Starting from this list of interpretation elements, a formal SeRQL query is dynamically created. In order to do this, references to instances are kept as they are, i.e. the instance URI is used directly in the query. References to classes are converted to query variables, with associated restrictions encoding the class membership. References to properties, either explicit or inferred, are used to create restrictions with regard to relations between the various query elements, either instance URIs or variables.

For example, we ran the system with the PROTON Ontology[2] populated with instances from the default knowledge base of the KIM system [12]. A textual query of "*Countries in Asia*" generated the following SeRQL formal query:

```
select c0, p1, i2
from
   {c0} rdf:type {<http://proton.semanticweb.org/2005/04/protonu#Country>},
   {c0} p1 {i2},
   {i2} rdf:type {<http://proton.semanticweb.org/2005/04/protonu#Continent>}
where
   p1=<http://proton.semanticweb.org/2005/04/protont#locatedIn> and
   i2=<http://www.ontotext.com/kim/2005/04/wkb#Continent_T.2>
```

The URI of "`http://www.ontotext.com/kim/2005/04/wkb#Continent_T.2`" is that of Asia.

For the provided query, the words "*Country*" and "*Asia*" were recognised as lexicalisations of objects in the ontology (a class and an instance respectively). In the context of the PROTON ontology, given the class "Country" as a

---

[2] The PROTON Ontology is a Base Upper-level Ontology developed within the Sekt project. More details about it can be found at its home page at http://proton.semanticweb.org/

domain and the class "Continent" (to which the instance "*Asia*" belongs) as a range, the list of candidate properties includes "partOf", "subRegionOf", and "locatedIn". The "partOf" property is defined at the very high level of the class "Entity", and so is very generic. The other two candidates have the same specificity, "locatedIn" being preferred in this case because of its closer textual similarity with the query due to the word "*in*" appearing the original input.

4 Evaluation – Coverage, Portability, Scalability

The first use case of QuestIO is in supporting developers working on open-source projects to find information from source code and user manuals, more efficiently via ontologies. More specifically, in the context of the TAO[3] project, we experimented first with learning semi-automatically a domain ontology [13] from the GATE source code and XML configuration files. It encodes the component model of GATE, the various plug-ins included in the default distribution, the types of modules included in each of the plug-ins, the parameters for for the different modules, their Java types, the associated comments, etc. The resulting ontology contains 42 classes, 23 object properties and 594 instances.

In order to evaluate QuestIO's ability to answer queries against a given ontology (i.e., its coverage and correctness), we started off by manually collecting a subset of 36 questions posted by GATE users to the project's mailing list in the past year, where they enquire about GATE plug-ins, modules and their parameters. Some example questions are: "Which PRs[4] take ontologies as a parameter?", "Which plugin is the VP Chunker in?", and "What is a processing resource?"

We divided these 36 randomly chosen postings into two groups: *answerable*, i.e., the GATE ontology contains the answer for this question; and *non-answerable*, i.e., the required information is missing in the ontology. We ran QuestIO on the subset of 22 answerable questions with the results as follows:

– The system interpreted *correctly* 12 questions and generated the appropriate SPARQL queries to get the required answer.
– *Partially-correct* answers were given to another 6 questions, where the system produced a query, but missed out one of the required constraints, so the answer was less focused. Partially correct answers are scored as 50% correct and 50% wrong.
– Lastly, the system *failed* to interpret 4 questions, so there was either no query generated or the generated query was not correct.

Overall, this means that 68% of the time QuestIO was able to interpret the question and generate the correct SPARQL queries. This result compares favourably to other ontology question-answering systems, e.g., in a similar experiment Aqua-Log was able to interpret around 50% of the questions posed.

---
[3] http://www.tao-project.eu
[4] PRs are a kind of GATE module for processing language.

Table 1. Knowledge Base Sizes

GATE Knowledge Base	
Classes	42
Object Properties	23
Instances	594
Total size (C + P + I)	659
Initialisation time	19 seconds

Travel Knowledge Base	
Classes	318
Object Properties	86
Instances	2790
Total size (C + P + I)	3194
Initialisation time	109 seconds

In another experiment, the scalability and portability of QuestIO were evaluated by using a completely different ontology with a larger knowledge base. The ontology used was Travel Guides Ontology[5] created to represent the travel guides domain, encoding geographical information for descriptions of locations as well as typical information from the tourism domain to do with hotels, room amenities, touristic attractions, etc. The ontology was created as an extension of the PROTON ontology, by adding 52 domain specific classes and 17 object properties to the existing 266 classes and 69 properties. The resulting ontology was then populated with 2790 instances [14]. A comparison between the sizes of the knowledge bases in the two experiments can be seen in Table 1. The last rows show the time spent during system initialisation in each case. It can be noted that, even for a sizeable knowledge base, this stays within reasonable limits, which is important even if it only occurs once in the system life-cycle.

Table 2 details the execution times for converting queries of varying degrees of complexity using the two knowledge bases. The complexity of the queries is measured in terms of the number of relations that are implied in the query – a query that mentions two concepts implies a relation between them, a query with three concepts implies two relations, etc. The queries shown in the last column are actual queries, as used during the experiments: note that they are mainly short fragment queries, with incomplete grammatical structure and with incorrect capitalisation. The last query used for the travel knowledge base makes use of the "`GlobalRegion`" class of PROTON, which is used to represent geographical areas smaller than a continent, e.g. "*North Asia*". This is somewhat unnatural but we used it in order to create an artificially longer query (the question having already been answered as a result of the previous query).

The two experiments show that the QuestIO system is capable of working without requiring any customisation on two knowledge bases that are quite different both semantically and in terms of size. As it can be seen from the figures in Table 2, the system scales well – the query conversion times remain within the range of a few seconds even for the more complex queries against the larger knowledge base.

Both knowledge bases used in the evaluation experiment were pre-existing ones and were in no way modified or customised for use with QuestIO. In a real application setting were the system is deployed, the results can be improved by

---

[5] http://goodoldai.org.yu/ns/tgproton.owl

Table 2. Query Execution Times

Query size (number of properties)	Execution time (seconds)	Number of results	Actual query
		GATE Knowledge Base	
1	0.148	15	"processing resources in ANNIE?"
2	0.234	37	"parameters for processing resources in ANNIE?"
3	0.298	37	"Java Class for parameters for processing resources in ANNIE?"
		Travel Knowledge Base	
1	1.013	52	"countries in asia"
2	2.030	52	"capitals of countries in asia"
3	3.307	52	"capitals of countries in global regions in asia"

ensuring a better lexicalisation of the ontology and knowledge base. This can be achieved by adding more labels or comments to the ontological resources which should help get better coverage of the domain vocabulary. Another option is to use a specialised domain dictionary that can provide synonyms for the domain terms. The use of general purpose dictionaries, such as WordNet [15,16], might not be advisable as, depending on the actual domain, they may have poor coverage or might actually introduce errors, in domains where the vocabulary is very precise.

The main purpose for developing the QuestIO system was to provide a user-friendly interface for interrogating knowledge bases. A qualitative indication of the success for this enterprise can be ascertained from comparing the textual query "*capitals of countries in global regions in asia*" with the equivalent formal query expressed in SeRQL:

```
select c0, p1, c2, p3, c4, p5, i6
from
  {c0} rdf:type{<http://proton.semanticweb.org/2005/04/protonu#Capital>},
  {c2} p1 [c0],
  {c2} rdf:type {<http://proton.../04/protonu#Country>},
  {c2} p3 {c4},
  {c4} rdf:type {<http://proton.../04/protonu#GlobalRegion>},
  {c4} p5 {i6},
  {i6} rdf:type {<http://proton.../04/protonu#Continent>}
where
  p1=<http://proton.semanticweb...04/protonu#hasCapital> and
  p3=<http://proton.semanticweb...04/protont#subRegionOf> and
  p5=<http://proton.semanticweb.../04/protont#subRegionOf> and
  i6=<http://www.ontotext.com/kim/2005/04/wkb#Continent_T.2>
```

5 Conclusion and Future Work

In this paper we presented the QuestIO system which converts a wide range of text queries into formal ones that can then be executed against a knowledge store. It works by leveraging the lexical information already present in the existing ontologies in the form of labels, comment and property values. By employing robust language processing techniques, string normalisation, and ontology-based disambiguation methods, the system is able to accept syntactically ill-formed queries or short fragments, which is what non-expert users have come to expect following their experience with popular search engines.

Work on QuestIO is continuing, and some of our plans for the future include moving from to the current single-shot query approach towards a session-based interaction, as also experimented with by AquaLog. This would make the interaction more user friendly in cases when the system can find no results for a query or the user is unhappy with the results. By making use of the session history combined with the information about the contents of the ontology, the system could guide the user toward the desired results, e.g. by suggesting relations that were missed due to the user choosing a different lexicalisation.

The current implementation already has the facility to track the process of transforming a textual query into a formal one, which was introduced to support better user interaction in the future. One example of using this would be to present, on request, a justification of how the current results are related to the query. The users who dislike black-box systems may use this to better understand how the query transformation works which, in turn, could improve the efficiency of them using the system.

We are also planning to perform a user satisfaction evaluation comparing our system with a forms-based interface, such that of KIM.

Acknowledgements. This research was partially supported by the EU Sixth Framework Program project TAO (FP6-026460).

References

1. Tablan, V., Polajnar, T., Cunningham, H., Bontcheva, K.: User-friendly ontology authoring using a controlled language. In: 5th Language Resources and Evaluation Conference (LREC), Genoa, Italy, ELRA (May 2006)
2. Funk, A., Tablan, V., Bontcheva, K., Cunningham, H., Davis, B., Handschuh, S.: Clone: Controlled language for ontology editing. In: Aberer, K., Choi, K.-S., Noy, N., Allemang, D., Lee, K.-I., Nixon, L., Golbeck, J., Mika, P., Maynard, D., Mizoguchi, R., Schreiber, G., Cudré-Mauroux, P. (eds.) ISWC 2007. LNCS, vol. 4825, Springer, Heidelberg (2007)
3. Kaufmann, E., Bernstein, A.: How useful are natural language interfaces to the semantic web for casual end-users? In: Franconi, E., Kifer, M., May, W. (eds.) ESWC 2007. LNCS, vol. 4519, Springer, Heidelberg (2007)
4. Noy, N., Sintek, M., Decker, S., Crubézy, M., Fergerson, R., Musen, M.: Creating Semantic Web Contents with Protégé-2000. IEEE Intelligent Systems 16(2), 60–71 (2001)

5. Popov, B., Kiryakov, A., Ognyanoff, D., Manov, D., Kirilov, A., Goranov, M.: Towards Semantic Web Information Extraction. In: Fensel, D., Sycara, K.P., Mylopoulos, J. (eds.) ISWC 2003. LNCS, vol. 2870, Springer, Heidelberg (2003)
6. Lei, Y., Uren, V., Motta, E.: Semsearch: a search engine for the semantic web. In: Managing Knowledge in a World of Networks, pp. 238–245. Springer, Heidelberg (2006)
7. Lopez, V., Motta, E.: Ontology driven question answering in AquaLog. In: Meziane, F., Métais, E. (eds.) NLDB 2004. LNCS, vol. 3136, Springer, Heidelberg (2004)
8. Cimiano, P., Haase, P., Heizmann, J.: Porting natural language interfaces between domains: an experimental user study with the orakel system. In: IUI 2007: Proceedings of the 12th international conference on Intelligent user interfaces, pp. 180–189. ACM, New York (2007)
9. Mithun, S., Kosseim, L., Haarslev, V.: Resolving quantifier and number restriction to question owl ontologies. In: Proceedings of The First International Workshop on Question Answering (QA 2007), Xian, China (October 2007)
10. Kaufmann, E., Bernstein, A., Zumstein, R.: Querix: A natural language interface to query ontologies based on clarification dialogs. In: Cruz, I., Decker, S., Allemang, D., Preist, C., Schwabe, D., Mika, P., Uschold, M., Aroyo, L.M. (eds.) ISWC 2006. LNCS, vol. 4273, pp. 980–981. Springer, Heidelberg (2006)
11. Damljanovic, D., Tablan, V., Bontcheva, K.: A text-based query interface to owl ontologies. In: 6th Language Resources and Evaluation Conference (LREC), Marrakech, Morocco, ELRA (May 2008)
12. Popov, B., Kiryakov, A., Kirilov, A., Manov, D., Ognyanoff, D., Goranov, M.: KIM – A semantic platform for information extraction and retrieval. Natural Language Engineering 10, 375–392 (2004)
13. Bontcheva, K., Sabou, M.: Learning Ontologies from Software Artifacts: Exploring and Combining Multiple Sources. In: Workshop on Semantic Web Enabled Software Engineering (SWESE), Athens, G.A., USA (November 2006)
14. Damljanovic, D., Devedzic, V.: Applying semantic web to e-tourism. In: Ma, Z. (ed.): The Semantic Web for Knowledge and Data Management: Technologies and Practices. IGI Global (2008)
15. Miller, G.A., Beckwith, R., Fellbaum, C., Gross, D., Miller, K.: Introduction to WordNet: On-line. In: Distributed with the WordNet Software (1993)
16. Fellbaum, C. (ed.): WordNet - An Electronic Lexical Database. MIT Press, Cambridge (1998)

Distinguishing between Instances and Classes in the Wikipedia Taxonomy

Cäcilia Zirn[1,2], Vivi Nastase[1], and Michael Strube[1]

[1] EML Research gGmbH, Heidelberg, Germany
{zirn,nastase,strube}@eml-research.de
http://www.eml-research.de/nlp
[2] Department of Computational Linguistics
University of Heidelberg, Heidelberg, Germany
zirn@cl.uni-heidelberg.de

Abstract. This paper presents an automatic method for differentiating between instances and classes in a large scale taxonomy induced from the Wikipedia category network. The method exploits characteristics of the category names and the structure of the network. The approach we present is the first attempt to make this distinction automatically in a large scale resource. In contrast, this distinction has been made in WordNet and Cyc based on manual annotations. The result of the process is evaluated against ResearchCyc. On the subnetwork shared by our taxonomy and ResearchCyc we report 84.52% accuracy.

1 Introduction

The World Wide Web (WWW) is a latent repository of multi-lingual and multi-faceted knowledge. This knowledge is hard to get at: for humans because of the overwhelming quantity; for computers because of noise (unedited texts, conflicting information) and lack of structure. This leads naturally to the desire for more structured web content, which can be accessed, used and shared among software agents for a wide range of activities. This is the desideratum of the Semantic Web endeavour [1]. Enhancing web pages with semantic annotations requires a large knowledge resource to serve as reference and to provide a portal to a large network of organized information and of reasoning capabilities.

Manually created resources, such as WordNet [2] and Cyc [3], have been in use since the beginning of the 90's. The Natural Language Processing (NLP) community has now much experience working and using them in applications. This has revealed to us their strengths and weaknesses, thus providing guidelines for the development of new and better knowledge resources [4]. One weakness is coverage. Manually built repositories cannot – and are not supposed to – cope with the extremely large number of entities to be captured, as they are edited by a small number of qualified people. The web has provided a collaborative medium through which this laborious task of creating knowledge resources can be distributed among a large number of contributors. The downside is that not all are computational linguists or lexicographers, and thus the task must be more flexible, and allow for the easy input of semi-structured knowledge. This is a very successful approach, as we have seen in the case of the online encyclopedia Wikipedia, which now covers more than 250 languages, 75 of which have more than 10,000 articles. It is therefore very appealing to try to induce from this semi-structured resource a large scale, multi-lingual ontology.

The first stage is to extract a taxonomy. Ponzetto and Strube [5] developed a method to accomplish this for the English Wikipedia, by inducing *isa* and *notisa* labels for the edges of the category network. An important feature of a useful taxonomy is differentiation between instances and classes – e.g. the authors of this paper are each an instance of the class PERSON[1]. This introduces a fundamental difference between elements in a resource:

- classes form the backbone of the network and form the actual ontology;
- classes are intensional descriptions of entities [6];
- in reasoning, instances are mapped to objects and classes to predicates;
- the class-class links are semantically different from the class-instance links;
- in texts, classes and instances have different syntactic behaviour – e.g. instances have no plural form, except in special cases when the speaker wants to emphasize some feature of the instance, as in the example: "I want it to be in a musically interesting catalogue (a label without compromises) so all the John Zorns and Mike Pattons of this globe are welcome to contact me."

The work presented in this paper focuses on developing methods to make the distinction between instances and classes automatically. In Section 2 we present a review of related work on distinguishing between instances and classes in large scale taxonomies. Section 3 reviews the methods for obtaining a taxonomy from the Wikipedia category network. Building upon this structure, we show in Section 4 our methods for labeling categories as instances or classes. The heuristics are evaluated against ResearchCyc, and the results are shown in Section 5.

2 Related Work

WordNet [2] is one of the most used lexical resources in NLP. It organizes open-class words in semantic networks: the nodes, *synsets*, represent senses, and contain a number of single or multi-word terms which have the same or very similar meaning; the edges represent different types of semantic relations, such as *hyponym-hypernym, meronym-holonym, antonymy, cause-effect, pertains to*. Among the members of a synset the *synonymy* relation holds. By far the most commonly used relation in WordNet is the *hyponym-hypernym*, which gives a taxonomic view of the resource. However, WordNet was not designed to be a taxonomy but rather a "map" of language. The organization has arisen as the links between synsets were added. Because of this, from the point of view of a

---

[1] We adopt the following notation conventions: Sans Serif for words and queries, *Italic* for relations, CAPITALS for Wikipedia pages and SMALL CAPS for concepts and Wikipedia categories.

formal taxonomy, WordNet is not perfect. One of the shortcomings is the lack of distinction between instances and classes [7,8].

Miller and Hristea [9,10] introduce manually the distinction between instances and classes in WordNet. The main features they used to identify instances are: (i) instances are proper nouns and therefore are capitalized; (ii) instances are unique, so they do not have instances themselves. Based on these criteria, the actual distinction was done manually by two judges. On 24,073 capitalized items, the agreement coefficient kappa was 0.75, indicating substantial, but not perfect, correspondence (for details on computing agreement statistics see [11]).

Cyc [3] is a large scale knowledge repository. It describes a wide variety of concepts and possible relations, and it is designed to be used for reasoning. Like WordNet, it was created manually, but there are functions to add non-atomic terms, e.g. "LiquidFN Nitrogen", automatically. From the beginning Cyc differentiated between instances and classes. There are two disjoint meta-concepts, #$Individual and #$SetOrCollection, based on which this distinction is made. All instances are linked through an *isa* chain to #$Individual, and all classes to #$SetOrCollection. Relations between two #$SetOrCollection items are labeled as *is generalized by (genls)*, whereas relations between an #$Individual item and a #$SetOrCollection item are labeled as *isa* (*isa*).

In both Cyc and WordNet the information about classes and instances is added manually, and is expensive both in terms of time and money.

Our literature review has not revealed any large scale resource in which this distinction is done automatically. The approach we present is based on the criteria identified by [9], and draws on the category structure of Wikipedia to assign automatically an instance or class label to categories and pages in this large network.

3 Wikipedia Taxonomy

Wikipedia is a free online encyclopedia, which grows through the collaborative efforts of volunteers over the Internet: anyone can contribute by writing or editing articles. As of March 2008, the English Wikipedia contains more than 2,300,000 articles[2]. The articles are organized in categories that can be created and edited as well. The categories themselves are organized into a hierarchy. Wikipedia's category and page network can be seen as a large semantic network. Categories in this network are connected by unlabeled links that represent different types of links: THUMB *isa* FINGER *part of* HAND.

Categories and pages linked with *isa* relations form a taxonomy. We show a sample in Figure 1. Ponzetto and Strube [5] describe methods to identify *isa* relations in Wikipedia's category network:

1. Filter out meta-categories used by Wikipedia for encyclopedia management using key words (e.g. template, user, portal).
2. Label as *is refined by* the relation between categories C_2 and C_1, if their names match the following patterns: C_1 = Y X and C_2 = X by Z – e.g. C_1 = MILES DAVIS ALBUMS and C_2 = ALBUMS BY ARTIST.

---
[2] http://en.wikipedia.org/wiki/Wikipedia:About#Related_versions_and_projects

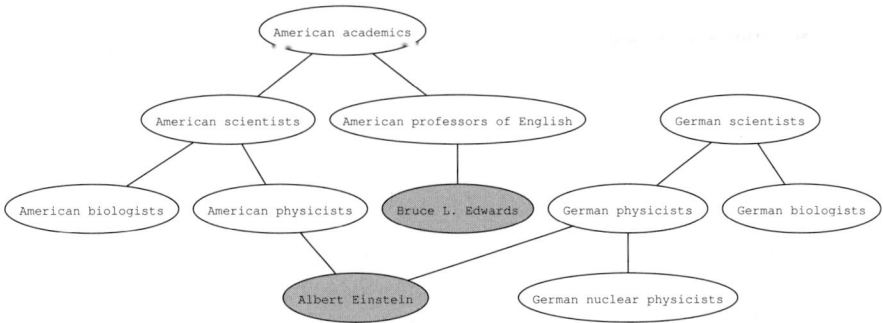

Fig. 1. Wikipedia Category Graph

3. Use two syntax-based methods:
   - Assign *isa* to the link between two categories if they share the same lexical head lemma – e.g. BRITISH COMPUTER SCIENTISTS and COMPUTER SCIENTISTS.
   - Assign *notisa* if one category contains the lemma of the lexical head of the other category in non-head position – e.g. CRIME COMICS and CRIME.
4a. Use structural information from the category network: for a category C, look for a page P with the same name. Take all of P's categories whose lexical heads are plural nouns $CP = \{C_1, C_2, ..., C_n\}$. Take all supercategories of $C_i, i = 1, ..., n$, $SC = \{SC_1, SC_2, ..., SC_k\}$. If the head lemma of one of C_i matches the head lemma of SC_j, label the relation between C and SC_j as *isa*. An example is provided in Figure 2. The category MICROSOFT has a homonymous page, categorized under COMPANIES LISTED ON NASDAQ which has the head lemma companies. MICROSOFT has a supercategory COMPUTER AND VIDEO GAME COMPANIES with the same head lemma. The link between MICROSOFT and COMPUTER AND VIDEO GAME COMPANIES is labeled as *isa*.

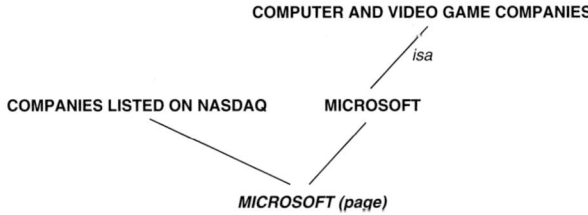

Fig. 2. Using structural information to induce *isa* links

4b. Assign *isa* label to the link between two categories if a page is redundantly categorized under both of them.

5. Use lexico-syntactic patterns in a corpus. This method uses two sets of patterns. One set is used to identify *isa* relations [12,13] – for example `such X as Y`, X and Y take the values of categories and their subcategories respectively. The second set is used to identify *notisa* relations. This last group includes patterns for finding meronymic, locative, temporal and other relations.

   These patterns are used with a corpus built from English Wikipedia articles, and separately with the Tipster corpus [14]. The label is assigned by majority voting between the frequency counts for the two types of patterns.
6. Assign *isa* labels to links based on transitive closures – all categories along an *isa* chain are connected to each other by *isa* links.

These methods lead to a fully automatically generated large scale taxonomy from the Wikipedia version of September 25th, 2006 – 127,311 nodes, 267,479 links of which 106,258 *isa* links. The work described in this paper enhances this taxonomy with `instance` and `class` information for each node.

4 Methods

In this section we describe our methods for distinguishing automatically between instances and classes in the Wikipedia taxonomy. It may seem intuitive that categories in Wikipedia are all classes, but that is not the case. For example, UNITED NATIONS, an instance of the class ORGANIZATIONS, appears both as a page and as a category in Wikipedia. The reason for this seems to be that it has many related concepts, which people naturally organize under this "umbrella". UNITED NATIONS SPECIALIZED AGENCIES, UNITED NATIONS CHARTER, INTERNATIONAL COURT OF JUSTICE, LEAGUE OF NATIONS are just some of the subcategories of the category UNITED NATIONS. This is not a rare situation, as many instances – organizations, individuals, locations, time intervals, etc. – about which Wikipedia contributors have extensive knowledge are structured in this way.

4.1 Structure-Based Method

Structure. The most important property of instances is that they are unique. As such, they do not have instances themselves [9]. Applying this to our taxonomy translates into finding categories which have no other subcategories or pages connected to them by *isa* relations. The reverse of this criterion is not true: if a category does not have other categories connected to it by *isa* relations, it is not necessarily an instance, it could also be a class that does not contain any other classes or instances. We use this criterion to determine categories that do have hyponyms, and which are therefore classes:

1. assign the `class` label to every category which has at least two hyponyms.

We adjust this rule to partially avoid erroneous *isa* labels introduced in the automatic *isa* links induction process, by introducing a second step:

2. assign the `class` label to every category which has exactly one hyponym, if this hyponym has more than one hyponym itself.

The remaining categories are processed based on clues from their names.

4.2 Category Name Analysis Methods

Named Entity Recognition (NER). Instances correspond to unique entities in the world, and are named entities. This fact is reflected in the Wikipedia category titles, which are accordingly capitalized. We then use an off-the-shelf named entity recognizer (NER), the CRFClassifier [15], and apply it to the category titles.

The category names consist of varied and complex noun phrase structures like MAIN KINGDOMS OF THE PURU CLAN or ... AND YOU WILL KNOW US BY THE TRAIL OF DEAD ALBUMS. We pass onto the NER only the lexical heads, which we extract using the Stanford Parser [16]. The parser may return several heads for a phrase, depending on the connectivity of the parse structure it produces. The NER tags the heads with one of the following labels: `Person`, `Location`, `Organization`, if they are recognized as named entities, `Other` otherwise. If a majority of the heads of a category are tagged as `Other`, the category is labeled as a class, else it is marked as an instance.

Capitalization. Bunescu and Paşca [17] have developed an approach to detect whether Wikipedia pages represent named entities: Following the Wikipedia naming conventions[3], all content words are capitalized if they constitute a part of a named entity. If not, they are lowercased. As the Wikipedia naming conventions for pages and the ones for categories[4] share these rules, we can apply this heuristic to the category titles. For example in the title ALL INDIA COUNCIL FOR TECHNICAL EDUCATION, all content words are capitalized, whereas in ALL AMERICA FOOTBALL CONFERENCE coaches the last word is lowercased, because it does not belong to a named entity. In Wikipedia category titles the first word is always capitalized; this introduces ambiguity for single-word categories.

Our method proceeds as follows:

1. preprocess the first word with the CRFClassifier mentioned in section 4.2: if it is a named entity keep the spelling, if not, lowercase the word.
2. filter out all function words (closed class words such as prepositions and determiners)[5].
3. analyze the remaining words in the title: if all of them are capitalized, the title is classified as an instance.

Plural. Instances are unique, therefore are mostly used in singular form. There are exceptions, as seen in the example: `The Millers are coming to our party`.

---

[3] http://en.wikipedia.org/wiki/Naming_conventions
[4] http://en.wikipedia.org/wiki/Wikipedia:Naming_conventions_(categories)
[5] http://www.marlodge.supanet.com/museum/funcword.html

This plural form is only used in particular situations. We therefore conclude that category titles that represent instances should be in singular. The grammatical number of the category title is the same as the number of its lexical head. To determine the heads and their numbers, we use the Stanford Parser and the part-of-speech tags assigned during parsing. If one of the category title phrase heads is marked as a plural noun (NNS, NNPS in the Penn Treebank Tagset used by this parser [18]), the category is labeled as `class`, otherwise as `instance`.

Page. Instructions for article authors in Wikipedia provide advice on creating categories: ARTICLES SHOULD BE PLACED IN CATEGORIES WITH THE SAME NAME. An article will therefore have a homonymous category. Because many articles refer to unique entities, the probability that a category containing a page with the same name is an instance is high. We use this as a heuristic to assign `instance` tags to categories.

5 Evaluation

The method presented was used with the Wikipedia version of September 25th, 2006. We use the same Wikipedia version as [5], whose work is briefly described in Section 3, as our research enhances the taxonomy they derived. The gold standard we use is ResearchCyc[6], in which the distinction between `#$Individual` and `#$SetOrCollection` is made for each entity in the repository.

Wikipedia and Cyc overlap on 7860 concepts, of which 44.35% (3486) are `#$Individual` and 55.65% (4374) are `#$SetOrCollection` (in Cyc). This constitutes our evaluation data set.

We first evaluated every method separately on the whole data set. Table 1 shows the results. We compute the reported scores as follows:

$$\text{Prec}_{indiv} = \frac{T_{indiv}}{T_{indiv}+F_{indiv}}$$

$$\text{Prec}_{coll} = \frac{T_{coll}}{T_{coll}+F_{coll}}$$

$$\text{Accuracy} = \frac{T_{indiv}+T_{coll}}{T_{indiv}+F_{indiv}+T_{coll}+F_{coll}}$$

where
- T_{indiv} is the number of nodes correctly classified as `instance` (they are classified as `#$Individual` in Cyc);
- F_{indiv} is the number of nodes incorrectly classified as `instance` (they are classified as `#$SetOrCollection` in Cyc);
- T_{coll} is the number of nodes correctly classified as `class` (they are classified as `#$SetOrCollection` in Cyc);
- F_{coll} is the number of nodes incorrectly classified as `class` (they are classified as `#$Individual` in Cyc).

---

[6] http://research.cyc.com/

Table 1. Results for separate evaluation of methods

Method	$Prec_{indiv}$ (%)	$Prec_{coll}$ (%)	Accuracy (%)
NER	85.23	76.84	79.69
Page	66.10	91.50	75.74
Capitalization	85.99	82.44	83.82
Plural	66.44	87.99	75.24
Structure	56.17	87.21	64.71

Using the results for individual methods, we have designed three different classification schemes, which we evaluate on 10 data sets – 5 rounds of binary random splits of the evaluation data, maintaining the #$Individual/#$SetOrCollection distribution – a form of cross-validation. We observe the standard deviation in performance over these splits, which will show the stability of performance.

A. Accuracy scheme:
   We pick the method with the highest accuracy, and this will constitute a baseline for the evaluation of the combination methods. The best performing method is Capitalization.
B. Precision scheme:
   We order the five methods according to their precision for correctly identifying instances respectively classes:
   1. Page – if the category does not have a corresponding page, it is classified as class.
   2. Plural – if the title is in the plural, the category is classified as class.
   3. Structure – if the category has hyponyms, it is classified as class.
   4. Capitalization – if the title is capitalized, the category is classified as instance.
   5. All remaining categories are assigned the class label.
C. Voting scheme:
   We chose the two methods with the highest precision for classifying instances and the two methods with the highest precision for classifying classes. We combine them to a voting scheme, the remaining categories are classified by all methods ordered by precision.
   1. Page & Plural – if a category has no corresponding page and if the title is in the plural, the category is classified as class.
   2. Capitalization & NER – if a category title is capitalized and if it is a named entity, the category is classified as instance.
   3. Page – if the page does not have a corresponding page, it is classified as class.
   4. Plural – if the title is in the plural, the category is classified as class.
   5. Structure – if the category has hyponyms, it is classified as class.
   6. Capitalization – if the title is capitalized, the category is classified as instance.
   7. All remaining categories are assigned the class label.

Table 2. Precision and accuracy ± standard deviation scores over 10 evaluation runs

Method	Precision$_{indiv}$ (%)	Precision$_{coll}$ (%)	Accuracy (%)
A.	85.99±0.54	82.44±0.63	82.82±0.5
B.	90.92±0.41	77.36±0.52	81.64±0.42
C.	89.21±0.46	81.82±0.52	84.52±0.34

The precision and accuracy averages and the standard deviation scores are presented in table 2.

5.1 Discussion

The methods used show very good results in distinguishing between classes and instances in Wikipedia category titles. The low standard deviation scores over the multiple runs indicate stable performance of the methods. This allows us to expect similar performance on the full Wikipedia category network.

It is interesting though to see where and why errors occur. A closer inspection of the erroneously classified categories reveal three main causes for these problems.

Preprocessing Errors. Some of our methods rely on output of NLP tools – lexical heads of noun phrases and part-of-speech tags. The category titles are quite complex, and some of them pose particular problems to the tagger and parser. Category titles like ALL INDIA COUNCIL FOR TECHNICAL EDUCATION are no challenge to be parsed, but it is expected too much to receive correct parsing results for titles like ...AND YOU WILL KNOW US BY THE TRAIL OF DEAD ALBUMS.

For an easier handling of the categories inside the system, all category titles are preprocessed by tokenization at the very beginning – ALL INDIA COUNCIL FOR TECHNICAL EDUCATION becomes [All] [India] [Council] [for] [Technical] [Education]. Punctuation marks are treated as separate tokens. So the name of a Japanese franchise-company .HACK becomes [.] [hack]. This not only leads to wrong parsing results, it also is not recognized as an instance by the method that checks the capitalization of the category titles.

Recognizing Named Entities. Named entity recognition is especially difficult when the components of the name are not named entities themselves. For example, BEE TRAIN is a Japanese animation studio, but the NER processes the two parts of the name separately, and neither **bee** nor **train** are tagged as named entities.

Concepts in Cyc. The gold standard we use, Cyc, is manually created. As agreement between human judges on assigning a class or instance label is not perfect, it is to be expected that there are some concepts in Cyc classified as instances respectively classes that do not match our definitions of instances and classes. As an example, we take the concept PHILOSOPHY. According to our definitions, PHILOSOPHY is a class, as it can contain subconcepts like ANALYTIC

PHILOSOPHY, APPLIED PHILOSOPHY, EPISTEMOLOGY, etc. In Cyc, it is classified as an instance.

5.2 Resource

We applied classification scheme C that performed best in our evaluation to the 127,124 categories in the Wikipedia taxonomy built by [5]. It classified 15,472 categories as instance and 111,652 categories as class. To make the results accessible we converted the obtained taxonomy into an RDF Schema file using the Jena Semantic Web Framework[7]. The RDF Schema file includes all of the 111,652 categories labeled as class and 13,258 of the categories labeled as instance. The reason for the missing instances is that the RDF Schema specification requires for each instance its corresponding class, otherwise the instance cannot be included. The structure of the taxonomy is not perfect, and there are categories labeled as instance not directly connected to a category labeled as class by an isa relation. The taxonomy converted to RDF Schema is available on our web page (http://www.eml-research.de/nlp/download/wikitaxonomy.php).

6 Conclusions

We have presented methods to distinguish automatically between instances and classes in a large scale taxonomy derived from the Wikipedia category network. Towards this end we exploited both structural information from the taxonomy, and naming characteristics and conventions. We have implemented such methods, and evaluated them separately, and then combined them to obtain even higher performance. Combining methods based on individual precision in identifying either instances or classes results in an algorithm that identifies instances with 90.92% precision. The most balanced approach uses voting from several methods, and gives the highest accuracy of 84.52%. From the five methods we presented one can generate a large number of combinations ($\sum_{i=1}^{n} \frac{n!}{i!}$ when each method is used at most once in a combination, n = 5 in our case). We have selected a few, based on individual performance of the methods. A more exhaustive analysis is left for future work.

The inter-judge agreement between human annotators for the task reported in [10] was $\kappa = 0.75$. This shows that the task is not easy, and that we have obtained very high performance with a fully automated approach.

There are multiple advantages of using Wikipedia: it is extremely up-to-date and it is multilingual. With slight modifications our methodology for distinguishing between instances and classes can be applied to other languages as well, provided a parser and a part-of-speech tagger are available.

Future work includes adding the Wikipedia article titles to the taxonomy and introducing the distinction between classes and instances at this level as well. Through the articles we can link to taxonomies in other languages, and ultimately create a huge, multi-lingual, knowledge base, to help users and software agents navigate through the World Wide Semantic Web.

---

[7] http://jena.sourceforge.net/index.html

Acknowledgements

We thank Simone Paolo Ponzetto for sharing his system for building the Wikipedia category network and for inducing the taxonomy, and for his feedback and input during various stages of this work. We thank the Klaus Tschira Foundation for financial support.

References

1. Berners-Lee, T., Hendler, J., Lassila, O.: The semantic web. Scientific American 284(5), 34–43 (2001)
2. Fellbaum, C. (ed.): WordNet: An Electronic Lexical Database. MIT Press, Cambridge (1998)
3. Lenat, D.B., Guha, R.V.: Building Large Knowledge-Based Systems: Representation and Inference in the CYC Project. Addison-Wesley, Reading (1990)
4. Fridman Noy, N., Hafner, C.D.: The state of the art in ontology design: A survey and comparative review. AI Magazine 18(3), 53–74 (1997)
5. Ponzetto, S.P., Strube, M.: Deriving a large scale taxonomy from Wikipedia. In: Proceedings of the 22nd National Conference on Artificial Intelligence, Vancouver, B.C., Canada, July 22–26, pp. 1440–1447 (2007)
6. Woods, W.A.: What's in a link: The semantics of semantic networks. In: Bobrow, D.G., Collins, A.M. (eds.) Representation and Understanding, pp. 35–79. Academic Press, New York (1975)
7. Gangemi, A., Guarino, N., Oltramari, A.: Conceptual analysis of lexical taxonomies: The case of WordNet top-level. In: Proceedings of the 2nd International Conference on Formal Ontology in Information Systems, Ogunquit, Maine, October 17-19, 2001, pp. 285–296 (2001)
8. Oltramari, A., Gangemi, A., Guarino, N., Masolo, C.: Restructuring WordNet's top-level: The OntoClean approach. In: Proceedings of the Workshop on Ontologies and Lexical Knowledge Bases at LREC 2002, Las Palmas, Spain, May 27, 2002, pp. 17–26 (2002)
9. Miller, G.A., Hristea, F.: WordNet nouns: Classes and instances. Computational Linguistics 32(1), 1–3 (2006)
10. Miller, G., Hristea, F.: Towards building a WordNet noun ontology. Revue Roumaine de Linguistique LI(3-4), 405–413 (2006)
11. Siegel, S., Castellan, N.J.: Nonparametric Statistics for the Behavioral Sciences, 2nd edn. McGraw-Hill, New York (1988)
12. Hearst, M.A.: Automatic acquisition of hyponyms from large text corpora. In: Proceedings of the 15th International Conference on Computational Linguistics, Nantes, France, August 23-28, 1992, pp. 539–545 (1992)
13. Caraballo, S.A.: Automatic construction of a hypernym-labeled noun hierarchy from text. In: Proceedings of the 37th Annual Meeting of the Association for Computational Linguistics, College Park, Md., June 20–26, 1999, pp. 120–126 (1999)
14. Harman, D., Liberman, M.: TIPSTER Complete. LDC93T3A, Linguistic Data Consortium, Philadelphia, Penn. (1993)
15. Finkel, J.R., Grenager, T., Manning, C.: Incorporating non-local information into information extraction systems by Gibbs sampling. In: Proceedings of the 43rd Annual Meeting of the Association for Computational Linguistics, Ann Arbor, Mich., June 25–30, 2005, pp. 363–370 (2005)

16. Klein, D., Manning, C.D.: Fast exact inference with a factored model for natural language parsing. In: Becker, S., Thrun, S., Obermayer, K. (eds.) Advances in Neural Information Processing Systems 15 (NIPS 2002), pp. 3–10. MIT Press, Cambridge (2003)
17. Bunescu, R., Paşca, M.: Using encyclopedic knowledge for named entity disambiguation. In: Proceedings of the 11th Conference of the European Chapter of the Association for Computational Linguistics, Trento, Italy, April 3–7, 2006, pp. 9–16 (2006)
18. Santorini, B.: Part of speech tagging guidelines for the Penn Treebank Project (1990), http://www.cis.upenn.edu/~treebank/home.html

Two Variations on Ontology Alignment Evaluation: Methodological Issues

Laura Hollink, Mark van Assem, Shenghui Wang, Antoine Isaac, and Guus Schreiber

Department of Computer Science
Vrije Universiteit Amsterdam
de Boelelaan 1081 HV
The Netherlands

Abstract. Evaluation of ontology alignments is in practice done in two ways: (1) assessing individual correspondences and (2) comparing the alignment to a reference alignment. However, this type of evaluation does not guarantee that an application which uses the alignment will perform well. In this paper, we contribute to the current ontology alignment evaluation practices by proposing two alternative evaluation methods that take into account some characteristics of a usage scenario without doing a full-fledged end-to-end evaluation. We compare different evaluation approaches in three case studies, focussing on methodological issues. Each case study considers an alignment between a different pair of ontologies, ranging from rich and well-structured to small and poorly structured. This enables us to conclude on the use of different evaluation approaches in different settings.

1 Introduction

The rise of the semantic web has lead to a large number of different and heterogeneous ontologies. This has created a need to interconnect these ontologies. Tools and algorithms have emerged that automate the task of matching two ontologies (see e.g. [18] or [13] for an overview). They create sets of correspondences between entities from different ontologies, which together are called alignments [6]. These correspondences can be used for various tasks, such as ontology merging, query answering, data translation, or for navigation on the semantic web[1]. Although the performance of ontology matching tools has improved in recent years [9], the quality of an alignment varies considerably depending on the tool and the features of the ontologies at hand.

The need for evaluation of ontology alignments has been recognised. Since 2004, the Ontology Alignment Evaluation Initiative (OAEI) organizes evaluation campaigns aimed at evaluating ontology matching technologies[2]. This has lead to the development of mature alignment evaluation methods. In this paper, we contribute to evaluation practices by investigating two alternative evaluation strategies.

---
[1] http://www.ontologymatching.org/
[2] http://oaei.ontologymatching.org/

Evaluation of alignments is commonly done in two ways: (1) assessing the alignment itself by judging the correctness of each correspondence and (2) comparing the alignment to a gold standard in the form of a reference alignment.

However, this type of evaluation does not guarantee that an application which uses the alignment will perform well. Evaluating the application that uses the alignment – commonly referred to as end-to-end evaluation – will provide a better indication of the value of the alignment [20]. In one of the seven test cases in the 2007 OAEI an end-to-end evaluation was performed; alignments were evaluated on the basis of their value for an annotation translation scenario [11].

Although many agree that end-to-end evaluation is desirable, it is hard to realize in practice. Even in a large scale initiative like OAEI it is time consuming. Another complicating factor is that real-world applications that use alignments are as yet scarce, and associated data on user behaviour and user satisfaction is even more rare. A more feasible alternative is to take into account some characteristics of a particular usage scenario without doing a full-fledged end-to-end evaluation [12]. The OAEI more and more incorporates usage scenarios in the evaluation. For example, in the Anatomy track of OAEI 2007, tools were asked to return a high-precision and a high-recall alignment, supporting the respective usage scenarios of fully automatic alignment creation and suggestion of candidate alignments to an expert [9]. Also, the number of tracks and test cases has increased every year [7, 8, 9], recognising the need for matching ontologies with different features, such as size, richness and types of relations (e.g. rdfs:subClassOf, part-of), depth of the hierarchy, etc.

Continuing this line of research, we investigate two evaluation methods that approximate end-to-end evaluation. Each strategy takes into account one characteristic of the targeted application context; the first takes into account the expected frequency of use of each correspondence and the second considers the expected effect of a particular misalignment on the performance of the application. We compare these two alternatives to the more common evaluation methods of assessing each individual correspondence and comparison to a reference alignment. We perform three case studies in which we evaluate an alignment using the four methods. This enables us to discuss not only the practicalities of each approach, but also the different conclusions that are the outcome of the approaches. Each case study compares a different pair of ontologies. Since the ontologies are structured differently, this allows us to discuss the effect of the features of the ontologies on the types of alignment errors that occur. In addition, we perform an end-to-end evaluation and compare this to the outcome of the four evaluation methods in a qualitative way.

The purpose of the paper is twofold. First, we intend to gain insight in methodological issues of evaluation methods. Second, we intend to give insight into the effect of the characteristics of the ontologies on the quality of the alignment, and on the best evaluation method to choose. It is not the purpose of this paper to evaluate a particular alignment or matching tool. In Section 2 we discuss the proposed methods, together with related work concerning these methods. In

Section 3 we present our case studies and in Section 4 the end-to-end evaluations. In Section 5 we discuss the results.

2 Alignment Evaluation Methods

Ideas have been put forward to find feasible alternatives to end-to-end evaluation. In this section we discuss related work on this topic, and describe two methods that each take into account one characteristic of an application that uses an alignment. The application scenario that we focus on is a query reformulation scenario, in which users pose a query in terms of one ontology in order to retrieve items that are annotated with concepts from another ontology. We assume that there is a partial alignment between the two ontologies, which is a realistic assumption given the state-of-the art of matching tools.

2.1 Evaluating Most Frequently Used Correspondences

If an alignment is large, evaluating all correspondences can be a time consuming process. A more cost-effective option is to evaluate a random sample of all correspondences, and generalize the results to get an estimate of the quality of the alignment as a whole.

An alternative to taking a random sample is purposefully selecting a sample. Van Hage et al. [20] note that in a particular application some correspondences affect the result (and thus user satisfaction) more than others. An end-to-end evaluation can take this into account but an evaluation of all (or a sample of) individual correspondences cannot. Evaluating the most important correspondences would better approximate the outcome of an end-to-end evaluation. The notion of 'importance' can mean different things in different application contexts. In this paper, we propose to use the estimated frequency of use of each correspondence as a weighting factor in the computation of performance measures. To this end, we divide all correspondences into two strata: infrequently and frequently used correspondences. As shown by Van Hage et al. [20], an intuitive way of stratified sampling is to aggregate the results of the strata 1 to L in the following manner:

$$\hat{P} = \sum_{h=1}^{L} \frac{N_h}{N} \hat{P_h} \quad (1)$$

where $\hat{P}$ is the estimated performance of the entire population, $\hat{P_h}$ is the estimated performance of stratum h, $\frac{N_h}{N}$ is a weighting factor based on the relative size of the stratum, where N_h is the size of the stratum, and N is the total population size.

Instead of weighting the strata based on their size, we propose to weight them based on their expected frequency of use:

$$\hat{P} = \sum_{h=1}^{L} \frac{\sum_{a \in H} \text{freq}(a)}{\sum_{a \in A} \text{freq}(a)} \hat{P_h} \quad (2)$$

where freq(a) is the frequency of use of correspondence a, H is the total set of correspondences in stratum h, and A is the total set of correspondences in the alignment.

Selecting the most frequently used correspondences for evaluation is beneficial in two situations. First, if there is a difference in quality between the frequently used correspondences and the infrequently used correspondences, the frequency-weighted precision will give a more reliable estimate of the performance of the application using the alignment. Second, if one intends a semi-automatic matching process in which suggested correspondences are manually checked and corrected by an expert, the frequency provides an ordering in which to check. This kind of scenario is targeted in the Anatomy track of OAEI 2007 [9] by asking participants to generate a high-recall alignment. Ehrig and Euzenat [4] consider the semi-automatic matching process by measuring the quality of an alignment by the effort it will take an expert to correct it. We argue that correction of a number of frequently used correspondences will positively affect the performance of the application more than correction of the same number of randomly selected correspondences.

Implementation of the Method. To estimate the frequency of use of each correspondence, we assume that each concept in source ontology X has an equal probability of being selected as a query by a user. For each query concept x, we determine the closest concept x' in X that has a correspondence to a concept y in ontology Y (the target ontology with which items are annotated). Closeness is determined by counting the number of steps in the (broader/narrower) hierarchy between x and x'. If a query concept x does not itself have a correspondence to Y, the correspondence of x' to ontology Y is used to answer the query, thus adding to the frequency count of correspondence $\{x', y\}$. Our estimation is biased, because in practice some query concepts are more often used than others. Logs of user and system behaviour can be used to determine more accurate prior probabilities of each query concept. However, logs are not always available.

2.2 Semantic Distance to a Reference Alignment

Comparing an alignment A to a reference alignment R gives precision as well as recall scores. Precision is the proportion of correspondences in A that are also found in reference alignment R, while recall is the proportion of the reference alignment R that is covered by A.

Incorrect correspondences negatively affect the performance of an application. However, this effect varies depending on how incorrect the correspondence is. Performance of an application will drop steeply if a correspondence links two completely unrelated concepts, while it may drop only slightly if a correspondence links two closely related concepts. We investigate the use of a semantic distance measure to capture this difference. More specifically, we use semantic distance to represent the distance between a correspondence in A and a correspondence in a reference alignment R. This allows us to distinguish between correspondences that cause incorrect results, and correspondences that are misaligned but still produce an acceptable result in the application.

The idea of a more nuanced precision and recall measure has been proposed before. Ehrig and Euzenat [4] propose to include a proximity measure in the evaluation of alignments. They suggest to use the effort needed by an expert to correct mistakes in an alignment as a measure of the quality of an alignment. In the same paper, they propose to use the proximity between two concepts as a quality measure. A very simple distance measure is used as an example.

In the current paper, we implement this idea by using the semantic distance measure of Leacock and Chodorow [15]. This measure scored well in a comparative study of five semantic distance measures by Budanitsky and Hirst [3], and has the pragmatic advantage that it does not need an external corpus. The measure by Leacock and Chodorow, sim_{LC}, actually measures semantic proximity:

$$sim_{LC} = -\log \frac{len_{(c_1,c_2)}}{2D}$$

where $len_{(c1,c1)}$ is the shortest path between concepts c1 and c2, which is defined as the number of nodes encountered when following the (broader/narrower) hierarchy from c1 to c2. D is the maximum depth of the hierarchy.

In our case studies, we compare each correspondence $\{x, y\}$ in A to a correspondence $\{x, y'\}$ in a reference alignment R. We use the semantic distance between y and y' as a relevance measure for the correspondence $\{x, y\}$. To calculate precision and recall, we normalize the semantic distance to a scale from 0 to 1.

A side effect of using a semantic distance measure is that the assessments are no longer dichotomous but are measured on an interval level. Common recall and precision measures are not suited for this scale. Therefore, we use *Generalised Precision* and *Generalized Recall* as proposed by Kekäläinen and Järvelin [14]:

$$gP = \sum_{a \in A} \frac{r(a)}{|A|} \qquad gR = \frac{\sum_{a \in A} r(a)}{\sum_{a \in R} r(a)} \qquad (3)$$

where $r(a)$ is the relevance of correspondence a, A is the set of all correspondences found by the matching tool, and R is the set of all correspondences in the reference alignment. A similar notion of this measure was later described by Euzenat [5]. The latter measure is more general since it is based on an overlap function between two alignments instead of distances between individual correspondences.

3 Case Studies

In this section we employ the four evaluation methods to evaluate alignments between three pairs of vocabularies. In three case studies, one for each alignment, we discuss the different outcomes of the evaluation methods.

In each case study we take a different source vocabulary: SVCN, WordNet and ARIA. The target vocabulary is the same in each case study: the Art and Architecture Thesaurus (AAT). Strictly speaking, some of these vocabularies are not ontologies. In practice, however, many vocabularies can be seen as ontologies with less formal semantics. They are widely used as annotation and search vocabularies in, for example, the cultural heritage field. Ontology alignment tools can be applied to these vocabularies, although the looser semantics may influence the quality of the alignment. We used the RDF representations of AAT, SVCN and ARIA provided by the E-Culture project [17], and WordNet's RDF representation by the W3C [19]. The three source vocabularies differ in size, granularity, structure, and topical overlap with the AAT, which allows us to investigate the role of vocabulary features in the evaluation methods.

The Getty Institute's AAT[3] is used by museums around the world for indexing works of art [16]. Its concepts have English labels and are arranged in seven facets including Styles and Periods, Agents and Activities. In our study we concentrate only on the Objects facet, which contains 16,436 concepts ranging from types of chairs to buildings and measuring devices, arranged in a monohierarchy with a maximum depth of 17. The broader/narrower hierarchy of this facet is ontologically clean. Concepts in the AAT typically have many labels, e.g. "armchair", "armchairs", "chairs, arm", "chaises á bras". The RDF version provided by the E-Culture project also has Dutch labels, obtained from a translation of AAT.[4]

The alignments were made with Falcon-AO, an automatic ontology matching tool [10]. We selected Falcon-AO because it was among the best performing tools in the OAEI 2006. We employed it as an off-the-shelf tool, because this paper does not focus on *tool* evaluation but on the characteristics of different methods of *alignment* evaluation.

3.1 Case 1: Alignment between AAT and SVCN

SVCN is a thesaurus developed and used by several Dutch ethnographic museums.[5] It has four facets, of which the Object facet has 4,200 concepts (making it four times smaller than the Object facet in the AAT). SVCN's Object facet was originally created by selecting AAT concepts and translating the labels to Dutch. However, over time intermediate and leaf concepts have been inserted and removed, resulting in a hierarchy with a maximum depth of 13. The broader/narrower hierarchy is well-designed, but contains more errors than AAT's.

Evaluating Individual Correspondences. Falcon produced 2,748 correspondences between SVCN and AAT. We estimated the frequency of use of each correspondence, as described in Section 2.1. Figure 1 displays cumulative percentages of these frequencies against cumulative percentages of the number of correspondences; all correspondences were ordered according to their frequency and displayed so that infrequent correspondences appear on the left side of the figure and the most frequent correspondences appear on the right. If each correspondence was used equally frequently, the graph would show a straight line from the origin to the top-right corner. SVCN does not deviate much from this straight line.

All correspondences were divided over two strata: frequently used and infrequently used correspondences. The size of the frequent stratum was set to 80

---
[3] See http://www.getty.edu/research/conducting_research/vocabularies/aat/
   The AAT is a licensed resource.
[4] http://www.aat-ned.nl/
[5] http://www.svcn.nl/

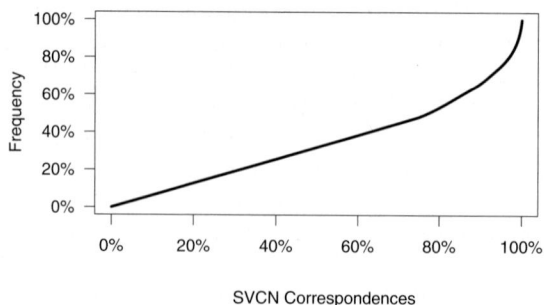

Fig. 1. Cumulative percentage of estimated use of SVCN-AAT correspondences in the application scenario. The total number of correspondences is 2,748.

(3% of all correspondences), which are responsible for 20% of the use in the application scenario (Figure 1). The choice for a size of 80 is pragmatic: it is a low number of correspondences that can be evaluated but still reflects a large frequency percentage. We evaluated all correspondences in the frequent stratum and a random sample of 200 from the infrequent stratum. Table 1 shows that the precision of the two strata differs, but not significantly so (0.93 and 0.89). We then weighted the outcomes of these evaluations in two ways: (1) according to the sizes of the strata (80 and 2,668) as in Equation 1 and (2) according to the frequency of use of the correspondences in the strata as in Equation 2. Both weighting schemes gave a precision of 0.89 (see Table 1).

Since in this use case the size of the population of all correspondences is large compared to the sample sizes, we used the binomial distribution to approximate the margins of error (shown in Table 1). The margin of error of a binomial distribution is given by:

$$\text{Margin of error} = 1.96\sqrt{\frac{p(1-p)}{n}} \qquad (4)$$

One reason for taking into account the frequency of use of correspondences is that it gives an order in which to manually check and correct the correspondences. We corrected all 80 correspondences in the frequent stratum and then

Table 1. Evaluation of the alignment between SVCN and AAT

Evaluation Type	Precision	Recall
Random sample of infrequent stratum	0.89±0.04	
Frequent stratum	0.93	
Weighted based on stratum size	0.89±0.03	
Weighted based on frequency of use	0.89±0.03	
After correction of frequent stratum	0.91±0.03	
After random correction	0.93±0.03	
Comparison to a reference alignment	0.84±0.07	0.80±0.08
Semantic distance to a reference alignment	0.90±0.06	0.86±0.07

recalculated the precision of the alignment, weighted by frequency of use. This gave a precision of 0.91, which is not a significant increase. After manual correction of a random sample of 80 correspondences the precision rises to 0.93, which is higher but again not a significant increase. A possible reason for the finding that random correction gives a better precision than correction of frequent correspondences, is the fact that there were more wrong correspondences in the random sample. Another factor is that the contribution to the total frequency of correspondences in the two strata is similar.

Comparison to a Reference Alignment. Reference alignment evaluation has the advantage that both precision and recall can be determined, but it is more costly because two vocabularies have to be aligned completely. Instead of aligning all concepts, we took a random sample of 100 concepts from SVCN and aligned those to AAT. Based on this partial reference alignment, Falcon's alignment has a precision of 0.84 and a recall of 0.80 (Table 1). As an alternative, we employ a semantic distance measure to compare the correspondences to the reference alignment; each correspondence $\{x,y'\}$ in the reference alignment is compared to a correspondence $\{x, y\}$ delivered by Falcon. We use the sim_{LC} measure between y and y', which results in a scaled value (0-1). Generalized precision and recall can then be calculated over these values (see Table 1). In the case of SVCN, the semantic distance based precision and recall are higher that the 'traditional' precision and recall, but the differences lie within the margins of error.

3.2 Case 2: Alignment between AAT and WordNet

WordNet is a freely available thesaurus of the English language developed by Princeton[6]. It has three top concepts: Physical entity, Abstraction and Thing. We only used the hierarchy below Physical entity ⊲– Physical object, which contains 31,547 concepts. Each concept has multiple synonymous terms. The main hierarchy is formed by the polyhierarchic hyponym relation which contains more ontological errors than AAT's hierarchy. The topical overlap with AAT is reasonable, depending on the part of the hierarchy. WordNet's Physical object hierarchy covers, for example, also biological concepts such as people, animals and plants while the Object facet of AAT does not. Other parts of WordNet are very similar to AAT. For example, the hierarchy from Furniture down to Chesterfield sofas is almost identical to that in AAT. The maximum depth of the Physical entity hierarchy is the same as AAT' Object Facet: 17 nodes.

Evaluating Individual Correspondences. Falcon produced 4.101 correspondences between WordNet and AAT. Applying our frequency estimation gives a distribution depicted in Figure 2. In this case the contribution of the most frequent correspondences is much greater, the top 20% of correspondences is already responsible for 70% of expected usage (reminiscent of Zipf's law).

We performed the same evaluation procedures as for the SVCN case, except that the size of the frequent stratum was set to 30 (0.7% of all correspondences).

---

[6] http://wordnet.princeton.edu/

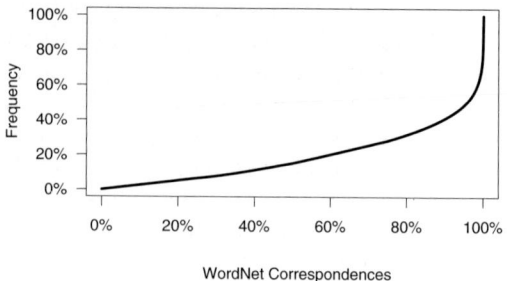

Fig. 2. Cumulative percentage of estimated use of WN-AAT correspondences in the application scenario. The total number of correspondences is 4.101.

Table 2. Evaluation of the alignment between WordNet and AAT

Evaluation Type	Precision	Recall
Random sample of infrequent stratum	0.72±0.06	
Frequent stratum	0.60	
Weighted based on stratum size	0.71±0.05	
Weighted based on frequency of use	0.68±0.04	
After correction of frequent stratum	0.81±0.04	
After random correction	0.72±0.04	
Comparison to a reference alignment	0.62±0.10	0.45±0.10
Semantic distance to a reference alignment	0.64±0.09	0.47±0.10

This is possible because here the contribution of the top correspondences is greater; the top 30 is responsible for 33% of total frequency. This reduction saves us a considerable evaluation effort. The results of the different evaluation methods are presented in Table 2. In the case of WordNet, weighting based on stratum size gives a slightly higher precision than weighting based on frequency (0.71 and 0.68, respectively).

Manual correction of all 30 frequent correspondences gives a higher precision than correcting 30 randomly selected correspondences from the complete set of correspondences (0.81 and 0.72, respectively, calculated by frequency-based weighting). This shows that in the WordNet case, it is sensible to prioritize correction of the most frequent correspondences.

Comparison to a Reference Alignment. We performed a sample reference alignment evaluation in the same manner as for the SVCN case (n=100). The results are much lower than those for SVCN, as can be seen in Table 2. The margins of error are somewhat higher because the sample size is smaller. The effect of applying semantic distance is smaller than the effect we saw for SVCN.

3.3 Case 3: Alignment between AAT and ARIA

ARIA is a thesaurus developed by the Dutch Rijksmuseum for a website that showcases some 750 masterpieces of the collection.[7] It contains 491 concepts

---

[7] http://www.rijksmuseum.nl/aria/

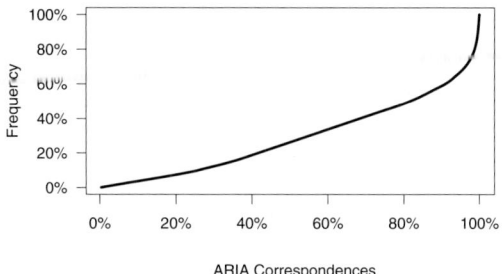

Fig. 3. Cumulative percentage of estimated use of ARIA-AAT correspondences in the application scenario. The total number of correspondences is 278.

which are all art-related object types. There are 26 top concepts such as Altarpieces, Household scenes and Clothing, only half of which have subconcepts. Each concept has one term in Dutch and one in English. Its hierarchy is at most 3 concepts deep and is arranged in a polyhierarchy; e.g. Retables is subordinate to Altarpieces and Religious paraphernalia. The broader/narrower relation used in ARIA can in many cases not be interpreted as rdfs:subClassOf. For example, Costumes and textiles has a grandchild Portable altars. ARIA is the smallest and most weakly structured of the three source vocabularies.

Evaluating Individual Correspondences. Falcon produced 278 correspondences between ARIA and AAT. Figure 3 shows the results of applying our frequency estimation. In this case the contribution of the most frequent correspondences is large; the top 20% of correspondences is responsible for 50% of expected usage.

Again we opted for a size of 30 for the frequent stratum (6% of all correspondences), which are responsible for 42% of the use in the application scenario. In this case, weighting according to the size of the stratum gave a precision of 0.74, while weighting according to the frequency of use gave a precision of 0.70.

Since the sample size is large compared to the size of the population of all correspondences in this case, we cannot approximate the margin of error with a binomial distribution. Instead, we used the following equation to compute the margin of error for a hypergeometric distribution [2]:

$$\text{Margin of error} = 1.96 \frac{N-n}{N-1} m(1 - \frac{m}{n}) \qquad (5)$$

where N is the size of the population, n is the size of the sample, and m is the number of correct correspondence found in the sample.

Manual correction of the most frequent stratum gives a precision of 0.85, which is again higher than random correction (0.74).

Comparison to a Reference Alignment. We performed a sample reference alignment evaluation in the same manner as for SVCN and WordNet (n=100). The recall and precision measures as shown in Table 3 are in between those for SVCN (highest) and WordNet (lowest). The effect of applying semantic distance is considerable.

Table 3. Evaluation of the alignment between Aria and AAT

Evaluation Type	Precision	Recall
Random sample of infrequent stratum	0.75±0.03	
Frequent stratum	0.63	
Weighted based on stratum size	0.74±0.03	
Weighted based on frequency of use	0.70±0.03	
After correction of frequent stratum	0.85±0.02	
After random correction	0.74±0.03	
Comparison to a reference alignment	0.66±0.09	0.63±0.09
Semantic distance to a reference alignment	0.80±0.08	0.76±0.08

4 End-to-End Evaluation

In this section we present an end-to-end evaluation performed using the three alignments from the case studies. The application scenario that we focus on is a query reformulation task for information retrieval: a user query for concept $x \in$ vocabulary X is transformed into a concept $y \in$ vocabulary Y. We queried a dataset of 15,723 art objects indexed with AAT provided by the E-Culture project. Objects annotated with concepts from Y are returned to the user and the relevance of these objects to the query x is rated. We used 20 randomly selected query concepts from each source vocabulary[8] and evaluated two different methods of reformulation for cases where a query $x \in X$ has no direct correspondence to Y: (1) find a concept x' in the hierarchy below x that has a correspondence to a concept $y \in Y$ (strategy "below"); or (2) find a concept x' above x with a correspondence to a concept $y \in Y$ (strategy "above").

The effectiveness of the reformulation was evaluated by assessing the relevance of objects annotated with concept y (or subconcepts of y) on on a six-point scale ranging from "very relevant" to "not relevant at all". Generalized precision and recall were calculated from these ordinal assessments (see Table 4). For comparison we also calculated precision and recall based on dichotomous (0/1) assessments by rescaling the ordinal values 0-2 to 0 and 3-5 to 1. Recall was calculated based on a recall pool[9].

We stress that the precision and recall figures presented in Table 4 refer to *relevancy* of the returned objects, instead of *correctness* of correspondences. This means that it is not possible to directly compare the results from reference alignment evaluations with the results from end-to-end evaluation. This is a general methodological difficulty when comparing evaluation methods, not only for the scenario presented in this paper.

---

[8] We excluded concepts that were too general such as **Physical object**
[9] A recall pool consists of the union of all objects returned by any of the systems; objects not in the pool are considered irrelevant. This method is regularly used in evaluation of text retrieval systems where evaluating all documents in the collection is practically infeasible.

Table 4. Precision and Recall of end-to-end evaluation for a six-point scale and a binary scale. Results are shown for two different query reformulation strategies.

Vocabulary	Strategy	Precision Binary Scale	Precision 6-point Scale	Recall Binary Scale	Recall 6-point Scale
ARIA	upward	0.27	0.37	0.83	0.88
	downward	0.70	0.66	0.49	0.43
SVCN	upward	0.46	0.48	0.93	0.96
	downward	0.79	0.76	0.42	0.36
WordNet	upward	0.46	0.48	0.80	0.81
	downward	0.63	0.67	0.18	0.18

5 Interpretation and Discussion of Results

The three case studies have illustrated differences between the evaluation methods and between the aligned vocabularies. The results show that the different evaluation methods stress different properties of an alignment.

A first observation is that SVCN outperforms WordNet and ARIA in all evaluations including the end-to-end evaluation. One exception is the result for recall in the downward strategy of the end-to-end evaluation; ARIA performs slightly better. The high scores of SVCN can be explained from its reasonably clean hierarchy and high similarity to the target vocabulary AAT. Evaluation of individual correspondences gives SVCN a precision of around 0.90 for all different weighting schemes. The different precision numbers lie around 0.70 for ARIA and WordNet. This suggests that a weakly structured, small vocabulary such as ARIA can be aligned with approximately the same precision as a large, richly structured vocabulary such as WordNet.

The variations in frequency of use of correspondences are most pronounced in WordNet. This can be explained from the fact that the proportion of WordNet concepts that has a correspondence to AAT is relatively small (13%). Queries for concepts without a correspondence to AAT will be reformulated to related concepts that do have a correspondence to AAT. This causes concepts that are central nodes in the hierarchy to get potentially high frequency counts. In line with this finding, correcting the most frequent correspondences gives a significantly higher precision than correcting randomly selected correspondences.

For ARIA, correcting frequent correspondences also showed a clear improvement of the results. This is not entirely expected, since ARIA has relatively many correspondences and ARIA's frequency distribution is less pronounced. The effect is partly due to the fact that the precision of the frequent stratum is lower than the precision of the infrequent stratum. The results suggest that for both WordNet and Aria, an evaluation that takes into account the frequency of use will result in a more realistic estimation of application performance than an evaluation that does not take this into account.

For SVCN, the frequency based weighting did not make a difference, nor did the correction of frequent correspondences. This lack of effect can be explained from two observations: (a) the precision of SVCN is already high and therefore correction will have less effect; and (b) the frequency distribution of SVCN correspondences is relatively gradual, so that the most frequent stratum has less influence than in the WordNet case. We conclude that in cases where only a small portion of a vocabulary can be aligned to a target vocabulary, for example when topical overlap is small, an estimation of the most frequently used correspondences gives a realistic image of application performance. In these cases it will be cost-effective to manually correct (only) the frequently used correspondences.

The comparison against a reference alignment produces a clear ordering of the three alignments, in both precision and recall: SVCN is best, followed by ARIA and finally WordNet. The alignment of WordNet has a low recall (0.45 and 0.47) compared to the other vocabularies. A possible cause is the size of WordNet and the relatively low number of correspondences that was found. Although we have no clear explanation, the effect is reflected in the end-to-end evaluation; WordNet has a remarkably low recall when using the downward strategy.

When comparing the 'traditional' precision and recall scores to those based on semantic distance, we see a clear difference between the two measures in the results of ARIA (an average difference of 7%). A difference is notable for SVCN (average of 4%) although less clear, and almost no difference is visible for WordNet (1%). This is mirrored in the end-to-end evaluation, where the differences between a binary scale and a 6-point scale show the same trend: large differences for ARIA (an average of 13%), small differences for SVCN (6%) and no differences for WordNet (2%). An explanation is that ARIA returns many results that are only moderately relevant, while WordNet returns mainly highly relevant results. For applications in which users expect to see also moderately relevant results, an evaluation based on semantic distance better reflects the quality of the alignment.

We see two practical consequences of our analyses. First, we conclude that two vocabularies that show a stark resemblance to each other with respect to structure and topical overlap, can be aligned with such high precision and recall that manual creation or correction of the alignment has little added value. This holds in particular for vocabularies that share a common source, such as SVCN and AAT. Second, a vocabulary with a weak structure is no impediment for a high-quality alignment. The ontological flaws of ARIA did not result in a worse alignment than the reasonable structure of WordNet.

Acknowledgements

The authors would like to thank Willem van Hage, Alistair Vardy, Tom Moons, Niels Schreiber and the members of the E-Culture project. The authors were supported by the NWO projects: CHIME, CHOICE and STITCH and the TELplus project.

References

[1] Ashpole, B., Ehrig, M., Euzenat, J., Stuckenschmidt, H. (eds.): Proceedings of the K-CAP 2005 Workshop on Integrating Ontologies (2005)
[2] Brink, W.P., van den Koele, P.: Statistiek, Boom, Amsterdam, The Netherlands, vol. 3 (2002) ISBN 90 5352 705 2

[3] Budanitsky, A., Hirst, G.: Semantic distance in wordnet: an experimental application oriented evaluation of five measures, In: Proceedings of the NACCL 2001 Workshop on WordNet and other lexical resources, Pittsburgh, PA, pp. 29–34 (2001)
[4] Ehrig, M., Euzenat, J.: Relaxed precision and recall for ontology matching. In: Ashpole, et al. [1]
[5] Euzenat, J.: Semantic precision and recall for ontology alignment evaluation. In: Veloso, M.M. (ed.) Proceedings of the International Joint Conferences on Artificial Intelligence, pp. 348–353 (2007)
[6] Euzenat, J., Shvaiko, P.: Ontology matching. Springer, Heidelberg (2007)
[7] Euzenat, J., Stuckenschmidt, H., Yatskevich, M.: Introduction to the ontology alignment evaluation 2005. In: Ashpole, et al. [1] (2005)
[8] Euzenat, J., Mochol, M., Shvaiko, P., Stuckenschmidt, H., Šváb, O., Svátek, V., van Hage, W.R., Yatskevich, M.: Results of the OAEI 2006. In: Ashpole, B., Ehrig, M., Euzenat, J., Stuckenschmidt, H. (eds.) Ontology Matching. CEUR Workshop Proceedings, vol. 225 (2006)
[9] Euzenat, J., Isaac, A., Meilicke, C., Shvaiko, P., Stuckenschmidt, H., Šváb, O., Svátek, V., van Hage, W.R., Yatskevich, M.: First results of the OAEI 2007. In: Ashpole, B., Ehrig, M., Euzenat, J., Stuckenschmidt, H. (eds.) Ontology Matching. CEUR Workshop Proc. (2007)
[10] Hu, W., Qu, Y.: Discovering simple mappings between relational database schemas and ontologies. In: Proceedings of the International Semantic Web Conference, pp. 225–238 (2007)
[11] Isaac, A., Wang, S.: Evaluation issues at the library testcase of oaei 2007. In: ESWC 2008 (accepted for publication, 2008)
[12] Isaac, A., Zinn, C., Matthezing, H., van der Meij, L., Schlobach, S., Wang, S.: The value of usage scenarios for thesaurus alignment in cultural heritage context. In: Proceedings of International Workshop on Cultural Heritage on the Semantic Web, ISWC 2007, Korea (2007)
[13] Kalfoglou, Y., Schorlemmer, M.: Ontology mapping: the state of the art. The Knowledge Engineering Review 18(1), 1–31 (2003)
[14] Kekäläinen, J., Järvelin, K.: Using graded relevance assessments in ir evaluation. Journal of the American Society for Information Science and Technology 53(13) (2002) ISSN 1532-2882
[15] Leacock, C., Chodorow, M.: Combining Local Context and WordNet Similarity for Word Sense Identification, ch. 11, pp. 265–285. MIT Press, Cambridge (1998)
[16] Peterson, T.: Introduction to the Art and Architecture Thesaurus. Oxford University Press, Oxford (1994)
[17] Schreiber, A.Th., Amin, A., van Assem, M., de Boer, V., Hardman, L., Hildebrand, M., Hollink, L., Huang, Z., van Kersen, J., de Niet, M., Omelayenko, B., van Ossenbruggen, J., Siebes, R., Taekema, J., Wielemaker, J., Wielinga, B.J.: Multimedian e-culture demonstrator. In: The Semantic Web Challenge at the Fifth International Semantic Web Conference, Athens, GA, USA (November 2006)
[18] Shvaiko, P., Euzenat, J.: A survey of schema-based matching approaches. Journal on Data Semantics 3730, 146–171 (2005)
[19] van Assem, M., Gangemi, A., Schreiber, G.: RDF/OWL Representation of WordNet. W3C Working Draft, World Wide Web Consortium (June 2006)
[20] van Hage, W.R., Isaac, A., Aleksovski, Z.: Sample evaluation of ontologymatching systems. In: Proceedings of the Fifth International Evaluation of Ontologies and Ontology-based Tools, Busan, Korea (2007)

Putting Ontology Alignment in Context: Usage Scenarios, Deployment and Evaluation in a Library Case

Antoine Isaac[1,2], Henk Matthezing[2], Lourens van der Meij[1,2], Stefan Schlobach[1], Shenghui Wang[1,2], and Claus Zinn[3]

[1] Vrije Universiteit Amsterdam
[2] Koninklijke Bibliotheek, Den Haag
[3] Max Planck Institute for Psycholinguistics, Nijmegen

Abstract. Thesaurus alignment plays an important role in realising efficient access to heterogeneous Cultural Heritage data. Current ontology alignment techniques, however, provide only limited value for such access as they consider little if any requirements from realistic use cases or application scenarios. In this paper, we focus on two real-world scenarios in a library context: thesaurus merging and book re-indexing. We identify their particular requirements and describe our approach of deploying and evaluating thesaurus alignment techniques in this context. We have applied our approach for the Ontology Alignment Evaluation Initiative, and report on the performance evaluation of participants' tools wrt. the application scenario at hand. It shows that evaluations of tools requires significant effort, but when done carefully, brings many benefits.

1 Introduction

Museums, libraries, and other cultural heritage institutions preserve, categorise, and make available a tremendous amount of human cultural heritage (CH). Many indexing schemes have been devised to describe and manage the heritage data. There are thesauri[1] specific to fields, disciplines, institutions, and even collections. With the advent of information technology and the desire to make available CH resources to the general public, there is an increasing need to facilitate interoperability across collections, institutions, and even disciplines.

By providing representational standards (such as SKOS [1]) as well as generic tool support [2], Semantic Web technology has recently taken a more prominent role in this facilitation. A technology that can help with some of the CH interoperability problems is ontology alignment [3]. Ontology alignment aims at aligning classes (and properties) from different ontologies, by creating sets of correspondences between these entities. Applied to CH vocabulary cases, this could help, for instance, to access a collection via thesauri it is not originally

---

[1] Here we use the word *thesaurus* to refer to all controlled vocabularies that are used in the Cultural Heritage field: classification schemes, subject heading lists *etc.* To denote the elements contained in these vocabularies, we will use the word *concept*.

indexed with, to interconnect distributed, differently annotated collections on the object level, or to merge two thesauri to rationalise thesaurus maintenance.

Unfortunately, experience shows that existing ontology alignment tools often under-perform in applications in the CH domain [4]. We believe that striving for generality of alignment technology is part of the problem. To this end, we argue that the *generation* and the *evaluation* of thesaurus alignments must take into account the application context. Current alignment research within the Semantic Web community, unfortunately, underestimates the importance of requirements from real applications. Evaluation efforts, such as those of the Ontology Alignment Evaluation Initiative[2] mostly favour "application-independent" settings, where, typically, manually-built gold standards are created and used. Such gold standards are actually biased towards – at best – one single usage scenario (*e.g.*, vocabulary merging), and can be of little use for other scenarios (*e.g.*, query reformulation). Efforts leading to an application-specific assessment are under way [5], but further work is required.

The following questions need to be answered to successfully deploy and evaluate alignment techniques:

- What kind of usage scenarios require alignment technology?
- For a given scenario,
  1. What is the meaning of an alignment, and how to exploit it?
  2. How to use current tools to produce the required type of alignment?
  3. How to evaluate an alignment appropriately?

Our aim is to illustrate how to answer these questions from a realistic application perspective. We focus on analysing application requirements and user needs as well as determining practical processes.

By answering these questions (both methodologically and empirically) we will validate two general hypotheses regarding the evaluation of ontology alignment:

- Evaluation results can depend on the evaluation strategy, even when applied to a same scenario.
- Evaluation results can depend on the scenario, even when the most appropriate evaluation strategy is applied.

These hypotheses, although quite obvious, have nevertheless been rather neglected. The ontology alignment community needs to better take on board requirements that come from a wide variety of real application contexts, and also evaluate the performance of their tools in the light of such requirements. Our aim is also to gain more insight with regard to the comparison between performances of different alignment techniques. The hypothesis, already formulated in alignment research, is that some techniques will be more or less interesting to pursue, depending on the application scenario at hand.

The next section introduces our application context, situated at the National Library of the Netherlands, where two thesauri need to be aligned for various scenarios.

---

[2] http://oaei.ontologymatching.org

2 The Need for Thesaurus Alignment at KB

The National Library of the Netherlands (KB) maintains two large collections of books. The *Deposit Collection* comprises all Dutch printed publications (one million items), and the *Scientific Collection* has about 1.4 million books on the history, language and culture of the Netherlands. Each collection is annotated – *indexed* – using its own controlled vocabulary. The Scientific Collection is described using the GTT thesaurus, a huge vocabulary containing 35,194 general concepts. The books in the Deposit Collection are mainly described against the Brinkman thesaurus, which contains a large set of headings (5,221) for describing the overall subjects of books. Currently, around 250,000 books are shared by both collections and therefore indexed with both GTT and Brinkman concepts.

The two thesauri have similar coverage (2,895 concepts actually have exactly the same label) but differ in granularity. Represented in SKOS [1] format, each concept has one preferred label, synonyms and other alternative labels, extra hidden labels and scope notes. Also, both thesauri are structured by *broader*, *narrower* and *related* relations between concepts, but this structural information is relatively poor. GTT (resp. Brinkman) contains only 15,746 (resp 4,572) hierarchical *broader* links and 6,980 (resp. 1,855) associative *related* links. On average, one can expect at most one parent per concept, for an average depth of 1 and 2, respectively. GTT has 19,752 root concepts.

The co-existence of these different systems, even if historically and practically justified, is not satisfactory. First, both thesauri are actively maintained but independently from each other, which doubles the management cost. Second, disconnected thesauri do not support unified access to both collections. Except the 250,000 dually indexed books, books can only be retrieved by concepts from the particular thesaurus they were originally indexed with.

In order to achieve better interoperability and reduce management cost, thesaurus alignment plays a crucial role, with regard to the following scenarios.

1. **Concept-based search:** support the retrieval of GTT-indexed books using Brinkman concepts, or *vice versa*.[3]
2. **Re-indexing:** support the indexing of GTT-indexed books with Brinkman concepts, or *vice versa*.
3. **Integration of one Thesaurus into the other:** support the integration of GTT concepts into the Brinkman thesaurus, or *vice versa*.
4. **Thesaurus Merging:** support the construction of a new thesaurus that encompasses both Brinkman and GTT.
5. **Free-text search:** support the search for books using free-text queries that match user search terms to GTT or Brinkman concepts.
6. **Navigation:** support users to browse both collections through a merged version of the two thesauri.

Different scenarios have different requirements with regard to the usages of the alignment. In the following sections, we will focus on two scenarios – thesaurus

---

[3] This is a simple version of query reformulation using links between thesauri.

merging and book re-indexing – and investigate their different impact on the general thesaurus alignment problem.

3 The Thesaurus Merging Scenario

To reduce thesaurus management and indexing costs, KB considers to merge their Brinkman and GTT thesauri into a single unified thesaurus. The question is whether the thesaurus merging task can be supported by ontology alignment tools, and how. Clearly, this application scenario requires ontology matchers to recognise semantic relations (in particular, equivalence) between the concepts of the two thesauri. These alignment links, together with the respective thesaurus-internal semantic relations, will constitute a semantic network that can then be exploited to create the unified thesaurus. This scenario can be likened to *ontology engineering* use cases for alignment, as presented in [3], chapter 1.

Until now, thesaurus merging scenarios have been rather neglected by the research community [6], since this task, like multilingual thesaurus building, can raise a multitude of languages-specific and cultural issues. GTT and Brinkman are both Dutch thesauri, and they describe similar and quite general domains. Therefore, such issues will not arise, and intuitively, alignments shall rather cover large parts on the input thesauri. The largest problem to address here is the different semantic granularity of the thesauri. GTT and Brinkman have different sizes for a similar subject coverage. Many concepts of one thesaurus, thus, will not have equivalent concepts in the other thesaurus. For instance, due to different indexing usages — *cf.* our discussion on post-coordination in section 4 — Brinkman and GTT both have the terms "gases" and "mechanics", but only Brinkman has the *compound* concept "Gases; mechanics".

3.1 Formulation of Thesaurus Merging Problem

In the thesaurus merging scenario, we define an alignment as a function that states whether two concepts are linked to each other by a semantic relation:

$$A_{merge} : \mathcal{G} \times \mathcal{B} \times \mathcal{SR} \to \{true, false\},$$

where $\mathcal{G}$ and $\mathcal{B}$ denote the sets of GTT and Brinkman concepts. $\mathcal{SR}$ denotes the set of semantic relations, containing the *equivalence* link that is used to merge concepts, as well as the *broader*, *narrower*, and *related* semantic links that are proposed in standard thesaurus building guidelines [7] and which are also used at the KB.

Alignments between *combinations of concepts* could help a thesaurus engineer to determine whether a complex subject from one thesaurus is covered by several simpler concepts from the other thesaurus. Note, however, that such alignments are not explicitly relevant for the task at hand. The GTT and Brinkman thesauri do not deal with complex concepts in their respective internal formats, and there is no reason why a unified thesaurus should entertain such structure.

3.2 Proceeding with Existing Alignments for Thesaurus Merging

Standard ontology matchers can give results that do not fit the function specified above. For instance, instead of typed relations, one tool can use only one generic symmetric similarity link, eventually coming with a certainty degree – typically in the $[0, 1]$ interval. Such certainty information could be computed, say, by counting the number of books that share Brinkman and GTT indices.

Such results would have to be re-interpreted and *post-processed*, *e.g.* by defining a threshold that filters out the weakest links, or validates some links as cases of *equivalence* and others as cases of mere *relatedness*. In some variants of this scenario, human experts can be involved, using the certainty information to accept or reject candidate links. Consider, for instance, the terms "making career" (denoting a series of actions) and "career development" (the result of these actions). A matcher might relate them via an equivalence link, but with a weak probability. But in some contexts, a thesaurus builder could nevertheless decide to merge them into a single concept, or to make one a specialisation of the other.

3.3 Evaluation Method

Alignments are evaluated in terms of their individual mappings as follows:

- If the mapping link is *related*, *broader* or *narrower*, then assess whether it would hold within one unified thesaurus, given both concepts were to be included in it;
- If the mapping link is *equivalence*, then assess whether the two concepts should be merged in such a thesaurus.

An evaluation must consider two aspects: (i) *correctness*: what is the proportion of correct (or acceptable) links in the results? (ii) *completeness*: what proportion of the links required by the scenario did the results contain? Two standard Information Retrieval measures, precision and recall, are normally used. But there are alternative options, which could fit well specific – supervised – scenario variants. For example, a semantic version of precision and recall, as proposed in [8], would help to discriminate between near misses and complete failures, when a human expert editing the proposed alignment links can transform these near misses into correct matches.

Ideally, the evaluation should be based on a complete reference alignment. If no complete gold standard is available, then absolute recall cannot be computed. In this case, especially valid when the focus is on comparing the relative performances of several alignments produced by different tools, one can measure *coverage*. For each alignment, we define the coverage as the proportion of all good mappings found by this alignment divided by the total number of distinct good mappings produced by all alignments. This coverage is proportional to the real recall, and in any case it provides an upper bound for it — as the correct mappings found by all participants give a lower bound for the total number of correct mappings.

The thesaurus merging evaluation actually resembles classical *alignment evaluation*, as presented, *e.g.*, in [9]. In the next section, we discuss the re-indexing scenario, where deployment and evaluation, formulated in terms of specific *information needs*, is more in line with the "end-to-end" approach described in [9].

4 The Book Re-indexing Scenario

To streamline the indexing of Dutch scientific books, currently described with both Brinkman and GTT, thesaurus alignment can be used as follows:

- Computer-supported book indexing with the following workflow: first, a new book is manually described with GTT by a human expert; subsequently, thesaurus alignment technology is asked to generate a Brinkman index, given its GTT annotation. In a supervised setting, the expert, not necessarily the same person, can then accept or adapt this suggestion.
- KB decides to terminate their use of GTT in favour of the Brinkman thesaurus. All books that have been indexed with GTT concepts shall be re-indexed with Brinkman using thesaurus alignment technology. Again, this re-indexing could be fully automatic or supervised. In the latter, a human expert takes a book's new Brinkman indexing as suggestion, possibly changing it by removing or adding Brinkman concepts.

This scenario is about *data migration*. Similar to the "catalogue integration" use case in [3], chapter 1, some tool transforms descriptions of objects — in our case book indices — from one vocabulary to the other.

Re-indexing books is fundamentally a non-trivial activity. Consider the following two books and their respective index in the GTT and in the Brinkman thesaurus:

- Book *Allergens from cats and dogs*
  - Brinkman: "allergie," (*allergy*) "katten," (*cats*) "honden" (*dogs*)
  - GTT: "allergenen," (*allergens*) "katten," "honden," "immunoglobulinen" (*immunoglobulins*)
- Book *Het verborgen leven van de kat*
  - Brinkman: "katten"
  - GTT: "diergedrag," (*animal behaviour*) "katten," "mens-dier-relatie" (*human-animal relation*)

As we can see, the same concept used in different indices should be jointly aligned to different sets of concepts. Some of these required alignments are obvious, while some are more complicated, sometimes even reflecting different analysis levels on a same book.

These phenomena are related to the use of *post-coordinate indexing*. As above examples show, when a book is annotated with several GTT subject concepts, these concepts are considered in combination, each being a factor of the subject of the whole book. The re-indexing function must therefore deal with more than just the (arbitrary) co-occurrence of concepts.

4.1 Formulation of the Book Re-indexing Problem

A book is usually indexed by a set of concepts; an alignment shall specify how to replace the concepts of a GTT book indexing with conceptually similar Brinkman concepts to yield a Brinkman indexing of the book:

$$A_{reindex} : 2^\mathcal{G} \rightarrow 2^\mathcal{B},$$

where $2^\mathcal{G}$ and $2^\mathcal{B}$ denote the powersets of the GTT and Brinkman concepts. Note that the sets of proposed Brinkman concepts would also be preferably small. Observation of usage reveals that 99.2% of the Deposit books are indexed with no more than 3 Brinkman concepts and that 98.4% of the GTT-indexed books have no more than 5 concepts.

The (informal) semantics of the required alignments can be determined the following way. First, consider the simple case – concerning 18.7% of KB's dually indexed books – where the GTT index of a given book consists of one GTT concept, and its Brinkman index book consists of one Brinkman concept. Here, our function needs to translate a single GTT concept g into a single Brinkman concept b. The re-indexing can be information-preserving if the concepts g and b are judged equivalent; the re-indexing can loose information if concept b is judged broader than concept g; and using a narrower concept or related concept B, additional information may be introduced, which could be wrong but not necessarily so. These cases correspond to well known mapping situations as described at the semantic level by [6] and given representation formats [1].

The simple case of one-to-one mappings can be generalised to many-to-many (set-to-set) mappings. A complex subject built from GTT concepts by means of post-coordination can be replaced by another complex subject built from Brinkman concepts (or a simple one) if these two complex subjects have equivalent meanings, or, to a lesser extent, if the meaning of the first subsumes the meaning of the second or if they have overlapping meaning.

4.2 Proceeding with Existing Alignments for Re-indexing Books

As mentioned before, off-the-shelves matchers usually produce only one-to-one mappings, possibly with a weight.[4] To meet the specific requirements of book re-indexing, a post-processing step is required to obtain multi-concept book indices. In our earlier research, we have presented a procedure – and several options – to do this [11]. The first step consists in grouping concepts based on the mappings they are involved in, so as to obtain translations rules of the form $\{g_0, g_1, \ldots, g_m\} \mapsto \{b_1, b_2, \ldots, b_n\}$. The set of GTT concepts attached to each book is then used to decide whether these rules are *fired* for this book. Given

---

[4] The *block matching* approach [10], which maps together sets of concepts, is an important exception. However, the nature of block matching prevents us from using such tools in our domain as it constructs sets of semantically close concepts rather than sets of semantically distinct concepts that can be used together. Furthermore, its computational complexity makes it difficult to apply to large datasets.

a book with a GTT annotation G_t, there are several conditions which can be tested for firing a given rule $G_r \mapsto B_r$: (1) $G_t = G_r$; (2) $G_t \supseteq G_r$; (3) $G_t \subseteq G_r$; (4) $G_t \cap G_r \neq \emptyset$. If several rules can be fired for a same book, several strategies can also be chosen for creating the final Brinkman annotation. The most simple one is to consider the union of the consequents of all the rules.

Note that the number of options may be further increased by considering scenarios where human experts are involved in the production of indices. Here, for instance, similarity weights, as given by alignment tools, could be used to generate probability of appropriateness for each candidate concept. An expert would validate the proposed indices using this information.

4.3 Evaluation Method

In the re-indexing scenario, evaluating an alignment's quality means assessing, for each book, the quality of its newly assigned Brinkman index. This assessment gives an indication of the quality of the original thesaurus alignment in the context of the $A_{reindex}$ function that was built from it. We can consider the following evaluation variants and refinements.

Evaluation Settings

Variant 1: Fully automatic evaluation. Reconsider the corpus of books that belong both to KB Scientific and Deposit collections. The corpus comprises 243,887 books that are already manually indexed against both GTT and Brinkman. In this variant, the existing Brinkman indices are taken as a gold standard that the evaluated re-indexing procedure must aim to match. That is, for each book in the given corpus, we compare its existing Brinkman index with the one that has been computed by applying $A_{reindex}$. The similarity between these two Brinkman concept sets can be computed, yielding a measure that indicates the general quality of $A_{reindex}$.

Variant 2: Manual evaluation In this variant, a human expert is asked to judge the correctness and completeness of candidate Brinkman indices for a sufficiently large set of books, hence producing a reference indexing. This assessment will vary depending on the scenario that defines how alignment technology is being deployed. Notice that in an unsupervised setting, strict notions of completeness and correctness apply. Here, instead of testing *e.g.* strict set equality, a human expert is likely to accept semantically close Brinkman concepts. The notions for correctness and completeness are thus different and possibly less strict. In this variant, experts are further asked to indicate the concepts which they may eventually use for indexing this book. If the proposed concepts are not part of their ideal choice, then experts can add those concepts. Ideally, this list should contain all the concepts that the human expert expects to describe a given book properly.

Having a human expert in the loop further helps dealing with three important evaluation issues:

- *Indexing variability.* Usually, there is more than one correct indexing of a given book, and two experts might index a given book in two different ways. Having an expert to complement a machine-produced Brinkman index with her own, might make explicit this variability. Also, asking a human expert on the *acceptability* – as opposed to strict validity – of a machine-generated index may increase completeness and correctness results, as human judgement is more flexible and open-minded than automatic measures.
- *Evaluation variability.* Along the same line, the assessment of a book index itself may vary among human evaluators. A manual evaluation allows us to compare several judgements on the same alignment results. One can attempt to address the reliability of the chosen evaluation measure, and then devise new approaches to compensate for the weaknesses that were found.
- *Evaluation set bias.* The corpus of dually indexed books that is needed for variant 1 might have some hidden specific features, while manual evaluation with human experts can be performed on any part of the KB collections.

Evaluation Measures. All evaluation variants depend on a test set of books indexed with GTT terms. Applying the re-indexing procedure for each book will then produce a set of Brinkman terms. These terms can then be compared against the reference set (or gold standard) that either stems from the existing Brinkman annotation or is set by human experts.

First, we measure how well the generated Brinkman book indices match the correct ones. Correctness and completeness of returned indices are assessed by *precision* and *recall* defined at the indexing level as follows:

$$P_a = \frac{\sum \frac{|\{b_1,...,b_n\} \cap A_{reindex}(\{g_1,...,g_m\})|}{|A_{reindex}(\{g_1,...,g_m\})|}}{\#books\_fired}, \quad R_a = \frac{\sum \frac{|\{b_1,...,b_n\} \cap A_{reindex}(\{g_1,...,g_m\})|}{|\{b_1,...,b_n\}|}}{\#books\_total},$$

where $\{b_1,\ldots,b_n\}$ (resp., $\{g_1,\ldots,g_m\}$) is the set of correct Brinkman (resp., existing GTT) concepts for the book; $\#books\_total$ is the number of books in the evaluation set; and $\#books\_fired$ is the number of books for which a re-indexing has been provided. We also use, as a combination of the precision and recall defined above, a Jaccard overlap measure between the produced annotation (possibly empty) and the correct one:

$$J_a = \frac{\sum \frac{|\{b_1,...,b_n\} \cap A_{reindex}(\{g_1,...,g_m\})|}{|\{b_1,...,b_n\} \cup A_{reindex}(\{g_1,...,g_m\})|}}{\#books\_total}$$

Second, we measure the performance of the re-indexing at a broader level, in terms of book retrieval. We consider that a book is retrievable when its correct and generated indices overlap, that is, $\{b_1,\ldots,b_n\} \cap A_{reindex}(\{g_1,\ldots,g_m\}) \neq \emptyset$ — we then call it a *matched* book. Here, *precision* is defined as the fraction of books which are considered as matches according to the previous definition over the number of books for which a new index was generated; and *recall* is defined by the fraction of the "matched" books over the total number of books:

$$P_b = \frac{\#books\_matched}{\#books\_fired}, \quad R_b = \frac{\#books\_matched}{\#books\_total}.$$

Note that in all these formulas, results are counted on a book and annotation basis, and not on a rule basis. This reflects the importance of different rules: a rule for a frequently used concept is more important for the application than a rule for a rarely used concept.

5 Implementing Application-Specific Evaluation for the OAEI Library Track

Since 2004, the Ontology Alignment Evaluation Initiative (OAEI) organises campaigns to review the performance of current state-of-the-art ontology alignment technologies in different cases. Among the six data sets of the 2007 campaign, the *Library* track[5] proposed to align the GTT and Brinkman thesauri, made available in the SKOS [1] and OWL [12] formats. Participants, who had no *a priori* knowledge of the evaluation procedures, were required to deliver SKOS mapping relations: `exactMatch`, `broadMatch` and `relatedMatch`. Three OAEI participants sent results for this track: **Falcon** [13] – 3,697 `exactMatch`, **DSSim** [14] – 9,467 `exactMatch` – and **Silas** [15] – 3,476 `exactMatch` and 10,391 `relatedMatch`.

5.1 Thesaurus Merging Evaluation

As there was no reference alignment available that maps the complete Brinkman thesaurus to the GTT thesaurus, we used coverage instead of recall for comparing the alignments, as presented in section 3.3. Moreover, to minimize the number of alignments that had to be evaluated, we decided to automatically construct a reference alignment based on a lexical procedure. The method compares labels with each other (literal string matching), but also exploits a Dutch morphology database to recognise variants of a word (*e.g.*, singular and plural). As a result, 3,659 correct equivalence links were obtained.

We only evaluated the `exactMatch` mappings, as only one participant provided another link type. For a representative sampling, the three sets of `exactMatch` mappings were partitioned into sections, one for each combination of the four considered sources (participant alignments plus reference set). For each of the resulting sections that were not in the lexical reference alignment, a sample of mappings was selected and evaluated manually. A total of 330 mappings were assessed by two Dutch native speakers.

From these assessments, precision and coverage were calculated with their 95% confidence intervals, taking into account sampling size and evaluator variability. The results are shown in Table 1.

Clearly, Falcon outperforms the other two systems. Falcon's high precision expresses in the following numbers: 3,493 links are common to Falcon's alignment and the reference alignment; Falcon's alignment has 204 mappings not in the reference alignment (of which 100 are judged correct); and the reference alignment has 166 mappings not in Falcon alignment.

---

[5] http://www.few.vu.nl/~aisaac/oaei2007

Table 1. Precision and Coverage for the thesaurus merging scenario

Participant	Precision	Coverage
Falcon	0.9725 ± 0.0033	0.870 ± 0.065
Silas	0.786 ± 0.044	0.661 ± 0.094
DSSim	0.134 ± 0.019	0.31 ± 0.19

Like Falcon, DSSim also uses a lexical approach for ontology alignment. However, its edit-distance-like approach is more prone to error: only between 20 and 400 of its 8,399 mappings not in the reference alignment were judged correct. In fact, given a selection of 86 mappings from the set of 8,363 mappings unique to DSSim, not a single one was evaluated as correct by the human evaluators. The Silas tool succeeds most in adding mappings to the reference alignment: 234 of its 976 "non-lexical" mappings are correct; nevertheless, it fails to reproduce one third of the reference mappings, and therefore, its coverage is relatively low.

5.2 Book Re-indexing Evaluation in OAEI 2007

Automatic Evaluation and Results. The automatic evaluation relies on comparing, for the dually indexed books, existing Brinkman indices with the ones that were generated using the alignment. Following the procedure of section 4.2, rules were generated to associate one GTT concept with a set of Brinkman concepts, using a simple grouping strategy. When considering exact matches only, this gives 3,618 rules for Falcon, 3,208 rules for Silas and 9,467 rules for DSSim. One rule is then fired on a given book if its GTT concept is contained in the GTT annotation of this book, *i.e.*, using the firing condition (4) introduced in Section 4.2. When several rules can be fired for a book, the union of their consequents forms the Brinkman re-indexing of the book, which can then be compared to the existing annotation.

Table 2 shows the evaluation results when only `exactMatch` mappings are exploited. Interestingly, comparing these results with Table 1, Silas performs as well as Falcon does here. The exploitation of the Falcon alignment resulted in at least one correct Brinkman term per book for nearly half of the test set. At the annotation level, half of the generated Brinkman concepts were judged incorrect, and more than 60% of the gold standard was not found. As mappings from

Table 2. Performance of book re-indexing generated from mappings

Participant	P_b	R_b	P_a	R_a	J_a
Falcon	65.32%	49.21%	52.63%	36.69%	30.76%
Silas	66.05%	47.48%	53.00%	35.12%	29.22%
DSSim	18.59%	14.34%	13.41%	9.43%	7.54%
Silas+related	69.23%	59.48%	34.20%	46.11%	24.24%

Falcon are mostly generated by lexical similarity, these figures clearly indicate that lexical approach is not sufficient for the book re-indexing scenario.

The results also confirms the sensitivity of mapping evaluation methods to certain application scenarios. Among the three participants, only Silas generated relatedMatch mappings. We combined these mappings with the exactMatch ones to generate a new set of 8,410 rules. As shown in the *Silas+related* row of Table 2, the use of relatedMatch mappings increases the chances of a book being given a correct annotation. However, unsurprisingly, the precision of annotations decreases as noisy results were introduced.

Manual Evaluation and Results

Evaluation process A sample of 96 books was randomly selected from the dually annotated books indexed by KB experts in 2006. For each of these books we applied the annotation translation rules derived from each participants' results, using only the exactMatch links. For each book, the results of these different procedures were merged into lists of candidate concept annotations. We also included the original annotations in the candidate lists. On average this procedure resulted in five candidate concepts per book.

To acquire experts' assessments of the candidate annotations, paper forms were created for each book in the sample. A form presented the book's cataloguing information — author, title, year of publication *etc.* — plus the candidate annotations found for this book.

Given a book's description and annotation, experts were then asked to judge the *acceptability*[6] for each and every annotation concept. The experts were also asked to select from the candidates the ones they would have *chosen as indices*. For this, experts had the opportunity to add terms to the candidate list they found most appropriate to describe the book.

A preliminary version of the evaluation form was tested with two professional indexers. The experts agreed with our notion of "acceptability" and also found the average number of candidate concepts adequate. Four professional book indexers from the Depot department at the KB, all native Dutch speakers, took part in the final evaluation. Each expert assessed the candidate annotation for every element of the sample set.

Results. Table 3 presents the results averaged over the four experts. Interestingly, these human assessments are significantly higher than the figures obtained from our automatic evaluation. It suggests that the chosen application context requires an evaluation that takes into account the indexing variability of human experts.

To assess *evaluation variability*, we computed the *Jaccard overlap* between evaluators' assessments. On average, two evaluators agreed on 60% of their

---

[6] As it was hinted in Section 4.3, this formulation aims to avoid too narrow judgements. The evaluator can here anticipate situations where other indexers might have selected indices different from hers, *e.g.* when the subject of the book is unclear, the thesaurus contains several concepts equally valid for the book.

Table 3. Performance of mappings as assessed by manual evaluation (left), compared to automatic evaluation results (right, from Table 2).

Participant	P_a	R_a	J_a	P_a	R_a	J_a
Falcon	74.95%	46.40%	42.16%	52.63%	36.69%	30.76%
Silas	70.35%	39.85%	35.46%	53.00%	35.12%	29.22%
DSSim	21.04%	12.31%	10.10%	13.41%	9.43%	7.54%

assessments. Using Krippendorff's α coefficient, a common measure for computational linguistics tasks [16], the overall agreement between two evaluators is $\alpha = 0.62$. According to standard interpretation, this indicates large variability.[7]

For measuring *indexing variability* between evaluators, we computed the average Jaccard overlap between their chosen indices as well as the α. Again, we have quite a low overall agreement value – 57% for the Jaccard, 0.59 for the α – which confirms the high intrinsic variability of the indexing task.

Additionally, evaluators assessed the original Brinkman indices for the books of the sample, which we had added in the candidate concepts to be evaluated. These concepts are the results of a careful selection of a human expert. Therefore, they cannot capture all the acceptable concepts for a book and recall is, unsurprisingly, relatively low (R_a=66.69%). More interestingly, almost one in five original index concept were judged not acceptable (P_a=81.60%), showing indeed that indexing variability matters considerably, even when the annotation selection criteria are made less selective.

6 Discussion and Conclusion

We reported on application scenarios that require the exploitation of thesaurus alignment. Existing off-the-shelves ontology alignment technology may be of limited practical use. It needs better characterisation for deployment and evaluation. We have studied these problems for the applications at hand, focusing on thesaurus merging and re-indexing scenarios.

All scenarios have some important common features, such as benefiting from alignments links that have thesaurus-inspired semantics. But there are also important differences, such as the emphasis on certain type of relations, or cardinality aspects. Furthermore, depending on the deployment strategy or the degree of human supervision, different levels of precision or recall can also be expected for alignments. Some cases actually hint at using less standard measures for correctness and completeness, or cautiously interpreting the evaluation results in the light of the specific characteristics of the application – *e.g.*, indexing variability.

The results we obtained for the OAEI Library track confirm the importance of considering applications when deploying and evaluating alignments. The practical usefulness of a certain alignment can vary from one scenario to the other,

---
[7] Although, the tasks usually analysed with this coefficient (part-of-speech tagging, for instance) are less variable than subject indexing.

and from one setting to another – even within one scenario. Evaluation needs to be done carefully.

Our approach is related to existing work on solving heterogeneity problems in thesaurus applications [6] or in wider controlled vocabulary contexts, including index translation and query reformulation, either from a general expert perspective [17] or with a strong emphasis on formalisation [18]. Yet, none of these efforts really study the gap between application requirements and alignments such as produced by state-of-the-art techniques. Our work started to investigate this problem, aiming at the alignment research community where application requirements have only recently come under consideration [19,5].

Our experiments show that the abovementioned gap is manifold. For instance, it is important to obtain asymmetric hierarchical alignment links – *e.g.* broader – instead of a plain similarity measure to address thesauri's different semantic granularity. Aligning sets of combined concepts instead of the standard one-to-one mappings will also be crucial for some data conversion scenarios. The lack of such capacities raises the need for a post-processing step. In such a step, decisions could be made that do not fit the assumptions guiding the computation of the alignment.

Current state-of-the-art alignment tools come with limited options with regard to the specific type of mapping that they can generate (limited mapping relations; 1-1 mappings). That makes it hard, if not impossible to use, evaluate, and deploy them in real-world contexts. We hope indeed that this paper will guide researchers from the Semantic Web domain to continue enhancing their existing tools, possibly by taking into account the diversity and richness of applications contexts and their requirements, some of which we reported here.

Our evaluations have also demonstrated that the compared usefulness of specific alignment strategies is dependent on the application scenario, confirming our last hypothesis. For the merging scenario, Falcon, which relies more on lexical matching when the structure of vocabularies is poor, outperforms the other two participants. While in the translation scenario, Silas, which detects links based on extensional information of concepts, performs as well as Falcon does. This is in line with the current trend in alignment research that investigates ways to perform case-specific selection of alignment strategies [19,20]. This also gives further reasons to keep up application-specific evaluation efforts in OAEI-like campaigns.

Evaluation when done carefully brings many benefits. We will therefore continue our own effort on determining deployment and evaluation contexts, including the cases we have only briefly mentioned here, as well as other cases outside the KB context. We also plan to investigate alignment methods that better match application requirements, extending for example our previous work on producing multi-concept alignment using instance-based similarity measures [11].

Acknowledgements

This work is funded by the CATCH programme of NWO, the Dutch Organisation for Scientific Research (STITCH project) and by the European Commission

(TELplus project). The evaluation at KB would not have been possible without the contribution of Yvonne van der Steen, Irene Wolters, Maarten van Schie, Erik Oltmans and Johan Stapels. We also thank the referees for their comments.

References

1. Isaac, A., Summers, E.: SKOS Simple Knowledge Organization System Primer. W3C Working Draft (2008)
2. Schreiber, G., et al.: Multimedian e-culture demonstrator. In: Cruz, I., Decker, S., Allemang, D., Preist, C., Schwabe, D., Mika, P., Uschold, M., Aroyo, L.M. (eds.) ISWC 2006. LNCS, vol. 4273, Springer, Heidelberg (2006)
3. Euzenat, J., Shvaiko, P.: Ontology Matching. Springer, Heidelberg (2007)
4. van Gendt, M., Isaac, A., van der Meij, L., Schlobach, S.: Semantic web techniques for multiple views on heterogeneous collections: a case study. In: Gonzalo, J., Thanos, C., Verdejo, M.F., Carrasco, R.C. (eds.) ECDL 2006. LNCS, vol. 4172, Springer, Heidelberg (2006)
5. Šváb, O., Svátek, V., Stuckenschmidt, H.: A study in empirical and casuistic analysis of ontology mapping results. In: Sure, Y., Domingue, J. (eds.) ESWC 2006. LNCS, vol. 4011, Springer, Heidelberg (2006)
6. Doerr, M.: Semantic problems of thesaurus mapping. Journal of Digital Information 1(8) (2001)
7. International Standards Organisation: ISO 2788-1986 Documentation - Guidelines for the establishment and development of monolingual thesauri (1986)
8. Euzenat, J.: Semantic precision and recall for ontology alignment evaluation. In: International Joint Conference on Artificial Intelligence (IJCAI 2007), Hyderabad, India (2007)
9. van Hage, W.R., Isaac, A., Aleksovski, Z.: Sample evaluation of ontology-matching systems. In: Fifth International Workshop on Evaluation of Ontologies and Ontology-based Tools, ISWC 2007, Busan, Korea. Springer, Heidelberg (2007)
10. Hu, W., Qu, Y.: Block matching for ontologies. In: Cruz, I., Decker, S., Allemang, D., Preist, C., Schwabe, D., Mika, P., Uschold, M., Aroyo, L.M. (eds.) ISWC 2006. LNCS, vol. 4273, Springer, Heidelberg (2006)
11. Wang, S., Isaac, A., van der Meij, L., Schlobach, S.: Multi-concept alignment and evaluation. In: Second International Workshop on Ontology Matching, ISWC 2007, Busan, Korea. Springer, Heidelberg (2007)
12. McGuinness, D.L., van Harmelen, F.: OWL Web Ontology Language Overview. W3C Recommendation (2004)
13. Hu, W., Zhao, Y., Li, D., Cheng, G., Wu, H., Qu, Y.: Falcon-AO: results for oaei 2007. In: Second International Workshop on Ontology Matching, ISWC 2007, Busan, Korea. Springer, Heidelberg (2007)
14. Nagy, M., Vargas-Vera, M., Motta, E.: DSSim – managing uncertainty on the semantic web. In: Second International Workshop on Ontology Matching, ISWC 2007, Busan, Korea. Springer, Heidelberg (2007)
15. Ossewaarde, R.: Simple library thesaurus alignment with SILAS. In: Second International Workshop on Ontology Matching, ISWC 2007, Busan, Korea. Springer, Heidelberg (2007)
16. Krippendorff, K.: Content Analysis: An Introduction to Its Methodology, 2nd edn. Sage, Thousand Oaks (2004)

17. British Standards Institution: Structured Vocabularies for Information Retrieval – Guide. Part 4: Interoperability between vocabularies. Working Draft (2006)
18. Miles, A.: Retrieval and the semantic web. Master's thesis, Oxford Brookes university (2006)
19. Euzenat, J., Ehrig, M., Jentzsch, A., Mochol, M., Shvaiko, P.: Case-based recommendation of matching tools and techniques. KnowledgeWeb Project deliverable D1.2.6 (2007)
20. Tan, H., Lambrix, P.: A method for recommending ontology alignment strategies. In: Aberer, K., Choi, K.-S., Noy, N., Allemang, D., Lee, K.-I., Nixon, L., Golbeck, J., Mika, P., Maynard, D., Mizoguchi, R., Schreiber, G., Cudré-Mauroux, P. (eds.) ISWC 2007. LNCS, vol. 4825, Springer, Heidelberg (2007)

CSR: Discovering Subsumption Relations for the Alignment of Ontologies

Vassilis Spiliopoulos[1,2], Alexandros G. Valarakos[1], and George A. Vouros[1]

[1] AI Lab, Information and Communication Systems Engineering Department, University of the Aegean, Samos, 83 200, Greece
{vspiliop,alexv,georgev}@aegean.gr
[2] Institution of Informatics and Telecommunications, NCSR "Demokritos", Greece

Abstract. For the effective alignment of ontologies, the computation of equivalence relations between elements of ontologies is not enough: Subsumption relations play a crucial role as well. In this paper we propose the "Classification-Based Learning of Subsumption Relations for the Alignment of Ontologies" (*CSR*) method. Given a pair of concepts from two ontologies, the objective of *CSR* is to identify patterns of concepts' features that provide evidence for the subsumption relation among them. This is achieved by means of a classification task, using state of the art supervised machine learning methods. The paper describes thoroughly the method, provides experimental results over an extended version of benchmarking series and discusses the potential of the method.

Keywords: ontology alignment, subsumption, supervised machine learning.

1 Introduction

In spite of the fact that ontologies provide a formal and unambiguous representation of domain conceptualizations, it is rather expectable to deal with different ontologies describing the same domain of knowledge, introducing heterogeneity to the conceptualization of the domain and difficulties in integrating information.

Although many efforts [1] aim to the automatic discovery of equivalence relations between the elements of ontologies, in this paper we conjecture that this is not enough: To deal effectively with the ontologies' alignment problem, we have to deal with the discovery of subsumption relations among ontology elements. This is particularly true, when we deal with ontologies whose conceptualizations are at different "granularity levels": In these cases, the elements (concepts and/or properties) of an ontology are more generic than the corresponding elements of another ontology. Although subsumption relations between the elements of two ontologies may be deduced by exploiting equivalence relations between other elements (e.g., a concept C_1 is subsumed by all subsumers of C_2, if C_1 is equivalent with a concept C_2), in the extreme cases where no equivalence relations exist, this can not be done. In any case, we conjecture that the discovery of subsumption relations between elements of different ontologies can enhance the discovery/filtering of equivalence relations, and vise-versa, augmenting the effectiveness of our ontology alignment and merging methods. This is of great importance when dealing with real-world ontologies, where, as it is also stated in the conclusions of the Consensus Track of OAEI 06 [2], current state of the art systems "confuse" subsumption relations with equivalence ones.

To make the above claims more concrete, let us consider the ontologies depicted in Fig. 1. These specify the concept Citation in O_1 (which is equivalent to the concept Reference in O_2), and Publication in O_2 (which is equivalent to the concept Work in O_1). Each of these ontologies elaborates on the specification of distinct concepts: O_2 elaborates on the concept Publication and O_1 on the concept Citation. Furthermore, as shown in Fig.1, concepts are related among themselves via object properties whose lexicalizations differ: For instance, in O_2, the concept Reference is related via the object property of with the concept Publication, while in O_1, the corresponding concept Citation is related via the object property to with the concept Work. Given these ontologies, and given that equivalent properties in the two ontologies do not have the same lexicalization, and that non-equivalent concepts do have the same lexicalization, we may distinguish two cases:

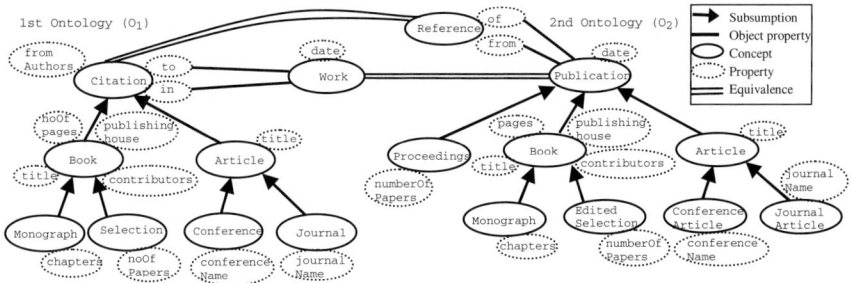

Fig. 1. Example ontologies for assessing the subsumption relation between concepts

In case that the equivalencies between the concepts of the two ontologies are not known, conclusions concerning subsumption relations between the concepts of the two ontologies cannot be drawn by a reasoning mechanism: This case clearly shows the need to discover both equivalence and subsumption relations between the concepts of the source ontologies.

In case the equivalences between the concepts of these two ontologies are known (the automatic discovery of these equivalencies is not a trivial task), one may deduce subsumption relations between the subsumees of these concepts. Specifically, a reasoning engine shall deduce that concepts that share the same lexicalization and the same properties are equivalent, which is wrong in our case: For instance, given that the properties of the concept Book of O_1 are pair wise equivalent to the properties of the concept Book of O_2, a reasoning service may wrongly assess that the concept Book of O_1 is equivalent to the concept Book of O_2: However Book from O_1 specifies book citations, while Book from O_2 specifies book publications. Therefore, it seems that although the discovery of subsumption relations among the elements of distinct ontologies must be done with respect to the known equivalence relations, a reasoning mechanism does not suffice for this purpose.

What is clearly needed is a method that shall discover subsumption relations between concept pairs of two distinct ontologies, separately from subsumptions and equivalencies that can be deduced by a reasoning mechanism. For instance, the concept Book of O_1 (respectively, O_2) must be assessed to be subsumed by the concept Reference (respectively, Work) of O_2 (respectively, O_1), without assessing that it is equivalent to the concept Book of this ontology, even if their properties and labels are identical. By admitting the later wrong mapping, numerous wrong subsumption relations can be deduced, e.g., that Book in O_1 is subsumed by the Publication in O_2 (Book specifies book citations, while Publication the publications themselves).

This paper deals with discovering subsumption relations between concepts of two distinct ontologies. This is done by using the "Classification-Based Learning of Subsumption Relations for the Alignment of Ontologies" (CSR) method. CSR computes subsumption relations between concept pairs of two distinct ontologies by means of a classification task, using state of the art supervised machine learning methods. The classification mechanisms proposed exploit two types of concepts' features: Concept properties, and terms extracted from labels, comments, properties and instances of concepts. Specifically, given a pair of concepts from the two source ontologies, the classification method "locates" a hypothesis concerning concepts' relation, which best fits to the training examples [3], generalizing beyond them. Concept pairs are represented as feature vectors of length equal to the number of the distinct features of source and target ontologies: In case features correspond to concept properties, properties that are equivalent (i.e., properties with equivalent meaning) correspond to the same vector component. In case features are terms, then terms with the same surface appearance correspond to the same vector component. It must be pointed that the examples for the training of the classifiers are being generated by exploiting the known subsumption and equivalence relations in both source ontologies, considering each source ontology in isolation.

The machine learning approach has been chosen since (a) there are no evident generic rules *directly* capturing the existence of a subsumption relation between ontology elements, and (b) concept pairs of the same ontology provide examples for the subsumption relation, making the method self-adapting to idiosyncrasies of specific domains, and non-dependant to external resources. The conjecture is that, if the supervised learning method generalizes successfully from the training examples, then the learned model shall capture the "patterns", in terms of the chosen features (i.e., properties or terms), for the discovery of subsumption relations that can not be deduced by a reasoning mechanism.

The rest of the paper is structured as follows: Section 2 states the problem and presents works that are most closely related to our approach. Section 3 provides background knowledge concerning the learning and classification methods used. Section 4 presents the proposed classification-based method for subsumption discovery. Section 5 presents and thoroughly discusses the experimental settings, as well as the results. Section 6 concludes the paper by pointing out the key points of our method and sketching further work for the improvement of the method.

2 Problem Statement and Related Work

2.1 Problem Statement

An ontology is a pair $O=(S, A)$, where S is the ontological signature describing the vocabulary (i.e., the terms that lexicalize ontology elements) and A is a set of ontological axioms, restricting the intended meaning of the terms included in the signature. In other words, A includes the formal definitions of ontology elements that are lexicalized by natural language terms in S. Subsumption relations are ontological axioms included in A. Distinguishing between concepts and properties, we consider a partition of S comprising the sets S_p and S_c, denoting the sets of terms lexicalizing ontology properties and ontology concepts, respectively. Let also T be the set of distinct terms that are in S, or that are extracted from labels, comments or instances of ontology elements.

Ontology mapping from a source ontology $O_1=(S_1,A_1)$ to a target ontology $O_2=(S_2,A_2)$ is a morphism $f:S_1 \rightarrow S_2$ of ontological signatures specifying elements' equivalences, such that $A_2 \vDash f(A_1)$, i.e., all interpretations that satisfy O_2's axioms also satisfy O_1's translated axioms. However, considering different types of relations between ontology elements, the ontology mapping problem can be stated as follows: Classify any pair (c^1,c^2) of elements of the input ontologies, such that c^i is a term in S_i, $i=1,2$, to any of the following relations, consistently: equivalence ($\equiv$), subsumption (inclusion) ($\sqsubseteq$), mismatch ($\bot$) and overlapping ($\sqcap$). By doing this, ontologies O_1 and O_2 can be aligned, resulting to a new consistent and coherent ontology.

In this paper we deal with the *subsumption computation problem* which, given the above generic problem, is as follows: Given (a) a source ontology $O_1=(S_1,A_1)$ and a target ontology $O_2=(S_2,A_2)$ such that $S_1=S_{1c} \cup S_{1p}$ and $S_2=S_{2c} \cup S_{2p}$, (b) the set $T_1 \cup T_2$ of distinct terms that appear in both ontologies (considering terms with the same surface appearance to be "equivalent" in meaning), and optionally (c) a morphism $f:S_{1p} \rightarrow S_{2p}$ from the lexicalizations of the properties of the source ontology to the lexicalizations of the properties of the target ontology (specifying properties' equivalences), classify each pair (c^1,c^2) of concepts, where c^1 is a term in S_{1c} and c^2 is a term in S_{2c}, to two distinct classes: To the "subsumption" ($\sqsubseteq$) class, or to the class "R". The latter class denotes pairs of concepts that are not known to be related via the subsumption[1] relation, or that are known to be related via the equivalence, mismatch or overlapping ones.

2.2 Related Work

Due to the evolving nature of ontologies, to the large number of elements that they comprise, and to the importance of the ontology alignment task, there are many research efforts towards automating this task. The majority of these methods focus on discovering equivalence relations between ontology elements [1] (e.g., concepts and properties). As a result, there has been a dramatic increase in the efficacy and efficiency of the methods that locate equivalences among ontology elements, while subsumption relations have not been thoroughly studied.

Concerning the computation of subsumption relations, related works have strong dependence on external resources, such as WordNet, domain ontologies or text corpora. A limitation that does not apply in the method proposed in this paper.

The method proposed in [4] transforms the mapping problem into a satisfiability problem, by taking into account the hierarchical relations between WordNet senses, along with the lexical and structural knowledge of the input ontologies.

Another related approach [5] introduces the WordNet Description Logics (WDL) language so as to align two different ontologies. WordNet is treated as an intermediate

---

[1] This means that a pair of concepts belonging to "R" may belong to the subsumption relation.

ontology. Similarly, in [6], [7] the authors propose the exploitation of background knowledge in the form of domain ontologies.

The authors in [8] loosen the formal constraints of the subsumption relation by exploiting hits returned by Google. Two more Google-based approaches [9], [10] exploit the so called Hearst patterns and test their validity by exploiting the returned hits.

Most machine learning based approaches aim to the discovery of equivalence relations between ontology elements and do not deal with subsumption relations as we do in this work. To the best of our knowledge, the most relevant machine learning technique is presented in [11]. Specifically, the authors propose a method based on Implication Intensity theory, which is a probabilistic model of deviation from statistical independence. The method takes as input a hierarchy of concepts and a set of documents, each one indexed under a specific concept. Then, the proposed model is applied in order to locate strong derivations between sets of terms that appear in the documents and as a consequence between their indexed concepts.

In this paper we consider the subsumption computation problem as a binary classification problem, where a classifier has to assess whether a pair of concepts belongs to the subsumption relation. As it will be explained, the semantics of the input ontologies are exploited in order the method to generate the appropriate examples for the training of the classifier. This makes the proposed method dependent only from the source ontologies and independent from any third/external domain resource.

3 Classification and Inductive Learning

Classification is one of the main problems addressed within the machine learning discipline. It concerns the classification of example cases into one of a discrete set of classes. When the number of classes is restricted to two, the problem is referred to as a binary classification problem. More accurately, the binary classification problem is defined as follows: Given a set of m examples (x^j, y^j), j=1, 2, ... , m (the training dataset) of vectors x^j sampled from some distribution D, the *output* is a function $c:R^n \rightarrow \{0,1\}$ (classifier) which classifies additional samples x^k sampled from the same distribution D to the classification classes $\{0,1\}$. It holds that $x^j \in R^n$ and $y^j \in \{0,1\}$. The i-th component of vector x^j is termed the *feature i*, X_i^j of the x^j sample.

In the context of studying the subsumption computation problem, we have used specific implementations of well studied classifiers: (a) *Probabilistic classifiers* specify the function c as a probabilistic function, assessing the probability $p(x^j, y^j)$ that the document x^j falls within a category y^j. From this category we have selected a Naïve Bayes (Nb) classifier. (b) *Memory-based classifiers* store the training data in memory and when a new instance is encountered, similar instances are retrieved from their memory and used for the instance classification. We used the k-nearest neighbor (Knn) with value of k=2. (c) *Support Vector Machines (SVMs)* based classifiers map input vectors to a higher dimensional space where a maximal separating hyperplane is constructed. The transformation of the data to the new space is made through functions called kernels. We have selected the libSVM [12] implementation with its default values and radial basis function as kernel. (d) *Decision Tree classifiers* exploit a tree structure in which each interior node corresponds to a feature. The branch from a node to a child represents a possible value of that feature, and a leaf node represents the classification class given the values of the features represented by the path from the root. Weka's j48 [3] is the implementation of the widely used state of the art C4.5 decision tree learning algorithm that we have used in this work.

4 The *CSR* Method

4.1 Description of the Overall *CSR* Method

The discrete steps of the *CSR* method, as depicted in Fig. 2, are the following:

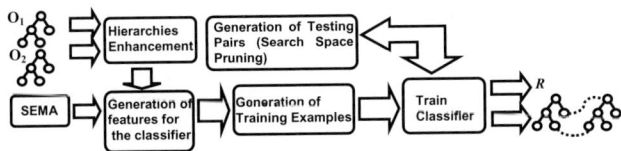

Fig. 2. Overview of the *CSR* method

- Reasoning services are being used for inferring all facts according to ontologies' specification semantics: The objective of this step is to compute implied subsumption and equivalence relations between existing ontology elements. This is a necessary step as it affects the generation of the training dataset (subsection 4.3).
- Currently *CSR* exploits two types of features: Concepts' properties and terms appearing in the "vicinity" of concepts. In both cases features are generated by gathering all discrete properties or terms from both ontologies. This is further detailed in the next subsection.
- The sets of training examples are being generated according to the rules defined in subsection 4.3. The balancing of the training dataset is an important issue that is being tackled in this step, as well.
- The classifier is being trained using the training dataset, and
- Concept pairs are being classified by the trained classifier, pruning the search space according to the method explained in subsection 4.4.

Studying the importance of concepts' properties to assessing the subsumption relation between concepts (a) appeals to our intuition concerning the importance of properties as distinguishing characteristics of concepts, (b) it provides the basis for a method considering only properties' equivalences. This basic method can be further enhanced with the computation of equivalences between other concepts' distinguishing features (e.g., concepts in a given vicinity), and can be further combined with other alignment methods. As far as the use of terms is concerned, (a) their use for describing the intended meaning of concepts appeals to our intuition, and (b) it does not necessitate the use of any method for the discovery of equivalence relations among ontology elements. This paper studies the potential of *CSR* with these two types of features, while leaving further enhancements and combinations for future work.

4.2 Features Generation

Each pair of concepts (c^1, c^2) is represented by a feature vector whose components' values are as follows:

- "0", if the corresponding feature does not appear neither in c^1 nor in c^2.
- "1", if the corresponding feature appears only in c^1.
- "2", if the corresponding feature appears only in c^2.
- "3", if the corresponding feature appears in both c^1 and c^2.

As it can be noticed, feature vectors are not identical for symmetrical pairs of concepts. This allows the computation of the direction of the subsumption relation. In the case where properties are being used as features of concept pairs, given the equivalences among ontology properties computed by a morphism f, equivalent properties correspond to the same component of concepts' feature vectors. To compute properties' equivalences, we have used the SEMA [13] mapping tool which has been evaluated in the OAEI 2007 contest [14]. Towards discovering mappings between properties SEMA exploits: (a) Lexical information concerning names, labels and comments of ontologies' properties. (b) Properties' domain, range and hierarchy for propagating similarities among properties. Therefore, we emphasize that by "property appearance" we do not mean the occurrence of the property's lexicalization, but the occurrence of property's meaning assessed by SEMA.

In the case where terms are being used for the representation of concepts' pairs, terms are being extracted from both ontologies. Each distinct term corresponds to a specific component of the feature vector and the length of the vector is equal to the total number of all distinct terms from both input ontologies. Specifically, for each concept of the input ontologies, terms are being extracted from its "vicinity", as it is specified by the following rule: Given a concept, the method extracts terms occurring in the local name, label and comments of this concept, from all of its properties (exploiting the properties' local names, labels and comments), as well as from all of its related concepts. Finally, terms from all instances of the corresponding concept are being extracted.

By exploiting the equivalence and disjoint relations between ontology elements, the conjunction and disjunction constructors, the appearance of a term in an element is determined by the following rules: (i) Given an ontology element, the method considers terms appearing in all its equivalent elements. (ii) If the corresponding element is defined as the disjunction (conjunction) of other elements, then the method unions (respectively, intersects) the sets of terms that appear in the constituent elements. (iii) If two elements are defined to be disjoint, then the method considers these terms that are not common in both elements.

During this step, tokenization, stemming and elimination of stop words is performed on the set of extracted words.

For example, according to the above, given the ontologies in Fig. 1 and the properties' equivalences of≡to, from≡in, pages≡noOfPages, numberOfPapers≡ noOfPapers provided by SEMA, the concept pair (Citation,Publication) is represented by the feature vector (0, 3, 3, 0, 0, 0, 0, 0, 0, 0, 0, 1). The features are date, of≡ to, from≡in, pages≡noOfPages, publishingHouse, numberOfPapers≡ noOfPapers, title, contributors, chapters, conferenceName, journalName and fromAuthors according to the order of their appearance. Concerning case where features are terms, the feature vector is (.., 1, 3, 1, 1, 1, 1, 3, 1, 1, 3, 3, 3, 3, 3, 3, 2, 2, 2, 2,..). The features in order of appearance are citation, from, authors, article, to, in, date, work, no, of, pages, publishing, house, book, title, contributors, publication, proceedings, number, and reference. All the vector components that are not shown correspond to terms that do not appear in the vicinity of any of the two concepts, so their value is set to 0.

4.3 Creating the Training Dataset

As it has been stated, training examples for classes "⊑" and R are being generated by exploiting the source and target ontologies, according to the semantics of specifications. The basic rules for the generation of the training examples for the class "⊑" are as follows:

Subsumption Relation. Include all concept pairs from both input ontologies that belong in the subsumption relation. The subsumption relation may or may not be direct. If more than one hierarchy is specified, then all hierarchies need to be exploited.

Equivalent concepts. Enrich the set of concept pairs generated by the above rule, by taking into account stated and inferred equivalence relations between concepts. In detail, for each concept pair (c^1,c^2) that belongs in the subsumption relation, and for each stated equivalence relation $c^i \equiv c^i_k$, $i \in \{1,2\}$, $k=1,2,\ldots$, then the pair (c^1, c^2_k) (or the pair (c^1_k, c^2)) belongs to the subsumption relation, as well.

Union of concepts. Enrich the set of pairs by exploiting the union construct in the definition of concepts: When one concept (e.g., the concept $C_4 \sqcup C_5$) is constructed as the union of others, and it is defined to be subsumed by another concept (e.g., by the concept C_2), then each concept in the union is subsumed by the more general one (i.e., it holds that $C_4 \sqsubseteq C_2$ and $C_5 \sqsubseteq C_2$). By taking into account also the equivalence rule (e.g., $C_4 \equiv C_3$), the concept C_4 can be substituted by its equivalent concept, and therefore, the pair (C_3, C_2) is included as well.

According to the open world semantics, we need to exploit the stated axioms for the generation of training examples: Therefore, in case there is not an axiom that specifies the subsumption relation between a pair of concepts (or in case this relation can not be inferred by exploiting the semantics of specifications), then this pair does not belong to the subsumption class and it is included in the generic class "R". The following cases summarize the rules for the generation of examples for the class "R":

Concepts belonging to different hierarchies. If two concepts belong to different hierarchies of the same ontology, then no explicit subsumption relation is defined among them. As a result, all pairs following this rule are characterized as training examples of the class "R". This set of pairs can be enriched by taking into account the stated equivalence and union relations between concepts, as explained in the case of class "⊑".

Siblings at the same hierarchy level. This includes pairs of concepts that are siblings (share the same subsumer) and that are not related via the subsumption relation. As a result, all possible pairs following this rule are characterized as training examples of the class "R". Similarly to the first category, this category can also be enriched by exploiting concepts' equivalences and unions.

Siblings at different hierarchy levels. If any concept that is in a pair belonging in the "siblings of the same hierarchy level" category is substituted by any of its subsumees, then new pair examples are recursively generated, until the leaf concepts of the ontology are reached. These examples constitute a new category called "siblings at different hierarchy levels". Similarly with the previous categories, this one also can be enriched by exploiting the union construct of concepts and the equivalence relation between concepts.

Concepts related thought a non-subsumption relation. This includes concepts that are related via an object property and are not related with a subsumption relation. As with the previous categories, this category may also be enriched by considering unions and equivalences between concepts.

Inverse pairs of class "⊑". All concepts pairs (C_2,C_1) such that C_1 subsumes C_2, but it cannot be inferred that C_2 subsumes C_1, constitute examples for the class "R".

As it is evidenced by the above, the number of training examples for the class "⊑" are much less than the ones for class "R". It is very important for the performance of the classifier that the training examples for both classes to be balanced in numbers.

Being balanced in numbers, we intend that the two classes are equally represented in the training dataset. This is referred as the *dataset imbalance* problem. In the context of the classification task, various techniques have been proposed towards its solution [15]. In this work, to tackle this problem, we have adopted two alternatives: The under-sampling and the over-sampling methods.

According to the under-sampling method, all different categories of class "R" are equally sampled (randomly), until the selected examples are equal in numbers with the ones of class "⊑". In the case of over-sampling, the method selects examples for the class "⊑" randomly, until the two classes have the same number of examples.

4.4 Pruning the Search Space

Taking into account the semantics of the subsumption relation, instead of generating all possible concept pairs from both ontologies, we prune the search space by excluding pairs of concepts for which a subsumption relation can not be assessed to hold, due to the existent and currently computed relations. First we provide two short definitions: A *root concept* is every concept of the ontology that does not have a subsumer. *Root concepts* may not have sub-concepts, hence are called *unit concepts*. We consider that an ontology may include more than one subsumption hierarchies for concepts.

In order to prune the search space, the proposed algorithm firstly checks all the concepts from the first ontology and unit/root concepts of the second ontology. If a pair is not classified in the class "⊑", then the hierarchy rooted by the corresponding concept of the second ontology is not being examined by the classifier. If a pair is assessed to belong to the class "⊑", then the concept of the first ontology is recursively being tested with the direct subsumees of the corresponding concept in the second ontology, until either a pair is assessed to belong in the class "R", or until the leaf concepts are reached.

5 Experimental Results and Discussion

5.1 The Dataset

The testing dataset has been derived from the benchmarking series of the OAEI 2006 contest [14]. As our method exploits the properties of concepts (in cases where properties are used as concept pairs' features), we do not include those OAEI 2006 ontologies where concepts have no properties. The compiled corpus is available at the URL http://www.icsd.aegean.gr/incosys/csr. For each pair of ontologies we have created the gold standard ontology, including subsumption relations among concepts.

All benchmarks (101-304) except those in categories *R1-R4* (real-world cases), define the second ontology of each pair as an alteration of the same first. The benchmarks can be categorized based on their common features as follows: (a) in categories *A1-A5* (101-210, 237, 238 and 249), elements' lexicalizations of the target ontologies are altered in various ways (e.g., uppercasing, underscore, foreign language, synonyms or random strings), (b) in categories *A6-A7* (225 and 230) restrictions are removed and/or properties are modeled in more detail and/or the hierarchy is flattened, (c) in categories *F1-F2* (222, 237, 251 and 258) the hierarchies are flattened and in *F2* also random lexicalizations of all elements are introduced, and (d) categories *E1-E2* (223, 238, 252 and 259) result from *F1-F2* with expanded hierarchies.

5.2 Experiments and Results

Results show the precision and recall of the proposed method as it is applied in the different types of ontology pairs specified in subsection 5.1. Precision is the ratio *#correct_pairs_computed/#pairs_computed* and recall is the ratio *#correct _pairs _ computed/#pairs_in_gold_standard*.

We have run experiments for the benchmark series specified using each of the classifiers: C4.5, Knn, NaiveBayes (Nb) and Svm. For each of the classifies we have run four experiments using terms or properties as features of concept pairs, in combination with the dataset balancing method: over and under-sampling. Subsequently, we denote each type of experiment with X+Y+Z, where X is the classifier, Y is the type of features used ("Props" for properties or "Terms" for terms) and Z is the type of dataset balancing method used ("over" and "under" for over- and under-sampling). For instance, the experiment type "C4.5+Props+Over" indicates the use of the C4.5 classifier in *CSR*, exploiting properties as features, with over-sampling for balancing the training dataset.

Furthermore, the results of our method are compared to the results of a baseline classifier, which is based on the Boolean Existential Model. This classifier does not perform any kind of generalization: In order to classify a testing concept pair, it consults the vectors of the training examples of the class "⊑", and selects the first exact match. The comparison with this classifier has been performed for showing how *CSR* classifiers generalize over the training examples, learning subsumption cases not present in the training examples. Here we have to point out that both *CSR* and the baseline classifier exploit the same information. As terms or properties are being used as features, two different types of experiments have been conducted using the baselines classifier: The one with properties (Baseline+Props) and the other with terms (Baseline+Terms).

To investigate whether, given a set of equivalence relations, a reasoning mechanism suffices for the purpose of computing subsumption relations among the elements of distinct ontologies we also compare *CSR* with a Description Logics' reasoning engine[2]. In order for the reasoner to be able to infer subsumption relations between concepts of the source ontologies we specify axioms concerning only properties' equivalencies

---

[2] We have used Pellet in our experiments (http://pellet.owldl.com).

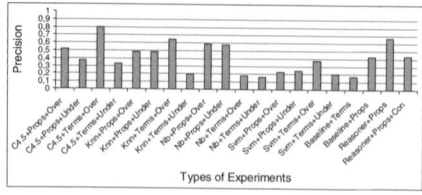

Fig. 3. Overall precision per experiment

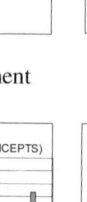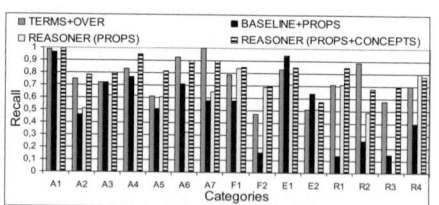

Fig. 4. Overall recall per experiment

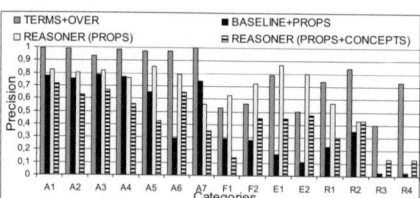

Fig. 5. Precision in all test categories

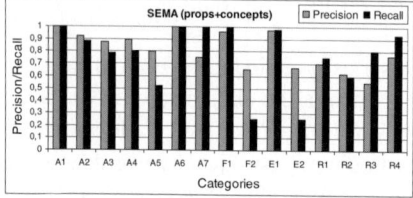

Fig. 6. Recall in all test categories

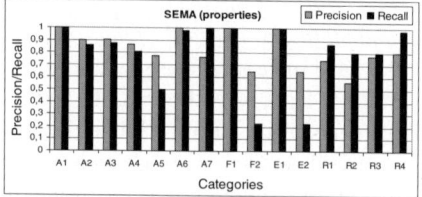

Fig. 7. SEMA's overall performance

Fig. 8. SEMA for properties

(Reasoner+Props), or alternatively, both properties' and concepts' equivalencies (Reasoner+Props+Con). At this point we must recall that when CSR exploits terms, then no equivalence mappings are required.

Fig. 3 and Fig. 4 depict the average precision and recall values in all types of experiments. The first important observation by looking at Fig. 3 and Fig. 4 is (as it was expected) that a reasoner cannot infer all the subsumption relations between concepts of the input ontologies. Especially, for the Reasoner+Props+Con type of experiments, the reasoner infers many false positives (precision: 41%). On the other hand, when only property mappings are exploited (Reasoner+Props) the reasoner achieves a low recall value (58%). Even in the case where the precision and recall of the equivalence mappings produced by SEMA are 100% (A1 category, Fig. 7 and Fig. 8) and the two ontologies are almost the same (only some minor axioms are suppressed in the second ontology), the reasoner achieves precision 82% and 71% depending of the type of equivalencies considered (Fig. 5). The same applies in Fig. 6 for the A1 ontologies. These results provide firm evidence that a reasoning mechanism does not suffice for the purpose of computing subsumption relations among the elements of distinct ontologies, even if equivalence relations are computed with high precision.

This conjecture is further evidenced by the results achieved in the real world cases (R1 to R4) shown in Fig. 5 and Fig. 6. Indeed, the reasoner, in both types of experiments in

R3 and R4 achieves low precision in a moderate recall. Especially, the case R4 is quite interesting, as the precision and recall of SEMA is quite high (Fig. 7 and Fig. 8). The same applies in cases F1 and A7: In F1 the precision and recall of SEMA is almost 100% (Fig. 7 and Fig. 8). In this case the reasoner exploiting both properties and concepts equivalences (Reasoner+Props+Con) achieves precision 13%, and recall 85%; the reasoner exploiting properties equivalences (Reasoner+Props) achieves precision 62% and recall 85%.

Furthermore, as Fig. 4 shows, nearly all types of experiments (with the exception of the SVM) contacted with the CSR method achieve a better recall than the baseline classifiers (Baseline+Props and Baseline+Terms). This means that classifiers do generalize, as they manage to locate subsumptions that are not in their training dataset. Moreover, for each of the classifiers there is a type of experiment in which CSR performs better than the baseline classifier in terms of precision (with the exception of the SVM). A fact which is very important, since it shows that in these cases overgeneralization does not take place.

As Fig. 3 shows, the C4.5 classifier exploiting terms with over-sampling (C4.5+Terms+Over) outperforms all classifiers in terms of precision. Also, as shown in Fig. 4, the same classifier achieves one of the highest recalls: Therefore, subsequently we shall focus on these type of experiments with C4.5.

The fact that C4.5+Terms+Over has the best overall performance in terms of precision and recall among all classifiers can be explained by the specific features of decision tree classifiers: (i) Disjunctive descriptions of cases, an inherent feature of decision trees, fits naturally to the subsumption computation problem. This is true since more than one features may indicate whether a specific concept pair belongs in the class "⊆". (ii) Decision trees are very tolerant to errors in the training set [3]. This is true as far as the training examples, as well as the values of vector components for the representation of examples are concerned. In our case, the values of vector components may not be correct as the task for the discovery of equivalencies among properties is erroneous.

If we compare the results achieved by the best CSR classifier that exploits terms as features (C4.5+Terms+Over) to the results achieved by the reasoner that exploits the least possible input (i.e., reasoner with properties' equivalencies), as shown in Fig.3 and Fig.4, CSR performs better in terms of recall and precision. Indeed, in this case C4.5 achieves the best balance between precision (80%) and recall (78%), than any other method in the experiments, although it does not exploit equivalence mappings.

By observing Fig. 5 and Fig. 6 we see that in cases where the source ontologies differ substantially (e.g., cases A7, R1-R4) CSR with C4.5 exploiting terms and over sampling (C4.5+Terms+Over) not only has the higher precision, but also is among the highest in terms of recall. In any case CSR achieves a good balance between recall and precision. Also, C4.5+Terms+Over performs better than the baseline classifier, generalizing beyond the training examples. Furthermore, in these cases it performsbetter than the reasoner, which means that it locates subsumptions that cannot be inferred by using the equivalence relations produced by SEMA (results in category A7 in Fig. 5 and Fig. 6 are very depictive).

Here we have to comment about categories E1, E2, F1, and F2: Although the conceptualizations of the target ontologies differ from the source, there is a special detail that highly favors the reasoner: In F1 and E1 the concepts' hierarchy is flattened or expanded, respectively, but the initial concepts (along with their properties and

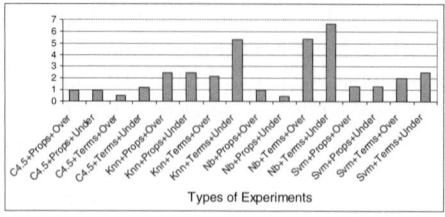 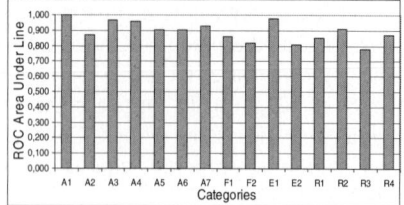

Fig. 9. Confused Equivalencies of CSR **Fig. 10.** ROC areas under line values

restrictions) defined in the first ontology remain almost unchanged in the second ontology. This means that the reasoner can relatively easily infer subsumptions through the equivalence relations returned by SEMA, in conjunction with the subsumption relations defined in each ontology hierarchy. Furthermore, the newly introduced concepts in E1 and E2 have no defined properties at all, a fact that lowers the discriminating ability of CSR when it uses properties as features. The same applies to categories F2 and E2, but now the lexical information is suppressed in the second ontology.

In Fig. 9 we present the number of relations computed, which, contrary to what CSR assesses, are equivalence rather than subsumption relations. As it is shown, C4.5 which is the best performing classifier has an average of one or less in all types of experiments. This is a really important feature of CSR, as it can perform a "filtering" in the results of any mapping system that locates equivalence relations [2].

To further assess the quality of the classifier with the best overall results, we performed ROC analysis (Fig. 10). In our case, ROC analysis indicates the "goodness" of the classifier in classifying testing examples in the distinct classes "⊑" and "R". It is generally accepted that values ranging in [0.5, 0.6] indicate a failure in the classification task, values ranging in [0.6, 0.7] indicate a poor classifier, values ranging in [0.7, 0.8] indicate a fair classifier, values ranging in [0.8, 0.9] indicate a good classifier and finally values in [0.9, 1.0] indicate an excellent classifier.

By examining the ROC area under line values of the CSR method with C4.5 in all types of experiments, it is obvious that the classifier is always "good" and in the majority of the test cases (8/15) can be characterized as "excellent". It must be stated that these values depict that, although the performance of the classifier in the class "R" is of no evident interest for the ontology alignment problem, as in these cases the classifier cannot decide, the CSR method performs even better there.

6 Conclusions and Future Work

In this paper we propose the CSR method. CSR aims to the computation of subsumption relations between concept pairs of two distinct ontologies by exploiting properties' equivalence mappings, as well as appearances of terms in concepts' vicinity. Towards this goal, CSR assesses whether concept pairs of the source ontologies belong to the subsumption relation by means of a classification task using state of the art machine learning methods. Given a pair of concepts from two ontologies, the objective of CSR is to identify patterns of concepts' features (properties or terms) that provide evidence for the subsumption relation among these concepts. For the learning of the classifiers, the proposed method generates training datasets from the source ontologies specifications, tackling also the problem of imbalanced training datasets.

Experimental results show the potential of the method: CSR generalizes effectively over the training examples, showing (a) the importance of both properties and terms to assessing the subsumption relation between concepts of discrete ontologies (b) the importance of incorporating more precise property mapping methods into the process, (c) the potential to further improve the method via the incorporation of more types of features, via the combination of different types of features or via its combination with other methods.

Lastly, it must be pointed that CSR manages to discriminate effectively between equivalence and subsumption relations. This is a really important feature of CSR, as it can be used for filtering equivalences computed by other alignment methods [2].

Acknowledgments. This research project is co-financed by E.U.-European Social Fund (75%) and the Greek Ministry of Development-GSRT (25%).

References

1. Shvaiko, P., Euzenat, J.: A survey of schema-based matching approaches. In: Spaccapietra, S. (ed.) Journal on Data Semantics IV. LNCS, vol. 3730, pp. 14–171. Springer, Heidelberg (2005)
2. Svab, O., Svatek, V., Stuckenschmidt, H.: A Study in Empirical and 'Casuistic' Analysis of Ontology Mapping Results. In: Franconi, E., Kifer, M., May, W. (eds.) ESWC 2007. LNCS, vol. 4519, Springer, Heidelberg (2007)
3. Mitchell, T.: Machine Learning. The McGraw-Hill Companies, Inc, New York (1997)
4. Giunchiglia, F., Yatskevich, M., Shvaiko, P.: Semantic Matching: Algorithms and implementation. Journal on Data Semantics IX (2007)
5. Bouquet, P., Serafini, L., Zanobini, S., Sceffer, S.: Bootstrapping semantics on the web: meaning elicitation from schemas. In: WWW, Edinburgh, Scotland (2006)
6. Aleksovski, Z., Klein, M., Kate, W., Harmelen, F.: Matching Unstructured Vocabularies Using a Background Ontology. In: Staab, S., Svátek, V. (eds.) EKAW 2006. LNCS (LNAI), vol. 4248, Springer, Heidelberg (2006)
7. Gracia, J., Lopez, V., D'Aquin, M., Sabou, M., Motta, E., Mena, E.: Solving Semantic Ambiguity to Improve Semantic Web based Ontology Matching. In: Ontology Matching Workshop, Busan, Korea (2007)
8. Risto, G., Zharko, A., Warner, K.: Using Google Distance to weight approximate ontology matches. In: WWW, Banff, Alberta, Canada (2007)
9. Hage, W.R., Van Katrenko, S., Schreiber, A., T.: A Method to Combine Linguistic Ontology Mapping Techniques. In: Gil, Y., Motta, E., Benjamins, V.R., Musen, M.A. (eds.) ISWC 2005. LNCS, vol. 3729, Springer, Heidelberg (2005)
10. Cimiano, P., Staab, S.: Learning by googling. In: SIGKDD Explor., Newsl., USA (2004)
11. Jerome, D., Fabrice, G., Regis, G., Henri, B.: An interactive, asymmetric and extensional method for matching conceptual hierarchies. In: EMOI – INTEROP Workshop, Luxembourg (2006)
12. Chang, C.-C., Lin, C.-J.: LIBSVM: a library for support vector machines (2001)
13. Spiliopoulos, V., Valarakos, A.G., Vouros, G.A., Karkaletsis, V.: SEMA: Results for the ontology alignment contest OAEI 2007. In: Ontology Matching Workshop, OAEI, Busan, Korea (2007)
14. Ontology Alignment Evaluation Initiative, http://oaei.ontologymatching.org/
15. Japkowicz, N.: The Class Imbalance Problem: Significance and Strategies. In: ICAI, Special Track on Inductive Learning, Las Vegas, Nevada (2000)

XSPARQL: Traveling between the XML and RDF Worlds – and Avoiding the XSLT Pilgrimage*

Waseem Akhtar[1], Jacek Kopecký[2], Thomas Krennwallner[1], and Axel Polleres[1]

[1] Digital Enterprise Research Institute, National University of Ireland, Galway
{firstname.lastname}@deri.org
[2] STI Innsbruck, University of Innsbruck, Austria
jacek.kopecky@uibk.ac.at

Abstract. With currently available tools and languages, translating between an existing XML format and RDF is a tedious and error-prone task. The importance of this problem is acknowledged by the W3C GRDDL working group who faces the issue of extracting RDF data out of existing HTML or XML files, as well as by the Web service community around SAWSDL, who need to perform lowering and lifting between RDF data from a semantic client and XML messages for a Web service. However, at the moment, both these groups rely solely on XSLT transformations between RDF/XML and the respective other XML format at hand. In this paper, we propose a more natural approach for such transformations based on merging XQuery and SPARQL into the novel language XSPARQL. We demonstrate that XSPARQL provides concise and intuitive solutions for mapping between XML and RDF in either direction, addressing both the use cases of GRDDL and SAWSDL. We also provide and describe an initial implementation of an XSPARQL engine, available for user evaluation.

1 Introduction

There is a gap within the Web of data: on one side, XML provides a popular format for data exchange with a rapidly increasing amount of semi-structured data available. On the other side, the Semantic Web builds on data represented in RDF, which is optimized for data interlinking and merging; the amount of RDF data published on the Web is also increasing, but not yet at the same pace. It would clearly be useful to enable reuse of XML data in the RDF world and vice versa. However, with currently available tools and languages, translating between XML and RDF is not a simple task.

The importance of this issue is currently being acknowledged within the W3C in several efforts. The Gleaning Resource Descriptions from Dialects of Languages [9] (GRDDL) working group faces the issue of extracting RDF data out of existing (X)HTML Web pages. In the Semantic Web Services community, RDF-based client software needs to communicate with XML-based Web services, thus it needs to perform transformations between its RDF data and the XML messages that are exchanged with

---

* This material is based upon works supported by the European FP6 projects inContext (IST-034718) and TripCom (IST-4-027324-STP), and by Science Foundation Ireland under Grant No. SFI/02/CE1/I131.

| ```
@prefix alice: <alice/> .
@prefix foaf: <...foaf/0.1/> .

alice.me a foaf:Person.
alice:me foaf:knows _:c.
_:c a foaf:Person.
_:c foaf:name "Charles".
``` | ```
<rdf:RDF xmlns:foaf="...foaf/0.1/"
  xmlns:rdf="...rdf-syntax-ns#">
<foaf:Person rdf:about="alice/me">
  <foaf:knows>
    <foaf:Person foaf:name="Charles"/>
  </foaf:knows>
</foaf:Person>
</rdf:RDF>
``` |
| (a) | (b) |
| ```
<rdf:RDF xmlns:foaf="...foaf/0.1/"
  xmlns:rdf="...rdf-syntax-ns#">
<rdf:Description rdf:nodeID="x">
  <rdf:type
      rdf:resource=".../Person"/>
  <foaf:name>Charles</foaf:name>
</rdf:Description>
<rdf:Description
     rdf:about="alice/me">
  <rdf:type
      rdf:resource=".../Person"/>
  <foaf:knows rdf:nodeID="x"/>
</rdf:Description>
</rdf:RDF>
``` | ```
<rdf:RDF xmlns:foaf="...foaf/0.1/"
  xmlns:rdf="...rdf-syntax-ns#">
<rdf:Description rdf:about="alice/me">
  <foaf:knows rdf:nodeID="x"/>
</rdf:Description>
<rdf:Description rdf:about="alice/me">
  <rdf:type rdf:resource=".../Person"/>
</rdf:Description>
<rdf:Description rdf:nodeID="x">
  <foaf:name>Charles</foaf:name>
</rdf:Description>
<rdf:Description rdf:nodeID="x">
  <rdf:type rdf:resource=".../Person"/>
</rdf:Description>
</rdf:RDF>
``` |
| (c) | (d) |

Fig. 1. Different representations of the same RDF graph

the Web services. The Semantic Annotations for WSDL (SAWSDL) working group calls these transformations *lifting* and *lowering* (see [12,14]). However, both these groups propose solutions which rely solely on XSL transformations (XSLT) [10] between RDF/XML [2] and the respective other XML format at hand. Using XSLT for handling RDF data is greatly complicated by the flexibility of the RDF/XML format. XSLT (and XPath) were optimized to handle XML data with a simple and known hierarchical structure, whereas RDF is conceptually different, abstracting away from fixed, tree-like structures. In fact, RDF/XML provides a lot of flexibility in how RDF graphs can be serialized. Thus, processors that handle RDF/XML as XML data (not as a set of triples) need to take different possible representations into account when looking for pieces of data. This is best illustrated by a concrete example: Fig. 1 shows four versions of the same FOAF (cf. http://www.foaf-project.org) data.[1] The first version uses Turtle [3], a simple and readable textual format for RDF, inaccessible to pure XML processing tools though; the other three versions are all RDF/XML, ranging from concise (b) to verbose (d).

The three RDF/XML variants look very different to XML tools, yet exactly the same to RDF tools. For any variant we could create simple XPath expressions that extract for instance the names of the persons known to Alice, but a single expression that would correctly work in all the possible variants would become more involved. Here is a list of particular features of the RDF data model and RDF/XML syntax that complicate XPath+XSLT processing:

– Elements denoting properties can directly contain value(s) as nested XML, or reference other descriptions via the `rdf:resource` or `rdf:nodeID` attributes.

---

[1] In listings and figures we often abbreviate well-known namespace URIs (http://www.w3.org/1999/02/22-rdf-syntax-ns#, http://xmlns.com/foaf/0.1/, etc.) with "...".

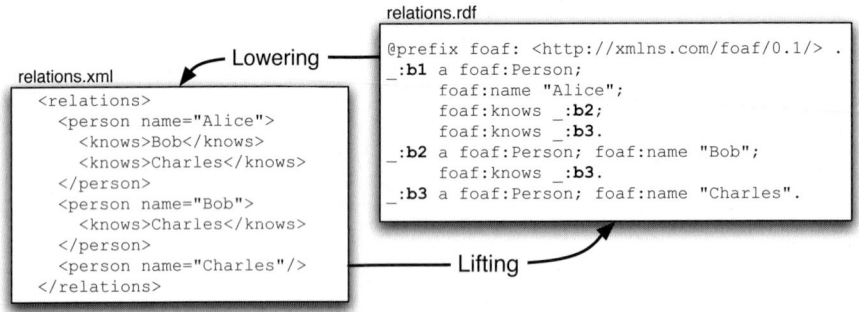

Fig. 2. From XML to RDF and back: "lifting" and "lowering"

- References to resources can be relative or absolute URIs.
- Container membership may be expressed as `rdf:li` or `rdf:_1`, `rdf:_2`, etc.
- Statements about the same subject do not need to be grouped in a single element.
- String-valued property values such as `foaf:name` in our example (and also values of `rdf:type`) may be represented by XML element content or as attribute values.
- The type of a resource can be represented directly as an XML element name, with an explicit `rdf:type` XML element, or even with an `rdf:type` attribute.

This is not even a complete list of the issues that complicate the formulation of adequate XPath expressions that cater for every possible alternative in how one and the same RDF data might be structured in its concrete RDF/XML representation.

Apart from that, simple reasoning (e.g., RDFS materialization) improves data queries when accessing RDF data. For instance, in FOAF, every Person (and Group and Organization etc.) is also an Agent, therefore we should be able to select all the instances of `foaf:Agent`. If we wanted to write such a query in XPath+XSLT, we literally would need to implement an RDFS inference engine within XSLT. Given the availability of RDF tools and engines, this seems to be a dispensable exercise.

Recently, two new languages have entered the stage for processing XML and RDF data: XQuery [5] is a W3C Recommendation since early last year and SPARQL [20] has finally received W3C's Recommendation stamp in January 2008. While both languages operate in their own worlds – SPARQL in the RDF- and XQuery in the XML-world – we show in this paper that the merge of both in the novel language XSPARQL has the potential to finally bring XML and RDF closer together. XSPARQL provides concise and intuitive solutions for mapping between XML and RDF in either direction, addressing both the use cases of GRDDL and SAWSDL. As a side effect, XSPARQL may also be used for RDF to RDF transformations beyond the capabilities of "pure" SPARQL. We also describe an implementation of XSPARQL, available for user evaluation.

In the following, we elaborate a bit more in depth on the use cases of lifting and lowering in the contexts of both GRDDL and SAWSDL in Section 2 and discuss how they can be addressed by XSLT alone. Next, in Section 3 we describe the two starting points for an improved lifting and lowering language – XQuery and SPARQL – before we announce their happy marriage to XSPARQL in Section 4. Particularly, we extend XQuery's **FLWOR** expressions with a way of iterating over SPARQL results.

```
<xsl:stylesheet
  xmlns:xsl="...XSL/Transform"
  xmlns:foaf="...foaf/0.1/"
  xmlns:rdf="...rdf-syntax-ns#"
  version="2.0">

<xsl:template match="/relations">
  <rdf:RDF>
    <xsl:apply-templates />
  </rdf:RDF>
</xsl:template>

<xsl:template match="person">
  <foaf:Person>
    <foaf:name>
      <xsl:value-of
          select="./@name"/>
    </foaf:name>
    <xsl:apply-templates/>
  </foaf:Person>
</xsl:template>

<xsl:template match="knows">
  <toaf:knows><foaf:Person>
    <foaf:name>
      <xsl:apply-templates/>
    </foaf:name>
  </foaf:Person></foaf:knows>
</xsl:template>

</xsl:stylesheet>
```

(a) mygrddl.xsl

```
<rdf:RDF xmlns:rdf="...rdf-syntax-ns#"
         xmlns:foaf="   foaf/0.1/">
<toaf:Person>
  <foaf:name>Alice</foaf:name>
  <foaf:knows><foaf:Person>
    <foaf:name>Bob</foaf:name>
  </foaf:Person></foaf:knows>
  <foaf:knows><foaf:Person>
    <foaf:name>Charles</foaf:name>
  </foaf:Person></foaf:knows>
</foaf:Person>
<foaf:Person>
  <foaf:name>Bob</foaf:name>
  <foaf:knows><foaf:Person>
    <foaf:name>Charles</foaf:name>
  </foaf:Person></foaf:knows>
</foaf:Person>
<foaf:Person>
  <foaf:name>Charles</foaf:name>
</foaf:Person>
</rdf:RDF>

@prefix toaf: <http://xmlns.com/foaf/0.1/>.
_:b1 a foaf:Person; foaf:name "Alice";
     foaf:knows _:b2; foaf:knows _:b3.
_:b2 a foaf:Person; foaf:name "Bob".
_:b3 a foaf:Person; foaf:name "Charles".
_:b4 a foaf:Person; foaf:name "Bob";
     foaf:knows _:b5 .
_:b5 a foaf:Person; foaf:name "Charles" .
_:b6 a foaf:Person; foaf:name "Charles".
```

(b) Result of the GRDDL transform
in RDF/XML (up) and Turtle (down)

Fig. 3. Lifting attempt by XSLT

We sketch the semantics of XSPARQL and describe a rewriting algorithm that translates XSPARQL to XQuery. By this we can show that XSPARQL is a conservative extension of both XQuery and SPARQL. A formal treatment of XSPARQL showing this correspondence is given in an extended version of this paper [1]. We wrap up the paper with an outlook to related and future works and conclusions to be drawn in Section 5 and 6.

2 Motivation – Lifting and Lowering

As a running example throughout this paper we use a mapping between FOAF data and a customized XML format as shown in Fig. 2. The task here in either direction is to extract for all persons the names of people they know. In order to keep things simple, we use element and attribute names corresponding to the respective classes and properties in the FOAF vocabulary (i.e., Person, knows, and name). We assume that names in our XML file uniquely identify a person which actually complicates the transformation from XML to RDF, since we need to create a unique, distinct blank node per name. The example data is a slight variant of the data from Fig. 1, where Alice knows both Bob and Charles, Bob knows Charles, and all parties are identified by blank nodes.

Because semantic data in RDF is on a higher level of abstraction than semi-structured XML data, the translation from XML to RDF is often called "lifting" while the opposite direction is called "lowering," as also shown in Fig. 2.

Lifting in GRDDL. The W3C Gleaning Resource Descriptions from Dialects of Languages (GRDDL) working group has the goal to complement the concrete RDF/XML

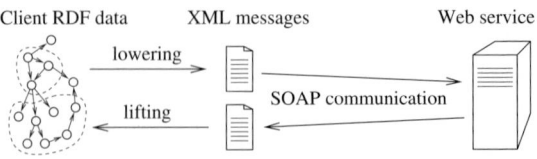

Fig. 4. RDF data lifting and lowering for WS communication

syntax with a mechanism to relate to other XML dialects (especially XHTML or "microformats") [9]. GRDDL focuses on the lifting task, i.e., extracting RDF from XML. To this end, the working group recently published a finished Recommendation which specifies how XML files or XML Schema namespace documents can reference transformations that are then processed by a GRDDL-aware application to extract RDF from the respective source file. Typically – due to its wide support – XSLT [10] is the language of choice to describe such transformations. However, writing XSLT can be cumbersome, since it is a general-purpose language for producing XML without special support for creating RDF. For our running example, the XSLT in Fig. 3(a) could be used to generate RDF/XML from the relations.xml file in Fig. 2 in an attempt to solve the lifting step. Using GRDDL, we can link this XSLT file mygrddl.xsl from relations.xml by changing the root element of the latter to:

```
<relations xmlns:grddl="http://www.w3.org/2003/g/data-view#"
           grddl:transformation="mygrddl.xsl"> ...
```

The RDF/XML result of the GRDDL transformation is shown in the upper part of Fig. 3(b). However, if we take a look at the Turtle version of this result in the lower part of Fig. 3(b) we see that this transformation creates too many blank nodes, since this simple XSLT does not merge equal names into the same blank nodes.

XSLT is a Turing-complete language, and theoretically any conceivable transformation can be programmed in XSLT; so, we could come up with a more involved stylesheet that creates unique blank node identifiers per name to solve the lifting task as intended. However, instead of attempting to repair the stylesheet from Fig. 3(a) let us rather ask ourselves whether XSLT is the right tool for such transformations. The claim we make is that specially tailored languages for RDF-XML transformations like XSPARQL which we present in this paper might be a more suitable alternative to alleviate the drawbacks of XSLT for the task that GRDDL addresses.

Lifting/Lowering in SAWSDL. While GRDDL is mainly concerned with lifting, in SAWSDL (Semantic Annotations for WSDL and XML Schema) there is a strong need for translations in the other direction as well, i.e., from RDF to arbitrary XML.

SAWSDL is the first standardized specification for semantic description of Web services. Semantic Web Services (SWS) research aims to automate tasks involved in the use of Web services, such as service discovery, composition and invocation. However, SAWSDL is only a first step, offering hooks for attaching semantics to WSDL components such as operations, inputs and outputs, etc. Eventually, SWS shall enable client software agents or services to automatically communicate with other services by means of semantic mediation on the RDF level. The communication requires both lowering

```
<xsl:stylesheet version="1.0" xmlns:rdf="...rdf-syntax-ns#"
  xmlns:foaf="...foaf/0.1/" xmlns:xsl="...XSL/Transform">
<xsl:template match="/rdf:RDF">
   <relations><xsl:apply-templates select=".//foaf:Person"/></relations>
</xsl:template>
<xsl:template match="foaf:Person"><person name="./@foaf:name">
   <xsl:apply-templates select="./foaf:knows"/>
</person></xsl:template>
<xsl:template match="foaf:knows[@rdf:nodeID]"><knows>
    <xsl:value-of select="//foaf:Person[@rdf:nodeID=./@rdf:nodeID]/@foaf:name"/>
</knows></xsl:template>
<xsl:template match="foaf:knows[foaf:Person]">
   <knows><xsl:value-of select="./foaf:Person/@foaf:name"/></knows>
</xsl:template>
</xsl:stylesheet>
```

Fig. 5. Lowering attempt by XSLT (mylowering.xsl)

and lifting transformations, as illustrated in Fig. 4. Lowering is used to create the request XML messages from the RDF data available to the client, and lifting extracts RDF from the incoming response messages.

As opposed to GRDDL, which provides hooks to link XSLT transformations on the level of whole XML or namespace documents, SAWSDL provides a more finegrained mechanism for "semantic adornments" of XML Schemas. In WSDL, schemata are used to describe the input and output messages of Web service operations, and SAWSDL can annotate messages or parts of them with pointers to relevant semantic concepts plus links to lifting and lowering transformations. These links are created using the `sawsdl:liftingSchemaMapping` and `sawsdl:loweringSchemaMapping` attributes which reference the transformations within XSL elements (`xsl:element`, `xsl:attribute`, etc.) describing the respective message parts.

SAWSDL's schema annotations for lifting and lowering are not only useful for communication with web services from an RDF-aware client, but for service mediation in general. This means that the output of a service S_1 uses a different message format than service S_2 expects as input, but it could still be used if service S_1 and S_2 provide lifting and lowering schema mappings, respectively, which map from/to the same ontology, or, respectively, ontologies that can be aligned via ontology mediation techniques (see [11]).

Lifting is analogous to the GRDDL situation – the client or an intermediate mediation service receives XML and needs to extract RDF from it –, but let us focus on RDF data lowering now. To stay within the boundaries of our running example, we assume a social network site with a Web service for querying and updating the list of a user's friends. The service accepts an XML format à la relations.xml (Fig. 2) as the message format for updating a user's (client) list of friends.

Assuming the client stores his FOAF data (relations.rdf in Fig. 2) in RDF/XML in the style of Fig. 1(b), the simple XSLT stylesheet mylowering.xsl in Fig. 5 would perform the lowering task. The service could advertise this transformation in its SAWSDL by linking mylowering.xsl in the `sawsdl:loweringSchemaMapping` attribute of the XML Schema definition of the `relations` element that conveys the message payload. However, this XSLT will break if the input RDF is in any other variant shown in Fig. 1. We could create a specific stylesheet for each of the presented variants, but creating one that handles all the possible RDF/XML forms would be much more complicated.

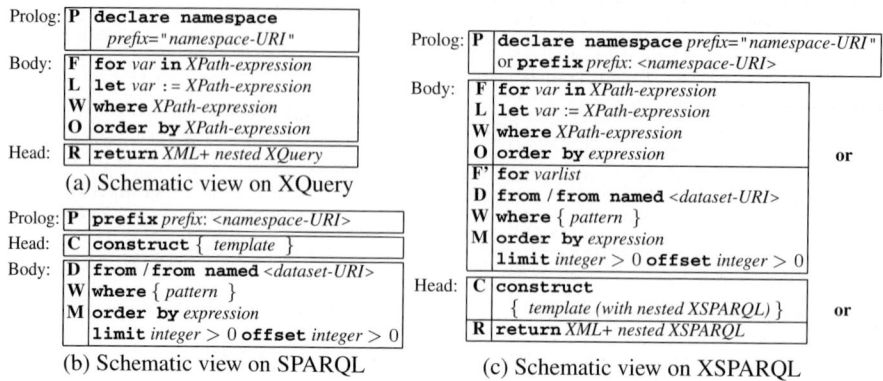

Fig. 6. An overview of XQuery, SPARQL, and XSPARQL

In recognition of this problem, SAWSDL contains a non-normative example which performs a lowering transformation as a sequence of a SPARQL query followed by an XSLT transformation on SPARQL's query results XML format [6]. Unlike XSLT or XPath, SPARQL treats all the RDF input data from Fig. 1 as equal. This approach makes a step in the right direction, combining SPARQL with XML technologies. The detour through SPARQL's XML query results format however seems to be an unnecessary burden. The XSPARQL language proposed in this paper solves this problem: it uses SPARQL pattern matching for selecting data as necessary, while allowing the construction of arbitrary XML (by using XQuery) for forming the resulting XML structures.

As more RDF data is becoming available on the Web which we want to integrate with existing XML-aware applications, SAWSDL will obviously not remain the only use case for lowering.

3 Starting Points: XQuery and SPARQL

In order to end up with a better suited language for specifying translations between XML and RDF addressing both the lifting and lowering use cases outlined above, we can build up on two main starting points: XQuery and SPARQL. Whereas the former allows a more convenient and often more concise syntax than XSLT for XML query processing and XML transformation in general, the latter is the standard for RDF querying and construction. Queries in each of the two languages can roughly be divided in two parts: (i) the retrieval part (*body*) and (ii) the result construction part (*head*). Our goal is to combine these components for both languages in a unified language, XSPARQL, where XQuery's and SPARQL's heads and bodies may be used interchangeably. Before we go into the details of this merge, let us elaborate a bit on the two constituent languages.

XQuery. As shown in Fig. 6(a) an XQuery starts with a (possibly empty) prolog (**P**) for namespace, library, function, and variable declarations, followed by so called **FLWOR** – or "flower" – expressions, denoting body (**FLWO**) and head (**R**) of the query. We only show namespace declarations in Fig. 6 for brevity.

```
 1  declare namespace foaf="...foaf/0.1/";
 2  declare namespace rdf="...-syntax-ns#";
 3  let $persons := //*[@name or ../knows]
 4  return
 5  <rdf:RDF>
 6  {
 7  for $p in $persons
 8  let $n := if( $p[@name] )
 9             then $p/@name else $p
10  let $id :=count($p/preceding::*)
11             +count($p/ancestor::*)
12  where
13    not(exists($p/following::*[
14        @name=$n or data(.)=$n]))
15  return
16  <foaf:Person rdf:nodeId="b{$id}">
17   <foaf:name>{data($n)}</foaf:name>
18   {
19   for $k in $persons
20   let $kn := if( $k[@name] )
21              then $k/@name else $k
22   let $kid :=count($k/preceding::*)
23               +count($k/ancestor::*)
24   where
25     $kn = data(//*[@name=$n]/knows) and
26     not(exists($kn/../following::*[
27        @name=$kn or data(.)=$kn]))
28   return
29    <foaf:knows>
30     <foaf:Person rdf:nodeID="b{$kid}"/>
31    </foaf:knows>
32   }
33  </foaf:Person>
34  }
35  </rdf:RDF>
```

(a) XQuery

```
declare namespace foaf="...foaf/0.1/";
declare namespace rdf="...-syntax-ns#";
let $persons := //*[@name or ../knows]
return


for $p in $persons
let $n := if( $p[@name] )
           then $p/@name else $p
let $id :=count($p/preceding::*)
           +count($p/ancestor::*)
where
  not(exists($p/following::*[
      @name=$n or data(.)=$n]))
construct {
_:b{$id} a foaf:Person;
         foaf:name {data($n)}.
{
for $k in $persons
let $kn := if( $k[@name] )
           then $k/@name else $k
let $kid :=count($k/preceding::*)
            +count($k/ancestor::*)
where
  $kn = data(//*[@name=$n]/knows) and
  not(exists($kn/../following::*[
     @name=$kn or data(.)=$kn]))
construct {
_:b{$id} foaf:knows _:b{$kid}.
_:b{$kid} a foaf:Person.
}
}
}
```

(b) XSPARQL

Fig. 7. Lifting using XQuery and XSPARQL

As for the body, **for** clauses (**F**) can be used to declare variables looping over the XML nodeset returned by an XPath expression. Alternatively, to bind the entire result of an XPath query to a variable, **let** assignments can be used. The **where** part (**W**) defines an XPath condition over the current variable bindings. Processing order of results of a **for** can be specified via a condition in the **order by** clause (**O**).

In the head (**R**) arbitrary well-formed XML is allowed following the **return** keyword, where variables scoped in an enclosing **for** or **let** as well as nested XQuery **FLWOR** expressions are allowed.

Any XPath expression in **FLWOR** expressions can again possibly involve variables defined in an enclosing **for** or **let**, or even nested XQuery **FLWOR** expressions. Together with a large catalogue of built-in functions [15], XQuery thus offers a flexible instrument for arbitrary transformations. For more details, we refer the reader to [5,7].

The lifting task of Fig. 2 can be solved with XQuery as shown in Fig. 7(a). The resulting query is quite involved, but completely addresses the lifting task, including unique blank node generation for each person: We first select all nodes containing person names from the original file for which a blank node needs to be created in variable $p (line 3). Looping over these nodes, we extract the actual names from either the value of the name attribute or from the knows element in variable $n. Finally, we compute the position in the original XML tree as blank node identifier in variable $id. The **where** clause (lines 12–14) filters out only the last name for duplicate occurrences of

```
prefix vc: <...vcard-rdf/3.0#>
prefix foaf: <...foaf/0.1/>
construct {$X foaf:name $FN.}
from <vc.rdf>
where { $X vc:FN $FN .}
```
(a)

```
prefix vc: <...vcard-rdf/3.0#>
prefix foaf: <...foaf/0.1/>
construct {_:b foaf:name
              {fn:concat("""",$N," ",$F,"""")}.}
from <vc.rdf>
where { $P vc:Given $N. $P vc:Family $F.}
```
(b)

Fig. 8. RDF-to-RDF mapping in SPARQL (a) and an enhanced mapping in XSPARQL (b)

```
<relations>{ for $Person $Name from <relations.rdf> where {$Person foaf:name $Name}
order by $Name return <person name="{$Name}">{
  for $FName from <relations.rdf> where
  {$Person foaf:knows $Friend. $Person foaf:name $Name. $Friend foaf:name $Fname}
  return <knows>{$FName}</knows> }</person> }</relations>
```

Fig. 9. Lowering using XSPARQL

the same name. The nested **for** (lines 19–31) to create nested foaf:knows elements again loops over persons, with the only differences that only those nodes are filtered out (line 25), which are known by the person with the name from the outer **for** loop.

While this is a valid solution for lifting, we still observe the following drawbacks: (1) We still have to build RDF/XML "manually" and cannot make use of the more readable and concise Turtle syntax; and (2) if we had to apply XQuery for the lowering task, we still would need to cater for all kinds of different RDF/XML representations. As we will see, both these drawbacks are alleviated by adding some SPARQL to XQuery.

SPARQL. Fig. 6(b) shows a schematic overview of the building blocks that SPARQL queries consist of. Again, we do not go into details of SPARQL here (see [20,17,18] for formal details), since we do not aim at modifying the language, but concentrate on the overall semantics of the parts we want to reuse. Like in XQuery, namespace prefixes can be specified in the Prolog (**P**). In analogy to **FLWOR** in XQuery, let us define so-called **DWMC** expressions for SPARQL.

The body (**DWM**) offers the following features. A *dataset* (**D**), i.e., the set of source RDF graphs, is specified in **from** or **from named** clauses. The **where** part (**W**) – unlike XQuery – allows to match parts of the dataset by specifying a graph *pattern* possibly involving variables, which we denote vars(*pattern*). This pattern is given in a Turtle-based syntax, in the simplest case by a set of triple patterns, i.e., triples with variables. More involved patterns allow unions of graph patterns, optional matching of parts of a graph, matching of named graphs, etc. Matching patterns on the conceptual level of RDF graphs rather than on a concrete XML syntax alleviates the pain of having to deal with different RDF/XML representations; SPARQL is agnostic to the actual XML representation of the underlying source graphs. Also the RDF merge of several source graphs specified in consecutive **from** clauses, which would involve renaming of blank nodes at the pure XML level, comes for free in SPARQL. Finally, variable bindings matching the **where** pattern in the source graphs can again be ordered, but also other solution modifiers (**M**) such as **limit** and **offset** are allowed to restrict the number of solutions considered in the result.

In the head, SPARQL's **construct** clause (**C**) offers convenient and XML-independent means to create an output RDF graph. Since we focus here on RDF construction, we omit the **ask** and **select** SPARQL query forms in Fig. 6(b) for brevity. A **construct** *template* consists of a list of triple patterns in Turtle syntax possibly involving variables, denoted by vars(*template*), that carry over bindings from the **where** part. SPARQL can be used as transformation language between different RDF formats, just like XSLT and XQuery for transforming between XML formats. A simple example for mapping full names from vCard/RDF (http://www.w3.org/TR/vcard-rdf) to foaf:name is given by the SPARQL query in Fig. 8(a).

Let us remark that SPARQL does not offer the generation of new values in the head which on the contrary comes for free in XQuery by offering the full range of XPath/XQuery built-in functions. For instance, the simple query in Fig. 8(b) which attempts to merge family names and given names into a single foaf:name is beyond SPARQL's capabilities. As we will see, XSPARQL will not only make reuse of SPARQL for transformations from and to RDF, but also aims at enhancing SPARQL itself for RDF-to-RDF transformations enabling queries like the one in Fig. 8(b).

4 XSPARQL

Conceptually, XSPARQL is a simple merge of SPARQL components into XQuery. In order to benefit from the more intuitive facilities of SPARQL in terms of RDF graph matching for retrieval of RDF data and the use of Turtle-like syntax for result construction, we syntactically add these facilities to XQuery. Fig. 6(c) shows the result of this "marriage." First of all, every native XQuery query is also an XSPARQL query. However we also allow the following modifications, extending XQuery's **FLWOR** expressions to what we call (slightly abusing nomenclature) **FLWOR'** expressions: (i) In the body we allow SPARQL-style **F'DWM** blocks alternatively to XQuery's **FLWO** blocks. The new **F' for** clause is very similar to XQuery's native **for** clause, but instead of assigning a single variable to the results of an XPath expression it allows the assignment of a whitespace separated list of variables (*varlist*) to the bindings for these variables obtained by evaluating the graph pattern of a SPARQL query of the form: **select** *varlist* **DWM**. (ii) In the head we allow to create RDF/Turtle directly using **construct** statements (**C**) alternatively to XQuery's native **return** (**R**).

These modifications allows us to reformulate the lifting query of Fig. 7(a) into its slightly more concise XSPARQL version of Fig. 7(b). The real power of XSPARQL in our example becomes apparent on the lowering part, where all of the other languages struggle. Fig. 9 shows the lowering query for our running example.

As a shortcut notation, we allow also to write "**for** *" in place of "**for** [*list of all variables appearing in the* **where** *clause*]"; this is also the default value for the **F'** clause whenever a SPARQL-style **where** clause is found and a **for** clause is missing. By this treatment, XSPARQL is also a syntactic superset of native SPARQL **construct** queries, since we additionally allow the following: (1) XQuery and SPARQL namespace declarations (**P**) may be used interchangeably; and (2) SPARQL-style **construct** result forms (**R**) may appear before the retrieval part; note that we allow this syntactic sugar only for queries consisting of a single **FLWOR'** expression,

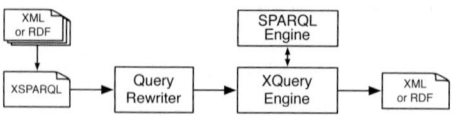

Fig. 10. XSPARQL architecture

with a single `construct` appearing right after the query prolog, as otherwise, syntactic ambiguities may arise. This feature is mainly added in order to encompass SPARQL style queries, but in principle, we expect the (**R**) part to appear in the end of a **FLWOR'** expression. This way, the queries of Fig. 8 are also syntactically valid for XSPARQL.

Semantics and Implementation. As we have seen above, XSPARQL syntactically subsumes both XQuery and SPARQL. Concerning semantics, XSPARQL equally builds on top of its constituent languages. In an earlier version of this paper [1], we have extended the formal semantics of XQuery [7] by additional rules which reduce each XSPARQL query to XQuery expressions; the resulting **FLWOR**s operate on the answers of SPARQL queries in the SPARQL XML result format [6]. Since we add only new reduction rules for SPARQL-like heads and bodies, it is easy to see that each native XQuery is treated in a semantically equivalent way in XSPARQL.

In order to convince the reader that the same holds for native SPARQL queries, we will illustrate our reduction in the following. We restrict ourselves here to a more abstract presentation of our rewriting algorithm, as we implemented it in a prototype.[2]

The main idea behind our implementation is translating XSPARQL queries to corresponding XQueries which possibly use interleaved calls to a SPARQL endpoint. The architecture of our prototype shown in Fig. 10 consists of three main components: (1) a query rewriter, which turns an XSPARQL query into an XQuery; (2) a SPARQL endpoint, for querying RDF from within the rewritten XQuery; and (3) an XQuery engine for computing the result document.

The rewriter (Alg. 1) takes as input a full XSPARQL *QueryBody* [7] q (i.e., a sequence of **FLWOR'** expressions), a set of bound variables b and a set of position variables p, which we explain below. For a **FL** (or **F'**, resp.) clause s, we denote by vars(s) the list of all newly declared variables (or the *varlist*, resp.) of s. For the sake of brevity, we only sketch the core rewriting function *rewrite*() here; additional machinery handling the prolog including function, variable, module, and namespace declarations is needed in the full implementation. The rewriting is initiated by invoking *rewrite*($q, \emptyset, \emptyset$) with empty bound and position variables, whose result is an XQuery. Fig. 11 shows the output of our translation for the `construct` query in Fig. 8(b) which illustrates both the lowering and lifting parts. Let us explain the algorithm using this sample output.

After generating the prolog (lines 1–9 of the output), the rewriting of the QueryBody is performed recursively following the syntax of XSPARQL. During the traversal of the nested **FLWOR'** expressions, SPARQL-like heads or bodies will be replaced by XQuery expressions, which handle our two tasks. The lowering part is processed first:

Lowering. The lowering part of XSPARQL, i.e., SPARQL-like **F'DWM** blocks, is "encoded" in XQuery with interleaved calls to an external SPARQL endpoint. To this end,

---
[2] http://www.polleres.net/xsparql/

```
1  declare vc = "http://www.w3.org/2001/vcard-rdf/3.0#";
2  declare foaf = "http://xmlns.com/foaf/0.1/";
3  declare sparql_result = "http://www.w3.org/2005/sparql-results#";
4  declare function local:rdf_term($NT as xs:string,$V as xs:string) as xs:string {
5   let $rdf_term := if($NT="literal") then fn:concat("""",$V,"""")
6     else if($NT="bnode") then fn:concat("_:",$V) else if($NT="uri")
7     then fn:concat("<",$V,">") else "" return $rdf_term };
8  declare variable $NS_1 := "prefix vc: <...vcard-rdf/3.0#> ";
9  declare variable $NS_2 := "prefix foaf: <...foaf/0.1/> ";
10 fn:concat("@",$NS_1,".","@",$NS_2,"."), let $aux_query := fn:concat(
11  "http://localhost:2020/sparql?query=", fn:encode-for-uri(fn:concat($NS_1, $NS_2,
12  "select $P $N $F from <vc.rdf> where {$P vc:Given $N. $P vc:Family $F.}")))
13 for $aux_result at $aux_result_pos in doc($aux_query)//sparql_result:result
14   let $P_Node := $aux_result/sparql_result:binding[@name="P"]
15   let $P := data($P_Node/*) let $P_NodeType := name($P_Node/*)
16   let $P_RDFTerm := local:rdf_term($P_NodeType,$P)
17   let $N_Node := $aux_result/sparql_result:binding[@name="N"]
18   let $N := data($N_Node/*) let $N_NodeType := name($N_Node/*)
19   let $N_RDFTerm := local:rdf_term($N_NodeType,$N)
20   let $F_Node := $aux_result/sparql_result:binding[@name="F"]
21   let $F := data($F_Node/*) let $F_NodeType := name($F_Node/*)
22   let $F_RDFTerm := local:rdf_term($F_NodeType,$F)
23 return ( fn:concat("_:b",$aux_result_pos," foaf:name "),
24        ( fn:concat("""",$N_RDFTerm," ",$F_RDFTerm,"""") ), "." )
```

Fig. 11. XQuery encoding of Example 8(b)

we translate **F'DWM** blocks into equivalent XQuery **FLWO** expressions which retrieve SPARQL result XML documents [6] from a SPARQL engine; i.e., we "push" each **F'DWM** body to the SPARQL side, by translating it to a native `select` query string. The auxiliary function *sparql*() in line 6 of our rewriter provides this functionality, transforming the **where** {*pattern*} part of **F'DWM** clauses to XQuery expressions which have all bound variables in vars(*pattern*) replaced by the values of the variables; "free" XSPARQL variables serve as binding variables for the SPARQL query result. The outcome of the *sparql*() function is a list of expressions, which is concatenated and URI-encoded using XQuery's XPath functions, and wrapped into a URI with `http` scheme pointing to the SPARQL query service (lines 10–12 of the output), cf. [6]. Then we create a new XQuery **for** loop over variable `$aux_result` to iterate over the query answers extracted from the SPARQL XML result returned by the SPARQL query processor (line 13). For each variable $x_i \in$ vars(*s*) (i.e., in the (**F'**) **for** clause of the original **F'DWM** body), new auxiliary variables are defined in separate `let`-expressions extracting its node, content, type (i.e., literal, uri, or blank), and RDF-Term (x_i_Node, x_i, x_i_NodeType, and x_i_RDFTerm, resp.) by appropriate XPath expressions (lines 14–22 of Fig. 11); the *auxvars*() helper in line 6 of Alg. 1 is responsible for this.

Lifting. For the lifting part, i.e., SPARQL-like **construct**s in the **R** part, the transformation process is straightforward. Before we rewrite the QueryBody q, we process the prolog (**P**) of the XSPARQL query and output every namespace declaration as Turtle string literals "@prefix ns: <URI>." (line 10 of the output). Then, Alg. 1 is called on q and recursively decorates every **for** $Var expression by fresh position variables (line 13 of our example output); ultimately, **construct** templates are rewritten to an assembled string of the pattern's constituents, filling in variable bindings and evaluated subexpressions (lines 23–24 of the output).

Algorithm 1. *rewrite*(q, b, p) : Rewrite XSPARQL q to an XQuery

Input: XSPARQL query q, set of bounded variables b, set of position variables p
Result: XQuery

1 **if** q *is of form* s_1 , ... , s_k **then**
2   **return** *rewrite*(s_1, b, p) , ... , *rewrite*(s_k, b, p)
3 **else if** q *is of form* **for** $\$x_1$ **in** $XPathExpr_1$, ... , $\$x_k$ **in** $XPathExpr_k$ s_1 **then**
4   **return for** $\$x_1$ **at** $\$x_{1\_pos}$ **in** $XPathExpr_1$, ... , $\$x_k$ **at** $\$x_{k\_pos}$ **in** $XPathExpr_k$ *rewrite*($s_1, b, p \cup \{\$x_{1\_pos}, \ldots, \$x_{k\_pos}\}$)
5 **else if** q *is of form* **for** $\$x_1 \cdots \$x_n$ **from** D **where** { *pattern* } M s_1 **then**
6   **return let** $\$aux\_query$:= *sparql*(D, $\{\$x_1, \ldots, \$x_n\}$, *pattern*, M, b) **for** $\$aux\_result$ **in** doc($\$aux\_query$)//sparql:result *auxvars*($\{\$x_1, \ldots, \$x_n\}$) *rewrite*($s_1, b \cup$ vars(q), p)
7 **else if** q *is of form* **construct** {*template*} **then**
8   **return return** (*rewrite-template*(*template*, b, p))
9 **else**
10   split q into its subexpressions $s_1, \ldots, s_n$
11   **for** $j := 1, \ldots, n$ **do** $b_j = b \cup \bigcup_{1 \leq i \leq j-1}$ vars(s_i)
12   **if** $n > 1$ **then return** $q\,[s_1/\textit{rewrite}(s_1, b_1, p), \ldots, s_n/\textit{rewrite}(s_n, b_n, p)]$
13   **else return** q
14 **end**

Blank nodes in **construct**s need special care, since, according to SPARQL's semantics, these must create new blank node identifiers for each solution binding. This is solved by "adorning" each blank node identifier in the **construct** part with the above-mentioned position variables from any enclosing **for**-loops, thus creating a new, unique blank node identifier in each loop (line 23 in the output). The auxiliary function *rewrite-template*() in line 8 of the algorithm provides this functionality by simply adding the list of all position variable p as expressions to each blank node string; if there are nested expressions in the supplied **construct** {*template*}, it will return a sequence of nested **FLWOR**s with each having *rewrite*() applied on these expressions with the in-scope bound and position variables.

Expressions involving **construct**s create Turtle output. Generating RDF/XML output from this Turtle is optionally done as a simple postprocessing step supported by using standard RDF processing tools.

5 Related Works

Albeit both XML and RDF are nearly a decade old, there has been no serious effort on developing a language for convenient transformations between the two data models. There are, however, a number of apparently abandoned projects that aim at making it easier to transform RDF data using XSLT. *RDF Twig* [21] suggests XSLT extension functions that provide various useful views on the "sub-trees" of an RDF graph. The main idea of RDF Twig is that while RDF/XML is hard to navigate using XPath, a subtree of an RDF graph can be serialized into a more useful form of RDF/XML. *Tree-Hugger*[3] makes it possible to navigate the graph structure of RDF both in XSLT and

---
[3] http://rdfweb.org/people/damian/treehugger/index.html

XQuery using XPath-like expressions, abstracting from the actual RDF/XML structure. *rdf2r3x*[4] uses an RDF processor and XSLT to transform RDF data into a predictable form of RDF/XML also catering for RSS. Carroll and Stickler take the approach of simplifying RDF/XML one step further, putting RDF into a simple *TriX* [4] format, using XSLT as an extensibility mechanism to provide human-friendly macros for this syntax.

These approaches rely on non-standard extensions or tools, providing implementations in some particular programming language, tied to specific versions of XPath or XSLT processors. In contrast, *RDFXSLT*[5] provides an XSLT preprocessing stylesheet and a set of helper functions, similar to RDF Twig and TreeHugger, yet implemented in pure XSLT 2.0, readily available for many platforms.

All these proposals focus on XPath or XSLT, by adding RDF-friendly extensions, or preprocessing the RDF data to ease the access with stock XPath expressions. It seems that XQuery and SPARQL were disregarded previously because XQuery was not standardized until 2007 and SPARQL – which we suggest to select relevant parts of RDF data instead of XPath – has only very recently received W3C's recommendation stamp.

As for the use of SPARQL, Droop et al. [8] suggest, orthogonal to our approach, to compile XPath queries into SPARQL. Similarly, encoding SPARQL completely into XSLT or XQuery [13] seems to be an interesting idea that would enable to compile down XSPARQL to pure XQuery without the use of a separate SPARQL engine. However, scalability results in [13] so far do not yet suggest that such an approach would scale better than the interleaved approach we took in our current implementation.

Finally, related to our discussion in Section 2, the SPARQL Annotations for WSDL (SPDL) project (http://www.w3.org/2005/11/SPDL/) suggests a direct integration of SPARQL queries into XML Schema, but is still work in progress. We expect SPDL to be subsumed by SAWSDL, with XSPARQL as the language of choice for lifting and lowering schema mappings.

6 Conclusion and Future Plans

We have elaborated on use cases for lifting and lowering, i.e., mapping back and forth between XML and RDF, in the contexts of GRDDL and SAWSDL. As we have seen, XSLT turned out to provide only partially satisfactory solutions for this task. XQuery and SPARQL, each in its own world, provide solutions for the problems we encountered, and we presented XSPARQL as a natural combination of the two as a proper solution for the lifting and lowering tasks. Moreover, we have seen that XSPARQL offers more than a handy tool for transformations between XML and RDF. Indeed, by accessing the full library of XPath/XQuery functions, XSPARQL opens up extensions such as value-generating built-ins or even aggregates in the construct part, which have been pointed out missing in SPARQL earlier [19].

As we have seen, XSPARQL is a conservative extension of both of its constituent languages, SPARQL and XQuery. The semantics of XSPARQL was defined as an extension of XQuery's formal semantics adding a few normalization mapping rules. We

---

[4] http://wasab.dk/morten/blog/archives/2004/05/30/transforming-rdfxml-with-xslt
[5] http://www.wsmo.org/TR/d24/d24.2/v0.1/20070412/rdfxslt.html

provide an implementation of this transformation which is based on reducing XSPARQL queries to XQuery with interleaved calls to a SPARQL engine via the SPARQL protocol. There are good reasons to abstract away from RDF/XML and rely on native SPARQL engines in our implementation. Although one could try to compile SPARQL entirely into an XQuery that caters for all different RDF/XML representations, that would not solve the use which we expect most common in the nearer future: many online RDF sources will most likely not be accessible as RDF/XML files, but rather via RDF stores that provide a standard SPARQL interface.

Our resulting XSPARQL preprocessor can be used with any available XQuery and SPARQL implementation, and is available for user evaluation along with all examples and an extended version [1] of this paper at http://www.polleres.net/xsparql/.

As mentioned briefly in the introduction, simple reasoning – which we have not yet incorporated – would significantly improve queries involving RDF data. SPARQL engines that provide (partial) RDFS support could immediately address this point and be plugged into our implementation. But we plan to go a step further: integrating XSPARQL with Semantic Web Pipes [16] or other SPARQL extensions such as SPARQL++ [19] shall allow more complex intermediate RDF processing than RDFS materialization.

We also plan to apply our results for retrieving metadata from context-aware services and for Semantic Web service communication, respectively, in the EU projects inContext (http://www.in-context.eu/) and TripCom (http://www.tripcom.org/).

References

1. Akthar, W., Kopecký, J., Krennwallner, T., Polleres, A.: XSPARQL: Traveling between the XML and RDF worlds – and avoiding the XSLT pilgrimage. Technical Report DERI-TR-2007-12-14, DERI Galway (December 2007)
2. Beckett, D., McBride, B. (eds.): RDF/XML syntax specification (revised). W3C Rec. (February 2004)
3. Beckett, D.: Turtle - Terse RDF Triple Language (November 2007)
4. Carroll, J., Stickler, P.: TriX: RDF Triples in XML. Tech. Report HPL-2004-56, HP (May 2004)
5. Chamberlin, D., Robie, J., Boag, S., Fernández, M.F., Siméon, J., Florescu, D. (eds.): XQuery 1.0: An XML Query Language, W3C Rec. (January 2007)
6. Clark, K., Feigenbaum, L., Torres, E.: SPARQL Protocol for RDF, W3C Prop. Rec. (November 2007)
7. Draper, D., Fankhauser, P., Fernández, M., Malhotra, A., Rose, K., Rys, M., Siméon, J., Wadler, P. (eds.): XQuery 1.0 and XPath 2.0 Formal Semantics, W3C Rec. (January 2007)
8. Droop, M., Flarer, M., Groppe, J., Groppe, S., Linnemann, V., Pinggera, J., Santner, F., Schier, M., Schöpf, F., Staffler, H., Zugal, S.: TranslatingXPath Queries into SPARQL-Queries. In: ODBASE 2007 (2007)
9. Connolly, D. (ed.): Gleaning Resource Descriptions from Dialects of Languages (GRDDL), W3C Rec. (September 2007)
10. Kay, M. (ed.): XSL Transformations (XSLT) Version 2.0, W3C Recommendation (January 2007)
11. Euzenat, J., Shvaiko, P.: Ontology matching. Springer, Heidelberg (2007)

12. Farrell, J., Lausen, H. (eds.): Semantic Annotations for WSDL and XML Schema. W3C Rec. (August 2007)
13. Groppe, S., Groppe, J., Linnemann, V., Kukulenz, D., Hoeller, N., Reinke, C.: Embedding SPARQL into XQuery/XSLT. In: SAC 2008 (March 2008) (to appear)
14. Kopecký, J., Vitvar, T., Bournez, C., Farrell, J.: SAWSDL: Semantic Annotations for WSDL and XML Schema. IEEE Internet Computing 11(6), 60–67 (2007)
15. Malhotra, A., Melton, J., Walsh, N. (eds.): XQuery 1.0 and XPath 2.0 Functions and Operators, W3C Rec. (January 2007)
16. Morbidoni, C., Polleres, A., Tummarello, G., Phuoc, D.L.: Semantic Web Pipes. Technical Report DERI-TR-2007-11-07, DERI Galway (December 2007)
17. Pérez, J., Arenas, M., Gutierrez, C.: Semantics and Complexity of SPARQL. In: Cruz, I., Decker, S., Allemang, D., Preist, C., Schwabe, D., Mika, P., Uschold, M., Aroyo, L.M. (eds.) ISWC 2006. LNCS, vol. 4273, Springer, Heidelberg (2006)
18. Polleres, A.: From SPARQL to Rules (and back). In: Proc. WWW 2007 (May 2007)
19. Polleres, A., Scharffe, F., Schindlauer, R.: SPARQL++ for mapping between RDF vocabularies. In: ODBASE 2007 (November 2007)
20. Prud'hommeaux, E., Seaborne, A. (eds.): SPARQL Query Language for RDF, W3C Rec. (January 2008)
21. Walsh, N.: RDF Twig: Accessing RDF Graphs in XSLT. In: Extreme Markup Languages (2003)

Streaming SPARQL - Extending SPARQL to Process Data Streams

Andre Bolles, Marco Grawunder, and Jonas Jacobi

University of Oldenburg, Germany,
Department for Computer Science, Information Systems and Databases
{andre.bolles,marco.grawunder,jonas.jacobi}@uni-oldenburg.de

Abstract. A lot of work has been done in the area of data stream processing. Most of the previous approaches regard only relational or XML based streams but do not cover semantically richer RDF based stream elements. In our work, we extend SPARQL, the W3C recommendation for an RDF query language, to process RDF data streams. To describe the semantics of our enhancement, we extended the logical SPARQL algebra for stream processing on the foundation of a temporal relational algebra based on multi-sets and provide an algorithm to transform SPARQL queries to the new extended algebra. For each logical algebra operator, we define executable physical counterparts. To show the feasibility of our approach, we implemented it within our ODYSSEUS framework in the context of wind power plant monitoring.

1 Introduction

Wind power plants are one promising way to reduce CO_2 emissions and by this the greenhouse effect. Large off-shore wind parks are capable of delivering electric power in a constant and reliable way. These parks need high initial investments, so keeping maintenance costs low is indispensable. Monitoring and early problem detection is essential. The current approach to monitor wind parks are proprietary SCADA (*S*upervisory *C*ontrol *a*nd *D*ata *A*cquisition) systems. This leads to the problem of missing interoperability between different wind power plant vendors and makes it difficult or even impossible for customers to extend the system. In analogy to the application of database management systems for many standard applications like personnel or warehouse management, we believe that data stream management systems (DSMS) [1] can help in reducing initial wind parks costs and providing higher interoperability. The International Electrotechnical Commission has proposed a standard to monitor and communicate with wind parks [2]. IEC 61400-25 contains a standard data model and a common information model. The standard determines no communication format. A common approach could be the usage of XML. In our work we analyze the use of RDF [3] as an alternative, because RDF can be used to model IEC 61850 family and substandards like 61400-25 in future. Furthermore a triple contains all information needed to understand the content therefore RDF is an ideal format for serialization and missing triples (e.g. because of network failure) do not necessary corrupt the whole stream.

In DSMS continuous queries are used to determine stream information. SPARQL is the W3C Recommendation for a query language over RDF. Unfortunately, this approach is not applicable directly to RDF data streams. A sensor produces typically unlimited datasets (e.g. an element every second). Some operators in SPARQL need the whole dataset to process, like distinct, join or sort. A common approach for this problem is the use of window queries. Instead of regarding the whole stream, only subparts, so called windows, are treated. Because recent data have typically higher relevance (e.g. average wind speed in the last 10 minutes) this approach is applicable. We regard different kinds of windows. Time based windows define the window by a unit of time (e.g. the last 10 minutes). These windows are called sliding windows, if they are moved for each new point in time. If the windows are moved only after a distinct period, these windows are called sliding δ windows (e.g. move the window every 5 minutes over the stream with the window size 10 minutes) [4]. Element based windows are defined by the number of elements in the window (e.g. the last 100 statements). We model these windows by defining temporal element validity, i.e. if an element is valid it is part of the window.

In query processing we follow the usual approach of translating a descriptive query into an internal representation (the logical plan) that can more easily be optimized and transformed. This logical plan will then be translated into a set of physical plans, containing executable physical algebra operators. Finally, one of these plans will be selected, e.g. by means of a cost model, and executed.

The contributions of this paper are:

- the extension of SPARQL to handle RDF based data streams,
- the precise definition of the semantics of this extension by extending the existing SPARQL algebra,
- a transformation algorithm to map the language extension to the extended SPARQL algebra and
- a set of physical algebra operators to execute the queries.

The remainder of this paper is as follows. The next section describes how we extended the SPARQL grammar to allow referencing data streams and defining windows over these streams. After that we give a clear formal definition of what our data streams are (Sec. 3) and, based on this definition, we show how we extended the SPARQL algebra (Sec. 4) and how SPARQL queries can be translated into that algebra (Sec. 5). Because algebra operators are not executable we defined physical counterparts of each logical operator (Sec. 6). Finally, we discuss related work (Sec. 7) and give an outlook on future work.

2 Extending the SPARQL Grammar

SPARQL as defined in [5] is not sufficient to define queries over data streams. One of our main goals is to preserve the known syntax and semantics of SPARQL as much as possible. We extend SPARQL with the capability to explicitly state data streams and to define windows over them. Windows can be defined in the FROM part of a SPARQL query to define a common window for the whole data stream. To allow a finer granularity we also allow windows in graph patterns.

Table 1 shows the extended grammar production rules in EBNF. First are listed those rules that are affected by our extensions[1]. Then new rules are listed. Extensions and new rules are shown in bold. The other rules from [5] do not need to be modified for processing data streams[2]. All nonterminals (underlined) that are not explained further are identical to [5].

Table 1. Extended SPARQL Grammar (extract)

SelectQuery	::=	'SELECT' ('DISTINCT' | 'REDUCED')? (Var | '*') (DatasetClause* | **DatastreamClause***) WhereClause SolutionModifier
NamedGraphClause	::=	'NAMED' SourceSelector
DatastreamClause	::=	**'FROM' (DefaultStreamClause | NamedStreamClause)**
DefaultStreamClause	::=	**'STREAM' SourceSelector Window**
NamedStreamClause	::=	**'NAMED' 'STREAM' SourceSelector Window**
SourceSelector	::=	IRIref
GroupGraphPattern	::=	'{' TriplesBlock? ((GraphPatternNotTriples | Filter) '.'? TriplesBlock?)* **(Window)?** '}'
Window	::=	**(SlidingDeltaWindow | SlidingTupelWindow | FixedWindow)**
SlidingDeltaWindow	::=	**'WINDOW' 'RANGE' ValSpec 'SLIDE' (ValSpec)?**
FixedWindow	::=	**'WINDOW' 'RANGE' ValSpec 'FIXED'**
SlidingTupelWindow	::=	**'WINDOW' 'ELEMS' INTEGER**
ValSpec	::=	**INTEGER Timeunit?**
INTEGER	::=	[0-9]+
Timeunit	::=	**('MS' | 'S' | 'MINUTE' | 'HOUR' | 'DAY' | 'WEEK')**

To state an input as a data stream the keyword STREAM followed by an IRI has to be used. We allow different window types, which require specific language extensions. If a window is defined in a query in the FROM and in the graph pattern parts, only the more special window of the graph pattern is evaluated. We provide time based windows and element based windows. Sliding δ-windows allow the definition of the window size with the RANGE keyword whereas we assume milliseconds to be the default timeunit. SLIDE defines the delay after which the window is moved. The value after SLIDE can be omitted. In that case the size is 1 of the unit defined in the RANGE-part. If SLIDE and RANGE contain the same values, we call this a fixed (or tumbling) window. As syntactic sugar we also allow the definition of this window using FIXED. Finally, ELEMS defines element based windows.

Listing 1.1 below gives an example of a query containing time and element based windows in the FROM and in the optional graph pattern part. The idea of the query is to return the number of starting a wind turbine in the last 30 minutes and, if available, the number of stopping a wind turbine that has to be reported in the last 1500 elements of the data stream.

---

[1] For space reasons we omitted Construct, Describe and Ask queries which are defined similar to Select.

[2] Of course there are possibly different semantics when defining queries over data streams.

```
PREFIX wtur: <http://iec.org/61400-25/root/ln/classes/WTUR#>
SELECT ?x ?y ?z
FROM STREAM <http://iec.org/61400-25/root/td.rdf>
    WINDOW RANGE 30 MINUTE SLIDE
WHERE { ?x wtur:StrCnt ?y .
        OPTIONAL { ?x wtur:StopCnt ?z .
                   WINDOW ELEMS 1500 }}
```

Listing 1.1. SPARQL query over an RDF stream with windows

3 Defining Data Streams

Before we present our extended algebra we define the logical base of our algebra, the data stream. Like [6] we distinguish between a raw data stream, representing the data received by the DSMS, a physical data stream which can be processed by the operators of the DSMS and finally a logical data stream over which the algebra and therefore the semantics of the extensions can be defined. Many of the following definitions are inspired by [6].

A raw data stream represents simply the data arriving at the DSMS. The data format could be in different formats like XML/RDF or Turtle representation. A special access operator can transform the statements to a unified representation. Because some sources in data stream applications typically produce timestamped information (e.g. a sensor for wind speed also transmits a timestamp for the measured value), we defined an RDF raw data stream with timestamps, too.

The physical RDF data stream contains prepared elements that can be used as input for physical data stream operators. This means especially that all elements are timestamped, defined either by the data source (raw data stream with timestamps) or by the DSMS (raw data streams without timestamps). Let $\mathbb{T} = (T; \leq)$ be a discrete time domain as proposed by [7][3]. Let $\mathbb{I} := \{[t_s, t_e) \in T \times T | t_s < t_e\}$ be the set of right open time intervals:

Definition 1 (Physical RDF data stream). *Let $\mathbb{S}^P_{RDF}$ be the set of all physical RDF data streams. A physical RDF data stream $S^P_{RDF} \in \mathbb{S}^P_{RDF}$ is defined as a pair $S^P_{RDF} = (M^P_{RDF}, \leq^P_t)$ where $M^P_{RDF} = [((s, p, o), [t_S, t_E)) | (s, p, o) \in \Omega, [t_S, t_E) \in \mathbb{I}]$ is an ordered possibly unlimited sequence of pairs, consisting of an RDF statement and a right open time interval. The order is defined by $\leq^P_t$:*

$$\forall x_i, x_j \in \Omega \times \mathbb{I}, i, j \in \mathbb{N}, i < j : x_i \leq^P_t x_j \Leftrightarrow x_i.t_S \leq x_j.t_S$$

In SPARQL only the first operators typically handle RDF based data. E.g. a basic graph pattern transforms RDF statements to SPARQL solutions μ which are the input to the following operators. To represent this in our algebra we define a set $\mathbb{S}^P_\mu$ of physical μ data streams analogous to $\mathbb{S}^P_{RDF}$.

---

[3] Without reducing generality, we allow only natural numbers for $\mathbb{T}$.

Logical algebra operators consume logical RDF streams. These logical streams are not defined as sequences but as multi-sets. Because no elements are processed on the logical level, it is only used for transformation and optimization; the order in the streams is not relevant. The great advantage of defining logical streams on multi-set semantics is to make it possible to base our algebra on the extended relational algebra [8]. Just like there, duplicates are allowed and expressed by the number of occurrences.

Definition 2 (Logical RDF data stream). *Let $\mathbb{S}^L_{RDF}$ be the set of all logical RDF data streams. A logical RDF data stream is a possibly unlimited multi-set of triples:* $S^L_{RDF} = \{((s,p,o),t,n)|(s,p,o) \in \Omega, t \in \mathbb{T}, n \in \mathbb{N}\backslash\{0\}\}$. *The triple $((s,p,o),t,n)$ expresses: The RDF statement (s,p,o) is valid at time t and occurs n-times at this point in time.* $\forall((s,p,o),t,n) \in S^L_{RDF} : \nexists((\hat{s},\hat{p},\hat{o}),\hat{t},\hat{n}) \in S^L_{RDF} : (s,p,o) \equiv_{RDF} (\hat{s},\hat{p},\hat{o}) \wedge t = \hat{t}. \equiv_{RDF}$ *means that subject, predicate and object of two statements are pairwise equal.*

Analogously, we define the set of logical μ data streams $\mathbb{S}^L_\mu$, because not all logical algebra operators can consume logical RDF data streams. We also define sequential logical RDF and sequential logical μ data streams to support sort operations. We will use $\mathbb{S}^L_x$ if no distinction between RDF and μ is necessary.

To relate physical and logical algebra operators, and thus defining the semantics of the physical operators, we need to define transformations between the different stream types. Thereby it is possible to assign plan transformations and optimizations on the logical query level also to the physical level. Due to space limitations we need to omit these transformations here.

4 Extending the SPARQL Algebra

Given the base definitions, we can now present some of the formal definitions of our new or extended SPARQL algebra operators.

The first operator we introduce is needed to define windows over the logical stream $\mathbb{S}^L_x$. It takes parameters $w \in \mathbb{T}$ defining the width of the window and $\delta \leq w$ defining the delay of window moving:

Definition 3 (Sliding δ-window). *Let $w \in \mathbb{T}$ be the width of a sliding window and $\delta \leq w \in \mathbb{T}$ be a time span, by which the window is moved. Let $\hat{t} \in \mathbb{T}$ be a point in time. A sliding δ-window is a function $\omega^{slide}_{w,\delta} : \mathbb{S}^L_x \times \mathbb{T} \times \mathbb{T} \times \mathbb{T} \to \mathbb{S}^L_x$, with:*

$$\omega^{slide}_{w,\delta}(S^L_x, \hat{t}) := \{(z,\hat{t},\hat{n})|\exists i \in \mathbb{N} : (i \cdot \delta) - w \leq \hat{t} < i \cdot \delta \wedge$$
$$\exists Y \subseteq S^L_x : Y \neq \emptyset \wedge$$
$$Y = \{(z,t,n)| \max\{(i \cdot \delta) - w, 0\} \leq t \leq \hat{t}\} \wedge$$
$$\hat{n} = \sum_{\{(z,t,n) \in Y\}} n\}$$

The special case $\delta = 1$ defines a sliding window, the case $\delta = w$ defines a tumbling window. Defining $\delta \leq w$ ensures not to miss elements by moving the window. The second kind of window, called a sliding tuple window, is defined over the element count in a stream. To define this operator we need to define a count function on logical data streams, that gives the number of elements in a logical RDF or μ stream (in the following represented by x) until a point in time t [6]:

Definition 4 (Count function of logical x data streams). *Let $t \in \mathbb{T}$ be a point in time. The function $m : \mathbb{S}_x^L \times \mathbb{T} \to \mathbb{N}$ calculates the number of all elements in the data stream until t:*

$$m(t, S_x^L) := \left|\{(z, \hat{t}, n) | \hat{t} \leq t\}\right|$$

This definition is only valid for $n \leq 1$, i.e. at every point in time only one element is added to the stream. With this function a sliding tuple window can be defined as follows:

Definition 5 (Sliding tuple window). *Let $c \in \mathbb{N}$ be the maximum size of a sliding tuple window and $\hat{t} \in \mathbb{T}$ a point in time. A sliding tuple window is a function ω_c^{count} : $\mathbb{S}_x^L \times \mathbb{N} \times \mathbb{T} \to \mathbb{S}_x^L$ with:*

$$\omega_c^{count}(S_x^L, \hat{t}) := \{(z, \hat{t}, \hat{n}) | \exists Y \subseteq S_x^L : Y \neq \emptyset \land \\ Y = \{(z, t, n) \in S_x^L | max\{m(\hat{t}, S_x^L) - c, 1\} \leq m(t, S_x^L) \leq m(\hat{t})\} \\ \land \hat{n} = \sum_{(z,t,n) \in Y} n\}$$

Triple Pattern Matching. In SPARQL [5] no operator for the transformation from RDF statements to SPARQL is defined. To be more general, we introduce the operator υ that filters statements from a (logical) RDF stream and transforms them to SPARQL solutions. To cope with the problem of blank nodes we need to extend the definition for an RDF instance mapping from [9]:

Definition 6 (Extended RDF instance mapping). *Let $\mathbb{E}$ be the set of all extended RDF instance mappings. An extended RDF instance mapping $\sigma_{ext} \in \mathbb{E}$ is a function σ_{ext} : RDF-T $\cup$ V $\to$ RDF-T $\cup$ V using a function f : RDF-B $\to$ RDF-T:*

$$\sigma_{ext}(x) := \begin{cases} f(x) \in \text{RDF-T}, & \text{if } x \in \text{RDF-B} \\ x, & \text{if } x \in I \cup \text{RDF-L} \cup V \end{cases}$$

We can then define the triple matching operator:

Definition 7 (Triple pattern matching). *Let $p = (x, y, z) \in \Lambda$ be a triple pattern in the set of all triple patterns, $S_{RDF}^L \in \mathbb{S}_{RDF}^L$ a logical data stream and σ_{ext} : RDF-T $\cup$ V $\to$ RDF-T $\cup$ V an extended RDF instance mapping. The triple pattern matching $\upsilon : \mathbb{S}_{RDF}^L \times \Lambda \to \mathbb{S}_\mu^L$ is defined as follows:*

$$\upsilon_p(S_{RDF}^L) := \{(\mu, t, n) | \exists ((s, p, o), t, n) \in S_{RDF}^L : \exists \sigma_{ext} \in \mathbb{E} : \\ \sigma_{ext}(x) \in \text{RDF-T} \Rightarrow \sigma_{ext}(x) = s \land \sigma_{ext}(y) \in \text{RDF-T} \Rightarrow \sigma_{ext}(y) = p \land \\ \sigma_{ext}(z) \in \text{RDF-T} \Rightarrow \sigma_{ext}(z) = o \land \sigma_{ext}(x) \in V \Rightarrow \mu(\sigma_{ext}(x)) = s \land \\ \sigma_{ext}(y) \in V \Rightarrow \mu(\sigma_{ext}(y)) = p \land \sigma_{ext}(z) \in V \Rightarrow \mu(\sigma_{ext}(z)) = o\}$$

The triple pattern matching operator transforms a logical RDF stream into a logical μ data stream, i.e. it changes the schema of the contained elements. All the following described operators consume logical μ streams. Therefore, the triple pattern matching operator must be placed in a plan before any of the following operators.

Filter. The filter operator evaluates the FILTER term of a SPARQL query. It selects from a logical μ stream those elements that satisfy a predicate $p(\mu, t) \in \mathbb{P}_s$ from the set of all SPARQL predicates.[4]

Definition 8 (Filter). *Let* $p : \Phi \times \mathbb{T} \to \{true, false\} \in \mathbb{P}_s$ *be a SPARQL predicate. A filter operator* f_p *is a function* $f : \mathbb{S}_\mu^L \times \mathbb{P} \to \mathbb{S}_\mu^L$, *defined as follows:*

$$f_p(S_\mu^L) := \{(\mu, t, n) | (\mu, t, n) \in S_\mu^L \land p(\mu, t)\}$$

Union. The union of two logical μ streams evaluates the UNION term of a SPARQL query.

Definition 9 (Union of logical μ data streams). *The union $\cup_+$ of two logical μ data streams is a function* $\cup_+ : \mathbb{S}_\mu^L \times \mathbb{S}_\mu^L \to \mathbb{S}_\mu^L$, *defined as follows:*

$$\cup_+(S_{\mu,1}^L, S_{\mu,2}^L) := \{(\mu, t, n_1 + n_2) | \\ (\exists (\mu, t, n_1) \in S_{\mu,1}^L \lor n_1 = 0) \land (\exists (\mu, t, n_2) \in S_{\mu,2}^L \lor n_2 = 0) \land \\ n_1 + n_2 > 0\}$$

The operator adds an element to the result stream if it occurs in one of the two streams. Because $\cup_+$ is a multi-set operator we need to sum the multiplicities [8].

Join. A join operator combines two compatible elements from two μ data streams. Two solutions are compatible [5] if:

$$\mu_1 \simeq \mu_2 : \forall v_1 \in dom(\mu_1), v_2 \in dom(\mu_2) : v_1 = v_2 \Rightarrow \mu_1(v_1) = \mu_2(v_2)$$

$\mu_1(v_1) = \mu_2(v_2)$ does also apply in this work, if v_1 in μ_1 and v_2 in μ_2 are unbound. Further let *merge* : $\Phi \times \Phi \to \Phi$ be a mapping corresponding to the merge function from [5] defined as follows:

$$merge(\mu_1, \mu_2) = \mu_{merge} :\Leftrightarrow \mu_1 \simeq \mu_2,$$
$$\text{with } dom(\mu_{merge}) = dom(\mu_1) \cup dom(\mu_2) \land$$
$$(\forall v \in dom(\mu_1) \backslash dom(\mu_2) : \mu_{merge}(v) = \mu_1(v)) \land$$
$$(\forall v \in dom(\mu_2) \backslash dom(\mu_1) : \mu_{merge}(v) = \mu_2(v)) \land$$
$$(\forall v \in dom(\mu_1) \cap dom(\mu_2) : \mu_{merge}(v) = \mu_1(v) = \mu_2(v))$$

We can then define the join as follows.

Definition 10 (Join). *The join of two logical μ data streams is a function* $\bowtie: \mathbb{S}_\mu^L \times \mathbb{S}_\mu^L \to \mathbb{S}_\mu^L$ *with:*

$$\bowtie (S_{\mu,1}^L, S_{\mu,2}^L) := \{(merge(\mu_1, \mu_2), t, n_1 \cdot n_2) | \exists (\mu_1, t, n_1) \in S_{\mu,1}^L \land \\ \exists (\mu_2, t, n_2) \in S_{\mu,2}^L \land \mu_1 \simeq \mu_2\}$$

Only elements that are valid at the same point in time can be joined to a new element.

---

[4] A SPARQL predicate is a boolean expression, consisting of operators from section 11.3 in [5] with boolean return values whose operands are variables or RDF terms.

Basic Graph Pattern Matching. This operator is defined in the SPARQL algebra, too, although it is a join over two or more triple pattern matchings. To assure that blank nodes are handled correctly, the included triple pattern matchings must use the same RDF instance mapping.

Definition 11 (Basic Graph Pattern matching). *Let $p_1, p_2 \in \Lambda$ be two triple patterns and $S^L_{RDF,1}, S^L_{RDF,2} \in \mathbb{S}^L_{RDF}$ be two logical RDF data streams. A basic graph pattern matching is the mapping $\chi : \mathbb{S}^L_{RDF} \times \mathbb{S}^L_{RDF} \to \mathbb{S}^L_\mu$, defined as follows:*

$$\chi(S^L_{RDF,1}, S^L_{RDF,2}) := \bowtie (\upsilon_{p_1}(S^L_{RDF,1}), \upsilon_{p_2}(S^L_{RDF,2})),$$

with $\sigma_{ext,1} = \sigma_{ext,2}$.

Left Join. In [5] a left join is used to evaluate an OPTIONAL term of a SPARQL query. We combine the left join directly with a filter predicate to preserve the correct OPTIONAL semantics. Unfortunately, the formal definition and its full description in [5] are inconsistent. According to the full description the left join does not deliver a result if there is no right element to potentially join with, so we use the following definition of a left join as opposed to the definition in [5]:

Definition 12 (W3C LeftJoin). *Let $p \in \mathbb{P}_s$ be a SPARQL predicate. A left join $\bowtie_{w3c}$ is a mapping $\bowtie_{w3c} : \Omega \times \Omega \to \Omega$, evaluated as follows:*

$$\bowtie_{w3c} (\Omega_1, \Omega_2) := \{merge(\mu_1,\mu_2) | \mu_1 \in \Omega_2 \wedge \mu_2 \in \Omega_2 \wedge \mu_1 \simeq \mu_2 \wedge p(merge(\mu_1,\mu_2))\}$$
$$\cup \{\mu_1 | \mu_1 \in \Omega_1 \wedge \nexists \mu_2 \in \Omega_2 : \mu_1 \simeq \mu_2 \wedge p(merge(\mu_1,\mu_2))\}$$

Based on this modified left join we give our definition of the stream based left join:

Definition 13 (LeftJoin). *Let $p \in \mathbb{P}_s$ be a SPARQL predicate. A left join $\bowtie$ is a mapping $\bowtie : \mathbb{S}^L_\mu \times \mathbb{S}^L_\mu \times \mathbb{P}_s \to \mathbb{S}^L_\mu$, defined as follows:*

$$\bowtie (S^L_{\mu,1}, S^L_{\mu,2}) := \{(merge(\mu_1,\mu_2), t, n_1 \cdot n_2) | \exists (\mu_1, t, n_1) \in S^L_{\mu,1} \wedge$$
$$\exists (\mu_1, t, n_2) \in S^L_{\mu,2} : \mu_1 \simeq \mu_2 \wedge p(merge(\mu_1,\mu_2))\} \cup_+$$
$$\{(\mu_1, t, n_1) | \exists (\mu_1, t, n_1) \in S^L_{\mu,1} \wedge \nexists (\mu_2, t, n_2) \in S^L_{\mu,2} : \mu_1 \simeq \mu_2$$
$$\wedge p(merge(\mu_1,\mu_2))\}$$

Other Operators. In addition to the described operators we also defined tolist, order by, duplicate elimination, duplicate reduction, slice, projection, construct, describe and ask.

5 Query Translation

We are now able to show how we can translate SPARQL queries over data streams into logical algebra plans. Most transformation rules from [5] can be applied directly, using our new algebra operators. Only the definition of windows over the data streams and the new triple pattern operator need special rules. The triple pattern matching operator transforms RDF statements into SPARQL solutions. Listing 1.2 shows a simple SPARQL query without data streams. The transformation of this query can be found in Figure 1.

```
SELECT ?w ?x ?y ?z
FROM <http://src.net/graph.rdf>
WHERE {?w my:name ?x } UNION { ?y my:power ?z }
```

Listing 1.2. Query 1

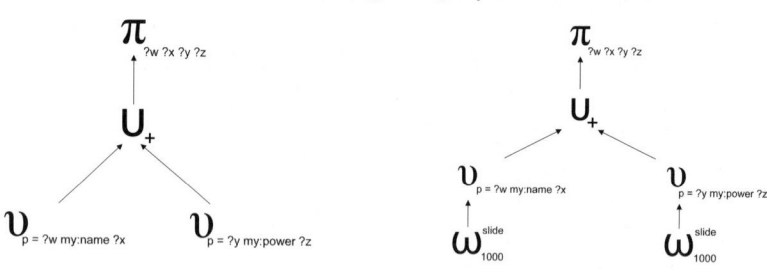

Fig. 1. Query plan for Query 1   **Fig. 2.** Query plan for Query 2

As you can see, the triple patterns {?w my:name ?x} and {?y my:power ?z} are both transformed to one triple pattern matching operator. By this, the RDF statements are filtered from the source and transformed into SPARQL solutions. Afterwards, the union and project operators can process the triple pattern matching results.

Window definitions in the FROM parts of a query are placed directly before the corresponding triple pattern matching operator as demonstrated in Query 2 in Listing 1.3 and Figure 2.

```
SELECT ?w ?x ?y ?z
FROM STREAM <http://src.net/graph.rdf> WINDOW RANGE 1000 SLIDE
WHERE {?w my:name ?x } UNION { ?y my:power ?z }
```

Listing 1.3. Query 2

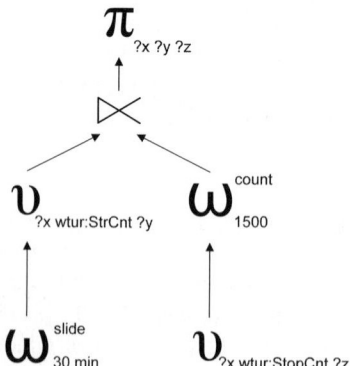

Fig. 3. Query plan for query 1.1

Algorithm 1. Method AlgebraOp createPlan(clause, actStream, namedStreams, win, graphVar)

AlgebraOp retVal = null;
List subplans = new List();
if *clause **instanceof** GroupGraphPattern* **then**
    win = clause.getWindow();
    foreach *Subclause sc: clause.getSubclauses()* **do**
        subplans.add(createPlan(sc, actStream, namedStreams, win, graphVar));
    retVal = createJoinHierarchyOverAll(subplans);
    setLeftJoinInHierarchyIfRightInputIsOptional(retVal);
else if *clause **instanceof** TriplesBlock* **then**
    foreach *Subclause sc: clause.getSubclauses()* **do**
        subplans.add(createPlan(sc, actStream, namedStreams, win, graphVar));
    retVal = createJoinHierarchyOverAll(subplans);
else if *clause **instanceof** GroupOrUnionGraphPattern* **then**
    foreach *Subclause sc: clause.getSubclauses()* **do**
        subplans.add(createPlan(sc, actStream, namedStreams, win, graphVar));
    retVal = createUnionHierarchyOverAll(subplans);
else if *clause **instanceof** Filter* **then**
    retVal = new Filter(clause.getExpression());
else if *clause **instanceof** GraphGraphPattern* **then**
    if *(uri = clause.getGraphClause().getURI()) ≠ null* **then**
        retVal = createPlan(clause.getSubclause(), namedStreams.get(uri),
        namedStreams, win, null);
    else
        foreach *s : namedStreams* **do**
            subplans.add(createPlan(clause.getSubclause(), s, namedStreams, win,
            clause.getGraphVar()));
        retVal = createUnionHierarchyOverAll(subplans);
else if *clause **instanceof** Triple* **then**
    TriplePatternMatching tpm = new TriplePatternMatching(clause, graphVar);
    if *win == null* **then**
        win = actStream.getWindow();
        retVal = tpm.withInput(win.withInput(createAccessOp(actStream.getURI())));
    else
        retVal = win.withInput(tpm.withInput(createAccessOp(actStream.getURI())));
return retVal;

Our complete transformation algorithm can be found in Algorithm 1. The basic idea of this algorithm is to recursively run through the query starting at the WHERE-clause and create operators for the serveral query parts (e. g. a left join operator for an optional graph pattern). With the parameter win it is possible to determine whether to put a window operator of a FROM-part or of a basic graph pattern part of a query into the queryplan.

Finally, we present the translation result of the query in Listing 1.1 in Figure 3 in our logical algebra.

6 Physical SPARQL Data Stream Algebra

The logical algebra is used to define the operator semantics and allow some static plan optimizations. A logical algebra cannot be used to execute queries. For this a physical algebra with executable operators is needed. In line with [6] we apply an interval based approach for the physical algebra.

The set of operators is divided into two groups. Stateless operators do not need any information about the stream history or stream future. Each stream element can be processed directly. Operators of this group are triple pattern matching, filter, union, project, construct and our time based window operators. The other group contains stateful operators. These operators need further information to process an element, e.g. a sort operator needs to know all elements before it can sort the set. Join, left join, basic graph pattern matching, duplicate elimination/reduction, orderBy, slice, ask and element based windows are stateful operators. They need a special data structure to process their input. As in [6] a so called sweep area is used as this data structure. It is an abstract datatype with methods to insert, replace or query elements. We will not further describe this data structure but refer to [6].

We will present our implementation of the sliding δ-window (Algorithm 2) first: A sliding δ window has a width $w \in \mathbb{T}$ and a step size $\delta \in \mathbb{T}$ to move the window over the data stream. In the following, x is used to either describe the set of RDF statements or the set of SPARQL solutions; z is an instance of this set.

Algorithm 2. Sliding δ window

Input: $S^P_{x,in} \in \mathbb{S}^P_x, w, \delta \in \mathbb{T}, \delta \leq w$
Output: $S^P_{x,out} \in \mathbb{S}^P_x$
$S^P_{x,out} \leftarrow \emptyset$;
foreach $e := (z, [t_S, t_E]) \hookleftarrow S_{RDF,in}$ **do**
$\quad e.t_E := \left(\left\lfloor \frac{t_S}{\delta} \right\rfloor \cdot \delta \right) + w;$
$\quad e \hookrightarrow S^P_{RDF,out};$

If a new element occurs in the stream, it will be read out and its end timestamp will be set according to $\left(\left\lfloor \frac{t_S}{\delta} \right\rfloor \cdot \delta \right) + w$. Afterwards, the element will be written into the output stream of the operator.

As stated above, a triple pattern matching transforms filtered statements of an RDF input stream to SPARQL solutions. This operator is stateless and can be implemented as in Algorithm 3.

The algorithm gets a triple pattern and an extended RDF instance mapping σ_{ext} as input. If the subject, predicate and object of the triple pattern match the subject, predicate and object of an RDF stream element, a SPARQL solution is generated, where variables of the triple pattern are mapped to the corresponding terms in the RDF stream statement. The SPARQL solution is written to the μ output stream.

We present our solution for the LeftJoin operator in Algorithm 4 and 5.

Algorithm 3. Triple Pattern Matching

Input: $S^P_{RDF,in} \in \mathbb{S}^P_\mu, p = (x, y, z) \in \Lambda, \sigma_{ext} \in \mathbb{E}$
Output: $S^P_{\mu,out} \in \mathbb{S}^P_\mu$
$S^P_{\mu,out} \leftarrow \emptyset$;
SPARQL Solution μ := new SPARQL Solution();
foreach $e := ((s, p, o), [t_S, t_E]) \hookleftarrow S_{RDF,in}$ **do**
  if $(x \in V \lor \sigma_{ext}(x) == e.s) \land (y \in V \lor \sigma_{ext}(y) == e.p) \land (z \in V \lor \sigma_{ext}(z) == e.o)$
  then
    if $x \in V$ **then**
      μ.add(x,s);
    if $y \in V$ **then**
      μ.add(y,p);
    if $z \in V$ **then**
      μ.add(z,o);
    $\mu \hookrightarrow S^P_{\mu,out}$;

Algorithm 4. LeftJoin

Input: $S^P_{\mu,in,1}, S^P_{\mu,in,2} \in \mathbb{S}^P_\mu$, SweepAreas SA_1, SA_2, Min-Heap $H, p \in \mathbb{P}_s$
Output: $S^P_{\mu,out} \in \mathbb{S}^P_\mu$
Let $S^P_{\mu,in,1}$ be the left and $S^P_{\mu,in,2}$ the right input stream;
foreach $e := (\mu, [t_S, t_E]) \hookleftarrow S^P_{\mu,in,j}, j \in \{1, 2\}$ **do**
  if $j == 1$ **then**
    /*Input from left input stream   */
    doLeft(this);
  else
    /*Input from right input stream   */
    doRight(this);
  $min_{t_S} \leftarrow min(\{t_S | (\mu, [t_S, t_E]) \in SA_1\})$;
  if $min_{t_S} \neq null$ **then**
    while $H \neq \emptyset$ **do**
      $top := (\tilde{\mu}, [\tilde{t}_S, \tilde{t}_E]) \leftarrow$ top element of H;
      if $top.\tilde{t}_S < min_{t_S}$ **then**
        $top \hookrightarrow S^P_{\mu,out}$;
        Remove top from H;
      else
        break;

7 Related Work

This work is based on the W3C Recommendation SPARQL [5], especially on the defined algebra and grammar. Also, possibilities for the serialization of RDF statements in data streams have been described. Our work is based on [6]. In that work a schema independent algebra for stream processing has been introduced. Because of some differences between the relational model and the RDF graph model, this algebra could not completely be reused for RDF stream processing. But some ideas like the time instant based approach for logical algebra operators and the interval based approach for physical algebra operators offer advantages for the definition of algebra operators in this work.

Algorithm 5. doLeft(LeftJoin op) and doRight(LeftJoin op) of a LeftJoin

doLeft(LeftJoin op);
SA_2.purgeElements(e);
$e.\tilde{t}_E = e.t_S$;
Iterator qualifies $\leftarrow SA_2$.query(e);
while *qualifies.hasNext()* **do**
  $\hat{e} := (\hat{\mu}, [\hat{t}_S, \hat{t}_E]) \leftarrow$ qualifies.next();
  if $e \bowtie \hat{e} \land p(merge(\mu, \hat{\mu}))$ **then**
    $e_{merge} := (merge(\mu, \hat{\mu}), intersection([t_S, t_E], [\hat{t}_S, \hat{t}_E]))$;
    if $e_{merge}.t_S > e.\tilde{t}_E$ **then**
      Insert $(\mu, [e.\tilde{t}_E, e_{merge}.t_S))$ into H;
    Insert e_{merge} into H;
    $e.\tilde{t}_E = e_{merge}.t_E$;
  else
    intersect := intersection$([t_S, t_E], [\hat{t}_S, \hat{t}_E])$;
    if *intersect*.$t_S > e.\tilde{t}_E$ **then**
      Insert $(\mu, [e.\tilde{t}_E, intersect.t_S))$ into H;
      $e.\tilde{t}_E = intersect.t_S$;

SA_1.insert(e);
doRight(LeftJoin op);
Iterator invalids $= SA_1$.extractElements(e);
while *invalids.hasNext()* **do**
  $\check{e} :=$ invalids.next();
  if $\check{e}.\tilde{t}_E < \check{e}.t_E$ **then**
    Insert $(\check{e}.\mu, [\check{e}.\tilde{t}_E, \check{e}.t_E))$ into H;

Iterator qualifies $\leftarrow SA_1$.query(e);
while *qualifies.hasNext()* **do**
  $\hat{e} := (\hat{\mu}, [\hat{t}_S, \hat{t}_E]) \leftarrow$ qualifies.next();
  if $e \bowtie \hat{e} \land p(merge(\mu, \hat{\mu}))$ **then**
    $e_{merge} := (merge(\mu, \hat{\mu}), intersection([t_S, t_E], [\hat{t}_S, \hat{t}_E]))$;
    if $e_{merge}.t_S > \hat{e}.\tilde{t}_E$ **then**
      Insert $(\hat{\mu}, [\hat{e}.\tilde{t}_E, e_{merge}.t_S))$ into H;
    Insert e_{merge} into H;
    $\hat{e}.\tilde{t}_E = e_{merge}.t_E$;
  else
    intersect = intersection$([t_S, t_E], [\hat{t}_S, \hat{t}_E])$;
    if *intersect*.$t_S > \hat{e}.\tilde{t}_E$ **then**
      Insert $(\hat{\mu}, [\hat{e}.\tilde{t}_E, intersect.t_S))$ into H;
      $\hat{e}.\tilde{t}_E := intersect.t_S$;

SA_2.insert(e);

Another approach for the expression of the validity of data stream elements is the positive-negative tuple approach introduced in [10]. This approach allows for using so called negative tuples, that mark the end of their positive counterparts. But this approach has the disadvantage that at least twice the elements have to be processed in comparison to the interval based approach of [6] and that all algebra operators have to distinguish between different types of elements.

There are other approaches like [11, 12] which also introduce stream processing languages. But the continuous query language CQL [11] uses the relational model in defining operators that transform a stream into a relation and vice versa. This is not useful for SPARQL stream processing because of the differences between the RDF and the relational models. It is the same with the extended XQuery language [12]. This language provides many constructs for handling XML data. Indeed RDF can be serialized as XML, but the extended XQuery constructs do not handle the RDF semantics correctly. So a real RDF query language has to be extended for stream processing.

Extending SPARQL for stream processing window operators had to be integrated into the SPARQL language. These operators have been introduced earlier (see [11, 12, 13, 14, 15]). Ideas for windows have been taken from these works, but also in this case the interval based approach in [6] offer advantages in implementing the corresponding physical algebra operators. So time based windows can be realized by calculating the end of a validity interval and tuple based windows can be realized by using a sweep area and setting the end of an earlier element to the start of a later element (see [6]).

8 Conclusions and Outlook

In this work we presented an extension to SPARQL to cope with window queries over data streams. We extended the SPARQL language to allow the definition of time and count based windows over data streams. We implemented the extended SPARQL processing in our ODYSSEUS system, which is an enhancement of our DynaQuest [16] framework with the ability to process data streams. For the translation of SPARQL we used the ARQ project[5] and slightly extended their query translation process. We added to this base our own query translation and execution framework and extended it with the described logical and physical algebra operators. We showed that query processing over RDF streams is possible. We now need to determine if this format is applicable for wind park monitoring. We will do this in the context of the Alpha Ventus[6] project, which creates an offshore test platform in the North Sea. Another application for our approach might be in the context of mobile information systems as they are developed in the C3World[7] research group.

There are multiple possible extensions for this work. Predicate based windows [17] define the boundaries using predicates over the stream content. We have already developed initial approaches but further research is necessary. Additionally, we are currently extending SPARQL to support group by clauses and aggregation, which are necessary in monitoring applications. Also alternatives to using windows like the positive-negative tuple approach [10] will be evaluated in future work.

References

1. Babu, S., Widom, J.: Streamon: An adaptive engine for stream query processing. In: Demo Track Session of ACM SIGMOD Conference (2004)
2. International Electrotechnical Commission Technical Commitee 88: 61400-25 communications for monitoring and control of wind power plants(version 61400-25-1_r0-4draftfdis_2006-05-31). Technical report, International Electrotechnical Commission (2006)
3. Manola, F., Miller, E., McBride, B.: RDF Primer. W3C Recommendation. In: World Wide Web Consortium (2004)
4. Tatbul, N., Zdonik, S.: Window-aware load shedding for aggregation queries over data streams. In: VLDB 2006: Proceedings of the 32nd international conference on Very largedata bases, VLDB Endowment, pp. 799–810 (2006)

---

[5] http://jena.sourceforge.net/ARQ/
[6] http://www.alpha-ventus.de/
[7] http://www.c3world.de/

5. Prud'hommeaux, E., Seaborne, A.: SPARQL Query Language for RDF. W3C Recommendation, World Wide Web Consortium (2007), http://www.w3.org/TR/rdf-sparql-query/
6. Krämer, J.: Continuous Queries over Data Streams - Semantics and Implementation. PhD thesis, University of Marburg (2007)
7. Bettini, C., Dyreson, C.E., Evans, W.S., Snodgrass, R.T., Wang, X.S.: A glossary of time granularity concepts. In: Temporal Databases, Dagstuhl, pp. 406–413 (1997)
8. Dayal, U., Goodman, N., Katz, R.H.: An extended relational algebra with control over duplicate elimination. In: PODS 1982: Proceedings of the 1st ACM SIGACT-SIGMOD symposium on Principles of database systems, pp. 117–123. ACM Press, New York (1982)
9. Hayes, P., McBride, B.: RDF Semantics. W3C Recommendation (2004)
10. Ghanem, T.M., Hammad, M.A., Mokbel, M.F., Aref, W.G., Elmagarmid, A.K.: Query processing using negative tuples in stream query engines. Technical report, Purdue University (2004)
11. Arasu, A., Babu, S., Widom, J.: An abstract semantics and concrete language for continuous queries overstreams and relations. In: 9th Interantional Workshop on Database Programming Languages, Stanford University, pp. 1–11 (2003)
12. Carabus, I., Fischer, P.M., Florescu, D., Kraska, D.K.T., Tamosevicius, R.: Extending xquery with window functions. Technical report, ETH Zürich (2006)
13. Sullivan, M.: Tribeca: A stream database manager for network traffic analysis. In: Vijayaraman, T.M., Buchmann, A.P., Mohan, C., Sarda, N.L. (eds.) VLDB 1996, Proceedings of 22th International Conference on Very Large DataBases, Mumbai (Bombay), India, September 3-6, p. 594. Morgan Kaufmann, San Francisco (1996)
14. Carney, D., Çetintemel, U., Cherniack, M., Lee, C.C.S., Seidman, G., Stonebraker, M., Zdonik, N.T.S.B.: Monitoring streams - a new class of data management applications. In: Bressan, S., Chaudhri, A.B., Li Lee, M., Yu, J.X., Lacroix, Z. (eds.) CAiSE 2002 and VLDB 2002. LNCS, vol. 2590, pp. 215–226. Springer, Heidelberg (2003)
15. Seshadri, P., Livny, M., Ramakrishnan, R.: Seq: A model for sequence databases. In: ICDE 1995: Proceedings of the Eleventh International Conference on Data Engineering, pp. 232–239. IEEE Computer Society, Washington, DC, USA (1995)
16. Grawunder, M.: DYNAQUEST: Dynamische und adaptive Anfrageverarbeitung in virtuellen Datenbanksystemen. PhD thesis, University of Oldenburg (2005)
17. Ghanem, T.M., Aref, W.G., Elmagarmid, A.K.: Exploiting predicate-window semantics over data streams. SIGMOD Rec. 35(1), 3–8 (2006)

The Creation and Evaluation of iSPARQL Strategies for Matchmaking

Christoph Kiefer and Abraham Bernstein

Department of Informatics, University of Zurich, Switzerland
{kiefer,bernstein}@ifi.uzh.ch

Abstract. This research explores a new method for Semantic Web service matchmaking based on iSPARQL strategies, which enables to query the Semantic Web with techniques from traditional information retrieval. The strategies for matchmaking that we developed and evaluated can make use of a plethora of similarity measures and combination functions from SimPack—our library of similarity measures. We show how our combination of structured and imprecise querying can be used to perform hybrid Semantic Web service matchmaking. We analyze our approach thoroughly on a large OWL-S service test collection and show how our initial strategies can be improved by applying machine learning algorithms to result in very effective strategies for matchmaking.

1 Introduction

Imagine the following situation: Kate, a young and successful Semantic Web researcher, is trying to find a set of services to invoke for her latest project. Unfortunately, her department does not offer any useful semantically annotated services. She decides to look for some suitable services in a large database of crawled OWL ontologies that her department has gathered. Two issues arise: (1) as the task requires searching, she probably wants to use SPARQL; and (2), because there does not seem to be an ultimate, widely-accepted ontology for the domain, she will also have to consider approximate matches to her queries—a task for which statistics-based techniques from traditional information retrieval (IR) are well-suited.

In this paper, we address the task of matching a given user request with available information in a particular knowledge base (KB) assuming the lack of a perfect domain model—one of the basic underlying assumptions in Semantic Web research. More precisely, for a given input service request (*i.e.*, Kate's preference criteria), we aim to find all the relevant services that are closest to the query.

While many matchmaking algorithms have been proposed [11,14,17,18,24], we suggest not to build a specialized matchmaker (algorithm) but to build a matchmaker out of off-the-shelf components that embeds both structured (*i.e.*, logical) as well as statistical elements. Our method is based on iSPARQL queries (*i.e.*, *iSPARQL strategies*) [13], which enable the user to query the Semantic Web with a SPARQL extension that incorporates the notion of similarity—an approach often used in traditional IR. The strategies for matchmaking that we developed and evaluate in this paper make use of a plethora of similarity measures and combination functions from SimPack—our library of similarity measures for the use in ontologies.[1]

We show how simple it is to find well-performing matchmaking strategies in an iterative and 'playful' procedure. Furthermore, we show how the performance of the initial matchmaking strategies can be greatly improved by applying standard machine learning algorithms such as regression, decision trees, or support vector machines (SVMs) to result in very effective strategies for matchmaking.

To evaluate our approach, we took a large collection of OWL-S services and compared the effectiveness of a number of simply constructed and learned/induced iSPARQL strategies in the search of the most effective strategies. We found that some simple-to-construct strategies performed surprisingly well, whilst a simple extension of those strategies based on machine learning outperformed even one of the most sophisticated matchmakers currently available.

The paper is structured as follows: next, we introduce the most important related work, before we give a brief overview of our iMatcher approach for matchmaking in Section 3. Section 4 discusses the performance measures and data sets used in the evaluation. The results of our experiments are presented in Section 5. We close the paper with a discussion of the results, our conclusions, and some insights into future work.

2 Related Work

In 1999, Sycara *et al.* [24] proposed LARKS—the Language for Advertisement and Request for Knowledge Sharing. The language defines a specification for the input and output parameters of services (among others) as well as for the constraints on these values. Their matchmaking process contains a *similarity filter* that computes the distances between service descriptions using TF-IDF [1] and other ontology-based similarity measures.

Three years later, Paolucci *et al.* [18] continued the work of [24] focusing on DAML-S.[2] One of the major differences between the LARKS and DAML-S approach is that the latter solely relies on logic and ontological reasoning.

Also related is the work of Di Noia *et al.* [17] who proposed a service matchmaking approach based on CLASSIC [4]. They discuss a purely logic-based implementation that matches service demands and supplies based on their explicit normal form (*i.e.*, demands and supplies are terminologically unfolded into their names, number restrictions, and universal role quantifications). The matchmaking algorithm then distinguishes between *potential* and *partial* matches of demands and supplies (*i.e.*, matches with no conflicts and matches with conflicting properties of demands and supplies).

In 2005, Jaeger *et al.* [11] presented the OWLSM approach for matching service inputs, service outputs, a service category, and user-defined service matchmaking criteria. The four individual matching scores are aggregated resulting in

---
[1] http://www.ifi.uzh.ch/ddis/simpack.html
[2] http://www.daml.org/

an overall matchmaking score. The result to the user is a ranked list of relevant services.

Recently, Klusch et al. [14] introduced the OWLS-MX matchmaker. As its name already implies, OWLS-MX focuses on the matching of services described in OWL-S, the successor of LARKS and DAML-S. The main focus of the matcher lies on a service profile's input and output (I/O) parameters. In comparison to the approaches proposed in [17,18], OWLS-MX uses both logic-based as well as IR-based matching criteria to identify services which match with a given query service. Specifically, OWLS-MX successively applies five different matchmaking filters: *exact*, *plug in*, *subsumes*, *subsumed-by*, and *nearest-neighbor*. Among those, the first three are purely logic-based, whereas *subsumed-by* and *nearest-neighbor* are hybrid as they incorporate a similarity measure to compute the syntactic similarity between query and service I/O concept terms. As such, the hybrid filters let some *syntactically similar*, but *logically disjoint* services be included in the answer set of a query, and thus, help improve query precision and recall. To compute the similarities, OWLS-MX particularly relies on similarity measures which have been shown to perform well in information retrieval [6].

A similar approach to [14] is described by Bianchini et al. [3], in which they presented the FC-MATCH hybrid matchmaking algorithm that is based on WordNet and the Dice coefficient [8].[3]

Finally, we would like to mention two studies that are similar to iSPARQL—the approach our matchmaker is based on. First, Corby et al. [7] introduced the conceptual graph-based Corese search engine that defines an RDF query language enabling both ontological and structural approximations.[4] Ontological approximation deals with distances between nodes in ontologies, whereas structural approximation is about structural divergences between the query and the queried RDF data set. Especially the latter allows the user to search for resources related by an arbitrary relations path between them. Second, Zhang et al. [27] presented the SPARQL-based Semplore system that combines structured querying with keyword searches using existing IR indexing structures and engines.

3 The iMatcher Approach

In this section we explain how our proposed matchmaking approach called *iMatcher* works. To that end, we first introduce the underlying iSPARQL technology and then explain the functionality of iMatcher itself.

3.1 Foundations: iSPARQL Matchmaking Strategies

The matchmaking approach we present in this paper is based on our previous work on iSPARQL [13]. We, therefore, succinctly review the most fundamental concepts. In iSPARQL we make use of ARQ property functions to apply similarity operators to SPARQL query elements.[5] The concept behind property

---
[3] WordNet lexical database of English, http://wordnet.princeton.edu/
[4] Conceptual graphs working draft, http://www.jfsowa.com/cg/cgstand.htm
[5] http://jena.sourceforge.net/ARQ/extension.html#propertyFunctions

```
 1  PREFIX isparql: <java:isparql.>
 2  PREFIX profile: <http://www.daml.org/services/owl-s/1.1/Profile.owl#>
 3
 4  SELECT ?p2 ?sim1 ?sim2 ?sim
 5  WHERE
 6    { # Basic graph pattern (BGP) matching part (lines 4-8)
 7        <P1> profile:name ?name1 ;
 8             profile:desc ?desc1 .
 9        ?p2  profile:name ?name2 ;
10             profile:desc ?desc2 .
11
12        # Virtual triple pattern matching part (lines 12-14)
13        IMPRECISE
14          { ?sim1 isparql:nameSimilarity ( ?name1 ?name2 ) .
15            ?sim2 isparql:textSimilarity ( ?desc1 ?desc2 ) .
16            ?sim  isparql:aggregate      ( 0.3 ?sim1 0.7 ?sim2 ) .
17          }
18    } ORDER BY DESC (?sim)
```

Listing 1.1. Example iSPARQL matchmaking strategy

functions is simple: whenever the predicate of a triple pattern is prefixed with a special name (*e.g.*, `isparql`), a call to an external similarity function is made and arguments are passed to the function (in our case by the object of the triple pattern). The similarity between the arguments is computed and bound to the subject variable of the triple pattern. As an example, consider the simple iSPARQL strategy in Listing 1.1 that matches the service profile *P1* to the following RDF data set (note that *P1*, *P2*, and *P3* stand for particular URLs, *e.g.*, http://example.org/grocerystore_food_service.owls#GROCERYSTORE_FOOD_PROFILE):

$$D = \{ \;(P1, \;\text{profile:name}, \;\text{``FoodService''}),$$
$$(P1, \;\text{profile:desc}, \;\text{``Returns food of grocery store''}),$$
$$(P2, \;\text{profile:name}, \;\text{``StoreService''}),$$
$$(P2, \;\text{profile:desc}, \;\text{``This service returns store food''}),$$
$$(P3, \;\text{profile:name}, \;\text{``GroceryFoodService''}),$$
$$(P3, \;\text{profile:desc}, \;\text{``A grocery service selling food''}) \;\}$$

We briefly explain the evaluation procedure of this strategy. Refer to [13] for a more elaborate description of this procedure. In a nutshell, the evaluation includes the following two steps: first, the basic graph pattern (BGP) matching part (lines 7–10) returns the Cartesian product of the sets of bindings from the defined variables; and second, the evaluation of the virtual triple patterns (lines 14–16) computes the similarities between the passed arguments and merges them with the solutions from BGP matching. Applying simple name and text similarity measures from SimPack—a library of similarity measures which iSPARQL uses—this could end up in the similarities $sc_{p_1,p_2} = 0.3 \cdot 0.571 + 0.7 \cdot 0.667 = 0.8095$ for *P1* and *P2*, and $sc_{p_1,p_3} = 0.3 \cdot 0.8 + 0.7 \cdot 0.189 = 0.372$ for *P1* and *P3*, being *P2* the most accurate service profile for *P1* (note that $sc_{p_1,p_1} = 1.0$).

It is important to note that the discovery phase of services is potentially very costly with our implementation as we do not optimize our iSPARQL strategies. Our queries likely result in the evaluation of expensive cross joins of triples as well as the (possibly) repeated computation of similarities in the evaluation of the virtual triple patterns. The first issue can be addressed with iSPARQL query optimization, which we investigated in [2,22]. The second issue, the optimization of virtual graph patterns inside an IMPRECISE clause, can be addressed with similarity indexes to cache repeated similarity computations—an issue which we have not addressed so far.

3.2 The iMatcher Procedure

Our proposed matchmaking system is called *iMatcher*.[6] The "i" stands for *imprecise* emphasizing its ability for approximate matching using techniques from information retrieval (IR). iMatcher performs hybrid Semantic Web service matchmaking. That is, it uses both logic- and IR-based techniques to search for suitable services, which makes it similar to other hybrid systems such as OWLS-MX [14] and FC-MATCH [3]. On the one hand, iMatcher uses logical inferencing in ontologies to reason about the services under consideration, and on the other hand, it computes statistics about the data as well as makes use of indexes and weighting schemes to determine the degree of match between queries and services.

Specifically, iMatcher computes similarities between the query and the services in the KB. The more similar a service to a query, the more likely it is to be a correct result. As parameters, iMatcher is given a set of similarity measures SM, a data set of queries and correct answers T, and a learning algorithm R. To *train iMatcher*, it first computes a similarity-based training set T^{sim} that contains the similarities between all the queries and services in T for all measures in SM and an indication if the combination of the service and the query is correct (*i.e.*, a true positive) or not. It then applies the algorithm R to T^{sim} to learn/induce an induction model M. To *use iMatcher*, it computes the similarities (by the measures SM) of a given query q to all services in a given KB, then uses M to predict the combined similarity (or likelihood) of a match, and returns the answers in decreasing order of similarity.

As iMatcher is based on iSPARQL, any similarity measure defined within SimPack can be use for SM. For the learning algorithms R, any induction algorithm implementing the Weka interface can be employed.[7] As a training set T, iMatcher expects a training data set consisting of a knowledge base with services, a list of queries, and a file specifying which services are the correct answers for any given query.

For the evaluation we ran the results through a simple statistics handler that knows of the set of relevant services (true positives) for a given input query service. This information is used to compute the precision vs. recall figures that we show throughout our evaluation (see Section 5). Furthermore, iMatcher implements the *SME2Plugin* interface to initialize the matchmaker with a particular service knowledge base and to run queries.[8]

---

[6] http://www.ifi.uzh.ch/ddis/imatcher.html
[7] http://www.cs.waikato.ac.nz/~ml/weka/
[8] Used at the 1st International Semantic Service Selection Contest (S3) at ISWC 2007 (http://www-ags.dfki.uni-sb.de/~klusch/s3/index.html).

4 Performance Evaluation

In this section we first briefly review the performance measures we use in our evaluation and then describe our detailed experimental procedure.[9] Note that both the performance measures and the procedure are slight extensions of the ones introduced and, hence, validated by Klusch [14].

4.1 Performance Measures for Matchmaking

Precision and Recall. Probably the most often used performance measures for matchmaking are precision and recall [1]. Given a query q, precision Pr is the fraction of the answer set A_q (retrieved by the matchmaker) that is relevant to the query, whereas recall Re is the fraction of relevant documents R_q which have been retrieved, i.e., $Pr_q = \frac{|Ra_q|}{|A_q|}$, $Re_q = \frac{|Ra_q|}{|R_q|}$, where Ra_q denotes the subset of relevant documents in A_q. As the evaluation of a single query is oftentimes not sufficient to make a statistically significant statement, many queries are involved and the averages of precision and recall computed.

Macro-Average. We are interested in *macro-averaging* precision and recall [20,21] over all queries, as it gives equal weight to each user query [16]. We introduce the set $L = \{0, 0.05, 0.1, \ldots, 1.0\}$ of 21 *standardized recall levels* [1] as our goal is to compute average precision vs. recall figures (see Section 5) for each investigated matchmaking strategy. Furthermore note that an interpolation procedure (aka *ceiling*) is necessary to compute precision and recall at each standardized recall level as each query likely has a different number of relevant services (i.e., different R_q values). For each query q_i, $i \in \{1, \ldots, n\}$, macro-averaging computes precision and recall separately at each level j, where $j \in L$, and then computes the mean of the resulting n values, i.e.,

$$Pr_j^M = \frac{1}{n} \times \sum_{i=1}^{n} \frac{|Ra_j^{q_i}|}{|A_j^{q_i}|} \qquad Re_j^M = \frac{1}{n} \times \sum_{i=1}^{n} \frac{|Ra_j^{q_i}|}{|R_{q_i}|}$$

R-Precision. R-Precision RPr [1] is used in this paper as single value summary of the performance of a particular matchmaking algorithm. It is the fraction of relevant services which have been retrieved when considering only the first $|R_q|$ services of the answer set, denoted Ra_q^*, i.e., $RPr_q = \frac{|Ra_q^*|}{|R_q|}$. We use R-Precision to compare two matchmaking algorithms A and B on a per-query basis, i.e.,

$$RPr_q^{A/B} = RPr_q^A - RPr_q^B$$

A positive result for $RPr_q^{A/B}$ highlights the fact that algorithm A is more effective than B for query q, and vice versa. A zero result denotes equal performance of both algorithms.

---

[9] For an elaborate treatment of these measures, refer to the textbook of Baeza-Yates and Ribeiro-Neto [1] as well as to the [16,21].

4.2 iMatcher Evaluation Procedure and Benchmarking Data Set

The evaluation procedure is the following: based on our experiences, we (1) create a set of iSPARQL strategies (see Section 3) that we assume will perform well; for each query, we (2) run our matchmaker to obtain a list of services ranked by their similarity to the query; given these ranked lists of services, we (3) compute macro-averages by considering the queries' relevance sets (their true answers) as explained in Section 4.1; finally, we (4) plot the results to visually examine the effectiveness of the created matchmaking strategies. By iteratively identifying and replacing weak parts of strategies with parts that improve the matchmaking performance, we are able to find well-performing strategies for the given task and data set.

For all our experiments we use *OWLS-TC v2*—an OWL-S Semantic Web service retrieval test collection as benchmarking data set.[10] The collection consists of 576 services from seven different domains. It specifies 28 queries with their relevance sets (true answers) allowing us to compute the aforementioned statistics for our matchmaker.

5 Experimental Results

The ultimate goal of our experiments was to demonstrate the ease of use of iMatcher to perform Semantic Web service matchmaking based on iSPARQL strategies. To that end, we evaluated three sets of strategies: (1) *primitive strategies*: evaluation of a set of simple, off-the-shelf strategies; (2) *induced strategies*: assessment of the quality of machine learning techniques for matchmaking; and (3) *customized strategies*: estimation of the improvements of iteratively improved, self-engineered strategies. We close our experiments with a short comparison of strategies from other systems.

5.1 Primitive Strategies

String-based Strategies. We start our evaluation with the comparison of simple string-based similarity strategies for matchmaking (see Figure 1(a)). In addition, the results for *OWLS-MX M4* (measure no. 6) are shown on this and every subsequent precision vs. recall figure presented in this work. OWLS-MX M4 is reported to be the best-performing matchmaker variant of the OWLS-MX matchmaker [14]. It uses the extended Jaccard similarity coefficient to compare two services based on their sets of unfolded input/output concepts.[11] Furthermore, Figure 1(a) illustrates the results of the TF-IDF measure (no. 5) that compares services based on their service descriptions [1].

As the results in Figure 1(a) show, TF-IDF clearly outperforms all other measures in terms of precision and recall. It also outperforms OWLS-MX M4

---

[10] http://projects.semwebcentral.org/projects/owls-tc/
[11] In all our experiments, we used nearest-neighbor as minimum degree of match and a value of 0.7 as syntactic similarity threshold for OWLS-MX M4. These values were suggested by the authors of OWLS-MX to obtain good results for OWLS-TC v2.

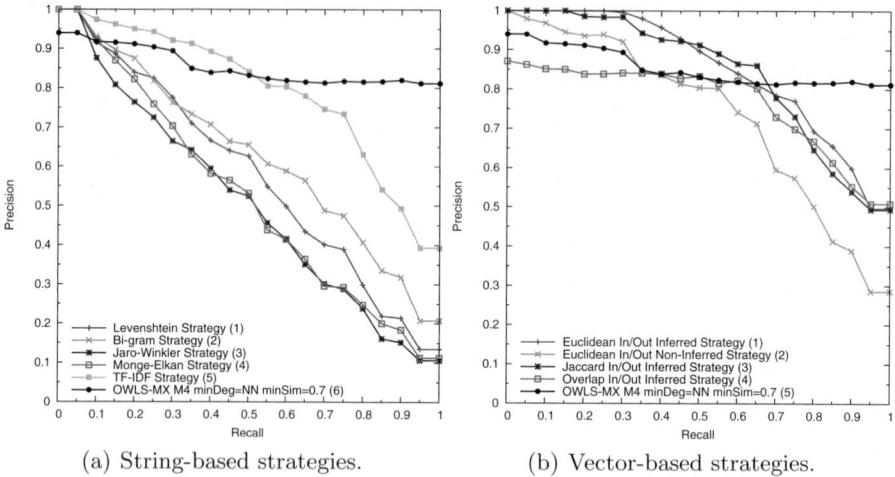

(a) String-based strategies.   (b) Vector-based strategies.

Fig. 1. Performance comparison of simple string- and vector-based strategies

until about half of the relevant services have been retrieved. Both strategies find all relevant services in the end (recall of 1.0), but OWLS-MX M4 with much higher, almost constant precision.

A clear performance trend is recognizable among the set of measures that syntactically compare the names of the services (measures no. 1–4). Among them, the Bi-gram string similarity measure (no. 2) performs slightly better than other prominent measures from this domain, such as the Levenshtein string similarity (no. 1) [15] and the Jaro measure (even with Winkler's reweighting scheme) [25].

Figure 2(a) additionally underscores the superiority of TF-IDF over the other strategies showing exemplarily a comparison with the Bi-gram measure on a per-query basis. For 19 out of 28 queries, comparing their textual descriptions with descriptions of services of the service KB turned out to be more efficient for matchmaking than comparing their service names. We speculate that these performance figures could be improved using even richer textual service descriptions.

We learned from this evaluation that the simple TF-IDF full-text similarity measure is very well suited for matchmaking in OWLS-TC v2 to achieve good results. We, therefore, will reuse it in subsequent strategies. Furthermore, to compare service names, we will use the Bi-gram measure.

Vector-based Strategies. Next, we compare a set of simple vector-based strategies for matchmaking. The results are depicted in Figure 1(b). Basically, these strategies take the sets of service I/O concepts, logically unfold them (using the Pellet reasoner), transform them to vectors, and measure the similarity between those vectors using one of the vector similarity measures from SimPack.

More formally, let n, m be the number of input concepts of two particular web services a and b. The set of all input concepts I_a, I_b can be represented as binary vectors $\mathbf{x}_a$, $\mathbf{x}_b$ of size $|I_a| + |I_b| - |I_a \cap I_b|$, where $|I_a| = n$, $|I_b| = m$.

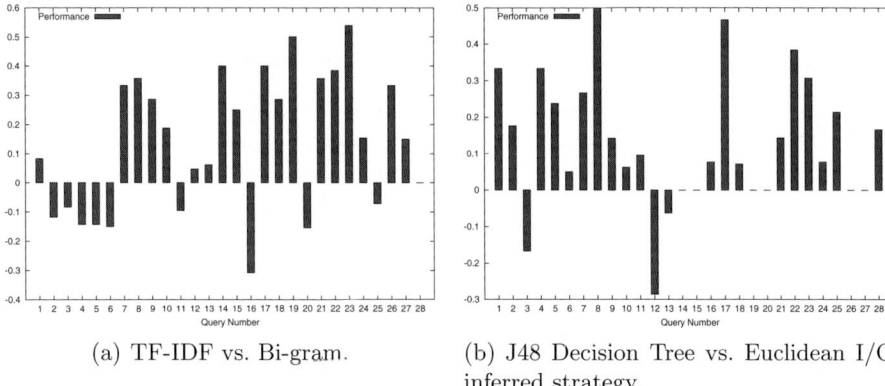

(a) TF-IDF vs. Bi-gram.

(b) J48 Decision Tree vs. Euclidean I/O inferred strategy.

Fig. 2. R-Precision comparison of selected matchmaking strategies

Consider, for example, the two services *surfing-beach-service* and *sports-town-service* from OWL-S TC v2 with their sets of unfolded input concepts $I_a = \{Surfing, Sports, Activity, Thing\}$ and $I_b = \{Sports, Activity, Thing\}$. The vector representation of these two sets is $x_a = [1, 1, 1, 1]^T$ and $x_b = [0, 1, 1, 1]^T$, that have, for instance, an extended Jaccard similarity of $\frac{|I_a \cap I_b|}{|I_a|+|I_b|-|I_a \cap I_b|} = 0.75$. Likewise, the similarity of the vectors representing the services' outputs is computed, which results in a value of 0.4. Averaging the two similarity scores, we obtain an overall similarity of 0.575 for the compared services a and b.

Figure 1(b) illustrates that simple measures such as the metric Euclidean vector distance are sufficient to achieve good results for I/O-based service matchmaking.[12] The extended Jaccard measure is only slightly outperformed by the simple Euclidean distance. The figure also shows that the Overlap measure fulfills the matchmaking task less precisely than the Euclidean and the extended Jaccard measure for about 65% of retrieved services. After that, its performance is comparable to the other measures.

It is interesting to observe the remarkable influence of ontological reasoning over the I/O concepts on the matchmaking task. Comparing the Euclidean measure with reasoning support turned on (measure no. 1 in Figure 1(b)) vs. the same measure without reasoning (no. 2) evidently shows that enabling reasoning boosts performance about 20% for the given matchmaking task. As a consequence, we will use reasoning in all subsequent experiments.

5.2 Machine-Learned Strategies

As we showed in our previous work [12], techniques from machine learning are well-suited for Semantic Web service discovery. Hence, we intend to also test their applicability for Semantic Web service matchmaking.

---
[12] The Euclidean distance d is converted into a similarity score by $\frac{1}{1+d}$.

Table 1. T^{sim} result table to learn a matchmaking strategy

Query	Service	Bi-gram	TF-IDF	EuclidIn	EuclidOut	TP(q,sv)
q_1	sv_1	$sc^{xs}_{q_1,sv_1}$	$sc^{tfidf}_{q_1,sv_1}$	$sc^{ei}_{q_1,sv_1}$	$sc^{eo}_{q_1,sv_1}$	1
...	...	...	...	...	...	...
q_1	sv_m	$sc^{xs}_{q_1,sv_m}$	$sc^{tfidf}_{q_1,sv_m}$	$sc^{ei}_{q_1,sv_m}$	$sc^{eo}_{q_1,sv_m}$	0
...	...	...	...	...	...	...
q_n	sv_1	$sc^{xs}_{q_n,sv_1}$	$sc^{tfidf}_{q_n,sv_1}$	$sc^{ei}_{q_n,sv_1}$	$sc^{eo}_{q_n,sv_1}$	1
...	...	...	...	...	...	...
q_n	sv_m	$sc^{xs}_{q_n,sv_m}$	$sc^{tfidf}_{q_n,sv_m}$	$sc^{ei}_{q_n,sv_m}$	$sc^{eo}_{q_n,sv_m}$	0

To that end, we induced four different machine learning models using algorithms from Weka[13] and LibSVM[14] as explained in 3.2. From Weka, we chose a linear regression, a logistic regression, and a J48 decision tree learner for the prediction if a service is a true answer to a query. From LibSVM, we chose the support vector regression model (ϵ-SVR) with an RBF kernel and cross-validation/grid-search done as recommended in [10] to search for the best parameter setting for the RBF kernel. For a more detailed discussion of these algorithms, refer to the textbook of Witten and Frank [26] for Weka and to Chang and Lin [5] for LibSVM.

For the four similarity measures *Bi-gram*, *TF-IDF*, *EuclidIn*, and *EuclidOut*, this resulted in T^{sim} shown in Table 1. From this table, we induced the models M, whilst ignoring the columns *Query* and *Service*, using the aforementioned algorithms from Weka and LibSVM.

The resulting iSPARQL strategy from Weka's linear regression model (its IMPRECISE clause) is shown in Listing 1.2. The model defines a weighted, linear combination of the input features (the similarity scores) for the prediction of the membership of a service to a query's relevance set. The weights learned for this particular model are given on the last two lines in Listing 1.2. The final similarity score sc is computed by aggregating/combining the individual scores, *i.e.*, $sc = \sum_{i \in SM} w_i \cdot sim_i$, where $SM = \{Bi\text{-}gram, TF\text{-}IDF, EuclidIn, EuclidOut\}$.

The results in Figure 3(a) show that the ϵ-SVR strategy outperforms all other strategies in terms of precision until about 65% of the relevant services have been retrieved on average. Then, the decision tree takes the lead until about 90% retrieved relevant services, before it gets again slightly outperformed by ϵ-SVR. Linear and logistic regression perform worse than the non-linear models.

The accuracy of the prediction for the J48 learner is 98.95% and 98.45% for the logistic regression learner. Note that the sole use of accuracies is, however, misleading, as they are heavily dependent on the prior of the data set. Therefore, Figure 3(b) graphs the Receiver Operating Characteristics (ROC) and the area under the ROC-curve (AUC; in legend) that both provide a prior-independent approach for comparing the quality of a predictor [19] for two of the chosen

---

[13] http://www.cs.waikato.ac.nz/~ml/weka/
[14] http://www.csie.ntu.edu.tw/~cjlin/libsvm/

```
IMPRECISE
    { ?sim1   isparql:bigram      (?serviceName ?queryName) .
      ?sim2   isparql:tfidf       (?serviceDescription ?queryDescription) .
      ?sim3   isparql:euclidIn    (?serviceProfile ?queryProfile) .
      ?sim4   isparql:euclidOut   (?serviceProfile ?queryProfile) .
      # learned combination of four different strategies
      ?sim    isparql:aggregate   (0.0443 ?sim1 0.785 ?sim2 0.481 ?sim3
                                   0.146 ?sim4 -0.158 1.0)
    }
```

Listing 1.2. Machine-learned strategy using a linear regression model

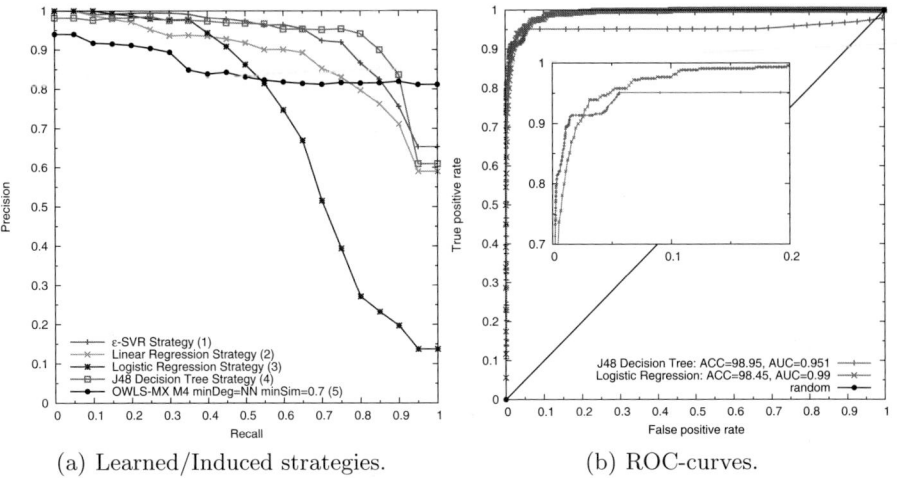

(a) Learned/Induced strategies.   (b) ROC-curves.

Fig. 3. Performance comparison of learned matchmaking strategies

methods. The x-axis shows the false positive rate and the y-axis the true positive rate. AUC is, typically, used as a summary number for the curve. A random class assignment (either YES or NO) is also shown as a line form the origin (0,0) to (1,1) and the ideal ROC-curve would be going from the origin straight up to (0,1) and then to (1,1).

Note that with this very simple approach we are able to clearly outperform OWLS-MX M4 in terms of precision and recall for almost 90% of relevant services. Also, the performance on a per-query basis illustrates the superiority of the machine learned strategies as plotted exemplarily in Figure 2(b) for the J48 decision tree strategy and the simpler vector-based Euclidean I/O strategy. We, therefore, conclude that the combination of logical deduction and statistical induction produces superior performance over logical inference only, which can be easily achieved with our approach.

5.3 Customized Strategies

As claimed in the introduction, iMatcher enables the creation of efficient matchmaking strategies in an iterative and 'playful' procedure. To point this out,

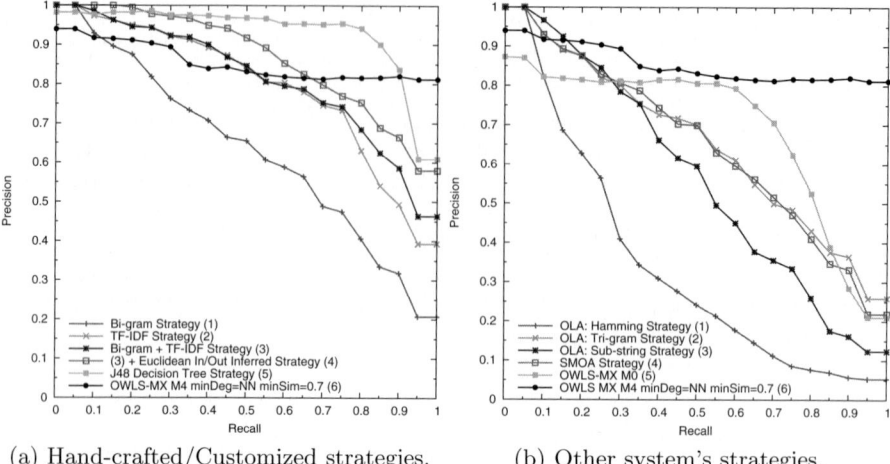

(a) Hand-crafted/Customized strategies. (b) Other system's strategies.

Fig. 4. Performance comparison of customized matchmaking strategies and strategies from other systems

Figure 4(a) summarizes the results for four iteratively improved strategies. Starting with the very simple name-comparing strategy Bi-gram (measure no. 1), we successively can create better strategies by first applying TF-IDF (no. 2), then combining 1 and 2 (= no. 3), followed by no. 4 that additionally takes service inputs and outputs into account, and last, by using machine learning to result in the best-performing strategy in this experiment.

5.4 Strategies from Other Systems

To close our experimental section, we succinctly compare the results of three other systems: (1) the ontology alignment tool OLA (OWL-Lite Alignment) [9]; (2) OWLS-MX [14]; and (3), the string metric for ontology alignment (SMOA) [23]. Note that two of these tools were initially created for ontology alignment rather than for matchmaking. However, as they also involve similarity measures to find correspondences in different ontologies, they can also be used by iMatcher.

The results in Figure 4(b) show that the Hamming strategy from OLA to compare the names of services is by far the most inaccurate approach for matchmaking. All of the other measures have much higher performance. The SMOA strategy behaves very similar to OLA's Tri-gram strategy. Finally, a comparison of OWL-S MX M0 that performs purely logic-based matchmaking with OWLS-MX M4 again illustrates the usefulness of taking into account simple methods from IR to improve precision and recall for Semantic Web service matchmaking.

6 Discussion and Limitations

Clearly, our empirical results are limited to an artificial data set, namely OWL-S TC v2. Therefore, the results must be taken with a pinch of salt. Given, however, the very low number of comparable test collections publicly available, this limitation is rather natural and must be accepted. Similarly, we note that focusing only on OWL-S as service description language is a limitation of our experimental setup and not of our approach itself.

Particularly interesting is the high influence of the IR techniques on matchmaking performance. The strategies that exploit textual information of the services turned out to be very effective, which underlines again the importance of the statistics-based approaches for matchmaking. On the other hand, as can be seen from Figure 1(b), strategies that involve reasoning over the input data can boost matchmaking performance by a factor of about 20%. It is interesting to observe that the combination of both approaches (see measure no. 4 in Figure 4(a)) is superior to each of the individual approaches. Whilst this is not a new finding, it emphasizes, nevertheless, the importance of combining statistical inference with logical deduction, which can also be concluded from our machine learning experiments in Section 5.2.

Of course, creating customized indexes for keyword search and document weighting (*e.g.*, for TF-IDF) involves a pre-processing step of the data, which might be costly depending on the size and dynamics of the data set.

As mentioned earlier, we did not yet consider iSPARQL optimization techniques. Besides the work achieved for SPARQL basic graph pattern optimization through selectivity estimation [2,22], we did, however, experiment with various similarity index structures, whose elaborate treatment we postpone to future work. This is especially important if our approach should be scalable and applicable to data sets which are much larger than the one used in this paper.

Another issue is the approximate matching procedure of iMatcher as opposed to a formal approach. Obviously, both approaches have their important role. Formal matching procedures ensure a correct match, which is especially important when, for example, automatically finding a service that shall be invoked without any adaptation. The formal approaches, however, might be too restrictive in open domains, where an exact match cannot be assumed [14]. Approximate matching is more suitable when an exact match cannot be found, when the formal approach provides too many answers, or when some adaptation procedure would adapt the "calling" code to the found service before invoking it. Also, approximate matching raises the issue of a threshold under which an answer should not be considered. In our evaluation we assumed that the caller would like an answer in any case and would like to examine the most suitable result. In other scenarios such a threshold would, however, be appropriate.

7 Conclusions and Future Work

We presented a new approach to perform Semantic Web service matchmaking based on iSPARQL strategies. We showed how our combination of structured and imprecise querying can be used to perform hybrid Semantic Web service matchmaking.

We evaluated our approach by analyzing a multitude of different matchmaking strategies that make use of the logical foundations of the Semantic Web, apply techniques from traditional information retrieval, involve machine learning, or employ a combination of the three to find very effective strategies for matchmaking. Our empirical results suggest that simple strategies are oftentimes sufficient to obtain good results. However, using weighted and learned combinations of simple measures boosts matchmaking performance even further.

It is left to future work to analyze iMatcher's behavior for different Semantic Web data sets, such as OWLS-S TC v2.2, and service description formats, such as WSMO[15] or SAWSDL.[16]

Finally, coming back to the introductory example, the question is, of course, what Kate would say to these results? Given our experimental results, iMatcher seems to be a practical and easy to use tool to help Kate solving the requirements of her latest project, ensuring she will receive good grades for her work.

Acknowledgment

We would like to thank the anonymous reviewers for their valuable comments, and Matthias Klusch for some interesting discussions regarding OWLS-MX and iMatcher.

References

1. Baeza-Yates, R., Ribeiro-Neto, B.: Modern Information Retrieval. Addison-Wesley, Reading (1999)
2. Bernstein, A., Kiefer, C., Stocker, M.: OptARQ: A SPARQL Optimization Approach Based on Triple Pattern Selectivity Estimation. Technical Report IFI-2007.02, Department of Informatics, University of Zurich (2007)
3. Bianchini, D., Antonellis, V.D., Melchiori, M., Salvi, D.: Semantic-Enriched Service Discovery. In: ICDEW, pp. 38–47 (2006)
4. Borgida, A., Brachman, R.J., McGuinness, D.L., Resnick, L.A.: CLASSIC: A Structural Data Model for Objects. In: SIGMOD, pp. 58–67 (1989)
5. Chang, C.-C., Lin, C.-J.: LIBSVM—A Library for Support Vector Machines (2001), Software available at http://www.csie.ntu.edu.tw/~cjlin/libsvm
6. Cohen, W.W., Ravikumar, P., Fienberg, S.: A Comparison of String Distance Metrics for Name-Matching Tasks. In: IJCAI Workshop, pp. 73–78 (2003)
7. Corby, O., Dieng-Kuntz, R., Gandon, F., Faron-Zucker, C.: Searching the Semantic Web: Approximate Query Processing Based on Ontologies. Intelligent Systems 21(1), 20–27 (2006)
8. Dice, L.R.: Measures of the Amount of Ecologic Association Between Species. Ecology 26(3), 297–302 (1945)
9. Euzenat, J., Loup, D., Touzani, M., Valtchev, P.: Ontology Alignment with OLA. In: EON, pp. 60–69 (2004)

---

[15] http://www.wsmo.org/
[16] http://www.w3.org/2002/ws/sawsdl/

10. Hsu, C.-W., Chang, C.-C., Lin, C.-J.: A Practical Guide to Support Vector Classification (2007)
11. Jaeger, M.C., Rojec-Goldmann, G., Mühl, G., Liebetruth, C., Geihs, K.: Ranked Matching for Service Descriptions using OWL-S. In: KiVS, pp. 91–102 (2005)
12. Kiefer, C., Bernstein, A., Lee, H.J., Klein, M., Stocker, M.: Semantic Process Retrieval with iSPARQL. In: Franconi, E., Kifer, M., May, W. (eds.) ESWC 2007. LNCS, vol. 4519, pp. 609–623. Springer, Heidelberg (2007)
13. Kiefer, C., Bernstein, A., Stocker, M.: The Fundamentals of iSPARQL: A Virtual Triple Approach for Similarity-Based Semantic Web Tasks. In: Aberer, K., Choi, K.-S., Noy, N., Allemang, D., Lee, K.-I., Nixon, L., Golbeck, J., Mika, P., Maynard, D., Mizoguchi, R., Schreiber, G., Cudré-Mauroux, P. (eds.) ISWC 2007. LNCS, vol. 4825, Springer, Heidelberg (2007)
14. Klusch, M., Fries, B., Sycara, K.: Automated Semantic Web Service Discovery with OWLS-MX. In: AAMAS, pp. 915–922 (2006)
15. Levenshtein, V.I.: Binary codes capable of correcting deletions, insertions and reversals. Soviet Physics Doklady 10(8), 707–710 (1966)
16. Lewis, D.D.: Evaluating Text Categorization. In: HLT Workshop, pp. 312–318 (1991)
17. Noia, T.D., Sciascio, E.D., Donini, F.M., Mongiello, M.: A System for Principled Matchmaking in an Electronic Marketplace. IJEC 8(4), 9–37 (2004)
18. Paolucci, M., Kawamura, T., Payne, T.R., Sycara, K.P.: Semantic Matching of Web Services Capabilities. In: Horrocks, I., Hendler, J. (eds.) ISWC 2002. LNCS, vol. 2342, pp. 333–347. Springer, Heidelberg (2002)
19. Provost, F., Fawcett, T.: Robust Classification for Imprecise Environments. Machine Learning 42(3), 203–231 (2001)
20. Raghavan, V.V., Bollmann, P., Jung, G.S.: Retrieval System Evaluation Using Recall and Precision: Problems and Answers. In: SIGIR, pp. 59–68 (1989)
21. Sebastiani, F.: Machine Learning in Automated Text Categorization. ACM Computing Surveys 34(1), 1–47 (2002)
22. Stocker, M., Seaborne, A., Bernstein, A., Kiefer, C., Reynolds, D.: SPARQL Basic Graph Pattern Optimization Using Selectivity Estimation. In: WWW (2008)
23. Stoilos, G., Stamou, G., Kollias, S.: A String Metric for Ontology Alignment. In: Gil, Y., Motta, E., Benjamins, V.R., Musen, M.A. (eds.) ISWC 2005. LNCS, vol. 3729, Springer, Heidelberg (2005)
24. Sycara, K., Klusch, M., Widoff, S., Lu, J.: Dynamic Service Matchmaking Among Agents in Open Information Environments. SIGMOD Rec. 28(1), 47–53 (1999)
25. Winkler, W.E., Thibaudeau, Y.: An Application of the Fellegi-Sunter Model of Record Linkage to The 1990 U.S. Decennial Census. Technical report, U.S. Bureau of the Census (1987)
26. Witten, I.H., Frank, E.: Data Mining: Practical Machine Learning Tools and Techniques. Morgan Kaufmann, San Francisco (2005)
27. Zhang, L., Liu, Q., Zhang, J., Wang, H., Pan, Y., Yu, Y.: Semplore: An IR Approach to Scalable Hybrid Query of Semantic Web Data. In: Aberer, K., Choi, K.-S., Noy, N., Allemang, D., Lee, K.-I., Nixon, L., Golbeck, J., Mika, P., Maynard, D., Mizoguchi, R., Schreiber, G., Cudré-Mauroux, P. (eds.) ISWC 2007. LNCS, vol. 4825, pp. 653–665. Springer, Heidelberg (2007)

Adding Data Mining Support to SPARQL Via Statistical Relational Learning Methods

Christoph Kiefer, Abraham Bernstein, and André Locher

Department of Informatics, University of Zurich, Switzerland
{kiefer,bernstein}@ifi.uzh.ch, andre@outerlimits.ch

Abstract. Exploiting the complex structure of relational data enables to build better models by taking into account the additional information provided by the links between objects. We extend this idea to the Semantic Web by introducing our novel SPARQL-ML approach to perform data mining for Semantic Web data. Our approach is based on traditional SPARQL and statistical relational learning methods, such as Relational Probability Trees and Relational Bayesian Classifiers.

We analyze our approach thoroughly conducting three sets of experiments on synthetic as well as real-world data sets. Our analytical results show that our approach can be used for any Semantic Web data set to perform instance-based learning and classification. A comparison to kernel methods used in Support Vector Machines shows that our approach is superior in terms of classification accuracy.

1 Introduction

The success of statistics-based techniques in almost every area of artificial intelligence and in practical applications on the Web challenges the traditional logic-based approach of the Semantic Web. We believe that we should treat statistical inference techniques as a complement to the existing Semantic Web infrastructure. Consequently, a big challenge for Semantic Web research is not if, but how to extend the existing Semantic Web techniques with statistical learning and inferencing capabilities. In this paper we (1) argue that the large and continuously growing amount of *interlinked* Semantic Web data is a perfect match for *statistical relational learning (SRL) methods* due to their focus on relations between objects in addition to features/attributes of objects of traditional, propositional data mining techniques; and (2) show two concrete Semantic Web research areas, where machine learning support is useful: service classification [6] and semantic data prediction [1]. Moreover, we think that the fact that companies such as Microsoft and Oracle have recently added data mining extensions to their relational database management systems underscores their importance, and calls for a similar solution for RDF stores and SPARQL respectively.

To support the integration of traditional Semantic Web techniques and machine learning-based, statistical inferencing, we developed an approach to create and work with data mining models in SPARQL. Our framework enables to predict/classify unseen data (or features) and relations in a new data set based on the results of a mining model. In particular, our approach enables the usage of SRL methods, which take the relations between objects into account. This allows us to induce statistical models without prior propositionalization (*i.e.*, translation to a single table) [2]—a cumbersome and error-prone task.

Our Contributions. We propose a novel extension to SPARQL called SPARQL-ML to support data mining tasks for knowledge discovery in the Semantic Web. Our extension introduces new keywords to the SPARQL syntax to facilitate the induction of models as well as the use of the model for prediction/classification. We, therefore, extend the SPARQL grammar with the CREATE MINING MODEL and the PREDICT statements (among others) as explained in Section 4.

To ensure the extensibility of SPARQL-ML with other learning methods, we created the *SPARQL Mining Ontology (SMO)* that enables the seamless integration of additional machine learning techniques (see Section 4). We show that SRLs—we use Relational Probability Trees (RPTs) and Relational Bayesian Classifiers (RBCs) [10,11]—are able to exploit the rich and complex heterogeneous relational structure of Semantic Web data (Section 5).

Experimental Results. To validate our approach, we perform three sets of experiments (all in Section 5): first, in the *project success experiment*, we show that, using a synthetic data set, the combination of statistical inference with logical deduction produces superior performance over statistical inference only; second, the *Semantic Web service domain prediction experiment* expands on these findings using a well-known Semantic Web benchmarking data set; last, the *SVM-benchmark experiment* shows the superiority of our approach compared to a state-of-the-art kernel-based Support Vector Machine (SVM) [1] using a real-world data set.

2 Related Work

Little work has been done so far on seamlessly integrating knowledge discovery capabilities into SPARQL. Recently, Kochut and Janik [9] presented SPARQLeR, an extension of SPARQL to perform semantic association discovery in RDF (*i.e.*, finding complex relations between resources). One of the main benefits of our work is that we are able to use a multitude of different, pluggable machine learning techniques to not only perform semantic association discovery, but also classification and clustering.

Similarily, Gilardoni *et al.* [4] argued that machine learning techniques are needed to build a semantic layer on top of the traditional Web. Therefore, the support from tools that are able to work autonomously is needed to add the required semantic annotations. We claim that our SPARQL-ML approach offers this support and, thus, facilitates the process of (semi-) automatic semantic annotation (through classification).

We are aware of two independent studies that focus on data mining techniques for Semantic Web data using Progol—an Inductive Logic Programming (ILP) system. Edwards *et al.* [3] conducted an empirical investigation of the quality of various machine learning methods for RDF data classification, whereas Hartmann [5] proposed the ARTEMIS system that provides data mining techniques to discover common patterns or properties in a given RDF data set. Our work extends their suggestions in extending the Semantic Web infrastructure in general with machine learning approaches, enabling the exploration of the suitability of a large range of machine learning techniques (as opposed to few ILP methods) to Semantic Web tasks without the tedious rewriting of RDF data sets into logic programming formalisms.

A number of studies relate to our experiments in Section 5. The study of Hess and Kushmerick [6] presented a machine learning approach for semi-automatic classification of web services. Their proposed application is able to determine the category of a WSDL web service and to recommend it to the user for further annotation. They treated the determination of a web service's category as a text classification problem and applied traditional data mining algorithms, such as Naive Bayes and Support Vector Machines. Our second experiment (see Section 5.2) is similar in that it employs OWL-S service descriptions. In contrast to [6], we employ SRL algorithms to perform service classification.

Last, Bloehdorn and Sure [1] explored an approach to classify ontological instances and properties using SVMs (*i.e.*, kernel methods). They presented a framework for designing such kernels that exploit the knowledge represented by the underlying ontologies. Inspired by their results, we conducted the same experiments using our proposed SPARQL-ML approach (see Section 5.3). Initial results show that we can outperform their results by a factor of about 10%.

3 Background—A Brief Introduction to SRL

We briefly review the two statistical relational learning (SRL) methods we use in this paper: Relational Bayesian Classifiers (RBCs) and Relational Probability Trees (RPTs). These methods have been shown to be very powerful for SRL, as they model not only the intrinsic attributes of objects, but also the extrinsic relations to other objects [2,10,11].

3.1 Relational Bayesian Classifiers (RBCs)

An RBC is a modification of the traditional Simple Bayesian Classifier (SBC) for relational data [11]. SBCs assume that the attributes of an instance are conditionally independent of its class C. Hence, the probability of the class given an example instance can be computed as the product of the probabilities of the example's attributes $A_1, \ldots, A_n$ given the class (*e.g.*, $P(C = YES \mid A_1, \ldots, A_n) = \alpha P(C = YES) \prod_{i=1}^{n} P(A_i \mid C = YES)$, where α is a scaling factor dependent only on the attributes $A_1, \ldots, A_n$). RBCs apply this independence assumption to relational data. Before being able to estimate probabilities, RBCs decompose (flatten) structured examples down to the attribute level.

Figure 1 shows an example instance (subgraph) to predict the success of a business project in a relational data set. This heterogeneous instance is transformed to homogenous sets of attributes shown in the table on the right. The

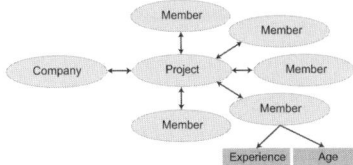

Success	Member's Age	Member's Exp.
YES	{23,25,37,32,41}	{2,4,3,1,12}
YES	{18,25}	{1,7}
NO	{17,26,45}	{1,3,17}
...	...	...

Fig. 1. Example relational data represented as subgraph [11] on the left, and decomposed by attributes on the right

attributes for the prediction of the project success include the age and experience distributions of the project members. Each row in the table stands for a subgraph, each column represents one of its attributes, and the cells contain the multisets (or distributions) of values of attributes.

Learning an RBC model then basically consists of estimating probabilities for each attribute. Such estimation methods include, but are not limited to, average-value and independent-value estimations [11]. As an example for average-value estimation, the probability of a business project B to be successful (i.e., $C = YES$) based only on the attribute "Member Age" (i.e., the project members age distribution) could be estimated by $P(C = YES \,|\, B) = \alpha P(Average_{Age} \geq 33 \,|\, C = YES)P(C = YES)$. For an elaborate treatment of this estimation techniques, please refer to the papers of Neville et al. [10,11].

3.2 Relational Probability Trees (RPTs)

RPTs extend standard probability estimation trees (also called *decision trees*) to a relational setting, in which data instances are heterogeneous and interdependent [10]. The algorithm first transforms the relational data to multisets of attributes (Figure 1). It then attempts to construct an RPT by searching over the space of possible binary splits of the data based on the relational features, until further processing no longer changes the class distributions significantly.

Figure 2 shows an example RPT for the task of predicting whether or not a business project is successful. The leaf nodes show the distribution of the training examples (that "reached the leaf") and the resulting class probabilities of the *isSuccessful* target label. Note that the probability estimates are smoothed by a Laplace correction (i.e., $\frac{p+1}{N+C}$, where p is the number of examples of a specific class, N the total number of examples, and C the number of classes). For the topmost left leaf this evaluates to $P(YES) = \frac{0+1}{117+2} = 0.008$ [13].

The features for splitting the data are created by mapping the multisets of values into single value summaries with the help of aggregation functions [10]. For our business project example in Figure 1 the root node splits the data (i.e., the projects) along the summary count(Non-Managers) into projects with less than 4 non-managers and those with four or more non-managers.

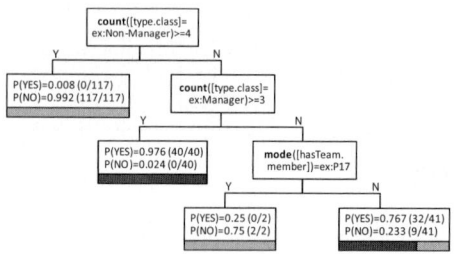

Fig. 2. Example RPT (pruned) to predict the successfulness of business projects

4 Our Approach: SPARQL-ML

SPARQL-ML (SPARQL Machine Learning) is an extension of SPARQL that extends the Semantic Web query language with knowledge discovery capabilities. Our extensions add new syntax elements and semantics to the official SPARQL grammar described in [15]. In a nutshell, SPARQL-ML facilitates the following two tasks on any Semantic Web data set: (1) train/learn/induce a model based on training data using the new CREATE MINING MODEL statement (Section 4.1); and (2), apply a model to make predictions via two new *property functions* (Section 4.2). The model created in the CREATE MINING MODEL step follows the definitions in our *SPARQL Mining Ontology (SMO)* presented in Section 4.1.

We implemented SPARQL-ML as an extension to ARQ—the SPARQL query engine for Jena.[1] Our current version of SPARQL-ML [2] supports, but is not limited to *Proximity*[3] and *Weka*[4] as data mining modules. Note that to preserve the readability of this paper we focus solely on supervised induction methods (in particular classification/regression), even though SPARQL-ML supports the whole breadth of machine learning methods provided by its data mining modules.

4.1 Learning a Model

SPARQL-ML enables to induce a classifier (model) on any Semantic Web training data using the new CREATE MINING MODEL statement. The chosen syntax was inspired by the Microsoft Data Mining Extension (DMX) that is an extension of SQL to create and work with data mining models in Microsoft SQL Server Analysis Services (SSAS) 2005.[5]

The extended SPARQL grammar is tabulated in Table 1 and Listing 1.1 shows a particular example query (used in Section 5.1). Note that we omit prefixes in all queries for compactness. Our approach adds the *CreateQuery* symbol to the official SPARQL grammar rule of *Query* [15]. The structure of *CreateQuery* resembles the one of *SelectQuery*, but has complete different semantics:

---

[1] http://jena.sourceforge.net/
[2] Available at http://www.ifi.uzh.ch/ddis/sparql-ml.html
[3] http://kdl.cs.umass.edu/proximity/index.html
[4] http://www.cs.waikato.ac.nz/ml/weka/
[5] http://technet.microsoft.com/en-us/library/ms132058.aspx

Table 1. Extended SPARQL grammar for the CREATE MINING MODEL statement

[1]	Query	::= Prologue(SelectQuery | ConstructQuery | DescribeQuery | AskQuery | CreateQuery)
[2]	CreateQuery	::= CREATE MINING MODEL' SourceSelector '{' Var 'RESOURCE' 'TARGET' (Var ('RESOURCE' | 'DISCRETE' | 'CONTINUOUS') 'PREDICT'?)+ '}' DatasetClause* WhereClause SolutionModifier UsingClause
[1.2]	UsingClause	::= 'USING' SourceSelector BracketedExpression

```
 1  CREATE MINING MODEL <http://www.example.org/projectSuccess>
 2    { ?project RESOURCE TARGET
 3      ?success DISCRETE PREDICT {'YES','NO'}
 4      ?member RESOURCE
 5      ?class RESOURCE
 6    }
 7  WHERE
 8    { # SPARQL Basic Graph Pattern (BGP) matching part (lines 9-11)
 9      ?project ex:isSuccess ?success .
10      ?project ex:hasTeam   ?member  .
11      ?member  rdf:type     ?class   .
12    } USING <http://kdl.cs.umass.edu/proximity/rpt>
```

Listing 1.1. SPARQL-ML CREATE MINING MODEL query

the *CreateQuery* expands to Rule 1.1 adding the new keywords CREATE MINING MODEL to the grammar followed by a *SourceSelector* to define the name of the trained model. In the body of *CreateQuery*, the variables (attributes) to train the model are listed. Each variable is specified with its content type, which is currently one of the following: RESOURCE—variable holds an RDF resource (IRI or blank node), DISCRETE—variable holds a discrete/nominal literal value, CONTINUOUS—variable holds a continuous literal value, and PREDICT—tells the learning algorithm that this feature should be predicted. The first attribute is additionally specified with the TARGET keyword to denote the resource for which a feature should be predicted (cf. Neville *et al.* [10]).

After the usual *DatasetClause*, *WhereClause*, and *SolutionModifier*, we introduced a new *UsingClause*. The *UsingClause* expands to Rule 1.2 that adds the new keyword USING followed by a *SourceSelector* to define the name and parameters of the learning algorithm.

Semantics of CREATE MINING MODEL Queries. According to Pérez *et al.* [12], a SPARQL query consists of three parts: the *pattern matching part*, the *solution modifiers*, and the *output*. In that sense, the semantics of the CREATE MINING MODEL queries is the construction of new triples describing the metadata of the trained model (*i.e.*, a new output type). An example of such metadata for a particular model is shown in Listing 1.2, which follows the definitions of our SPARQL Mining Ontology (SMO) in Figure 3. The ontology enables to permanently save the parameters of a learned model, which is needed by the predict queries (see next section). The ontology includes the model name, the used learning algorithm, all variables/features being used to train the classifier, as well as additional information, such as where to find the generated model file. In Listing

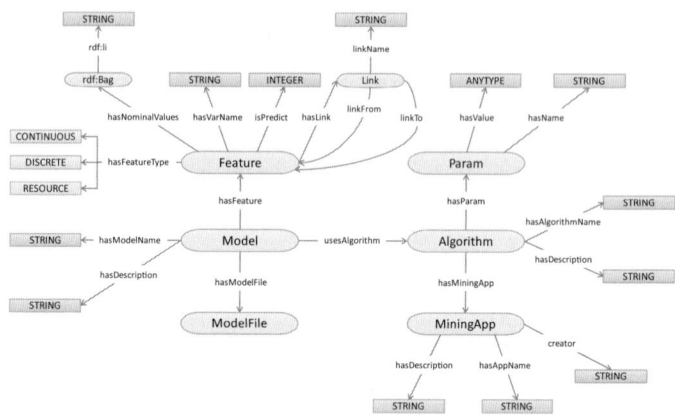

Fig. 3. SPARQL-ML Mining Ontology (SMO)

1.2, lines 1–7 show the constructed triples of a model with name *projectSuccess*, while lines 9–15 show the metadata for a particular feature of the model.

4.2 Making Predictions

After inducing a model with CREATE MINING MODEL, SPARQL-ML allows the user to make predictions with the model via two new property functions (*i.e.*, sml:predict and sml:mappedPredict).[6] The concept behind property functions is simple: whenever the predicate of a triple pattern is prefixed with a special name (*i.e.*, sml), a call to an external function is made and arguments are passed to the function (in our case by the object of the triple pattern).

The example query in Listing 1.3 explains the usage of sml:mappedPredict (line 7). As arguments, the function takes the identifier of the previously learned model (Section 4.1) and the instance as defined by the parameters used whilst training the model (in our case specified by the variables ?project, ?success, ?member, and ?class). In Listing 1.1, we induced a classifier to predict the value for the variable ?success on the *training data*. This classifier is then used on line 7 in Listing 1.3 to predict the value for ?award on the *test data*. The result of the prediction, either 'YES' or 'NO', and its probability are finally bound on line 6 to the variables ?prediction and ?probability respectively.

One benefit of the chosen approach is that we are able to use a number of different models in the same query. In some sense, the property functions introduce *virtual triples* [8] (*i.e.*, for the predictions and probabilities) into the query result set that are computed at run-time rather than present in the knowledge base.

Semantics of Predict Queries. The semantics of our predict queries is basically that of a *prediction join*:[7] (1) mappedPredict maps the variables in the

---

[6] http://jena.sourceforge.net/ARQ/extension.html#propertyFunctions
[7] http://msdn2.microsoft.com/en-us/library/ms132031.aspx

```
1  <http://www.example.org/projectSuccess>
2      smo:hasModelFile <http://www.example.org/projectSuccess/model_RPT.xml> ;
3      smo:hasFeature <http://www.example.org/projectSuccess#project> ;
4      smo:hasFeature <http://www.example.org/projectSuccess#success> ;
5      smo:usesAlgorithm <http://kdl.cs.umass.edu/proximity/rpt> ;
6      smo:hasModelName "projectSuccess" ;
7      a          smo:Model .
8
9  <http://www.example.org/projectSuccess#success>
10     smo:hasVarName "success" ;
11     smo:isPredict "1" ;
12     smo:hasFeatureType "DISCRETE" ;
13     smo:hasLink <http://www.example.org/projectSuccess/link/isSuccessful> ;
14     smo:hasNominalValues _:b1 ;
15     a          smo:Feature .
16
17 <http://www.example.org/projectSuccess/link/isSuccessful>
18     smo:linkName "isSuccess" ;
19     smo:linkFrom <http://www.example.org/projectSuccess#project> ;
20     a          smo:Link .
21
22 _:b1   rdf:li "NO" ;
23        rdf:li "YES" ;
24     a          rdf:Bag .
```

Listing 1.2. Example metadata for a learned classifier

```
1  SELECT DISTINCT ?person ?award ?prediction ?probability
2  WHERE
3    { ?person ex:hasAward    ?award .
4      ?person ex:hasFriend ?friend .
5      ?friend rdf:type       ?class .
6      ( ?prediction ?probability )
7              sml:mappedPredict ( <http://www.example.org/projectSuccess>
8                                  '?project = ?person' '?success = ?award'
9                                  '?member = ?friend' '?class = ?class' )
10   }
```

Listing 1.3. SPARQL-ML example predict query 1: apply the model on a data set with a different ontology structure

Basic Graph Pattern (BGP) to the features in the specified model, which allows us to apply a model on a data set with a different ontology structure; (2) mappedPredict creates instances out of the mappings according to the induced model; (3) the model is used to classify an instance as defined in the CREATE MINING MODEL query (Listing 1.1); and (4), the values of the prediction and its probability are bound to variables in the predict query (line 6 in Listing 1.3). Note that we also defined a shorter version for mappedPredict in case the model is used on the same data set (*i.e.,* sml:predict; Listing 1.4).

5 Experimental Analysis

The goal of our experiments was to show the usefulness and the simplicity of the integration of machine learning methods with the existing Semantic Web infrastructure. Furthermore, we wanted to show the advantage that the com-

```
1  SELECT DISTINCT ?project ?success ?prediction ?probability
2  WHERE
3    { ?project   ex:isSuccess  ?success .
4      ?project   ex:hasTeam    ?member .
5      ?member    rdf:type      ?class .
6      ( ?prediction ?probability )
7                    sml:predict ( <http://www.example.org/projectSuccess>
8                                  ?project ?success ?member ?class )
9    }
```

Listing 1.4. SPARQL-ML example predict query 2: apply the model on a data set with the same ontology structure

bination of logic deduction and statistical induction holds over induction only. Last, we wanted to compare the performance of the SRL algorithms that we integrated into SPARQL-ML with another state-of-the-art approach. To that end, we conducted three experiments. The *project success experiment* and the *Semantic Web service domain prediction experiment* both show the ease of use as well as the advantage of the combination of induction and deduction. The *SVM-benchmark experiment* compares the prediction performance of our approach to another state-of-the-art method. In the following, we present each of the experiments in detail describing our experimental setup and discussing the empirical results.

5.1 The Project Success Experiment

Evaluation Procedure and Data Set. In order to show the ease of use and predictive capability of our system, we put together a proof of concept setting with a small, artificially created data set. We chose an artificial data set to better understand the results and to reduce any experimental noise. The constructed *business project data set* consists of different business projects and the employees of an imaginary company. The company has 40 employees each of which having one out of 8 different occupations. Figure 4 shows part of the created ontology in more detail. 13 employees belong to the superclass Manager, whereas 27 employees belong to the superclass Non-Manager.

We then created business projects and randomly assigned up to 6 employees to each project. The resulting teams consist of 4 to 6 members. Finally, we randomly defined each project to be successful or not, with a bias for projects being more successful, if more than three team members are of type Manager. The resulting data set contains 400 projects with different teams. The prior probability of a project being successful is 35%. We did a 50:50 split of the data and followed a single holdout procedure, swapping the roles of the testing and training set and averaged the results.

Experimental Results. Listing 1.1 shows the CREATE MINING MODEL query that we used in the model learning process. We tested different learning algorithms with and without the support of inferencing. With the reasoner disabled, the last triple pattern in the WHERE clause (line 11) matches only the direct type of the received employee instance (*i.e.*, if an employee is a 'direct' instance of

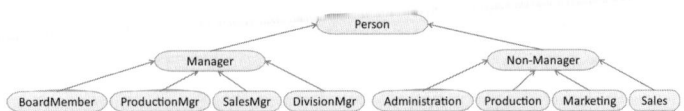

Fig. 4. Example business ontology

class `Manager`). This is the typical situation in relational databases without the support of inheritance. With inferencing enabled, the last triple pattern also matches all inferred types, indicating if an employee is a `Manager` or not.

Given the bias in the artificial data set, it is to be expected that the ability to infer if a team member is a `Manager` or not is central to the success of the induction procedure. Consequently, we would expect that models induced on the inferred model should exhibit a superior performance. The results shown in Figure 5 confirm our expectations. The Figure shows the results in terms of prediction accuracy (ACC; in legend), Receiver Operating Characteristics (ROC; graphed), and the area under the ROC-curve (AUC; also in legend). The ROC-curve graphs the true positive rate (y-axis) against the false positive rate (x-axis), where an ideal curve would go from the origin to the top left (0,1) corner, before proceeding to the top right (1,1) one [14]. It has the advantage to show the prediction quality of a classifier independent of the distribution (and, hence, prior) of the underlying data set. The area under the ROC-curve is, typically, used as a summary number for the curve. Note that a random assignment whether a project is successful or not is also shown as a line form the origin (0,0) to (1,1). The learning algorithms shown are a Relational Probability Tree (RPT), a Relational Bayes Classifier (RBC), both with and without inferencing, and, as a baseline, a k-nearest neighbor learning algorithm (k-NN) with inferencing and $k = 9$ using a maximum common subgraph isomorphism metric [16] to compute the closeness to neighbors.

As the Figure shows, the relational methods clearly dominate the baseline k-NN approach. As expected, both RPT and RBC with inferencing outperform the respective models without inferencing. It is interesting to note, however, that RPTs seem to degrade more with the loss of inferencing than RBCs. Actually, the lift of an RBC with inferencing over an RBC without inferencing is only small. These results support our assumption that the combination of induction and deduction should outperform pure induction. The major limitation of this finding is the artificial nature of the data set. We, therefore, decided to conduct a second experiment with the same goal using a real-world data set, which we present next.

5.2 The Semantic Web Service Domain Prediction Experiment

Evaluation Procedure and Data Set. Our first experiment showed that SPARQL-ML can easily induce and apply accurate models that rely on inferred data. The goal of our second set of experiments was to provide further support for these findings using a real-world test-collection in a non-binary classifica-

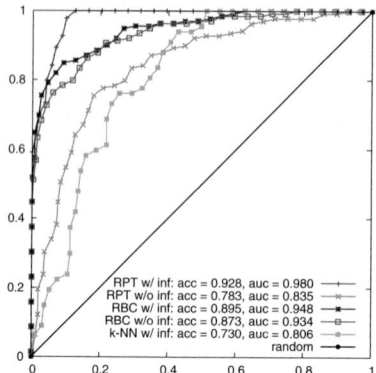

Fig. 5. ROC-Curves of business project success prediction

tion task. Specifically, we used the OWLS-TC v2.1 service retrieval test collection that contains 578 OWL-S Semantic Web service descriptions.[8] OWLS-TC contains OWL-S services in seven service categories: *communication, economy, education, food, medical, travel,* and *weapon*. The prior distribution of the services is $communication = 5.02\%$, $economy = 35.63\%$, $education = 23.36\%$, $food = 4.33\%$, $medical = 8.99\%$, $travel = 18.34\%$, and $weapon = 4.33\%$ (*i.e., economy* is the domain with most services).

For our experiments, we induced an RPT to predict the service category of a service based on its input and output concepts. We limited our investigations to the I/O parameters as we believe that they are most informative for this task. Again, we ran the experiment once on the asserted and once on the (logically) inferred model. Furthermore, we performed a 10-fold cross validation, where 90% of the data was used to learn a classification model and the remaining 10% to test the effectiveness of the learned model. This approach is standard practice in machine learning.

Experimental Results. Listing 1.5 shows the `CREATE MINING MODEL` query that we used in the model learning step. By using `OPTIONAL` patterns, we enable the inclusion of services with no outputs or inputs. The additional `OPTIONAL` pattern for the `rdfs:subClassOf` triple enables us to run the same query on the asserted and the inferred data.

The averaged classification accuracy of the results of the 10 runs is 0.5102 on the asserted and 0.8288 on the inferred model. Hence, the combination of logical deduction with induction improves the accuracy by 0.3186 over pure induction. The detailed results in Table 2 further confirm the results for all seven domains by listing the typical data mining measures false positive rate (FP Rate), Precision, Recall, and F-measure for all categories. As the results of the t-test show, the differences for Recall and F-measure are (highly) significant. The results for Precision just barely misses significance at the 95% level.

---

[8] http://projects.semwebcentral.org/projects/owls-tc/

```
1  CREATE MINING MODEL <http://www.ifi.uzh.ch/ddis/services>
2    { ?service RESOURCE TARGET
3      ?domain DISCRETE PREDICT
4        { 'communication','economy','education','food','medical','travel',
5          'weapon'
6        }
7      ?profile RESOURCE
8      ?output RESOURCE
9      ?outputType RESOURCE
10     ?outputSuper RESOURCE
11     ?input RESOURCE
12     ?inputType RESOURCE
13     ?inputSuper RESOURCE
14   }
15 WHERE
16   { ?service service:presents   ?profile .
17     ?service service:hasDomain ?domain .
18     OPTIONAL
19       { ?profile profile:hasOutput     ?output .
20         ?output  process:parameterType ?outputType .
21         OPTIONAL
22           { ?outputType rdfs:subClassOf ?outputSuper . }
23       }
24     OPTIONAL
25       { ?profile profile:hasInput      ?input .
26         ?input   process:parameterType ?inputType .
27         OPTIONAL
28           { ?inputType rdfs:subClassOf ?inputSuper . }
29       }
30   } USING <http://kdl.cs.umass.edu/proximity/rpt> ('maxDepth' = 6)
```

Listing 1.5. CREATE MINING MODEL query for service classification

Table 2. Detailed results for the Semantic Web service classification experiments

Domain	FP Rate		Precision		Recall		F-measure	
	w/o inf	w/ inf	w/o inf	w/ inf	w/o inf	w/ inf	w/o inf	w/ inf
communication	0.007	0.004	0.819	0.900	0.600	0.600	0.693	0.720
economy	0.081	0.018	0.810	0.964	0.644	0.889	0.718	0.925
education	0.538	0.090	0.311	0.716	0.904	0.869	0.463	0.786
food	0	0.002	0	0.960	0	0.800	0	0.873
medical	0.006	0.030	0	0.688	0	0.550	0	0.611
travel	0	0.069	1	0.744	0.245	0.873	0.394	0.803
weapon	0.002	0.002	0.917	0.964	0.367	0.900	0.524	0.931
average	0.091	0.031	0.551	0.848	0.394	0.783	0.399	0.807
t-test (paired, one-tailed)	p=0.201		p=0.0534		p=0.00945		p=0.0038	

When investigating the structure of the relational probability trees, the trees induced on the inferred model clearly exploit inheritance relations using the rdfs:subClassOf predicate, indicating that the access to the newly inferred triples improves the determination of a service's category. These observations further support our finding that a combination of deduction and induction is useful for Semantic Web tasks and can be easily achieved with SPARQL-ML.

5.3 The SVM-Benchmark Experiment

Evaluation Procedure and Data Set. With our third set of experiments, we aimed to show possible advantages of SPARQL-ML over another state-of-the-art method. Specifically, we compared the off-the-shelf performance of a simple 19-lines SPARQL-ML statement (see Listing 1.6) with a Support Vector Machine (SVM) based approach proposed by Bloehdorn and Sure [1] following exactly their evaluation procedure.[9] In their work, they introduced a framework for the design and evaluation of kernel methods that are used in Support Vector Machines, such as SVM^{light} [7]. The framework provides various kernels for the comparison of classes as well as datatype and object properties of instances. Moreover, it is possible to build customized, weighted combinations of such kernels. Their evaluations include two tasks: (1) prediction of the affiliation a person belongs to (*person2affiliation*), and (2) prediction of the affiliation a publication is related to (*publication2affiliation*).

As a dataset they used the SWRC ontology—a collection of OWL annotations for persons, publications, and projects, and their relations from the University of Karlsruhe.[10] The data contains 177 instances of type `Person`, 1155 of type `Publication`, as well as some instances of types `Topic`, `Project`, and `ResearchGroup`.

For the person2affiliation task, we used the `worksAtProject` and `worksAtTopic` (essentially `workedOnBy`) object properties as well as properties pointing to the publications of a given person (`publication`). For the publication2affiliation task, we used the `isAbout` and `author` object properties pointing to associated topics and authors respectively. In order to predict the affiliation a publication is related to, we defined the affiliation to be the research group, where the major part of the authors of this publication belong to.

Experimental Results. Table 3 summarizes the macro-averaged results that were estimated via Leave-One-Out Cross-Validation (LOOCV). We applied both, an RBC and an RPT learning algorithm to both tasks. The table also reports the best-performing SVM results from Bloehdorn and Sure's experiments. The RBC clearly outperformed the RPT in both predictions, hence, we report only on the results given by the RBC. For both tasks the performance of the inferred model is not very different from the one induced on the asserted model. When consulting Listing 1.6 (for person2affiliation) it is plausible to conclude that the only inferred properties (types of persons and publications) do not help to classify a person's or a publication's affiliation with an organizational unit.

As Table 3 also shows, our method clearly outperforms the kernel-based approach in terms of prediction error, recall, and F-Measure, while having an only slightly lower precision. The slightly lower precision could be a result of the limitation to just a few properties used by an off-the-shelf approach without a single parameter setting, whereas the SVM approach is the result of extensive testing and tuning of the kernel method's properties and parameters.

We conclude from this experiment, that writing a SPARQL-ML query is a simple task for everyone familiar with the data and the SPARQL-ML syntax. Kernels, on the other hand, have the major disadvantage that the user has to

---

[9] We would like to thank them for sharing the exact data set used in their paper.
[10] http://ontoware.org/projects/swrc/

```
1  CREATE MINING MODEL <http://www.ifi.uzh.ch/ddis/affiliations>
2    { ?person RESOURCE TARGET
3      ?group RESOURCE PREDICT
4      ?project RESOURCE
5      ?topic RESOURCE
6      ?publication RESOURCE
7      ?pertype RESOURCE
8      ?pubtype RESOURCE
9    }
10 WHERE
11   { ?person swrc:affiliation ?group .
12     ?person rdf:type            ?personType .
13     OPTIONAL { ?person swrc:worksAtProject ?project . }
14     OPTIONAL { ?person swrc:worksAtTopic ?topic . }
15     OPTIONAL
16       { ?person       swrc:publication ?publication .
17         ?publication rdf:type           ?publicationType .
18       }
19   } USING <http://kdl.cs.umass.edu/proximity/rbc>
```

Listing 1.6. CREATE MINING MODEL query for the person2affiliation task

Table 3. LOOCV results for the *person2affiliation* and *publication2affiliation* tasks

person2affiliation					publication2affiliation				
algorithm	*err*	*prec*	*rec*	F_1	*algorithm*	*err*	*prec*	*rec*	F_1
sim-ctpp-pc, c=1	4.49	95.83	58.13	72.37	sim-cta-p, c=10	0.63	99.74	95.22	97.43
RBC w/o inf	3.53	87.09	80.52	83.68	RBC w/o inf	0.09	98.83	99.61	99.22
RBC w/ inf	3.67	85.72	80.18	82.86	RBC w/ inf	0.15	97.90	99.25	98.57

choose from various kernels, kernel modifiers, and parameters. This constitutes a major problem for users not familiar with kernels and SVM algorithms.

6 Conclusions, Limitations, and Future Work

We have presented a novel approach we call SPARQL-ML that extends traditional SPARQL with data mining support to perform knowledge discovery in the Semantic Web. We showed how our framework enables to predict/classify unseen data in a new data set based on the results of a mining model. In particular, we demonstrated how models trained by statistical relational learning (SRL) methods outperform models not taking into account additional deduced (or inferred) information about the links between objects.

We fully analyzed SPARQL-ML on synthetic and real-world data sets to show its excellent prediction/classification quality as well as its superiority to other related approaches, such as kernel methods used in Support Vector Machines. Our approach is extensible in terms of the supported machine learning algorithms and generic as it is applicable for any Semantic Web data set.

We note that the performance loss when mining on inferred knowledge bases (see Table 3) is a limitation of the employed ontologies (or data sets) and not of our approach and the used techniques themselves. We speculate that the loss could be eliminated by using more comprehensive ontologies.

Future work will evaluate further relational learning methods such as the ones proposed by NetKit[11] or Alchemy.[12] Finally, given the usefulness and the ease of use of our novel approach, we believe that SPARQL-ML could serve as a standardized approach for data mining tasks on Semantic Web data.

References

1. Bloehdorn, S., Sure, Y.: Kernel Methods for Mining Instance Data in Ontologies. In: ISWC, pp. 58–71 (2007)
2. Dzeroski, S.: Multi-Relational Data Mining: An Introduction. SIGKDD 5(1), 1–16 (2003)
3. Edwards, P., Grimnes, G.A., Preece, A.: An Empirical Investigation of Learning from the Semantic Web. In: ECML/PKDD Workshop, pp. 71–89 (2002)
4. Gilardoni, L., Biasuzzi, C., Ferraro, M., Fonti, R., Slavazza, P.: Machine Learning for the Semantic Web: Putting the user into the cycle. In: Dagstuhl Seminar (2005)
5. Hartmann, J.: A Knowledge Discovery Workbench for the Semantic Web. In: SIGKDD Workshop (2004)
6. Hess, A., Kushmerick, N.: Machine Learning for Annotating Semantic Web Services. In: AAAI (2004)
7. Joachims, T.: SVM light—Support Vector Machine (2004), Software, available at http://svmlight.joachims.org/
8. Kiefer, C., Bernstein, A., Stocker, M.: The Fundamentals of iSPARQL: A Virtual Triple Approach for Similarity-Based Semantic Web Tasks. In: ISWC, pp. 295–309 (2007)
9. Kochut, K., Janik, M.: SPARQLeR: Extended SPARQL for Semantic Association Discovery. In: ESWC, pp. 145–159 (2007)
10. Neville, J., Jensen, D., Friedland, L., Hay, M.: Learning Relational Probability Trees. In: KDD, pp. 625–630 (2003)
11. Neville, J., Jensen, D., Gallagher, B.: Simple Estimators for Relational Bayesian Classifiers. In: ICDM, pp. 609–612 (2003)
12. Pérez, J., Arenas, M., Gutierrez, C.: The Semantics and Complexity of SPARQL. In: ISWC, pp. 30–43 (2006)
13. Provost, F., Domingos, P.: Well-Trained PETs: Improving Probability Estimation Trees. CeDER Working Paper #IS-00-04 (2000)
14. Provost, F.J., Fawcett, T.: Robust Classification for Imprecise Environments. Machine Learning 42(3), 203–231 (2001)
15. Prud'hommeaux, E., Seaborne, A.: SPARQL Query Language for RDF. Technical report, W3C Recommendation (January 15, 2008)
16. Valiente, G.: Algorithms on Trees and Graphs. Springer, Berlin (2002)

---

[11] http://www.research.rutgers.edu/~sofmac/NetKit.html
[12] http://alchemy.cs.washington.edu/

A Semantic Web Middleware for Virtual Data Integration on the Web

Andreas Langegger, Wolfram Wöß, and Martin Blöchl

Institute of Applied Knowledge Processing
Johannes Kepler University Linz
Altenberger Straße 69, 4040 Linz, Austria
{al,martin.bloechl,wolfram.woess}@jku.at

Abstract. In this contribution a system is presented, which provides access to distributed data sources using Semantic Web technology. While it was primarily designed for data sharing and scientific collaboration, it is regarded as a base technology useful for many other Semantic Web applications. The proposed system allows to retrieve data using SPARQL queries, data sources can register and abandon freely, and all RDF Schema or OWL vocabularies can be used to describe their data, as long as they are accessible on the Web. Data heterogeneity is addressed by RDF-wrappers like D2R-Server placed on top of local information systems. A query does not directly refer to actual endpoints, instead it contains graph patterns adhering to a virtual data set. A mediator finally pulls and joins RDF data from different endpoints providing a transparent on-the-fly view to the end-user.

The SPARQL protocol has been defined to enable systematic data access to remote endpoints. However, remote SPARQL queries require the explicit notion of endpoint URIs. The presented system allows users to execute queries without the need to specify target endpoints. Additionally, it is possible to execute join and union operations across different remote endpoints. The optimization of such distributed operations is a key factor concerning the performance of the overall system. Therefore, proven concepts from database research can be applied.

1 Introduction

One of the best use cases for Semantic Web technology is probably large-scale data integration across institutional and national boundaries. Compared to traditional approaches based on relational database systems, there are several aspects of the Semantic Web, which makes it well suited for the integration of data from globally distributed, heterogeneous, and autonomous data sources. In short, these are: a simple but powerful and extensible data model which is the Resource Description Framework (RDF), URIs (or IRIs) used for global naming, and the possibility of reasoning based on Description Logic.

In this paper a system is presented which is developed primarily for sharing data in scientific communities. More precisely, the system is developed as part of the Austrian Grid project to enable transparent access to distributed data for scientific collaboration. The main project is called *Semantic Data Access Middleware for Grids* (G-SDAM)

[16]. Because of the generic architecture of the mediator component, which is responsible for processing queries, and because of its supposed relevance for the Semantic Web community, it has been detached from the rest of the Grid middleware. It is expected that other Semantic Web applications will benefit from this *Semantic Web Integrator and Query Engine* (SemWIQ). The system cannot only be used for other data integration applications such as library repositories, directories, or various scientific archives and knowledge bases, it could also be used to complement Semantic Web search engines and Linked Data browsers (explained in Section 6).

The architecture of the data integration system is based on the following findings: scientific data is usually structured[1], regional or globally distributed, stored in heterogeneous formats, and sometimes access is restricted to authorized people. To provide a transparent access to such kinds of data, a mediator-wrapper architecture is commonly used [29]. It basically consists of a mediator which is accepting queries from clients and then collects data translated by a number of wrappers attached to local data sources. These wrappers use mappings to translate data from the underlying information systems into a common schema which can be processed by the mediator. In the case of virtual data integration, mappings are used for on-the-fly translation of data during query processing. In the next section some related work will be discussed.

The remainder of the paper is structured as follows. After the related work section, the concept of the system and its architecture are described (Section 3). Details about query federation and the implementation of the mediator are presented (Section 4). A sample scenario with sample queries and results are presented (Section 5). Optimization concepts for distributed SPARQL queries and future work are discussed (Section 6) and finally, the conclusion can be found in Section 7.

2 Related Work

Related work can be divided into four main categories: (a) schema integration using ontologies, (b) schema mapping, and translating source data to RDF, (c) distributed query processing, and (d) similar projects addressing ontology-based data integration.

While schema integration is not a new topic in general (an overview can be found in [4]), ontology matching and alignment is rather new[2] [25]. The presented system uses OWL DL ontologies for the global data model and an extended version of SPARQL for processing queries. When integrating data sources of a specific (scientific) domain, it is required to create global ontologies that are expressive enough to describe all data that will be provided. Because data is usually not stored as RDF graphs originally, schema integration has to be done over arbitrary data models. This is a difficult task which is related to a fairly new discipline called *model management* [18]. On the other hand, ontologies have already been developed over the past years for several domains (e.g. medicine, biology, chemistry, environmental science). These can be reused and

---

[1] Apart from data which is shared before being analyzed, as for instance data collected at CERN which is distributed across Europe. But this is for scalability reasons, there is no data integration taking place.

[2] Starting with 2004, there is an annual workshop as part of the International Semantic Web Conference (ISWC): http://oaei.ontologymatching.org

combined freely because of the modular nature of RDF which is based on namespaces identified by globally unique IRIs. Thus, the current process is finding associations between concepts of local data models and existing ontologies. For this process tools can help but for the creation of meaningful mappings experts will be needed.

Concerning the second topic (b), only mapping frameworks that allow manual or (semi-)automatic mapping from arbitrary data models to RDF are relevant. And since SPARQL is used for global queries at the mediator, these mappings have to support on-the-fly data access based on SPARQL. There are a few frameworks that support this [21,22,7][3]. Because the mapping language used by D2R-Server [7] is most powerful, it was chosen as a wrapper for relational database systems. For other information systems or protocols like LDAP, FTP, IMAP, etc. or CSV files and spreadsheets stored in file systems, it is possible to adapt the D2R wrapping engine. This has successfully been done for the library protocol standard OAI-PMH (Open Archives Initiative Protocol for Metadata Harvesting) in the context of a library integration system developed at the University of Vienna [12].

A powerful capability of the presented system is the possibility to execute binary operations across distributed data sources. These operations are involved when matching multiple basic graph patterns, optional patterns, or alternatives. It is possible because the system is completely based on a pipelined query processing workflow. However, optimization of distributed query plans is a pre-condition in order to supply results as fast as possible. It seems that the discussion about federating SPARQL queries and query optimization in general is gaining importance in the community. Because shipping data over the internet is more expensive than most of the local operations, other policies and algorithms have to be used than for local queries. In [23] blocking of remote subqueries has been described reflecting a common approach in distributed query processing which is called row blocking [9]. The re-ordering of binary operations (especially joins) is one of the most complex tasks in a distributed scenario. Some early conceptual ideas have been presented in [15], but further research and experiments will be required (Section 6).

Lessons learned from the development of the D2R wrapper showed that people are demanding for real data integration: "Mapping to RDF is not enough" [6]. Recently, several approaches addressing ontology-based data integration have been proposed. Some of them were initiated in an inter-disciplinary setting (e.g., biomedical [20]). Because of the large number of application scenarios related work will concentrate on implementations and systems. In [24] a concept-based data integration system is described which was developed to integrate data about cultural assets. Although RDF ontologies are used to describe these assets, query processing is based on XML and a special query language called CQuery. Web services are used as a common gateway between the mediator and data sources and XSLT is used to map XML data to the global structure. The authors also presented several optimization techniques like push-down of selections or cashing of results at the mediator. Two other projects based on RDF and SPARQL are *FeDeRate* [22] and *DARQ* [3]. The first one, FeDeRate, just adds support to SPARQL for remote sub-queries using named graph patterns. Thus, it can be seen as a

---

[3] A more comprehensive list is collected at
http://esw.w3.org/topic/RdfAndSql (Dec10, 2007)

multi-database language because endpoints have to be specified explicitly. DARQ, which is an acronym for *Distributed ARQ*, provides access to multiple, distributed service endpoints and is probably related most closely to the presented system. However, there are some important differences. Setting up DARQ requires the user to explicitly supply a configuration file which includes endpoint descriptions. These descriptions include: capabilities in the form of lists of RDF property IRIs which are used at the endpoint for data descriptions, cardinalities for instances, and selectivities. But these statistics are no longer representative when the corresponding data is changed. The system presented here uses a concept-based approach based on DL ontologies, however the idea is similar: data source selection is based on type information instead of RDF properties only. Furthermore, a dynamic catalog is used storing meta data and statistics about available endpoints. These statistics are automatically gathered by a monitoring service. It has to be said that DARQ was only the prototype of a rather small project at HP Labs which unfortunately has not been continued yet.

Virtuoso [21], a comprehensive data integration software developed by OpenLink Software, is also capable of processing distributed queries. Because Virtuoso is also a native quad store, the strength of this software is its scalability and performance. Beside the commercial edition, there is also an open source version available. A relatively new application also provided by OpenLink is the *OpenLink Data Spaces* platform, which is promoted as being able to integrate numerous heterogeneous data from distributed endpoints. Finally, there is also a commercial-only software package, called *Semantic Discovery System* [8], claiming to be able to integrate all kinds of data from distributed sources based on SPARQL with optimal performance. Because no resource materials could be found about the internals the software has not been tested further.

3 Concept and Architecture

For the implementation of SemWIQ the Jena2 RDF library developed at HP Labs [14] is used. Since the system is based on the mediator-wrapper approach [29], its architecture depicted in Fig. 1 looks similar to other mediator-based systems. Clients establish a connection to the mediator and request data by submitting SPARQL queries (1). Patterns in such *global* queries adhere to a virtual graph which refers to classes and properties from arbitrary RDFS or OWL vocabularies. As long as these vocabularies are accessible on the Web according to the *Best Practice Recipes for Publishing RDF Vocabularies* [19], they can be used by data sources to describe provided data. The parser (2), which is part of Jena (ARQ2), calculates a canonical query plan which is modified by the federator/optimizer component (3). Query federation and optimization is tightly coupled and will be described in Section 4. The federator analyzes the query and scans the catalog for relevant registered data sources. The output of the federator/optimizer is an optimized global query plan which is forwarded to the query execution engine (4). The query execution engine processes the global plan which includes remote sub-plans executed at wrapper endpoints (5). Currently, the SPARQL protocol is used between mediator and wrapper endpoints and sub-plans are therefore serialized back into lexical queries. This is a first approach, but it is limited and does not allow for sophisticated optimization strategies as will be discussed later in Section 6.

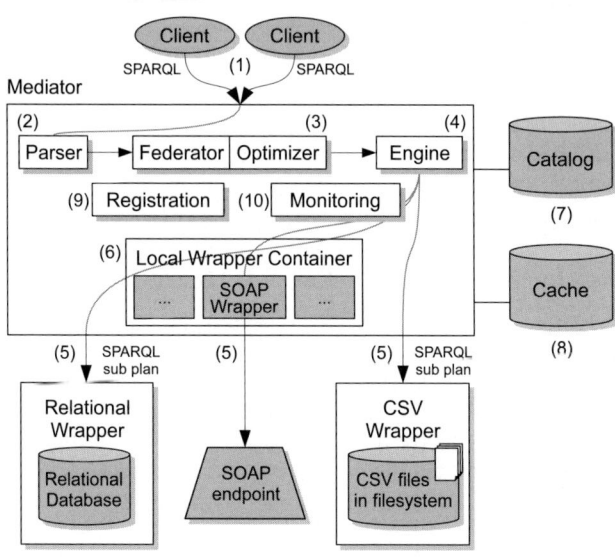

Fig. 1. Mediator-Wrapper architecture of SemWIQ

Any data source must use a SPARQL-capable wrapper to provide local data unless it is a native RDF data source supporting the SPARQL protocol. As explained earlier there are several wrappers available for relational database systems of which D2R has been chosen because of its powerful mapping language. Usually, a wrapper is placed as close as possible to the local information system to be able to benefit from local query processing capabilities. However, it may occur that a data source cannot be extended by a wrapper (e.g. web service endpoints or generally, when there is no control over the remote system). For such cases, the mediator provides a local wrapper container (6). A wrapper inside the container may process a sub-plan and issue native access operations to the actual remote endpoint. The catalog (7) stores descriptions and statistics about registered data sources and a local RDF cache (8) will be used in future to cache triples for global join operations and recurring queries. The registration component is currently a simple REST-based service. A data source can register itself at the mediator by sending a HTTP POST request with an RDF document attached. It specifies the endpoint URI and meta data as for example the providing *Virtual Organization* (VO) and a contact person[4]. De-registration is currently not implemented. If the endpoint becomes unavailable, it is automatically removed from the catalog so it has to register again. The extension of the registration service is one part of future work. The monitoring component (9) is periodically fetching statistics from registered data sources.

---

[4] G-SDAM, mentioned in the introduction, uses the *Grid Security Infrastructure* (GSI) to authenticate requests and Virtual Organizations. Within SemWIQ virtually anybody can register a data source.

3.1 Catalog and Monitoring

The catalog is implemented as a fast in-memory Jena graph. It is currently persisted into a file when the mediator shuts down. However, since the Jena assembler API is used, it is just a matter of changing a configuration file to persist the whole catalog into a database. Beside the endpoint URI and description, it stores statistics gathered by the monitoring component. Because the optimization concepts explained later are not fully implemented in the current prototype, the statistics are rather simple. They are also used for data source selection by the federator. The currently used catalog vocabulary is described by `http://semwiq.faw.uni-linz.ac.at/core/2007-10-24/catalog.owl`.

The monitoring service is designed for simplicity and easy adoption. No configuration is required at remote data sources. They just need to register and the mediator will do the rest. It fetches data using an extended SPARQL syntax implemented in ARQ2 which allows for aggregate queries. The following pseudo-algorithm and queries are used to fetch a list of classes and the number of instances a data sources provides for each class as well as a list of properties and their occurrences:

```
foreach (DataSource as ds) {
  SELECT DISTINCT ?c WHERE { [] a ?c }
  foreach (solution mapping for ?c as c)
    addConceptCount(ds, c, (SELECT COUNT(*) WHERE { ?s a <c> }))

  SELECT DISTINCT ?p WHERE { [] ?p [] }
  foreach (solution mapping for ?p as p)
    addPropertyCount(ds, p, (SELECT COUNT(*) WHERE { ?s <p> ?o }))
}
```

The COUNT(*) statement is currently not part of the SPARQL recommendation and thus, there are a few different proprietary syntaxes around for aggregate functions. It would be better to use distinct counting, however this would break compatibility with Virtuoso and also DBpedia [2] as a consequence.

3.2 Wrapping Data to RDF

Most structured data is currently stored in relational database systems (RDBMS). However, especially in the scientific community file-based data formats are also very popular because sharing and interchange is often easier when having a file. While newer file-based data formats are often based on XML, there are several legacy formats around like FITS for instance, the *Flexible Image Transport System* endorsed by NASA and the International Astronomical Union. It can be compared to JPEG files including EXIF data. In terms of a RDBMS, this would be a record with several fields next to the BLOB or file path/URI where the image is stored. Because often tools already exist which enable the import from legacy data like FITS files into a RDBMS, a wrapper for these information systems is most important. D2R-Server [7] has been chosen because it uses a powerful mapping language. Later on, additional wrappers will be added for other information systems, protocols like LDAP, FTP, IMAP, etc. and CSV files or spreadsheets stored in file systems. D2R-Server also provides a good basis for the implementation of further wrappers.

The mapping language used by D2R can be seen as a global-as-view approach. Starting with global classes and properties defined using RDF Schema or OWL, a view is defined which maps tables and attributes from the database to the global concepts. Regarding query optimization, D2R relies on the optimizer of the local database management system. At the scale of the complete mediator-based data integration system, the wrapper more or less breaks the query processing pipeline. The optimizer at the mediator can currently not take into account special indices of local database systems. To benefit from these – which would decrease response time and lower memory consumption – it will be required to introduce a special protocol instead of the SPARQL protocol which is able to communicate capabilities and estimated costs during plan optimization. Functional mappings as well as translation tables are both supported by D2R. Scientific data sets often use varying scales and metrics. A functional mapping can transform data accordingly. A translation table can be used for varying nominal scales or naming schemes.

3.3 Vocabularies

Depending on the data an endpoint provides, different vocabularies to describe them may be used. There are many vocabularies around created for social software like *Friend-of-a-Friend* (foaf), *Description-of-a-Project* (doap), *Semantically-Interlinked Online Communities* (sioc), etc. and also several scientific communities have created more or less suitable ontologies which can be used to describe data. If there are no applicable vocabularies available for a specific domain, usually a small group of people will introduce a new one and publish them according to [19]. For instance, an ontology for solar observation which is used by the prototype scenario developed together with the Kanzelhöhe Solar Observatory.

A major design goal is to remain flexible and keep the setup process as simple as possible. Initially it was assumed to introduce a collaborative management system for globally used ontologies including versioning support. However, finding a consensus on collaboratively developed ontologies is a difficult, time-consuming, and sometimes frustrating process. It is often better to design vocabularies bottom-up than rounding up as many people as possible and following a top-down approach. Now, everybody can re-use and publish new vocabularies as long as they are accessible on the Web. The registered SPARQL endpoint will be queried by the monitoring service as described and automatically fetch required vocabularies from the Web.

3.4 Authentication and Data Provenance

Scientific data is often only shared inside a community where people know each other. To restrict access to a data source the Grid-enabled middleware G-SDAM uses the *Grid Security Infrastructure*. It is based on a PKI (*Private Key Infrastructure*) and each person using the service requires a Grid certificate issued by a central or regional authority. Data sources are usually provided by *Virtual Organizations*, a concept introduced by the Grid community [11]. For granting access on a data source to specific persons, tools supplied by current Grid middleware like Globus Toolkit or gLite can be used. Because there are also components available for billing of used Grid resources, it may be possible in future to buy and sell scientific data over G-SDAM. Regarding the SemWIQ project, there is currently no support for fine-grained access control. Only the mediator may be secured by SSL and HTTP Basic Authentication which is actually up to the system administrator.

Another important aspect when sharing scientific data – which also may become more and more an issue when the Semantic Web takes off – is data provenance. For any data integration software, it is very important to know where data is coming from. Within SemWIQ this can be achieved by a convention. Upon registration the data source has to supply an instance of type `cat:DataSource`[5]. This instance must be identified by a fully-qualified IRI representing the data source. For each instance returned by the data source, a link to this IRI must be created using the `cat:origin`-property. The mediator will always bind this value to the magic variable `?_origin`. If this variable is added in the projection list, it will appear in the result set. If it is not specified or if the asterisk wildcard is used, it will not occur in the result set.

4 Federating SPARQL Queries

With the SPARQL Working Draft of March 2007 a definition of SPARQL [28, Section 12] was added which introduces a common algebra and semantics for query processing. Query plans can now be written using a prefix syntax similar to that of LISP. In this section the federator will be described. Since the mediator is based on Jena, it also uses ARQ for parsing global SPARQL queries. The idea of concept-based data integration is that every data item has at least one asserted type. For a global query, the mediator requires type information for each subject variable in order to be able to retrieve instances. This implies that there are several restrictions to standard SPARQL:

– All subjects must be variables and for each subject variable its type must be explicitly or implicitly (through DL constraints) defined. A BGP like {?s :p ?o} is not allowed unless there is another triple telling the type for ?s. For instance, the BGP {?s :p ?o ; rdf:type <some-type>} is valid. In a future version, when DESCRIBE-queries are supported, it may become valid to constraint the subject term to an IRI.
– For virtual data integration it is not required to have multiple graphs. A query may only contain the default graph pattern. Furthermore, federation is done by the mediator and not explicitly by the user (this can be done with *FeDeRate* [22]).
– Currently only SELECT-queries are supported, but support for DESCRIBE is planned. Supporting DESCRIBE may be useful to get further information to an already known record, however the mediator has to go through all data sources.

All other features of SPARQL like matching group graph patterns, optional graph patterns, alternative graph patterns, filters, and solution modifiers are supported but they still require further optimization. To demonstrate the federation algorithm, Query 2 shown in Fig. 2 is taken as an example. Firstly, ARQ parses the query string and generates the canonical plan:

---

[5] The prefix cat is generally used for the namespace
http://semwiq.faw.uni-linz.ac.at/core/2007-10-24/catalog.owl#

```
(project (?dt ?groups ?spots ?r ?fn ?ln)
  (filter (&& (= ?fn "Wolfgang") (= ?ln "Otruba"))
    (BGP
      (triple ?s rdf:type sobs:SunspotRelativeNumbers)
      (triple ?s sobs:dateTime ?dt)
      (triple ?s sobs:groups ?groups)
      (triple ?s sobs:spots ?spots)
      (triple ?s sobs:rValue ?r)
      (triple ?s obs:byObserver ?obs)
      (triple ?obs rdf:type obs:Observer)
      (triple ?obs person:firstName ?fn)
      (triple ?obs person:lastName ?ln)
    )))
```

For better readability namespace prefixes are still used for printing plans. Next, the federator transforms this plan according to the following Java-like algorithm:

```
visit(opBGP) by visiting plan bottom-up {
  Hashtable<Var, BasicPattern> sg = createSubjectGroups(opBGP); // group by subjs
  Op prev = null;

  // iterate over subject groups
  Iterator<Node> i = sg.keySet().iterator();
  while (i.hasNext()) {
    Node subj = i.next();
    String type = getType(sg, subj); //+does static caching of already known types
    OpBGP newBGP = new OpBGP((BasicPattern)sg.get(subj));
    Op prevU = null;

    // look for sites storing instances of determined type...
    Iterator<DataSource> dit = catalog.getAvailable(type);
    while(dit.hasNext()) {
      DataSource ds = dit.next();
        Op newDS = new OpService(Node.createURI(
        ds.getServiceEndpoint()), newBGP.copy());
      prevU = (prevU==null) ? newDS : new OpUnion(prevU, newDS);
    }
    prev = (prev==null) ? prevU : OpJoin.create(prev, prevU);
  }
  if (prev == null) return opBGP; // no data sources found
    else return prev;
}
```

Because there are two registered data sources providing instances of obs:Observer and sobs:SunspotRelativeNumbers the resulting (un-optimized) query plan is:

```
(project (?dt ?groups ?spots ?r ?fn ?ln)
  (filter (&& (= ?fn "Wolfgang") (= ?ln "Otruba"))
    (join
      (union
        (service <http://keas.kso.ac.at:8002/sparql>
          (BGP
            (triple ?s rdf:type sobs:SunspotRelativeNumbers)
            (triple ?s sobs:dateTime ?dt)
            (triple ?s sobs:groups ?groups)
            (triple ?s sobs:spots ?spots)
            (triple ?s sobs:rValue ?r)
            (triple ?s obs:byObserver ?obs)
          ))
        (service <http://solarscience.msfc.nasa.gov:8004/sparql>
          (BGP
            (triple ?s rdf:type sobs:SunspotRelativeNumbers)
            (triple ?s sobs:dateTime ?dt)
            (triple ?s sobs:groups ?groups)
```

```
              (triple ?s sobs:spots ?spots)
              (triple ?s sobs:rValue ?r)
              (triple ?s obs:byObserver ?obs)
           ))
     )
     (union
       (service <http://keas.kso.ac.at:8002/sparql>
           (BGP
              (triple ?obs rdf:type obs:Observer)
              (triple ?obs person:firstName ?fn)
              (triple ?obs person:lastName ?ln)
           ))
       (service <http://solarscience.msfc.nasa.gov:8004/sparql>
           (BGP
              (triple ?obs rdf:type obs:Observer)
              (triple ?obs person:firstName ?fn)
              (triple ?obs person:lastName ?ln)
           ))
     )
  )))
```

It can be seen that there is a new operator which is not part of official SPARQL: service[6]. The implementation of this operator in the query execution engine serializes sub-plans back into SPARQL and uses the protocol to execute it on the endpoint specified by the first parameter. In the next step, the query plan is handed on to the optimizer for which currently only the concepts discussed in Section 6 exist.

5 Sample Queries and Results

For the following sample queries, real-world data of sunspot observations recorded at Kanzelhöhe Solar Observatory (KSO) have been used. The observatory is also a partner in the Austrian Grid project. In the future, the system presented should replace the current archive CESAR (*Central European Solar ARchives*), which is a collaboration between KSO, Hvar Observatory Zagreb, Croatia, and the Astronomical Observatory Trieste, Italy. Because of the flexible architecture other observation sites can easily take part in future. The tests where performed with the following setup: the mediator (and also the test client) where running on a 2.16 GHz Intel Core 2 Duo with 2 GB memory and a 2 MBit link to the remote endpoints. All endpoints where simulated on the same physical host running two AMD Opteron CPUs at 1.6 GHz and 2 GB memory. Local host entries were used to simulate the SPARQL endpoints described in Table 1. The NASA endpoint is imaginary. The table only shows registered data sources which are relevant for the sample queries. The statistics shown were collected by the monitoring service.

The queries are shown in Fig. 2. Query 1 retrieves the first name, the last name, and optionally the e-mail address of scientists who have done observations. Query 2 retrieves all observations ever recorded by Mr. Otruba. This query is used to show a distributed join and filter. Query 3 retrieves all sunspot observations recorded in March 1969. Query 4 shows how the mediator's catalog can be accessed. It will list all data sources currently available, the *Virtual Organization*, and the contact person. The plan transformation for Query 2 has been described in the previous section. In Table 2 the

---

[6] Special thanks to Andy Seaborne, who implemented this extension after an e-mail discussion about federation of queries in July 2007.

Table 1. Endpoints registered when processing samples queries

Sunspot observations at KSO	instances
endpoint: <http://keas.kso.ac.at:8002/sparql>	
sobs:SunspotRelativeNumbers	9973
sobs:SunExposure	288
sobs:SolarObservationInstrument	7
sobs:Detector	9
obs:Observer	17
Sunspot observations by NASA (imaginary)	**instances**
endpoint: <http://solarscience.msfc.nasa.gov:8004/sparql>	
sobs:SunspotRelativeNumbers	89
obs:Observer	1

response time for the 1st solution, the total execution time (median of 10 samples), and the returned solution mappings are shown. The improvement of the performance is currently on top of the agenda. The results presented in this paper were generated from a prototype not using any optimization of distributed query plans.

Table 2. Test results for Query 1–4 and catalog status depicted in Table 1

Query #	1st solution	total time	solutions
1	162 ms	1.4 s	10
2	290 ms	3.8 s	1,272
3	1,216 ms	37.8 s	43
4	74 ms	0.2 s	6

Because ARQ is using a pipelining concept the response time is very good, even when data has to be retrieved from a remote data source. The reasons why Query 2 and especially Query 3 come off so badly will be discussed in the next section.

6 Optimizations and Future Work

For a mediator, minimizing response time is usually more important than maximizing throughput. Because ARQ is pipelined, response time is very good. However, shipping data is costly, so another goal is the minimization of the amount of data transfered. When using the REST-based SPARQL protocol a second requirement is to minimize the number of required requests. Query 2 and 3 show bad performance mainly because of bad join ordering and not pushing down filters to local sub-plans.

6.1 Optimization of Distributed Query Plans

The following optimization concepts are currently[7] being implemented: push-down of filter expressions, push-down of optional group patterns (becoming left-joins),

---
[7] i.e. at the time of writing this contribution – results are expected to be available for the conference in June 2008 and will be published later on.

```
SELECT  ?obs ?fname ?lname ?em
WHERE
  { ?obs a obs:Observer ;
         person:lastName     ?lname ;
         person:firstName    ?fname .
    OPTIONAL
      { ?obs   person:email  ?em .}
  }
```
Query 1

```
SELECT  ?dt ?groups ?spots ?r ?fn ?ln
WHERE
  { ?s a sobs:SunspotRelativeNumbers;
       sobs:dateTime       ?dt ;
       sobs:groups         ?groups ;
       sobs:spots          ?spots ;
       sobs:rValue         ?r ;
       obs:byObserver      ?obs .
    ?obs a obs:Observer ;
         person:firstName  ?fn ;
         person:lastName   ?ln .
    FILTER ( ?fn = "Wolfgang" &&
             ?ln = "Otruba" )
  }
```
Query 2

```
SELECT *
WHERE
  { ?s a sobs:SunspotRelativeNumbers;
       sobs:groups        ?groups ;
       sobs:spots         ?spots ;
       sobs:rValue        ?r ;
       obs:description    ?desc;
       obs:byObserver     ?obs ;
       sobs:dateTime      ?dt .
    FILTER ( ?dt >=
"1969-03-01T00:00:00"^^xsd:dateTime
         && ?dt <
"1969-04-01T00:00:00"^^xsd:dateTime )
  }
```
Query 3

```
SELECT * WHERE {
  ?ds a cat:DataSource .
  ?ds cat:maintainedBy ?vom .
  ?vom a cat:VOMember .
  ?vom person:firstName ?fn .
  ?vom person:lastName ?ln .
  OPTIONAL { ?vom person:email ?e } .
  ?ds cat:providedBy ?vo . }
```
Query 4

Fig. 2. Sample queries

push-down of local joins whenever possible, and optimization through global join and union re-ordering which is a rather complex task. A holistic approach for finding optimal plans based on *Iterative Dynamic Programming* (IDP) [10] will require heavy modifications to ARQ which should also be discussed in the future. By contrast to implementing static optimization algorithms based on general assumptions, IDP systematically enumerates all possible (equivalent) plans and prunes those with high cost as early as possible during the iteration. At a second stage the implementation of the new service-operator as part of the query execution engine will be extended to support row blocking to reduce the amount of HTTP requests. Some of the algorithms proposed by the database community can be re-used for SPARQL query processing. It is expected that query federation and optimization of distributed SPARQL queries will become more important in future to be able to manage large distributed data stores. Discussions about the *Billion Triples Challenge 2008* [1] indicate a need for scalable base technology which is not limited to local RDF data management.

In a highly optimized mediator-wrapper system like Garlic [17] or Disco [26], each wrapper provides several operators reflecting local data access and processing capabilities. For instance, an endpoint could support a distributed join operation with another remote node and joins could even be executed in parallel. Exploiting local capabilities would require heavy changes to the current query execution process.

6.2 Future Work

Other future work will be the support for DESCRIBE-queries and IRIs as subjects. In future the mediator should also use a OWL-DL reasoner to infer additional types for subject nodes specified in the query pattern. Currently, types have to be explicitly specified for each BGP (more precisely for the first occurrence: the algorithm caches already known types). OWL-DL constraints like for example a qualified cardinality restriction on obs:byObserver with owl:allValuesFrom obs:Observer would allow the mediator to deduce types of other nodes in the query pattern. Supporting subsumption queries like { ?p a p:Person } returning all resources that have a sub-type of p:Person may considerably inflate the global query plan. Such queries should be supported in future, when global plan optimization has been implemented.

6.3 The Role of Mediators in the Web of Data

As mentioned in the introduction, mediators like SemWIQ can be used to complement Semantic Web search engines and the web of Linked Data [5]. While the traditional hypertext web is separated into the *Surface Web* and the – for search engines hardly reachable – *Deep Web* [13], this separation disappears when browsing through Linked Data. The problem which remains is the fact that the so-called Deep Web is huge. Endpoints in the Web of data may expose large data stores and archives to the public and it will be hard for search engines to index all of these data efficiently. Sindice [27], for instance, is a Semantic Web crawler which is also able to crawl over SPARQL endpoints. However, their authors admitted that this feature is not used at the moment, because of the danger of getting lost in such *black holes*. Fig. 3 shows how SemWIQ can be embedded between a cloud of SPARQL endpoints. The registration component could be extended by a crawler which is autonomously registering new SPARQL endpoints which use RDF Schema or OWL vocabularies to describe their data (e.g. endpoints with D2R). A vocabulary browser which is visualizing all the vocabularies used by registered SPARQL endpoints including freetext search for concepts can be provided to users to examine the virtual data space. Compared to a search engine like Sindice, the mediator

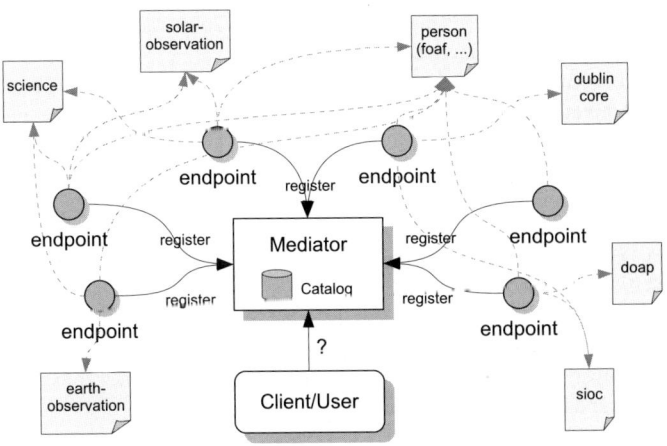

Fig. 3. SemWIQ

allows declarative queries over the complete data space. To maintain its scalibility, it is also possible to use multiple mediators at different locations which may synchronize catalog metadata and statistics over a peer-to-peer overlay network.

7 Conclusion

In this contribution a mediator-based system for virtual data integration based on Semantic Web technology has been presented. The system is primarily developed for sharing scientific data, but because of its generic architecture, it is supposed to be used for many other Semantic Web applications. In this paper query federation based on SPARQL and Jena/ARQ has been demonstrated in detail and several concepts for query optimization which is currently on the agenda have been discussed. Additional contributions can be expected after the implementation of additional features mentioned before.

Acknowledgements

The work is supported by the *Austrian Grid Project*, funded by the Austrian BMBWK (*Federal Ministry for Education, Science and Culture*), contract GZ 4003/2-VI/4c/2004.

References

1. The Billion Triples Challenge (mailing list archive at Yahoo!) (2007) (last visit December 12, 2007), http://tech.groups.yahoo.com/group/billiontriples/
2. Auer, S., Bizer, C., Lehmann, J., Kobilarov, G., Cyganiak, R., Ives, Z.: DBpedia: A nucleus for a web of open data. In: Aberer, K., Choi, K.-S., Noy, N., Allemang, D., Lee, K.-I., Nixon, L., Golbeck, J., Mika, P., Maynard, D., Mizoguchi, R., Schreiber, G., Cudré-Mauroux, P. (eds.) ISWC 2007. LNCS, vol. 4825, pp. 715–728. Springer, Heidelberg (2007)
3. Quilitz, B.: DARQ – Federated Queries with SPARQL (2006) (last visit December 12, 2007), http://darq.sourceforge.net/
4. Batini, C., Lenzerini, M., Navathe, S.B.: A comparative analysis of methodologies for database schema integration. ACM Comput. Surv. 18(4), 323–364 (1986)
5. Berners-Lee, T., Chen, Y., Chilton, L., Connolly, D., et al.: Tabulator: Exploring and analyzing linked data on the semantic web. In: Proceedings of the ISWC Workshop on Semantic Web User Interaction (2006)
6. Bizer, C., Cyganiak, R.: D2RQ – lessons learned. In: The W3C Workshop on RDF Access to Relational Databases (October 2007), http://www.w3.org/2007/03/RdfRDB/papers/d2rq-positionpaper/
7. Bizer, C., Cyganiak, R.: D2R Server – Publishing Relational Databases on the Semantic Web. In: 5th International Semantic Web Conference (2006)
8. In Silico Discovery. Semantic discovery system (2007) (last visit December 12, 2007), http://www.insilicodiscovery.com
9. Kossmann, D.: The State of the Art in Distributed Query Processing. ACM Comput. Surv. 32(4), 422–469 (2000)
10. Kossmann, D., Stocker, K.: Iterative dynamic programming: a new class of query optimization algorithms. ACM Trans. Database Syst. 25(1), 43–82 (2000)

11. Foster, I., Kesselman, C., Tuecke, S.: The Anatomy of the Grid: Enabling Scalable Virtual Organizations. In: Sakellariou, R., Keane, J.A., Gurd, J.R., Freeman, L. (eds.) Euro-Par 2001. LNCS, vol. 2150, Springer, Heidelberg (2001)
12. Haslhofer, B.: Mediaspaces (2007), http://www.mediaspaces.info/
13. He, B., Patel, M., Zhang, Z., Chang, K.C.-C.: Accessing the deep web. Commun. ACM 50(5), 94–101 (2007)
14. UK HP Labs, Bristol. Jena – A Semantic Web Framework for Java (last visit March 2007), http://jena.sourceforge.net/
15. Langegger, A., Blöchl, M., Wöß, W.: Sharing data on the grid using ontologies and distributed SPARQL queries. In: Wagner, R., Revell, N., Pernul, G. (eds.) DEXA 2007. LNCS, vol. 4653, pp. 450–454. Springer, Heidelberg (2007)
16. Langegger, A., Wöß, W., Blöchl, M.: Semantic data access middleware for grids (last visit December 2007), http://gsdam.sourceforge.net
17. Haas, L.M., Kossmann, D., Wimmers, E.L., Yang, J.: Optimizing Queries Across Diverse Data Sources. In: Proceedings of the 23th International Conference on Very Large Databases, Athens, VLDB Endowment, Saratoga, Calif, pp. 276–285 (1997)
18. Melnik, S.: Generic Model Management: Concepts And Algorithms. LNCS. Springer, New York (2004)
19. Miles, A., Baker, T., Swick, R.: Best practice recipes for publishing RDF vocabularies (2006) (last visit December 12, 2007), http://www.w3.org/TR/swbp-vocab-pub/
20. Noy, N.F., Rubin, D.L., Musen, M.A.: Making biomedical ontologies and ontology repositories work. Intelligent Systems 19(6), 78–81 (2004)
21. OpenLink Software. OpenLink Virtuoso (last visit March 2007) http://www.openlinksw.com/virtuoso/
22. Prud'hommeaux, E.: Optimal RDF access to relational databases (April 2004), http://www.w3.org/2004/04/30-RDF-RDB-access/
23. Prud'hommeaux, E.: Federated SPARQL (May 2007), http://www.w3.org/2007/05/SPARQLfed/
24. Sattler, K.-U., Geist, I., Schallehn, E.: Concept-based querying in mediator systems. The VLDB Journal 14(1), 97–111 (2005)
25. Tan, H., Lambrix, P.: A method for recommending ontology alignment strategies. In: Aberer, K., Choi, K.-S., Noy, N., Allemang, D., Lee, K.-I., Nixon, L., Golbeck, J., Mika, P., Maynard, D., Mizoguchi, R., Schreiber, G., Cudré-Mauroux, P. (eds.) ISWC 2007. LNCS, vol. 4825, pp. 491–504. Springer, Heidelberg (2007)
26. Tomasic, A., Raschid, L., Valduriez, P.: Scaling heterogeneous databases and the design of disco. ICDCS 00, 449 (1996)
27. Tummarello, G., Delbru, R., Oren, E.: Sindice.com: Weaving the open linked data. In: Proceedings of the 6th International Semantic Web Conference (ISWC) (November 2007)
28. W3C. SPARQL Query Language for RDF, W3C Proposed Recommendation (last visit May 2007), http://www.w3.org/TR/rdf-sparql-query/
29. Wiederhold, G.: Mediators in the architecture of future information systems. Computer 25(3), 38–49 (1992)

Graph Summaries for Subgraph Frequency Estimation

Angela Maduko[1], Kemafor Anyanwu[2], Amit Sheth[3], and Paul Schliekelman[4]

[1] Department of Computer Science, University of Georgia
[2] Department of Computer Science, North Carolina State University
[3] Kno.e.sis Center, Wright State University
[4] Department of Statistics, University of Georgia
maduko@cs.uga.edu, kogan@ncsu.edu, amit.sheth@wright.edu,
pdschlie@stat.uga.edu

Abstract. A fundamental problem related to graph structured databases is searching for substructures. One issue with respect to optimizing such searches is the ability to estimate the frequency of substructures within a query graph. In this work, we present and evaluate two techniques for estimating the frequency of subgraphs from a summary of the data graph. In the first technique, we assume that edge occurrences on edge sequences are position independent and summarize only the most informative dependencies. In the second technique, we prune small subgraphs using a valuation scheme that blends information about their importance and estimation power. In both techniques, we assume conditional independence to estimate the frequencies of larger subgraphs. We validate the effectiveness of our techniques through experiments on real and synthetic datasets.

Keywords: Frequency estimation, Graph summaries, Data summaries.

1 Introduction

Graphs are increasingly used to model data on the Web, the emerging Semantic Web and complex biochemical structures such as proteins and chemical compounds. They offer a representation amenable to analysis and knowledge extraction. Structure search that matches a query graph over a graph database, is a common technique for retrieving information from graphs. In biochemistry, search for common features in large sets of molecules is used for drug discovery and drug design studies. These searches, which return all graphs that contain the query graph, can be computationally challenging. As demonstrated in [12] and [15], path or subgraph indexes help to cope with this difficulty. The idea is to use indexed fragments of the query graph to retrieve a set of data graphs, from which those containing the whole query graph are found using subgraph isomorphism tests. The frequency of the fragments in the query graph plays a crucial role in optimizing these searches. As an illustration, the left part of Figure 1 shows a database of graphs, with the subgraphs in Figure 1h and Figure 1i indexed. Given the query graph of Figure 1g, the graphs in Figure 1b–1f that contain the indexed fragments are retrieved. With subgraph isomorphism tests, only Figure 1b and 1f are found to contain the entire query graph. One strategy for optimizing this process is to reduce the number of isomorphism tests performed, which depends on the frequency of the indexed subgraphs. Figure 1h and Figure 1i have 9 and at least 18 occurrences in the selected graphs respectively. Thus the isomorphism tests can be performed by matching Figure 1h first, then expanding to the rest of the query graph.

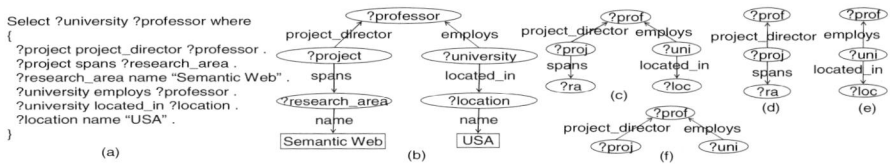

Fig. 1. A Sample Graph Database, Its Indexed Fragments and a Query Graph

Most proposed query languages for querying graph representations of semi-structured data, such as RDF [5], support structure search as a primary query paradigm. In storage schemes for graph databases that organize graphs as edge relations, the embeddings of a given query graph in the database are computed using join operations. For such scenarios, subgraph frequency estimates are needed to determine the cardinality of intermediate join results during query optimization.

Fig. 2. A SPARQL Query, Its Graph and Subgraph Patterns

To illustrate, the SPARQL [10] query (namespaces are omitted) of Figure 2a (with graph pattern shown in Figure 2b) asks for professors employed in universities in the United States who direct a Semantic Web project. Three join operations (shown in Figure 2c) are needed to process this query if edge relations is the storage scheme for the data. To optimize this query, the optimizer will need accurate estimates of the frequency of subpatterns of Figure 2c, such as those shown in Figure 2d, e and f.

In this work, we focus on efficient techniques for estimating the frequency of subgraph patterns in a graph database. Noting that (1) the number of possible subgraphs in a graph database could be exponential and (2) it is more expedient if the estimates are computed without disk accesses, since they are needed at optimization time, we focus on techniques that summarize subgraphs in the graph database so as to fit in the available memory. Obviously, such a summary will be useful only if it captures the correlations among subgraphs. Our choice of subgraphs is strengthened by the observations in [13], [15] and [16], where it is shown that subtrees and subgraphs perform better than paths in capturing correlations among graphs and trees, respectively. However, the number of unique subgraphs greatly exceeds the number of paths in a graph. It is thus infeasible to examine all subgraphs, and efficient pruning techniques are needed. We propose two summaries that differ in their pruning techniques. The Maximal Dependence Tree (MD-Tree) and the Pattern Tree (P-Tree).

The pruning technique of the MD-Tree is based on the observation that high-order statistical dependencies often exist among subgraphs. It may be prohibitive to keep all such dependencies; thus we attempt to capture the most informative dependencies in the given space. The pruning technique of the P-Tree is based on two insights: (1) the frequency of a graph may be close to that of a function of its subgraph; and (2) information about the importance of subgraphs could lead to characterizing some as more important than others. For example, frequent subgraphs from a query workload are more important than infrequent ones for tuning purposes. We prune the P-Tree by blending the significance of patterns for estimation and for tuning purposes.

This paper is structured as follows: Section 2 formally defines the problem we address in this work and briefly discusses background work. Section 3 discusses the proposed summaries while Section 4 presents the experimental evaluation of the summaries. In Section 5 we discuss related work and conclude the paper in Section 6.

2 Preliminaries

Data Model. A collection of connected graphs or a graph database can be viewed as a large graph with several connected components. We use the term "graph" to refer to such a large graph. We focus on a directed labeled graph model that represents named binary relationships (edges labeled with names) between entities (nodes) such. Such named relationships can be viewed as triples (entityA relationship1 entityB).

Definition 1. Let L and T be finite sets of labels. We define a graph G as a 4-tuple (V, E, λ, τ). V and E are sets of nodes and edges of G respectively, $\lambda : (V \cup E) \to L$ is a many-to-one function that maps nodes/edges of G to labels in L and $\tau : 2^V \to T$ is a multivalued type function that maps sets of nodes of G to labels in T, (i.e. $L \cap T \neq \emptyset$) so that for a node v, $\tau(v)$ may be perceived as a conceptual entity to which v belongs.

Note that, both τ and λ map nodes not perceived as members of a conceptual entity to the same label so that for such a node v, $\tau(v) = \lambda(v)$. Our graph model captures many semantic data models, particularly the RDF model[5].

Definition 2. Given graphs G = (V, E, λ, τ) and G' = (V', E', λ', τ'), we say that G' is embedded in G if there is an injective function $f : V' \to V$ such that:
- $\tau'(v) = \tau(f(v)), \forall v \in V'$
- $\forall (u, v) \in E', (f(u), f(v)) \in E$ and $\lambda'(u, v) = \lambda(f(u), f(v))$

Problem Definition. Given graphs G and G', the frequency of G' in G is the number of unique embeddings it has in G. The problem we address is stated succinctly as:
Given a graph G and a space budget B, create a summary of size at most B, for obtaining accurate estimates of the frequencies of graphs embedded in G.

2.1 Background

With minimal modifications, efficient pattern-mining algorithms such as gSpan [14] can be used to discover subgraphs and count their frequencies. This technique uses a canonical label for a graph for computing subgraphs frequencies. We now briefly review the *minimum DFS code* [14] canonical label of a graph, adopted in this work.

DFS Coding. This technique transforms a graph into an edge sequence called DFS code, by a DFS traversal. Each edge (u, v) in the graph is represented by a 5-tuple <i, j, l_i, $l_{(i, j)}$, l_j>, where integers i and j denote the DFS discovery times of nodes u and v, l_i, l_j and $l_{(i, j)}$ are the labels of u, v, and the edge (u, v), respectively. The edges are ordered by listing those in the DFS tree (tree edges) in order of their discovery. The rest are inserted into the ordered list as follows: For a tree edge (u, v), all non-tree edges from v come immediately after (u, v); if (u_i, v_j) and (u_i, v_k) are two non-tree edges, (u_i, v_j) is listed before (u_i, v_k) only if j < k. A graph may have many DFS codes, so the minimum, based on a linear ordering of all its DFS codes, is its canonical label. Details of the DFS coding and gSpan algorithm can be found in [14].

3 Approach

In this section, we present our proposed summaries. We discuss the Maximal Dependence Tree in section 0, then the Pattern Tree in section 0. To create each summary, we generate and count the frequencies of all subgraphs of length at most maxL, using a slight modification of gSpan and input graph G = (V, E, λ, τ). We represent each edge e = (u, v) in G by a 5-tuple <i, j, λ(e), τ(u), τ(v)>, where integers i and j are the DFS discovery times of nodes u and v and λ and τ are the functions defined in Definition 1, with λ(.) and τ(.) (i.e., the ranges of λ and τ) mapped to unique integers. The sequence of edges/quintuples obtained after the algorithm is run, represents the structure of subgraphs of G; thus, given any two edges e_1 = (u_1, v_1) and e_2 = (u_2, v_2), if v_1 = v_2, it follows that τ(v_1) = τ(v_2). [14] discusses the minimum DFS code in the context of undirected labeled graphs. For directed labeled graphs, we ignore edge directions during the DFS traversal so as to maintain the connectivity of the graph. However the directions are kept implicitly in the quintuples. We use the term pattern ambiguously to refer to a subgraph as well as to its minimal DFS code.

Example 1. *Figure 3*a shows a directed labeled graph of conference paper information. The same graph is shown in *Figure 3c, with nodes and edges assigned integer ids as shown* in *Figure 3b*. To obtain the edge sequence for the subgraph au1 authorOf pub1, pub1 submittedTo conf1, pc1 pcMember of conf1, we begin DFS with the edge authorOf as it is lexicographically the smallest label. DFS proceeds as indicated by the boxed discovery ids to yield (1,2,5,1,4) (2,3,7,4,3) (4,3,6,3,2). Note that the direction of the edge labeled "pcMemberOf" is implicit in the sequence. *Figure 3d shows all* patterns of length at most 3 and their frequencies.

Note that, in our directed graph model, an edge, for example (5, 1, 4) in Figure 3, may appear in a pattern of length at least two, in one of three possible directions: forward, as in (1, 2, 5, 1, 4); backward, as in (3, 2, 5, 1, 4); or self-loop, as in (2, 2, 5, 1, 4).

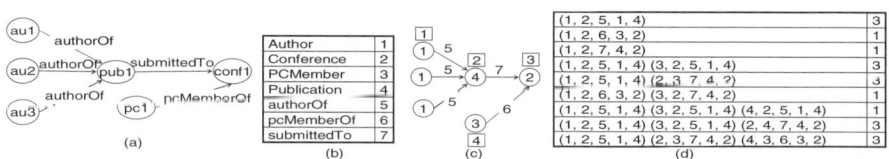

Fig. 3. Unique Edge Sequences for Subgraphs

3.1 Maximal Dependence Tree (MD-Tree)

To motivate the MD-Tree, we observe that edges in certain positions in patterns may largely determine their probabilities. For example, the three patterns of length 3 in Figure 3d have the edge in the first position in common so that the edge in the second position rather than the first, will exert a greater influence in their probabilities. Our MD-Tree approach exploits the existence of such edge positions. If none exists, we assume edges occur independently at each position. We construct the MD-Tree through an adaptation of the Maximal Dependence Decomposition [2] technique.

Notations. Let N_E denote the number of unique edge labels in a graph $G = (V, E, \lambda, \tau)$ that is N_E is size of the mapping $(\lambda(e), \tau(u), \tau(v))$, for each edge $e = (u, v)$ in E. Further, let β be an integer in the range [1, 3] such that β is: (a) 1 if all edges in E are forward edges, (b) 2 if E also contains backward edges and (c) 3 if in addition to forward edges, E also contains self loop edges or both backward and self loop edges. We denote a set of patterns of length k by P_K and define **freq** as a function with domain the power set of patterns of length at most maxL and range the set of positive integers. If X is a single pattern, freq(X) maps to the frequency of the pattern but if X contains more than one pattern, freq(X) maps to the sum of the frequencies of patterns in X. We denote the probability with which edge x occurs at position y by $\mathbf{Pr(x_{(y)})}$.

We begin the discussion of the MD-Tree by introducing *weight matrices* for a set of pattern. A weight matrix WM_K for P_K is a $\beta N_E \times k$ matrix whose rows represent the possible edge patterns that may appear in patterns in P_K and columns represent positions in which the edge types may occur. Cell (i,j) contains the probability that edge i occurs at position j. Given P_K and the assumption that edges occur independently at any position in the patterns, a weight matrix for P_K suffices for estimating the frequency of any pattern $p = (e_1, e_2, ..., e_k)$ in P_K as $freq(p) = freq(P_K) \times Pr(e_{1(1)}) \times Pr(e_{2(2)}) \times ... \times Pr(e_{k(k)})$. To construct a weight matrix WM_K for P_K, we obtain the row indices by assigning unique integer ids to the possible edge patterns that may appear in patterns of length at least two, in multiples of β. We assign integer x to edge type y such that if x modulo β is 0, 1 or 2 then x identifies y in the forward or backward directions or self-loop, respectively. The column indices are the k positions in which edge patterns may occur.

<table>
<tr><td>

(1, 2, 5, 1, 4) (3, 2, 5, 1, 4)	3
(1, 2, 5, 1, 4) (2, 3, 7, 4, 2)	3
(1, 2, 6, 3, 2) (3, 2, 7, 4, 2)	1

(a)

</td><td>

1	(5, 1, 4)
2	(5, 1, 4)
3	(6, 3, 2)
4	(6, 3, 2)
5	(7, 4, 2)
6	(7, 4, 2)

(b)

</td><td>

	1	2
1	6/7	0
2	0	3/7
3	1/7	0
4	0	0
5	0	3/7
6	0	1/7

(c)

</td></tr>
</table>

Fig. 4. A Weight Matrix for Patterns of Length 2

Example 2. To construct WM_2 with dimension 6×2 (i.e. β is 2, since there are no self loop edges) for the patterns in *Figure 4*a, we assign integer ids to the edge types (5, 1, 4), (6, 3, 2) and (7, 4, 2) as shown in *Figure 4*b. Next, we compute the entries for each cell (i, j) in WM_2 as shown in *Figure 4*c. Thus, cell (2, 1) holds the probability that edge (5, 1, 4) occurs in a backward direction at position 2 etc. Under the independence assumption, the frequency of (1, 2, 5, 1, 4)(3, 2, 5, 1, 4) is estimated as 7(6/7)(3/7) i.e. 18/7, which rounds to 3.

Definition 3 Base MD-Tree. Given the sets $P_1, P_2, ..., P_{maxL}$ of patterns of length at most maxL, a base MD-Tree for the patterns in P_i $1 \leq i \leq$ maxL is a triple (R_T, V_T, E_T) where $R_T \in V_T$ is the root of the tree and V_T and E_T are the sets of nodes and edges of the tree, such that $|V_T - R_T| = |E_T| =$ maxL. All nodes in $V_T - R_T$ are ordered children of R_T such that child i is associated with the weight matrix WM_i, for patterns in P_i. Each edge (R_T, i) is labeled with freq(P_i), the total frequency of all patterns in P_i.

If the independence assumption does not hold, a refinement process on the base MD-Tree is required to capture edge dependencies. Given P_K, if it is known that the occurrence of an edge at position i, $1 \leq i \leq k$, $i \neq m$, depends on the edge at position m, we estimate the frequency of a pattern $p = (e_1, e_2, ..., e_k)$ in P_K as:

$$\text{freq}(P_K) \times \Pr(e_{m(m)}) \times \prod_{i=1, i \neq m}^{k} \Pr(e_{i(i)} | e_{m(m)}), \quad (1)$$

where $\Pr(e_{i(i)} | e_{m(m)})$ is the conditional probability that e_i occurs at position i given that e_m occurred at position m. The base MD-Tree is modified to reflect this dependence. We refer to the modified tree as a *refined MD-Tree*.

Definition 4 Refined MD-Tree. A refined MD-Tree is a triple (R_T, V_T, E_T) where $R_T \in V_T$ is the root and V_T and E_T are the sets of its nodes and edges respectively. V_T can be partitioned into two disjoint non-empty sets V_{Tleaf} and $V_{Tnon-leaf}$ such that for $v \in V_{Tleaf}$ or $v \in V_{Tnon-leaf}$, v is a leaf node or non-leaf node, respectively. Weight matrices are associated only with leaf nodes and every non-leaf node has βN_E ordered children except the root, which has maxL children.

For ease of exposition, we illustrate the refinement process with an example.

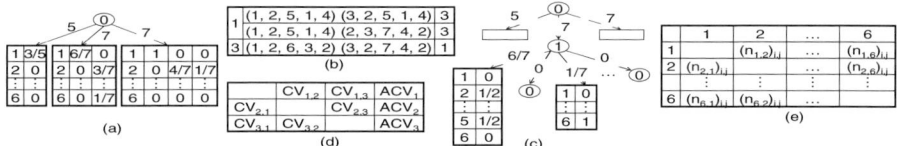

Fig. 5. Refining the Base MD-Tree of patterns of length 2, at position 1

Example 3. *Suppose that edge types at position 1 influence those at position 2, for patterns of length 2. We refine the second child (which we denote as v_2) of the root of the base MD-Tree of Figure 5a as follows. First, we create βN_E (in this case 6) ordered children nodes for node v_2, one for each edge type. Next, we obtain the length 2 patterns used to create the weight matrix associated with v_2 i.e., the patterns in Figure 4a. We then partition these patterns with respect to the occurrence of the 6 edge types at position 1. As shown in Figure 5b, only the partitions for edge types 1 and 3 are non-empty. Using the patterns in these partitions, we create two new 6×1 weight matrices for child nodes 1 and 3 of v_2 respectively. We label the edges to these nodes 6/7 and 1/7 i.e. the probabilities that the edge types with ids "1" and "3" occur at position 1 in patterns in their respective partitions. Finally, we delete v_2's weight matrix and assign it the split position 1. Figure 5c shows the refined MD-Tree.*

Given P_K, we refine its base MD-Tree by finding the position in its patterns that most influences others, by chi-square association tests for edge types at all pairs of positions i and j, $1 \leq i,j \leq k, i \neq j$. The test statistic is given by:

$$\sum_{m=1}^{\beta N_E} \sum_{n=1}^{\beta N_E} \frac{(O_{m,n} - E_{m,n})^2}{E_{m,n}} \text{ where } E_{m,n} = \left(\sum_{a=1}^{\beta N_F} O_{m,a} \sum_{b=1}^{\beta N_F} O_{b,n} \right) \bigg/ \sum_{a=1}^{\beta N_E} \sum_{b=1}^{\beta N_E} U_{a,b}. \quad (2)$$

$O_{m,n}$ is the sum of the frequency of patterns in P_K for which edge types m and n occur at positions i and j respectively. We clarify this with an illustration.

Example 4. *To find a position of maximal dependence for a set of patterns of length 3, first, we create a 3×4 matrix as shown in Figure 5d. To compute $CV_{1,2}$ i.e. the chi-square value for cell (1,2) for instance, we create a 6×6 matrix as shown in Figure 5e, where cell i,j contains the number of times the edge types with integer ids i and j occur at positions 1 and 2 in the patterns respectively. $CV_{1,2}$ is the value of the test statistic for this 6×6 matrix. Next, we obtain the aggregate chi-square value (ACV) stored in the fourth column of each row by summing the chi-square values in each row of the 3×4 matrix. Then we find the maximum ACV over the three rows. Suppose it is ACV_2, we then conclude that position 2 has the greatest influence on others but only if at least one of $CV_{2,j}$ is statistically significant.*

If v is a leaf node of a base or refined MD-Tree T, we say that v is significant if there is a position j of maximal dependence in the set of patterns used to create the weight matrix of v. We denote an MD-Tree that is completely refined (has no significant nodes) as a *Complete MD-Tree*. A complete MD-Tree is ideal for estimating the frequency of patterns but its size may exceed the budget. Our *optimal MD-Tree* then is a refined MD-Tree that fits the budget and that gives the best estimates of pattern frequencies. We now formalize the problem of finding the optimal MD-Tree.

Given a complete MD-Tree (R_T, V_T, E_T), let $T' = (V', E')$ be a tree such that $V' \subseteq V_T$, where $V' = \{R_T, v_1, v_2, ..., v_m\}$ contains all significant nodes of V_T and every edge (u, v) in E' is an edge in E_T. Also, let $S = (0, s_{v1}, s_{v2}, ..., s_{vm})$ be the size increment induced on the MD-Tree when v_i is refined and let $I = (0, i_{v1}, i_{v2}, ... i_{vm})$ be the impact of node v_i, given by $\max(ACV)/C_v$, rounded to the nearest integer. C_v is the number of columns of the weight matrix associated with v. The problem is to find a tree $T'' = (V'', E'')$, $T'' \subseteq T'$ rooted at R_T, such that $\sum_j (S_{vj}) \leq B$ and $\sum_j (i_{vj})$ is maximized. This problem is an instance of the Tree Knapsack Problem, which is known to be NP-hard. Given x_j, an indicator variable with value 1 if v_j is selected as part of the optimal solution or 0 otherwise, TKP is formulated as:

$$\text{Maximize } \sum_j^m i_{vj} x_j \text{ constrained on } \sum_j^m s_{vj} x_j \leq B, \; x_{\text{pred}(j)} \geq x_j, \quad (3)$$

where pred(j) is the predecessor of j in T'. With this reformulation, we employ a greedy approximation with $O(|V'|^2)$ running time. Given $T' = (V', E')$, vectors S and I and the size budget B, our greedy approximation creates the tree $T'' = (V'', E'')$ by keeping maximal impact subtrees of T' that fit the budget.

Frequency Estimation Using the MD-Tree. Given an optimal MD-Tree (R_T, V_T, E_T), let λ_V map nodes in V_T to the integers or weight matrices they are associated with and and λ_E map edges in E_T to integers or real numbers they are labeled with. Let the function id on edge patterns return the integer id of its edge type. Let $p' = e_1, e_2, ..., e_k$ be the edge sequence of a graph $G' = (V', E', \lambda, \tau)$ of length k. To estimate the frequency of P', we first check that the structure of p' exists in the structural summary given in Definition 1, which we keep along with the MD-Tree. If so, beginning from the kth child v of R_T, we estimate freq(p') as:

$$\lambda_E(R_T, v) \left(\prod_{i=1}^{j} \lambda_E(v^i, v^{i+1})_{\lambda_V(V^i)} \right) \left(\prod_{r=1, r \notin S}^{k} (\lambda_V(v^{j+1}))_{(id(e_r), r)} \right). \quad (4)$$

In this product, the subscript r of an edge $(v, v')_r$ denotes the rth edge of node v. The integer j is the number of edges of the optimal MD-Tree found on the path from the root to a leaf node as defined by the subscripts on the edges, so that the node v^{j+1} is a leaf. The subcripts (r, r') are integers indices for accessing cell (r, r') of the weight matrix associated with node $\lambda_V(v^{j+1})$. The set S holds labels of all nodes on the path from R_T to v^{j+1} so that at v^{j+1}, any integer in the range [1, maxL] not in the set S did not label any node on this path. The depth of the MD-Tree is at most maxL; thus the time complexity for estimating pattern frequencies is O(maxLlog(maxL)).

Example 5. *To estimate the frequency of the pattern $p = e_1, e_2$ given by (1,2,5,1,4) (3,2,5,1,4) from Figure 5c, we first access v_1, the second child of the root. Since $\lambda_V(v_1)$ = 1, we insert 1 into set S and set the frequency of p (freq(p)) to freq(p) which is 7. Recall from Figure 4b that $id(e_1) = id(1, 2, 5, 1, 4)$ is 1. So, we access v_2, the node on which the first edge of v_1 is incident. Next, we multiply freq(p) by $\lambda_E(v_1, v_2)$ given by 6/7, resulting in 6. Then we obtain $\lambda_V(v_2)$ i.e. the weight matrix WM_2 of v_2. S contains the integer 1, so the lone column of WM_2 must index position 2 of patterns in P_2. Further, $id(e_2) = id(3,2,5,1,4)$ is 2; thus we access the cell that represents the index (2, 2) in WM_2 to obtain 1/2. We then multiply freq(p) by 1/2 to obtain 3.*

3.2 Pattern Tree (P-Tree)

The idea of our P-Tree approach is to identify sets of patterns with almost the same edge patterns, such that for a set P_K, the frequencies of patterns in P_K are within δ of that of at least one pattern in P_K say p. Given p, the frequencies of patterns in P_K can be estimated within δ error thus we can safely eliminate all patterns but p from the summary. Given $P = P_1, P_2, ..., P_{maxL}$ a set of patterns of length at most maxL, the unpruned P-Tree for P is a prefix tree of patterns in P. Its nodes are labeled with edge patterns so that a pattern in P is obtained by concatenating node labels on a path from the root. Also, each node is associated with the frequency of the pattern it represents.

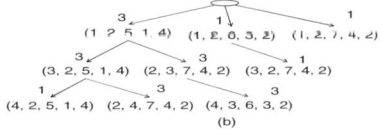

Fig. 6. Pattern Tree

If the size of the P-Tree exceeds the budget, it must be systematically pruned so as to avoid a large increase in its overall estimation error. The question is *which nodes are to be pruned and in what order?* We answer this question with the concepts of *observed* and *estimation* values of patterns. We begin by introducing some notations.

Notations. In addition to the notations introduced in section 0, we define the function **children** whose domain and range are the set of patterns of length at most maxL and the power set of patterns of length at most maxL respectively, such that for a pattern p, children(p) maps to the set of children of p in the P-Tree.

Definition 5 Observed Value of a Pattern. Let $P = (p_1, p_2, \ldots, p_m)$ be a set of patterns with frequencies $(freq(p_1), freq(p_2), \ldots, freq(p_m))$. Let T, a positive integer, be an importance threshold T and let $P_{OI} = (p_{OI1}, p_{OI2}, \ldots, p_{OIm})$ be a vector such that $0 \leq p_{OIi} \leq 1$ for every $p_{OIi} \in P_{OI}$ and p_{OIi} defines the importance of pattern p_i. We define the observed value of p_i (p_{OVi}) as the number of patterns that are less important than p_i that is the number of patterns p_j in P such that $p_{OIi} > p_{OIj}$.

We do not assume any particular technique for computing the importance of a pattern. However, for the purpose of tuning the summary to favour frequent patterns, it can simply be computed as the ratio of its frequency to that of the most frequent.

To motivate the estimation value of patterns, we note that if there is a match for a pattern $p = e_1, e_2, \ldots, e_k$ in the tree, its frequency freq(p) is the integer associated with the matched node labeled e_k. If e_k is contracted, we guess freq(p) as $p'_{GR} \times freq(p')$, where $p' = e_1, e_2, \ldots, e_{k-1}$ is the parent of p and p'_{GR} is the growth rate of p', under the assumption that children of p' have a uniform frequency distribution. When the children of p' are to be contracted, we keep its growth rate given by $N/(m \times freq(p'))$, where m is the number of children of p' and N is their total frequency. Thus the frequency of each child is estimated as N/m. We keep the growth rate and not N/m, for ease of propagation as we will discuss later. We keep one growth rate for p' for all its children, to avoid overly increasing the size of the tree as patterns are pruned. To validate our uniformity assumption, we prune the P-Tree by deleting the children of patterns that are uniformly or nearly uniformly distributed. To do this, we let the random variable Y define the occurrence of a child of p'; then we measure the evenness of the probability distribution of Y using its *entropy* [10] $H(Pr_Y)$, given by $-\sum_j Pr_Y(p_j)\log_2(Pr_Y(p_j))$. We compute the probability of the occurrence of any child p of a pattern p' as its proportion to the total frequency of children of p' (i.e., freq(p)/N, where N is the sum of $freq(p_j)$ for all p_j in children(p')). The entropy of a probability distribution is maximized if the distribution is uniform. Thus to measure the uniformity of Y, we normalize $H(Pr_Y)$ by a division by its maximum entropy. We denote this ratio as p'_{ENT} for pattern p' in the tree. If p' has one child, we set p'_{ENT} to 1.

Definition 6 Estimation Value of a Pattern. Given a set of patterns $P = (p_1, p_2, \ldots, p_m)$ with frequencies $(freq(p_1), freq(p_2), \ldots, freq(p_m))$ and some $\varepsilon \geq 0$, the estimation value of p_i (p_{EVi}) is given by:

$$p_{ENTi} \times \frac{\left(|\{p_j \mid p_j \in P_{Ci} \text{ and } |(freq(p_i) \times p_{GRi}) - freq(p_j)| \leq \varepsilon\}|\right)^h}{|P_{Ci}|}. \qquad (5)$$

By definition, p_{ENT} is at most 1. It is 1, if the distribution of the children of p is uniform. If the exponent h is 1, the second term of the product measures how closely p estimates all its children within ε error. Thus if p_i and p_j both have three children and p_i estimates only two within ε while p_j estimates just one within ε, this value will be higher for p_i (2/3) than for p_j (1/3). However, if p_i has six children and estimates only two within ε, then the value will be 1/3 for both p_i and p_j, although p_i estimates more children outside ε than p_j. We set h to 1.5 to prevent the numerator from overly dominating the denominator. To find the optimal ε, beginning with exponent 0 and base 2, we recursively increment the exponent until we get to 2^i, such that enough patterns can be pruned to meet the budget. Then, we search between 2^{i-1} and 2^i for the value that allows for pruning the fewest patterns. We now combine the observed and estimation values of patterns to obtain a single value for pruning the P-Tree.

Definition 7 Let $P = (p_1, p_2, ..., p_m)$ be the set of patterns in the P-Tree and p_{EVmax}, the maximum expected value of patterns in P. Given a constant $c > 0$, the value of a pattern p_j is given by:

$$p_{Vj} = (1 + p_{EVj})(1 + p_{OVj}) + ip_{EVmax} . \qquad (6)$$

where i is an indicator variable whose value is 1 if $p_{OVj} \geq c$ and 0 otherwise. The additive constants ensure that the value of a pattern is non-zero when either its observed or its estimation value is zero. The second term allows for tuning the P-Tree by boosting the values of important patterns, to delay their contraction.

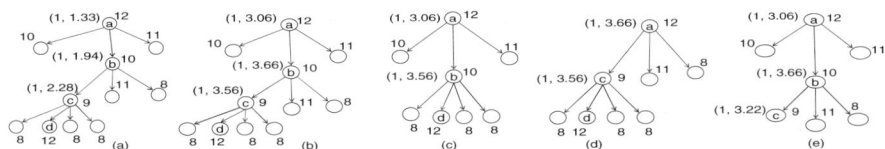

Fig. 7. Contracting Nodes of the Pattern Tree

Example 6. *Figure 7a shows a subtree of a P-Tree. The 2-tuple (p_Z, p_V) for each internal node is its growth rate and its value, computed at ε = 1 and c = 0, with no importance information (i.e. p_{OVi} is zero for all patterns). The growth rate of node "a" is given by (10+10+11)/(3×12) ≈ 1, so its frequency (12) estimates that of one child (11). With exponent 1.5, the second term of the equation in Definition 7 is 0.333. The entropy of the frequency distribution of its children is 10/31 × $\log_2$(31/10) + 10/31 × $\log_2$(31/10) + 11/31× $\log_2$(31/11) = 1.583, maximized at ($\log_2$(3)) = 1.585 with ratio 1.583/1.585 = 0.999. Its estimation value is 0.999(0.333) = 0.333, so its value is 1+0.333 = 1.333. In Figure 7b, the values are computed at ε = 2. In Figure 7c, d, and e, the children of nodes b, a, and c have been contracted at ε = 2, with total estimation errors of 4, 5, and 6. Our technique will result in the contraction of Figure 7c since node b has the largest value.*

When children(p_i) are to be contracted, if the children of any node in children(p_i) have been contracted, the average of the growth rates of such nodes are computed and associated with p_i. Thus, a node in the pruned P-Tree may have at most maxL growth rates, ordered in increasing order of the original depths of their sources in the P-Tree.

Frequency Estimation Using the Pattern Tree. Given $G' = (V', E', \lambda, \tau)$, we obtain its edge sequence $p = e_1, e_2, \ldots, e_k$ and check that its structure exists in the structural summary given in Definition 1, which we keep along with the P-Tree. If so, we match p against the P-Tree. If we find a complete match for p, we return the frequency of the matched node e_k in the P-Tree. If we find a partial match, we consider the last matched node v_j in the P-Tree. If it matches e_k, we return its frequency, which is the exact frequency of p if no descendant of v_j was contracted. If it matches e_i $i < k$, we use its frequency to estimate that of the contracted node that originally matched e_k. Estimating the frequency of e_k requires estimating and propagating those of its k-i-1 immediate contracted ancestors. If $\xi_1, \xi_2, \ldots, \xi_r$, $k \leq r \leq maxL$ are the growth rates kept with v_j and $freq(p_j)$ is the frequency of v_j, we estimate the frequency of e_k as:

$$freq(p_j) \times \prod_{r=1}^{k-i} \xi_r . \qquad (7)$$

Example 7. *To estimate the frequency of (a, b, c, d) from Figure 7c, we find the partial match (a, b, _, d) and we return 12 since d is matched. With the P-Tree of Figure 7e, we find the matches (a, b, c, _). We return 9, since the growth rate of c is 1.*

3.3 Estimating the Frequency of Large Patterns

Given a subgraph $G' = (V', E', \lambda, \tau)$ with $|E'| > maxL$, as always, we check that the structure of $p = e_1, e_2, \ldots, e_k$ exists in the structural summary given in Definition 1. If so, we partition G' into $G'_1, G'_2, \ldots, G'_{|E|-maxL+1}$ non-disjoint connected subgraphs such that G'_i intersects G'_{i-1} in all but one edge. Let G''_i denote the intersecting edges of G'_i and G'_{i-1}. Next, we obtain the edge sequences $p'_1, p'_2, \ldots, p'_{|E|-maxL+1}$ and $p''_2, p''_3, \ldots, p''_{|E|-maxL+1}$ for the subgraphs $G'_1, G'_2, \ldots, G'_{|E|-maxL+1}$ and $G''_2, G''_3, \ldots, G''_{|E|-maxL+1}$, respectively. As in [13], we assume conditional independence to estimate the frequency of G' as follows:

$$freq(p') = freq(p'_1) \times \prod_{r=2}^{|E|+maxL-1} \frac{freq(p'_r)}{freq(p''_r)} . \qquad (8)$$

Since G' may be partitioned into $G'_1, G'_2, \ldots, G'_{|E|-maxL+1}$ in several different ways, we select the partition for which frequency estimates of the patterns $P'_1, P'_2, \ldots, P'_{|E|-maxL+1}$ are obtained along the deepest paths, that is paths with the maximum total split nodes in the MD-Tree or along paths with the fewest contracted nodes in the P-Tree.

4 Experimental Evaluation

In this section, we (1) show the efficiency of the proposed techniques in terms of the accuracy of the estimates and (2) evaluate the situations in which one technique may be preferred over the other.

Datasets. We used part of the SwetoDBLP [18] dataset, which follows a Zipfian distribution. We also experimented on a synthetic graph generated from TOntoGen [6] RDF graph generator, in which edge types/labels are uniformly distributed across their corresponding source and destination node types.

Table 1. Dataset Properties

	SwetoDBLP	TOntoGen
# Nodes	1037856	200001
# Edges	848839	749825
# Unique edge labels	87	9
Avg node degree	1	7

Implementation Details. Implementation is in C++ with experiments performed on a 1.8GHz Dual AMD Opteron processors and 10GB RAM. We created sparse matrices using sparseLib++ [17] libraries and used BRAHMS [18] to parse the graphs.

Summaries. The unsummarized size of all patterns of length three for the SwetoDBLP dataset is 6036340 bytes and the size of the unpruned P-Tree and MD-Tree are 245000 and 259200 bytes, giving a 95% reduction in size. We summarized the P-Tree and MD-Tree, constructing two sets of summaries with budgets 10KB, 25KB and 50KB, with one set tuned for frequent patterns. On the other hand, the TOntoGen dataset has fewer unique edge labels, so it had fewer unique patterns of lengths at most three with an unsummarized size of 10890 bytes. Its unsummarized P-Tree and MD-Tree are 4916 and 7554 bytes, with at least a 30% reduction. We summarized the P-Tree and MD-Tree, constructing two sets of summaries of sizes 1000 bytes, 1250 bytes, and 1500 bytes, with one set tuned for frequent patterns. For both datasets, we used a 5% significance level and a β value of 3 for constructing the MDTree summaries.

Time Analysis. The time for discovering all patterns of length maxL is the most time-consuming part of our approach. Fortunately, it is a preprocessing step and depends on the connectedness of the dataset but will typically take a couple of hours. The time needed for constructing the P-Tree and MD-Tree summaries is in the order of tenths of seconds whereas the estimation time is negligible, running in tens of milliseconds.

Query Workloads. We used two sets of three workloads: (1) the positive workload has patterns with non-zero frequencies; (2) the frequent workload has those with frequencies of at least 500; (3) the negative workload has patterns with zero frequencies. One set contains patterns of length at most three and the other has patterns of length at least four and at most six. All workloads contain 500 randomly selected patterns as appropriate for the workload.

Error metrics. We used the absolute error metric |freq(p) − freq(p^)| to measure the estimation error (freq(p) and freq(p^) are the true and estimated frequencies of p). In our charts, we measure the overall estimation error by cumulating on the y-axis, the percentage of patterns estimated with at most ε error, for $\varepsilon \in [0, E]$ and E, the observed maximum error. The error values (in logscale) are shown on the x-axis.

Accuracy of Positive Queries. Figure 8a – c show the accuracy of estimates from summaries of the SwetoDBLP dataset using 10KB, 25KB and 50KB space. With the 10KB summary, about 20% of queries in the workload are estimated with very high accuracy (with error value 0 or 1). With the 50KB summary, at least 50% of queries are estimated with high accuracy. Recall that the 10KB and 50KB summaries can only hold about 4% and 20% of the patterns in the unsummarized P-Tree (MD-Tree). Thus, their performances are encouraging. The P-Tree performs better than the MD-Tree, since the accuracy of the base MD-Tree assumes edge patterns occur independently while that of the refined MD-Tree depends on the existence of single points of dependence among edge patterns. Although these assumptions do not hold in all sets of patterns for this dataset, the MD-Tree still exhibits an encouraging performance. Figure 8g shows the accuracy of the estimates obtained from the summaries of the TOntogen dataset. For lack of space, we show only the summary constructed at 1500 bytes space which can hold only about 30% of the patterns in the original unsummarized P-Tree (MD-Tree). The P-Tree that estimates 40% of the patterns with very high accuracy exhibits an encouraging performance. Although the dataset was created by assigning edge types/labels to node types in a uniform manner, our assumption of uniform growth rate of patterns does not necessarily hold. The MD-Tree, on the other hand, does not perform as well, because the optimal MD-Tree constructed is the base MD-Tree and the assumptions of independence of edge patterns upon which the base MD-Tree rests, does not hold for this dataset.

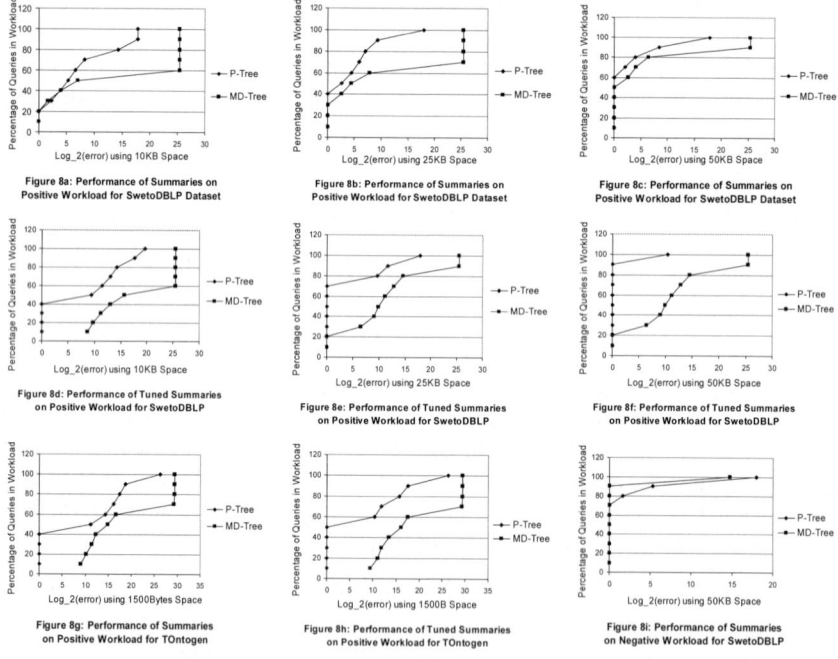

Fig. 8. Comparing the Accuracy of Estimates Obtained from the Summaries

Accuracy of Frequent Queries. Figure 8d–f show the accuracy of the estimates for the summaries of the SwetoDBLP dataset, tuned to favour frequent patterns. As expected, the accuracy of the tuned P-Tree surpasses that of the untuned P-Tree. We tuned the MD-Tree by increasing the impact of significant nodes that also had more frequent patterns over others. However, as the performance of the MD-Tree shows, the TKP greedy algorithm may yet prune such a node if its subtree does not fit the budget. Figure 8h shows the results from the tuned summaries of the TOntogen dataset. For lack of space, we only show the 1500 byte tuned summary. The tuned P-Tree performed better while the performance of the MD-Tree remained unchanged since the optimal MD-Tree constructed is the base MD-Tree and the independence assumption does not hold for this dataset.

Accuracy of Negative Queries. Figure 8i show the accuracy of the estimates from the summaries of the SwetoDBLP dataset, using 50KB space on the negative workloads. For of lack of space, we show only the result of this largest summary, which represents the worst case scenario for negative queries. Since we encode the structure of the graph in the patterns, we are mostly able to detect patterns whose structures do not exist. Thus both the P-Tree and MD-Tree exhibit very good performances. However, some non-zero estimates are obtained due to spurious paths and cycles that may be introduced in the structure.

For lack of space, we do not show the results of the query workload of longer patterns. However, the results are consistent with those of the shorter patterns albeit with larger estimation errors.

5 Related Work

To the best of our knowledge, ours is the first work on summarizing graphs for subgraph frequency estimation. Work most closely related are techniques that summarize XML data for selectivity estimation for path expressions [1] and twigs[8][9][13]. A fundamental difference between these techniques and ours is the data and query models. All the techniques except [9] assume tree-structured XML data. More important, all the techniques are proposed for either path or twig queries so that it is unclear how they apply to arbitrary graph-structured queries. Our estimation value for patterns is similar in spirit to the notion of δ-derivable twigs introduced in [13] for pruning twigs whose estimated frequencies are within δ error of their true frequencies. However, the technique of [13] may blindly prune a pattern that, if left unpruned, may have caused more twigs to be pruned, thereby reducing the summary size further. In contrast, our value-based approach makes a more informed choice of patterns to be pruned.

Several efforts have been made in using graph-indexing schemes to reduce the cost of processing graph containment queries, over a collection of many disconnected graphs. In these approaches, a graph containment query is processed in two steps. The first step retrieves a candidate set of graphs that contain the indexed features of the query graph. The second step uses subgraph isomorphism to validate each candidate graph. GraphGrep [12] uses a path-based indexing approach that selects all paths of up to length l_p as the indexing feature. The size of the candidate set obtained in the first step could be large since paths do not keep graph structure. To cope with this, GIndex [15] uses frequent graph fragments as the indexing feature. To reduce the large (potentially exponential) number of frequent fragments, only discriminative frequent fragments are kept. Noting that the set of frequent graph fragments contain many more tree than non-tree structures, Tree+Δ [16] indexes frequent trees, reducing the large index construction time of GIndex due to graph-mining. On demand, Tree+Δ further reduces the size of the candidate set by selecting a small portion of discriminative non-tree features related to query graphs only. In complement, our work allows allows for optimizing the subgraph isomorphism tests, in the second step, using estimates of the cardinalities of both indexed and non-indexed fragments of the query. In addition, our technique can also be applied to a large connected graph.

6 Conclusions and Future Work

Structure querying is important for eliciting information from graphs. Optimizing structure queries requires estimating the frequency of subgraphs in a query graph. In this work, we presented two techniques for summarizing the structure of graphs in limited space. The Pattern Tree is relatively stable for all datasets but performs best when graph patterns that share a common sub-graph pattern co-occur. The MD-Tree performs best when single points of dependence exist among subgraphs. As our experiments showed, the untuned MD-Trees had more encouraging results for the SwetoDBLP dataset than the tuned MD-Trees for the same dataset. This is mostly because our current representation of sparse matrices as a sparse vector and two one-dimensional arrays is feasible only when the sparse matrix is at most half-filled, otherwise, the space overhead results in the pruning of deep maximal impact subtrees. In the future, we will explore more compact alternatives that reduce this space overhead and characterize the performance of the MD-Tree for datasets that have multiple points of dependence. We will also provide a comprehensive evaluation of the benefits of our summaries in terms of speeding up structure queries. In addition, we will look into estimating patterns in graphs such as RDF graphs, which may have subsumption hierarchies on the edges. Further we will investigate techniques for gracefully accommodating updates to the data graph into our summaries.

Acknowledgments. This work is funded by NSF-ITR-IDM Award #0325464 and #071444.

References

1. Aboulnaga, A., Alameldeen, A., Naughton, J.: Estimating the Selectivity of XML Path Expressions for Internet Scale Applications. In VLDB, 2001.
2. Burge, C. Identification of Complete Gene Structures in Human Genomic DNA. Ph.D. Thesis, Stanford University, Stanford, CA. 1997
3. Dehaspe, L., Toivonen, H., King, R. D.: Finding Frequent Substructures in Chemical Compounds. In KDD, 1998.
4. Desphande, M. Kuramochi, M. Wale, N.: Frequent Substructure-Based Approaches for Classifying Chemical Compounds. In TKDE. Vol. 17, No. 8. Aug. 05.
5. Klyne, G., Carroll, J. J.: RDF Concepts and Abstract Syntax. W3C Recommendation. (Revised) February 2004. http://www.w3.org/TR/rdf-syntax-grammar/
6. Perry, M. TOntoGen: A Synthetic Data Set Generator for Semantic Web Applications. In SIGSEMIS Bulletin.
7. Pei, J., Dong, G., Zou, W., Han, J.: On Computing Condensed Frequent Pattern Bases. In ICDM, 2002.

8. Polyzotis, N., Garofalakis, M., Ioannidis, Y.: Selectivity Estimation for XML Twigs. In ICDE, 2004.
9. Polyzotis, N., Garofalakis, M.: Statistical Synopses for Graph-Structured XML Databases. In SIGMOD, 2002.
10. Prud'hommeaux, E., Seaborne, A.: SPARQL Query Language for RDF. W3C Working Draft. 19th April 2005. http://www.w3.org/TR/rdf-sparql-query/
11. Shannon, C.E. A Mathematical Theory of Communication, Bell Syst. Tech. Journal 27, 379-423, 623-656. 1948.
12. Shasha, D., Wang, J. T. L., Giugno, R. Algorithmics and Applications of Tree and Graph Searching. In PODS, 2002
13. Wang, C., Parthasarathy, S., Jin, R.: A Decomposition-Based Probabilistic Framework for Estimating the Selectivity of XML Twig Queries. In EDBT, 2006.
14. Yan, X., Han, J.: gSpan: Graph-Based Substructure Pattern Mining. In ICDM, 2002.
15. Yan, X., Yu, P. S., Han, J. Graph Indexing: A Frequent Structure-based Approach. In SIGMOD, 2004.
16. Zhao, P., Yu, J. X., Yu, P. S.: Graph Indexing: Tree + Delta >= Graph. In VLDB, 2007.
17. http://math.nist.gov/sparselib++/
18. http://lsdis.cs.uga.edu/projects/semdis/brahms
19. http://lsdis.cs.uga.edu/projects/semdis/swetodblp

Querying Distributed RDF Data Sources with SPARQL

Bastian Quilitz and Ulf Leser

Humboldt-Universität zu Berlin
{quilitz,leser}@informatik.hu-berlin.de

Abstract. Integrated access to multiple distributed and autonomous RDF data sources is a key challenge for many semantic web applications. As a reaction to this challenge, SPARQL, the W3C Recommendation for an RDF query language, supports querying of multiple RDF graphs. However, the current standard does not provide transparent query federation, which makes query formulation hard and lengthy. Furthermore, current implementations of SPARQL load all RDF graphs mentioned in a query to the local machine. This usually incurs a large overhead in network traffic, and sometimes is simply impossible for technical or legal reasons. To overcome these problems we present DARQ, an engine for federated SPARQL queries. DARQ provides transparent query access to multiple SPARQL services, i.e., it gives the user the impression to query one single RDF graph despite the real data being distributed on the web. A service description language enables the query engine to decompose a query into sub-queries, each of which can be answered by an individual service. DARQ also uses query rewriting and cost-based query optimization to speed-up query execution. Experiments show that these optimizations significantly improve query performance even when only a very limited amount of statistical information is available. DARQ is available under GPL License at *http://darq.sf.net/*.

1 Introduction

Many semantic web applications require the integration of data from distributed, autonomous data sources. Until recently it was rather difficult to access and query data in such a setting because there was no standard query language or interface. With SPARQL [1], a W3C Recommendation for an RDF query language and protocol, this situation has changed. It is now possible to make RDF data available through a standard interface and query it using a standard query language. The data does not need be stored in RDF but can be created on the fly, e.g. from a relational databases or other non-RDF data sources (see D2R Server[1] and SquirrelRDF[2]). We expect that more and more content provider will make their data available via a SPARQL endpoint. Nevertheless, it is still difficult to integrate data from multiple data sources. RDF data integration is

---

[1] D2R Server: *http://www.wiwiss.fu-berlin.de/suhl/bizer /d2r-server/*
[2] SquirrelRDF: *http://jena.sf.net/SquirrelRDF/*

often done by loading all data into a single repository and querying the merged data locally. In many cases this will not be feasible for legal or technical reasons. Often it will not be allowed to create copies of the whole data source due to copyright issues. Possible technical reasons are that local copies are not up-to-date if the data sources change frequently, that data sources are too big, or that the RDF instances are created on-the-fly from non-RDF data, like relational databases, web services, or even websites. This clearly shows the need for virtual integration of RDF datasets.

In this paper, we present DARQ[3], a query engine for federated SPARQL queries. It provides transparent query access to multiple, distributed endpoints as if querying a single RDF graph. We introduce *service descriptions* that describe the capabilities of SPARQL endpoints and a *query optimization algorithm* that builds a cost-effective query plan considering limitations on access patterns [2]. Sources with limited access patterns require some some variables in a query to be bound or fail to answer the query.

Related work. Data integration has been a research topic in the field of database systems for a long time. Systems providing a single interface to many underlying data sources are generally called *federated information systems* [3]. Solutions range from multi-database query languages (MDBQL) such as SchemaSQL [4] to federated databases [5] to mediator based information systems (MBIS) [6]. While multi-database query languages require that the user explicitly specifies the used data sources in the query MBIS hide the federation from the user by providing a single, unified schema. In this notation, SPARQL currently can be considered as a MDBQL for RDF allowing the user to specify the graphs to be used in the query. In contrast, DARQ offers source transparency to the user, but unlike MBIS it does not assume an integrated schema.

In [7] Stuckenschmidt et. al theoretically describe how to extend the Sesame RDF repository to support distributed SeRQL queries over multiple Sesame RDF repositories. They use a special index structure to determine the relevant sources for a query. To this end, they restricted themselves to path queries. In [8] the authors describe a system for SPARQL queries over multiple relational databases. To our best knowledge there exists no system that supports SPARQL query federation for multiple regular SPARQL endpoints. Also, none of the described systems uses service descriptions to declaratively describe the data sources nor do they support limitations on access patterns. A special characteristic of DARQ is that it strongly relies on standards and is compatible with any endpoint that supports the SPARQL standards. There is no other need for cooperation except of the support of the SPARQL protocol.

Research on query optimization for SPARQL includes query rewriting [9] or basic reordering of triple patterns based on their selectivity [10]. Optimization for queries on local repositories has also focused on the use of specialized indices for RDF or efficient storage in relational databases, e.g. [11,12]. However, none of the approaches targets SPARQL queries across multiple sources. There has been

---

[3] Distributed ARQ, as an extension to ARQ (*http://jena.sourceforge.net/ARQ/*)

```
SELECT ?name ?mbox WHERE   {
  ?x foaf:name ?name .
  ?x foaf:mbox ?mbox .
  FILTER regex(?name, "^Tim") && regex(?mbox, "w3c")
} ORDER BY ?name LIMIT 5
```

Listing 1.1. Example SPARQL Query

a lot of research on query optimization in the context of databases and federated information systems. An excellent overview of distributed query processing techniques can be found in [13]. In this paper, we show that existing techniques from relational systems, such as query rewriting and cost based optimization for join ordering can be adopted to federated SPARQL. We also propose a way to estimate the result sizes of SPARQL queries with only very few statistical information.

Structure of this paper. The rest of the paper is structured as follows. Section 2 gives a brief introduction to the SPARQL query language. In Section 3 we show the architecture of DARQ, give an introduce service descriptions and describe the used query planning and optimization algorithms we use in our current implementation. We show initial results of the evaluation of the system in Section 4 and conclude and discuss future directions in Section 5.

2 Preliminaries

Before we describe our work on federated queries we give a short introduction to the SPARQL query language and the operators of a SPARQL query that are considered for this report. For a more detailed introduction to RDF and SPARQL we refer the interested reader to [14,1,15]. In the following we use the definitions from the SPARQL Recommendation in [1].

A SPARQL query Q is defined as tuple $Q = (E, DS, R)$. Basis of SPARQL query is an algebra expression E that is evaluated with respect to a RDF graph in a dataset DS. The results of the matching process are processed according to the definitions of the *result form* R (SELECT, CONSTRUCT, DESCRIBE, ASK). The algebra expression E is build from different *graph patterns* and can also also include *solution modifiers*, such as PROJECTON, DISTINCT, LIMIT, or ORDER BY.

The simplest graph pattern defined for SPARQL is the *triple pattern*. A triple pattern t is similar to a RDF triple but allows the usage of variables for subject, predicate, and object:

$$t \in TP = (\textit{RDF-T} \cup V) \times (I \cup V) \times (\textit{RDF-T} \cup V)$$

with $RDF\text{-}T$ being the set of RDF Terms (RDF Literals and Blank Nodes), I being a set of all IRIs, and V a set of all variables [1].

A *basic graph pattern* BGP is defined as a set of triple patterns $BGP = \{t_1..t_n\}$ with $t_1..t_n \in TP$. It matches a subgraph if all contained triple patterns match. Basic graph patterns can be mixed with value constraints (FILTER) and other graph patterns. The evaluation of basic graph patterns and value constraints is order independent. This means, a structure of two basic graph patterns BGP_1 and BGP_2 separated by a constraint C can be transformed into one equivalent basic graph pattern followed by the constraint. We refer to a basic graph pattern followed by one or more constraints as *filtered basic graph pattern* (*FBGP*).

Example 1. Listing 1.1 shows a SPARQL query with one filtered basic graph pattern that retrieves the names and email addresses of persons whose name start with *"Tim"* and email address contains *"w3c"*. The results are ordered by the name, the number of results is limited to five.

SPARQL furthermore defines other types of graph patterns such as GRAPH, UNION, or OPTIONAL. We omit these patterns here, because DARQ works on basic graph patterns as we will see in Section 3.2. Note however, that the engine is able to process all other patterns by distributing the FBGPs contained in these patterns and doing local post-processing. This means that the FBGPs in every of these pattens are handled separately, i.e. the scope for distribution and cost-based optimization is always limited to one FBGP. DARQ correctly handles the order-dependent OPTIONAL pattern, but may waste resources transferring unnecessary results when OPTIONAL is used to express negation as failure.

3 DARQ: Federated SPARQL Queries

To provide transparent query access to multiple data sources we adopt an architecture of mediator based information systems [6] as shown in Figure 1. The DARQ query engine has the role of the mediator component. Non-RDF data sources can be wrapped with tools such as D2R and SquirrelRDF. A DARQ query engine itself can work as SPARQL endpoint and may be integrated by another instance of DARQ. Data sources are described by service descriptions (see Section 3.1). The query engine uses this information for query planning and optimization. In contrast to MBIS the schema is not fixed and does not need to be specified, but is determined by the underlying data sources.

A query is processed in 4 stages:

1. **Parsing.** In the first stage the query string is parsed into a tree model of SPARQL. The DARQ query engine reuses the parser shipped with ARQ.
2. **Query Planning.** In the second stage the query engine decomposes the query and builds multiple sub-queries according to the information in the service descriptions, each of which can be answered by one known data source (see Section 3.2).
3. **Optimization.** In the third stage, the query optimizer takes the sub-queries and builds an optimized query execution plan (see Section 3.3).
4. **Query Execution.** In the fourth stage, the query execution plan is executed. The sub-queries are sent to the data sources and the results are integrated.

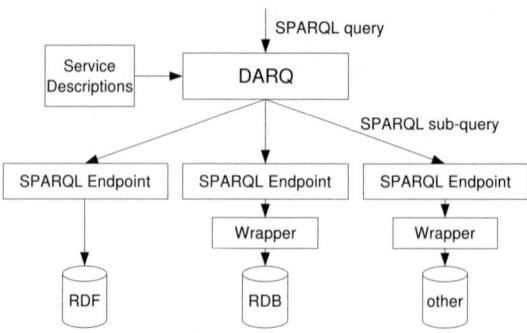

Fig. 1. DARQ - integration architecture

3.1 Service Descriptions

To find the relevant information sources for the different triples in a query and to decompose the query into sub-queries the query engine needs information about the data sources. To this end, we introduce *service descriptions* which provide a declarative description of the data available from an endpoint and allow the definition of limitations on access patterns. Furthermore, service descriptions can include statistical information used for query optimization. Service Descriptions are represented in RDF.

Data Description. A service description describes the data available from a data source in form of capabilities. Capabilities define what kind of triple patterns can be answered by the data source. The definition of capabilities is based on predicates. The capabilities of a data source D are a set C_D of tuples $c = (p, r) \in C_D$, where p is a predicate existing in D and r is a constraint on subjects and objects. This constraint is a regular SPARQL filter expression that enables a more precise source selection, e.g. we can express that a data source only stores data about specific types of resources. We denote the constraint as function $r(subject, object)$ with $r : (\textit{RDF-T} \cup V) \times (\textit{RDF-T} \cup V) \to \{true, false\}$. For example, the constraints can be used for horizontal partitioning. It is possible to define a constraint that says that a Service A can only answer queries for names starting with a letter from A to R, whereas another service can answer queries for names starting with a letter from Q to Z.

Limitations on Access Patterns. Some data sources have limitations on access patterns [2]. For example, a wrapper that transforms results from a web form into RDF may require some input values that can be entered into the form to compute the results. Another example is a wrapper for an LDAP server may require that the name of a person or their email address is always included in the query because the server owner does not allow other queries.

DARQ supports the definition of limitations on access patterns in the service descriptions in form of patterns that must be included in a query. Because predicates must be bound we use them as basis for the pattern definition. Let L_D be a set of limitations on access patterns for data source D and $(S,O) \in L_D$ be one pattern with S and O being sets of predicates that must have bound subject (S) or bound objects (O).

Source D could contribute to the query answer of a query with graph pattern P if it satisfies at least one of the defined access patterns for D. Let $bound(x)$ be a function that returns $false$ if x is a variable and $true$ otherwise. An access pattern (S,O) is satisfied if

$$(\forall p_s \in S \backslash O : \exists (s,p_s,o) \in P : bound(s))$$
$$\wedge (\forall p_o \in O \backslash S : \exists (s,p_o,o) \in P : bound(o))$$
$$\wedge (\forall p_b \in S \cap O : \exists (s,p_b,o) \in P : bound(s) \wedge bound(o))$$

Example 2. To come back to the example of the LDAP server, the service description in this example would contain two access patterns, (S_1, O_1) and (S_2, O_2), with $S_1 = S_2 = \emptyset$ and $O_1 = \{foaf : name\}$ $O_2 = \{foaf : mbox\}$

Statistical Information. Defining statistical information about the data available from a data source helps the query optimizer to find a cost-effective query execution plan. Service descriptions include the total number of triples N_s in data source D and optionally information for each capability $(p,r) \in C_D$: (1) The number of triples $n_D(p)$ with the predicate p in D, (2) the selectivity $ssel_D(p)$ of a triple pattern with predicate p if the subject is bound (default=$\frac{1}{n_D(p)}$), and (3) the selectivity $osel_D(p)$ of a triple pattern with predicate p if the object is bound (default=1). We deliberately use only these simple statistics because we expect every data source to be able to provide them, or at least rough estimations. More precise statistics such as histograms would be preferable but will not be available from many sources. Future work should explore what other statistics are required for more complex cost-models and how they can be estimated. In this context, aggregate functions, such as *count*, could be a valuable addition to future SPARQL version.

RDF Representation. Service Descriptions are represented in RDF. Listing 1.2 shows an example service description for a FOAF data source, e.g. an LDAP Server. The data source defined in the example can answer queries for foaf:name, foaf:mbox and foaf:weblog. Objects for a triple with predicate *foaf:name* must always start with a letter from A to R. In total it stores 112 triples. The data source has limitations on access patters, i.e. a query must at least contain a triple pattern with predicate *foaf:name* or *foaf:mbox* with a bound object. More detailed examples of service descriptions can be found at *http://darq.sf.net/*.

3.2 Query Planning

When querying multiple data sources it is necessary to decide which data source can contribute to answer a query. The process of finding relevant sources and

```
[]   a             sd:Service ;
     sd:capability  [ sd:predicate   foaf:name ;
                      sd:objectFilter "REGEX(?object,"^[A-R]")";
                      sd:triples     51 ] ;
     sd:capability  [ sd:predicate   foaf:mbox ;
                      sd:triples     51 ] ;
     sd:capability  [ sd:predicate   foaf:weblog ;
                      sd:triples     10 ] ;
     sd:totalTriples     "112" ;
     sd:url              "EndpointURL" ;
     sd:requiredBindings [ sd:objectBinding foaf:name ] ;
     sd:requiredBindings [ sd:objectBinding foaf:mbox ] .
```

Listing 1.2. Example Service Description

feasible sub-queries is referred to as query planning. In this section we describe the query planning algorithm used by DARQ. Query planning is based on the information provided in the service descriptions. In the following let $R = \{(d_1, C_1), ..., (d_n, C_n)\}$ be a set of data sources $d_1..d_n$ and their capabilities $C_1..C_n$, where $C_i = \{(p_{i,1}, r_{i,1})..(p_{i,m}, r_{i,m})\}$.

Source Selection. A SPARQL query contains one or more filtered basic graph patterns each containing the actual triple patterns. Query Planning is performed separately for each filtered basic graph pattern. The algorithm for finding the relevant data sources for a query simply matches all triple patterns against the capabilities of the data sources. The matching compares the predicate in a triple pattern with the predicate defined for a capability and evaluated the constraint for subject and object. Because matching is based on predicates, DARQ currently only supports queries with bound predicates.

Let BGP be a set of triple patterns in a filtered basic graph pattern. The result of the source selection is a set of data sources D_j for each triple pattern $t_j = (s_j, p_j, o_j) \in BGP$ with

$$D_j = \{d | (d, C) \in R \wedge \exists (p_j, r) \in C : r(s_j, o_j) = true\}$$

Building Sub-Queries. The results from source selection are used to build sub-queries that can be answered by the data sources. Sub-queries consist of one filtered basic graph pattern per data source. We represent a sub-query as triple (T, C, d), where T is a set of triple patterns, C is a set of value constraints and d is the data source that can answer the sub-query. Algorithm 1 shows how the sub-queries are generated. If a triple pattern matches exactly one data source ($D_i = \{d\}$) the triple will be added to the set of a sub-query for this data source. All triples in this set can later be sent to the data source in one sub-query. If a triple matches multiple data sources the triple must be sent individually to all matching data sources in separate sub-queries.

Example 3. Let data source A and B be two data sources with the capabilities (name,*true*) and (mbox,*true*). A stores the triple (a, name,"Tim"), B stores the triple (a ,mbox,"Tim@x.y"). The query shown in Listing 1.1 will return no results if sent to A and B with both triple patterns or the correct result if triple patterns are sent in separate sub-queries and the results are joined afterwards.

Algorithm 1. Sub-query generation

Require: $T = \{t_1, .., t_n\}$, // set of triple patterns
 $D = \{D_1, .., D_n\}$ // sets of data sources matching to the triple patterns
1: $queries = \emptyset$, $separateQueries = \emptyset$
2: **for** each $t_i \in T$ **do**
3:    **if** $D_i = \{d\}$ **then**
4:       $q = queries.getQuery(d)$
5:       **if** q not null **then**
6:          $q.T = q.T + t_i$
7:       **else**
8:          $queries = queries + (\{t_i\}, \{\}, d)$
9:       **end if**
10:   **else**
11:      **for** each $d_j \in D_i$ **do**
12:         $separateQueries = separateQueries \cup (\{t_i\}, \{\}, d_j)$
13:      **end for**
14:   **end if**
15: **end for**
16: **return** $queries \cup seperateQueries$ // Return all queries

3.3 Optimization

After query planning the query plan consists of multiple sub-queries. The task of the query optimizer is to build a feasible and cost-effective query execution plan considering limitations on the access patterns. To build the plan we use logical and physical query optimization.

Logical Optimization. Logical query optimization uses equalities of query expressions to transform a logical query plan into an equivalent query plan that is likely to be executed faster or with less costs. The current implementation of DARQ uses logical query optimization in two ways. First, we use rules based on the results in [15] to rewrite the original query before query planning so that basic graph patterns are merged whenever possible and variable are replaced by constants from filter expressions.

Example 4. Listing 1.3 shows the original query submitted by the user. There are two separate Basic Graph Patterns, each with one triple pattern. In the rewritten query that is shown in Listing 1.4 the two patterns are merged. Also, variables that occur in filters with an *equal* operator are substituted. In our example, `?name` is substituted by `"Tim"`.

```
SELECT ?mbox WHERE {
  { ?x foaf:name ?name . }
  FILTER (?name = "Tim")
    && regex(?mbox, "w3c")
  { ?x foaf:mbox ?mbox . }
}
```

```
SELECT ?mbox WHERE {
  ?x foaf:name "Tim"
  ?x foaf:mbox ?mbox .
  FILTER regex(?mbox, "w3c")
}
```

Listing 1.3. Query before rewriting

Listing 1.4. Query after rewriting

Second, we move possible value constraints into the sub-queries to reduce the size of intermediate results as early as possible. Let $Q = (T, C, d)$ be a sub-query and $FGP = (T', C')$ a filtered basic graph pattern. The value constraint C' can be moved to the sub-query if all variables in the constraint are also used in the triple patterns in the sub-query. Filters that contain variables from more than one sub-query and that cannot be split using a limited set of rules are applied locally inside the DARQ query engine.

Example 5. Listing 1.1 shows a query with a conjunctive filter on two attributes. Let us assume that the two triple patterns are split into two sub-queries for services A and B. In this case, the single filter cannot be moved into the sub-queries because one of the variables would be unbound. However, to benefit from filtering at the remote site the conjunction can be split into two filters `FILTER regex(?name, "^Tim")` and `FILTER regex(?mbox, "w3c")` that can then be moved into the sub-queries. If the optimizer is not able to split a filter using its limited set of rewriting rules, it will apply the filter locally, inside DARQ, as soon as all used variables are bound.

Physical Optimization. Physical query optimization has the goal to find the 'best' query execution plan among all possible plans and uses a cost model to compare different plans. In case of federated queries with distributed sources network latency and bandwidth have the highest influence on query execution time. Thus, the main goal in our system is to reduce the amount of transferred data and to reduce the number of transmissions, which will lead to less transfer costs and faster query execution. We use the expected result size as the cost factor of sub-queries.

We use iterative dynamic programming for optimization considering limitations on access patterns. Currently, we support two join implementations:

- **nested-loop join** ($\bowtie$) The nested-loop join is the simplest join implementation. For every binding in the outer relation, we scan the inner relation and add the bindings that match the join condition to the result set.
- **bind join** ($\bowtie_B$) The bind join was introduced in [16]. Basically it is a nested loop join where intermediate results from the outer relation are passed to the inner to be used as filter. This means that DARQ sends out the sub-query for the inner relation multiple times with the join variables bound. We use the bind join for data sources with limitations on access patterns. Furthermore, it can help to drastically reduce the transfer costs if the unbound query would return a large result set.

We calculate the result size of joins with

$$|R(q_1 \bowtie q_2)| = |R(q_1)| \, |R(q_2)| \, sel_{12}$$

where q_1 and q_2 are the joined query plan elements, i.e sub-query or join, $|R(q)|$ is the result size of q, and sel_{12} is a selectivity factor for the join attributes. For DARQ, we currently set $sel_{12} = 0.5$ because the current statistics in the service descriptions do not provide enough information for a better estimation.

The (transfer) costs of a nested loop join is estimated as

$$C(q_1 \bowtie q_2) = |R(q_1)| c_t + |R(q_2)| c_t + 2c_r$$

while the costs of a bind join are estimated as

$$C(q_1 \bowtie_B q_2) = |R(q_1)| c_t + |R(q_1)| c_r + |R(q_2')| c_t$$

with c_t and c_r being the transfer costs for one result tuple[4] and one query, respectively, and q_2' being the query with variables bound with values of a result tuple from q_1.

Query result size estimation The result size estimation for a sub-query is based on the statistics provided in the service descriptions. Currently, service descriptions include for each capability $(p, r) \in C_d$ of service d: (1) the number of triples $n_d(p)$ with the predicate p in data source d, (2) the average selectivity $ssel_d(p)$ if the subject is bound, and (3) the average selectivity $osel_d(p)$ if the object is bound. With this information we estimate the result size of a query with a single triple pattern (s, p, o) that is sent to a service d using the function $cost_d : TP \times V \to N$ with

$$costs_d((s,p,o),b) = \begin{cases} n_d(p) & \text{if } \neg bound(s,b) \land \neg bound(o,b), \\ n_d(p) * osel_d(p) & \text{if } \neg bound(s,b) \land bound(o,b), \\ n_d(p) * ssel_d(p) & \text{if } bound(s,b) \land \neg bound(o,b), \\ 0.5 & \text{if } bound(s,b) \land bound(o,b). \end{cases}$$

where b is a set of previously bound variables and $bound(x, b)$ is a function that returns *true* if x is bound given the bound variables in b and *false* otherwise.

Estimating the result size of a combination of two or more triple patterns is more complex. Note that adding a triple pattern to a query can restrict the result size or introduce new results because of a join. Adding more triple patterns with the same subject will not introduce new results, but rather reduce the result size. In contrast, adding triple pattern with another subject potentially increases the result size. Thus, we start with estimating the result size for all triple patterns with the same subject or subject variable. Let $T = \{t_1, ..., t_n\}$ be a set of triple patterns where $t_1, ... t_n$ all have the same subject. Triple patterns with a bound object restrict the possible solutions. We use the minimum function over all triple patters with a bound object to estimate an upper bound for the number of subjects. Note that this is different from the *attribute independence assumption* that is widely used in SQL query optimization [17]. Triple patterns with an unbound object can introduce new bindings for the used object variable. The overall result size for the set of triple patterns is the product of number of subjects and the result sizes of all triple patterns with unbound object. Using the cost function for a single triple pattern we estimate the result size of as follows:

---

[4] For simplicity, we currently disregard the specific tuple size.

$$costs_d(T,b) = \min_{v \in T_{bound}} (costs_d(v,b)) * \prod_{u \in T_{unbound}} costs_d(u,b)$$

with

$$T_{bound} = \{t | t = (s,p,o) \in T \land bound(o)\} \text{ and}$$
$$T_{unbound} = \{t | t = (s,p,o) \in T \land \neg bound(o)\}$$

Finally, we must combine the groups of triples with one subject to to compute the estimated result size for the complete sub-query. The result sizes of the single triple groups strongly depend of the already bound variables. Algorithm 2 builds groups of triple patterns with same subjects and then incrementally selects the group with the minimal result size considering the variables bound by the previously selected groups. We calculate the overall costs of the query as the product of the result sizes of all groups.

Algorithm 2. Result size estimation for a general basic graph pattern
---
Require: $T = \{t_1, ..., t_n\}$ // basic graph pattern
1: $result = 1$, $bindings = \emptyset$, $groups = \{g_1, ..., g_m\} = buildGroups(T)$
2: **while** $groups \neq \emptyset$ **do**
3:    $g = null$, $costs = positiveInfinity$
4:    **for** each $g_i \in groups$ **do**
5:       $c = costs(g_i, bindings)$
6:       **if** $c < costs$ **then**
7:          $g = g_i$, $costs = c$
8:       **end if**
9:    **end for**
10:   $groups = groups - \{g\}$, $bindings = bindings \cup var(g)$
11:   $result = result * costs$
12: **end while**
13: **return** $result$
---

4 Evaluation

In this section we evaluate the performance of the DARQ query engine. The prototype was implemented in Java as an extension to ARQ[5]. We used a subset of DBpedia[6]. DBpedia contains RDF information extracted from Wikipedia. The dataset is offered in different parts. The names of the parts we used can be found in the description column of Table 1(a).

The dataset has about 31.5 million triples in total. For our experiments we split the dataset into multiple parts located at different endpoints as shown in Table 1(a). To make sure that the endpoints are not a bottleneck in our setup we split all data over two Sun-Fire-880 machines (8x sparcv9 CPU, 1050Mhz, 16GB

---
[5] http://jena.sf.net/ARQ/
[6] http://dbpedia.org (Version 2.0)

RAM) running SunOS 5.10. The SPARQL endpoints were provided using Virtuoso Server 5.0.3[7] with an allowed memory usage of 8GB . Note that, although we use only two physical servers, there were five logical SPARQL endpoints. DARQ was running on Sun Java 1.6.0 on a Linux system with Intel Core Duo CPUs, 2.13 GHz and 4GB RAM. The machines were connected over a standard 100Mbit network connection.

Table 1. Overview on data sources and queries

(a) data sources

No.	Description	#triples
S1	Articles	7.6M
S2	Categories	6.4M
S3	Yago	2M
S4	Infoboxes	14.6M
S5	Persons	0.6M
	Total	31.5M

(b) queries

No.	Used sources	#results
Q1	S4, S5	452
Q2	S4, S5	452
Q3	S2, S4, S5	6
Q4	S1, S3, S4, S5	1166

We run four example queries and evaluated the runtime with and without optimization. For queries without optimization we used the ARQ 1.5 default execution strategies without any changes, i.e. bind joins of all sub-queries in order of appearance. The queries can be found in Listings 1.5-1.8. The queries use different numbers of sources and have different result sizes. An overview is given in Table 1(b). For all queries we had a timeout of 10 minutes. Q1 and Q2 demonstrate the effect of pushing filters into the sub-queries. Q3 and Q4 are rather complex queries, involving three to four sources, but have very different result sizes. The results shown in the following are the average values over four runs.

```
/* Find all movies of actors
born in Paris */

SELECT ?p ?m WHERE {
?p dbpedia2:birthPlace :Paris .
?p foaf:name ?name .
?m dbpedia2:starring ?p.
}
```
Listing 1.5. Q1

```
/* Find all movies of actors
born in Paris */
SELECT ?p ?m WHERE {
?p foaf:name ?name ..
?p dbpedia2:birthPlace ?paris.
?m dbpedia2:starring ?p.
FILTER (?paris=
<http://dbpedia.org/resource/
 Paris>)
}}
```
Listing 1.6. Q2

```
/*Find name, birthday and image of
german musicians born in Berlin*/
SELECT ?n ?b ?p ?img WHERE {
?p foaf:name ?n .
?p dbpedia2:birth ?b .
?p dbpedia2:birthPlace :Berlin .
?p skos:subject
         cat:German_musicians .
OPTIONAL { ?p foaf:img ?img }}
```
Listing 1.7. Q3

```
/* Find all Movies with actors
born in London with an image */
SELECT * WHERE { ?n rdf:type
  yago:MotionPictureFilm103789400.
?n dbpedia2:starring ?p.
?p dbpedia2:birthPlace :London.
?p foaf:name ?name .
?n rdfs:label ?label.
?n foaf:depiction ?img.
FILTER (LANG(?label) = 'en') .}
```
Listing 1.8. Q4

---
[7] http://virtuoso.openlinksw.com/

Figure 2(a) shows the query execution times. The experiments show that our optimizations significantly improve query evaluation performance. For query Q1 the execution times of optimized and unoptimized execution are almost the same. This is due to the fact that the query plans for both cases are the same and bind joins of all sub-queries in order of appearance is exact the right strategy. For queries Q2–Q4 the unoptimized queries took longer than 10 min to answer and timed out, whereas the execution time of the optimized queries is quiet reasonable. The optimized execution of Q1 and Q2 takes almost the same time because Q2 is rewritten into Q1.

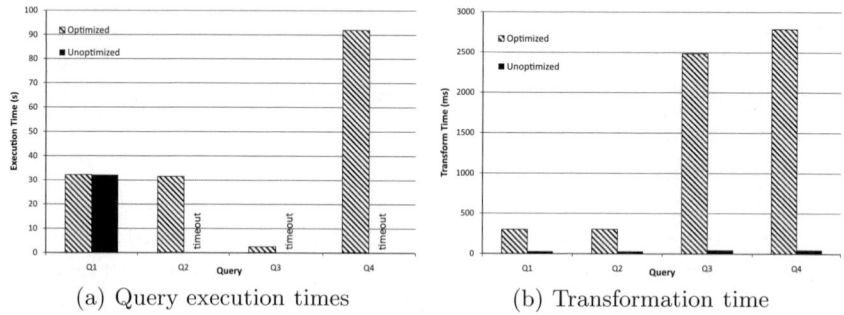

(a) Query execution times (b) Transformation time

Fig. 2. Benchmark Results

Figure 2(b) shows the time needed for query planning and optimization (transformation time). We can see that the transformation times for optimized queries increase with query complexity from around 300 ms to 2800ms. Compared to this transformation times for unoptimized queries (query planning only) are negligible, around 30–40 ms. However, in comparison to the performance gains for query execution, the transformation times including optimization still remain very small.

Our evaluations show that even with a very limited amount of statistical information it is possible to generate query plans that perform relatively well. All queries where answered within less than one and a half minutes. Of course it would be possible for the user to write down triple patterns in exact the right order for ARQ, but this is in conflict with the declarative nature of SPARQL. Note that optimized queries in DARQ will be less performant if all the sub-queries are very unselective, e.g. contain no values for subject and object or a very unselective filter. In this case DARQ has few possibilities to improve performance by optimization.

5 Conclusion and Future Work

DARQ offers a single interface for querying multiple, distributed SPARQL endpoints and makes query federation transparent to the client. One key feature of

DARQ is that it solely relies on the SPARQL standard and therefore is compatible to any SPARQL endpoint implementing this standard. Using service descriptions provides a powerful way to dynamically add and remove endpoints to the query engine in a manner that is completely transparent to the user. To reduce execution costs we introduced basic query optimization for SPARQL queries. Our experiments show that the optimization algorithm can drastically improve query performance and allow distributed answering of SPARQL queries over distributed sources in reasonable time. Because the algorithm only relies on a very small amount of statistical information we expect that further improvements are possible using techniques as described in [16,13].

An important issue when dealing with data from multiple data sources are differences in the used vocabularies and the representation of information. In further work, we plan to work on mapping and translation rules between the vocabularies used by different SPARQL endpoints. Also, we will investigate generalizing the query patterns that can be handled and blank nodes and identity relationships across graphs.

Acknowledgments. Major parts of this work were done at HP Labs Bristol. Bastian Quilitz is grateful to Andy Seaborne and the other members of the HP Semantic Web Team for insightful discussions, their support, and their work on Jena/ARQ. This research was supported by the German Research Foundation (DFG) through the Graduiertenkolleg METRIK, grant no. GRK 1324.

References

1. Prud'hommeaux, E., Seaborne, A.: SPARQL Query Language for RDF. W3C Recommendation (January 2008), *http://www.w3.org/TR/rdf-sparql-query/*
2. Florescu, D., Levy, A., Manolescu, I., Suciu, D.: Query optimization in the presence of limited access patterns. In: International conference on Management of data (SIGMOD), pp. 311–322. ACM, New York (1999)
3. Busse, S., Kutsche, R.D., Leser, U., Weber, H.: Federated information systems: Concepts, terminology and architectures. Technical Report Forschungsberichte des Fachbereichs Informatik 99-9, Technische Universität Berlin (1999)
4. Lakshmanan, L.V.S., Sadri, F., Subramanian, I.N.: SchemaSQL - a language for interoperability in relational multi-database systems. In: Vijayaraman, T.M., Buchmann, A.P., Mohan, C., Sarda, N.L. (eds.) 22th International Conference on Very Large Data Bases (VLDB), Mumbai (Bombay), India, September 1996, pp. 239–250 (1996)
5. Sheth, A.P., Larson, J.A.: Federated database systems for managing distributed, heterogeneous, and autonomous databases. ACM Comput. Surv. 22(3), 183–236 (1990)
6. Wiederhold, G.: Mediators in the architecture of future information systems. Computer 25(3), 38–49 (1992)
7. Stuckenschmidt, H., Vdovjak, R., Houben, G.-J., Broekstra, J.: Index structures and algorithms for querying distributed rdf repositories. In: WWW 2004 (2004)

8. Chen, H., Wang, Y., Wang, H., Mao, Y., Tang, J., Zhou, C., Yin, A., Wu, Z.: Towards a semantic web of relational databases: a practical semantic toolkit and an in-use case from traditional chinese medicine. In: 4th International Semantic Web Conference (ISWC), Athens, USA. LNCS, pp. 750–763. Springor, Heidelberg (2006)
9. Hartig, O., Heese, R.: The sparql query graph model for query optimization. In: 4th European Semantic Web Conference (ESWC), pp. 564–578 (2007)
10. Bernstein, A., Christoph Kiefer, M.S.: OptARQ: A SPARQL Optimization Approach based on Triple Pattern Selectivity Estimation. Technical Report ifi-2007.03, Department of Informatics, University of Zurich (2007)
11. Harth, A., Decker, S.: Optimized index structures for querying rdf from the web. In: Third Latin American Web Congress (LA-WEB), Washington, DC, USA, p. 71. IEEE Computer Society, Los Alamitos (2005)
12. Harris, S., Gibbins, N.: 3store: Efficient bulk rdf storage. In: PSSS - Practical and Scalable Semantic Systems (2003)
13. Kossmann, D.: The state of the art in distributed query processing. ACM Comput. Surv. 32(4), 422–469 (2000)
14. Manola, F., Miller, E.: RDF Primer, W3C Recommendation (2004), http://www.w3.org/TR/rdf-primer/
15. Pérez, J., Arenas, M., Gutierrez, C.: Semantics and Complexity of SPARQL. In: 4th International Semantic Web Conference (ISWC), Athens, GA, USA, pp. 30–43 (November 2006)
16. Haas, L.M., Kossmann, D., Wimmers, E.L., Yang, J.: Optimizing queries across diverse data sources. In: 23rd Int. Conference on Very Large Data Bases (VLDB), pp. 276–285. Morgan Kaufmann Publishers Inc., San Francisco (1997)
17. Selinger, P.G., Astrahan, M.M., Chamberlin, D.D., Lorie, R.A., Price, T.G.: Access path selection in a relational database management system. In: International conference on Management of data (SIGMOD), pp. 23–34. ACM Press, New York (1979)

Improving Interoperability Using Query Interpretation in Semantic Vector Spaces

Anthony Ventresque[1], Sylvie Cazalens[1], Philippe Lamarre[1], and Patrick Valduriez[2]

LINA, University of Nantes
FirstName.LastName@univ-nantes.fr
INRIA and LINA, University of Nantes
Patrick.Valduriez@inria.fr

Abstract. In semantic web applications where query initiators and information providers do not necessarily share the same ontology, semantic interoperability generally relies on ontology matching or schema mappings. Information exchange is then not only enabled by the established correspondences (the "shared" parts of the ontologies) but, in some sense, limited to them. Then, how the "unshared" parts can also contribute to and improve information exchange ? In this paper, we address this question by considering a system where documents and queries are represented by semantic vectors. We propose a specific query expansion step at the query initiator's side and a query interpretation step at the document provider's. Through these steps, unshared concepts contribute to evaluate the relevance of documents wrt. a given query. Our experiments show an important improvement of retrieval relevance when concepts of documents and queries are not shared. Even if the concepts of the initial query are not shared by the document provider, our method still ensures 90% of the precision and recall obtained when the concepts are shared.

1 Introduction

In semantic web applications where query initiators and information providers do not necessarily share the same ontology, semantic interoperability generally relies on ontology matching or schema mappings. Several works in this domain focus on what (*i.e.* the concepts and relations) the peers share [9,18]. This is quite important because, obviously if nothing is shared between the ontologies of two peers, there is a little chance for them to understand the meaning of the information exchanged. However, no matter how the shared part is obtained (through consensus or mapping), there might be concepts (and relations) that are not consensual, and thus not shared. The question is then to know whether the unshared parts can still be useful for information exchange.

In this paper, we focus on semantic interoperability and information exchange between a query initiator p_1 and a document provider p_2, which use different ontologies but share some common concepts. The problem we address is to *find documents which are relevant to a given query although the documents and the query may be both represented with concepts that are not shared.* This problem is very important because in semantic web applications with high numbers of participants, the ontology (or ontologies) is rarely entirely shared. Most often, participants agree on some part of a reference ontology to exchange information and internally, keep working with their own ontology [18,22].

We represent documents and queries by *semantic vectors* [25], a model based on the vector space model [1] using concepts instead of terms. Although there exist other, richer representations (conceptual graphs for example), semantic vectors are a common way to represent unstructured documents in information retrieval. Each concept of the ontology is weighted according to its representiveness of the document. The same is done for the query. The resulting vector represents the document (respectively, the query) in the n-dimensional space formed by the n concepts of the ontology. Then the relevance of a document with respect to a query corresponds to the proximity of the vectors in the space.

In order to improve information exchange beyond the "shared part" of the ontologies, we promote both *query expansion* (at the query initiator's side) and *query interpretation* (at the document provider's side). Query expansion may contribute to weight linked shared concepts, thus improving the document provider's understanding of the query. Similarly, by interpreting an expanded query with respect to its own ontology (*i.e.* by weighting additional concepts of its own ontology), the document provider may find additional related documents for the query initiator that would not be found by only using the matching concepts in the query and the documents. Although the basic idea of query expansion and interpretation is simple, query interpretation is very difficult because it requires to precisely weight additional concepts given some weighted shared ones, while the whole space (i.e. the ontology) and similarity measures change.

In this context, our contributions are the following. First, we propose a specific query expansion method. Its property is to keep separate the results of the propagation from each central concept of the query, thus limiting the noise due to inaccurate expansion. Second, given this expansion, we define the relevance of a document. Its main, original characteristic is to require the document vector to be requalified with respect to the expanded query, the result being called *image* of the document. Third, a main contribution is the definition of query interpretation which enables the expanded query to be expressed with respect to the provider's ontology. Fourth, we provide two series of experiments with still very good results although few concepts are shared.

This paper is organized as follows. Section 2 gives preliminary definitions. Section 3 presents our query expansion method and the image based relevance of a document. For simplicity, we assume a context of shared ontology. This assumption is relaxed after in Section 4, where we consider the case where the query initiator and the document provider use different ontologies and present the query interpretation. Section 5 discusses the experiments and their results. The two last sections are respectively devoted to related work and conclusion.

2   Preliminary Definitions

We define an ontology as a set of concepts together with a set of relations between these concepts. In our experiments, we consider an ontology with only one relation: the is-a relation (specialization link). This does not restrict the generality of our relevance computation. Indeed, the presence of several relations only affects the definition of the similarity of a concept wrt. another. A *semantic vector* $\overrightarrow{v_\Omega}$ is an application defined on the set of concepts $\mathcal{C}_\Omega$ of the ontology Ω : $\forall c \in \mathcal{C}_\Omega, \overrightarrow{v_\Omega} : c \to [0..1]$. A popular way to compute the relevance of a document is to use the cosine-based proximity of the document and query vectors in the space [19]. The problem with cosine is the independence of dimensions : a query on concept c_i and a document on concept c_j very close from c_i could not match. Query expansion is generally used to express these links between concepts, by propagating initial weights on other linked concepts. To define a query expansion, we need a *similarity function* [11] which expresses how much a concept is similar to another within the ontology : $sim_c: \mathcal{C}_\Omega \to [0, 1]$, is a similarity function iff $sim_c(c) = 1$ and $0 \leq sim_c(c_j) < 1$ for all $c_j \neq c$ in $\mathcal{C}_\Omega$. Then, propagation from a central concept c of weight v assigns a weight to every value of similarity with c.

Definition 1 (Propagation function). *Let c be a concept of Ω valued by v; and let sim_c be a similarity function.*

A function $\mathcal{P}f_c : \begin{vmatrix} [0..1] & \mapsto & [0..1] \\ sim_c(c') & \to & \mathcal{P}f_c(sim_c(c')) \end{vmatrix}$ *is a propagation function from c iff*

- $\mathcal{P}f_c(sim_c(c)) = v$, *and*
- $\forall c_k, c_l \in \mathcal{C}_\Omega\ sim_c(c_k) \leq sim_c(c_l) \Rightarrow \mathcal{P}f_c(sim_c(c_k)) \leq \mathcal{P}f_c(sim_c(c_l))$

Among different types of propagation functions those inspired by the membership functions used in fuzzy logic work fine (see Figure 1) in our experiments. It is defined by three parameters v (weight of the central concept), l_1 (similarity value until which concepts have the same weight : v) and l_2 (similarity value until which concepts have non zero weight) such that, $\forall x = sim_c(c'), c' \in c_\Omega$:

$$\mathcal{P}f_c(x) = f_{v,l_1,l_2}(x) = \begin{vmatrix} v & if\ x \geq l_1 \\ \frac{v}{l_1-l_2}x + \frac{l_2 \times v}{l_1-l_2} & if\ l_1 > x > l_2 \\ 0 & if\ l_2 \geq x \end{vmatrix}$$

3   Query Expansion and Image Based Relevance

In this section, we present our method to compute the relevance of a document wrt a query. For the sake of simplicity, we assume that the query initiator and the document provider use the same ontology. However, they can still differ on the similarity measures and the propagation functions. First, we compute a *query expansion*, and then an *image of a document vector* to compute the relevance of the document wrt. a query in a single space.

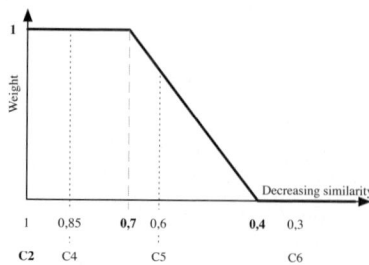

Fig. 1. Example of a propagation function $f_{1,0.7,0.4}$ with central concept c_2

To our knowledge, most query expansion methods propagate the weight of each weighted concept in *the same vector*, thus directly adding the expanded terms in the original vector [13]. When a concept is involved in several propagations conducted from different central concepts, an aggregation function (e.g. the maximum) is used. We call this kind of method "rough" propagation. Although its results are not bad, such a propagation has some drawbacks among which a possible unbalance of the relative importance of the initial concepts [16]. First, let us denote by $\mathcal{C}_{\vec{q}}$ the set of the *central concepts* of query $\vec{q}$, i.e. those weighted concepts which represent the query. To keep separate the effects of different propagations, each central concept of $\mathcal{C}_{\vec{q}}$ is *semantically enriched* by propagation, in a separate vector.

Definition 2 (Semantically Enriched Dimension). *Let $\vec{q}$ be a query vector and let c be a concept in $\mathcal{C}_{\vec{q}}$. A semantic vector $\vec{sed_c}$ is a semantically enriched dimension, iff $\forall c' \in \mathcal{C}_\Omega, \vec{sed_c}[c'] \leq \vec{sed_c}[c]$.*

Definition 3 (Expansion of a query). *Let $\vec{q}$ be a query vector. An expansion of $\vec{q}$, noted $\mathcal{E}_{\vec{q}}$ is a set defined by:*
$$\mathcal{E}_{\vec{q}} = \{\vec{sed_c} : c \in \mathcal{C}_{\vec{q}}, \forall c' \in \mathcal{C}_\Omega, \vec{sed_c}[c'] = \mathcal{P}f_c(c')\}$$

Figure 2 illustrates the expansion of a query $\vec{q}$ with two weighted concepts c_4 and c_7. It contains two semantically enriched dimensions. In dimension $\vec{sed_{c_7}}$, concept c_7 has the same value as in the query. The weight of c_7 has been propagated on c_3, c_{11} and c_6 according to their similarity with c_7. The other dimension is obtained from c_4 in the same way.

The expanded query is composed of several semantic vectors (the SEDs). Our aim is then to transform the semantic vector of a document, $\vec{d}$, in an *image* through the expanded query, i.e. to characterize the document wrt. each central concept c (dimension) of the query, as far as it has concepts related to c, in particular even if c is not initially weighted in $\vec{d}$. Given a SED $\vec{sed_c}$, we aim at valuating c in the image of the document $\vec{d}$ according to the relevance of $\vec{d}$ to $\vec{sed_c}$. To evaluate the impact of $\vec{sed_c}$ on $\vec{d}$ we consider the product of the respective values of each concept in $\vec{sed_c}$ and $\vec{d}$. Intuitively, all the concepts

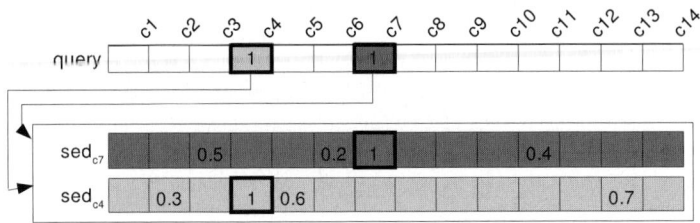

Fig. 2. A query expansion composed of 2 semantically enriched dimensions

of the document which are linked to c through $\overrightarrow{sed_c}$ have a nonnull value. The image of $\overrightarrow{d}$ keeps track of the best value assigned to one of the linked concepts if it is better than $\overrightarrow{d}[c]$, which is the initial value of c. This process is repeated for each SED of the query. Algorithm 1.1 gives the computation of the image of document $\overrightarrow{d}$, noted $\overrightarrow{i}_d$. This algorithm ensures that all the central concepts of the initial query vector are also weighted in the image of the document as far as the document is related to them. Wrt. the query, the image of the document is more accurate because it enforces the documents characterization over each dimension of the query. However, in the image, we keep unchanged the weights of the concepts which are not linked to any concept of the query (*i.e.* which are not weighted in any SED). The example of Figure 3 illustrates how the image of a document is computed.

Algorithm 1.1. Image of a document wrt a query

(* *Input*   : *a semantic vector* $\overrightarrow{d}$ *on an ontology* Ω;
               an expanded query $\mathcal{E}_{\overrightarrow{q}}$ *)
(* *Output*: *a semantic vector* $\overrightarrow{i}_d$, *image of* $\overrightarrow{d}$. *)
begin
    for $c \in \mathcal{C}_{\overrightarrow{q}}$ **do**
        for $c' : \overrightarrow{sed_c}[c'] \neq 0$ **do**
            $\overrightarrow{i}_d[c] \leftarrow \max(\overrightarrow{d}[c'] \times \overrightarrow{sed_c}[c'], \overrightarrow{i}_d[c])$;

    for $c \notin \mathcal{C}_{\overrightarrow{q}}$ **do if** $\exists c' \in \mathcal{C}_{\overrightarrow{q}} : \overrightarrow{sed_{c'}}[c] \neq 0$ **then** $\overrightarrow{i}_d[c] \leftarrow 0$
                                                                                       else $\overrightarrow{i}_d[c] \leftarrow \overrightarrow{d}[c]$;

    return $\overrightarrow{i}_d$
end;

We define the relevance of $\overrightarrow{d}$ wrt. $\overrightarrow{q}$ by $\cos(\overrightarrow{i}_d, \overrightarrow{q})$. Considering the image enables to take into account the documents that have concepts linked to those of the query. Using a cosine, and thus the norm of the vectors, assigns a lower importance to the documents with an important norm, which are often very general.

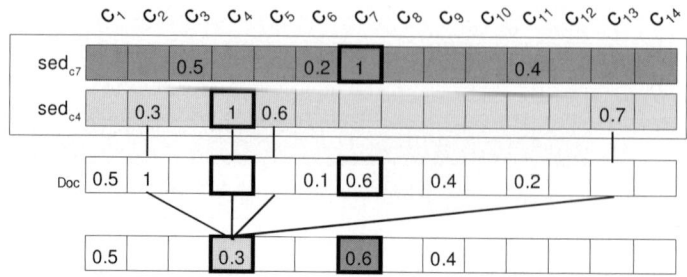

Fig. 3. Obtaining the image of a document

4 Relevance in the Context of Unshared Concepts

In this section, we assume that the query initiator and the document provider do not use the same ontology. We follow the approach adopted in Section 3, using a query expansion at the query initiator's side and the computation of the image of the document at the provider's side. But things get complicated by the fact that the query initiator and the document provider do not use the same vector space. An additional step is needed in order to evaluate relevance in a same and single space. Thus, we introduce a *query interpretation* step at the provider's side.

4.1 Computing Relevance: Overview

As shown in Figure 4, the query initiator, denoted by p_1, works within the context of ontology Ω_1, while the document provider, noted p_2, works with ontology Ω_2. Through its semantic indexing module, the query initiator (respectively the document provider) produces the query vector (respectively the document vector), which is expressed on Ω_1 (respectively Ω_2). Both p_1 and p_2 also have their own way of computing both the similarity and the propagation.

We assume that the query initiator and the document provider *share* some common concepts, meaning that each of them regularly, although may be not often, runs an ontology matching algorithm. Ontology matching results in an *alignment* between two ontologies, which is composed of a (non empty) set of correspondences with some cardinality and, possibly some meta-data [4]. A *correspondence* establishes a relation (equivalence, subsumption, disjointness...) between some entities (in our case, concepts), with some confidence measure. Each correspondence has an identifier. In this paper, we only consider the equivalence relation between concepts and those couples of equivalent concepts of which confidence measure is above some threshold. We call them the *shared* concepts. For simplicity, when there is an equivalence, we make no distinction between the name of the given concept at p_1's, its name at p_2's, and the identifier of the correspondence, which all refer to the same concept. Hence, the set of shared concepts is denoted by $\mathcal{C}_{\Omega_1} \cap \mathcal{C}_{\Omega_2}$.

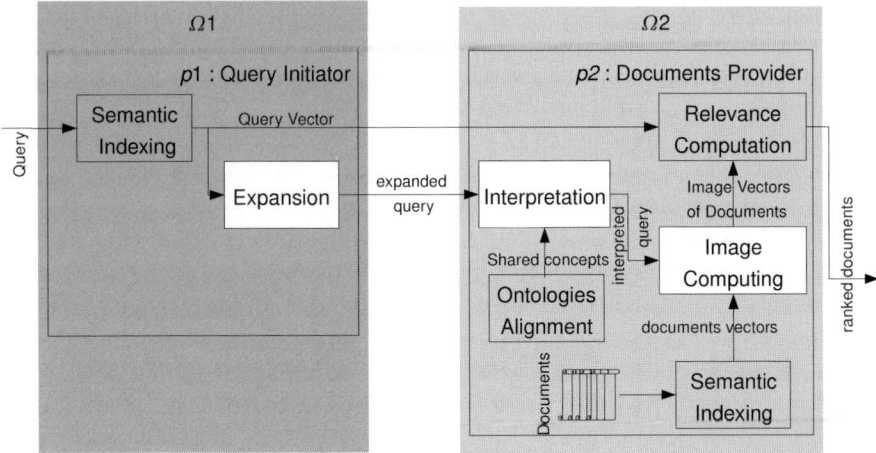

Fig. 4. Overview of relevance computation

Given these assumptions, computing relevance requires the following steps:

Query Expansion. It remains unchanged. The query initiator p_1 computes an *expansion* of its query, which results in a set of SEDs. Each SED is expressed on the set $\mathcal{C}_{\Omega_1}$, no matter the ontology used by p_2. Then, the expanded query is sent to p_2, together with the initial query.

Query Interpretation. Query interpretation by p_2 provides a set of interpreted SEDs on the set $\mathcal{C}_{\Omega_2}$ and an interpreted query. Each SED of the expanded query is interpreted separately. Interpretation of a SED $\overrightarrow{sed_c}$ is decomposed in two problems, which we address in the next subsections:

- The first problem is to find a concept in $\mathcal{C}_{\Omega_2}$ that corresponds to c, noted c̃. This is difficult when the central concept is not shared. In this case, we use the weights of the shared concepts to guide the search. Of course, this is only a "contextual" correspondence as opposed to one that would be obtained through matching.
- The second problem is to attribute weights to shared and unshared concepts of $\mathcal{C}_{\Omega_2}$ which are linked to $\overrightarrow{sed_c}$. This amounts to interpret the SED.

Image of the Document and Cosine Computation. They remain unchanged. Provider p_2 computes the image of its documents wrt. the interpreted SEDs and then, their cosine based relevance wrt. the interpreted query, no matter the ontology used by p_1.

In the following, we describe the steps involved in the interpretation of a given SED.

4.2 Finding a Corresponding Concept

The interpretation of a given SED $\overrightarrow{sed_c}$ leads to a major problem: finding a concept in $\mathcal{C}_{\Omega_2}$ which corresponds to the central concept c. This corresponding concept is noted $\tilde{c}$ and will play the role of the central concept in the interpretation of $\overrightarrow{sed_c}$, noted $\overrightarrow{sed_{\tilde{c}}}$. If c is shared, we just keep it as the central concept of the interpreted SED. When c is not shared we have to find a concept which seems to best respect the "flavor" of the initial SED.

Theoretically, all the concepts of $\mathcal{C}_{\Omega_2}$ should be considered. Several criterias can apply to choose one which seems to best correspond. We propose to define the notion of *interpretation function* which is relative to a SED $\overrightarrow{sed_c}$ and a candidate concept $\tilde{c}$ and which assigns a weight to each value of silmilarity wrt. $\tilde{c}$. Definition 4 consists of four points. The first one requires the interpretation function to assign the value of $\overrightarrow{sed_c}[c]$ to the similarity value 1, which corresponds to $\tilde{c}$. In the second point, we use the weights assigned by $\overrightarrow{sed_c}$ to the shared concepts (c_1, c_2, c_3 and c_6 in figure 5) and the ranking of concepts in function of $sim_{\tilde{c}}$. However, there might be several shared concepts that have the same similarity value wrt. $\tilde{c}$, but have a different weight according to $\overrightarrow{sed_c}$. Thus, we require function $f_i^{\overrightarrow{sed_c},\tilde{c}}$ to assign the minimum of these values to the corresponding similarity value. This is a pessimistic choice and we could either take the maximum or a combination of these weights. As for the third point, let us call c_{min}, the shared concept with the lowest similarity value (c_6 in Figure 5 (a) and c_3 in Figure 5 (b)). We consider that we have not enough information to weight the similarity values lower than $sim_{\tilde{c}}(c_{min})$. Thus we assign them the zero value. The fourth point is just a mathematical expression which ensures that the segments of the affine function are only those defined by the previous points.

Definition 4 (Interpretation function). *Given a SED $\overrightarrow{sed_c}$ and a concept $\tilde{c}$, $f_i^{\overrightarrow{sed_c},\tilde{c}} : [0..1] \to [0..1]$, noted f_i if no ambiguity, is an* interpretation function *iff it is a piecewise affine function and:*

- $f_i(1) = \overrightarrow{sed_c}[c]$;
- $\forall c' \in \mathcal{C}_{\Omega_1} \cap \mathcal{C}_{\Omega_2}$, $f_i(sim_{\tilde{c}}(c')) = \min_{\substack{c'' \in \mathcal{C}_{\Omega_1} \cap \mathcal{C}_{\Omega_2} \\ sim_{\tilde{c}}(c') = sim_{\tilde{c}}(c'')}} (\overrightarrow{sed_c}[c''])$;
- $\forall x \in [0..1]$, $x < sim_{\tilde{c}}(c_{min}) \Rightarrow f_i(x) = 0$;
- $Seg = \|\{x : \exists c' \in \mathcal{C}_{\Omega_1} \cap \mathcal{C}_{\Omega_2}, c' \neq \tilde{c}$ and $sim_{\tilde{c}}(c') = x\}\| + 1$ *where Seg is the number of segments of f_i.*

Intuitively, the criterias for choosing a corresponding concept among all the possible concepts can be expressed in terms of the properties of the piecewise affine function f_i. Of course, there are as many different function f_i as candidate concepts. But the general idea is to choose the function f_i wich resembles the more to a propagation function. Let us consider the example of Figure 5 (a) and (b) where c_1, c_2, c_3 and c_6 are shared. The function in Figure 5 (a) is obtained considering c'_1 as the corresponding concept (and thus ranking the other concepts in function of their similarity with $c'1$). The function in Figure 5 (b) is obtained similarly, considering c'_2. Having to choose between c'_1 and c'_2 we would prefer

c'_1 because function $f_i^{\overrightarrow{sed_c},c'_1}$ is monotonically decreasing whereas $f_i^{\overrightarrow{sed_c},c'_2}$ shows a higher "disorder" wrt. the general curve of a propagation function.

Several characteristics of the interpretation function can be considered to evaluate "disorder". For example, one could choose the function which minimizes the number of local minima (thus minimizing the number of times the sign of the derived function changes). Another example is to choose the function which minimizes the variations of weight between local minima and their next local maximum (thus penalizing the functions which do not decrease monotonically). A third could combine these criteria.

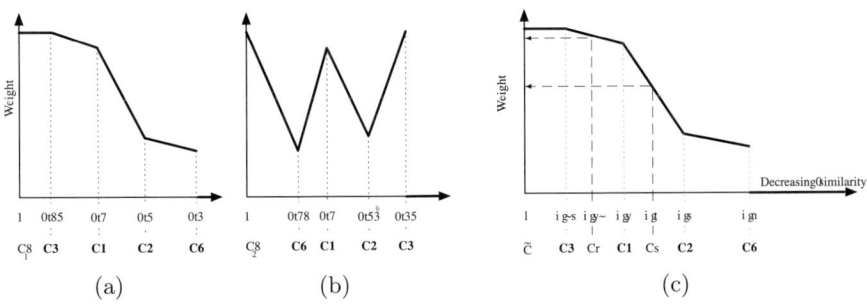

Fig. 5. Two steps of the interpretation of a SED : (a) f_i for candidate concept c'_1, (b) f_i for candidate concept c'_2 and (c) weighting the unshared concepts.

4.3 Interpreting a SED

We define the interpretation of a given SED $\overrightarrow{sed_c}$ as another SED, with central concept $\tilde{c}$ which has been computed at the previous step. We keep their original weight to all the shared concepts. The unshared concepts are weighted using an interpretation function as defined above.

Definition 5 (Interpretation of a SED). Let $\overrightarrow{sed_c}$ be a SED on $\mathcal{C}_{\Omega_1}$ and let $\tilde{c}$ be the concept corresponding to c in $\mathcal{C}_{\Omega_2}$. Let $sim_{\tilde{c}}$ be a similarity function and let $f_i^{\overrightarrow{sed_c},\tilde{c}}$, noted f_i, be an interpretation function. Then SED $\overrightarrow{sed_{\tilde{c}}}$ is an interpretation of $\overrightarrow{sed_c}$ iff:

- $\overrightarrow{sed_{\tilde{c}}}[\tilde{c}] = f_i(1)$;
- $\forall c' \in \mathcal{C}_{\Omega_1} \cap \mathcal{C}_{\Omega_2}$, $\overrightarrow{sed_{\tilde{c}}}[c'] = \overrightarrow{sed_c}[c']$;
- $\forall c' \in \mathcal{C}_{\Omega_2} \setminus \mathcal{C}_{\Omega_1}$, $\overrightarrow{sed_{\tilde{c}}}[c'] = f_i(sim_{\tilde{c}}(c'))$;

Figure 5 (c) illustrates this definition. Document provider $p2$ ranks its own concepts in function of $sim_{\tilde{c}}$. Among these concepts, some are shared ones for which the initial SED $\overrightarrow{sed_c}$ provides a given weight. This is the case for c_1, c_2, c_3 and c_6 which are in bold face in the figure. The unshared concepts are assigned the weight they obtain by function f_i (through their similarity to $\tilde{c}$). This is illustrated for concepts c_4 and c_5 by a dotted arrow.

5 Experimental Validation

In this section, we use our approach based on *image based relevance* to find documents which are the most relevant to given queries. We compare our results with those obtained by the *cosine based method* and the *rough propagation method*. In the former method, relevance is defined by the cosine between the query and document vectors. In the latter, the effects of propagating weights from different concepts are mixed in a single vector; then relevance is obtained using the cosine.

5.1 General Setup for the Experiments

We use the Cranfield corpus, a testing corpus consisting of 1400 documents and 225 queries in natural language, all related to aeronautical engineering. For each query, each document is scored by humans as relevant or not relevant (boolean relevance). Our ontology is lightweight, in the meaning of [7], *i.e.* an ontology composed of a taxonomy of concepts : WordNet [5]. In Information Retrieval, there was a debate whether WordNet is suitable for experimentation (see the discussion in [24]). However, more recent works show that it is possible to use WordNet, and sometimes other resources, and still get good results [8]. Semantic indexing [20] is the process which can compute the semantic vectors from documents or queries in natural language. The aim is to find the most representative concepts for documents or queries. We use a program made in our lab : RIIO [3], which is based on the selection of synsets from WordNet. Although it is not the best indexing module, one of its advantages is that there is no human intervention in the process. The semantic similarity function we use is that of [2], because it has good properties and results which are discussed in Section 6. We slightly modified that function due to normalization considerations. Following the framework of membership functions presented in Section 2 we can define many propagation functions. We tested three different types of functions : "square" (of type f_{v,l_1,l_1}), "sloppy" (of type $f_{v,1,l_2}$), or hybrid (of type f_{v,l_1,l_2} with $l_1 = 2 \times l_2$). Our experiments show no important difference, but sloppy propagation has slightly better results. So we use only this propagation function, adding ten concepts in average for a given central concept.

In order to evaluate whether our solution is robust, we would need ontologies which agree on different percentages of concepts : 90%, 80%, 70%, ..., 10%. This is very difficult to obtain. We could build artificial ontologies, but this would force us to give up the experiments on a real corpus. Thus, we decided to stick to WordNet and simulate semantic heterogeneity. Both the query initiator and the provider use WordNet, but we make so that they are not able to understand each other on some concepts (a given percentage of them). To do so, we remove some mappings between the two ontologies. Thus it simulates the case where the query intiator and the document provider use the same ontology but are not aware of it. It is then no more possible to compare queries and documents on those concepts. The aim is to evaluate how the answers to queries expressed with removed matchings, change. Note that the case with no removed matching reduces to a single ontology.

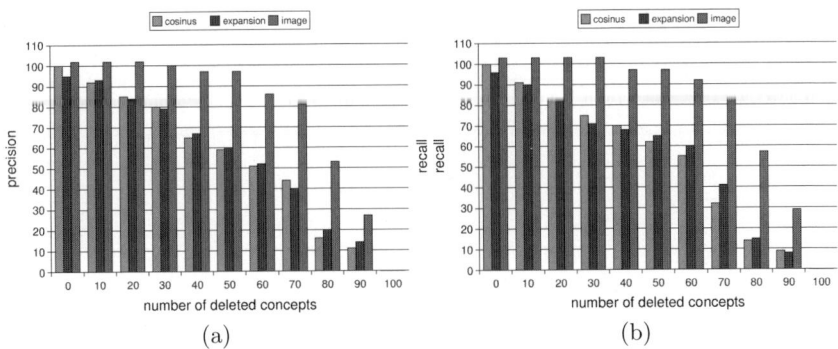

Fig. 6. Evolution of (a) precision and (b) recall in function of the random removal percentage of mappings

In a first experiment, we progressively reduce the number of mappings, thus increasing the percentage of removed mappings (10%, 20%, ... until 90%). The progressive reduction in their common knowledge is done randomly. In a second experiment, we remove the mappings concerning the central concepts of the queries in the ontology of the document manager. This is now an intentional removing, which is the worst case for most of the techniques in IR : removing only the elements that match. For both experiments, we take into account the results obtained with the 225 queries of the corpus.

5.2 Results

Figure 6 shows the results obtained in average for the all 225 queries of the testing corpus. The reference method is the cosine one when no matching is removed, which gives a given reference precision and recall. Then, for each method and each percentage of removed matching, we compute the ratio of the precision obtained (respectively recall) by the reference precision. When the percentage of randomly removed matchings increases, precision (Figure 6 (a)) and recall (Figure 6 (b)) decrease *i.e.* the results are less and less relevant. However, our "image and interpretation based" solution shows much better results. When the percentage of removed matchings is under 70%, we still get 80% or more of the answers obtained in the reference case.

In the second experiment, we consider that the document manager does not understand (*i.e.* share with the query initiator) the central concepts of the query (see Figure 7). With the cosine method, there is no more matching between concepts in queries and concepts in documents. Thus no relevant document could be retrieved. With the query expansion, some of the added concepts in the query allow to match with concepts in documents that are close to the central concepts of the query. This leads to precision and recall at almost 10%. Our image-based retrieving method has more than 90% of precision and recall in the retrieval. This is also an important result. Obviously, as we have the same ontology and

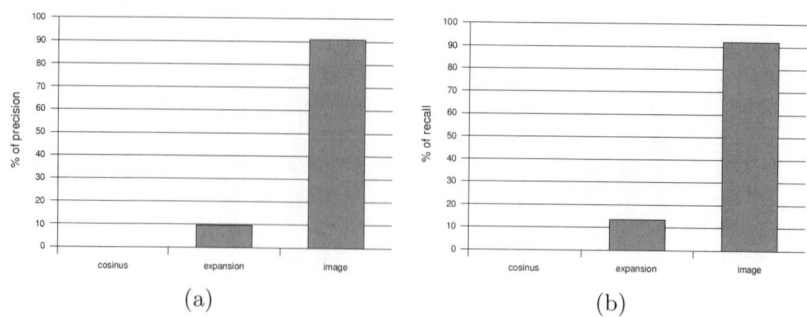

Fig. 7. Precision (a) and recall (b) when the central concepts of the query are unshared

the same similarity function, the interpretation can retrieve most of the central concepts of the query. But the case presented here is hard for most of the classical techniques (concepts of the query unshared) and we obtain a very important improvement.

6 Related Work

The similarity that we use in our experiments is the result of a thorough study of the properties of different similarity measures. We looked for a similarity which is not a distance (does not satisfy similarity nor triangle inequality), based on the result of [23]. Hence we use one classical benchmark of this domain : the work of Miller and Charles [15] on the human assessments of similarity between concepts. Thirty eight students were asked to mark how similar thirty couples of concepts were. We have implemented four similarity measures: [26,21,12,2], respectively noted *Wu and P.*, *Seco*, *Lin* and *Bidault* in table 1. Correlation is the ratio between those measures on the human results. The results show that only Bidault's measure does not meet symmetry nor triangle inequality. Moreover, it obtains a slightly better correlation. Hence, it was preferred to rank the concepts according to their (dis)similarity with a central concept.

The idea of query expansion is shared by several fields. It was already used in the late 1980's in Cooperative Answering Systems [6]. Some of the suggested techniques expanded SQL queries considering a taxonomy. In this paper, we do not consider SQL queries, and we use more recent results about ontologies and

Table 1. Comparison of similarity measures

	Wu & P.	Seco	Lin	Bidault
symmetry	yes	yes	yes	**no**
triangle inequality	no	no	no	no
correlation	0.74	0.77	0.80	0.82

their interoperability. Expansion of query vectors is used for instance in [17,24]. However, this expansion produces a single semantic vector only. This amounts to mix the effects of the propagations from different concepts of the query. Although this method avoids some silence, it often generates too much noise, without any highly accurate sense disambiguation [24]. Consequently, the results can be worse than in the classical vector space model [1]. Our major difference with this approach is that (1) the propagations from the concepts of the query are kept separate and that (2) they are not directly compared with the document. Rather, they are used to modify its semantic vector. In our experiments, our method gives better results. Also, we join [16] on their criticism of the propagation in a single vector, but our solutions are different.

Our approach also relies on the correspondences resulting from the matching of the two ontologies. Several existing matching algorithms could be used in our case [4]. In the interpretation step, we provide a very general algorithm to find the concept corresponding to the central concept of a SED. In case the concept is not shared, one could wonder whether matching algorithms could be used. In the solution we propose, the problem is quite different because the *weights of the concepts are also used* to find the corresponding concept (through the interpretation function). This is not the case in traditional ontology matching, which aim is to find general correspondences. In our case, one can see the problem as finding a "contextual" matching, the results of which cannot be used in other contexts. Because it is difficult to compute all the interpretation functions, one can use an *approximation algorithm* (for example, taking the least common ancestor as we did in our experiments). In that case, existing proposals can fit like [10,14]. But it is clear that they do not find the best solution every time.

Finally, the word *interpretation* is used very often and reflects very different problems. However, to the best of our knowledge, it never refers to the case of interpreting a query expressed on some ontology, within the space of another ontology, by considering the weights of the concepts.

7 Conclusion

The main contribution of this paper is a proposal improving information exchange between a query initiator and a document provider that use different ontologies, in a context where semantic vectors are used to represent documents and queries. The approach only requires the initiator and the provider to share some concepts and also uses the unshared ones to find additional relevant documents. To our knowledge, the problem has never been addressed before and our approach is a first, encouraging solution. In short, when performing query expansion, the query initiator makes more precise the concepts of the query by associating an expansion to each of them (SED). The expansion depends on the initiator's characteristics: ontology, similarity, propagation function. However, as far as shared concepts appear in a SED, expansion helps the document provider interpreting what the initiator wants, especially when the central concept is not shared. Interpretation by the document provider is not easy because the peers do not share the same vector space. Given its own ontology and similarity function, it first finds out a correspondent concept for the central concept of each SED, and then interprets the whole SED. The interpreted SEDs are used to compute an image of the documents and their relevance. This is only possible because the central concepts are expanded separately. Indeed if the effects of propagations from different central concepts were mixed in a single vector, the document provider wouldn't be able to interpret the query as precisely.

Although our approach builds on several notions (ontology, ontology matching, concept similarity, semantic indexing, relevance of a document wrt a query...) it is not stuck to a specific definition or implementation of them and seems compatible with many instantiations of them. It is important to notice that there is no human intervention at all in our experiments, in particular for semantic indexing. Clearly, in absolute, precision and recall could benefit from human interventions at different steps like indexation or the definition of the SEDs. Results show that our approach significantly improves the information exchange, finding up to 90% of the documents that would be found if all the concepts were shared.

As future work, we plan to test our approach in several different contexts in order to verify its robustness. Many different parameters can be changed: similarity and propagation functions, ontologies, indexing methods, corpus... Complexity is another point that should be considered carefully. Indeed, naive implementations would lead to unacceptable execution time. Although an implementation is running for the experiments within admissible times, it could benefit from a more thorough study of theoretical complexity.

References

1. Berry, M.W., Drmac, Z., Jessup, E.R.: Matrices, vector spaces, and information retrieval. SIAM Rev. 41(2) (1999)
2. Bidault, A., Froidevaux, C., Safar, B.: Repairing queries in a mediator approach. In: ECAI (2000)
3. Desmontils, E., Jacquin, C.: Indexing a web site with a terminology oriented ontology. In: The Emerging Semantic Web (2002)
4. Euzenat, J., Shvaiko, P.: Ontology matching. Springer, Heidelberg (DE) (2007)
5. Fellbaum, C.: WordNet: an electronic lexical database (1998)
6. Gaasterland, T., Godfrey, P., Minker, J.: An overview of cooperative answering. J. of Intelligent Information Systems 1(2), 123–157 (1992)
7. Gómez-Pérez, A., Fernández, M., Corcho, O.: Ontological Engineering. Springer, London (2004)
8. Gonzalo, J., Verdejo, F., Chugur, I., Cigarran, J.: Indexing with wordnet synsets can improve text retrieval. In: COLING/ACL 1998 Workshop on Usage of WordNet for NLP (1998)
9. Ives, Z.G., Halevy, A.Y., Mork, P., Tatarinov, I.: Piazza: mediation and integration infrastructure for semantic web data. Journal of Web Semantics (2003)
10. Jiang, G., Cybenko, G., Kashyap, V., Hendler, J.A.: Semantic interoperability and information fluidity. Int. J. of cooperative Information Systems 15(1), 1–21 (2006)

11. Jiang, J., Conrath, D.: Semantic similarity based on corpus statistics. In: International Conference on Research in Computational Linguistics (1997)
12. Lin, D.: An information-theoretic definition of similarity. In: International Conf. on Machine Learning (1998)
13. Manning, C.D., Schtze, H.: Foundations of statistical natural language processing. MIT Press, Cambridge (1999)
14. Mena, E., Illaramendi, A., Kashyap, V., Sheth, A.: Observer: An approach for query processing in global information sytems based on interoperation across preexisting ontologies. Int. J. distributed and Parallel Databases 8(2), 223–271 (2000)
15. Miller, G.A., Charles, W.G.: Contextual correlates of semantic similarity. Language and Cognitive Processes (1991)
16. Nie, J.-Y., Jin, F.: Integrating logical operators in query expansion invector space model. In: SIGIR workshop on Mathematical and Formal methods in Information Retrieval (2002)
17. Qiu, Y., Frei, H.P.: Concept based query expansion. In: SIGIR (1993)
18. Rousset, M.-C.: Small can be beautiful in the semantic web. In: International Semantic Web Conference, pp. 6–16 (2004)
19. Salton, G., MacGill, M.J.: Introduction to Modern Information Retrieval. McGraw-Hill, New York (1983)
20. Sanderson, M.: Retrieving with good sense. Information Retrieval (2000)
21. Seco, N., Veale, T., Hayes, J.: An intrinsic information content metric for semantic similarity in wordnet. In: ECAI (2004)
22. Tempich, C., Pinto, H.S., Staab, S.: Ontology engineering revisited: An iterative case study. In: ESWC, pp. 110–124 (2006)
23. Tversky, A.: Features of similarity. Psychological Review 84(4) (1977)
24. Voorhees, E.M.: Query expansion using lexical-semantic relations. In: SIGIR, Dublin (1994)
25. Woods, W.: Conceptual indexing: A better way to organize knowledge. Technical report, Sun Microsystems Laboratories (1997)
26. Wu, Z., Palmer, M.: Verb semantics and lexical selection. In: ACL (1994)

Hybrid Search: Effectively Combining Keywords and Semantic Searches

Ravish Bhagdev[1], Sam Chapman[1], Fabio Ciravegna[1], Vitaveska Lanfranchi[1] and Daniela Petrelli[2]

[1] Department of Computer Science
[2] Department of Information Studies
University of Sheffield, Regent Court, 211 Portobello Street,
S1 4DP Sheffield, United Kingdom
{N.Surname}@shef.ac.uk

Abstract. This paper describes hybrid search, a search method supporting both document and knowledge retrieval via the flexible combination of ontology-based search and keyword-based matching. Hybrid search smoothly copes with lack of semantic coverage of document content, which is one of the main limitations of current semantic search methods. In this paper we define hybrid search formally, discuss its compatibility with the current semantic trends and present a reference implementation: K-Search. We then show how the method outperforms both keyword-based search and pure semantic search in terms of precision and recall in a set of experiments performed on a collection of about 18.000 technical documents. Experiments carried out with professional users show that users understand the paradigm and consider it very powerful and reliable. K-Search has been ported to two applications released at Rolls-Royce plc for searching technical documentation about jet engines.

Keywords: Semantic search, Semantic Web in use.

1 Introduction

The Semantic Web (SW) is a creative mix of metadata designed according to multiple ontologies and unstructured documents (e.g. classic Web documents). The assumption that the SW is not a Web of documents, but a Web of relations between resources denoting real world objects [4] is too restrictive of the true nature of the SW. There are a number of applications and situations where coexistence of documents and metadata is actually required. One example is the legal scenario, where access to documents is the main focus and the available metadata is the means to reach a specific set of documents [5]. However it may well happen that the available metadata does not cover parts of the document that are of interest to some users because: (i) the ontology used for annotation has a different focus and does not model that part of the content or (ii) annotations can be incomplete, whether user or system provided. A human annotator may miss some or provide spurious ones; in the same way automated means such as Information Extraction from texts (IE) may be unable to reliably extract the information required. This is because IE is a technology that performs very well on simple tasks (such as named entity recognition), but poorly on more complex tasks such as event capture [8]. Therefore, some metadata modelled by an ontology may be impossible to capture with IE thus preventing any future operation (e.g. retrieval) via that metadata.

In this paper, we focus on searching the SW as a collection of both documents and metadata, with the aim of accommodating different user tasks: document retrieval and/or knowledge retrieval. A document retrieval task implies searching for documents using concepts or keywords of interest; a knowledge retrieval task concerns retrieving facts from a knowledge base (i.e. triples). Differently from previous literature [1, 2, 3, 4, 9], we consider the issue of working in a complex environment where metadata only partially covers the user information needs. We therefore propose to use a strategy (called Hybrid Search, (HS)) where a mix of keyword-based and metadata-based strategies are used. We formally define the approach and describe how to organise a HS architecture. We then describe K-Search, a reference implementation of HS. In implementing an approach, a number of decisions are made: methodological (e.g. we selected a form based approach [1]), and technical (e.g. on the expressivity of covered language and architecture design). We discuss how these choices impact the HS mechanism. Then we present two experiments performed using a K-Search application:

- *in vitro:* K-search was applied to a large corpus of legacy documents; an evaluation of the resulting application shows HS outperforming both keyword based searching and semantic searching;
- *in vivo:* the application was evaluated with real users; the results show that users appreciate the full power of the HS concept.

Finally we compare our work to the state of the art, we discuss how it is possible to extend the currently available semantic search paradigms to cope with HS, draw conclusions and highlight future work.

2 Hybrid Search

HS combines the flexibility of keyword-based retrieval (as in traditional search engines) with the ability to query and reason on metadata typical of semantic search systems. Metadata is information associated to a document describing both its context (e.g. author, title, etc.) and its content (as provided by RDF triples annotating portions of the documents with respect to an ontology, e.g. *<"installed_part" upon "engine_type">*). In concrete terms, HS is defined as:

- the application of semantic (metadata-based) search for the parts of the user queries where metadata is available;
- the application of keyword-based search for the parts not covered by metadata.

Three types of queries are possible with HS: (i) pure semantic search via unique identification of concepts/relations/instances (e.g. via *URIs* or unique identifiers); (ii) keyword-based search on the whole document and (iii) keyword-in-context search. Keyword-in-context searches the keywords only within the portion of the document annotated with a specific concept or relation; for example in the aerospace domain, it enables searching for the string "fuel" but only in the context of all the text portions annotated with the concept *affected-engine-part*. The keyword-in-context mechanism was the core of the mechanism proposed in [14].

It is important to stress that differently from other approaches (e.g. [9]), in HS conditions on metadata and keywords coexist. For example consider an application in the aerospace diagnostic domain where metadata is associated to documents for events described (e.g. discoloration) and the affected component (e.g. a high pressure blade) but not for the part of the component affected (e.g. the trailing edge). An example of hybrid query could be:

> ∀z /(discoloration y) & (component x) & (located-on y x) &
> (provenance-text-contains x "blade") & (document z) &&
> (provenance y z) & (contains z "trailing edge")

This can be read as: retrieve all documents *(document z)* that contain the string "trailing edge" *(contains z "trailing edge")* with associated metadata *(provenance y z)* involving:

- an instance of discoloration – *(discoloration y)*
- an instance of component where the provenance text contains the word "blade" *(component x) & (provenance-text-contains x "blade")*
- the component is affected by the discoloration *(located-on y x)*

To our knowledge, no other approaches allows such flexibility in qiuerying, as most of them just allow queries based on metadata [1, 2, 3, 4, 9].

2.1 Hybrid Search for Document Retrieval

The most commonly used method for document retrieval is keyword-based search (KS). KS effectiveness is often affected by two main issues, ambiguity and synonymity. Ambiguity arises in traditional keyword search systems because keywords can be polysemous, i.e. they can have multiple meanings. A search containing ambiguous terms will return spurious documents (low precision). Synonymity is found when an object can be identified by multiple equivalent terms. When searching documents using just one of the terms, the documents containing other synonym are not retrieved (low recall). Semantic search as metadata-based search defined according to an ontology, enables overcoming both issues because annotations are unambiguous and do not suffer from synonymity.

Nonetheless when pure Semantic Search is applied to a document retrieval task, it can fail to encompass the user information needs (either because of limitations in the ontology or because the metadata is unavailable for a specific document), as it would restrict the types of queries users can perform (low recall).

HS combines the disambiguation capabilities of semantic search (when metadata is available) with the generality and extensibility of keyword-based search (for the other cases). The expected result is that:

- precision and recall are increased with respect to the standard keyword-based search because ambiguity and synonymity are dealt with by semantic search when available;

- the use of keywords where metadata is missing enables to answer otherwise impossible queries (increased recall with respect to semantic search). As keywords are combined with metadata in the same query, the context given by the available metadata helps in disambiguating keywords as well (higher precision than keyword-based search).

2.2 Hybrid Search for Knowledge Retrieval

In addition to document retrieval, HS can provide highly effective knowledge retrieval by using keywords as "context" of the metadata, hence enabling to further focus the results in a way that is impossible with semantic search.
In the aerospace example, it will be possible to retrieve

```
∀y,x /(discoloration y) & (component x) & (contains-
    keyword x "blade") & (located-on y x) & (document
    z) && (provenance y z) & (contain-keyword z
    "trailing edge")
```

Searching on metadata only allows a query like

```
∀y,x /(discoloration y) & (component x) & (located-on y x)
```

which would return a large amount of spurious results (low precision). The results of the hybrid query would still be sub-optimal (i.e. not equivalent to a semantic search where all the metadata is available) because the keyword search part will still suffer problems of synonymity and polysemy, but it would be far more high quality than the pure semantic search which will miss essential conditions. Also, it is expected that the matching of metadata will help reducing the issue of polysemy because it will work as context for the keywords.

3 Architecture for Hybrid Search

This section discusses a generic architecture for HS, while the next one presents an actual implementation.

At indexing time, documents are indexed using a standard keyword–based engine such as SolR[1]. Annotations (e.g. generated by an IE system) are stored in a Knowledge Base (e.g. a triple store like Sesame[2]) in the form of RDF triples. Provenance of facts must be recorded, for example in the form of triples connecting the facts' URIs and those of the document of origin, as well as the original strings used in the documents.

At retrieval time, HS performs the following steps:

- the query is parsed and the different components (keywords, keywords-in-context and metadata-based) identified;
  - keyword matches are sent to the traditional information retrieval system;

---

[1] http://lucene.apache.org/solr/
[2] http://www.openrdf.org/

- metadata searches are translated into a query language like SPARQL[3] and sent to a triple store;
- keywords-in-context queries are matched with the provenance of annotations in documents (again using SPARQL and a triple store);
- finally, the results of the different queries are merged, ranked and displayed.

Merging of results. A direct matching between keyword and semantic results is not straightforward as their results are incompatible. Keyword matching returns a set of *URIs* of documents (KSDocUriSet) of size n.

$$\text{KSDocUriSet} \subset \text{URIs, where KSDocUriSet} = \begin{vmatrix} uri1, \\ uri2, \\ ... \\ urin \end{vmatrix}$$

while a semantic search performed on a knowledge base returns an <u>unordered</u> set *rSet* (size m) of individual assertions $< subj, rel, obj >$[4]

```
OSTripleSet = all triples ∈ R that satisfy the Ontology-based Query
```

Using the provenance information associated to each triple, it is possible to compute the set of documents that contain the information retrieved from the RDF store.

```
OSDocUriSet = Union of Provenance(triple^i) for all i where triple^i ∈
              OSTripleSet
```

In order to provide the answer for users interested in **document retrieval**, the list of *URIs* of documents generated using provenance information is now directly compatible with the output of keyword matching. The result of the query is given by the <u>intersection</u> of the two sets of document *URIs*.

```
            HybridSearchUriSet = KSDocUriSet ∩ OSDocUriSet
```

In order to provide answers to users interested in **knowledge retrieval**, a list of triples must be returned. In this case, the list of triples is filtered so to remove those whose provenance does not point to any of the documents returned by the keyword-based search engine. Formally:

$$\text{HSTripleSet} = \begin{vmatrix} \text{All triples} \in \text{OSTripleSet} \\ \text{Where Provenance(triple}^i) \in \text{KSDocUriSet} \end{vmatrix}$$

Ranking. Effective ranking (i.e. the ability to return relevant documents first) is extremely important for a positive user experience. The results returned by the different modalities provide material for orthogonal ranking methods:

---

[3] http://www.w3.org/TR/rdf-sparql-query/
[4] Both ontology-based and keyword in context queries are covered here.

- **keyword-based systems** like Lucene enable ranking of documents according to (1) their ability to match the keyword-based query; (2) the keywords used in anchor links (i.e. the text associated to hyperlinks pointing to a specific document) and (3) the document popularity measured as function of the weight of the links referring to the document itself;
- **semantic search** ranks according to the presence and quality of metadata.

Different ranking solutions can be adopted accordingly to the use case. The most natural one is to adopt the ranking provided by the keyword based search, as it is based on solidly proven methods, especially the use of anchor texts and hyperlinking However more sophisticated strategies can be designed, especially for organisational repositories where such interlinking is generally not present [14].

Presentation of results. Depending on the task (i.e. document retrieval Vs knowledge retrieval), results can be presented in different ways: as a list of ranked documents, as aggregated metadata (e.g. via graphs or charts) with associated provenance, etc.

4 K-Search: Putting Hybrid Search into Practice

K-Search is an implementation of the HS paradigm. In realising HS in a real world system, a number of choices need to be made in order to:

- create an interface that communicates to the user the optimal strategy to mix metadata and keywords for the task at hand, so to maximize effectiveness and efficiency of searches;
- decide what strategies to adopt for ranking, visualisation, annotation, etc.

We have chosen to model our search interface on a form data entry paradigm. The interface (Fig 1) works in a standard browser and enables the definition of complex hybrid queries in an intuitive way. Keywords can be inserted into a default

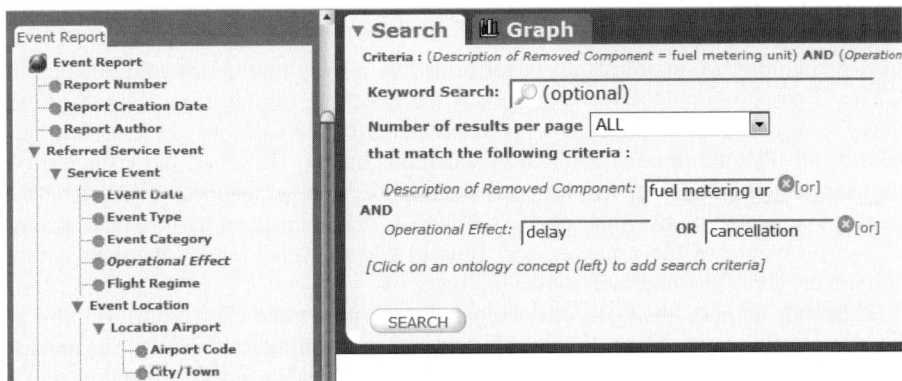

Fig. 1. Interface detail: the query form. Clicking a concept on the ontology creates a form item enabling inserting restrictions on metadata. Disjunctions are easily introduced by clicking [or].

form field in a way similar to that required by search engines; Boolean operators *OR* and *AND* can be used in their combination. Conditions on metadata can be added to the query by clicking on the ontology tree (left side of interface in Figure 2). This creates a form item to insert conditions on the specific concept. As multiple constraints can be added to the query, the **logical language** is restricted to provide a simple and intuitive interface: only common Boolean combinations are supported. This decision was supported by the observation that in carrying out their tasks, users adopted strategies that do not require the full logical language; furthermore research done in human-computer interaction shows that graphical representation of the whole Boolean logic is not understood by most users [10].

AND constructs are allowed among conditions checking different concepts in the ontology. So for example, *contains(removed-component, "fuel") AND contains(jet-engine-name, "engineA")* is acceptable, but *contains(removed-component, "fuel") AND contains(removed-component, "meter")* is not. The latter is acceptable if formulated as *contains (removed-component, "fuel meter")*. Conditions in AND are displayed on different lines in the interface (Figure. 3 shows an example of a combination of *removed-component AND operational-effect*). The expressivity restrictions are motivated by the results of our user studies, which showed which types of queries the users wanted to make.

OR constructs are acceptable only if between conditions on the same concept. So *contains(removed-component, "fuel") OR contains(removed-component, "meter")* is accepted, but *contains(removed-component, "fuel") OR contains(jet-engine-name, "engineA")* is not. The latter must be split into two different queries. Again, these restrictions are motivated by results of our user studies.

Figure 1 shows how the query *retrieve all events where removal of a fuel meter unit caused delay or cancellation"* - logically translated in *(contains(removed-component "fuel meter unit"))* AND equal*(operational-effect (delay OR cancellation))* - appears at the interface level: two concepts (*removed-component* and *operational-effect*) have been selected; removed-component has been specified with a single option (fuel meter unit) while *operational-effect* covers two alternatives (delay or cancellation).

4.1 Ranking and Presentation of Results

In K-Search the ranking of results is performed by relying on the keyword ranking, in this case based upon TF/IDF, because - as the matching on the metadata part of the query is strict (i.e. only the documents that match *all* the conditions are returned) - all the documents tend to be equivalent in semantic content. However, the visualisation interface enables the user to change the ranking by focusing on specific metadata values. For example, given the query in Figure 1, documents can be sorted according to e.g. the value of the removed part (this is done by clicking on the appropriate column header of the interface shown in Figure 2).

K-Search supports the tasks of document retrieval and knowledge retrieval also at the presentation level, by providing different views on the search results. The default use of K-Search is for document retrieval. Therefore when a query is fired, a set of ranked documents displayed as a list are presented (see mid-right panel of the interface in Figure 2). Each item in the list is identified by the title (or file name) of

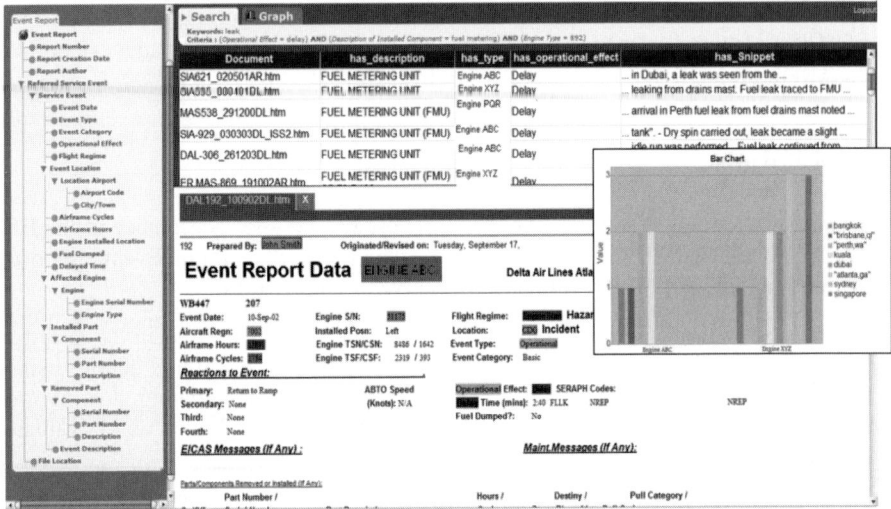

Fig. 2. The interface showing the list of documents returned (centre top), an annotated document and a graph produced from the results (image modified to protect confidential data)

the document and the values in the metadata that satisfy the semantic search. Clicking on one item in the list opens the corresponding document on the bottom right. The document is presented in its original layout with added annotations via colour highlighting; advanced features or services are associated to annotations [12, 13], including refining the current query. Multiple documents can be opened simultaneously in different tabs.

As for knowledge retrieval, K-Search provides two ways of inspecting the returned metadata. On the one hand the triples extracted are visible in the document list, because the values that satisfy the semantic query are listed for each document (middle panel). This enables an exhaustive and user-friendly inspection of the content of the triples and re-ranking of results according to these values. On the other hand, K-Search enables the creation of bi-dimensional graphs via selection of style (pie or bar chart) and variables to plot (e.g. engine Vs affected component). The graph in Figurre 2 plots the results of the previous query by location and engine type. Each graphic item (each bar in the example) is active and can be clicked to focus on the subset of documents that contains that specific occurrence. All retrieved triples can be exported in RDF or in CSV format for further statistical processing.

4.2 Indexing and Annotation

In order to make available document metadata and indexes, K-Search uses: (i) SolR for indexing documents and (ii) a generic semantic annotation plugin . Plugins currently exist for AktiveMedia (manual and semi-automatic annotation [6]) and some information extraction tools (T-Rex, an ontology-based IE tool [15] and Saxon, a rule-based extraction system[5]). Extracted information (ontology-based annotations)

---

[5] http://nlp.shef.ac.uk/wig/tools/saxon/

is stored in the form of RDF triples according to OWL or RDF ontologies into a triple store. K-Search provides plugins for Sesame and 3store; query languages supported are SPARQL and SeRQL.

5 Evaluation

Tests were carried out to evaluate the effectiveness and the user acceptance of the HS paradigm. The evaluation was performed using the K-Search Event Reports application (developed for Rolls-Royce plc) in two separate steps:

- *in vitro*: First of all the precision and recall of the IE system used in the specific case were evaluated; then 21 user-defined topics were translated into queries using three options: keyword-based searching, ontology-based searching and hybrid searching and the performances were recorded; these tests enabled us to evaluate the effectiveness of the method in principle;
- *in vivo:* 32 Rolls-Royce plc employees were involved in a usability test and commented on efficiency, effectiveness, and satisfaction; this evaluation enabled measuring the extent to which users understand the HS paradigm and feel that it returns appropriate results.

5.1 *In Vitro* Evaluation

The *in vitro* evaluation is composed by two parts, one to evaluate the effectiveness of the IE, the second to compare HS to keyword-based and semantic search.

IE evaluation

We analyzed a corpus of 18,097 reports on operational conditions of jet engines provided by Rolls-Royce plc. They are semi-structured Word documents containing tables and free text. As these documents are generated as part of the same management process, they all contain broadly the same relevant information but tables are user defined, so in principle each document can contain different types of table. However, some regularity occurs in tables across documents as users tend to reuse previously generated documents as template. The documents were converted into XML and HTML then indexed using SolR and metadata were generated using T-Rex. The ontology included concepts like the location where the event occurred, installed component(s), removed component(s), event details, what was the operational effect on the flight (delay, cancellation etc.), location, author, etc. The evaluation of the IE system was performed in order to understand which metadata were recognisable with an acceptable accuracy. Information in tables tends to be captured reliably by the IE system. This is because, although tables are irregular (e.g. sometimes the semantics is on the rows, sometimes on the columns, sometimes the information is spread over multiple cells, sometimes multiple information is compressed in one single cell), they roughly contain the same information and derive from evolution of common tables. T-Rex's learning curve assumed an asymptotic shape after learning from about 200 manually annotated documents. The combined evaluation results on all fields obtained in a two-cross folder test using 400 documents were Precision=98%,

Recall=99%, (harmonic) F-Measure=98%. Information in tables contained most of the metadata required in the ontology with the exception of the event cause.

As for the information contained in the free text (which was mainly describing the event cause), instead, accuracy was not at a level adequate to the user expectations (which was – according to our studies very close to 100% for recall and >90% for precision) therefore it was not made available to semantic search; it was however still available for searching via keywords.

Hybrid Search Comparative Evaluation
The goal of the comparative evaluation was to show that HS can provide better results than the pure keyword based or semantic search, by combining their reciprocal strengths. The evaluation was done considering a set of 21 topics generated on the basis of observed tasks, sequences of user queries recorded in the event corporate database or as elaboration of direct input from users (i.e. examples of their recent searches). Each topic represents a realistic information-seeking task of Rolls-Royce engineers, which typically could be previously answered only via repeated searches and extensive manual work. As it turned out, some topics, like *"How many events happened during maintenance in 2003"*, can be answered using pure semantic search (because all the relevant metadata was captured by the IE system), others, like *"What events happened during maintenance in 2003 due to control units?"* can only be answered by combining annotations and keywords (in this case due to the lack of metadata about the cause of the event). Finally one topic could only be answered using keyword-based search, as no parts of it are covered by metadata.

During evaluation, topics were transformed into queries by manually translating the topics into semantic, hybrid and keyword-based queries. An example of hybrid query is ((flight-regime maintenance) AND (event-date year-2003)) + (keywords-contained "control unit" OR "control" OR "unit").

Precision and Recall were computed on the first 20 and 50 documents returned by each modality. We used standard Precision and Recall measures.

$$Precision = \frac{Correct\ System\ Answers}{System\ Answers} \qquad Recall = \frac{Correct\ System\ Answers}{Expected\ Answers}$$

Evaluation of results on such a large amount of documents is quite a difficult issue. The problem comes in computing recall's *Expected Answers* without manually matching all the 18,097 documents against all 21 topics. Therefore, we decided to approximate *Expected Answers* with the cardinality of the set of all the relevant documents returned by any of the three modalities. We believe that this measure is enough for the purpose of this evaluation because our goal is to demonstrate that HS outperforms the other two approaches in terms of precision and recall in returning relevant documents in the first 20 and 50 returned results; this means HS must show the ability:

(i) not to omit relevant documents identified by the other methodologies when intersecting the two sets (high recall)
(ii) to rank the relevant documents in the first 20 and 50 respectively (high precision)

Query	POS	Keyword 20			Ontology 20			Hybrid 20 General		
		COR	ACT	EXP	COR	ACT	EXP	COR	ACT	EXP
Q1	84	16	20	20	20	20	20	20	20	20
Q2	22	16	20	20	0	0	20	16	20	20
Q3	25	1	20	20	11	20	20	11	20	20
Q4	63	19	20	20	19	20	20	19	20	20
Q5	27	9	20	20	12	20	20	12	20	20
Q6	5	4	8	5	0	0	5	4	8	5
Q7	7	6	6	7	0	0	7	6	6	7
Q8	1	1	1	1	0	0	1	1	1	1
Q9	5	3	3	5	0	0	5	5	5	5
Q10	83	12	20	20	0	0	20	20	20	20
Q11	2	1	1	2	0	0	2	1	1	2
Q12	3	3	3	3	0	0	3	3	3	3
Q13	7	6	6	7	0	0	7	6	6	7
Q14	145	19	20	20	19	20	20	20	20	20
Q15	40	8	20	20	0	0	20	20	20	20
Q16	11	1	16	11	11	11	11	11	11	11
Q17	13	3	20	13	0	0	13	4	4	13
Q18	7	1	4	7	0	0	7	4	20	7
Q19	25	10	17	20	0	0	20	11	11	20
Q20	53	3	20	20	20	20	20	20	20	20
Q21	37	18	20	20	0	0	20	20	20	20
TOTAL	665	160	285	281	112	131	281	234	276	281
		PREC	REC	F-MEAS	PREC	REC	F-MEAS	PREC	REC	F-MEAS
		0.56	0.57	0.57	0.85	0.40	0.54	0.85	0.83	0.84

Fig. 3. Comparative Evaluation of KS, pure semantic search, and HS on 20 queries. POS are the possible correct answers. COR is the number of correct answers returned by the system. ACT is the number of total answers provided by the system. EXP=min(POS, 20) and is the number of correct answers expected, given the limitation to 20 hits.

Moreover, consider that HS is defined as ranked intersection of results from the other two methodologies; therefore it cannot discover more documents than the other two methods. So its recall is a direct function of the recall of the other two modalities.

The experiments showed that semantic search has very high precision, but the lowest recall in identifying relevant documents in the first 20 returned results (Figure. 3).

This is because the metadata did not cover completely 6 of the topics. Keyword-based search has the lowest precision and fair recall in the same task. Hybrid Search reports very high precision (same as OS, +51% with respect to KS), and the highest recall (+46% with respect to keywords and +109% with respect to ontology-based search). (weighted harmonic) F-Measure is +49% with respect to keywords and +55% with respect to ontology-based. In conclusion, in our experiment HS outperforms the other methods in ranking relevant documents within the first 20 results. Experimental results for the first 50 returned documents are largely equivalent.

5.2 In-Vivo Evaluation

A user evaluation was carried on with 32 users (recruited from a number of departments of Rolls-Royce plc) that individually tested the system. The goal was to evaluate K-Search as a means for both document retrieval and knowledge retrieval. The individual sessions lasted an average of 90 minutes. After a short introduction to the system participants were asked to carry out an assisted training task to familiarise with the features of K-Search and the idea of HS. Then they were asked to carry out a second task without assistance. Finally they were asked to propose and carry out a

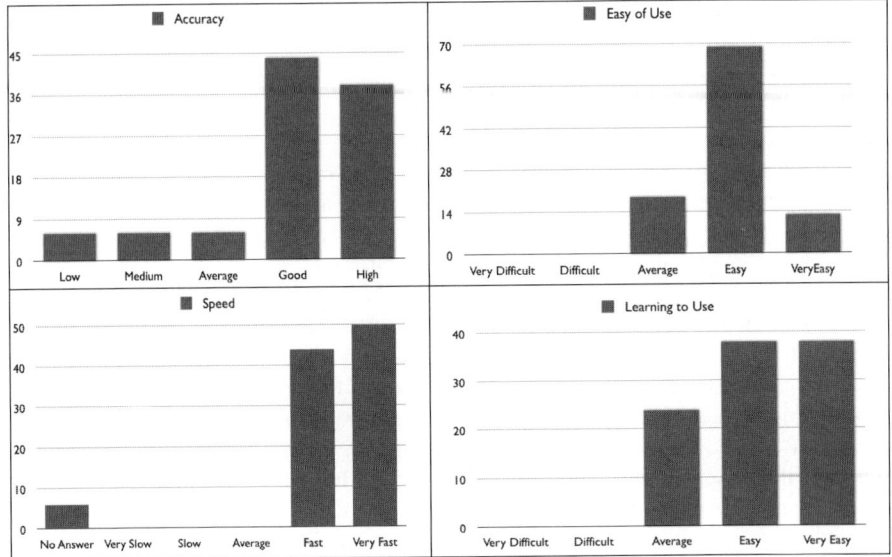

Fig. 4. Results of evaluation of K-Search by 32 users (values are in %)

task that reflected their work experience and interests. A user satisfaction questionnaire was filled in at the end of the test; a short interview on the experience closed the session.

The data collected allows assessing the validity of the HS paradigm as well as the usability of K-Search (Figure 4):

- **Use of HS:** all users appeared to have grasped the concept of HS. Users adopted different strategies: some started querying using keywords and added conditions on metadata in a second iteration; others instead composed conditions on metadata and keywords in a single search; others used metadata search initially and added keywords later to refine the task. This means that different user's searching strategies can be accommodated within the framework.
- **Learning:** 75% of users found easy or very easy to learn to use the system, 25% found it average.
- **System accuracy** (system reliability in retrieving relevant documents): 82% of the users judged K-Search reliable or highly reliable; although this could seem a feature of the system rather than of HS, in our view the comment refers to the fact that with HS the searches were effective.
- **Searching experience:** 82% of users found K-Search easy or very easy to use; the ease of use was often commented about in the interviews;
- **System Speed:** the system was judged fast or very fast in executing the queries allowing a quick task completion by 98% of users.

6 Comparison with the State of the Art

Most of the methodologies proposed for semantic search [1, 2, 3, 4, 9] consider accessing documents using metadata only and do not consider the cases where metadata is unavailable. However, the HS idea can in principle be implemented with all the main types of semantic search described in literature. In [1] Uren et al classify the approaches according to: keywords-based approaches, view-based approaches, natural language approaches and form-based approaches. **Keyword-based approaches** [3, 16] are based on the interpretation of keywords according to an underlying ontology. They require translating all the keywords in order to perform the query. These methods could implement HS by replacing keywords in the query with concepts in the ontology when possible while leaving the rest for pure keyword-based searching. A **View-based approach** [17] is based on querying by building visual graphs exploring the ontology. This is quite effective and appealing method to query; however as [2] noted, experimentally this is one of the least preferred methods for querying. These approaches could easily support HS by just adding a new arc labelled e.g. *document-contains* that sends the query to the indexer rather than to the knowledge base engine. A **natural language approach** [18] addresses querying the knowledge base using natural language. A parser and a semantic analyser interpret the query and transform it into formal queries to the knowledge base. These methods are quite appealing to users, but generally suffer from limitations in the expressiveness of the underlying supported language [2]. They could implement the approach quite naturally by recognising expressions like ("and the document contains..."). Finally, we have seen HS can be implemented in a **form-based approach.** The model could be easily built into other models of this type presented in literature [19]. We have chosen to implement a form-based approach in K-Search because user analysis for the use cases at hand showed that this was the way of interaction preferred by our users.

Concerning other hybrid models, Rocha et al [14] presented a hybrid approach where users input a set of keywords which are sent to a search engine. The results of the search engine are re-ranked using semantic information associated to fields in the ontology. For example, they use the (generally long) provenance text of some annotations to decide that some documents are semantically more relevant than others. They have a spreading mechanism to reach also concepts not explicitly mentioned in the document. This method is similar in spirit to the keyword-based semantic searches such as [3] but it allows retrieval of the cases where there is no metadata available. So it centres on some of our initial objectives. However, the method is equivalent to the use of our keyword searching plus keyword-in-context searching. There is no way to address unique concepts and relations directly as in our model. Also, the keywords-first approach does not solve the issue of synonymity mentioned for keyword-based searches.

KIM [9] provides keyword-based search and ontology-based search as alternative options, i.e. a query is either based on keywords or on metadata but it does not enable mixing them. This is quite reductive with respect to the full HS mechanism.

LKMS [5] enables integration of keyword-based search and ontology-based search, but the actual functionality, the way the combination is performed, the expressive power of the formalism used and a number of details are unclear in the literature. It appears that their annotation is limited to named entities and that their form of HS reduces to searching for the presence of a concept in a document or in the metadata. They do not seem to provide any facility for Boolean queries. Also, even if the system has dozens of real world users, it is unclear how and to what extent they actually use the HS mechanism.

7 Conclusions and Future Work

In this paper we have proposed HS, a mixed approach to searching based on a combination of keyword-based and semantic search. We believe that hybrid search is interesting because it overcomes an implicit limitation of most of the current literature that is that semantic search must rely on metadata only. We have given a formal definition of the method and we have shown experimentally that HS outperforms both keyword-based search and pure semantic search in a real case scenario. We have also shown how the strategy is compatible with most of the current models presented in literature. We believe that this is because:

- Hybrid search performs equally well as pure semantic search when metadata is fully available for a specific query;
- When metadata does not cover the whole information need, HS reaches higher recall than pure semantic search via the use of keywords for sections not covered by IE. Recall is boosted with limited loss in precision;
- HS outperforms keyword-based search in terms of both precision and recall. Higher precision is obtained by the use of metadata when available. Higher recall is obtained thanks to better ranking capabilities due to the use of metadata.
- In cases where the metadata is unavailable, HS is equivalent to keyword-based search.

HS has been implemented in a working system, K-Search, and two real world applications have been developed for Rolls-Royce plc. Such applications are currently deployed for real users to (i) access event reports and (ii) retrieve document and knowledge about requests of product technical variances. User studies carried out before the launch of the applications have shown appreciation for the system and the hybrid approach in general. More applications are being planned. A University spin-out company has been created to exploit the HS approach and its applications.

Future work will clarify some outstanding issues. The major issue concerns the use of IE in tasks where it does not perform at a very high standard. In those cases, the findings could change; because it could be no longer true that semantic search provides high precision. All the findings above are based on this important aspect. With lower precision, the strategy of designing hybrid search as applying semantic search when possible and resorting to keyword for the uncovered parts could actually prove to be not the most effective strategy. Experiments have to be carried out to understand the consequences of reduced precision and recall in the annotation process.

Acknowledgments. This work was supported by IPAS, a project jointly funded by the UK DTI (Ref. TP/2/IC/6/I/10292) and Rolls-Royce plc and by X-Media (www.x-media-project.org), an Integrated Project on large scale knowledge management across media, funded by the European Commission under the IST programme, (IST-FP6-026978). Thanks to Colin

Cadas (Rolls-Royce) for the constant support in the past two years. Thanks to all the users for their very positive attitude and the helpful feedback.

References

1. Uren, V., Lei, Y., Lopez, V., Liu, H., Motta, E., Giordanino, M.: The usability of semantic search tools: a review, Knowledge Engineering Review (in press)
2. Kaufmann, E., Bernstein, A.: How Useful are Natural Language Interfaces to the Semantic Web for Casual End-users? In: Proceedings of the 6th International Semantic Web Conference and the 2nd Asian Semantic Web Conference, Busan, Korea (November 2007)
3. Lei, Y., Uren, V., Motta, E.: SemSearch: A Search Engine for the Semantic Web. In: Staab, S., Svátek, V. (eds.) EKAW 2006. LNCS (LNAI), vol. 4248, Springer, Heidelberg (2006)
4. Guha, R., McCool, R., Miller, E.: Semantic Search. In: 12th International Conference on World Wide Web (2003)
5. Gilardoni, L., Biasuzzi, C., Ferraro, M., Fonti, R., Slavazza, P.: LKMS – A Legal Knowledge Management System exploiting Semantic Web technologies. In: Proceedings of the 4th International Conference on the Semantic Web (ISWC), Galway (November 2005)
6. Chakravarthy, A., Lanfranchi, V., Ciravegna, F.: Cross-media Document Annotation and Enrichment. In: Proceedings of the 1st Semantic Authoring and Annotation Workshop, 5th International Semantic Web Conference (ISWC 2006), Athens, GA, USA (2006)
7. Ireson, N., Ciravegna, F., Califf, M.E., Freitag, D., Kushmerick, N., Lavelli, A.: Evaluating Machine Learning for Information Extraction. In: Proceedings of the 22nd International Conference on Machine Learning (ICML 2005), Bonn, Germany (2005)
8. Kiryakov, A., Popov, B., Terziev, I., Manov, D., Ognyanoff, D.: Semantic annotation, indexing, and retrieval. Journal of Web Semantics 2(1), 49–79
9. Shneiderman, B.: Designing the User Interface, 3rd edn. Addison-Wesley, Reading (1997)
10. Dzbor, M., Domingue, J.B., Motta, E.: Magpie - towards a semantic web browser. In: Fensel, D., Sycara, K.P., Mylopoulos, J. (eds.) ISWC 2003. LNCS, vol. 2870, Springer, Heidelberg (2003)
11. Lanfranchi, V., Ciravegna, F., Petrelli, D.: Semantic Web-based Document: Editing and Browsing in AktiveDoc. In: Proceedings of the 2nd European Semantic Web Conference, Heraklion, Greece (2005)
12. Rocha, R., Schwabe, D., Poggi de Aragão, M.: A Hybrid Approach for Searching in the Semantic Web. In: The 2004 International World Wide Web Conference, New York, May 17-22 (2004)
13. Iria, J., Ciravegna, F.: A Methodology and Tool for Representing Language Resources for Information Extraction. In: Proc. of LREC 2006, Genoa, Italy (May 2006)
14. Tran, T., Cimiano, P., Rudolph, R., Studer, R.: Ontology-based Interpretation of Keywords for Semantic Search. In: Proceedings of the 6th International Semantic Web Conference and the 2nd Asian Semantic Web Conference, Busan, Korea (November 2007)
15. Catarci, T., Di Mascio, T., Franconi, E., Santucci, G., Tessaris, S.: An Ontology Based Visual Tool for Query Formulation Support. In: 16th European Conference on Artificial Intelligence (ECAI 2004), Valencia, Spain (2004)
16. Kaufmann, E., Bernstein, A., Zumstein, R.: Querix: A natural language interface to query ontologies based on clarification dialogs. In: 5th ISWC, Athens, GA, pp. 980–981 (2006)
17. Corby, O., Dieng-Kuntz, R., Faron-Zucker, C., Gandon, F.: Searching the Semantic Web: Approximate Query Processing Based on Ontologies. IEEE Intelligent Systems 21(1) (2006)

Combining Fact and Document Retrieval with Spreading Activation for Semantic Desktop Search

Kinga Schumacher, Michael Sintek, and Leo Sauermann

Knowledge Management Department
German Research Center for Artificial Intelligence (DFKI) GmbH,
Kaiserslautern, Germany
{firstname.surname}@dfki.de

Abstract. The Semantic Desktop is a means to support users in Personal Information Management (PIM). It provides an excellent test bed for Semantic Web technology: resources (*e. g.*, persons, projects, messages, documents) are distributed amongst multiple systems, ontologies are used to link and annotate them. Finding information is a core element in PIM. For the end user, the search interface has to be intuitive to use, natural language queries provide a simple mean to express requests. State of the art semantic search engines focus on fact retrieval or on semantic document retrieval. We combine both approaches to search the Semantic Desktop exploiting all available information. Our semantic search engine, built on *semantic teleporting* and *spreading activation*, is able to answer natural language queries with facts, *e. g.*, a specific phone number, and/or relevant documents. We evaluated our approach on ESWC 2007 data in comparison with Google site search.

1 Introduction

The Semantic Desktop [20,7] is a means for personal knowledge management. It transfers the Semantic Web to desktop computers by consistent application of Semantic Web standards like the Resource Description Framework (RDF) and RDF Schema (RDFS). Documents, e-mails, contacts are identified by URIs, across application borders. The user is able to annotate, classify, and relate these resources, expressing his view in a *Personal Information Model* (PIMO) [21]. On a full-featured Semantic Desktop many data sources are integrated and searchable: a personal categorization system of topics, projects, contacts, *etc.* is established. The text and metadata of all documents is indexed and categorized. Together with the collected facts about people, organizations, events and processes, a critical amount of information is available to the user.

To use all information, we propose to use a Semantic Desktop search engine which is able to find structured information, to focus on unstructured information, and combine them providing a comprehensive search solution for the user.

To achieve this goal, the search engine automatically explores the knowledge base to find facts which answer the query and accomplish enhanced document retrieval embracing the found facts including the metadata of the documents. The engine should also facilitate not only the searching for documents but also *semantic teleporting*. Teleporting is a search strategy where the user tries to jump directly to the information target, *e. g.*, the query 'phone number of the DFKI KM-Group secretary' delivers the document which contains the wanted phone number [23]. Semantic teleporting does not deliver the document which contains the wanted phone number but the phone number itself. Another basic point is to enable *natural language queries* to keep knowledge overhead of special query languages away from the user. Free-text queries are usually dealt with NLP technology which, among other things, has to resolve syntactic ambiguity, *i. e.*, words with multiple meanings and structural ambiguity[1], *i. e.*, the ambiguity of the underlying structure of complex expressions [10]. For this reason we focus on semantic search approaches based on natural language queries which support the resolution of ambiguity.

Recently, several semantic search approaches with diverse application areas have been published. There are two main research thrusts: semantic search engines to retrieve documents which are enriched with semantics (*semantic document retrieval*) and engines which are deployed to search within ontologies (*fact retrieval*). Semantic document retrieval augments traditional keyword search with semantic techniques. Such search engines often use *thesauri* for query expansion and/or apply *graph traversal* algorithms to the available ontology for generalization, specification of the query terms, or to match predefined categories [10,15]. Fact retrieval approaches apply three kinds of core search techniques: *reasoning*, *triple based* (also referred to as statements based), *i. e.*, structural interpretation of the query guided by semantic relations, and *graph traversal* [10].

According to the requirements of searching the Semantic Desktop, our engine combines fact retrieval with semantic document retrieval using a triple-based algorithm and graph traversal. We selected these two approaches for the following reasons: Triple-based search provides the resolving of syntactic and structural ambiguity since the existing triples can constrict the possibilities of the interpretation of a concrete query (see Sect. 3.1). The reason for choosing a graph traversal algorithm, especially *spreading activation* [6], is that this approach enables an effective combination of fact retrieval and document retrieval (see Sect. 3.3). The goal of our approach is to provide a simple to use, yet powerful search functionality to users of the Semantic Desktop, through which all information from a PIMO can be retrieved, *i. e.*, both facts like a special phone number and relevant documents. We supports queries like 'literature by Sintek', 'semantic search papers', 'persons interested in document retrieval', 'phone number of Nepomuk's project manager', 'abstract and authors of papers about Semantic Desktop applications', 'when and where is the welcome reception of the ESWC 2007', etc..

---

[1] For example, the sentence 'I saw the man with the telescope' has two underlying structures; it is not clear who is using the telescope [11].

The rest of this paper is organized as follows. The next section gives a state of the art overview on semantic search. Section 3 explains our approach, where Sect. 3.1 describes the fact retrieval, *i. e.*, the *semantic teleporting*, Sect. 3.2 the semantic document retrieval and Sect. 3.3 the way of their combination. The experimental evaluation including method, data, and results is dealt with in Sect. 4. Section 5 concludes the paper.

2   State of the Art

In recent years, several semantic search engines have been developed. There are some main publications which give a good overview on this research field.

Hildebrand *et al.* consider in [10] 35 existing systems and examine them according to the three main steps of the search process, *i. e.*, query construction, the core search process, and the representation of the results. They identify the general approaches for the core search process. C. Mangold focuses in [16] on semantic document retrieval. He introduces a classification scheme and compares 22 semantic document retrieval engines by means of the defined criteria.

Lei, Uren and Motta investigate how current semantic search approaches address user support without making restrictions on considered approaches [14]. This work points out the problem of knowledge overhead and the lack of support for answering complex queries. Search engines which allow the user to specify the query by choosing ontologies, classes, properties, values, and query language front-end engines require that the user possesses knowledge about the back-end knowledge base or copes with a special query language. In contrast, keyword-based semantic search is user-friendly but breeds the problem of supporting complex queries since allowing free-text queries causes syntactic and structural ambiguity [14,10].

Given that we aspire to provide natural language queries, we focus upon semantic search approaches which enable to resolve these ambiguities. Syntactic ambiguity can be solved by pre-query disambiguation, *e. g.*, in Squiggle [2], through user interaction, *e. g.*, MIT's Semantic Desktop Haystack [13]. A further possibility is to include the user context, *e. g.*, TAP [9]. In terms of resolving structural ambiguity, one such approach is a triple-based algorithm which defines and combines query templates to translate the free-text query into a formal query, introduced in [14]. Another triple-based approach uses keyword search to identify concepts of the ontology and forms RDF queries with one or two variables based on matched properties and non-properties, *i. e.*, subjects and objects of a triple [8].

F. Crestani gives in [6] a good introduction on the application of spreading activation techniques in semantic networks in information retrieval. Rocha *et al* applies spreading activation as graph traversal algorithm for semantic document retrieval using weights assigned to links based on certain properties of the ontology [19]. Berger *et al.* [1] use spreading activation and take additionally the relatedness of terms into account.

Current approaches especially to Semantic Desktop Search concentrate on semantic document retrieval. Beagle^{++} [4,12,5] enhances document retrieval on the desktop by exploiting activity-based meta data, *e. g.*, a saved email attachment is annotated with the sender, date, subject, body text and status of the email. The engine executes a keyword search on the document index and on available meta data and ranks the found documents with an enhanced ranking system. The ranks are computed based on ontological relation weights and on ranks computed by considering external sources. Further approach for semantic document retrieval on the Semantic Desktop applies spreading activation to find resources which are sparsely annotated with semantic information [22]. The underlying semantic network (see Sect. 3.2) is initially set up with a path length based semantic similarity measure of concepts.

3 Searching the Semantic Desktop

Figure 1 depicts the architecture of our semantic search approach. The search engine is composed of a fact retrieval engine, a semantic document retrieval engine and a component which controls both in order to exploit all available data of the Semantic Desktop. The engines are linked among each other and they are connected to the knowledge base by the use of an interface which can be used to define and apply views and inferences. Our knowledge base, the PIMO, consists of the ontology and its instances. Native structures, *i. e.*, folders in file system and in email client are mapped to ontological concepts. Files and other information objects, *i. e.*, text documents, emails, address book entries are mapped to instances of the ontology. Documents also refer to the knowledge base, they are semantically tagged.

3.1 Fact Retrieval Approach

Our fact retrieval approach is based on the triple-based search algorithm by Goldschmidt and Krishnamoorthy [8]. Their semantic search engine processes natural language queries. Syntactic matching is done by string matching against the labels (`rdfs:label`), comments (`rdfs:comment`), and literals (`rdfs:literal`) in the knowledge base. Results are the URIs of found properties and non-properties (subjects and objects of a triple), weighted according to their frequency. The semantic matching is carried out by generating RDF queries as combinations of potential properties (p_i) and non-properties (n_j) with one or two variables, *e. g.*, (($n_1\ p_1$?), (? p_1 ?), (? $p_1\ n_2$)). The process of creating and applying queries is repeated for each new resource detected but limited to only two hops in the knowledge base in order to avoid flooding with irrelevant inferences [8].

We adopted and extended this basic idea in order to enhance results by improving the syntactic matching and by providing more hops in the knowledge base conforming with the query. The *syntactic matching* of the query against the knowledge base step enacts, after removing stop words, a keyword search over the textual content of the knowledge base, *i. e.*, over all kinds of labels,

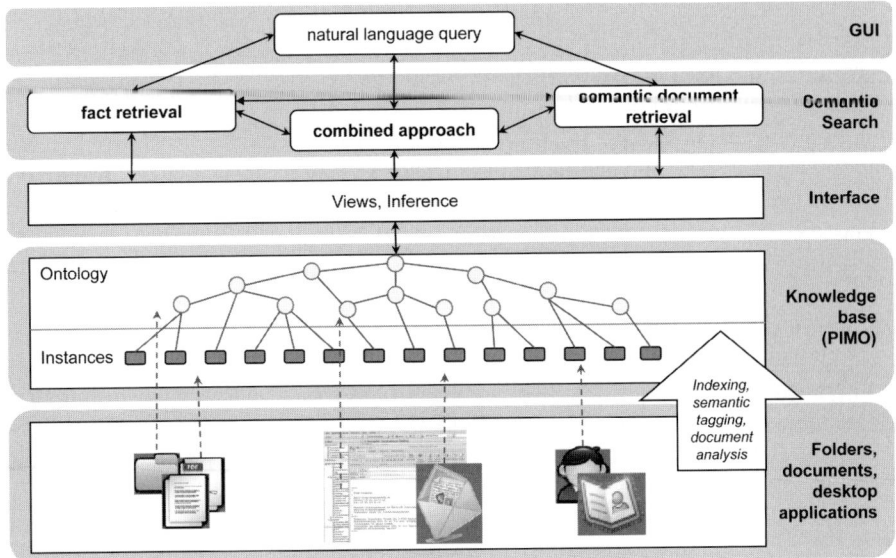

Fig. 1. Architecture

literals, local names of URIs and comments if available. Our knowledge base supports the search engine with synonyms/alternative names of the knowledge base elements (classes, properties, and instances). We use elements of the SKOS[2] formal language to define them, `<skos:prefLabel>` contains the labels which are used for visualizing and textual output, `<skos:altLabel>` stands for other alternative designators. Thus we perform query rewriting by augmentation [16], i.e., we derive further terms from the ontological context of the query term. However, this part of the search engine focuses on the knowledge base in order to find the *perfect* answer and it is optimized for the Semantic Desktop where in addition to the use of thesauri the user is enabled to assign alternative names of ontological elements himself. Therefore, we only use the SKOS labels of the matched elements for this step, i.e., without an explicit generalizing and specializing through hypernyms, meronyms, or adjacent concepts. This feature is customizable according to the application area. We use the n-gram method to compute the syntactic similarity of two strings[3]. The matches are also weighted with the n-gram value of the query term and label, literal or comment using

---

[2] Simple Knowledge Organisation Systems, http://www.w3.org/2004/02/skos/

[3] The N-gram method divides a sequence in subsequences of n items, e.g., a string in subsequences of n characters. The similarity of two terms can then be determined by mapping the words to vectors containing the number of occurrences of several n-grams and computing the distance/similarity of them. More different ways are used for the n-gram decomposition and the computation of the similarity of two sequences. For example, the term 'basic' can be decomposed in the 2-grams ('ba', 'as', 'si', 'ic') and the Dice distance can be used as similarity measure [17]. We use 2-grams combined with 3-grams.

predefined thresholds for the selection of potential elements. Additionally we support phrase matching by determining possible phrases and comparing their n-gram weights with the weights of the single terms. For further steps, we differentiate between matched classes, properties and instances for each query term.

The *semantic matching* is simple in the case of one query term. When instances are matched, the engine returns them. For matched properties it delivers the associated triples of instances, in case of classes their instances. When the query is composed of multiple query terms we start a search process over three levels in order to detect relevant triples from the instance base.

1st level. At first, we consider all pairs of terms side by side, e. g., if a query is composed of 3 terms we consider the pairs (t_1, t_2) and (t_2, t_3). We determine and execute the possible RDF queries based on the matched ontological elements of the terms and gathering the result bindings. For example, if the term t_1 matches the instance $inst_j$, t_2 matches the property $prop_i$ and the class $class_k$, the possible RDF queries are:

$$(inst_j \; ? \; class_k), \; (? \; prop_i \; inst_j), \; (inst_j \; prop_i \; ?) \tag{1}$$

For matched classes, we consider the class itself to find class-instance relations and if it returns no results with the adjacent terms, we replace the class by its instances. That means, for the example above, if neither the RDF query $(inst_j \; ? \; class_k)$ nor the RDF queries created based on t_2 and t_3 match existing triples, we replace $class_k$ by its instances and apply the possible RDF queries. If at least one of the executed RDF queries based on a term pair (t_a, t_b) results in existing triples from the knowledge base, t_a and t_b are marked as matched.

2nd level. At the second level, the found triples and until now unmatched query terms build the base for generating RDF queries along the same way as in the first level. This step is iterated until all new results and all still unmatched terms have been processed. Example: Assuming that the query $(inst_j \; prop_i \; ?)$ results in the triple $(inst_j \; prop_i \; inst_n)$ and the property $prop_l$ of t_3 is not marked as matched, the query $(inst_n \; prop_l \; ?)$ is executed. Embedded in this process, we discover and handle enumerations of instances, properties or classes, since we iterate over unmatched terms and found triples following the order of the query terms but considering the results of all terms in a set of found triples. This feature enables to answer queries like 'mail address, phone number and affiliation of the NEPOMUK project members.'

3rd level. The third level combines the found triples if possible. This includes as well the identification of connected triples, thus triples which build a coherent subgraph, as the identification of subgraphs which are seen as single results. In our example, we assume that the 1st and the 2nd level results are the triples:

$$(inst_j \; prop_i \; inst_n), \; (inst_n \; prop_l \; inst_m), \; (inst_j \; \texttt{rdf:type} \; class_k),$$

$$(inst_q \; prop_i \; inst_r), \; (inst_r \; prop_l \; inst_s), \; (inst_q \; \texttt{rdf:type} \; inst_r).$$

The first two triples have a joint instance, they belong together. The subject of the 3rd triple $inst_j$ is same as the subject of the 1st triple, i. e., the three triples build a subgraph. The next three triples build a subgraph, too. There is a link between the two subgraphs since both instances $inst_j$ and $inst_q$ have the same class but we identify the both subgraphs as different results, the engine also outputs the two subgraphs.

Figure 2 demonstrates this process by means of the query 'phone number of the KM-Group secretary.' The numbers of edges state the order of executed steps.

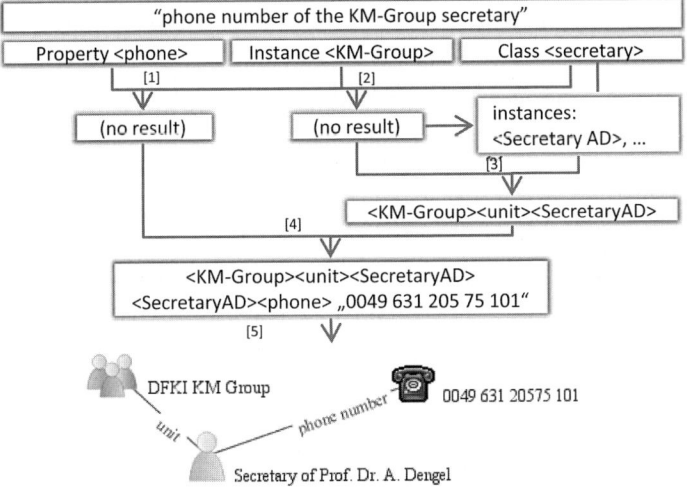

Fig. 2. Example for Semantic Teleporting

The process stops either when all query terms are matched or when there is no possibility to include all terms given that some term pairs do not lead to existing triples. The result consists of a set of instances and a set of triples which constitute the whole answer including semantic relations which explain why the result should be relevant to the query.

The ranking is based on the n-gram values, which result from the syntactic matching. Since our goal is to find the 'perfect answer', the number of involved query terms is used in the ranking function. Starting weight is the computed n-gram value of a query term and textual knowledge base content relating to the matched element. The weight of a matched triple is the sum of the n-gram weights of participating ontological elements. Each added triple increases the weight of the partial result by the appropriate n-gram value. Class-instance information is included by adding the weight of the class and assigning the appropriate query term to the partial result. Finally, the rank of a result is the sum of participating elements' n-gram weights divided by the number of query terms.

Resolving structural ambiguity is supported by the step by step, i.e., triple to triple processing. Since we do not directly transform the natural language query to an RDF query, the triples found step by step lead to possible ways. Syntactic ambiguity is resolved in most cases by the same process. The ontological elements, which lead to existing triples, are higher ranked than elements that cannot be combined with other matched elements.

Compared to [8], our algorithm enables to make as many hops as required by the query and stops when all query terms are matched or there is no possibility to include further triples since no existing ones corresponds to the query. Furthermore, we include information about classes and involve their instances if the class itself does not lead to further results. The recognition of enumerations enables to answer queries which ask for multiple properties of multiple resources.

3.2 Semantic Document Retrieval Approach

The semantic document retrieval approach applies spreading activation to enhance the results by exploiting available domain knowledge. Before starting the description of our solution, it is helpful to explain spreading activation in general.

The basic idea behind spreading activation is to find more relevant information based on retrieved relevant information items (e.g., results of traditional keyword search on the document index) by exploiting associations represented by semantic networks [6]. Spreading activation considers the knowledge base as a network structure, where the ontological concepts are the nodes and the properties are the edges. The edges are usually directed and weighted. The processing technique is an iteration of two main steps: pulses and termination check. A pulse is the incoming activation of a node, which propagates from a node to connected nodes along the edges by computing the outgoing activation, i.e., the activation spreads through the network. The basic formula to compute the input I_j of the node j is:

$$I_j = \sum_i O_i w_{ij}$$

where O_i is the output of node i and w_{ij} is the assigned weight of the edge between node i and node j, thus the strength of the association between node i and node j. Usually, the activation level is the output of the node, determined by an output function of the input value:

$$O_i = f(I_j)$$

Commonly used functions for f are the threshold function, linear function, step function, or sigmoid function. The spreading activation results in the activation level of each node at the termination time. To avoid an uncontrolled flooding of the network, the application of constraints is needed, e.g., distance constraint to limit the number of hops from the initially activated nodes or fan-out constraint to stop spreading at nodes with high connectivity since they often have a broad semantic meaning [6].

Spreading activation requires coupling the documents with the knowledge base, i. e., the metadata of documents refers explicitly to instances of the knowledge base. Furthermore, the configuration of the network by choosing useful relations and corresponding weights (w_{ij}), defining adequate constraints, input and output function, requires domain knowledge.

We design our semantic network as follows: a default weight is assigned to each kind of relation w_{ij} and the graph is represented as a similarity matrix. Rows and columns are dedicated to the ontological concepts, w_{ij} to the weight of the connecting edge. The activation level of a node i is recorded in w_{ii}.

We use a Lucene[4] index for keyword matching on the document corpus. The query is expanded with the alternative names of knowledge base elements which are found by the meta data search. Alternatively, it is possible to index the textual content of the ontology with Lucene. In this case, the query modification is carried out implicitly during the spreading activation. Lucene delivers a ranked set of documents which match one or more terms of the expanded query. These documents are the initial activation points for spreading, inputs are the appropriate ranks. For this simple spreading activation, we extended the basic activation function with the factor α, so-called 'loss of energy' [19]:

$$I_j = \sum_i O_i w_{ij}(1-\alpha)$$

Parameter α can be seen as an attenuation factor. It decreases the activation by factor α by each propagation from one node to a connected one. This feature is combined with an activation constraint, which stops spreading at a node when its activation level does not exceed a defined threshold. A fan-out constraint averts the danger of a too wide spreading through nodes with high connectivity, thus to become noise in results. At the same time, these constraints define the stop condition. It is important to assure that each edge is processed only once (in case of directed edges each direction is considered as one edge). The process stops when no more nodes have an activation level above the defined threshold or the nodes above the threshold have no pending edges. The activation level of nodes determines their ranks.

3.3 Combined Approach

In this section, we describe the combination of the fact retrieval and the document retrieval with spreading activation, shown in Fig. 3.

First, the fact retrieval is processed. If it results in nothing we process the semantic document retrieval (described in Sect. 3.2). If the user attempts teleporting and the engine delivers the 'perfect answer' no further steps are needed. We define the 'perfect answer' for a query with multiple terms/phrases as a set of results with a rank close to 1.0 where each result is composed of triples and the triples of one result match all query terms. In case of one term matching

---
[4] Lucene is a high-performance full text search engine library, cf. http://lucene.apache.org/java/docs/

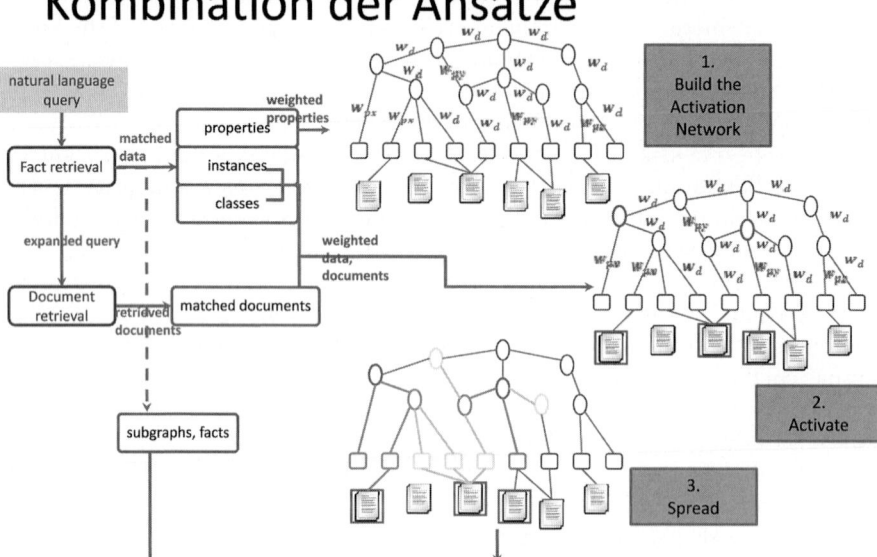

Fig. 3. Overview of the combined approach

a particular instance in the knowledge base, we assume that the spreading activation delivers helpful contextual information to the wanted resource. When no 'perfect answer' is found, we extract the following information from the fact retrieval results:

- instances, which are components of the results,
- alternative names, synonyms of the matched instances,
- matched properties.

The query is expanded with the alternative names of found instances. We chose this way of query expansion since it enables better to specify which documents are relevant. A graph-based query expansion would spread all resources associated with an activated instance which is suited for thesauri. Domain ontologies, which do not mainly describe linguistic relations between terms but support domain-specific relations between resources (instances, documents), require a very specific configuration of the spreading activation process. For example, the query 'spreading activation' on the Lucene index delivers the document 'Application of Spreading Activation Techniques in Information Retrieval' by Crestani. This document is categorized with the topics 'spreading activation', 'information retrieval' and 'semantic networks'. The activation of these topics results in documents which are on 'information retrieval' or on 'semantic networks', but not on 'spreading activation', the precision decreases. To avoid this noise, we should define additional constraints for categorization topics. Query expansion by found facts before spreading is also more effective.

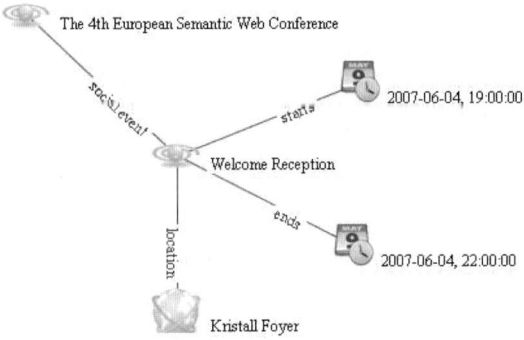

Fig. 4. Result graph of the query 'when and where is the welcome reception of the ESWC 2007'

Fig. 5. Presentation of a result composed of the first name, affiliation and paper of a person

After querying the document index we have all required information to create the network model and to start spreading activation. The network model includes all classes, instances and documents and involves all properties between instances, i. e., all properties which link instances but not instances with literal values. These properties are marked as spreadable in the knowledge base and receive a low initial weight in the network model. We assign to the properties which are delivered by the fact retrieval their n-gram weight. We also assign to each edge the according weight, where some rules are carried out:

- directed relation from node i to node j, which has an inverse relation from node j to node i (e. g., 'hasAuthor' – 'authorOf'): assigning the weight from i to j
- directed relation without an inverse: assign the weight in both directions
- instance-class relation (rdf:type): assign the weight from class to instance

The second point makes sure that relations without an inverse are involved if one of the connected instances is activated. The third condition helps to avoid spreading from one instance to all other instances of a particular type but enables to identify all instances of a class, if a class is initially activated.

The initial activation points are the instances, classes of the fact retrieval results and documents found by keyword search. The strength of activation corresponds to the weights and ranks which the elements have. The spreading activation process complies with the one described in Sect. 3.2 using an activation constraint and a fan-out constraint. It results in a ranked set composed of instances of the knowledge base and related documents.

We visualize the 'perfect answer' as a graph[5] with associated symbols for persons, projects, email, phone, event, location, date *etc.*, in order to enable the user to recognize the result immediately. Furthermore, a graph shows at same time the explanation of the result, see Fig. 4. Other results are shown as items in a list according to their ranks and enriched with the requested information. Properties which are matched by the fact retrieval and which have a literal value are added to the appropriate results. The matched properties which connect two instances are used to group the instances and documents which belongs together. Figure 5 shows an example.

4 Evaluation

As there is no standardized and annotated test data set for Semantic Desktop evaluations yet [3], we used the ESWC 2007 [18] data for our evaluations which has a data structure similar to the PIMO. The ESWC conference knowledge base describes the conference including information about people, talks, papers, posters, conference events, it also involves abstract concepts and references to documents like the Semantic Desktop.

We extended the knowledge base with some synonyms of the ontological elements and instances and created the index of the document set using Lucene. The evaluation is based on 11 queries carried out with our semantic search engine and Google site search using '$<query>$ site::://www.eswc2007.org/'. We computed the precision of the results, shown in Fig. 6. The last column states the results as 'perfect answer' or not, *i.e.*, subgraph(s) which accomplish the conditions of a 'perfect answer' (see Sect. 3.3) found by the fact retrieval engine. Since the semantic search engine applies additional information as ontological elements and instances it is not possible to build a ground truth for this evaluation. For instance, a session is included in the knowledge base but there is no document on the ESWC 2007 web sites about a single session. For this reason, recall, f-measure of the results cannot be determined.

The results demonstrate clearly the power of our combined approach, it returns precise results to extensive queries by exploiting additional information from the knowledge base. The additional results for 'RDF' compared with the Google site results are caused partly by the expansion of the query with 'Resource Description Framework' which enhance the results of the document retrieval and by the found instances through our combined approach, *e.g.*, sessions with topic 'RDF'. In case of simple queries like 'social networks', the precision of the semantic search engine is lower since the engine results documents which

---

[5] The graph visualization was developed by Björn Forcher (DFKI) and is based upon the Java Universal Network/Graph Framework and the Batik SVG Toolkit, http://jung.sourceforge.net/ and http://xmlgraphics.apache.org/batik/

QUERY	Google site search		Semantic desktop search		
	No. of results	precision	No. of results	precision	Perfect answer
RDF	55	1.0	77	1.0	
Social networks	13	1.0	24	0.58	
Sintek	11	1.0	15	0.8	
System descriptions	0	0.0	10	1.0	yes
Organizer of workshop 5	0	0.0	4	1.0	yes
Invited talks	0	0.0	4	1.0	yes
Semantic search demos	13	0.077	2	1.0	yes
Papers from Voelker	2	0.5	2	1.0	yes
Authors and abstracts of reasoning papers	21	1.0	1	1.0	yes
DFKI members at the ESWC 2007	1	0.0	6	1.0	yes
When and where is the welcome reception	2	0.5	1	1.0	yes
Average precision		0.4615		0.9436	

Fig. 6. Table of results

were presented in the same session as the relevant documents. The additionally found documents are related since the sessions are organized based on topics, but not relevant to the query. This result demonstrates the failing of spreading activation: if there are no certain properties matched against the query and/or only documents are matched, the spreading activation propagates to all adjacent nodes with the same intensity. For some queries it is helpful for the user to have this additional information based on the ontological context of found resources. In all further evaluated cases, the semantic search engine delivered precise answers. The query 'authors and abstract of reasoning papers' is, compared to the Google site results, a special case. It results in one document with the keyword 'reasoning' incl. its abstract and authors. Since it is a 'perfect answer', no document retrieval is processed and so all other documents containing the term but not focused on 'reasoning' are not returned. Since the engine is configured to search the semantic desktop and to response as precise answers as possible, we consider this result as actually the 'perfect answer' but we enable the user to inquire more information, thus the results of the complete search process.

The benefit of our semantic search is that it supports the user with precise information (facts) to extensive queries. Furthermore, the user receives useful additional information to the found resources, e. g., not only the title of a paper and the appropriate text snippet, but also the authors and keywords. Furthermore, the ranking which combines the ranks from the fact retrieval and the ranks from document retrieval delivers a good order of the results, since the found facts support more precise information about documents and instances of the knowledge base. The example query 'RDF' returns at the top of the result list papers and sessions with this topic, i. e., with the keyword 'RDF', thus papers which focus on 'RDF'. A document retrieval just delivers documents that contain this term and ranks them based on the number of the query term's frequency.

5 Conclusions and Future Work

This paper describes a semantic search approach which combines fact retrieval and document retrieval with spreading activation in order to exploit all available information on a Semantic Desktop. The developed semantic search engine works with natural language queries and supports not only a semantic document search but also *semantic teleporting*. We evaluated our approach based on the ESWC 2007 knowledge base and documents comparing the results of our engine with the results of Google Site search since there is no standardized test data set for Semantic Desktop or Semantic Desktop search evaluations. The evaluation results demonstrate the power of our combined approach; the engine returns precise results to extensive queries by exploiting facts, metadata of documents from the knowledge base, *i. e.*, it is appropriate to search for information on the Semantic Desktop in a goal-directed way.

For future work, we plan to extend our approach with machine learning technology in order to learn the network model's edge-weights by exploiting user feedback. We assume that the user is often interested in specific information about an instance or in a specific aspect of a document. Using relevance feedback, both explicitly by the user and implicitly by user observation, allows the adaption of the weights according to the user's needs, *i. e.*, to personalize the search on the Semantic Desktop.

Acknowledgements. Part of this work has been supported by the Rheinland-Pfalz cluster of excellence "Dependable adaptive systems and mathematical modeling" DASMOD, project ADIB,[6] and by the European Union IST fund (Grant FP6-027705, Project NEPOMUK[7]). We also want to thank the various developers of Gnowsis[8], Aperture and Nepomuk.

References

1. Berger, H., Dittenbach, M., Merkl, D.: An adaptive information retrieval system based on associative networks. In: Proc. of the first Asian-Pacific Conference on Conceptual Modelling, pp. 27–36 (2004)
2. Celino, I., Turati, A., Valle, E.D., Cerizza, D.: Squiggle – a semantic search engine at work. In: Proc. of the 4th European Semantic Web Conference (2007)
3. Chernov, S., Serdyukov, P., Chirita, P.-A., Demartini, G., Nejdl, W.: Building a desktop search test-bed. In: Proc. of the 29th European Conference on Information Retrieval (2007)
4. Chirita, P.-A., Costache, S., Nejdl, W., Paiu, R.: Beagle^{++}: Semantically enhanced searching and ranking on the desktop. In: Proc. of the 3rd European Semantic Web Conference, pp. 348–362 (2006)
5. Chirita, P.-A., Gavriloaie, R., Ghita, S., Nejdl, W., Paiu, R.: Activity based metadata for semantic desktop search. In: Proc. of the 2nd European Semantic Web Conference, pp. 439–454 (2005)

---

[6] http://www.dasmod.de/twiki/bin/view/DASMOD/ADIB
[7] http://nepomuk.semanticdesktop.org/
[8] http://www.gnowsis.org/

6. Crestani, F.: Application of spreading activation techniques in information retrieval. Artificial Intelligence Review 11(6), 453–482 (1997)
7. Decker, S., Frank, M.R.: The networked semantic desktop. In: WWW Workshop on Application Design, Development and Implementation Issues in the Semantic Web (2004)
8. Goldschmidt, D.E., Krishnamoorthy, M.: Architecting a search engine for the semantic web. In: Proc. of the AAAI Workshop on Contexts and Ontologies: Theory, Practice and Applications (2005)
9. Guha, R., McCool, R., Miller, E.: Semantic search. In: Proc. of the 12th International Conference on Word Wide Web (2003)
10. Hildebrand, M., Ossenbruggen, J., van Hardman, L.: An analysis of search-based user interaction on the semantic web. Report, CWI, Amsterdam, Holland (2007)
11. Hindle, D., Rooth, M.: Structural ambiguity and lexical relations. Computational Linguistics 19(6), 103–120 (1993)
12. Iofciu, T., Kohlschütter, C., Nejdl, W., Paiu, R.: Keywords and rdf fragments: Integrating metadata and full-text search in beagle++. In: Proc. of Semantic Desktop Workshop at the International Semantic Web Conference, vol. 175 (2005)
13. Karger, D.R., Bakshi, K., Huynh, D., Quan, D., Sinha, V.: Haystack: A customizable general-purpose information management tool for end users of semistructured data. In: Proc. of the 2nd Conference on Innovative Data Systems Research (2005)
14. Lei, Y., Uren, V.S., Motta, E.: Semsearch: A search engine for the semantic web. In: Proc. of the 15th International Conference on Knowledge Engineering and Knowledge Management, pp. 238–245 (2006)
15. Mäkelä, E.: Survey of semantic search research. In: Proc. of the Seminar on Knowledge Management on the Semantic Web (2005)
16. Mangold, C.: A survey and classification of semantic search approaches. International Journal of Metadata, Semantics and Ontologies 2(1), 23–34 (2007)
17. Manning, C.D., Schütze, H.: Foundations of statistical natural language processing. MIT Press, Cambridge (1999)
18. Möller, K., Heath, T., Handschuh, S., Domingue, J.: Recipes for semantic web dog food - the eswc and iswc metadata projects. In: Aberer, K., Choi, K.-S., Noy, N., Allemang, D., Lee, K.-I., Nixon, L., Golbeck, J., Mika, P., Maynard, D., Mizoguchi, R., Schreiber, G., Cudré-Mauroux, P. (eds.) ISWC 2007. LNCS, vol. 4825, pp. 802–815. Springer, Heidelberg (2007)
19. Rocha, C., Schwabe, D., Aragao, M.P.: A hybrid approach for searching in the semantic web. In: Proc. of the 13th International Conference on World Wide Web, pp. 374–383 (2004)
20. Sauermann, L., Bernardi, A., Dengel, A.: Overview and outlook on the semantic desktop. In: Proc. of the Semantic Desktop Workshop at the 4th International Semantic Web Conference, vol. 175 (2005)
21. Sauermann, L., van Elst, L., Dengel, A.: Pimo – a framework for representing personal information models. In: Proc. of the I-SEMANTICS 2007, pp. 270–277 (2007)
22. Scheir, P., Ghidini, C., Lindstaedt, S.N.: Improving search on the semantic desktop using associative retrieval techniques. In: Proc. of the I-Semantics 2007, pp. 415–422 (2007)
23. Teevan, J., Alvarado, C., Ackerman, M.S., Karger, D.R.: The perfect search engine is not enough: a study of orienteering behavior in directed search. In: Proc. of the SIGCHI conference on Human factors in computing systems, pp. 415–422 (2004)

Q2Semantic: A Lightweight Keyword Interface to Semantic Search

Haofen Wang[1], Kang Zhang[1], Qiaoling Liu[1], Thanh Tran[2], and Yong Yu[1]

[1] Department of Computer Science & Engineering
Shanghai Jiao Tong University, Shanghai, 200240, China
{whfcarter,jobo,lql,yyu}@apex.sjtu.edu.cn
[2] Institute AIFB, Universität Karlsruhe, Germany
{dtr}@aifb.uni-karlsruhe.de

Abstract. The increasing amount of data on the Semantic Web offers opportunities for semantic search. However, formal query hinders the casual users in expressing their information need as they might be not familiar with the query's syntax or the underlying ontology. Because keyword interfaces are easier to handle for casual users, many approaches aim to translate keywords to formal queries. However, these approaches yet feature only very basic query ranking and do not scale to large repositories. We tackle the scalability problem by proposing a novel clustered-graph structure that corresponds to only a summary of the original ontology. The so reduced data space is then used in the exploration for the computation of top-k queries. Additionally, we adopt several mechanisms for query ranking, which can consider many factors such as the query length, the relevance of ontology elements w.r.t. the query and the importance of ontology elements. The experimental results performed against our implemented system Q2Semantic show that we achieve good performance on many datasets of different sizes.

1 Introduction

The Semantic Web can be seen as an ever growing web of structured and interlinked data. Examples for large repositories of such data available in RDF are DBpedia[1], TAP[2] and DBLP[3]. A snippet of RDF data contained in TAP is shown in Fig. 1. It describes the entity SVGMobile (a W3CSpecification) in terms of its relations to the other entities and its attribute values.

The increasing availability of this semantic data offers opportunities for semantic search engines, which can support more expressive queries that address complex information needs [1]. However, query interfaces in current semantic search engines [2,3] only support formal queries e.g. SPARQL[4]. For example, when a person wants to find specifications about "SVG" whose author's name is "Capin", he

---
[1] http://dbpedia.org
[2] http://tap.stanford.edu
[3] http://dblp.uni-trier.de/
[4] http://www.w3.org/TR/rdf-sparql-query

```
<rdf:Description rdf:about="SVGMobile">        PREFIX   tap: <http://tap.stanford.edu/tap#>
<rdf:type> W3CSpecification</rdf:type>          SELECT   ?spec
<tap:hasAuthor rdf:resource="Capin,_Tolga"/>    WHERE    {
<rdfs:label  xml:lang="en">Mobile   SVG             ?spec     tap:hasAuthor    ?person.
Profiles: SVG Tiny and SVG Basic</rdfs:label>       ?spec     tap:label        "SVG".
</rdf:Description>                                  ?person   tap:name         "Capin".
                                                }
```

Fig. 1. a) Sample RDF snippet. b) Sample SPARQL query

needs to type in the SPARQL query shown in Fig.1. The user thus needs to learn the complex syntax of the formal query. Moreover, the user also needs to know the underlying schema and the literals expressed in the RDF data.

Keyword interfaces is one solution to this problem. User's are very familiar with these interfaces due to their widespread usage. Compared with formal queries, keyword queries have the following advantages: (1) *Simple Syntax*: they are simply lists of keyword phrases (2) *Open Vocabularies*: the users can use their own words when expressing their information needs. In the above example, the user would have to type in only "Capin" and "SVG".

Since keyword interfaces seem to be suitable for casual users, many studies have been carried out to bridge the gap between keyword queries and formal queries, notably in the information retrieval and database communities [4,5,6,7]. There also exist approaches that specifically deal with keywords interfaces for semantic search engines. The template-based approach discussed in [8] fixes the possible interpretations and thus, cannot always capture the meaning intended by the users. This problem has been tackled recently by [9,10]. In [10], a more generic graph-based approach has been proposed to explore the connections between nodes that correspond to keywords in the query. This way, all interpretations that can be derived from the underlying RDF graph can be computed.

However, three main challenges still remain: (1) How to deal with keyword phrases which are expressed in the user's own words which do not appear in the RDF data? (2) How to find the relevant query when keywords are ambiguous (ranking)? For instance, [10] exploits only the query length for ranking. (3) How to return the relevant queries as quickly as possible (scalability)? Both [9,10] require the exploration of a possibly large amount of RDF data, and thus, cannot efficiently deal with large repositories.

In this paper, we address the above challenges as follows:

- (1) We leverage terms extracted from Wikipedia to enrich literals described in the original RDF data. This way, users need not use keywords that exactly match the RDF data.
- (2) We adopt several mechanisms for query ranking, which can consider many relevant factors such as the query length, the relevance of ontology elements w.r.t. the query as well as the importance of ontology elements.
- (3) We propose an exploration algorithm and a novel graph data structure called clustered graph, which represents only a summary of the original RDF data. This improves scalability particularly because the data space relevant

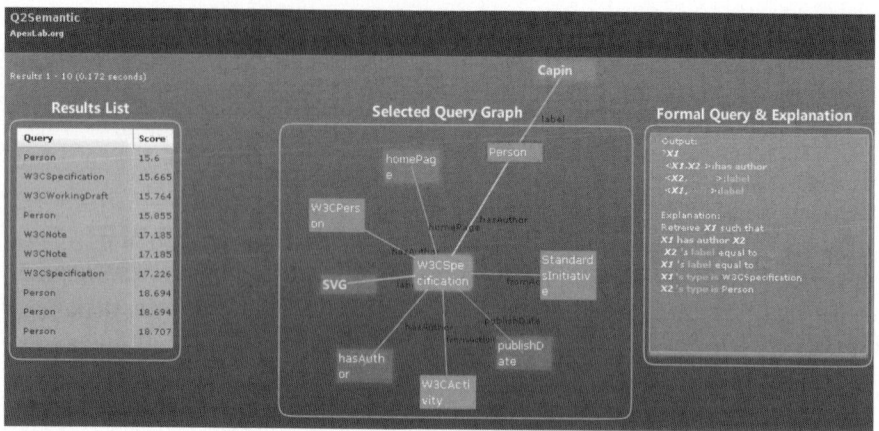

Fig. 2. The result view of Q2Semantic

for exploration becomes smaller in size. Additionally, the exploration algorithm also allows for the construction of the top-k queries, which can help to terminate the interpretation process more quickly.

Also, we have implemented a keyword interface called Q2Semantic to evaluate our approach. The experiments performed on several large datasets show that our solution achieves high effectiveness and efficiency.

The rest of the paper is organized as follows. We will start in section 2 with an overview of Q2Semantic. Section 3 shows how the underlying data models are preprocessed. Section 4 elaborates on how these models are used in the main steps involved in the translation process. Section 5 presents several mechanisms for query ranking. The experimental results are given in section 6. Section 7 contains information on related work. Finally, we conclude the paper with discussions of current limitations and future work in section 8.

2 Q2Semantic

2.1 Feature Overview of Q2Semantic

Q2Semantic is equipped with a keywords-based search interface. In order to facilitate usage, this interface supports auto-completion. This feature exploits the underlying RDF literals enriched with Wiki terms to assist the user in typing keywords. This is extended to "phrase completion" such that when the first keyword has been entered, Q2Semantic will automatically generate a list of phrases containing these keywords from which the user can choose from.

After submitting the keyword query, the user sees the results as shown in the screenshot of our AJAX interface in Fig. 2 (corresponds to our example query "Capin" and "SVG"). On the left, the query results are listed in an ascending order according to the ranking scores of their corresponding queries. For the

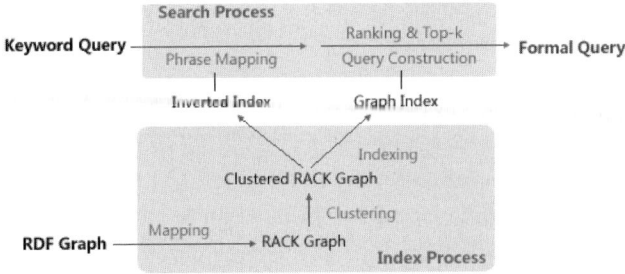

Fig. 3. Workflow of Q2Semantic.

selected result, the corresponding formal query and its natural language explanation are presented on the right. In the middle, the data space that is explored to compute the queries is visualized for the user to understand and explore the queries. For the selected query, the relevant path in this data space is highlighted (in yellow and green). The user explores the data by double clicking on a node to see (further) neighbors. These and other features such as query refinement can be tested at http://q2semantic.apexlab.org/UI.html.

2.2 Query Translation in Q2Semantic

Q2Semantic supports the translation of keyword queries to formal queries. In particular, a keyword query K is composed of keyword phrases $\{k_1, k_2, \cdots, k_n\}$. Each phrase k_i has correspondence (i.e. can be mapped) to literals contained in the underlying RDF graph. A formal query F can be represented as a tree of the form $\langle r, \{p_1, p_2, \cdots, p_n\}\rangle$, where r is the root node of F and p_i is a path in F, which starts from r and ends at leaf nodes that correspond to k_i. The root node of F represents the target variable of the query. So basically, we restrict our definition of formal queries to a particular type of tree-shaped conjunctive queries [11] where the leaf nodes correspond to keywords entered by the user. In our example, K includes k_1 = "Capin" and k_2 = "SVG", and $F = \langle r, \{p_1, p_2\}\rangle$, where r = W3CSpecification, $p_1 = \langle x1, label, SVG \rangle$ and $p_2 = \langle x1, hasAuthor, x2, name, Capin \rangle$. Since SPARQL is essentially, conjunctive query plus additional features, our formal query can be directly rewritten as triple patterns to obtain a SPARQL query like the one presented in section 1.

The translation process is illustrated in Fig. 3, which includes two main steps: (1) *Phrase Mapping*: Retrieve terms stored in an inverted index using the keyword phrases entered by the user (2) *Query Construction and Ranking*: Search the clustered graph to construct potential formal queries and assign costs to them. Meanwhile, top-k queries are returned based on the costs. Note that these online activities are performed on the inverted and the graph index. There is more pre-processing required to build these two data structures, including mapping, clustering and indexing. We will continue with a detailed elaboration on these pre-processing steps and then, discuss the online activities required to translate the keywords.

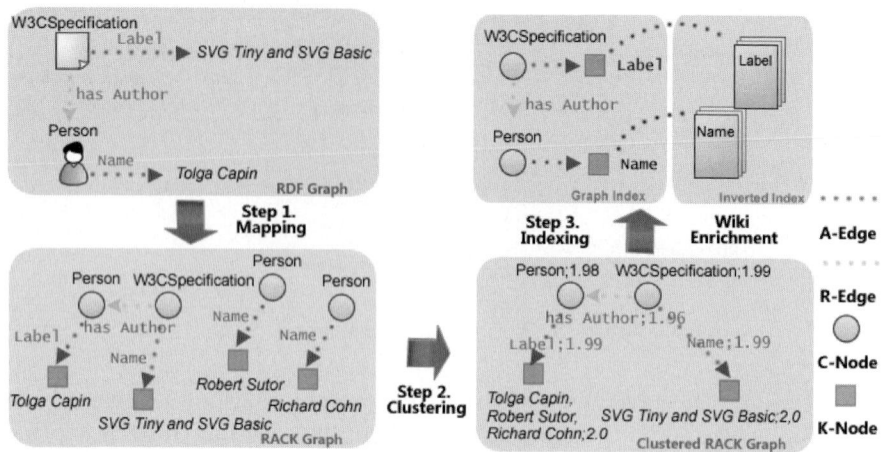

Fig. 4. Index process

3 Data Pre-processing in Q2Semantic

3.1 Graph Construction Via Mapping and Clustering

Graph exploration as done in other approaches is expensive due to the large size of the A-Box (RDF graph) [9,10]. As observed in [12], similar instances always share similar attributes and relations. Adopting this idea, we propose a clustered RACK graph which corresponds to a summary of the original RDF graph. This reduction in size enables faster query construction and ranking especially for RDF graph containing a large number of instances. In the following, we will describe our notion of RACK graph and the rules for clustering.

A RACK graph consists of the four elements R-Edge, A-Edge, C-Node and K-Node, obtained from the original RDF graph through the following mappings:

- Every instance of the RDF graph is mapped to a *C-Node* labelled by the concept name that the instance belongs to.
- Every attribute value is mapped to a *K-Node* labelled by the value literal.
- Every relation is mapped to a *R-Edge* that is labelled by the relation name and connects two C-Nodes.
- Every attribute is mapped to an *A-Edge* that is labelled by the attribute name and connects a C-Node with a K-Node.

As shown in Fig. 4, the mapping process results in a RACK graph. Note that each instance is mapped to the most special concepts if it belongs to multiple concepts. We also do not consider any axioms (e.g. subsumption between concepts) in the RACK graph as it does not support reasoning capability for query interpretation. A clustered RACK graph can be further obtained by the iterative application of the following four rules.

- Two C-Nodes are clustered to one if they have the same label.

- Two R-Edges are clustered to one if they have the same label and connect the same pair of C-Nodes.
- Two A-Edges are clustered to one if they have the same label and is connected to the same C Node.
- Two K-Nodes are clustered to one if they are connected to the same A-Edge. The resulting node inherits the labels of both these K-Nodes.

For each clustered node, we track and store the number of original nodes that collapsed to it during the clustering. Also, for each clustered edge, we store the number of node pairs that were connected by the original edges collapsed to it. These numbers stored in nodes and edges are used to compute their costs on the basis of cost functions discussed in section 5. The costs are shown in Fig. 4. They will be used later in the construction and ranking of the query.

3.2 Clustered Graph Indexing

The clustered RACK graph computed in the previous step can be stored in a graph index as discussed in [13]. In our current experiments, we directly load the clustered RACK graph model into the memory for fast query construction since it is very small. However, the graph index will be used when the clustered graph is too big to be loaded into memory.

3.3 Phrase Indexing

We make use of an inverted index to store the labels of K-Nodes. This index is used to locate relevant K-Nodes for a given keyword phrase faster. In particular, we create a document for each K-Node and take its labels as the document content. This document is further enriched with terms extracted from Wikipedia.

This enrichment is performed to support keywords that are expressed in the user's own words that do not match the literals of RDF data. In fact, we adopt the idea in [14] to leverage Wikipedia. Namely, for each K-Node label, we search the Wikipedia database to see whether it matches the title of any article. If so, several semantic features of the article as introduced in [15] are added as additional labels of the K-Node. These features include the title, the anchor texts that link to the article, and the titles of other articles that redirect to the article. Therefore, user keywords might be mapped to the actual labels of the K-Nodes or any of these extracted features added to the K-Nodes.

4 Query Interpretation in Q2Semantic

The query interpretation begins with the mapping of user keywords to the labels of K-Nodes in the inverted index. Starting from the matched K-Nodes, an exploration on the clustered graph is performed, which is similar to the single-level search algorithm discussed in [16]. It expands the current nodes to their neighbors iteratively until reaching a common root. In this process, the edge with the lowest cost is selected for traversal. The process terminates until the top-k queries have been found. In the following subsections, we will describe these steps in detail.

4.1 Phrase Mapping

Each keyword phrase k_i in K entered by the user is submitted as a query to the index, resulting in hits that represent the matching K-Nodes. They are returned in a ranked list as $KL_i = \{k\text{-}node_{i1}, k\text{-}node_{i2}, \cdots, k\text{-}node_{im_i}\}$, associated by the retrieval engine with the matching score $S_i = \{s_{i1}, s_{i2}, \cdots, s_{im_i}\}$. Each s_{ij} is used as the dynamic weight of the respective $k\text{-}node_{ij}$ with respect to k_i. For instance, KL_1 contains one K-Node that matches "Capin" while KL_2 contains three K-Nodes matching "SVG", as illustrated in Fig. 5.

4.2 Query Construction

After obtaining these K-Nodes, we construct the potential queries by exploring the clustered RACK graph. The process is as follows: For each keyword phrase, we create a thread. Then we do traversal in these threads until all the threads converge at a same node. This way, the traversal paths correspond to a tree, from which we construct a tree-shaped formal query. In the following, we first define the thread and the expansion operations required to traverse the graph. Then we will present the detailed algorithm.

A *thread* maintains cursors that haven't been expanded yet. A *cursor* is defined on a node, which traces the expansion track in a thread. Each cursor has four fields $(c; n; p; k)$, where c represents the cost for the track, n is the node where the cursor locates in, p is the parent cursor of the current cursor, and k is the keyword phrase corresponding to the thread that the cursor is in. Note that all cursors in the same thread share the same keyword phrase.

Given a thread, a *thread expansion* (*T-Expansion*) selects a cursor in it, executes cursor expansion, and then removes the cursor from it. Given a cursor C_{cur}, a *cursor expansion* (*C-Expansion*) includes a validation step and an exploration step. In the validation, we check whether a new formal query rooted at the node $C_{cur}.n$ has been found. It is accomplished by checking whether cursors in other threads have arrived at this node. In the exploration, new cursors (e.g. C_{new}) are created for all neighbors of the node $C_{cur}.n$ and added to the current thread, i.e. $C_{new}.k = C_{cur}.k$. The current cursor then becomes parent cursor of these new cursors, i.e. $C_{new}.p = C_{cur}$. The costs of the new cursors are calculated

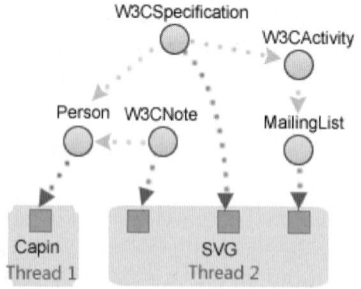
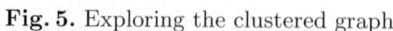
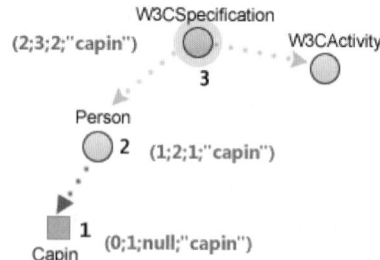

Fig. 5. Exploring the clustered graph

Fig. 6. Example on repeated expansion

Input: $K = \{k_1, k_2, ..., k_n\}$, where k_i hits the K-Nodes
$KL_i = \{k\text{-}node_{i1}, k\text{-}node_{i2}, \cdots, k\text{-}node_{im_i}\}$ with the matching relevance
as $S_i = \{s_{i1}, s_{i2}, \cdots, s_{im_i}\}$;
Output: A: result set, initially $\emptyset$;
Data: τ_{prune}: pruning threshold, initially τ_0;

1 **for** $i \in [1, n]$ **do**
2   t_i = new $Thread()$;
3   **for** $j \in [1, m_i]$ **do**
4        $t_i.add($ new $Cursor(s_{ij}, k\text{-}node_{ij}, NULL, k_i))$;
5   **end**
6 **end**
7 **while** $\exists i \in [1, n]\ :\ t_i.peekCost() \neq \infty$ **do**
8     $j \leftarrow$ pick from $[1, n]$ in a round-robin fashion;
9     $c \leftarrow t_j.popMin()$;
10    $C\text{-}Expansion(c)$; // A and τ_{prune} will be updated here;
11   **if** $t_j.peekCost() > \tau_{prune}$ **then**
12   Output the top k answers in A;
13   **end**
14 **end**

Algorithm 1. Query Interpretation Process

using the formula $C_{new}.c = C_{cur}.c + dist(C_{cur}.n, C_{new}.n)$, where $dist()$ is a distance function between two nodes in the graph. By default, it is the cost of the edge which connects the two nodes.

The sequence of doing T-Expansions has an impact on the speed of query construction. This speed is also influenced by the sequence of C-Expansions performed during the T-Expansions. Considering that, we use the following two strategies when choosing what to expand next: 1) *Intra-Thread Strategy*: In a T-Expansion, we choose the cursor with the lowest cost for the next C-Expansion. 2) *Inter-Thread Strategy*: Within different threads, we choose the thread with the lowest number of expanded cursors for the next T-Expansion in order for a round robin fasion. These two strategies have been proven optimal in the single-level search algorithm [16].

This query construction process is described in Algorithm 1. We first initialize thread t_i for each keyword phrase k_i in K (Line 1), and fill t_i with cursors for the K-Nodes in KL_i (Line 1). Then we do T-Expansions on the threads in a round-robin fashion (Line 1). In each T-Expansion, we do C-Expansion on the cursor which has the lowest cost (Line 1). Note that for each thread, $popMin()$ pops out the cursor with the minimal cost, whereas $peekCost()$ just returns the minimal cost. Line 1 to Line 1 is the optimization for top-k termination, which will be discussed in the next subsections.

As shown in Fig. 5, after the initialization, t_1's cursor locates in the K-Node labelled "Capin", and t_2's cursors point to three K-Nodes. When we expand the cursor in t_1 to the C-Node Person, and assuming cursors in t_2 have already reached this node (e.g. a cursor starts from "SVG", expands through W3CNote and reaches Person, we get a formal query rooted at Person. One path of the query is from Person to the K-Node labelled "Capin", and the other is from Person to the most left K-Node labelled "SVG".

4.3 Optimization for Top-k Termination

In order to find out the top-k queries only, we maintain a pruning threshold called τ_{prune}, which is the current kth minimal cost of the already computed queries. τ_{prune} will be initialized to τ_0. When we find a valid formal query in C-Expansion, the cost of the query is calculated by the ranking mechanism, which will be discussed in Section 5. For a new formal query to be in a top k position, its cost should be no greater than τ_{prune}. When such a query is found, it will be added to the answer set A and τ_{prune} will be updated accordingly. Since a cursor actually indicates a path in query, if all cursors' costs are larger than τ_{prune}, new queries including these paths will have even larger costs. Therefore, we can stop the query interpretation process and output the top-k formal queries.

4.4 Optimization for Repeated Expansion

We assume that it rarely happens for people to propose a query which contains the same relations several times (e.g. "find Tom's friends' friends' friend, who is Spanish"). Based on this assumption, we adopt a mechanism to avoid redundant exploration of the same elements, which can speed up the construction process. Namely, we add penalty to the cursor whose track contains repeated nodes. This is done by using a different $dist()$ function for C-Expansion, namely

$$dist^*(n_1, n_2) = \begin{cases} P & \text{If } n_2 \text{ has been visited} \\ dist(n_1, n_2) \end{cases} \quad (1)$$

where P is set to a large number as the predefined penalty parameter.

In Fig. 6, there is a cursor on W3CSpecification. Its track is indicated by 1, 2 and 3. Assume that the cost of the current cursor is two, every edge has one as weight, and P is set to five. Then the cost of the new cursor on W3CActivity gets three, while the one on Person gets seven as it has been visited already at 2. This way, repeated expansion on Person is still allowed but with a higher cost.

5 Query Ranking in Q2Semantic

Since the query construction process can result in many queries, i.e. possible interpretations of the keywords, a ranking scheme is required to return the queries that most likely match the user intended meaning. Ranking has been dealt with in other approaches. For ranking ontologies, [17] returns the relevant ontologies based on the matching score of the keywords w.r.t. the ontology elements. It also considers the importance of nodes and edges in the ontology graph as a static score similar to Google's PageRank. For ranking complex relationships, [18,19] employ the length of the relation paths. Besides these approaches for ranking ontology (answers) and relations, work has been done for ranking queries. [10] uses the length of the formal query and [9] considers also the keywords' matching score.

We define three ranking schemes from simple to complex, which adopt ideas from other approaches mentioned above, to extend existing work on ranking queries. They compute the cost for a query. The most complex scheme leverages all the above factors including the query length, the keyword matching score and the importance of nodes and edges.

Path Only: The basic ranking scheme R_1 considers the query length only, which is as follows:

$$R_1 = \sum_{1 \leq i \leq n} (\sum_{e \in p_i} 1) \qquad (2)$$

This formula computes the total length of paths in the formal query, where p_i is a path and e is an edge in p_i. Each p_i represents a connection between the root of the formal query and a matched K-Node. Lower cost queries are preferred over higher cost queries. Since the cost of every edge is defaulted to exactly one, in effect, shorter queries are preferred over longer ones. As discussed in [10], shorter queries tend to capture stronger connections between keyword phrases.

Adding matching relevance: When further considering the matching distance between the user's keyword phrases and the literals in the RDF graph, a ranking scheme R_2 can be defined as

$$R_2 = \sum_{1 \leq i \leq n} (\frac{1}{D_i} \sum_{e \in p_i} 1) \qquad (3)$$

where D_i is the score stored in the p_i's starting K-Node, which has been computed in the phrase mapping. In this case, R_2 prefers shorter queries with higher matching score of keyword phrases w.r.t. K-Nodes labels.

Adding Importance of Edges and Nodes: This ranking scheme assumes that users prefer to find entities with types and relations that are more "important". Ranking scheme R_3 considers also the importance of query elements. In particular, specific cost functions are defined for nodes and edges, which reflect their importance for the RDF graph. R_3 and these cost functions are defined as

$$R_3 = cost_r \sum_{1 \leq i \leq n} (\frac{1}{D_i} \sum_{e \in p_i} cost_e) \qquad (4)$$

$$cost_{node} = 2 - \log_2(\frac{|node|}{N} + 1) \qquad (5)$$

$$cost_{edge} = 2 - \log_2(\frac{|edge|}{M} + 1) \qquad (6)$$

where N is the total number of nodes in the original RACK graph, $|node|$ is the number of original nodes clustered to the node (as discussed for clustering in section 3), M is the total number of edges in the original RACK graph, $|edge|$ is the number of original edges clustered to the edge, and $cost_r$ is the cost function of the query root. The cost functions guarantee the cost of each node and edge to be in the interval (1,2). Since both local frequencies, i.e. the number of original elements clustered to an element, and total number of nodes and edges are incorporated, these function compute the importance of nodes and edges in a manner similar to TF/IDF used in information retrieval. Note that the higher its frequency is, the more important a node is considered to be

Table 1. Table of TAP sample queries

Query	Keywords	Potential information need
Q3	Supergirl	Who is called "supergirl"
Q5	Strip, Las Vegas	What is the well-known "Strip" in Las Vegas
Q9	Web Accessibility Initiative, www-rdf-perllib	Find persons who work for Web Accessibility Initiative and involve in the activity with mailing list "www-rdf-perllib"

because it will obtain a lower cost. As the cost is lower for elements with high importance, they have more positive impact on the rank of the query.

6 Evaluation

6.1 Experiment Setup

As there is yet no standard benchmark for evaluating the performance of translating keyword queries to formal queries, we use TAP, DBLP and LUBM [20] for the experiment. For TAP, we manually construct nine scenarios where the keywords and the corresponding potential information needs are listed in table 1 for three scenarios. The experiments are conducted on a Intel PC with 2.6GHz Pentium processor and 2GB memory. Note that the following presentation will focus on results performed on TAP. The proposed queries and their results for DBLP and LUBM as well as the extended presentation of our experiments can be found in the technical report [21] at http://q2semantic.apexlab.org/Pub/Q2Semantic-TR.pdf.

6.2 Effectiveness Evaluation

For ranking query, precision and recall as applied for information retrieval can not be used directly because only one of the computed query matches the meaning of the keywords intended by the user. Hence, we introduce a new metric called *Target Query Position (TQP)* to evaluate the effectiveness of query ranking. Namely, $TQP = 11 - P_{target}$, where P_{target} means the position of the intended query in the ranked list. Note the higher the rank of the intended query, the higher its TQP score. If the rank of a query is greater than ten, its TQP is simply 0. Thus, the TQP score of a query range from 0 to 10.

Since this metric is sensitive to the query rank, it can be used to evaluate our approach for query construction and the different ranking schemes. For this, We invite twelve graduate students to identify the query from a ranked list computed by Q2Semantic, which corresponds to their interpretations of the given keyword query. For each keyword query, we compute the final TQP score as an average of the scores obtained for each participant. Fig. 7 illustrates results of our experiments performed on TAP using the three different ranking schemes.

Note that the performance of R_1 is relatively good for Q1-Q4. This is because keywords in these queries have little ambiguity, i.e. can be mapped exactly one or two K-Nodes (e.g. "supergirl" in Q3). In these cases, the query length is very effective in ranking the queries. When applying R_2, significant improvements

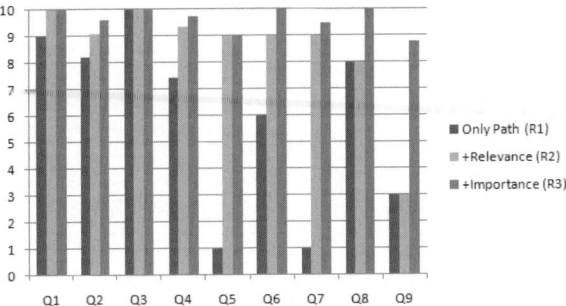

Fig. 7. TQPs of different ranking schemes on TAP

Table 2. RACK graph versus Clustered RACK graph

	R-Edge		A-Edge		C-Node		K-Node	
TAP	41914	158	87796	666	167656	314	87796	666
LUBM(1,0)	41763	43	30230	39	16221	13	30230	39
LUBM(20,0)	1127823	43	815511	39	411815	13	815511	39
LUBM(50,0)	2788382	43	2015672	39	1018501	13	2015672	39
DBLP	5619110	19	12129200	23	1366535	5	12129200	23

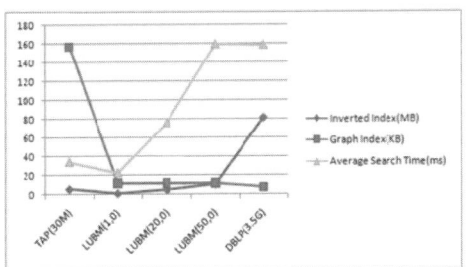

Fig. 8. Index size and search time on different datasets

can be obtained for Q5-Q7. Keyword phrases in these queries are ambiguous, i.e. mapped to many K-Nodes, resulting in a lot of queries having the same path length. As R_2 also considers the matching score of the keyword phrases to K-Nodes, it helps to resolve this ambiguity. For instance, the query containing K-Nodes with highest matching score for "Strip" and "Las Vegas" (Q5) is indeed the one intended by the user. Finally, another improvement is obtained for Q8 and Q9 when using R_3 to consider also the importance of nodes and edges. This improvement comes from the usage of costs for nodes and edges, which guide the traversal and the selection of the root note. Note that elements with higher importance are preferred during expansions. For instance, in Q9, the author is preferred over the book because it has higher importance (lower cost).

In summary, the results show that our approach offers high quality translation of keywords to formal queries, especially when using R_3 as the ranking scheme.

Our technical report also shows that the overall performance on all keyword queries for LUBM is promising and the average TQP reaches 9.125 by using R_3.

6.3 Efficiency Evaluation

Table 2 compares the statistical information of the original RACK graphs and the clustered RACK graphs (Bold numbers). The number of K-Node is the same as that of A-Edge according to the fourth clustering rule. It is observed that the sizes are largely reduced after clustering. That is, the relevant data space to be explored in the query interpretation process is much smaller, which leads faster query construction. This is indicated by the average time in Fig. 8, which also shows the size of the inverted and the graph index for TAP, LUBM(1,0), LUBM(20,0), LUBM(50,0) and DBLP. The average time ranges from 20ms to 160ms on all datasets. Since no evaluation has been carried out to measure performance in previous approaches, we cannot make any comparative analysis. However, the reduction in data space must have a positive effect on performance.

In this regard, we found out that the size of the clustered graph index depends heavily on the schema structure of the original RDF graph. Namely, the simpler the schema (number of T-Box axioms), the smaller the index size. For example, the graph index size of DBLP and LUBM is smaller than that of TAP as TAP contain much more concepts. Also the performance depends on the size of the inverted index. This mainly depends on the number of literals in the ontology.

In summary, the experiments show promising performance. Besides the reduction of the original RDF graph, top-k query answering helps to terminate even more quickly, i.e. avoid the calculation of all possible queries. Our technical report provides more details on the impacts of the ranking mechanism, the top-k parameter, and the penalty parameter on the performance.

7 Related Work

Translating keywords to formal queries is a line of research that has been carried out in both the information retrieval and the database communities. Notably, [4,5,22] support keyword queries over databases while [6,7] specifically tackle XML data by translating keyword queries to XQuery expressions. However, none of them can be directly applied to semantic search on RDF data since the underlying data model is a graph rather than relational or tree-shaped XML data.

[8] represents an attempt that specifically deals with keyword queries in semantic search engines. There, keywords are map to elements of triple patterns of predefined query templates. These templates fix the structure of the resulting queries a priori. However, only some but not all interpretations of the keywords can be captured by such templates. Also, since queries with more than two keywords lead to a combinatorial explosion of different possible interpretations, a large number of templates would be needed. These problems have been tackled recently by [9,10]. In [10], a more generic graph-based approach has been proposed to explore all possible connections between nodes that correspond to

keywords in the query. This way, all interpretations that can be derived from the underlying RDF graph can be computed.

With respect to these recent works [9,10], our approach is distinct in three aspects. Firstly, we enrich RDF data with terms extracted from Wikipedia. Thus, users can also use their own words because keywords can map also to Wikipedia terms. Secondly, we extend the ranking mechanism in [10] to a more general framework for query ranking, which can incorporate many factors besides the query length. Most importantly, query construction has been relied on a large number of A-Box queries that are performed on the original RDF graph. Our approach greatly reduces this space to a summary graph, and thus scales to large repositories. The additional support for top-k queries can further help to terminate the translation even more quickly.

8 Conclusions and Future Work

In this paper, we propose a solution to translate keyword queries to formal queries that can address drawbacks in current approaches. RDF Data is enriched with terms from Wikipedia to support also keywords specified in the user own words. The RDF graph used for exploration is clustered down to a summary graph. Combined with top-k query answering, this increases scalability and efficiency of the translation process. To improve effectiveness, a more general ranking scheme is proposed that considers the query length, the element matching score and the importance of the elements. Evaluation of the implemented system Q2Semantic shows high quality translation of keyword queries processed against datasets of different sizes and domains.

Currently, our approach support keywords that match literals and concepts contained in the RDF data (where concepts are treated as special K-Nodes in the current implementation). We will extend the current query capability to support also keywords in the form of relations and attributes. Another future work is to integrate query interpretation with query answering in a unified graph index as one still need to use the original graph instead of the clustered RACK graph for answering the translated queries from keywords.

References

1. Tran, D.T., Bloehdorn, S., Cimiano, P., Haase, P.: Expressive resource descriptions for ontology-based information retrieval. In: Proceedings of the 1st International Conference on the Theory of Information Retrieval (ICTIR 2007), Budapest, Hungary, October 18- 20, pp. 55–68 (2007)
2. Broekstra, J., Kampman, A., van Harmelen, F.: Sesame: A generic architecture for storing and querying rdf and rdf schema. In: ISWC, pp. 54–68 (2002)
3. Lu, J., Ma, L., Zhang, L., Brunner, J.S., Wang, C., Pan, Y., Yu, Y.: Sor: A practical system for ontology storage, reasoning and search. In: VLDB, pp. 1402–1405 (2007)
4. Hristidis, V., Papakonstantinou, Y.: Discover: Keyword search in relational databases. In: VLDB, pp. 670–681 (2002)
5. Bhalotia, G., Hulgeri, A., Nakhe, C., Chakrabarti, S., Sudarshan, S.: Keyword searching and browsing in databases using banks. In: ICDE, pp. 431–440 (2002)

6. Hristidis, V., Papakonstantinou, Y., Balmin, A.: Keyword proximity search on xml graphs. In: ICDE, pp. 367–378 (2003)
7. Guo, L., Shao, F., Botev, C., Shanmugasundaram, J.: Xrank: Ranked keyword search over xml documents. In: SIGMOD Conference, pp. 16–27 (2003)
8. Lei, Y., Uren, V.S., Motta, E.: Semsearch: A search engine for the semantic web. In: Staab, S., Svátek, V. (eds.) EKAW 2006. LNCS (LNAI), vol. 4248, pp. 238–245. Springer, Heidelberg (2006)
9. Zhou, Q., Wang, C., Xiong, M., Wang, H., Yu, Y.: Spark: Adapting keyword query to semantic search. In: Aberer, K., Choi, K.-S., Noy, N., Allemang, D., Lee, K.-I., Nixon, L., Golbeck, J., Mika, P., Maynard, D., Mizoguchi, R., Schreiber, G., Cudré-Mauroux, P. (eds.) ISWC 2007. LNCS, vol. 4825, pp. 694–707. Springer, Heidelberg (2007)
10. Tran, T., Cimiano, P., Rudolph, S., Studer, R.: Ontology-based interpretation of keywords for semantic search. In: Aberer, K., Choi, K.-S., Noy, N., Allemang, D., Lee, K.-I., Nixon, L., Golbeck, J., Mika, P., Maynard, D., Mizoguchi, R., Schreiber, G., Cudré-Mauroux, P. (eds.) ISWC 2007. LNCS, vol. 4825, pp. 523–536. Springer, Heidelberg (2007)
11. Horrocks, I., Tessaris, S.: Querying the semantic web: a formal approach. In: Horrocks, I., Hendler, J. (eds.) ISWC 2002. LNCS, vol. 2342, pp. 177–191. Springer, Heidelberg (2002)
12. Fokoue, A., Kershenbaum, A., Ma, L., Schonberg, E., Srinivas, K.: The summary abox: Cutting ontologies down to size. In: Cruz, I., Decker, S., Allemang, D., Preist, C., Schwabe, D., Mika, P., Uschold, M., Aroyo, L.M. (eds.) ISWC 2006. LNCS, vol. 4273, pp. 343–356. Springer, Heidelberg (2006)
13. Zhang, L., Liu, Q., Zhang, J., Wang, H., Pan, Y., Yu, Y.: Semplore: An ir approach to scalable hybrid query of semantic web data. In: Aberer, K., Choi, K.-S., Noy, N., Allemang, D., Lee, K.-I., Nixon, L., Golbeck, J., Mika, P., Maynard, D., Mizoguchi, R., Schreiber, G., Cudré-Mauroux, P. (eds.) ISWC 2007. LNCS, vol. 4825, pp. 652–665. Springer, Heidelberg (2007)
14. Milne, D., Witten, I.H., Nichols, D.: A knowledge-based search engine powered by wikipedia. In: Proc. of CIKM (2007)
15. Fu, L., Wang, H., Zhu, H., Zhang, H., Wang, Y., Yu, Y.: Making more wikipedians: Facilitating semantics reuse for wikipedia authoring. In: Aberer, K., Choi, K.-S., Noy, N., Allemang, D., Lee, K.-I., Nixon, L., Golbeck, J., Mika, P., Maynard, D., Mizoguchi, R., Schreiber, G., Cudré-Mauroux, P. (eds.) ISWC 2007. LNCS, vol. 4825, pp. 128–141. Springer, Heidelberg (2007)
16. He, H., Wang, H., Yang, J., Yu, P.S.: Blinks: ranked keyword searches on graphs. In: SIGMOD Conference, pp. 305–316 (2007)
17. Ding, L., Pan, R., Finin, T.W., Joshi, A., Peng, Y., Kolari, P.: Finding and ranking knowledge on the semantic web. In: Gil, Y., Motta, E., Benjamins, V.R., Musen, M.A. (eds.) ISWC 2005. LNCS, vol. 3729, pp. 156–170. Springer, Heidelberg (2005)
18. Anyanwu, K., Maduko, A., Sheth, A.P.: Semrank: ranking complex relationship search results on the semantic web. In: WWW, pp. 117–127 (2005)
19. Lehmann, J., Schüppel, J., Auer, S.: Discovering unknown connections - the dbpedia relationship finder. In: Proc. of the 1st SABRE Conference on Social Semantic Web (CSSW) (2007)
20. Guo, Y., Pan, Z., Heflin, J.: Lubm: A benchmark for owl knowledge base systems. J. Web Sem. 3(2-3), 158–182 (2005)
21. Wang, H., Zhang, K., Liu, Q., Yu, Y.: Q2semantic: Adapting keywords to semantic search. Technical report, APEX Data & Knowledge Management Lab (2007)
22. Balmin, A., Hristidis, V., Papakonstantinou, Y.: Objectrank: Authority-based keyword search in databases. In: VLDB, pp. 564–575 (2004)

Conceptual Situation Spaces for
Semantic Situation-Driven Processes

Stefan Dietze, Alessio Gugliotta, and John Domingue

Knowledge Media Institute,
Open University,
MK7 6AA, Milton Keynes, UK
{s.dietze,a.gugliotta,j.b.domingue}@open.ac.uk

Abstract. Context-awareness is a highly desired feature across several application domains. Semantic Web Services (SWS) technologies address context-adaptation by enabling the automatic discovery of distributed Web services for a given task based on comprehensive semantic representations. Whereas SWS technology supports the allocation of resources based on semantics, it does not entail the discovery of appropriate SWS representations for a given situation. Describing the complex notion of a situation in all its facets through symbolic SWS representation facilities is a costly task which may never lead to semantic completeness and introduces ambiguity issues. Moreover, even though not any real-world situation completely equals another, it has to be matched to a finite set of parameter descriptions within SWS representations to enable context-adaptability. To overcome these issues, we propose Conceptual Situation Spaces (CSS) to facilitate the description of situation characteristics as members in geometrical vector spaces following the idea of Conceptual Spaces. CSS enable fuzzy similarity-based matchmaking between real-world situation characteristics and predefined situation descriptions. Following our vision, the latter are part of semantic Situation-Driven Process (SDP) descriptions, which define a composition of SWS Goals suitable to support the course of an evolving situation. Particularly, we refer to the WSMO approach for SWS. Consequently, our approach extends the expressiveness of WSMO by enabling the automatic discovery, composition and execution of achievable goals for a given situation. To prove the feasibility, we apply our approach to the domain of eLearning and provide a proof-of-concept prototype.

Keywords: Semantic Web, Conceptual Spaces, Semantic Web Services, WSMO.

1 Introduction

Context is a highly important aspect in information systems (IS) and has been subject to intensive research across a wide variety of application domains throughout the last decade [5] [14] [26]. Context-adaptation can be defined as the ability of IS to adapt their behavior to multiple possible situations. A *situation* is a complex combination of features; i.e. its parameters. Usually, situations evolve throughout the course of a *process*. The latter, therefore, can be best perceived as a sequence of intermediate situations, leading from an initial to a final situation.

To consider the relation between situations and processes, we introduce the notion of *Situation-Driven Processes (SDP)*, which consider a process context from two perspectives: the user and the system perspective. Whereas the user perspective is concerned with user situations and goals throughout the course of a process, the system perspective takes into account the resources - data and services - which are required to support each user goal in a given situation.

Semantic Web Services (SWS) [9] address the system perspective of a process since they enable the automatic discovery and selection of distributed resources - services and data exposed via Web services - for a particular goal. Current results of SWS research are available in terms of reference ontologies, such as OWL-S [22] and WSMO [28], as well as comprehensive frameworks (see DIP project[1] results). However, the definition and discovery of the most appropriate goal for a given process situation – i. e. the user perspective of a process - remains a challenging task. Current SWS approaches do neither explicitly consider the notion of a situation nor do they facilitate the grounding of purely symbolic SWS representations to a conceptual level to fully support semantic meaningfulness [3][21]. Therefore, we claim that fuzzy matchmaking methodologies are crucially required, to match a possibly infinite number of (real-world) situation characteristics to a finite set of predefined parameter instance representations as part of SDP and consequently SWS descriptions.

Conceptual Spaces (CS), as introduced by Gärdenfors [10] [11], follow a theory of describing entities at the conceptual level in terms of their natural characteristics, similar to natural human cognition to avoid the symbol grounding issue. CS enable the representation of objects as vector spaces within a geometrical space which is defined by a set of quality dimensions. For instance, a particular color may be defined as point described by vectors measuring the quality dimensions hue, saturation, and brightness. Describing instances as vector spaces where each vector follows a specific metric enables the automatic calculation of their semantic similarity in terms of their Euclidean distance in contrast to the costly description of such knowledge through symbolic representations. Even though several criticisms have to be taken into account when utilizing CS (Section 6), they are considered to be a viable option for knowledge representation.

To enable the use of SWS technology as part of SDP, we propose *Conceptual Situation Spaces (CSS)*, which are mapped to the established SWS framework WSMO and enable the discovery of appropriate SWS representations capable to achieve a given task within a particular situation. Extending merely symbolic SWS descriptions based on WSMO with context information at a conceptual level through CSS enables fuzzy, similarity-based matchmaking between real-world situation characteristics and predefined SWS representations. Whereas similarity between situation parameters, as described within a CSS, is indicated by the Euclidean distance between them, real-world situation parameters are classified along predefined prototypical parameters which are implicit elements of a SWS description. Consequently, the expressiveness of SWS facilities, respectively WSMO, is extended through CSS in order to enable fuzzy matchmaking mechanisms when allocating resources for a given situation.

Since a situation always occurs within a specific domain setting, and can be described based on domain-context specific entities only, the CSS metamodel

---

[1] DIP Project: http://dip.semanticweb.org

considers domain-specific derivations. In this paper, we exemplarily refer to the e-Learning domain. To prove the feasibility of our approach, we provide a proof-of-concept prototype that uses CSS to describe learning styles, following the Felder-Silverman Learning Style theory [6], as particular learning situation parameter utilized within comprehensive SDP models.

The remainder of the paper is organized as follows: Section 2 introduces our vision of SDP as motivations for CSS, which are introduced in Section 3. Section 4 illustrates the application of CSS to the eLearning domain as utilized within a prototype application which is explained in Section 5. Finally, we discuss and conclude our work in Section 6.

2 Motivation: Situation-Driven Processes

Following our vision, a SDP consists of *Situations* (S) and *Goals* (G), where a Goal represents a particular activity within a process from a user perspective and links two situations leading from an initial situation to a desired situation. Each Goal is supported by a set of *Brokered Goals* (BG) which are achievable and brokered via a SWS Execution Environment. Since we refer to WSMO as SWS reference implementation, BG are derived WSMO Goals (Figure 1).

BG and SWS support the system perspective of a process, since they are linked (via mediators) to semantic descriptions of available Web services that, once discovered and selected, will provide the appropriate resources needed to progress a situation. For

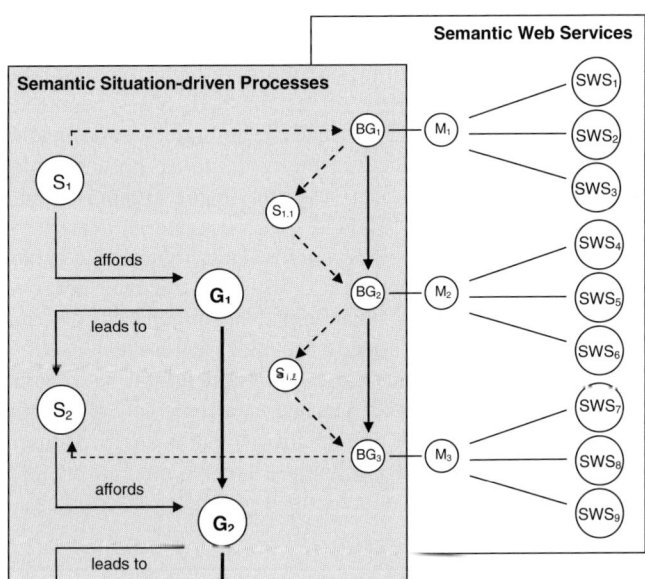

Fig. 1. Utilizing SWS to support Situation-Driven Processes

instance, one BG could be aimed at providing required information out of specific databases whereas another aims at computing a specific calculation, such as the current stock of a specific article. Note that the achievement of BG subsequently modifies the actual situation until the desired final situation is reached (Figure 1).

Utilizing the OCML knowledge modelling language [20], the SDP metamodel has been formalized into an ontology *(SDPO)*, which is aligned to an established and well known foundational ontology: the Descriptive Ontology for Linguistic and Cognitive Engineering (DOLCE) [13] and, in particular, its module Descriptions and Situations (D&S) [12]. WSMO is adopted to enable the description of SWS, as well as data resources in terms of its main elements – goals, Web services, ontologies, and mediators. As a result, the SDP metamodel extends the expressiveness of WSMO representation facilities by incorporating WSMO Goals into meaningful situation-based process context descriptions. It is important to note that process situations are highly dependent on the domain and nature of a process, since each domain emphasizes different situation parameters. Therefore, we foresee multiple domain-specific derivations of the introduced meta-model.

However, to fully enable situation-aware discovery of resources through SDP and SWS, the following shortcomings have to be considered:

I1. *Lack of explicit notion of context*: Current SWS technology does not entirely specify how to represent domain contexts. For example, WSMO [28] addresses the idea of context: Goal and web service represent the user and provider local views, respectively; the domain ontologies define the terminologies used in each view; and the mediators are the semantic bridges among such distinct views. However, WSMO does not specify what a context description should define and how the context elements should be used.

I2. *Symbolic Semantic Web representations lack grounding to conceptual level*: the symbolic approach, i.e. describing symbols by using other symbols, without a grounding in the real world, of established SWS, and Semantic Web representation standards in general, leads to ambiguity issues and does not entail semantic meaningfulness, since meaning requires both the definition of a terminology in terms of a logical structure (using symbols) and grounding of symbols to a conceptual level [3][21].

I3. *Lack of fuzzy matchmaking methodologies*: Describing the complex notion of a specific situation in all its facets is a costly task and may never reach semantic completeness. Whereas not any situation and situation parameter completely equals another, the number of (predefined) semantic representations of situations and situation parameters is finite. Therefore, a possibly infinite set of given (real-world) situation characteristics has to be matched to a finite set of predefined parameter instance representations which are described within an IS. Consequently, fuzzy classification and matchmaking techniques are required to classify a real-world situation based on a limited set of predefined parameter descriptions.

3 Conceptual Situation Spaces

To address I1 – I3 introduced in Section 2, we propose *Conceptual Situation Spaces (CSS)* applying CS to represent situations as part of SDP.

3.1 CSS Metamodel

CSS enable the description of a particular situation as a member of a dedicated CS. Referring to [11],[17],[24], we define a CSS (*css:Conceptual Situation Space* in Figure 2) as a vector space:

$$C^n = \{(c_1, c_2, ..., c_n) | c_i \in C\}$$

with c_i being the quality dimensions (*css:Quality Dimension*) of C. Please note, that we do not differentiate between domains, as sets of integral dimensions [11], but enable dimensions to be detailed further in terms of subspaces. Hence, a dimension within one space may be defined through another conceptual space by using further dimensions [24]. In such a case, the particular quality dimension c_j is described by a set of further quality dimensions with

$$c_j = D^n = \{(d_1, d_2, ..., d_n) | d_k \in D\}.$$

In this way, a CSS may be composed of several subspaces and consequently, the description granularity of a specific situation can be refined gradually. This aspect of CSS corresponds to the approach Dolce D&S [12], utilized within SDPO, to gradually refine a particular description by using parameters where each parameter can be described by an additional description. To reflect the impact of a specific quality dimension, we consider a prominence value p (*css:Prominence*) for each dimension. Therefore, a conceptual space is defined by

$$C^n = \{(p_1 c_1, p_2 c_2, ..., p_n c_n) | c_i \in C, p_i \in P\}$$

where P is the set of real numbers. However, the usage context, respectively the domain, of a particular CSS strongly influences the ranking of its quality dimensions. For instance, within a learning situation the competencies of a particular learner may be more important whereas in a business situation, the costs of a particular task may be weighted higher. This clearly supports our position of describing distinct CSS explicitly for specific domains only.

Particular members (*css:Member*) in the CSS are described through a set of valued dimension vectors (*css:Valued Dimension Vectors*). Moreover, referring to [11] we consider prototypes which represent specific prototypical members (*css:Prototypical Member*) within a particular space. Prototypical members are utilised to categorize a specific CSS member as they enable the classification of any arbitrary member m within the same space, by simply calculating the Euclidean distances between m and all prototypical members to identify the closest neighbours of m. For instance, given a CS to describe apples based on their shape, taste and colour, a green apple with a strong and fruity taste may be close to a prototypical member representing the typical characteristics of the Granny Smith species. Figure 2 depicts the CSS metamodel.

The metamodel introduced above has been formalized into a *Conceptual Situation Space Ontology (CSSO)*, utilizing OCML [20]. In particular, each of the depicted entities is represented as a concept within CSSO whereas associations are reflected as their properties in most cases. The correlation relationship between several quality dimensions indicates whether two dimensions are correlated or not. For instance,

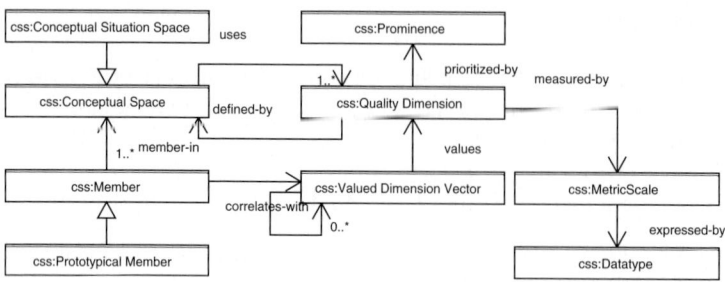

Fig. 2. The CSS metamodel

when describing an apple the quality dimension describing its sugar content may be correlated with the taste dimension. Information about correlation is expressed within the CSSO through axioms related to a specific quality dimension instance.

Semantic similarity between two members of a space can be perceived as a function of the Euclidean distance between the points representing each of the members. Applying a formalization of CS proposed in [24] to our definition of a CSS, we formalize the Euclidean distance between two members in a CSS as follows. Given a CSS definition C and two members represented by two vector sets V and U, defined by vectors $v_0, v_1, ..., v_n$ and $u_1, u_2, ..., u_n$ within C, the distance between V and U can be calculated as:

$$|d(u,v)|^2 = \sum_{i=1}^{n}(z(u_i) - z(v_i))^2$$

where $z(u_i)$ is the so-called Z-transformation or standardization [4][24] from u_i. Z-transformation facilitates the standardization of distinct measurement scales which are utilized by different quality dimensions in order to enable the calculation of distances in a multi-dimensional and multi-metric space. The z-score of a particular observation u_i in a dataset is to be calculated as follows:

$$z(u_i) = \frac{u_i - \bar{u}}{s_u}$$

where $\bar{u}$ is the mean of a dataset U and s_u is the standard deviation from U. Considering prominence values p_i for each quality dimension i, the Euclidean distance $d(u,v)$ indicating the semantic similarity between two members described by vector sets V and U can be calculated as follows:

$$d(u,v) = \sqrt{\sum_{i=1}^{n} p_i ((\frac{u_i - \bar{u}}{s_u}) - (\frac{v_i - \bar{v}}{s_v}))^2}$$

3.2 Aligning CSS and SWS

CSS are aligned to WSMO to support the automatic discovery of the most appropriate goal representation for a specific process situation. Figure 3 depicts the main relationships between CSS and WSMO.

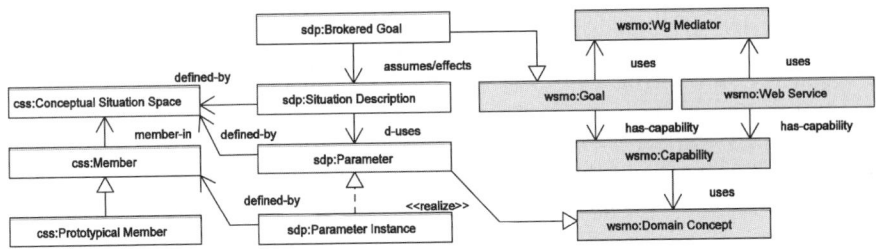

Fig. 3. Alignment of CSS and WSMO

Grey colored concepts in Figure 3 represent concepts of WSMO [28]. A goal description (*wsmo:Goal*, *sdp:Brokered Goal*) utilizes particular domain concepts (*wsmo:Domain Concept*, *sdp:Parameter*) to semantically describe its capabilities, i.e. its assumptions, effects, preconditions and postconditions. A WSMO runtime reasoning engine utilizes capability descriptions to identify SWS (*wsmo:Web Service*) which suit a given Goal. In contrast, the preliminary selection of the most appropriate goal description for a given situation is addressed by classification of situation parameters through CSS. For instance, given a set of real-world situation parameters, described as members in a CSS, their semantic similarity with predefined prototypical situation descriptions (*css:Prototypical Member*) is calculated. Given such a classification of a particular real-world situation, a goal representation which assumes matching prototypical parameter instances can be selected and achieved through the reasoning engine.

4 Spanning a Conceptual Learning Situation Space

As Gärdenfors states in [11], the prioritization of certain quality dimensions within a CS is highly dependent on the context of the space. The same applies to situations which are described within a CSS. In order to validate the applicability of our approach, we defined a CSS for the eLearning domain, a *Conceptual Learning Situation Space (CLSS)*.

Since situation parameters usually are complex theoretical constructs, each parameter itself is described as a CSS subspace (Section 0). In this Section we focus exemplarily on the representation of one parameter, which is of particular interest within the eLearning domain: the learning style of a learner. A learning style is defined as an individual set of skills and preferences on how a person perceives, gathers, and processes learning materials [19]. Whereas each individual has his/her distinct learning style, it affects the learning process.

We refer to the *Felder-Silverman Learning Style Theory (FSLST)* [6], which is supposed to be suitable to describe learning styles within computer-aided educational environments. It is important to note that distinct theories can be applied to describe each situation parameter, and FSLST just serves as example to illustrate the application of CSS in this paper. Following FSLST, a learning style can be described by four quality dimensions [6]. In short, the *Active-Reflective* dimension describes whether or not a learner prefers to interact with learning material, whereas the *Sensing-Intuitive* dimension, describes whether a learner tends to focus on facts and details (Sensing) rather than abstract theories (Intuitive). The *Visual-Verbal* dimension obviously covers, whether a learner prefers visual rather than verbal learning material, while the *Global-Sequential* dimension describes, whether a learner tends to learn gradually in small steps (Sequential) rather than following a holistic learning process marked by large learning leaps. Literature shows [8][15][27], that these dimensions can be assumed to be virtually linearly independent. Consequently, we define a *CLSS L* with 4 quality dimensions l_i:

$$L^4 = \{(l_1, l_2, l_3, l_4) | l_i \in L\}$$

Figure 4 depicts the key concepts describing *L* as subspace (*clss:FSLST Space*) within CLSS representing FSLST.

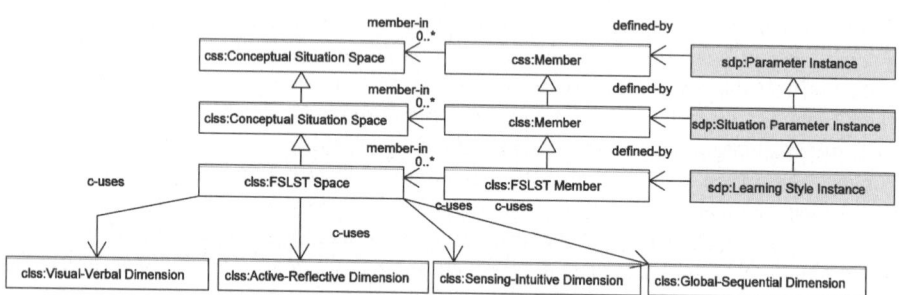

Fig. 4. Key concepts of an ontology representing the FSLST as particular subspace

Moreover, Figure 4 depicts the alignment of subspace *L* (*clss:FSLST Space*) with SDP (Section 0), represented via grey-colored concepts. Symbolic representations of parameter instances within the SDP are defined by members within CLSS-based representations (*clss:Member*), where a particular learning style is a specific SDP parameter instance (*sdp:Learning Style Instance*) and is defined by a particular member (*clss:FSLST Member*). The metric scale, datatype, value range and prominence value for each dimension l_i are presented in Table 1:

Table 1. Quality dimensions $l_1 - l_4$ of CLSS *L* describing learning styles following FSLST

	Quality Dimension	Metric Scale	Datatype	Range	Prominence
l_1	Active-Reflective	Interval	Integer	-11..+11	1.5
l_2	Sensing-Intuitive	Interval	Integer	-11..+11	1
l_3	Visual-Verbal	Interval	Integer	-11..+11	1.5
l_4	Global-Sequential	Interval	Integer	-11..+11	1

As depicted in Table 1, each quality dimension is ranked on an interval scale with its value range being an integer between -11 and +11. This particular measurement scale was derived from an established assessment method, the Index of Learning Styles (ILS) questionnaire defined by Felder and Soloman [7], aimed at identifying and rating a particular learning style of an individual.

The authors would like to highlight, that prominence values have been assigned which rank the first (l_1) and the third dimension (l_3) higher than the other two, since these have a higher impact on the context of the learning situation, which is focused on the aim to deliver appropriate learning material to the learner. Since dimensions l_1 and l_3 are highly critical for context-adaptation and SWS discovery (Section 5), a higher prominence value was assigned. It is obvious, that the assignment of prominence values is a highly subjective process, strongly dependent on the purpose, context and individual preferences. Therefore, future work is aimed at enabling learners to assign rankings of quality dimensions themselves individually.

To classify an individual learning style (*clss:FSLST Member*), we define prototypical members (*clss:FSLST Prototypical Member*) in the FSLST-based vector space *L*. To identify appropriate prototypes, we utilized existing knowledge about typical correlations between the FSLST dimensions, as identified throughout research studies such as [8][27]. Particularly, we refer to correlation coefficients identified in [27] which led to the description of the following five prototypical members and their characteristic vectors:

5 Contexts-Adaptive Composition and Accomplishment of SDP

To prove the feasibility of our approach, a proof-of-concept prototype application has been provided which applies CSS (Section 3) to the domain of eLearning[2]. The lifecycle of a SDP instance consists of the following three stages:

S1. Classification of a situation given a domain-specific CSS;
S2. Composition of SDP as sequence of BG which satisfy a given situation;
S3. Runtime execution of SDP, in terms of BG achievements.

Figure 5 depicts the architecture used to support reasoning on SDP and CSS through a Semantic Execution Environment (SEE) which is implemented through IRS-III [2] in our case.

Table 2. Prototypical learning styles defined as prototypical members in the CLSS ontology

Prototype	Act/Ref	Sen/Int	Vis/Ver	Seq/Glo
P1: Active-Visual	-11	-11	-11	+11
P2: Reflective	+11	-11	-11	0
P3: Sensing-Sequential	-11	-11	-11	-11
P4: Intuitive-Global	-11	+11	-11	+11
P5: Verbal	-11	+11	+11	+11

---

[2] The application and ontologies have been developed in the context of the EU FP6 project LUISA [18].

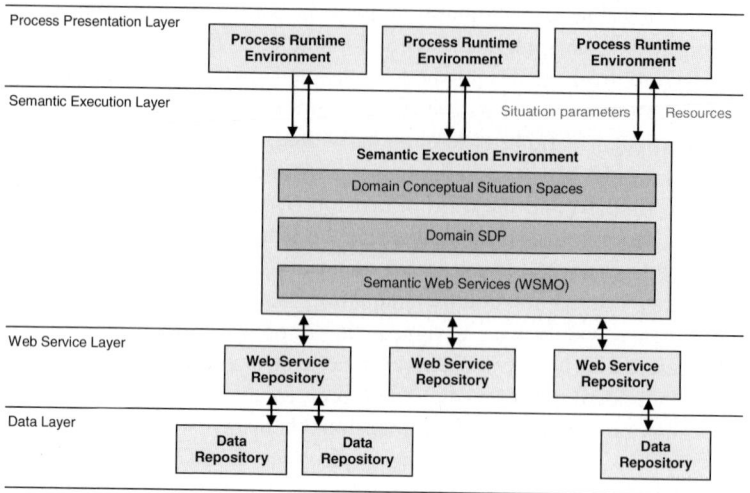

Fig. 5. Architecture to support runtime reasoning on SDP/CSS

SEE makes use of WSMO-based SWS descriptions, semantic representations of the SDP and CSS metamodels, and particularly their derivations for the eLearning domain. Multiple runtime environments interact with SEE to provide information about the current real-world situation on the one hand and to present and accomplish SDP-based processes on the other hand.

Starting from a set of real-world situation parameters, their semantic similarity with a set of prototypical situation parameters (Section 4) is calculated to support *S1*. Given such a classification of an identified prototypical situation, the appropriate SDP (Section 2) is composed as a sequence of Goals and supportive Brokered Goals (BG) to accomplish *S2*. Finally, SEE accomplishes *S3* by discovering and orchestrating appropriate Web services, which show the capabilities to suit the Brokered Goals associated with the user Goals within the SDP.

5.1 Context Classification through CSS

To enable a description of a situation, a learner is first authenticated in order to retrieve information about his/her actual preferences. During the following situation refinement, learners are enabled to define situation-specific parameters, such as his/her current learning aim or the available learning time, while other parameters are calculated automatically, such as the competency gap between the current learner situation and the desired aim. Referring to the learning style as particular situation parameter defined by the CLSS subspace L in Section 4, its semantic similarity with each of the prototypical members is indicated by their Euclidean distance. Since we utilize a CSS described by dimensions which each use the same metric scale (ordinal scale), the distance between two members U and V can be calculated disregarding a Z-transformation (Section 3) for each vector as follows:

$$d(u,v) = \sqrt{\sum_{l-1}^{n} p_i (u_i - v_i)^2}$$

The calculation of similarities using the formula shown above is performed by a standard Web service that is exposed as SWS. Given a particular CSS description, a member (representing a specific parameter instance) as well as a set of prototypical members, similarities are calculated at runtime in order to classify a given situation parameter. For instance, a particular situation description includes a learner profile indicating a learning style parameter which is defined by a member U in the specific CLSS subspace L (*clss:FSLST Space*) with the following vectors:

$$U = \{(u_1 = -5, u_2 = -5, u_3 = -9, u_4 = 3) | u_i \in L\}$$

Learning styles, such as the one above, could be introduced to individual learners by utilizing the ILS Questionnaire [7] as assessment method. Calculating distances between U and each of the prototypes described in Table 2 of Section 4 led to the following results:

Table 3. Euclidean distances between U and prototypical learning styles

Prototype	Euclidean Distance
P1: Active-Visual	12.649110640673518
P2: Reflective	20.85665361461421
P3: Sensing-Sequential	17.08800749063506
P4: Intuitive-Global	19.493588689617926
P5: Verbal	31.20897306865447

As depicted in Table 3, the lowest Euclidean distance applies to *P1*, indicating a rather active and visual learning style described as in Table 2 of Section 4. Utilizing such similarity-based classifications enables the gradual refinement of learning situation description and fuzzy matchmaking between real-world situation parameters, such as U, and prototypical parameters such as *P1*.

5.2 Situation-Driven Composition and Accomplishment of SDP

The authors would like to highlight, that not only processes (SDP) but also entire application scenarios are accomplished by achieving BG at runtime. Given a classified situation description, the SEE first identifies semantic Goal representations (*sdp:Brokered Goal*) and finally discovers and orchestrates Web services to suit the given runtime situation. Given the alignment of CSS and WSMO (Section 3.2) we address the issue of discovering the most appropriate BG representation for a given situation by enabling a classification of an individual situation based on semantic similarity calculations as described in the previous sections.

For instance, referring to Section 5.1, given a classified learning style, together with classifications of all further situation parameters, a BG representation (*sdp:Brokered Goal*) which assumes equivalent situation parameter instances (*sdp:Parameter Instance*) is achieved at runtime through SEE. Consequently, following the alignment of CSS and WSMO, context-aware SWS applications are enabled which automatically discover not only Web services for a given task but also SWS Goal descriptions for a given situation.

Following the lifecycle stages *S1-S3*, the application supports the automatic composition of SDP *(S2)*, their transformation into non-semantic process metadata manifestations for two metadata standards - ADL SCORM [1] and IMS LD [16] - and the accomplishment of the SDP-based process *(S3)*. Process composition is accomplished by a Web service which composes a SDP as a set of Goals and Brokered Goals, which show the appropriate effects and assumptions to progress from an initial situation S_i to the final situation S_f. The service takes into account the situation parameters of S_i, for instance the available learning time of the learner or his/her learning style (Section 4). A particular parameter, the actual aim of a learner, is linked to a set of desired competencies which consequently are part of the final situation S_f. Composition functionalities could be provided by different Web services following distinct composition strategies. At runtime, the most appropriate composition and transformation service for a given situation is selected and invoked automatically by the SEE. The achievement of BG at runtime considers the actual learning situation parameters, and enables a more fine-grain adaptation to the actual learning context. At runtime, a process is executed *(S3)* whether through a metadata standard specific runtime environment or a runtime environment dedicated to interpret semantic SDP models. Whereas a web-based user interface is utilized to interpret and present semantic process instances, software clients of the RELOAD-project [25] are utilized to present dynamically transformed XML-manifestations following the IMS LD and ADL SCORM standard.

6 Discussion and Conclusions

We proposed the CSS approach aimed at the classification of situations to enable the automatic discovery, composition and accomplishment of SWS-based Goal representations to suit a given process situation. We introduced the notion of a Situation-Driven Process (SDP), which perceives a process from two perspectives, the user perspective which describes a process in terms of user Goals and situations and the system perspective, which links user Goals to automatically achievable SWS Goals. Whereas current SWS frameworks such as WSMO and OWL-S support the system perspective of a process by enabling the allocation of distributed services for a given (semantically) well-described task, the CSS approach particularly addresses the similarity-based discovery of the most appropriate SWS task representation for a given situation. Consequently, by aligning CSS to established SWS frameworks, the expressiveness of symbolic SWS standards is extended with context information on a conceptual level to enable fuzzy context aware delivery of resources at runtime. Deriving the CSS metamodel for specific domains enables the composition of domain-specific SDP at runtime while considering the specific characteristics of the runtime situation. Based on mappings with domain-specific process metadata standards, a semantic SDP based process model can be transformed into non-semantic metadata standards to support interoperability of a process.

To prove the feasibility of our approach, a proof-of-concept prototype was described, which applies CSS to the eLearning domain. Whereas the Felder-Silverman Learning Style Theory (FSLST) was exemplarily represented as CSS, the authors would like to highlight that distinct theories could be applied to represent situation parameters. In this paper, FSLST just serves the purpose to illustrate the application of CSS but is not explicitly supported by the authors. The application supports the automatic classification of a situation based on similarity calculation within CSS, the situation-driven composition of semantic process instances following the SDP approach at runtime, their transformation into non-semantic metadata standards and the runtime execution of a SDP-based model by automatically achieving SWS to retrieve resources which suit the actual process situation.

However, even though our approach aims at solving SWS-related issues such as the symbol grounding problem, several criticisms still have to be taken into account. Whereas defining objects, respectively "instances" within a given CSS appears to be a straightforward process of assigning specific values to each quality dimension of a CSS, the definition of a CSS itself is not trivial at all and strongly dependent on individual perspectives and appraisals. Whereas semantics of an object are grounded to metrics in geometrical vector spaces within a CSS, the quality dimensions themselves are subject to ones perspective what may lead to ambiguity issues. Consequently, the approach of CSS does not appear to completely solve the symbol grounding issue but to shift it from the process of describing instances to the definition of a CSS. Apart from that, whereas the size and resolution of a CSS is indefinite, defining a reasonable CSS for a specific context may become a challenging task. Moreover, distance calculation as major contribution of the CSS approach always relies on the fact, that objects are described in the same geometrical space. Consequently, CSS may be perceived as step forward but do not fully solve the issues related to symbolic Semantic Web (Services)-based knowledge representations.

Future work has to deal with the aforementioned issues. For instance, we foresee to enable adjustment of prominence values to quality dimensions of a specific CSS to be accomplished by a user him/herself, in order to most appropriately suit his/her specific preferences, since prioritization of dimensions is a highly individual and subjective process. Besides that, we consider the representation of further situation parameters as CSS subspaces.

The authors are aware, that the current prototype applies SDP/CSS to one domain only. However, we are strongly convinced that applying the idea of SDP in further process domains is feasible, since processes across several domains share similar notions, as well as concepts and have to deal with related issues, such as process design, process resource allocation, and context-sensitivity. Therefore, further research will be concerned with the application of our approach to further domains. It is apparent, that our approach requires an initial effort to produce domain models following SDP/CSS which can be instantiated at runtime. However, once these derivations are available, these can be reused across multiple application settings to enable context-aware runtime composition and accomplishment of processes.

References

[1] Advanced Distributed Learning (ADL) SCORM 2004 Specification, http://www.adlnet.org
[2] Cabral, L., Domingue, J., Galizia, S., Gugliotta, A., Norton, B., Tanasescu, V., Pedrinaci, C.: IRS-III: A Broker for Semantic Web Services based Applications. In: Cruz, I., Decker, S., Allemang, D., Preist, C., Schwabe, D., Mika, P., Uschold, M., Aroyo, L.M. (eds.) ISWC 2006. LNCS, vol. 4273, Springer, Heidelberg (2006)
[3] Cregan, A.: Symbol Grounding for the Semantic Web. In: 4th European Semantic Web Conference 2007, Innsbruck, Austria (2007)
[4] Devore, J., Peck, R.: Statistics - The Exploration and Analysis of Data, 4th edn. Duxbury, Pacific Grove (2001)
[5] Dietze, S., Gugliotta, A., Domingue, J.: A Semantic Web Services-based Infrastructure for Context-Adaptive Process Support. In: Proceedings of IEEE 2007 International Conference on Web Services (ICWS), Salt Lake City, Utah, USA (2007)
[6] Felder, R.M., Silverman, L.K.: Learning and Teaching Styles in Engineering Education. Engineering Education 78, 674–681 (1988), Preceded by a preface in 2002 http://www.ncsu.edu/felderpublic/Papers/LS-1988.pdf
[7] Felder, R.M., Soloman, L.K.: Index of Learning Styles Questionnaire. (1997) (retrieved, October 2007), Online version http://www.engr.ncsu.edu/learningstyles/ilsweb.html
[8] Felder, R.M., Spurlin, J.: Applications, Reliability and Validity of the Index of Learning Styles. International Journal on Engineering Education 21(1), 103–112 (2005)
[9] Fensel, D., Lausen, H., Polleres, A., de Bruijn, J., Stollberg, M., Roman, D., Domingue, J.: Enabling Semantic Web Services – The Web service Modelling Ontology. Springer, Heidelberg (2006)
[10] Gärdenfors, P.: Conceptual Spaces - The Geometry of Thought. MIT Press, Cambridge (2000)
[11] Gärdenfors, P.: How to make the semantic web more semantic. In: Vieu, A.C., Varzi, L. (eds.) Formal Ontology in Information Systems, pp. 19–36. IOS Press, Amsterdam (2004)
[12] Gangemi, A., Mika, P.: Understanding the Semantic Web through Descriptions and Situations. In: Meersman, R., Tari, Z., et al. (eds.) Proceedings of the On The Move Federated Conferences (OTM 2003). LNCS, Springer, Heidelberg (2003)
[13] Gangemi, A., Guarino, N., Masolo, C., Oltramari, A., Schneider, L.: Sweetening Ontologies with DOLCE. In: Gómez-Pérez, A., Benjamins, V.R. (eds.) EKAW 2002. LNCS (LNAI), vol. 2473, Springer, Heidelberg (2002)
[14] Gellersen, H.-W., Schmidt, A., Beigl, M.: Multi-Sensor Context-Awareness in Mobile Devices and Smart Artefacts. ACM journal Mobile Networks and Applications (MONET) 7(5) (October 2002)
[15] Graf, S., Viola, S.R., Kinshuk, Leo, T.: Representative Characteristics of Felder-Silverman Learning Styles: An Empirical Model. In: Proceedings of the IADIS International Conference on Cognition and Exploratory Learning in Digital Age (CELDA 2006), Barcelona, Spain, December 8-10 (2006)
[16] IMS Learning Design Specification, http://www.imsglobal.org
[17] Keßler, C.: Conceptual Spaces for Data Descriptions. In: The Cognitive Approach to Modeling Environments (CAME), Workshop at GIScience 2006, Münster, Germany, pp. 29–35 (2006), SFB/TR 8 Report No. 009-08/2006
[18] LUISA Project - Learning Content Management System Using Innovative Semantic Web Services Architecture, http://www.luisa-project.eu/www/

[19] Johnson, C., Orwig, C.: What is Learning Style? (1998), http://www.sil.org/LinguaLinks/LanguageLearning/OtherResources/YorLrnngStylAndLnggLrnng/WhatIsALearningStyle.htm
[20] Motta, E.: An Overview of the OCML Modelling Language. In: The 8th Workshop on Methods and Languages (1998)
[21] Nosofsky, R.: Similarity, scaling and cognitive process models. Annual Review of Psychology 43, 25–53 (1992)
[22] Object Management Group: Business Process Modelling Notation Specification, http://www.omg.org/docs/dtc/06-02-01.pdf
[23] OWL-S 1.0 Release, http://www.daml.org/services/owl-s/1.0/
[24] Raubal, M.: Formalizing Conceptual Spaces. In: Varzi, A., Vieu, L. (eds.) Formal Ontology in Information Systems, Proceedings of the Third International Conference (FOIS 2004). Frontiers in Artificial Intelligence and Applications, vol. 114, pp. 153–164. IOS Press, Amsterdam (2004)
[25] Reload Project, http://www.reload.ac.uk/
[26] Schmidt, A., Winterhalter, C.: User Context Aware Delivery of E-Learning Material: Approach and Architecture. Journal of Universal Computer Science (JUCS) 10(1) (January 2004)
[27] Viola, S.R., Graf, S., Kinshuk, Leo, T.: Investigating Relationships within the Index of Learning Styles: A Data-Driven Approach. International Journal of Interactive Technology and Smart Education 4(1), 7–18 (2007)
[28] WSMO Working Group, D2v1.0: Web service Modeling Ontology (WSMO). WSMO Working Draft (2004), http://www.wsmo.org/2004/d2/v1.0/

Combining SAWSDL, OWL-DL and UDDI for Semantically Enhanced Web Service Discovery

Dimitrios Kourtesis and Iraklis Paraskakis

South East European Research Centre (SEERC),
Research Centre of the University of Sheffield and CITY College
Mitropoleos 17, 54624 Thessaloniki, Greece
{dkourtesis,iparaskakis}@seerc.org

Abstract. UDDI registries are included as a standard offering within the product suite of any major SOA vendor, serving as the foundation for establishing design-time and run-time SOA governance. Despite the success of the UDDI specification and its rapid uptake by the industry, the capabilities of its offered service discovery facilities are rather limited. The lack of machine-understandable semantics in the technical specifications and classification schemes used for retrieving services, prevent UDDI registries from supporting fully automated and thus truly effective service discovery. This paper presents the implementation of a semantically-enhanced registry that builds on the UDDI specification and augments its service publication and discovery facilities to overcome the aforementioned limitations. The proposed solution combines the use of SAWSDL for creating semantically annotated descriptions of service interfaces and the use of OWL-DL for modelling service capabilities and for performing matchmaking via DL reasoning.

Keywords: Semantic Web Services, Web Service Discovery, Universal Description Discovery and Integration (UDDI), Semantic Annotations for WSDL (SAWSDL), Web Ontology Language (OWL).

1 Introduction

The Universal Description, Discovery and Integration (UDDI) service registry specification [1] is currently one of the core standards in the Web service technology stack and an integral part of every major SOA vendor's technology strategy. The UDDI specification defines an XML-based data model for storing descriptive information about Web services and their providers, and a Web service-based programmatic interface for publishing this type of information to the registry and performing inquiries. A UDDI service registry may be deployed and used within a private corporate network, a restricted network of business partners, or even made available over the Web. In any of the three settings, a UDDI registry can serve as the basis for establishing service lifecycle management and is one of the fundamental building blocks for realising design-time and run-time SOA governance.

A UDDI registry allows service providers to describe the functionality of their advertised services by means of references to externally maintained technical specifications or classification schemes developed by service providers or third-parties. The UDDI specification is generic and does not prescribe the use of any specific method, formal or informal, for creating such specifications. Their definition and interpretation are beyond the scope of a UDDI registry and are left to the discretion of service providers and service consumers. A UDDI service advertisement may reference numerous such descriptions in order to represent different aspects of a Web service's functional and non-functional properties. This allows services advertised in UDDI registries to be searched for and discovered by service consumers based on their declared conformance to some technical specification, or their attributed categorisation within a classification system. In addition, service advertisements can be retrieved through a text-based search for keywords contained in service names, although the effectiveness of this mechanism is clearly rather limited.

The fundamental problem with the externally maintained specifications that service advertisements refer to, like WSDL [2] documents, is that even if they are machine-processable, they still lack the formal rigour and machine-understandable semantics that would make them amenable to logic-based reasoning and automated processing. As a result, UDDI registries cannot support fine-grained matchmaking based on the actual definitions of these technical specifications or classification systems, and effectively, cannot support truly automated service discovery. In a typical service discovery scenario a developer still needs to retrieve the WSDL document and any additional specification documents referenced by a UDDI service advertisement and inspect them manually, in order to assert that the advertised service is fully interoperable with other services assembled in a service composition. Semantic Web Services research aims at addressing this problem by bringing semantics into the realm of service specifications, such that service capabilities can be explicated in an unambiguous and machine-interpretable manner that not only allows for fully automated discovery in service registries, but enables the automation of a broad range of design-time and run-time activities in SOA.

This paper presents the implementation of the FUSION Semantic Registry, a semantically-enhanced service registry developed within the FUSION project[1] and released as open source software[2]. FUSION is an EU-funded research project aiming to promote business process integration and interoperability within and across enterprises, through a semantics-based approach for Enterprise Application Integration (EAI) in service-oriented business applications. Semantically-enhanced service discovery based on widely accepted standards is an essential requirement for the theoretical and technological approach that FUSION puts forward. The FUSION Semantic Registry relies on a combination of three standards from the domain of Web service and Semantic Web technologies to achieve its objectives: UDDI, for storing and retrieving syntactic and semantic information about services and service providers, SAWSDL [3], for creating semantically annotated descriptions of service interfaces, and OWL [4], for modelling service characteristics and performing fine-grained service matchmaking via DL reasoning.

The remaining of this paper is organised as follows. Section 2 presents the service discovery requirements that the FUSION Semantic Registry addresses, and the

---

[1] http://www.fusion-strep.eu/
[2] http://www.seerc.org/fusion/semanticregistry/

approach put forward in the FUSION project for describing service advertisements and service requests in a semantically-enriched manner. Section 3 presents an overview of the FUSION Semantic Registry architecture and its interfaces. Section 4 provides a walkthrough of the core activities performed during service publication, while section 5 provides a walkthrough of the activities performed during service discovery. Section 6 examines related research works that focus on the enrichment of the discovery facilities that UDDI registries offer through semantic enhancements, and section 7 concludes the paper by summarising the main points of the presented work.

2 Service Discovery Requirements in the Scope of FUSION

The FUSION project aims at delivering a reference framework, a supporting methodology, and a set of tools for realising Enterprise Application Integration (EAI) through Semantic Web Service technology. Semantically-enhanced publication and discovery of services is central to the approach that FUSION puts forward, and encompasses two objectives in order to be sufficiently supported by the FUSION Semantic Registry.

Firstly, describing service advertisements and service requests in a machine-understandable form that captures their salient characteristics and allows for comparing them in a fully automated way. Secondly, augmenting the typical functions supported by UDDI registries (i.e. storing syntactic metadata about services and their providers) with the addition of a mechanism for semantic service matchmaking and indexing.

The latter, i.e. the technical means employed to augment UDDI-based service registries with semantic matchmaking extensions is discussed in subsequent sections. This section of the paper discusses the first objective, and more specifically, describes (i) *what* are the salient service characteristics that should be captured for the purpose of matchmaking in the scope of FUSION, and (ii) *how* these characteristics should be captured in a suitable semantic representation formalism.

2.1 Service Characteristics Considered for Matchmaking

The Semantic Web Services research literature features an abundance of different approaches for service matchmaking, each of them addressing a different set of requirements and therefore focusing on a different set of service properties, functional or non-functional ones. The service characteristics that the FUSION Semantic Registry considers during matchmaking are a combination of functional and non-functional properties. In the following we describe these matchmaking requirements in detail, as a means to allow for comparisons among the FUSION Semantic Registry and other implementations or specifications of semantically-enhanced service registries in the literature.

Functional Properties. The majority of research works in the literature focus on functional properties of services, and more specifically, on approaches for matchmaking among descriptions of service inputs, outputs, preconditions and effects (IOPE). IO and PE descriptions are means to represent two different aspects of a service's functional properties: the information transformation that a service produces through the inputs it consumes and the outputs it generates, and the state-wise conditions that need to hold before a service can be invoked (preconditions), or will eventually hold after the service's invocation (effects). The matchmaking requirements that the FUSION Semantic Registry addresses concern only the first aspect, i.e. the data semantics of a Web service. Extending the capabilities of the Semantic Registry to include matchmaking based on behavioural semantics is a subject of future work, beyond the scope of the FUSION project.

More specifically, the matchmaking that the FUSION Semantic Registry performs among the inputs or outputs of a service advertisement and a service request should be able to detect if *data-level interoperability* can be guaranteed among an advertised service and its prospective consumer. This requires evaluating the degree to which the consumer provides all input data that the advertised service expects to receive when invoked, and the degree to which the advertised service produces all output data that the consumer expects to obtain after execution. As discussed in more detail in [5], this is an essential requirement for guaranteeing flawless communication among the systems participating in a collaborative business process.

From a purely practical perspective, this poses some important requirements on: (i) the expressivity of the semantic representation formalism that is employed for describing Web service inputs and outputs (it should facilitate modelling of arbitrarily complex XSD schemata), and (ii) the sensitivity of the matchmaking mechanism that is employed for comparing the above semantic representations of inputs or outputs (it should be able to detect mismatches at a fine level of granularity). As presented in detail in [5], the FUSION Semantic Registry must be able to detect mismatches among inputs or outputs of a service advertisement and a service request at two distinct levels:

- *Message-level*: The goal here is to determine the degree to which a service can produce the set of data parameters that the requestor wants to obtain, and the degree to which the requestor can provide the set of data parameters that a service expects to receive when invoked. In the case of atomic, non-transactional Web service operations, this so-called set of data parameters corresponds trivially to an operation's request or response message. In the case of complex, transactional Web services that involve the invocation of numerous operations to fulfil one goal, the set of input data parameters corresponds to the superset of all sets of input data parameters exchanged as part of request messages for the operations involved, while the equivalent holds for output data parameters. Comparing sets of input or output data parameters rather than request and response messages allows us to abstract from the differences among complex and atomic Web services and support discovery for both.
- *Schema-level:* The goal here is to determine the degree to which the schema of some data parameter that is produced or consumed by an advertised service contains all attributes specified in the schema of the corresponding data parameter at the service consumer's end. This type of matching is meaningful in cases where an advertised service and a service consumer share a data model specification as a basis for exchanging interoperable business objects or electronic documents, but are not obliged to instantiate or make use of all schema attributes for every entity defined in that model. As a result, the case may arise where the developers of different applications have chosen to instantiate the schema attributes of a base entity

(e.g. address) in different ways, thus arriving to only partially overlapping and effectively incompatible definitions of data parameters that however share a common name.

Non-functional properties. Non-functional properties also play an important role in service discovery, and are increasingly attracting the interest of the Semantic Web Services research community. Non-functional properties may relate to quality of service (QoS), policy compliance, adherence to technical standards or protocols, or categorisation within a classification scheme. The only type of non-functional property that is taken into account for matchmaking by the FUSION Semantic Registry is the latter, i.e. the categorisation of a service advertisement with regard to some semantically represented classification scheme, in order to designate the functionality of that service and assist in simple tasks like coarse-grained filtering of services during matchmaking and browsing. Extending the capabilities of the Semantic Registry to include matchmaking based on additional non-functional properties will be a subject of future work.

The end goal in the categorisation-level matching that the FUSION Semantic Registry should support, is determining if the semantic categorisation class attributed to a service request is equivalent, more specific, or more generic than the one specified in some service advertisement. In order to have a positive match, the classification concept associated with a request must subsume the classification concept of an advertisement (i.e. the first must be equivalent or more generic than the second). As an example, consider the case of a service request classified under Supply Chain Management services, and some advertisement classified under Freight Costing services, a subcategory of Transportation services that is itself classified under Supply Chain Management services. A semantically represented service taxonomy and a suitable matchmaking mechanism should allow detecting that the service advertisement can satisfy the request, since the category of Supply Chain Management services subsumes the Freight Costing services category.

2.2 Semantic Representation of Service Characteristics

By using a semantic representation formalism to express the characteristics of Web services offered or needed, providers and requestors can create definitions of service capabilities that are automatically processable through reasoning and logic-based inference. In turn, this can facilitate high-precision retrieval for services that address the matchmaking requirements presented above. Evidently, the extent to which this can be achieved depends on the semantic representation formalism that is adopted for this purpose. The recent years have seen numerous Semantic Web Service frameworks being proposed and promoted for standardisation through W3C member submissions. The most prominent ones are OWL-S [6], WSMO [7], WSDL-S [8], and more recently the W3C Recommendation of SAWSDL [3], which evolved from the WSDL-S specification.

Although the FUSION reference framework does not prescribe the use of any specific Semantic Web Service description framework, the tools that comprise the reference implementation of the FUSION System, including the FUSION Semantic Registry, build on SAWSDL. In contrast to developing Web service descriptions at a high conceptual level and then linking these specifications to concrete Web service interfaces that are described in WSDL (as proposed in OWL-S and WSMO), the approach that SAWSDL puts forward is bottom-up: the WSDL documents themselves are to be enriched with annotations that capture machine processable semantics by pointing to concepts defined in externally maintained semantic models. This approach has numerous advantages, but the most important one is that SAWSDL can be agnostic to the knowledge representation formalism one adopts for representing service characteristics.

The semantic model that serves as the basis for creating, storing, and reasoning upon representations of service characteristics in the FUSION project is the FUSION Ontology [9]. Its multi-faceted structure reflects different types of concepts necessary for modelling a service: the data structures a service exchanges through input and output messages (data semantics), the functionality categorisation of a service with regard to a taxonomy (classification semantics), and the behaviour it may expose within a complex and stateful process execution (behavioural semantics). As already discussed, behavioural semantics are not in the range of matchmaking requirements that the FUSION Semantic Registry addresses.

The FUSION Ontology is encoded in OWL-DL, a sublanguage of the Web Ontology Language (OWL) W3C standard that has been so named due to its correspondence with description logics. OWL-DL strikes a satisfactory balance between expressiveness and computational completeness [4] and facilitates decidable reasoning with the help of DL reasoning engines. The expressivity of the DL sublanguage is a prerequisite for modelling Web service inputs and outputs at a sufficient degree of complexity that preserves the semantics of XSD schemata defined in a WSDL document. The DL sublanguage adds a number of OWL modelling constructs to those offered by OWL-Lite that are essential for this purpose, such as the ability to define enumerated classes, Boolean combinations of classes (intersectionOf, unionOf, complementOf), disjoint classes, and also place restrictions on the values that properties may have, or on their cardinality.

In order to represent the functional and non-functional service properties that are of interest for matchmaking in the FUSION Semantic Registry, one needs to create a Functional Profile, and define its key attributes in terms of references to the abovementioned FUSION Ontology. As presented in [5], a Functional Profile is expressed as a named OWL class that is attributed a set of three different OWL object properties:

1. hasCategory: Associates a FunctionalProfile with exactly one TaxonomyEntity concept from the service classification taxonomy that is part of the FUSION Ontology, to represent the service's categorisation.
2. hasInput: Associates a FunctionalProfile with an InputDataSet concept, in order to represent the set of data parameters that a service expects to receive and consume. The cardinality of this property is zero in the case of an *out-only* Message Exchange Pattern (MEP), or one, in the case of an *in-out* MEP.
3. hasOutput: Associates a FunctionalProfile with an OutputDataSet concept, in order to represent the set of data parameters that a service will produce if invoked. The cardinality of this property is zero in the case of an *in-only* MEP, or one, in the case of an *in-out* MEP.

Finally, each InputDataSet and OutputDataSet concept is associated with one or more DataFacetEntity concepts through a hasDataParameter object property, in order to represent the data parameters exchanged. Depending on the perspective from which the Functional Profile is viewed, that of the provider or the requestor, we can differentiate among Advertisement Functional Profiles (AFPs) and Request Functional Profiles (RFPs). The first are created automatically by the FUSION Semantic registry at the time of service publication, while the latter are created by the service requestor at the time of discovery (or even at an earlier stage to be used as service request templates).

To allow for the automated construction of Advertisement Functional Profiles (AFPs) in the FUSION Semantic Registry, service providers need to augment the WSDL interfaces of their provided services with semantic annotations, as per the SAWSDL specification. According to the SAWSDL annotation conventions that are applied in the context of FUSION, the semantics of a Web service's input and output data should be captured by adding `modelReference` annotations to the appropriate `<xs:element>` entities under `<wsdl:types>`, while functionality categorisation semantics should be captured via `modelReference` annotations on `<wsdl:portType>` entities.

3 FUSION Semantic Registry Architecture

A UDDI-based service registry supporting semantically-enhanced publication and discovery can be realised in a multitude of ways. A number of relevant attempts, each addressing a different set of requirements, are reviewed in the related work section of this paper. This section provides an overview of the architecture employed in the development of the FUSION Semantic Registry, and the programmatic interfaces that it exposes.

A distinctive characteristic of the FUSION Semantic Registry architecture that is discussed in this section is that it augments the purely syntactic search facilities that a UDDI registry can offer with semantic matchmaking support, without requiring any modifications to the implementation of the UDDI server or the UDDI specification API. This is considered an important advantage compared to other approaches, as it allows adopters of this solution to use their existing or preferred UDDI server implementation without performing any changes, thus encouraging uptake of such technology by the industry.

As illustrated in Figure 1, we propose an architecture where the UDDI server stands independently to the semantically-enabled service registry modules. The FUSION Semantic Registry exposes two specialised Web service APIs to the client for publication and discovery functions, and is responsible for performing the associated SAWSDL parsing, OWL ontology processing, and DL reasoning operations. Approaches based on this principle of accommodating semantic processing functions without imposing any changes to the UDDI server implementation or interface have been also proposed in [10], [11], and [12].

The UDDI module that is depicted in Figure 1 can be any UDDI server implementation that complies with the UDDI v2 or v3 specification as ratified by OASIS [1], although the FUSION Semantic Registry has been developed and tested using Apache

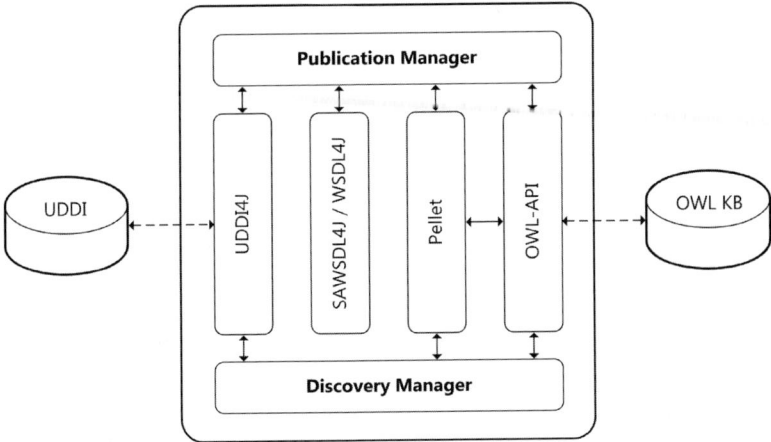

Fig. 1. FUSION Semantic Registry Architecture

jUDDI[3]. The OWL KB module is a typical OWL ontology with RDF/XML serialisation that the Semantic Registry uses for storing the Advertisement Functional Profiles it generates at the time of service publication, as will be explained in the next section of the paper. In the centre of the figure is the actual FUSION Semantic Registry, a Web Application that complies with the Java Servlet 2.4 specification and can be deployed on any compatible container implementation, such as Apache Tomcat.

The Publication Manager module of the FUSION Semantic Registry provides a Web service interface to the user for adding, removing, or updating Web service advertisements, as well as adding, removing, or updating descriptions of service providers. The Discovery Manager module provides a Web service interface for retrieving a specific service advertisement or service provider record via its key, discovering a set of services or service providers through keyword-based for terms contained in their names, and most importantly, discovering a set of services based on a Request Functional Profile. The dependencies that these two manager modules have on the third-party components that are depicted in the centre of the figure are examined in the following sections, along with the overviews of the semantic service publication and discovery processes.

4 The Publication Process

As already mentioned, the Publication Manager Module provides a Web service interface to the user for adding, removing, or updating descriptions of Web services, as well as adding, removing, or updating descriptions of service providers. This section of the paper focuses on the most important of these functions, the process of publishing a semantically-enhanced service description.

The publication query that initiates the publication process comprises: (i) the service provider ID (every service advertisement is associated to exactly one service

---

[3] http://ws.apache.org/juddi/

provider that is identified by a UUID key), (ii) a URL pointing to the SAWSDL document that describes the service, (iii) an optional service name, and (iv) an optional free text description. The process that follows based on this input comprises a number of phases that are presented in the following subsections.

4.1 Parsing of SAWSDL Document

The first step that the Publication Manager performs is to retrieve the SAWSDL document from the specified URL and parse it to extract the semantic annotations it contains. As discussed in section 2, WSDL interfaces are augmented with potentially multiple `modelReference` annotations on `<xs:element>` entities, in order to capture the data semantics of the service (consumed inputs or produced outputs), and a single `modelReference` annotation on `<wsdl:portType>` entities to capture its functionality categorisation semantics. At the time of this writing the current implementation of the Semantic Registry SAWSDL parser relies on the WSDL4J[4] and SAWSDL4J[5] libraries to create an in-memory representation of the SAWSDL document and extract the URIs of the ontological concepts to which the `modelReference` annotations point.

4.2 Construction of UDDI Advertisement

The next step in the publication process is to map the information that was provided as part of the publication query (i.e. the service name, free text description, and service provider's UUID) and the information that was extracted by parsing the SAWSDL document (i.e. input, output, and category annotation URIs), into a UDDI service advertisement. Communication between the FUSION Semantic Registry and the UDDI server for this purpose is facilitated by UDDI4J[6].

As illustrated in Figure 2, this mapping requires creating a `uddi:businessService` entity and instantiating the values of its `uddi:name`, `uddi:description`, and `uddi:businessKey` attributes, as well as a `uddi:categoryBag` that includes one `uddi:keyedReference` entity for every extracted annotation URI. The FUSION Semantic Registry makes use of so-called canonical tModels for representing the different types of semantic annotations that can be placed on SAWSDL documents (input, output, or category annotations). Depending on the type of semantic information being modelled, each `uddi:keyedReference` entity should point to the appropriate canonical tModel (Input Annotation tModel, Output Annotation tModel, or Category Annotation tModel). As depicted in Figure 2, an additional canonical tModel is used for indexing service advertisements with respect to the Request Functional Profiles that they can readily satisfy (Semantic Indexing tModel), but the `uddi:keyedReference` entities which point to this tModel are created at a later stage in the publication process.

4.3 Generation of Functional Profile and Publication-Time Matchmaking

The next step in the process is to create an Advertisement Functional Profile (AFP) based on the extracted semantic annotations and add it to the registry's internal OWL

---

[4] http://sourceforge.net/projects/wsdl4j
[5] http://knoesis.wright.edu/opensource/sawsdl4j/
[6] http://uddi4j.sourceforge.net/

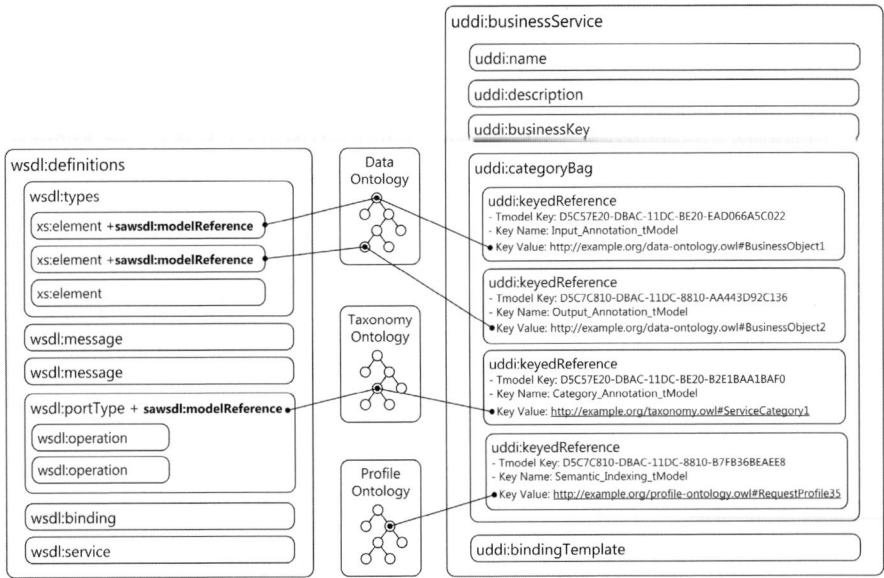

Fig. 2. SAWSDL to UDDI mapping methodology

Knowledge Base (KB) with the help of the OWL API library[7]. The construction of the AFP follows the modelling conventions analysed in section 2. Once the AFP has been constructed, the Pellet DL reasoner[8] is used for performing an "eager" semantic classification of the new AFP against all known Request Functional Profiles (RFPs). The purpose of this classification procedure is to identify RFPs representing service requests that the newly added service advertisement can readily satisfy.

We refer to this classification procedure as "eager" since it takes place at publication-time. In contrast, a "lazy" classification procedure would not have taken place before the actual need for matchmaking arises during discovery-time. This approach may be placing an overhead on the time required to complete the publication of a service advertisement, but it substantially reduces the time required to perform matchmaking at discovery-time, so it is considered particularly beneficial.

Three conditions must hold in order to claim that the new service advertisement can satisfy a service request: (i) the InputDataSet concept associated with the RFP must be subsumed by the InputDataSet of the AFP, (ii) the OutputDataSet of the RFP must subsume the OutputDataSet of the AFP, and (iii) the TaxonomyEntity concept associated with the RFP must subsume the TaxonomyEntity of the AFP.

4.4 Indexing of Semantic Matching Results in UDDI

The last step in the publication process is to map the semantic matchmaking information that resulted from the publication-time matchmaking algorithm described above

---

[7] http://owlapi.sourceforge.net/
[8] http://pellet.owldl.com/

into the UDDI service advertisement. This requires retrieving the advertised `uddi:businessService` entity and its associated `uddi:categoryBag` from the UDDI server, and creating one `uddi:keyedReference` for every RFP that the service matches with. What this essentially achieves is indexing the service advertisement with respect to all service requests it can readily satisfy. As depicted in Figure 2, `uddi:keyedReference` entities should be made to point to the canonical tModel used for this purpose (the Semantic Indexing tModel), and the URI of each RFP should be specified as the Key Value of the `uddi:keyedReference`. When this step is completed, a new semantic service advertisement has been created, registered with the UDDI registry, and is available for discovery.

5 The Discovery Process

The Discovery Manager module provides a Web service interface for retrieving service advertisements or service provider records via their unique keys, discovering sets of services or service provider records through keyword-based search, and most importantly, discovering sets of services based on a Request Functional Profile that represents the requirements of the service consumer. This latter type of semantic matchmaking functionality is the focus of this section.

The discovery query that initiates the semantic matchmaking process comprises two elements: (i) a URI pointing to some Request Functional Profile (RFP), and (ii) an optional UUID designating the preferred service provider, i.e. the company, business unit, or specific business application that should expose the service. The RFP that the URI points to may be defined within an ontology that is shared by service providers and service requestors alike (i.e. be a reusable RFP defined in the FUSION Ontology), or within some third-party ontology that imports and extends the shared ontology (i.e. be a custom-built and non-shared RFP). Depending on which of the two cases holds, the algorithm would follow a different discovery path. Resolving the location of the ontology in which the RFP is identified is therefore the first step in the discovery process.

If the RFP is defined in the shared FUSION Ontology the Discovery Manager will look for service advertisements indexed in UDDI with a reference to that RFP. This means looking for services with AFPs that have matched the requested RFP during the "eager" publication-time classification. To retrieve such advertisements the Discovery Manager places a simple syntactic matchmaking query to the UDDI server, looking for `uddi:businessService` entities having a `uddi:categoryBag` that contains a `uddi:keyedReference` which points to the Semantic Indexing tModel, and moreover, has a Key Value that is equal to the URI of the RFP.

Since the matchmaking and indexing process is repeated every time a new RFP is created and added to the shared ontology, the UDDI server's semantic matching index is bound to always be accurate and up to date. This means that if some service advertisement matches some RFP which is defined in the shared ontology, the registry is guaranteed to have this association indexed in the UDDI server, and be able to instantly retrieve the advertised service.

Due to the shared ontology assumption that is valid in the context of FUSION, this is the most typical type of discovery querying envisaged for the FUSION Semantic

Registry, and is also the simplest and fastest type of matchmaking possible. Since the time-consuming process of subsumption reasoning and hierarchy classification has been already performed at publication-time, the computational complexity of discovery-time matchmaking for RFPs defined in a shared ontology is essentially as low as that of a conventional UDDI server. In other words, the use of semantics does not impose any noteworthy overhead compared to syntactic matchmaking.

If the RFP is defined in a non-shared ontology the Discovery Manager would need to load that ontology into memory and perform a complete semantic matchmaking process among the specified RFP and all AFPs stored in the OWL-KB. The conditions that need to be checked in order to assert that a service advertisement can satisfy the request are the same as the ones defined for publication-time matchmaking (see section 4.3).

The result of the discovery process, regardless of the ontology in which the RFP is defined, is a list of UUID keys corresponding to advertisements of services that comply with the matchmaking criteria modelled in the RFP. If a service provider UUID has been also specified in the discovery query, the UDDI server will restrict the result set to only those services offered by the specified provider.

6 Related Work

The use of semantics for representing service characteristics and facilitating semantically-enhanced matchmaking in UDDI-based registries has been the focus of numerous works in recent years. Largely due to the fact that OWL-S was the first Semantic Web Service framework to be developed, most of the relevant approaches in the research literature rely on OWL-S.

In [13] the authors propose that discovery in UDDI registries should be achieved through semantic matchmaking among service capability descriptions expressed as DAML-S Profiles. To that end, they propose a matchmaking algorithm that can recognise various degrees of match among the inputs and outputs of advertisements and requests described in DAML-S, and also propose the incorporation of a matching engine inside the UDDI registry. In [14] the authors expand on the work introduced in [13] and define a mapping among DAML-S Profiles and UDDI data structures, such that semantic information can be recorded in UDDI. Subsequent work in [15] proposes a revised mapping between OWL-S Profiles and the UDDI data model, and also an improved version of the matchmaking algorithm from [13].

A research work by a different group that expands on the approach firstly introduced in [13] and [14] is presented in [16]. The authors in [16] present a method to improve the effectiveness of service discovery in UDDI based on a two-stage service discovery process, combining syntactic and semantic search for service inputs and outputs. They also propose an extension to the specification of the UDDI inquiry API in order to support automatic service composition based on DAML-S semantics.

Another approach for developing OWL-S-based semantically-extended UDDI registries is presented in [12]. The key feature of the proposed solution is that hierarchical relationships among ontology concepts are resolved at the time of publication and indexed in UDDI in a way that allows processing in a purely syntactic manner at the time of discovery. The modules for publishing and query processing are placed on the client-side and as a result no modifications to the UDDI server implementation or interface are mandated.

An approach that utilises the WSDL-S specification is introduced in [17] and elaborated in [18]. In [17] the authors present a theoretical approach for publishing Web service descriptions that have been semantically annotated with references to concepts defined in an OWL ontology. Their proposed approach includes a WSDL-S to UDDI mapping for storing these semantic annotations, and facilitating subsequent discovery of Web service operations based upon them. In a subsequent work [18] the authors describe the way in which Web service descriptions can be annotated, published and discovered using Radiant and Lumina, a pair of graphical tools that are integrated with the METEOR-S Web Services Discovery Infrastructure (MWSDI). Discovery with the proposed system is performed based on a semantic request template that specifies the desired functionality, inputs, and outputs, by references to ontological concepts. A number of research prototypes that support the WSDL-S specification and were developed in the context of METEOR-S are currently undergoing a process of harmonisation with the SAWSDL standard, but it seems that a UDDI-based service registry supporting the standard has not yet been realised.

In contrast to our matchmaking desiderata, the above described approaches do not seem to consider the problem of schema-level mismatch among inputs and outputs as an important use case, and thus do not seem to address the need for matchmaking at a fine level of granularity. With the exception of [17] and [18], classification-based matchmaking is not addressed either. Another distinction among the approach that the FUSION Semantic Registry puts forward and the other reviewed approaches, with the exception of [12], is in the proposed architecture for incorporating semantic matchmaking capabilities in UDDI. While most of the approaches necessitate some form of modification to the UDDI server's programmatic interface or internal business logic, the approach that we suggest in this paper does not mandate any such changes.

7 Conclusions

Despite its indisputable success and wide-spread adoption by the industry, the UDDI specification features a service discovery mechanism with some important limitations. The services advertised in a UDDI registry are currently being described and discovered by means of references to externally maintained technical specifications or classification schemes that lack the machine-understandable semantics that would be necessary to support fully-automated service discovery. This paper presents the implementation of the FUSION Semantic Registry, a semantically-enhanced service registry that builds on the UDDI specification and augments its service publication and discovery facilities in order to address this challenge and meet the requirements that the FUSION project puts forward for service discovery. We have presented a theoretical and technological solution that relies on a combination of three standards from the domain of Web service and Semantic Web technologies to achieve its objectives: UDDI, for storing and retrieving syntactic and semantic information about services and service providers, SAWSDL, for creating semantically annotated descriptions of service interfaces, and OWL-DL, for modelling service characteristics and performing fine-grained service matchmaking via DL reasoning. To the best of our knowledge the presented work represents the first attempt to combine these standards into a comprehensive and self-contained system. The proposed solution has been fully implemented and tested within the FUSION project, and is contributed to the community as open source software.

Acknowledgements. Research project FUSION (Business process fusion based on semantically-enabled service-oriented business applications) is funded by the European Commission's 6th Framework Programme for Research and Technology Development under contract FP6-IST-2004-170835 (http://www.fusion-strep.eu/). We would like to acknowledge the valuable contribution, to this work, of our colleagues from the FUSION consortium, Andreas Friesen (SAP Research, Germany) and Panagiotis Gouvas (Institute of Communications and Computer Systems, Greece).

References

1. Clement, L., Hately, A., von Riegen, C., Rogers, T. (eds.): UDDI Version 3.0.2 Specification, UDDI Specification Committee (October 2004)
2. Chinnici, R., Moreau, J.J., Ryman, A., Weerawarana, S. (eds.): Web Services Description Language (WSDL) Version 2.0. W3C Recommendation (June 2007)
3. Farrell, J., Lausen, H. (eds.).: Semantic Annotations for WSDL and XML Schema. W3C Recommendation (August 2007)
4. McGuinness, D.L., van Harmelen, F.: OWL Web Ontology Language Overview, W3C Recommendation (February 2004)
5. Kourtesis, D., Paraskakis, I.: Web Service Discovery in the FUSION Semantic Registry. In: Abramowicz, W., Fensel, D. (eds.) BIS 2008. LNBIP, vol. 7, Springer, Heidelberg (2008)
6. Martin, D., Burstein, M., Hobbs, J., Lassila, O., McDermott, D., McIlraith, S., Narayanan, S., Paolucci, M., Parsia, B., Payne, T., Sirin, E., Srinivasan, N., Sycara, K.: OWL-S: Semantic Markup for Web Services. W3C Member Submission (November 2004)
7. Bruijn, J.d., Bussler, C., Domingue, J., Fensel, D., Hepp, M., Keller, U., Kifer, M., Konig-Ries, B., Kopecky, J., Lara, R., Lausen, H., Oren, E., Polleres, A., Roman, D., Scicluna, J., Stollberg, M.: Web Service Modeling Ontology (WSMO). W3C Member Submission (June 2005)
8. Akkiraju, R., Farrell, J., Miller, J., Nagarajan, M., Schmidt, M.T., Sheth, A., Verma, K.: Web Service Semantics - WSDL-S. W3C Member Submission (November 2005)
9. Bouras, A., Gouvas, P., Mentzas, G.: ENIO: An Enterprise Application Integration Ontology. In: 1st International Workshop on Semantic Web Architectures For Enterprises, 18th International Conference on Database and Expert Systems Applications, Regensburg, Germany, September 3-7 (2007)
10. Colgrave, J., Akkiraju, R., Goodwin, R.: External Matching in UDDI. In: Proceedings of the 2004 IEEE International Conference on Web Services (ICWS 2004), San Diego, USA (July 2004)
11. Pokraev, S., Koolwaaij, J., Wibbels, W.: Extending UDDI with Context Aware Features based on Semantic Service Descriptions. In: Proceedings of the 2003 International Conference on Web Services (ICWS 2003), Las Vegas, USA (June 2003)
12. Luo, J., Montrose, B., Kim, A., Khashnobish, A., Kang, M.: Adding OWL-S Support to the Existing UDDI Infrastructure. In: Proceedings of the 2006 IEEE International Conference on Web Services (ICWS 2006), Chicago, USA (September 2006), ISBN 0-7695-2669-1

13. Paolucci, M., Kawamura, T., Payne, T.R., Sycara, K.: Semantic Matching of Web Service Capabilities. In: Horrocks, I., Hendler, J. (eds.) ISWC 2002. LNCS, vol. 2342, Springer, Heidelberg (2002)
14. Paolucci, M., Kawamura, T., Payne, T.R., Sycara, K.: Importing the Semantic Web in UDDI. In: Proceedings of Web Services, E-Business and Semantic Web Workshop, Toronto, Canada, May 2002, pp. 225–236 (2002)
15. Srinivasan, N., Paolucci, M., Sycara, K.: Adding OWL-S to UDDI, Implementation and Throughput. In: Cardoso, J., Sheth, A.P. (eds.) SWSWPC 2004. LNCS, vol. 3387, Springer, Heidelberg (2005)
16. Akkiraju, R., Goodwin, R., Doshi, P., Roeder, S.: A method for semantically enhancing the service discovery capabilities of UDDI. In: Proceedings of the Workshop on Information Integration on the Web (IIWeb 2003), Acapulco, Mexico (August 2003)
17. Sivashanmugam, K., Verma, K., Sheth, A., Miller, J.: Adding Semantics to Web Services Standards. In: Proceedings of the 2003 International Conference on Web Services (ICWS 2003), Las Vegas, USA (June 2003)
18. Li, K., Verma, K., Mulye, R., Rabbani, R., Miller, J., Sheth, A.: Designing Semantic Web Processes: The WSDL-S Approach. In: Cardoso, J., Sheth, A. (eds.) Semantic Web Services, Processes and Applications, pp. 163–198. Springer, Heidelberg (2006)

Web Service Composition with User Preferences

Naiwen Lin[1], Ugur Kuter[1], and Evren Sirin[2]

[1] Department of Computer Science and Institute for Advanced Computer Studies,
University of Maryland, College Park, MD 20742, USA
{nwlin,ukuter}@cs.umd.edu
[2] Clark & Parsia, LLC, 926 N Street, NW Rear, Studio # 1,
Washington, DC 20001, USA
evren@clarkparsia.com

Abstract. In Web Service Composition (WSC) problems, the composition process generates a composition (i.e., a plan) of atomic services, whose execution achieves some objectives on the Web. Existing research on Web service composition generally assumed that these objectives are absolute; i.e., the service-composition algorithms must achieve all of them in order to generate successful outcomes; otherwise, the composition process fails altogether. The most straightforward example is the use of OWL-S process models that specifically tell a composition algorithm how to achieve a functionality on the Web. However, in many WSC problems, it is also desirable to achieve users' preferences that are not absolute objectives; instead, a solution composition generated by a WSC algorithm must satisfy those preferences as much as possible. In this paper, we first describe a way to augment Web Service Composition process, where services are described as OWL-S process models, with qualitative user preferences. We achieve this by mapping a given set of process models and preferences into a planning language for representing Hierarchical Task Networks (HTNs). We then present SCUP, our new WSC planning algorithm that performs a best-first search over the possible HTN-style task decompositions, by heuristically scoring those decompositions based on ontological reasoning over the input preferences. Finally, we discuss our experimental results on SCUP.

1 Introduction

Web Service Composition (WSC), dynamic integration of multiple Web Services to fulfill the requirements of the task at hand, is one of the most important area in the Web Services research. Most research on composing Web services has been focused on developing new algorithms based on existing AI planning techniques. Examples of this approach include [1,2,3,4,5,6]. All of these works generally assumed that the composition process generates a solution composition (i.e., a plan) of atomic services, whose execution achieves some objectives on the Web. Those objectives are almost always required to be *absolute*; i.e., a service-composition algorithm must achieve all of them in order to generate successful outcomes; otherwise, the composition process fails altogether. As an example, in WSC planning systems such as [2,4,6], such absolute objectives and constraints specify structural ways of decomposing higher-level process models in OWL-S [7] into an atomic-level composition that solves the input WSC problem.

In many real-life WSC scenarios, on the other hand, the aim is to generate a solution that not only achieves some absolute goals and any constraints associated with them, but also is desirable with respect to some user-provided preferences. For example, there are abundant number of travel Web sites providing transportation and accommodation services. Possibly, there will be several combinations of these services that will allow one to plan for a trip from a source location to a destination. However, an agent making such trip arrangements will need to take into account user preferences such as a desired range on the total cost of the trip, preferences on particular transportation companies/hotels, and certain times/dates for the trip. None of these are necessarily absolute goals in the sense that, if violated, they do not affect the correctness of the compositions but specify that certain compositions are more preferable than others.

In this paper, we describe a new WSC technique in which the Web services are described as OWL-S process models and the composition process is augmented with user preferences. Our contributions are as follows:

- We first describe a planning formalism that is based on the previous languages developed for *Hierarchical Task Networks (HTNs)*. Then, we describe a way to take OWL-S service descriptions and preferences described in the PDDL3 language [8], and translate them into our formalism. This enables us to investigate the semantics of user preferences in the context of HTNs and provides a clear, unifying framework for augmenting service composition process with user preferences.
- We describe our new automated planning algorithm, called SCUP (for *Service Composition with User Preferences*). SCUP combines HTN planning with best-first search that uses a heuristic selection mechanism based on ontological reasoning over the input user preferences, state of the world, and the HTNs. In this context, the ontological reasoning provides a way to compute the heuristic cost of a method before decomposing it. Using best-first search, SCUP generates compositions for WSC problems with minimal cost of violations of the user preferences.
- We present an experimental evaluation, demonstrating that our approach is a promising one. On two benchmark problem suites that were used in the latest International Planning Competition (IPC-5) [9], SCUP generated solutions that had substantially better quality than the ones generated by the planning algorithm SGPlan [10,11], the winner of the IPC-5 in the"Planning with Qualitative Preferences" track.

2 Preliminaries

Our formalism is based on the several previous HTN-planning formalisms; in particular, we use similar definitions as for the *Universal-Method Composition Planner (UMCP)* [12,13,14] and the *HTN-DL* WSC system [4,15].

We assume the traditional definitions for logical constant and variable symbols, atoms, literals, and ontologies. A *state* is a set of ontological assertions describing the world. A state s is *consistent* if there is an interpretation μ that satisfies all axioms and assertions in s; otherwise we say the state is *inconsistent*.

We assume the existence of finite set of symbols, called *task symbols*. A *task* is an expression of the form $t = \langle head, \text{Inp}, \text{Out}, \text{Pre}, \text{Eff} \rangle$. $head$, the name of this task, is a task symbol. Inp and Out, the set of input and output parameters respectively, are finite sets of variable symbols. Pre and Eff are the preconditions and the effects of the task, respectively, both of which are in the form of conjunctive ontological assertions.

A *task network* is of the form $((n_1 : t_1, n_2 : t_2, \ldots, n_k : t_k), \Delta)$ such that each n_i is a unique label symbol, each t_i is a task, and Δ is a set of *task-network constraints*. Label symbols in a task network are used to differentiate with different occurrences of a particular task in the network. We define two kinds of task-network constraints: *task-ordering* and *state constraints*. A *task-ordering constraint* is an expression of the form $(n \prec n')$, where n and n' are the labels associated with two tasks t and t' in the current task network, respectively. Intuitively, $n \prec n'$ means that the task denoted by the label n must be accomplished before the task denoted by n'.

A *state constraint* is either of the following forms: (n, ϕ), (ϕ, n), or (n, ϕ, n'). Here, ϕ is an arbitrary logical formula, and n, n' are label symbols. Intuitively, state constraints (n, ϕ) and (ϕ, n) mean that the literal ϕ must be true in the state immediately after or before the task labeled with n, respectively. These constraints are used to specify the effects and preconditions of a task labeled by n. A state constraint of the form (n, ϕ, n') means that ϕ is true in every state between n and n'.

If we want a state constraint to hold for every occurrence of the task regardless of the label symbols, we simply use the task name in the constraint, e.g. (ϕ, t). We also use *task variables* in constraint expressions, e.g. $(\phi, ?t)$ and $(?t_1 \prec ?t_2)$, in order to represent an existential constraint. Intuitively, a task variable will be bound (only once) to a task that exists in the task network such that the constraint will be satisfied. If no such binding is possible throughout the decomposition process we say that the constraint is violated. In addition, we allow a *task-binding constraint* in the form $(?t = t)$ and $(?t \neq t)$ to restrict the tasks that a task variable can be bound to.

Let s be a state, w be a task network, and x be a task-network constraint in w. Then, we define a *violation-cost function*, χ, as the partial function:

$$\chi(s, w, x) = \begin{cases} c, & \text{if } x \text{ does not hold in } w \text{ given } s, \\ 0, & \text{otherwise} \end{cases} \quad (1)$$

where c is either a positive number or it is ∞. The *overall violation cost* of a task network $w = (\tau, \Delta)$ in a state s is

$$\Xi(s, w, \Delta) = \sum_{x \in \Delta} \chi(s, w, x). \quad (2)$$

A *WSC planning operator* is an expression of the form $o = (task, \text{Inp}, \text{Out}, \text{Pre}, \text{Eff})$ where $task$ is a task symbol that specifies the task of this operator, Inp is the set of input parameters, Out is the set of output parameters, Pre is the preconditions of the operator, and Eff is the effects of the operator. Pre and Eff are in the form of conjunctive ontological assertions. An *action* is a ground operator instance which can be executed directly on the Web. A *plan* (or equivalently, a *service composition*) is a sequence of actions. We describe the state change from s to s' when applying an action a by the *state-transition function* γ: that is, $s' = \gamma(s, a)$

A WSC planning operator o is *applicable* to a task t in a state s, if there is a variable substitution σ such that $\sigma(o)$ is ground, the preconditions of $\sigma(o)$ hold in the state s, o has the same task symbol as in t, and $\sigma(o)$ entails the effects of $\sigma(t)$.

A *task-decomposition method* is a tuple $m = (task, \mathsf{Inp}, \mathsf{Out}, \mathsf{Local}, \mathsf{Pre}, \mathsf{Eff}, \Gamma)$, where $task$, Inp, Out, Pre, and Eff are defined the same as in the above definition of a planning operator. Γ is a task network. Local is a set of local variables used in P and Γ. Local variables are those variables in m that does not appear in the inputs, but are bound in the preconditions of the method or of a task network. A method m is *applicable* to a task t in a state s, if there is a variable substitution σ such that $\sigma(m)$ is ground, the preconditions of $\sigma(m)$ are satisfied in the state s, m has the same task symbol as in t, and the effects of the $\sigma(m)$ entails the desired effects of $\sigma(t)$.

A *WSC planning domain* D is a triple of (O, M, T_{ont}), where O is the set of operator descriptions, M is the set of methods, and T_{ont} is the task ontology. We assume that the task ontology T_{ont} is an OWL ontology [16] that describes the set of all possible tasks in the domain. A *WSC planning problem* is a triple $P = (s_0, w, D, \Xi)$, where s_0 is the initial state, w is the initial task network, D is the WSC planning domain, and Ξ is the overall violation-cost function. A *solution* for P is a task network $w^* = (\tau, \Delta)$ such that (1) all of the tasks in τ are primitive; (2) there exists a total-ordering of the tasks in τ (i.e., a solution plan $\tau = \langle a_1, a_2, ..., a_k \rangle$) that has a minimum overall cost $\Xi(s_k, w^*, \Delta) < \infty$, where s_k is the final state generated by applying the actions in τ in s_0. Note that a plan that has a violation-cost of ∞ is not a solution even if it somehow generates the final state, when applied in the initial state s_0.

We remark that our formalism above is not intended to depend on a particular WSC system and can be implemented as the input languages of any of the existing HTN-based WSC systems such as [4,15,6]. The rationale behind this formalism is that it provides a clear semantics for augmenting a Web Service composition process, where the services are described as OWL-S process models, and user preferences in a single unifying framework, as we will describe and discuss in the rest of this paper.

3 Modeling User Preferences

Our preference language is largely based on the recent language **PDDL3** [8] that allows to incorporate user preferences in AI planning problems. In **PDDL3**, preferences are described as logical assertions over states and state trajectories by defining *basic preferences* and *temporal preferences*. A *basic preference* (BP) is a logical formula ϕ, or a formula built with logical connectives $\neg, \wedge, \vee$ from other basic preferences. Let ϕ be a basic preference. Then, a *temporal preference* (TP) is any of the following:

$$\langle TP \rangle := always(\langle BP \rangle) \mid sometime(\langle BP \rangle) \mid at\_most\_once(\langle BP \rangle) \mid final(\langle BP \rangle) \mid$$
$$sometime\_after(\langle BP \rangle, \langle BP \rangle) \mid sometime\_before(\langle BP \rangle, \langle BP \rangle)$$

As the above definition demonstrates, we do not allow the nesting of temporal constraints in this paper. We assume that each temporal preference has a unique label; e.g., $p_1 : (final\ \phi)$, where p_1 is the unique label and ϕ is a basic preference.

The satisfaction of temporal constraints is defined with respect to a plan π and its state trajectory $S = \langle s_0, \ldots s_n \rangle$ in an initial state s_0 as follows.

- If ϕ is a ground atom then $\langle \pi, S, s_i \rangle \models \phi$ iff $s_i \models \phi$;
- Quantifiers and logical connectives have the same meaning as in first-order logic. For example, $\langle \pi, S, s_i \rangle \models \phi \wedge \phi'$ iff $\langle \pi, S, s_i \rangle \models \phi$ and $\langle \pi, S, s_i \rangle \models \phi'$;

- $\langle \pi, S, s_i \rangle \models always(\phi)$ iff $\forall j : i \leq j \leq n \cdot \langle \pi, S, s_j \rangle \models \phi$;
- $\langle \pi, S, s_i \rangle \models sometime(\phi)$ iff $\exists j : i \leq j \leq n \cdot \langle \pi, S, s_j \rangle \models \phi$;
- $\langle \pi, S, s_i \rangle \models final(\phi)$ iff $\langle \pi, S, s_n \rangle \models \phi$;
- $\langle \pi, S, s_i \rangle \models at\_most\_once(\phi)$ iff (1) $\exists j : i \leq j \cdot \langle \pi, S, s_j \rangle \models \phi$ and $\not\exists k : i \leq k \cdot \langle \pi, S, s_k \rangle \models \phi$, or (2) there is no $i \leq j$ such that $\langle \pi, S, s_j \rangle \models \phi$;
- $\langle \pi, S, s_i \rangle \models sometime\_after(\phi, \phi')$ iff (1) $\exists j : i \leq j \cdot \langle \pi, S, s_j \rangle \models \phi$ and $\exists k : j \leq k \cdot \langle \pi, S, s_k \rangle \models \phi'$, or (2) true, if $\not\exists j : i \leq j \cdot \langle \pi, S, s_j \rangle \models \phi$;
- $\langle \pi, S, s_i \rangle \models sometime\_before(\phi, \phi')$ iff (1) $\exists j : i \leq j \cdot \langle \pi, S, s_j \rangle \models \phi$ and $\exists k : i \leq k \leq j \cdot \langle \pi, S, s_k \rangle \models \phi'$, or (2) true, $\not\exists j : i \leq j \cdot \langle \pi, S, s_j \rangle \models \phi$;

PDDL3 also allows for representing preferences with universal and existential quantification. As an example, consider the following two different kinds of quantified preferences specified in **PDDL3** for a transportation domain:

```
p1A: sometime (forall (?t - truck) at(?t, loc1))
     forall (?t - truck) (p1B: (sometime at(?t, loc1)))
```

The preference p1A suggests all trucks stay at location $loc1$ sometime in the plan trajectory. If a truck does not satisfy the condition, the preference $p1A$ is violated. The preference p1B defines a group of preferences with different truck instantiation: e.g., if there are three trucks in the problem defintion, $p1B$ defines three independent preferences. Violation of one preference will not affect the satisfaction of the others.

4 Augmenting Web Service Composition with User Preferences

In this section, we describe a translation methodology that takes **OWL-S** descriptions of Semantic Web services and **PDDL3**-style preferences as above, and maps them into our planning formalism described in Section 2. The rationale behind this methodology is to represent both the Web services and the preferences in a single unifying framework, in which a planning algorithm could work to generate compositions of services while satisfying the user preferences as much as possible.

From OWL-S to HTN-Based Constraints. Our translation method for encoding **OWL-S** in our HTN-based constraint language is very similar to the technique described in [4] for encoding **OWL-S** in the input language of the **SHOP2** planner [17].

Suppose K is a collection of **OWL-S** process models. First of all, for every process C in K, we create a task of the form $(C\ \boldsymbol{v}\ \boldsymbol{u}\ \boldsymbol{p}\ \boldsymbol{e})$ where $\boldsymbol{v}$ and $\boldsymbol{u}$ are the list of input and output parameters defined in D_C, $\boldsymbol{p}$ and $\boldsymbol{e}$ are the preconditions and effects of D_C, and a corresponding label n_C for that task.

We translate each atomic process in K into a corresponding planning operator in the same way as in [4]. The translation of a composite process C in K with a **sequence** control construct is as follows. Let D_C be the **OWL-S** definition of the process C. Next, we define an HTN method m in our constraint language whose head is C and whose inputs, outputs, preconditions, and effects are $\boldsymbol{v}, \boldsymbol{u}, \boldsymbol{p}$ and $\boldsymbol{e}$, respectively. The task network specified by the method m is translated from the process C as follows:

1. For each precondition p defined in D_C, we create a state-constraint of the form (p, n_C) where n_C is the label of the task corresponding to the process C.
2. For the sequence $C_1, C_2, \ldots, C_k$ of composite process defined in the **sequence** control construct of C, we do the following.
   - We define a task $t_i = (C_i\ v_i\ u_i\ p_i\ e_i)$ as above, and a corresponding task label n_i. The task network in the method m contains the tasks defined as $((n_i : t_i))_{i=1}^{k}$.
   - For each $1 \leq i < k$, we define an ordering constraint $(n_i \prec n_{i+1})$ where n_i and n_{i+1} are task labels for the tasks that correspond to the processes C_i and C_{i+1}.

Both sets of constraints above go into the constraint definition Δ of the task network associated with the method m.

The translation of a process C with an **if-then-else** control construct is as follows. Let P_{if} be the conditions for the **if** clause as defined in D_C. We define two HTN methods m_{if} and m_{else} to encode C such that the heads of both methods are the same task $(C v)$. Then, we do the following translation for the constraint part of those methods:

- For m_{if}:
  1. For each precondition p either defined in D_C or in P_{if}, we define a state-constraint (p, n_C) where n_C is the label of the task corresponding to C.
  2. For the process C_{then} defined for the **then** construct in D_C, we define a task $t = (C_{then}\ v_{then}\ u_{then}\ p_{then}\ e_{then})$ as above, and a corresponding label n. The task list of the task network associated with the method m_{if} is $((n : t))$.
- For m_{else}:
  1. For each precondition p defined in D_C, we define a state-constraint of the form (p, n_C) where n_C is the label of the task corresponding to the process C.
  2. For each precondition p defined in P_{if}, we define a state-constraint of the form $(\neg p, n_C)$ where n_C is the label of the task corresponding to the process C.
  3. For the process C_{else} defined for the **else** construct in D_C, we define a task $t = (C_{else}\ v_{else}\ u_{else}\ p_{else}\ e_{else})$ as above, and a corresponding label n. The task list of the task network associated with the method m_{else} is $((n : t))$.

The translations for **repeat-while**, **repeat-until**, **choice**, and **unordered** constructs that may appear in the processes in K are similar to the ones above.

For every HTN task-network constraint generated with our translation, we assign a violation-cost of ∞, independent of the state and task network that the constraints might be evaluated in. By doing so, we ensure that the translation generates a bijection between the **OWL-S** process models (where such constraints must be satisfied with absolute certainty) and the translated HTN constructs.

Proposition 1. *The translation from* **OWL-S** *service process models into HTN constraints is correct, i.e. for each* **OWL-S** *process construct as described above, there is a corresponding set of HTN constructs that specify the same process-model semantics.*

Translating Preferences into HTN Constraints. We will now describe how to translate a preference ψ to one compound task-network constraint $\Gamma(\psi)$. With this translation, whenever we augment a task network w with $\Gamma(\psi)$, we have a direct method to evaluate if a state trajectory (i.e., a plan) satisfies or violates the preference.

In our translation, we use two special task symbols t_{start} and t_{end} as the start and end tasks that will be added to every initial task network w with labels n_{start} and n_{end}. We also add the ordering constraint $n_{start} \prec n$ and $n \prec n_{end}$ for every $n : t \in w$.

We use the special symbol $\top$ to denote a trivially satisfiable HTN constraint and $\bot$ to denote an unsatisfiable HTN constraint with the usual semantics that $\neg \top = \bot$, $\neg \bot = \top$, $\top \wedge \phi = \psi$, $\top \vee \psi = \top$, $\bot \wedge \psi = \bot$ and $\bot \vee \psi = \psi$. If $\Gamma(\psi) = \top$, then every plan satisfies ψ and if $\Gamma(\psi) = \bot$ there is no plan that can satisfy ψ.

Now we describe how we construct $\Gamma(\psi)$ by analyzing the temporal preferences separately:

- $\Gamma(\psi) \equiv (n_{start}, \phi, n_{end})$, if $\psi \equiv always(\phi)$, where ϕ is a basic preference.
- $\Gamma(\psi) \equiv (?t, \phi)$, if $\psi \equiv sometime(\phi)$, where ϕ is a basic preference.
- $\Gamma(\psi) \equiv (\phi, n_{end})$, if $\psi \equiv final(\phi)$, where ϕ is a basic preference.
- $\Gamma(\psi) = (?t_1, \phi) \wedge (?t_2, \phi') \wedge (?t_1 \prec ?t_2)$, if $\psi \equiv sometime\_after(\phi, \phi')$.
- $\Gamma(\psi) = (?t_1, \phi) \wedge (?t_2, \phi') \wedge (?t_2 \prec ?t_1)$, if $\psi \equiv sometime\_before(\phi, \phi')$.
- $\Gamma(\psi) = (\phi, ?t)$ and for every task t, $(\phi, t) \implies ?t = t$, if $\psi \equiv at\_most\_once(\phi)$.

We translate the universally- and existentially-quantified preferences as follows:

- If $\psi \equiv \forall (?x) \cdot \phi$, where $?x$ is a variable symbol, then we apply the following rules:
  1. If ϕ is a basic preference, then $\Gamma(\psi) \equiv \Gamma(\phi(?x = x_1) \wedge \phi(?x = x_2) \wedge \ldots \wedge \phi(?x = x_n))$, for all possible instantiations x_i of $?x$, $i = 1, \ldots, n$.
  2. If ϕ is a temporal preference, then $\Gamma(\psi) \equiv \{\Gamma(\phi(?x = x_1)), \Gamma(\phi(?x = x_2)), \ldots, \Gamma(\phi(?x = x_n))\}$, for all possible instantiations x_i of $?x$, $i = 1, \ldots, n$.
- We have $\psi \equiv \exists(?x)(\phi)$ as $\Gamma(\psi) \equiv \phi$ since existential quantification has the same logical meaning for both basic and temporal preferences.

Proposition 2. *The translation from preferences into HTN constraints is correct: if $\pi = \langle a_1, a_2, \ldots, a_k \rangle$ is a solution plan to a WSC planning problem (s_0, w, D, Ξ) with a violation cost of v, where the user preferences ψ are translated into HTN constraints $\Gamma(\psi)$ in the task network w, then $v = \Xi(s_k, w^*, \Gamma(\psi))$, where $w^* = (\pi, \Delta)$, s_k is the final state generated by applying π in s_0 and $\Gamma(\psi) \subseteq \Delta$.*

5 SCUP: Service Composition with User Preferences

We are now ready to describe our new WSC planning algorithm, SCUP, for generating service compositions while satisfying user preferences as much as possible. Figure 1 shows the abstract description of the SCUP algorithm. SCUP takes five inputs s_0, w_0, D, ψ, and Ξ. s_0 is the initial state and w_0 is the initial task network. D is the domain knowledge including the task-ontology definitions, HTN methods, and planning operators. ψ is the set of user preferences and Ξ is the overall violation cost function.

The algorithm first generates the task network w, where the tasks in $w_0 = (\tau, \Delta)$ are rearranged based on the ordering constraints in w_0 and the ordering constraints in the translation $\Gamma(\psi)$ of the input preferences in ψ. The Preprocess subroutine is responsible for this operation (Line 2 of the pseudocode in Figure 1). Preprocess first scans

```
Procedure SCUP($s_0, w_0, D, \psi, \Xi$)
  1.  $\pi \leftarrow \emptyset$
  2.  $w = (\tau, \Delta) \leftarrow$ Preprocess($w_0, D, R$)
  3.  $OPEN \leftarrow \{(s_0, w, \pi, \Xi(s_0, w, \Delta))\}$
  4.  while $OPEN \neq \emptyset$ do
  5.    select the first node $x = (s, w, \pi, v)$ from $OPEN$ and remove it
  6.    if $\pi$ is a solution then return $\pi$
  7.    $Succ \leftarrow$ Decompose($s, w, \pi, v, D$)
  8.    $OPEN \leftarrow OPEN \cup$
              $\{(s', w', \pi', \Xi(s', w', \Delta')) \mid (s', w' = (\tau', \Delta'), \pi') \in Succ\}$
  9.    sort $OPEN$ in the ascending order based on the cost function $\Xi$
  10. return $failure$
```

Fig. 1. An abstract description of the SCUP algorithm

all preferences and see if there are final, sometime_before, sometime_after, and sometime preferences. If there are no such preferences in ψ, the algorithm simply returns the original task network w_0 with $\Delta \cup \Gamma(\psi)$. Otherwise, Preprocess checks all of the preferences of the above types and verify if additional tasks will be added in w_0 in order to guarantee the satisfaction of these preferences in ψ as much as possible.

The rationale behind the preprocessing step is as follows:

- **Preprocessing (sometime ϕ).** If no task in τ can achieve the effect ϕ, there might be still other tasks in the input task ontology, when/if decomposed into a primitive HTN successfully, can achieve ϕ. Thus, Preprocess searches for such an additional task t whose effect entails ϕ. If there is such a task in the input task ontology, Preprocess adds t in τ and the state constraint (t, ϕ) into Δ. The state-constraint (t, ϕ) ensures that the condition ϕ is supposed to be achieved after the decomposition of t is done. As a result, if the task t is accomplished successfully during planning, then the preference ϕ will be satisfied as well.

- **Preprocessing (sometime_before ϕ ϕ').** In this case, Preprocess generates two sets of tasks T_1 and T_2 that can accomplish the effect ϕ and ϕ', respectively. For each task t in T_1 with a label n, the subroutine chooses a task t' in T_2 with a label n', and adds the task-network constraints $(n \prec n')$, (n, ϕ), and (n', ϕ') to Δ.

The preprocessing of final and sometime_after preferences are are similar to sometime and sometime_before as described above.

After adding all task-ordering constraints into the task network w_0 based on the preferences in ψ, the set of ordering and state constraints in w_0 may not be satisfiable. For example, in a transportation-services domain as in [6], we may have both $(n_1 : load(truck1, package1) \prec n_2 : drive(truck1))$ and $(n_2 : drive(truck1) \prec n_1 : load(truck1, package1))$ in ψ. As another example, we may also have a cyclic constraint such as $(n_1 : drive(truck1) \prec n_2 : load(truck1, package1))$, $(n_2 : load(truck1, package1) \prec n_3 : deliver(package1))$ and $(n_3 : deliver(package1) \prec n_1 : drive(truck1))$ in w_0. There is no solution plan which can satisfy these HTN constraints during the planning process. Thus, Preprocess, in such cases, greedily removes the constraint with less weight from w_0 in order to satisfy more important preferences (i.e., preferences with smaller costs).

A couple of remarks on **Preprocess** are in order. First, it is important to note that the additional HTN constraints that **Preprocess** adds to Δ is due to the translation Γ of those preferences based on the additional tasks from the task ontology. Second, we only work with `final`, `sometime_before`, `sometime_after`, and `sometime` preferences in the preprocessing phase. The reason is that we can improve the cost value by adding additional tasks for `sometime` and `final`, and rearrange task orderings for `sometime_before`, `sometime_after`, and `sometime`. As for other preferences like `always` and `at_most_once`, we can only check their satisfaction in decomposition and choose the best tuple in $OPEN$ list by using best first search.

After rearranging the task network $w = (\tau, \Delta)$, the algorithm computes the overall violation-cost value $v = \Xi(s_0, w, \Delta)$ and puts the tuple (s_0, w, π, v) in the $OPEN$ list (see Line 3), where π is the current partial plan. As we mentioned above, SCUP is an HTN task-decomposition planner that is based on a best-first search procedure. At each iteration of the loop, SCUP first takes the first tuple $(s, w, \pi, v) \in OPEN$ and removes it. If π is a solution composition (i.e., plan), then SCUP returns π. Otherwise, the algorithm decomposes the task-network w according to the task-decomposition methods in D, generates the set $Succ$ of successor state and task-network pairs, computes the violation costs for those new pairs in $Succ$, and puts them back in $OPEN$ (see Line 9). The tuples in $OPEN$ are then sorted in the ascending order of their violation costs. This process continues until SCUP finds a solution or fails to generate a plan.

In Line 8 of Figure 1, the **Decompose** subroutine first nondeterministically chooses a task t in w with no predecessors and finds the set A of all applicable operators and methods for task t. If it cannot find any applicable operators or methods, it simply returns failure. Otherwise, for each item u in A, **Decompose** does the following:

- If u is an operator, **Decompose** first generates the action a by applying input bindings to u. Then, it computes the next state $s' = \gamma(s, a)$, where s is the current state, and the successor task-network w' by removing t from w. It appends a to the partial plan π.
- If u is a method, the subtasks w'' will replace task t in w. The effects of each task in w'' is then used to check if any preference is satisfied or violated in the current state, and thus, to update the overall cost. This helps us foresee the violation costs without further decomposition, and decide whether we should continue on this task network. **Decompose** adds in $Succ$ the tuple with the same state s, the updated task nework w' by replacing t with w'' in w, and the same partial plan π.

After the loop over all applicable operators and methods, **Decompose** checks and removes duplicate tuples in $Succ$ that are those which have the same task network and partial plan, in order not to repeat the same decompositions during planning.

An example of how **Decompose** uses the task effects to compute the heuristic violation costs is in order. First, recall that the preprocessing phase generates the postcondition state-constraint (t, ϕ) for each task t that is inserted into the task network due to a preference. Once the planner completely decomposes t into a primitive task network, **Decompose** checks if the post-condition ϕ is satisfied in the current state.

For example, suppose we have a task $SendPackage(package1, location3)$ with an post-condition $delivered(truck3, package1)$ due to a preference (p1 : sometime $delivered(truck3, package1)$). The following is an abstract description of an **OWL-S** service for $SendPackage$:

Composite Service: SendPackage
  Input Parameters: (?pac ?loc)
  Local Parameters: (?truck)
  Preconditions: (available(?truck))
  ComposedOf: (Load(?truck, ?package), Deliver(?truck, ?pac, ?loc))
  Results: (delivered(?truck, ?pac))

Note that $truck3$ is not the input to $SendPackage$, and the HTN method that corresponds to the above service uses a variable $?truck$ to match any available trucks near $package1$. Suppose the planner finds three trucks $truck1, truck2, truck3$ in the current state. Then, it will create three tuples in $Succ$ each possible truck. However, since $SendPackage$ has the post-condition $delivered(truck3, package1)$, only the tuple with the truck variable binding $truck3$ will satisfy the preference $p1$ and the others will violate it, increasing their cost values. Thus, when the tuples in $Succ$ are merged into the $OPEN$ list and sorted based on the costs of its tuples, **SCUP** will always consider the tuple that satisfies the preference $p1$.

6 Implementation and Preliminary Experiments

We have implemented a prototype of the **SCUP** algorithm. Our current prototype is built on our previous WSC system HTN-DL [15]. HTN-DL does not directly implement our planning language as we described in Section 2. Thus, we assumed that the translation from OWL-S to HTN-DL has been done automatically as described in [15] and we implemented the translation of user preferences from **PDDL3** into HTN-DL's input language. Furthermore, we used the interface functionality between HTN-DL and Pellet, an OWL DL reasoner [18], that we used in our **SCUP** prototype for knowledge inference and ontological reasoning.

During planning, the size of **SCUP**'s $OPEN$ list may get very large and this may induce serious performance drawbacks for the **SCUP** algorithm. For that reason, we implemented an additional input parameter k that bounds the size of the $OPEN$, which, as a trade-off, may affect the solution plan quality. If k is too small, a tuple in $OPEN$ with prospective solution may be discarded in the first few decompositions. This happens when the local optimal search node is not the global optimal search node. In our experiments described below, we used $k = 20$ as the size of the $OPEN$ list and this was sufficient to generate optimal (or in some cases, near-optimal) solutions.

We have designed the following preliminary experiments with **SCUP** in order to evaluate our planner's performances with compared to other approaches. Since, to the best of our knowledge, the only preference-based WSC techniques are based on AI planning algorithms, we have chosen an AI planning algorithm, SGPlan [10,11], that was participated in the "Planning with Qualitative Preferences" track of the most recent International Planning Competition (IPC-5) (http://zeus.ing.unibs.it/ipc-5/), and won the 1st place; hence, it is considered as the state-of-the-art

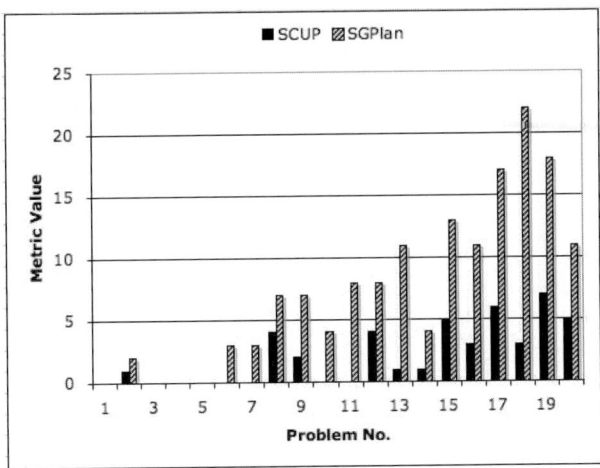

Fig. 2. The violation costs of the solutions generated by SCUP and SGPlan on the Trucks problems

technique for preference-based planning. We have not run the SGPlan ourselves for our experiments; instead, we used the published results for it from the IPC-5 [9]. Note that this does not affect our results since we are comparing SCUP with SGPlan on the costs of their solutions and neither of the planners had any memory issues in our experiments.

We have conducted experiments with two benchmark planning domains used in IPC-5: Trucks and Rovers. In the Trucks domain, the goal is to move packages between locations by using trucks. There are only four operators *load*, *drive*, *unload*, and *deliver*. Each truck may have multiple truck areas so as to carry packages, but there are constraints and penalties on the ordering of loading packages and the input user preferences model choices over the trucks based on those constraints. Generally, loading multiple packages will result in preference violations, and on the other hand, delivering only a single package at a time may delay the delivering deadlines of other packages.

In the Rovers domain, the goal is to navigate the rovers on a planet surface, collect scientific data such as rock and soil samples, and send them back to a lander. These tasks need to be achieved by considering the input spatial constraints and temporal preferences on the operation of a rover. Each rover has only a limited storage capacity for the collected samples and it is only capable of sampling either soil, rock, image, or some combination of them specified in the problem descriptions. A rover can only travel between certain waypoints only when the path from source to destination is visible.

We used the exact planning domain and problem descriptions that were used in IPC-5. For each planning domain, we have 20 problems with increasing number of instances, goals and preferences. In our experiments, we compared the overall violation cost values of the solutions generated by SCUP and SGPlan.

Figure 2 shows our results in the Trucks domain. SCUP has generated solutions that have cost values that are substantially less than SGPlan in all 20 problems, where SCUP satisfied all preferences in 8 of the experimental problems. The average cost value for SCUP is 2.00, less than one-third of SGPlan's average 7.45. Most violations in SCUP

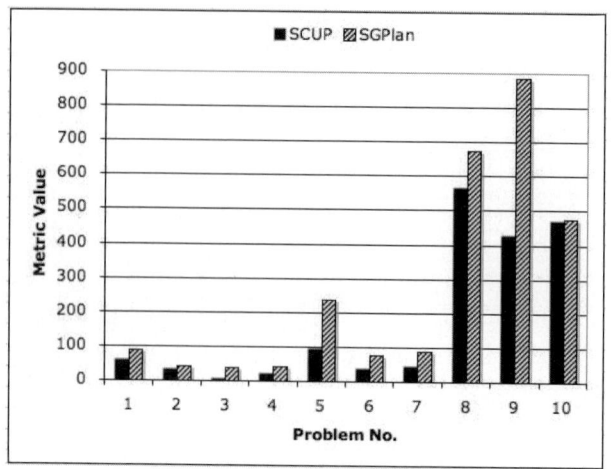

Fig. 3. The violation costs of the solutions generated by SCUP and SGPlan on the smaller Rover problems

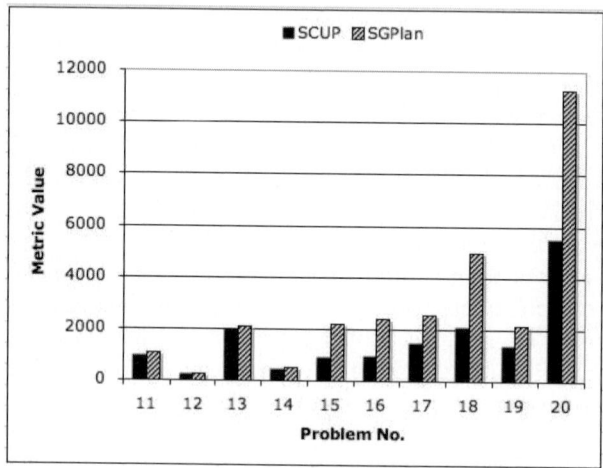

Fig. 4. The violation costs of the solutions generated by SCUP and SGPlan on the larger Rover problems

solutions result from the conflicts between preferences; e.g., delivering multiple packages on a particular truck is necessary to satisfy delivery deadline preferences that had larger costs, and this unavoidably causes violations of using multiple truck areas.

Figures 3 and 4 illustrate how SCUP and SGPlan performed in the Rovers domain. SCUP has outperformed SGPlan in 19 out of 20 problems. In these experiments, we found that as the size of the problems is increased, SCUP discarded possible solution tuples due to the $OPEN$ list size limitation. However, generally SCUP has higher quality solutions with average cost value **888.645**, compared to SGPlan's **1608.378**.

7 Related Work

There has been many advances in WSC planning in the recent years [2,4,6,19,20,21]. Probably the first work in this research area is the one described in [2], where the states of the world and the world-altering actions are modeled as Golog programs, and the information-providing services are modeled as external functions calls made within those programs. The goal is stated as a Prolog-like query and the answer to that query is a sequence of world-altering actions that achieves the goal, when executed in the initial state of the world. During the composition process, however, it is assumed that no world-altering services are executed. Instead, their effects are simulated in order to keep track of the state transitions that will occur when they are actually executed.

In [4], the WSC procedure is based on the relationship between the OWL-S process ontology [7] used for describing Web services and *Hierarchical Task Networks* as in HTN Planning [17]. OWL-S processes are translated into tasks to be achieved by the SHOP2 planner [17], and SHOP2 generates a collection of atomic process instances that achieves the desired functionality. [6] extended this work to cope better with the fact that information-providing Web services may not return the needed information immediately, or at all. The ENQUIRER algorithm presented there does not cease the search process while waiting answers to some of its queries, but keeps searching for alternative compositions that do not depend on answering those specific queries.

In all works above, search for desirable solutions have been incorporated into the service-composition process as *hard constraints*; i.e., constraints that must be satisfied by all of the solutions. Recently, several different approaches have been developed for planning with preferences; i.e., *soft constraints* that are *preferably* but *not necessarily* satisfied by a plan. There are various different approaches for integrating user preferences in the planning process. Examples include [22,23,10,24].

In our experimental study here, we considered one of those state-of-the-art planners, SGPlan [10,11], that won the recent International Planning Competition in the "Planning with Qualitative Preferences" track. SGPlan is a planning algorithm that uses a divide-and-conquer approach: the planner serializes a large planning problem into sub-problems with subgoals, solves the sub-problems, merges the solutions to those sub-problems, and tries to remove conflicts between them. It has been demonstrated in experimental studies that this approach largely reduces the search space compared to the state-space of the original problem.

In another work that is very related to our paper, Baier et al. proposed the heuristic-search planner HPlan-P in [24]. They used a best-first search algorithm with a goal distance function to find the first solution, and tried to improve solution by using other heuristics. Unfortunately, we were not able to get the HPlan-P planner for experimentation due to its distribution-licensing problems at the time we personally communicated with the authors of that planner, and since SGPlan significantly outperformed HPlan-P in IPC-5, we have not included its results here.

8 Conclusions and Future Work

In many interesting Web Service Composition (WSC) problems, the goal is to generate desirable solutions with respect to some user preferences that are not necessarily absolute objectives or constraints, but a composition algorithm needs to try to satisfy them as much as possible. In this paper, we have described a novel approach for incorporating user preferences in planning for Web Service Composition. We have first described a way to take **OWL-S** service descriptions and **PDDL3**-style preferences, and translate them into a planning language for Hierarchical Task Networks (HTNs). Based on this translation, we have then described an HTN planning algorithm, called **SCUP**, that combines HTN planning and best-first search for WSC planning with preferences.

Our preliminary experiments demonstrated that **SCUP** is a promising approach: our prototype implementation of **SCUP** was able to generate solutions that satisfy more user preferences and those preferences that have more value (i.e., less cost) than the state-of-the-art planner SGPlan, the winner of the latest International Planning Competition in the "Planning with Qualitative Preferences" track.

In the near future, we are planning to conduct extensive theoretical and experimental evaluation of our approach. In parallel to that evaluation, there are a couple of future research direction based on our work described in this paper. One possible direction is to develop ways to augment user preferences directly into **OWL-S** process models by representing them in OWL. This will enable us to use encode and reason about user preferences in real Semantic Web services on the Web.

In another direction, we have just started a new research project based on **SCUP** that will focus on reasoning about user preferences in a social setting; in this project, our objective is to develop ways to represent user preferences based on social factors such as trust and confidence in social networks. As Web-based social networks is a very important research area, this will enable us to investigate Web services in the existing social settings and how users generate service compositions to achieve their objectives based on social factors.

Acknowledgments. This work was supported in part by DARPA's Integrated Learning Program. The opinions in this paper are those of the author and do not necessarily reflect the opinions of the funders.

References

1. McDermott, D.: Estimated-regression planning for interactions with web services. In: AIPS, pp. 204–211 (2002)
2. McIlraith, S., Son, T.: Adapting Golog for composition of semantic web services. In: KR-2002, Toulouse, France (April 2002)
3. Martinez, E., Lespérance, Y.: Web service composition as a planning task: Experiments using knowledge-based planning. In: Proceedings of the ICAPS-2004 Workshop on Planning and Scheduling for Web and Grid Services, June 2004, pp. 62–69 (2004)
4. Sirin, E., Parsia, B., Wu, D., Hendler, J., Nau, D.: HTN planning for web service composition using SHOP2. Journal of Web Semantics 1(4), 377–396 (2004)
5. Traverso, P., Pistore, M.: Automated composition of semantic web services into executable processes. In: International Semantic Web Conference, pp. 380–394 (2004)
6. Kuter, U., Sirin, E., Nau, D., Parsia, B., Hendler, J.: Information gathering during planning for web services composition. In: McIlraith, S.A., Plexousakis, D., van Harmelen, F. (eds.) ISWC 2004. LNCS, vol. 3298, pp. 335–349. Springer, Heidelberg (2004)

7. W3C: Owl-based web service ontology, www.daml.org/services/owl-s/
8. Gerevini, A., Long, D.: Plan constraints and preferences in pddl3 (2006)
9. Gerevini, A., Long, D.: The fifth international planning competition (2006), http://ipc5.ing.unibs.it/
10. Hsu, C.W., Wah, B.W., Huang, R., Chen, Y.X.: Handling soft constraints and preferences in sgplan. In: ICAPS Workshop on Preferences and Soft Constraints in Planning (June 2006)
11. Hsu, C.W., Wah, B.W., Huang, R., Chen, Y.X.: Constraint partitioning for solving planning problems with trajectory constraints and goal preferences. In: IJCAI, pp. 1924–1929 (2007)
12. Erol, K., Hendler, J., Nau, D.S.: UMCP: A sound and complete procedure for hierarchical task-network planning. In: Proceedings of the International Conference on AI Planning Systems (AIPS), June 1994, pp. 249–254 (1994)
13. Erol, K., Hendler, J., Nau, D.S., Tsuneto, R.: A critical look at critics in HTN planning. In: Proceedings of the International Joint Conference on Artificial Intelligence (IJCAI) (1995)
14. Erol, K., Hendler, J., Nau, D.S.: Complexity results for hierarchical task-network planning. Annals of Mathematics and Artificial Intelligence 18, 69–93 (1996)
15. Sirin, E.: Combining Description Logic Reasoning with AI Planning for Composition of Web Services. PhD thesis, Department of Computer Science, University of Maryland (2006)
16. W3C: Web ontology language (owl) (2004), www.w3c.org/2004/OWL/
17. Nau, D., Au, T.C., Ilghami, O., Kuter, U., Murdock, W., Wu, D., Yaman, F.: SHOP2: An HTN planning system. JAIR 20, 379–404 (2003)
18. Sirin, E.: Pellet: The open source owl dl reasoner (2006), http://pellet.owldl.com/
19. Martinez, E., Lespérance, Y.: Web service composition as a planning task: Experiments using knowledge-based planning. In: ICAPS-2004 Workshop on Planning and Scheduling for Web and Grid Services (June 2004)
20. Pistore, M., Barbon, F., Bertoli, P., Shaparau, D., Traverso, P.: Planning and monitoring web service composition. In: Bussler, C.J., Fensel, D. (eds.) AIMSA 2004. LNCS (LNAI), vol. 3192, pp. 106–115. Springer, Heidelberg (2004)
21. Traverso, P., Pistore, M.: Automated composition of semantic web services into executable processes. In: McIlraith, S.A., Plexousakis, D., van Harmelen, F. (eds.) ISWC 2004. LNCS, vol. 3298, pp. 380–394. Springer, Heidelberg (2004)
22. Son, T., Pontelli, E.: Planning with preferences using logic programming. In: Lifschitz, V., Niemelä, I. (eds.) LPNMR 2004. LNCS (LNAI), vol. 2923, pp. 247–260. Springer, Heidelberg (2003)
23. Bienvenu, M., Fritz, C., McIlraith, S.: Planning with qualitative temporal preferences. In: Proceedings of the 10th International Conference on Principles of Knowledge Representation and Reasoning (KR 2006) (2006)
24. Baier, J.A., Bacchus, F., McIlraith, S.A.: A heuristic search approach to planning with temporally extended preferences. In: IJCAI (2007)

Enhancing Workflow with a Semantic Description of Scientific Intent

Edoardo Pignotti[1], Peter Edwards[1], Alun Preece[2], Nick Gotts[3], and Gary Polhill[3]

[1] School of Natural & Computing Sciences, University of Aberdeen
Aberdeen, AB24 5UE, UK
{e.pignotti,p.edwards}@abdn.ac.uk
[2] School of Computer Science, Cardiff University
Cardiff, CF24 3AA, UK
A.D.preece@cs.cf.ac.uk
[3] The Macaulay Institute
Craigiebuckler, Aberdeen, AB15 8QH, UK
{n.gotts,g.polhill}@macaulay.ac.uk

Abstract. In the e-Science context, workflow technologies provide a problem-solving environment for researchers by facilitating the creation and execution of experiments from a pool of available services. In this paper we will show how Semantic Web technologies can be used to overcome a limitation of current workflow languages by capturing experimental constraints and goals, which we term *scientist's intent*. We propose an ontology driven framework for capturing such intent based on workflow metadata combined with SWRL rules. Through the use of an example we will present the key benefits of the proposed framework in terms of enriching workflow output, assisting workflow execution and provenance support. We conclude with a discussion of the issues arising from application of this approach to the domain of social simulation.

Keywords: eScience, semantic grid, workflow, SWRL, constraints, goals.

1 Introduction

In recent years there has been a proliferation of scientific resources available through the internet, including datasets and computational modelling services. Scientists are becoming more and more dependent on these resources, which are changing the way they conduct their research activities (with increasing emphasis on 'in silico' experiments as a computational means to test a hypothesis). Scientific workflow technologies [1] have emerged as a problem-solving tool for researchers by facilitating the creation and execution of experiments given a pool of available services.

As part of the PolicyGrid[1] project we are investigating the use of semantic workflow tools to facilitate the design, execution, analysis and interpretation of simulation experiments and exploratory studies, while generating appropriate metadata automatically. The project involves collaboration between computer scientists and social scientists at the University of Aberdeen, the Macaulay Institute (Aberdeen) and elsewhere in the

---
[1] http://www.policygrid.org

UK. The project aims to support policy-related research activities within social science by developing appropriate Semantic Grid [2] tools which meet the requirements of social science practitioners. Where Grid technologies [3] provide an infrastructure to manage distributed computational resources, the vision of the Semantic Grid is based upon the adoption of metadata and ontologies to describe resources (services and data sources) in order to promote enhanced forms of collaboration among the research community. The PolicyGrid project is developing a range of services to support social scientists with mixed-method data analysis (involving both qualitative and quantitative data sources) together with the use of social simulation techniques. Issues surrounding usability of tools are also a key feature of PolicyGrid, with activities encompassing workflow support and natural language presentation of metadata [4].

The main benefit of current workflow technologies is that they provide a user-friendly environment for both the design and enactment of experiments without the need for researchers to learn how to program. Many different workflow languages exist including: MoML (Modelling Markup Language) [5], BPEL (Business Process Execution Language) [6], Scufl (Simple conceptual unified flow language) [7]. All these languages are designed to capture the flow of information between services (e.g. service addresses and relations between inputs and outputs).

As more computational and data services become available and researchers begin to share their workflows and results, there will be an increasing need to capture provenance associated with such workflows. Provenance (also referred to as lineage or heritage) aims to provide additional documentation about the processes that lead to some resource [8]. Goble [9] expands on the Zachman Framework [10] by presenting the '7 W's of Provenance': *Who, What, Where, Why, When, Which, & (W)How*. While some progress has been made in terms of documenting processes [11] (*Who, What, Where, When, Which, & (W)How*), little effort has been devoted to the *Why* aspect of research methodology. This is particularly important in the context of policy appraisal [12].

A typical experimental research activity [13] involves the following steps: observation, hypothesis, prediction (under specified constraints), experiment, analysis and write-up. While workflow technologies provide support for a researcher to define an experiment, there is no support for capturing the conditions under which the experiment is conducted, therefore making it difficult to situate the experiment in context. While existing provenance frameworks can provide information about an experiment by documenting the process, we argue that in order to fully characterise scientific analysis we need to go beyond such low-level descriptions by capturing the experimental conditions. The aim here is to make the constraints and goals of the experiment, which we describe as the *scientist's intent*, transparent.

PolicyGrid aims to provide an appropriate provenance framework to support evidence-based policy assessment where the focus is on how a particular piece of evidence was derived. To date the project has developed a `resource`[2] and a `task`[3] ontology to capture such provenance information. The `resource` ontology describes the type of resources used by social scientists (e.g. `Questionnaire`, `SimulationModel`, `InterviewTranscript`). The `task` ontology describes activities associated with the creation

---
[2] http://www.policygrid.org/ResourceV4.owl
[3] http://www.policygrid.org/TaskV3.owl

of resources (e.g. SimulationDataAnalysis, SimulationParameterExploration). These ontologies together provide the underlying framework which defines the provenance for a piece of evidence. Moreover, our work on capturing *scientist's intent* provides additional information on how experiments were conducted, giving an improved insight into the evidence creation process.

This paper is organized as follows. Section 2 introduces our motivation through the use of a workflow example, section 3 presents an ontology for capturing scientist's intent. In section 4 we discuss the requirements for a semantic workflow infrastructure supporting our scientist's intent ontology. In section 5 we present some examples of how scientist's intent can enrich workflows. In section 6 we discuss issues arising from application of this approach to the domain of social simulation, and finally in section 7 and 8 we discuss related work and our conclusions.

2 Motivation and Example

Recent activities in the field of social simulation [14] indicate the need to improve the scientific rigour of agent-based modelling. One of the important aspects of any scientific activity is that work should be repeatable and verifiable, yet results gathered from possibly hundreds of thousands of simulation runs cannot be reproduced conveniently in a journal publication. Equally, the source code of the simulation model, and full details of the model parameters used are also not journal publication material. We have identified activities that are relevant to such situations. These are:

- Being able to access the results, to check that the authors' claims based on those results are justifiable.
- Being able to re-run the experiments to check that they produce broadly the same results.
- Being able to manipulate the simulation model parameters and re-run the experiments to check that there is no undue sensitivity of the results to certain parameter settings.
- Being able to understand the conditions under which the experiment was carried out.

In a previous project, FEARLUS-G [15], we tried to meet the needs of agent based modelling using Semantic Grid technologies [2]. FEARLUS-G provided scientists interested in land-use phenomena with a means to run much larger-scale experiments than is possible on standalone PCs, and also gave them a Web-based environment in which to share simulation results. The FEARLUS-G project developed an ontology which centred on the tasks and entities involved in simulation work, such as experiments, hypotheses, parameters, simulation runs, and statistical procedures. We demonstrated that it is possible to capture the context in which a simulation experiment is performed making collaboration between scientists easier. However, FEARLUS-G was not designed to be a flexible problem-solving environment as the experimental methodology was hard-coded into the system. We feel that, in this context, workflow technologies can facilitate the design, execution, analysis and interpretation of simulation experiments and exploratory studies. However, we argue that current workflow technologies can only capture the method and not the scientist's intent which we feel is essential to make such experiments truly transparent.

We have identified a number of scenarios through interaction with collaborators from the social simulation community. We now present a simple example using a virus model developed in NetLogo[4]: an agent-based model that simulates the transmission and perpetuation of a virus in a human population. An experiment using this model might involve studying the differences between different types of virus in a specific environment. A researcher wishing to test the hypothesis 'Smallpox is more infectious than bird flu in environment A' might run a set of simulations using different random seeds. If in this set of simulations, Smallpox outperformed Bird Flu in a significant number of simulation runs, the experimental results could be used to support the hypothesis.

Figure 1(bottom) shows a workflow built using the Kepler editor tool [16] that uses available services to perform the experiment described above. The VirusSimulationModel generates simulation results based on a set of parameters loaded at input from a data repository; the experiment definition is selected by Experiment ID. These simulation results are aggregated and fed into the Significance Test component which outputs the results of the test. The hypothesis is tested by looking at the result of the significance test; if one virus that we are considering (e.g. smallpox) significantly outperforms another, we can use this result to support our hypothesis.

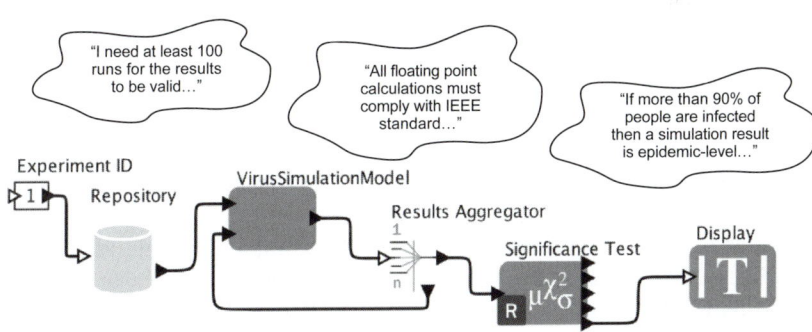

Fig. 1. Example of Simulation Workflow

The experimental workflow outlined in Figure 1(bottom) has some limitations as it is not able to capture the scientist's goals and constraints (scientist's intent) as illustrated in Figure 1(top). For example, the goal of this experiment is to obtain significant simulation results that support the hypothesis. Imagine that the researcher knows that the simulation model could generate out-of-bounds results and these results cannot be used in the significance test. For this reason, we do not know a priori how many simulation runs per comparison we need to do. Too few runs will mean that the experiment will return inconclusive data, while too many runs will waste computing resources executing unnecessary simulations. There may also be constraints associated with the workflow (or specific activities within the workflow) depending upon the intent of the scientist.

---

[4] http://ccl.northwestern.edu/netlogo/

For example, a researcher may be concerned about floating point support on different operating systems; if the Significance Test activity runs on a platform not compatible with IEEE 754 specifications, the results of the simulation could be compromised. A researcher might also be interested in detecting and recording special conditions (e.g. a particularly virulent virus) during the execution of the workflow to support the analysis of the results. Existing workflow languages are unable to explicitly associate such information with their workflow descriptions.

The main challenges we face are to represent scientist's intent in such a way that:

- it is meaningful to the researcher, e.g. providing information about the context in which an experiment has been conducted so that the results can be interpreted;
- it can be reasoned about by a software application, e.g. an application can make use of the intent information to control, monitor or annotate the execution of a workflow;
- it can be re-used across different workflows, e.g. the same high-level intent may apply to different workflows;
- it can be used as provenance (documenting the process that led to some result).

3 Scientist's Intent

As part of our approach we have developed an ontology for capturing the scientist's intent based upon goals and constraints. Before discussing this ontology we need to specify some of the concepts and properties associated with workflows:

- **Workflow model** - The representation of the flow of data between tasks needed to complete a certain (in-silico) experiment.
- **Workflow activity** - A basic task in the workflow or a sub-workflow. Properties associated with a workflow activity are of two types: a) properties describing the activity itself (e.g. Name, Type, Location) b) properties describing the status of the activity at run-time.
- **Abstract workflow activity** - An abstract view of workflow activity that does not map to a specific task but its instantiation is decided at run-time.
- **Workflow links** - Indicate the temporal relationship between workflow activities e.g. the pipeline between workflow activities. This relationship is established by combining workflow activities' inputs and outputs. A typical property is the datatype of workflow inputs and outputs.

This leads us to the definition of goals and constraints associated with a workflow experiment:

- **Constraints** - A formal specification of a restriction on the properties of workflow activity (single task or sub-workflow), workflow activity at run-time, and workflow links (inputs and outputs).
- **Goal** - A formal specification of a desired state which is defined by a sub-set of workflow activity (single task or sub-workflow), workflow activity at run-time, and workflow links (inputs and outputs).

Based on the definitions of goal and constraint given above we propose an ontology for capturing scientist's intent associated with a workflow experiment, as shown in Figure 2. We begin by defining a WorkflowExperiment which represents a specific instance of a workflow model used to conduct a scientific experiment. A Workflow Experiment is designed to automate one or more tasks defined in the PolicyGrid task ontology (e.g. DataAnalysisTask, DataCollectionTask, etc.). A WorkflowExperiment contains one or more ComputationalResource instances which define the computational services (Grid, Web or local) associated with a workflow activity. Each ComputationalResource might have an associated ontology describing the resource as an entity but also describing properties of the resource at run-time. The metadata associated with ComputationalResource instances during the execution of the workflow provides information about the WorkflowState. A WorkflowExperiment has one or more instances of a WorkflowState capturing the temporal changes of workflow metadata. This is based on the idea of *Abstract State Spaces* [17] where a particular execution of services denotes a sequence of state transitions $\tau = (s_0, \ldots, s_m)$ [18]. A WorkflowExperiment is performed by a WorkflowEngine which characterizes a specific software implementation, e.g. Kepler[16]. Each WorkflowEngine implementation supports zero or more WorkflowActions, e.g. stop workflow, pause workflow, show message.

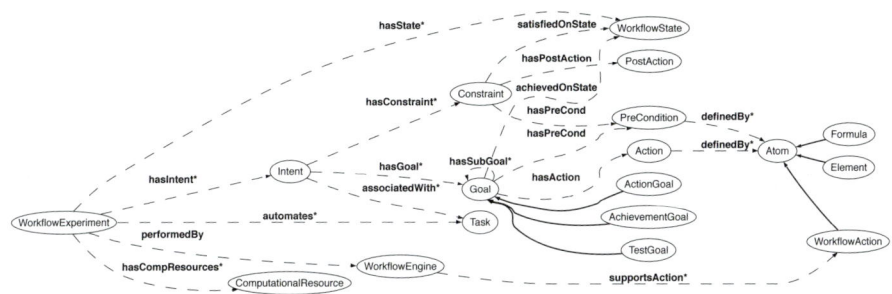

Fig. 2. Scientist's Intent Ontology

Central to our approach is the concept of Intent which is characterized by a set of Goal and Constraint statements. A WorkflowExperiment might have zero or more Intent instances. Although, from the definition above, Goal and Constraint are conceptually different, they share similar properties as they both have a PreCondition and a Action. Both properties are based on their constituent Atoms which can take the form of a metadata Element, a Formula or a WorkflowAction. As in SWRL[5] a condition is a conjunction of its Atoms. A PreCondition is a specific condition on the WorkflowState that can satisfy the Constraint or is achieved by the Goal [18]. An Action is an artifact that we use to trigger actions where the workflow engine is not able to reason itself about formal goals and constraints. Ideally a workflow engine would be fully aware of the goals and constraints defined in the *scientist's intent* and therefore be

---
[5] http://www.w3.org/Submission/SWRL/

able to reason about them but unfortunately this is not the case for most of the workflow engines currently available. In a *scientist's intent* aware workflow engine the planning and scheduling of the workflow execution can be optimized based on goals and constraints. Therefore, the concept of `PreCondition` is sufficient to represent both `Goal` and `Constraint` instances. However, most of the available workflow engines cannot be made fully compatible with scientist's intent without a major re-implementation, and therefore the concept of `Action` is required to overcome such limitations by providing additional metadata about the workflow state when goals are achieved and constraints are satisfied. Examples of such goals and constraints and their use will be presented later in this paper. However, to illustrate our approach, the scientific intent reflected in the example in Figure 1 can be represented as a combination of goals and constraints as follows:

- **Goal:** Run enough simulations to provide valid results to support the hypothesis. (valid-run > 100)
- **Constraint:** Significance Test has to run on a platform compatible with IEEE 754. (platform = IEEE 754).

In our view details of the intent need to be kept separate from the operational workflow as embedding constraints and goals directly into the workflow representation would make it overly complex (e.g. with a large number of conditionals) and would limit potential for sharing and re-use. Such a workflow would be fit for only one purpose and addition of new constraints would require it to be substantially re-engineered. Using the support for scientific intent proposed here, a new experiment might be created just by changing the rules but not the underlying operational workflow.

4 Semantic Workflow Infrastructure

In this section we present a semantic workflow infrastructure solution based on the scientist's intent ontology described above, highlighting the requirements for the various components. We base our solution on open workflow frameworks (e.g. Kepler) that allow the creation and execution of workflows based on local, Grid or Web services. A key part of this infrastructure is the workflow metadata support which provides information about the workflow components, inputs and outputs, and their execution. We also require a scientist's intent framework that manages goals and constraints of the experiment based on the workflow metadata.

Open workflow frameworks are the core of our solution as they provide the tools and systems to model and execute workflow. Different workflow frameworks may take different approaches; in this section we highlight the core functionality necessary to provide support for our solution. An important element of a workflow framework is the modelling tool (or editor) that allows researchers to design a workflow from available services. The key requirement here is that the editor is capable of working with both local and Grid services and that the resulting workflow is represented in a portable and machine processable language (e.g. XML). Workflow frameworks also provide the execution environment necessary to enact the workflow. Usually the execution environment provides a monitoring tool which allows the scientist to inspect the status of the

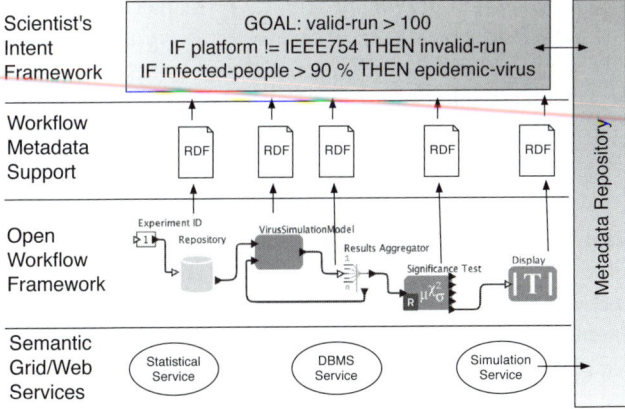

Fig. 3. Scientist's Intent Framework

execution. An important requirement is the ability to monitor and control the workflow execution through the use of APIs from external applications. This will provide the appropriate software in which the scientist's intent framework can operate.

A crucial aspect of our framework is that the workflow must have supporting ontologies and should produce metadata that can be used against scientific intent to reason about the workflow. We have identified the following possible sources of metadata:

- metadata about the result(s) generated upon completion of the workflow (e.g. a significance test);
- metadata about the data generated at the end of an activity within the workflow or sub-workflow (e.g. simulation model run);
- metadata about the status of an activity over time, for example while the workflow is running (e.g. infected people, immune people).

Central to our idea of capturing intent is the use of Semantic Grid services to perform the activities defined in the workflow. The main benefit of using such metadata enriched services is that they provide supporting information so that shared terms (e.g. virus, experiment, simulation model, floating point standard, etc.) can be used in the context of scientist's intent.

We have identified SWRL[6] (Semantic Web Rule Language) as a language for capturing rules associated with scientist's intent. SWRL enables Horn like rules to be combined with metadata. The rules take the form of an implication between an antecedent (PreCondition) and consequent (Action). This formalism is suitable for capturing scientist's intent, as the rules can capture the logic behind goals and constraints, while the ontology and metadata about the workflow provide the 'knowledge base' upon which the rules can operate. We selected the Bossam rule engine[7] for use within our architecture as it seamlessly integrates SWRL, OWL ontologies and RDF instances.

---

[6] http://www.w3.org/Submission/SWRL/
[7] http://bossam.wordpress.com/

Semantic grid services can provide different type of metadata: metadata about the service itself or metadata about the service execution at runtime. The latter involve many thousands of triples, and therefore a repository component is required to store such data. The Scientist's Intent Framework can then make use of the repository to extract metadata necessary to validate the rules but also to store any additional metadata (e.g. inferred statements).

5 Scientist's Intent & Workflow

We will now present some examples of goals and constraints to illustrate the benefits of scientist's intent in term of enriching workflow output, assisting workflow execution and provenance support.

5.1 Scientist's Intent to Assist Workflow Execution

As mentioned earlier, `Action` statements are used to add additional metadata to the workflow state if the workflow engine is unable to reason directly about goals and constraints. The example constraint below is used to check *if the significance test activity is running on a platform compatible with IEEE 754* as otherwise it will produce invalid results.

```
PreCondition:
  significanceTest( ?x1 ) ∧
  platform( ?x2, ''IEEE754'' ) ∧
  runsOnPlatform( ?x1,?x2 ) ∧
  hasResult( ?x1, ?x3)
Action:
  hasValidresult( ?x1, ?x3 )
```

If the significance test activity (see Figure 1) was defined as an abstract activity and the workflow engine was capable to interpret constraints directly, the selection of an appropriate significance test service could be made based on the pre-condition above.

The goal described below is used to specify the main goal for an experiment, i.e. to *run more that 100 valid simulation runs*. However we do not know a-priori how many simulation runs will be invalid.

```
PreCondition:
  significanceTest( ?x1 ) ∧
  hasResult( ?x1, ?x2 ) ∧
  hasValidresult( ?x1, ?x3 ) ∧
  more-than ( ?x3, 100)
```

The constraint below is used to check if the results of a particular simulation are invalid, specifically to determine *if the number of infected people (in a particular run) is greater than the number of not immune people (in the entire population)*.

```
PreCondition:
  population( ?x1 ) ∧
  virusMode( ?x2 ) ∧
  testPopulation( ?x2, ?x1 ) ∧
  hasModelRun( ?x2, ?x3 ) ∧
  notImmunePeople( ?x1, ?x4 ) ∧
  infectedPeople( ?x3, ?x5 ) ∧
  more-than( ?x5, ?x4 )
Action:
  hasInvalidRun( ?x2, ?x3) ∧
  stop(?x2) (Workflow Action)
```

Actions based on scientist's intent (e.g. `stop(?x2)`) will depend on the ability of the workflow framework to detect events from the scientist's intent framework and execute an action. We are currently extending the Kepler workflow tool to operate with our scientist's intent framework by registering the events that it is capable to detect and perform. These include: stop and pause the workflow, exit from a loop, show warning and error messages, prompt the user for information or intervention, display activity status.

5.2 Scientist's Intent to Enrich Workflow Output

Using the scientist's intent formalism it is possible to capture special kinds of constraints whose purpose is to enrich workflow outputs. While the previous goals and constraints support the verification and execution of the workflow by identifying invalid results or simulation runs, the constraints defined below aim to facilitate the analysis of results by enriching them with additional metadata. For example: *if the number of infected people in a simulation run is more than 90%, the virus tested is epidemic.*

```
PreCondition:
  virus( ?x1 ) ∧
  virusModel( ?x2 ) ∧
  testVirus( ?x2, ?x1 ) ∧
  hasModelRun( ?x2, ?x3 ) ∧
  infectedPeople( ?x3, ?x4 ) ∧
  more-than ( ?x4, 90% )
Action:
  isEpidemic( ?x1 )
```

The new metadata resulting from the application of this constraint (`isEpidemic( ?x1 )`) can be used as part of a *PreCondition* on other goals and constraints or as an annotation about the workflow outputs to facilitate analysis of the experimental results.

5.3 Provenance Support

As explained earlier in this paper, provenance is important for documenting the process that leads to a particular resource. We established that traditional provenance frameworks are not sufficient for all applications (e.g. policy appraisal) as it is very important to understand *why* particular steps in the process have been selected. We think that scientist's intent can be used to provide the important *why* context. For example, consider some of the constraint examples presented earlier. When looking back at the provenance of a simulation experiment it would be possible to determine *why* a particular statistical service had been selected (`platform (?x2,''IEEE754'')`) or *why* a particular simulation result was invalid (`notImmunePeople < infectedPeople`);

6 Case Study Discussion

We are exploring the use of workflow technologies in combination with our scientist's intent framework with a group of simulation modellers. Their work focuses on simulation of rural land use change in the Grampian region of Scotland over the past few decades, and on likely future responses to climate change, and to regulatory and market responses to it. The work is being supported by the Scottish Government through the research programme "Environment - land use and rural stewardship", and by the European Commission through the CAVES project[8]. It is in planning sequences of simulation runs, and associated statistical testing, required to validate, refine and use the models that it is planned to use workflow technologies.

Although a full evaluation of the scientist's intent framework in this real case-study environment has not yet been carried out, a number of issues about enabling agent-based models to work with our framework have been raised.

As mentioned earlier, one of the key issues for agent-based modelling has been the question of repeatability [19]. Authors reporting replication of agent-based modelling work have often commented that considerable interaction with the developers of the original model was necessary [20]. Using workflow technologies with the scientist's intent framework, it will be possible to record metadata about activities undertaken using a piece of modelling software and goals and constraints associated with it. This means that if one has access to the software from which conclusions were derived, it is possible to reconstruct the simulation output basis on which the conclusions were reached. This is also a timely contribution in the context of increasing demands from funding bodies for recognised standards to audit traceability of scientific results (in some cases, under the auspices of ISO9001).

A full replication of a piece of agent-based modelling work would ideally involve a reimplementation of the model, without any code reuse from the original software. Workflow metadata and scientist's intent is also useful here, as the re-implemented model can then presumably undergo the same processes used to derive conclusions as were used with the original model. However, deeper ontological support covering the structure of the model itself would facilitate the reimplementation process, and (related to this) provide a basis for verifying the similarity of the original and reimplemented models. Whilst some such information may be covered in accompanying documentation if available (often in the form of UML diagrams), ontological support can capture meanings in software representations not covered by the semantics of the implementing programming language [21], as well as providing a resource with which automated reasoning can be used.

---

[8] http://cfpm.org/caves/

7 Related Research

Many of the concepts underlying today's eScience workflow technologies originated from business workflows. These typically describe the automation of a business process, usually related to a flow of documents. Scientific workflow on the other hand is about the composition of structured activities (e.g. database queries, simulations, data analysis activities, etc.) that arise in scientific problem solving [16]. However, the underlying representation of the workflow remains the same (data and control flow). For example the language BPEL [6], originally designed for business, has been adapted for scientific workflow use. BPEL4WS is an extension of BPEL and provides a language for the formal specification of processes by extending the Web services interaction model to enable support for business transactions. The workflow is executed in terms of blocks of sequential service invocations. The main limitation of BPEL is that it does not support the use of semantic metadata to describe the workflow components and their interaction but instead relies entirely on Web services described by WSDL (Web Service Description Language). This type of language in not the best fit for our solution as we need rich metadata support for the workflow to describe not only service related information (e.g. platform, inputs and outputs) but also high level concepts (e.g. virus, population and model).

XScufl is a simple workflow orchestration language for Web services which can handle WSDL based web service invocation. The main difference from BPEL is that XScufl, in association with a tool like Taverna [7] allows programmers to write extension plug-ins (e.g. any kind of Java executable process) that can be used as part of the workflow. Taverna is a tool developed by the myGrid[9] project to support 'in silico' experimentation in biology, which interacts with arbitrary services that can be wrapped around Web services. It provides an editor tool for the creation of workflows and the facility to locate services from a service directory with an ontology-driven search facility. The semantic support in Taverna allows the description of workflow activities but is limited to facilitating the discovery of suitable services during the design of a workflow. Our scientist's intent framework relies not only on metadata about the activity, but also on metadata generated during the execution of the workflow.

MoML [5] is a language for building models as clustered graphs of entities with inputs and outputs. Like Taverna with XScufl, Kepler [16], is a workflow tool based on the MoML language where Web and Grid services, Globus Grid jobs, and GridFTP can be used as components in the workflow. Kepler extends the MoML language by using Directors which define execution models and monitor the execution of the workflow. Kepler also supports the use of ontologies to describe actors' inputs and outputs, enabling it to support automatic discovery of services and facilitate the composition of workflows. Like other workflow tools, Kepler does not allow the use of metadata at runtime. However, the Director component and the integration of ontologies with workflow activities provide an ideal interface within which our framework can operate.

Gil et al. [22] present some interesting work on generating and validating large workflows by reasoning on the semantic representation of workflow. Their approach relies on semantic descriptions of workflow templates and workflow instances. This description

---

[9] www.mygrid.org.uk

includes requirements, constraints and data products which are represented in ontologies. This information is used to support the validation of the workflow but also to incrementally generate workflow instances. Although in our research we are not focusing on assisted workflow composition, we do share the same interest in the benefit of enhanced semantics in workflow representation. While both our approaches rely on logical statements that apply to workflow metadata, we are taking a more user-centred approach by capturing higher level methodological information related to scientist's intent, e.g. `valid simulation result`, `epidemic virus`, etc.

Also relevant to our work is the model of provenance in autonomous systems presented by Miles et al. [23]. This model combines a description of goal-oriented aspects of agency with existing provenance frameworks in service-oriented architectures.

8 Discussion

Our evaluation strategy involves assessing the usability of the enhanced workflow representation using real workflows from the case-studies identified with our collaborators. We are using Kepler as a design tool and Grid services that we have developed over time as workflow activities (e.g. various simulation models). User scientists are central to the evaluation process, as they will use the tools and then supply different types of feedback via questionnaire, interview or through direct observation.

Lack of space prevents us discussing the evaluation plan in detail, but we will now present our key evaluation criteria:

- **Expressiveness of the intent formalism:** Is the formalism sufficient to capture real examples of intent? Were certain constraints impossible to express? Were some constraints difficult to express?
- **Reusability:** Can an intent definition be reused - either in its entirety or in fragments? Does our framework facilitate reusability?
- **Workflow execution:** Does the inclusion of intent information affect the computational resources required during the execution of a workflow? (This type of evaluation will be carried out in simulated conditions by running workflow samples with and without scientist's intent support and measuring the Grid resources used and the time involved.)

From a user perspective, creating and utilizing metadata is a non-trivial task; the use of a rule language to capture scientist's intent does of course provide additional challenges in this regard. Although we have not currently addressed these issues in this research, other work ongoing within the PolicyGrid project may provide a possible solution. Hielkema et al. [4] describe a tool which provides access to metadata (create, browse and query) using natural language. The tool can operate with different underlying ontologies, and we are sure that it could be extended to work with SWRL rules - creating a natural language interface for defining and exploring scientist's intent.

In conclusion, we aim to provide a closer connection between experimental workflows and the goals and constraints of the researcher, thus making experiments more transparent. While the scientist's intent provides context for the experiment, its use should also facilitate improved management of workflow execution. We have the underlying provenance framework to capture metadata about resources and tasks. The scientist's intent framework provides additional metadata about goals and constraints associated with a task (or set of tasks). Moreover through the use of Action the scientist's intent framework can also provide additional provenance generated from goals and constraints. However, we acknowledge that to truly understand the intent of the scientist a meta-level interpretation of all the above sources of provenance is necessary and this is beyond the current scope of our work.

Acknowledgments. This research is funded by the Aberdeen Centre for Environmental Sustainability (ACES) and conducted within the PolicyGrid project involving the Macaulay Institute and the University of Aberdeen; PolicyGrid is funded by the UK Economic & Social Research Council, award reference RES-149-25-1027.

References

1. Pennington, D.: Supporting large-scale science with workflows. In: Proceedings of the 2nd workshop on Workflows in support of large-scale science, High Performance Distributed Computing (2007)
2. Roure, D.D., Jennings, N., Shadbolt, N.: The semantic grid: a future e-science infrastructure. Grid Computing: Making the Global Infrastructure a Reality (2003)
3. Foster, I., Kesselman, C., Nick, J., Tuecke, S.: Grid services for distributed system integration. Morgan Kaufmann, San Francisco (2002)
4. Hielkema, F., Edwards, P., Mellish, C., Farrington, J.: A flexible interface to community-driven metadata. In: Proceedings of the eSocial Science conference 2007, Ann Arbor, Michigan (2007)
5. Lee, A., Neuendorffer, S.: Moml — a modeling markup language in xml —version 0.4. Technical report, University of California at Berkeley (2000)
6. Andrews, T.: Business process execution for web services, version 1.1 (2003), ftp://www6.software.ibm.com/software/developer/library/ws-bpel.pdf
7. Oinn, T., Addis, M., Ferris, J., Marvin, D., Senger, M., Greenwood, M., Carver, T., Glover, K., Pocock, M., Wipat, A., Li, P.: Taverna: a tool for the composition and enactment of bioinformatics workflows. Bioinformatics Journal 20(17), 3045–3054 (2004)
8. Groth, P., Jiang, S., Miles, S., Munroe, S., Tan, V., Tsasakou, S., Moreau, L.: An architecture for provenance systems. ECS, University of Southampton (2006)
9. Goble, C.: Position statement: Musings on provenance, workflow and (semantic web) annotation for bioinformatics. In: Workshop on Data Derivation and Provenance, Chicago (2002)
10. Zachman, J.A.: A framework for information systems architecture. IBM Syst. J. 26(3), 276–292 (1987)
11. Greenwood, M., Goble, C., Stevens, R., Zhao, J., Addis, M., Marvin, D., Moreau, L., Oinn, T.: Provenance of e-science experiments. In: Proceedings of the UK OST e-Science 2nd AHM (2003)
12. Chorley, A., Edwards, P., Preece, A., Farrington, J.: Tools for tracing evidence in social science. In: Proceedings of the Third International Conference on eSocial Science (2007)
13. Wilson, E.: An introduction to scientific research. McGraw-Hill, New York (1990)
14. Polhill, J., Pignotti, E., Gotts, N., Edwards, P., Preece, A.: A semantic grid service for experimentation with an agent-based model of land-use change. Journal of Artificial Societies and Social Simulation 10(2)2 (2007)

15. Pignotti, E., Edwards, P., Preece, A., Polhill, G., Gotts, N.: Semantic support for computational land-use modelling. In: Proceedings of the Fifth IEEE International Symposium on Cluster Computing and Grid (CCGrid) 2005, vol. 2, pp. 840–847. IEEE Press, Los Alamitos (2005)
16. Ludäscher, B., Altintas, I., Berkley, C., Higgins, D., Jeager, E., Jones, M., Lee, E., Tao, J.: Scientific workflow management and the kepler system. Concurrency and Computation: Practice and Experience (2005)
17. Keller, U., Lausen, H., Stollberg, M.: On the semantics of functional descriptions of web services. In: Sure, Y., Domingue, J. (eds.) ESWC 2006. LNCS, vol. 4011, pp. 605–619. Springer, Heidelberg (2006)
18. Stollberg, M., Hepp, M.: A refined goal model for semantic web services. In: Proceedings of the Second International Conference ion Internet and Web Applications and Services. ICIW apps 2007, pp. 17–23 (2007)
19. Hales, D., Rouchier, J., Edmonds, B.: Model-to-model analysis. Journal of Artificial Societies and Social Simulation 6(4) (2003)
20. Axtell, R., Axelrod, R., Epstein, J., Cohen, M.D.: Aligning simulation models: A case study and results. Computational and Mathematical Organization Theory 1(2), 123–141 (1996)
21. Polhill, J.G., Gotts, N.M.: Evaluating a prototype self-description feature in an agent-based model of land use change. In: Amblard, F. (ed.) Proceedings of the Fourth Conference of the European Social Simulation Association, Toulouse, France, September 10-14, pp. 711–718 (2007)
22. Gil, Y., Ratnakar, V., Deelman, E., Spraragen, M., Kim, J.: Wings for pegasus: A semantic approach to creating very large scientific workflows. In: Proceedings of the 19th Annual Conference on Innovative Applications of Artificial Intelligence (IAAI), Vancouver, British Columbia, Canada (2006)
23. Miles, S., Munroe, S., Luck, M., Moreau, L.: Modelling the provenance of data in autonomous systems. In: AAMAS 2007: Proceedings of the 6th international joint conference on Autonomous agents and multiagent systems, pp. 1–8. ACM, New York (2007)

WSMO Choreography: From Abstract State Machines to Concurrent Transaction Logic

Dumitru Roman[1], Michael Kifer[2], and Dieter Fensel[1]

[1] STI Innsbruck, Austria
[2] State University of New York at Stony Brook, USA

Abstract. Several approaches to semantic Web services, including OWL-S, SWSF, and WSMO, have been proposed in the literature with the aim to enable automation of various tasks related to Web services, including discovery, contracting, enactment, monitoring, and mediation. The ability to specify processes and to reason about them is central to these initiatives. In this paper we analyze the WSMO choreography model, which is based on Abstract State Machines (ASMs), and propose a methodology for generating WSMO choreography from visual specifications. We point out the limitations of the current WSMO model and propose a faithful extension that is based on Concurrent Transaction Logic (CTR). The advantage of a CTR-based model is that it uniformly captures a number of aspects that previously required separate mechanisms or were not captured at all. These include process specification, contracting for services, service enactment, and reasoning.

1 Introduction

The field of Semantic Web services marries the technology of delivering services over the Web with the Semantic Web. This brings up a variety of issues, which range from service-specific and domain ontologies to service discovery, service choreography (i.e., specification of how autonomous client agents interact with services), automated contracting for services, service enactment, execution monitoring, and others. A number of past and ongoing projects proposed solutions to some of these problems. These include OWL-S,[1] SWSL,[2] WSMO,[3] and WS-CDL.[4] WSMO is one of the more comprehensive approaches to Semantic Web Services, as it covers virtually all of the aforementioned areas. In this paper, we focus on one particular aspect of WSMO—its support for service choreography, which is based on the formalism of Abstract State Machines (ASMs) [3].

Although ASMs have been used to design and verify software in the past, this formalism is too general and leaves Web service designers to their own devices.

---

[1] http://www.daml.org/services/owl-s/
[2] http://www.w3.org/Submission/SWSF-SWSL/
[3] http://www.wsmo.org/
[4] http://www.w3.org/TR/ws-cdl-10/

To the best of our knowledge, there has been little work to help guide Web service designers through the process of actual building of ASM-based choreographies. Our first contribution is to show that control flow graphs, which are commonly used to design workflows (and thus can be used as visual design tools for choreographies), have a modular translation into ASMs and can be used to specify WSMO choreography interfaces.

Next, we discuss some of the drawbacks of using ASMs in WSMO. We argue that the problem of choreography is inextricably related to the problem of contracting for Web services. We define one very useful instance of the problem of contracting and show that ASMs are inadequate to deal with this problem. We then propose a solution grounded in Concurrent Transaction Logic (CTR) [2]. Based on our previous results [15], we show that both service choreography and service contracts have natural representation in CTR: the former as deductive rules and the latter as constraints. In addition, CTR enables us to reason about consistency of the contracts with the choreographies, about service enactment, and other issues. In this way, CTR-based WSMO choreography can be seen as a natural extension of the ASM-based WSMO choreography.

An introduction to WSMO choreography is given in Section 2. Section 3 proposes a more intuitive, visual language for specifying service interactions, and develops a methodology for automatic generation of WSMO choreography transition rules. Section 4 defines the problem of service contracting as choreography under the constraints specified as service policies and customers requirements and highlights the limitations of the current WSMO choreography in this area. It then proposes CTR as a formalism that overcomes those problems. Section 5 presents related work, and Section 6 concludes this paper.

2 WSMO Choreography: An Abstract State Machine Model

WSMO choreography [17] is a state-based model inspired by the methodology of *Abstract State Machine* (ASMs) [3]. It provides basic mechanisms for modeling interactions between service providers and clients at an abstract level. The use of an ASM-based model has several benefits:

- *Minimality*: ASMs are based on a small assortment of modeling primitives.
- *Expressivity*: ASMs can model arbitrary computations.
- *Formality*: ASMs provide a formal framework to express dynamics.

WSMO choreography borrows its basic mechanisms from ASMs. A *signature* defines predicates and functions to be used in the description. *Ground facts* specify the underlying database states. State changes are described using *transition rules*, which specify how the states change by falsifying (deleting) some previously true facts and inserting (making true) some other facts.

In WSMO, signatures are defined using *ontologies*. The ground facts that populate database states are instances of concepts and relations defined by the ontologies. State changes are described in terms of creation of new instances or changes to attribute values of objects.

The transition rules used in WSMO have one of the following forms:

- **if** *Condition* **then** *Rules*
- **forall** *Variables* **with** *Condition* **do** *Rules*
- **choose** *Variables* **with** *Condition* **do** *Rules*

The *Condition* part (also called *guard*) is a logical expression, as defined by WSML.[5] The *Rules* part is a set of ASM rules, which can be primitive state changes, like *add*, *delete*, or *update* (modify) a fact. More complex transition rules can be defined with the help of *if-then*, *forall* and *choose* rules. As usual with ASMs, WSMO Choreography rules are evaluated in parallel and the updates are executed in parallel as well. When all rules of an ASM are executed in this way, the ASM makes a *transition* from one database state to another.

A *run* of a WSMO Choreography is a finite or infinite sequence of states, $s_0, s_1, ...,$ where s_0 is an initial state of the choreography and, for each $n \geq 0$, the choreography can make a transition from state s_n to s_{n+1}.

In the following section we look at how a WSMO Choreography can be generated from graphical representations of interactions.

3 From Visual Modeling to WSMO Choreography Rules

Although the state-based model of WSMO choreography is appealing for modeling interaction with services, the area of process aware information systems [7] has been dominated by various graphical notations and visual languages for process modeling. Examples of such notations and languages include: UML 2.0 Activity Diagram (AD), Business Process Definition Metamodel (BPDM), Business Process Modeling Notation (BPMN), Business Process Modeling Language (BPML), Event Driven Process Chain (EPC), Petri Nets, etc. Although these tools differ in their expressivity, they share core visual elements for specifying sequential, conditional, and parallel executions. One common way of depicting such core elements is thorough AND/OR graphs, which, in the context of business processes, are referred to as *control flow graphs*.

To illustrate, consider an example from [15] in Figure 1. The figure shows a fairly complex pattern of interaction with a service that sells high ticket items. It includes provisions for optionally giving rebates to customers who fulfill certain requirements as well as a possibility that customers might return the ordered items and receive partial refund. Payment is allowed by credit cards or cheques, and in some cases the service might require those payments to be secured by a credit card (if the available limit exceeds the price) or by providing a guarantor for the payment. Under certain circumstances, the client may get a rebate. The figure represents a pattern of interaction with the service using an AND/OR *control flow graph*. Each node in the graph represents an action. In practice every action is usually invoked through a message.

---

[5] http://www.wsmo.org/TR/d16/d16.1/v0.3/

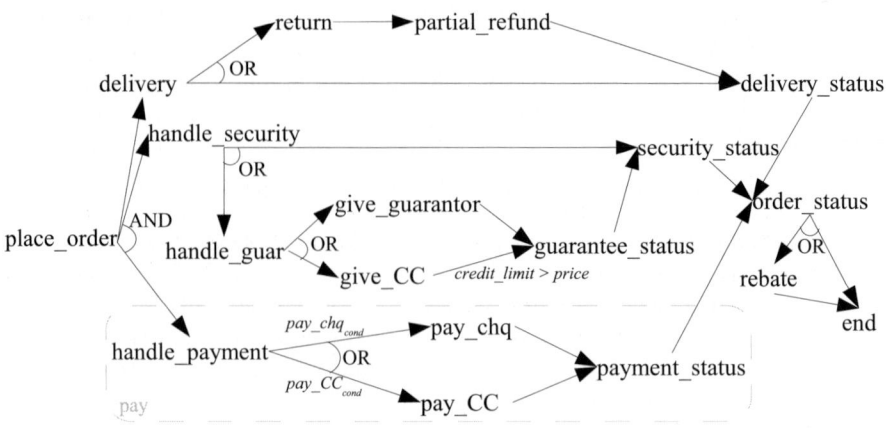

Fig. 1. A process specification

Control flow graphs are typically used to specify local execution dependencies among the interactions with the service; it is a good way to visualize the overall flow of control. Such a graph has an initial and a final *interaction task*, the successor-interaction for each interaction in the graph (except the final interaction), and a notation that tells whether the successors of an interaction must *all* be executed concurrently (represented by **AND**-split nodes), or whether only one of the alternative branches needs to be executed non-deterministically (represented by **OR**-nodes).

In Figure 1, all successors of the initial interaction **place_order** must be executed, since **place_order** is an **AND**-node. The branches that correspond to these successors eventually join in an **AND**-node called **order_status**. On the other hand, the successors of **place_order** are **OR**-nodes. For example, the successor called **handle_payment** is an **OR**-node, so only one of *its* successors, **pay_chq** or **pay_CC**, is supposed to be executed. The node **delivery** is also an **OR**-split. The upper branch hanging from this node represent a situation where a customer accepts delivery but then returns the purchased item. The lower branch, however, has no interactions—it connects **delivery** with the final node on the upper branch, the node **delivery_status**. This means that the upper branch is *optional*: the customer may or may not return the item. Similarly, the segment growing out of the node **order_status** indicates that **rebate** is an optional interaction.

Two more things about the control flow graph in Figure 1 should be noted: the shaded box labeled with **pay**, and the conditions on arcs such as $credit\_limit > price$ attached to the arc leaving the node **give_CC**, and $pay\_chq_{cond}, pay\_CC_{cond}$ on the arcs adjacent to the node **handle_payment**. Shaded boxes delineate control flow corresponding to *complex interactions* (such as payments) that are composed of several sub-interactions. If a control graph represents a process then these complex interactions are referred to as complex tasks and correspond to *subprocesses*.

The condition $credit\_limit > price$ attached to an arc is called a *transition condition*. It says that in order for the next interaction with the service to take place the condition must be satisfied. The parameters $credit\_limit$ and $price$ are obtained by querying the current state of the service or they may be passed as parameters from one interaction to another—the actual method depends on the details of the representation. Conditions pay_chq_{cond} and pay_CC_{cond} have a similar meaning, but since they are on arcs that start at an **OR**-node, it is assumed that these conditions are mutually exclusive. In general, transition conditions are Boolean expressions attached to the arcs in control flow graphs. Only the arcs whose conditions evaluate to true can be followed at run time. Sometimes a control flow graph may contain additional elements, such as loops, but we do not deal with these here.

In order to represent this example using the ASM-based WSMO Choreography model, we develop a methodology for translating control-flow graphs to WSMO Choreography transition rules. For the purpose of this paper we make the following pragmatic assumptions:

- The signature of a WSMO Choreography consists of:
  - A set of Boolean propositions, where each proposition represents a task in the interaction process;
  - The *cond* predicate which takes two task names as parameters and represents the conditions attached to the arc connecting these two tasks; and
  - The *chosen* predicate which takes a task name as a parameter and is used to model nondeterminism.

- The state of a WSMO Choreography is determined by the truth values of:
  - The propositions corresponding to the tasks in the choreography process. If such a proposition is true, it means that the task has already been executed; if it is false then the task has not yet been executed.
  - The instances of the predicate *cond*. An instance $cond(task_1, task_2)$ evaluates to true if there is an arc from $task_1$ to $task_2$ and the condition attached to that arc evaluates to true. It evaluates to "false" otherwise. If the condition attached to the arc connecting $task_1$ to $task_2$ is not specified explicitly then $cond(task_1, task_2)$ is assumed to be true.
  - The instances of the predicate *chosen*. For each OR-split node, t, with successors $t_1, \ldots, t_n$, it is assumed that exactly one of the predicates $chosen(t_1), \ldots, chosen(t_n)$ evaluates to true. How exactly this evaluation happens is left unspecified. This can be due to an external action or due to the execution of the task t.

Now we can define the translation $\mathfrak{T}$, which transforms a control graph, G, into a set of WSMO Choreography transition rules. The update rules of the form $add(b)$ are used to change the truth value of the proposition b to $true$. When b becomes true, it means that the task represented by b has finished its execution.

$$\mathfrak{T}(G) = \begin{cases} \textbf{if } a \wedge \neg b \wedge cond(a,b) \wedge chosen(b) \textbf{ then } add(b), \\ \quad \text{if there is an arc from } a \text{ to } b \\ \quad \text{and } a \text{ is an OR-split node} \\ \textbf{if } a_1 \wedge \cdots \wedge a_n \wedge \neg b \wedge cond(a_1,b) \wedge \cdots \wedge cond(a_n,b) \textbf{ then } add(b), \\ \quad \text{if there is an arc from each } a_1 \ldots a_n \text{ to } b \\ \quad \text{and } b \text{ is an AND-join node} \\ \textbf{if } a \wedge \neg b \wedge cond(a,b) \textbf{ then } add(b), \\ \quad \text{if there is an arc from } a \text{ to } b \\ \quad \text{and } a \text{ is not OR-split node} \\ \quad \text{and } b \text{ is not AND-join node} \end{cases}$$

Applying this transformation to the example in Figure 1 yields the following. For the arcs originating from **OR**-split nodes we have:

if $delivery \wedge \neg return \wedge chosen(return)$ **then** $add(return)$
if $delivery \wedge \neg delivery\_status \wedge chosen(delivery\_status)$ **then** $add(delivery\_status)$
if $handle\_security \wedge \neg security\_status \wedge chosen(security\_status)$
   then $add(security\_status)$
if $handle\_security \wedge \neg handle\_guar \wedge chosen(handle\_guar)$ **then** $add(handle\_guar)$
if $handle\_guar \wedge \neg give\_guarantor \wedge chosen(give\_guarantor)$
   then $add(give\_guarantor)$
if $handle\_guar \wedge \neg give\_CC \wedge chosen(give\_CC)$ **then** $add(give\_CC)$
if $handle\_payment \wedge \neg pay\_chq \wedge pay\_chq_{cond} \wedge chosen(pay\_chq)$ **then** $add(pay\_chq)$
if $handle\_payment \wedge \neg pay\_CC \wedge pay\_CC_{cond} \wedge chosen(pay\_CC)$ **then** $add(pay\_CC)$
if $order\_status \wedge \neg rebate \wedge chosen(rebate)$ **then** $add(rebate)$
if $order\_status \wedge \neg end \wedge chosen(end)$ **then** $add(end)$

(1)

Here we omit those instances of the predicate $cond(t_1,t_2)$, which are not specified explicitly and thus are *true* according to our assumptions. For all the arcs ending in the **AND**-join node, $delivery\_status$, we have:

if $delivery\_status \wedge payment\_status \wedge security\_status \wedge \neg order\_status$
   then $add(order\_status)$

For the remaining arcs, we have:

if $place\_order \wedge \neg delivery$ **then** $add(delivery)$
if $place\_order \wedge \neg handle\_security$ **then** $add(handle\_security)$
if $place\_order \wedge \neg handle\_payment$ **then** $add(handle\_payment)$
if $return \wedge \neg partial\_refund$ **then** $add(partial\_refund)$
if $partial\_refund \wedge \neg delivery\_status$ **then** $add(delivery\_status)$
if $give\_guarantor \wedge \neg guarantee\_status$ **then** $add(guarantee\_status)$
if $give\_CC \wedge \neg guarantee\_status \wedge (credit\_limit > price)$
   then $add(guarantee\_status)$
if $pay\_chq \wedge \neg payment\_status$ **then** $add(payment\_status)$
if $pay\_CC \wedge \neg payment\_status$ **then** $add(payment\_status)$
if $rebate \wedge \neg end$ **then** $add(end)$

The above WSMO Choreography rules capture the behavior of the process depicted in Figure 1. It is interesting to note that the parallelism represented in the control flow graph is simulated using concurrent, interleaved execution of tasks.

4 Extending WSMO Choreography with Services Policies and Client Requirements

In the context of service-orientation service providers and requester represent autonomous entities and need to agree on how to interact. Specifications, like the one in Figure 1 provide a mechanism to describe potential interactions between providers and requesters, but service providers and their clients might want to impose additional policies and constraints on those interactions. In this section we explain how such policies can be represented and formulate the problem of agreement between providers and requesters, which is a form of service *contracting*. Then we point out an inadequacy in the current WSMO choreography for representing policies and propose Concurrent Transaction Logic as an extension which allows not only to model service interaction and policies, but also perform certain types of reasoning.

4.1 The Problem of Contracting for Services

In order for service providers and requesters to engage in a business deal, they need to agree on the terms of the engagement, or on a *contract*. Since contract requirements can vary and might depend on the client, control-flow graphs are too rigid for representing variable contract terms. These terms often take the form of global temporal and causality constraints among the tasks performed by the services and/or clients. An example of a temporal constraint is a statement that a set of tasks must be executed in a particular temporal order. A causality constraint is a statement of the form "if some particular task is executed, some other tasks must not execute or be executed in a particular temporal relation with the former task." Such constraints coming from the service provider side are called *service policies*, and the constraints coming from the client are called *client contract requirements*. To illustrate, we will again use the choreography depicted in Figure 1. The constraints appropriate for that example are shown in Figure 2.

Generally, the terms that may go into a contract can be very complex, but we will focus on temporal and causality constraints, such as the requirement that certain tasks must all be performed in the course of the execution of a service; that certain tasks must *not* occur together in the same execution; or that some tasks require that certain other tasks execute before or after a given task. We refer the reader to the language of constraints defined in [15] for the specifics of the language used for policy specification.

Once we have a way of describing both the choreographies and constraints, we need a technique for automated support for tasks, such as contracting and enactment, which informally can be defined as follows:

Service policy:
1. If **pay_CC** (paying by credit card) takes place after accepting **delivery** then **guarantee_status** (giving security) must *precede* **delivery**
2. If **pay_chq** takes place after accepting **delivery** then **pay_chq** (paying by cheque) *immediately* follows **delivery**
3. If **rebate** is given then **pay** must precede accepting **delivery**

Client contract requirement:
4. The interaction of accepting **delivery** must precede **pay_chq**

Fig. 2. Service policies and client contract requirements

1. **Contracting:** Given a control-flow graph, G, with a set of service policies $C_{service}$ and client requirements C_{client}, the question of determining if a contract between the sides is possible is the problem of finding out if there is an execution of G such that $C_{service}$ and C_{client} are satisfied (i.e. finding out if $G \wedge C_{service} \wedge C_{client}$ is true).
2. **Enactment:** Given a control-flow graph G, with a set of service policies $C_{service}$ and client requirements C_{client}, the problem of enactment is that of generating a sequence of task executions (from G) such that $C_{service}$ and C_{client} are satisfied.

Before we describe a solution to these problems, we need to spend some time looking at the limitations of the current WSMO choreography.

4.2 Limitations of WSMO Choreography

The methodology proposed in Section 3 offers a way of translating control-flow graphs to WSMO choreography rules. To better understand how WSMO will cope with contract requirements, consider the first constraint from Figure 2:

If **pay_CC** (paying by credit card) takes place after accepting **delivery** then **guarantee_status** (giving security) must *precede* **delivery**

To compile this constraint into WSMO choreography, as defined by the rules given in Section 3, the rule that enables *pay_CC* (the 8th rule in (1)) needs to be replaced with the following two rules:

if $handle\_payment \wedge \neg pay\_CC \wedge pay\_CC_{cond} \wedge chosen(pay\_CC) \wedge \neg delivery$ **then** $add(pay\_CC)$

if $handle\_payment \wedge \neg pay\_CC \wedge pay\_CC_{cond} \wedge chosen(pay\_CC) \wedge delivery$
$\wedge\, after(guarantee\_status, delivery)$
  then $add(pay\_CC)$

The first rule above ensures that if *pay_CC* was chosen to execute and *delivery* was not executed, the constraint does not apply and thus *pay_CC* can execute. The second rule says that if *pay_CC* has to execute and *delivery* was previously executed, then *pay_CC* may execute only if *delivery* was executed after *guarantee_status* (i.e., if $after(guarantee\_status, delivery)$ is true). To ensure the latter, an additional rule is needed:

if $place\_order \land \neg delivery \land guarantee\_status$
then $add(after(guarantee\_status, delivery))$

Let us now consider the second constraint from Figure 2:

If **pay_chq** takes place after accepting **delivery**
then **pay_chq** (paying by cheque) *immediately* follows **delivery**

To capture this constraint, the WSMO Choreography rule that enables *pay_chq* (the 9th rule in (1)) need to be replaced with two rules as follows:

if $handle\_payment \land \neg pay\_chq \land pay\_chq_{cond} \land chosen(pay\_chq) \land \neg delivery$ **then** $add(pay\_chq)$

if $handle\_payment \land \neg pay\_chq \land pay\_chq_{cond} \land chosen(pay\_chq) \land delivery \land last\text{-}task = delivery$
then $add(pay\_chq)$

The first rule above ensures that if *pay_chq* was chosen to execute and *delivery* was not executed, the constraint does not apply and thus *pay_chq* can execute. The second rule says that if *pay_chq* has to execute and *delivery* was previously executed, then *pay_chq* must execute only if *delivery* was executed immediately prior to that (i.e. $last\text{-}task = delivery$ is true). Now, to ensure that *last-task* always records the last task which executed, all the rules need to be extended such that whenever one executes, it records that it is the last task which executed. That is, whenever we have $add(task)$, we need to add $last\text{-}task = task$, for example

if $return \land \neg partial\_refund$
then $add(partial\_refund), add(last\text{-}task = partial\_refund)$

The above exercise should give an idea how much constraints might complicate specification of a WSMO choreography. Even a simple constraint, like the first constraint above, may cause an increase in the number of rules as well as in the number of conditions in the guards of these rules. Adding or deleting constraints will require a recompilation of the entire choreography specification, and such a scenario may quickly become unmanageable. Slightly more complex constraints, like the second constraint above, may lead to further complications, such as the requirement to log the execution of the choreography tasks using auxiliary predicates, such as *last-task*. Additional techniques might need to be developed to handle other constraints described in [15]. All this suggests that constraints must be kept separate from WSMO choreography specifications, if the latter are to be represented as ASMs.

The solution we propose is to use Concurrent Transaction Logic (CTR) [2] as a unifying mechanism for representing both WSMO choreography rules (derived using the methodology in Section 3) and constraints. An added benefit is that CTR supports certain forms of reasoning about such choreographies.

4.3 CTR to the Rescue

Concurrent Transaction Logic (CTR) [2] is a formalism for declarative specification, analysis, and execution of transactional processes. It has been successfully applied to modeling and reasoning about workflows and services [5,19,6,15]. In this paper we show how CTR can model and extend WSMO choreography, including the ability to specify contract requirements. We first give an informal introduction to CTR and then illustrate its use for the examples in Figures 1 and 2.

The atomic formulas of CTR are identical to those of the classical logic, *i.e.*, they are expressions of the form $p(t_1, \ldots, t_n)$, where p is a predicate symbol and the t_i's are function terms. Apart from the classical connectives $\vee, \wedge, \neg, \forall$, and $\exists$, CTR has two new connectives, $\otimes$ (*serial conjunction*) and $|$ (*concurrent conjunction*), plus one modal operator $\odot$ (*isolated execution*). The intended meaning of the CTR connectives can be summarized as follows:

- $\phi \otimes \psi$ *means*: execute ϕ then execute ψ. In terms of control flow graphs (cf. Figure 1), this connective can be used to formalize arcs connecting adjacent activities.
- $\phi \mid \psi$ *means*: ϕ and ψ must both execute concurrently, in an interleaved fashion. This connective corresponds to the "AND"-nodes in control flow graphs.
- $\phi \wedge \psi$ *means*: ϕ and ψ must both execute along the *same* path. In practical terms, this is best understood in terms of *constraints* on the execution. For instance, ϕ can be thought of as a transaction and ψ as a constraint on the execution of ϕ. It is this feature of the logic that lets us specify constraints as part of process specifications.
- $\phi \vee \psi$ *means*: execute ϕ *or* execute ψ non-deterministically. This connective corresponds to the "OR"-nodes in control flow graphs.
- $\neg \phi$ *means*: execute in any way, provided that this will *not* be a valid execution of ϕ. Negation is an important ingredient in the specifications of temporal constraints.
- $\odot \phi$ *means*: execute ϕ in isolation, *i.e.*, without interleaving with other concurrently running activities.

Implication $p \leftarrow q$ is defined as $p \vee \neg q$. The purpose of the implication in CTR is similar to that of Datalog: p can be thought of as the name of a procedure and q as the definition of that procedure. However, unlike Datalog, both p and q assume truth values on execution paths, not at states. More precisely, $p \leftarrow q$ means: if q can execute along a path $\langle s_1, ..., s_n \rangle$, then so can p. If p is viewed as a subroutine name, then the meaning can be rephrased as: one way to execute p is to execute its definition, q.

The semantics of CTR is based on the idea of *paths*, which are finite sequences of database states. For the purpose of this paper, the reader can think of these states as just relational databases. For instance, if $s_1, s_2, ..., s_n$ are database states, then $\langle s_1 \rangle$, $\langle s_1, s_2 \rangle$, and $\langle s_1, s_2, ..., s_n \rangle$ are paths of length 1, 2, and n, respectively. Just as in classical logic, CTR formulas are assigned truth values.

However, *unlike* classical logic, the truth of CTR formulas is determined over paths, *not* at states. If a formula, ϕ, is true over a path, $\langle s_1, ..., s_n \rangle$, it means that ϕ can *execute* starting at state s_1. During the execution, the current state will change to $s_2, s_3, ...,$ etc., and the execution terminates at state s_n. Details of the model theory of CTR can be found in [2].

The control flow of a service choreography is represented as a *concurrent-Horn goal*. Subworkflows are defined using *concurrent Horn rules*. We define these notions now.

- A *concurrent Horn goal* is defined recursively as follows:
  - any atomic formula is a concurrent-Horn goal;
  - $\phi \otimes \psi$, $\phi \mid \psi$, and $\phi \vee \psi$ are concurrent-Horn goals, if so are ϕ and ψ;
  - $\odot \phi$ is a concurrent-Horn goals, if so is ϕ.
- A *concurrent-Horn rule* is a CTR formula of the form *head* $\leftarrow$ *body*, where *head* is an atomic formula and *body* is a concurrent-Horn goal.

A service choreography is represented in CTR as a concurrent Horn goal which is built out of atomic formulas that represent interaction tasks or subprocesses. The difference between interaction tasks and subprocesses is that the former are considered atomic primitive tasks, while subprocesses are defined using concurrent-Horn rules. Now we can specify the choreography depicted in Figure 1 as a concurrent Horn goal as follows:

$$place\_order \otimes ((delivery \otimes (refund \vee \texttt{state})) \mid (security \vee \texttt{state}) \mid pay) \otimes (rebate \vee \texttt{state}) \otimes end$$

where *security*, *pay*, and *refund* are subprocesses defined using concurrent-Horn rules as follows:

$$security \leftarrow \\ (give\_guarantor \vee (give\_CC \otimes credit\_limit \otimes Limit > Price)) \\ \otimes guarantee\_status$$
$$pay \leftarrow (pay\_chq_{cond} \otimes pay\_chq) \vee (pay\_CC_{cond} \otimes pay\_CC)$$
$$refund \leftarrow return \otimes partial\_refund$$

Here we should point out the use of a special proposition $\texttt{state}$, which is true only on paths of length 1, that is, on database states. It is used in the above to indicate optional actions. Another propositional constant that we will use in the representation of constraints is $\texttt{path}$, defined as $\texttt{state} \vee \neg \texttt{state}$, which is true on every path.

The above representation is much more compact and modular than the corresponding WSMO choreography rules described in Section 3. First, the definitions of *security* and *refund* are significantly simpler and clearer. Second, because of the non-deterministic nature of disjunction in CTR, we can abstract from the actual representations of conditions on the arcs originating from **OR**-nodes and do not need to specify them explicitly. We no longer need nodes such as *delivery_status*, *security_status*, or *order_status*, which were required in the WSMO choreography rules.

CTR can also naturally represent the constraints used for service contracts. For instance, the CTR representation of the constraints from Figure 2 is as follows:

1. $(\nabla delivery \otimes \nabla pay\_CC) \rightarrow (\nabla guarantee\_status \otimes \nabla delivery)$
2. $(\nabla delivery \otimes \nabla pay\_chq) \rightarrow \nabla \odot (delivery \otimes pay\_chq)$
3. $\nabla rebate \rightarrow (\nabla pay \otimes \nabla delivery)$
4. $\nabla delivery \otimes \nabla pay\_chq$

Here $\nabla task$ is a short cut for the formula $\texttt{path} \otimes task \otimes \texttt{path}$, which means that $task$ must happen sometime during the execution of the choreography. The formula $\texttt{path} \otimes \odot (task_1 \otimes \cdots \otimes task_n) \otimes \texttt{path}$ means that tasks $task_1, ..., task_n$ must execute next to each other with no other events in-between. Observe that although representing such constraints complicates WSMO choreography rules even further (and is not modular), their representation in CTR is simple and natural.

Now, to specify that a choreography specification must obey the terms of a contract, we must simply build a conjunction of the choreography (which is specified as a concurrent Horn goal) and the constraints that form the service contract specification, i.e., $G \wedge C_{service} \wedge C_{client}$, where G is a choreography specification, $C_{service}$ are the service policies, and C_{client} are the client contract requirements.

A technique for proving formulas of the form $ConcurrentHornGoal \wedge Constraints$ and to constructively find a sequence of states on which $ConcurrentHornGoal \wedge Constraints$ is true (i.e., enactment of the choreography subject to the contract requirements) was initially proposed in [5] and further developed in [15].

In this section we pointed put the limitations of the current ASMs-based WSMO choreography model and showed a way to overcome these problems with the help of CTR. Another advantage of CTR over the ASMs-based model is that any ASM can be mapped into a CTR formula and thus using CTR instead of ASMs does not limit the generality of WSMO choreography. This line of research is pursued in [16].

5 Related Work

To the best of our knowledge, this paper is the first to introduce a methodology that enables reasoning about WSMO choreography. More specifically, together with the results developed in [15], the present work paves the way to automatic contracting and enactment of WSMO services.

Automatic service contracting and enactment have been identified as core tasks in the context of semantic Web services (see e.g. [10]), however no approaches have been proposed that directly deal with such tasks. Several reasoning and verification techniques have been developed for the OWL-S [20] Process Model—the counterpart of WSMO choreography in OWL-S. For example, [13] proposed a Petri Net-based operational semantics for the OWL-S Process Model, and applied existing Petri Net analysis techniques to verify process models for various properties such as reachability, deadlocks, etc. In [1], a verification technique of OWL-S Process Models using model checking is proposed. A planning technique for automated composition of Web services, described in OWL-S process models, is presented in [21]. Although these works are related, they are complementary to the task of service contracting. The Semantic Web Services Framework [18] introduces a rich behavioral process model based on the Process Specification Language PSL [11]. Although PSL is defined in first order logic, which in theory makes behavioral specifications in PSL amenable to automated reasoning, we are not aware of specific works that deal with service contracting in PSL.

In the broader area of Web services, reasoning and verification about service choreographies (especially in the context of WS-CDL[6]) have been investigated in various works, such as [4] and [14]. As in the case of previous works on the OWL-S Process Model, such approaches deal with verification and analysis of various properites of service interactions, but leave the problem of service contracting out. Compared to such approaches, the work presented in this paper deals with a specific choreography model developed in the context of WSMO and proposes a concrete mechanism for service contracting using this model. From the modeling point of view, WS-CDL distinguishes between global and local views on a multi-party interaction. This is not an issue for our approach since each party can describe its own control flow graph and constraints in CTR. Contracting is then a problem of finding executions of all control flow graphs such that all constraints are satisfied. Thus, our approach extends straightforwardly to multi-party interactions.

Web service policies is another related area. WS-Policy [8] provides a general purpose model to describe the policies of entities in a Web services-based system. Although WS-Policy is widely adopted in industry, it is primarily used as a syntactic mechanism for policy specification. Instead, our approach works at a semantic level and, although focused on a specific choreography model, it enables automatic service contracting and enactment. Other Semantic Web-based policy frameworks, such as KAoS [22] or Rein [12], focus on authentication and authorization policies, whereas our approach targets service choreographies.

Last but not least, our contribution can be considered in the context of the emerging area of compliance checking between business processes and business contracts [9].

6 Conclusions

Semantic Web services (SWS) promise to bring high degree of automation to service discovery, contracting, enactment, monitoring, and mediation. WSMO is a major initiative in this area, which is supported by a number of academic institutions and companies. As with other SWS approaches (e.g. OWL-S, SWSF), modeling and reasoning about service behavior are central to the

---

[6] http://www.w3.org/TR/ws-cdl-10/

WSMO framework. While other parts of WSMO are based on a logic, WSMO choreography relies on Abstract State Machines (ASMs) for its formalization. ASMs have many advantages, but they are hard to reason about and there are no methodologies for building ASM-based process specifications that choreograph the interactions with Web services. To overcome these problems, this paper developed a methodology, based on control-flow graphs, for designing WSMO choreographies. We highlighted the limitations of the current WSMO choreography model and showed a way to remedy these problems with the help of Concurrent Transaction Logic. In this way, we enabled reasoning about service contracts and enactment in WSMO and opened up several new opportunities for future work. For example, integration of WSMO Web service discovery[7] with service contracting and enactment is one such opportunity. Verification and optimization of choreography interfaces is another promising direction.

Acknowledgments. Part of this work was done while Michael Kifer was visiting Free University of Bozen-Bolzano, Italy. His work was partially supported by the BIT Institute, NSF grant IIS-0534419, and by US Army Research Office under a subcontract from BNL. Dumitru Roman was partly funded by the BIT Institute, the SUPER project (FP6-026850), and the Knowledge Web project (FP6-507482).

Refererences

1. Ankolekar, A., Paolucci, M., Sycara, K.P.: Towards a formal verification of owl-s process models. In: International Semantic Web Conference, pp. 37–51 (2005)
2. Bonner, A.J., Kifer, M.: Concurrency and Communication in Transaction Logic. In: Joint International Conference and Symposium on Logic Programming (1996)
3. Börger, E., Stärk, R.F.: Abstract State Machines—A Method for High-Level System Design and Analysis. Springer, Heidelberg (2003)
4. Carbone, M., Honda, K., Yoshida, N.: Structured global programming for communication behaviour. In: De Nicola, R. (ed.) ESOP 2007. LNCS, vol. 4421, Springer, Heidelberg (2007)
5. Davulcu, H., Kifer, M., Ramakrishnan, C.R., Ramakrishnan, I.V.: Logic Based Modeling and Analysis of Workflows. In: PODS, pp. 25–33 (1998)
6. Davulcu, H., Kifer, M., Ramakrishnan, I.V.: CTR–S: A Logic for Specifying Contracts in Semantic Web Services. In: WWW2004, p. 144+ (2004)
7. Dumas, M., van der Aalst, W.M., ter Hofstede, A.H.: Process-aware information systems: bridging people and software through process technology. John Wiley & Sons, Inc., Chichester (2005)
8. Vedamuthu, A., et al. (eds.): Web Services Policy 1.5 – Framework (WS-Policy), W3c recommendation, W3C (September 2007),
http://www.w3.org/TR/ws-policy
9. Governatori, G., Milosevic, Z., Sadiq, S.: Compliance checking between business processes and business contracts. In: EDOC 2006, pp. 221–232. IEEE Computer Society, Los Alamitos (2006)

---

[7] http://www.wsmo.org/TR/d5/d5.1/

10. Grosof, B., Gruninger, M., Kifer, M., Martin, D., McGuinness, D., Parsia, B., Payne, T., Tate, A.: Semantic Web Services Language Requirements Version 1. Working draft, SWSI Language Committee (2005), http://www.daml.org/services/swsl//requirements/swsl-requirements.shtml
11. Grüninger, M., Menzel, C.: The process specification language (PSL) theory and applications. AI Mag. 24(3), 63–74 (2003)
12. Kagal, L., Berners-Lee, T., Connolly, D., Weitzner, D.J.: Using Semantic Web Technologies for Policy Management on the Web. In: AAAI (2006)
13. Narayanan, S., McIlraith, S.: Simulation, verification and automated composition of web services. In: Proceedings of the 11th International World Wide Web Conference (WWW 2002), Honolulu, Hawaii (May 2002)
14. Qiu, Z., Zhao, X., Cai, C., Yang, H.: Towards the theoretical foundation of choreography. In: WWW, pp. 973–982 (2007)
15. Roman, D., Kifer, M.: Reasoning about the Behavior of Semantic Web Services with Concurrent Transaction Logic. In: VLDB (2007)
16. Roman, D., Kifer, M.: Simulating Abstract State Machines (ASMs) with Concurrent Transaction Logic (CTR). WSMO Deliverable D14.2v0.1. Technical report, DERI Innsbruck (2007), http://www.wsmo.org/TR/d14/d14.2/v0.1/d14.2v01.pdf
17. Roman, D., Scicluna, J., Nitzsche, J. (eds.): Ontology-based Choreography. WSMO Deliverable D14v0.1. Technical report, DERI Innsbruck (2007), http://www.wsmo.org/TR/d14/v1.0/
18. Semantic Web Services Framework. SWSF Version 1.0 (2005), http://www.daml.org/services/swsf/1.0/
19. Senkul, P., Kifer, M., Toroslu, I.: A Logical Framework for Scheduling Workflows under Resource Allocation Constraints. In: VLDB 2002, pp. 694–705 (2002)
20. The OWL Services Coalition. OWL-S 1.1 beta release (July 2004), http://www.daml.org/services/owl-s/1.1B/
21. Traverso, P., Pistore, M.: Automated composition of semantic web services into executable processes. In: International Semantic Web Conference, pp. 380–394 (2004)
22. Uszok, A., Bradshaw, J.M., Jeffers, R., Tate, A., Dalton, J.: Applying KAoS Services to Ensure Policy Compliance for Semantic Web Services Workflow Composition and Enactment. In: International Semantic Web Conference (2004)

WSMO-Lite Annotations for Web Services*

Tomas Vitvar[1], Jacek Kopecký[1], Jana Viskova[2], and Dieter Fensel[1]

[1] Semantic Technology Institute (STI),
University of Innsbruck, Austria
{firstname.lastname}@uibk.ac.at
[2] Department of Information Networks,
University of Zilina, Slovakia
viskova@kis.fri.utc.sk

Abstract. Current efforts in Semantic Web Services do not sufficiently address the industrial developments of SOA technology in regards to bottom-up modeling of services, that is, building incremental layers on top of existing service descriptions. An important step in this direction has been made in the W3C by the SAWSDL WG proposing a framework for annotating WSDL services with arbitrary semantic descriptions. We build on the SAWSDL layer and define WSMO-Lite service ontology, narrowing down the use of SAWSDL as an annotation mechanism for WSMO-Lite. Ultimately, our goal is to allow incremental steps on top of existing service descriptions, enhancing existing SOA capabilities with intelligent and automated integration.

1 Introduction

Current efforts in enhancing Web service technology with semantics, such as WSMO and OWL-S, adopt the top-down approach to modeling of services. They define complete frameworks for describing semantics for services while they assume that a service engineer first models the semantics (usually as ontologies, functional, non-functional, and behavioral descriptions) before grounding them in service invocation and communication technologies (e.g. WSDL and SOAP). This approach, however, does not fit well with industrial developments of SOA technology, such as WSDL and REST, where thousands of services are already available within and outside enterprises (i.e., on the Web). In other words, it is hard to use the semantic frameworks in a bottom-up fashion, that is, for building increments on top of existing services while at the same time enhancing SOA capabilities with intelligent and automated integration.

In 2007, the W3C finished its work on Semantic Annotations for WSDL and XML Schema (SAWSDL)[1]. SAWSDL defines simple extensions for WSDL and XML Schema used to link WSDL components with arbitrary semantic descriptions. It thus provides the grounds for a bottom-up approach to service modeling: it supports the idea of adding small increments (and complexity) on top of WSDL, allowing to adopt results from various existing approaches. As the basis for bottom-up modeling, SAWSDL is independent

---

* This work is supported by the EU project SOA4ALL (FP7-215219).
[1] http://www.w3.org/TR/sawsdl/

of any particular semantic technology, i.e., it does not define any types, forms or languages for semantic descriptions.

In this paper we describe the WSMO-lite service ontology, created as a recent result of the community effort in the Conceptual Models for Services working group (CMS WG)[2]. WSMO-Lite is the next evolutionary step after SAWSDL, filling the SAWSDL annotations with concrete semantic service descriptions and thus embodying the semantic layer of the Semantic Service Stack. With the ultimatate goal to support realistic real-world challenges in intelligent service integration, WSMO-Lite addresses the following requirements:

- Identify the types and a simple vocabulary for semantic descriptions of services (a service ontology) as well as languages used to define these descriptions;
- Define an annotation mechanism for WSDL using this service ontology;
- Provide the bridge between WSDL, SAWSDL and (existing) domain-specific ontologies such as classification schemas, domain ontology models, etc.

The rest of this paper is structured as follows. In Section 2, we introduce the Semantic Service Stack along with the state-of-the-art technologies for services and Semantic Web languages used in the stack. In Section 3, we describe the WSMO-Lite Service Ontology and summarize the resolution of the major points from its development. In Section 4, we describe the WSMO-Lite semantic annotations for WSDL, and in Section 5 we describe some relevant aspects for WSMO-Lite applications.

2 Semantic Service Stack

As depicted in Figure 1, there are two levels in the Semantic Service Stack, namely *semantic* and *non-semantic* level. In addition, there are two types of stakeholders in the stack, namely a *service engineer* (human being) and a *client* (software agent). The service engineer uses Web services through the client, with particular tasks such as service discovery, selection, mediation, composition and invocation. Through these tasks the client or service engineer (depending on the level of automation) decide whether to bind with the service or not. In order to faciliate such decisions, services should describe their offers using so called *service contracts*. The Semantic Service Stack adopts the following general types of service contracts:

- *Information Model* defines the data model for input, output and fault messages.
- *Functional Descriptions* define service functionality, that is, what a service can offer to its clients when it is invoked.
- *Non-Functional Descriptions* define any incidental details specific to the implementation or running environment of a service.
- *Behavioral Descriptions* define external (public choreography) and internal (private workflow) behavior.
- *Technical Descriptions* define messaging details, such as message serializations, communication protocols, and physical service access points.

In the following sections, we show how the Semantic Service Stack represents the above general description types for service contracts at the two different levels.

---
[2] http://cms-wg.sti2.org

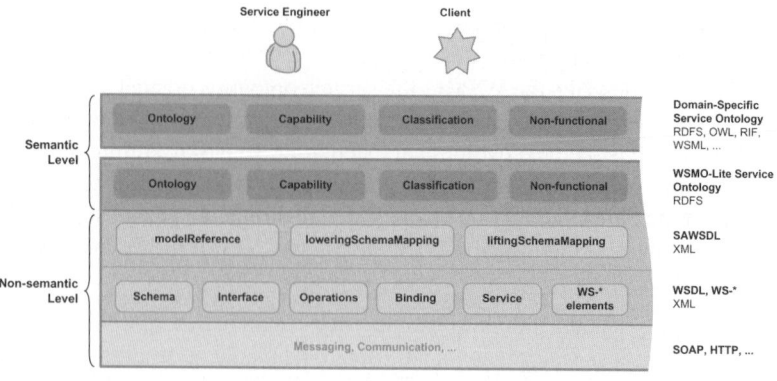

Fig. 1. Semantic Service Stack

2.1 Non-Semantic Level

In regard to SOA technology developments today, the Semantic Service Stack represents service contracts at the non-semantic level using the existing de-facto and de-jure standards: WSDL, SAWSDL, and related WS-* specifications. They all use XML as a common flexible data exchange format. Service contracts are represented as follows:

- *Information Model* is represented using XML Schema.
- *Functional Description* is represented using a WSDL Interface and its operations.
- *Non-Functional Description* is represented using various WS-* specifications, such as WS-Policy, WS-Reliability, WS-Security, etc.
- *Behavioral Description* is represented using the WS-* specifications of WS-BPEL[3] (for the workflow) and WS-CDL[4] (for the choreography).
- *Technical Description* is represented using WSDL Binding for message serializations and underlying communication protocols, such as SOAP, HTTP; and using WSDL Service for physical endpoint information.

In addition, while SAWSDL does not fall into any of the service contract descriptions, it is an essential part of the non-semantic level of the stack, providing the ground for the semantic layer. SAWSDL defines a simple extension layer that allows WSDL components to be annotated with semantics, using three extension attributes:

- *modelReference* for pointing to concepts that describe a WSDL component,
- *loweringSchemaMapping* and *liftingSchemaMapping* for specifying the mappings between the XML data and the semantic information model.

2.2 Semantic Level

The Semantic Service Stack represents service contracts at the semantic level using the WSMO-Lite service ontology as follows (see Section 3 for a detailed description of WSMO-Lite):

---

[3] http://docs.oasis-open.org/wsbpel/2.0/wsbpel-v2.0.pdf
[4] Choreography Description Language, http://www.w3.org/TR/ws-cdl-10/

- *Information Model* is represented using a domain ontology.
- *Functional Descriptions* are represented as *capabilities* and/or functionality *classifications*. A capability defines *conditions* which must hold in a state before a client can invoke the service, and *effects* which hold in a state after the service invocation. Classifications define the service functionality using some classification ontology (i.e., a hierarchy of categories).
- *Non-Functional Descriptions* are represented using an ontology, semantically representing some policy or other non-functional properties.
- *Behavioral Descriptions* are not represented explicitly in WSMO-Lite. In Section 5.1 we show how the public part of the behavioral description of a Web service may be derived from the functional descriptions of its operations.
- *Technical Descriptions* are not represented semantically in the service ontology, as they are sufficiently covered by the non-semantic description in WSDL.

In order to create or reuse domain-specific service ontologies on top of the Semantic Service Stack, a service engineer can use any W3C-compliant language with an RDF syntax. This preserves the choice of language expressivity according to domain-specific requirements. Such languages may include RDF Schema (RDFS), Web Ontology Language (OWL) [5], Rule Interchange Format (RIF)[5] or Web Service Modeling Language (WSML)[6] [10].

3 WSMO-Lite Service Ontology

Listing 1.1 shows the WSMO-Lite service ontology in RDFS, serialized in Notation 3[7]. Below, we explain the semantics of the WSMO-Lite elements:

```
1   @prefix rdfs: <http://www.w3.org/2000/01/rdf-schema#> .
2   @prefix rdf: <http://www.w3.org/1999/02/22-rdf-syntax-ns#> .
3   @prefix owl: <http://www.w3.org/2002/07/owl#> .
4   @prefix wl: <http://www.wsmo.org/ns/wsmo-lite#> .
5
6   wl:Ontology rdf:type rdfs:Class;
7       rdfs:subClassOf owl:Ontology.
8   wl:ClassificationRoot rdfs:subClassOf rdfs:Class.
9   wl:NonFunctionalParameter rdf:type rdfs:Class.
10  wl:Condition rdfs:subClassOf wl:Axiom.
11  wl:Effect rdfs:subClassOf wl:Axiom.
12  wl:Axiom rdf:type rdfs:Class.
```

Listing 1.1. WSMO-Lite Service Ontology

- *wl:Ontology* (lines 6–7) defines a container for a collection of assertions about the information model of a service. Same as *owl:Ontology*, *wl:Ontology* allows for meta-data such as comments, version control and inclusion of other ontologies. *wl:Ontology* is a subclass of *owl:Ontology* since as we already mentioned, it has a special meaning of the ontology used as the service information model.

---

[5] http://www.w3.org/2005/rules/
[6] For details on compliance of WSML with W3C languages see [3].
[7] http://www.w3.org/DesignIssues/Notation3.html

- *wl:ClassificationRoot* (line 8) marks a class that is a root of a classification which also includes all the RDFS subclasses of the root class. A classification (taxonomy) of service functionalities can be used for functional description of a service.
- *wl:NonFunctionalParameter* (line 9) specifies a placeholder for a concrete domain-specific non-functional property.
- *wl:Condition* and *wl:Effect* (lines 10–12) together form a *capability* in functional service description. They are both subclasses of a general *wl:Axiom* class through which a concrete language can be used to describe the logical expressions for conditions and effects. We illustrate this on an example in Listing 1.2 (lines 26–42).

Below, we describe the resolutions of major points that came up while WSMO-Lite was under development in the CMS WG.

Relation of WSMO-Lite to WSMO. WSMO-Lite has been created due to a need for lightweight service ontology which would directly build on the newest W3C standards and allow bottom-up modeling of services. On the other hand, WSMO is an established framework for Semantic Web Services representing a top-down model identifying semantics useful in a semantics-first environment. WSMO-Lite adopts the WSMO model and makes its semantics lighter in the following major aspects:

- WSMO defines formal user goals and mediators, while WSMO-Lite treats mediators as infrastructure elements, and specifications for user goals as dependent on the particular discovery mechanism used. They both can be adopted in the running environment in combination with WSMO-Lite.
- WSMO-Lite only defines semantics for the information model, functional and non-functional descriptions (as WSMO Service does) and only implicit behavior semantics (see below). If needed, an application can extend WSMO-Lite with its own explicit behavioral descriptions, or it can adopt other existing technologies.
- While WSMO uses the WSML language for describing domain-specific semantic models, WSMO-Lite allows the use of any ontology language with an RDF syntax (see Section 2.2 for more details).

Concrete semantics for conditions and effects. To work with conditions and effects, it is necessary to define the environment in which these axioms are evaluated. Such an environment depends on the particular logical language in which the axioms are expressed. WSMO-Lite does not prescribe any concrete language for functional service semantics, and therefore it cannot define semantics for conditions and effects as they are language-dependent.

4 WSMO-Lite Annotations for WSDL

Section 2.2 mentions briefly how the WSMO-Lite ontology is used for the semantic descriptions. In this section, we formally define the particular types of annotations supported by WSMO-Lite.

4.1 Definitions

Ontology. The fundamental building block for all types of semantic descriptions offered by WSMO-Lite is the ontology. We use a general definition of the ontology

$$\Omega = (C, R, E, I) \quad (1)$$

where the sets C, R, E, I in turn denote classes (unary predicates), relations (binary and higher-arity predicates[8]), explicit instances (extensional definition), and axioms (intensional definition) which describe how new instances are inferred.

A particular axiom common in I is the *subclass* relationship: if c_1 is subclass of c_2 (written as $c_1 \sqsubset c_2$), every instance of c_1 is also an instance of c_2. We call this axiom out because it is necessary for Definition 2 below.

We distinguish several sub-types of ontologies: we denote an information model ontology as $O_I \equiv \Omega$; a functionality classification ontology with root $r \in C$ as $O_F(r) \equiv \Omega$; and an ontology for non-functional descriptions as $O_N \equiv \Omega$.

Capability. Functional description of a service as a capability is defined here as

$$K = (\Sigma, \phi^{pre}, \phi^{eff}), \quad (2)$$

where $\Sigma \subseteq (\{x\} \cup C \cup R \cup E)$ is the signature of symbols, i.e., identifiers of elements from C, R, E of some ontology O_I complemented with variable names $\{x\}$; ϕ^{pre} is a condition which must hold in a state before the service can be invoked, and ϕ^{eff} is the effect, a condition which must hold in a state after the successful invocation. Conditions and effects are defined as statements in logic $\mathcal{L}(\Sigma)$.

In Definition 1 below, we specify a *restriction* relationship (partial ordering $\leq$) between capabilities, and in Definition 2 we define an analogous relationship between categories in a functionality classification. Practically, if a capability/category K_1 is a restriction of another capability/category K_2, any discovery algorithm that discovers K_1 as a suitable capability/category for some goal would also discover K_2 as such.

Definition 1 (capability restriction). A capability $K_1 = (\Sigma, \phi_1^{pre}, \phi_1^{eff})$ is a restriction of $K_2 = (\Sigma, \phi_2^{pre}, \phi_2^{eff})$ (written as $K_1 \leq K_2$) if the condition ϕ_1^{pre} only holds in states (denoted as s) where also ϕ_2^{pre} holds, and if the same is true for the effects:

$$K_1 \leq K_2 \iff \forall s : (holds(\phi_1^{pre}, s) \Rightarrow holds(\phi_2^{pre}, s)) \land$$
$$(holds(\phi_1^{eff}, s) \Rightarrow holds(\phi_2^{eff}, s)) \quad (3)$$

Definition 2 (category restriction). For two functionality categories K_1 and K_2 from classification $O_F(r)$, K_1 is a restriction of K_2 (written as $K_1 \leq K_2$) if $K_1 \sqsubset K_2$:

$$K_1 \leq K_2 \iff K_1 \sqsubset K_2 \quad (4)$$

---

[8] Note that a minimal definition would combine the sets of classes and relations as a set of predicates, but we choose to split them, due to familiarity and also reuse in further definitions.

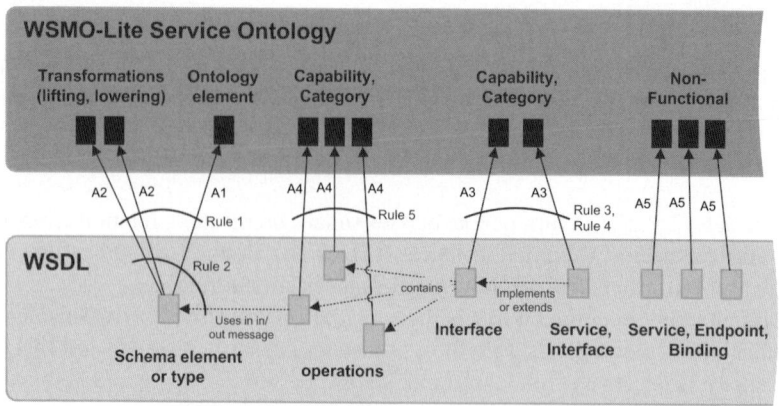

Fig. 2. Illustration of Annotations and Rules

WSDL. We denote an XML Schema in WSDL as S, a WSDL interface as I and a service as W. Further, we denote $\{x\}_S$ as the set of all element declarations and type definitions of S, and $\{op\}_I$ as the set of all operations of I. Each operation $op \in \{op\}_I$ may have one input message element $m \in \{x\}_S$ and one output message element $n \in \{x\}_S$ and a corresponding MEP[9] denoted here as $op.mep$.

Annotations. According to SAWSDL, we distinguish two types of annotations, namely *reference annotations* and *transformation annotations*. A reference annotation points from a WSDL component to a semantic concept. This is denoted as the binary relation $ref(x,s)$ where $x \in (\{x\}_S \cup \{I\} \cup \{op\}_I)$ — any WSDL or Schema component; $s \in (C \cup R \cup E \cup \{K\})$ — an ontology element or a capability. SAWSDL represents ref using *modelReference* extension attribute on the WSDL or XML Schema component).

A transformation annotation specifies a data transformation called *lifting* from a component of schema S to an element of ontology O_I; and a reverse transformation (from ontology to XML) called *lowering*. We denote these annotations as the binary relations $lower(m, f(c_1))$ and $lift(n, g(n))$, where $m, n \in \{x\}_S$. The function $f(c_1) = m$, where $c_1 \in (C \cup R)$, is a lowering function transforming data described semantically by c_1 to the XML message described by schema m (SAWSDL represents this annotation using *loweringSchemaMapping* extension attribute on m). Analogously, function $g(n) = c_2$, where $c_2 \in (C \cup R)$, is a lifting function transforming XML data from the message n to semantic data described by c_2 (SAWSDL represents this annotation using *liftingSchemaMapping* extension attribute on n).

4.2 Annotations and Rules

Figure 2 illustrates a set of annotations (marked *A1...A5*) and their associated rules (marked *Rule 1...Rule 5*). The rules have been refined from [13] to conform to the latest WSMO-Lite service ontology specification. The purpose of the rules is to ensure that the annotations are:

---
[9] Message Exchange Pattern, http://www.w3.org/TR/wsdl20-adjuncts/#meps

```
1   // namespaces and prefixes
2   @prefix rdfs: <http://www.w3.org/2000/01/rdf-schema#> .
3   @prefix rdf: <http://www.w3.org/1999/02/22-rdf-syntax-ns#> .
4   @prefix wl: <http://www.wsmo.org/ns/wsmo-lite#> .
5   @prefix ex: <http://example.org/onto#> .
6   @prefix xs: <http://www.w3.org/2001/XMLSchema#> .
7   @prefix wsml: <http://www.wsmo.org/wsml/wsml-syntax#> .
8
9   // ontology example
10  <> rdf:type wl:Ontology.
11
12  ex:Customer rdf:type rdfs:Class .
13  ex:hasService rdf:type rdf:Property ;
14      rdfs:domain ex:Customer ;
15      rdfs:range ex:Service .
16  ex:Service rdf:type rdfs:Class .
17  ex:hasConnection rdf:type rdf:Property ;
18      rdfs:domain ex:Customer ;
19      rdfs:range ex:NetworkConnection .
20  ex:NetworkConnection rdf:type rdfs:Class .
21  ex:providesBandwidth rdf:type rdf:Property ;
22      rdfs:domain ex:NetworkConnection ;
23      rdfs:range xs:integer .
24  ex:VideoOnDemandService rdfs:subClassOf ex:Service .
25
26  // capability description example
27  ex:VideoOnDemandSubscriptionPrecondition rdf:type wl:Condition ;
28      rdf:value "
29          ?customer[hasConnection hasValue ?connection]
30              memberOf Customer and
31          ?connection[providesBandwidth hasValue ?y]
32              memberOf NetworkConnection and
33          ?y > 1000
34      "^^wsml:AxiomLiteral .
35
36  ex:VideoOnDemandSubscriptionEffect rdf:type wl:Effect ;
37      rdf:value "
38          ?customer[hasService hasValue ?service]
39      "^^wsml:AxiomLiteral .
40
41  // definition of the axiom for WSML language
42  wsml:AxiomLiteral rdf:type rdfs:Datatype .
43
44  // non-functional property example
45  ex:PriceSpecification rdfs:subClassOf wl:NonFunctionalParameter .
46  ex:VideoOnDemandPrice rdf:type ex:PriceSpecification ;
47      ex:pricePerChange "30"^^ex:euroAmount ;
48      ex:installationPrice "49"^^ex:euroAmount .
49
50  // classification example
51  ex:SubscriptionService rdf:type wl:ClassificationRoot .
52  ex:VideoSubscriptionService rdfs:subClassOf ex:SubscriptionService .
53  ex:NewsSubscriptionService rdfs:subClassOf ex:SubscriptionService .
```

Listing 1.2. Example of domain-specific service ontology

- *complete*, that is, no gaps are left in the semantic annotations, so that the client can see all the parts of the service description; for instance, all the operations should be semantically annotated so that they are reachable to automatic discovery.
- *consistent*, that is, no related annotations are contradictory; for instance the schema annotations by model reference need to point to concepts that are the outputs of the lifting schema mapping transformation, or inputs of the lowering one.

Listing 1.2 shows an example ontology we use to illustrate annotations. It defines a simple ontology for a telecommunication service (lines 9–24); the capability for a concrete Video on Demand subscription service (lines 26–39) (the condition says that the customer must have a network connection with some minimal bandwidth, the effect says that the customer is subscribed to the service); a non-functional property describing the pricing (lines 44–48); and a simple functionality classification (lines 50–53). We also define the *wsml:AxiomLiteral* data type (line 42) for WSML-Rule axioms so that a client can correctly process them according to the WSML specification.

A1: Annotations of XML Schema (ontology). The schema used in WSDL to describe messages, i.e., the element declarations and type definitions, can carry reference annotations linking to classes from the service information model ontology.

A2: Annotations of XML Schema (transformations). To be able to communicate with a service, the client needs to transform data between its semantic model and the service-specific XML message structures. The schema may contain transformation annotations which specify the appropriate mappings.

```
1  <xs:element name="NetworkConnection" type="NetworkConnectionType"
2     sawsdl:modelReference="http://example.org/onto#NetworkConnection"
3     sawsdl:loweringSchemaMapping="http://example.org/NetCn.xslt"/>
```

Listing 1.3. Example of annotations A1 and A2

Listing 1.3 shows an example of annotations *A1* and *A2* (the lowering transformation is omitted for brevity). Below, Rule 1 defines consistency of *A1* and *A2* annotations on schema components; Rule 2 defines completeness of these annotations on element declarations used as operation input and output messages.

Rule 1 (consistency). Let S be a schema, and O_I be an ontology. If for any $m \in \{x\}_S$ there exist the annotations $ref(m, c_1)$ (*A1*) and $lower(m, f(c_1))$ (*A2*), then it must hold that $f(c_1) = m$. Analogously, if for any $n \in \{x\}_S$ there exist the annotations $ref(n, c_2)$ (*A1*) and $lift(n, g(n))$ (*A2*), then it must hold that $g(n) = c_2$.

Rule 2 (completeness). Let S be a schema, and I be an interface. For each $m \in \{x\}_S$ where m is the input message element of any operation in $\{op\}_I$, the element must have consistent annotations $ref(m, c_1)$ (*A1*) and $lower(m, f(c_1))$ (*A2*). Analogously, for each $n \in \{x\}_S$ where n is the output message element of any operation in $\{op\}_I$, the element must have consistent annotations $ref(n, c_2)$ (*A1*) and $lift(n, g(n))$ (*A2*).

A3: Annotations of WSDL Interface and Service (functional). Functional descriptions (both capabilities and categories) apply both to concrete web services and to the reusable and abstract interfaces. A reference annotation points from a service or an interface to its appropriate functional description. Listing 1.4 shows an example of multiple *A3* annotations:

```
1  <wsdl:interface name="NetworkSubscription"
2      sawsdl:modelReference="http://example.org/onto#VideoSubscriptionService
3             http://example.org/onto#VideoOnDemandSubscriptionPrecondition
4             http://example.org/onto#VideoOnDemandSubscriptionEffect" >
```

Listing 1.4. Example of annotations A3

Please note that a WSDL interface may be shared by multiple services, therefore the functional description of the interface should be general. A concrete functional description attached to the service then refines the functional description of the interface. Additionally, aggregate interfaces or services (i.e., those that combine multiple potentially independent functionalities) may be annotated with multiple functional descriptions. Rule 3 defines consistency for *A3* annotations: each functionality of a service must be a restriction of some functionality of the service's interface (see Definition 1 and Definition 2). This allows discovery to first find appropriate interfaces and then only check services that implement these interfaces. Rule 4 is analogous to Rule 3 with the difference that it applies to interface extension[10], ensuring that functionality cannot be lost through WSDL interface extension.

Rule 3 (consistency). Let W be a service and I be an interface such that W implements I. Then, for each annotation $ref(W, F)$ (*A3*) there must exist an annotation $ref(I, G)$ (*A3*) such that $F \leq G$.

Rule 4 (consistency). Let I and J be some interfaces such that I extends J. Then, for each annotation $ref(I, F)$ (*A3*) there must exist an annotation $ref(J, G)$ (*A3*) such that $G \leq F$.

A4: Annotations of WSDL Interface operations (functional). Functional descriptions (both capabilities and categories) apply also to interface operations, to indicate their particular functionalities. A reference annotation points from an operation to its appropriate functional description.

Functional annotation of interface operations can be used for services whose interfaces are simply collections of standalone operations. For example, a network subscription service may offer independent operations for subscription to a bundle, cancellation of a subscription, or price inquiry. A client will generally only want to use one or two of these operations, not all three. This shows that service discovery can, in such cases, become operation discovery. Also, operation annotations can be used for defining the order in which the operations should be invoked (see Section 5.1).

Rule 5 defines completeness for *A4* annotations: all operations within an interface must be annotated with a functional description. This rule ensures that no operation is left invisible to the automated clients.

Rule 5 (completeness). For all $o \in \{op\}_I$ there must exist some functional description F (capability or category) such that $ref(o, F)$ is defined.

Please note that annotations *A3* and *A4* apply to both types of functional descriptions, i.e., a capability or a category from some functional classification. It is even possible to combine them for a service, interface and its operations.

A5: Annotations of WSDL Service, Endpoints, and Binding (non-functional). Non-functional descriptions apply to a concrete instance of a Web service, that is, a Service,

---

[10] Interface extension is a feature of WSDL 2.0.

its Endpoints, or its Binding. A reference annotation can point from any of these components to a non-functional property. Listing 1.5 shows an example of annotation A5:

```
1  <wsdl:service name="ExampleCommLtd"
2      interface="NetworkSubscription"
3      sawsdl:modelReference="http://example.org/onto#VideoOnDemandPrice">
4      <wsdl:endpoint ...
5  </wsdl:service>
```

Listing 1.5. Example of annotation A5

Please note that non-functional descriptions are always specific to a concrete service, therefore, annotating interfaces or interface operations with non-functional properties is not defined. In case non-functional properties need to be specified on the operations (for example, different operations may have different invocation micropayment prices), a WSDL binding operation components (which mirror the operations of some interface) may be used to capture these properties. Due to the domain-specific nature of non-functional properties, WSMO-Lite cannot formulate any consistency or completeness rules for non-functional descriptions.

5 On Top of WSMO-Lite Annotations

WSMO-Lite annotations for Web services allow additional tasks on top: in particular we show implicit representation of a service choreography and illustrate the overall use of WSMO-Lite annotations for various SWS tasks essential for the client's automated decisions about services.

5.1 Implicit Choreography

In this section we show how WSMO-Lite interface operation annotations implicitly represent a choreography, understood according to [9] as a protocol from a single service's point of view[11], and formalized as an Abstract State Machine (ASM, [2]) as

$$X = (\Sigma, L) \qquad (5)$$

where $\Sigma \subseteq (\{x\} \cup C \cup R \cup E)$ is the signature of symbols, i.e., variable names $\{x\}$ or identifiers of elements from C, R, E of some ontology O_I; and L is a set of rules. Further, we denote by Σ_I and Σ_O the input and output symbols of the choreography (subsets of $C \cup R \cup E$), corresponding to the input data sent to the service and the returned output data. Each rule $r \in L$ defines a state transition $r : r^{cond} \rightarrow r^{eff}$ where r^{cond} is an expression in logic $\mathcal{L}(\Sigma)$ which must hold in a state before the transition is executed; r^{eff} is an expression in logic $\mathcal{L}(\Sigma)$ describing a condition which holds in a state after the execution. And finally, we use ontology elements as conditions (as in $c_1 \in O_I : c_1 \wedge \phi^{pre}$ within the algorithm), by which we mean that there exists an entity in the knowledge

---

[11] WS-CDL defines a different type of a choreography, i.e., as a common behavior of collaborating parties. The relationship of WSMO-Lite to WS-CDL is an open research question.

Table 1. MEPs, Rules and WSDL operations

MEP and Rule	WSDL Operation

in-out:
if $c_1 \wedge cnd1$ then $c_2 \wedge eff1$
$c_1 \in \Sigma_I$, $ref(msg1, c_1)$
$c_2 \in \Sigma_O$, $ref(msg2, c_2)$
	`<operation name="op1" pattern="w:in-out"`
	`    sawsdl:modelReference="ex:cnd1 ex:eff1">`
	`  <input  element="msg1"/>`
	`  <output element="msg2"/>`
	`</operation>`

in-only:
if $c_3 \wedge cnd2$ then $eff2$
$c_3 \in \Sigma_I$, $ref(msg3, c_3)$
	`<operation name="op2" pattern="w:in-only">`
	`    sawsdl:modelReference="ex:cnd2 ex:eff2">`
	`  <input  element="msg3"/>`
	`</operation>`

out-only:
if $cnd3$ then $c_4 \wedge eff3$
$c_4 \in \Sigma_O$, $ref(msg4, c_4)$
	`<operation name="op3" pattern="w:out-only">`
	`    sawsdl:modelReference="ex:cnd3 ex:eff3">`
	`  <output element="msg4"/>`
	`</operation>`

out-in:
if $cnd4$ then c_5
if $c_5 \wedge c_6$ then $eff4$
$c_5 \in \Sigma_O$, $ref(msg5, c_5)$
$c_6 \in \Sigma_I$, $ref(msg6, c_6)$
	`<operation name="op4" pattern="w:out-in">`
	`    sawsdl:modelReference="ex:cnd4 ex:eff4">`
	`  <output element="msg5"/>`
	`  <input  element="msg6"/>`
	`</operation>`

base which fits the description of the ontology element; for example, if the ontology element c_1 is a class, the knowledge base contains an instance of this class.

We construct the choreography from capability annotations of interface operations, according to the following algorithm.

Inputs:

- An interface I with operations $\{op\}_I$, ontology O_I, and a set of capabilities $\{K\}$.
- Complete *A4* annotations using only capabilities from $\{K\}$ for all $\{op\}_I$.
- Consistent and complete *A1* and *A2* annotations using O_I for all input and output messages of operations $\{op\}_I$.

Output:

- Choreography X with Σ_I, Σ_O and L.

Algorithm:

1: **for all** $ref(op, K)$, $op \in \{op\}_I$, $K = (\phi^{pre}, \phi^{eff}) \in \{K\}$ **do**
2:    get $ref(m, c_1)$ where m is the input message of op, $c_1 \in O_I$; add c_1 to Σ_I.
3:    get $ref(n, c_2)$ where n is the output message of op, $c_2 \in O_I$; add c_2 to Σ_O.
4:    **if** $op.mep$ in $\{in\text{-}out, in\text{-}only, out\text{-}only\}$ **then**
5:      create the rule r: $r^{cond} = c_1 \wedge \phi^{pre}$, $r^{eff} = c_2 \wedge \phi^{eff}$; add r to L.
6:    **else if** $op.mep$ in $\{out\text{-}in\}$ **then**
7:      create the rule r_1: $r_1^{cond} = \phi^{pre}$; $r_1^{eff} = c_2$; add r_1 to L.
8:      create the rule r_2: $r_2^{cond} = c_1 \wedge c_2$; $r_2^{eff} = \phi^{eff}$; add r_2 to L.
9:    **end if**
10: **end for**

The algorithm creates the sets of choreography input and output symbols from the semantic representations of the input and output messages of all the operations (lines 2–3). In addition, it creates choreography rules where the conditions contain assertions about the input messages and the effects contain assertions about output messages of operations. The algorithm creates one rule for operations with the *in-out, in-only* or *out-only* MEPs (lines 4–5). Since the ASM rules always represent an *in-out* interaction, two rules need to be created for operations with the *out-in* MEP: one representing the output and one the following input interaction. In order to further illustrate the results of the algorithm, Table 1 shows the resulting rules for the four MEPs (please note, that we do not currently cover fault messages). Here, a transition rule $r^{cond} \rightarrow r^{eff}$ is represented as if r^{cond} then r^{eff}; the symbols msg1...msg6 refer to schema elements used for input/output messages of operations; the symbols $c_1 \ldots c_6$ refer to identifiers of semantic descriptions of these messages; $ref(m,c)$ denotes the *A1* annotation, w: is a shortening for the URI http://www.w3.org/ns/wsdl/ and ex: for some application URI http://example.org/onto#. With a choreography constructed according to this algorithm, the client is able to automatically invoke a service, i.e., its operations in the correct and expected order.

5.2 Service Use Tasks

Not all annotations described in Section 4.2 are always needed, only those required by the tasks at hand in a particular domain-specific setting. Table 2 provides a summary, with *A1...A5* denoting the annotations and *R1...R5* denoting the rules. The symbol • marks the annotations and rules required to automate a given task, and the symbol ○ marks rules that are helpful but not absolutely required.

- *Service Discovery*, operating on functional descriptions (capabilities or categories), requires annotations *A3*. Rule 3 and Rule 4 help improve the scalability of the discovery through narrowing down a set of interfaces and services to be searched. If the discovery mechanism determines that an interface is not suitable, all the services implementing it and all the interfaces extended by it can immediately be discarded from further consideration.
- *Operation Discovery*, operating on functional descriptions of individual operations, requires annotations *A4*. Operation discovery might be useful with interfaces that

Table 2. Service Tasks, Annotations and Rules

Service Task	A1	A2	A3	A4	A5	R1	R2	R3	R4	R5
Service Discovery			•					○	○	
Operation Discovery				•						○
Composition			•							
Ranking and Selection					•					
Operation Invocation	•	•				•	•			
Service Invocation	•	•		•		•	•			○
Data Mediation	•	•					•			
Process Mediation	•	•		•		•	•			○

are collections of standalone, independent operations. Rule 5 ensures that no operation is left invisible to this discovery process.
- *Composition* uses capability descriptions, i.e., annotations *A3* restricted to capabilities, to put together multiple services to achieve a complex goal.
- *Ranking and Selection* processes non-functional descriptions, i.e., annotations *A5*, to select the service that most suits some particular requirements.
- *Operation Invocation* is the invocation of a single operation, requiring data transformations between the semantic model on the client and the service's XML message structure. This requires *A1* and *A2* annotation, kept consistent by Rule 1. Rule 2 ensures that all operation messages have these annotations.
- *Service Invocation* requires the operations of the service to be invoked in a proper order. This task therefore uses the implicit interface choreography (Section 5.1) and requires annotations *A4*. Rule 5 ensures that no operation is omitted from the choreography.
- *Data Mediation* uses data annotations *(A1* and *A2)* — assuming two different schemas correspond to a single shared ontology, the *A1* annotations make it possible to discover such a correspondence, and the *A2* annotations then enable data mapping transformations: lifting from one schema and lowering to the other.
- *Process Mediation* combines data mediation and choreography processing and thus requires the combined annotations *A1, A2* and *A4* . As described in [4], process mediation is applied during conversation between two services mediating their choreographies and messages.

This provides certain modularity to WSMO-Lite, enabling different environments using this service ontology to mix and match the annotations as necessary for the required tasks. On top of already being light-weight, WSMO-Lite provides value even if only parts of it are used.

6 Related Work

The major stream of related work is in the frameworks for Semantic Web Services (SWS), including WSMO [10], Semantic Markup for Web Services (OWL-S [12]) and Web Service Semantics (WSDL-S [1]). WSMO is a top-down conceptual model for SWS that defines four top-level components: ontologies, mediators, goals and web services. As we already mentioned, WSMO was the major input for WSMO-Lite. On the other hand, OWL-S was the first major ontology for SWS defining three interlinked ontologies: Service Profile (for the functional and non-functional descriptions), Service Model (for the behavioral descriptions), and Service Grounding (for physical Web service access). There are also recent works on OWL-S grounding that uses SAWSDL [8,7]. In comparison with that work, WSMO-Lite takes the additional step of simplifying the annotations into a lightweight ontology. WSDL-S was created in the METEOR-S[12] project as a specification of how WSDL can be annotated with semantic information. WSDL-S itself does not provide a concrete model for SWS, instead it makes the assumption that the concrete model will be expressible as annotations

---

[12] http://lsdis.cs.uga.edu/projects/meteor-s/

in WSDL and XML Schema documents. WSDL-S was taken as the basis for SAWSDL. In addition, there is a major orthogonal work to WSMO-Lite called SA-REST [11], aiming to enrich the informal descriptions of RESTful services, usually available in HTML, with RDFa[13] annotations. This work is complementary to WSMO-Lite as it could serve as an additional annotation mechanism for WSMO-Lite service ontology used for RESTful services. We will work on such integration within the W3C SWS-Testbed XG.

7 Conclusion and Future Work

In this paper, we describe the latest results from the development of WSMO-Lite, a minimal lightweight ontology for Semantic Web Services, building on the newest W3C standards. WSMO-Lite fills in SAWSDL annotations, and thus enables the Semantic Service Stack, open for various customizations according to domain-specific requirements, languages of required expressivity and domain-specific ontologies. WSMO-Lite supports the idea of incremental enhancements of SAWSDL as Amit Sheth points out in [6]: "Rather than look for a clear winner among various SWS approaches, I believe that in the post-SAWSDL context, significant contributions by each of the major approaches will likely influence how we incrementally enhance SAWSDL. Incrementally adding features (and hence complexity) when it makes sense, by borrowing from approaches offered by various researchers, will raise the chance that SAWSDL can present itself as the primary option for using semantics for real-world and industry-strength challenges involving Web services."

In our future work we plan to work on validation of WSMO-Lite annotations, together with a compiler for WSMO-Lite descriptions. We also plan to integrate WSMO-Lite with other research efforts within the W3C SWS-Testbed XG (such as already mentioned SA-REST) and to support service mashups with the WSMO-Lite ontology. In addition, we plan to integrate the WSMO-Lite ontology with the results of the semantic business processes research.

References

1. Akkiraju, R., et al.: Web Service Semantics - WSDL-S, Tech. rep., LSDIS Lab. (2005), http://lsdis.cs.uga.edu/projects/meteor-s/wsdl-s/
2. Börger, E., Stärk, R.: Abstract State Machines: A Method for High-Level System Design and Analysis. Springer, Heidelberg (2003)
3. de Bruijn, J., Fensel, D., Lausen, H.: D34v0.1: The Web Compliance of WSML. Technical report, DERI (2007), http://www.wsmo.org/TR/d34/v0.1/
4. Haselwanter, T., et al.: WSMX: A Semantic Service Oriented Middleware for b2b Integration. In: ICSOC, pp. 477–483 (2006)
5. Horrocks, I.: Owl: A description logic based ontology language. In: van Beek, P. (ed.) CP 2005. LNCS, vol. 3709, pp. 5–8. Springer, Heidelberg (2005)
6. Martin, D., Domingue, J.: Semantic web services: Past, present and possible futures (systems trends and controversies). IEEE Intelligent Systems 22(6) (2007)

---

[13] http://www.w3.org/TR/xhtml-rdfa-primer/

7. Martin, D., Paolucci, M., Wagner, M.: Bringing Semantic Annotations to Web Services: OWL-S from the SAWSDL Perspective. In: Aberer, K., Choi, K.-S., Noy, N., Allemang, D., Lee, K.-I., Nixon, L., Golbeck, J., Mika, P., Maynard, D., Mizoguchi, R., Schreiber, G., Cudré-Mauroux, P. (eds.) ISWC 2007. LNCS, vol. 4825, pp. 340–352. Springer, Heidelberg (2007)
8. Paolucci, M., Wagner, M., Martin, D.: Grounding OWL-S in SAWSDL. In: Krämer, B.J., Lin, K.-J., Narasimhan, P. (eds.) ICSOC 2007. LNCS, vol. 4749, pp. 416–421. Springer, Heidelberg (2007)
9. Roman, D., Scicluna, J.: Ontology-based choreography of wsmo services. Wsmo d14 final draft v0.3, DERI (2006), http://www.wsmo.org/TR/d14/v0.3/
10. Roman, D., et al.: Web Service Modeling Ontology. Applied Ontology 1(1), 77–106 (2005)
11. Sheth, A.P., Gomadam, K., Lathem, J.: SA-REST: Semantically Interoperable and Easier-to-Use Services and Mashups. IEEE Internet Computing 11(6), 91–94 (2007)
12. The OWL Services Coalition. OWL-S 1.1 Release (November 2004), http://www.daml.org/services/owl-s/1.1/
13. Vitvar, T., Kopecky, J., Fensel, D.: WSMO-Lite: Lightweight Semantic Descriptions for Services on the Web. In: ECOWS (2007)

Semantic Sitemaps: Efficient and Flexible Access to Datasets on the Semantic Web

Richard Cyganiak, Holger Stenzhorn, Renaud Delbru,
Stefan Decker, and Giovanni Tummarello

Digital Enterprise Research Institute (DERI),
National University Ireland, Galway

Abstract. Increasing amounts of RDF data are available on the Web for consumption by Semantic Web browsers and indexing by Semantic Web search engines. Current Semantic Web publishing practices, however, do not directly support efficient discovery and high-performance retrieval by clients and search engines. We propose an extension to the Sitemaps protocol which provides a simple and effective solution: Data publishers create Semantic Sitemaps to announce and describe their data so that clients can choose the most appropriate access method. We show how this protocol enables an extended notion of authoritative information across different access methods.

Keywords: Sitemaps, Datasets, RDF Publishing, Crawling, Web, Search, Linked Data, SPARQL.

1 Introduction

Data on the Semantic Web can be made available and consumed in many different ways. For example, an online database might be published as one single RDF dump. Alternatively, the recently proposed Linked Data paradigm is based on using resolvable URIs as identifiers to offer access to individual descriptions of resources within a database by simply resolving the address of the resources itself [1]. Other databases might offer access to its data via a SPARQL endpoint that allows clients to submit queries using the SPARQL RDF query language and protocol.

If several of these options are offered simultaneously for the same database, the choice of access method can have significant effects on the amount of networking and computing resources consumed on the client and the server side.

For example, a Semantic Web search engine that wants to index an entire database might prefer to download the single dump file, instead of crawling the data piecemeal by fetching individual Linked Data URIs. A client interested in the definition of only a few DBpedia [2] resources would, on the other hand, be well-advised to simply resolve their URIs. But if the client wants to execute queries over the resources, it would be better to use the available SPARQL service.

Such choices can have serious implications: For example, on February the 2nd 2007, Geonames[1] was hit by what appeared to be the first distributed denial of service attack against a Semantic Web site[2].

What happened, however, was not a malicious attack but rather a Semantic Web crawler bringing down the site by rapid-firing requests to Geonames' servers up to the point where the site's infrastructure could not keep up. One could simply judge this as a case of inconsiderate crawler behavior but, it highlights an important issue: What crawling rate should a crawler have on a Semantic Web site and would this be compatible with the size of the datasets? Considering that results are generated from semantic queries, it might be sensible to limit the query rate to one document per second, for example. It has to be noted however that crawling Geonames' 6.4M RDF documents would take 2.5 months under this condition, so that periodic recrawls to detect changes would take the same unsatisfactory amount of time and it would be impossible to have data in a fresh state.

Conceptually, the Geonames incident could have been avoided: Geonames offers a complete RDF dump of their entire database so this could have been bulk imported instead of crawling. But how is an automated crawler able to know about this dump and to know that the dump contains the entire Geonames database?

This paper describes a methodology with the primary goal to allow publishers to provide exactly such information and thus enable the smart selection of data access methods by clients and crawlers alike.

The methodology is based on extending the existing Sitemap Protocol (section 2) by introducing several new XML tags for announcing the presence of RDF data and to deal with specific RDF publishing needs (section 3).

2 The Sitemap Protocol and robots.txt

The Sitemap Protocol[3] defines a straighforward XML file for automatic agents (e.g. crawlers) that holds a list of URLs they should index. This is possible through tags that describe the location of each crawlable resource along with meta-information such as, for example, the expected rate of change for each individual URL or the date when this was last modified. An example of a sitemap is shown in the following listing:

```
<?xml version="1.0" encoding="UTF-8"?>
<urlset xmlns="http://www.sitemaps.org/schemas/sitemap/0.9">
  <url>
    <loc>http://www.example.com/</loc>
    <lastmod>2005-01-01</lastmod>
```

---

[1] http://geonames.org/
[2] http://geonames.wordpress.com/2007/02/03/friendly-fire-semantic-web-crawler-ddos/
[3] http://www.sitemaps.org/protocol.php

```
      <changefreq>monthly</changefreq>
      <priority>0.8</priority>
   </url>
</urlset>
```

Once a sitemap has been created, it must be saved in a file on the server. The protocol defines a way to extend the `robot.txt` file, so that a robot can find the location of the sitemap file on a given site.

3 The State of RDF Publishing

Even though there is no universal agreement on how information should be best published on the Semantic Web, we can characterize some of the options that are in widespread use today; list some of the limitations imposed by those options; and look at existing proposals to address these limitations.

3.1 Access Methods to RDF Data

Several different access methods to RDF data are in widespread use today. They vary widely along several dimensions. Two dimensions warrant closer attention, namely discoverability and cost of access.

The RDF Web and Linked Data. The traditional World Wide Web can be characterized as the HTML Web. It consists of a large number of HTML documents made available via the HTTP protocol and connected by hyperlinks. By analogy, the RDF Web consists of the RDF documents that are available via HTTP. Each RDF document can be parsed into an RDF graph. Often, different documents share URIs, and therefore merging the documents results in a connected graph. Unlike HTML hyperlinks, statements in these RDF graphs usually do not connect the URIs of documents, but instead they connect the URIs of resources described in those documents.

RDF uses URIs to identify resources of interest, but does not prescribe any particular way of choosing URIs to name resources. Experience shows that it is a good idea to choose URIs so that a Web lookup (an HTTP GET operation) on the URI results in fetching the RDF document that describes them. This effect is typically achieved by either appending a fragment identifier (e.g. `#me`) to the document URI, or by employing HTTP's `303 See Other` status code to redirect from the resource's URI to the actual document.

Datasets on the RDF Web are typically served so that resolving the URI of a resource will return only the RDF statements closely describing the resource, rather than resolving to a file containing the entire dataset. This is usually achieved by using a properly configured server which creates such descriptions on demand.

The cost of accessing this publishing method is usually relatively low. Descriptions of a single resource are usually much smaller than the description of an entire knowledge base. Similarly, the computational complexity for generating resource descriptions is limited, albeit non negligible.

RDF dumps. Like documents on the RDF Web, RDF dumps are serializations of an RDF graph made available on the Web. But they usually contain descriptions of a large number of resources in a single file, the resource identifiers usually do not resolve to the dump itself, and the file is often compressed. Many RDF datasets are published in this way on the Web and they cannot be easily browsed, crawled or linked into. It should be noticed, however, that:

- dumps are obviously useful to provide the entire knowledge behind a dataset at a single location (e.g. the Geonames dump as a single file to process it directly)
- RDF datasets that are not crawlable or do not contain resolvable resources may exist for perfectly valid technical reasons.
- Producing RDF dumps is technically less challenging than operating a SPARQL endpoint or serving Linked Data which makes RDF dumps a popular option.

Again, the cost of accessing this publishing method is computationally very low, as dumps are usually precomputed and can be easily cached. Since dumps represent the entire knowledge base, they can be however expensive in terms of network traffic and memory requirements for processing them.

SPARQL endpoints. SPARQL endpoints can provide descriptions of the resources described in a knowledge base and can further answer relational queries according to the specification of the SPARQL query language.

While this is the most flexible option for accessing RDF data, the cost of this publishing method is high:

- For simple tasks such as obtaining a single resource description, it is more involved than a simply resolving a URI. A query needs to be written according to the correct SPARQL syntax and semantics, the query needs to be encoded, the results need to be parsed into a useful form.
- It leaves a Semantic Web database open to potential denials of service due to queries with excessive execution complexity.

Multiple access methods. It is a common scheme for publishers to provide large RDF datasets through more than one access method. This might be partially explained by the early state of the Semantic Web, where there is no agreement on the best access method; but it also reflects the fact that the methods' different characteristics enable fundamentally different applications of the same data. Our proposal embraces and formalizes exactly this approach.

3.2 Current Limitations

In this section we list some limitations imposed by current RDF publishing practices which we want to address with the Semantic Sitemaps proposal.

Crawling performance. Crawling large Linked Datasets takes a long time and is potentially very expensive in terms of computing resources of the remote server.

Disconnected datasets. Not all datasets form a fully connected RDF graph. Crawling such datasets can therefore result in an incomplete reproduction of the dataset on the client side.

Scattered RDF files. To find scattered, poorly linked RDF documents, an exhaustive crawl of the HTML Web is necessary. This is likely not cost-effective for clients primarily interested in RDF, leading to missed documents.

Cataloging SPARQL endpoints. Even a full HTML Web crawl will not reveal SPARQL endpoints because support for the SPARQL protocol is not advertised in any way when requests are made to a specific SPARQL endpoint URI.

Discovering a SPARQL endpoint for a given resource. If all we have is a URI then we can resolve it to reveal some data about it. But how can we discover a potentially existing SPARQL endpoint for asking more complex queries about the resource?

Provenance. Provenance is a built-in feature of the Web, thanks to the grounding of HTTP URIs in the DNS. But control over parts of a domain is often delegated to other authorities. This delegation is not visible to the outside world.

Identifying RDF dumps. An HTML Web crawl will reveal links to many compressed archive files. Some of them may contain RDF data when uncompressed, but most of them will most likely contain other kinds of files. Downloading and uncompressing all of those dumps is likely not cost-effective.

Closed-world queries about self-contained data. RDF semantics is based on the open-world assumption. When consuming RDF documents from the Web, we can never be sure to have complete knowledge and therefore cannot with certainty give a negative answer to questions like "Does Example, Inc. have an employee named Eric?"

This list will serve a double purpose: First, it motivates several requirements for Semantic Sitemaps. Second, it serves as the basis for the evaluation of our proposal, as discussed in section 5.

3.3 Related Work

Most of the problems listed above have emerged just recently due to an increase in the amount and diversity of data available on the Semantic Web. The emergence of generic protocols for accessing RDF, especially the SPARQL protocol and Linked Data, enables the development of generic clients such as Tabulator [1] letting the user explore RDF documents anywhere on the Web. Equipped with a SPARQL API and a working knowledge of the query language, developers can interrogate any SPARQL endpoint available on the Web. As a result of these standardized protocols and tools, talking to data sources has become much easier. By contrast, efficient discovery and indexing of Semantic Web resources has become an important bottleneck that needs to be addressed quickly. A number of protocols and ideas have been evaluated and considered to address this.

It is clear that the problem of indexing datasets is tightly linked to the problem of indexing the Deep Web. Our Semantic Sitemaps proposal "piggy backs" on the Sitemap protocol, a successful existing approach to indexing the Deep Web.

Semantic Web Clients such as Disco[4] and Tabulator [1] are also somewhat related since they could ideally make use of this specification, e.g., to locate SPARQL endpoints. Use cases to serve search engines such as Sindice [3], SWSE [4] and Swoogle [5] have been of primary importance in the development of these specifications.

There many examples of services which index or provide reference to collections of RDF documents using diverse methodologies. PingTheSemanticWeb[5], for example, works by direct notifications from data producers of every file put online or updated. SchemaWeb.info[6] offers a large repository of RDF documents submitted by users. Even Google provides a good amount of RDF files when asked specifically with a `filetype:rdf` query.

Related proposals include POWDER[7], a method for stating assertions about sets of resources. A POWDER declaration can assert that all resources starting with a certain URI prefix are `dc:published` by a specific authority. In fact, a GRDDL transform could be used to turn a Semantic Sitemap into a POWDER declaration.

Finally, the Named Graph proposal provides a vocabulary for assertions and quotations with functionalities which can be used in conjunctions with these specifications [6].

4 The Semantic Sitemaps Extension

This section introduces the Semantic Sitemaps proposal. We will start by clarifying the notion of a *dataset*, then list the key pieces of information that can be provided in a Semantic Sitemap, and finally look at two specific issues: obtaining individual resource descriptions from a large dump; and the topic of authority on the Semantic Web.

4.1 Datasets

A dataset represents a knowledge base which is a very useful notion for the following reasons:

– Different access methods, such as those mentioned above, can and should indeed be simply thought of as only "access methods" to the same knowledge base.
– The knowledge base is more than just the sum of its parts: by itself, it is of informational value whether a piece of information belongs to a dataset

---

[4] http://sites.wiwiss.fu-berlin.de/suhl/bizer/ng4j/disco/
[5] http://pingthesemanticweb.com/
[6] http://schemaweb.info
[7] http://www.w3.org/2007/powder/

or not and queries using aggregates can make this explicit. For example, the question "Is `Berlin` the *largest* city mentioned in DBpedia?" cannot be answered by only getting the description of the resource itself (in case of Linked data though resolving their `Berlin` URI). On the contrary, such queries can be answered only the entire content of the dataset is available.

The Semantic Sitemap extension has the concept of dataset at its core: Datasets are well defined entities which can have one or more access methods. It is well defined what properties apply to a certain access method and what properties apply to a given dataset. Therefore properties that apply to that dataset will be directly related to all the data that can be obtained, independently from the access method.

This statement of existence of an underlying dataset implies that in case of multiple offered access methods, then they will provide information consistent with each other. The specification mandate that this in fact must be the case, with the exceptions only limited to:

- operational issues such as delays in the publication of dumps.
- information that pertains only to a certain access method, such as `rdfs:seeAlso` statements that link together the documents of a linked data deployment.

A publisher can host multiple datasets on the same site and can describe them independently using different sections of the Semantic Sitemap. While there is nothing that prevents information overlap or contradictions between different datasets, it is expected that this is not the case.

4.2 Adding Dataset Descriptions to the Sitemap Protocol

The Semantic Sitemap extension allows the description of a dataset via the tag `<sc:dataset>`, to be used at the same level as `<url>` tags in a regular sitemap. Access options for the datasets are given by additional tags such as `<sc:dataDump>`, `<sc:sparqlEndpoint>` and `<sc:linkedDataPrefix>`. If a sitemap contains several dataset definitions which are treated independently. The following example shows a sitemap file applying the extension.

```
<?xml version="1.0" encoding="UTF-8"?>
<urlset xmlns="http://www.sitemaps.org/schemas/sitemap/0.9"
        xmlns:sc="http://sw.deri.org/2007/07/sitemapextension">
  <sc:dataset>
    <sc:datasetLabel>
      Example Corp. Product Catalog
    </sc:datasetLabel>
    <sc:datasetURI>
       http://example.com/catalog.rdf#catalog</sc:datasetURI>
    <sc:linkedDataPrefix sc:slicing="subject-object">
       http://example.com/products/</sc:linkedDataPrefix>
```

```
      <sc:sampleURI>
          http://example.com/products/widgets/X42</sc:sampleURI>
      <sc:sampleURI>
          http://example.com/products/categories/all</sc:sampleURI>
      <sc:sparqlEndpoint sc:slicing="subject-object">
          http://example.com/sparql</sc:sparqlEndpoint>
      <sc:dataDump>
          http://example.com/data/catalogdump.rdf.gz</sc:dataDump>
      <sc:dataDump>
          http://example.org/data/catalog_archive.rdf.gz</sc:dataDump>
      <changefreq>weekly</changefreq>
  </sc:dataset>
</urlset>
```

The dataset is labeled as the *Example Corp. Product Catalog* and identified by `http://example.com/catalog.rdf#catalog`. Hence it is reasonable to expect further RDF annotations about the dataset `http://example.com/catalog.rdf`.

The "things" described in the dataset all have identifiers starting with `http://example.com/products/`, and their descriptions are served as Linked Data. A dump of the entire dataset is available, split into two parts and the publisher states that dataset updates can be expected weekly.

RDF dataset dumps can be provided in formats such as RDF/XML, N-Triples and N-Quads (same as N-Triples with a fourth element specifying the URI of the RDF document containing the triple; the same triple might be contained in several different documents). Optionally, dump files may be compressed in GZIP, ZIP, or BZIP2 format.

4.3 Other Elements

Other interesting elements in the specifications include:

`<sc:sparqlGraphName>`. If this optional tag is present, then it specifies the URI of a named graph within the SPARQL endpoint. This named graph is assumed to contain the data of this dataset. This tag must be used only if `<sc:sparqlEndpointLocation>` is also present, and there must be at most one `<sc:sparqlGraphName>` per dataset. If the data is distributed over multiple named graphs in the endpoint, then the publisher should either use a value of * for this tag, or create separate datasets for each named graph. If the tag is omitted, the dataset is assumed to be available via the endpoint's default graph.

`<sc:datasetURI>`. An optional URI that identifies the current dataset. Resolving this URI may yield further information, possibly in RDF, about the dataset, but this is not required.

`<sc:datasetLabel>`. An optional label that provides the name of the dataset.

`<sc:sampleURI>`. This tag can be used to point to a URI within the dataset which can be considered a representative sample. This is useful for Semantic Web clients to provide starting points for human exploration of the dataset. There can be any number of sample URIs.

4.4 Defining the DESCRIBE Operator

Publishing an RDF dataset as Linked Data involves creating many smaller RDF documents, each served as a response when accessing the URI itself via HTTP.

This is usually the result of a SPARQL DESCRIBE query performed giving as argument the resource's identifier. But the SPARQL specification does not specify how DESCRIBE queries are actually answered: it is up to the data provider to choose the amount of data to return as a description.

It is important that the Semantic Sitemap provides a way to specify how this description process is most likely to happen. Knowing this enables a process that we call "slicing" of the dataset, i.e. turning a single large data dump into many individual RDF models served online as description of the resource identifiers. This process is fundamental for creating large indexes of online RDF documents descriptions, as illustrated in section 5.1.

For this, the `<sc:linkedDataPrefix>` and `<sc:sparqlEndpointLocation>` tags can have an optional `sc:slicing` attribute taking a value from the list of slicing methods below and which mean that the description of a resource X includes:

- `subject`: All triples whose subject is X.
- `subject-object`: All triples whose subject or object is X.
- CBD: The Concise Bounded Description [7] of X.
- SCBD: The Symmetric Concise Bounded Description [7] of X.
- MSG: All the Minimal Self-Contained Graphs [8] involving X.

Publishers that want to use a slicing method that is not in the list should pick the value that most closely matches their desired method, or they may omit the `sc:slicing` attribute. If the slicing method is very different from any in the list, it is recommended to publish a dump in N-Quads format.

4.5 Sitemaps and Authority

While there is no official definition of authoritative information specifically for the Semantic Web, many currently agree on extending the standard definition given in *Architecture of the World Wide Web, Volume One* [9], which links authority to the ownership of the domain name in which the URIs are minted.

For example, a piece of information coming from the owner of the domain `example.com` would be considered authoritative about the URI `http://example.com/resource/JohnDoe`. Following this definition, any information obtained by resolving the URI of information resources served as Linked Data can be considered authoritative, as it comes from the same domain where the URI is hosted.

However, no mechanism is currently defined to get information about such an authority: The only possibility for this is via DNS domain records, which are outside the actual Web architecture.

For this reason the Semantic Sitemap extension proposes an `<sc:authority>` element, which is used at the top level of a sitemap file to specify a URI identifying the person, organization or other entity responsible for the sitemap's URI space. This is only useful if the URI is resolvable and yields additional information about the authority. The authority URI has to be within the sitemap's URI space, which makes the authority responsible for providing their own description.

The semantics of `<sc:authority>` is such that, given an authority URI a, for any document d within the sitemap's URI space, there is an implied RDF statement d `dc:publisher` a. For example, if a sitemap file in the root directory of `http://example.com/` declares an `sc:authority` of `http://example.com/foaf.rdf#me`, then the entity denoted by that URI is considered to be the publisher of all documents whose URI starts with `http://example.com/`. It has to be kept in mind that publication does not necessarily imply endorsement; it merely means that a is responsible for making d available. To express a propositional attitude, like assertion or quotation, additional means are necessary, such as the Semantic Web Publishing Vocabulary [6].

Delegation of authority. The boundaries of DNS domains do not always match the social boundaries of URI ownership. An authority sometimes delegates responsibility for chunks of URI space to another social entity. The Semantic Sitemap specification accounts for this pattern. If another sitemap file is placed into a sitemap's URI space, then the second sitemap's URI space is considered to be not under the authority of the first. For example, publishing a sitemap file at `http://example/~alice/sitemap.xml` delegates the subspace rooted at `http://example.com/~alice/`.

Furthermore, `robots.txt` must link to a sitemap index file that in turn links to both of the sitemap files. This ensures that all visitors get identical pictures of the site's URI space. The approach is limited: It only allows the delegation of subspaces whose URIs begin with the same URI as the secondary sitemap file.

This feature proves to be very useful in properly assigning authority to Semantic Web information published in popular URI spaces like `purl.org`.

Joining URI spaces. Besides partitioning of URI spaces, it is sometimes necessary to join URI spaces by delegating responsibility of additional URIs to an existing sitemap. This is particularly useful when a single dataset spans over multiple domains, for example, if the SPARQL endpoint is located on a different server. The joining of URI spaces is achieved by placing a sitemap file into each of the spaces and having the `<sc:subSitemap>` element point to the other's URI. The sub-sitemap also needs a reciprocating `<sc:parentSitemap>` element. This prevents fraudulent appropriation of URI spaces.

Authoritative descriptions from SPARQL endpoints and dumps. An RDF description of a URI u is considered authoritative if resolving u yields an RDF document containing that description. An authoritative description is known to originate directly from the party owning the URI u, and thus can be considered to be definitive with regard to the meaning of u.

Providing authoritative information thus requires the publication of RDF in the "RDF Web" style. Descriptions originating from SPARQL endpoints or RDF dumps cannot be considered authoritative, because it cannot be assumed that information about u from a SPARQL endpoint at URI e is indeed authorized by the owner of u. The presence of a Semantic Sitemap, however, changes this. If e and u and d are in the same sitemap's URI space, then they originate from the same authority, and therefore we can treat information from the SPARQL endpoint as authoritative with respect to u even if u is not resolvable. This has two benefits. First, it allows RDF publishers to provide definitive descriptions of their URIs even if they choose not to publish RDF Web-style documents. Second, it allows RDF consumers to discover authoritative descriptions of URIs that are not resolvable, by asking the appropriate SPARQL endpoint or by extracting from the RDF dump.

5 Evaluation

We now revisit the challenges identified in section 3.2 and will show how the Semantic Sitemaps proposal can address each of them.

Crawling performance. Crawling large Linked Data deployments takes a long time and is potentially very expensive in terms of computing resources of the remote server. An RDF publisher can make a dump of the merged Linked Data documents available and announce both the `<sc:linkedDataPrefix>` and `<sc:dataDump>` in a sitemap file. Clients can now discover and download the dump instead of crawling the site, thus dramatically reducing the required time.

Disconnected datasets. Not all datasets form a fully connected RDF graph. Crawling such datasets can result in an incomplete reproduction of the dataset on the client side. Sitemaps address this in two ways. Firstly, clients can again choose to download a dump if it is made available. Secondly, by listing at least one ¡sc:sampleURI¿ in each component of the graph, the publisher can help to ensure a complete crawl even if no dump is provided.

Scattered RDF files. To find scattered, poorly linked RDF documents, an exhaustive crawl of the HTML Web is necessary. This is likely not cost-effective for clients primarily interested in RDF, leading to missed documents. Site operators can provide an exhaustive list of those documents using `<sc:dataDump>` or `<sc:sampleURI>`.

Cataloging SPARQL endpoints. Even a full HTML Web crawl will not reveal SPARQL endpoints, because support for the SPARQL protocol is not advertised in any way when requests are made to a SPARQL endpoint URI. However, if a sitemap has been provided, the crawler can discover the services via `<sc:sparqlEndpoint>`.

Discovering a SPARQL endpoint for a given resource. If all we have is a URI, we can resolve it to reveal some data about it. But to discover a potentially existing SPARQL endpoint for asking more complex queries we can look for a sitemap on the URI's domain and inspect it for `<sc:sparqlEndpoint>` elements.

Provenance. Provenance is a built-in feature of the Web, thanks to the grounding of HTTP URIs in the DNS. But control over parts of a domain is often delegated to other authorities. Semantic Sitemaps allow site operators to make this delegation visible, by appropriate placement of multiple sitemap files, and also by employing `<sc:subSitemap>` and `<sc:parentSitemap>` elements.

Identifying RDF dumps. An HTML Web crawl will reveal links to many compressed archive files. Some of them may contain RDF data when uncompressed, but most of them will most likely contain other kinds of files. Downloading and uncompressing all of those dumps is likely not cost-effective. Semantic Sitemap files explicitly list the locations of RDF dumps in the `<sc:dataDump>` element and therefore allow crawlers to avoid the cost of inspecting dumps that turn out not to contain any RDF.

Closed-world queries about self-contained data. RDF semantics is based on the open-world assumption and hence, when consuming RDF documents from the Web, we can never be sure to have complete knowledge. So we cannot with certainty give a negative answer to the question "Does Example, Inc. have an employee named Eric?". Datasets defined in Semantic Sitemaps are natural candidates for applying local closed-world semantics. Drawing on `<sc:datasetLabel>`, this allows us to give a stronger answer: "Example, Inc. has not given information about such an employee in the dataset labeled *Example, Inc. Employee List.*"

5.1 Processing of Large RDF Dumps

A major motivation for our proposal is its potential to reduce the cost, both in raw time and computing resources, of indexing the largest RDF datasets currently available on the RDF Web. These datasets are usually also available as RDF dumps, and Semantic Sitemaps enable clients to download the dump and avoid crawling. This section presents the results of some experiments into quantifying the benefits of this approach and showing the general feasibility of processing very large RDF dumps. We employ the Hadoop framework[8] for parallel data processing on a "mini" cluster of low cost machines (single core, 4gb ram).

The issue with crawling large collections of RDF documents from a single host, such as the UniProt datasets available on `http://purl.uniprot.org`, is that a crawler has to space its request by a reasonable amount of time in order to avoid overloading the server infrastructure. Semantic Web servers are likely

---

[8] `http://lucene.apache.org/hadoop/`

to be experimental and not optimized for high load. Therefore submitting more than one request per second seems to be unadvisable for RDF crawlers.

The UniProt dataset[9] contains approximately 14M RDF documents. A full crawl at this rate would take more than five months, and a full re-crawl to detect updated documents would take another five months.

UniProt provides a dump (over 10 GB in size, compressed). The time for gathering the data can thus be reduced to downloading of the dump files, plus subsequent local data processing to slice the dumps into parts that are equivalent to the individual documents. Each part describes a single resource. The parts can be passed on to our indexing system grouped in .TAR.GZ files each containing 10000 individual RDF files.

The first step in our processing was to convert the dumps from the RDF/XML format to N-Triples. The line-based N-Triples format can be more easily processed with Hadoop's off-the-shelf classes. We observed processing rates of about 60k triples/s for this format conversion. We did not attempt to further optimize or parallelize this step.

For the actual slicing of the dumps we used Hadoop's MapReduce [10] facility. In the mapping phase, each triple is sorted into buckets based on the URIs mentioned in the subject and object position. In the reduce phase, the triples in each bucket are written to a separate file.

To gather some proof-of-concept results, we first processed the file `uniref.nt` (270M triples) on a single machine, then two machines and then four. We then ran a test on the complete UniProt dataset (1.4B triples) on four machines.

# Machines	1	2	4	4
# Triples	272M	272M	272M	1.456M
Conversion	1:17:01	1:19:30	1:19:10	6:19:28
MapReduce	10:04:35	5:18:55	2:56:54	16:36:26

These preliminary results show that processing of a very large RDF dump to obtain slices equivalent to the documents that are being served as description of the resources using the Linked Data paradigm is feasible. The processing task can be parallelized, and the results indicate that, unlike with the crawling approach, the data retrieval step is unlikely to be a bottleneck.

6 Adoption

Currently, the Sitemap Specification is currently available in its 5th release[10]. The specification creation process involved a great deal of interaction with data producers and developers alike. Most of this interaction happened in specialized Mailing Lists and trough technical workshop dissemination [11].

---

[9] Available from
  ftp://ftp.uniprot.org/pub/databases/uniprot/current_release/rdf
[10] http://sw.deri.org/2007/07/sitemapextension/

The process has so far been very interactive, with data producers proposing features and clarifications they considered important such as, for example split datasets for DBpedia, and details on how to use a sitemap on shared domains like purl.org.

Thanks to this direct interaction, adoption at Data Producer level has so far been very satisfactory. Most large datasets provide a Semantic Sitemap and in general we report that data producers have been very keen to add one when requested given the very low overhead and the perceived lack of negative consequences.

At consumer level and as discussed earlier, the Sindice Semantic Web indexing engine adopts the protocol [3] and thanks to it has indexed, as today, more than 26 million RDF documents.

We can report that the SWSE Semantic Web Search Engine [4] will also soon be serving data obtained thanks to dumps downloaded using this extension.

7 Future Work

While it is unlikely that the current specifications will change profoundly, we envision that future versions of the Semantic Sitemaps will address issues such as: Data published in formats such as RDFa and using GRDDL transformations, datasets available in multiple versions, enhance SPARQL endpoint descriptions, mirrored datasets and copyright and legal related information.

8 Conclusion

In this paper we have discussed an extension to the original Sitemap Protocol to deal with large Semantic Web datasets. We have highlighted several challenges to the way Semantic Web data have been previously published and subsequently showed how these can be addressed by applying the proposed specifications, thus allowing both clients and servers alike to provide and important novel functionalities.

We have further verified the feasibility of an important use case: server side "dump splitting" to efficiently process and index millions of documents from the Semantic Web. We have showm that this task can be performed efficiently and in a scalable fashion even with modest hardware infrastructures.

Acknowledgements

Many people have provided valuable feedback and comments about Semantic Sitemaps including Chris Bizer (Free University Berlin), Andreas Harth (DERI Galway), Aidan Hogan (DERI Galway), Leandro Lopez (independent), Stefano Mazzocchi (SIMILE - MIT), Christian Morbidoni (SEMEDIA - Universita' Politecnica delle Marche), Michele Nucci (SEMEDIA - Universita' Politecnica delle Marche), Eyal Oren (DERI Galway), Leo Sauermann (DFKI)

References

1. Berners-Lee, T.: Tabulator: Exploring and Analyzing linked data on the Semantic Web. In: Procedings of the The 3rd International Semantic Web User Interaction Workshop (2006)
2. Auer, S., Bizer, C., Kobilarov, G., Lehmann, J., Cyganiak, R., Ives, Z.: Dbpedia: A nucleus for a web of open data. In: Aberer, K., Choi, K.-S., Noy, N., Allemang, D., Lee, K.-I., Nixon, L., Golbeck, J., Mika, P., Maynard, D., Mizoguchi, R., Schreiber, G., Cudré-Mauroux, P. (eds.) ISWC 2007. LNCS, vol. 4825, pp. 722–735. Springer, Heidelberg (2007)
3. Tummarello, G., Oren, E., Delbru, R.: Sindice.com: Weaving the open linked data. In: Aberer, K., Choi, K.-S., Noy, N., Allemang, D., Lee, K.-I., Nixon, L., Golbeck, J., Mika, P., Maynard, D., Mizoguchi, R., Schreiber, G., Cudré-Mauroux, P. (eds.) ISWC 2007. LNCS, vol. 4825, pp. 547–560. Springer, Heidelberg (2007)
4. Harth, A., Hogan, A., Delbru, R., Umbrich, J., O'Riain, S., Decker, S.: Swse: Answers before links! In: Semantic Web Challenge 2007, 6th International Semantic Web Conference (2007)
5. Ding, L., Finin, T., Joshi, A., Pan, R., Cost, R.S., Peng, Y., Reddivari, P., Doshi, V., Sachs, J.: Swoogle: a search and metadata engine for the semantic web. In: CIKM 2004: Proceedings of the thirteenth ACM international conference on Information and knowledge management, pp. 652–659. ACM, New York (2004)
6. Carroll, J., Bizer, C., Hayes, P., Stickler, P.: Named Graphs. Journal of Web Semantics 3(4), 247–267 (2005)
7. Stickler, P.: CBD - Concise Bounded Description (2005) (retrieved 09/25/2006), http://www.w3.org/Submission/CBD/
8. Tummarello, G., Morbidoni, C., Puliti, P., Piazza, F.: Signing individual fragments of an rdf graph. In: WWW 2005: Special interest tracks and posters of the 14th international conference on World Wide Web, pp. 1020–1021. ACM, New York (2005)
9. Jacobs, I., Walsh, N.: Architecture of the World Wide Web, Volume One - W3C Recommendation (2004) Retrieved 09/25/2006, http://www.w3.org/TR/webarch/
10. Dean, J., Ghemawat, S.: Mapreduce: Simplified data processing on large clusters. In: OSDI 2004: Proceedings of the 6th conference on Symposium on Operating Systems Design and Implementation, pp. 137–150 (2004)
11. Tummarello, G.: A sitemap extension to enable efficient interaction with large quantity of linked data. In: Presented at W3C Workshop on RDF Access to Relational Databases (2007)

On Storage Policies for Semantic Web Repositories That Support Versioning

Yannis Tzitzikas, Yannis Theoharis, and Dimitris Andreou

Computer Science Department, University of Crete, GREECE, and
Institute of Computer Science, FORTH-ICS, Greece
{tzitzik,theohari,andreou}@ics.forth.gr

Abstract. This paper concerns versioning services over Semantic Web (SW) repositories. We propose a novel storage index (based on partial orders), called POI, that exploits the fact that RDF Knowledge Bases (KB) have not a unique serialization (as it happens with texts). POI can be used for storing several (version-related or not) SW KBs. We discuss the benefits and drawbacks of this approach in terms of storage space and efficiency both analytically and experimentally in comparison with the existing approaches (including the change-based approach). For the latter case we report experimental results over synthetic data sets. POI offers notable space saving as well as efficiency in various cross version operations. It is equipped with an efficient version insertion algorithm and could be also exploited in cases where the set of KBs does not fit in main memory.

1 Introduction and Motivation

The provision of versioning services is an important requirement of several modern applications of the Semantic Web, e.g. for digital information preservation [3,14], and for e-learning applications [13]. Two key performance aspects of a version management system is the *storage space* and the *time* needed for creating (resp. retrieving) a new (resp. existing) version. In the Semantic Web (SW) there exist only limited support of versioning services. Most of the related works [9,16,10,11] propose high level services for manipulating versions but none of these have so far focused on the performance aspect of these services (they mainly overlook the storage space perspective). In general, we could identify the following approaches:

(a) Keep stored each version as an independent triple store. This is actually the approach adopted in all previous works [16,11,9] on versioning for SW repositories. The obvious drawback of this approach is the excessive storage space requirements.
(b) Keep stored only the deltas between two consecutive versions [12,2]. To construct the contents of a particular version one has to execute a (potentially long) sequence of deltas (which could be computationally expensive). However, appropriate comparison functions and change operation semantics can result in smaller in size deltas as described in [17]. The extreme case where only the change log is kept stored and marked positions (on that log) are used to indicate versions, is elaborated on [15] which also provides methods for reducing the size of a sequence of deltas.

In this paper we propose and analyze an approach (actually a storage index structure) that stands between the above two extremes. It aims at exploiting the fact that it is expected to have several versions (not necessarily consecutive) whose contents overlap. In addition, it exploits the fact that RDF graphs have not a unique serialization (as it happens with text), and this allows us to explore directions that have not been elaborated by the classical versioning systems for texts (e.g. [12,2]). Specifically, we view an RDF KB as a *set* of triples. In a nutshell, the main contribution of this work is the introduction of a storage data structure, called POI, based on partial orders that exploits the expected overlap between versions' contents in order to reduce the storage space, also equipped with algorithms (including an auxiliary caching technique) for efficient version insertion and retrieval. It is important to note that the structure (and occupied space) of POI is independent of the version history. Knowledge of version history can be exploited just for speeding up some operations (specifically the insertion of versions that are defined by combining existing versions). It follows that the benefits from adopting POI are not limited to versioning. It could also be exploited for building repositories appropriate for collaborative applications, e.g. for applications that require keeping personal and shared spaces (KBs) as it is the case of modern e-learning applications (e.g. see trialogical e-learning [13]). In such cases, we need to store a set of KBs that are not historically connected and are expected to overlap.

In comparison to the change-based approaches (proposed in the context of the SW [17] or not [4,5]) we could say that POI stores explicitly only the versions with the minimal (with respect to set containment) contents. All the rest versions are stored in a positively incremental way, specifically positive deltas are organizing as a partially ordered set (that is history-independent) aiming at minimizing the total storage space. It follows that POI occupies less space than the change-based approach in cases where there are several versions (or KBs in general) not necessarily consecutive (they could be even in parallel evolution tracks) whose contents are related by set inclusion ($\subseteq$).

Regarding version retrieval time, the cost of retrieving the contents of a version in the change-based approach is analogous to the distance from the closest stored snapshot (either first or last version, according to the delta policy adopted [5]). Having a POI the cost is independent of any kind of history, but it depends on the contents of the particular version, specifically on the depth of the corresponding node in the POI graph.

In comparison with other works on versioning (not in the SW context), [5] focuses on providing fast access to the current (or recent) versions, while we focus on minimizing the storage space, in an attempt to support applications with vast amounts of overlapping versions. [6] focuses on composite versioned objects, i.e. objects composed of other versioned objects.

In general, POI is an advantageous approach for archiving set-based data, especially good for inclusion-related and "oscillating"[1] data. This is verified analytically and experimentally. The remainder of this paper is organized as follows: Section 2 introduces basic notions and notations. Section 3 introduces POI. Section 4 elaborates on the storage requirements of POI by providing analytical and experimental results. Section 5 provides version insertion algorithms and reports experimental results. Section 6

---

[1] Suppose, for instance, a movie database describing movies, theaters that showtimes. Unlike movie descriptions, which rarely change, theaters and showtimes change daily.

discusses other operations that can be performed efficiently with a POI. Finally, Section 7 summarizes and identifies issues for future research. Due to space limitations, proofs and other details are available in the extended version of this paper[2].

2 Framework and Notations

Def 1. If $\mathcal{T}$ is the set of all possible RDF triples (or quadruples if we consider graph spaces [7]), then :

- a *knowledge base* (for short KB) is a (finite) subset S of $\mathcal{T}$,
- a *named knowledge base* (for short NKB) is a pair (S, i) where i is an identifier and S is a KB,
- a *multi knowledge base* (for short MKB) is a set of NKBs each having a distinct identifier, and
- a *versioned knowledge base* (for short VKB) is a MKB plus an acyclic *subsequent* relation over the identifiers of the NKBs that participate to MKB.

Let Id be the set of all possible identifiers (e.g. the set of natural numbers). If $v = (S, i)$ is a NKB (i.e. $S \subseteq \mathcal{T}, i \in Id$) then we will say that i is the identifier of v, and S is the content of v (we shall write $id(v) = i$ and $D(i) = S$ respectively).

Let $V = \{v_1, \ldots, v_k\}$ be a MKB. We shall use $id(V)$ to denote the identifiers of v_i, i.e. $id(V) = \{ id(v) \mid v \in V \}$, and $D(V)$ to denote their KBs, i.e. $D(V) = \{ D(v) \mid v \in V \}$. We shall use T_V denote the set of all distinct triples of V, i.e. $T_V = \cup_{v \in V} D(v)$. So $D(V)$ is a family of subsets of T_V.

A *subsequent* (version) relation over a MKB V is any function of the form $next : id(V) \to \mathcal{P}(id(V))$, where $\mathcal{P}(\cdot)$ denotes powerset. For instance, suppose that $next(i) = \{j, k\}$. In this case we will say that i is a direct previous version of j and k, and that j and k are direct next versions of i. If $next(i) = \emptyset$, then we will call version i *leaf* version of V. We call a version id i the *root* version of V, if there does not exist any version id j, such that $i \in next(j)$. A pair $(V, next)$ is a VKB if the graph $(id(V), next)$ is acyclic.

Table 1 presents some version management services and their semantics in the form of conditions that should hold before their call and after their run. In particular, *insert* adds a new version with id i and version content S. Additionally, *merge* differs from *insert* service in that the content of the new version (with identifier h) is the result of the application of a set operator, denoted by $\odot$, on the contents of two (or more) existing versions (having identifiers i and j). In this way we could model operators like those proposed in [8].

3 The Partial Order Index (POI)

Consider a versioned knowledge base $(V, next)$. Specifically consider a V comprising four versions with: $D(1) = \{a, b\}, D(2) = \{a, b, c\}, D(3) = \{a, c\}$ and $D(4) = \{a, b, c\}$, where a, b, c denote triples. This means that $V = \{(\{a, b\}, 1), (\{a, b, c\}, 2),$

---

[2] http://www.csd.uoc.gr/~theohari/pub/eswc2008.pdf

Table 1. Version creation services

	Service	Pre-condition	Post-Condition
1	$insert(S,i)$	$i \notin id(V)$	$V' = V \cup \{(S,i)\}$
2	$merge(i,j,h,\odot)$	$\{i,j\} \subseteq id(V), h \notin id(V)$	$V' = V \cup (D(i) \odot D(j), h),$ $next'(i) = next(i) \cup \{h\},$ $next'(j) = next(j) \cup \{h\}$

Individual Copies	Change based	Hasse Diagram	Partial Order Index
{a, b} ← 1	Δ^+ Δ^-	1	1
{a, b, c} ← 2	$\Delta(\to 1)$: {a, b} ∅	{a, b} {a, c} ← 2	{a, b} {a, c} ← 2
{a, c} ← 3	$\Delta(1\to 2)$: {c} ∅		
{a, b, c} ← 4	$\Delta(2\to 3)$: ∅ {b}	{a, b, c} ← 3	∅ ← 3
	$\Delta(3\to 4)$: {b} ∅	← 4	← 4

Fig. 1. Storage Example when the partial order index is used (right) or not (left)

$(\{a, c\}, 3), (\{a, b, c\}, 4)\}$. Below we present methods for storing V (we will not discuss the storage of $next$ as this is trivial and of minor importance). To aid understanding, Figure 1 sketches the storage policies that we will investigate in the sequel. One trivial approach, which is adopted by current SW versioning tools [16,11,9], is to store each individual NKB independently and entirely. Another approach [17,4,5] is to store the initial NKB and the deltas of every other version with respect to its previous version. Since versions usually contain overlapping triples, we propose the use of an index in order to reduce the number of triple copies. For instance, in the example of Figure 1 four copies of triple a are stored in the trivial case (see the left part of Figure 1). However, with the use of the index we propose, only two a copies are stored (see the right part of Figure 1). To describe this index, we first introduce some preliminary background material and definitions.

Given two subsets S, S' of $\mathcal{T}$, we shall say that S is narrower than S', denoted by $S \leq S'$, if $S \supseteq S'$. So, $\emptyset$ is the top element of $\leq$, and the infinite set $\mathcal{T}$ is the bottom element. Clearly, $(\mathcal{P}(\mathcal{T}), \leq)$ is a partially ordered set (poset).

We can define the *partial order of a versioned KB* by restricting $\leq$, on the elements of $D(V)$. For brevity, we shall use the symbol $\sqsubseteq$ to denote $\leq_{|D(V)}$.

In our running example, we have $D(V) = \{\{a, b\}, \{a, b, c\}, \{a, c\}, \{a, b, c\}\}$ and the third diagram of Figure 1 shows the Hasse diagram of the partially ordered set $(D(V), \sqsubseteq)$. This diagram actually illustrates the structure of the so-called *storage graph* that we introduce below.

A *storage graph* Υ is any pair $\langle \Gamma, stored \rangle$ where $\Gamma = (N, R)$ is a directed acyclic graph and *stored* is a function from the set of nodes N to $\mathcal{P}(\mathcal{T})$. If $(a, b) \in R$, i.e. it is an edge, we will also write $a \to b$.

For a node $n \in N$, $stored(n)$ is actually a set of triples, so it is the storage space associated with node n. Table 2 shows all notations (relating to storage graphs) that will be used.

Table 2. Notations for Storage Graphs

Notation	Definition	Equiv. Notation
R	a binary relation over the set of nodes N	$\to$
R^t	the transitive closure of the relation R	$\to^t$
R^r	the reflexive and transitive reduction of the relation R	$\to^r$
$Up(n)$	$= \{n' \mid (n,n') \in R\} = \{n' \mid n \to n'\}$	
$Down(n)$	$= \{n' \mid (n',n) \in R\} = \{n' \mid n' \to n\}$	
$Up^t(n)$	$= \{n' \mid (n,n') \in R^t\} = \{n' \mid n \to^t n'\}$	
$Down^t(n)$	$= \{n' \mid (n',n) \in R^t\} = \{n' \mid n' \to^t n\}$	
$content(n)$	$= \cup\{ stored(n') \mid n' \in Up^t(n)\}$	

For each node n of a storage graph we can define its *content*, denoted dy $content(n)$, by exploiting the structure of the graph and the function *stored*. Specifically we define:

$$content(n) = \cup\{ stored(n') \mid n \to^t n'\} \tag{1}$$

so it is the union of the sets of triples that are stored in all nodes from n to the top elements of Γ. We should stress that $content(n)$ is not stored, instead it is computed whenever it is necessary.

The *Partial-Order Index*, for short POI, is a storage graph whose structure is that of $(D(V), \sqsubseteq^r)$. Note that $\sqsubseteq^r$ denotes the reflexive and transitive reduction of $\sqsubseteq$. Consider a NKB $(D(i), i)$. Each version id i is associated with a node n_i whose storage space is defined as:

$$\begin{aligned} stored(n_i) &= D(i) \setminus \{ D(j) \mid D(j) \sqsubseteq D(i) \} \\ &= D(i) \setminus \{ stored(n_j) \mid n_i \to^t n_j \} \\ &= D(i) \setminus \{ stored(n_j) \mid n_j \in Up^t(n_i) \} \end{aligned}$$

It follows easily that if $n_i \to^t n_j$ then it holds $stored(n_i) \cap stored(n_j) = \emptyset$. The fourth diagram of Figure 1 illustrates the storage graph of this policy for our running example. For each node n_i the elements of $stored(n_i)$ are shown at the internal part of that node. Although we have 4 versions, Γ contains only 3 nodes. As an example, $D(4) = \{a,b\} \cup \{a,c\} = \{a,b,c\}$.

4 Analyzing Storage Space Requirements

Let IC (from "individual copies") denote the policy where each individual version is stored independently, POI denotes the case where a POI is adopted and CB (from Changed Based) the case where deltas are stored. Below we compare these policies with respect to *storage space*.

If Z denotes a policy, we shall use $space_t(Z)$ to denote the number of triples that are stored according to policy Z. It is not hard to see that

$$|T_V| \leq space_t(\text{POI}) \leq space_t(\text{IC}) = \sum_{v_i \in V} |D(v_i)|$$

$$|T_V| \leq space_t(\text{CB}) \leq 2 \sum_{v_i \in V} |D(v_i)|$$

Regarding the first formula, strict inequality holds, i.e. $space_t(\text{POI}) < space_t(\text{IC})$ if there is a version whose content is a subset of the content of another version. Specifically, the worst case for POI is when all nodes of the storage graph are leaves (except for the root). That case leads to space requirements equal to those of IC. On the other hand, the best case for POI, is when the content of every version is a subset of the content of every version with greater content cardinality. In that case every triple is stored only once in the storage graph. Regarding CB, in the worst case it stores $2 \times space_t(\text{IC})$ triples, while in the best case stores every triple only once and thus coincides with the best case of POI.

Regarding graph size, for IC and CB no index structure has to be kept as we store explicitly the entire of every version. Concerning POI, the number of nodes of the storage graph is $|D(V)|$. Notice that $|D(V)| \leq |id(V)|$, i.e. less than or equal to the number of versions[3]. The number of edges coincides with the size of the relation $\sqsubseteq^r$. This relation can have at most $\frac{N^2}{4}$ relationships. This value is obtained when $\varGamma$ is a bipartite graph, whose $\frac{N}{2}$ nodes are connected with all other $\frac{N}{2}$ nodes. More on the overall comparison of these policies are described below.

4.1 Experimental Evaluation

In order to measure the storage space required by each policy we created a testbed comprising from 100 to 1000 versions, each having 10,000 triples on average, where the size of each triple is 100 bytes (a typical triple size). As in real case scenarios, a new version is commonly produced by modifying an existing version, in order to generate the content of a new version, we first choose at random a parent version and then we either *add* or *delete* triples from the parent contents. The difference in triples with respect to the parent content is 10%, i.e. 1000 triples. We have an additional parameter d that defines the probability to choose triple additions (so with probability $1 - d$ we subtract triples). In this respect, we create versions whose contents are either supersets or subsets of the contents of existing versions. We experimented with d in the range of [0.5, 1.0] (we ignored values smaller than 0.5 as subtractions usually do not exceed additions). For additions, we assumed that the 25% of the additional triples are triples which already exist in the KB (in the content of a different than the parent version), while the rest 75% are brand new triples. This is motivated by the fact that in a versioning system it is more rare to re-add a triple which exists in an old version and was removed in one of the subsequent versions, than to add new triples. Clearly, as d approaches 1, more new triples are created and less are deleted (so the total number of distinct triples increases). The minimum (resp. maximum) sized version contains 6000 (resp. 13000) triples, while $|T_V| = 43000$. We should also stress that, if we only create versions by adding triples, then the resulting storage graph resembles a tree. The higher the probability of deleting triples is, the higher the probability of having nodes with more than one fathers becomes. In the later case, we have increased number of edges. The more edges we have the higher the probability of having nodes with overlapping stored contents.

We compared the three approaches IC, CB and POI using a PC with a Pentium IV 3.4GHz processor and 2 GB of main memory, over Windows XP. For the

---

[3] In our running example $|D(V)| = 3$ while $|id(V)| = 4$.

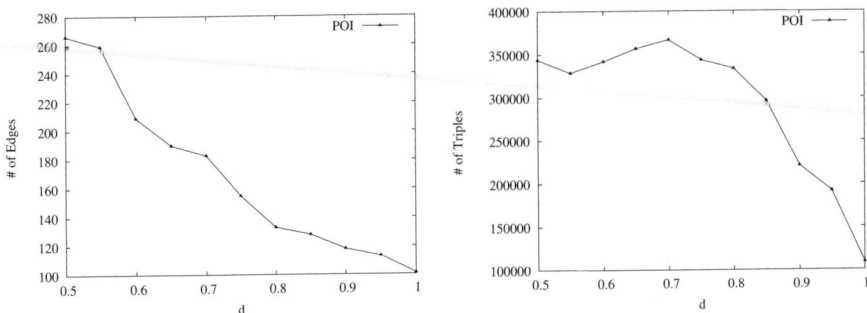

Fig. 2. # of graph edges (left) and triples (right) for POI

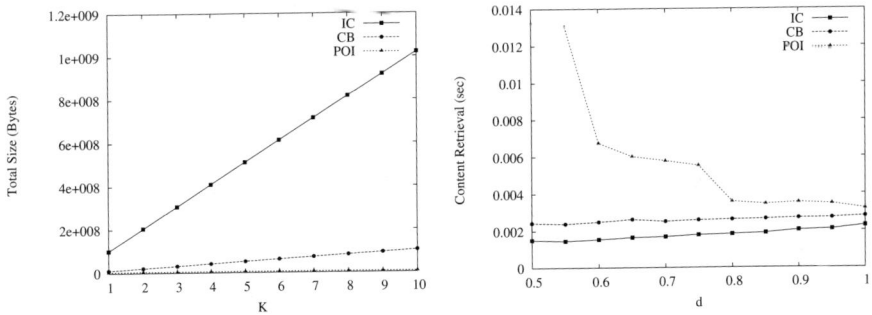

Fig. 3. Storage space requirements (left) and retrieval costs (right) of IC, CB and POI

change-based approach (CB), we compute and store the delta Δ_e [17] between two consecutive versions. Regarding, the storage graph of POI, we assume node size equal to 20 bytes and edge size equal to 12 bytes. To further reduce the storage space, we used a table listing all triples each associated with a unique identifier. The contents of each version (specifically $n.stored$) is represented as a set of identifiers (rather than triples).

We first show the characteristics of POI and then we proceed with its comparison with the rest two policies. The left part of Figure 2 shows the number of graph edges of POI. As d increases, the number of edges (and consequently of paths) decreases and POI gains advantage by its invariant to store every triple only once in a single path. The right part of Figure 2 shows the number of triples stored in POI. Notice that as d increases, the number of stored triples tends to decrease (because duplicates decrease).

To show that POI saves space in cases where there are versions with equal or inclusion-related content, we used datasets that guarantee that each distinct triple set is content of K versions. The left part of Figure 3 shows the total size of IC, POI, and CB for $d = 0.5$ and various values of K (specifically $1 \leq K \leq 10$)[4]. Notice that CB and POI are much better than IC (9 and 18 times better respectively), and the

---

[4] The total number of versions is 100*K.

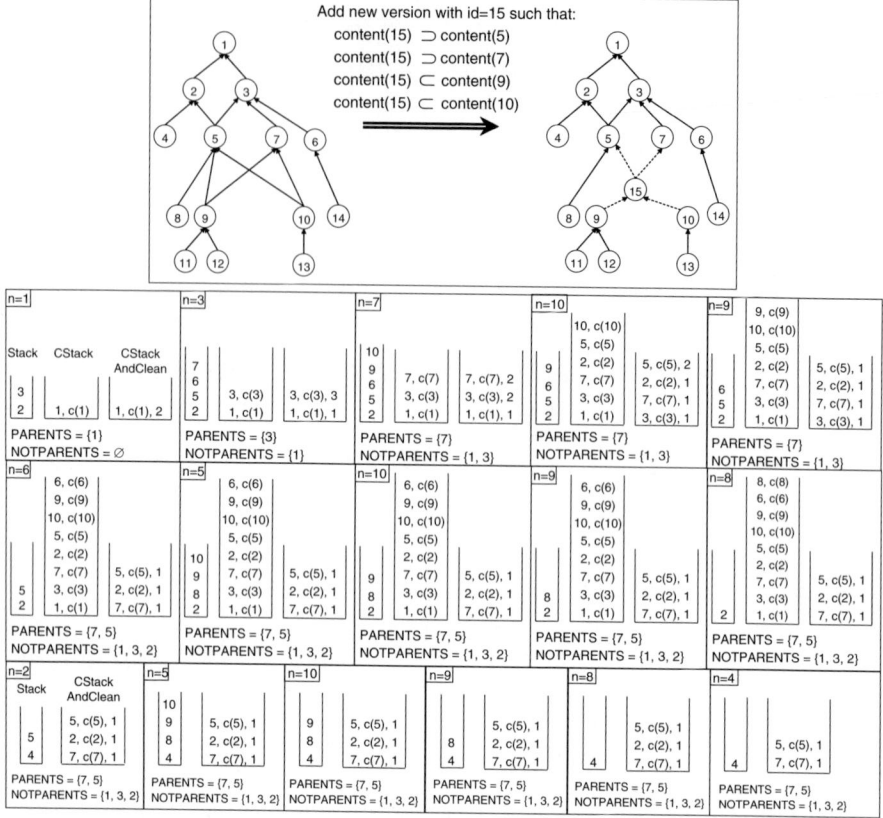

Fig. 4. Example of adding a new version using Algorithm *Insert POIds*

greater the value of K is, the better POI than CB is. For $K = 1$ POI is roughly 2 times better than CB, while for $K = 10$, POI is roughly 18 times better than CB.

Regarding version retrieval times, we measured the time to retrieve the contents of all versions and report the average. The results are shown in the right part of Figure 3 for various values of d and $K = 1$. Obviously, IC is the best regarding content retrieval, since no structure should be traversed, as every node is assigned to its content. In contrast, POI needs to traverse all the ancestors of the given node. CB slightly outperforms POI, but we have to note that the history paths in our datasets are very small. In a realistic setting CB would be much slower than POI. Once again, the decrease in number of edges as long as d increases, results in shorter traversals for POI and as a consequence its difference with IC tends to decrease. In any case, POI content retrieval time is acceptable (i.e., max. of 0.013 sec).

5 Version Insertion Algorithms

The insertion algorithms exploit the structure and the semantics of the storage graph. Intuitively, we have to check whether the new version is subset or superset of one of the existing versions. To this end, we start from the root(s) of the storage graph and we descend. To be more specific, let $\mathcal{B}$ be a family of subsets of $\mathcal{T}$, let $<$ be the cover relation over these and let $<^r$ be its transitive reduction. Suppose that we want to insert a new subset A (i.e. $A \notin \mathcal{B}$) and update accordingly the relation $<^r$. We define:

$$Parents(A) = \min_{<}\{ B \mid A < B \} = \{ B \mid A <^r B \}$$

$$Children(A) = \max_{<}\{ B \mid B < A \} = \{ B \mid B <^r A \}$$

To update $<^r$ we have to add the relationships $A <^r p$ for every $p \in Parents(A)$, and

---

Algorithm 1. InsertInPoset(A)

Input: a set of triples A
Output: updated cover relation of the poset so that to contain a node corresponding to A (if there is no such node already)

==FIND PARENTS================================
(1)   Stack = new STACK(); PARENTS = new Set(); NOTPARENTS = new Set();
(2)   push (Stack, {root(Γ)})
(3)   while not(isEmpty(Stack))
(4) n = pop(Stack)
(5)   if (n $\in$ NOTPARENTS)
(6)   push(Stack, $\{x \in Down(n) \mid x \notin PARENTS\}$)
(7)   else if n.contents $= A$ // a node with contents A already exists
(8)   break
(9)   else if n.contents $\subset A$
(10)   PARENTS = (PARENTS $\cup \{n\}$) \ $Up^t(n)$ // all upper nodes of n are certainly
                                                           // not parents
(11)   NOTPARENTS = NOTPARENTS $\cup Up^t(n)$
(12)   push(Stack, $Down(n)$)
==FIND CHILDREN=================================
(13) Stack = new STACK(PARENTS) // a new stack with initial contents the set PARENTS
(14) CHILDREN = new Set()
(15) while not(isEmpty(Stack))
(16) n = pop(Stack)
(17)   if n.contents $\supset A$
(18)   CHILDREN = CHILDREN $\cup \{n\}$
(19)   else
(20)   push(Stack, $Down(n)$)
==CONNECT A=====================================
(21) nA = new node(A)
(22) For each $c \in$ CHILDREN Add($c \to nA$)   // i.e. Add($c <^r A$)
(23) For each $p \in$ PARENTS Add($nA \to p$)   // i.e. Add($A <^r p$)
==ELIMINATE REDUNDANCIES=========================
(24) for each $c \in$ CHILDREN
(25)   for each $p \in$ PARENTS
(26)   if $c \to p$ then Delete($c \to p$)   // i.e. if $c <^r p$ then Delete($c <^r p$)

$c <^r A$ for every $c \in Children(A)$. In addition we have to eliminate redundant relationships (that may exist between $Parent(A)$ and $Children(A)$, specifically we have to eliminate all $c <^r p$ relationships where $c \in Children(A)$ and $p \in Parents(A)$). Let now see how we could find the children and the parents of a set A and update appropriately the relation $<^r$. Returning to the problem at hand, this scenario corresponds to the case where each node n of a storage graph had explicitly stored $contents(n)$. Algorithm 1 sketches the crux of the algorithm in pseudocode. It is based on a Stack and two sets called PARENTS and NOTPARENTS. The root of the storage is denoted by $root(\Gamma)$ and every storage graph has a single root corresponding to a dummy version with an empty content. Note that the relation $\rightarrow$ of the storage graph corresponds to the relation $<^r$, i.e. $a \rightarrow b \Rightarrow content(b) \subset content(a)$. An indicative example of version addition is illustrated in Figure 4. The stack contents are shown in every step.

To implement version insertion we could use Alg. 1. The only difference is that in a storage graph $n.contents$ (where n is a node) is not explicitly stored (instead only $n.stored$ is stored). One naive approach would be to compute $n.contents$ by taking the union of the stored triples of its (direct and indirect) broader nodes, i.e. to use the formula (1). Each such computation would require $\mathcal{O}(d(n))$ set union operations where $d(n)$ is the depth of the node n multiplied to the average number of parents. If the storage graph is a tree, then all set union operations would actually be concatenations (and thus faster). In case of DAG, we have to perform set union operations only for nodes that have more than one father. We will hereafter call this algorithm POI-*plain* insertion (for short Insert POI$_p$) algorithm.

5.1 *Insert POI-DoubleStack* (Insert POI$_{ds}$) **Insertion Algorithm**

To reduce the number of set union operations that are issued by the Insert POI$_p$ algorithm for computing $content(n)$ for a node n, here we present a more time efficient algorithm which employs a second stack (actually it is a cache) for keeping stored (and thus reusing) results of operations that have already been computed. The second stack, called *CStack* (where $'C'$ comes from contents), stores elements comprising of two components: a version id and its content (i.e. a set of triples). The extension of Alg. 1 with the second stack is Algorithm 4.

Algorithm 2. TSContent(n)

Input: a node id
Output: the set of triples comprising the content of node n (the stack CStack is updated, if necessary).

(1) if (e=**lookup**(CStack, n))
(2)   return e.content //returns the contents of element e
(3) else
(4)   Res = stored(n)
(5)   for each $n' \in Up(n)$
(6)   Res = Res $\cup$ **TSContent(n')** // this is a concatenation if $|Up(n)| = 1|$
(7)   **push**(CStack, (n, Res)) //pushes a pair (key, content) to the stack
(8) return Res

To compute the contents of a node n it uses the function **TSContent** (Algorithm 2) which accesses the second stack. If the storage graph were a tree then we would be sure that all broader nodes of a node are in the stack (in both *Stack* and *CStack* stacks). However the storage graph is a DAG in the general case, so this is not always true. That's why TSContent uses a lookup and if the sought element is not in *CStack* it creates and stores it to *CStack*. We will hereafter call this algorithm *Insert POI-DoubleStack* (for short Insert POI$_{ds}$) insertion algorithm.

Algorithm 3. TSContentAndClean(n, PARENTS, initialN)

Input: a node id, the ids of its parents, a node id (in the root call of the routine n = initialN)
Output: the sets of triples comprising the content of node n. The stack is updated (addition, deletion) if necessary.

(1)   if (e = **lookup**(CStack, n))
(2)   Res = getElemContent(CStack, n)
(3)   if ((e.Y == 1) & (n $\notin$ PARENTS))
(4)   **delElem**(CStack, n)
(5)   else
(6)   **editElem**(CStack, (n, Res, Y), (n, Res, Y-1)) //decreases the 3rd component
                                                //of the stack element
(7)   else
(8)   Res = stored(n)
(9)   for each $n' \in Up(n)$
(10)   Res = Res $\cup$ **TSContentAndClean**(n', PARENTS, initialN) //concatanation
(11)   **push**(CStack, (n, Res, $|Down(n)|$))
             //pushes a triple (key, content, number of children) to the stack
(12) return Res

To reduce the total space needed by *CStack*, Alg. 4 actually uses a different implementation of Alg. 2 called **TSContentAndClean** (Alg. 3) that frees the contents of those versions that are not needed any more. Specifically, an element of *CStack* should be removed if one of the following two conditions holds:

(a) its content is not a subset of A, so the traversal of the storage graph will not continue to its descendants and therefore its contents are not needed any more,
(b) it is not a (definite) parent of the node to be inserted and all its children are already in *CStack*. Specifically, if all its children are already in *CStack* the version content is not needed because it is a subset of the content of every child of it.

To this end we extend the structure of each *CStack* element with a third component, denoted by Y, which is actually a variable initialized to the number of children (of the corresponding node) that is decreased by one whenever lookup finds and fetches that element. When it reaches 0, the element (if not a parent) should be removed because that means that the contents of all its children are already stored in *CStack*. We will hereafter call this algorithm *Insert POI-DoubleStack(Clean)* (for short Insert POI$_{dsc}$).

Figure 4 shows these stacks in our running example. Specifically, the left stack is the *Stack*, the center stack is the *CStack* as employed by *TSContent* algorithm, while the right one is the *CStack* as employed by *TSContentAndClean* algorithm. After the PARENTS have been computed, *CSTack* remains the same and therefore we show only *Stack*. Notice how shorter *CStack* is according to *TSContentAndClean*.

Algorithm 4. Insert $\text{POI}_{\text{dsc}}(A)$

Input: a set of triples A
Output: updated storage graph and CStack

==FIND PARENTS=============================
(1) Stack = new STACK(); PARENTS = new Set(); NOTPARENTS = new Set()
(2) push (Stack, {root($\varGamma$)})
(3) while not(isEmpty(Stack))
(4) n = pop(Stack)
(5)   if ($n \in$ NOTPARENTS)
(6)   push(Stack, $\{x \in Down(n) \mid x \notin \text{PARENTS})\}$)
(7)   else if n.contents $= A$
(8)   break
(9)   else if **TSContentAndClean**(n, PARENTS, n) $\subset A$
(10)   PARENTS = (PARENTS $\cup \{n\}) \setminus Up^t(n)$
(11)   NOTPARENTS = NOTPARENTS $\cup Up^t(n)$
(12)   **delElems**(CStack, $Up^t(n)$)
(13)   push(Stack, $Down(n)$)
(14)   else
(15)   **delElem**(CStack, n)
==FIND CHILDREN===========================
(16) Stack = new STACK(PARENTS)
(17) CHILDREN = new Set()
(18) while not(isEmpty(Stack))
(19) n = pop(Stack)
(20)   if **TSContentAndClean**(n, PARENTS, n) $\supset A$
(21)   CHILDREN = CHILDREN $\cup \{n\}$
(22)   else
(23)   push(Stack, $Down(n)$)
==CONNECT A==============================
(24) nA = new node(A)
(25) nA.stored = $A \setminus \cup \{$**TSContentAndClean**(n') $\mid n' \in$ Parents$\}$
(26) For each $c \in$ CHILDREN Add($c \rightarrow nA$)   // i.e. Add($c <^r A$)
(27) For each $p \in$ PARENTS Add($nA \rightarrow p$)   // i.e. Add($A <^r p$)
==ELIMINATE REDUNDANCIES=================
(28) ... as in Alg. 1
==UPDATE THE STORED CONTENTS OF THE CHILDREN NODES =====
(29) for each $c \in$ CHILDREN
(30) c.stored = c.stored - A

5.2 Experimental Evaluation

To compare the Insert POI implementations we employed the same testbed of 100 versions, of 10,000 triples on average, that has been presented in Section 4.1. The left (resp. right) part of Figure 5 illustrates the average version insertion time (resp. main memory required) in log scale. Of course, the size of the storage graph is the same irrespective of

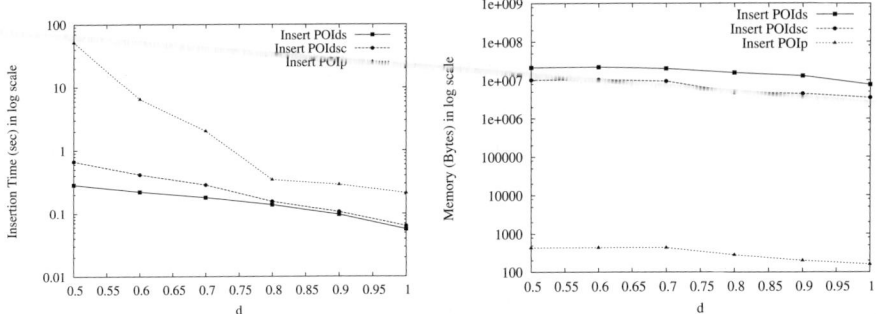

Fig. 5. Comparison of Insert-POI implementations

Fig. 6. Comparison of Version Insertion Algorithms

the employed insertion algorithm. As one would expect, Insert POI_p is the slowest but the less main memory consuming implementation of POI. On the other edge of the time-space tradeoff, lies Insert POI_{ds}, which is the fastest but the most main memory consuming implementation. In particular, Insert POI_p is from 181 (for d=0.5) to 3.7 (for d=1) times slower than Insert POI_{ds}, but Insert POI_{ds} needs 4 orders of magnitude more main memory. POI_{dsc} is 2-3 times less main memory consuming than Insert POI_{ds} and from 2.3 (for $d = 0.5$) to 1.1 (for $d = 1$) slower than Insert POI_{ds}. Figure 6 summarizes the above results.

5.3 History-Based Version Insertion Speedup

We have provided a general method for inserting versions and recall that the storage space requirements of POI are independent of the evolution history. However the knowledge of the evolution history can speed up version insertion, especially in the case we insert versions that are defined through set operations over existed versions. For instance, consider the case we want to insert a new version $v_3 = v_1 \cup v_2$. In that case, the search for parents begins with v_1 and v_2, instead of the root, because $v_1 \subset v_3$ and $v_2 \subset v_3$. Nodes v_1 and v_2 will be the parents of v_3, unless there exists a descendant v_4 of v_1 or v_2, such that $v_4 \subset v_3$. Additionally, the case where $v_3 = v_1 \cap v_2$ is in a sense dual to the previous one, as the children of v_3 will be v_1, v_2 or some nodes above them.

6 Other Applications and Extensions

Below we discuss a number of operations that can be performed efficiently if a POI is available.

Cross version operations can take advantage from the existence of a POI. For instance, inclusion checking can clearly benefit from a POI. To decide whether $D(i) \subseteq D(j)$ one could pose a reachability query on the storage graph (no need to access the contents of the versions). Moreover, by adopting a labeling scheme [1] for the storage graph Γ we could decide inclusion in $\mathcal{O}(1)$.

Additionally, let S be a set of triples. Suppose we want to find all versions i such that $S \subseteq D(i)$ (or $D(i) \subseteq S$). Such queries would be very expensive in the IC or in the CB approach. By employing a POI we can use the insertion algorithm to insert S to the storage graph. Let n be the inserted node. The sought versions are those that point in nodes of $Nr^t(n)$ (resp. $Br^t(n)$).

We should also mention, that, if the storage graph $\langle \Gamma, stored \rangle$ does not fit in main memory, then we could keep only Γ in main memory, while the function $stored$ could be kept in a relational storage with schema Stored(vid,tid) where tid is the triple identifier. To retrieve the contents of a version i, we need to compute $Br^t(i)$ using Γ and then send to the db a disjunctive query (with all ids in $Br^t(i)$).

7 Concluding Remarks and Further Research

To the best of our knowledge, this is the first work that focuses on the storage aspect of SW repositories that support versioning. We proposed an index called POI, we verified the space gains of this index experimentally and we provided an efficient version insertion algorithm with acceptable main memory space requirements. From our experiments, POI can be 180 times more space economical compared to IC and 18 times compared to CB for parallel version tracks. Moreover, POI allows performing efficiently various cross-version operations.

It is worth mentioning that we have experimented also with storage graphs that have a *semi-lattice* structure, specifically with graphs that contain a node for each intersection of version contents. Such graphs guarantee that each triple is stored at most once. However, the storage gains obtained are compensated by the space required to keep the excessive number of edges of the storage graph. In future, we plan to compare POI with the *inverse* POI, i.e. with storage graphs that store explicitly the maximal elements and the internal nodes are negative deltas. We also plan to experiment with real data sets (currently we did not manage to find long version histories of SW data). Last, we could explore possible combinations of POI with change-based storage policies for enabling more sophisticated policies.

Acknowledgment

This work was partially supported by the EU projects CASPAR (FP6-2005-IST-033572) and KP-Lab (FP6-2004-IST-4).

References

1. Agrawal, R., Borgida, A., Jagadish, H.V.: Efficient Management of Transitive Relationships in Large Data and Knowledge Bases. SIGMOD Records 18(2), 253–262 (1989)
2. Berliner, B.: CVS II: Parallelizing software development. In: Procs. of the USENIX Winter 1990 Technical Conf., Berkeley, CA, pp. 341–352. USENIX Association (1990)
3. Cheney, J., Lagoze, C., Botticelli, P.: Towards a Theory of Information Preservation. In: Constantopoulos, P., Sølvberg, I.T. (eds.) ECDL 2001. LNCS, vol. 2163, pp. 340–351. Springer, Heidelberg (2001)
4. Chien, S., Tsotras, V.J., Zaniolo, C.: Version Management of XML Documents. In: Selected papers from the Third Intern. Workshop WebDB 2000 on The World Wide Web and Databases, London, UK, pp. 184–200. Springer, Heidelberg (2001)
5. Dadam, P., Lum, V.Y., Werner, H.D.: Integration of Time Versions into a Relational Database System. In: VLDB 1984: Procs of the 10th Intern. Conf. on Very Large Data Bases, San Francisco, CA, USA, pp. 509–522 (1984)
6. Gançarski, S., Jomier, G.: A Framework for Programming Multiversion Databases. Data & Knowledge Engineering 36(1), 29–53 (2001)
7. Gutierrez, C., Hurtado, C., Mendelzon, A.: Foundations of Semantic Web Databases. In: Procs of the 23th ACM Symp. on Principles of Database Systems (PODS) (2004)
8. Kaoudi, Z., Dalamagas, T., Sellis, T.: RDFSculpt: Managing RDF Schemas Under Set-Like Semantics. In: Gómez-Pérez, A., Euzenat, J. (eds.) ESWC 2005. LNCS, vol. 3532, pp. 123–137. Springer, Heidelberg (2005)
9. Klein, M., Fensel, D., Kiryakov, A., Ognyanov, D.: Ontology versioning and change detection on the web. In: Gómez-Pérez, A., Benjamins, V.R. (eds.) EKAW 2002. LNCS (LNAI), vol. 2473, pp. 197–212. Springer, Heidelberg (2002)
10. Noy, N.F., Kunnatur, S., Klein, M., Musen, M.A.: Tracking Changes During Ontology Evolution. In: Procs of the 3rd Intern. Conf. on the Semantic Web (ISWC-2004), Japan (2004)
11. Noy, N.F., Musen, M.A.: Ontology versioning in an ontology management framework. IEEE Intelligent Systems 19(4), 6–13 (2004)
12. Tichy, W.F.: RCS-a system for version control. Software Practice & Experience 15(7), 637–654 (1985)
13. Tzitzikas, Y., Christophides, V., Flouris, G., Kotzinos, D., Markkanen, H., Plexousakis, D., Spyratos, N.: Emergent Knowledge Artifacts for Supporting Trialogical E-Learning. Journal of Web-based Learning and Teaching Technologies (IJWLTT) 2(3), 16–38
14. Tzitzikas, Y., Flouris, G.: Mind the (Intelligibily) Gap. In: Kovács, L., Fuhr, N., Meghini, C. (eds.) ECDL 2007. LNCS, vol. 4675, Springer, Heidelberg (2007)
15. Tzitzikas, Y., Kotzinos, D. (Semantic Web) Evolution through Change Logs: Problems and Solutions. In: Procs. of the Artificial Intelligence and Applications, AIA 2007, Innsbruck, Austria (February 2007)
16. Volkel, M., Winkler, W., Sure, Y., Kruk, S.R., Synak, M.: SemVersion: A Versioning System for RDF and Ontologies. In: Procs. of the 2nd European Semantic Web Conf., ESWC 2005, Heraklion, Crete, May 29–June 1 (2005)
17. Zeginis, D., Tzitzikas, Y., Christophides, V.: On the Foundations of Computing Deltas Between RDF Models. In: Aberer, K., Choi, K.-S., Noy, N., Allemang, D., Lee, K.-I., Nixon, L., Golbeck, J., Mika, P., Maynard, D., Mizoguchi, R., Schreiber, G., Cudré-Mauroux, P. (eds.) ISWC 2007. LNCS, vol. 4825, pp. 637–651. Springer, Heidelberg (2007)

Semantic Reasoning: A Path to New Possibilities of Personalization*

Yolanda Blanco-Fernández, José J. Pazos-Arias, Alberto Gil-Solla,
Manuel Ramos-Cabrer, and Martín López-Nores

Department of Telematics Engineering, University of Vigo, 36310, Spain
{yolanda,jose,agil,mramos,mlnores}@det.uvigo.es

Abstract. Recommender systems face up to current information overload by selecting automatically items that match the personal preferences of each user. The so-called content-based recommenders suggest items similar to those the user liked in the past, by resorting to syntactic matching mechanisms. The rigid nature of such mechanisms leads to recommend only items that bear a strong resemblance to those the user already knows. In this paper, we propose a novel content-based strategy that diversifies the offered recommendations by employing reasoning mechanisms borrowed from the Semantic Web. These mechanisms discover extra knowledge about the user's preferences, thus favoring more accurate and flexible personalization processes. Our approach is generic enough to be used in a wide variety of personalization applications and services, in diverse domains and recommender systems. The proposed reasoning-based strategy has been empirically evaluated with a set of real users. The obtained results evidence computational feasibility and significant increases in recommendation accuracy w.r.t. existing approaches where our reasoning capabilities are disregarded.

1 Introduction

Recommender systems provide personalized advice to users about items or services they might be interested in. Currently, these tools are gaining momentum in the Digital Revolution, helping people efficiently manage content overload and reducing complexity when searching for relevant information.

To fulfill these personalization needs, three main components are required in a recommender system: (i) a database where the available items are stored, (ii) personal profiles where the users' preferences are modeled, and (iii) recommendation strategies aimed at selecting personalized suggestions for each individual. The first such strategy was the so-called *content-based filtering*, which suggests to a user items similar to those he/she liked in the past. In spite of its accuracy, this technique is limited due to the employed similarity metrics. These metrics are based on rigid syntactic approaches

---

* Work funded by the Ministerio de Educación y Ciencia (Gobierno de España) research project TSI2007-61599, by the Consellería de Educación e Ordenación Universitaria (Xunta de Galicia) incentives file 2007/000016-0, and by the Programa de Promoción Xeral da Investigación de la Consellería de Innovación, Industria e Comercio (Xunta de Galicia) PGIDIT05PXIC32204PN.

that only detect similarity between items that share all or some of their attributes [1]. Consequently, traditional content-based approaches lead to *overspecialized suggestions* including only items that bear a strong resemblance to those the user already knows (i.e. items with attributes defined in his/her profile).

To fight overspecialization, researchers devised a new strategy named *collaborative filtering*, based on offering to each user items that were appealing to others with similar preferences (named *neighbors*). Collaborative filtering reduces the effects of overspecialization by considering other users' interests, but it also causes new limitations, such as scalability problems, difficulties to select each user's neighborhood when the available preferences are sparse (commonly named *sparsity problem*), and privacy concerns related to the confidentiality of the users' personal data (see [1] for details).

Bearing in mind the severe drawbacks of the collaborative solutions, we propose a novel *content-based strategy* that exploits the main strengths of this personalization paradigm and overcomes the overspecialized nature of its recommendations. For that purpose, our strategy diversifies the offered suggestions without resorting to other users' preferences, thus protecting their privacy. Specifically, we fight syntactic limitations of the existing content-based approaches by employing two reasoning techniques borrowed from the Semantic Web field: the so-called *semantic associations* [3] and *Spreading Activation techniques* (henceforth, SA techniques) [7]. Instead of using the traditional syntactic similarity metrics, these associations trace semantic bonds between the user's preferences and the items available in the recommender system, which are previously formalized in a domain ontology along with their semantic annotations. Next, SA techniques efficiently explore these semantic relationships and discover new knowledge related to the users' interests. This knowledge permits our strategy to compare in a more flexible way the user's preferences with the available items, thus offering more accurate recommendations. Although the adopted reasoning mechanisms have been widely used in the Semantic Web [3,14,15], their internals must be adapted to fulfill personalization requirements of a recommender system. So, these mechanisms must allow to: (i) learn automatically new knowledge about the users' preferences from their feedback, and (ii) adapt dynamically the strategy as these preferences evolve.

In spite of the generality of our reasoning-based approach, in this paper we have adopted a specific context with the goal of describing in detail its use in a domain where the information overload is noticeable. Specifically, we have exploited the reasoning capabilities of our content-based strategy in order to enhance the recommendations offered to viewers of the Interactive Digital TV (IDTV). Today, TV viewers are exposed to overwhelming amounts of information, and challenged by the plethora of interactive functionality provided by the current digital receivers. As there are hundreds of channels with an abundance of programs available, it is likely that appealing TV programs go unnoticed. To assist these viewers, it is possible to take advantage of the personalization capabilities provided by a TV recommender system, which sifts through the myriad of programs available in the digital stream and selects those that match the viewers' preferences by using our reasoning based strategy

This paper is organized as follows: Sect. 2 describes the two key elements in our reasoning framework: (i) the ontology where the domain knowledge is formalized, including the available TV programs and their semantic descriptions, and (ii) the user modeling approach employed to create the users' profiles. Next, Sect. 3 describes how the semantic associations and SA techniques are exploited in our content-based strategy. Then, a sample example where a set of TV programs are suggested to a given viewer is presented in Sect. 4. The tests carried out to validate our reasoning-based approach are explained in detail in Sect. 5. Finally, Sect. 6 draws some conclusions and points out possible lines of further work.

2 Domain Ontology and User Modeling

2.1 The Domain Ontology

Two elements are needed to formalize the IDTV domain by an ontology: (i) the semantic descriptions of the TV programs that can be suggested, and (ii) a language expressive enough to represent the concepts (i.e. classes and their instances) and relationships (i.e. hierarchical links and properties) identified in the domain. In our approach, the semantic descriptions have been extracted from TV-Anytime metadata specifications [6], whereas the OWL (DL) language has been selected due to its expressive capability, which allows to formalize concepts and expressions not supported in RDF and RDFS.

Starting from TV-Anytime metadata, we have defined and included in our OWL ontology several hierarchies of classes and properties, as well as specific instances of them, as shown in the TV ontology depicted in Fig. 1. The considered TV programs (identified by unique IDs) have been automatically extracted from the Internet Movie DataBase (IMDB) and the BBC web server[1], and are represented as specific instances belonging to a hierarchy of genres organized in several levels (e.g. *fiction*, *leisure*, *romance*, etc.), as shown at top of Fig. 1. The main attributes of these programs (e.g. involved credits, topics and places, intended audience, intention, etc.) are also instances related to them by labeled properties. These attributes also belong to hierarchically organized classes. As some of these classes are already defined in existing conceptualizations, we have imported ontologies about different domains such as sports, geographical information, credits involved in TV programs (e.g. actors), among others[2].

2.2 Our User Modeling Approach

Our approach models the user's profiles by reusing the knowledge available in the domain ontology, that is why we named them *ontology-profiles*. Specifically, we propose a semantic model for each user that gives information about: (i) the TV programs that were appealing or uninteresting for him/her (named *positive* and *negative preferences*, respectively), (ii) their main attributes, and (iii) the genres under which these programs are classified in the TV ontology (see at the top of Fig. 1). This user modeling approach provides a formal representation of the users' preferences, permitting to reason about them and discover additional knowledge about their interests. Such knowledge permits

---

[1] See *http://www.imdb.com* and *http://backstage.bbc.co.uk/data/7DayListingData* for details.
[2] These ontologies were extracted from the DAML repository located in *www.daml.org/ ontologies* and converted to the OWL language by means of a tool developed by the MINDSWAP Research Group (see *http://www.mindswap.org/2002/owl.shtml* for details).

Semantic Reasoning: A Path to New Possibilities of Personalization 723

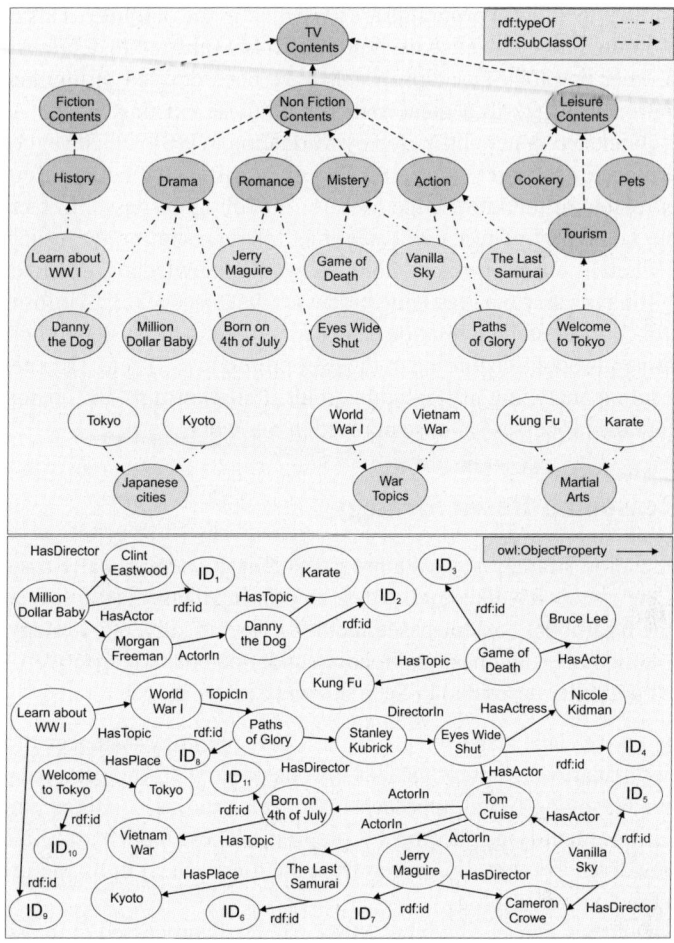

Fig. 1. Excerpt from classes, properties and instances in our TV ontology

to compare, in a more effective way, the users' preferences with the available items, thus leading to personalization processes more accurate than the traditional syntactic approaches [1]. In this regard, note that our ontology profiles greatly improve other flat lists-based approaches which are not well structured to favor the discovery of new knowledge (see [4] for details).

Fulfilling the goals of our personalization strategy requires identifying the interest of the user in both TV programs defined in his/her profile and their attributes and genres. Specifically, these *Degrees Of Interest* (named DOI indexes and belonging to [-1,1]) can be explicitly stated by the user or automatically inferred from his/her viewing behavior (e.g. programs accepted or rejected after recommendations, viewing time for each suggested program, etc.). Once the DOI indexes of each program in the user's profile have been established, we compute the indexes corresponding to their attributes and to the genres under which these programs are classified in the ontology. This computation mechanism -omitted here due to space limitations- is explained in detail in [5].

Although other ontology-based proposals have been devised in literature, our user modeling approach differs to a great extent from these existing works. As an example, note the Quickstep system proposed by Middleton in [10], which suggests research papers according to the users' interests. The main difference between our work and Quickstep is related to the knowledge used for modeling purposes. In fact, Quickstep uses a simple taxonomy of research categories for representing the papers each user appreciates, whereas our proposal exploits the whole knowledge formalized in the ontology, permitting to carry out reasoning processes that discover extra information about users' preferences. The same limitation can be identified in the system proposed in [16], which recommends books according to the user preferences. There, the knowledge discovery is based on analyzing just only hierarchical relationships, thus hampering more complex inference processes as those pursued in our work.

3 Our Reasoning-Based Strategy

Our personalization strategy suggests programs that are semantically associated with the contents the viewer has liked in the past, improving the syntactic similarity metrics adopted in the traditional content-based methods. Specifically, our strategy consists of two stages –named filtering phase and recommendation phase, respectively–, which are sketched next and fully described in Sect. 3.1 and Sect. 3.2.

- *Filtering phase.* This stage selects in the OWL ontology instances of classes and properties that are relevant for the user, by considering his/her personal preferences. Next, our reasoning-based approach infers semantic associations among the selected entities identifying specific TV programs. These hidden associations –which we borrow from [3]– are discovered from the hierarchical links and properties defined in the domain ontology.
- *Recommendation phase.* The inferred knowledge is processed in the second phase by employing SA techniques. This intelligent mechanism works as concept explorer, as it detects concepts that are closely related to the user's preferences by exploring the entities and semantic associations inferred during the filtering phase.

3.1 Filtering Phase

Firstly, our strategy locates in the domain ontology the programs that were (un)appealing to the user (defined in his/her profile). Next, it traverses successively the properties bound to these programs until reaching new class instances (nodes referred to programs, actors, topics...) in the ontology. To guarantee the computational feasibility, we have developed a controlled inference mechanism that works as follows. As new nodes are reached from a given instance, our approach firstly quantifies their relevance for the user. Then, the nodes whose *relevance indexes* are lower than a specific threshold[3] are ignored, in a such

---

[3] The value of this threshold depends on both the domain ontology and the recommender system that adopts our content-based strategy. In our tests in DTV field, we have used values around 0.65.

way that our inference mechanism continues traversing successively only the properties that permit to reach new nodes from those that are significant for the user (according to his/her profile). Consequently, our strategy explores solely entities of interest for the user, thus filtering those that probably do not provide knowledge useful for the personalization process.

In our filtering mechanism, the more significant the relationship between a given node and the user's preferences (either positive or negative preferences), the more relevant this node. In order to measure this relevance value, we have developed a technique that takes into account diverse ontology-dependent *filtering criteria*. Some of these criteria –described in detail in [5]– are sketched next.

1. *Length of the chain of properties that enables to reach the considered node starting from the user's preferences.* Specifically, the longer this property sequence[4], the lower the relevance index of the node, as its relationship to the user's preferences is less significant due to the presence of many intermediate nodes.

    Example: Let us consider that a user has enjoyed the documentary *Learn about WW I* shown in Fig. 1. Here, it is possible to find the property sequence *Learn about WW I - World War I - Paths of Glory - Stanley Kubrick - Eyes Wide Shut*. In this case, the relevance index of the program *Paths of Glory* is greater than the index of *Eyes Wide Shut*, as the relationship between *Paths of Glory* and *Learn about WW I* is more significant than the relationship between this documentary and *Eyes Wide Shut*. In other words, the relation "movie about the World War I" is more relevant than "movie whose director has directed movies about the World War I".

2. *Existence of hierarchical relationships between the node and the user's preferences.* The relevance of a node is increased when it is possible to find a common ancestor between it and the user' preferences in the ontology hierarchies.

    Example: Let us consider again the user who has liked the war documentary mentioned in the previous example, whose topic is represented in our ontology by the instance *World War I*. In this case, the filtering phase increases the relevance index of other instances that share the common ancestor *War Topics* with the class instance *World War I* (e.g. *Vietnam War*).

3. *Existence of implicit relationships between the node and the user's preferences detected by concepts from graph theory.* In graph theory [8], the betweenness among three nodes is high when in the most of paths existing between the first and the second node, the third node is also included. So, from a high value of betweenness, it follows that the involved nodes are strongly related. In our approach, these nodes are the user's preferences and the class instance whose relevance is measured.

    Example: Let us consider a user who has liked the movies *Vanilla Sky* and *Jerry Maguire* with *Tom Cruise* as leading actor. In this case, the relevance index of the instance *Born on the 4th of July* gets higher, as this movie is closely related to the user's preferences. In fact, as shown in Fig. 1, the node *Tom Cruise* is included in all the paths established between *Born on the 4th of July* and the two movies defined in the user's profile.

---

[4] The length of a sequence is defined as the number of properties included in it.

Once the nodes related to the user's interests (and the properties linking them to each other) have been selected, our strategy infers semantic associations between the instances referred to TV programs. Specifically, we adopt three associations that have been defined by Anyanwu and Sheth in [3]:

- ρ-*path association*. In our approach, two programs are ρ-*pathAssociated* when they are linked by a chain or sequence of properties in the ontology. For instance, in Fig. 1, it is possible to trace a sequence between the documentary *Learn about WW I* and the movie *Paths of Glory* by means of the *World War I* instance.
- ρ-*join association*. Two programs are ρ-*joinAssociated* when their respective attributes belong to the same class in the domain ontology. For instance, in Fig. 1 there exists a ρ-*join* association between the documentary *Welcome to Tokyo* and the movie *Last Samurai*, as both programs are bound to different cities in Japan (*Tokyo* and *Kyoto*, respectively, which are classified as *Japanese cities* in Fig. 1).
- ρ-*cp association*. Two programs are ρ-*cpAssociated* when they share a common ancestor in the genre hierarchy defined in the ontology. For instance, note that all the movies depicted at the top of Fig. 1 are ρ-*cpAssociated* by the ancestor *Non Fiction Contents*.

By means of the filtering and knowledge inference processes, our approach has built a network for the user, whose nodes are the instances of classes selected during the filtering phase, and whose links are both the properties joining these instances in the ontology and the semantic associations inferred from it. The knowledge represented in this network is explored during the second phase of the strategy by exploiting the inference capabilities provided by SA techniques, which are one of the most-used processing frameworks for semantic networks.

3.2 Recommendation Phase

We emphasize the use of SA techniques as a computational mechanism able to: (i) explore efficiently the relationships among the nodes interconnected in the user's network (henceforth *SA network*), and (ii) infer from them knowledge useful for the recommendation process by detecting concepts closely related to the user's preferences. According to the guidelines established in [7], these techniques work as follows:

- The nodes of the network have an implicit relevance, named *activation level*. Besides, each link joining two nodes has a *weight*, in a such way that the stronger the relationship between both nodes, the higher the assigned weight. Initially, a set of nodes are selected and their activation levels are spread until reaching the nodes connected to them by links (named *neighbor nodes*).
- The activation level of a reached node is computed by considering the levels of its neighbors and the weights assigned to the links that join them to each other. Consequently, the more relevant the neighbors of a given node (i.e. higher their activation levels), and the stronger the relationship between the node and its neighbors (i.e. higher the weights of the links between them), the more relevant this node.

- This spreading process is repeated successively until reaching all the nodes of the network. Finally, the highest activation levels correspond to the nodes that are closest related to those initially selected.

The SA techniques have been widely used in the fields of searching and information retrieval [15,12,9]. However, to combine their inferential capabilities with the personalization requirements of a recommender system, it is necessary to extend the existing approaches. The modifications we propose affect mainly to two issues: (i) the kind of links traditionally modeled in the SA network, and (ii) their weighting process.

- On the one hand, the links considered in traditional approaches model only simple and direct relationships[5], thus disregarding during the spreading activation process a huge amount of knowledge hidden behind more complex relationships. In order to fight this limitation, our SA network models both the properties defined in the ontology and the semantic associations inferred in the filtering phase. This way, the links corresponding to the associations allow to spread the relevance of the user's preferences until reaching programs appealing to him/her, which would go unnoticed in traditional SA-based approaches.
- On the other hand, as weights of the links depends only on the strength of the relationship between the connected nodes, these values remain static in existing SA approaches. Bearing in mind the purposes of a recommender system, our SA-based strategy must also consider the user's preferences during the weighting process, in a such way that this process adapts to changes in his/her interests.

In our approach, the proposed strategy activates in the user's network the nodes referred to the programs defined in his/her profile, and assigns them an initial activation level equal to their respective DOI indexes. Next, it is necessary to weight conveniently the links of the network, which represent both the explicit knowledge formalized in the ontology (i.e. properties), and the implicit knowledge discovered from it (i.e. semantic associations). As we mentioned in the previous section, our strategy adjusts dynamically these values as the user's preferences evolve over time. In fact, the weight of a link joining two nodes in our user's network is computed by considering their respective relevance indexes, which are measured during the filtering phase. As a result, our approach leads to a highly positive weight when the two linked nodes and the user's positive preferences are strongly related, and to a negative weight when the relationship is established to his/her negative preferences. This way, as the user's preferences change, the weights of the links in his/her SA network are conveniently modified and, hence, the elaborated recommendations are also updated.

According to the traditional SA techniques, once the propagation process has reached all the nodes in the user's network, the highest activation levels correspond to TV programs satisfying two conditions: (i) their neighbor nodes are also relevant for the user (that is why their high activation levels), and (ii) they are closely related to the user's preferences (that is why the high weight of the links). For that reason, these nodes identify the TV programs finally suggested by our content-based strategy.

---

[5] For instance, in information retrieval, a link between two nodes referred to terms indicates their co-occurrence in a document (see [7] for details).

As a conclusion, note that the spreading process employed in our reasoning-based approach has three main advantages in the personalization field:

- Firstly, our strategy is able to discover that a TV program is appealing to the user even when its attributes are not defined in his/her profile. Thanks to the SA techniques, this program is relevant if it is semantically associated with the user's preferences. Consequently, our reasoning-based strategy offers diverse recommendations, beyond the overspecialized suggestions offered by the traditional syntactic content-based techniques.
- Secondly, our reasoning mechanisms consider both the positive and negative preferences of the user. Whereas the interests help to identify contents appealing to the user, the negative preferences decrease the activation levels of the nodes to which are related (either explicitly by means of properties, or implicitly by semantic associations). This way, our strategy prevents from suggesting programs associated with those the user did not like.
- Lastly, note that our approach not only favors the knowledge reusing, but also permits the user's network to adapt easily to changes in his/her preferences. This way, as these interests evolve, the filtering phase selects new nodes, properties and semantic associations, and incorporates them into the current network of the user.

4 A Sample Scenario

The example described in this section shows the differences between our reasoning-based recommendations and those offered by traditional content-based strategies. Due to space limitations, we consider only the brief excerpt from our TV ontology shown in Fig. 1. However, this restriction does not prevent from highlighting the associations used by our SA techniques for the selection of personalized recommendations.

In this scenario, we consider that the target user is U, who has enjoyed the documentaries *Welcome to Tokyo* and *Learn about World War I*, and the movies *Vanilla Sky* and *Jerry Maguire* starring *Tom Cruise*. As for U's negative preferences, we assume that this user liked neither *Morgan Freeman* (supporting actor in *Million Dollar Baby*), and *Game of Death* (a movie about *martial arts* with *Bruce Lee* as leading actor).

Filtering Phase: Selecting Instances Relevant for U
Firstly, our strategy selects in the domain ontology instances that are relevant for the user U by considering his/her personal preferences. For that purpose, the strategy locates in the ontology the nodes referred to the programs defined in U's profile, and explores successively the nodes joined to them by properties. For instance, from the node identifying U's favorite actor (*Tom Cruise* in Fig. 1), it is possible to reach the instances *Born on the 4th of July* and *The Last Samurai*. Considering the second filtering criterion, our strategy selects the two reached instances, as they share common ancestors with U's positive preferences. In fact, as shown in the hierarchy at the top of Fig. 1, *Born on the 4th of July* and *Jerry Maguire* are *Drama* movies, and *The Last Samurai* and *Vanilla Sky* are classified as *Action* movies.

Next, our filtering phase continues exploring the instances linked to the two previously selected nodes (i.e. *Vietnam War* from the node *Born on the 4th of July*, and *Kyoto* from *The Last Samurai*). In this case, both instances are relevant for U, as this user has appreciate other instances belonging to their classes in the ontology (specifically, *World War I* belonging to the *War Topics* class, and *Tokyo* belonging to *Japanese cities*).

We also search for class instances related to the user's negative preferences. These instances are relevant during the personalization processes, as they help to identify programs the user will probably not enjoy. Among such nodes, note *Danny the Dog*. As shown in Fig. 1, this movie not only involves an actor U does not like (*Morgan Freeman*), but also it is about *Karate*, a topic that seems to be unappealing to this user, who has not liked the movie about martial arts entitled *Game of Death*. This knowledge will be used by our SA-based reasoning to elaborate recommendations for U.

Filtering Phase: Inferring Semantic Associations Between TV Programs

Once the instances relevant for U have been identified, our strategy infers associations between the programs included in the selected property sequences (see Table 1).

Table 1. Property Sequences and Semantic Associations inferred for the user U

Sequence Properties	Semantic Associations
Learn about WW I - World War I - Paths of Glory	ρ-path (Jerry Maguire, Born on 4th July)
Jerry Maguire - Tom Cruise - Born on 4th July - Vietnam War	ρ-join (Welcome to Tokyo, The Last Samurai)
Vanilla Sky - Tom Cruise - The Last Samurai - Kyoto	ρ-join (Learn about WW I, Born on 4th July)
Welcome to Tokyo - Tokyo	ρ-cp (Vanilla Sky, The Last Samurai)
Danny the Dog - Karate	ρ-cp (Danny the Dog, Game of Death)
Game of Death - Kung Fu	ρ-path (Danny the Dog, Million Dollar Baby)

The next step is to build the user U's SA network, by including as nodes the class instances selected by the filtering process, and as links both the properties that join these instances to each other in the ontology, and the associations inferred from it. Once the links have been weighted, U's network (in Fig. 2) is processed by SA techniques, which reason about the represented knowledge to select the personalized recommendations.

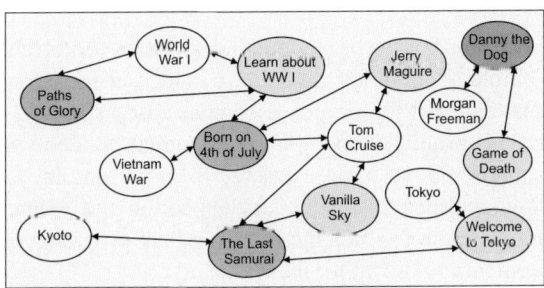

Fig. 2. Network used by SA techniques to select content-based recommendations for U

Recommendation Phase: Suggesting TV Programs by Means of SA Techniques

After spreading the activation levels of U's preferences until reaching all the nodes in the SA network, our strategy suggests the TV programs with the highest levels. According to what we explained in Sect. 3.2, these programs receive links from other contents which are appealing to the user U. This way, our content-based approach suggests to U the movies *Paths of Glory*, *Born on the 4th of July* and *The Last Samurai*, by employing the associations inferred between these contents and his/her positive preferences.

- **Paths of Glory:** The activation level corresponding to this movie is increased thanks to the weight spread from the program *Learn about WW I*. The link between the two movies in the user's SA network is due to the fact that both contents are about the *World War I* in which U is interested. For that reason, the relevance of *Learn about WW I* reaches the suggested movie *Paths of Glory*.
- **Born on the 4th of July:** The activation level of this movie gets higher thanks to the links from two programs relevant for U: *Learn about WW I* and *Jerry Maguire*. The war topic turns the documentary into a program appealing to the user, whereas the movie *Jerry Maguire* is relevant as it involves U's favorite actor (*Tom Cruise*).
- **The Last Samurai:** As shown in Fig. 2, *Vanilla Sky* and *Welcome to Tokyo* inject positive weights in *The Last Samurai* node. Both programs are specially relevant for U, thus increasing the activation level of *The Last Samurai*, a movie the user will appreciate due to two reasons: (i) his/her favorite actor takes part in it, and (ii) the movie is set in a city of Japan, a country that seems to be interesting for U in view of the documentary about history and customs in Tokyo this user has liked.

Our SA network permits also to discover programs unappealing to U, which are identified by the low activation levels obtained after the spreading process. For instance, note the movie *Danny the Dog*, whose level in decreased by the negative weights received from the nodes *Morgan Freeman* and *Game of Death* (U's negative preferences).

In conclusion, our content-based approach suggests not only programs with the *same* attributes defined in U's profile, but also contents that are significantly related to his/her preferences. So, out of the available movies starring U's favorite actor, we suggest those bound to war topic and Japan, two issues especially relevant for U according to the programs *Learn about WW I* and *Welcome to Tokyo* in his/her profile.

5 Testing

We have implemented a prototype of recommender system for testing purposes including: (i) the OWL ontology about TV domain, (ii) our user modeling technique based on *ontology-profiles*, and (iii) the proposed content-based strategy. Our experimental evaluation involved this prototype and 400 undergraduate students from University of Vigo. These users rated 400 programs extracted from our ontology, by assigning values in [-1,1] (positive and negative values for appealing and uninteresting programs, respectively). A brief description was included for each program, so that all the users could judge even contents unknown for them.

Our goals were: (i) to evaluate the accuracy of our reasoning-based recommendations, and (ii) to compare this approach (henceforth *Asso-SA*) with other existing techniques that are devoid of our semantic inference capabilities. Due to space limitations, here we include just only two well-known approaches defined in the field of personalization. The first one –proposed by O'Sullivan *et al.* in [13]– combines content-based and collaborative filtering with discovery of association rules to measure similarity between programs; the second approach –proposed by Mobasher *et al.* in [11]– devises a collaborative strategy enhanced by the semantic descriptions of the suggested items.

5.1 Approach Based on Association Rules

This approach (henceforth *Rules*) mixes a collaborative phase with a content-based stage to select personalized recommendations for a set of users. The collaborative phase compares (program to program) each user's preferences against the profiles of the remaining users. Then, it extracts his/her k most similar neighbors. The similarity between contents is based on identifying association rules between them. These rules are automatically extracted by the Apriori algorithm [2], which works on a set of training profiles containing several programs. So, given two programs A and B, a rule $A \Rightarrow B$ means that if a profile contains A, it is likely to contain the program B as well. In fact, each rule $A \Rightarrow B$ has a confidence value interpretable as the conditional probability p(B/A). This way, the similarity between A and B is computed as the confidence of the rule involving both contents, as long as this value is greater than a threshold confidence.

Once the k most similar profiles to the user's one have been identified, the approach *Rules* applies the content-based phase. Here, the programs in these neighbors' profiles (and unknown for the user) are ranked according to their relevance for this user and the top-N_1 are returned for recommendation. A program is more relevant when: (i) it is very similar to those contained in the user's profile, (ii) it occurs in many neighbors' profiles, and (iii) these neighbors' profiles are strongly correlated to the user' preferences.

5.2 Semantically Enhanced Collaborative Filtering

The goal in the approach of Mobasher *et al.* (henceforth *Sem-CF*) is to predict the rating of the user U on a given item t (in our case, a TV program). For that purpose, the *Sem-CF* approach selects the N_2 contents in U's profile which are most similar to the program t. Next, the level of interest of the user U in t is predicted as the sum of the ratings given by U on the N_2 contents more similar to t. Each rating is weighted by the corresponding similarity between t and the N_2 selected programs.

The similarity between two items i_p and i_q is computed by combining linearly two components: (i) the ratings of the users who have rated both items in their profiles (denoted by $RateSim(i_p, i_q)$), and (ii) the semantic attributes of the compared items (denoted by $SemSim(i_p, i_q)$). In order to compute $SemSim$ component, Mobasher *et al.* build a matrix S where each row refers to each one of the n items available in the recommendation process, and each column corresponds to the values of their respective semantic attributes. Once this matrix has been created, the *Sem-CF* approach applies techniques based on Latent Semantic Indexing (LSI) to reduce its dimension, resulting in a much less sparse matrix S'. Finally, Mobasher *et al.* apply the mathematical techniques described in [11] and obtain from S' a $n \times n$ square matrix in which an entry i, j corresponds to the semantic similarity between the items i and j ($SemSim(i,j)$).

5.3 Test Data

The preferences of the 400 users were divided into two groups: *training* and *test users*.

Training users (40%). The programs rated by these users with positive DOI indexes were employed to build the so-called *training profiles*. From these profiles, the similarity between two specific contents is computed in both the approach *Rules* and *Sem-CF*. In the rule-based work, these profiles are used to train the Apriori algorithm, whose task is to discover association rules between the programs contained in them and to quantify their similarity. Regarding *Sem-CF*, note that the $RateSim$ component for two given programs is measured by selecting the training users who have rated both contents in their profiles, and computing the correlation Pearson-r between their respective ratings.

Test users (60%). The three evaluated strategies were executed by considering the preferences of these users (defined in the so-called *test profiles*). To identify these preferences, we proceeded as follows. Out of the initially rated 400 programs, we selected for each user those 10 ones most appealing to him/her and those 10 he/she found less interesting (according to their ratings), and we built their respective profiles. In this process, a low overlapping between the programs rated by these users was obtained, leading to a great sparsity level (89%). The remaining 240 programs rated by these users and their DOI indexes (named hereafter *evaluation data*) were hidden in order to be compared to the programs offered by each approach.

5.4 Methodology and Accuracy Metrics

Our evaluation is organized as follows. Firstly, the evaluated strategies were executed for each test user. Each strategy suggested 20 TV programs and as configuration parameters, we have used $k = 30$ neighbors and $N_1 = 20$ programs in *Rules*, $\alpha = 0.6$ and $N_2 = 10$ programs in *Sem-CF*, and finally, a filtering threshold of 0.65 in *Asso-SA*[6].

Next, we compute for each user the values of the employed accuracy metrics: recall and precision. The recall of a strategy for a given user is the ratio of programs rated as interesting in his/her evaluation data (i.e. with positive DOI index) that were suggested by this strategy. The precision is the ratio of contents recommended by this technique that have a positive DOI in the user's evaluation data. Lastly, the average and variance of the recall and precision were computed over the 240 test users, to detect possible abrupt dispersions between the values measured for each user and the average value.

5.5 Discussion on Experimental Results

The results from Fig. 3 indicate that the reasoning capabilities of our approach increase greatly the recommendation accuracy, offering diverse suggestions the users appreciate and excluding the typical drawbacks of the two evaluated collaborative approaches (*Rules* and *Sem-CF*). In spite of high sparsity level in the used data (89%), our approach selected, in average, the greatest number of programs interesting to the test users (highest recall), and the lowest account of contents they found unappealing (highest precision). This improvement is due to the combination of semantic associations and SA

---

[6] These values were chosen after several experiments because they offered the most accurate recommendations in each evaluated approach.

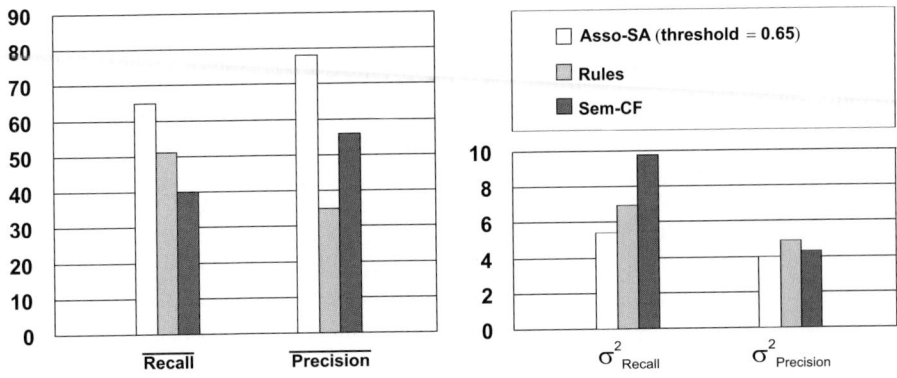

Fig. 3. Average (left) and variance (right) of recall and precision for the evaluated approaches

techniques as reasoning mechanisms. As shown in Fig. 3, these techniques discover programs of interest for the test users, which go unnoticed in the other two evaluated approaches devoid of reasoning capabilities.

Instead of reasoning about the knowledge of a domain ontology, the approach *Rules* extracts association rules between programs by considering just only the occurrence patterns detected by Apriori from the training profiles. In our tests, there was a low overlap among the programs rated by the training users, thus obtaining a reduced number of significant rules. Consequently, it was difficult to select accurately the k most similar neighbors for each user, thus offering recommendations less precise than ours.

Something similar happened in the approach *Sem-CF*, where many programs appealing to the test users could not be suggested due to both the intrinsic limitations of collaborative solutions and the absence of semantic reasoning processes. On the one hand, the low overlap among the programs rated by the test users affected to the computation of the $RateSim$ component, hampering the creation of the users' neighborhoods in an accurate way. On the other side, the $SemSim$ component only detected similarity between programs with the same attributes, thus disregarding the semantic associations established between them. Besides, according to what we described in Sect. 5.1 and Sect. 5.2, we checked that both *Rules* and *Sem-CF* suggested only programs defined in the test users' profiles. This fact caused the lower recall values of the two approaches w.r.t. our strategy, since many programs could not be suggested as they had not been rated by the users involved in our experiments.

6 Conclusions and Further Work

In this paper, we have proposed a personalization strategy that overcomes limitations unresolved in traditional content-based recommender systems, by resorting to reasoning mechanisms borrowed from the Semantic Web. Specifically, our strategy is based on: (i) filtering from a domain ontology instances that are irrelevant for the user by a threshold-based process, (ii) inferring semantic associations between his/her preferences and the resulting instances, and (ii) exploring efficiently the discovered knowledge by Spreading Activation techniques to select the personalized recommendations. Our strategy is clearly promising for its generality, as it is not exclusively bound to a specific application domain. For that reason, the proposed technique is flexible enough to be reused in other recommendation domains, becoming a good starting point to implement diverse personalization services for the Semantic Web users.

Our content-based strategy has been implemented in a prototype of TV recommender system and evaluated in a scenario with real users. The obtained results show computational feasibility and significant increases in the accuracy of our suggestions w.r.t. traditional syntactic approaches, which are devoid of our reasoning capabilities.

Our further work is focused on two aspects. On the one hand, we are developing a mechanism to weight automatically the filtering threshold used in our strategy. Broadly speaking, this threshold will be dynamically adjusted as the user's preferences change, and its value will depend on both the domain ontology and the feedback provided by the user regarding the past recommendations. On the other hand, we also plan to continue the experimental evaluation of our strategy by involving the 80000 subscribers of the cable networks of Spanish operator R[7], over which a recommender system using our reasoning-based strategy is being deployed.

References

1. Adomavicius, G., Tuzhilin, A.: Towards the next generation of recommender systems: a survey of the state-of-the-art and possible extensions. IEEE Transactions on Knowledge and Data Engineering 17(6), 739–749 (2005)
2. Agrawal, R., Srikant, R.: Fast Algorithms for Mining Association Rules. In: 20th International Conference on Very Large Data Bases, pp. 487–499 (1994)
3. Anyanwu, K., Sheth, A.: ρ-Queries: Enabling Querying for Semantic Associations on the Semantic Web. In: 12th International World Wide Web Conference, pp. 115–125 (2003)
4. Ardissono, L., Kobsa, A., Maybury, M.: Personalized Digital Television: Targeting Programs to Individual Viewers. Kluwer Academic Publishers, Dordrecht (2004)
5. Blanco, Y., Pazos, J.J., Gil, A., Ramos, M., López, M.: A Flexible semantic inference methodology to reason about user preferences in knowledge-based recommender systems. Knowledge-based Systems (in press), http://dx.doi.org/10.1016/j.knosys.2007.07.004
6. Broadcast and On-line Services: Search, select, and rightful use of content on personal storage systems. ETSI TS 102 822 series (2003)
7. Crestani, F.: Application of Spreading Activation Techniques in Information Retrieval. Artificial Intelligence Review 11(6), 453–482 (1997)
8. Diestel, R.: Graph Theory. Springer, Heidelberg (2000)
9. Huang, Z., Chen, H.: Applying associative retrieval techniques to alleviate sparsity in collaborative filtering. ACM Transactions on Information Systems 22(1), 116–142 (2004)
10. Middleton, S.: Capturing Knowledge of User Preferences with Recommender Systems. PhD thesis, University of Southampton (2003)
11. Mobasher, B., Jin, X., Zhou, Y.: Semantically Enhanced Collaborative Filtering on the Web. In: Web Mining: applications and techniques, pp. 57–76. IDEA Group Publishing (2004)

---

[7] http://www.mundo-r.com

12. O'Hara, K., Alani, H., Shadbolt, N.: Identifying Communities of Practices: Analyzing Ontologies as Networks to Support Community Recognition. Information Systems, 89–102 (2002)
13. O'Sullivan, D., Smyth, B., Wilson, D., McDonald, K.: Improving the Quality of the Personalized EPG. User Modeling and User-Adapted Interaction 14, 5–36 (2004)
14. Perry, M., Janik, M., Ramakrishnan, C., Arpinar, B., Sheth, A.: Peer-to-Peer Discovery of Semantic Associations. In: Workshop on P2P Knowledge Management, pp. 1–12 (2005)
15. Rocha, C., Schawabe, D., Poggi, M.: A Hybrid Approach for Searching in the Semantic Web. In: 13th International World Wide Web Conference, pp. 374–383 (2004)
16. Ziegler, C., Lausen, G.: Exploting Semantic Product Descriptions for Recommender Systems. In: 2nd Semantic Web and Information Retrieval Workshop, pp. 25–29 (2004)

An User Interface Adaptation Architecture for Rich Internet Applications

Kay-Uwe Schmidt[1], Jörg Dörflinger[1], Tirdad Rahmani[1], Mehdi Sahbi[1],
Ljiljana Stojanovic[2], and Susan Marie Thomas[1]

[1] SAP AG, Research, Vincenz-Prienitz-Straße 1, 76131 Karlsruhe
˜http://www.sap.com
[2] FZI Forschungszentrum Informatik, Haid-und-Neu-Straße 10-14, 76131 Karlsruhe
˜http://www.fzi.de

Abstract. The need for adaptive and personalized Rich Internet Application puts a new dimension to already existing approaches of Adaptive Hypermedia Systems. Instead of computing the adaptation steps at the server, Rich Internet Applications need a client-side approach that can react immediately on user input. In this paper we present a novel approach that holistically combines page annotations, semantic Web usage mining, user modeling, ontologies and rules to adapt AJAX pages. The focus of our pater is the conceptual introduction of the autonomous client. An autonomous client directly executes all necessary adaptation steps based on a user model, without requesting any logic on the server. In order to realize this, we use ontologies to annotate Rich Internet Applications and to describe the user model as well as semantic Web usage mining for detecting adaptation rules. Additionally, we provide a detailed overview and evaluation of how we moved resource-intensive ontology processing and rules execution from the server to the client.

1 Introduction

In Adaptive Hypermedia Systems (AHSs) adaptation strategies have been intensively studied [1] and are well understood for conventional Web applications adhering to the Web page paradigm[1]. With conventional techniques, the tracking of user clicks, the user modeling, as well as the adaptation take place on the server. This limits the possibilities of user tracking to the user requests seen by the server [2], which is actually a subset of the user clicks. Furthermore adaptation can only take place when a user requests a new page, which then is adapted to his/her needs. On-the-fly adaptation, without reloading the whole page, is not obtainable. Recently, with the rise of AJAX [3], new possibilities appeared for user tracking and user interface adaptation. With AJAX the look and feel of Web pages are transformed to that of desktop applications and users are accustomed to highly responsive user interfaces. State of the art Web applications

---
[1] The Web paradigm determines that very Web page in a series of pages is downloaded separately.

obtain a responsive user interface by encoding the adaptation logic in static script languages like JavaScript.

In this paper we present a novel solution for on-the-fly adaptation of Rich Internet Applications (RIAs) applications. We introduce a holistic framework covering the whole adaptation cycle, from obtaining rules, over ad-hoc user modeling, to on-the-fly user interface adaptation. Compared to common AHSs our approach goes two steps beyond, as we introduce not only ontologies, for capturing and storing the user model, and declarative logical rules, for carrying out the adaptation, but also advance the state of the art through the client-side realization of user modeling and portal adaptation. We consider the evaluation and execution of the adaptation rules on the client-side as the major contribution of this paper, as it is the prerequisite to responsive user interfaces for RIAs. Client-side rule processing has several advantages, such as reduction of client-server communication to a minimum, and in-time response to user interactions. The acquisition of adaptation rules is an indispensable pre-requisite to adaptation. Adaptation rules declaratively encode adaptation logic based on the user's behavior, which means based on the recorded user interactions with the RIAs. Adaptation rules can be gained by mining user access log data. With the help of annotations, added to the Web application in advance, we present a semantic approach of Web usage mining in order to find common Web usage patterns. In turn, the most useful patterns can be directly modeled as adaptation rules, guiding the user while interacting with the RIA. The user interactions are stored at run-time in a user model residing together with the application rules on the client-side. We do not assume that the system has any previous information about a user like explicitly specified user preferences or user roles determined by log-in information. In each user session the user model has to be acquired from scratch. The user model, the RIA and the rules are modeled in ontologies that are transformed at design-time into executable AJAX snippets. The adaptation ontologies as a backbone of our adaptive solution were already comprehensively described in [4] and are not the focus of this paper.

The rest of the paper is structured as follows. Section 2 an example is presented in order to motivate our work. In Section 3 we give an overview of the logical system architecture and illustrate the adaptation loop. The following two Sections 4 and 5 go into the details of the design-time and run-time architecture accordingly. An evaluation of our approach is given in Section 6, and in Section 7 we discuss related work. In Section 8 acknowledgments are given, and, finally, the paper closes with conclusions and prospects for future work.

2 Motivating Example

Searching for the right form is an widespread problem in portals especially in e-Government portals. The average user of an e-Government portal is usually not an expert but rather a novice regarding the use of online forms. E-Government Web applications are designed for end users without special training. Two major requirements for e-Government Web applications are: Citizen-centric services and ease of use [5]. To meet these requirements a form of non-intrusive user guidance can be provided.

In our first motivating example we pick up the idea of user guidance. We want to recommend links related to the forms the user already filled in. Lets consider the following use case: Building application. The citizen officially has to apply for building permission at the local department of housing and urban development. The building application can be filled in online and consists of several forms like the main building application form, building license form, building description form, start of construction form, to mention just a few. We assume no pre-defined workflow determining the number and order of forms the citizen has to fill in. After filling in the main form the user wants to know which form to fill in next. This can be accomplished by suggesting related forms based on the forms filled in by the current user compared to the collaboratively filtered Web usage behavior of past users.

3 Logical System Architecture: The Adaptation Loop

Our user interface adaptation architecture for RIAs, as depicted in Figure 1, is a two-stage approach consisting of three cycles forming the adaptation loop. The design-time stage and the run-time stage logically divide the components of our architecture into off-line and online components respectively. That is, the stages refer to the invocation time of the components comprising our architecture. The three cycles, the modeling cycle on the left, the adaptation cycle on the right and the larger transfer cycle in the middle illustrate the self-adaptive character of our architecture.

The modeling cycle stands for the design time components in charge of constructing the adaptation rules. At design time, an indispensable prerequisite for our RIA adaptation approach, the portal must be annotated by using a portal annotation tool. After annotating the structure and content of the RIA, user access log data, collected in the past, can be mined for useful Web usage patterns. This is done by the semantic Web usage mining component. Once useful patterns are found, they can be formulated as adaptation rules by using a rule design tool. The adaptation rules are stored in an ontology format. After designing the adaptation rules on a conceptual level, based on the annotated RIA, they are translated at design-time by the rule transformer into a client-readable format like JavaScript.

The rightmost cycle in Figure 1, the adaptation cycle, is executed, like the transfer cycle in the middle of the figure, at run-time. The aim of the adaptation cycle is to adapt the RIA based on the predefined adaptation rules and the current user model. The adaptation rules are obtained from the modeling cycle via the transfer adaptation rules component of the transfer cycle. The user model is built up by the user tracking component, which records the user's interactions with the RIA. Based on the user model, which is constructed on-the-fly, the adaptation rules are evaluated and, if the condition part holds, are fired. It is the task of the rule evaluation component and the rule execution component to

An User Interface Adaptation Architecture for Rich Internet Applications

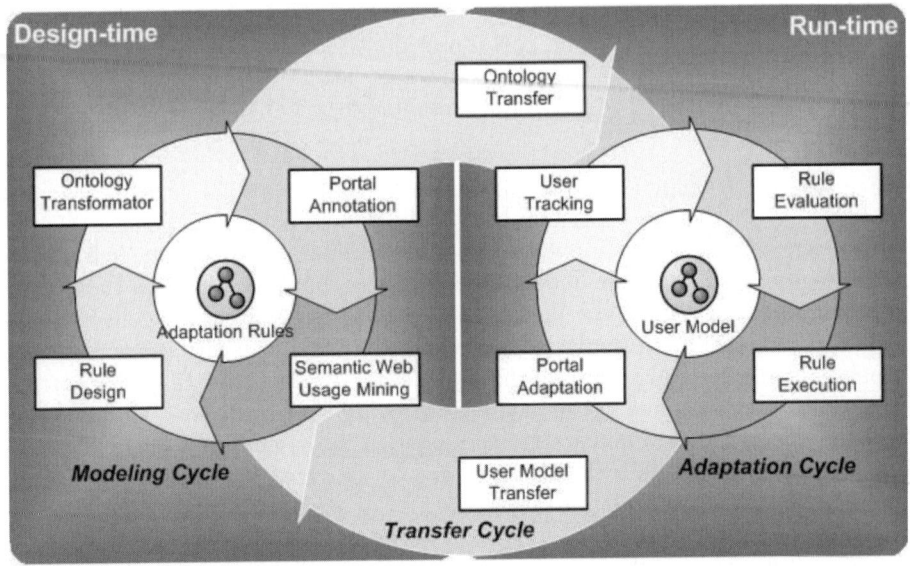

Fig. 1. Logical System Architecture: The Adaptation Loop

carry out rule processing. If a rule fires the corresponding actions are executed and the RIA adapts itself directly on the client-side without server requests. At the end of the session the tracked user model is sent back to the server. On the server-side all user models are collected and fed back into the modeling cycle in order to mine new behavioral patterns. Thus, the adaptation loops starts again.

4 Design-Time Architecture

The design-time architecture consists of the components constituting the modeling cycle as depicted in Figure 1. These tools and components are executed off-line during the annotation, mining, design and transformation phases.

4.1 Ontology Creation and Portal Annotation

Suitable ontologies are crucial for our RIA adaptation approach. We developed an approach amalgamating ontology learning [6], ontology refinement [7] and annotating RIAs into one coherent tool. The main ideas behind this approach are described in detail in [8]. Based on this approach we developed a domain-specific e-Government ontology, as well as an annotation knowledge-base linking concepts of the domain ontology to contents of the RIA. Additionally, a RIA ontology, describing the structural aspects, and a user model ontology were developed using standard ontology development tools like Protege[2]. An overview

---
[2] http://protege.stanford.edu/

and description of the developed ontologies are given in [4]. As a proof of concept we annotated our internal demo portal and the e-Government portal of the city of Vöcklabruck[3] with our ontolgies under supervision of e-Government experts. All ontologies are described using the Web Ontology Language (OWL) [9].

4.2 Semantic Web Usage Mining

Web Usage Mining is the application of data mining algorithms on Web server access logs to gain a better understanding of user behavior. Besides the access logs, metadata describing the Web resources and their content are conceptually helpful for data-mining analysis. The utilization of the metadata is strongly dependent on its organization and the way it can be combined with the log entries [10]. In recent years the research areas semantic Web and Web mining have become more important and are merged together into a new research field called semantic Web mining which has been deeply analyzed [11,12]. In particular, semantic Web Usage Mining as a subcategory of semantic Web mining enables tracking of user behavior at a conceptual level.

Figure 2 shows the different stages of our semantic Web usage mining architecture. The first stage is the preprocessing stage. Here the ontologies are designed, the Web resources are annotated and the user sessions are reconstructed. Reconstructing user sessions is a difficult task, because of the lack of log-in data. The user sessions are reconstructed using common reconstruction methods as detailed in [13]. The available data sources for the second stage, the data mining step, comprise the reconstructed sessions, the annotated Web resources and the domain ontology. The last two items form a knowledge-base of available

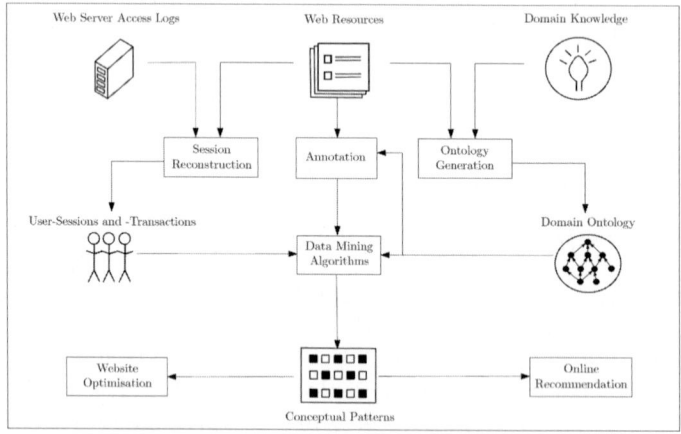

Fig. 2. Semantic Web usage mining architecture

---

[3] http://www.voecklabruck.at/

metadata. The Web usage mining algorithms used for this purpose are association rules, sequential rules, and multi level rules based on a concept hierarchy. Moreover, clustering approaches which consider the user behavior as well as semantic contents are considered. More technical details are given in [14].

The results of the data mining step can be used for website optimization or online recommendation based on the user behavior and semantic content. The new item problem nicely illustrates the benefits of using semantics in Web usage mining. New items can be recommended directly after their annotation. That is possible, because our data mining approach, as well as our recommendation engine, work on concepts rather than concrete URLs or IDs. As an example the following rule is considered: $C1 \wedge C2 \rightarrow C3$. This rule states that if pages annotated with $C1$ and $C2$ are visited, all pages annotated with concept $C3$ are candidates to be recommended. If now a new Web resource annotated with $C3$ is introduced, it can be added to these candidates, because the rules are on the conceptual level.

4.3 Design of Adaptation Rules

After applying semantic Web usage mining to access-log files, the discovered patterns need to be analyzed by an e-Government expert. The domain expert has to judge, whether the patterns are useful or not. Patterns, which have been judged useful, are then encoded into a rule language as adaptation rules by an ontology engineer with the help of customary ontology and rule editors. So, for instance, we discovered the following rule after evaluating the patterns found in the annotated access log file of the city of Vöcklabruck: 85% of all users that filled in the marriage certificate form and the wedding day form also filled in the birth certificate form[4].

Example 1 (Adaptation rule in SWRL)
   portal:Form(?a) ∧ portal:isVisited(?a, true) ∧
   domain:WeddingDay(?b) ∧ portal:isAnnotated(?a, ?b) ∧
   portal:Form((?c)) ∧ portal:isVisited(?c, true) ∧
   domain:MarriageCertificate(?d) ∧ portal:isAnnotated(?c, ?d) ∧
   domain:BirthCertificate(?e) → portal:showLink(?e)

We are using the Semantic Web Rule Language (SWRL) [15] because it nicely fits to our OWL ontologies. Example 1 shows how an ontology and rule engineer could formulate the rule described above in SWRL. Translated to English the rule states, that whenever a form annotated with WeddingDay and a form annotated with MarriageCertificate were visited show all links to forms annotated with BirthCertificate as link recommendations. WeddingDay, DateOfWedding and BirthCertificate are concepts taken from the domain ontology. The functionality of displaying recommended links is realized as a SWRL built-in. The built-in finds all forms annotated with BirthCertificate, reads the link from the appropriate property and, finally, recommends these links.

---
[4] Support: 0,01; confidence: 0,85.

4.4 Ontology Transformer

Having the ontologies, annotations and adaptation rules in place, the last step in the modeling cycle is still the transformation of all of these parts into a client-readable format that can be executed by a browser's JavaScript engine. As an Internet browser on a client machine has only limited processing power and main memory capacity, both ontologies and rules must be translated beforehand in an easy-to-parse format that can be effortlessly executed on the client-side. Due to the lack of a client-side reasoner we materialize all ontologies at the server-side. By using an OWL reasoner we check the consistency of the ontology at design-time, classify the instances and infer the class hierarchy. There are two possibilities to represent ontologies on the client-side: XML or JSON (JavaScript Object Notation) [16]. XML is very verbose and adds additional overhead to the payload. Furthermore, XML rules encoded in XML cannot be executed directly on the client, but have to be parsed, an added expense. Therefore, we decided to represent ontologies, annotations and rules in the compact and directly executable data interchange format JSON.

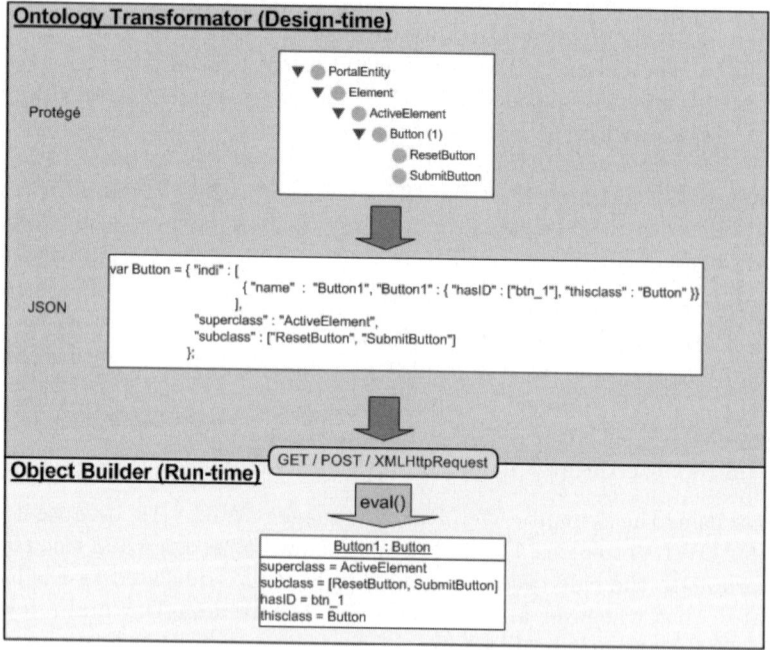

Fig. 3. Transformation of ontologies

Figure 3 shows what the JSON format looks like, after transforming the concept Button. As depicted in the concept hierarchy of Protege the class Button has several super and sub concepts. Additionally, there exist one instance of

Button in the ontology. Not shown are the properties of the Button class, which in fact are: *hasID* and *thisclass*. A class is represented as object in JSON and its instances are collected in a property of type Array called *indi*. This array contains all instances as objects whereas the objects in turn hold their properties as attributes. The last attribute *thisclass* is a reference to the instantiated class and is automatically added during the translation. The class hierarchy is stored directly in the JSON object representing the class as Array attributes: superclass and subclass. SWRL rules are also encoded as objects consisting of a condition and action part. SWRL build-ins are encoded manually as JavaScript functions beforehand. A call to these external functions is placed into the JSON translation whenever the transformer detects a built-in in the original adaptation rules format. At run-time the object builder generates real objects from the JSON string, as depicted in the lower part on Figure 3.

5 Run-Time Architecture

The core responsibility of the run-time architecture is to ensure the user-centric adaptiveness of the RIA. The run-time architecture is constituted by the adaptation and transfer cycle as depicted in Figure 1. After transforming the ontologies, annotations and adaptation rules into a client-readable format at design-time, they can be transmitted as JavaScript code in answer to a client request at any point in time. When a user requests the RIA, not only content and layout data are send to the client, but also the JSON representation of the ontlogies, annotations and rules. On the client-side the user model is built up and the portal is adapted by tracking user interactions and executing adaptation rules. Figure 4 shows the interplay of the constituent run-time components.

In a Web browser HTML pages are internally represented as a DOM (Document Object Model). Whenever a user interacts with the Web page the DOM fires appropriate events which can be caught by the event handler component. In order

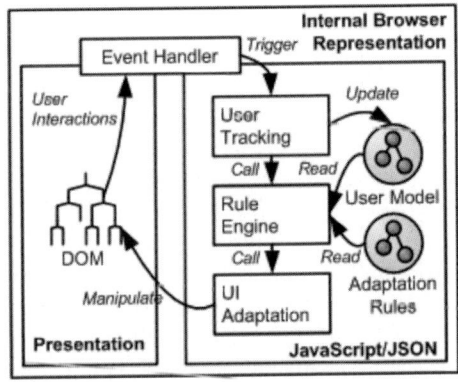

Fig. 4. Logical client-side run-time architecture

to catch events the event handler has to register first to specific event types. In our current implementation this is done manually. The Web programmer has to explicitly specify which kinds of events shall be tracked. Each recognized event results in a call of the user tracking component and, in a second step, the invocation of the rule engine. The user tracking component resolves the relationships between the JavaScript events, the user interface elements and their annotations. Furthermore it records the events to the user model. In this way the user model materializes the browsing history of the current user on the level of JavaScript events.

Based on the Web usage data stored in the user model the rule engine evaluates the adaptation rules. We implemented a stateless rule evaluation based on the sequential algorithm [17]. The rationale behind this approach is that each of the independent rules fires once its conditions hold. There is no agenda to resolve any eventual conflicts caused by executing rules. Furthermore, this approach implies that the rules do not affect each other. Despite the disadvantages of this approach we chose the sequential algorithm because of its simple loop-like implementation. Once a rule has fired, the rule body, in most of the cases translated SWRL built-ins, is executed by the UI adaptation component and the user interface is manipulated.

At the end of the session the user model is sent back to the server using the asynchronous communication facility of AJAX. The accumulated user models form the basis for a further modeling cycle.

6 Evaluation

The implementation of the conceptual framework was realized using Java libraries for the design-time code generation and AJAX for the run-time components. We evaluated our prototypical implementation at different levels. First, we looked at the time consumption of rebuilding the ontologies and executing the adaptation rules at runtime on the client-side. Secondly, we evaluated our approach theoretically. That means we looked at the JSON format representing ontologies and rules and we also examined restrictions imposed by our rule execution algorithm.

6.1 Computational Evaluation

The design-time modeling of ontologies and rules, as well as the subsequent translation into JSON is the non-time critical part of the application. The transformation from RDF/XML syntax into JSON rules and the mapping of OWL concepts, instances and relations into JSON occurs at design-time. But already at this stage optimization is a crucial issue. The preparation of the JSON file for later usage on client-side requires an effective mapping-method to keep the amount of data represented on the client to a minimum. The compressed JSON format, as the result of translating ontologies, annotations and rules, lets the file size shrink to 50% of it original size. The file size decreased from 42,1 KB (RDF/XML) to 20,3 KB (JSON).

The more time critical issues are the run-time tasks, like the initial loading and creation of OWL concepts on client-side as well as the execution of rules and user interface adaptation on the client side. At the time of accessing the RIA the ontolgies, annotations and rules have to be up-loaded to the client in a first step. The JSON file is executed using the JavaScript function *eval()* and concepts, instances and rules are represented as JSON objects on the client-side. An evaluation of this initial client-side concept creation is depicted in the following Figure 5 a). The initial transfer and construction of the rules and ontologies does not affect the usability too much, since it takes place within the first couple of seconds a user accesses a new page which is a time period of almost no interaction. As the diagram shows, the time consumption is below 200 ms for up to 10000 concepts, which is not recognized by the user when loading the page.

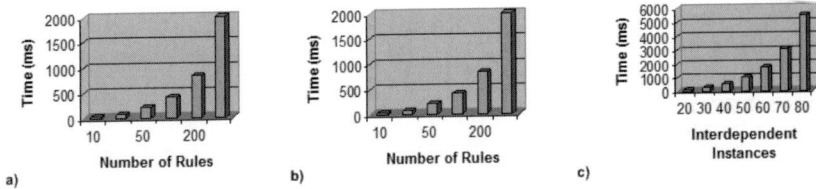

Fig. 5. Evaluation: a) Client side initial JSON concept creation; b) Rule execution time evaluation; c) Interdependent instances

During run-time the most important task is the execution of rules and the subsequent adaption of the user interface. The client-side rule engine is implemented as a stateless sequential algorithm. The rules are mapped to nested *IF/THEN* statements and executed in the order they are received. There is no conflict solving, complex event processing or any kind of inference during the execution since rules cannot be triggered by another rule. Each rule is evaluated and, if all conditions hold, the body (action) of the rule is executed. In Figure 5 b) the performance of the rule execution is evaluated. As an evaluation constraint we let each rule fire, that means that all conditions of our evaluation rule set hold. Each rule manipulates the user interface of our exemplary RIA. The evaluation of each rule starts with loading the rule from the JSON structure into a local variable. This loading process is realized with a non-optimized algorithm and needs refinement. A big system performance improvement will be reached by revising this algorithm which occupies most of the execution time. Future development will also include the evaluation and integration of more matured rule execution algorithms like Rete [18] to achieve better performance.

The number of instances of ontology concepts have a considerable bearing on the execution time of the application rules on the client-side. The most time-intensive parts of instance-handling are *N-to-M* relations between instances. The rationale behind this is: If there is one instance Ia of concept A and one instance Ib of concept B it is easy to utilize a rule because there is only one relation between the two instances. But if there are several instances $Ia - n$ for concept A and several instances $Ib - n$ for concept B the rule engine has to evaluate the Cartesian product of the instances, which leads to an exponential time consumption. Figure 5 c) depicts the time consumption of rules coping with interdependent instances in more detail. However, in our motivating examples we did not have to deal with that, because of carefully designed adaptation rules.

During the performance evaluation a slight distinction in the measurement results between the two tested web browsers (Internet Explorer[5] and Mozilla Firefox[6]) has been determined. The diagrams are based on the average measurement results of both web browsers.

6.2 Theoretical Evaluation

First we evaluated the JSON format we created in order to represent OWL ontologies. Our JSON ontology serialization is conceived to minimize time of accessing instances at the client-side. The format puts some limitations on the representation of ontologies. On the other side, these limitations have no restrictive effects to the overall approach of client-side adaptation of RIAs as we solely rely on the concept taxonomy in our adaptation rules. By computing the subsumption hierarchy at design-time we can construct the entire class hierarchy graph. However, in doing so we lose all informations regarding OWL class axioms like equivalent classes and class descriptions like union or intersection. All axioms and descriptions are mapped to a simple sub class relation. Also all information about relations are lost. Relations only appear as attributes in objects and are no longer represented as discrete entities. Only individuals are transformed without any information loss. But as already mentions, as consistency checks are performed at design-time and the adaptation rules only rely on instances, their relations, and class hierarchy, this puts practically no constrains to our approach.

In a second step we evaluated the gains and losses of the sequential algorithm implemented by our client-side rule engine. One advantage of the sequential algorithm is its simplicity. It is quick to implement and easy to maintain. Furthermore, in the computational evaluation, we showed the feasibility of our approach even with a simple rule evaluation strategy. But the sequential algorithm does not come for free. So, we have exponential time consumption when evaluating the Cartesian product of class instances. This could be reduced to by using the forward-chaining Rete algorithm for the evaluation of the adaptation rules. Rete as efficient pattern matching algorithm would introduce real inferencing and a stateful rule evaluations. Our current implementation of the client-side rule engine imposes some restrictions to the design of the adaptation rules. So, the rule engine can only evaluate 2-ary predicates which have a model consisting exactly of pairs of the form $(a, b_i)_{i \in I}$ or conversely $(a_i, b)_{i \in I}$, where I

---

[5] http://www.microsoft.com/windows/products/winfamily/ie/default.mspx
[6] http://www.mozilla.com/en-US/firefox/

is an arbitrary finite index set. In the case of $(a_i, b_j)_{i,j \in I}$ the outcome is currently undefined. We are working on a solution to that. Furthermore all variables have to be explicitly introduced as we can not conclude the type of a variable for its occurrence in an 2-ary predicate. This means that in the Example 1 the atoms portal:Form(?a), domain:WeddingDay(?b), domain:MarriageCertificate(?d) and domain:BirthCertificate(?e) are mandatory. Also the order matters. These atoms have to occur before the variables are used in an arbitrary 2-ary relation.

One of our goals is the autonomous working of the application (without client-server communication). It means that our main instrument for the detection of events is JavaScript. This fact constitutes a restriction with respect to complex events. Suppose that some complex event CE is modelled as the conjunction of two other not necessarily atomic events E_1 and E_2. We must have a rule $E_1 \wedge E_2 \rightarrow CE$ expressing this situation. It is possible to create a new instance for the complex event CE^7, however the application remains unaware of the latter, since it was not detected by JavaScript. We have solved this issue by coding the occurrence of such a complex event in some attributes of the existing instances, and then replacing the above rule by a checking of the attributes. This fact remains, however, a handicap toward an elegant and natural modeling of the rules. The above evaluation confirms our presentiment in [4], SWRL is in fact not sufficient for our purposes. It does not constitute a suitable framework for the modeling of complex rules. The simulation of the *seq* operator does not allow an efficient processing of the events and rules. A recent investigation has shown that Event-Condition-Action (ECA) rules[8] may be more appropriate.

7 Related Work

In [19] the integration of semantics in Web usage mining techniques is shown applied to a movie website. On the basis of a movie ontology and the user behaviour, user profiles were constructed, which are used for online recommendations. In the center of our aproach are e-Government websites consisting of forms, services and information.

Comparing our work with standard models for adaptive hypermedia systems like e.g. AHAM [20], we observe that they use several models like conceptual, navigational, adaptational, teacher and learner models. Compared to our approach, these models correspond to ontologies presented in Section 4, but miss their formal representation. Moreover, we express adaptation functionalities as encapsulated and reusable OWL-DL rules, while the adaptation model in AHA uses a rule based language encoded into XML.

The Personal Reader [21] provides a framework for designing, implementing and maintaining Web content readers, which provide personalized enrichment of Web content for each individual user. The adaptive local context of a learning

---

[7] This situation is even impossible with respect to the ontology. However, we make use of SWRL bultins to achieve the addition of new instances to the ontology.

[8] ECA rules are event driven rules in the form of: *ON* (event) *IF* (condition) *DO* (action).

resource is generated by applying methods from adaptive educational hypermedia in a semantic Web setting. Similarly [22] focuses on content adaptation, or, more precisely, on personalizing the presentation of hypermedia content to the user. However, both approaches do not focus on the on-line discovery of the profile of the current user that is one of the main features of our approach. Another difference would be the self-adaptivity.

In [23] the authors suggest the use of ontologies and rules in order to find related content on the Web, based on the content currently displayed to the user. We enhance this work by not only adapting the content based on concept similarity but rather based on accumulated Web usage data. Furthermore we show a way how to link semantics and content. Still the main difference remains the introduction of the autonomous client, as we are dealing with Rich Internet Applications and not with common dynamic Web applications executed on a Web server.

8 Conclusions and Future Work

In this paper we presented a novel approach that holistically combines page annotations, semantic Web usage mining, user modeling, ontologies and rules to adapt AJAX pages. We showed and evaluated how our concept of an autonomous client works. With our prototypical implementation we demonstrated the proof of concept and our motivating example taken form the e-Government domain emphasized the practical relevance of our work. Currently, we are investigating the Rete algorithm with respect to its adoption for client-side rule processing. We expect a better run-time performance by substituting the sequential algorithm with the Rete algorithm. Furthermore we plan to extend SWRL from deduction rules to event ECA rules to allow complex event processing directly on the client side. Then the transformer component of the design-time cycle can automatically generate event handlers based on the ECA rules.

Acknowledgements

The work is based on research done within the FIT project - Fostering self-adaptive e-Government service improvement using semantic technologies. The FIT project is co-funded by the European Commission under the "Information Society Technologies" Sixth Framework Program (2002-2006).

References

1. Brusilovsky, P.: Methods and techniques of adaptive hypermedia. User Model. User-Adapt. Interact. 6(2-3), 87–129 (1996)
2. Mobasher, B., Cooley, R., Srivastava, J.: Automatic personalization based on web usage mining. Commun. ACM 43(8), 142–151 (2000)
3. Garrett, J.J.: Ajax: A new approach to web applications (2005), http://www.adaptivepath.com/publications/essays/archives/000385.php

4. Schmidt, K.-U., Stojanovic, L., Stojanovic, N., Thomas, S.: On enriching ajax with semantics: The web personalization use case. In: Franconi, E., Kifer, M., May, W. (eds.) ESWC 2007. LNCS, vol. 4519, pp. 686–700. Springer, Heidelberg (2007)
5. Thomas, S., Schmidt, K.-U.: D4: Identification of typical problems in e-government portals. Technical report, FIT consortium (July 2006), http://www.fit-project.org/Documents/D4.pdf
6. Maedche, A., Staab, S.: Semi-automatic engineering of ontologies from text. In: Proceedings of the 12th International Conference on Software Engineering and Knowledge Engineering (2000)
7. Stojanovic, L., Ma, J., Stojanovic, N.: D9: Methods and tools for semi-automatic learning of a domain ontology that models the content of a front office. Technical report, FIT consortium (January 2007), http://www.fit-project.org/Documents/D9.pdf
8. Stojanovic, L., Stojanovic, N., Ma, J.: An approach for combining ontology learning and semantic tagging in the ontology development process: egovernment use case. In: Benatallah, B., Casati, F., Georgakopoulos, D., Bartolini, C., Sadiq, W., Godart, C. (eds.) WISE 2007. LNCS, vol. 4831, pp. 249–260. Springer, Heidelberg (2007)
9. Bechhofeer, S., van Harmelen, F., Hendler, J., Horrocks, I., McGuinness, D., Patel-Schneider, P., Stein, L.A.: Owl - web ontology language reference. Recommendation, W3C, February 10 (2004)
10. Dai, H., Mobasher, B.: Using ontologies to discover domain-level web usage profiles. In: 2nd Semantic Web Mining Workshop at ECML/PKDD-2002 (2002)
11. Berendt, B., Hotho, A., Mladenic, D., van Someren, M., Spiliopoulou, M., Stumme, G.: A roadmap for web mining: From web to semantic web. In: Berendt, B., Hotho, A., Mladenič, D., van Someren, M., Spiliopoulou, M., Stumme, G. (eds.) EWMF 2003. LNCS (LNAI), vol. 3209, pp. 1–22. Springer, Heidelberg (2004)
12. Stumme, G., Hotho, A., Berendt, B.: Semantic web mining: State of the art and future directions. Semantic Grid –The Convergence of Technologies 4(2), 124–143 (2006)
13. Spiliopoulou, M., Mobasher, B., Berendt, B., Nakagawa, M.: A framework for the evaluation of session reconstruction heuristics in web usage analysis. INFORMS Journal of Computing, Special Issue on Mining Web-Based Data for E-Business Applications 15 (2003)
14. Stojanovic, L., Ma, J., Yu, J., Stojanovic, N., Schmidt, K.-U., Thomas, S., Rahmani, T.: D18: Methods and tools for mining the log data by taking into account background knowledge. Technical report, FIT consortium (July 2007), http://www.fit-project.org/Documents/D18.pdf
15. Horrocks, I., Patel-Schneider, P.F., Boley, H., Tabet, S., Grosof, B., Dean, M.: Swrl: A semantic web rule language combining owl and ruleml. Technical report, W3C Member submission, May 21 (2004)
16. Crockford, D.: Rfc4627: Javascript object notation. Technical report, IETF (2006)
17. Berstel, B., Bonnard, P., Bry, F., Eckert, M., Patranjan, P.-L.: Reactive rules on the web. In: Antoniou, G., Aßmann, U., Baroglio, C., Decker, S., Henze, N., Patranjan, P.-L., Tolksdorf, R. (eds.) Reasoning Web. LNCS, vol. 4636, pp. 183–239. Springer, Heidelberg (2007)
18. Forgy, C.L.: Rete: a fast algorithm for the many pattern/many object pattern match problem. Artificial Intelligence 19, 17–37 (1982)
19. Dai, H., Mobasher, B. (eds.): Using Ontologies to Discover Domain-Level Web Usage Profiles (2002)

20. Romero, C., Ventura, S., Herváas Martinez, C., De Bra, P.: In: Proceedings of the Fifth International Conference on Human System Learning, ICHSL, Europia (November 2005)
21. Dolog, P., Henze, N., Nejdl, W., Sintek, M.: The personal reader: Personalizing and enriching learning resources using semantic web technologies. In: De Bra, P.M.E., Nejdl, W. (eds.) AH 2004. LNCS, vol. 3137, pp. 85–94. Springer, Heidelberg (2004)
22. Frasincar, F., Houben, G.-J.: Hypermedia presentation adaptation on the semantic web. In: De Bra, P., Brusilovsky, P., Conejo, R. (eds.) AH 2002. LNCS, vol. 2347, pp. 133–142. Springer, Heidelberg (2002)
23. Ankolekar, A., Tran, D.T., Cimiano, P.: Rules for an ontology-based approach to adaptation. In: 1st International Workshop on Semantic Media Adaptation and Personalization, Athen, Greece (December 2006)

OntoGame: Weaving the Semantic Web by Online Games

Katharina Siorpaes[1] and Martin Hepp[1,2]

[1] SEBIS, Semantic Technology Institute (STI), University of Innsbruck, Austria
[2] Chair of General Management and E-Business, Bundeswehr University Munich, Germany
katharina.siorpaes@sti2.at, mhepp@computer.org

Abstract. Most of the challenges faced when building the Semantic Web require a substantial amount of human labor and intelligence. Despite significant advancement in ontology learning and human language technology, the tasks of ontology construction, semantic annotation, and establishing alignments between multiple ontologies remain highly dependent on human intelligence. This means that individuals need to contribute time and sometimes other resources. Unfortunately, we observe a serious lack of user involvement in the aforementioned tasks, which may be due to the absence of motivations for people who contribute. As a novel solution, we (1) propose to masquerade the core tasks of weaving the Semantic Web behind online, multi-player game scenarios, in order to create proper incentives for human users to get involved. Doing so, we adopt the findings from the already famous "games with a purpose" by von Ahn, who has shown that presenting a useful task, which requires human intelligence, in the form of an online game can motivate a large amount of people to work heavily on this task, and this for free. Then, we (2) describe our generic OntoGame platform, and (3) several gaming scenarios for various tasks plus our respective prototypes. Based on the analysis of user data and interviews with players, we provide preliminary evidence that users (4) enjoy the games and are willing to dedicate their time to those games, (5) are able to produce high-quality conceptual choices. Eventually we show how users entertaining themselves by online games can unknowingly help weave and maintain the Semantic Web.

1 Introduction

A pre-requisite for the Semantic Web to become a reality is the broad availability of ontologies and annotation data. However, the knowledge acquisition bottleneck [1] strikes the Semantic Web as it struck other endeavors in the past. Despite significant advancement in tools and semi-automatic approaches, we still need a significant amount of human labor and intelligence for the construction of ontologies, for the annotation of data in various modalities and formats, and for aligning the conceptual elements in multiple ontologies. Making the Semantic Web a reality requires an increase of available metadata by orders of magnitude as compared to the current state. However, we observe that it is hard to motivate people to dedicate their time to those three tasks. At the same time, the amount of Web content in complex modalities (like images, videos, sounds, or Flash applets) and services exposed on the Web is increasing; such is even harder to annotate without the aid of human intelligence.

Obviously, there are still many tasks that most humans can solve easily but state of the art computers cannot [2, 3]. A famous example for such tasks are CAPTCHAs [3]: challenges related to image analysis that can be used to test whether the user is a human being or a computer agent. Those challenges are employed by many Web applications to block access by unwanted bots and scripts.

Similar to CAPTCHAs, most of the tasks for lifting the current Web to a semantic level remain dependent on human intelligence. Now – why would people want to invest time in building ontologies or annotating content? Clearly, we can observe a sharp contrast in user interest in two branches of Web activity – the "Web 2.0" movement lives from an unprecedented amount of contributions from Web users, while the work on the Semantic Web side is hampered by a substantial lack of user involvement in the aforementioned tasks. In our opinion, this is mainly because Web 2.0 environments provide direct rewards for user involvement, mostly in the form of improved access to Web content [4-6]: Users who tag objects in collaborative tagging systems immediately improve their own access to those objects, while at the same time improving the shared metadata. As for the Semantic Web, many important tasks come without a proper reward for the contributing humans: Building an ontology is a fairly abstract task and thus pretty much decoupled from immediate rewards. Also, heavyweight annotations often require a lot more time from a single skilled individual than this individual will ever save by means of the improved access.

This leaves us with two options for overcoming the lack of ontologies, annotations, and alignments: Either we make a leap in technology so that humans can be eliminated from those tasks. Or we fix the broken incentive scheme for the Semantic Web, i.e., create proper rewards for contributing humans. Luis van Ahn has demonstrated with his already famous games [2, 7-10] that one can exploit computer gaming scenarios for having people contribute human intelligence to actual problems. We adopt his approach for overcoming the key bottlenecks to building the Semantic Web: the lack of people actually dedicating intelligence and judgment for building and maintaining it.

1.1. Related Work

The most popular games with a purpose have been described by **Von Ahn** and colleagues, who have also coined the term *"human computation"*: The ESP game [8] aims at labeling images on the Web. Two players, who do not know each other, have to come up with identical tags describing an image. Peekaboom [7] is a related game for locating objects within images. Verbosity [10] is a game for collecting common sense facts. Phetch [9] is a computer game that collects explanatory descriptions of images in order to improve accessibility of the Web for the visually impaired. Only very recently, Law, von Ahn, and colleagues [11] also came up with a game called Tagatune for music and sound annotation based on tags. **Lieberman** and colleagues describe the game Common Consensus [12], which aims at collecting human goals in order to recognize goals from user actions and conclude a sequence of actions from these goals. Another approach to collecting common sense knowledge is the FACTory Game[1] published by

---

[1] http://game.cyc.com

Cycorp[2]: FACTory is a single-player online game that randomly chooses facts from the Cyc knowledge base [13] and presents them to the players. The player has to say whether the statement is true, false, doesn't make sense, or whether the user does not know. The answers are scored depending on accordance with the majority of answers. Apart from Verbosity, Common Consensus, and FACTory, we do not know of any other work that uses computer game scenarios for the collection of knowledge, and none of those is directly linked to the Semantic Web.

1.2. Contribution and Overview

In this paper, we (1) propose to masquerade the core tasks of weaving the Semantic Web behind online, multi-player game scenarios, in order to create proper incentives for humans to get involved, (2) describe our generic OntoGame platform, and (3) multiple gaming scenarios for various task plus respective prototypes. Based on the analysis of user data and interviews with players, we provide preliminary evidence that users (4) enjoy the games and are willing to dedicate their time to those games, (5) are able to produce high-quality conceptual choices, and show (6) how they may unknowingly help weave the Semantic Web by doing so. Please check our project Web page at http://www.ontogame.org for the first fully-fledged public game and other prototypes. This paper extends our very first overview of experiments described in [14], in which we asked humans to judge whether a particular Wikipedia page primarily describes a set of objects (i.e. a class) or an individual (i.e. an instance).

2 Multi-player Games for Weaving the Semantic Web

In the following, we describe multi-player games for subtasks in ontology construction, ontology alignment, and ontology population (annotation).

2.1 Games for Ontology Construction

Ontology construction involves the following five tasks that are hard to delegate to computers:

Collecting named entities: Relevant conceptual elements of the domain of discourse must be identified and a unique key assigned.

Typing named entities according to the ontology meta-model: The type of conceptual element according to the distinctions of the applicable ontology meta-model must be determined for each named entity. For example, many popular ontology meta-models support *classes*, *properties*, and *individuals* as core types.

Adding taxonomic and non-taxonomic relations: A flat collection of ontological elements can be enriched by adding taxonomic and non-taxonomic relations. The most prominent form of this task is arranging the concepts into a subsumption hierarchy by introducing subClassOf relations.

---

[2] http://www.cyc.com

Modularization: Depending on the domain of discourse, it is often useful to define groups of concepts - either based on their ontological nature or by target applications, since such may be more manageable.

Lexical enrichment: Ontology engineering methodologies tend to focus on formal means for specifying ontologies. In order to describe the *intended* semantics of ontology elements, informal means, like natural language labels or synonyms are albeit also needed. However, relating a conceptual element to terms or synonym sets requires careful human judgment, since otherwise, inconsistencies between the informal part and the formal part of the ontology may result.

In the following, we describe some game scenarios for those tasks.

Table 1. Games for Ontology Construction

Task	Input		Output
	Computational Side	Human Side	
Collecting and typing named entities	Users are presented with a class definition.	The players have to come up with and agree upon a label for an attribute its range.	Attributes and their ranges
Typing Named Entities	Users are shown a conceptual entity (e.g. a Wikipedia article).	The players have to agree whether the respective entity represents a class, a property, or an individual.	Meta-model classification of input entities
Adding taxonomic and non-taxonomic relations	Users are shown two classes.	The players have to judge whether one class subsumes the other or to come up and agree upon a label of a relationship between the classes.	Taxonomic relations and labels for other relationship types
Adding taxonomic relations	Users are shown a class.	Users have to come up and agree upon a label for a super-class, i.e. an abstraction.	Classes, taxonomic relations
Lexical Enrichment	Users are presented with one element from an ontology as well as a lexical resource (e.g. WordNet) including the possibility to browse the resource.	The players have to select an entity from the lexical resource, such as a synonym of the class label or a translation.	Links to terms
Modularization	Users are presented with a domain name (from a list of relevant domains) as well as a set of ontological elements.	The players have to define a subset of relevant ontological entities for that domain (and agree on this assignment).	Domain ontology modules

2.2. Games for Ontology Alignment

In an open environment such as the Web, it is likely that multiple, partly overlapping ontologies evolve and are being used. For improved access of the related information, the elements of overlapping ontologies must be aligned to each other; and since ontologies evolve due to conceptual dynamics in domain and advancement of our understanding of the world, such is a continuous effort rather than a one-time task. It's burdensome and never done. Euzenat and Shvaiko [15] distinguish four different techniques of ontology matching: (1) terminological techniques that rely on lexical resources within the ontology, (2) structural techniques that focus on the relations between entities, i.e. ontology elements, (3) extensional techniques comparing extensions of entities, and finally (4) semantic techniques that exploit formalized knowledge.

Despite significant advancement towards automatic matching of ontologies without human intervention, current systems are often not able to perform reliable automatic matching on real-world ontologies yet. The less formal the input ontologies are, the less likely it is that a machine will ever be able to reliably determine the proper semantic relationships between elements from two different ontologies.

In this paper, we focus on semantic relationships between classes, individuals, relations, and data types. Between such entities, there are different possible types of correspondence, of which the most relevant set-theoretic relations are equivalence (=), more general ($\supseteq$), disjointness ($\perp$), and subsumption ($\subseteq$) as described in [15]. We think that the following tasks are particularly suited for the representation as game scenarios:

Equivalence of classes, relations, attributes: Indicating whether two classes or properties are equivalent, based on the label, a description, and additional lexical resources.

Subsumption between classes: Indicating whether a class is a sub-class of another class.

The tasks in ontology matching were outlined in the previous section. In literature, equivalence (=), subsumption ($\subseteq$), and disjointness ($\perp$) are described as the most important matching relations. Thus, we do not only want to know from our players whether two classes are the same but we want to know the kind of relation that exists between them. In our games (Table 2) we let players choose from a set of possible relations. Furthermore, one has to keep in mind that our goal is to attract as many users as possible to play in order to create a wealth of data, even if only lightweight. Therefore, we decided to make use of SKOS [16] relations: SKOS (Simple Knowledge Organization System) core is a lightweight meta-model that describes just the minimal set of classes and properties that are necessary to express knowledge in simple structures. We have preliminary evidence that players are able to understand the meaning of SKOS relations, such as broader or narrower, more easily than the precise meaning of subClassOf [17]. Thus, we use the following relations for

Table 2. Games for Ontology Matching

Scenario (Task)	Input		Output
	Computational Side	Human Side	
Matching classes	Players are faced with the two concepts c1 from ontology A and c2 from ontology B and a set of possible mapping relations.	Players have to select and agree on the most appropriate relation between the concepts.	Alignments
Matching classes	Players are presented with concepts c1 from ontology A and the complete subsumption hierarchy of ontology B plus the set mapping relations.	Players have to select the most specific corresponding class in ontology B, the appropriate relation between the concepts, and agree on both choices.	Alignments

matching ontologies: (1) equivalent (=), (2) broader: a concept that is more general in meaning, (3) narrower: a concept that is semantically narrower in some sense, (4) related: a concept with which there is an associative semantic relationship, (5) partly overlapping with: there is an overlap in meaning between these concepts, (6) strict subClassOf; this relation is intended only for expert games, (7) Not related: disjointness ($\perp$).

2.3. Games for Semantic Annotation

Generally, all annotation scenarios require (1) a resource, e.g. a Wikipedia article or a media object, and (2) an ontology, e.g. the Proton ontology. The players are then asked to annotate the resource using the given ontology (Table 3). For each consensual aspect, both players will earn points. In many cases it will be necessary to hide the ontology behind a graphical user interface or natural language patterns in order to increase the game fun as well as the comprehensibility of the task. Candidate resources that are vastly available on the Web are textual resources, images, videos, sounds, software, and Web services. The (semi-) automatic annotation of multimedia content is especially challenging for a machine; however, this is a task that can often be easily done by a human actor. Thus, we see an especially large potential in turning multimedia content annotation into games. Additionally, games that involve music, pictures, or videos are more enjoyable for players. Another potential application area, which will not be addressed in this paper, is the annotation of Web services.

For annotation games, we depend on the availability of sufficiently detailed (domain) ontologies, which can be a bottleneck as of today. This is why we aim at interweaving games for annotation with games for ontology construction.

Table 3. Games for Semantic Annotation

Scenario (Task)	Input		Output
	Computational Side	Human Side	
Annotation	Players are shown a resource, which can be text or multimedia content, and a suitable (domain) ontology.	Players have to select and agree on the appropriate annotation of the resource.	Semantic Annotations

3 OntoGame: A Generic Game Infrastructure

In order to keep up interest, the set of available games should be changed or the games being updated frequently. Also, the resulting data from past games should be stored in a generic format so that we can run statistical analyses when deriving ontologies, annotations, or mappings from consensual games. Note that the games do not directly return the correct modeling; moreover, we will use an appropriate threshold of consensual, matching rounds that must confirm a particular modeling choice before it is assumed to be correct.

The heart of our OntoGame is a generic game infrastructure that allows to plug-in various scenarios with minimal modifications. All user inputs and results are stored in RDF for simple analysis and reuse. The user interface is designed in a way that in can be easily adjusted to a new scenario.

3.1 System Description

Each OntoGame is an online, multi-user game where players play in teams of two: these teams are selected randomly and anonymously. The players have no means to communicate with or identify the counterpart. This is important in order to avoid cheating or false input, which will be discussed in detail in a later section.

In all game scenarios, users are faced with a task, e.g. matching two classes or finding a suitable abstraction of a Wikipedia article in a given ontology. The players have to reach consensus on their choice in order to earn credits. After each choice, both players get feedback about what their partner's choice was, regardless of whether they reached consensus or not.

Before using the system, each user has to register. It is desirable to have users login with the same username every time they play because of two reasons. First, **competition:** users can build a reputation in the system and work on their rank, which constitutes an additional incentive to play games [6]. Second, **reliability:** if users have a history of good, meaningful game rounds, their judgment is more reliable than that of others. This can be exploited when deriving formal content and when to spot cheating. Upon pressing a "play!" button, the user is randomly paired with another player and the game starts. In case there is not an even number of users on-line, a single-player mode is started; this alternative remains invisible to the user, though. In single-player mode, users play pre-recorded challenges as if playing with a real partner.

Players can skip a step (Fig. 2) and abort the current challenge in the games; the team will then proceed with a new challenge. At the moment, this feature follows the principle of consensus as well: only when both players decide to skip, they are taken to the next round. Skipping is an important feature, because it is possible that poor or incomprehensible challenges are given, for which players may simply be unable to produce consensual solutions. Instead of encouraging random guesses, we rather motivate users to proceed to a new challenge.

3.2 Implementation

The OntoGame platform (Fig. 1) is a client-server infrastructure based on Java. The game server runs on a Apache Tomcat 5.5[3] server together with the RDF repository Sesame[4] [18] and servlets. The game server connects to the repository via a database connector and runs the servlets. The servlets connect to the client via an object stream over an HTTP tunnel. The controller runs the graphical user interface. The game

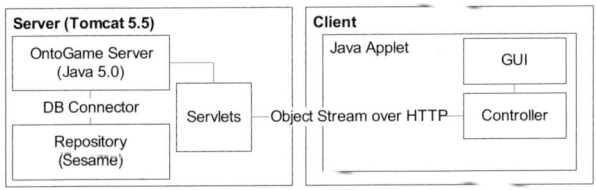

Fig. 1. OntoGame Platform

---

[3] http://tomcat.apache.org/
[4] http://www.openrdf.org/

server implements the singleton pattern, which is used to restrict instantiation of a class to one object, because in OntoGame exactly one object is needed to coordinate actions across the system including the games, discovering matches, etc. Four different servlets perform the following tasks: login, communication flows for the game, handling user input, matching, and skipping.

3.3 Cheating

One may argue that cheating and other forms of destructive user behavior endanger the quality of the game output. However, von Ahn has already shown that the impact of cheating can be minimized yet by several simple mechanisms. We follow his suggestions and use the following techniques: First, the players are paired anonymously and have no way to communicate with each other. Second, we check whether the IP addresses of partners are different, so one cannot simply run the game on the same machine multiple times and hope for being teamed up with oneself. Third, simple cheating strategies like always choosing the first option or enter pre-agreed words as text input can be detected rather easily by having them play one challenge for which the correct result is known. If the consensual solution to such a challenge is different from the set of known solution, user input from both players will be ignored when deriving formal content. Also, one can monitor the response times and assume bots when they are significantly lower than the average.

We are also considering more sophisticated reputation mechanisms for future releases.

4 Four Cool OntoGame Scenarios

In the following, we describe four game scenarios for weaving the Semantic Web that we consider most promising and that address real-world problems, such as searching videos or product search in e-Bay. The first two ones are already released to the general public. The two others are design studies for which the implementation is underway.

4.1 Turning Wikipedia into a Huge Domain Ontology with Proton Grounding

In this game, we show the first paragraph from a randomly selected Wikipedia page. By Wikipedia convention, this is almost always a reliable excerpt of the page content. Then, we ask the user to select whether this Wikipedia entry rather describes a set of objects (i.e., a class) or a significant single object (i.e., an individual), see Fig. 2^5). If both players agree on that choice, they proceed to the next level. In this level, they have to agree upon the most specific class of the Proton ontology [19] of which the Wikipedia entry is a subclass or instance (see Fig. 3). The use of Proton is mainly motivated by two factors. First, we needed a general-purpose ontology that would make sense as an upper-level ontology above all Wikipedia entries. This ontology should already contain sufficient specializations so that the difference in the level of abstraction as compared to Wikipedia URIs was appropriate for average users. In the future, we will also consider upper ontologies such as DOLCE[6] or SUMO[7].

---

[5] Larger screenshots are available at http://www.ontogame.org
[6] http://www.loa-cnr.it/DOLCE.html
[7] http://www.ontologyportal.org/

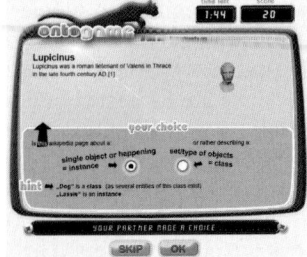 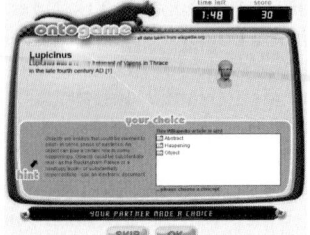

Fig. 2. Ontologizing Wikipedia: Step 1   **Fig. 3.** Ontologizing Wikipedia: Step 2

The deeper the teams manage to go into the hierarchy, the more Wikipedia articles they play, and the more Proton abstractions they find within 2 minutes, the more points they are awarded. For the moment, we do not make use of the Wikipedia category system due to its diverging and unstructured nature, but may use this in the future for suggesting suitable Proton choices.

The motivation for this game is that the URIs of the more than 1.8 Million Wikipedia entries are reliable identifiers for countless useful conceptual entities [20]. For example, Wikipedia contains more than 220,000 URIs for types of products and services and is thus eight times larger than eCl@ss or UNSPSC, the two largest categories for products and services. If we are able to ground those 1.8 Million conceptual elements properly in the Proton ontology, we will create the largest general interest ontology for annotating Web resources – 1.8 Million identifiers for anything from artists to high schools, from products to organizations. This game is online for playing by the general public at http://www.ontogame.org.

Example

Alice and Bob play the game: they both see an excerpt of the Wikipedia article about Lupicinus[8]. They first have to agree on whether the most important ontological role of Lupicinus is to be a class or an instance. Alice and Bob agree on instance (because it is an instance of Person), get 20 points and are taken to the next step. Here they are shown the first level of the Proton ontology, which divides things into *abstracts, happenings,* or *objects*. Alice and Bob both agree on *object*, get 10 points and are taken into the object branch of Proton. Here they agree on *agent* and are awarded 20 points and are taken even deeper in the Proton hierarchy. Our players both choose *person* in the next level, get 30 points, and finally agree on the Proton class *man*, receiving 40 points. The round ends here and they are taken to the next randomly chosen article. This continues until the time of 2 minutes is over.

4.2 Annotating YouTube Videos

The objective of this scenario (see Fig. 4 and 5) is to annotate YouTube[9] videos. It is inspired by Jim Hendler's comment at last year's ISWC's panel discussion that search in YouTube (and videos in general) was a key application of semantic search.

---

[8] http://en.wikipedia.org/wiki/Lupicinus
[9] http://www.youtube.com

In order to annotate YouTube videos in games, we specified a simple domain ontology that describes the content of videos. The relevant standard for the description of multimedia is MPEG-7[10]. We also took IMDB (Internet Movie DataBase)[11] into consideration as input. IMDB has a huge user base and we are interested in what users are searching for when they search for videos. Therefore, we had a close look at the search interface of IMDB in order to model a very simple video content ontology. Our approach to annotating YouTube videos is to start with a very lightweight conceptual model, which will be extended once the game will have generated a wealth of data. From the ontology, we derived a set of challenges that are posed to users. This game is online for playing by the general public at http://www.ontogame.org. We are currently integrating the ontology produced by the first scenario in this game. We are also considering how we could exploit the existing YouTube tags.

Describing Video Content

Both players are shown the first ten seconds of a randomly chosen YouTube video with the option to view further parts of the video. Then they are presented with challenges: each time the players agree on an answer based on a predefined ontology, they are taken to the next level. Again, the number of points players can earn increases with the number of mastered levels. Also, the total amount of time available is two minutes.

Fig. 4. Annotating YouTube: Level 1   **Fig. 5.** Annotating YouTube: Level 2

The set of challenges for each video is:

1. The video is: Non Fiction or Fiction.
2. The video's color is: black/white or color.
3. The video's genre can be best described as: {list of 27 genres ranging from action over drama western as used by IMDB}
4. Generally, the video is about: {set of topics; the players can take multiple guesses}
5. The language of the video is: {list of languages including option "no language"}
6. The location of the video is: {set of countries and locations; the players can take multiple guesses}
7. The time period the video plays in is: {users enter the earliest and latest covered year or decade}
8. The video was produced: by a private person or by a company.

---

[10] http://www.chiariglione.org/MPEG/standards/mpeg-7/mpeg-7.htm
[11] http://www.imdb.com

Example

The first video presented to Alice and Bob is a video where Tim Berners-Lee is speaking about the Semantic Web. They agree on that the video is non-fiction (+10 pts.). On the next level, they quickly agree that the video is color (+10 pts.). Next, they consensually choose "Scientific Talk" as genre (+30pts.). On the following level, Alice selects that the video is about Tim Berners-Lee while Bob selects "Web" (none of them can see the other's guesses). Next, they both enter "Semantic Web" and get 40 points. After specifying the language of the video as English, they can not reach agreement on the location and thus decide to skip and go to the next video. (This continues until the time of two minutes is up.)

4.3 Mapping UNSPSC and eCl@ss

UNSPSC and eCl@ss are the two most important categorization standards for products and services, and establishing mappings between them for achieving data interoperability is one of the long-lasting target applications of semantic technology [21]. In this game scenario (Fig. 6), we have humans weave a net of semantic alignments between classes in both standards. Players are faced with a randomly chosen class from UNSPSC, as well as a set of possible relations, and the eCl@ss tree. In each step, the players have to agree on a class from eCl@ss and the kind of relation between the UNSPSC class the eCl@ss one. Before choosing a branch in eCl@ss, players can open the branch and see sub-classes in order to get a better understanding of the branch they are choosing. Choosing multiple classes is allowed. As described in an earlier section, we use the matching relations "same as", "narrower than", and "partly overlapping with".

Fig. 6. Mapping UNSPSC and eCl@ss   **Fig. 7.** Annotating eBay with eCl@ss

4.4 Annotating eBay with eClassOWL

The objective of this game (see Fig. 7) is to annotate offerings in eBay auctions with the product categories and product properties in eClassOWL [22]. For this purpose, we randomly select eBay auctions and present them to the players. The players are provided with a tree view of the eClassOWL ontology. Similar to the scenario where Wikipedia articles where annotated with the Proton ontology, the players have to choose a class from eClassOWL and reach agreement on this choice. In most cases, classes on a high level of eClassOWL will have several sub-classes, where the first step is repeated: the deeper the players manage to get in the hierarchy, the more points they are awarded. In many cases, it will increase gaming fun and quality of results when the system comes up with a suggestion for a branch of eClassOWL that is likely to fit. Therefore, we will investigate in how far we can (1) use matching algorithms in the background and (2) make use of the eBay category system. Also, the attributes of eClassOWL may be considered in future games.

5 Evaluation

While the last three scenarios are currently prototypes with still limited scalability, scenario 1 has been made available to the general public recently. In the following, we summarize our evaluation of the approach based in this scenario. First, we checked the data produced by the game for ontological correctness. Second, we conducted interviews among all participants who played the game in order to find out about the fun factor of the game.

5.1 Methodology

We invited 33 individuals in 5 groups with different backgrounds and asked them to play OntoGame for a duration of between 10 and 20 minutes. We asked each group to play at the same time to ensure that there were enough players to play OntoGame. Each individual of one group was asked to play separately in order to evaluate the single player mode and to verify the results of already played games. Only very few had experience with building ontologies due to their professional background (research). During most experiments, the participants were in different rooms and did not communicate with each other during playing. In two cases, the groups were in one room. However, we supervised the experiments and made sure they did not communicate with each other. The game was explained to the participants briefly before playing it online.

All of the games were logged. After the game, we interviewed participants about their experiences with the game and analyzed the output of the recorded games. They were asked the following questions:

1. Were the rules of the game hard to understand?
2. How do you rate the challenge of the game? ("OK", "too easy", "too hard")
3. Was it fun to play the game?
4. What did you especially like/dislike about the game?
5. Would you play it again?
6. General feedback.

5.2 Results

The results of our preliminary evaluation are encouraging, as summarized below.

Quality of Results: 27 individuals actually played the game. 170 Wikipedia articles were played by different players in 825 games, i.e. some pages were played multiple times. Players decided to skip directly and proceed to the next article in about 11% of the games. We took the remaining 733 games as a basis for our evaluation and

Table 4. Summary of Results

	Criterion	Number	Percent
1	General		
1.1	Number of Wikipedia pages that were played at least once	170	-
1.2	Total number of challenges played	825	-
1.2	Challenges that were not skipped and actually played	733	88.85%
2	Consensus		
2.1	Challenges in which only the first task was completed consensually	147 of 733	20.05%
2.2	Challenges in which both tasks where completed consensually	586 of 733	79.95%
2.3	Challenges in which both tasks were completed consensually, and the consensus was at the leaf level of Proton	405 of 733	55.25%
3	Conceptual Quality of the Consensual Solutions		
3.1	Amount and ratio of challenges in 2.1 of which the consensual choice for only task 1 was correct	142 of 147	96.60%
3.2	Amount and ratio of challenges in 2.2 of which the consensual choice for tasks 1 AND 2 were correct	581 of 586	99.15%
3.3	Amount and ratio of challenges in 2.3 of which the consensual choice for tasks 1 AND 2 were correct	404 of 405	99.75%
4	Mistakes		
4.1	Total of wrong choices	10	-
4.2	Wrong judgment of ontological nature	5	-
4.3	Wrong abstraction	5	-

analyzed (1) how many were correct regarding the choice class vs. instance and (2) regarding the abstraction in Proton, and (3) how many and (4) which mistakes were made (Table 4). For this purpose, we manually analyzed the data generated by the games.

Excluding those challenges that were skipped immediately (n=92, 11%), our players were able to agree on both the ontological nature and a Proton class in almost 80% of the cases (n=586). Of these tasks that were completed consensually more than 99% (n=581) were semantically correct. Of the challenges for which the player agreed on class vs. instance only (n=147 of 733), 99% of choices (n=142) were correct. In a nutshell, we can see that if consensus is reached, it largely represents correct choices.

Only a marginal amount of the consensual choices games were conceptually wrong. The following mistakes were made:

Class vs. of instance: In one case, players classified an article as a class while it was an instance (a person). Four Wikipedia articles were categorized as instances while they were classes. We are aware that the judgment whether the dominant ontological role of a conceptual entity is a class or an instance is sometimes subjective.

Wrong abstraction in Proton: In the remaining cases, the teams chose wrong abstractions in Proton, i.e. a park was classified as *abstract* while it is a *location* or a bank classified as a *service* while it is an *organization*.

While this tentative assessment is encouraging, it is currently a very preliminary evaluation. In particular, the extremely high conceptual reliability may have been caused by a substantial amount of single-player games which used recorded game-scenarios. Since the amount of recorded game-scenarios was initially small, the share of correct solutions based on us researchers playing the game may have been higher than in a large-scale deployment. However, a more comprehensive analysis is already in preparation and in principle confirms the first assessment.

Fun Factor: We received very positive feedback from the participants: surprisingly, those without any background in computer science enjoyed playing the game especially. In earlier experiments many participants experienced problems to grasp the distinction between class and instance caused troubles. Therefore, we changed the descriptions in the game to make it more understandable. Almost all participants confirmed that the rules of the game were easy to grasp.

More than 80% found the game challenging enough, all of them described the time pressure and the variety of concepts in the Proton branches as challenging. Four participants found it too easy. Six participants mentioned that in the beginning the game was too hard when one does not know the Proton ontology. Furthermore, they indicated that abstract Wikipedia articles were hard to classify. However, they also indicated that they enjoyed learning the Proton ontology and hence increased their playing pace. 21 players liked the game and said it was enjoyable to play. Six found it neither especially exciting nor especially boring. Two said that they found it boring. 19 stated that they would play the game again. Seven participants mentioned that they liked making sense of a rather short excerpt of the Wikipedia article. The majority described the second step of the game, i.e. matching the article to a Proton class, as the most fun part of the game.

Almost all participants enjoyed playing with a human counterpart and liked the consensus component of the game. Two mentioned that they would have liked to know who they were playing with. We are therefore working on a functionality that gives an additional reward to the players in form of information that is revealed about their partner (e.g. gender or nationality). Fifteen participants perceived the ranking of players displayed in the beginning of each game as a motivation to further improve their abilities and thus status in the system.

6 Conclusion and Outlook

In this, paper we proposed to masquerade the core tasks of weaving the Semantic Web behind online, multi-player game scenarios, in order to create proper incentives for humans to get involved. We presented game scenarios that in combination have the potential to increase the amount of ontologies, annotations, and alignment data in the Semantic Web substantially. If only 1,000 individuals in the world will play our games 1 hour per day for three months, this will mean 90,000 hours of volunteer work; something that would otherwise cost about a million euro at an hourly rate of 11 euro – and few experts in the Semantic Web will work for 11 euro per hour.

Based on the analysis of user data and interviews with players, we provide preliminary evidence that users enjoy the games and are willing to dedicate their time to those games and are able to produce high-quality conceptual choices.

Each of the scenarios addresses a real-world problem: Annotating Wikipedia does not only help to learn Proton and learn new topics from randomly selected Wikipedia pages, but it will allow help extend Proton to make it one of the biggest domain ontologies in the world. Annotating video content will make the vast amount of content for entertainment and education available at video portals such as YouTube accessible to search at the semantic level. Using games for creating alignments between eCl@ss and UNSPSC has the potential to mitigate one of the most substantial data interoperability problems in the product data domain. Annotating eBay offerings with references to eClassOWL will help make the vision of Semantic Web-based e-commerce a reality. Please play OntoGame @ www.ontogame.org, and help weave the Semantic Web!

Acknowledgments. We would like to thank Werner Huber, Michael Waltl, and Roberta Hart-Hilber. The work presented has been funded by the Austrian BMVIT/FFG under the FIT-IT Semantic Systems project myOntology (grant no. 812515/9284).

References

[1] Wagner, C.: Breaking the Knowledge Acquisition Bottleneck Through Conversational Knowledge Management. Information Resources Management Journal 19(1), 70–83 (2006)
[2] Von Ahn, L.: Games with a Purpose. IEEE Computer 29(6), 92–94 (2006)
[3] Von Ahn, L., Blum, M., Hopper, N., Langford, J.: CAPTCHA: Using Hard AI Problems for Security. In: International Conference on the Theory and Applications of Cryptographic Techniques, Warsaw, Poland. LNCS, Springer, Heidelberg (2003)
[4] Hotho, A., Jaeschke, R., Schmitz, C., Stumme, G.: BibSonomy: A Social Bookmark and Publication Sharing System. In: Conceptual Structures Tool Interoperability Workshop at the Conference on Conceptual Structures, Aalborg University Press (2006)
[5] Mika, P.: Ontologies are us: A unified model of social networks and semantics. In: Gil, Y., Motta, E., Benjamins, V.R., Musen, M.A. (eds.) ISWC 2005. LNCS, vol. 3729, pp. 522–536. Springer, Heidelberg (2005)
[6] Marlow, C., Naaman, M., Boyd, D., Davis, M.: Tagging, Taxonomy, Flickr, Article,ToRead. In: World Wide Web Conference (WWW 2006), ACM, Edinburgh, Scotland (2006)
[7] Von Ahn, L.: Peekaboom: A Game for Locating Objects in Images. In: Conference on Human Factors in Computing Systems (CHI 2006), ACM, Montreal, Canada (2006)
[8] Von Ahn, L., Dabbish, L.: Labeling Images with a Computer Game. In: Conference on Human Factors in Computing Systems (CHI 2004), ACM, New York (2004)
[9] Von Ahn, L., Ginosar, S., Kedia, M., Liu, R., Blum, M.: Improving Accessibility of the Web with a Computer Game. In: Conference on Human Factors in Computing Systems (CHI 2006), ACM Press, Montreal, Canada (2006)
[10] Von Ahn, L., Kedia, M., Blum, M.: Verbosity: a game for collecting common sense facts. In: Conference on Human Factors in Computing Systems (CHI 2006), ACM, Montreal, Canada (2006)
[11] Law, E., von Ahn, L., Dannenberg, R., Crawford, M.: Tagatune. In: International Conference on Music Information Retrieval (ISMIR 2007), Vienna, Austria (2007)
[12] Lieberman, H., Smith, D., Teeters, A.: Common Consensus: A Web-based Game for Collecting Commonsense Goals. In: Workshop on Common Sense for Intelligent Interfaces, ACM Conference on Intelligent User Interfaces (IUI 2007), Honolulu (2007)
[13] Lenat, D.B., Guha, R.V.: Building Large Knowledge-based Systems: Representation and Inference in the Cyc Project. Addison-Wesley, Boston (1990)
[14] Siorpaes, K., Hepp, M.: OntoGame: Towards Overcoming the Incentive Bottleneck in Ontology Building. In: International IFIP Workshop On Semantic Web & Web Semantics (SWWS 2007), OTM conferences. LNCS, Springer, Vilamoura, Portugal (2007)

[15] Euzenat, J., Shvaiko, P.: Ontology Matching. Springer, Heidelberg (2007)
[16] W3C, Simple Knowledge Organisation System (SKOS), http://w3.org/2004/02/skos/
[17] Hepp, M.: Possible Ontologies: How Reality Constrains the Development of Relevant Ontologies. IEEE Internet Computing 11(7), 96–102 (2007)
[18] Broekstra, J., Kampman, A., Van Harmelen, F.: Sesame: A Generic Architecture for Storing and Querying RDF and RDF Schema. In: Horrocks, I., Hendler, J. (eds.) ISWC 2002. LNCS, vol. 2342, Springer, Heidelberg (2002)
[19] SEKT Consortium, PROTON Ontology, http://proton.semanticweb.org
[20] Hepp, M., Siorpaes, K., Bachlechner, D.: Harvesting Wiki Consensus: Using Wikipedia Entries as Vocabulary for Knowledge Management. IEEE Internet Computing 11(5), 54–65 (2007)
[21] Schulten, E., Akkermans, H., Guarino, N., Botquin, G., Lopes, N., Doerr, M., Sadeh, N.: The E-Commerce Product Classification Challenge. IEEE Intelligent Systems 16(4), 86–89 (2001)
[22] Hepp, M.: eCl@ssOWL, http://www.heppnetz.de/eclassowl/

SWING: An Integrated Environment for Geospatial Semantic Web Services

Mihai Andrei[1], Arne Berre[2], Luis Costa[2], Philippe Duchesne[3], Daniel Fitzner[4], Miha Grcar[5], Jörg Hoffmann[6], Eva Klien[4], Joel Langlois[7], Andreas Limyr[2], Patrick Maue[4], Sven Schade[4], Nathalie Steinmetz[6], Francois Tertre[7], Laurentiu Vasiliu[1], Raluca Zaharia[1], and Nicolas Zastavni[3]

[1] DERI Galway, Ireland
⟨first⟩.⟨last⟩@deri.org
[2] SINTEF Group, Oslo, Norway
⟨first⟩.⟨last⟩@sintef.no
[3] IONIC Software, Liege, Belgium
⟨first⟩.⟨last⟩@ionicsoft.com
[4] University of Münster, Germany
⟨first⟩.⟨last⟩@uni-muenster.de
[5] Jozef Stefan Institute, Ljubljana, Slovenia   miha.grcar@ijs.si
[6] STI Innsbruck, Austria
⟨first⟩.⟨last⟩@sti2.at
[7] BRGM, Orleans, France
⟨first-initial⟩.⟨last⟩@brgm.fr

Abstract. Geospatial Web services allow to access and to process Geospatial data. Despite significant standardisation efforts, severe heterogeneity and interoperability problems remain. The SWING environment[1] leverages the Semantic Web Services (SWS) paradigm to address these problems. The environment supports the entire life-cycle of Geospatial SWS. To this end, it integrates a genuine end-user tool, a tool for developers of new Geospatial Web services, a commercial service Catalogue, the Web Service Execution Environment platform (WSMX), as well as an annotation tool. The demonstration includes three usage scenarios of increasing complexity, involving the semantic annotation of a legacy service, the semantic discovery of a Geospatial SWS, as well as the composition of a new Geospatial SWS.

1 Introduction

Geospatial Web services provide access to, and processing functions for, Geospatial data. The need for sharing and processing such data on a large scale has lead to significant standardisation efforts by the *Open Geospatial Consortium (OGC)* and the Technical Committee 211 of the International Organisation for Standardisation (ISO/TC 211). The OGC drives standardisation efforts regarding Geospatial languages such as the Geographic Markup Language (GML)[1], GeoRSS, or the Keyhole Markup Language (KML)[2]. Also, standardised interfaces for certain types of Geospatial Web services have been

---

[1] Developed in the SWING project (Semantic Web services INteroperability for Geospatial decision making), funded by the European Commission (FP6-26514).

fixed, such as the *Web Feature Service (WFS)* which provides access to a Geospatial database. However, these specifications are purely syntactic, and so severe heterogeneity and interoperability problems remain, with little support for resolving them. The WFS protocol is for example syntactic in the way that the optional meta-information that can be provided to describe the geo-information (features) is not formalized. The SWING environment leverages the *Semantic Web Services (SWS)* paradigm to address these problems.

The SWING environment supports the entire life-cycle of Geospatial SWS; to this end, it integrates: a genuine end-user tool (called *MiMS*); a tool (*DEV*) supporting the composition of new Geospatial Web services from existing ones; a commercial service Catalogue (*CAT*); a state-of-the-art SWS platform (*WSMX*); and a tool (*ANNOT*) that helps with the creation of the semantic annotations. Interconnecting all these disparate components is a challenge in itself. Our solution shows how all the issues regarding the required interplays can be resolved, and hence how Geospatial SWS can be realized. Also, various contributions are made regarding base technologies, such as Geospatial SWS ontologies, and particular methods for semantic annotation and discovery.

Section 2 overviews the SWING environment and outlines our technical contributions. Section 3 explains what will be demonstrated at ESWC'08. Section 4 wraps up.

2 The SWING Environment

Figure 1 illustrates the SWING environment, in terms of its components and their intuitive relations [3]. MiMS is the environment for the Geospatial domain expert, i.e., the end-user who will need access to Geospatial data; MiMS is a genuine tool used at the French Geological Survey (BRGM), the leading institution in France for Geospatial decision making. DEV is a UML modelling tool adapted for the Geospatial domain, enabling human developers to conveniently compose new Geospatial Web services from existing ones. Both MiMS and DEV require the ability to discover existing Geospatial SWS, and to semantically annotate services (legacy services in MiMS, new composed services in DEV) for later use. These functionalities are supported by the CAT, WSMX, and ANNOT components; they interface to MiMS and DEV via the query annotation GUI (*Query Annot*) for discovery queries, and via the service annotation GUI (*Service Annot*) for annotating services. Query Annot accesses ANNOT for helping to construct the semantic part of the discovery query; Service Annot does the same for the semantic annotation. ANNOT is a tool developed especially for the SWING environment, employing term matching techniques and ontology structure analysis to help map a natural language description of a query/service into a semantic query/annotation. CAT is an adaptation of a commercial Geospatial Web service Catalogue, handling service storage and discovery. For executing a discovery query, CAT performs the spatial part of the query itself, and then calls WSMX [4] for semantic matching. New Geospatial Web services composed in DEV are automatically exported into the orchestration format understood by WSMX; these orchestrations are registered and discovered in CAT just like any other Geospatial SWS; when executing an orchestration, the *WFS Wrapper* component sits between WSMX and MiMS so that, from the end-user point of view, the composed service exposes a standard OGC interface. In this way, WSMX technology can be seamlessly integrated into existing Geospatial Web service environments.

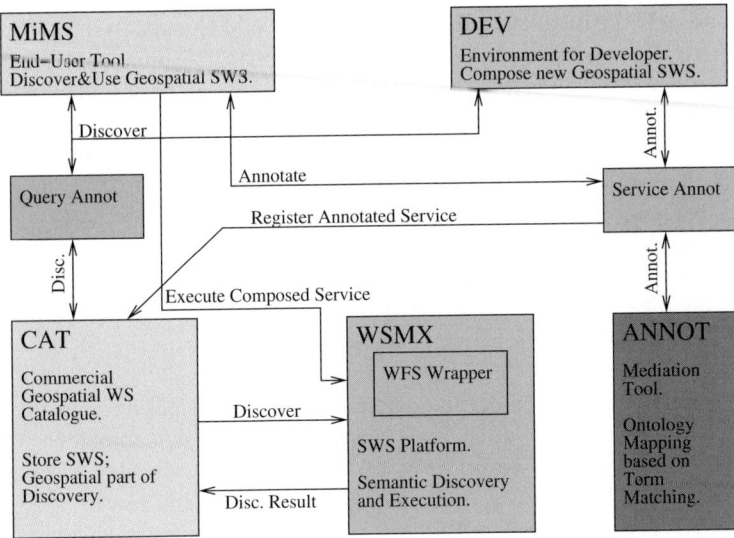

Fig. 1. A schematic overview of the SWING environment

It should be clear that the integration of such a diverse set of components involves significant engineering issues. Also, the construction of such a complex environment involves numerous design decisions, which we cannot describe in detail for lack of space.

Apart from the methodological and software contributions made by the SWING environment, many of its elements are technical contributions in themselves; in particular:

– **Geospatial ontologies.** We created WSML [5] ontologies formalising many aspects of (the exploitation of) mineral resources. The necessary domain knowledge was acquired from domain experts at BRGM [6].
– **Semantic annotation of WFS and WPS.** We designed new strategies to annotate and discover WFS and *Web Processing Services (WPS)*; In difference to previous approaches [7,8] our strategies are light-weight in that they use query containment in Datalog (in WSML-Flight [5], to be precise) for the required matching and keep the annotations simple. In addition, we designed techniques to automatically suggest semantic annotations by mapping natural language statements into our ontologies [9]. A GUI supports the selection of a corresponding ontology sub-graph. It is automatically transformed into the required WSML-Flight expression.
– **Enhancement of geospatial catalogue discovery capabilities.** The catalogue interface is used in an alternate way by exchanging the keyword set for the thematic search with more structured WSML goals [10]. Furthermore the functionality of the catalogue is enhanced by delegating the thematic search to a WSMX server.
– **Convenient UML modelling of composed Geospatial Web services.** DEV allows to orchestrate existing Web services into new ones at UML level [3]. Support is provided, e.g., for integration issues regarding WFS: being an entry point into a database, every WFS provides a rich set of possible outputs, with non-uniform structure and naming. We resolve this using semantic annotations.

- **Exporting UML into semantic orchestrations.** WSMX supports semantic orchestrations by *Abstract State Machines (ASM)*, which operate on semantic data and are very flexible in orchestration specification and execution. We designed an automatic transformation from UML orchestrations into ASMs.
- **Executing semantic orchestrations.** ASM execution is a very powerful and flexible mechanism which relies heavily on reasoning and hence pays a computational prize. By a number of enhancements to ASM execution in WSMX, we obtained speed-ups of several orders of magnitude.

3 Demonstration

The demonstration focuses on the "big picture" of the integrated environment, showing how users interact with it, in three usage scenarios of increasing complexity; technical issues are highlighted on the way. The first scenario shows how a MiMS user annotates a legacy Geospatial Web service via an interface to ANNOT, and how the service is registered in CAT. As an example, a WFS that provides information on quarry boundaries and aggregate production rates of quarries is annotated. The second scenario shows how a MiMS user creates a semantic discovery query supported by ANNOT, and how that query is executed in an interplay of CAT and WSMX. In this step, the previously annotated quarry WSF is discovered. The third scenario shows how a DEV user composes a new WFS from existing Web services, with a graphical modelling tool; the new WFS is annotated, registered, and exported into an ASM for execution by WSMX; the MiMS user discovers the new WFS, and the WSMX execution is hidden behind a standard OGC interface. The quarry WFS is used as part of the composition.

Consider the third and most complex scenario. A MiMS user needs a WFS providing aggregated mineral resource consumption/production data for a particular area in France [11]. This is a classical business use case including complex systems with complex Geospatial data models. No WFS can be found that delivers the desired data; so a developer in DEV must create such a WFS. The developer uses Query Annot, the semantic discovery interface, to discover several WFS delivering either consumption or production data. The previously mentioned quarry WFS is used to serve production data. In each discovery step, a short natural language statement suffices to select concepts and relations from the domain ontology, from which the required WSML-flight query is automatically created. After discovering the services, the developer combines them. A particular detail highlighted by the demonstration is the following. A major source of complications when integrating several WFS is the need to integrate data from diverse WFS with non-uniform output structure and naming. This issue is addressed by means of semantic annotations: the DEV user identifies, only once,the desired domain ontology concept; the correct output of each WFS is then automatically selected based on its annotation. For non-annotated WFS, a term matching tool can directly be used to find the output attribute that is the best match for the desired concept.

Once the graphical model is complete, the developer pushes a button to export it into an ASM. The service is then annotated using Service Annot, selecting concepts and relations from the domain ontology similarly as in the discovery interaction explained above. Now the new WFS is discovered in MiMS, and the MiMS user invokes it just like any other WFS. The invocation is handled by the WFS Wrapper component. Towards MiMS, this acts like a standard OGC interface; it communicates to WSMX the parameter values needed for the ASM execution, and it extracts from the outcome of the execution the data desired by MiMS. The demonstration shows the WFS Wrapper at work and, if desired, the trace of the ASM execution that is behind it.

4 Summary

The SWING environment comprises support for the entire life-cycle of Geospatial semantic Web services. From a Semantic Web perspective, this is a detailed case study of putting semantics to use; the case study is important due to the omni-presence of Geospatial data in human society. The SWING environment shows how the diverse involved components can be integrated. This is best presented, by far, in the form of a system demonstration illustrating the use of the environment as a whole.

In contrast to previous affords in connecting Geospatial Web services with Semantic Web technology like [12,13], SWING aims at an integrated framework for annotation, development, discovery and execution of Geospatial SWS. The demonstration represents a related best practice. A second major difference to other approaches is the use of WSML, specifically of the variant WSML-Flight. WSML-Flight offers alternate ways for service annotation, ontology formalisation and discovery

References

1. OGC: Open Geospatial Consortium, Inc., Geography Markup Language (GML) Encoding Specification (GML) Version 2.1.2 (2002)
2. OGC: Open Geospatial Consortium, Inc., KML 2.2: An OGC Best Practice. Best Practice Paper (2007)
3. Hoff, H., et al.: D6.1 The Architecture of the Development Environment. Deliverable of the SWING Project (2006)
4. Fensel, D., et al.: Enabling Semantic Web Services: The Web Service Modeling Ontology. Springer, Heidelberg (2006)
5. de Bruijn, J., Lausen, H., Polleres, A., Fensel, D.: The web service modeling language: An overview. In: Sure, Y., Domingue, J. (eds.) ESWC 2006. LNCS, vol. 4011, pp. 590–604. Springer, Heidelberg (2006)
6. Klien, E., Schade, S., Hoffmann, J.: D3.1 Ontologies in the SWING Application - Requirement Specification. Deliverable of the SWING Project (2007)
7. Lutz, M., Klien, E.: Ontology-based retrieval of geographic information. International Journal of Geographic Information Science 20(3), 233–260 (2005)
8. Lutz, M.: Ontology-based descriptions for semantic discovery and composition of geoprocessing services. Geoinformatica (2006)
9. Grcar, M., et al.: D4.1 Representational language for Web-service annotation models. Deliverable of the SWING Project (2006)
10. Duchesne, P., Zastavni, N.: D5.1 Online services. Deliverable of the SWING Project (2007)
11. Langlois, J., et al.: D1.1 Use Case Definition and I&T Requirements. Deliverable of the SWING Project (2007)
12. Klopfer, M., Kanellopoulos, I.: Orchestra - an open service architecture for risk management (2008)
13. Lemmens, R.: Semantic interoperability of distributed geo - services. PhD thesis, Nederlandse Commissie voor Geodesie (NCG), Delft (2006)

Semantic Annotation and Composition of Business Processes with Maestro*

Matthias Born[1], Jörg Hoffmann[2], Tomasz Kaczmarek[3], Marek Kowalkiewicz[1], Ivan Markovic[1], James Scicluna[2], Ingo Weber[1], and Xuan Zhou[1]

[1] SAP Research, Karlsruhe, Germany
{mat.born,marek.kowalkiewicz,ivan.markovic,ingo.weber,xuan.zhou}@sap.com
[2] STI Innsbruck, Austria
{joerg.hoffmann,james.scicluna}@sti2.at
[3] Poznan University of Economics, Poland
t.kaczmarek@kie.ae.poznan.pl

Abstract. One of the main problems when creating execution-level process models is finding implementations for process activities. Carrying out this activity manually can be time consuming, since it involves searching in large service repositories. We present Maestro for BPMN, a tool that allows to annotate and automatically compose activities within business processes. We explain the main assumptions and algorithms underlying the tool, and we overview what will be demonstrated at ESWC.

1 Introduction

One of the biggest challenges within Service Oriented Architectures (SOA) is the composition of different Web services, achieving a higher utility. While in the realm of Web services such a combination is usually presented as a new service, we focus on how such compositions can be used as partial implementations of business processes, which is one of the key aspects of the SUPER project.

Web services need to be formally annotated in order for tools to automatically compose them into an orchestration (defining the control flow between them). We tackle the composition problem as a key part of the question on how to implement a business process with a set of given Web services. Business processes are often modeled as a set of activities (or tasks) together with their control flow. In order to make a process executable, e.g., in a workflow execution engine, all tasks in the process have to be carried out manually or automatically by Web services. Our aim is to semantically annotate such tasks and to automatically discover or compose (if needed) the services which collectively implement the required functionality. A business process modeling tool serves as a main user interface for performing these activities. This article focuses on the extensions made to *Maestro for BPMN*, a modeling tool from SAP Research. These

---

* This work is partly funded by the EU 6th Framework Programme, within Information Society Technologies (IST) under the SUPER project (http://www.ip-super.org). Thanks also to Alina Dima, Florian Dörr, and Mario Karrenbrock for their coding support and Christian Drumm and Christian Brelage for their advice.

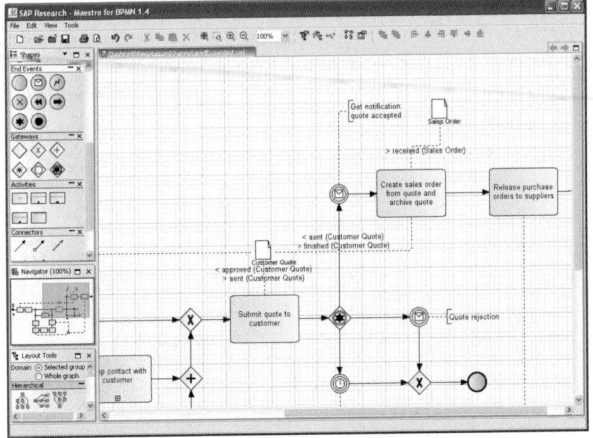

Fig. 1. A fragment of a process model represented in Maestro for BPMN

extensions enable semantic annotation of the business process and automatic discovery and composition of Web services for this process.

In the last years, Business Process Modeling Notation (BPMN) has received wide attention as a graphical representation of business process models. Within SUPER, business processes are stored as ontology descriptions. The *sBPMN* ontology[1][1] serves as a meta model for BPMN process models, featuring the concepts, relations and attributes for standard BPMN. This ontology has been extended, mainly featuring the ability to define a state of the process before and after execution of successive activities. With these extensions we can derive semantic goal descriptions for activities – i.e., formal descriptions of the functionality which an implementation of a particular task needs to perform. This is in line with most popular approaches to Semantic Web Service description[2] where Web services can be annotated with pre and postconditions.

Handling semantic descriptions for activities was one of the features added to Maestro for BPMN. We also equipped it with the ability to call a composition tool with the annotated tasks as input[3] and integrated its output back to the user interface provided by the modeling tool.

Within the composer, we tackle some of the interesting opening issues in the Semantic Web Service Composition (WSC) area. We define a formal framework for WSC, inspired by A.I. Planning methodologies [9,3]. We consider plug-in matches, where services do not have to match exactly, but have to be able to connect in all possible situations. In particular, we take the background ontology into account during the composition process; in contrast, many existing works assume exact matches (of concept names). The distinguishing feature of our work on composition is that we explore restrictions on the background ontology in order to find a solution (i.e., a composition) efficiently.

---

[1] sBPMN is written in WSML (http://www.wsmo.org/TR/d16/d16.1/v0.21/)
[2] Followed for example in WSMO (http://www.wsmo.org)
[3] Note that the annotated tasks will in effect be equivalent to a WSMO Goal.

2 Process Modeling

From the graphical point of view, Maestro for BPMN follows BPMN. However, it makes use of the sBPMN ontology, by creating on-the-fly a set of instances for sBPMN classes. If a new BPMN task is created on the drawing pane, an instance of the concept *Task* is created in the in-memory working ontology. This enables supportive reasoning over the working ontology. The underlying conceptual work on the ontology and design choices are documented in [2].

The main goal of the tool extensions is to allow a user-friendly semantic annotation of process models. This is achieved by allowing to link semantically expressed process activities to a domain ontology. We focus on how process activities manipulate business objects in terms of their life cycles. E.g., a task "Send offer" sets the status of the object "Offer" to the state "sent". For this purpose, the domain ontology needs to specify the business objects of interest together with their life cycles[2]. This technique enables the user to define formal pre and postconditions of tasks in a human-friendly way.

For creating such links, we implemented matchmaking methods that filter the domain ontology based on the process context and rank the concepts to include the pre/postconditions. The textual descriptions of tasks (or other elements) are matched against the entities of interest in the domain ontology using linguistic methods, such as the edit distance between strings. E.g. if a task has label "Send offers", then the object "offer" from the domain ontology may be suggested as a top match. Another way to restrict the set of matches is by employing the process structure, e.g., by not suggesting the same activity twice or by comparing the process control flow to the object life cycle. The extensions made are conceptually independent of the tool chosen, and could be ported to other modeling notations.

3 Task Discovery and Composition

As a first step in finding process task implementations, we try to discover a single Semantic Web Service (SWS) for each annotated task. To achieve this, we check if the concept from the domain ontology describing a SWS matches the concept used for annotating the task. We follow a matching technique proposed in [8], analysing intersection of ontological elements in service descriptions and rating two descriptions as relevant whenever they specify an overlapping functionality. For that, we use standard reasoning task of *concept satisfiability* of a conjunction between the concepts taken from the task and Web service descriptions [7].

If a Web service cannot be found, WSC is performed. This is computationally hard and has two main sources of complexity: (i) combinatorial explosion of possible compositions, and (ii) worst-case exponential reasoning. We tackle (i) using heuristic search - a well known technique for dealing with combinatorial search spaces. We address (ii) by trading off expressivity of the background ontology against efficiency, i.e., we investigate restricted classes of ontologies allowing reasoning to be performed in polynomial time. Problem (ii) is closely related to the notion of "belief updates" in A.I. We define a clear formal model that combines these notions and those from planning techniques, following recent formalisations of WSC [6,4] and use heuristic methods for efficient searching [5]. One of our

$s_0 := \textit{reasoning-startstate}(); (h, H) := \textit{heuristic-function}(s_0);$ open-list $:= \langle(s_0, h, H)\rangle;$
while TRUE **do**
    $(s, h, H) := \textit{remove-front}(\text{open-list});$
    if $\textit{is-solution}(s)$ **then return** path leading to s;
    for all applicable calls a of SWS in H **do**
        $s' := \textit{reasoning-resultstate}(s, a);$
        $(h', H') := \textit{heuristic-function}(s');$
        insert-ordered-by-increasing-h(open-list,s',h',H');

Fig. 2. The main loop of our WSC algorithm

results is a polynomial time reasoning over background ontologies with Binary Clauses. For Horn Ontologies, we use an approximate update-reasoning technique that still runs efficiently but sacrifices some precision, preserving either soundness or completeness. The restricted ontologies allow to describe (amongst others) subsumption hierarchies, cardinality bounds and image type restrictions. Other features (such as QoS) are part of our ongoing work.

The main algorithm of the composer is shown in Fig. 2. The algorithm performs a forward search in a space of states s corresponding to different situations during the execution of the various possible compositions. The key elements are the *reasoning-startstate*, *reasoning-resultstate*, and *is-solution* procedures - maintaining the search states and detecting solutions - and the *heuristic-function* procedure - taking a state and returning a solution distance estimate h as well as a set H of promising Web services by solving a relaxed version of the problem. The states are ordered by increasing h, which is a standard method called "best-first-search". The set H is used for *filtering* the explored SWS calls. Filtering is widely perceived to be essential in WSC since it "forces" to check the most promising services first, leading to a considerable speed-up of the search.

The performance of our WSC tool was tested on two testbeds: the Telekomunikacja Polska (TPSA) which defines how a service (e.g. VoIP) is created for a new customer and the Virtual Traveling Agency (VTA) whereby the user specifies the kind of services that she/he would like for a trip (such as flight and hotel). The composer was set up in different configurations: *Blind* uses neither h nor H; *Heuristic* uses only h; *Filtering* uses only H; *Full* uses both. The results are plotted in Fig. 3, showing how runtime scales over the number of available services N; N was increased by generating additional services through randomized modifications of the original services.

4 Demo Scope

The demo will show an example of a realistic Business Process (Fig. 1). We will first demonstrate how data objects are associated with tasks (annotation) and how the states of these objects can be attached to the pre and postconditions of the task. Discovery is then used to find a Web service that fulfils that particular task. We will also show a task for which composition is required (rather than discovery). The annotated task will serve as the input goal to the composer. The component is run in the background and once a solution is found, the task in the

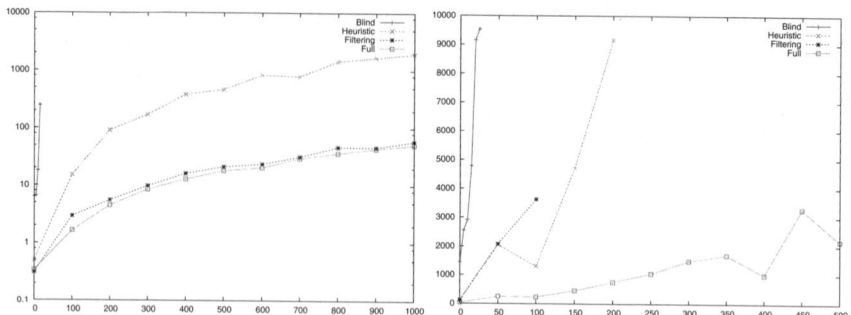

Fig. 3. Results for TPSA (left) and VTA (right), plotted as Seconds (y-axis) against N (x-axis)

process is replaced with the sequence of Web services found by the composer. The search for a solution will be performed within a large number of Web services such that the audience can clearly see the scalability of our approach.

5 Conclusion

We have presented extensions to Maestro for BPMN, which demonstrate tool support for semantic annotation of the process models, as well as task discovery and composition. This enables more agile business process development and deployment. We showed how business analysts can easily annotate process elements (in particular—tasks) and automatically find Web services that fulfil them using discovery. If the latter fails, an efficient composer tool that we developed can be used to find a chain of Web services that can adequately fulfil the task.

References

1. Abramowicz, W., Filipowska, A., Kaczmarek, M., Kaczmarek, T.: Semantically enhanced business process modelling notation. In: SBPM Workshop (2007)
2. Born, M., Dörr, F., Weber, I.: User-friendly semantic annotation in business process modeling. In: Hf-SDDM Workshop (December 2007)
3. Eiter, T., Faber, W., Leone, N., Pfeifer, G., Polleres, A.: A logic programming approach to knowledge-state planning: Semantics and complexity. Transactions on Computational Logic 5(2), 206–263 (2004)
4. De Giacomo, G., Lenzerini, M., Poggi, A., Rosati, R.: On the approximation of instance level update and erasure in description logics. In: AAAI (2007)
5. Hoffmann, J., Nebel, B.: The FF planning system: Fast plan generation through heuristic search. J. AI Research 14, 253–302 (2001)
6. Lutz, C., Sattler, U.: A proposal for describing services with DLs. In: DL (2002)
7. Markovic, I., Karrenbrock, M.: Semantic web service discovery for business process models. In: Hf-SDDM Workshop (December 2007)
8. Trastour, D., Bartolini, C., Preist, C.: Semantic web support for the business-to-business e-commerce lifecycle. In: WWW, pp. 89–98 (2002)
9. Winslett, M.: Reasoning about action using a possible models approach. In: AAAI, pp. 89–93 (1988)

Learning Highly Structured Semantic Repositories from Relational Databases: The RDBToOnto Tool

Farid Cerbah

Dassault Aviation
DPR/ESA
78, quai Marcel Dassault 92552 Saint-Cloud – France
farid.cerbah@dassault-aviation.fr

Abstract. Relational databases are valuable sources for ontology learning. Methods and tools have been proposed to generate ontologies from such structured input. However, a major persisting limitation is the derivation of ontologies with flat structure that simply mirror the schema of the source databases. In this paper, we show how the RDBToOnto tool can be used to derive accurate ontologies by taking advantage of both the database schema and the data, and more specifically through identification of taxonomies hidden in the data. This extensible tool supports an iterative approach that allows progressive refinement of the learning process through user-defined constraints.

1 Motivation

Ontology learning from relational databases is not a new research issue. Several methods and tools have been developed to deal with such structured input (e.g. [1–3]). However, a major persisting limitation of the existing methods is the derivation of ontologies with flat structure that simply mirror the schema of the source databases. For example, the DataMaster Protégé plugin [3] is a convenient tool that allows to import schema definition and data into Protégé, but the target populated models are simply based on ontologies of the relational model (such as Relational.OWL [4]). Such tools can significantly ease the transitioning task by automatically expressing legacy data into ontology representation formats. However, the results might not fully meet the expectations of users that are primarily attracted by the rich expressive power of semantic web formalisms and that could hardly be satisfied with target knowledge repositories that look like their source relational databases. A natural expectation is to get at the end of the learning process ontologies that better capture the underlying conceptual structure of the stored data.

Ontologies with flat structure is the typical result of learning techniques that exclusively exploit information from the schema without considering the data. One of the main motivations behind the RDBToOnto tool is to implement a process that allows to learn populated ontologies with rich taxonomies by exploiting both the schema and the data in the identification of the ontology structure.

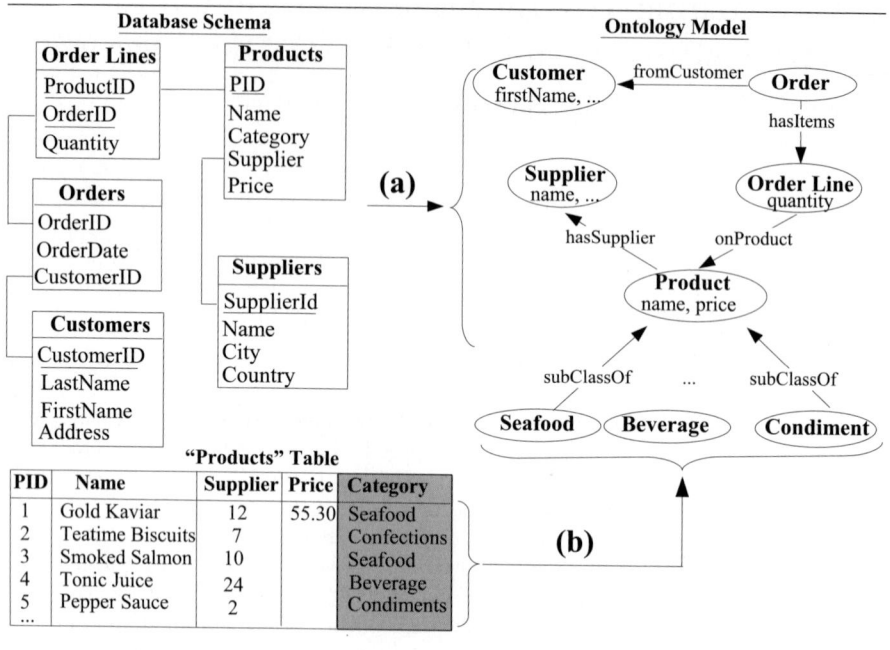

Fig. 1. Ontology model built by exploiting both the schema and the data

Additionally, a second major objective is to provide support for an iterative approach that allows progressive refinement of the learning process through user-defined constraints.

To give an illustration of how both schema definition and data can be exploited as input, let us start by depicting the typical transitioning process on an academic example. Figure 1 shows the input and the potential output of such of a process when applied on a sample database.

The derivations applied to get the target ontology can be divided in two parts. The first part, named **(a)** in the figure, includes derivations that are motivated by the identification of patterns from the database schema. In this example, each relation (or table) definition from the schema is the source of a class in the ontology. Such simple mappings from relations to classes are often relevant (though some exceptions need to be handled). Datatype properties are derived from some of the relation attributes and binary key-based associations between tables are the most reliable source for linking classes through object properties.

The derivations applied to obtain this upper part of the ontology are well covered by current methods and, if applied on this database example, most of the methods would provide the result of the **(a)** derivations as final output. However, by looking closer at the data, we can notice that the process can go further. In the Product table, additional structuring patterns can be exploited to make the ontology more accurate. More particularly, the **(b)** part of the derivations shows how the Product class can be refined with subclasses derived from the values of Category column in the source Products table. In the same vein, the Supplier class can be extended with a two-level hierarchy by interpreting the values in both Country and City columns of the corresponding table (resulting in subclasses Sweden Supplier ⟶ Stockholm Supplier, Göteborg Supplier, etc).

These are typical examples of subsumption relations that can be discovered by mining the database content. One of the key issues addressed in this work is the identification of relation attributes that may serve as good *categorisation sources*. In our ontology learning approach, it is assumed that these attributes can be revealed by combining identification of lexical cues in attribute names and entropy-based estimation of data redundancy in attribute extensions.

We give in other publications a formal description of our comprehensive learning method which takes advantage of both the database schema and the data. It is the main method implemented in the RDBToOnto tool described below.

2 The RDBToOnto Tool

RDBToOnto[1] is a highly configurable tool that ease the design and implementation of methods for ontology learning from relational databases. It is also a user oriented tool that supports the complete transitioning process from access to the input databases to generation of populated ontologies. The settings of the learning parameters and control of the process are performed through a full-fledged dedicated interface (figure 2).

A basic principle in the design of RDBToOnto is to allow the derivation of an exploitable ontology in a fully automated way. By using the tool with its default configuration, a user can get a populated ontology by simply providing as input the uri of the input database. However, it should also allow the user to iteratively refine the result. This is performed by adding local constraints. Several types of constraints are pre-defined while allowing (experienced) users to define new ones. As briefly discussed in previous section, the main learning method implemented in RDBToOnto includes data-driven mechanisms to automatically mine categorisation patterns in the database content. To further refine the ontology structure, the user can add local constraints to specify categorisation patterns that have been missed by the automated mechanisms (i.e. by selecting relevant categorisation attributes through the interface). Constraints on instance naming are also highly useful when building fine-tuned ontologies. Instead of letting the system assign arbitrary names to instances, it is possible to specify through local constraints attached to source relations how names should be derived from attribute values (e.g., to an Employees source relation, it is possible to attach a constraint specifying that instance names should be formed by combining values of FirstName and LastName attributes).

Another key improvement over existing tools is the inclusion of a database normalisation step in the supported process. It is often assumed that the ontology

---

[1] http://www.tao-project.eu/researchanddevelopment/demosanddownloads/RDBToOnto.html

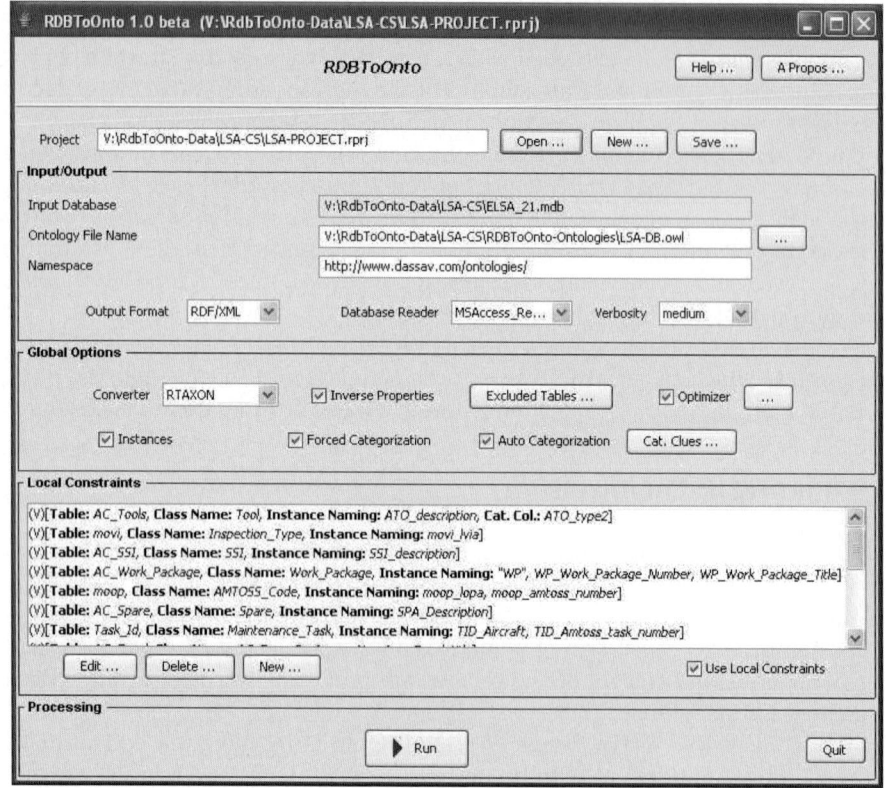

Fig. 2. The user interface of RDBToOnto. This extensible tool is designed to allow the integration of several learning methods (each method is implemented by a "converter"). A method can have its specific types of global options and local constraints.

learning process starts with well-designed databases. While theoretically acceptable, this assumption has some drawbacks in practice as many databases that are relevant for ontology learning suffer from redundancy problems. Without a proper integrated support for database normalisation, users might be tempted to directly take the databases as input even if badly designed. In RDBToOnto, main effect of the normalisation step is to eliminate data duplication in the source tables (through the interpretation of inclusion dependencies defined by the user). The model transformation performed to eliminate redundancy ultimately results in the introduction of inter-class relations (i.e. object properties).

A set of reusable components can be directly exploited to implement new methods. More particularly, database readers for some of the most common database formats are included in the tool and new ones can be integrated. Additionally, the database normalisation task is supported by a reusable component. The user interface can be extended to handle the specific constraints of new methods.

3 Evaluation

RDBToOnto has been evaluated on a set of 50 databases from different domains. One of the representative transitioning experiments performed with this tool has been conducted in the context of the TAO project[2]. In this significant case study, the input is a complex database in the domain of aircraft maintenance that includes technical descriptions of aircraft parts and all logistic resources involved in maintenance operations (spares, tools, manpower, ...). For this project, a thorough specification of the learning process has been performed resulting in 70 constraints (mostly, inclusion dependencies to optimise the model and naming constraints for classes and instances). The process produced an ontology of 600K triples corresponding to 70 classes populated with 50K instances (50 object properties with 40K instances and 120 datatype properties with 350K assigned values). The ten class hierarchies that have been discovered appeared to be relevant and the variety of some prominent concepts (such as tools and spares) are captured in these hierarchies.

4 Conclusion

We described in this paper the functionalities of RDBToOnto, a tool that implements a novel approach to ontology learning from relational databases. The prominent features of the supported approach are:

- A method that takes advantage of both database schema and content, and that can identify reliable categorisation patterns hidden in the data.
- A fully automated learning process that can be influenced through user-defined local constraints of various types.
- A database normalisation step incorporated in the implemented process that can reduce the redundancy of the source databases before ontology learning.
- A framework that eases the implementation of new methods.

References

1. Stojanovic, L., Stojanovic, N., Volz, R.: Migrating data-intensive web sites into the semantic web. In: Proc. of ACM Symp. on Applied Computing, Madrid (2002)
2. Astrova, I.: Reverse engineering of relational databases to ontologies. In: Bussler, C.J., Davies, J., Fensel, D., Studer, R. (eds.) ESWS 2004. LNCS, vol. 3053, Springer, Heidelberg (2004)
3. Nyulas, C., O'Connor, M., Tu, S.: Datamaster - a plug-in for importing schemas and data from relational databases into protégé. In: 10th Intl. Protégé Conference, Budapest (2007)
4. de Laborda, C.P., Conrad, S.: Relational.OWL: a data and schema representation format based on OWL. In: APCCM 2005: Proc. of the 2nd Asia-Pacific conference on Conceptual modelling, Darlinghurst, Australian Computer Society, Inc. (2005)

---

[2] http://www.tao-project.eu/

Cicero: Tracking Design Rationale in Collaborative Ontology Engineering

Klaas Dellschaft, Hendrik Engelbrecht, José Monte Barreto,
Sascha Rutenbeck, and Steffen Staab

Universität Koblenz-Landau, ISWeb Working Group
Universitätsstr. 1, 56070 Koblenz, Germany
{klaasd,engelbrecht,monte,srutenbeck,staab}@uni-koblenz.de,
http://isweb.uni-koblenz.de

Abstract. Creating and designing an ontology is a complex task requiring discussions between domain and ontology engineering experts as well as the users of an ontology. We present the Cicero tool, that facilitates efficient discussions and accelerates the convergence to decisions. Furthermore, by integrating it with an ontology editor, it helps to improve the documentation of an ontology.

1 Introduction

Creating and designing an ontology is a complex task that requires the collaboration of domain and ontology engineering experts. For coming to a consensual model of a domain that is expressed by an ontology, the participants in the engineering process must discuss their different viewpoints in an efficient manner. Thus, discussions are an important part of collaborative ontology engineering.

In the following, we will present the Cicero tool that has been developed in the context of the NeOn project.[1] It facilitates an asynchronous discussion and decision taking process between participants of an ontology engineering project. Two main objectives of capturing discussions in Cicero can be distinguished:

- *Higher efficiency*: Cicero supports its users in discussing the design rationale of ontologies. The whole discussion including the pro and contra arguments is recorded, leading to fewer redundancies in disputes. It has been shown that the applied discussion methodology facilitates efficient discussions and accelerates convergence to a solution.
- *Enhanced documentation*: The captured discussions reflect the design rationale of an ontology. By attaching a discussion to the entities in the ontology, it is possible later to understand why certain elements are modeled as they are. Furthermore, prior discussions can easily be resumed if e. g. new requirements have to be taken into account.

---

[1] http://www.neon-project.org/. This work has been supported by the European project *Lifecycle Support for Networked Ontologies* (NeOn, IST-2006-027595).

The first objective is accomplished by the Cicero tool itself.[2] Its underlying argumentation model and discussion workflow are described in section 3 and 4. The second objective is accomplished by integrating Cicero with an ontology editor, which is described in section 5.

2 Use Case

Examples for collaborative ontology engineering are the development processes of the AGROVOC thesaurus[3] or the Gene Ontology[4]. They are both maintained by teams of ontology engineers and domain experts that are coordinated by a central organization like FAO or the Gene Ontology Consortium respectively.

The world-wide users of the publicly available ontologies can influence their further development by e.g. proposing new terms or definitions for inclusion or by suggesting the reorganization of sections of the ontologies. The issues are then discussed between the members of the maintenance team. During the discussion, it may be necessary to ask the issue creator for further clarification or for commenting on the proposed solution(s).

3 Cicero Argumentation Model

The argumentation model, that is underlying the Cicero tool, is based on the DILIGENT argumentation framework [1] and the Potts and Bruns model [2]. They are both extensions of the idea of the Issue Based Information Systems [3,4]. All these models help in structuring an *issue* or problem and to simultaneously derive possible solutions with the help of discussions.

On the one hand, the Potts and Bruns model extends the IBIS approach to discussions that lead to the creation or change of a concrete artifact (e.g. elements in an ontology). The discussion serves as a connection between the old and the new version of the changed artifact, thus documenting its design rationale or provenance. On the other hand, the DILIGENT argumentation framework amongst others introduces certain argument types to IBIS, that accelerates issue resolution. In [1], it has been shown that participants of a discussion, who mainly used the identified argument types, needed less time for coming to a successful conclusion of the discussion.

The Cicero argumentation model combines the general structure for representing discussions from the DILIGENT argumentation framework with the idea of annotating ontology elements and changes with the corresponding discussions. In Cicero, a discussion always starts with an issue that is raised by either of the participants (i.e. ontology/knowledge engineers or users). Subsequently, solutions are proposed and discussed. After some time, it has to be decided which of the proposed solutions will be implemented in the ontology (see Fig. 1).

---

[2] It can be downloaded from http://isweb.uni-koblenz.de/Research/Cicero/
[3] http://www.fao.org/agrovoc/
[4] http://www.geneontology.org/

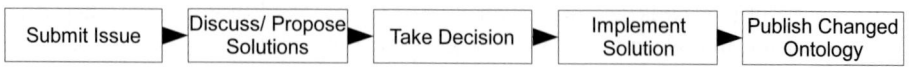

Fig. 1. Prototypical workflow for discussions about the design rationale

4 Using the Model in Cicero

Cicero is an extension of the Semantic MediaWiki. On the one hand, this has several advantages: For example, one can also use the Wiki for maintaining documents related to the collaborative ontology engineering, e. g. requirements documents. Furthermore, the Semantic MediaWiki allows for accessing its contents as RDF streams which is useful for integrating it with other tools like the NeOn toolkit. But on the other hand, it has to be extended with functionality that supports the discussion workflow presented in the previous section.

4.1 Discussions

In Cicero exists for each discussion thread an overview page summarizing the issue and all proposed solutions (see Fig. 2). The page is automatically generated as soon as a new issue is created. The actual discussion consisting of solution proposals and arguments then takes place on the talk page of the overview page.

During the discussion, a user can propose solutions of the issue. The different solution proposals can then be supported or objected by all discussion

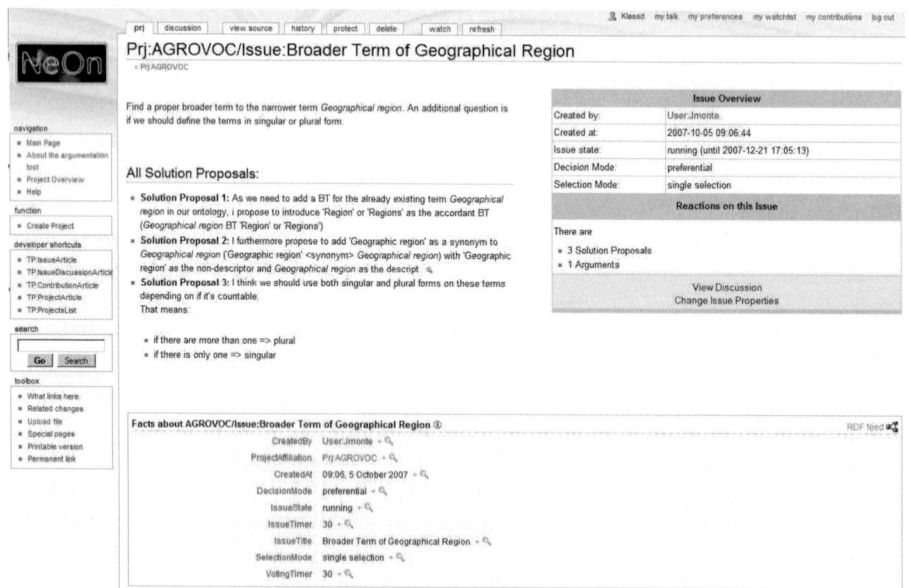

Fig. 2. Overview page of an issue in Cicero

participants. Three different argument types are supported by Cicero, that have been identified in [1] as accelerating issue resolution.:

- **Example:** An supporting or objecting example corresponds to a pattern that should or should not be imitated. Examples are used for illustrating similar cases that may serve as a model for the solution to which they reply.
- **Evaluation:** An evaluation gives criteria which help to assess the strengths and weaknesses of a solution proposal.
- **Justification:** A justification describes the relevant circumstances that help to understand why a certain solution is supported or objected by a user.

4.2 Taking a Decision

After some time of discussion, the decision taking procedure may be started in Cicero. During decision taking, no new solution proposals or arguments can be added to the discussion and users with the corresponding access right can cast their ballot. It depends on the settings of an issue whether users can cast their ballot for one solution proposal only or for several of the solution proposals.

Two decision taking modes can be distinguished: During *preferential voting mode*, all users with the corresponding access right can cast their ballot. The solution proposal with the most votes is subsequently marked as the decided solution. During *dictator mode*, only a single user with the corresponding access right decides on the solution of the issue.

5 Integration with NeOn Toolkit

The NeOn toolkit[5] is an ontology engineering environment that can be easily extended. Currently, a plugin is being developed that integrates the functionality of Cicero into the NeOn toolkit.[6] The plugin will allow for an easy creation of new discussions and for searching existing discussions that are related to specific ontology entities.

For example, if a user of the NeOn toolkit wants to create a new discussion he may select the ontology entities in the environment that should be discussed, i. e. the issue gets related to ontology entities. Before actually creating the new issue, a list of already existing issues related to the selected ontology entities can then be presented to the user. Thus, he can check whether it is really necessary to create a new issue or whether it is better to participate in the discussion of one of the already existing issues.

After a decision is taken, the NeOn toolkit is used for implementing the selected solution proposal. All changed, deleted or added ontology entities get also related to the discussion, i. e. a provenance relationship between ontology entities and their design rationale in form of the discussions is established.

---

[5] http://www.neon-toolkit.org/
[6] http://www.neon-toolkit.org/wiki/index.php/Cicero

6 Conclusions and Related Work

Aspects of Cicero and its integration with the NeOn toolkit can be compared with Collaborative Protégé [5], Tadzebao and WebOnto [6] as well as Compendium [7,8]. Because of lack of space, the comparison in Tab. [6] is concentrated on the aspects of Cicero highlighted in this paper.

Table 1. Comparison of Cicero with related work

	Collaborative Protégé	Tadzbao & WebOnto	Compendium	Cicero
IBIS-based Arg. Framework	no	no	yes	yes
Integration with Ontology Editor	yes	yes	no	yes
Establishing Provenance Links	yes	no	no	yes

Altogether, Cicero helps with its IBIS-based argumentation framework to have better structured and more efficient discussions. Furthermore, the integration with an ontology editor leads to a better support of the ontology engineering lifecycle in which discussions about the design rationale of ontology elements play an inherent role. Finally, Cicero and its integration with the NeOn toolkit reduces the required effort for establishing the provenance links between design rationale discussions and the affected ontology elements.

References

1. Pinto, H.S., Staab, S., Tempich, C.: DILIGENT: Towards a fine-grained methodology for Distributed, Loosely-controlled and evolving Engineering of oNTologies. In: Proc. of ECAI (2004)
2. Potts, C., Bruns, G.: Recording the reasons for design decisions. In: ICSE, pp. 418–427 (1988)
3. Kunz, W., Rittel, H.: Issues as elements of information systems. WP 131, Institute of Urban and Regional Development, University of California, Berkeley (1970)
4. Rittel, H.W.J., Webber, M.M.: Dilemmas in a general theory of planning. Policy Sciences 4(2), 155–169 (1973)
5. Tudorache, T., Noy, N.: Collaborative Protégé. In: Social and Collaborative Construction of Structured Knowledge Workshop held at WWW 2007 (2007)
6. Domingue, J.: Tadzebao and WebOnto: Discussing, browsing and editing ontologies on the web. In: 11th Knowledge Acquisition Workshop (1998)
7. Buckingham Shum, S., Selvin, A., Sierhuis, M., Conklin, J., Haley, C., Nuseibeh, B.: Hypermedia Support for Argumentation-Based Rationale: 15 Years on from gIBIS and QOC. In: Dutoit, A., McCall, R., Mistrk, I., Paech, B. (eds.) Rationale Management in Software Engineering, pp. 111–132. Springer, Heidelberg (2006)
8. Buckingham Shum, S., Motta, E., Domingue, J.: Augmenting design deliberation with compendium: The case of collaborative ontology design. In: Workshop on Facilitating Hypertext-Augmented Collaborative Modeling at the ACM Hypertext Conference (2002)

xOperator – An Extensible Semantic Agent for Instant Messaging Networks

Sebastian Dietzold[1], Jörg Unbehauen[2], and Sören Auer[1]

[1] Universität Leipzig, Department of Computer Science
Johannisgasse 26, D-04103 Leipzig, Germany
{dietzold,auer}@informatik.uni-leipzig.de
[2] Leuphana - University of Lneburg, Faculty III Environmental Sciences and Engineering, Volgershall 1, D-21339 Lneburg
joerg@unbehauen.net

Abstract. Instant Messaging is in addition to Web and Email the most popular service on the Internet[1]. With xOperator we demonstrate the implementation of a strategy which deeply integrates Instant Messaging networks with the Semantic Web. The xOperator concept is based on the idea of creating an overlay network of collaborative information agents on top of social IM networks. It can be queried using a controlled and easily extensible language based on AIML templates. Such a deep integration of semantic technologies and Instant Messaging bears a number of advantages and benefits for users when compared to the separated use of Semantic Web technologies and IM, the most important ones being context awareness as well as provenance and trust. Our demonstration showcases how the xOperator approach naturally facilitates enterprise and personal information management as well as access to large scale heterogeneous information sources.

1 Background and Application Context

With estimated more than 500 Million users Instant Messaging (IM) is in addition to Web and Email the most popular service on the Internet. IM is used to maintain a list of close contacts (such as friends or co-workers), to synchronously communicate with those, exchange files or meet in groups for discussions. Examples of IM networks are ICQ, Skype, AIM or the Jabber protocol and network[2]. The latter is an open standard and the basis for many other IM networks such as Google Talk, Meebo and Gizmo.

The xOperator concept is based on the idea of additionally equipping an users' IM identity with a number of information sources this user owns or trusts (e.g. his FOAF profile, iCal calendar etc.). Thus the social IM network is overlaid with a network of trusted knowledge sources. An IM user can query his local knowledge sources using a controlled (but easily extensible) language based on

---

[1] According to a sum up available at:
http://en.wikipedia.org/wiki/Instant_messaging#User_base
[2] http://www.jabber.org/

Artificial Intelligence Markup Language (AIML) templates[6]. The AIML component translates natural language into SPARQL queries according to predefined templates. In order to pass the generated machine interpretable queries to other xOperator agents of friends in the social IM network xOperator makes use of the standard message exchange mechanisms provided by the IM network. After evaluation of the query by the neighbouring xOperator agents results are transferred back, filtered, aggregated and presented to the querying user.

Such a deep integration of semantic technologies and IM bears a number of advantages and benefits for users when compared to the separated use of Semantic Web technologies and IM. From our point of view the two most crucial ones are:

- **Context awareness.** Users are not required to world wide uniquely identify entities, when it is clear what/who is meant from the context of their social network neighbourhood. When asked for the current whereabout of Sebastian, for example, xOperator can easily identify which person in my social network has the name Sebastian and can answer my query without the need for further clarification.
- **Provenance and trust.** IM networks represent carefully balanced networks of trust. People only admit friends and colleagues to their contact list, who they trust seeing their online presence, not being bothered by SPAM and sharing contact details with. Overlaying such a social network with a network for semantic knowledge sharing and querying naturally solves many issues of provenance and trust. Future versions of xOperator will allow more fine grained access control mechanisms, based upon group and individual policies.

2   Communication Scenarios

This section describes the three demonstrated agent communication settings for the xOperator system demonstration. Figure 1 shows a schematic depiction of the communication scenarios. The figure is divided vertically into four layers.

The first two layers represent the World Wide Web. Mutually interlinked RDF documents (such as FOAF documents) reference each other using relations such as `rdf:seeAlso`. These RDF documents could have been generated manually, exported from databases or could be generated from other information sources. For the system demonstration we use the Semantic Web Conference Corpus[3] together with an instance of our semantic Wiki OntoWiki [1], a number of FOAF profile documents and private and public calendars[4].

The lower two layers in Figure 1 represent the Jabber Network. Here users are interacting synchronously with each other, as well as users with artificial agents (such as xOperator) and agents with each other. A user can pose queries in natural language to an agent and the agent transforms the query into one or multiple SPARQL queries. Thus generated SPARQL queries can be forwarded either to

---

[3] http://data.semanticweb.org/
[4] We use Masahide Kanzaki's ical2rdf service for the live conversion of iCal calendars (e.g. from the Google's calendar service).

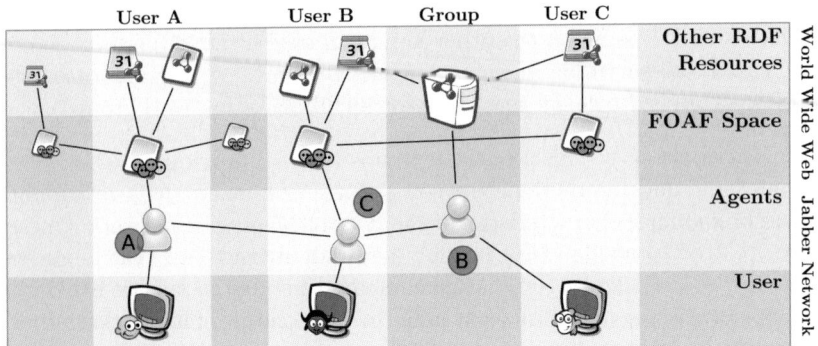

Fig. 1. Agent communication scenarios: (a) personal agent, (b) group agent, (c) agent network

a SPARQL endpoint or neighbouring agents via the IM networks transport protocol (XMPP in the case of Jabber). Queries are forwarded to all neighboring agents, but not beyond to prevent flooding of the network. SPARQL endpoints evaluate the query using a local knowledge base, dynamically load RDF documents from the Web or convert Web accessible information sources into RDF. The results of SPARQL endpoints or other agents are collected, aggregated, filtered and presented to the user depending on the query as list, table or natural language response.

The different communication scenarios of the demo are presented in the remainder of this section.

Personal Agent (A). A user of an Instant Messaging network installs his own personal agent and configures information sources he owns or trusts. Information sources can be for example a FOAF profile of the user containing personal information about the user and about relationships to other people he knows and where to find further information about these. Additionally the FOAF profile can link to other RDF documents which contain more information about the user and his activities (e.g. the users' iCal calendar). Such links span a network of information sources as depicted in Figure 1. Each user maintains his own information and links to information sources of his acquaintances. Depending on the query, the agent will access the respective resources. The following example queries are possible, when FOAF profiles are known to the agent: Give me the phone / homepage / ... of Frank! What is the birthday of Michael? Where is Dave now? Who knows Alex?

Group Agent (B). This communication scenario differs from the Personal Agent scenario in that multiple users gain access to the same agent. The agent does not only access remote documents but can also use triple stores for answering queries. For agents themselves, however, the distinction between RDF sources on the Web and information contained in a local triple store is not relevant. When used within a corporate setting this triple store can for example contain a directory with information about employees or customers. A group agent accessing DBpedia on the other hand can enrich the xOperator query answering capabilities with background knowledge. This, e.g., enables queries such as: Which airports are easily reachable for members of my workgroup?

Agent Network (C). This scenario extends the two previous ones by allowing communication and interaction between agents. The rationale is to exploit the trust and provenance characteristics of the Instant Messaging network: Questions about or related to acquaintances in my network of trust can best be answered by their respective agents. Hence, agents should be able to talk to other agents on the IM network. A personal agent uses the IM account of its respective owner, has access to his contact list and is thus a part of its owner's social network. The agent is able to recognise other personal agents of acquaintances in the contact list and it is possible for agents to communicate without interfering with the communication of their owners. After other agents are identified it is possible to forward SPARQL queries (originating from a user question) to these agents, collect their answers and present them to the user. This enables, e.g., queries such as: "What is the next possible meeting date and place for members of my workgroup?" (based on the evaluation of workgroup members' calendars).

3 Demonstration Overview

The xOperator concept was implemented in Java and is available as open-source software together with an online demo from the AKSW group website[5]. Figure 2 shows the communication with the xOperator agent by means of an ordinary

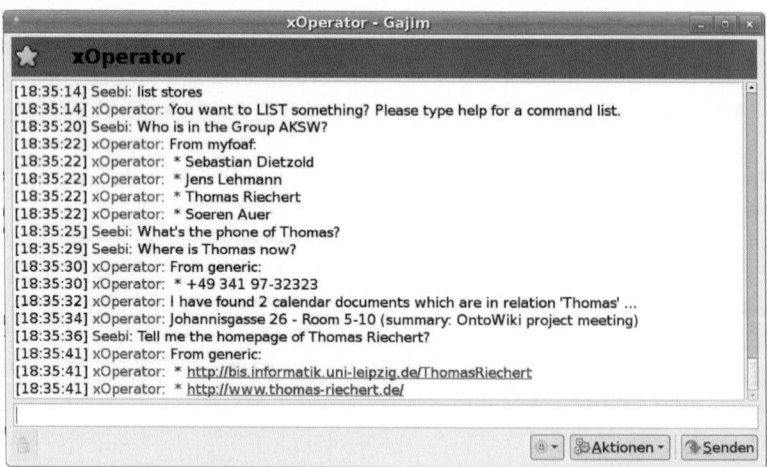

Fig. 2. Communication with xOperator by means of an ordinary Jabber client

---

[5] http://aksw.org/Projects/xOperator

Jabber client, which facilitates the instant evaluation by interested users. In order to instantly try xOperator the AKSW maintains a publicly available xOperator instance with the xOperator account `xoperator-demo@aksw.org`, which can be simply added to a Jabber contact list. Example queries can be run on this demo instance, such as:

- `tell me (the) * of *` e.g. the phone of Sebastian
- `where is * now` e.g. where is Sebastian now

On the `help` xOperator will respond with information about the usage. Some data stores such as the DBpedia endpoint and the AKSW group OntoWiki are already pre-configured for the use within the xOperator demo. Additional ones can be easily added using the keywords `add ds`. The keyword `add template` allows xOperator users to add AIML templates on-the-fly. Parameters are the AIML pattern (using * as a placeholder) and a SPARQL query with %%n%%-references to the value of the nth placeholder in the AIML pattern.

4 Related Work

Agent software is a rapidly developing area of research. In [4] the author proposed a typology of agents. According to this typology xOperator is an interface agent for the Semantic Web. Proposals and first prototypes which are closely related to xOperator and inspired its development are Dan Brickley's JQbus[6] and Chris Schmidt's SPARQL over XMPP[7]. However, both works are limited to the pure transportation of SPARQL queries over XMPP. Quite different but the xOperator approach nicely complementing are works regarding the semantic annotation of IM messages [5,2]. Finally, in [3] the author enhanced AIML bots by generating AIML categories from RDF models. Different to xOperator, these categories are static and represent only a fixed set of statements.

References

1. Auer, S., Dietzold, S., Riechert, T.: OntoWiki - A Tool for Social, Semantic Collaboration. In: Cruz, I., Decker, S., Allemang, D., Preist, C., Schwabe, D., Mika, P., Uschold, M., Aroyo, L.M. (eds.) ISWC 2006. LNCS, vol. 4273, pp. 736–749. Springer, Heidelberg (2006)
2. Franz, T., Staab, S.: SAM: Semantics Aware Instant Messaging for the Networked Semantic Desktop. In: Semantic Desktop Workshop at the ISWC (2005)
3. Freese, E.: Enhancing AIML Bots using Semantic Web Technologies. In: Proc of Extreme Markup Languages (2007)
4. Nwana, H.S.: Software Agents: An Overview. Knowledge Engineering Review 11(3), 205–244 (1996)
5. Osterfeld, F., Kiesel, M., Schwarz, S.: Nabu - A Semantic Archive for XMPP Instant Messaging. In: Semantic Desktop Workshop at the ISWC (2005)
6. Wallace, R.: Artificial Intelligence Markup Language (AIML). Working draft, A.L.I.C.E. AI Foundation, February 18 (2005)

---

[6] http://svn.foaf-project.org/foaftown/jqbus/intro.html
[7] http://crschmidt.net/semweb/sparqlxmpp/

LabelTranslator - A Tool to Automatically Localize an Ontology

Mauricio Espinoza[1], Asunción Gómez-Pérez[1], and Eduardo Mena[2]

[1] UPM, Laboratorio de Inteligencia Artificial, 28660 Boadilla del Monte, Spain
asun@fi.upm.es, mespinoza@delicias.dia.fi.upm.es
[2] IIS Department, Univ. of Zaragoza, María de Luna 1, 50018 Zaragoza, Spain
emena@unizar.es

Abstract. This demo proposal briefly presents LabelTranslator, a system that suggests translations of ontology labels, with the purpose of localizing ontologies. LabelTranslator takes as input an ontology whose labels are described in a source natural language and obtains the most probable translation of each ontology label into a target natural language. Our main contribution is the automatization of this process, which reduces human efforts to localize manually the ontology.

Keywords: Ontology localization, Multilingual ontologies.

1 Introduction

Typically, the names assigned to ontology terms appear in a specific natural language. Thus, in order to achieve a more generally applicable ontology, it is necessary to guarantee that the same knowledge be recognizable in different natural languages. Moreover, some organizations working in a multilingual environment demand multilingual ontologies.

To cope with these problems we present LabelTranslator, a system that automatically localizes ontologies in English, Spanish, and German. The Ontology Localization Activity (OLA) consists in adapting an ontology to a concrete language and culture community, as defined in [3]. The technological background of the system comprises: 1) the extraction of possible single-lemma translations from semantic and translation resources, 2) the disambiguation of the translation senses and 3) the ranking of the translations.

Some details on the aspects above enumerated are given in the following section. For a more comprehensive description of our system, we refer to [2].

2 System Description

The current version of LabelTranslator has been implemented as a Neon[1] plug-in, but it can easily become an independent module, for example a web service. Figure 1 illustrates the main steps given by our system and their executing

---
[1] http://www.neon-toolkit.org/

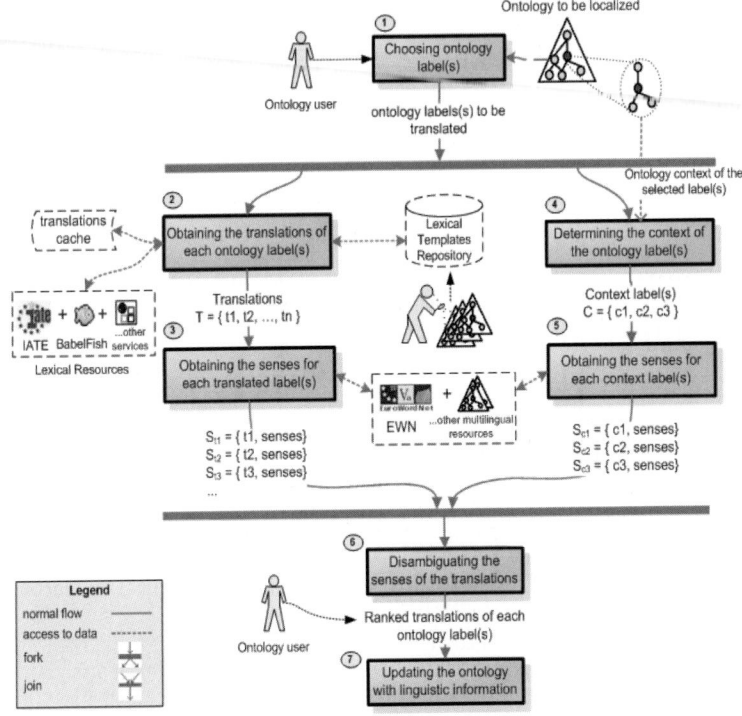

Fig. 1. Main steps of LabelTranslator

sequence to localize an ontology among different natural languages. Notice that, the steps two and three are executed sequentially, then, the system enforces that the steps four and five be finished to continue its execution.

1. *Choosing ontology terms(s) to localize.* The system starts with an ontology (described) in OWL-DL, F-Logic, or RDF provided by the ontology user. The system uses some views of the Neon ToolKit to load the ontology and store the multilingual results, respectively. In Figure 2, we show a screenshot of both the *Ontology Navigator* and the *Entity Properties* view with information related to our sample ontology[2].

    When the system imports a new ontology, it offers a perspective over ontological data using the NeOn ToolKit-style. Then, the user chooses the label of the ontology term(s) to be translated. In our example, the concept *planta* (plant) By right clicking on a frame (concept, attribute, or relation), the *Translate* action performs the translation of an ontology label (see Figure 2).

2. *Obtaining the translations of each ontology label(s).* To obtain the translations of the selected ontology label(s), the system relies on different linguistic

---

[2] The sample ontology used here belongs to biosphere domain and it needs to be localized from Spanish into English.

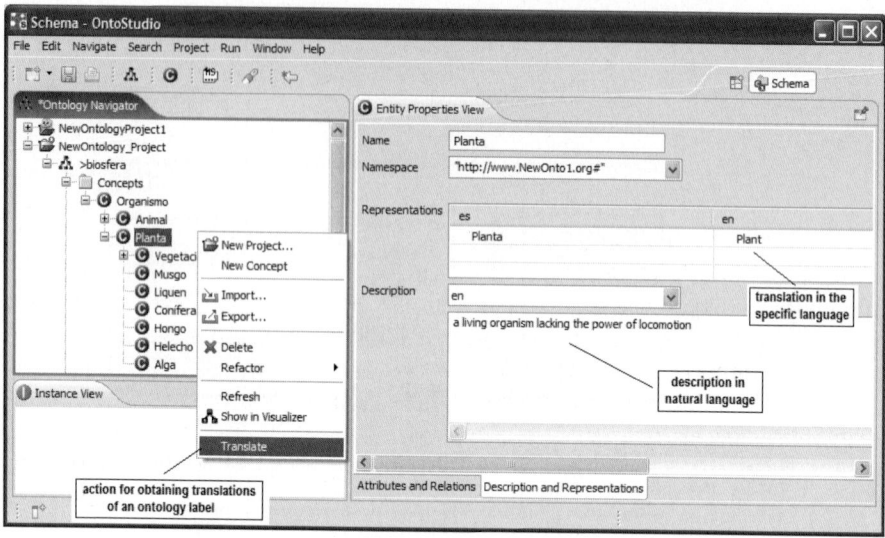

Fig. 2. A screenshot of the NeOn Toolkit views used by the LabelTranslator plug-in

resources: 1) multilingual translation web services such as Babelfish[3], GoogleTranslate[4], etc., and 2) other lexical resources as IATE[5]. The current prototype supports translations from/into English, Spanish, and German. A *cache* stores previously translations to avoid accessing the same remote data twice. In our example the system retrieves 12 translations for the selected label "planta". Thus, $T = \{$ sole, flora, plant, mill, story, level, plant life, ..., factory, manufactory $\}$.

In the case of compound labels where no entry is present in the translation resources used, the label is split into its components; the individual components are translated and then combined into a compound label the target language. The original order of each component is considered to combine components respecting the word order of the target language using lexical templates[6]. The learning of the lexical templates was semi-automatically derived from the patterns found in the labels used in ontologies of different domain and described in English, Spanish, and German. A more detailed description of this process can be found in [2].

3. *Obtaining the senses for each translated label(s)*. For each translated label the system retrieves a list of semantic senses using the approach proposed in a previous work [6]. Our system takes as input a list of words (each translated label), discovers their semantics in run-time and obtains a list of senses

---

[3] http://babelfish.altavista.com/
[4] http://www.google.com/translate_t
[5] http://iate.europa.eu/iatediff/SearchByQueryLoad.do?method=load
[6] The notion of lexical template proposed in this paper refers to text correlations found between a pair of languages.

extracted from different ontology pools; it deals with the possible semantic overlapping among senses. These senses are used by the *disambiguation method* (see step 6) to sort the different translations of an ontology label according to similarity with its lexical and semantic context.
4. *Determining the context of the ontology label(s).* In order to determine the context of an ontology label, the system retrieves the set of labels associated with the label under consideration. The list of context labels comprises a set of names which can be direct label names and/or attributes label names, depending on the type of term that is being translated. The number of context labels is limited to those labels with the higher values of similarity according to the Normalized Google Distance [1] (NGD). NGD measures the semantic relatedness between any two terms, considering the relative frequency in which two terms appear in the Web within the same documents. Those labels that have the higher values of similarity with the label under consideration are chosen. In our example the system selects only three context labels, out of eight previously found. Thus, $C = \{$ *organismo(organism), liquen(lichen), conífera(conifer)* $\}$.
5. *Obtaining the senses for each context label(s).* The senses of each context label are discovered with the same process used to discover the senses of each translated label (see step 3).
6. *Disambiguating the senses of the translations.* A disambiguation process is needed to sort the list of translations according to similarity with the context of the label to translate. This method takes as input the set of senses of each translated label (step 3) and the senses of their context labels (step 5). From this set of senses, the method relies on a relatedness measure [4] based on glosses to disambiguate the translations. Candidate translations are then ranked according to the similarity with their context in the ontology, and the ranked list is used to either present the user the best candidates first, or to use the highest-scoring candidate to translate the label automatically.

Figure 3 shows a screenshot of the translation dialog with the translations of the sample ontology label "planta". In our example, the system correctly suggests "plant" in the sense of "a living organism lacking of power of locomotion" as first translation of the selected ontology label.
7. *Updating the ontology with linguistic information.* Once the right sense has been selected, the system updates the linguistic information of the ontological term. Our system supports the linguistic model [5] designed for the representation and structuration of multilingual information in ontologies. In the current version, the system fills in runtime the fields *representations* and *description* of the Entity Properties view (shown in Figure 2), according to the most probable translations proposed by the disambiguation method. These fields represent the link between the conceptual knowledge and the linguistic information discovered. In a future release (current under development), we will add support to the new linguistic model which captures all the relevant linguistic/terminological information associated with concepts.

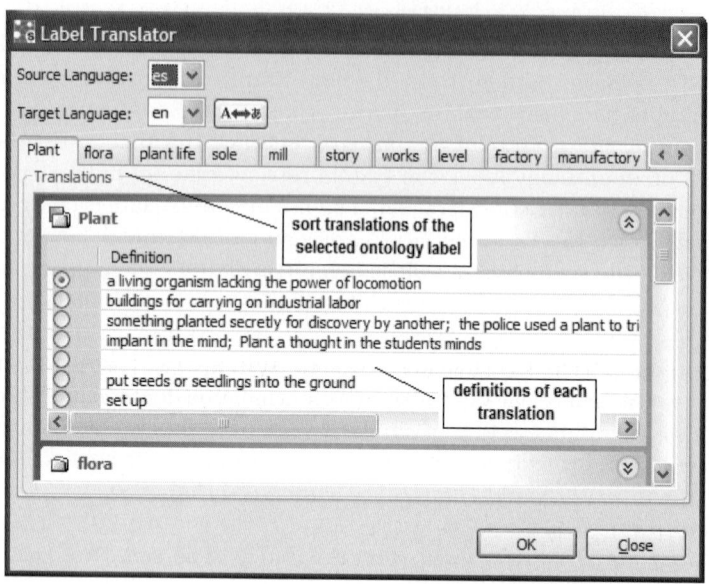

Fig. 3. Translations into English of the spanish ontology label "planta"

Acknowledgements

This work is supported by the European Commission's Sixth Framework Program under the project name: Lyfecycle support for networked ontologies (NeOn) (FP6-027595), the National Project "GeoBuddies" (TSI2007-65677C02), and the spanish CICYT project TIN2007-68091-C02-02.

References

1. Cilibrasi, R.L., Vitnyi, P.M.: The Google Similarity Distance. IEEE Transactions on Knowledge and Data Engineering 19(3), 370–383 (2007)
2. Espinoza, M., Gómez-Pérez, A., Mena, E.: Enriching an ontology with multilingual information. In: Proc. of 5th European Semantic Web Conference (ESWC 2008), Tenerife (Spain) (June 2008)
3. Suarez-Figueroa, M.C.: NeOn Development Process and Ontology Life Cycle. NeOn Project Deliverable 5.3.1 (2007)
4. Pedersen, T., Banerjee, S., Patwardhan, S.: Maximizing Semantic Relatedness to Perform Word Sense Disambiguation. Research Report UMSI 2005/25, University of Minnesota Supercomputing Institute (March 2005)
5. Peters, W., Montiel-Ponsoda, E., Aguado de Cea, G.: Localizing ontologies in owl. In: OntoLex 2007 (2007)
6. Trillo, R., Gracia, J., Espinoza, M., Mena, E.: Discovering the semantics of user keywords. Journal on Universal Computer Science. Special Issue: Ontologies and their Applications (2007) ISSN 0948-695X

RKBExplorer.com: A Knowledge Driven Infrastructure for Linked Data Providers

Hugh Glaser, Ian C. Millard, and Afraz Jaffri

School of Electronics and Computer Science
University of Southampton, UK
{hg,icm,a.o.jaffri}@ecs.soton.ac.uk

Abstract. RKB Explorer is a Semantic Web application that is able to present unified views of a significant number of heterogeneous data sources. We have developed an underlying information infrastructure which is mediated by ontologies and consists of many independent triple-stores, each publicly available through both SPARQL endpoints and resolvable URIs. To realise this synergy of disparate information sources, we have deployed tools to identify co-referent URIs, and devised an architecture to allow the information to be represented and used. This paper provides a brief overview of the system including the underlying infrastructure, and a number of associated tools for both knowledge acquisition and publishing.

1 Introduction

The Linking Open Data Initiative (http://linkeddata.org/) has encouraged the widespread creation and deployment of RDF data using URIs that are both dereferenceable and linked to URIs from other data sources. This initiative led to the creation of DBpedia (http://wiki.dbpedia.org/Datasets), an RDF version of Wikipedia which itself established a base from which other data sets could be linked. There are now several hundred thousand links from DBpedia to other RDF sources and Linked Data sites.

There have been guidelines laid down for the creation of Linked Data [1] but for users not familiar with Semantic Web technology, creating and maintaining a Linked Data site requires a serious amount of time and effort. Such a scenario is encountered in our own ReSIST project, an EU funded Network of Excellence in Resilient Systems. From its early conceptions, it was proposed that the entire project would be supported by a semantically-enabled knowledge infrastructure. The vision was of sets of services and applications for both acquiring and publishing knowledge, working together as a unified coherent resource. Project members and others would be able to explore the knowledge created and acquired from distributed and heterogeneous resources, enabling them to identify relationships and resources that may not previously have been evident.

Many of the participants in the project have very little or no experience or knowledge of the Semantic Web. The aim of www.rkbexplorer.com was to create an infrastructure where knowledge acquired from different data sources could become part of 'The Semantic Web' and not just used as a closed part of a static system. The system that has been built provides data providers with a SPARQL endpoint, dereferenceable URIs, data browser, data explorer, consistent reference services and keyword search. The rest of this paper describes these services and highlights the minimal cost of use for non-expert users.

2 Related Work

There have been a number of efforts recently to facilitate the management of linked data. Most noticeable is D2R Server [2] which was produced so that data providers who held their data in relational databases could expose the database via a SPARQL endpoint and HTML browser. This method has the convenience of not having to convert any of the existing data into RDF which involves the execution of custom extraction scripts. A disadvantage of this approach is that additional functionality that can only be performed on RDF data cannot be performed on data exposed through the D2R server. An example of this is the co-reference analysis that is described in Section 4.2.

The individual systems made for exposing and exploring semantic data have been produced on a stand alone basis. A true semantic web site should possess all of the features that are required for Linking Data, and more, and also have the added advantage of being simple in creation and management. The architecture described in the rest of this paper provides just such an infrastructure that is a live system being used daily by many users.

3 Knowledge Acquisition

Ideally, in an active Semantic Web world, we would have simply been able to use existing knowledge sources. These sources would publish their contents against well-known ontologies, both as SPARQL endpoints and resolvable URIs. We would then use them, possibly needing some ontology translation on the way. Unfortunately, this is not yet the case. When the project started in January 2006, there were few such citizens of the Semantic Web, and so we resolved to undertake the bootstrap process ourselves.

We therefore harvested the information from the places we identified, and made it openly available as both resolvable URIs and SPARQL endpoints, against our AKT ontology (http://www.aktors.org/publications/ontology/). Since this also involved minting all new URIs, each data source is held in a separate triplestore, on a separate domain, not necessarily running on the same machine.

At present we have acquired RDF data on people, publications and institutions from the 18 partners in the ReSIST project. People and publication data have also been harvested from a number of major metadata resources. We chose the major publishers and aggregators in Computer Science, and have to date harvested some 50 million triples from Citeseer, the ACM, DBLP, NSF and selected IEEE conferences.

The way in which we have structured the system that collects and organises this knowledge means that knowledge from any new data provider can easily be brought into the existing infrastructure, automatically giving all of the benefits that have been built into the system. For example, if a new partner was to join the ReSIST project, information from their site would be converted into RDF. Putting this in a triplestore on the rkbexplorer.com site means that their knowledge has been exposed via SPARQL, tabulator-style browsing, graphical interaction and keyword based searching. The knowledge will also be analysed for any coreferences amongst people and papers and this information will also be stored in a separate knowledge base. All the functionality of Linked Data will have been given to the user for a small investment in knowledge acquisition. They can thus have a fully-functional Linked Data site up and being linked into other sites in minutes or longer, depending primarily on the RDF provided.

4 Information Infrastructure

4.1 Triplestores

We use 3Store [3] as our base repository, with a separate knowledge base representing each data source which we have acquired. The separate knowledge bases facilitate the system scalability, and help to provide the high query performance needed for an application of this sort, while allowing the assertion of the volumes of data (many tens of millions of triples) that we have. Each repository is complemented by a number of services and interfaces, which are openly accessible at http://<repository>.rkbexplorer.com/. These resources are additionally being indexed by semantic search services such as http://sindice.com/.

4.2 Consistent Reference Service (CRS)

The way in which the RKB explorer and other applications give a unified view of tens of triplestores (knowledge bases) with tens of millions of triples, requires a well-founded method of allowing URIs to bridge between the triplestores, when they are considered to refer to the same concept.

The ReSIST activity embraces this. It includes in its architecture the deployment of a number of CRSes, which are knowledge bases of URI equivalences for the application being considered, according to appropriate criteria.

We chose to keep this knowledge separately from the main data. One reason is simply that of good engineering practice. It is easier to maintain knowledge that is being created by the CRS builder separate from the knowledge that is being created by the information provider. Indeed, different CRS providers will exist for the same information in an open Semantic Web world. A second reason is that a CRS is designed for a purpose, or set of purposes. Some applications might wish to consider that two concepts are the same, while this may not be the case for another application over the same knowledge. For example, in undertaking citation analysis, a paper with the same title and text that appeared both as a journal article and technical report should be considered as two separate papers. In an application such as ours, where we are considering who works with whom on what topic, it might well be more appropriate to consider that they should be treated as one resource, while still representing the separate details in a consistent fashion. As information providers of the basic information, we include a `coref:hasCRS` to the associated CRS in the RDF for a resource, so that it can be easily found, although there can be more than one CRS, corresponding to different policies.

Thus, the CRS is essentially an open service, which gives a view of URI equivalence: when presented with a URI, it returns all the URIs it considers equivalent. Note that it aims to avoid the mistake of creating a new URI; such an action would simply add further to the problem by being a new authority. It was also decided not to use `owl:sameAs`, since this is a much stronger assertion than the CRS is making. Of course, the knowledge bases themselves may still be using it where appropriate.

To generate knowledge for the CRS, the system uses the expected heuristic of string similarity (very conservatively), confirming the identification with the other relational uses, such as publication place (for papers), funding body (projects) and place of work (people), where these have already been the subject of equivalence identification. This means that to begin the generation of knowledge for the CRS, there is a 'cold start' problem, as there are almost no real URIs that are in common. This is achieved by string analysis of the titles of publication, and hence spreading to authors.

4.3 The RKB Explorer

For the user interface for exploring the knowledge bases we settled on a simple and very static presentation, with one window. This was primarily because a previous attempt based on a more dynamic style, for example allowing a choice of topics, was criticised as being non-intuitive by many of our users, few of whom had any knowledge of Semantic Web technologies.

A coherent and unified view of the underlying data sources is presented to the user, through the application of co-reference resolution algorithms and community of practise analysis to consolidate duplicate references and to identify related resources. Users may search and browse through the information available based around the four core themes which are present within the Explorer interface, namely People, Resilience Mechanisms, Publications and Projects. At any one time the top half of the interface window details a given instance of one of these types of resource, while the lower half lists those resources of each type which are related to the currently selected item.

This enables 'opportunistic' browsing, allowing a user to discover further information related to that which they are viewing, with associations determined by detailed domain specific analysis over the underlying knowledge bases. Utilising this interface, end users may be presented with data that is of interest or related to their field, potentially including work from areas or communities of which they were previously unaware.

Since the general problem of distributed queries remains unsolved, the system has to implement querying as appropriate for its environment. To gather all the

RDF related to a particular URI, it firstly resolves the URI (which includes any `owl:sameAs` in that store). It then looks up the URI in the associated CRS, which can be identified from the `coref:hasCRS` that was provided, and finds other, equivalent URIs. These can now be looked up in their CRSes, and the process continues, essentially to the fixed point. There is also the provision for other CRSes which are not directly associated with information sources to be consulted. It would be possible to consult all CRSes, but we consider this unnecessary. The CRSes we choose to trust for equivalence in these applications are either the original information providers, or ones we have chosen ourselves.

5 Conclusion

In order to ensure data providers take advantage of the benefits of linked data, the creation and maintenance processes that accompany the publishing of such data must not be a hindrance to those not willing to invest much time or effort.

We have presented a real-world Semantic Web application that is based on large-scale information from independent sources, using an ontology to mediate between them and rank resources when presenting consolidated results to users.

It provides a number of related applications, including the RKB Explorer, which gives an accessible and functional user interface. This, along with the usefulness of the knowledge resources have been extensively validated by the ReSIST Project partners, as reported in [4].

Since the system does not function by harvesting information into a common store, it is thus truly web-based. By employing resolvable URIs and distributed repositories to which queries can be fielded, we have created a real-world and scalable solution. Our system empowers linked data providers by providing a complete infrastructure for the curation and integration of large sets of RDF. The system will be shown to work with the existing sources of data that we have acquired and the various features that have been described in this paper will also be demonstrated. In addition, we will be happy to be provided with interesting RDF resources and use the system to provide a Linked Data site for them.

This work is supported by the ReSIST Network of Excellence, funded by FP6 under contract IST 4 026764 NOE.

References

[1] Bizer, C., Cyganiak, R., Heath, T.: How to publish linked data on the web (2007)
[2] Bizer, C., Cyganiak, R.: D2r server - publishing relational databases on the web. In: Proceedings of the 5th International Semantic Web Conference (2006)
[3] Harris, S., Gibbins, N.: 3Store: Efficient bulk RDF storage. In: Proceedings of the 1st International Workshop on Practical and Scalable Semantic Systems (2003)
[4] Glaser, H., Millard, I.C., Anderson, T., Randell, B.: ReSIST Project Deliverable D10: Prototype knowledge base. Tech. Rept., University of Southampton. Technical report (2007)

Semantic Browsing with PowerMagpie

Laurian Gridinoc, Marta Sabou, Mathieu d'Aquin,
Martin Dzbor, and Enrico Motta

Knowledge Media Institute (KMi), The Open University, Milton Keynes, UK
{l.gridinoc,r.m.sabou,m.daquin,m.dzbor,e.motta}@open.ac.uk

Abstract. PowerMagpie is a tool that brings semantic interpretation to classical web pages by dynamically—*i.e.* during browsing—selecting and making use of a wide range of online available ontologies. We introduce the idea of extending browsing through semantic, ontology-based interpretation. Then, we provide a brief description of the architecture. In the end we underline which aspects of the available online semantic data are demonstrated, what the user may learn and which are the future directions.

1 Introduction

The web supports a large array of interactions and tasks by providing a scalable infrastructure which makes it possible to locate, explore and exchange knowledge. However, not much support is available for the interpretation of web pages [2], beyond what is provided by the author of each particular resource. Generally speaking the web offers two mechanisms to help users make sense of a resource. The first is hypertext linking, that is normally used to indicate additional explicative material, which can help the user in understanding web content. The second is the *Related Pages* feature provided by several search engines and directories. While these mechanisms can be useful in many cases, they nevertheless suffer from limitations.

Hypertext linking is provided by the author of a web page, who may have a particular audience in mind or may not be aware of other relevant material. This mechanism cannot be very robust with respect to the generation of new information, which may in principle be useful to interpret earlier resources as few resource providers will keep resources up to date. The *Related Pages* facility improves on the static hypertext linking, as it dynamically identifies interesting relevant pages and therefore supports a user in exploring a domain and even in moving across domains. However, from the point of view of supporting sensemaking, this feature is not very precise and effective, given that the linking is based on lexical similarity and that typically the search space the user needs to explore explodes very quickly, in sharp contrast to the targeted manner in which page designers insert hyperlinks in web resources.

1.1 Motivation

Semantic layering (the realisation of *semantic* links, in addition to the existing *syntactic* ones) is a notion that was introduced in connection with browsing and navigation in early papers about Magpie [3]. Magpie finds and highlights the entities from a particular ontology in the current web page. A pre-condition of a successful parsing within the Magpie-enabled browser is the specification of an ontology-derived lexicon which is used to automatically associating a semantic layer to a web resource. This Magpie-mediated association between an ontology and a web resource provides an *interpretative viewpoint* or *context* over the resource in question. While Magpie has been successfully used in a number of applications in the domains of Climate Science, Agriculture, and Academic Research, it requires the user to select a priori which ontology has to be used. Hence, it suffers from the brittleness that typically affects knowledge-based systems: while it works well within a well-defined domain, as soon as the browsing session moves away from the original domain, the support declines dramatically.

A number of other approaches can be found in the literature. However, most of them, when not extracting data from the pages and offering it out of context in a different interface (PiggyBank [4]), tend to deal with structured content, *e.g.* Sifter [5], and pay no attention to the unstructured web documents. Other tools, which bring in the same interface semantic data within the web page, rely mostly on collaborative annotation, like Trailblazer [7]. While this approach can work for well defined communities in well defined domains, it is nevertheless brittle—the quality of the interpretation support relies on the availability and quality of annotation. Hence, there is the need for a tool which is able to dynamically bring relevant semantic information into a browsing session, without being limited to one particular ontology. The goal of PowerMagpie is to bring to the user, opportunistically, the appropriate semantic information relevant to his current information needs, in principle from any ontology available on the web.

2 Architecture

From a user perspective, PowerMagpie is an extension of a classical web browser, but the main functionalities it provides are actually realized by server-side software that can be seen as a back-end for the whole architecture. In this way, PowerMagpie is particularly easy to install, as it does not require particular software except a JavaScript enabled web browser.

The PowerMagpie server is the central element of the architecture: it is in charge of realising term extraction, ranking and ontology selection and to transmit the results to the browser extension. The term extraction and ranking are done via a hybrid TF*IDF algorithm [6] which weights each term against its popularity in a search engine (Google or Yahoo). The top ranked terms represents a lexical document signature [8] and those will be used to bootstrap ontology selection. Watson [1]—an ontology search engine—which makes it possible to access large scale semantic content with the degree of efficiency that is required

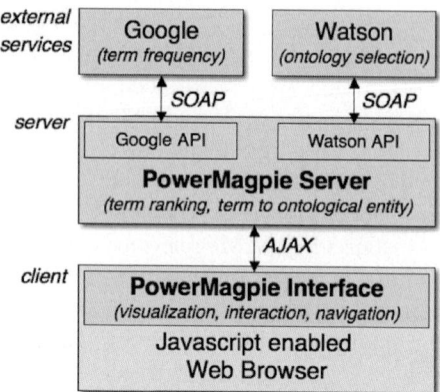

Fig. 1. PowerMagpie architecture

for open semantic browsing. Finally, the interface for the user is in charge of realising the navigation of the semantic and textual information. It is the only part of the architecture that is required to be installed on the client-side, *i.e.* on the user's computer. The advantage of relying on a common server is that the information computed by the PowerMagpie server can be cached, improving the performance for all the users, based on each individual use of PowerMagpie.

3 Demonstration

In our demonstration we will show various functionalities of the tool (*i.e.*, extracting semantic data, navigating the semantic data through the two panels provided—ontology and entity view), applied on web pages of various content while underlying the role of each component within the architecture.

From a user perspective, PowerMagpie is an extension of a classical web browser and takes the form of a narrow, vertical widget displayed on top of the currently browsed web page. This widget provides several functionalities that allow exploring relevant semantic information. A first view, displayed under the *Entities* panel, lists the key terms from the web page, ordered according to their importance to the document. Each term can be extended to display the semantic information (*i.e.*, ontological entities) it is attached to and allows navigating to the text snippet from where it was extracted. A second, ontology centric view is shown in the *Ontologies* panel and displays all the ontologies that were identified as relevant for the page. Each ontology can be extended to explore its semantic entities that are relevant for the web page. While the navigation from the ontological entities to the text is a powerful tool for apprehending the document from a semantic perspective, navigating from text to ontologies is important in a situation in which the user needs a semantic definition of a term, either because he is unfamiliar with it or because additional information is needed. This functionality is provided as follows: whenever a word in the text

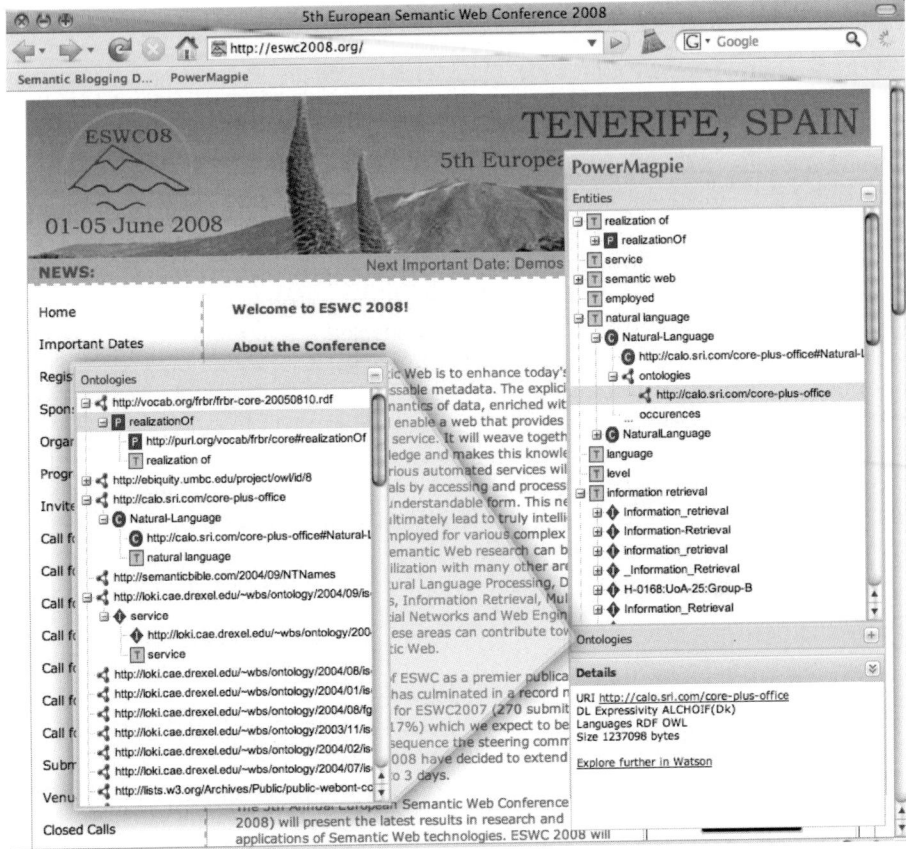

Fig. 2. PowerMagpie screenshot with entity panel in view, ontology panel detailed

of the web page is double-clicked, the PowerMagpie interface moves to the entity panel, to display the semantic description of this term. If the given word has not been already recognised as a key term, then it is automatically added, and the relevant ontologies and ontological entities dynamically selected and explored.

The user will learn—by watching the demo and using the tool himself—how to acquire relevant semantic information and how to make use of it in order to support the sense making process. Because the semantic information is brought in from online available ontologies (via Watson), the user will get a glimpse of the quality of online available data. Further, he will have the chance to use one of the first applications that exploit the semantic web at runtime and experience the added value of such a direct interaction.

A demo of PowerMagpie and a demonstrative video are available at http://powermagpie.open.ac.uk/.

4 Future Work

As the semantic information we can align with the document may be too abstract and of little value to a user which is not a knowledge engineer, we plan to use this information to expand the computed lexical signature of the document to a semantic one—using it as a query expansion approach with a classical lexical search engine (Google or Yahoo) in order to discover semantically related pages. Since query expansion yields high recall and low precision, we will have to filter those results by using the already discovered ontologies in order to validate which of the discovered documents are relevant with the current domain.

We also have to address the presentation of the related semantic information which is cryptic to the regular user by emphasising relations between entities and group them by provenance rather than just displaying URIs as they were intended to be opaque. Regarding the semantically related pages upcoming feature, we intend to address from the previewing of such information to the extent that this information existence may decide which semantic information will be shown to the user, as ontologies that will just define and not 'lead' anywhere would be of little value in semantic browsing.

References

1. d'Aquin, M., Sabou, M., Dzbor, M., Baldassarre, C., Gridinoc, L., Angeletou, S., Motta, E.: Watson: A Gateway for the Semantic Web. In: Proc. of European Semantic Web Conference, ESWC, Poster Session (2007)
2. Domingue, J., Dzbor, M.: Magpie: supporting browsing and navigation on the semantic web. In: Proceedings of the 9th international conference on Intelligent user interface, pp. 191–197 (2004)
3. Domingue, J., Dzbor, M., Motta, E.: Semantic Layering with Magpie. In: Staab, S., Studer, R. (eds.) Handbook on Ontologies in Information Systems, Springer, Heidelberg (2003)
4. Huynh, D., Mazzocchi, S., Karger, D.: Piggy bank: Experience the semantic web inside your web browser (2005)
5. Huynh, D.F., Miller, R.C., Karger, D.R.: Enabling web browsers to augment web sites' filtering and sorting functionalities. In: Proceedings of the 19th annual ACM symposium on User interface software and technology, pp. 125–134 (2006)
6. Manning, C.D., Raghavan, P., Schütze, H.: Introduction to Information Retrieval. Cambridge University Press, Cambridge (to appear, 2008)
7. Johnston, A., Croke, P., Tighe, K.: Using named entities as a basis to share associative trails between semantic desktops. In: Semantic Desktop Workshop (2005)
8. Phelps, T.A., Wilensky, R.: Robust Hyperlinks Cost Just Five Words Each. University of California, Berkeley, Computer Science Division (2000)

Tagster - Tagging-Based Distributed Content Sharing

Olaf Görlitz, Sergej Sizov, and Steffen Staab

ISWeb, University of Koblenz-Landau, Germany
{goerlitz,sizov,staab}@uni-koblenz.de,
http://isweb.uni-koblenz.de

Abstract. Collaborative tagging systems like Flickr and del.icio.us provide centralized content annotation and sharing which is simple to use and attracts many people. A combination of tagging with peer-to-peer systems overcomes typical limitations of centralized systems, however, decentralization also hampers the efficient computation of global statistics facilitating user navigation. We present Tagster, a peer-to-peer based tagging system that provides a solution to this challenge. We describe a typical scenario that demonstrates the advantages of distributed content sharing with Tagster.

1 Introduction

Collaborative Tagging systems have become quite popular in recent years. Their success comes from their ease of use for collaboratively annotating and sharing information objects, e.g. photos in Flickr, videos in YouTube, or bookmarks in del.icio.us. However, centralized systems have a number of serious drawbacks, including limited resource allocation (users are charged for additional contents beyond the strict limitation of free space), vulnerability to denial-of-service attacks with possible temporal unavailability of the service, and the need to sign up with multiple services (which all need to be trusted) when different types of resources shall be shared.

Modern peer-to-peer (P2P) systems instead are self-organizing, decentralized infrastructures which can handle huge amounts of available resources. To this end, they are highly attractive for content sharing applications with collaborative annotation (tagging) of contents. In this demo paper we describe *Tagster*, a distributed content sharing application with embedded tagging functionality. It comes as a small Java client program that, similar to a normal file browser, allows the owner to navigate through his locally stored resources (e.g. media files) and to assign arbitrary text labels (tags) to them. Additionally, the tagging information becomes instantly available for all other users that are connected to the Tagster Peer-to-Peer (P2P) overlay network. A distributed index structure ensures that tagging metadata is always available to all other users even if not all peers are permanently online. Furthermore, based on this index structure, Tagster incorporates a tag-based user characterization that takes into account the global tag statistics for better navigation and ranking of resources.

2 Using Tagster

As a motivation example for Tagster, we consider the common scenario of conference participation for computer science researchers.

Our sample user Tom is attending the ESWC 2008 conference. He has submitted two contributions, downloaded three other papers he's interested in, and he has already taken 200 photos and 10 video clips at several conference locations. All that data is stored on his laptop and Tom is looking for an easy way to organize it and share the resources with colleagues and friends.

Using Tagster, Tom annotates his resources with 'eswc' and '2008' by selecting the respective files and folders and typing in these tags. He also adds annotations 'paper' and 'publication' to the papers and some contextual annotations like names (e.g. 'Bill', 'George') and the location (e.g. 'Tenerife') to his photos and videos. Additional media-specific information like photo resolution, format, or creation timestamp is automatically extracted by Tagster from the corresponding media files and added as further tags to their annotation.

After organizing all files, Tom likes to get an overview over the annotations of his resources. The annotation summary is represented by a so-called 'tag cloud' which visualizes aggregated statistics of the tag usage. Tom realizes that his preferable tags for the ESWC trip are 'eswc', '2008' and 'photo'. The click on a tag in the tag cloud opens an overview of associated resources. Multiple tag selections narrow down the set of displayed items.

In the next step, Tom wants to share his resources with other ESWC participants and also see the resources of other colleagues. With Tagster it does not require any additional effort as all tagging metadata is automatically published in the corresponding peer-to-peer overlay network. Thus, searching for 'eswc' and 'photo' will not only return Tom's own locally stored media files but also a list of all resources tagged with those two tags by all other users connected to the overlay network.

Assuming that Tom's search returns a number of photos taken by Bill. Tom can select/download particular resources of Bill and also display their annotations. Furthermore, he can list all resources offered by this user. For better navigation, Tagster displays Bill's profile including an overview of his shared resources, and the corresponding user-characteristic tag cloud.

Finally, Tom aims to find in the network other users with similar interests. A common way for similarity search is the analysis of user-characteristic tag clouds. To return the ranked list of most relevant recommendations, Tagster internally maps the user-specific tag clouds onto vectors in a multi-dimensional feature space. The values of particular features are constructed with respect to local frequencies of tag occurrences in the user's own data, and global tag statistics, analogously to tf*idf feature weighting known from text IR. An important point is the accurate approximation of global tag statistics on particular peers in a decentralized overlay network, which aims to avoid unnecessary high communication overhead.

3 Design Choices and Related Work

One drawback of centralized tagging platforms is the limitation to a single media type. Some exceptions exist, like BibSonomy[1] (bookmarks + bibtex), sevenload[2] (pictures + video), or technorati[3] (blogs + video). But still they are far from being a comprehensive platform for organizing all types of personal data. MyTag[4] partially overcomes this limitation by providing a single interface to retrieve combined results from Flickr, YouTube, and del.icio.us. However, manipulating the data is not possible. Therefore, Tagster is designed as a distributed application for organizing any type of locally stored data and globally sharing the tagging metadata.

Peer-to-peer systems have drawn a lot of attention in the last decade. Two major types of peer-to-peer systems exist: unstructured and structured ones.

Unstructured peer-to-peer networks are, e.g., Gnutella[5] and its successors. For example, Bibster[3] is a peer-to-peer application based on an unstructured network and that allows for sharing bibliographic data. The bibliographic description and search is based on an ontology. Query propagation is done by semantic routing based on content similarity. However, the centrally-defined ontology impedes user annotation and distributed tagging is hampered as it is difficult to determine global tag use - leading to problems with search and browsing scalability.

Structured peer-to-peer networks, like distributed hashtables (DHT) as Chord[8] or P-Grid[1], have the advantage that every piece of distributed information can be located with low overhead, usually within $\log n$ hops where n is the number of peers in the network. There are also sophisticated replication mechanisms available to cope with offline or frequently leaving peers.

The recent popularity of tagging systems has also increased research efforts in order to understand tagging behavior and semantics, or to improve access to data found in such tagging systems. For instance, folkrank[5] is a PageRank like mechanism for recommending resources.

For Tagster, we have focused on approaches based on the vector space model that are suitable for DHT-based P2P systems. An important problem in peer-to-peer systems is the cardinality estimation for item sets, which is necessary for constructing feature weights in our approach. It is possible to directly exploit the underlying network structure [4], as with DHTs, or use a gossip-based approach [6]. However, tracking a huge number of cardinalities cannot be efficiently implemented in such a way. Therefore, we have developed and integrated PINTS, an original dynamic algorithm for computing and updating such feature vectors. The explanation of PINTS is beyond the scope of this demo description and we refer the reader to [2].

---

[1] http://bibsonomy.org
[2] http://sevenload.com
[3] http://technorati.com
[4] http://mytag.uni-koblenz.de
[5] http://gnutella.org

4 System Architecture

All tagging data, i.e. the tag assignments relation between user, tag, and resource, is represented as RDF[6] triples conforming to a tagging ontology. With the possibility to integrate more semantic tagging information and having different tagging ontologies for different users, this provides the necessary flexibility and scalability for future development. Sesame[7] is used as a persistent RDF data store. Queries against the repository are done with the SPARQL[8] query language which allows for flexible and complex queries on the tagging data.

Tagster implements a distributed index structure that holds user-tag, user-resource, and resource-tag relations. For example, accessing a resource ID in the index will return all tags assigned to that resource and all users who tagged it. The index has been realized with Bamboo[9], a distributed hashtable (DHT) implementation similar to Pastry[7] which allows for efficient access to the stored data. DHTs generally ensure that within a given id space, evenly distributed among all peers, all key/value pairs get assigned to exactly one responsible peer and can be stored and retrieved with at most $log(n)$ routing hops. When selecting a tag in Tagster, both the local repository and the distributed index are queried to retrieve the respective resources associated with the tag as well as all users using the tag.

5 Tagging Statistics

The knowledge of global statistics about tagging data is useful for different purposes. The frequency of tags, for example, as seen in the whole system or of different users, together with the tags global popularity, can be used to find similar resources or users with similar interests. Moreover, information about frequently co-occurring tags is helpful for discovering semantic relations in the tagging data. The latter can then be used to identify concepts and construct concept hierarchies or simply for clustering tagging data.

The big challenge in a distributed tagging system, however, is to efficiently gather all collection statistics. The main problem is that the user's statistics are kept at the user peer while a collection's cardinality information is stored at the index peer and both are updated independently of each other. In the naive case, an update would require to contact all peers in the network to see if the current statistics is still correct. Obviously, this is not feasible in terms of scalability and message complexity. Instead, we need to estimate these collection cardinalities. Since tagging systems evolve and new tags and resources are continuously added it also has to be flexible enough to accommodate to these changes.

We have developed and implemented such a dynamic algorithm[2] for Tagster. It predicts the future collection cardinality development and automatically

---
[6] http://www.w3.org/RDF/
[7] http://openrdf.org/
[8] http://www.w3.org/TR/rdf-sparql-query/
[9] http://bamboo-dht.org

updates the respective feature vector entries if the actual deviation violates a predefined error margin. This ensures consistent user statistics in the whole network while keeping the peer-to-peer networks message complexity as low as possible.

6 Outlook

The next steps in the development of Tagster will include (i) the collection of experience data from Tagster use, (ii) sophisticated means for recommending resources from the distributed peers, and (iii) additional wrappers for facilitating data collection from further semantic and non-semantic sources. With such support Tagster may offer a viable open alternative to closed, centralized systems.

Our long-term objective is the efficient and effective infrastructure for decentralized, self-organizing Web 2.0 applications which allows for scalable sharing, annotation, searching, and browsing of relevant resources.

Acknowledgements. This work has been supported by the EU FP6 research project Tagora - Semiotic Dynamics in Online Social Communities (IST-2006-34721, http://tagora-project.eu).

References

1. Aberer, K., Cudré-Mauroux, P., Datta, A., Despotovic, Z., Hauswirth, M., Punceva, M., Schmidt, R.: P-Grid: A Self-organizing Structured P2P System. SIGMOD Record 32(3), 29–33 (2003)
2. Görlitz, O., Sizov, S., Staab, S.: PINTS: Peer-to-peer infrastructure for tagging systems. In: Proceedings of the Seventh International Workshop on Peer-to-Peer Systems, IPTPS 2008, Tampa Bay, USA (February 2008)
3. Haase, P., Broekstra, J., Ehrig, M., Menken, M., Mika, P., Plechawski, M., Pyszlak, P., Schnizler, B., Siebes, R., Staab, S., Tempich, C.: Bibster - a semantics-based bibliographic peer-to-peer system. In: McIlraith, S.A., Plexousakis, D., van Harmelen, F. (eds.) ISWC 2004. LNCS, vol. 3298, pp. 122–136. Springer, Heidelberg (2004)
4. Horowitz, K., Malkhi, D.: Estimating network size from local information. Inf. Process. Lett. 88(5), 237–243 (2003)
5. Hotho, A., Jäschke, R., Schmitz, C., Stumme, G.: Information retrieval in folksonomies: Search and ranking. In: Sure, Y., Domingue, J. (eds.) ESWC 2006. LNCS, vol. 4011, pp. 411–426. Springer, Heidelberg (2006)
6. Jelasity, M., Montresor, A.: Epidemic-style proactive aggregationin large overlay networks. In: Proceedings of The 24th International Conference on Distributed ComputingSystems (ICDCS 2004), Tokyo, Japan, pp. 102–109. IEEE Computer Society, Los Alamitos (2004)
7. Rowstron, A.I.T., Druschel, P.: Pastry: Scalable, decentralized object location, and routing for large-scale peer-to-peer systems. In: Guerraoui, R. (ed.) Middleware 2001. LNCS, vol. 2218, pp. 329–350. Springer, Heidelberg (2001)
8. Stoica, I., Morris, R., Karger, D.R., Kaashoek, M.F., Balakrishnan, H.: Chord: A scalable peer-to-peer lookup protocol for internet applications. In: Proc. of the ACM SIGCOMM, San Diego, August 2001, pp. 149–160 (2001)

The Web Service Modeling Toolkit

Mick Kerrigan and Adrian Mocan

Semantic Technology Institute (STI) Innsbruck,
Universität Innsbruck, Austria
firstname.lastname@sti2.at

Abstract. The development of software is not an easy task and the availability of adequate tool support is an important step towards reducing the effort that a developer must put into the Software Development Cycle. As an emerging technology, it is vital that Semantic Web Services can be quickly and easily created by developers to ensure that this new technology can be easily adopted. In this demo the process of developing Semantic Web Service descriptions, through the WSMO paradigm, using the Web Service Modeling Toolkit (WSMT) will be presented.

1 Introduction

Web services are quickly becoming one of the most important technologies for business to business integration; however some of the promises claimed by Service Oriented Architectures in terms of loose decoupling and easy reusability have failed to be met, as the ability to dynamically find and link Web services at runtime cannot be realistically performed with syntactic technologies. Semantic Web Services are the extension of ontologies to describe Web services in such a way that they can be dynamically *discovered*, *composed*, *ranked*, *selected*, *mediated* and *invoked* at runtime. Such functionality reduces the amount of effort that a developer must spend building an application using a Service Oriented Architecture and improves the overall quality of that application as new services, which may be cheaper or have better performance, become immediately available to the application once they are published by the provider.

The life of the Semantic Web Service developer is not such an easy one though. As noted in 1994 by the Standish group chaos report[10], 31% of all software development projects in their survey failed, i.e. were canceled at some point during the development cycle, another 53% where completed but over-budget, over the estimate time or provided fewer functions and features than originally intended, thus leaving only 16% of all software projects surveyed as completing successfully. Given that the software projects were using well establish technologies like Java, C# or Visual Basic with good tool support provided by development environments equivalent to the Eclipse Java Development Toolkit[1], SharpDevelop[2] and Visual Studio[3], it would be surprising if the Semantic Web Service developer could do much better without any tool support at all.

---

[1] http://www.eclipse.org/jdt/
[2] http://www.icsharpcode.net/
[3] http://msdn.microsoft.com/vstudio/

The Web Service Modeling Toolkit (WSMT)[6,7] is an integrated development environment for Semantic Web Services that enables developers to develop Ontologies, Web Services, Goals and Mediators through the Web Service Modeling Ontology (WSMO)[3] formalism. The WSMT is implemented as a collection of plug-ins for the Eclipse[4] framework such that it can be integrated with other toolkits like the Java Development Toolkit JDT or the Web Tools Platform (WTP)[5] so that a developer can develop his java code, Web services and Semantic Web Services side by side in the one application. The main aim of the WSMT is to support the developer through the full Software Development Cycle of his Semantic Web Service from requirements, through design, implementation, testing, and deployment such that the process of developing Semantic Web Services can become cheaper to perform and remove many of the tedious activities that the developer must currently perform.

2 The Web Service Modeling Toolkit

The WSMT has been under development since early 2005 and is made up of three main areas of functionality:

- **Creation and Management of WSML Artifacts:** The ability to quickly and cheaply create and test WSMO Ontologies, Web Services, Goals and Mediators through the Web Service Modeling Language (WSML)[8] is key to the successful creation of Semantic Web Services. The WSMT provides the **WSML perspective** with multiple editors[5] for creating and testing WSMO descriptions, conversion tools to and from RDF and OWL, embedded reasoners for testing the behavior of ontologies in their target environment, and embedded discovery engines for ensuring that Goals and Web Services match each other as expected.
- **Creation and Management of Mediation Mappings:** One of the key challenges in semantics is the interoperability of ontologies. In the Semantic Web Service field this becomes even more important when the service requester and service provider use different ontologies to describe the same domain. The WSMT provides the **Mapping Perspective**[1] within which mediation mappings between two or more ontologies can be created at design time, such that they can later be executed at runtime. The tools in this perspective guide the developer through the process of creating mappings using visual cues, suggestion algorithms and embedded testing functionality.
- **Interfacing with Semantic Execution Environments:** Crucially once all the artifacts related to a Semantic Web Service have been created these artifacts need to be deployed the execution environment within which they will be used. The **SEE Perspective** provides functionality for interfacing with Semantic Execution Environments like the Web Service Execution Environment (WSMX)[4] and IRSIII[2]. Artifacts can be stored to and retrieved from

---

[4] http://www.eclipse.org
[5] http://www.eclipse.org/webtools/

these environments, or can used to invoke the functionality of the Semantic Execution Environments, for example discovering services that match, or fully invoking the best service that matches, a provided Goal.

The WSMT development team is also in the process of branching out into new areas including the encoding of business processes semantically in WSML through the many projects we are involved in. These new tools will form a number of new key WSMT perspectives that will support the developer in creating Semantically Enabled Service Oriented Architectures.

3 Contents of the Demo

The demo of the Web Service Modeling Toolkit takes the user through the different tools that are available for developing those artifacts related to Semantic Web Services and how these tools interact with each other. The demo begins by introducing the user to the different editors available for creating WSML Ontologies, Web Services, Mediators and Goals based on the expertise of the user.

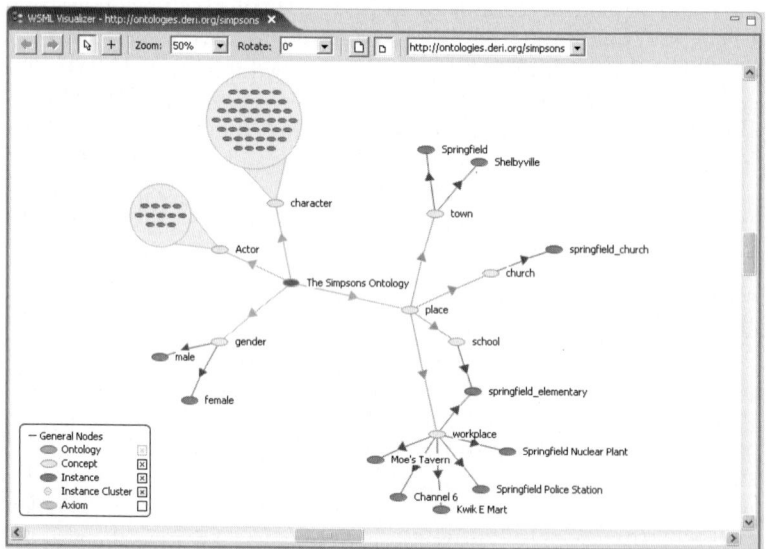

Fig. 1. An Ontology in the WSML Visualizer

The WSML Visualizer[5], as can be seen in figure 1, can be used by novice users who know very little about ontologies to get an overview of WSMO descriptions or by expert users to dig deeper into the intricacies of the relationship between elements in the semantic description. Crucially the WSML Visualizer also has fully embedded editing support allowing the user to directly manipulate the semantic description vis the graph representation. The WSML Form Based Editor[7] provides a more functional view of the semantic description providing custom forms to the user which can filled in to create and maintain semantic descriptions. Finally the WSML Text Editor[7] is available to the user who is more experienced with the Human-readable syntax of WSML, it provides functionality like syntax highlighting, syntax and content assistance, code folding, and bracket highlighting. At this stage the user is also introduced to the validation support available in the WSMT, which is used to give immediate feedback to the developer whenever they make mistakes.

Now that the user is familiar with the different types of editing support available for WSML descriptions, and has an understanding of how to create and maintain them, they are introduced to the WSML Reasoner View[7] within which queries can be executed over the currently visible WSMO Ontology. The reasoner view gives the developer the opportunity to execute WSML queries over WSML-Flight and WSML-Rule ontologies using the IRIS, MINS or KAON2 reasoners and over WSML-DL ontologies. The user is introduced to the Query template preferences, that enables the creation of complex logic programming queries by a domain expert and are later displayed to the end user as natural language queries, where drop downs and text fields can be used for configuring them. These query templates enable those that are not familiar with logics to access the power of reasoning over ontologies.

Having introduced ontologies as the basis upon which Semantic Web Services are built, the user is brought back to the concept of Goals and Web Services with the WSML Discovery View[6]. The WSML Discovery View provides access to a number of underlying discovery engines with which the developer can ensure that a specified Goal matches the expected set of Semantic Web Services. This is especially important as a Semantic Web Service that does not match the expected Goal is essentially unfindable by service requesters and thus will never be used in applications. The discovery view can also give a competitive advantage to new service providers on the market, allowing them to test that their new Semantic Web Services match the sample Goals provided by existing providers on the market. Thus they can attempt to take business from its competitors by providing services with better quality of service or lower cost.

As already mentioned the importance of the ability to mediate between different ontologies is necessary in the heterogeneous world of the web. Thus the user is introduced to the WSMT Mapping perspective and particularly the View Based Editor[9] for the semi-automatic creation of ontology to ontology mappings for runtime instance transformation[1] with the WSMX[4] environment. At runtime these mappings are used within WSMX to transform instances provided by the service requester into the terms expected by the service provider and vice versa. Alongside the view based editor the user is presented with the MUnit view, which can be used for creating unit tests for mediation mappings. With this view the developer can define tests consisting of a set of source instances and a set of target instances. The tests can be executed whenever the ontologies or mappings evolve in order to be sure that the mappings are still valid.

Acknowledgements

The work is funded by the European Commission under the projects ASG, DIP, enIRaF, InfraWebs, Knowledge Web, Musing, Salero, SEKT, SEEMP, SemanticGov, Super, SHAPE, SWING and TripCom; by the FFG (Österreichische ForschungsFörderungsGeselleschaft mbH) under the projects Grisino, RW, SemNetMan, SeNSE, TSC, OnTourism.

References

1. Mocan, A., Cimpian, E.: An Ontology-based Data Mediation Framework for Semantic Environments. International Journal on Semantic Web and Information Systems (IJSWIS) 3(2), 66–95 (2007)
2. Cabral, L., Domingue, J., Galizia, S., Gugliotta, A., Norton, B., Tanasescu, V., Pedrinaci, C.: IRS-III: A Broker for Semantic Web Services Based Applications. In: Proceedings of the 5th International Semantic Web Conference (ISWC 2006), Athens, Georgia, USA (2006)
3. Fensel, D., Lausen, H., Polleres, A., de Bruijn, J., Stollberg, M., Roman, D., Domingue, J.: Enabling Semantic Web Services – The Web Service Modeling Ontology. Springer, Heidelberg (2006)
4. Haller, A., Cimpian, E., Mocan, A., Oren, E., Bussler, C.: WSMX - A Semantic Service-Oriented Architecture. In: Proceedings of the International Conference on Web Services (ICWS 2005), Orlando, Florida, USA (July 2005)
5. Kerrigan, M.: WSMOViz: An Ontology Visualization Approach for WSMO. In: Proceedings of the 10th International Conference on Information Visualization (IV 2006), London, England (July 2006)
6. Kerrigan, M., Mocan, A., Tanler, M., Bliem, W.: Creating Semantic Web Services with the Web Service Modeling Toolkit (WSMT). In: Proceedings of the workshop on Making Semantics Work For Business (MSWFB2007) at the 1st European Semantic Technology Conference (ESTC2007), Vienna, Austria (May 2007)
7. Kerrigan, M., Mocan, A., Tanler, M., Fensel, D.: The Web Service Modeling Toolkit - An Integrated Development Environment for Semantic Web Services (System Description). In: Proceedings of the 4th European Semantic Web Conference (ESWC2007), Innsbruck, Austria (June 2007)
8. Lausen, H., de Bruijn, J., Polleres, A., Fensel, D.: WSML - A Language Framework for Semantic Web Services. In: Proceedings of the W3C Workshop on Rule Languages for Interoperability (April 2005)
9. Mocan, A., Cimpian, E.: Mapping creation using a view based approach. In: 1st International Workshop on Mediation in Semantic Web (2005)
10. Standish Group. The CHAOS Report. Technical report, Standish Group (1994)

Mymory: Enhancing a Semantic Wiki with Context Annotations

Malte Kiesel, Sven Schwarz, Ludger van Elst, and Georg Buscher

Knowledge Management Department
German Research Center for Artificial Intelligence DFKI GmbH,
Trippstadter Straße 122, 67663 Kaiserslautern, Germany
{firstname.lastname}@dfki.de

Abstract. For document-centric work, meta-information in form of annotations has proven useful to enhance search and other retrieval tasks.

The Mymory project[1] uses a web-based workbench based on the semantic wiki *Kaukolu* that allows annotating texts both with concepts modeled in the user's personal information model and other ontologies in a flexible way. Annotations get enriched with contextual information gathered by a context elicitation component. Reading annotations are created with the help of an eyetracker.

In the demonstration, we use contextualized annotations and semantic search using annotations in order to support knowledge workers in the domain of software licenses.

1 An Introduction to the Mymory Scenario

Knowledge-intensive work often means working with multiple sources of information in parallel, reading large amounts of text, skimming even more text, creating digests, (subjectively) rating information, and many other things. Some of these texts may be thoroughly read and understood very closely, with the respective amount of cognitive processing and resulting in artifacts like annotations, summaries, citations in other texts, etc. For this demonstration, we take the domain of software licenses as an example: License texts are a typical complex texts that are difficult to read completely, that must be interpreted in multiple contexts, and that benefit a lot from both generic and personal annotations: Law experts and knowledge engineers read software license texts and create fine-grained annotations that formalize license texts. Experts and engineers can also use simple highlighting and rating annotations as personal notes useful during their formalization work. People needing to choose a software license for their project can use search and annotations created before to find licenses of interest. License text passages read by people in context of their current project get associated automatically with that project through contextualized reading annotations.

---

[1] http://www.dfki.uni-kl.de/mymory

2 Key Technologies Used in Mymory

The Mymory workbench is a comprehensive environment for document-centric work, especially for reading and writing text documents, annotating them, searching in the document pool, etc. In the following we will present some of the (technical) approaches elaborated in the Mymory project and how these technical components handle the above mentioned issues of our scenario. Especially, we deal with

- a *Semantic Wiki* as a central hub for reading, writing, and annotating,
- *automated annotation* based on a user's reading behaviour as well as on context elicitation technology,
- *manual annotation support* with concepts from the user's Personal Information Model and other ontologies, and
- an *ontology-based search* interface which allows for dynamic assembly of faceted queries.

The Mymory workbench is implemented by Kaukolu[2], a semantic wiki research prototype. Its annotation features are demonstrated in Figure 1, depicting a software license text and its annotations. A Kaukolu annotation is an RDF resource (typically an instance of a subclass of an `Annotation` class) associated with a part of wiki text. Contrary to most other semantic wikis that typically allow annotation of links and complete pages only, any text part can get annotated here[3].

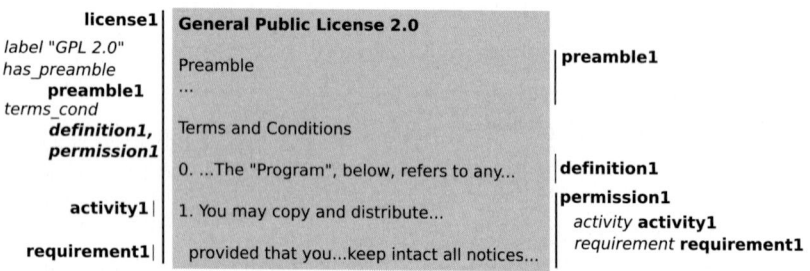

Fig. 1. Structure of software license anntations in Kaukolu

Creating annotations manually is done using a form-based approach that fetches possible annotation classes from the Mymory RDF repository. Additionally, an **eyetracker**[4] can create reading annotations automatically. For example, these annotations allow searching for individual text passages that have

---

[2] http://kaukoluwiki.opendfki.de/
[3] Character offsets of the annotated text are stored and updated on markup edits by a modified text diff algorithm.
[4] We use a third-party eyetracker integrated in a display—no glasses or other additional equiment is needed.

been read closely. For further information about the actual eyetracker's workings, see [1]. Whenever an annotation is created, information about the current user context is attached to it. This allows filtering annotations when doing search. User context consists of author and creation time of the annotation plus the most relevant PIMO concepts currently active. **PIMO** is the *Personal Information Model (Ontology)*, a layered approach to Personal Information Models [2]. A user's PIMO contains persons, topics, companies, and projects of importance to her as well as relationships between those entities. **Automatic user context capturing** keeps track of the *attention distribution* of PIMO concepts in form of a set of PIMO concepts along with activation values. This set is created by an additional component that hooks into key applications on the desktop. It exploits usage information and analyses documents currently used to detect PIMO concepts relevant in current context. A data structure for that (dynamic) distribution is kept together with a unique identifier (URI) for that context. If the user switches to another context, Mymory detects this and changes the current context URI as well as PIMO activation values accordingly. As the PIMO attention distribution is kept separately for each context, one context's distribution is not "polluted" by other context's attentions. Finally, annotations are used in **semantic search** that returns wiki text paragraphs satisfying search critera as results. Search criteria can be various types of metadata; searching for authors of texts is supported as well as searching for text annotated using specified annotations and their context. In the end, this allows searching for text passages read in a context of a certain project, or searching for text passages annotated with a domain-specific statement such as "Here, the term *derivative work* is defined" for the software license domain.

3 System Demonstration

The live demonstration uses a document set consisting of software license texts, project websites, and wikipedia articles. It uses a software license ontology (for formalization of software license semantics), a personal ontology (containing projects and people the user knows and that is used for modeling context), and a generic annotation ontology (facilitating simple annotations such as rating texts or highlighting). In the following, we give example use cases of the features used in the demonstration.

Example 1: We want to annotate a software license text using instances according to a license ontology. Kaukolu's form-based annotation feature is used for this task. In the end, we get fine-grained annotations in the license text that on one hand decompose the text into subparts (sections containing term definitions, etc.) and on the other hand relate statements in the text semantically with each other (*Here, copying of the software is allowed under the condition Y*).

Example 2: Imagine Arthur, a Nepomuk member, given the task to identify potential collaboration with the Mymory team. He has been informed that collaboration has taken place recently. As both projects use the Kaukolu wiki, he

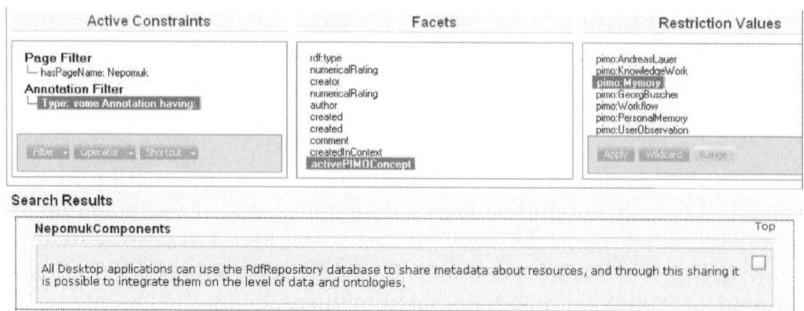

Fig. 2. Searching for text passages in Nepomuk pages with Mymory context

decides to "ask" the wiki about text passages about Nepomuk with Mymory context (and vice versa). In other words, he searches the wiki using two filters: i) passage is part of a page with a name containing `Nepomuk`; ii) passage has an annotion with `activePimoConcept Mymory` in its context. Figure 2 shows a text passage found in a Nepomuk page annotated with Mymory context.

Example 3: We want to take a look at text passages in license texts that we have read some time ago. Reading annotations shown in Figure 3 generated by the eyetracker are used for this during search.

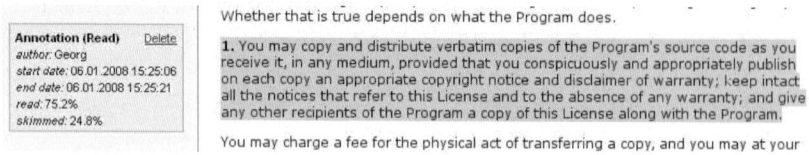

Fig. 3. A reading annotation created by the eyetracker

Example 4: We want to get an overview of the individual software licenses used in components of the Nepomuk project. Therefore, we want to retrieve text passages read in context of the project *Nepomuk* and the topic *Software License*. This can be done by searching for text paragraphs that have been annotated with the according PIMO concepts during reading.

Example 5: After a while of using the annotation feature, we get a very large number of annotations per document. This is especially true in a multi-user environment and when using the eyetracker—then, for every user every paragraph read gets annotated with a *Read* annotation automatically. Therefore, just showing every annotation as highlighted text passages does not scale, rendering documents unreadable. In this example we use a filtering feature that shows only annotations relevant in the current user's context. Also, we can show annotations not only as direct highlighting but also as less obtrusive icons in the document's margin which also helps keeping the system usable.

4 Conclusion

In the demonstration, we have shown that *annotations enhance the understanding and management of documents*, especially when the documents are used by more than one person and for more than one context. Annotations allow additional filters to be applied to search, which can *enhance retrieval precision*—this holds particularly for contextual annotations, respectively contextual filtering. The possibility to annotate whole wiki pages as well as small text passages with PIMO concepts allows fine-grained semantic markup of such text and enables *semantic search for information elements*. Automatic elicitation of the user's context, based on user observation providing a stream of contextual evidence, allows automatic contextualizations of annotations or even *automatic generation of attentional annotations* (e.g., annotating that a text passage has been read/skimmed by the user). We have demonstrated how storing the user's context as part of the annotations enables *contextual search/filtering* in retrieval scenarios.

Acknowledgement

The Mymory project is funded by the Bundesministerium für Bildung und Forschung (Federal Ministry of Education and Research) under grant 01 IW F01. For this work, many ideas and components from the NEPOMUK project have been used. NEPOMUK is funded by the IST Programme of the European Union under grant FP6-027705.

References

1. Buscher, G.: Attention-based information retrieval. In: SIGIR 2007: Proceedings of the 30th annual international ACM SIGIR conference on research and development in information retrieval (doctoral consortium) (2007)
2. Sauermann, L., van Elst, L., Dengel, A.: Pimo - a framework for representing personal information models. In: Pellegrini, T., Schaffert, S. (eds.) Proceedings of I-MEDIA 2007 and I-SEMANTICS 2007 International Conferences on New Media Technology and Semantic Systems as part of TRIPLE-I 2007, J. UCS, pp. 270–277. Know-Center, Austria (2007)

Pronto: A Non-monotonic Probabilistic Description Logic Reasoner

Pavel Klinov

The University of Manchester, Manchester, M13 9PL, UK
pklinov@cs.man.ac.uk

Abstract. The demonstration presents Pronto - a prototype of a non-monotonic probabilistic reasoner for very expressive Description Logics. Pronto is built on top of the OWL DL reasoner Pellet, and is capable of performing default probabilistic reasoning in the Semantic Web. It can handle uncertainty in terminological and assertional DL axioms. The demonstration covers Pronto's features and capabilities as well as current challenges and limitations. It describes how an involved realistic problem of breast cancer risk assessment can be formalized in terms of probabilistic reasoning in Pronto. As an important outcome, it is anticipated that attendees should learn and better understand the potential of ontology based approaches to modeling problems involving reasoning under uncertainty.

1 Introduction

One of the limitations of current Description Logic (DL) reasoners is the inability to handle uncertain knowledge. It is a serious obstacle to the expansion of the Semantic Web because many domains of human interest contain knowledge that cannot be represented with absolute certainty. One example of an uncertain domain is medicine, in particular, disease diagnosing. Symptoms, causes and consequences of many diseases are uncertain which complicates conceptualization of such domains in formal ontologies and thus restricts machine understanding.

This demonstration presents Pronto - a probabilistic DL reasoner prototype [1]. Pronto is an attempt to provide reasoning services for P-$\mathcal{SHIN}$(D) - a very expressive formalism that is a probabilistic generalization of OWL with the exception of nominals [2]. In addition to presenting the Pronto's features and capabilities, the demo displays how Pronto can aid in the probabilistic modeling of a realistic medical problem by providing representation and reasoning services for constructing and using a probabilistic ontology for breast cancer risk assessment.

2 Research Background

Pronto provides means for the representation of uncertain ontological statements and offers a collection of reasoning services. It uses the syntax of Lukasiewicz's conditional constraints to express uncertain OWL axioms [2] and Lehmann's lexicographic entailment to perform default probabilistic reasoning [3] [2].

2.1 Representation of Uncertain Knowledge

Pronto represents uncertain ontological knowledge using P-$\mathcal{SHIN}$(D) formalism [2] that is a probabilistic generalization of the very expressive DLs $\mathcal{SHIN}$(D). It is heavily based on the earlier developed approach to default probabilistic reasoning with *conditional constraints* [4]. In the context of P-$\mathcal{SHIN}$(D), conditional constraints are expressions of the form $(D|C)[l, u]$ where C and D are arbitrary $\mathcal{SHIN}$(D) concepts and $[l, u]$ is a closed interval within $[0, 1]$.

Using conditional constraints Pronto is capable of representing both, generic and individual uncertainty. For example, a generic constraint $(Fly|Bird)[0.9, 0.95]$ where Fly and $Bird$ are DL concepts, can express the uncertainty that randomly picked bird can fly. Similarly, a constraint $(Penguin|\top)[0.7, 0.8]$ for individual $Tweety$ captures the uncertainty in $Tweety$ being an instance of $Penguin$.

The important feature of P-$\mathcal{SHIN}$(D) supported by Pronto is the ability to capture *default* terminological knowledge. That is, TBox constraints are default in the sense that they represent statements that are generally true but might fail for some specific individuals. They can also be overridden by more specific statements. The demo will exemplify this possibility.

2.2 Default Probabilistic Reasoning

The main reasoning task in P-$\mathcal{SHIN}$(D) is entailing new conditional constraints, both, terminological and assertional, from probabilistic knowledge bases. Given that standard notion of logical entailment is too weak [4], it has been proposed to use Lehmann's lexicographic entailment that obeys desirable non-monotonic properties [3]. In particular, it allows to resolve conflicts between conditional constraints by preferring some constraints to others [4].

Computing probabilistic entailments involves other reasoning procedures, such as probabilistic satisfiability and probabilistic consistency. Satisfiability is defined in a traditional way, i.e., as a problem of determining whether given probabilistic knowledge base has a model. Consistency has no analogue in monotonic DL reasoning. Knowledge base is *consistent* iff given a probabilistic entailment relation, it is possible to resolve all conflicts during reasoning.

All aforementioned reasoning procedures are implemented in Pronto.

3 Technological Basis

One of the principal requirements for Pronto was that the uncertainty could be gradually introduced into existing OWL ontologies and that the existing OWL reasoning services should be retained. To meet that requirement, Pronto was designed on top of the OWL reasoner Pellet [5] that performs reasoning with the classical part of ontologies and provides routines for higher level probabilistic reasoning procedures, e.g., lexicographic entailment.

Pronto also employs methods for solving linear optimization and related subproblems. Following Lukasiewicz's approaches to probabilistic default reasoning and probabilistic logic programming, Pronto reduces the probabilistic satisfiability (PSAT) to the problem of solvability of a corresponding linear system. In addition, lexicographic entailment is reduced to a number of logical entailments each of which is computed by performing linear optimization. Pronto makes use of an LP solver to perform these tasks [1].

Finally, Pronto uses OWL 1.1 axiom annotations to associate probability intervals with uncertain OWL axioms.

4 Demonstration

It has been chosen to use Life Sciences domain, in particular, medical informatics to demonstrate Pronto's capabilities. Medical informatics has a successful history of using ontologies for modeling domain knowledge [6]. At the same time, medical domain knowledge is often uncertain especially when it is required to diagnose diseases or estimate risks of developing them in the future.

One relevant medical problem is assessing women risk of developing breast cancer. Given that neither all the relevant risk factors are known nor their impact is sufficiently investigated, the assessment cannot be done with absolute certainty. There have been proposed few statistical approaches to probabilistically estimate the breast cancer risk, for example, Gail model used by the National Cancer Institute (NCI) risk calculator[2] [7].

The demo aims to show how Pronto might aid in approaching the same problem by incorporating the statistical knowledge into a cancer ontology. It will first present how the problem can be modeled using a probabilistic ontology, and second, how risk assessment can be performed on the basis of probabilistic default reasoning. Extended version of web based ontology browser OWLSight [3] will serve as the interface to Pronto.

4.1 Probabilistic Model Demonstration

The model of breast cancer risk assessment consists of two major parts - classical OWL ontology and the probabilistic part that represents domain uncertainty. The classical part of the ontology models two types of risk of developing breast cancer. First is *absolute* risk, i.e., the risk that can be measured without the reference to other categories of women. Statements like "*an average woman has up to 13.2% of developing breast cancer in her lifetime*" are the examples of the absolute risk. Second, the ontology models relative breast cancer risk, i.e., the risk comparably to an average woman. Statements like "*having BRCA1 gene mutation increases the risk of developing breast cancer by a factor of four*" are the examples of the relative risk.

The ontology defines risk factors that are relevant to breast cancer. It makes the distinction between the factors that should be known to a woman, e.g., age,

---

[1] Operations Research library, available at: http://opsresearch.com/OR-Objects/
[2] http://www.cancer.gov/bcrisktool
[3] Standard version is available at: http://pellet.owldl.com/ontology-browser/

family cancer history, breastfeeding, and those that can only be inferred on the basis of other factors or by the examination, e.g., BRCA gene mutation, breast and bone densities, etc.

Following the assumption that the subjective probabilities representing risk factors for a certain individual can be combined with objective probabilities representing the statistical knowledge, the probabilistic part combines a set of uncertain ABox and TBox axioms [4]. ABox axioms define risk factors that are relevant to a particular individual. TBox axioms model generic probabilistic relationships between the risk factors and classes of women.

It will be shown how to express various dependencies between risk factors. One possibility is to represent how the presence of one risk factor allows to guess on the presence of others. This is the principal way to use *inferred* risk factors, i.e., those unknown to a woman. For example, it is known that Ashkenazi Jews are more likely to develop BRCA gene mutation.

It will be demonstrated how to capture the impact of combining risk factors, i.e., if they are known to strengthen or weaken each other. Classical part of the ontology provides classes that are combinations of multiple risk factors. The model can define the risk for women having multiple risk factors to be higher (or lower) than if they had just one of the factors. This is possible using the *overriding* feature of the default probabilistic reasoning.

Finally, the ontology contains a number of ABox axioms that represent risk factors for specific individuals. The motivation is that while the generic probabilistic model that provides all the necessary statistics can be developed and maintained by a central cancer research institute, individual women can supply the knowledge about the risk factors that are known to them, e.g., age. It will also be shown how to express uncertainty in having some particular risk factor.

4.2 Probabilistic Reasoning Demonstration

The modeling described above is necessary to reduce the problem of assessing breast cancer risks to the standard lexicographic entailment implemented in Pronto. Risk assessment for a particular woman corresponds to the entailment of an ABox constraint. For example, $(WomanWithBRCInLongTerm|\top)[0.6, 0.8]$ implies that some woman's risk of developing cancer in life time is 60%-80%. The reasoning will be demonstrated on a number of test probabilistic individuals.

It will also be presented how Pronto justifies the results of the risk assessment by generating the *explanations* for the entailments. In particular, it can retrieve exactly those risk factors and generic statistical axioms that caused the inference for a particular woman and filter out all the irrelevant risk factors. In addition to being useful for end users, this capability can aid the model developers in testing the accuracy and adequacy of their model.

Finally, the demo will reveal some pitfalls of default probabilistic reasoning by presenting seemingly unobvious, yet sound entailments. This will provide a better insight into the nature of probabilistic reasoning and also demonstrate the need of explanations.

---

[4] Available at: http://www2.cs.man.ac.uk/~klinovp/pronto/brc/cancer_cc.owl

5 Discussion

Although the demo does not pretend to cover all the aspects of default probabilistic reasoning in the Semantic Web, it is expected to help the attendees learn the following important things:

- Features and capabilities of Pronto. The demo will present Pronto's fundamental features, e.g., reasoning services and representational capabilities.
- How Pronto can help in modeling real life problems that involve uncertain knowledge. It is important to learn that the limits of applicability of ontologies will be pushed forward once they become capable of representing uncertain knowledge. The breast cancer risk assessment is one of the problems that can be given a probabilistic ontological model and then approached using a Semantic Web reasoner.
- How current OWL ontologies can be reused in probabilistic models.
- Current challenges and limitations of probabilistic reasoning as well as future research directions.

Acknowledgment. Pronto was developed when the first author was an intern at Clark & Parsia, LLC (www.clarkparsia.com) during the summer 2007. Pronto is a property of Clark & Parsia and is available under the AGPL license at: http://pellet.owldl.com/pronto. The authors are thankful to Bijan Parsia, Evren Sirin, Michael Smith and Kendall Clark for numerous useful suggestions and other support regarding the nature of this work.

References

1. Klinov, P., Parsia, B., Mazlack, M.J.: Pronto - a non-monotonic probabilistic description logic reasoner. In: The European Semantic Web Conference (submitted, 2008)
2. Lukasiewicz, T.: Probabilistic description logics for the semantic web. Technical Report Nr. 1843-06-05, Institut fur Informationssysteme, Technische Universitat Wien (2007)
3. Lehmann, D.: Another perspective on default reasoning. Annals of Mathematics and Artificial Intelligence 15(1), 61–82 (1995)
4. Lukasiewicz, T.: Probabilistic default reasoning with conditional constraints. Annals of Mathematics and Artificial Intelligence 34(1-3), 35–88 (2002)
5. Sirin, S., Parsia, B., Grau, B.C., Kalyanpur, A., Katz, Y.: Pellet: A practical OWL-DL reasoner. Technical Report CS 4766, University of Maryland, College Park, MD (2005)
6. Golbeck, J., Fragoso, G., Hartel, F.W., Hendler, J.A., Oberthaler, J., Parsia, B.: The national cancer institute's thesaurus and ontology. Journal of Web Semantics 1(1), 75–80 (2003)
7. Gail, M.H., Brinton, L.A., Byar, D.P., Corle, D.K., Green, S.B., Shairer, C., Mulvihill, J.J.: Projecting individualized probabilities of developing breast cancer for white females who are being examined annually. Journal of the National Cancer Institute 81(25), 1879–1886 (1989)

User Profiling for Semantic Browsing in Medical Digital Libraries

Patty Kostkova, Gayo Diallo, and Gawesh Jawaheer

City ehalth Research Centre, City University, Northampton Square, London,
EC1V 0HB, UK
patty@soi.city.ac.uk,
{Gayo.Diallo.1,Gawesh.Jawaheer.1}@city.ac.uk

Abstract. Semantic Browsing provides contextualized dynamically generated Web content customizing the knowledge to better meet user expectations. The real-world medical digital library, the National electronic Library of Infection (NeLI, www.neli.org.uk), enriched with an infection domain ontology enables new semantic services to be developed qualitatively. In this paper, we will address the use of group profiling to customize semantic browsing by integrating distributed knowledge sources. The service is evaluated by web server logs analysis, dynamically enhancing the profiles and by qualitative feedback from real users of the NeLI portal.

1 Introduction and Background

The Semantic Web leverages the knowledge integration on the Web to new levels. Despite the efforts put into the technical and research issues, there are few applications actually deploying and evaluating semantic web with real users. Semantic web can only deliver if it is driven by user needs, context or profiles to seamlessly integrate the knowledge on the web to really provide desirable content.

This is in particular relevant in the medical domain. The Internet enabled patients and healthcare professionals to access vast amount of available information but this often results in the inability to find what is needed and when it is needed [1]. Medical sites need to support a profile-based semantic search and a contextualized browsing to integrate knowledge from other medical portals needed by particular medical users or patients.

Context and customisation are some of the key factors for accurate, effective relevant information access in Internet digital libraries and in general – in the Semantic Web. Allan et al. [2] define contextual retrieval as a general framework combining search technologies and knowledge about a query and so called "user context" into a single framework in order to provide the most appropriate results for users' information needs. The context of the user may include his/her level of expertise and domains of interest. An user profile is a record of user specific data that define users interests, his/her level of expertise, and his/her context.

2 Semantics Profiling in NeLI

2.1 Overview

The work presented in this paper demonstrates using profiles to customise user access in the UK based National electronic Library in Infection (NeLI) and to semantically integrate NeLI with other medical portals. Since 2000, NeLI has provided a single access point to the best available evidence around all aspects of infection and is currently being used by more than 15 000 real-world medical professionals a month. Portal users range from members of the public, General Practitioners (GPs), nurses, consultants communicable disease control (CCDCs) to senior Primary Care Trust executives. NeLI has been gathering user and group profiles providing contextual information essential for support of customised services. The development of the new version of NeLI portal takes into consideration the emerging Semantic Web where ontologies are one of the essential components [3].This is particularly the case in the (bio)medical domain where substantial efforts have been made to develop standards, medical terminologies and coding systems (SNOMED, MeSH and UMLS, which integrates more than 100 most relevant vocabulary sources in medicine[1]), thereby, providing knowledge bases for encoding medical evidence. As there is no standard ontology meeting our specific need, the NeLI ontology has been developed representing the infection medical domain with several hundred concepts.

Resources in the library are indexed using a NeLI ontology that has been created form a pruned sub-tree of the MeSH[2] (Medical Subject Headings) vocabulary and a customised classification for infection control and public health. The development of the NeLI ontology has been carried out by 3 NeLI content managers, in a close collaboration with the project Advisory Board members and infection experts in the UK.

Profiling for recommending research papers based on ontologies was investigated by Middleton [4]. Dynamic profiling applied to information retrieval on the web has been widely applied [5]. However, a very common application of profiles is the vision of customised content relevant to the particular profile [6].

In this section, we will look into the spectrum of NeLI users and discuss how they could be categorised in terms of profiles. NeLI users come from different professional backgrounds and specialties which determine their medical interests, information needs and type of questions they are asking on the portal. In addition to their professions, they have particular specialities and are likely to have particular interests in treatment or investigation of a particular disease or a group of diseases. Due to the size limitation of this demo paper, personal user profiles are outside the remit of this publication and can be found in [7]. Based on personal profiles, which have been inserted manually by NeLI users, we have an understanding of their professional backgrounds and personal interested. Users have provided this information to us when subscribing to project updates at conferences or online. We use the personal profiles as a base for semi-automated development of professional group profiles.

---

[1] http:// umlsks.nlm.nih.gov/
[2] www.nlm.nih.gov/mesh/2002/index.html

2.2 Professional Group Profiles

The professional group profile is defined as a tuple: $N=\{p,s,e,o,t\}$ where $p \, \varepsilon \, P$, $s \, \varepsilon \, P$, $e \, \varepsilon \, O$, $o \, \varepsilon \, O$ and $t \, \varepsilon \, T$, where O is the set of NeLI ontology concepts, P is the list of NeLI recognised professions and T is a list of NeLI-recommended external knowledge web portals in a form of a URL (called Targets). The list P of NeLI recognised professions include: clinical scientists, nurse, consultant, environmental health officer, general practitioner, lecturer, microbiologist. An example of a user is illustrated in Table 1.

Table 1. Example of a speciality group user profile

Profession	Nurse
Speciality	Infection Control
Expertise Area	TB
Other Area of Interest	Hand washing Antibiotic Resistance National Policy
Targets	Department of Health WHO

Currently, we are looking into a dynamic web server log-based generation of the profiles (the Other Area of Interest and the Targets fields being dynamically populated by frequently searched keywords from the NeLI ontology). The aim is to dynamically enhance the profiles by evaluation of the search keywords and navigation terms from the web server logs. Users in a group may have different interests, but also sufficiently similar interests for the development of group default preferences. The need for a combination of professional, speciality and topic comes from different questions asked by different users. For example, all users search for TB (tuberculosis) but the information they actually need varies: a clinician will ask about latest TB guidelines provided by the Department of Health (target), an infection control nurse about isolating patients with TB, a public health office is more concerned about high-risk TB populations and outreach to them, while a GP might need to check the latest diagnosis and treatment recommendations for TB. A case scenario around TB is shown in Table 2. Based on the user group (profession), different information is to be provided (targets – portals with further knowledge).

Table 2. User profiles information needs example based on a TB scenario

Professional group	Targets
GPs	BNF (British National Formulary) NeLI treatment pages
clinicians	PubMed Clinical Evidence
nurse	Department of Health
public	Wikipedia HPA Public leaflets Nathnac

3 Semantic Browsing and Semantic Knowledge Integration

Semantic browsing provides users with dynamically selected concepts or links from an ontology – enriched by the profile-based customization this selects and integrates web portals by working as a "semantic recommender" system. NeLI semantic browsing is being developed in the context of the SeaLife project [8] enabling users to semantically browse the Web by highlighting ontology concepts and providing dynamic access to Web servers or knowledge portals semantically related to the Web content retrieved (targets).

To facilitate this, NeLI has integrated the Conceptual Open Hypermedia Service (COHSE) system [9] developed at University of Manchester. COHSE automatically inserts hyperlinks on web pages by recognizing terms contained in background knowledge, based on an uploaded ontology, and presents the user with search web services linking to relevant targets, see Table 2. For any term from the ontology, resources are provided for broader, narrower and related terms based on ontological taxonomical and non-taxonomical relationships. Selecting the relevant set of targets for a particular profile improves the contextualization of the search and recommender function.

Demonstrator

Unlike many other demos serving just as a proof of concept, NeLI is a real-world Internet medical library with over 15 000 unique users a month. In the demonstration, we will show the ontology-based selection of targets based on the user group profiles. A medical scenario semantically integrating the following targets based on ontology-based relationships will be shown for two profiles: public and nurses.

A "public" user visits NeLI to search for tropical diseases. A list of search results based on taxonomical relationships in the NeLI ontology is displayed containing (among others) malaria resources. NeLI ontology terms on the page are highlighted by the NeLI-COHSE system and ontological relationships to malaria are shown in a pop-up box. Among non-taxonomical relationship *"is caused"* is shown giving *Plasmodium*. When selected, the targets for public profile show the Nathnac web site http://www.nathnac.org giving travelers information on malaria for public.

A "nurse" searches for healthcare associated infection. Results contain documents on MRSA, TB etc., provided by the NeLI ontology. When a TB concept is highlighted a pop-up box providing the ontology relationships shows a non-taxonomical "transmission mode" giving *airborne*. The HPA (www.hpa.org.uk) and Department of Health (www.dh.gov.uk) targets, provided for a nurse profile, are searched to give policy resources on management of healthcare acquired TB – in particular, respiratory isolation and respiratory protection.

These two case scenarios demonstrate the semantic search, use of taxonomical and non-taxonomical relationships in the NeLI ontology and the new concept of selection of targets for different professional profiles and the construction of searches on those external sites to provide contextualized information. These are new features enhancing the COHSE system developed by the NELI team to support profile-based customization. A screen shot cand be found in Figure 1.

The visitors will see a unique working example of domain ontology-driven medical portal in use by practising clinicians and a semantically integrated knowledge from other portals based on professional group profiles.

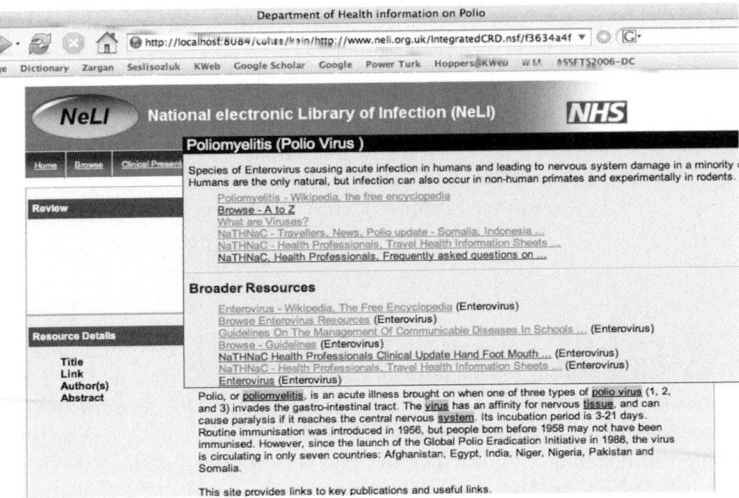

Fig. 1. NeLI-COHSE Semantic Browsing and Recommender Case Scenario

Acknowledgments. We acknowledge the EC for providing funding for the SeaLife project (IST-2006-027269). We thank the University of Manchester for providing the COHSE System used in this research.

References

1. Muir Gray, J.A., de Lustignan, S.: National electronic Library for Health (NeLH). BMJ 319, 1476–1479 (1999)
2. Allan, J., et al.: Challenges in Information Retrieval and Language Modelling: Report of a workshop held at the centre for intelligent information retrieval, University of Massachusetts Amherst, September 2002 (2003). ACM SIGIR Forum 37(1), 31–47
3. Berners-Lee, T., Hendler, Lassila, O.: The Semantic Web. Scientific Am., 34–43 (May 2001)
4. Middleton, S.E., Shadbolt, N.R., De Roure, D.C.: Ontological User Profiling in Recommender Systems. ACM Trans on Information Systems (TOIS) 22(1), 54–88 (2004)
5. Danilowicz, C., Indyka-Piasecka, A.: Dynamic User Profiles Based on Boolean Formulas. In: Orchard, B., Yang, C., Ali, M. (eds.) IEA/AIE 2004. LNCS (LNAI), vol. 3029, pp. 779–787. Springer, Heidelberg (2004)
6. Wærn, A.: User Involvement in Automatic Filtering: An Experimental Study. User Modeling and User-Adapted Interaction 14(2-3) (June 2004)
7. Kostkova, P., Diallo, G., Jawaheer, G.: Application of User Profiling on Ontology Module Extraction for Medical portals. In: The Proceedings of the MedSemWed 2007 Workshop, held in conjunction with the MedInfo 2007 Conference, Brisbane, Australia (August 2007)
8. Schroeder, M., Burger, A., Kostkova, P., Stevens, R., Habermann, B., Dieng-Kuntz, R.: From a Service-based eScience Infrastructure to a Semantic Web for the Life Sciences: The Sealife Project. In: Workshop on Network Tools and Applications in Biology, NETTAB 2006, Italy (2006)
9. Bechhofer, S., Yesilada, Y., Horan, B.: COHSE: Knowledge-Driven Hyperlinks. In: The Semantic Web Challenge at the International Semantic Web Conference (ISWC 2006) (2006)

SWiM – A Semantic Wiki for Mathematical Knowledge Management

Christoph Lange

Computer Science, Jacobs University Bremen
ch.lange@jacobs-university.de

Abstract. SWiM is a semantic wiki for collaboratively building, editing and browsing mathematical knowledge represented in the domain-specific structural semantic markup language OMDoc. It motivates users to contribute to collections of mathematical knowledge by instantly sharing the benefits of knowledge-powered services with them. SWiM is currently being used for authoring content dictionaries, i. e. collections of uniquely identified mathematical symbols, and prepared for managing a large-scale proof formalisation effort.

1 Research Background and Application Context: Mathematical Knowledge Management

A great deal of scientific work consists of collaboratively authoring *documents*— taking down first hypotheses, commenting on results of experiments, circulating informal drafts inside a working group, and structuring, annotating, or re-organising existing items of knowledge, finally leading to the publication of a well-structured article or book. Here, we particularly focus on the domain of mathematics and on tools that support collaborative authoring by utilizing the knowledge contained in the documents. In recent years, several *semantic markup* languages have been developed to represent the clearly defined and hierarchical structures of mathematics. The XML languages MathML [9], OpenMath [11], and OMDoc [3] particularly aim at exchanging mathematical knowledge on the web. OMDoc, employing Content MathML or OpenMath representing the functional structure of mathematical *formulæ*—as opposed to their visual appearance—and adding support for mathematical *statements* (like symbol declarations or axioms) and *theories*, has many applications in publishing, education, research, and data exchange [3, chap. 26]. The main challenge is *acquiring* a large collection of OMDoc-formalised knowledge that can power such added-value services. In an open, collaborative environment, the workload can be distributed among many authors, but as semantic markup make fine-grained structures explicit, it is tedious to author. As the community can only benefit from added-value services after a substantial initial investment (writing, annotating and linking) on the author's part, we sought for motivating the author into action by offering "elaborate [...] services for the concrete situation" they are in [2].

2 Key Technology: Semantic Wiki and Ontologies

Our research is motivated by the assumption that in this context a semantic wiki comes in handy. OMDoc supports all levels of formalisation, from human-readable texts to fully formal representations for automated theorem proving, and semantic wikis have been found appropriate for collaboratively refining knowledge models (cf. [13]). User motivation in semantic wikis by instant gratification has been investigated in earlier works [1]. The ultimate goal of our work is to achieve a feedback loop where users are supported to contribute well-structured knowledge, which is then exploited to offer services, which in turn facilitate editing and motivate new contributions [5].

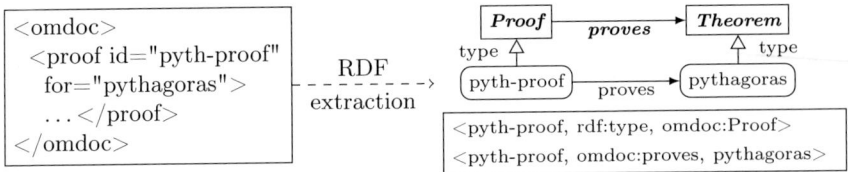

Fig. 1. RDF extraction from OMDoc markup in a wiki page

Semantic markup has deep structures: an OMDoc document can contain theories containing statements that contain formulæ referring to symbols defined in other theories. This is uncommon for most semantic wikis, where the structures are rather flat and one aims at small pages to prevent editing conflicts and to facilitate search and navigation. So to adapt OMDoc's model of knowledge to a semantic wiki, we had to choose an appropriate granularity of wiki pages and arrived at one page holding one mathematical statement or one theory. To make knowledge from OMDoc documents usable on the semantic web, information about the resources represented by pages and their interrelations (e.g. "a *proof for* the Pythagorean theorem") are extracted to RDF. As a vocabulary for this, we modeled OMDoc's structures explicitly in a *document ontology* [5] in OWL-DL. This ontology contains e.g. the information that both theorems and proofs are specialisations of a general "mathematical statement", and that a proof can prove a theorem (Fig. 1). Moreover, generic transitive dependency and containment relations have been modeled. For example, having one theory import another theory (and reusing symbols defined there) establishes a dependency. One theory logically contains its statements; similarly, statements can contain sub-statements, as in the case of a proof that consists of multiple steps.

3 The SWiM 0.2 Prototype: IkeWiki + OMDoc

As a base system for the implementation, we chose IkeWiki [12]. Among the systems evaluated, it offered the richest XML infrastructure—a key requirement

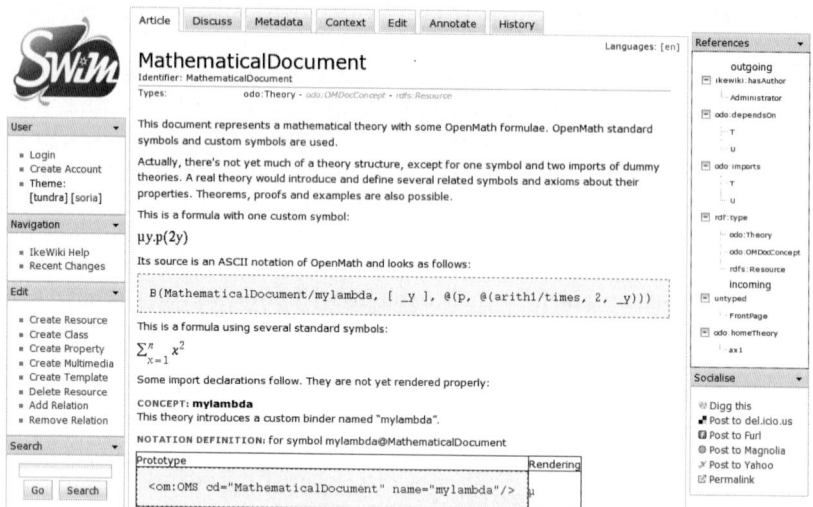

Fig. 2. A mathematical document in SWiM

for adding OMDoc support—and was found to be most extensible [4]. Its backend consists of a PostgreSQL database for the page contents, a Jena RDF store for the RDF graph and the ontologies. Additional ontologies can easily be imported. The frontend heavily relies on the Dojo Ajax toolkit.

Technically, the extension of IkeWiki to SWiM required supporting OMDoc in addition to the HTML-like wiki page format. To foster stepwise formalisation of informal text, we chose to mix OMDoc fragments with wiki markup. Thus we could still rely on IkeWiki's WYSIWYG HTML editor, which just had to be enhanced by support for OMDoc XML elements. Moreover, this choice allowed for an easier maintenance of the OMDoc-related enhancements to the SWiM code base and avoided changes to the underlying database schema. The document ontology is preloaded into the RDF store. RDF triples are extracted from the OMDoc markup upon saving a page or importing an OMDoc file. Additional XSLT template rules care for rendering embedded OMDoc fragments. In order to render mathematical formulæ, there is a *notation definition* for every semantic symbol. These notation definitions can be imported and edited right in the wiki, as parts of OMDoc documents [6]. An efficient, specialised renderer supporting the upcoming MathML 3 standard [10,9] applies them to the symbols in the formulæ. In the editing view, statement- and theory-level structures of OMDoc are made accessible as special HTML tables, whereas mathematical formulæ given in semantic markup are made accessible in a simplified ASCII notation of OpenMath. OMDoc documents are browsable via inline links manually set in the informal parts, via links from occurrences of symbols in formulæ to the place of their declaration, set by the formula renderer, and via RDF links, displayed in a separate box by IkeWiki. The latter comprise those triples that are extracted from the markup (cf. Fig. 1), as well as triples inferred by a reasoner[1].

SWiM also relies on the ontology for reacting on changes to notation definitions. When an author changes a notation definition n for a symbol s, exactly those wiki pages that contain a formula using s or that include other pages containing such formulæ need to be re-rendered. Looking up the symbol s rendered by n, the formulæ f_i using s, or pages (transitively) including the f_i would be clumsy in the OMDoc XML sources, but is easy in the RDF graph, as this information is extracted from the documents and represented using ontology properties such as *NotationDefinition–renders–Symbol* and *Statement–contains–Formula; Formula–uses–Symbol*. This service allows for instant visual debugging of notation definitions [6]. For upcoming releases, more ontology-powered services are planned, including more general change management, learning assistance, and editing facilitations like editing of subsections and auto-completion of link targets [7]. There is some evidence that many services can be based on the most generic relations of dependency and (physical or logical) containment [5]. With scientists and knowledge engineers in mind, we envisage SWiM as a development environment that conveniently supports refactorings of knowledge[2].

4 Use Cases and Applications

Now that viewing, browsing, editing, importing and exporting mathematical documents basically works, we are evaluating SWiM in practical settings. The **Flyspeck** project is about large-scale formalisation of a proof of the Kepler conjecture. We are starting to support this effort by "crowdsourcing" the knowledge compiled so far (hundreds of proof sketches that are not yet machine-verifiable) on a SWiM site [8]. The main challenge is giving an interested visitor an impression of the extent of the project and, using appropriate SPARQL queries, showing him where work needs to be done. Currently we are investigating how the original LaTeX sources can be utilised by automatically converting them to HTML with MathML, then to informal OMDoc, breaking that into wiki pages, and letting the users formalise them stepwisely. For the upcoming **OpenMath 3** standard, SWiM is currently being extended to an editor for OpenMath Content Dictionaries [6], which could be regarded as flat OMDoc theories that just define symbols and do not import anything. There, mainly editing Dublin Core metadata and notation definitions is of interest.

---

[1] The ontology is prepared for DL reasoning, but currently only the RDFS reasoner built into Jena is used.
[2] This is common in mathematics, e.g. in algebra: If one just needs groups, they can be defined by a theory with the four well-known axioms. For explicitly modeling related structures as well, one would break this into smaller theories—*semigroup* just defining an associative operation on a set, *monoid* importing this and extending it by an identity element, and finally the refactored *group*, adding inverse elements.

5 Conclusion and Related Work

SWiM makes mathematical documents editable collaboratively and particularly facilitates browsing them by exploiting the knowledge they contain. Domain-specific services are powered by an ontology that models structures of documents—an advantage over generic semantic wikis, which would not be able to offer additional services for mathematical knowledge. Competing non-semantic approaches like the math encyclopædia *PlanetMath* (evaluated in [4]) are less flexible, as they cannot exploit the structures of their presentation-oriented LaTeX formulæ and rely on a fixed set of metadata. Most services for editing and browsing need to be hard-coded, which potentially restricts the scale of knowledge managment tasks the systems can be applied to. The SWiM approach of integrating a semantic markup language into a wiki by choosing an appropriate page granularity, modeling a document ontology, and extracting relevant facts from the markup into RDF has successfully been applied to OMDoc and the closely related but syntactically different OpenMath [6] and is likely to be portable to other domains as well, e. g. for the chemical markup language CML.

References

1. Aumüller, D., Auer, S.: Towards a semantic wiki experience – desktop integration and interactivity in WikSAR. In: 1st Workshop on The Semantic Desktop (2005)
2. Kohlhase, A., Müller, N.: Added-Value: Getting People into Semantic Work Environments. In: Rech, J., Decker, B., Ras, E. (eds.) Emerging Technologies for Semantic Work Environments: Techniques, Methods, and Applications, Idea Group (in press, 2008)
3. Kohlhase, M.: OMDoc – An Open Markup Format for Mathematical Documents [version 1.2]. LNCS (LNAI), vol. 4180. Springer, Heidelberg (2006)
4. Lange, C.: SWiM – a semantic wiki for mathematical knowledge management. Technical Report 5, Jacobs University (2007), http://kwarc.info/projects/swim/pubs/tr-swim.pdf
5. Lange, C.: Towards scientific collaboration in a semantic wiki. In: Hotho, A., Hoser, B. (eds.) Bridging the Gap between Semantic Web and Web 2.0 (2007)
6. Lange, C.: Mathematical Semantic Markup in a Wiki: The Roles of Symbols and Notations. In: The 3rd Semantic Wiki Workshop at ESWC (submitted, 2008), http://kwarc.info/projects/swim/pubs/semwiki08-notation-semantics.pdf
7. Lange, C.: SWIM development roadmap (2008), https://trac.kwarc.info/swim/roadmap/
8. Lange, C., McLaughlin, S., Rabe, F.: Flyspeck in a semantic wiki, 2008. In: The 3rd Semantic Wiki Workshop at ESWC 2008 (submitted, 2008), http://kwarc.info/projects/swim/pubs/flyspeck-wiki-eswc08.pdf
9. Mathematical Markup Language (MathML) version 3.0. W3C working draft, World Wide Web Consortium (2007), http://www.w3.org/TR/MathML3
10. Müller, C., Müller, N., Kohlhase, M.: A library for transforming Content MathML/OpenMath into Presentation MathML (2008), http://kwarc.info/projects/mmlkit/

11. The Open Math standard, version 2.0. Technical report, The Open Math Society (2004), http://www.openmath.org/standard/om20
12. Schaffert, S.: IkeWiki: A semantic wiki for collaborative knowledge management. In: 1st International Workshop on Semantic Technologies in Collaborative Applications (STICA) (2006)
13. Schaffert, S.: Semantic social software. In: Sure, Y., Schaffert, S. (eds.) Semantics (2006)

Integrating Open Sources and Relational Data with SPARQL

Orri Erling and Ivan Mikhailov

OpenLink Software, 10 Burlington Mall Road Suite 265 Burlington, MA 01803 U.S.A
{oerling,imikhailov}@openlinksw.com,
http://www.openlinksw.com

Abstract. We believe that the possibility to use SPARQL as a front end to heterogeneous data without significant cost in performance or expressive power is key to RDF taking its rightful place as the lingua franca of data integration. To this effect, we demonstrate how RDF and SPARQL can tackle a mix of standard relational workload and data mining in public data sources.

We discuss extending SPARQL for business intelligence (BI) workloads and relate experiences on running SPARQL against relational and native RDF databases. We use the well known TPC H benchmark as our reference schema and workload. We define a mapping of the TPC H schema to RDF and restate the queries as BI extended SPARQL. To this effect, we define aggregation and nested queries for SPARQL.

We demonstrate that it is possible to perform the TPC H workload restated in SPARQL against an existing RDBMS without loss of performance or expressivity and without changes to the RDBMS.

Finally, we demonstrate how to combine TPC-H or XBRL financial reports with RDF data from CIA factbook and DBpedia.

1 Introduction and Motivation

RDF promises to be a top level representation for data extracted or accessed or demand from any conceivable source. Thus, chief promise of RDF is in the field of information integration, analysis and discovery. Yet it is difficult to imagine any business reporting, let alone more complex information integration task that would not involve aggregating and grouping.

As a data access and data integration vendor, OpenLink has a natural interest in seeing SPARQL succeed as a top level language for answering business questions on data mapped from any present day data warehouse or other repository.

For SPARQL to deliver on this potential, several extension and scalability issues have to be addressed.

These include:

- Expressive power of SPARQL must be at least on par with SQL. As of the present, the SPARQL recommendation is lacking aggregation, grouping, expressions in result sets, nested subqueries and full text support, to name

a few. All these are either part of SQL or universally available in RDBMS, as in the case of full text. The baseline business intelligence benchmark, TPC H, relies on these all, except for full text.
- Efficient mapping of SPARQL to relational queries against one or more relational databases. SPARQL's promise is greatest in combining data from diverse sources. Still, in cases where a straightforward translation of SPARQL to SQL is possible, the performance should not be much less than that of the relational back-end when accessed through SQL.
- Scalability of RDF storage. Parallelization and clustering are needed for scaling into the tens of billions of triples and beyond.

We intend to demonstrate how we address all these questions with our Virtuoso product.

2 The Data and Queries

We draw on a combination of real-world and synthetic data sets for the demonstration. In specific, we use the following:

- DBpedia;
- US Census;
- real world XBRL financial data mapped into RDF;
- various Linking Open Data sets, such as Geonames and the CIA Factbook;
- TPC H benchmark data, a scalable industry standard benchmark data set.

The TPC H data is stored in relational form as well as as RDF triples. We demonstrate queries combining these data in novel ways. For example:

- comparing sales figures from the TPC H data with population and GDP figures from the CIA Factbook;
- combining XBRL financial results with geography and DBpedia information on the same companies;
- comparing two TPC H data sets, one as a relational database and one in RDF form.

In addition to aggregate queries such as the above, we show navigation by following data links between these sets.

We also present loading and query times for data sets such as the LUBM benchmark data and the Uniprot data set.

The complete source code of the queries and data definitions and mappings is published at the OpenLink web site at the time of the demonstration (hboxhttp://demo.openlinksw.com/tpc-h/). The data itself is either linked open data or synthetic data that can be generated with generally available tools. Thus the things demonstrated are readily reproducible.

2.1 SPARQL Extensions

We show how we have extended SPARQL with the following:

- Subqueries and derived tables.
- Aggregates, grouping and expressions in results.
- Syntax sugar for following chains of references, as in region of country of customer of order X.

The below is the SPARQL version of Q2 from the TPC H queries.

```
prefix tpcd: <http://www.openlinksw.com/schemas/tpcd#>
select
  ?supp+>tpcd:acctbal   ?supp+>tpcd:name
  ?supp+>tpcd:has_nation+>tpcd:name as ?nation_name
  ?part+>tpcd:partkey   ?part+>tpcd:mfgr
  ?supp+>tpcd:address   ?supp+>tpcd:phone   ?supp+>tpcd:comment
from <http://example.com/tpcd>
where {
  ?ps a tpcd:partsupp ; tpcd:has_supplier ?supp ; tpcd:has_part ?part .
  ?supp+>tpcd:has_nation+>tpcd:has_region tpcd:name 'EUROPE' .
  ?part tpcd:size 15 .
  ?ps tpcd:supplycost ?minsc .
  { select ?part min(?ps+>tpcd:supplycost) as ?minsc
     where {
        ?ps a tpcd:partsupp ;
          tpcd:has_part ?part ; tpcd:has_supplier ?ms .
        ?ms+>tpcd:has_nation+>tpcd:has_region tpcd:name 'EUROPE' .
     } }
     filter (?part+>tpcd:type like '%BRASS') }
order by
  desc (?supp+>tpcd:acctbal)
  ?supp+>tpcd:has_nation+>tpcd:name
  ?supp+>tpcd:name
  ?part+>tpcd:partkey
```

We notice a subquery used for determining the lowest supply cost for a part. We also notice the pattern

```
{ ?ms+>tpcd:has_nation+>tpcd:has_region tpcd:name 'EUROPE' }
```

which is a shorthand for

```
{ ?ms tpcd:has_nation ?t1 . ?t1 tpcd:has_region ?t2 .
  ?t2 tpcd:name "EUROPE" }
```

The notation with +> differs from a join path expressed with [] in that these are allowed in expressions and that common subpaths are guaranteed to be included only once in the evaluation. Thus

```
sum ( ?c+>has_order+>has_line+>l_extendedprice *
  (1 - ?c+>has_order+>has_line->l_discount) )
```

evaluates to the sum of each line's extendedprice multiplied by the line's discount whereas

```
sum ( ?extprice * (1 - ?discount))
...
?c has_order [ has_line [ l_extendedprice ?extprice ] ] .
?c has_order [ has_line [ l_discount ?discount ] ]
```

would mean the sum of every price times every discount.

For brevity we have omitted the declarations for mapping the TPC H schema to its RDF equivalent. The mapping is straightforward, with each column mapping to a predicate and each table to a class.

2.2 Linked Data

Virtuoso has an integrated HTTP server used for providing web services end points and web app hosting. For presenting the TPC H data as linked data, we have added a virtual collection which presents the data as dereferenceable URI's, redirecting the dereference to a describe query against the SPARQL end point.

2.3 Performance of Mapping

As a baseline, we take the performance of Virtuoso executing TPC H queries in SQL against Oracle. There Virtuoso parses the SQL query, makes a distributed execution plan, finds out the whole query can go to Oracle and finally rewrites the query as a single Oracle SQL query. This takes an average of 7 ms per query, including time to send and retrieve results. The rest of the real time is spent by Oracle.

Adding the SPARQL to SQL layer on top of this adds another 9 ms to each query. The cost of SPARQL is negligible in the cases where the resulting SQL query passes as a single unit to Oracle.

We note that the single most important factor in any distributed query performance as opposed to local query performance is the number of synchronous round trips between the processes involved.

Some SPARQL queries make a suboptimal SQL that does not pass as a unit to Oracle (even if it should), so the execution is divided between Virtuoso and Oracle and there is significant cost from message latency. Fixing this is a current work in progress.

3 System Demonstrated

The demonstration databases run on a cluster of X86-64 servers either at our offices or Amazon's EC2. Smaller scale local demonstration can be run on laptops with the same software but less data.

The software demonstrated includes:

- Virtuoso 6.0 RDBMS and triple store.
- Oracle 10G RDBMS accessed both directly and through Virtuoso's RDF to relational mapping.
- Diverse RDF browsers (Tabulator, OpenLink RDF Browser and Zitgist).

4 Conclusions

Mapping of relational data to RDF has existed for a long time [6][7]. The work shown here represents its coming of age. We can tackle a standard SQL workload without loss of performance or added complexity. Basically, we can bring any data warehouse to the world of linked data, giving dereferenceable URI's and SPARQL while retaining the performace of SQL.

We would point out that bringing SPARQL on par with SQL for decision support queries is not aimed at replacing SQL but at making SPARQL capable of fulfilling its role as a language for integration.

Indeed, we retain all of SPARQL's and RDF's flexibility for uniquely identifying entities, for abstracting away different naming conventions, layouts and types of primary and foreign keys and so forth.

In the context of mapping relational data to RDF, we can map several instances of comparable but different schemes to the common terminology and couch all our queries within this terminology. Further, we can join from this world of mapped data to native RDF data, such as the data in the Linking Open Data project.

Once we have demonstrated that performance or expressivity barriers do not cripple SPARQL when performing traditional SQL tasks, we have removed a significant barrier from enterprise adoption of RDF and open data.

References

1. W3C RDF Data Access Working Group: SPARQL Query Language for RDF, http://www.w3.org/TR/rdf-sparql-query/
2. Transaction Processing Performance Council: TPC-H – a Decision Support Benchmark, http://www.tpc.org/tpch/
3. Linking Open Data Project, http://linkeddata.org/
4. DBpedia – A Community Effort to Extract Structured Information From Wikipedia, http://dbpedia.org/
5. XBRL - Extensible Business Reporting Language, http://www.xbrl.org/Home/
6. Seaborn, A.: Counting and GROUP BY in ARQ, http://seaborne.blogspot.com/2007/09/counting-and-group-by.html
7. Weiske, C., Auer, S.: Implementing SPARQL Support for Relational Databases and Possible Enhancements. In: Proceedings of the 1st Conference on Social Semantic Web. Leipzig (CSSW 2007), SABRE. LNI 113 GI 2007, Bonner Kollen Verlag (2007), http://www.informatik.uni-leipzig.de/~auer/publication/sparql-enhancements.pdf, ISBN 978-3-88579-207-9
8. Erling, O., Mikhailov, I.: Adapting an ORDBMS for RDF Storage and Mapping. In: Proceedings of the 1st Conference on Social Semantic Web. Leipzig (CSSW 2007), SABRE. LNI 113 GI 2007, Bonner Kollen Verlag (2007) ISBN 978-3-88579-207-9

Previewing Semantic Web Pipes ⋆

Christian Morbidoni[2], Danh Le Phuoc[1], Axel Polleres[1], Matthias Samwald[1], and Giovanni Tummarello[1]

[1] DERI Galway, National University of Ireland, Galway
{firstname.lastname}@deri.org
[2] SeMedia Group, Universita' Politecnica delle Marche, Ancona, Italy
christian@deit.univpm.it

Abstract. In this demo we present a first implementation of Semantic Web Pipes, a powerful tool to build RDF-based mashups. Semantic Web pipes are defined in XML and when executed they fetch RDF graphs on the Web, operate on them, and produce an RDF output which is itself accessible via a stable URL. Humans can also use pipes directly thanks to HTML wrapping of the pipe parameters and outputs. The implementation we will demo includes an online AJAX pipe editor and execution engine. Pipes can be published and combined thus fostering collaborative editing and reuse of data mashups.

1 Introduction

Making effective use of RDF data published online (e.g. in sources as RDF DBLP, DBPEDIA etc) is, in practice, all but straightforward: data might be fragmented or incomplete so that multiple sources needs to be joined, different identifiers (URIs) are usually employed for the same entities, ontologies need alignment, certain information might be need to be "patched", etc. The only approach available to these problems so far has been custom programming such transformations for the specific task to be performed in a Semantic Web application. In this paper we present a paradigm for creating and reusing such transformation in a easy way: a Web based Software Pipeline for the Semantic Web.

A similar metaphor has been implemented in Yahoo Web Pipes[1], which allows to implement customized services and information streams by processing and combining Web sources (usually RSS feeds) using a cascade of simple operators. Since Web pipes are themselves HTTP retrievable data sources, they can be reused and combined to form other pipes. Also, Web pipes are "live": they are computed on demand at each HTTP invocation, thus reflect the current status of the original data sources.

Unfortunately Yahoo Web Pipes are engineered to operate using fundamentally the RSS paradigm (item list) which does not map well at all with the graph based data model of RDF. For this purpose Semantic Web Pipes have been written from the start to

---
⋆ This work has been supported by the European FP6 project inContext (IST-034718), by Science Foundation Ireland under the Lion project (SFI/02/CE1/I131), and by the European project DISCOVERY(ECP-2005-CULT-038206).

[1] http://pipes.yahoo.com/

operate also on Semantic Web data, offering specialized operators to perform the most important data aggregation and transformation tasks.

When a pipe is invoked, simply fetching the pipe URL, the external sources are fetched dynamically and transformed transparently and thus the Semantic Web pipe will reflect the most up to date data available online.

2 Basic Operators

A *Semantic Web pipe* implements a predefined workflow that, given a set of RDF sources (resolvable URLs), processes them by means of special purpose operators. Unlike fully-fledged workflow models, our current pipes model is a simple construction kit that consists of linked *operators* for data processing. Each operator allow a set of unordered inputs in different yformats (to make them distinguishable) as well as a list of optional ordered inputs, and exactly one output.

Figure 1(b) shows a set of base operators which we implemented so far and which we will shortly explain below. *The ⊎-Operator: RDF Merge:* This operator takes a list

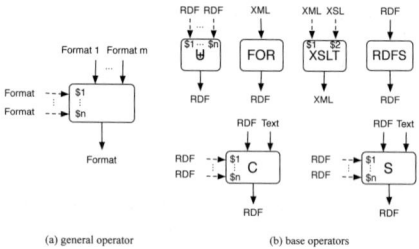

Fig. 1. Semantic Web pipe operators

of RDF graphs as inputs, expressed in RDF/XML, N3 or Turtle format, and produces an RDF graph that is composed by the merge of its inputs. The standard implementation of the ⊎-operator simply standardizes blank nodes apart, according to RDF merge definition in [2], thus possibly generating non-lean graphs.

The C- and S-Operators: CONSTRUCT and SELECT: The C-operator outputs the result of a SPARQL [4] CONSTRUCT query given as textual input performed on the standard input RDF graphs. Similarly, the S-Operator performs a SELECT query and outputs the result in the SPARQL-Result XML format.

The RDFS-Operator: This operator basically performs materialization of the RDFS closure of the input graph by applying RDFS inference rules. We currently implement this using OWLIM.

The FOR-Operator: It works by taking a SPARQL XML result list (i.e. the output from the S operator) and binding each result with temporary variables which are then used as parameters in a subpipe which can be embedded inside it. The FOR operator is fundamental to enable many useful processing which involve discovering and using open data on the Semantic Web.

The XSLT-Operator: Finally, the XSLT-Adapter performs an XML transformation on a generic input XML document. This operator is particularly handy when custom XML output formats are needed or when an input source in a custom XML format shall be transformed to RDF/XML.

Examples for all operators can be found at http://pipes.deri.org.

2.1 A Semantic Web Pipe Example: About TBL

Pipes enable flexible aggregation of RDF data from various sources, here we present a simple example that show them in action. Data about Tim Berners-Lee is available on various sources on the Semantic Web, e.g. his FOAF file, his RDF record of the DBLP scientific publication listing service and from DBPedia. This data cannot simply be merged directly as all three sources use different identifiers for Tim. Since we prefer using his self-chosen identifier from Tim's FOAF file, we will create a pipe as an aggregation of components that will convert the indentifiers used in DBLP and DBPedia. This is performed by using the C-operator with a SPARQL [4] query as shown below for DBLP:

```
CONSTRUCT {<http://www.w3.org/People/Berners-Lee/card#i> ?p ?o.
           ?s2 ?p2 <http://www.w3.org/People/Berners-Lee/card#i>}
   WHERE {{<http://dblp.l3s.de/d2r/.../Tim_Berners-Lee> ?p ?o}
    UNION {?s2 ?p2 <http://dblp.l3s.de/d2r/.../Tim_Berners-Lee>} }
```

A similar query is done to perform fix the identifier for Tim's DBPedia entry.[2] The whole use case is then easily adressed by the pipe shown in Figure 2: URIs are normalized via the C-operators and then joined with Tim's FOAF file.

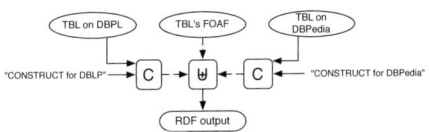

Fig. 2. A pipe that combines a Foaf file with DBLP and DBPedia entries

For lack of space we do not discuss more compex examples here. However it would be simple to perform more interesting operations, such as fetching the 10 top hints from the Sindice[3] search engine (e.g. querying for TBL's URI or even for his email address as an IFP) and using a FOR block to merge them with the end results (possibly after a proper transformation or filtering).

3 Implementation

An open-source implementation is available online at http://pipes.deri.org and is composed by an execution engine and an AJAX based pipe editor. The engine

---
[2] http://dbpedia.org/resource/Tim_Berners-Lee
[3] http://sindice.com

supports the basic operators from Figure 1 plus more advanced ones which provide support for patching RDF graphs, or smushing URIs based on *owl:sameAs* relations. As the output of a pipe is an HTTP-retrievable RDF model or XML file, simple pipes can work as sources for more complex pipes. Additional functionalities are also available, such as "parametric pipes" which inject extra parameters via HTTP GET query string, allowing pipes to act within other pipes not only as sources but as full featured operators. Pipes are written in a simple XML language.[4] The following XML code show two pipes: a simple mix between two RDF sources (M-operator) and the one shown in Figure 2.

```
<mix>
  <source><fetch><location>
    http://www.w3.org/People/Berners-Lee/card
  </location></fetch></source>
  <source><fetch><location>
    http://glo.net/foaf.rdf
  </location></fetch></source>
</mix>
```

```
<mix>
  <source><fetch><location>
    http://www.w3.org/People/Berners-Lee/card#i
  </location></fetch></source>
  <source><construct>
    <source><fetch><location>
        http://dblp.l3s.de/.../Tim_Berners-Lee
    </location></fetch></source>
    <query> <![CDATA[ CQ1 ]]> </query>
  </construct></source>
  <source><construct>
    <source><fetch><location>
        http://dbpedia.org/.../Tim_Berners-Lee
    </location></fetch></source>
    <query> <![CDATA[ CQ2 ]]> </query>
  </construct></source>
</mix>
```

Here, CQ1 and CQ2 stand for CONSTRUCT queries such as the ones previously shown. While it would be possible to implement pipe descriptions themselves in RDF, our current ad hoc XML language is more terse and legible. If an RDF representation will be later needed, it will be possible to obtain it via GRDDL.

HTTP-compliant caching is performed to avoid to recompute a pipe output if the sources have not changed. Whenever content is fetched it is hashed to detect changes. When no changes are detected the cached result is returned.

Circular invocations of the same pipe, which could create denial of services, can be easily detected within the same pipe engine, but not when different engines are involved. In this cases our solution relies on extra HTTP headers: whenever a model is fetched coming from an another pipe engine, an HTTP GET is performed putting an extra *PipeTTL* (Time To Live) header. The TTL number is decremented at each subsequent invocation. A pipe engine refuses to fetch more sources if the PipeTTL header is ≤ 1.

The AJAX pipe editor provides inline operator documentation when inserting a component. It presents a list of available pipes, fostering pipe reuse and composition. While normal runtime behavior is very accommodating to network errors (using copies of pre-

---

[4] A grahical editor following the notation in Section 2 is currently under development.

vious files on network timeouts or treating malformed input as empty sources), a debug mode is available, which highlights execution errors.

Finally, thanks to HTTP content negotiation, humans can use Semantic Web Pipes directly. Pipes parameters can be inputted directly in HTML boxes and the results will be shown by the use of the Simile Exhibit data browser[5].

4 Related Works

Semantic Web pipes as described in this paper are similar, in sense, to UNIX pipes[6], but they allow to connect outputs to multiple inputs of other operators so that there can be multiple branches executed at the same time.

Cascaded XML transformations are sometimes referred to as XML pipelines and have been successfully employed in projects like Apache Cocoon.[7]

The Yahoo Web Pipes framework was greatly inspiring our work, but lacks in functionality to address our desired use cases. Yahoo pipes provides an easy to use and powerful Web based graphic composer for pipes.

Concerning the Semantic Web world, the need for a cascade of operators to process RDF repositories is also addressed in the SIMILE Banach project[8], that enhance the Sesame triplestore by implementing pipelined stack of operators (implemented as SAILS). These can both process data and rewrite queries.

5 Conclusions and Future Works

Semantic Web pipes were also shown to be a paradigm that can do more than data harmonization alone: they implement workflows which can be used to model data flow scenarios that also include collaborative aspects. Most importantly, Semantic Web pipes are based on the union of functional operators specific to Semantic Web with the HTTP REST paradigm. Such combination fosters clean implementations, and promotes reuse of data sources as well as pipes themselves.

As we mentioned, a number of additional operators can then be imagined to aid ontology and data alignment when SPARQL CONSTRUCT queries are inconvenient or do not have the required features. Also, it will be interesting to consider how to achieve interaction between advanced RSS feed processing tools like Yahoo Pipes and Semantic Web operators. The SPARQL SELECT operator, producing XML, together with an XSLT transforms could provide a base for this. Many technical solutions can also be put in place to achieve scalability. These range from smart pipe execution strategies, advancing those explained in the previous sections, to others such as, for example, differential updates of the local copy of large remote RDF graphs [5].

Finally, while Semantic Web pipes (like Web pipes and Unix pipes) are certainly a tool for expert users, it is undeniable that the overall engine will be much more useful

---

[5] http://simile.mit.edu/exhibit/
[6] http://www.linfo.org/pipe.html
[7] http://cocoon.apache.org/
[8] http://simile.mit.edu/wiki/Banach

once a visual pipe editor is availale. A grahical editor for our XML format following the notation in Section 2 is currently under development.

References

1. Brickley, D., Miller, L.: FOAF Vocabulary Spec. (July 2005)
2. Hayes, P.: RDF semantics, W3C Rec. (February 2004)
3. Morbidoni, C., Polleres, A., Tummarello, G., Le Phuoc, D.: Semantic Web Pipes. Technical Report (November 2007), http://pipes.deri.org/
4. Prud'hommeaux, E., Seaborne, A.: SPARQL Query Language for RDF, W3C Cand. Rec. (June 2007)
5. Tummarello, G., Morbidoni, C., Bachmann-Gmur, R., Erling, O.: RDFSync: efficient remote synchronization of RDF models. In: 6th Int.l Semantic Web Conf (ISWC 2007) (2007)

Demo: Visual Programming for the Semantic Desktop with *Konduit*

Knud Möller[1], Siegfried Handschuh[1], Sebastian Trüg[2], Laura Josan[1], and Stefan Decker[1]

[1] Digital Enterprise Research Institute, National University of Ireland, Galway
{knud.moeller,siegfried.handschuh,laura.josan,stefan.decker}@deri.org
[2] Mandriva S.A., France
strueg@mandriva.com

Abstract. In this demo description, we present *Konduit*, a desktop-based platform for visual programming with RDF data. Based on the idea of the semantic desktop, non-technical users can create, manipulate and mash-up RDF data with Konduit, and thus generate simple applications or workflows, which are aimed to simplify their everyday work by automating repetitive tasks. The platform allows to combine data from both Web and desktop and integrate it with existing desktop functionality, thus bringing us closer to a convergence of Web and desktop.

1 Introduction

With the Semantic Web gaining momentum, more and more structured data becomes available online. The majority of applications that use this data today are concerned with aspects like search and browsing. However, a greater benefit of structured Web data is its potential for reuse: being able to integrate existing Web data in a workflow relieves users from the investment of creating this data themselves[1]. When it comes to working with data, users still rely on desktop-based applications (exceptions such as Google Docs only serve to support this rule), which are embedded in a familiar environment, using familiar UI metaphors. Web-based applications either simply don't exist, or have shortcomings in terms of usability. What's more, web-based applications can only access Web data, and do not integrate with data that a user might already have on their own desktop, let alone with other applications on the user's desktop. Even considering that it may be benefical for users to publish some desktop data on the Web, releasing all their data to the Web is not an option, since this may raise significant privacy issues. Instead, what is needed is a way of accessing structured Web data from the desktop, integrate it with existing desktop-data and applications and work with both in a unified way.

The advent of the Semantic Desktop [1] through projects such as *NEPOMUK* [2] now opens up new possibilities of solving this problem of integrating

---

[1] Nicely illustrated in the idea of *TCO - Total Cost of Ontologies*, e.g., http://www.w3.org/2005/Talks/1110-iswc-tbl/ (12/12/2007)

data and functionality from both Web and desktop. On the Semantic Desktop, data is lifted from application-specific formats to a universal format (RDF) in such a way that it can be interlinked across application boundaries: emails can be linked to calendar events, address book contacts to pictures or PDF documents, electronic plane tickets to tasks in a task management system. This allows new ways of organizing data, but also new views on and uses of arbitrary desktop data. To use a Web 2.0 term, data can now be *mashed-up*. What is more, because desktop data is now available in a Web format, it can also be interlinked and processed together with genuine Web data: e.g., a book editor could now query a SW-enabled conference website for the contact details of all authors who wrote papers about a specific topic (assuming that this data is available, such as e.g. on `http://data.semanticweb.org`), mash this data locally with a mail template and use his preferred mail client to send a call for contribution to all those authors. While the unified data model makes this scenario easier than it previously was, implementing it would ordinarily still require an experienced developer, who would use a full-fledged programming language to create applications that manipulate and visualize RDF data. With current tools, casual or naïve users would not be able to perform such tasks.

In this demo, we will present *Konduit*, a software that allows users to build simple applications and workflows which can create, manipulate, mash-up and visualize RDF data. Konduit is based on the ideas of a semantic desktop, combined with the principles of UNIX pipes, which has been a central part of UNIX and its derivatives since 1973, when it was introduced by M. Doug McIlroy. In order to allow non-expert users to work with RDF, Konduit is realized as a form of visual programming, meaning that it is *"a computer system whose execution can be specified without scripting"* [3][2]. In a sense, Konduit and similar systems are also related to the concept of data-flow programming [4], in which a program consists of a series of components with inputs and outputs, which become active when all of their inputs are valid.

We will outline an example scenario which illustrates the motivation behind building Konduit in Sect. 2, followed by a presentation of Konduit itself in Sect. 3 and a discussion of what will be contained in the actual demonstration in Sect. 4.

2 Motivating Example

The following example is inspired from a real-life scenario which recently occurred at our institute (names have been changed) and illustrates what Konduit can do for the user: The SemBar Institute needs to prepare a report for its funding agency, which has to show how well the SemBar researchers are connected to their research community. Among other things, co-authorship is a good indicator for this. For this reason, SemBar's scientific director Mary asked secretary Jim to do the following:

---

[2] Of course, that does not mean that the user is prohibited from extending the system with scripting, e.g., in the form of custom SPARQL queries.

- *Compile a list of all researchers.* To get this information, Jim accesses his electronic address book, in which he has cleverly organized all SemBar employees into groups like researchers, administrative staff, technical staff, etc.
- *For each researcher, compile a list of recent publications.* Unfortunately, SemBar does not yet have an internal list of publications (it's Jim's task to compile such a list, after all), so Jim has to resort to a web-based service like http://data.semanticweb.org.
- *For each publication, compile a list of co-authors.* Jim will have to spend a while to manually create these lists.
- *Organize all co-authors into one list and send it to Mary by email.* Jim has to manually remove duplicate and send the list using his mail client.

This scenario highlights a number of important aspects that our approach addresses, and illustrates how a tool such a Konduit can be used by Jim to aid him in his work:

- **Accessing and processing desktop data:** Since Jim is using a semantic desktop, his address book data is available as RDF and can therefore be processed by *Konduit*.
- **Accessing and processing Web data:** Service such as http://data.semanticweb.org expose their data as RDF, which means that our system can use it in the same way as desktop data.
- **Merging desktop and Web data:** Since both kinds of data sources use a unified data model, Konduit can simply mash both together.
- **Using desktop functionality:** Since our approach is desktop-based, we can easily access and integrate the functionality of arbitrary desktop applications (such as Jim's preferred email client).

3 Konduit

Konduit is a desktop-based visual programming environment that presents the user with a working environment not unlike a drawing application, and is as such similar to systems like Yahoo Pipes[3]. However, where Yahoo Pipes is restricted to web-based data in news feed-like form, Konduit works with any kind of RDF source, both from the Web and the user's desktop.

Within the Konduit environment, users can choose from a set of ready-made building blocks, which are organized into various groups (e.g., data-sources or operators). Each component has a number of inputs and zero or more outputs. The user can drag the components onto the "drawing" area, move them around, connect outputs to inputs, set parameters and in this way build simple applications. Konduit defines two basic element types: *Source* and *Sink*. Source elements have an arbitrary number of sockets on which they provide data. Sink elements define plugs which can be plugged into the sockets to define the flow of the data. An element can also be a source and a sink at the same time. This is especially

---

[3] http://pipes.yahoo.com/ (12/12/2007)

useful for elements that modify a stream of data (filter it, combine two streams into one, split the data, enrich the data with meta information, etc.). Based on these two simple types, a number of other components are available, such as various application adapters, SPARQL-based filter components, etc.

A Konduit workflow always has a start and an end point[4], represented by a source and sink element, respectively. The actual flow of data is activated from the end point, by clicking the activation button on a sink component (e.g., the "Send" button in Fig. 1). The connectors of each building block within Konduit are restricted to allow input or output of RDF — in this way, any group of components which have been combined to a workflow will also always have RDF input or output. As a result, combinations of components can be combined into complex elements and become part of the library of components the user can choose from.

An example of how Konduit looks like is given in Fig. 1. The screenshot shows a simple Konduit project, which will send mails to all members of a specific project group, combining desktop sources such as the address book and project data, a SPARQL construct query and functionality of a mail client in one workflow.

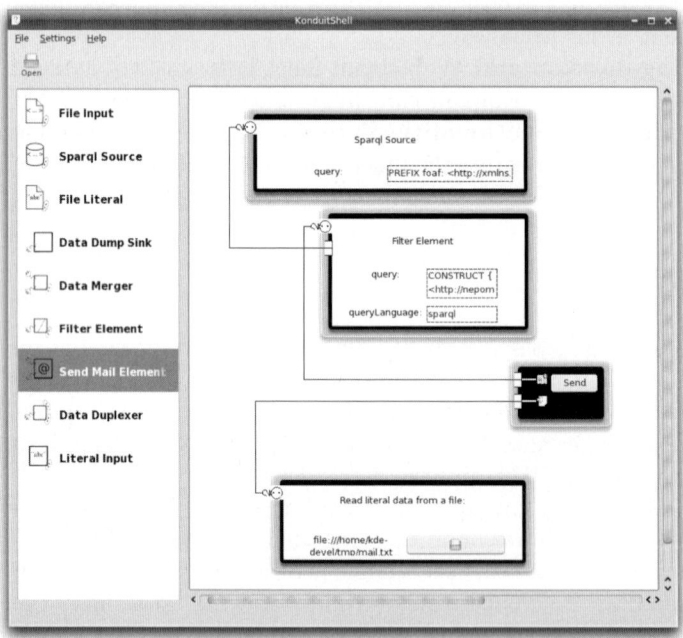

Fig. 1. A workflow for sending mails to all members of a project in *Konduit*

---

[4] Since the flow of data can be split, and different data flows can be combined, a Konduit workflow can also have multiple start and end points.

3.1 Implementation

Konduit is implemented as a desktop-based application for KDE 4 Linux, and is based on the Plasma engine[5]. Each component is realized as a plugin into the Konduit platform. Since components are also Plasma applets, designing and implementing new plugins for Konduit is quite straightforward (from the point of view of a KDE developer). The RDF functionality of Konduit makes use of the semantic desktop features that come as part of Nepomuk-KDE[6] implementation in the upcoming KDE 4, especially the Soprano RDF framework[7].

4 The Demo

During the demo, the audience will have the opportunity to get a live, hands-on experience of Konduit. The demo will consist of a running Nepomuk KDE installation from a fictitious user (one of the *personas* from the NEPOMUK project[8]) with various datasets that ordinary desktop users would find on their desktop, i.e. a contact database, calendar events, emails, textual documents, etc. A number of example scenarios will be prepared (such as the mailer example in the screenshot, or the more complex example in Sect. 2) to showcase what can be done with Konduit. However, visitors will be invited to experiment with and test the system in any way they like: access other data, re-order and re-connect components, interact with other application functionality adaptors, etc.

Acknowledgements

The work presented in this paper was supported (in part) by the Líon project supported by Science Foundation Ireland under Grant No. SFI/02/CE1/I131 and (in part) by the European project NEPOMUK No FP6-027705.

References

1. Decker, S., Frank, M.R.: The networked semantic desktop. In: WWW Workshop on Application Design, Development and Implementation Issues in the Semantic Web (2004)
2. Groza, T., Handschuh, S., Möller, K., Grimnes, G., Sauermann, L., Minack, E., Mesnage, C., Jazayeri, M., Reif, G., Gudjonsdottir, R.: The NEPOMUK project - on the way to the social semantic desktop. In: Pellegrini, T., Schaffert, S. (eds.) Proceedings of I-Semantics 2007, pp. 201–211. JUCS (2007)
3. Menzies, T.: Visual programming, knowledge engineering, and software engineering. In: Proc. 8th Int. Conf. Software Engineering and Knowledge Engineering, SEKE, ACM Press, New York (1996)
4. Orman, L.: A multilevel design architecture for decision support systems. SIGMIS Database 15(3), 3–10 (1984)

---

[5] http://plasma.kde.org/ (10/01/2008)
[6] http://nepomuk-kde.semanticdesktop.org/ (12/12/2007)
[7] http://soprano.sourceforge.net/ (10/01/2008)
[8] http://nepomuk.semanticdesktop.org/ (10/01/2008)

SCARLET: SemantiC RelAtion DiscoveRy by Harvesting OnLinE OnTologies

Marta Sabou, Mathieu d'Aquin, and Enrico Motta

Knowledge Media Institute (KMi), The Open University, Milton Keynes,
{r.m.sabou,m.daquin,e.motta}@open.ac.uk

Abstract. We present a demo of SCARLET, a technique for discovering relations between two concepts by *harvesting the Semantic Web*, i.e., *automatically* finding and exploring *multiple* and *heterogeneous* online ontologies. While we have primarily used SCARLET's relation discovery functionality to support ontology matching and enrichment tasks, it is also available as a stand alone component that can potentially be integrated in a wide range of applications. This demo will focus on presenting SCARLET's functionality and its different parametric settings that can influence the trade-off between its accuracy and time performance.

1 Introduction

A novel trend of harvesting the Semantic Web, i.e., *automatically* finding and exploring *multiple* and *heterogeneous* online knowledge sources, has been favored by the recent growth of online semantic data and the increased interest in building gateways that allow quick exploration of this data[1] (e.g., Watson [4]). For example, Alani proposes a method for ontology learning that relies on cutting and pasting ontology modules from online ontologies relevant to keywords from a user query [1]. Then, in [5] the authors describe a multi-ontology based method that exploits the Semantic Web rather than WordNet to disambiguate the senses of keywords that are given as a query to a search engine (e.g., *star* is used in its sense of celestial body in [*astronomy, start, planet*]).

SCARLET [2] follows this paradigm of automatically selecting and exploring online ontologies *to discover relations between two given concepts*. For example, when relating two concepts labeled *Researcher* and *AcademicStaff*, SCARLET 1) identifies (at run-time) online ontologies that can provide information about how these two concepts inter-relate and then 2) combines this information to infer their relation. We describe two increasingly sophisticated strategies to discover and exploit online ontologies for relation discovery. The first strategy derives a relation between two concepts if this relation is defined within a single online ontology, e.g., stating that *Researcher* ⊑ *AcademicStaff* (Section 2.1). The second strategy (Section 2.2) addresses those cases when no single online ontology states

---
[1] http://esw.w3.org/topic/TaskForces/CommunityProjects/LinkingOpenData/SemanticWebSearchEngines
[2] http://scarlet.open.ac.uk/

the relation between the two concepts by combining relevant information which is spread over two or more ontologies (e.g., that *Researcher* $\sqsubseteq$ *ResearchStaff* in one ontology and that *ResearchStaff* $\sqsubseteq$ *AcademicStaff* in another). The DBPedia Relation Finder [6] provides a similar functionality (it discovers connections between two objects in the DBPedia data set), but it relies on semantically weaker methods (e.g., graph clustering instead of reasoning).

SCARLET originates from earlier work in the field of ontology matching, from the design of a matcher that exploits the entire Semantic Web as a source of background knowledge [7]. In essence, this matcher discovers semantic relations (mappings) between the elements of two ontologies by using the methods described above. A large-scale evaluation of this matcher lead to precision values of over 70% [8]. SCARLET's relation discovery functionality has also been used to semantically enrich folksonomy tagsets [2]. Given a set of implicitly related tags, we used SCARLET to identify relations between these tags and then merged them into a new knowledge structure (ontology).

While we have used SCARLET to support two different tasks, we also provide it as a stand-alone component as its functionality could be useful when integrated in a variety of other tools. To cater for applications with different needs, SCARLET can be used with a variety of parametric settings that regulate the trade-off between its accuracy and time performance. In this demo we provide an insight in the internal working of SCARLET and its different settings.

2 SCARLET: Technology Overview

In this section we present an overview of SCARLET's two strategies (as described in [7]) and its main parameters. The parameters can be fine-tuned from the command line, and soon through a Web based interface. Each strategy is presented as a procedure that takes two candidate concepts (denoted as A and B) as an input and returns the discovered relation between them. The corresponding concepts to A and B in an online ontology O_i are A'_i and B'_i ("anchor terms"). We rely on the description logic syntax for semantic relations occurring between concepts in an online ontology O_i (e.g., $A'_i \sqsubseteq B'_i$) and on a C-OWL like notation [3] for the returned relations (e.g., $A \xrightarrow{\sqsubseteq} B$). Note that the current version of SCARLET only explores taxonomic and disjoint relations.

2.1 Strategy S1: Relation Discovery Within One Ontology

Strategy S1 consists of finding ontologies containing concepts similar with the candidate concepts (e.g., by relying on Watson) and then deriving a relation from their relations in the selected ontologies. Figure 1 (a) illustrates this strategy with an example where three ontologies are discovered (O_1, O_2, O_3) containing the concepts A' and B' corresponding to A and B. The first ontology contains no relation between the anchor concepts, while the other two ontologies declare a subsumption relation. The concrete steps of this strategy are:

1. Anchor A and B to corresponding concepts A' and B' in online ontologies;
2. Select ontologies containing A' and B';
3. For a given ontology (O_i) apply the following rules:
   - if $A'_i \equiv B'_i$ then derive $A \xRightarrow{} B$;
   - if $A'_i \sqsubseteq B'_i$ then derive $A \xrightarrow{\sqsubseteq} B$;
   - if $A'_i \sqsupseteq B'_i$ then derive $A \xrightarrow{\sqsupseteq} B$;
   - if $A'_i \perp B'_i$ then derive $A \xrightarrow{\perp} B$;
4. Combine all relations derived from the considered ontologies.

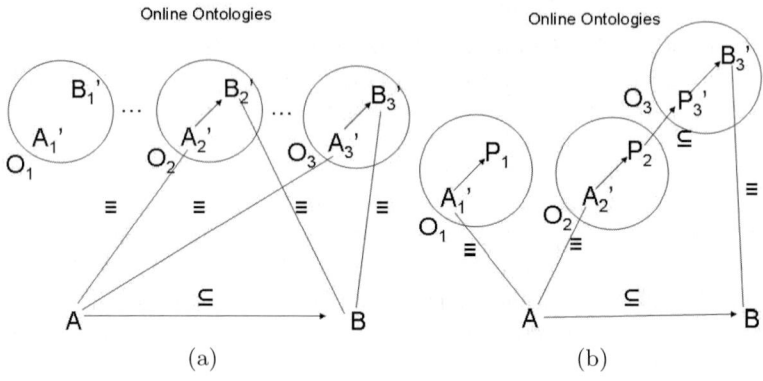

Fig. 1. Relation discovery (a) within one ontology (S1) and (b) across ontologies (S2)

For example, when matching two concepts labeled *Drinking Water* and *tap_water*, appropriate anchor terms are discovered in the TAP ontology and the following subsumption chain in the external ontology is used to deduce the relation: *DrinkingWater* $\sqsubseteq$ *FlatDrinkingWater* $\sqsubseteq$ *TapWater*.

2.2 Strategy S2: Cross-Ontology Relation Discovery

The previous strategy assumes that a relation between the candidate concepts can be discovered in a single ontology. However, some relations could be distributed over several ontologies. Therefore, if no ontology is found that relates both candidate concepts, then the relation should be derived from two (or more) ontologies. In this strategy, relation discovery is a recursive task where two concepts can be matched because the concepts they relate to in some ontologies are themselves matched. Figure 1 (b) illustrates this strategy where no ontology is available that contains anchor terms for both A and B, but where one of the parents (P_2) of the anchor term A'_2 can be matched to B in the context of a third ontology (O_3). For example, a relation between *Cabbage* and *Meat* can be derived by taking into account that *Cabbage* $\sqsubseteq$ *Vegetable*[3] and then discovering that *Vegetable* $\perp$ *Meat*[4] through another discovery step. The concrete steps are:

---

[3] http://139.91.183.30:9090/RDF/VRP/Examples/tap.rdf
[4] http://www.co-ode.org/resources/ontologies/Pizzademostep1.owl

1. Anchor A and B to corresponding concepts A' and B' in online ontologies;
2. If no ontologies are found that contain both A' and B' then select all ontologies containing A';
3. For a given ontology O_i apply the following rules:
   (a) for each P_i such that $A'_i \sqsubseteq P_i$, search for relations between P_i and B;
   (b) for each C_i such that $A'_i \sqsupseteq C_i$, search for relations between C_i and B;
   (c) derive relations using the following rules:
   - (r1) if $A'_i \sqsubseteq P_i$ and $P_i \xrightarrow{\sqsubseteq} B$ then $A \xrightarrow{\sqsubseteq} B$
   - (r2) if $A'_i \sqsubseteq P_i$ and $P_i \xrightarrow{\equiv} B$ then $A \xrightarrow{\sqsubseteq} B$
   - (r3) if $A'_i \sqsubseteq P_i$ and $P_i \xrightarrow{\perp} B$ then $A \xrightarrow{\perp} B$
   - (r4) if $A'_i \sqsupseteq C_i$ and $C_i \xrightarrow{\sqsupseteq} B$ then $A \xrightarrow{\sqsupseteq} B$
   - (r5) if $A'_i \sqsupseteq C_i$ and $C_i \xrightarrow{\equiv} B$ then $A \xrightarrow{\sqsupseteq} B$
4. Combine all relations derived from the considered ontologies.

2.3 Main Parameters

There are several parameters that can influence the way SCARLET works.

Ontology gateway. Currently, SCARLET can rely on either Swoogle or Watson to access online available ontologies.

Strategy. SCARLET can function according to two different strategies: S1, when the relation is derived from within a single ontology and S2, when information spread over several ontologies is combined to derive a relation.

Number of derived relations. The anchoring step of both strategies (step 1), identifies all the ontologies that possibly contain a relation between the input concepts. An important decision is the number of ontologies used to derive relations. On the one hand, using a single ontology is the easiest way to deal with the multiple returned ontologies but it assumes that the discovered relation can be trusted and there is no need to inspect the other ontologies as well. On the other hand, using a subset (or all) of the returned ontologies is computationally more expensive but it has a higher accuracy by taking into account all the information that can be possibly derived. In these cases a relation is derived from each ontology and then these are combined into a final relation (using a method selected by the next parameter).

Method for combining multiple relations. If all possible relations between the terms are derived, then it is important to specify what will actually be returned by SCARLET. By default, SCARLET returns all the derived relations. Another option is to return a relation only if all the derived relations were the same (i.e., all ontologies stated the same relation). A final possibility is to return the relation that was most frequently derived, i.e., on which most sources agree.

S2: Depth of hierarchy considered. The number of parent/child classes in the anchor ontology used in S2 have an influence on the complexity of the algorithm. In the simplest case, only the direct parents/children are considered. In the most complex case, S2 will investigate all parents/children. The larger the considered depth, the longer the processing times and (possibly) higher the number of discovered relations.

3 Demonstration Plan

The demo will focus on showcasing the functionality of SCARLET as well as the use of its various parameters. Additional information will be provided about the paradigm on which this technique relies, the case studies in which it was used and its internal functioning.

We expect that several aspects of this demo will be of interest to the visitors. First, the visitor will learn about the new paradigm of harvesting the Semantic Web and experience through SCARLET one of its concrete implementations. Second, he will get to know the functionality of SCARLET and the types of tasks in which it has already been used. This information could help him to understand whether SCARLET could be useful in his own work. Finally, visitors interested in the internal working of SCARLET, will be explained how the software works and how it can be fine-tuned through its various parameters. Those visitors that intend to use this functionality in their own work could already determine what settings suit their needs best.

References

1. Alani, H.: Position Paper: Ontology Construction from Online Ontologies. In: Proc. of WWW (2006)
2. Angeletou, S., Sabou, M., Specia, L., Motta, E.: Bridging the Gap Between Folksonomies and the Semantic Web: An Experience Report. In: Proc. of the ESWC Workshop on Bridging the Gap between Semantic Web and Web 2.0 (2007)
3. Bouquet, P., Giunchiglia, F., van Harmelen, F., Serafini, L., Stuckenschmidt, H.: Contextualizing ontologies. Journal of Web Semantics 1(4), 24 (2004)
4. d'Aquin, M., Baldassarre, C., Gridinoc, L., Sabou, M., Angeletou, S., Motta, E.: Watson: Supporting Next Generation Semantic Web Applications. In: Proc. of WWW/Internet conference, Vila Real, Spain (2007)
5. Gracia, J., Trillo, R., Espinoza, M., Mena, E.: Querying the Web: A Multiontology Disambiguation Method. In: Proc. of ICWE (2006)
6. Lehmann, J., Schuppel, J., Auer, S.: Discovering Unknown Connections - the DBpedia Relationship Finder. In: Proc. of CSSW (2007)
7. Sabou, M., d'Aquin, M., Motta, E.: Using the Semantic Web as Background Knowledge for Ontology Mapping. In: Proc. of the Ontology Matching WS (2006)
8. Sabou, M., Gracia, J., Angeletou, S., d'Aquin, M., Motta, E.: Evaluating the Semantic Web: A Task-based Approach. In: Proc. of ASWC/ISWC (2007)

ODEWiki: A Semantic Wiki That Interoperates with the ODESeW Semantic Portal

Adrián Siles, Angel López-Cima, Oscar Corcho, and Asunción Gómez-Pérez

Facultad de Informática, Universidad Politécnica de Madrid,
Campus de Montegancedo, sn. 28660 Boadilla del Monte (Madrid, Spain)
asiles@delicias.dia.fi.upm.es, {alopez,ocorcho,asun}@fi.upm.es

Abstract. We present ODEWiki, a technology for the development of Semantic Wikis, which has a combined set of added-value features over other existing semantic wikis in the state of the art. Namely, ODEWiki interoperates with an existing semantic portal technology (ODESeW), it manages inconsistencies raised because of the distributed nature of knowledge base development and maintenance, it uses RDFa for the annotation of the resulting wiki pages, it follows a WYSIWYG approach, and it allows decoupling wiki pages and ontology instances, that is, a wiki page may contain one or several ontology instances. Although some of these features appear in some of the state-of-the-art semantic wikis, but they are not combined together in a single solution.

Keywords: Semantic Wiki, ODEWiki.

1 Introduction and Background

A wiki is software that allows users to create, edit, and link web pages easily[1]. Wikis are often used to create collaborative websites and to power community websites. As it happens with other Web sites, wikis are mainly focused on the provision of content for human users, and lack from a clear semantic description of their content. Semantic wikis have appeared in the last years as a technology that builds on top of existing wikis, providing the same functionalities as these plus the possibility of adding semantic annotations (normally in the form of RDF triples) to the wiki pages that are generated.

In the last years, there has been a huge growth in the number of semantic wikis that have been developed and made available to the community. For instance, the OntoWorld site provides a non-exhaustive list of approximately 30 semantic wikis[2].In addition, there have been several workshops whose only topic has been about semantic wikis. In our analysis, we have focused on some of the most popular ones: Makna [1], Rhizome [2], Semantic Mediawiki [3], SweetWiki [4] and IkeWiki [5].

Some of the common characteristics of all these technologies are related to the following aspects:

---

[1] http://en.wikipedia.org/wiki/Wiki
[2] http://ontoworld.org/wiki/Semantic_Wiki_State_Of_The_Art

- Annotation. This is the mechanism by which semantic annotations are related to the information provided in a wiki page (normally in a textual form). These semantic annotations may be related to existing ontologies or only be plain RDF, which is used mainly as a syntax for providing additional information.
- Edition. This is the mechanism by which users can add semantic information not only to the wiki pages that they are editing (this is the annotation process aforementioned), but also the knowledge base that acts as a knowledge repository for those wiki pages.
- Semantic search. This is the mechanism by which users may look for wiki pages taking into account their semantic annotations.
- Visualization. This is the mechanism by which semantic wikis visualize the semantic information about ontologies and their corresponding instances in the knowledge base.
- Navigation. This is the mechanism by which users can navigate between wiki pages taking into account the semantic relationships that are established by the annotations associated to them.

Although these are common features for semantic wikis, not all the existing semantic wiki technologies offer all these services, and there are many differences among different technologies with respect to the degree of complexity and functionality in each of these categories.

After an analysis of the state of the art in semantic wiki technologies, we have discovered several limitations of all of them:

- None of the existing semantic wikis is used together with a knowledge portal. This means that, although these applications are similar (semantic wikis and knowledge portals allow creating and maintaining knowledge bases), their strengths are not combined in any single solution. This could facilitate processes of collaborative knowledge update and different publishing options for knowledge portals, and knowledge edition and curation workflows for semantic wikis.
- Only one of the analyzed semantic wiki technologies (SweetWiki) makes use of the latest developments in the use of XHTML tags to include RDF triples (RDFa[3]). This fact is very important for automatic data discovery using same wiki page, that is, persons and computers can be extract same information from this page.
- Only one of the analyzed semantic wiki technologies (SweetWiki) follows a WYSIWYG (What You See Is What You Get) approach, which facilitates the annotation of wiki pages for non-experts. The rest of technologies require users to learn a new set of tags or syntax to include these annotations, and also require users to know the ontologies to be used in advance, or to have them opened in a different part of the web site.
- In most cases, each wiki page is associated to only one instance of an ontology. This means that each page may only describe an individual, which makes it difficult, for instance, to create and annotate more complex pages which contain and declare several instances (e.g., a page that contains information about a set of persons that belong to an institution, where we want to have more information about them than simply their name).

---

[3] http://www.w3.org/TR/xhtml-rdfa-primer/

Taking into account the aforementioned limitations, we have designed and implemented ODEWiki, which is a semantic wiki that is integrated with the ODESeW semantic portal [6] [7] [8]. This semantic wiki provides the most common functionalities that are provided by most of the existing semantic wikis: annotation, edition, search, navigation and visualization. And besides being integrated with a semantic portal, which acts as a knowledge repository for the wiki plus as an alternative knowledge publishing system, ODEWiki combines features that are only found in some of the state-of-the-art semantic wikis, such as the use of RDFa for the annotation of the resulting wiki pages, a WYSIWYG approach, and the decoupling of wiki pages and ontology instances, that is, a wiki page may contain one or several ontology instances.

In this demo, we will demonstrate how to edit semantically enabled wiki pages with this technology, together with the advantages of the integration with a semantic portal, so that the visitor will know how to operate with this technology.

2 Annotation and Edition in ODEWiki

Most of the limitations identified in the state of art are related to characteristics of the annotation and edition functionalities offered by wikis. In annotation and edition, users can use a WYSIWYG user interface to annotate wiki pages with semantic data, which is included in the wiki page using the RDFa format. This wiki page, and the semantic data included into it, is not modified until the user decides to modify it. This fact sometimes causes the semantic data to have external inconsistencies with the underlying knowledge base, and therefore with the ODESeW semantic portal or other wiki pages. ODEWiki provides a mechanism to automatically detect and manually correct these external inconsistencies. We understand by external inconsistencies those that are related to the following situations:

- The wiki page contains an instance that is not present in the knowledge base (e.g., "onto:AdriánSiles rdf:type onto:Person"). This may have happened because another wiki page or an external system has deleted that instance from the knowledge base.
- The wiki page contains annotations of an instance that exists in the knowledge base. However, the annotation itself (in other words, the RDF triple) does not exist in the knowledge base[4] (e.g. "onto:AdriánSiles onto:belongsTo onto:UPM", where onto:belongsTo has been removed from the ontology, onto:UPM has disappeared, this person has stopped belonging to onto:UPM, etc.). To some extent, this may be seen as a generalization of the previous case.

These external inconsistencies are checked whenever a wiki page is going to be visualized or when the user is going to start editing it. The user is notified of the

---

[4] Taking into account the Open World Assumption (OWA), this situation should not be considered as a problem, since we can always have annotations about an ontology instance that has not been defined elsewhere, but since this system is connected to a knowledge portal where we consider the Closed World Assumption (CWA), we must assume here this behavior. This behavior, however, could be changed accordingly in case that the OWA.

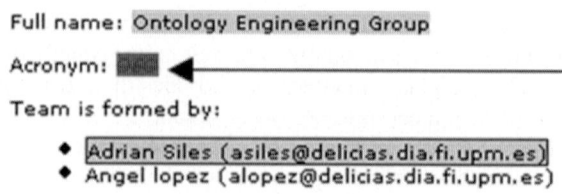

Fig. 1. Notification of inconsistencies in a wiki page

inconsistencies that have been found (as shown in figure 1) and can take any of the following decisions: delete the corresponding RDF triple from the wiki page or from the knowledge base, or include the RDF triple into the knowledge base.

Users can annotate wiki pages in two ways, using semantic data from the knowledge base or adding new semantic data. Semantic data added to a wiki page is automatically added into the knowledge base. Not only can users add semantic data into the underlying knowledge base, but also can they delete, using a special action attribute in RDFa triple, semantic data form the knowledge base and from the wiki page. Users are responsible to take specific actions to avoid deleting relevant semantic data. In fact, ODEWiki is not recommended to be used in open environments, that is, ODEWiki is specifically designed to be used inside companies or organizations.

ODEWiki also provides a mechanism to automatically detect and manually correct internal inconsistencies. We understand by internal inconsistencies those that are related to the following situation:

- A wiki page contains annotations of an instance that is, at the same time, removed from that page ("onto:AdriánSiles onto:belongsTo onto:UPM" is added to the knowledge base from the wiki page, but at the same time the wiki page contains an action to remove the triple "onto:AdriánSiles rdf:type onto:Person" and there is no other "onto:AdriánSiles rdf:type X" or "onto:AdriánSiles rdf:label X" triple in the knowledge base, which causes the removal of that instance).

3 Conclusions and Future Work

In this paper, we have shown some of the limitations of the most relevant semantic wiki technologies, which are mainly, related to their annotation and edition functionalities, especially in what respects to collaborative and distributed edition of semantic information. To overcome these limitations, we have built ODEWiki, a semantic wiki that is interconnected with a knowledge portal, which acts both as a knowledge repository and as an alternative publication and knowledge edition platform.

As part of our future work in this technology, we will focus on the provision of an improved search functionality that goes beyond the current state of the art in semantic search in existing semantic wikis (which is based on the SPARQL protocol). And in an improved navigation model that inherits some of the characteristics of the navigation model of ODESeW [8].

Acknowledgement

This work has been funded by the UPM-funded project "Extensiones a portales semánticos con wikis semántico"[5] (CCG06-UPM/INF-284).

References

1. Nixon, L.J.B., Paslaru, E., Simperl, B.: Makna and multimakna: towards semantic and multimedia capability in wikis for the emerging web. In: Proceedings of the Semantics 2006, Vienna, Austria (November 2006)
2. Souzis, A.: Bringing the "wiki-way" to the semantic web with rhizome. In: Völkel, M., Schaert, S. (eds.) Proceedings of the First Workshop on Semantic Wikis From Wiki To Semantics, Workshop on SemanticWikis (ESWC2006), Budva, Montenegro (June 2006)
3. Millard, I., Jaffri, A., Glaser, H., Rodriguez-Castro, B.: Using a semantic mediawiki to interact with a knowledge based infrustructure. In: 15th International Conference on Knowledge Engineering and Knowledge Management (EKAW 2006), Podebrady, Czech Republic (October 2006)
4. Buffa, M., Gandon, F.: Sweetwiki: semantic web enabled technologies in wiki. In: Proceedings of the international symposium on Symposium on Wikis (WikiSym 2006), New York, NY, USA, pp. 69–78 (2006)
5. Schaffert, S.: Ikewiki: A semantic wiki for collaborative knowledge management. In: 1st International Workshop on Semantic Technologies in Collaborative Applications (STICA 2006), Manchester, UK (June 2006)
6. López-Cima, Á., de Figueroa, M.d.C.S., Gómez-Pérez, A.: The ODESeW platform as a tool for managing EU projects: the Knowledge Web case study. In: Managing Knowledge in a World of Networks. 15th Internation Conference (EKAW 2006), Podêbrady, Czech Republic (October 2006)
7. López-Cima, Á., Gómez-Pérez, A.: Rapid ontology-based Web application development with JSTL. In: Scripring for the Semantc Web 2007 from European Semantic Web Conference (ESWC 2007), Innsbruck, Austria (June 2007)
8. López-Cima, A., Corcho, O., Gómez-Pérez, A.: A Platform for the Development of Semantic Web portals. In: 6th International Conference on Web Engineering (ICWE 2006), Stanford, CA (July 2006)

---

[5] Extensions to semantic portals with semantic wikis.

Simplifying Access to Large-Scale Health Care and Life Sciences Datasets

Holger Stenzhorn[1,2], Kavitha Srinivas[3],
Matthias Samwald[2,4], and Alan Ruttenberg[5]

[1] Department of Medical Informatics, University Medical Center Freiburg, Germany
[2] Digital Enterprise Research Institute (DERI), National University of Ireland, Galway, Ireland
[3] IBM T.J. Watson Research Center, Yorktown Heights, New York, USA
[4] Section on Medical Expert and Knowledge-Based Systems, Medical University of Vienna, Austria
[5] Science Commons, Cambridge, Massachusetts, USA

1 Introduction

Within the health care and life sciences (HCLS) domain, a plethora of phenomena exists that range across the whole "vertical scale" of biomedicine. To accurately research and describe those phenomena, a tremendous amount of highly heterogeneous data have been produced and collected with various research methodologies encompassing the genetic, molecular, tissue, and organ level. An initial step to provide researchers with access to this data has been through creating integrated views on existing and open biomedical datasets published on the Web. In order to make the next step, we need to now create easy-to-use yet powerful applications that enable researchers to efficiently query, integrate and analyze those datasets.

One effort in that direction is currently carried out by the World Wide Web Consortium's Semantic Web Health Care and Life Sciences Interest Group (HCLSIG)[1]. It is intended as a bridge between the Semantic Web community's technology and expertise and the information challenges and experiences in the HCLS communities [8]. It brings together scientists, medical researchers, science writers, and informaticians working on new approaches to support biomedical research. Participants come from both academia, government, non-profit organizations as well as health care, pharmaceuticals, and industry vendors.

In the following we show some results of this effort by describing two demonstrations of our approach on preparing and applying biomedical information on the Semantic Web. (All demonstration materials can be freely downloaded[2].)

2 Data Modeling, Storage and Provision

As starting point for our activities, we have re-modeled several biomedical datasets in OWL in order to take advantage of that language's well-defined semantics.

---

[1] http://www.w3.org/2001/sw/hcls
[2] http://esw.w3.org/topic/HCLS/Banff2007Demo

Those datasets include[3] PubMed[4], the Gene Ontology Annotations (GOA)[5], Entrez Gene[6], the Medical Subject Headings (MeSH)[7], the Foundational Model of Anatomy (FMA)[8], and the Allen Brain Atlas (ABA)[9].

2.1 Modeling Principles

Great care has been taken in modeling the dataset, e.g., to clearly distinguish between database records and real world statements such as about proteins in cells. We have used (and extended) the OBO Foundry design principles [5] to create interoperability between the information sources and the OBO ontologies[10].

As a proof for the viability of this approach, we have successfully aligned a principled representation of GOA with two new representations of neuroscience databases (NeuronDB[11] and the Brain Architecture Management System (BAMS)[12]). Additionally, to create a specific anatomical view over the resources we have created mappings from MeSH to the FMA using UMLS[13].

2.2 Storage and Provision

We have created two demonstrations based on the OWL dataset representations:

In the first one, we have used the Openlink Virtuoso[14] RDF triple store to save more than 300 million triples (made publicly accessible through a SPARQL endpoint[15]). Although it is possible to store OWL in triple stores, Virtuoso does not support native OWL inference. In order to support more expressive queries, we have provided limited inference support in this implementation by using a combination of Virtuoso's native transitive closure support, simple rules based on their implementation of SPARQL-Update (SPARUL)[16] and loading partonomy relationships[17] pre-computed by the Pellet OWL Reasoner [6].

In the second one, we have used the SHER OWL management and inference system from IBM Research [2]. In this system also about 300 million triples have been stored, providing OWL inferencing over both GO[18] and FMA [4].

---

[3] The full dataset list is available at http://sw.neurocommons.org/2007/kb-sources
[4] http://www.ncbi.nlm.nih.gov/pubmed
[5] http://www.geneontology.org/GO.annotation.shtml
[6] http://www.ncbi.nlm.nih.gov/sites/entrez?db=gene
[7] http://www.nlm.nih.gov/mesh
[8] http://sig.biostr.washington.edu/projects/fm
[9] http://www.brainatlas.org
[10] http://www.obofoundry.org
[11] http://senselab.med.yale.edu/senselab/NeuronDB
[12] http://brancusi.usc.edu/bkms
[13] http://umlsinfo.nlm.nih.gov
[14] http://www.openlinksw.com/virtuoso
[15] http://hcls.deri.ie/demo, http://sparql.neurocommons.org:8890/nsparql
[16] http://jena.hpl.hp.com/~afs/SPARQL-Update.html
[17] http://esw.w3.org/topic/HCLS/PartOfInference
[18] http://www.geneontology.org

2.3 Identifiers

We have also assured to employ a URI scheme to uniquely name resources and biological entities based on the `purl.org` resolver: Stable URIs have been given both to existing resources where providers do not currently have any stable identification scheme as well as for newly defined classes and instances. To avoid resolver redirection overhead for large query numbers, Semantic Web agents can query the resolver once to retrieve rewrite rules and implement those in their application to then access the actual resource directly.

3 Query Capabilities

Our two implementations highlight two different approaches to inference and reasoning. But both of them aim at retrieving precise answers to narrow queries.

Fig. 1 shows an example of a SPARQL query against our triple store querying for *genes associated with CA1 Pyramidal Neurons* (as defined by MeSH) and *signal transduction processes* (as defined by GO), returning 40 pairings of gene and process, compared to about 175,000 returned by the query against Google for *genes involved in pyramidal neuron signal transduction*.

```
prefix go: <http://purl.org/obo/owl/GO#>
prefix mesh: <http://purl.org/commons/record/mesh/>
prefix owl: <http://www.w3.org/2002/07/owl#>
prefix rdfs: <http://www.w3.org/2000/01/rdf-schema#>
prefix sc: <http://purl.org/science/owl/sciencecommons/>
prefix ro: <http://www.obofoundry.org/ro/ro.owl#>

SELECT DISTINCT ?gene ?process WHERE {
  graph <http://purl.org/commons/hcls/pubmesh>
  { ?pubmedrecord ?p mesh:D017966.
    ?article sc:identified_by_pmid ?pubmedrecord.
    ?generecord sc:describes_gene_or_gene_product_mentioned_by ?article. }
  graph <http://purl.org/commons/hcls/goa>
  { ?protein rdfs:subClassOf ?res.
    ?res owl:onProperty ro:has_function.
    ?res owl:someValuesFrom ?res2.
    ?res2 owl:onProperty ro:realized_as.
    ?res2 owl:someValuesFrom ?process.
    graph <http://purl.org/commons/hcls/20070416/classrelations>
      {{ ?process <http://purl.org/obo/owl/obo#part_of> go:GO_0007166. }
        union { ?process rdfs:subClassOf go:GO_0007166. }}
    ?protein rdfs:subClassOf ?parent.
    ?parent owl:equivalentClass ?res3.
    ?res3 owl:hasValue ?generecord. }}}
```

Fig. 1. Complex SPARQL query to retrieve all genes which are associated with both *CA1 Pyramidal Neurons* and the *signal transduction processes*

The query works by linking MeSH associated with Pubmed records to genes via Entrez Gene, narrowing the genes by GO associations narrowed to signal transduction processes or parts of those processes.

On the other hand, SHER provides reasoning capabilities over FMA and GO. Reasoning on FMA is well known to be problematic for current reasoners due to the fact that FMA represents a deep mereological hierarchy in which both *part-of* as well as its inverse *has-part* relations are employed [1]. This occurs partly to work around modeling constraints found in OWL. Even though mereological hierarchies are better modeled as description graphs [3] this would require a complete re-modeling of FMA, and hence, as a workaround, we reasoned only over the *part-of* relations in FMA with the SHER OWL reasoner. Supported sample queries are of the following form (marked concepts are taken from FMA and GO respectively and require inferencing): *Find the genes known to be involved in Alzheimer's disease, in the **hippocampal** region that have a role in **dendrite development**.* This expands the search not only to the hippocampus but also to its sub-parts, such as the CA1 region as well as to processes that are part of dendrite development such as dendrite morphogenesis.

4 User Interface

In order to demonstrate the capabilities of the developed system we have created two different browser interfaces using freely available tools:

In the first one, we have combined query results with data made available by the Alan Brain Institute and presented the combined data using Exhibit [7]. It shows images of mouse brain slices stained for expressed genes with each gene's details, and visualizing its transcript regions and genomic context (cf. Fig. 2).

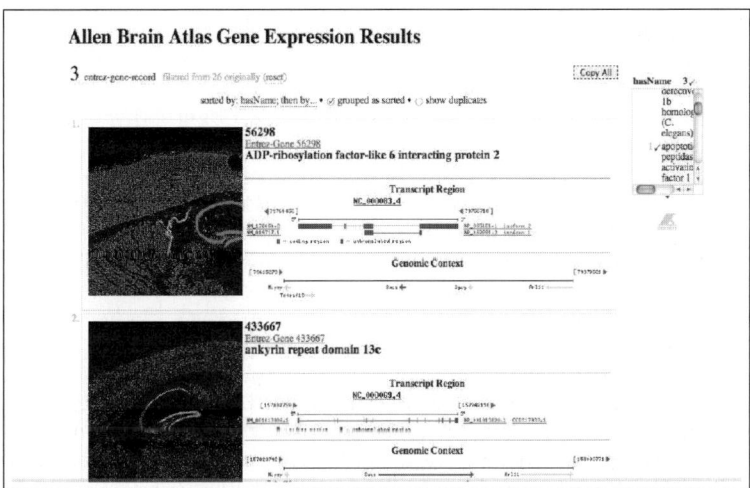

Fig. 2. Screenshot of a gene query result in Exhibit, showing expressions with images from the Allen Brain Atlas combined with transcripts from Entrez Gene

In the second interface, we have provided an intuitive keyword interface to search medical literature with keywords being internally converted into a logical query. The example sentence would be translated into (y is the selected variable) aboutGene(x,y) ⊓ hasFunction(x,m) ⊓ rdf:type(m,GO:dendrite_development) ⊓ evidence(x,z) ⊓ source(z,p) ⊓ hasPubMedID(p,q) ⊓ hasAsMesh(q, Alzheimer's_disease) ⊓ hasAsMesh(q,r)⊓ rdf:type(r, FMA:hippocampus).

As a next step, we envisage to develop a further interface for simplifying the creation and maintainance of complex SPARQL queries for the first system.

Acknowledgments

This work is the product of many participants in the HCLSIG and includes (besides the authors) John Barkley, Olivier Bodenreider, Bill Bug, Huajun Chen, Paolo Ciccarese, Kei Cheung, Tim Clark, Don Doherty, Julian Dolby, Kerstin Forsberg, Achille Fokoue, Ray Hookaway, Aditya Kalyanpur, Vipul Kashyap, June Kinoshita, Joanne Luciano, Li Ma, Scott Marshall, Chris Mungall, Eric Neumann, Chintan Patel, Eric Prud'hommeaux, Jonathan Rees, Edith Schonberg, Mike Travers, Gwen Wong and Elizabeth Wu. Susie Stephens coordinated the BioRDF subgroup of the HCLSIG in which this work was developed.

References

1. Dameron, O., Rubin, D., Musen, M.: Converting the Foundational Model of Anatomy into OWL. In: Proc. AMIA Symp 2005, Washington, DC (2005)
2. Dolby, J., Fokoue, A., Kalyanpur, A., Kershenbaum, A., Ma, L., Schonberg, E., Srinivas, K.: Scalable Semantic Retrieval Through Summarization and Refinement. In: Proc. AAAI 2007, Vancouver, Canada (2007)
3. Motik, B., Cuenca Grau, B., Sattler, U.: Structured Objects in OWL: Representation and Reasoning. Technical Report, University of Oxford, UK (2007)
4. Fokoue, A., Kershenbaum, A., Ma, L., Schonberg, E., Srinivas, K.: The Summary Abox: Cutting Ontologies Down to Size. In: Proc. 5th International Semantic Web Conference, Athens, GA, USA (2006)
5. Smith, B., Ashburner, M., Rosse, C., Bard, J., Bug, W., Ceusters, W., Goldberg, L., Eilbeck, K., Ireland, A., Mungall, C., Leontis, N., Rocca-Serra, P., Ruttenberg, A., Sansone, S., Scheuermann, R., Shah, N., Whetzel, P., Lewis, S.: The OBO Foundry: Coordinated Evolution of Ontologies to Support Biomedical Data Integration. Nature Biotechnology 25, 1251–1255 (2007)
6. Sirin, E., Parsia, B., Cuenca Grau, B., Kalyanpur, A., Katz, Y.: Pellet: A practical OWL-DL reasoner. UMIACS Technical Report, 2005-68 (2005)
7. Huynh, D., Karger, D., Miller, R.: Exhibit: Lightweight Structured Data Publishing. In: Proc. of the World Wide Web 2007 Conference, Banff, Canada (2007)
8. Ruttenberg, A., Clark, T., Bug, W., Samwald, M., Bodenreider, O., Chen, H., Doherty, D., Forsberg, K., Gao, Y., Kashyap, V., Kinoshita, J., Luciano, J., Marshall, M., Ogbuji, C., Rees, J., Stephens, S., Wong, G., Wu, E., Zaccagnini, D., Hongsermeier, T., Neumann, E., Herman, I., Cheung, K.: Advancing translational research with the Semantic Web. BMC Bioinformatics 8 (2007)

GRISINO - An Integrated Infrastructure for Semantic Web Services, Grid Computing and Intelligent Objects

Ioan Toma[1], Tobias Bürger[1,2], Omair Shafiq[1], and Daniel Döegl[3]

[1] Semantic Technology Institute - STI Innsbruck, University of Innsbruck, Austria
{ioan.toma,omair.shafiq,tobias.buerger}@sti2.at
[2] Salzburg Research Forschungsgesellschaft mbH, Salzburg, Austria
tobias.buerger@salzburgresearch.at
[3] Uma Information Technology GmbH, Vienna, Austria
daniel.doegl@uma.at

Abstract. Future information, knowledge and content infrastructures which provide highly automated support in fulfilling users goals will most likely rely on some form of GRID computing. In combination with Semantic Web technologies and Semantic Web Services, such infrastructure will be much enhanced to form the Semantic Grid. Last but not least, the content, which can range from multimedia data to intelligent objects, can and must be interpreted by the services of the Semantic Grid. In this demo we will detail the GRISINO Common Infrastructure, an integrated infrastructure for Semantic Web Services, Intelligent Content Objects and Grid, that are brought together in the search for an emerging solution for next generation distributed applications.

1 Introduction

The aim of GRISINO [6] is to combine three leading edge technologies which complement each other, for the definition of intelligent and dynamic business processes:

1. **Semantic Web Services (SWS)** as the future standard for the declaration of web-based semantic processes.
2. **Intelligent Content Objects** as the unit of value which can be manipulated by SWS, and
3. **Grid Computing** as a pervasive service distribution infrastructure for a future, ambient intelligence space.

Semantic Web Services: The vision of Semantic Web Services (SWS) is to describe the various aspects of a Web Service using explicit, machine-understandable semantics, that can enable automatic discovery or composition of complex Web Services or facilitate seamless interoperation between different Web Services.

Intelligent Content: The term Intelligent Content (IC) [1] is a notation for content containing information with explicit semantic descriptions of its properties which might include technical metadata, subject matter information, how the content is typically used or what rights apply to it.

Grid Computing: A grid as defined in [3] is a system that coordinates resources that are not subject to centralized control, using standard, open, general-purpose protocols and interfaces. Grid Computing is a means of multiple independent computing clusters offer online computation and storage services for the creation of virtual companies.

Each of the three technologies plays a crucial role and has a defined scope in the integrated GRISINO platform as follows:

- The **scope of SWS technology** in GRISINO is to offer a business infrastructure which exposes and manages functionalities as services. Using SWS technology, the application development of GRISINO applications becomes compliant with the Service Oriented Architectures (SOA) - paradigm. Services are expected to use the computational and storage power provided by the Grid in order to provide their business value. Additionally the information they are going to exchange will be semantically annotated thus having a meaningful content.
- The **scope of (Semantic) Grid** is to provide a computational/ organizational infrastructure on which SWS are hosted. On one hand the (Semantic) Grid will provide the resource backbone of GRISINO in terms of computational and storage power.
- The **scope of Intelligent Content Objects** it to provide a content infrastructure for GRISINO. Intelligent Content is content that carries semantic information that can (and must) be interpreted by the services of the (Semantic) Grid. This might include descriptions about properties of the content or descriptions on how to handle the content. Intelligent Content Objects provide the intelligent data paradigm that can be seen as an intelligence and metadata wrapper for (rich) media content ranging from text files to multimedia objects.

GRISINO aims to develop an experimental test-bed combining advanced prototypes of each of the three technologies: Knowledge Content Objects (KCOs) as a model for the unit of value together with its management framework KCCA (Knowledge Content Carrier Architecture)[1] that can deal with KCOs, WSMO/L/X ([5],[2],[4]) as a framework for the description and execution of SWS and Globus[2] as the Grid infrastructure for managing resources and hosting services.

---

[1] http://metokis.salzburgresearch.at
[2] http://www.globus.org/toolkit/

2 System Description

The GRISINO system architecture is shown in Figure 1. It provides a set of APIs and the implementation of these APIs that shall ease the handling and development of applications which intend to use the three technologies together:

- the GRISINO API providing application developers easy access to the combined functionality of the three technologies.
- the Transformer API including protocol transformations between the technologies.
- the Selector API issuing calls to Transformer or Foundation API.
- the Foundation API is an abstracted view of the APIs of core components.

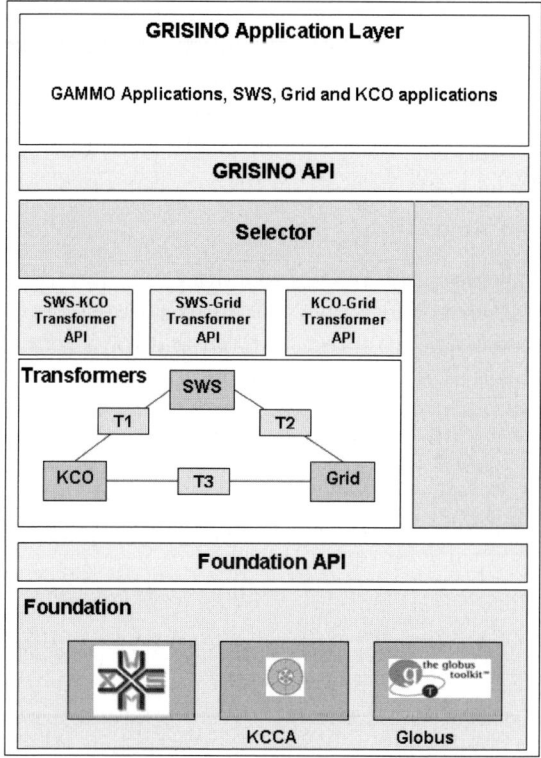

Fig. 1. GRISINO Architecture

The rest of this paper focuses on the integration of the technologies which is mainly done through the transformer components. The three technologies are integrated in a service oriented manner and the communication is realized via Web services. Furthermore the integration is supported and enhanced by using semantic technologies.

2.1 SWS-Grid Integration

The main intention of the integration of SWS and Grid technologies is to enhance Grid technology with semantics to achieve the vision of the Semantic Grid, in which Grid services can be automatically discovered, invoked and composed. To integrate the two SWS and Grid systems, namely WSMX and KCCA, we have taken a service-based integration approach. We have deployed a set of statefull Grid services using the Globust toolkit. For each of these services semantic annotations in WSML has been provided. The integration approach is as follows:

1. A formal WSML goal description is submitted to WSMX.
2. The WSML description is translated to an XML format and a SOAP message is created with data from the goal.
3. The SOAP message is sent to the endpoint specified in the grounding part of the WSML service description.
4. The grid service deployed in the Globus Toolkit is invoked and a SOAP response is created and sent back to WSMX.
5. A lifting mechanism (based on XSLT transformation) is used to construct out of the SOAP message a WSML description containing the response.

2.2 SWS-KCO Integration

The aim of the integration of WSMX and KCCA is to enable automatic discovery of KCCA services and to be able to execute KCO plans through the invocation of Web services. To integrate SWS and Intelligent Objects systems, namely WSMX and KCCA, we follow a service-based integration approach extended with semantic features. The integration approach is as follows:

1. A formal SWS request in the form of a WSML goal description is submitted to WSMX
2. The WSML description is translated to an XML format and a SOAP messages are created with data from the goal.
3. The SOAP message is sent to the endpoint specified in the grounding part of the WSML service description.
4. The KCCA Web service is invoked and a SOAP response is created and sent back to WSMX.
5. A lifting mechanism (based on XSLT transformation) is used to construct out of the SOAP message a semantic representation in WSML of the KCCA response.

3 Demonstration

The proposed system demo will present the basic interaction between Semantic Web services, Grid Computing and Intelligent Objects technologies.

We use the exhibition generation scenario to demonstrate the functionality of the system. The scenario involves multimedia objects, multimedia processing services and computational resources for video and image processing. The first use case of this scenario will be demonstrated during the system demo. It shows the process of a digital catalog creation which involves the use of various services: (1) getting artifacts from different sources (artifacts specifying), (2) building catalogs based on Semantic Web Services technology (digital catalog producing) and (3) registering intelligent objects (digital catalog registering). These services are deployed as Grid services using a Globus toolkit container. For each of these services a Semantic Web service has been created and deployed in WSMX. Last but not least simple Intelligent Objects are exchanged between above mentioned services.

The visitor interested in the demo will have the chance to see how the three fundamental technologies for distributed computing can be integrated in a service oriented manner. We will show, using the previous mentioned scenario, how calls are made between the three systems: WSMX, Globus and KCCA, what kind of data is passed between them and how semantics enables and supports this integration.

Acknowledgements

This work is funded by the FIT-IT (Forschung, Innovation, Technologie - Informationstechnologie) under the project GRISINO[3] - Grid semantics and intelligent objects. The authors would like to thank all the people who are involved in GRISINO project and the funding support from Austrian Government.

References

1. Behrendt, W., Gangemi, A., Maass, W., Westenthaler, R.: Towards an Ontology Based Distributed Architecture for Paid Content. In: Gómez-Pérez, A., Euzenat, J. (eds.) ESWC 2005. LNCS, vol. 3532, pp. 257–271. Springer, Heidelberg (2005)
2. de Bruijn, J. (ed.): The Web Service Modeling Language WSML, Technical report, WSML, WSML Final Draft D16.1v0.21. (2005),
   http://www.wsmo.org/TR/d16/d16.1/v0.21/
3. Foster, I., Kesselman, C., Tuecke, S.: The Anatomy of the Grid: Enabling Scalable Virtual Organizations. In: Sakellariou, R., Keane, J.A., Gurd, J.R., Freeman, L. (eds.) Euro-Par 2001. LNCS, vol. 2150, pp. 1–26. Springer, Heidelberg (2001)
4. Haller, A., Cimpian, E., Mocan, A., Oren, E., Bussler, C.: WSMX - A Semantic Service-Oriented Architecture. In: Proceedings of International Conference on Web Services (ICWS 2005), Orlando, Florida, USA (2005)
5. Roman, D., Lausen, H., Keller, U.: Web Service Modeling Ontology (WSMO), Working Draft D2v1.4, WSMO (2007), http://www.wsmo.org/TR/d2/v1.4/
6. Toma, I., Bürger, T., Shafiq, O., Doegl, D., Behrendt, W., Fensel, D.: Grisino: Combining semantic web services, intelligent content objects and grid computing. e-science 0, 39 (2006)

---

[3] http://www.grisino.at/

SemSearch: Refining Semantic Search

Victoria Uren, Yuangui Lei, and Enrico Motta

Knowledge Media Institute, The Open University,
Milton Keynes, MK7 6AA, UK
{y.lei,e.motta,v.s.uren}@open.ac.uk

Abstract. We demonstrate results presentation and query refinement functions of the SemSearch engine for semantic web portals and intranets.

Keywords: Semantic search, query refinement, semantic intranet.

1 Introduction

SemSearch is a search engine for RDF knowledge bases [1] [2]. The driving factor in its design is to make the formulation of semantic queries straightforward for users who may not know the details of the ontology underlying the knowledge base. To achieve this, it has a query translation engine which takes keyword input and translates it into formal semantic queries.

The issue of the usability of semantic search systems is being addressed actively. In a recent review [3], we identified four main query modes: keyword, form, visual and natural language systems. We have taken keyword systems as the stepping off point because we believe they are the most familiar to users. This requires a "translation" process that converts keyword input to formal, semantic queries. Our approach to automatic query formulation from keywords is closest to those proposed in [4] and [5]. However, in many cases, users need to refine their searches to get closer to the results they want, a fact widely acknowledged by IR researchers, e.g. [6] and [7]. To achieve this, they need both clear presentations of the results they have so far and mechanisms to constrict, or alternatively broaden, their search. Our hypothesis is that different query modes come into their own at different stages of the search refinement cycle. In this demo, we focus on recent developments in the results presentation and query refinement facilities of SemSearch that exploit several query formulation modes.

SemSearch is intended for two kinds of scenario. The first is the relatively familiar scenario of semantic web portals; the example in figure 1 is taken from the KMi Semantic Web Portal[1]. The second scenario could be called "semantic intranets", i.e., intranets in which organizations employ semantic web technology to give access to heterogeneous resources which may be scattered across an organization's servers on different sites. Both scenarios would be expected to employ a relatively small number of known and trusted ontologies. However, in the case of semantic intranets the volume of resources covered may be very large scale.

---

[1] http://semanticweb.kmi.open.ac.uk:8080/ksw/index.html

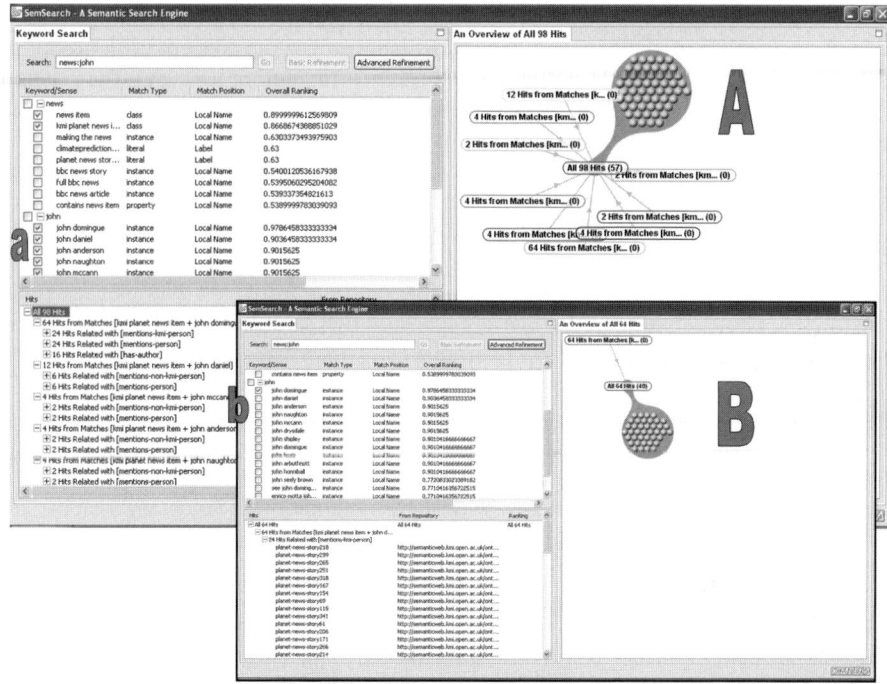

Fig. 1. Results presentation in SemSearch for the search "news: john" showing: a) the list of entities automatically selected by SemSearch, A) ClusterMap results visualisation for the search, at a glance it is messy, too many Johns have been selected, b) the form interface is used to reduce the list to just john-domingue, B) with a clear change to the visualisation the user can judge whether the search refinement has had the desired effect.

2 System Details

Implementation. SemSearch is a Java application implemented using the Eclipse[2] rich client platform. The query translation function is also implemented as a web service that can be built into web portals. Lucene[3] is used to make an index of the textual parts of the RDF knowledge base. This is exploited by the query interpretation engine, which generates formal SeRQL queries that are sent to the Sesame[4] query engine. Results are visualized using the Aduna Cluster Map[5] tool.

Query Translation. The query translation engine finds out the semantic meanings of the keywords specified in user queries and composes multiple senses into formal queries. This process is described in detail elsewhere [1],[2]. Here, we provide a summary to assist understanding of other parts of the description.

---

[2] Eclipse http://www.eclipse.org/
[3] Lucene http://lucene.apache.org/
[4] Sesame http://www.openrdf.org/
[5] Cluster Map http://www.aduna-software.com/technologies/clustermap/overview.view

From the semantic point of view, a keyword entered by the user may match i) general concepts (e.g., the keyword "news" which matches the concept news-item), ii) semantic relations between concepts, (e.g. the keyword "author" matches the relation has-author), or iii) instance entities (e.g., the keyword "Yuangui" which matches the instance Yuangui-Lei). The system exploits the Lucene text search to match the keyword against the indexes of local names, labels, and short literal values built with Lucene. This process can produce a number of matches for each input keyword. These matches must be assembled into SeRQL queries. In the example shown in Figure 1, the keyword "news" has eight possible matches and the keyword "phd students" has three, giving twenty four possible pairings for which queries need to be constructed. SemSearch ranks the queries and initially searches the high ranking ones. In the example, "news" matches two classes, which are selected, ignoring the instances. But "john" matches only instances, ten of which rank high enough to be selected. The current ranking process is described in detail elsewhere [1].

Presenting Results. Semantic search systems provide access both to the RDF data itself, as triples, and to documents with RDF annotations. Summary results are particularly helpful at the search refinement stage when the user needs to determine whether and in what way to modify their query. The most basic summary result presentation is the ranked listing. SemSearch provides such a listing using the same ranking method it applies for selecting queries. The RDF results can also be summarized using Cluster Map (shown on the right in figure 1). This generates a graphical representation that shows how many occurrences of each entity have been found and whether they are linked to other entities found in the search. The resulting visualizations give rapid insight into which entities dominate the results. In particular, it can indicate that "something is wrong" much faster than scanning a results listing or reading through documents. In our example, it indicates that too many instances have been included in searches. Finally, the user may need to see the original documents from which RDF annotations were derived. Again, a ranking is provided to let them see the documents that best represent each entity first.

By examining the results presented in these ways the user can determine whether they have found a satisfactory answer to their query. If they are not satisfied, the presentations should provide them with the clues they need to proceed to the next step: query refinement.

Query Refinement. One way a user can refine their SemSearch query is by changing the selection of matching entities using a simple form. The close matches are presented as a list, with the entities automatically selected by SemSearch ticked (shown in figure 1 in the top left panel of the interface). By selecting and deselecting terms in the list a different selection of semantic queries can be run. Motivations for this kind of refinement include removing spurious matches, which contain the same string as the query but are not interesting (the extra "john" instances in our example), and including matches that were not selected but look promising to the user.

Another means of query refinement tackles the case in which a user's initial query has been pitched at the wrong conceptual level and they need to narrow down or broaden the search scope. This is a case where the hierarchical organization of ontologies gives semantic search systems a real advantage over pure text search in which the user has to keep guessing at appropriate new keywords until they get the results they want. A standard approach is to use an ontology browser. However, we wished to avoid this in SemSearch because 1) it can be more information than the user requires, 2) we plan to extend SemSearch for use with multiple ontologies and so need an approach that will scale up. Therefore, we have developed a visual query formulation function in which the user is shown only the immediate neighbourhood of an entity as an interactive graph. Working from this fragment the user can navigate up or down through the hierarchy to find the right level.

3 Work in Progress

We are continually improving SemSearch and adding new functionality. One open issue currently under investigation is semantic ranking. This topic is being actively researched by the semantic web community. Methods are being developed for ranking whole ontologies or RDF/OWL documents, e.g. [8], [9] and [10], for associations between semantic entities, e.g., [11] and [12], and for query results, e.g. [13]. For SemSearch, we are presently investigating alternative algorithms for ranking matches to semantic entities and combinations of matches in formal queries generated by the query translation engine. Currently, the matches are ranked using an algorithm that exploits the similarity, the domain context, and the query context factor. Improved rankings are also needed for annotations found as results, and we are investigating methods for this.

We have demonstrated how SemSearch makes keyword, form and visual search modes available at different stages of the query refinement process. Future work will extend the use of multiple, complementary search modes demonstrated in the current prototype.

Acknowledgement

This work funded by the X-Media project (www.x-media-project.org) sponsored by the European Commission as part of the Information Society Technologies (IST) programme under EC grant number IST-FP6-026978.

References

1. Lei, Y., Uren, V., Motta, E.: SemSearch: a search engine for the semantic web. In: Staab, S., Svátek, V. (eds.) EKAW 2006. LNCS (LNAI), vol. 4248, pp. 238–245. Springer, Heidelberg (2006)
2. Lei, Y., Lopez, V., Motta, E., Uren, V.: An Infrastructure for Building Semantic Web Portals. Journal of Web Engineering 6(4), 283–308 (2007)
3. Uren, V., Lei, Y., Lopez, V., Liu, H., Motta, E., Giordanino, M.: The usability of semantic search tools: a review. Knowledge Engineering Review 22, 361–377 (2007)
4. Tran, T., Cimiano, P., Rudolph, S., Studer, R.: Ontology-Based Interpretation of Keywords for Semantic Search. In: Aberer, K., Choi, K.-S., Noy, N., Allemang, D., Lee, K.-I., Nixon, L., Golbeck, J., Mika, P., Maynard, D., Mizoguchi, R., Schreiber, G., Cudré-Mauroux, P. (eds.) ISWC 2007. LNCS, vol. 4825, pp. 523–536. Springer, Heidelberg (2007)

5. Zhou, Q., Wang, C., Xiong, M., Wang, H., Yu, Y.: SPARK: Adapting Keyword Query to Semantic Search. In: Aberer, K., Choi, K.-S., Noy, N., Allemang, D., Lee, K.-I., Nixon, L., Golbeck, J., Mika, P., Maynard, D., Mizoguchi, R., Schreiber, G., Cudré-Mauroux, P. (eds.) ISWC 2007. LNCS, vol. 4825, pp. 694–707. Springer, Heidelberg (2007)
6. Belkin, N.J., Cool, C., Kelly, D., Lin, S.-J., Park, S.Y., Perez-Carballo, J., Sikora, C.: Iterative exploration, design and evaluation of support for query reformulation in interactive information retrieval. Information Processing and Management 37(3), 403–434 (2005)
7. White, R.W., Kules, B., Bederson, B.: Exploratory Search Interfaces: Categorization, Clustering and Beyond. Report on the XSI 2005 Workshop at the Human-Computer Interaction Laboratory, University of Maryland. ACM SIGIR Forum 39(2), 52–56 (2005)
8. Ding, L., Pan, R., Finin, T., Joshi, A., Peng, Y., Kolari, P.: Finding and ranking knowledge on the semantic web. In: Gil, Y., Motta, E., Benjamins, V.R., Musen, M.A. (eds.) ISWC 2005. LNCS, vol. 3729, pp. 156–170. Springer, Heidelberg (2005)
9. Alani, H., Brewster, C.: Ontology Ranking based on the analysis of concept structures. In: Proceedings of the 3rd International Conference on Knowledge Capture 2005 (K-CAP 2005), pp. 51–58. ACM Press, New York (2005)
10. Hogan, A., Harth, A., Decker, S.: ReConRank: A Scalable Ranking Method for Semantic Web Data with Context. In: 2nd International Workshop on Scalable Semantic Web Knowledge Base Systems (SSWS 2006), Athens, GA, USA, November 5 (2006)
11. Aleman-Meza, B., Halaschek-Wiener, C., Arpinar, I.B., Ramakrishnan, C., Sheth, A.P.: Ranking Complex Relationships on the Semantic Web. IEEE Internet Computing 9(3), 37–44 (2005)
12. Anyanwu, K., Maduko, A., Sheth, A.P.: SemRank: ranking complex relationship search results on the semantic web. In: Proceedings of the 14th international conference on World Wide Web, pp. 117–127. ACM Press, New York (2005)
13. Stojanovic, N., Maedche, A., Staab, S., Studer, R., Sure, Y.: SEAL: a framework for developing SEmantic PortALs. In: Proceedings of the 1st International Conference on Knowledge Capture (K-CAP 2001), pp. 155–162. ACM Press, New York (2001)

The Combination of Techniques for Automatic Semantic Image Annotation Generation in the IMAGINATION Application*

Andreas Walter[1] and Gabor Nagypal[2]

[1] FZI Research Center for Information Technologies, Information Process Engineering,
Haid-und-Neu-Straße 10-14, 76131 Karlsruhe, Germany
Andreas.Walter@fzi.de
[2] disy Informationssysteme GmbH, Erbprinzenstr. 4-12, Eingang B, 76133 Karlsruhe, Germany
nagypal@disy.net

Abstract. The IMAGINATION project provides image-based navigation for digital cultural and scientific resources. Users can click on parts of an image to find other, interesting images to a given context. In this paper, we present the core parts of the IMAGINATION application. To allow the navigation through images, this application automatically generates high quality semantic metadata. Therefore it combines automated processes for person and object detection, face detection and identification in images together with text mining techniques that exploit domain specific ontologies.

1 Introduction

State of the art systems for the navigation through images either assume that users manually create semantic image annotations or they use automated processes for image annotation creation in isolation. SemSpace [1] uses domain ontologies and allows the semantic annotation of image parts. This application does not use automated processes at all. Thus, the annotation of images is a time consuming task in this system. Riya [2] uses a face detection algorithm. This reduces the annotation time for users, but leads to new problems. The sole usage of face detection algorithm leads to the problem of incorrectly generated annotations. E.g., so-called "phantom faces" may be generated.

The goal of IMAGINATION is to minimize the human effort to create high-quality image annotations. First, the annotation time for the creation of semantic images has to be reduced by automatic generation. Second, these generated artifacts must be of high-quality to reduce the time needed for manual corrections. To achieve these goals, IMAGINATION combines different automated processes. Moreover, background knowledge in a common domain ontology is also exploited to achieve best result.

In Section 2, we present the automated processes that are used in the IMAGINATION application. In Section 3, we present a sample scenario to demonstrate the interaction of the processes and how this interaction helps increase to quality of annotations.

---

* This work was co-funded by the European Commission within the project IMAGINATION.

2 Components of the IMAGINATION Application

In this section, we present the components of the IMAGINATION application. They work together to automatically generate semantic image annotations of the highest possible quality.

2.1 ImageNotion - Collaborative Generation of Domain Ontologies

Domain ontologies used in IMAGINATION base on the *ImageNotion* concept [3]. Ontologies according to ImageNotion consist of imagenotions. An *imagenotion* (formed from the words image and notion) graphically represents a semantic notion through an image. Furthermore, similarly to many existing ontology formalisms, it is possible to associate descriptive information, e.g. textual labels and date information with an imagenotion. Further, it is possible to add links to related web pages for an imagenotion. Links can help text mining algorithms to gather background information from web pages. In addition, relations between imagenotions are also supported. To achieve maximal understandability, ImageNotion makes no distinction between concepts and instances. Based on the ImageNotion methodology, users of the IMAGINATION system can collaboratively generate required domain ontologies. Also, it is possible to import and extend existing ontologies, such as CIDOC-CRM ([4]). Semantic annotations can be created either for the whole image or for image parts. Fig. 1 shows an the imagenotion for Manuel Barroso, the current president of the EU commission.

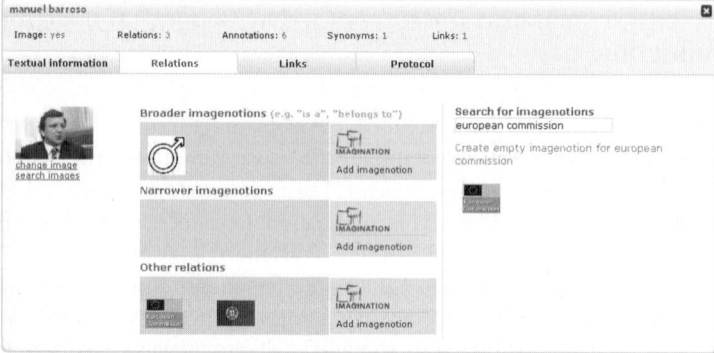

Fig. 1. Imagenotion for Manuel Barroso

2.2 Automated Processes

Most images contain text base image annotations. The *text mining* algorithms of JSI[1] [5] allow the detection of semantic elements in such text based image annotations. They can also new ontology elements. Especially for this task, the background texts that are stored at imagenotions as links to external web pages are very useful.

---

[1] Links to partner web pages can be found on the project web site: http://www.imagination-project.org

Face detection and identification algorithms are provided by Fraunhofer IIS ([6]). The face detection algorithm can detect parts on images that display faces. In addition, gender classification of a detected person is also possible. The face identification algorithm aims to detect the relevant person for a detected face. It returns a list of proposals for the person and their corresponding relevance.

Person and object detection algorithms are provided by NTUA. Person detection finds the image area showing a person or a part of it. Object detection algorithms can identify objects, e.g. tanks, airplanes or cars.

2.3 Controller for the Generation of Automatically Generated Image Annotations

The controller for the generation of automatically generated image annotations gets an image as input. First, it loads available textual and manually created semantic annotations for the image. Then, it iteratively invokes the available automated processes one by one. Each of the automated processes can read and change the available annotations and image regions, and add new ones. This allows the correction or the refinement of existing result. E.g., face detection algorithms may thoroughly examine areas for faces that were detected as persons by the person detection algorithms. In addition, the text mining results may help to eliminate wrong suggestions in the face identification step. The controller stops the annotation process, when there are no further changes in an iteration.

3 An Example Scenario for Using the IMAGINATION Application

In this section we present an example scenario for using the IMAGINATION application. The input is an image showing the current president of the EU commission, Manuel Barroso, together with the former president of the EU commission, Romano Prodi.

First, the user uploads the image. Then, it is possible to create manual annotations. In Fig. 2, a user has added the imagenotion "Manuel Barrosso" for the complete image with a high rating.

In the next step, the user can start the automated annotation process. Fig. 3 shows the result of the automated processes. In the first iteration, the text mining algorithm has created the semantic annotations "Romano Prodi" and "Manuel Barroso", based on the textual title of the image: "EU president Barroso meets Prodi". The person and object detection algorithms have created two image annotations for the shapes of the two persons. The face detection algorithm has created two image annotations for the detected faces. The face identification has identified "Manuel Barroso" with a score of 80 percent and "Guenther Verheugen" with a score of 20 percent. Also, the gender of the second face was detected as "male". The controller now initiates another annotation round. In this round, the person identification can use the results of the other automated processes. For the second face, it creates a new image annotation "Romano Prodi" for Prodi's face with a score of 100 percent and sets the score of "Guenther Verheugen" to zero, since the information of the text mining algorithm and the manual annotation

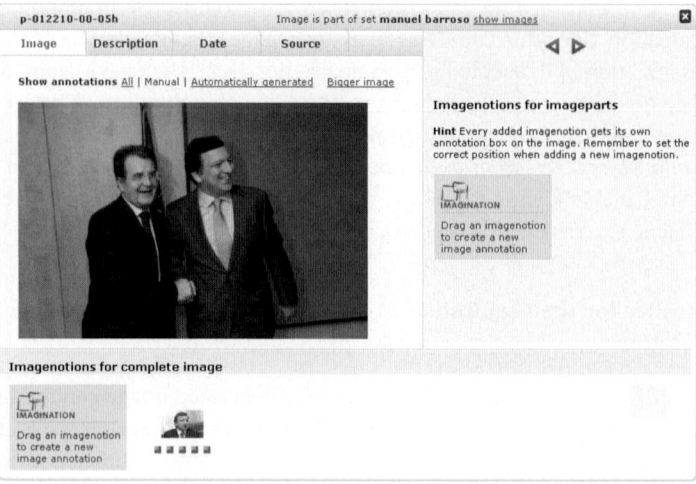

Fig. 2. Manual annotation of an image

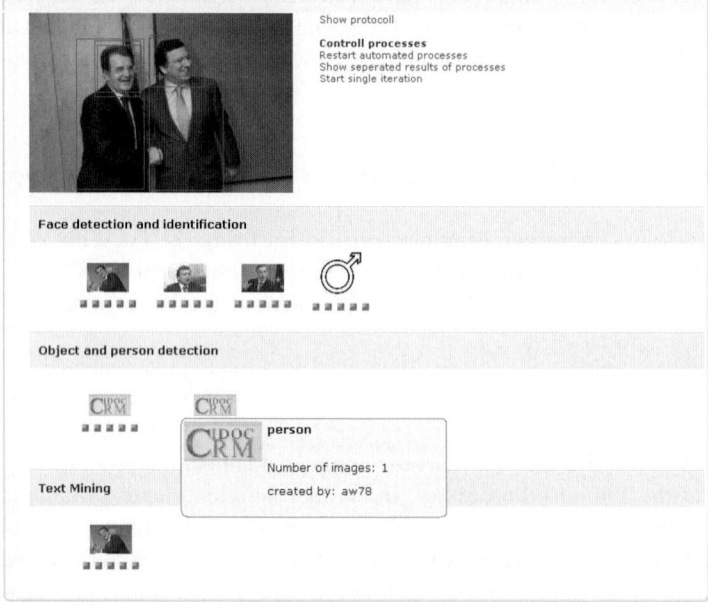

Fig. 3. Results of the automated processes

states, that there must be the person "Manuel Barroso". Also, it sets the score of the detected areas for the persons to zero, since two faces were detected[2]. In the third iteration round, there are no more changes and the controller returns the result.

---

[2] Our users prefer faces to body contours, when the size of the face is big enough.

Fig. 4. Resulting semantic image annotations for the image

Fig. 2 shows the final result of the annotation when a user searches for the image. The result contains the detected annotation boxes of "Manuel Barroso" and "Romano Prodi". Thus, the combination of the automated processes with domain ontologies lead to automatically created image annotations with a higher quality than using the automated processes in isolation.

The application is accessible at www.imagenotion.com

References

1. van Ossenbruggen, J., Troncy, R., Stamou, G., Pan, J.Z.: Image Annotation on the Semantic Web. W3C working draft, W3C (2006)
2. Riya: Riya - Visual search (2007) (accessed 2007-12-09), http://www.riya.com/
3. Walter, A., Nagypal, G.: Imagenotion - methodology, tool support and evaluation. In: GADA/DOA/CoopIS/ODBASE 2007 Confederated International Conferences DOA, CoopIS and ODBASE, Proceedings. LNCS, Springer, Heidelberg (2007)
4. Crofts, N., Doerr, M., Gill, T., Stead, S., Stiff, M.: Definition of the cidoc conceptual reference model version 4.2. In: CIDOC CRM Special Interest Group (2005)
5. Fortuna, B., Grobelnik, M., Mladenic, D.: Semi-automatic data-driven ontology construction system. In: Proc. of the 9th International multi-conference Information Society IS-2006, Ljubljana (2006)
6. Küblbeck, C., Ernst, A.: Face detection and tracking in video sequences using the modified census transformation. In: Baker, K.D. (ed.) Image and Vision Computing, vol. (6), Elsevier, Amsterdam (2006)

WSMX: A Solution for B2B Mediation and Discovery Scenarios[*]

Maciej Zaremba[1] and Tomas Vitvar[2]

[1] Digital Enterprise Research Institute (DERI),
National University of Ireland, Galway
maciej.zaremba@deri.org
[2] Semantic Technology Institute (STI2),
University of Innsbruck, Austria
tomas.vitvar@sti2.at

Abstract. We demonstrate Web Service Execution Environment (WSMX), a semantic middleware platform for runtime service discovery, mediation and execution, applied to SWS Challenge scenarios. We show the modelling principles as well as execution aspects of the WSMX semantic technology addressing the real-world requirements.

1 Introduction

Semantic Web Services (SWS) technologies offer promising potential to enable integration and discovery that is more flexible and adaptive to changes occurring over a software system's lifetime. However, there remains very few publicly available, realistic, implemented scenarios that showcase the benefits of semantics for services. In this respect we develop the WSMX[1] – a middleware system that operates on semantic description of services and facilitates automation in service integration and discovery. We demonstrate the value of the WSMX on real-world scenarios from SWS Challenge[2], a community-driven initiative that provides a set of scenarios with real Web services along with a methodology for the evaluation of solutions. We contribute to the SWS Challenge by implementing solutions based on the WSMX showing how this technology can be used to facilitate dynamic discovery and mediation in B2B integration. Users of our demonstration are able to learn how existing, non-semantic Web services can be semantically enabled and what are the benefits of semantics in the context of B2B integration and service discovery.

2 SWS Challenge Scenarios and WSMX

We build our solutions on the SWS framework developed in DERI including conceptual model for SWS (Web Service Modeling Ontology, WSMO[2]), language

---

[*] This work is supported by the Science Foundation Ireland Grant No. SFI/02/CE1/I131, and the EU projects SUPER (FP6-026850), and SemanticGov (FP-027517).
[1] WSMX is an open-source project, see http://sourceforge.net/projects/wsmx
[2] http://www.sws-challenge.org

for service modeling (Web Service Modeling Language, WSML[2]), middleware system (Web Service Execution Environment, WSMX[4]), and modelling framework (Web Service Modelling Toolkit, WSMT[3]). In order to model the scenario, we use WSMO for modeling of services and goals (i.e. required and offered capabilities) as well as ontologies (i.e. information models on which services and goals are defined) all expressed in the WSML-Flight ontology language. WSML-Flight provides a Datalog expressivity extended with inequality and stratified negation that is sufficient for addressing requirements of SWS Challenge scenarios. We use KAON2 reasoner[4] for the inference over WSML-Flight ontologies. In addition, we use Java SWING[5] based monitoring facility to display WSMX components' progress of use-case executions.

Figure 1 depicts solution architectures for SWS Challenge discovery (part A) and mediation (part B) scenarios. **Discovery** (A) defines tasks for identifying and locating Business Services. **Selection** (A) selects most appropriate service according to user's preferences. **Orchestration** (A and B) executes the composite business process. **Mediation** (B) resolves data and process heterogeneities. **Reasoning** (A, B) performs logical reasoning over semantic descriptions of services.

2.1 Mediation Scenario

The mediation scenario (Figure 1, part B) describes a data and process mediation of a trading company called Moon. Moon uses two back-end systems to manage its order processing, namely a Customer Relationship Management system (CRM) and an Order Management System (OMS). The SWS Challenge provides access to both of these systems through public Web services. The scenario describes how Moon interacts with its partner company called Blue using RosettaNet PIP 3A4 purchase order specification[6]. Using the WSMT data mapping tool we map the Blue RosettaNet PIP 3A4 message to messages of the Moon back-end systems. We then apply the WSMX data and process mediation components to resolve incompatibilities of message exchanges defined by the RosettaNet PIP 3A4 process and those defined in the Moon back-end systems.

Our major contributions to the mediation scenario shows:

- how flat XML schema of RosettaNet purchase-order and other proprietary messaging schema used by different partners could be semantically enriched using the WSML ontology language as Listing 1.1 shows,
- how services provided by partners could be semantically described as WSMO services and built on top of existing systems,
- how conversation between partners and their services can be facilitated by the WSMX integration middleware enabling semantic integration, and

---

[3] http://sourceforge.net/projects/wsmt
[4] http://kaon2.semanticweb.org
[5] http://java.sun.com/docs/books/tutorial/uiswing
[6] RosettaNet is the B2B integration standard and PIP (Partner Interface Process) define various interactions patterns and vocabularies for business integration.

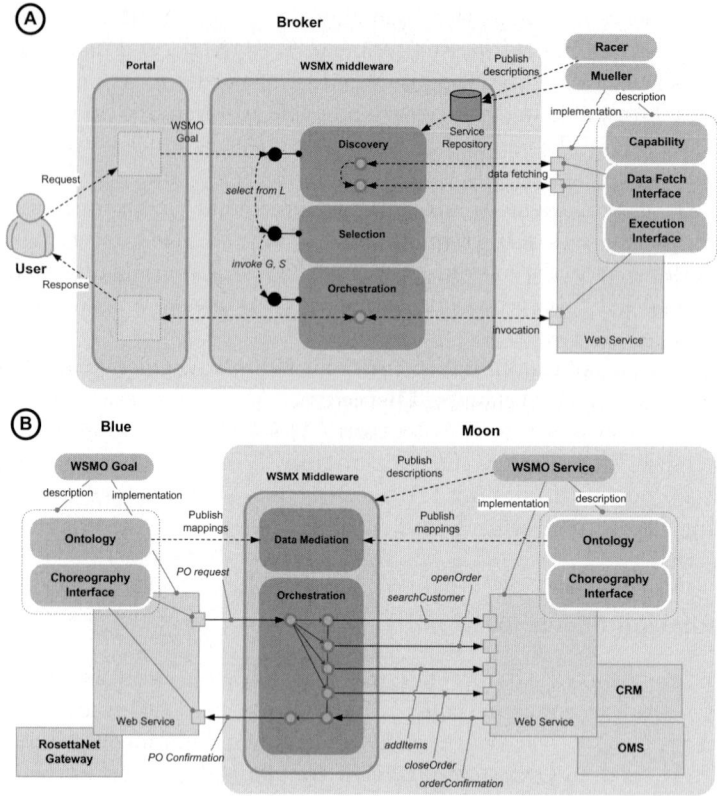

Fig. 1. Solution Architectures for SWS Challenge Scenarios

- how data and process mediation can be applied between heterogeneous services within the integration process.

```
1   /* XSLT Extract of lifting rules from XML message to WSML */
2   instance PurchaseOrderUID memberOf por#purchaseOrder
3     por#globalPurchaseOrderTypeCode hasValue " <xsl:value−of select="dict:GlobalCode"/>"
4     por#isDropShip hasValue IsDropShipPo
5       <xsl:for−each select="po:ProductLineItem">
6         por#productLineItem hasValue ProductLineItem<xsl:value−of select="position()"/>
7       </xsl:for−each>
8   ...
9   /* message in WSML after transformation */
10  instance PurchaseOrderUID memberOf por#purchaseOrder
11    por#globalPurchaseOrderTypeCode hasValue "Packaged product"
12    por#isDropShip hasValue IsDropShipPo
13  ...
```

Listing 1.1. Lifting in XSLT and resulting WSML message

Since the core WSMX functionality operates on semantic descriptions of messages, WSMX needs to also facilitate transformations between semantic and non-semantic messages through so called grounding descriptions (i.e. lifting and lowering). For the modelling phase we demonstrate how WSMT toolkit can be applied to modeling of both semantic and grounding definitions for the Moon and Blue companies, that is, how we model service orchestrations, domain ontologies, lifting/lowering groundings. Data Mediation is based on declarative, rule-based mappings between source and target ontologies. We show how these mappings are created during the design-time and how they are executed on the messages (instance level) using KAON2 reasoner during the execution time. We demonstrate details of semantic B2B integration focusing on types of data and process heterogeneities that WSMX is able to handle. We also show how changes in utilized ontologies and processes are handled. In case of source or target ontology changes, adjustments of their declarative mappings using WSMT are required. Changes on the service public process level are handled automatically by Process Mediator without any additional design time steps.

2.2 Discovery Scenario

The discovery scenario (Figure 1, part A) describes a user who uses a third-party company (broker or e-hub) in order to buy certain products with shipment to certain location. A number of shippers allow to ship products with different shipment conditions (places of shipment, price, etc.). Our approach to discovery is to match a WSMO Goal with a WSMO Web service through their semantic descriptions as well as to use additional data not available in the semantic descriptions (e.g., shipment price). The WSMX fetches this information during runtime through a specific Web service data-fetching interface. In [3] we define a conceptual framework supporting integration of dynamically fetched data into the discovery context.

```
1   /* general abstract definition of the axiom in the common ontology */
2   relation isShipped(ofType sop#ShipmentOrderReq)
3
4   /* specification of the axiom in the Mueller ontology */
5   axiom isShippedDef definedBy
6       ?sOrder[sop#to hasValue ?temp, sop#package hasValue ?p] memberOf sop#SOrderReq and
7       ?temp[so#address hasValue ?to] and ?to[so#city hasValue ?city] and
8       isShippedContinents(?city, so#Europe, so#Asia, so#NorthAmerica, so#Africa) and
9       ( (?p [so#weight hasValue ?weight] memberOf so#Package) and (?weight =< 50) )
10      implies
11      sop#isShipped(?sOrder).
```

Listing 1.2. isShipped relation declared in the common and Mueller ontologies

We semantically describe shipment capabilities offered by different companies using common shipment ontology. We take the advantage of the shared ontology when defining "abstract" axioms and their specialization in the concrete shipment service ontology (e.g., *isShipped* axiom as Listing 1.2 shows). The axiom is shared by both the shipping services and the goals (representing service requester) and provides an interface-like mechanism[7] to define a common

---

[7] An analogy are interfaces in programming languages like Java. The interface declares some functionality but does not say how this should be implemented.

evaluation criteria for service discovery. Requestor does not need to know how *isShipped* is specified by the service, but it can use it in its request to check whether given service is able to ship for a specified input (i.e., source and target location, package weight, dimension, etc.). There is a context Knowledge Base (KB) created for every instance of matchmaking between goal and a service. Logic query provided in the goal is evaluated against this KB by KAON2 reasoner and depending on the evaluation and variable binding, a service is assigned to the given matching category and is further ranked for runtime selection purposes.

Our solution demonstrates how domain ontologies, shipment goals and services are semantically described (concepts, instances, relationships, rules) as well as how service discovery works. We demonstrate Web services and Goals modelling principles and we show how extra information (e.g., shipping price) can be dynamically provided into the discovery context by utilizing data-fetching service interface.

3 Related Work

Our work can be compared to other solutions of SWS Challenge scenarios like Diane or SWE-ET as described in [1]. Diane solution provides a solution based on the language supporting fuzzy sets modelling, although without rule support. On the other hand, SWE-ET is based on combination of software modelling workflow-based methods with Flora-2 utilized for semantic descriptions.

4 Conclusion

With our contribution to the SWS Challenge we proved the value of the WSMX semantic technology in the context of B2B integration. Our solutions have been evaluated, by peer-review, according to the evaluation methodology of the SWS Challenge[8]. The evaluation criteria targets the adaptivity of the solutions, that is, solutions should handle introduced changes by modification of declarative descriptions rather than code-changes. Success level 0 indicates a minimal satisfiability level, where messages between middleware and back-end systems are properly exchanged. Success level 1 is assigned when changes introduced in the scenario require code changes and recompilation. Success level 2 indicates that introduced changes did not entail any code modifications and only declarative parts had to be changed. Finally, success level 3 is assigned when the system is able to automatically adapt to the new conditions. WSMX proved to deliver a generic solution[9] scoring level 2 as there were no changes required in WSMX code when addressing new scenarios but it sufficed to adapt or provide a new semantic descriptions of involved services and service requestors.

---

[8] http://sws-challenge.org/wiki/index.php/SWS_Challenge_Levels
[9] http://sws-challenge.org/wiki/index.php/Workshop_Innsbruck

References

1. Kuster, U., et al.: Service Discovery with SWE-ET and DIANE - A Comparative Evaluation By Means of Solutions to a Common Scenario. In: 9th International Conference on Enterprise Information Systems (ICEIS 2007), Funchal, Madeira-Portugal (June 2007)
2. Roman, D., et al.: Web Service Modeling Ontology. Applied Ontologies 1(1), 77–106 (2005)
3. Vitvar, T., Zaremba, M., Moran, M.: Dynamic Service Discovery through Meta-Interactions with Service Providers. In: Franconi, E., Kifer, M., May, W. (eds.) ESWC 2007. LNCS, vol. 4519, Springer, Heidelberg (2007)
4. Vitvar, T., et al.: Semantically-enabled service oriented architecture: Concepts, technology and application. Service Oriented Computing and Applications 1(2) (2007)

Conceptual Spaces in ViCoS

Claus Zinn

Max Planck Institute for Psycholinguistics
Wundtlaan 1, 6525 XD Nijmegen, The Netherlands
Claus.Zinn@mpi.nl

Abstract. We describe ViCoS, a tool for constructing and visualising conceptual spaces in the area of language documentation. ViCoS allows users to enrich existing lexical information about the words of a language with conceptual knowledge. Their work towards language-based, informal ontology building must be supported by easy-to-use workflows and supporting software, which we will demonstrate.

1 Research Background and Application Context

Language documentation aims at the creation of a representative and long lasting, multipurpose record of natural languages. Such documentation contributes to maintain, consolidate or revitalize endangered languages, and inherently, also contributes to the description of cultural elements of a language community [1]. Our aim is to increase this cultural aspect by allowing users to complement linguistic information — defined mainly by lexica and annotated media recordings — with ontological information. Our approach in centered around the notion of knowledge spaces, where users model and manipulate a world of concepts and their interrelations rather than just lexical entries.

In language documentation, both linguistic and ontological resources are targeted for *human consumption*. By and large, there are two main user groups: scientists such as linguists and anthropologists, and members of the language community. Scientists may contribute to and exploit resources to study the language and culture of a community, or to compare them to the ones of other communities. This user group requires technology for intelligent search, potentially across lexica with their diverse structure and metadata, and thus, a machinery that can cope with data heterogeneity. The focus for the second user group is a different one; the empowerment of community members to actively participate in describing their language and culture and to learn from such resources. For community members, words are keys to access and describe relevant parts of their life and cultural traditions such as food preparation, house building, medicine, ceremonies, legends *etc*. For them, words are of foremost practical rather than theoretical nature; and their understanding of words is best described by the various associations they evoke rather than in terms of any formal and abstract theory of meaning. This user group requires effective technology for the construction of knowledge spaces, a sort of *informal* ontology of fuzzily-defined concepts and relationships, that enables community members to anchor the words of a linguistic resource, or the objects of a multimedia archive, to such associations.

2 Related Work

Given existing lexical and multimedia resources, users shall be empowered to represent, communicate, and thus preserve complex facts about their language and culture. Resources may help eliciting their tacit knowledge, but tools are required that help transform it into tangible representations that can be viewed and manipulated by others. Resulting conceptual structures may then facilitate the revival of a shared understanding within the community. Mind mapping and concept mapping software are a good starting point to support this task.

Mind maps form a tree structure as there is only a single central concept to which other concepts can be related to. Usually, the nature of these relations is left underspecified. *Concept maps* are more expressive. First, there is no single central concept so that the hierarchy of concepts form a graph structure. Moreover, it is possible to name the relations that hold between concepts, and links between concepts can be directed. There is, however, no formal definition of node types (classes, instances), or arc types (defining the relation types). *Ontologies* can be seen as a formalisation of concept maps as they are based on using formal languages, and thus, concept and relation types have a well-defined semantics.

Existing tools for ontology engineering (*e.g.*, Protégé) have only limited relevance for a community-driven knowledge engineering effort as their proper and effective use requires a considerable amount of expertise in knowledge representation formalisms. Moreover, the use of existing ontologies, in addition to the need to know their content and structure, is problematic as they induce a significant (and usually Westernized) bias of how the world should be modelled. Existing tools for constructing mind/concept maps are much easier to use but, for our purpose, lack integration with lexical resources and multimedia archives.

Our ViCoS tool is based on the notion of *conceptual spaces* (CS), where a concept must be anchored to its lexical resource(s), and where it can be illustrated with archived material (images, sounds, videos). In contrast to concept maps, we require users to define relation types before their first use. Their definition is semi-formal but help addressing some interoperability issues when doing studies across lexica and conceptual spaces. In contrast to ontology builders, there is, thus, no formal semantics of concepts or relations. In this sense, conceptual spaces are closer to topic maps [6] rather than RDF/OWL-based approaches. Conceptual Spaces also bear much resemblance to the representational aspect of ConceptNet [7], a semantic network where concept nodes (semi-structured English fragments) are interrelated by twenty rather pragmatic semantic relations.

3 ViCoS

ViCoS is a simple but effective tool that aims at empowering a broad base of users to describe and link together those words (and the concepts they denote), which they feel are important in their language and culture. ViCoS is coupled with Lexus [2], our tool for building and maintaining lexical resources, and has access to an archive that hosts multimedia objects and their annotations [3]. ViCoS is a web-based application requiring a Javascript-enabled web browser.

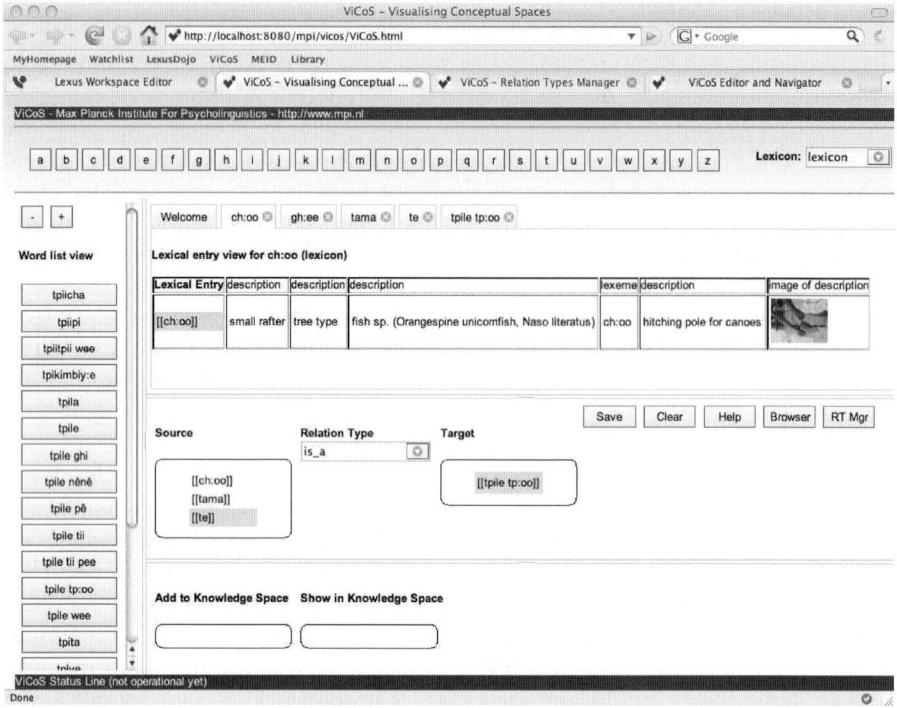

Fig. 1. ViCoS Main Window, showing the Yélî Dnye example lexicon)

Fig. 1 depicts ViCoS' main window consisting of: the Lexicon Selection and Start Letter Selection frame; the Word List frame where lexical entries are represented by their lexeme; the Lexical Entry frame where selected lexical entries are displayed; and the Conceptual Workspace frame where relations can be created.

In the screenshot, we have selected the Yélî Dnye demo lexicon, and clicked on the letter "t" to obtain (from Lexus) all lexemes that start with this letter. We then selected "te" (Engl. "fish") from the Word List frame to display its lexical entry. We then copied the complete lexical entry "[[te]]" (via drag&drop) to become the target of a "is-a" relation. In a similar way, be obtained the two fish species "ch:oo" and "ghee" as sources. Likewise, any relation between any pair of lexical entries can be created. With a tab mechanism that maintains access to prior selections of lexical entries, links can also be created across lexica.

ViCoS offers a few standard pre-defined relation types such as hypnomy, meronymy, holonymy, synonymy, and antonymy. Users may also define new relation types. To view the conceptual space for a lexical entry, the user drags this lexical entry from the Lexical Entry frame into the "Show in Knowledge Space" option of the Conceptual Space frame. The lexical entry's corresponding concept will take center stage in the conceptual space browser, surrounded by all concepts that are directly related to it (see Fig. 2).

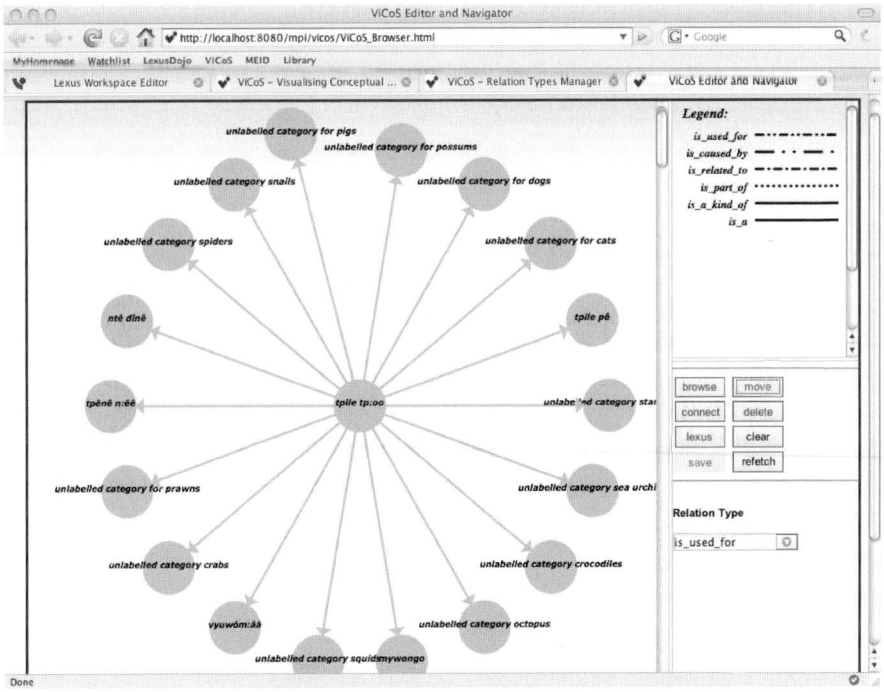

Fig. 2. ViCoS Browser Window, showing a few classes and their interrelations

The graphical editor and browser offers various modes: in Browse mode, when clicking on a concept node, additional information that stems from the lexical entry is displayed; and in Connect mode, new relations between concepts can be drawn. A double-click on a node will show a CS with the chosen node as centre, together with all concepts it is directly related to. There is also functionality for moving around nodes, or deleting concepts and relations. In Lexus mode, clicking on the node with open Lexus to show the corresponding lexical entry in linguistic context, thus strongly linking together lexical and ontological spaces.

On the back-end, ViCoS uses the OWL knowledge representation language [4]. Conceptual spaces are stored within the JENA framework, which provides a programmatic environment for OWL constructs as well as a rule-based inference engine for retrieving implicit information [5]. Once a query user interface is in place, users can then ask ViCoS, for instance, whether a given speech community categorises a dolphin as a type of fish, or whether dolphin meat is used as ingredient for food recipes. The reasoning engine will exploit class-subclass hierarchies and the characteristics of relation types, for instance, their domain and range, and whether they are symmetric, transitive, or functional.

ViCoS has entered a test phase with end-users. First feedback from two documentation teams has been positive. We are currently implementing new functionality that aims at supporting users in the creation of conceptual spaces: an interface to Lexus' search engine where search results can be easily entered into the CS space; and a mechanism that suggests, for each noun, a CS that contains its corresponding concept as well as candidate concepts to which it may be related to. These candidates will stem from the lexicon, exploiting, i.e., "semantic" data categories, or from word co-occurrences within example sentences.

4 Discussion

Language documentation projects aim at preserving endangered cultural heritage for future generations, ideally within the communities themselves. Thus an active involvement of community members is crucial for such projects to succeed. To overcome the limitations of a purely linguistic approach to language documentation, we use knowledge engineering methods that allow members of indigenous communities to play an active role in the documentation process. This emphasises that a language is so much more than a list of lexical entries and their scientific description in linguistic parlour. Our approach turns words into culturally relevant concepts and places them in relation to other concepts. It attempts to engage and inspire community members to explore and to extend the resulting knowledge space. Because our design preserves the relationship between lexical and ontological space, users can browse them more or less simultaneously and can thus gain a richer experience of the language and culture being documented. In a way, our approach bridges scientific resources (lexica constructed by linguists; and multimedia assets annotated by experts) with indigenous knowledge resources (knowledge spaces constructed by community members).

Ensuring that emerging CSs stay manageable instead of becoming chaotic and hard to interpret will be a main challenge. One could hope that the users themselves organise the spaces they build. In particular, relation types will be an issue as it is unclear whether the built-in relation types will be accepted and properly used by community members. It is also to be seen how their set grows, and how well they will be documented and reused; regular interventions of moderators may prove necessary to clarify their semantics.

Acknowledgements. I would like to thank Jacquelijn Ringersma (MPI) and Gaby Cablitz (University of Kiel) for testing ViCoS, and for reporting their feedback.

References

1. Documentation of Endangered Languages (DoBeS), http://www.mpi.nl/dobes
2. Lexus, http://www.lat-mpi.eu/tools/lexus
3. Archive for linguistic resources, http://corpus1.mpi.nl
4. OWL Web Ontology Language Overview. W3C Recommendation from (February 10, 2004), http://www.w3.org/TR/owl-features
5. JENA, http://jena.sourceforge.net/index.html
6. Topic Maps, http://www.ontopia.net/topicmaps/materials/tao.html
7. ConceptNet, http://web.media.mit.edu/~hugo/conceptnet

Author Index

Akhtar, Waseem 432
Alves de Medeiros, Ana Karla 49
Andrei, Mihai 767
Andreou, Dimitris 705
Anyanwu, Kemafor 508
Auer, Sören 19, 787

Baeza-Yates, Ricardo 1
Barnard, Andries 273
Barreto, José Monte 782
Bazzanella, Barbara 258
Berges, Idoia 5
Berlanga, Rafael 185
Bermúdez, Jesús 5
Bernstein, Abraham 463, 478
Berre, Arne 767
Bhagdev, Ravish 554
Blöchl, Martin 493
Blanco-Fernández, Yolanda 720
Bolles, Andre 448
Bontcheva, Kalina 361
Born, Matthias 772
Bouquet, Paolo 258
Bräuer, Matthias 34
Bürger, Tobias 869
Buscher, Georg 817

Castano, Silvana 170
Cazalens, Sylvie 539
Cerbah, Farid 777
Chapman, Sam 554
Cho, Heeryon 65
Ciravegna, Fabio 554
Corcho, Oscar 859
Costa, Luis 767
Cyganiak, Richard 690

d'Amato, Claudia 288, 318
d'Aquin, Mathieu 802, 854
Damljanovic, Danica 361
Decker, Stefan 80, 124, 139, 690, 849
Delbru, Renaud 690
Dellschaft, Klaas 782
Diallo, Gayo 827
Dietze, Stefan 599

Dietzold, Sebastian 19, 787
Döegl, Daniel 869
Dolbear, Catherine 348
Domingue, John 49, 599
Dörflinger, Jörg 736
Duchesne, Philippe 767
Dzbor, Martin 802

Edwards, Peter 303, 644
Engelbrecht, Hendrik 782
Erling, Orri 838
Espinoza, Mauricio 333, 792
Esposito, Floriana 288, 318

Fanizzi, Nicola 288, 318
Fensel, Dieter 659, 674
Ferrara, Alfio 170
Fitzner, Daniel 767

Gerber, Aurona 273
Gil-Solla, Alberto 720
Glaser, Hugh 797
Gómez-Pérez, Asunción 859
Goñi, Alfredo 5
Görlitz, Olaf 807
Gotts, Nick 644
Grau, Bernardo Cuenca 185
Grawunder, Marco 448
Grcar, Miha 767
Gridinoc, Laurian 802
Grimnes, Gunnar AAstrand 303
Groza, Tudor 80
Gugliotta, Alessio 599
Gutierrez, Claudio 3

Handschuh, Siegfried 80, 124, 139, 849
Hart, Glen 348
Hepp, Martin 751
Hoffmann, Jörg 767, 772
Hollink, Laura 388
Hyvönen, Eero 95, 110

Illarramendi, Arantza 5
Isaac, Antoine 388, 402
Ishida, Toru 65

Jacobi, Jonas   448
Jaffri, Afraz   797
Jawaheer, Gawesh   827
Jiménez-Ruiz, Ernesto   185
Johnson, Martina   348
Josan, Laura   849

Kaczmarek, Tomasz   772
Kauppinen, Tomi   110
Kerrigan, Mick   812
Kiefer, Christoph   463, 478
Kiesel, Malte   817
Kifer, Michael   659
Klien, Eva   767
Klinov, Pavel   822
Klyne, Graham   154
Kopecký, Jacek   432, 674
Kostkova, Patty   827
Kourtesis, Dimitrios   614
Kowalkiewicz, Marek   772
Krennwallner, Thomas   432
Kuter, Ugur   629

Lamarre, Philippe   539
Lanfranchi, Vitaveska   554
Lange, Christoph   832
Langegger, Andreas   493
Langlois, Joel   767
Le Phuoc, Danh   843
Lei, Yuangui   874
Leser, Ulf   524
Limyr, Andreas   767
Lin, Naiwen   629
Liu, Qiaoling   584
Locher, André   478
Lochmann, Henrik   34
López-Cima, Angel   859
López-Nores, Martín   720
Lorusso, Davide   170

Maduko, Angela   508
Markovic, Ivan   772
Matthezing, Henk   402
Maue, Patrick   767
Mena, Eduardo   333, 792
Mikhailov, Ivan   838
Millard, Ian C.   797
Mocan, Adrian   812
Möller, Knud   80, 849
Möller, Ralf   170

Morbidoni, Christian   843
Motta, Enrico   802, 854, 874

Näth, Tobias Henrik   170
Nagypal, Gabor   879
Nastase, Vivi   376

Oyama, Satoshi   65

Pan, Jeff Z.   245
Paraskakis, Iraklis   614
Pazos-Arias, José J.   720
Pedrinaci, Carlos   49
Pérez, Asunción Gómez   333, 792
Petrelli, Daniela   554
Pichler, Reinhard   200
Pignotti, Edoardo   644
Polhill, Gary   644
Polleres, Axel   200, 432, 843
Preece, Alun   303, 644

Quilitz, Bastian   524

Rahmani, Tirdad   736
Ramos-Cabrer, Manuel   720
Roman, Dumitru   659
Rosati, Riccardo   215
Rutenbeck, Sascha   782
Ruttenberg, Alan   864

Sabou, Marta   802, 854
Sahbi, Mehdi   736
Samwald, Matthias   843, 864
Sattler, Ulrike   185
Sauermann, Leo   569
Scerri, Simon   124
Schade, Sven   767
Schliekelman, Paul   508
Schlobach, Stefan   402
Schmidt, Kay-Uwe   736
Schneider, Thomas   185
Schreiber, Guus   388
Schumacher, Kinga   569
Schwarz, Sven   817
Scicluna, James   772
Seppälä, Katri   95
Shadbolt, Nigel   4
Shafiq, Omair   869
Sheth, Amit   508
Shotton, David   154
Siles, Adrián   859

Sintek, Michael   569
Siorpaes, Katharina   751
Şirin, Evren   629
Sizov, Sergej   807
Spiliopoulos, Vassilis   418
Srinivas, Kavitha   864
Staab, Steffen   782, 807
Steinmetz, Nathalie   767
Stenzhorn, Holger   690, 864
Stoermer, Heiko   258
Stojanovic, Ljiljana   736
Strube, Michael   376
Suntisrivaraporn, Boontawee   230

Tablan, Valentin   361
Takasaki, Toshiyuki   65
Tertre, Francois   767
Thai, VinhTuan   139
Theoharis, Yannis   705
Thomas, Susan Marie   736
Toma, Ioan   869
Topor, Rodney   245
Trüg, Sebastian   849
Tran, Thanh   584
Tummarello, Giovanni   690, 843
Tuominen, Jouni   95
Tzitzikas, Yannis   705

Unbehauen, Jörg   19, 787
Uren, Victoria   874

Väätäinen, Jari   110
Valarakos, Alexandros G.   418

Valduriez, Patrick   539
van Assem, Mark   388
van der Meij, Lourens   402
van der Merwe, Alta   273
van Elst, Ludger   817
Vasiliu, Laurentiu   767
Ventresque, Anthony   539
Viljanen, Kim   95
Viskova, Jana   674
Vitvar, Tomas   674, 884
Vouros, George A.   418

Walter, Andreas   879
Wang, Haofen   584
Wang, Kewen   245
Wang, Shenghui   388, 402
Wang, Zhe   245
Weber, Ingo   772
Wei, Fang   200
Wöß, Wolfram   493
Woltran, Stefan   200

Yu, Yong   584

Zaharia, Raluca   767
Zaremba, Maciej   884
Zastavni, Nicolas   767
Zhang, Kang   584
Zhao, Jun   154
Zhou, Xuan   772
Zinn, Claus   402, 890
Zirn, Cäcilia   376

Printing: Mercedes-Druck, Berlin
Binding: Stein+Lehmann, Berlin

Lecture Notes in Computer Science

Sublibrary 3: Information Systems and Application, incl. Internet/Web and HCI

For information about Vols. 1– 4587
please contact your bookseller or Springer

Vol. 5021: S. Bechhofer, M. Hauswirth, J. Hoffmann, M. Koubarakis (Eds.), The Semantic Web: Research and Applications. XIX, 897 pages. 2008.

Vol. 5017: T. Nanya, F. Maruyama, A. Pataricza, M. Malek (Eds.), Service Availability. XII, 225 pages. 2008.

Vol. 5013: J. Indulska, D.J. Patterson, T. Rodden, M. Ott (Eds.), Pervasive Computing. XIV, 315 pages. 2008.

Vol. 5006: R. Kowalczyk, M. Huhns, M. Klusch, Z. Maamar, Q.B. Vo (Eds.), Service-Oriented Computing: Agents, Semantics, and Engineering. X, 154 pages. 2008.

Vol. 4997: B. Monien, U.-P. Schroeder (Eds.), Algorithmic Game Theory. XI, 363 pages. 2008.

Vol. 4976: Y. Zhang, G. Yu, E. Bertino, G. Xu (Eds.), Progress in WWW Research and Development. XVIII, 699 pages. 2008.

Vol. 4956: C. Macdonald, I. Ounis, V. Plachouras, I. Ruthven, R.W. White (Eds.), Advances in Information Retrieval. XXI, 719 pages. 2008.

Vol. 4952: C. Floerkemeier, M. Langheinrich, E. Fleisch, F. Mattern, S.E. Sarma (Eds.), The Internet of Things. XIII, 378 pages. 2008.

Vol. 4947: J.R. Haritsa, R. Kotagiri, V. Pudi (Eds.), Database Systems for Advanced Applications. XXII, 713 pages. 2008.

Vol. 4936: W. Aiello, A. Broder, J. Janssen, E.. Milios (Eds.), Algorithms and Models for the Web-Graph. X, 167 pages. 2008.

Vol. 4932: S. Hartmann, G. Kern-Isberner (Eds.), Foundations of Information and Knowledge Systems. XII, 397 pages. 2008.

Vol. 4928: A.H.M. ter Hofstede, B. Benatallah, H.-Y. Paik (Eds.), Business Process Management Workshops. XIII, 518 pages. 2008.

Vol. 4903: S. Satoh, F. Nack, M. Etoh (Eds.), Advances in Multimedia Modeling. XIX, 510 pages. 2008.

Vol. 4900: S. Spaccapietra (Ed.), Journal on Data Semantics X. XIII, 265 pages. 2008.

Vol. 4892: A. Popescu-Belis, S. Renals, H. Bourlard (Eds.), Machine Learning for Multimodal Interaction. XI, 308 pages. 2008.

Vol. 4882: T. Janowski, H. Mohanty (Eds.), Distributed Computing and Internet Technology. XIII, 346 pages. 2007.

Vol. 4881: H. Yin, P. Tino, E. Corchado, W. Byrne, X. Yao (Eds.), Intelligent Data Engineering and Automated Learning - IDEAL 2007. XX, 1174 pages. 2007.

Vol. 4877: C. Thanos, F. Borri, L. Candela (Eds.), Digital Libraries: Research and Development. XII, 350 pages. 2007.

Vol. 4872: D. Mery, L. Rueda (Eds.), Advances in Image and Video Technology. XXI, 961 pages. 2007.

Vol. 4871: M. Cavazza, S. Donikian (Eds.), Virtual Storytelling. XIII, 219 pages. 2007.

Vol. 4858: X. Deng, F.C. Graham (Eds.), Internet and Network Economics. XVI, 598 pages. 2007.

Vol. 4857: J.M. Ware, G.E. Taylor (Eds.), Web and Wireless Geographical Information Systems. XI, 293 pages. 2007.

Vol. 4853: F. Fonseca, M.A. Rodríguez, S. Levashkin (Eds.), GeoSpatial Semantics. X, 289 pages. 2007.

Vol. 4836: H. Ichikawa, W.-D. Cho, I. Satoh, H.Y. Youn (Eds.), Ubiquitous Computing Systems. XIII, 307 pages. 2007.

Vol. 4832: M. Weske, M.-S. Hacid, C. Godart (Eds.), Web Information Systems Engineering – WISE 2007 Workshops. XV, 518 pages. 2007.

Vol. 4831: B. Benatallah, F. Casati, D. Georgakopoulos, C. Bartolini, W. Sadiq, C. Godart (Eds.), Web Information Systems Engineering – WISE 2007. XVI, 675 pages. 2007.

Vol. 4825: K. Aberer, K.-S. Choi, N. Noy, D. Allemang, K.-I. Lee, L. Nixon, J. Golbeck, P. Mika, D. Maynard, R. Mizoguchi, G. Schreiber, P. Cudré-Mauroux (Eds.), The Semantic Web. XXVII, 973 pages. 2007.

Vol. 4823: H. Leung, F. Li, R. Lau, Q. Li (Eds.), Advances in Web Based Learning – ICWL 2007. XIV, 654 pages. 2008.

Vol. 4822: D.H.-L. Goh, T.H. Cao, I.T. Sølvberg, E. Rasmussen (Eds.), Asian Digital Libraries. XVII, 519 pages. 2007.

Vol. 4820: T.G. Wyeld, S. Kenderdine, M. Docherty (Eds.), Virtual Systems and Multimedia. XII, 215 pages. 2008.

Vol. 4816: B. Falcidieno, M. Spagnuolo, Y. Avrithis, I. Kompatsiaris, P. Buitelaar (Eds.), Semantic Multimedia. XII, 306 pages. 2007.

Vol. 4813: I. Oakley, S.A. Brewster (Eds.), Haptic and Audio Interaction Design. XIV, 145 pages. 2007.

Vol. 4810: H.H.-S. Ip, O.C. Au, H. Leung, M.-T. Sun, W.-Y. Ma, S.-M. Hu (Eds.), Advances in Multimedia Information Processing – PCM 2007. XXI, 834 pages. 2007.

Vol. 4809: M.K. Denko, C.-s. Shih, K.-C. Li, S.-L. Tsao, Q.-A. Zeng, S.H. Park, Y.-B. Ko, S.-H. Hung, J.-H. Park (Eds.), Emerging Directions in Embedded and Ubiquitous Computing. XXXV, 823 pages. 2007.

Vol. 4808: T.-W. Kuo, E. Sha, M. Guo, L.T. Yang, Z. Shao (Eds.), Embedded and Ubiquitous Computing. XXI, 769 pages. 2007.

Vol. 4806: R. Meersman, Z. Tari, P. Herrero (Eds.), On the Move to Meaningful Internet Systems 2007: OTM 2007 Workshops, Part II. XXXIV, 611 pages. 2007.

Vol. 4805: R. Meersman, Z. Tari, P. Herrero (Eds.), On the Move to Meaningful Internet Systems 2007: OTM 2007 Workshops, Part I. XXXIV, 757 pages. 2007.

Vol. 4804: R. Meersman, Z. Tari (Eds.), On the Move to Meaningful Internet Systems 2007: CoopIS, DOA, ODBASE, GADA, and IS, Part II. XXIX, 683 pages. 2007.

Vol. 4803: R. Meersman, Z. Tari (Eds.), On the Move to Meaningful Internet Systems 2007: CoopIS, DOA, ODBASE, GADA, and IS, Part I. XXIX, 1173 pages. 2007.

Vol. 4802: J.-L. Hainaut, E.A. Rundensteiner, M. Kirchberg, M. Bertolotto, M. Brochhausen, Y.-P.P. Chen, S.S.-S. Cherfi, M. Doerr, H. Han, S. Hartmann, J. Parsons, G. Poels, C. Rolland, J. Trujillo, E. Yu, E. Zimányie (Eds.), Advances in Conceptual Modeling – Foundations and Applications. XIX, 420 pages. 2007.

Vol. 4801: C. Parent, K.-D. Schewe, V.C. Storey, B. Thalheim (Eds.), Conceptual Modeling - ER 2007. XVI, 616 pages. 2007.

Vol. 4797: M. Arenas, M.I. Schwartzbach (Eds.), Database Programming Languages. VIII, 261 pages. 2007.

Vol. 4796: M. Lew, N. Sebe, T.S. Huang, E.M. Bakker (Eds.), Human–Computer Interaction. X, 157 pages. 2007.

Vol. 4794: B. Schiele, A.K. Dey, H. Gellersen, B. de Ruyter, M. Tscheligi, R. Wichert, E. Aarts, A. Buchmann (Eds.), Ambient Intelligence. XV, 375 pages. 2007.

Vol. 4777: S. Bhalla (Ed.), Databases in Networked Information Systems. X, 329 pages. 2007.

Vol. 4761: R. Obermaisser, Y. Nah, P. Puschner, F.J. Rammig (Eds.), Software Technologies for Embedded and Ubiquitous Systems. XIV, 563 pages. 2007.

Vol. 4747: S. Džeroski, J. Struyf (Eds.), Knowledge Discovery in Inductive Databases. X, 301 pages. 2007.

Vol. 4744: Y. de Kort, W. IJsselsteijn, C. Midden, B. Eggen, B.J. Fogg (Eds.), Persuasive Technology. XIV, 316 pages. 2007.

Vol. 4740: L. Ma, M. Rauterberg, R. Nakatsu (Eds.), Entertainment Computing – ICEC 2007. XXX, 480 pages. 2007.

Vol. 4730: C. Peters, P. Clough, F.C. Gey, J. Karlgren, B. Magnini, D.W. Oard, M. de Rijke, M. Stempfhuber (Eds.), Evaluation of Multilingual and Multi-modal Information Retrieval. XXIV, 998 pages. 2007.

Vol. 4723: M. R. Berthold, J. Shawe-Taylor, N. Lavrač (Eds.), Advances in Intelligent Data Analysis VII. XIV, 380 pages. 2007.

Vol. 4721: W. Jonker, M. Petković (Eds.), Secure Data Management. X, 213 pages. 2007.

Vol. 4718: J. Hightower, B. Schiele, T. Strang (Eds.), Location- and Context-Awareness. X, 297 pages. 2007.

Vol. 4717: J. Krumm, G.D. Abowd, A. Seneviratne, T. Strang (Eds.), UbiComp 2007: Ubiquitous Computing. XIX, 520 pages. 2007.

Vol. 4715: J.M. Haake, S.F. Ochoa, A. Cechich (Eds.), Groupware: Design, Implementation, and Use. XIII, 355 pages. 2007.

Vol. 4714: G. Alonso, P. Dadam, M. Rosemann (Eds.), Business Process Management. XIII, 418 pages. 2007.

Vol. 4704: D. Barbosa, A. Bonifati, Z. Bellahsène, E. Hunt, R. Unland (Eds.), Database and XML Technologies. X, 141 pages. 2007.

Vol. 4690: Y. Ioannidis, B. Novikov, B. Rachev (Eds.), Advances in Databases and Information Systems. XIII, 377 pages. 2007.

Vol. 4675: L. Kovács, N. Fuhr, C. Meghini (Eds.), Research and Advanced Technology for Digital Libraries. XVII, 585 pages. 2007.

Vol. 4674: Y. Luo (Ed.), Cooperative Design, Visualization, and Engineering. XIII, 431 pages. 2007.

Vol. 4663: C. Baranauskas, P. Palanque, J. Abascal, S.D.J. Barbosa (Eds.), Human-Computer Interaction – INTERACT 2007, Part II. XXXIII, 735 pages. 2007.

Vol. 4662: C. Baranauskas, P. Palanque, J. Abascal, S.D.J. Barbosa (Eds.), Human-Computer Interaction – INTERACT 2007, Part I. XXXIII, 637 pages. 2007.

Vol. 4658: T. Enokido, L. Barolli, M. Takizawa (Eds.), Network-Based Information Systems. XIII, 544 pages. 2007.

Vol. 4656: M.A. Wimmer, J. Scholl, Å. Grönlund (Eds.), Electronic Government. XIV, 450 pages. 2007.

Vol. 4655: G. Psaila, R. Wagner (Eds.), E-Commerce and Web Technologies. VII, 229 pages. 2007.

Vol. 4654: I.-Y. Song, J. Eder, T.M. Nguyen (Eds.), Data Warehousing and Knowledge Discovery. XVI, 482 pages. 2007.

Vol. 4653: R. Wagner, N. Revell, G. Pernul (Eds.), Database and Expert Systems Applications. XXII, 907 pages. 2007.

Vol. 4636: G. Antoniou, U. Aßmann, C. Baroglio, S. Decker, N. Henze, P.-L. Patranjan, R. Tolksdorf (Eds.), Reasoning Web. IX, 345 pages. 2007.

Vol. 4611: J. Indulska, J. Ma, L.T. Yang, T. Ungerer, J. Cao (Eds.), Ubiquitous Intelligence and Computing. XXIII, 1257 pages. 2007.

Vol. 4607: L. Baresi, P. Fraternali, G.-J. Houben (Eds.), Web Engineering. XVI, 576 pages. 2007.

Vol. 4606: A. Pras, M. van Sinderen (Eds.), Dependable and Adaptable Networks and Services. XIV, 149 pages. 2007.

Vol. 4605: D. Papadias, D. Zhang, G. Kollios (Eds.), Advances in Spatial and Temporal Databases. X, 479 pages. 2007.

Vol. 4602: S. Barker, G.-J. Ahn (Eds.), Data and Applications Security XXI. X, 291 pages. 2007.

Vol. 4601: S. Spaccapietra, P. Atzeni, F. Fages, M.-S. Hacid, M. Kifer, J. Mylopoulos, B. Pernici, P. Shvaiko, J. Trujillo, I. Zaihrayeu (Eds.), Journal on Data Semantics IX. XV, 197 pages. 2007.

Vol. 4592: Z. Kedad, N. Lammari, E. Métais, F. Meziane, Y. Rezgui (Eds.), Natural Language Processing and Information Systems. XIV, 442 pages. 2007.